To Marjorie Hoy
– with the authors' very
best wishes & personal regards

Dave

# SPECIES OF *APHYTIS* OF THE WORLD

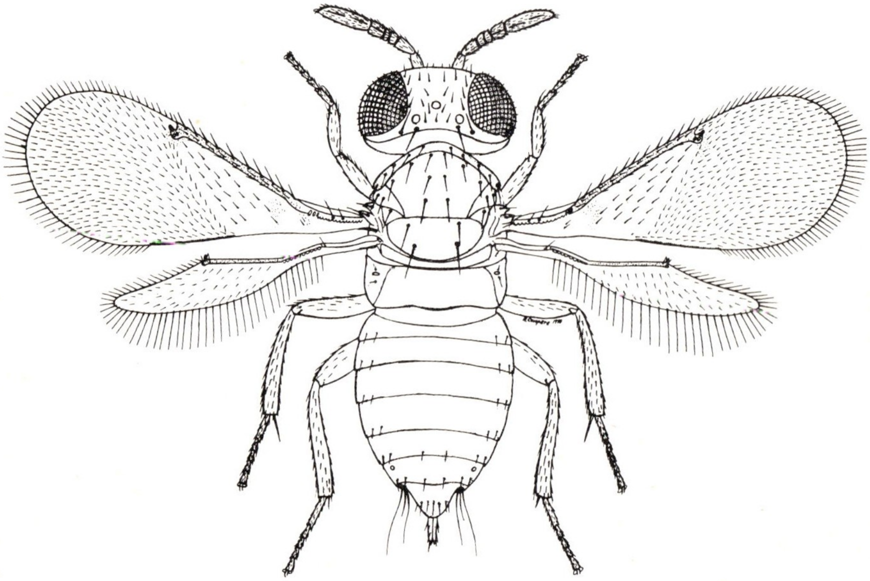

Frontispiece: *Aphytis chrysomphali* (Mercet); drawn by H. Compere, 1933.

# SPECIES OF *APHYTIS* OF THE WORLD

## (HYMENOPTERA: APHELINIDAE)

by

David Rosen and Paul DeBach

ISRAEL UNIVERSITIES PRESS, JERUSALEM

DR. W. JUNK BV PUBLISHERS, THE HAGUE—BOSTON—LONDON 1979

Published in December 1978 by
Israel Universities Press,
an imprint of
Keter Publishing House Jerusalem, Ltd.,
Givat Shaul, Jerusalem, Israel

This title is distributed worldwide, except Israel by:
Dr. W. Junk BV, Publishers, The Hague—Boston—London

Cover design: Shlomo Yehudayan

ISBN 90 6193 127 4
cat. no. 222125

Manufactured by KeterPress, Jerusalem

Printed in Israel

We take real pleasure in dedicating this work to our colleague at the University of California, Riverside, P.H. Timberlake. He, in 1924, resurrected the genus *Aphytis*, which had lain completely dormant since its description in 1900, by separating the coccid-feeding *Aphytis* spp. from the aphid-feeding *Aphelinus* spp. under which *Aphytis* had previously been included. However, this represents but a minuscule part of his contributions as a taxonomist. Timberlake ideally reflects our picture of the prototype classical dedicated, meticulous systematist. After more than sixty-seven working years, taxonomy is still his entire life and consuming passion. At the age of ninety-three he still works in his laboratory from 8 to 5 daily without interruption and then continues on in the evening at home. He is an inspiration to all who know him, and his dedication emphasizes the fascination of systematics that grips many of us.

THE AUTHORS

**David Rosen** is Professor of Entomology at the Hebrew University, Faculty of Agriculture, Rehovot, Israel.

**Paul DeBach** is Professor of Biological Control and Entomologist at the University of California, Riverside.

# CONTENTS

## PART I. GENERAL

## PART II. SPECIFIC DESCRIPTIONS

# APPENDICES

# PART I

# GENERAL

# INTRODUCTION

The great importance of accurate systematics for biological control has been pointed out and discussed time and again by numerous authors (see Clausen, 1942; Sabrosky, 1955; Schlinger and Doutt, 1964; Compere, 1969; Rosen and DeBach, 1973; DeBach, 1974; Delucchi, Rosen and Schlinger, 1976; and others) and does not require any further elaboration here. Suffice it to say that when natural enemies are being sought, or are transferred from one region to another in order to bring about biological control of an arthropod pest, correct identification of both the target pests and their natural enemies is an essential prerequisite for ultimate success. Failure in biological control has often resulted from inadequate systematics. This monograph, presenting a biosystematic revision of an important group of natural enemies, is therefore intended as a contribution to biological control as well as to basic science.

The species of the genus *Aphytis* Howard (Hymenoptera: Chalcidoidea: Aphelinidae) are minute, yellow or grayish wasps usually not exceeding one millimeter in length (see Figure 1). They develop exclusively as primary ectoparasites of armored scale insects (Homoptera: Coccoidea: Diaspididae) and are the most important natural enemies of these serious pests.

Armored scale insects are usually free beneath the hard covering scale or shield. The adult *Aphytis* female pierces the shield with the ovipositor and lays one to several eggs on the body of the scale insect. Protected by the covering scale, the *Aphytis* larvae feed externally by sucking the body fluids from the scale insect. The fully grown larvae then excrete characteristic meconial pellets and pupate underneath the empty scale. The adult parasites usually emerge through a hole they chew in the covering scale, but they may merely push their way out under the edge. In addition to the hosts killed by parasitism, numerous scale insects may be destroyed by predatory host-feeding by the adult *Aphytis* female.

As natural enemies, the species of *Aphytis* are generally superior in effectiveness to the endoparasites and predators of armored scale insects. Several species have been successfully employed in biological control projects directed against economically important armored scale insect pests, and such projects have played a significant role in the development of biological control as a science. Of still greater general importance are the naturally occurring species of *Aphytis* that are instrumental in keeping the populations of many potential pests at subeconomic levels without human intervention. The species of *Aphytis* are usually the most abundant—and often appear to be the most effective—natural enemies of their respective hosts.

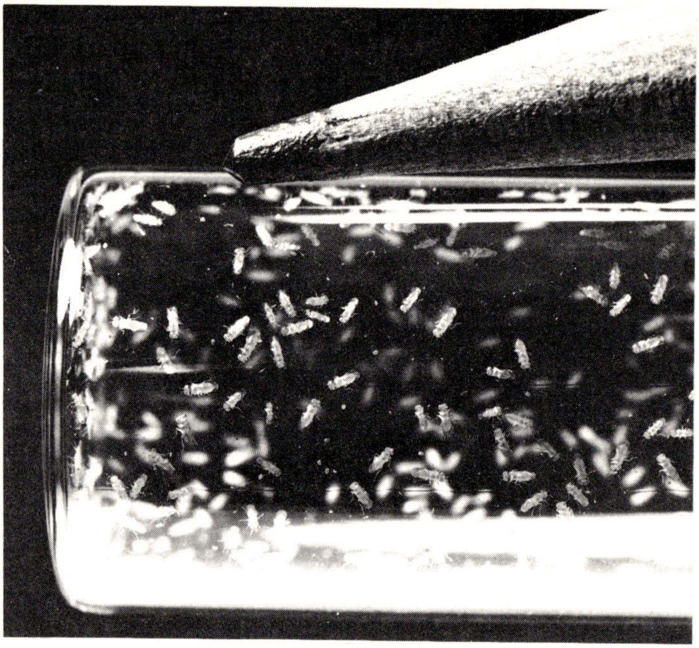

Figure 1. *Aphytis lingnanensis* individuals inside a glass vial. The pencil point at the top illustrates the minute size of these parasites. (From DeBach and Sundby, 1963.)

Unfortunately, in spite of the undisputed economic importance of *Aphytis*, systematic knowledge of this large genus has remained rather fragmentary. Identification and separation of the species of *Aphytis* are extremely difficult, due mainly to their minute size, the relative scarcity of reliable distinguishing characters, the common occurrence of sibling species, as well as to the fact that a significant proportion are uniparental (thelytokous). In the latter cases males are lacking or so scarce as to preclude hybridization tests to determine species status. As just intimated, several complexes of sibling species or biological races have recently been shown to exist and, while their separation may be of prime importance to the researcher in practical biological control, it is often impossible to tell them apart by known morphological criteria. Great confusion also formerly existed regarding the correct identification of several old species of *Aphytis* which were originally based on unreliable characters and described from poorly mounted specimens. Many new species were suspected to exist, but their identity could not be established until all known species were thoroughly revised and redescribed. Thus, until recently, misidentification of *Aphytis* species was probably the rule rather than the exception, and many potentially effective natural enemies were not recognized as distinct species. In fact, the inadequate state of the systematics of *Aphytis* had caused major setbacks in important biological control projects.

In view of this, a worldwide, biosystematic revision of the species of *Aphytis* was undertaken by the authors at the Division of Biological Control, University of California, Riverside, as part of an international project on the biological control of armored scale insects carried out under the auspices of the International Biological Program (IBP) and largely supported by the U.S. National Science Foundation. This revision was based mainly on the enormous *Aphytis* collection and numerous live cultures kept at the Division. The collection, accumulated over a period of several decades, now comprises about 150,000 cleared, slide-mounted specimens representing most of the known species and is the largest collection of *Aphytis* in the world. Possibly two thirds of these have been added during the past twenty years by the second author in connection with extensive foreign exploration for new parasites of armored scale insects as well as through cooperative exchange under the IBP program. In addition, types, other dead specimens and often live material of various species were obtained from collaborators at the Plant Protection Research Institute, Pretoria; the Queensland Museum, Brisbane; the U.S. National Museum, Washington; the British Museum (Natural History), London; the Hope Department of Entomology, Oxford; the USSR Academy of Sciences, Leningrad; the ORSTOM Laboratory, New Caledonia; the Benaki Phytopathological Institute, Athens; the Muséum National d'Histoire Naturelle, Paris; and many other institutions. The types and ample material of most described species, as well as numerous undescribed ones, were thus available in our laboratory for study.

The morphology of all available species was studied in detail. Large series of cleared, properly mounted specimens are essential for correct identification of *Aphytis*. Permission was often sought—and sometimes granted—to remount borrowed type specimens that had been inadequately preserved on slides or mounted on points. Slide-mounted specimens were studied under stereoscopic and phase-contrast microscopes, measurements and counts were made of all significant characters, and numerous photomicrographs were taken. Scanning electron microscopy was also employed in an attempt to clarify various morphological aspects. Whenever possible, results of these comparative morphological studies were corroborated by various biosystematic investigations. On the other hand, biosystematic studies sometimes provided the first and perhaps ultimate means of species separation.

Our studies have led to a redefinition of the genus in relation to closely allied aphelinid genera, especially *Marietta* Motschulsky and *Marlattiella* Howard, and to a re-evaluation of the validity of specific diagnostic characters. At this point in our studies, the genus *Aphytis* comprises 90 distinct species, including 14 that have been transferred from other genera and 34 that were recognized as new to science during the course of this investigation. The confusion that had existed regarding the identity of several species has been largely clarified, some difficult species complexes have been untangled, quite a few synonymies and homonymies have been established, and numerous misidentifications corrected. Several old, long-forgotten species have been resurrected and redefined, and a number of new potential natural enemies of injurious armored scale insect pests have been recognized. All these may be regarded as new weapons added to the arsenal of biological control.

Preliminary accounts of some of these results were reported by Rosen and DeBach (1970, 1976, 1977a) and by DeBach and Rosen (1976a, 1976b).

## ACKNOWLEDGEMENTS

Studies that led to this monograph were supported by the following grants from the U.S. National Science Foundation, awarded to Paul DeBach: GB20870 (1962–1965); GB7444 (1968–1970); GB17829 (1970–1975). Publication was facilitated by special grants from the University of California and from the Hebrew University of Jerusalem.

Our studies were based on extensive collaboration with numerous colleagues, students and associates, and would not have been possible without their invaluable cooperation and assistance. We are especially grateful to David P. Annecke, Pretoria, South Africa; Oscar Beingolea, Lima, Peru; C. Benassy, Antibes, France; Barnard D. Burks, Washington, D.C.; S. K. Cheng, Hong Kong; Paul Cochereau, Noumea, New Caledonia; Irma de Crouzel, Buenos Aires, Argentina; E. C. Dahms, Brisbane, Australia; Luis De Santis, La Plata, Argentina; Michael Kosztarab, Blacksburg, Virginia; H. E. Martin, Riyadh, Saudi Arabia; T. Sankaran, Bangalore, India; Geoffrey J. Snowball, Sydney, Australia; Arturo Teran, Tucuman, Argentina; V. A. Yasnosh, Tbilisi, USSR; Keizo Yasumatsu, Fukuoka, Japan; and many others.

Staff Research Associates involved with the various *Aphytis* projects at the Division of Biological Control, University of California, Riverside, include Chris Elder, Marilyn Huftile, John Landi, Harry Maltby, John LaSalle, Mike Rose, Richard Warkentin, Stanley C. Warner, Ernest B. White and Walter White.

Graduate students who conducted research on the biosystematics and ecology of *Aphytis* spp. at Riverside include John Baker, Gordon Gordh, Hanif Gulmahamad, Syed Khasimuddin, Dale Meyerdirk, Sudha (Rao) Nagarkatti, and Yoram Rössler. Several of them continued their collaboration with us long after having completed their graduate research.

"Post-Docs" who did research in our Riverside laboratory on various aspects of *Aphytis* biology and ecology include Loukia C. Argyriou of Athens, Greece; Gerard Fabres of Noumea, New Caledonia; Kaichi Furuhashi of Shizuoka, Japan; Pelagia Sisojević of Belgrade, Yugoslavia; Ragnhild A. Sundby of Vollebekk, Norway; Mine Tunçyürek of Bornova, Turkey; and J. C. Van Lenteren of Leiden, Holland.

Hilary Hardin and Yaacov Zoref of Israel Universities Press have spared no efforts in the preparation of this volume.

Most of the 1342 figures illustrating this monograph are original. The scanning electron microscope (SEM) micrographs were taken at Applied Space Products, Anaheim, California and at the University of California, Riverside. All the prints were made at the Photo Lab of the University of California, Riverside. The sources of all other figures are acknowledged in the captions.


# HISTORICAL REVIEW

The early history of *Aphytis* was rather confused. Although the genus was not established until 1900, several of the species that are now included in it were described during the nineteenth century, mostly in the older genus *Aphelinus* Dalman. However, descriptions at that time were generally inadequate by modern taxonomic standards, and quite a few of those early species are at present unrecognizable. Subsequently, even after *Aphytis* was formally described, a quarter of a century passed before it came to be correctly interpreted and used.

Apparently the first to describe a species that may be referable to *Aphytis* was Nees ab Esenbeck (1834). His "*?Eulophus flavus*" (see p. 733) is unrecognizable and of questionable generic position, but the possibility cannot be discounted that it was an *Aphytis*. Three years later, Westwood (1837) described the species *Agonioneurus albidus*. The genus *Agonioneurus* Westwood now stands as a synonym of *Aphelinus*, whereas *albidus* is a synonym of *Aphytis mytilaspidis* (Le Baron) (see p. 467).

Walker (1839) described another European species, *Aphelinus proclia*, which is recognized here as a valid member of *Aphytis*. Thirteen years later, Ratzeburg (1852) described the genus *Coccobius*, comprising 5 species including *luteus*, another bona fide representative of *Aphytis*, and *pallidus*, which was listed by several subsequent authors as a synonym of *flavus* Nees and is at present unrecognizable. Later, Gahan and Fagan (1923) designated another of Ratzeburg's species, *annulicornis*, as generotype of *Coccobius*; this genus is now placed in the Encyrtidae.

The center of activity shifted early from Europe to the New World. Le Baron (1870) described the first American species of *Aphytis*, which he named *Aphelinus mytilaspidis*, and was the first to note the economic importance of this effective parasite. Ashmead (1880) described *Trichogramma flavus*, an unrecognizable species which may have been yet another member of *Aphytis* (see p. 732). Then Howard (1881) described 3 additional species: *Aphelinus abnormis*, *A. diaspidis* and *A. fuscipennis*. Of these, only *diaspidis* is currently recognized as a valid member of *Aphytis*, whereas *abnormis* is an unrecognizable representative and *fuscipennis* is a synonym of *diaspidis*. It is noteworthy that although Howard included *Eriophilus mali* Haldeman, a parasite of aphids, in the genus *Aphelinus*, he regarded this genus as mainly comprising scale-insect parasites.

The first two Oriental species now included in *Aphytis* were described during the last decade of the nineteenth century: Cameron (1891) described *Aphelinus theae*, an Indian species that has recently been resurrected by Rosen and DeBach (1977) as a bona fide member of *Aphytis*, and Zehntner (1897) described *Aphelinus simplex*, an unrecognizable species from Java.

Thus, at the turn of the century, 5 valid species now included in *Aphytis* were already known, along with several others that are now regarded as unrecognizable or doubtful members of this genus. All these (except *simplex*) were listed by Dalla Torre (1898) in his catalogue, among 41 species that he referred to *Aphelinus*.

Howard (1900) described a South American species, *chilensis*, for which he erected the new genus *Aphytis*. His original generic description speaks for itself and is presented here in full:

APHYTIS—New Genus.

*Female.*—Resembles Aphelinus in the oblique hairless line extending from the stigmal vein transversely to base of wing. It differs principally from Aphelinus in the antennae, which are only 5-jointed, the first ring-joint apparently being absent. The pedicel is nearly cylindrical; the first funicle joint cubical; the second funicle joint long = oval, wider than the first and more than twice as long; club long, elliptical, longer than pedicel and funicle together. The mesonotal sclerites resemble those of Aphelinus,

> but the ovipositor is exserted to about one-third the length of the abdomen, as with Centrodora; hind thighs somewhat swollen; stigmal vein is short and knobbed and the post-marginal vein is absent; the mandibles are tridentate; the ocelli large and placed in the form of an oblique angled triangle.

Obviously, Howard considered *Aphytis* to be closely related to *Aphelinus*, but with 5 rather than 6 segments in the antennae. This, however, was based on an error. *A. chilensis* was known to him from a single, poorly preserved specimen (the female holotype), which is uncleared and mounted on a slide in such a way that both antennae are folded and obscured by the head, and the funicular segments cannot be seen clearly. In fact, the antennae of *chilensis* are 6-segmented, but the first funicular segment is much smaller than in most other species and was overlooked by Howard (see p. 349 for a redescription of *chilensis*).

Because of Howard's misinterpretation of the generotype of *Aphytis* as having 5 antennal segments, his new genus was largely ignored by his contemporaries. Even Howard himself evidently considered *Aphytis* to be a freak, and did not refer to it in any of his subsequent publications on the Aphelinidae. Thus, there is no mention of *Aphytis* in his revised table of aphelinid genera (Howard, 1907). Incidentally, in his 1907 paper Howard described two species from the Australian region, *Perissopterus capillatus* and *P. noumeaensis*. The genus *Perissopterus* Howard is currently listed as a synonym of *Marietta* Motschulsky, but both *capillatus* and *noumeaensis* are recognized here as bona fide members of *Aphytis*.

A year later Howard (1908), still regarding *Aphelinus* as comprising mainly scale-insect parasites, suggested the advisability of erecting a new genus for the aphid-inhabiting species. However, he never put that suggestion into practice.

The number of known species eventually assigned to *Aphytis* was more than doubled during the following decade. In Europe, Masi (1911) described *Aphelinus maculicornis*, a valid member of *Aphytis*, whereas Mercet (1911a) described *Aphelinus longiclavae*, a synonym of *chilensis*, and in the same year (Mercet, 1911b) added *Aphelinus aonidiae*, another valid species. Then Mercet (1912a) described *Aphelinus chrysomphali*, and in his monograph on the "Afelininos" (Mercet, 1912b) he redescribed all the known species of *Aphelinus* and presented two additional valid species, *Aphelinus maculicornis* var. *hispanica* (now known as *Aphytis hispanicus*) and *A. opuntiae*. He divided *Aphelinus* into two "sections," comprising predominantly black parasites of aphids and predominantly yellow parasites of coccids, respectively, but did not follow Howard's suggestion to erect a new genus for the aphid parasites.

De Gregorio (1914) described a new subgenus of *Aphelinus*, *Prospaphelinus*, based on his new species *silvestrii*, which in the following year (De Gregorio, 1915) he reduced to a variety of *chrysomphali*. Then Silvestri (1915) described *Aphelinus erythraeus*, now regarded as a valid species of *Aphytis*, from Africa. In the same year, Rust (1915) described three species: *Aphelinus capitis* from California, a synonym of *chilensis*; *A. quaylei* from Peru, a synonym of *chrysomphali*; and *A. limonus* from Hawaii, an unrecognizable species of *Aphytis*.

Meanwhile in South America Brèthes (1913) described the species *Trichogrammatoidea signiphoroides*. The genus *Trichogrammatoidea* Girault now stands as a synonym of *Trichogramma* Westwood in the Trichogrammatidae, but *signiphoroides* is a synonym of *chilensis*. Brèthes (1916) described another species, *Aphelinus argentinus*, which is recognized here as a synonym of *hispanicus*. *Aphelinus bovelli*, described by Malenotti (1918b) from Barbados, is another synonym of *hispanicus*. Incidentally, Malenotti was the first to utilize propodeal crenulae, one of the most useful diagnostic characters, for the identification of *Aphytis* species.

The first detailed account of the developmental history of an *Aphytis* was presented by Imms (1916) for *A. mytilaspidis*.

In Australia, Dodd (1917) described *Aphelinus ciliatus*, a valid species of *Aphytis*. A. A. Girault described one Indian and 19 Australian members of this genus in 9 papers published over a period of 20 years, 1913–1933 (5 of these papers were privately published pamphlets). Unfortunately, quite a few of Girault's species were based on single specimens, captured on window panes or "in forest," his types were usually poorly preserved, and his descriptions were often extremely brief. It is a sad commentary on Girault's work that only 4 of his 20 species are recognized here as valid members of *Aphytis*, all the rest being synonyms or unrecognizable species.

Girault (1913) synonymized the genus *Paraphelinus* Perkins (now regarded as a synonym of *Centrodora* Foerster) with *Aphelinus*, but maintained *Perissopterus* Howard as a separate genus. He described *Aphelinus grotiusi*, *A. minutissimus* and *A. newtoni*, all of which are listed here as unrecognizable species of *Aphytis*, and presented a key to the Australian species of *Aphelinus*. He also redescribed *Perissopterus capillatus*. Two years later, Girault (1915) synonymized *Perissopterus* with *Aphelinus*, and described two additional unrecognizable species of *Aphytis* under the names *Aphelinus perissoptroides* and *A. ruskini*. Then Girault (1917a) described *Marietta maculatipes*, another unrecognizable member of *Aphytis*, and *M. novicapillata*, a synonym of *capillatus*, and transferred *Aphelinus perissoptroides* to the genus *Marietta* Motschulsky. But in the same year (1917b) he described *punctaticorpus*, yet another unrecognizable *Aphytis*, in the genus *Perissopterus*. Two years later, Girault (1919) described *Aphelinus ovidii*, a synonym of *diaspidis*.

Two additional synonyms of *capillatus*—*Perissopterus cowperi* and *P. emersoni*—were published by Girault (1923); in the following year, Girault (1924) described *Aphelinus wallumbillae*, a valid species of *Aphytis*.

Seven species that are now referable to *Aphytis* were included in a small pamphlet published by Girault (1932): *Aphytis ulianovi*, a valid member of *Aphytis* with 5-segmented antennae, that was presumably referred by Girault to this genus on the basis of Howard's original description; *Aphelinus columbi*, another valid species of *Aphytis*; *Marietta peculiaris*, an Indian species recognized here as yet another valid *Aphytis*; *M. distonota*, a synonym of *columbi* (Girault actually described the female and male of the same species in two different genera, both on the same page of his pamphlet); *M. lessingi* and *M. romae*, two additional synonyms of *capillatus*; and *M. angeloni*, an unrecognizable species of *Aphytis*. Finally, Girault (1933) described *Aphelinus stellaris*, yet another synonym of *columbi*.

Needless to say, it is systematists like Girault, who unfortunately cannot be ignored, that have actually hindered the progress of systematic research and have placed the heaviest burdens on modern workers.

Meanwhile Mercet (1921) concluded that the genus *Aphytis* had no reason to exist, and synonymized it with *Aphelinus*. He also drew attention to the "extraordinary" similarity between *A. chilensis* and *A. longiclavae* but did not synonymize them.

Timberlake (1924) finally resurrected *Aphytis*, and gave this genus its modern interpretation, when he decided to follow Howard's suggestion to segregate the aphid-feeding species of *Aphelinus* from the coccid-feeding species. Since by that time Kurdjumov (1913) had clearly shown that the generotype of *Aphelinus*, *A. abdominalis* (Dalman), was an aphid parasite, it became apparent that the coccid parasites were the ones to be segregated. In view of the similarity between *chilensis* and *longiclavae*, Timberlake correctly assumed that Howard's original description of *Aphytis*, with 5-segmented antennae, was erroneous, and stated: "I believe it is safe to conclude that *Aphytis chilensis* really has three funicle joints, with the first one very small and overlooked by Dr. Howard, and I therefore adopt this name for the group of the old genus *Aphelinus* containing the coccid-feeding species." Timberlake's decision was soon accepted by Mercet (1928), as well as by practically all subsequent workers.

Compere (1925) described a new genus and species from the Orient, *Paraphytis vittata*, which he later (Compere, 1936) transferred to *Marietta*. We consider *Paraphytis* to be synonymous with *Aphytis*.

Mercet (1930, 1932) presented keys to the known European species of *Aphytis*. In his 1930 paper he listed *Prospaphelinus* De Gregorio as a synonym of *Aphytis*, but in 1932 he decided to split the genus *Aphytis* into two subgenera, *Aphytis* Howard sensu stricto and *Prospaphelinus* De Gregorio. Mercet's subgeneric classification was adopted by De Santis (1946, 1948) but not by other workers, and eventually was formally rejected by Compere (1955). (See p. 108 here for notes on subgenera.)

Four additional bona fide species of *Aphytis* have been described in the genus *Marietta* Motschulsky: *M. maculatipennis*, described by Dozier (1933) from Haiti; *M. nigripes*, by Compere (1936) from Australia; *M. costalimai*, by Gomes (1942) from Brazil; and *M. haywardi*, by Blanchard (in De Santis, 1948) from Argentina. Compere (1936) included in his revision of *Marietta* 12 species that are now referred to *Aphytis*; he also described *Marlattiella secunda*, an African species recognized here as another member of *Aphytis*.

De Santis (1948) presented extensive bibliographies for species of *Aphytis* occurring in Argentina, and described *Aphytis dubius* and *A. dubius* var. *intermedia*, both of which are regarded here as synonyms of *aonidiae*.

In Madagascar, Risbec (1952) described two species: *Aphytis opuntiae* in the subfamily Aphelininae and *Prospaphelinus madagascariensis* in the subfamily Coccophaginae. Both were rather inadequately described and figured. Risbec's *opuntiae* was, of course, a homonym of *opuntiae* Mercet, and was renamed *risbeci* by Annecke and Insley (1971). At any rate, both of Risbec's species are synonymous with *diaspidis*. In the same year, Nikol'skaya (1952) presented a key to the species of *Aphytis* occurring in the USSR, which was essentially adapted from Mercet's (1932) key.

Compere's (1955) excellent revision of *Aphytis* opened a new era in the systematic study of this genus. Compere reconfirmed Timberlake's interpretation of *Aphytis* and clarified the morphological distinctions between *Aphytis* and *Aphelinus*. He revised, redescribed and discussed all the known species, cleared up much of the confusion surrounding some of them, and indicated clearly the areas in which knowledge was insufficient. He established 3 of the main species groups of *Aphytis*, and described 13 additional species from practically all over the world: *angustus*, *anomalus*, *cercinus*, *citrinus*, *cylindratus*, *dealbatus*, *funicularis*, *ignotus*, *immaculatus*, *lepidosaphes*, *lingnanensis*, *melanostictus* and *merceti*. All of these, with the possible exception of *citrinus*, are recognized here as valid species. Altogether, Compere listed 33 species of *Aphytis*. Our present revision of *Aphytis* is based primarily on the solid foundation laid by Harold Compere.

Alam (1956a) recorded 3 species of *Aphytis* from England. Both his *A. zonatus* and "*A. diaspidis* (Silwood material)" are synonyms of *proclia*, whereas his *A. variolosum* is an unrecognizable species.

In the USSR, Chumakova (1957) described *Aphytis diaspidioti*, considered here as unrecognizable (see p. 731). A few years later (Chumakova, 1961) she added *A. testaceus*, a valid species. Then Chumakova (1964) described *A. caucasicus*, yet another unrecognizable species.

DeBach (1959) recorded several species and strains of *Aphytis* from the Orient and described *A. fisheri* and *A. melinus*. He was the first to employ reciprocal crossing tests for the separation of closely related species of *Aphytis*. A year later, DeBach (1960) described two additional Oriental species, *A. coheni* and *A. holoxanthus*, and then DeBach and Azim (1962) described *A. japonicus* from Japan.

Compere and Annecke (1961) described *Aphytis spiniferus*, a South American species that was shortly after placed by DeBach (1963) in synonymy with *costalimai*.

Azim (1963b) recorded 10 species of *Aphytis* occurring in Japan. He described two species, *A. debachi* and *A. yasumatsui*, both of which happen to be valid although they cannot be identified according to his confused key and inadequate original descriptions.

Peck (1963) listed 14 species of *Aphytis*, along with extensive bibliographies, in his catalogue of Nearctic Chalcidoidea. In the same year, Annecke (1963) described *faurei*, an aberrant African species with 4-segmented antennae, which he correctly assigned to the genus *Aphytis*.

Quednau (1964b) presented a quite detailed account of 19 indigenous and introduced species of *Aphytis* occurring in South Africa, including 5 that he described as new, namely *africanus*, *ciliatus*, *flavus*, *griseus* and *taylori*. His *ciliatus* and *flavus* being homonyms, they were subsequently renamed *setosus* and *rolaspidis*, respectively, by DeBach and Rosen (1976b). In spite of some conspicuous misidentifications, Quednau's descriptions and figures are usually quite reliable.

DeBach (1964) discussed some of the taxonomic problems involved in the study of *Aphytis* and gave keys to 11 species and 7 forms occurring in Greece, including 4 introduced species. Erdös (1964) recorded 8 species of *Aphytis* from Hungary. At about the same time, Ferrière (1965) included 11 species of *Aphytis* in his monograph on the Aphelinidae of Europe and the Mediterranean Basin, whereas Nikol'skaya and Yasnosh (1966) recorded 12 species from the European part of the USSR and the Caucasus, including one, *Aphytis moldavicus*, that they described as new. Rosen (1966) presented an illustrated key to the species of *Aphytis* attacking citrus pests in Israel.

Delucchi (1964) described *Aphytis riadi*, a synonym of *chilensis*, from Lebanon, whereas Agarwal (1964) described an unrecognizable species, *A. alami*, which he recorded mistakenly as endoparasitic in a mealybug in India (see p. 730). Then De Santis (1965) added a valid species, *Aphytis notialis*, from South America.

Traboulsi (1969) gave a fairly detailed account of the morphology of *Aphytis*. He recorded several species and forms from Lebanon, and described a new species, *A. libanicus*. Another valid species, *A. neuter*, was described by Yasnosh and Myartseva (1971) from Soviet Central Asia. Yasnosh (1972) then presented biosystematic keys to the species of *Aphytis* recorded from the USSR. She described another species, *A. sugonjaevi*, which is regarded here as a synonym of *proclia*. Annecke and Insley (1971) catalogued the species of *Aphytis* occurring in the Ethiopian region. Finally, DeBach and Gordh (1974) described *Aphytis roseni*, an African species introduced into South America.

During the course of the present revision, Rosen and DeBach (1970, 1976, 1977a) redefined the genus *Aphytis* and the main species groups, and DeBach and Rosen (1976b) presented brief descriptions of 20 new species and several new combinations. Fourteen additional species are described as new in the present monograph, bringing the total number of valid *Aphytis* species to 90.

Biosystematic studies have indicated that additional species may exist in the *lingnanensis*, *maculicornis* and *mytilaspidis* complexes, and probably in most other groups of *Aphytis* as well (see Rao and DeBach, 1969a; Rössler and DeBach, 1972a; Khasimuddin and DeBach, 1976b, 1976c). These species have not been formally described as yet, pending the accumulation of further biosystematic information.

## SOME NOMENCLATURAL CONSIDERATIONS

### THE GENDER OF *APHYTIS*

The name *Aphytis* was derived from a locality in Ancient Greece. It was presumably chosen by Howard (1900) because of its proximity to the Chalcidice, an area in northern Greece (see Figure 2) from which the name of the superfamily Chalcidoidea was derived. The place no longer exists and does not appear on modern maps of Greece. The word Aphytis does not have any meaning in Greek, and therefore does not connote any particular gender.

According to the International Code of Zoological Nomenclature (ICZN, 1961), Article 30(b)(ii), "A genus-group name consisting of a word that is neither Greek, Latin, nor modern Indo-European, or that is an arbitrary combination of letters, takes the gender expressly attributed to it by its author, or implied by an originally associated species-group name. If no gender was assigned or implied, the name is to be treated as masculine, except that if the ending is clearly a natural classical feminine or neuter one, the gender is that appropriate to the ending."

Although the ending of *Aphytis* seems to suggest a feminine gender, this is by no means obvious. Since Howard did not expressly attribute any particular gender to his genus, and since no gender is implied by the only species-group name that he originally associated with it (*chilensis*), we prefer to use the masculine gender for species-group

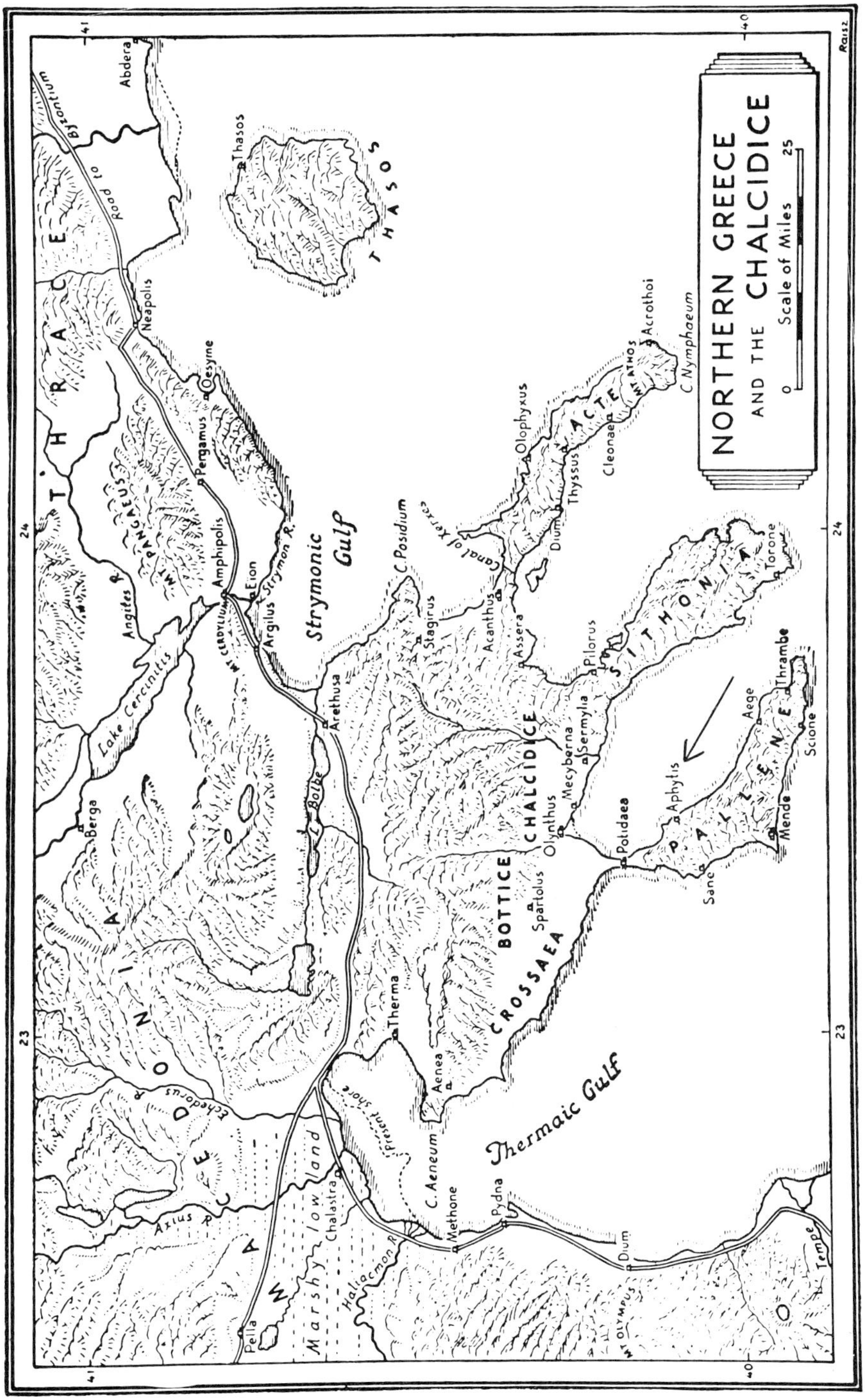


Figure 2. Map of northern Greece in ancient times. Aphytis is indicated by an arrow.

names that are combined with it. This agrees with the usage of most earlier authors (e.g., Timberlake, 1924; Compere, 1955; and others). We therefore cannot accept the change to the feminine gender adopted without any explanation by Nikol'skaya and Yasnosh (1968).

### TRANSLITERATION OF RUSSIAN NAMES

The transliteration of Russian author names into the Latin alphabet presented a minor problem, for which no easy solution could be found. Following the author's own example would have been, of course, advisable—provided that the Russian authors themselves were consistent in transliterating their own names. Unfortunately, such was not the case, as is illustrated by the following example. In one paper, Chumakova (1964) had her own name transliterated in no less than four different ways, *viz.*: Chumakova, Tchumacova, Tshumakova, and Tshumakowa.

We have therefore decided to adopt the following policy: Russian names are phonetically transliterated throughout the text and bibliography, but are retained as the authors originally wrote them in the names of species. Hence, Chumakova (1961) but *Aphytis testaceus* Tshumakova, Yasnosh (1972) but *Aphytis sugonjaevi* Jasnosh, etc. Likewise, the titles of Russian papers have been freely translated in the bibliography, even if this may have somewhat differed from a translation provided in the original paper.

## COLLECTING AND MOUNTING SPECIMENS

Rearing *Aphytis* specimens from their scale-insect hosts is definitely preferable to collecting them on vegetation or elsewhere. General sweeping, beating or suction techniques may, of course, at times yield identifiable specimens that may serve to indicate the presence of an *Aphytis* species in a given area. However, in the absence of host records such data are usually of little value compared to information obtained from material reared from accurately identified hosts.

Samples of armored scale insects are best collected with the infested plant parts, such as twigs, bark, leaves or fruit. These can then be kept in the laboratory for several weeks for parasite emergence. Almost any kind of container will do for this purpose, as long as it is adequately ventilated: cloth-covered glass or plastic jars, plastic boxes, even cloth or paper bags or envelopes. Regular brown paper bags, folded several times at the top and stapled, make convenient, inexpensive containers. The use of plastic bags should be avoided, as plant material tends to mold if kept in them for any prolonged period. Although some of the parasites present in such a sample will undoubtedly perish with their dying hosts, a good proportion will usually be able to complete their development and emerge. A blackened container, with a lighted compartment or a glass vial attached to it, may be useful for separating the emergent parasites from the sample, especially if live specimens are desired (adult *Aphytis* wasps are positively phototropic and will be attracted to the light). Otherwise, the emergent parasites may be left to die in the sample. Eventually, the entire contents of the container can be studied under a dissecting microscope, and the dead, dry wasps may be removed with the aid of a fine brush, forceps, or an entomological pin.

Mixed populations of scale insects are commonly encountered in nature. Special care should be exercised to avoid the inclusion of several host species in a given sample. Erroneous host records have often resulted from heterogenous samples of scale insects. Final examination of the host material under a binocular microscope will usually reveal whether or not more than one host species is involved and, if so, the true host can usually be determined by finding dead unemerged *Aphytis* adults or diagnostic pupae or cast skins in association with a particular host.

*Aphytis* specimens are best stored dry, in small glass vials or gelatin capsules, with some loose cotton to prevent their bouncing around and breaking. Storage in alcohol should be avoided, as it usually results in insufficient clearing and inadequate slide mounts.

Dried specimens of *Aphytis* shrivel, and important diagnostic characters cannot be seen. Mounting them on points is therefore not recommended. Cleared, properly prepared slide mounts are essential for meaningful study and identification of *Aphytis*.

**Preparation of Slide Mounts.** Dead, dried specimens make the best study mounts for general taxonomic purposes. Live specimens should be killed and then dried in an oven at 45°C prior to mounting. If material preserved in alcohol must be used, it should be similarly dried before mounting.

The main purpose is to clear the specimens to transparency and to restore the normal body turgidity before mounting. The following procedure (DeBach 1959, 1964) has consistently given the best results with *Aphytis*: Thoroughly dried specimens are soaked in glacial acetic acid (7 drops) mixed with lactophenol or chloral phenol (5 drops) in small watch glasses (U.S. Bureau of Plant Industry model). After one to three days the specimens should be satisfactorily cleared and the body shape restored to normal. The process may be speeded up by very gentle heating, but this is not necessary. The cleared specimens are then mounted directly in Hoyer's medium, which does not cause shrinkage or collapse of the specimens and usually results in additional clearing. (Hoyer's medium is prepared by slowly dissolving 12 g gum arabic in 20 cc warm distilled water, then adding 80 g chloral hydrate and 20 g glycerine, and filtering through filter paper with the aid of a suction pump.) After drying for about two weeks under room conditions, the slide cover is ringed with a suitable sealer in order to insure a permanent mount. Double ringing with Zut® is usually quite satisfactory. Occasionally the ringing material will eventually separate from the slide, permitting the Hoyer medium to dry out and the mount to deteriorate. However, this does not spoil the specimens. Such slides can be reconstituted by removing the ringing material with a dissecting needle, placing the slide in a humidity chamber until the Hoyer medium is again quite liquid and then removing the specimens on a point and remounting them on a new slide.

An alternative method, commonly employed for larger parasites, is to clear the specimens in a ten percent potassium hydroxide solution. If heat is used complete clearing can be accomplished in minutes. The specimens are then transferred through glacial acetic acid, an alcohol series and xylol, and are mounted in Canada balsam which is a much more permanent mounting medium than Hoyer's and does not require ringing. However, the KOH method will disfigure the wings and sometimes crumple the body and appendages, and, most important, it will eliminate melanization in *Aphytis*, a very important diagnostic character. It is also much more cumbersome than the glacial acetic acid–lactophenol method. If anything, its best usage may lie with dissected individual parts.

Phase-contrast microscopy is essential for the satisfactory detailed taxonomic study of *Aphytis*. With phase contrast, no staining is necessary. Specimens mounted in Hoyer's medium definitely provide a better resolution and reveal more details under phase contrast than do balsam mounts.

The balsam on old slides can be dissolved in xylol, and the specimens may then be cleared and remounted in Hoyer's medium. With some experience and when apropos, as with certain type slides, a drop of xylol may be run under the cover slip, and a specimen can be teased out without disturbing the rest of the specimens on the slide.

**Preparation of Developmental Stages.** The eggs, larvae and pupae of *Aphytis* are made readily available by simply turning over the covering scale of the parasitized host with a fine forceps or dissecting needle. Eggs and first-instar larvae can be mounted directly in Hoyer's medium, whereas larger larvae and pupae should be cleared by soaking them in a mixture of glacial acetic acid and lactophenol or chloral phenol, as described earlier for adult specimens. Third-instar larvae should first be punctured with an entomological micro-pin, to eliminate the opaque gut contents. Addition of small amounts of iodine and potassium iodide to the standard Hoyer medium tends to improve the resolution somewhat, besides making it easier to locate the transparent specimens on the slide (Rosen and Eliraz, 1978).

**Preparation for Scanning Electron Microscopy.** The recent advent of the scanning electron microscope has placed a powerful tool in the hands of morphologists, providing resolution and depth of focus unequaled by conventional light microscopy. Special techniques were developed by our colleague Mike Rose for preparing *Aphytis* specimens for study and micrography under the scanning electron microscope. Live adult specimens can be used, but these shrivel after 10–15 minutes of exposure in the vacuum chamber of the microscope. Dried, gold-coated specimens are best suited for this purpose, but specimens of *Aphytis* normally shrink and are badly distorted when dried. This has been avoided by adopting the following freeze-drying procedure: Live specimens are anesthetized with ether, then soaked in an alcohol series, from 70 percent to absolute ethanol. Drops of alcohol containing the specimens are subsequently released into a lyophilization flask immersed in liquid nitrogen that has already attained a temperature of $-141.7°C$. After several minutes at this temperature, the flask and specimens are transferred to a bath of acetone and dry ice, and lyophilized in a Virte freeze drier at 75 microns of mercury and $-47.8°C$ for 24 hours. The dried specimens are then mounted in various positions on aluminum disks with double-sticky tape, and coated with gold.

No special treatment is required for the developmental stages of *Aphytis*. Eggs, larvae and pupae can be studied and micrographed alive on the body of the scale insect host, after removal of the covering scale. When excessive movement of larvae is likely to distort the scanning micrograph, the specimens can be anesthetized with ether.

# DEVELOPMENTAL HISTORY

Quite detailed accounts of developmental history were presented by Imms (1916) and Griswold (1925) for *Aphytis mytilaspidis* and by DeBach and Landi (1961) for *A. lepidosaphes*. Descriptions of developmental stages were also presented by Quayle (1910) for *A. chrysomphali* (erroneously recorded as *A. diaspidis*), by Parker (1924) for *A. chilensis* (recorded as *A. longiclavae*, a synonym), by Taylor (1935) for *A. chrysomphali* (but apparently more than one species was included in that study), and by Azim (1963a, 1963c) for several species of *Aphytis* occurring in Japan.

All the known species of *Aphytis* develop as primary ectoparasites of armored scale insects. As far as we know, records of other insects, such as soft scales, mealybugs or asterolecaniids, as hosts of *Aphytis* are all erroneous, resulting from misidentification of rearing samples. Field-collected samples of various coccoids frequently include a few cryptic specimens of armored scale insects, from which species of *Aphytis* may issue. An inexperienced entomologist may list such species as parasites of the insect constituting the bulk of the sample, and the error may subsequently be perpetuated by cataloguers. All such host records should therefore be carefully verified.

Recently, Rosen and Eliraz (1978) presented a detailed account of the developmental history and the various developmental stages of *A. chilensis*, and pointed out some inaccuracies of previous treatments. The following descriptions are based mostly on their paper.

## THE EGG (Figures 3–12)

**The ovarian egg** of *Aphytis* is of the double-bodied type common in the Encyrtidae, Miscogasteridae and some Aphelinidae (Hagen, 1964), comprising the distal egg proper and a proximal bulb, connected by a narrower neck (see, for instance, Figure 924, p. 576).

The eggs are deposited on the dorsal or ventral aspect of the body of the scale insect host, underneath the covering scale (see p. 44 for a discussion of ovipositional behavior). They adhere to the surface of the host by means of a small adhesive pad (see Figure 3). During oviposition, the bulb collapses and its contents are forced into the egg proper, which is the first part to travel through the ovipositor. The resulting **deposited egg** is stalked (Figures 5, 6). It is whitish, with a smooth, semitransparent chorion that appears to be somewhat thicker at the distal end (i.e., opposite the stalk end). There appears to be a distinct border separating the egg proper from the stalk; however, when pressure is applied to a newly deposited egg, some of the contents of the egg proper may flow back into the collapsed bulb (Figure 8). Male and female eggs are apparently identical in *Aphytis* (see Eliraz and Rosen, 1978).

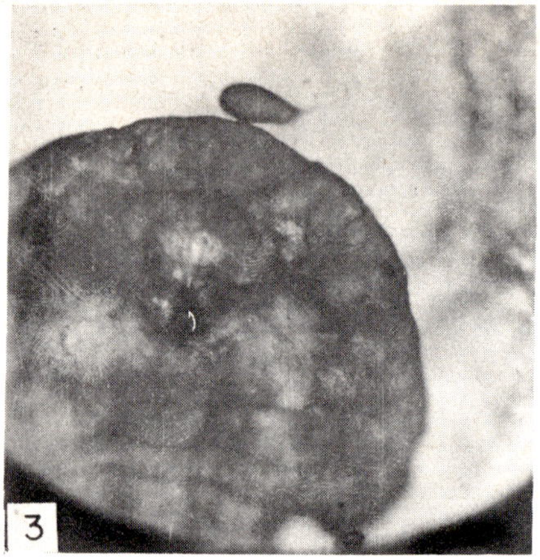

Figure 3. *Aphytis lingnanensis* Compere: egg affixed to California red scale by adhesive pad. (From Fisher, 1952.)

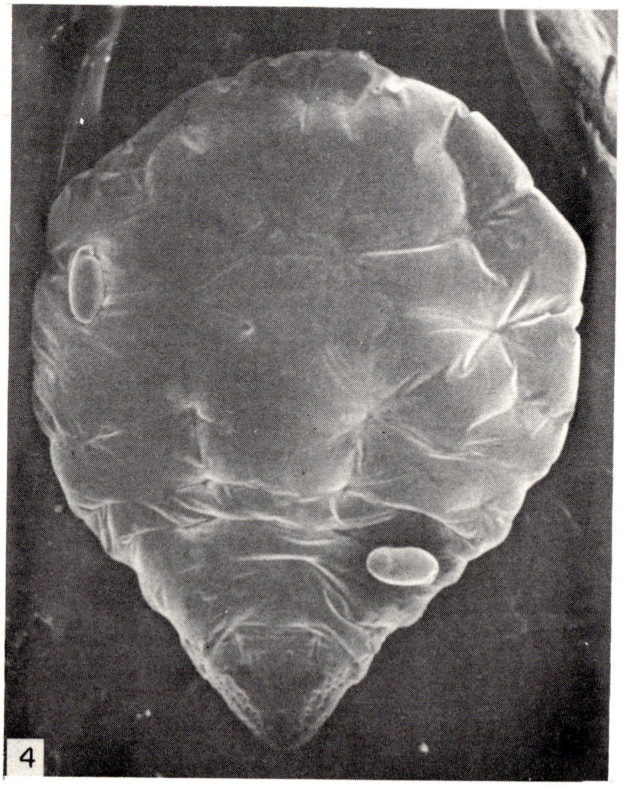

Figure 4. *Aphytis melinus* DeBach: two eggs on oleander scale. (From Rosen and DeBach, 1976.)

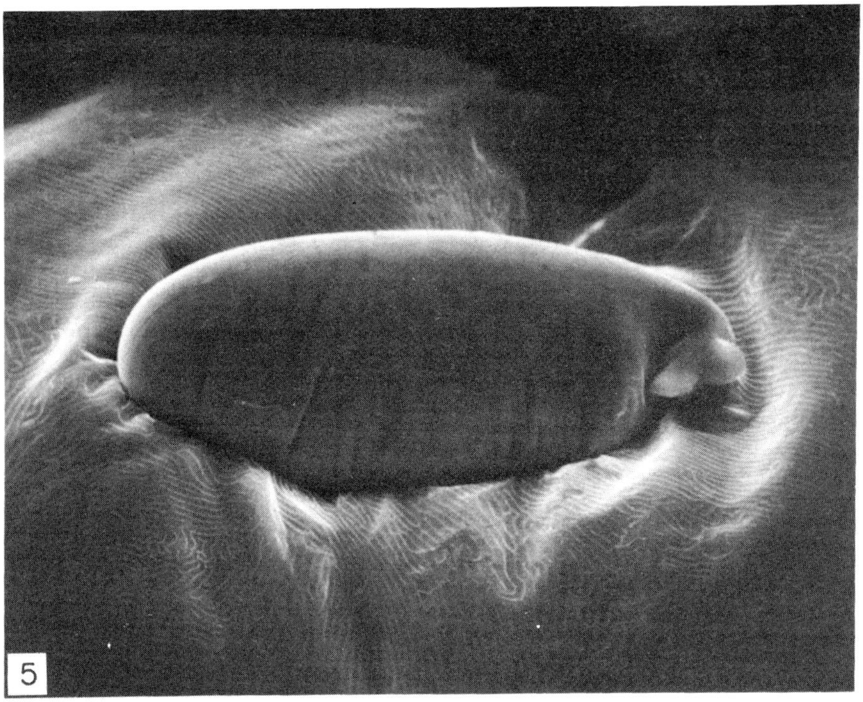

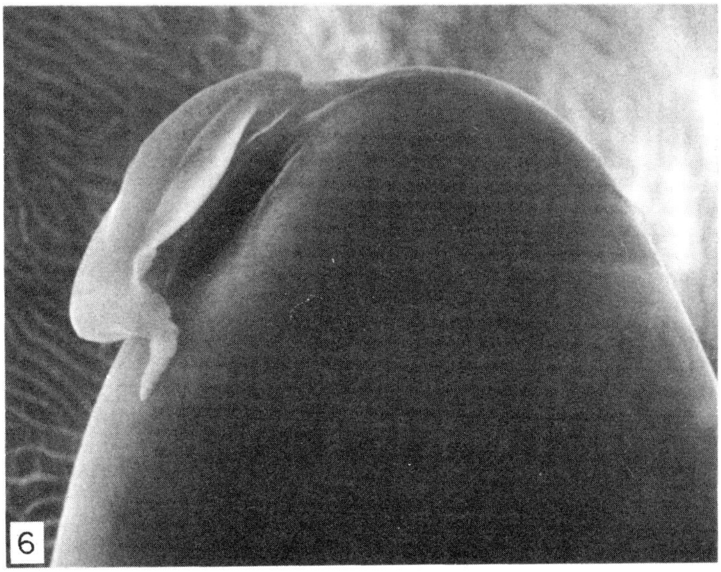

Figures 5, 6. *Aphytis melinus* DeBach
5. One of the eggs in Figure 4, enlarged. (From Rosen and DeBach, 1976.)
6. Proximal end of egg, showing stalk.

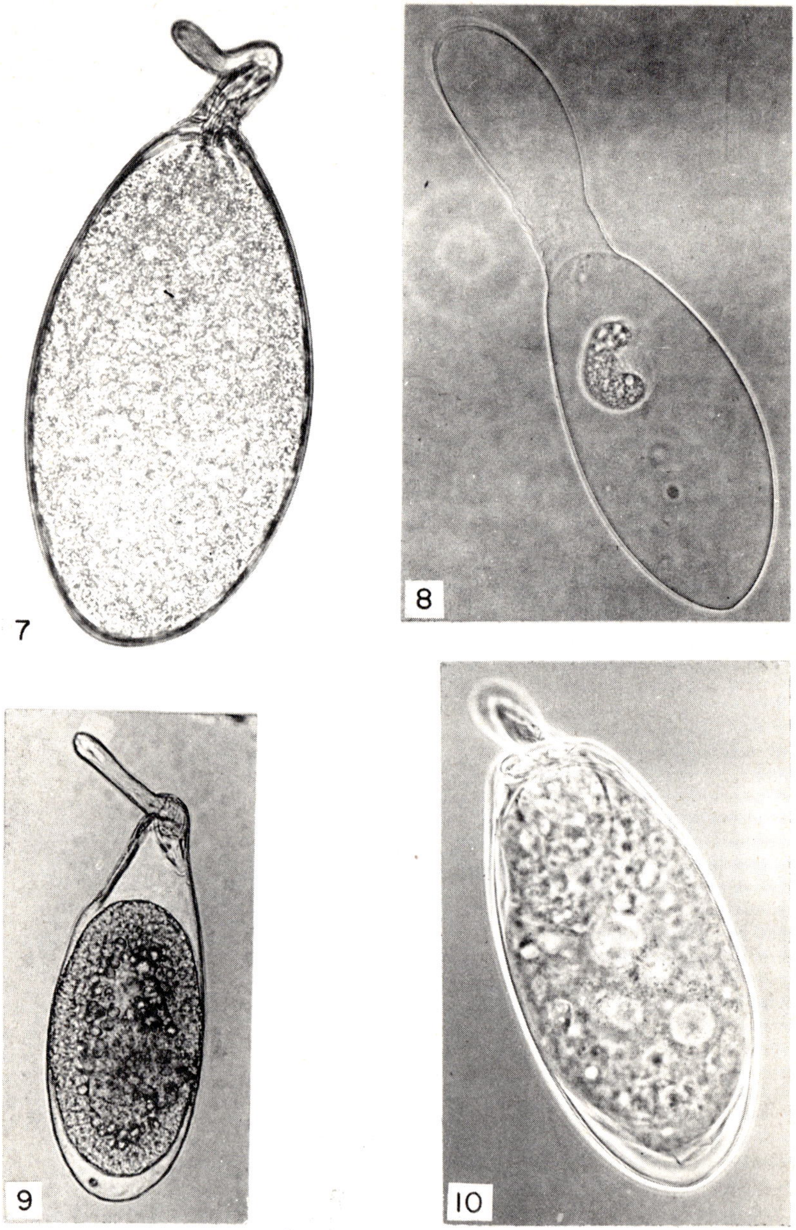

Figures 7–10. *Aphytis chilensis* Howard: eggs, showing embryonic development.
7. Newly deposited egg. 8. Early blastoderm formation (the pressure of the cover slip has forced some of the contents of the egg proper back into the bulb). 9. Double-layered embryo. 10. Fully-formed embryo; note head capsule facing stalk end of egg.
(From Rosen and Eliraz, 1978.)

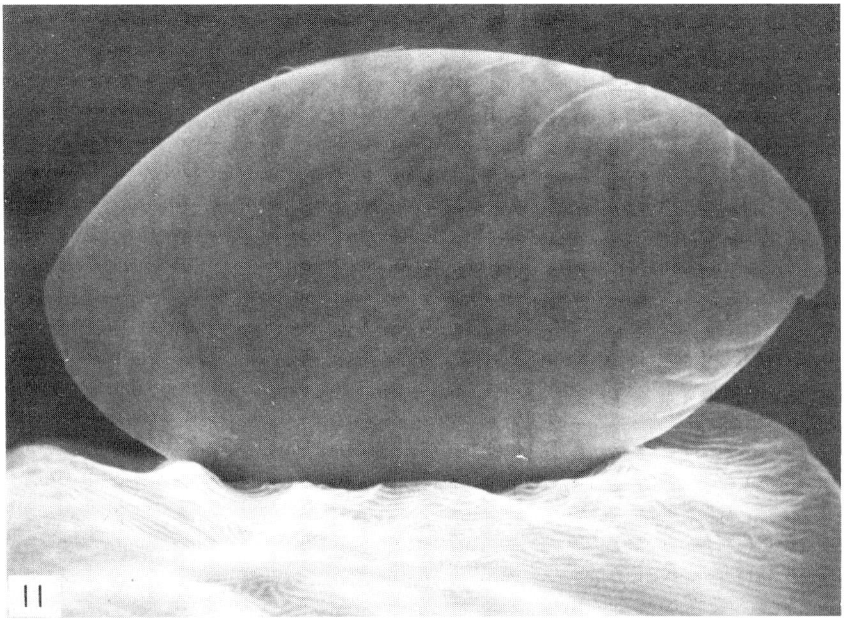

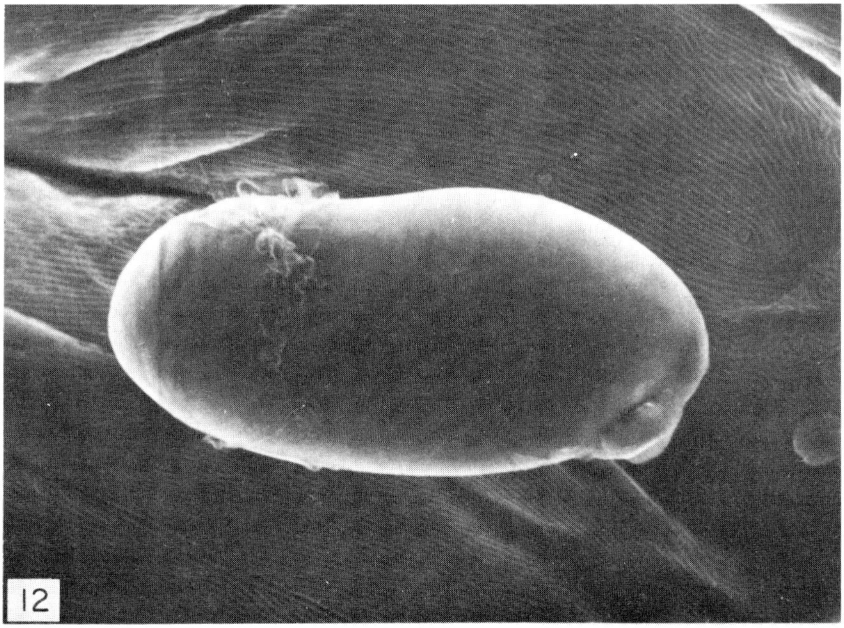

Figures 11, 12. *Aphytis melinus* DeBach
11. Egg on oleander scale, apparently ready to hatch.
12. Egg on oleander scale, hatching; note fragments of chorion.

Stalked eggs are common among the parasitic Hymenoptera. In the Aphelinidae, this egg type has been recorded in the genera *Aphytis*, *Marietta* and *Centrodora*, which are closely related, and also in *Aspidiotiphagus* (Hagen, 1964).

During early blastoderm formation, the developing embryo appears as a small group of cells at the proximal third of the egg (Figure 8). After some time the embryo nearly fills the lumen of the egg, but still encompasses a mass of undigested yolk (Figure 9). The head capsule of the fully formed embryo (Figure 10) is always directed toward the stalk end of the egg. During the incubation period the egg increases considerably in size (Figure 11). Prior to hatching, the embryo performs pendulumlike movements to the right and left of the longitudinal axis of the egg, accompanied by vigorous movements of the mandibles. The thin egg chorion becomes fragmented during hatching (Figure 12).

## THE LARVA (Figures 13–31)

All aphelinid genera apparently have three larval instars (Nikol'skaya and Yasnosh, 1966), and *Aphytis* is no exception. The larval exuvia are very thin (Figures 17, 18) and usually cannot be detected with conventional light microscopy prior to pupation, but the three instars are readily recognizable by the different size and shape of their mandibles (Figure 13). The first instar also differs from the others in the number of open spiracles. Male and female larvae cannot be separated by any known morphological criteria.

As the larva feeds and grows, the host becomes visibly depleted of body fluids and gradually shrinks (Figures 15, 19, 21), until only its crumpled, shriveled integument remains.

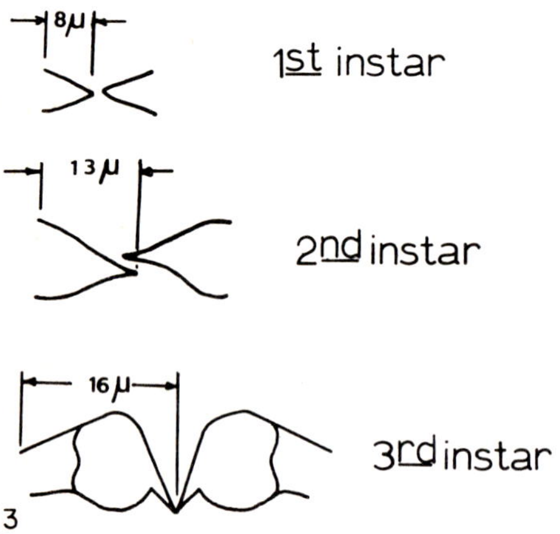


13

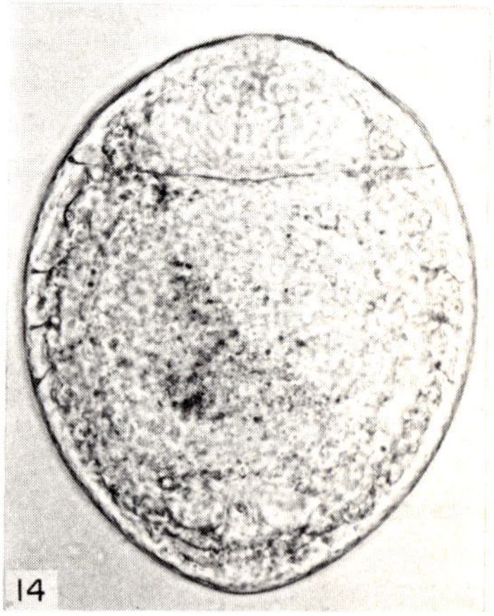

14

Figure 13. *Aphytis chilensis* Horward: mandibles of the 3 larval instars, semidiagrammatic. (From Rosen and Eliraz, 1978.)

Figure 14. *Aphytis melinus* DeBach: first-instar larva; note 4 pairs of spiracles.

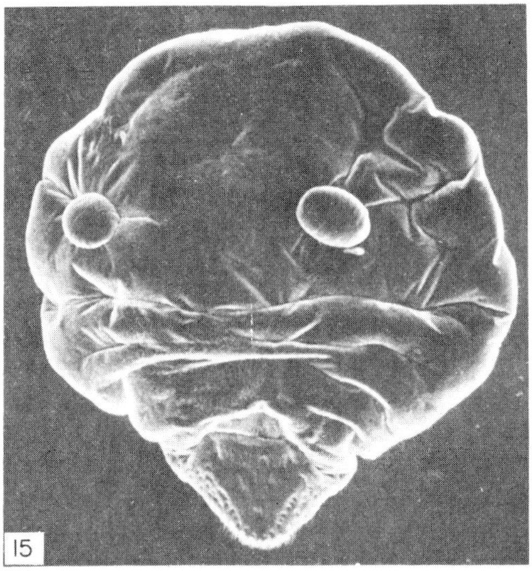

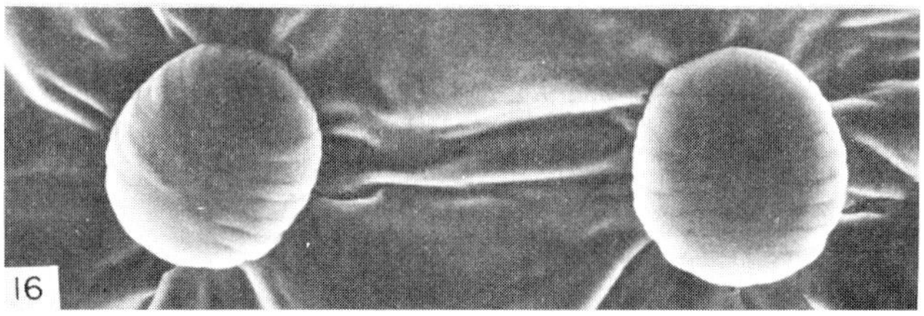

Figures 15, 16. *Aphytis melinus* DeBach: two young larvae on oleander scale. (Figure 15 from Rosen and DeBach, 1976.)

**First instar.** The newly hatched larva of *Aphytis* is ovoid in shape. Segmentation is not clearly visible, but in some specimens the head and 12 body segments are evident.

The first-instar larva of *Aphytis* possesses 4 pairs of spiracles: one pair in the mesothoracic segment and one in each of the first 3 abdominal segments (Figure 14). The cephalic skeleton is readily visible; the mandibles (Figure 13) are minute, triangular. Antennal disks, setae, or any other cuticular formations cannot be detected in this instar.

**Second instar.** The second instar differs markedly from the first in having 8 pairs of functioning spiracles: one pair in the mesothoracic segment and one pair in each of the first 7 abdominal segments. Segmentation is much more pronounced in this instar, with the head and 13 body segments clearly evident. The mandibles (Figure 13) are longer and more acutely pointed than in the first instar. Antennal disks, setae and cuticular tubercles are evident in the second-instar larva.

Except for the differences in body size and in the shape of the mandibles, the second instar is virtually identical to the third instar, which is described in greater detail below.

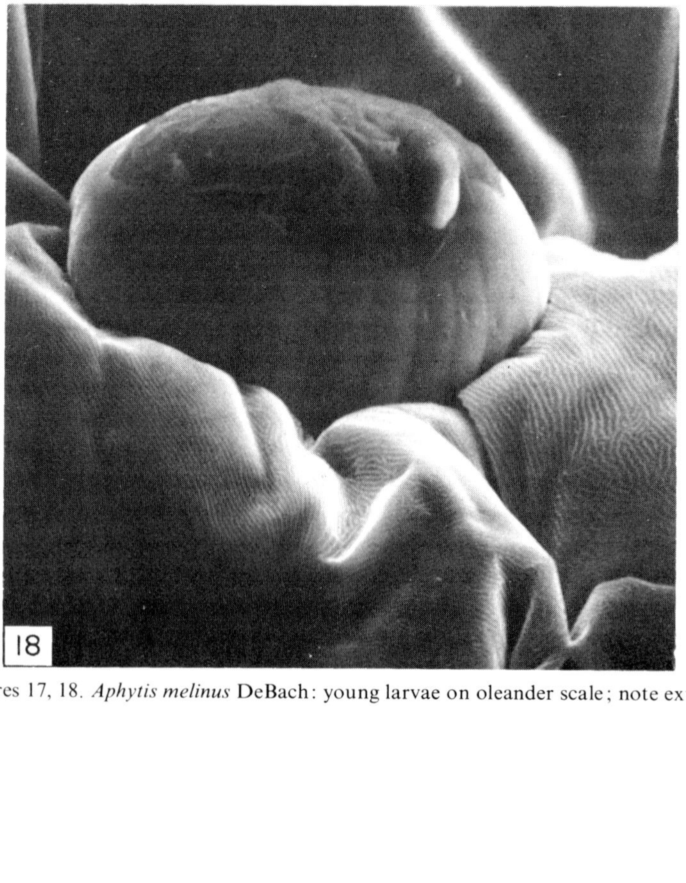


Figures 17, 18. *Aphytis melinus* DeBach: young larvae on oleander scale; note exuvia.

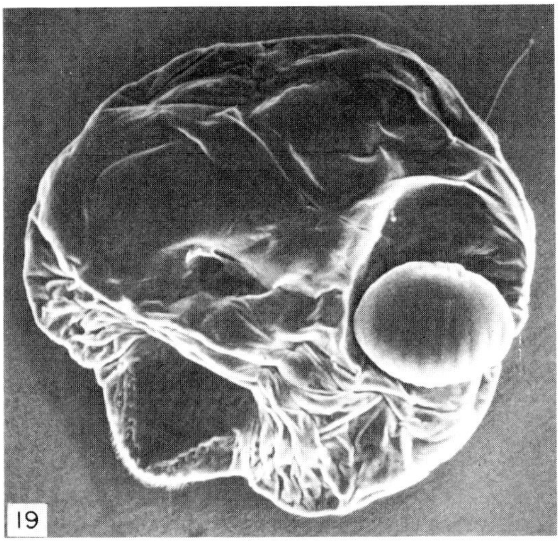
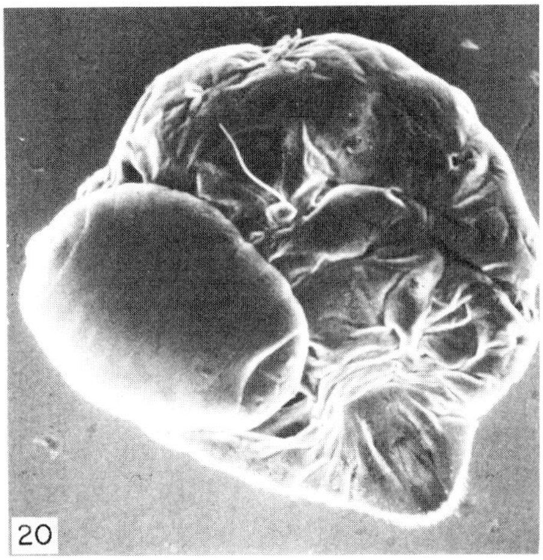
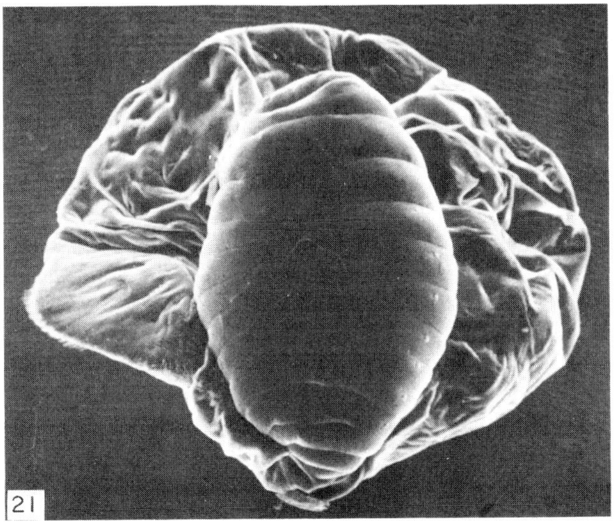

Figures 19–21. *Aphytis melinus* DeBach: larvae on oleander scale.
19. Young larva on dorsal aspect of host; note depleted area.
20. Older larva on ventral aspect of host.
21. Large larva on shrunken host.
(Figures 19, 20 from Rosen and DeBach, 1976.)

**Third instar.** The third, or final, instar is considerably larger than the second, elongate, rounded anteriorly and somewhat narrower posteriorly. Segmentation is quite pronounced, with the head and 13 body segments rather clearly demarcated. The opaque, yellow or brown midgut occupies about $\frac{3}{4}$ of the length of the body; its peristaltic movements are clearly visible in live specimens.

The *cephalic skeleton* (Figure 22) is rather simple, the epistomal, pleurostomal, hypostomal and tentorial sclerites forming a continuous ring. The mandibles (Figures 13, 22, 24) are larger than in the preceding instar, with a distinct, pointed denticle. The 1-segmented antennae are represented by minute disks (Figures 22, 23).

The *tracheal system* (Figure 25) consists of a pair of longitudinal trunks passing laterad of the midgut, connected by transverse commissures in the mesothorax and in the ninth abdominal segment, thus forming a complete ring. Ten short trunks extend dorso-laterally from the longitudinal trunk on each side of the body, 8 of them leading to open spiracles: one pair in the mesothorax and one pair in each of the first 7 abdominal segments (see also Figure 26). The lateral trunks in the metathorax and in the eighth abdominal segment appear to be more slender and are sometimes twisted into tiny loops; they are slightly thickened apically but do not end in functional spiracles. Two additional branches depart from the longitudinal trunk at the junction of each of the 10 lateral trunks: one directed ventro-laterally toward the integument, the other toward the midgut and branching profusely around it. The anterior transverse commissure sends out a pair of dichotomous branches toward the head, and another pair,

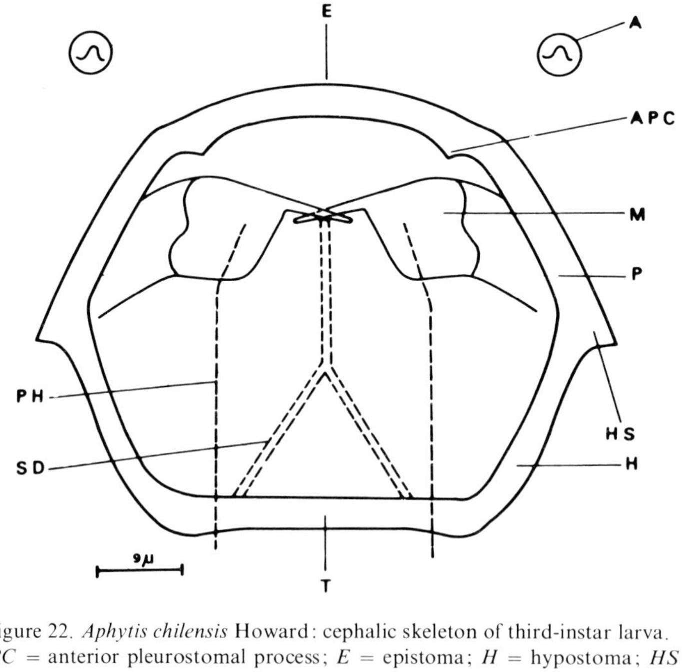


Figure 22. *Aphytis chilensis* Howard: cephalic skeleton of third-instar larva. *A* = antenna; *APC* = anterior pleurostomal process; *E* = epistoma; *H* = hypostoma; *HS* = hypostomal spur; *M* = mandible; *P* = pleurostoma; *PH* = pharynx; *SD* = salivary duct; *T* = tentorium. (From Rosen and Eliraz, 1978.)

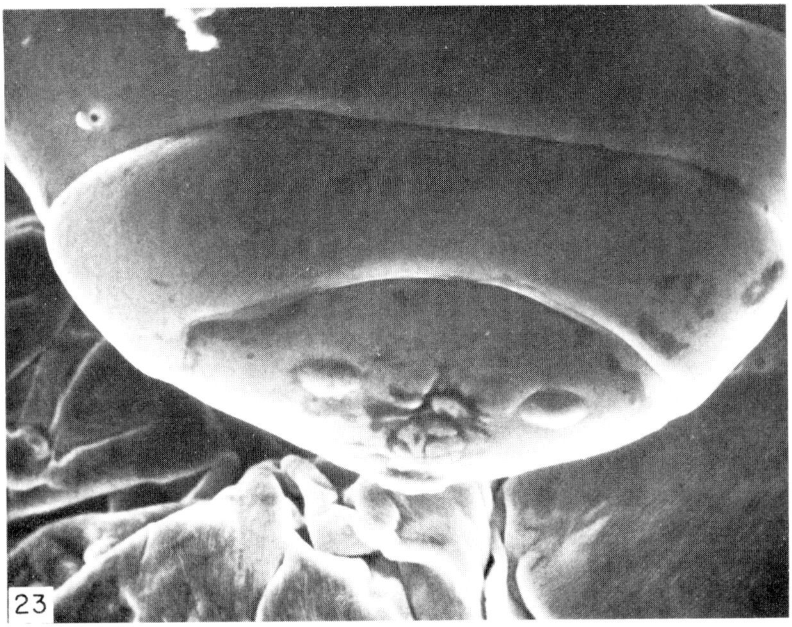


Figure 23. *Aphytis melinus* DeBach: head of larva in Figure 20; note antennal disks.

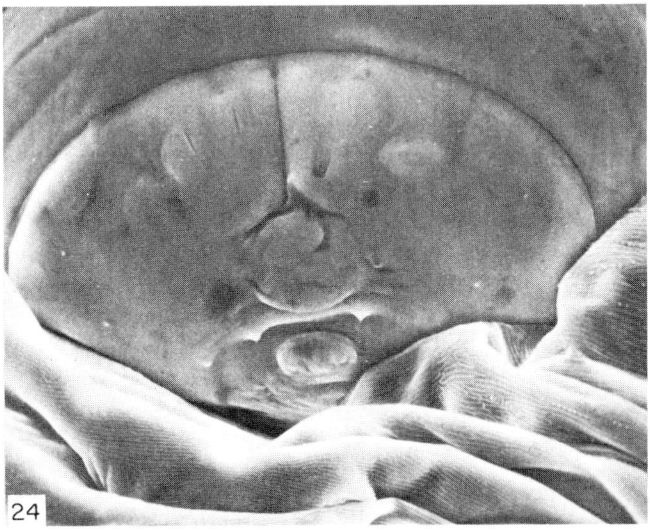


Figure 24. *Aphytis melinus* DeBach: head capsule of larva in frontal view, showing mandibles. (From Rosen and DeBach, 1976.)

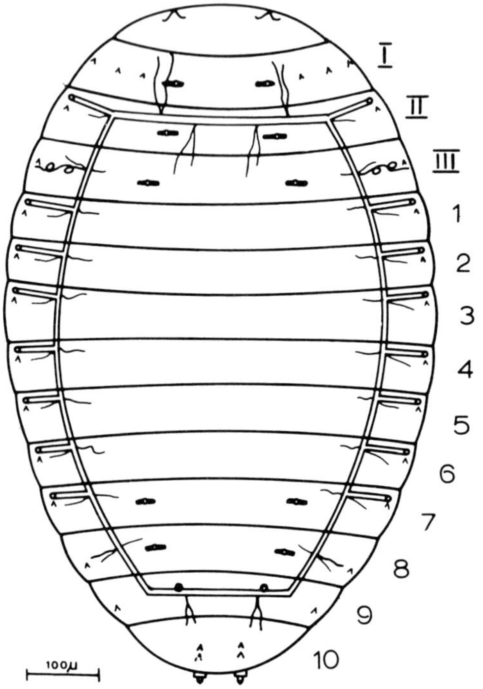


Figure 25. *Aphytis chilensis* Howard: third-instar larva, showing respiratory system and some integumentary formations, semidiagrammatic. (From Rosen and Eliraz, 1978.)

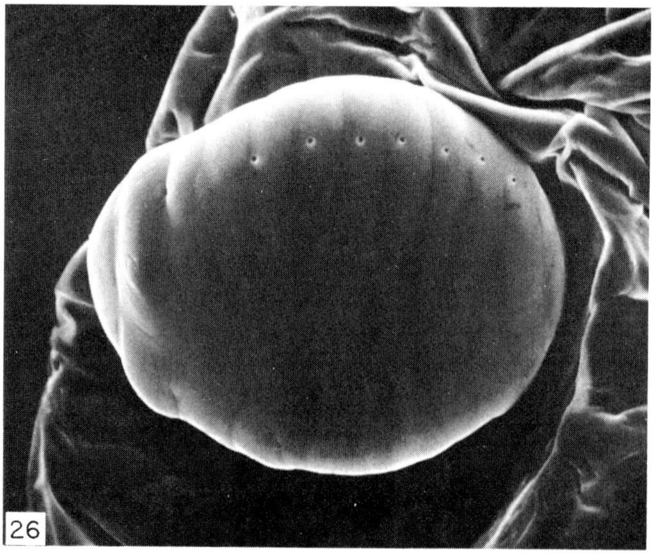


Figure 26. *Aphytis melinus* DeBach: larva on oleander scale, showing 8 spiracles.

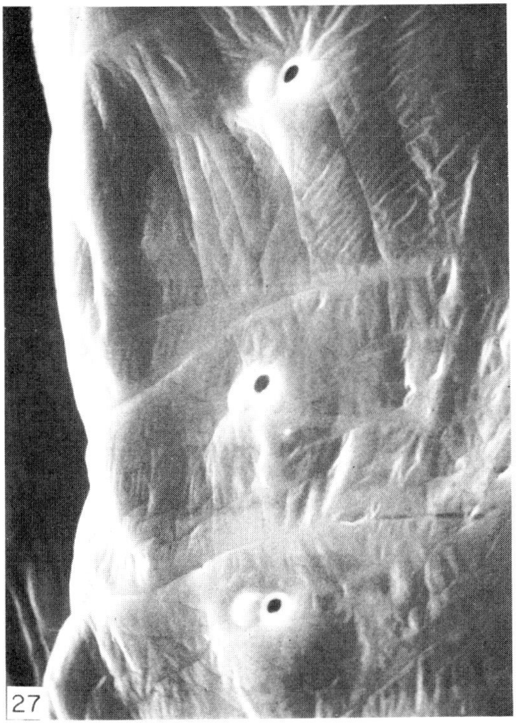

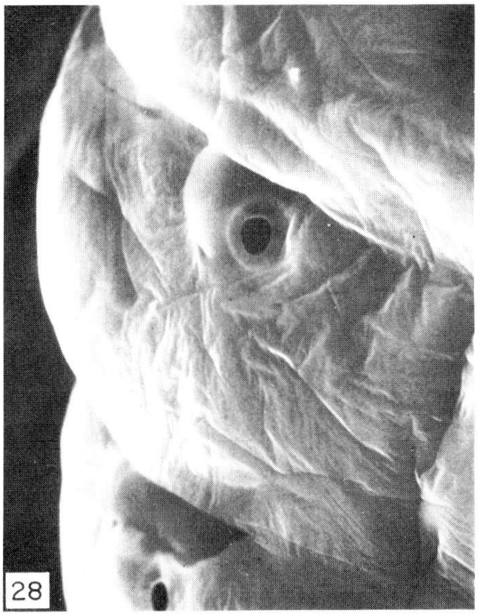

Figures 27, 28. *Aphytis melinus* DeBach: spiracles of third-instar larvae; note cuticular plate.

mesad of the latter, toward the midgut. The posterior transverse commissure sends a similar pair of dichotomous branches toward the caudal end of the body.

As pointed out by Rosen and Eliraz (1978), final-instar larvae of most aculeate Hymenoptera have 10 pairs of open spiracles, whereas those of most parasitic Hymenoptera usually have 9 pairs. In *Aphytis* only 8 pairs are present, the lateral tracheal trunks in the metathorax and in the eighth abdominal segment apparently representing vestigial spiracles that have been lost in evolution.

The *spiracles* (Figures 27, 28) are very simple, consisting of a spherical atrium followed by a series of chambers leading to the trachea. A closing mechanism is not evident. Adjacent to each spiracle is a minute sclerotized plate or tubercle.

The *integument* is smooth or delicately wrinkled. Minute cuticular tubercles are present on all segments: 5–10 pairs on the head, between the mouthparts and antennal disks (some of these may in fact be minute pores rather than tubercles; see Figures 29, 30, also p. 198), 3 in a dorsal transverse row on each side of the prothorax, and one pair of dorso-lateral tubercles on each of the two posterior thoracic segments and first 9 abdominal segments, adjacent to a spiracle if present. The tenth abdominal segment bears 2 pairs of submedian tubercles—one dorsal and one ventral—and a pair of minute, 3-segmented caudal sensoria on each side of the anus (Figure 25).

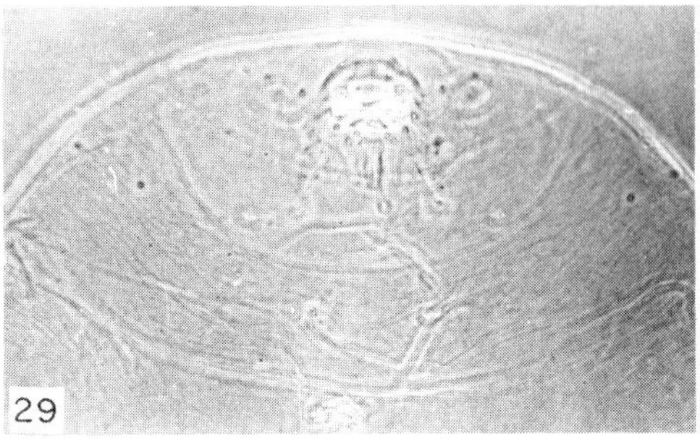


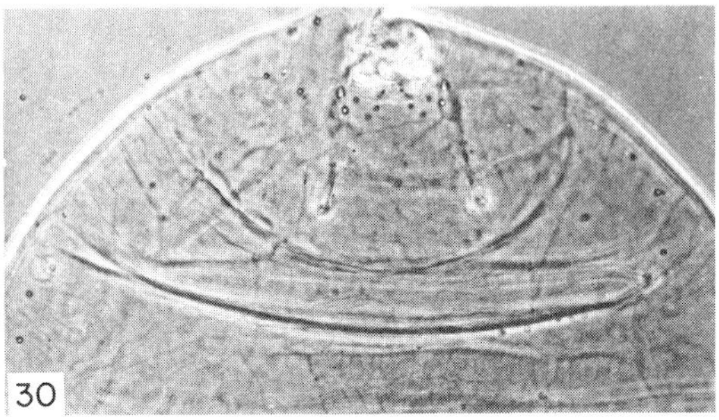


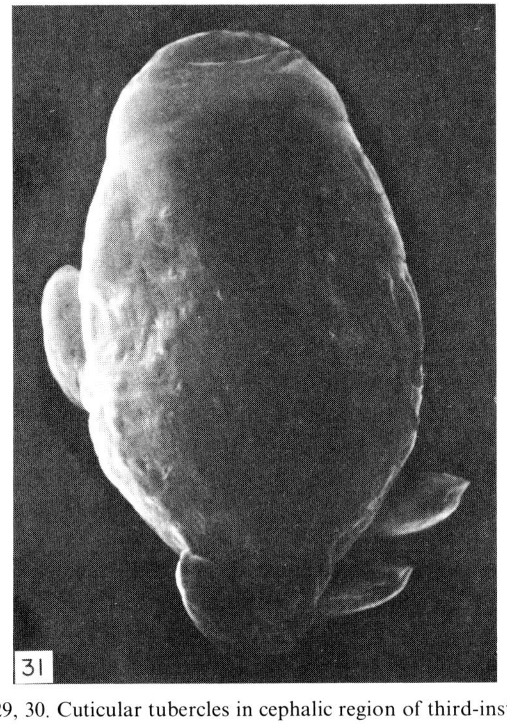


Figures 29, 30. Cuticular tubercles in cephalic region of third-instar larva.
29. *Aphytis hispanicus* (Mercet). 30. *Aphytis chrysomphali* (Mercet). (From Eliraz and Rosen, 1978.)
Figure 31. *Aphytis melinus* DeBach: prepupa and meconia, dorsal view.
(From Rosen and DeBach, 1976.)

A pair of transverse, "bacilliform" rod formations, each containing a "sporelike" discoid structure, are present dorsally on each of the 3 thoracic segments and on the seventh, eighth and ninth abdominal segments. Apparently sclerotized, their arrangement is fairly constant, whereas their form in any given species is quite variable, with the discoid structure positioned on each side of the center of the rod; on the ninth abdominal segment, often only the discoid structures are present (Figure 25). The function of these peculiar formations is at present unknown.

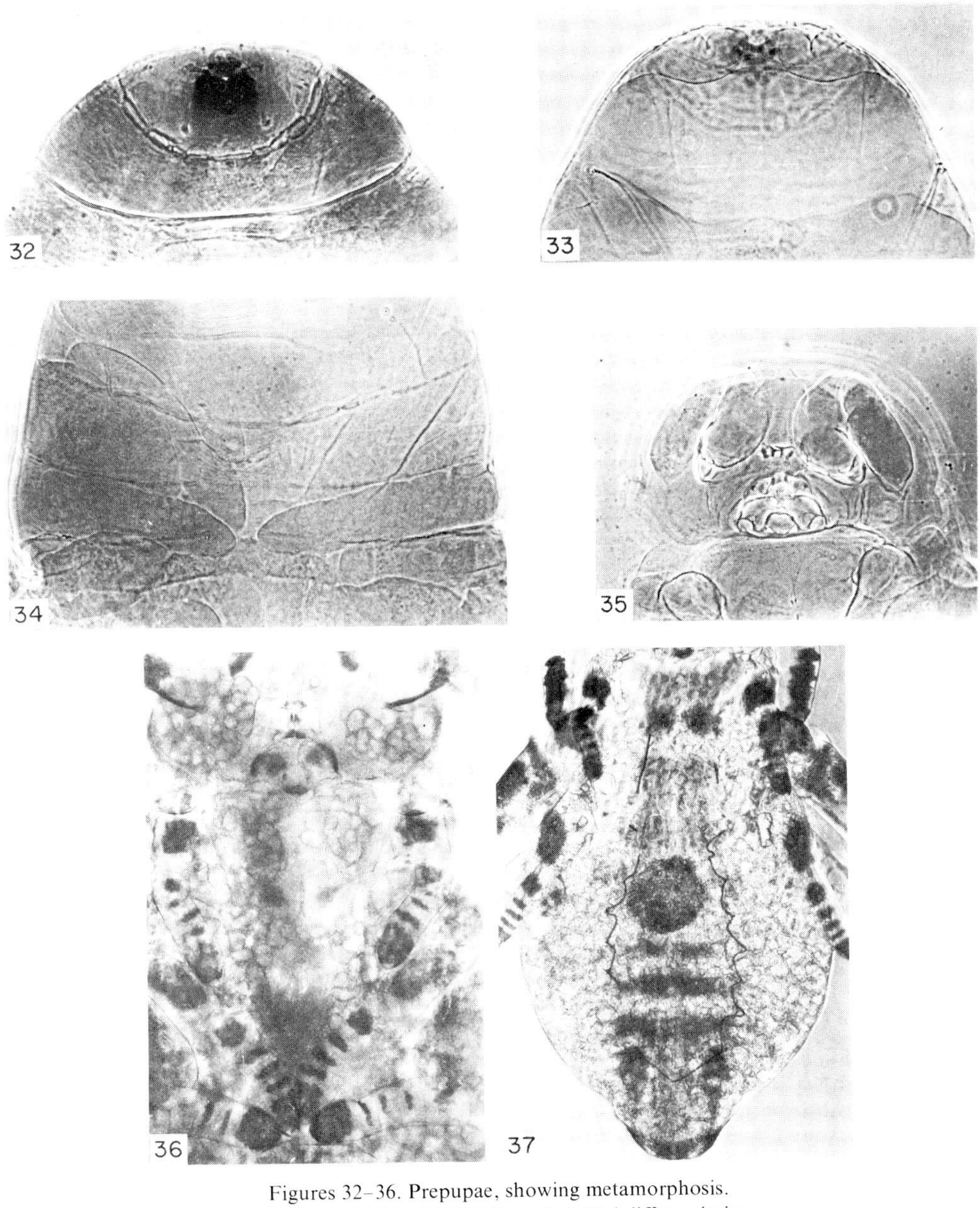


Figures 32–36. Prepupae, showing metamorphosis.
32. *Aphytis mytilaspidis* (Le Baron): initial differentiation.
33–35. *Aphytis melinus* DeBach: 33. Early formation of antennal cases. 34. Early formation of leg and wing cases. 35. Antennal and mouthpart cases; note larval mandibles in prepupal integument.
36. *Aphytis chilensis* Howard: late prepupa; the leg cases are still directed toward the midline. (From Rosen and Eliraz, 1978.)

Figure 37. *Aphytis chilensis* Howard: newly formed pupa; the appendages have migrated to the sides of the body. (From Rosen and Eliraz, 1978.)

Toward the end of the larval period, the third-instar larva enters a short **prepupal stage**. Inasmuch as in the parasitic Hymenoptera this stage is not marked by apolysis and ecdysis, it cannot be considered a distinct instar. All feeding ceases at this point, and the hind gut becomes linked with the midgut. The larva turns over, with its ventral aspect facing the covering scale of the host, and excretes fecal material in the form of several distinctive, brown or black meconial pellets (see Figure 31, also p. 200). It then enters a resting period, during which rapid metamorphosis takes place.

The prepupa is usually milky white; it is more elongate, and its caudal end is more distinctly pointed, than in earlier larval stages (Figure 31). Larval structures, including the mandibles, the cephalic skeleton, the tracheal system and all cuticular formations, are still present throughout this stage (see, for instance, Figure 35). The developing pupal organs are readily visible in cleared, slide-mounted specimens.

Differentiation of pupal structures first becomes evident with the appearance of a transverse row of rectangular cells along the intersegmental line separating head from prothorax (Figure 32). This is followed by development of the pupal antennal cases (Figure 33). Next, the pupal mouthpart, leg and wing cases develop simultaneously. The leg and wing cases are directed toward the midline of the prepupa (Figures 34, 36). The constriction between the pupal head and thorax is now apparent (Figure 35); the abdomen appears to be the last part of the body to become differentiated.

During the process of pupation (i.e., the ecdysis marking the onset of the pupal stage), the leg and wing cases migrate to the sides of the body (Figure 37). The crumpled larval exuvium often adheres to the tip of the abdomen of the pupa.

## THE PUPA (Figures 38–50)

As in other Hymenoptera, the pupa of *Aphytis* is exarate. It is distinctly flattened dorso-ventrally, considerably wider than thick. It invariably lies on its dorsum, with its ventral aspect (and mouthparts) facing the covering scale of the host.

On the ventral aspect of the pupa (Figures 39, 40), the geniculate antennal cases (*Ant*) are present on each side of the head. Between their bases are the mouthpart cases (Figures 39–41): the labrum (*Lr*) is small, transverse; immediately below it are the quadrate mandibles (*Md*); the pupal maxillae (*Mx*) are 2-segmented, whereas the labium (*Lb*) is composed of 3 parts. The prosternum (*Ps*) and mesosternum (*Ms*) are large, bilobed plates, whereas the metasternum (*Mt*) is represented by 2 triangular plates. Six sternal plates are evident in the abdomen. The cases of the forelegs (*Fl*) are situated mesad and caudad of the antennal cases; they are followed by the cases of the middle (*Ml*) and hind legs (*Hl*). A triangular process on the middle leg case marks the site of the developing mid-tibial spur. The forewing cases (*Fw*) are situated laterad of the middle legs, with the tips of the hind wing cases (*Hw*) protruding from underneath their caudal edge. The integument of the pupa is generally smooth, except for sublateral patches of minute spines on the abdominal sternites (Figures 40, 42, 43).

Whereas the sexes are not distinguishable in any of the preceding developmental stages, they are readily recognizable in the pupal stage. In the female pupa of *Aphytis*, a pair of minute subrectangular plates is present ventrally near the tip of the abdomen

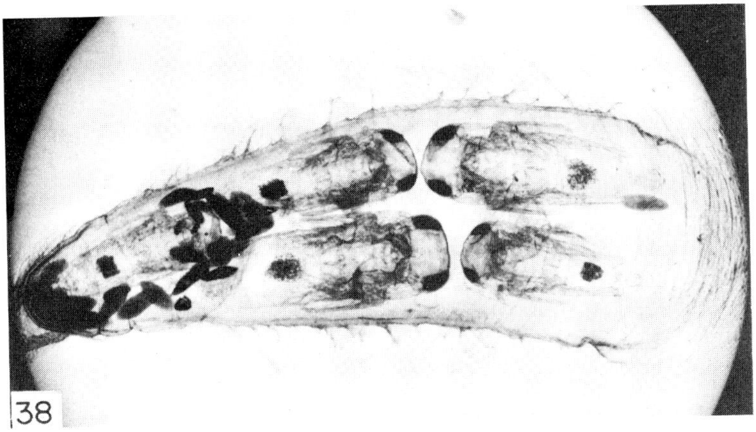

38

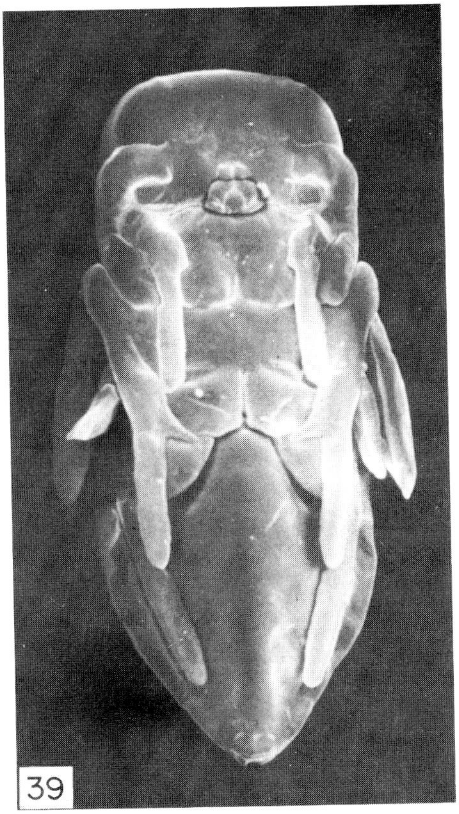

39

Figure 38. *Aphytis lepidosaphes* Compere: pupae under cover of purple scale. (From DeBach and Landi, 1961.)

Figure 39. *Aphytis melinus* DeBach: female pupa, ventral view. (From Rosen and DeBach, 1976.)

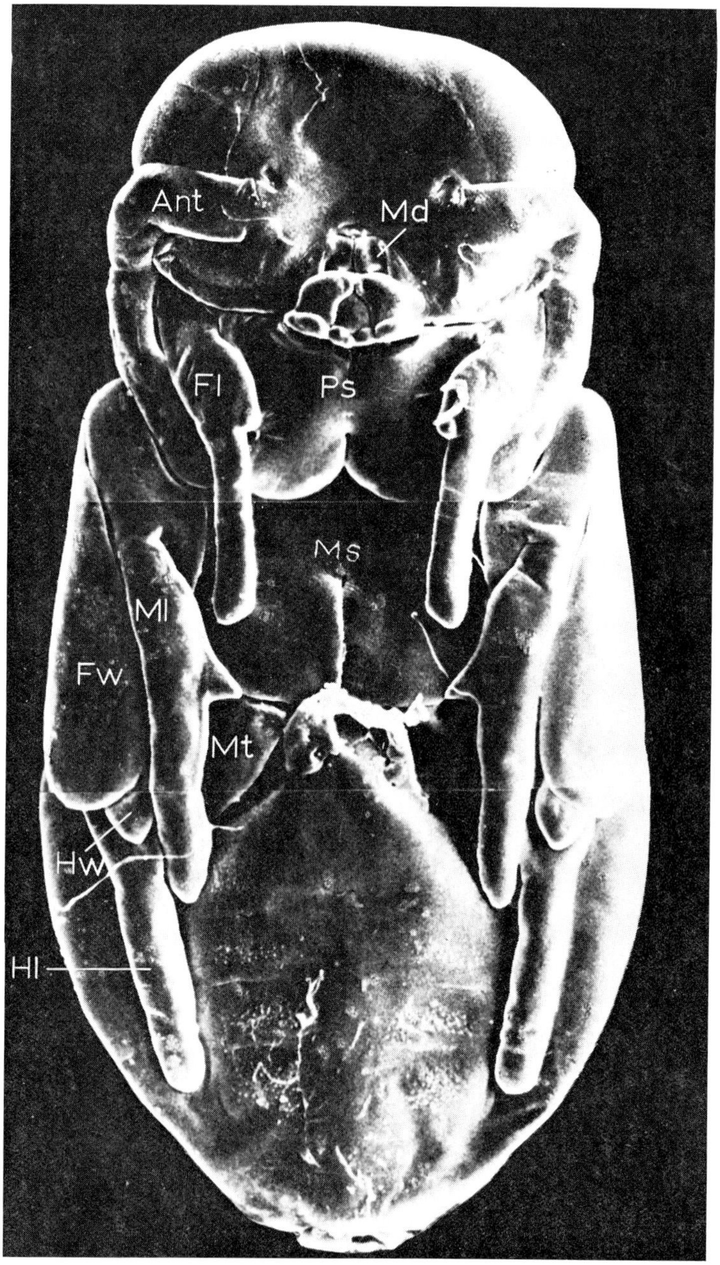


Figure 40. *Aphytis chilensis* Howard: female pupa, ventral view (a composite scanning electron micrograph). *Ant* = antennal case; *Fl* = foreleg case; *Fw* = forewing case; *Hl* = hind leg case; *Hw* = hind wing case; *Md* = mandible; *Ml* = middle leg case; *Ms* = mesosternum; *Mt* = metasternum; *Ps* = prosternum. (From Rosen and Eliraz, 1978.)

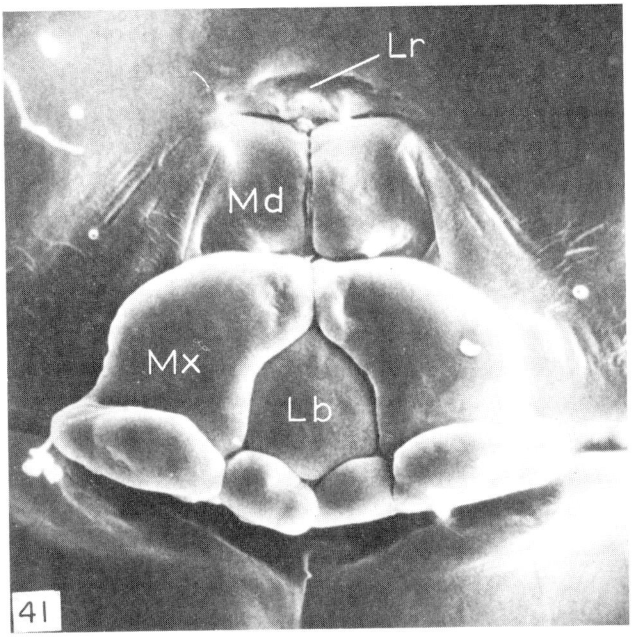


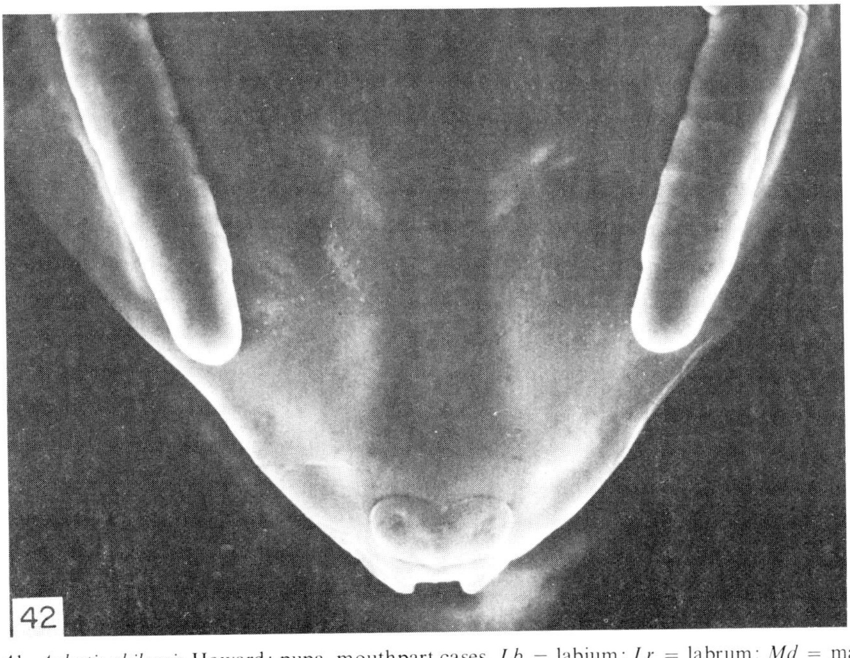


Figure 41. *Aphytis chilensis* Howard: pupa, mouthpart cases. *Lb* = labium; *Lr* = labrum; *Md* = mandible; *Mx* = maxilla. (From Rosen and Eliraz, 1978.)

Figure 42. *Aphytis melinus* DeBach: female pupa, ventral view, showing hind leg cases, stippling, and plates near tip of abdomen.

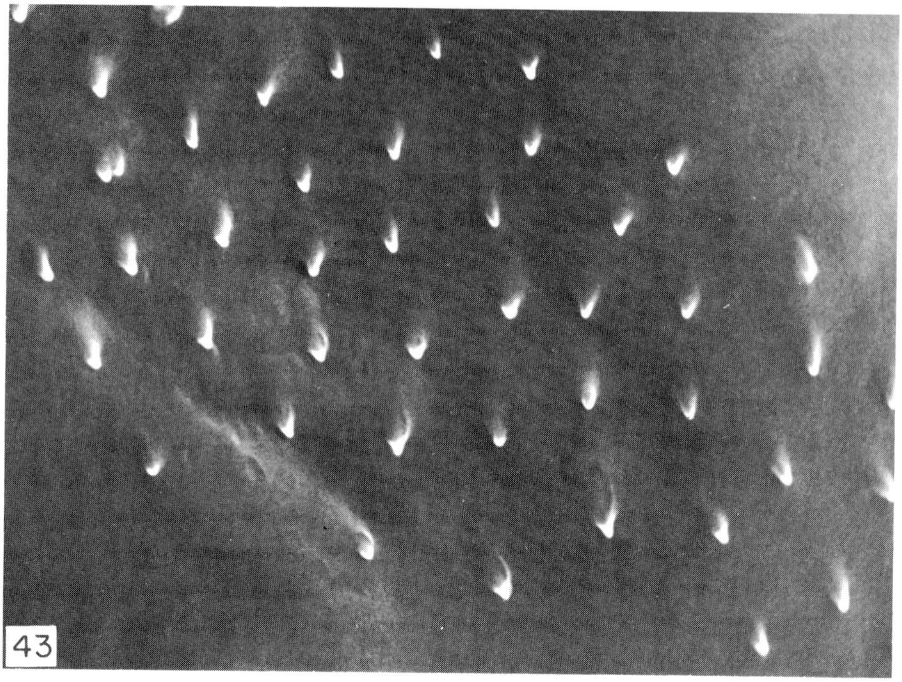

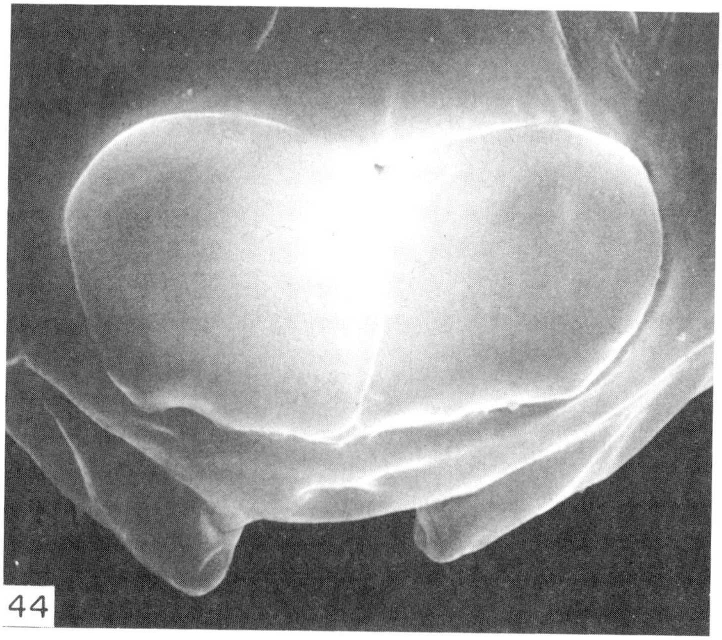

Figures 43, 44. *Aphytis melinus* DeBach: female pupa, ventral view.
43. Patch of spines on abdominal sternite. 44. Plates near tip of abdomen.

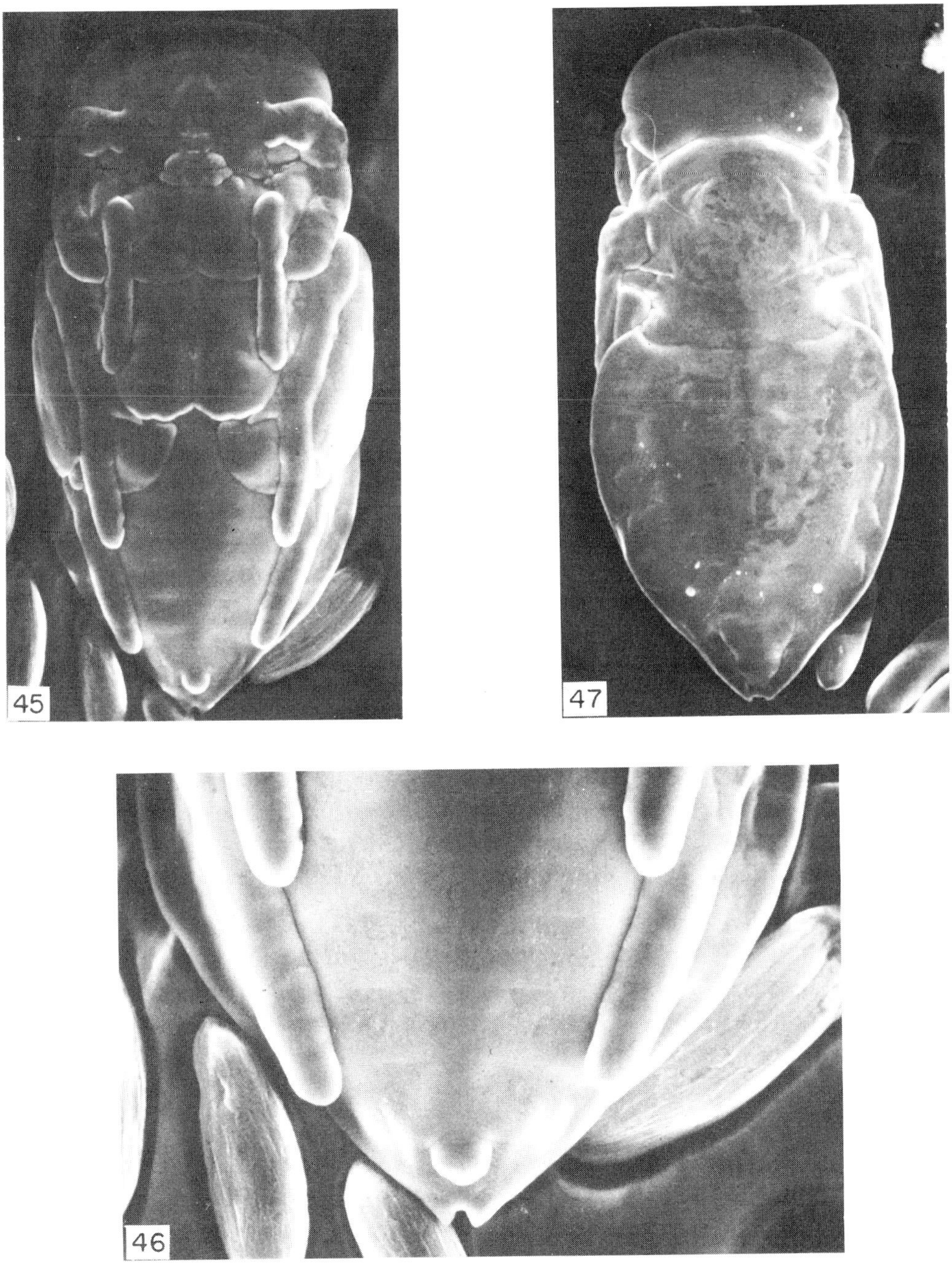


Figures 45–47. *Aphytis melinus* DeBach
45, 46. Male pupa, ventral view; note subapical plate and notch at tip of abdomen.
47. Female pupa, dorsal view. (From Rosen and DeBach, 1976.)

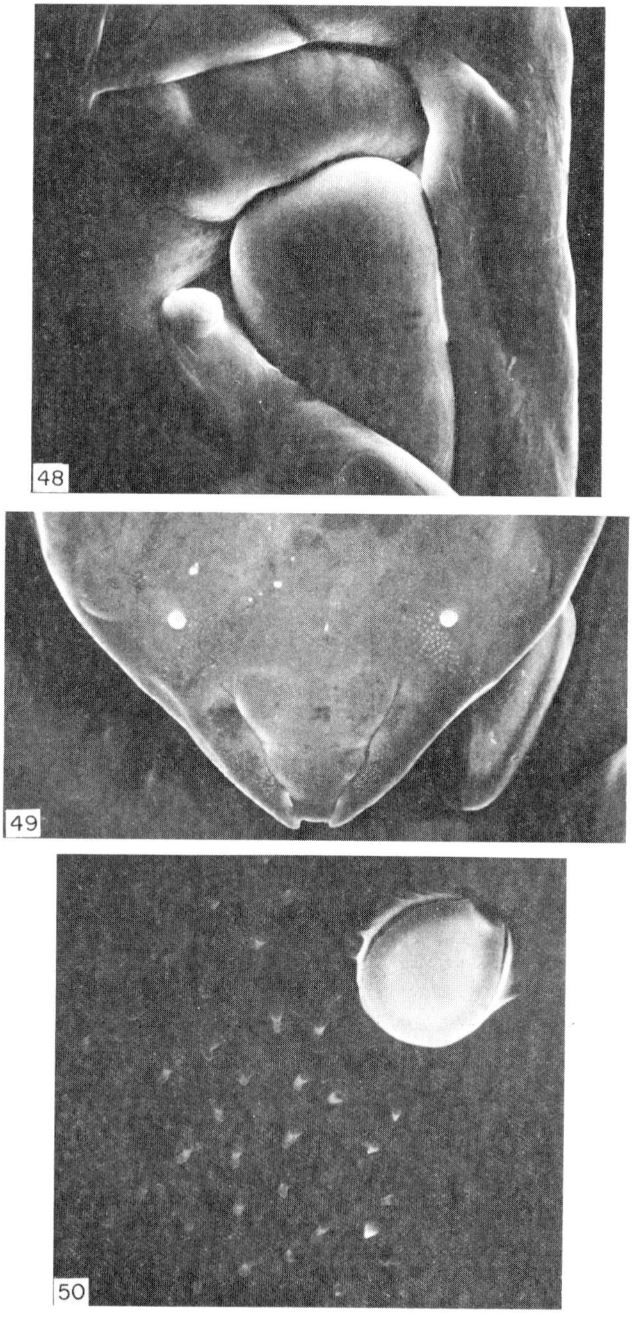


Figures 48–50. *Aphytis melinus* DeBach: female pupa, dorsal view. 48. Propodeal spiracle. 49. Tip of abdomen, showing spiracles. 50. Spiracle and patch of spines on eighth abdominal tergite.

(Figures 39, 42, 44). In the male pupa (Figures 45, 46), a single subapical plate is present ventrally, and the tip of the abdomen is distinctly notched.

The dorsal aspect of the pupa (Figure 47) reveals relatively few morphological characters. The antennal and wing cases are visible also from this side. The dorsum of the thorax appears to be undivided, whereas the abdominal region consists of 10 poorly defined tergites, each marked by a pair of sublateral patches of minute spines. The ninth and tenth abdominal tergites form a semitriangular median lobe, flanked on the sides by a pair of elongate lateral lobes (Figures 47, 49). Three pairs of spiracles are present: one concealed on the pronotum, one on the propodeum (Figure 48), and one on the eighth abdominal tergite (Figures 49, 50). The opening of the posterior spiracle is covered by a protective cuticular operculum (Figure 50).

The eyes of the developing pupa gradually turn from colorless to red to green, thus providing an indication of the age of the pupa.

## EMERGENCE OF THE ADULT

Eclosion occurs underneath the covering scale of the host. The pupal skin is broken into 3 parts. The cephalic exuvium (Figure 51), including the antennal cases, and the thoracic exuvium (Figure 52), including the leg and wing cases, are rather strongly sclerotized and are usually recognizable after emergence, whereas the abdominal exuvium is considerably thinner and much more fragile, and is often torn and fragmented.

Emergence from the host is usually effected by chewing an exit hole through the covering scale, although in some cases the emerging adult wasps may merely push

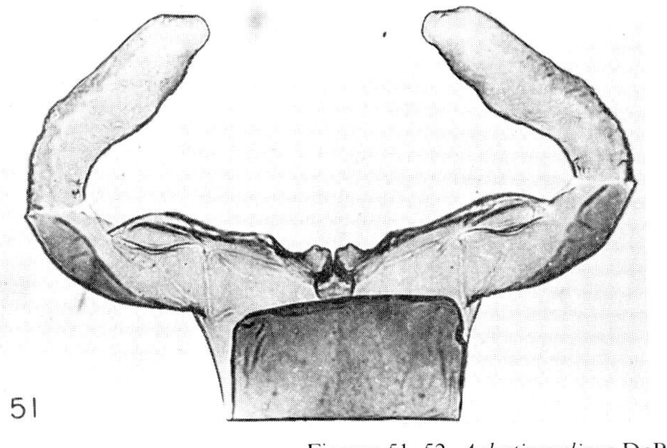


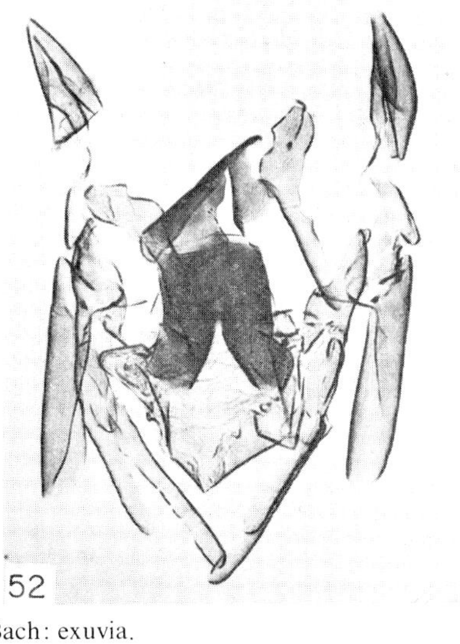


Figures 51, 52. *Aphytis melinus* DeBach: exuvia.
51. Cephalic exuvium. 52. Thoracic exuvium.

their way out under the edge of the scale. Like the pupa, the emerging wasp lies on its dorsum, with its ventral aspect—and mouthparts—facing the covering scale. It starts by gnawing a small opening in the scale, then extends one mandible through it and proceeds to chew, in scissorlike fashion, an oval hole. From time to time the wasp stops gnawing and retracts its head, presumably to discard pieces of cut material. As soon as the exit hole is large enough, the wasp rotates its head through it, one eye at a time. It then slowly extricates itself, until the forelegs become free and help pull the rest of the body out. At room temperature, the entire process may take about 30 minutes. If more than one wasp have developed on the same host, they all invariably emerge through one exit hole.

Long after emergence, the distinctive meconia and characteristic exuvia, as well as the oval exit hole when present, are the unmistakable tell-tale signs attesting to the fact that a dead armored scale insect had earlier been parasitized by *Aphytis*.

## DURATION OF DEVELOPMENT

The developmental period of *Aphytis* is usually short, unless interrupted by diapause. At 28°C, for instance, *A. chilensis* completes its entire development from oviposition to emergence in 18.6 days (Rosen and Eliraz, 1978), *A. hispanicus* in 16 days (Gerson, 1968), and *A. coheni* in only 12 days (Avidov, Balshin and Gerson, 1970). As can be seen from Figure 53, egg incubation and the prepupal period are relatively short, whereas pupal development takes about one third of the total.

Abdelrahman (1974c) noted that when *A. melinus* eggs were laid on the same day in different California red scale hosts on the same lemon, their total developmental period varied considerably. By contrast, parasites developing on the same scale always emerged within a few minutes of each other. Pupation, pupal development as indicated

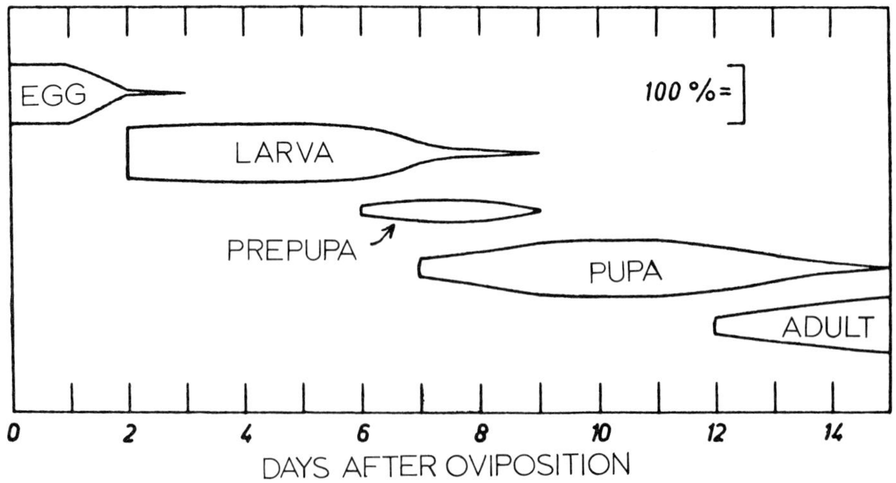


Figure 53. *Aphytis holoxanthus* DeBach: duration of development at 27°C.

by eye color, and adult emergence were synchronized in parasites on the same host, irrespective of their sex or size. Synchronization was observed also in *A. chrysomphali*.

Most species of *Aphytis* are multivoltine, develop continuously throughout the year, and have 2 or 3 generations to one of their host. However, winter diapause is known to occur in the larval stage, especially in species developing on hosts that overwinter as eggs, such as *A. mytilaspidis* on the oystershell scale, *Lepidosaphes ulmi* (L.). In such cases, the parasite may be bivoltine or even possibly univoltine (see Le Baron, 1870; Imms, 1916; DeBach, 1964; Yasnosh, 1972). Since the same species may develop uninterruptedly on multivoltine hosts, it may be assumed that diapause in *Aphytis* is induced by the physiological state of the host on which it develops. Our field observations of *A. mytilaspidis* on the oystershell scale in Greece indicate that the breaking of diapause is not due to increasing photoperiod but to rising temperatures following a certain summation of low winter temperatures.

## REPRODUCTIVE BIOLOGY

Like the majority of other Hymenoptera, most known species of *Aphytis* are biparental and reproduce sexually. However, a relatively high proportion—about one fourth of all species whose sexuality is known—are uniparental.

The biparental species exhibit arrhenotokous reproduction, unfertilized eggs developing into males and fertilized eggs into females. Unmated females produce male progeny only, whereas mated females produce both male and female progeny. Males are common in such species, although the sex ratio is rarely 1:1, females usually constituting 60–70% of the populations.

The uniparental species of *Aphytis* exhibit thelytokous parthenogenesis, females producing female progeny without fertilization by males. However, males are produced regularly in these species, usually at a low rate of 1–5%. Until recently, such males were usually considered to be nonfunctional, and their occurrence in uniparental populations was interpreted as a genetic mistake, of no significant consequence to the survival of the species. White (1964), for instance, made the following statement regarding thelytoky: "In this genetic system fertilization is entirely suppressed. Males are absent from the population or are only produced as occasional, nonfunctional anomalies, frequently intersexual." DeBach (1969b) concurred and added: "I know of no hard data to dispute this statement."

Recent studies by Rössler and DeBach (1972a, 1973) have, however, demonstrated that males of thelytokous species are not always nonfunctional. A thelytokous "form" of *A. mytilaspidis* was found to be apomictic, i.e., to produce eggs that undergo meiosis. Restoration of diploidy appears to be achieved by terminal fusion, i.e., by the fusion of two sister nuclei produced by the second meiotic division. The males have viable haploid sperm in the testes, and whenever they are present, some inseminated females are found in the population. Furthermore, the thelytokous females can be inseminated and fertilized by the males of a closely related arrhenotokous "form", although some prezygotic barriers exist in the recognition and copulation phases. Likewise, the uniparental males are capable of inseminating and fertilizing the arrhenotokous females. By utilizing genetic markers, Rössler and DeBach (1972a, 1973) have shown that when thelytokous females are inseminated, they produce more than 50% diploid female progeny that originate from fertilized eggs. In other words, when a sperm enters a "thelytokous" egg, it apparently competes successfully with the second polar body for conjugation with the egg nucleus. Thus, at least in some uniparental populations of *Aphytis*, sexual processes appear to play a much more significant role in enhancing genetic variability than has been hitherto realized.

Environmental conditions may affect the rate at which males occur in thelytokous populations. However, contrary to some sweeping statements found in the entomological literature (e.g., Flanders, 1945, 1965), in *Aphytis* we do not know of any authenticated record of an actual change in the mode of reproduction (from thelytoky to arrhenotoky, or vice versa) being affected by changes in the environment. As pointed out by DeBach (1969b), such apparent "changes" in insectary cultures are often due to the accidental contamination, and eventual competitive displacement, of one species by another. We individually tested 100,000 unmated individual females of arrhenotokous *A. lingnanensis* to ascertain whether occasional thelytokous females might be produced. None were—only male progeny resulted.

# ADULT BEHAVIOR

Upon emergence from the host, the adult *Aphytis* wasp rests for a while, then proceeds to groom itself. This was described by Rosen and Eliraz (1978) in *A. chilensis* as follows: "Grooming is usually performed in clockwise order, from front to rear. First, the right foreleg is used to clean the eyes, the head and the right antenna. The middle leg then cleans the foreleg. The hind leg cleans the middle leg, subsequently grooms the abdomen and wings. The wasp then rubs both hind legs and proceeds to groom the other side of the body. The grooming process is repeated several times before the wasp finally moves away."

Active locomotion of adult *Aphytis* in the presence of a suitable host population involves mainly running. Flight commonly occurs on warm sunny days. Long-distance dispersal is thought to occur under sunny conditions and is probably aided by low-speed air movement. Higher-velocity wind causes adults to cease movement and cling to the substrate.

All known species of *Aphytis* exhibit strong negative geotaxis. They are all attracted to light (positive phototaxis), and our preliminary studies have indicated that attraction is to the green-yellow region ($5.0–6.0 - 10^3$ Å) of the electromagnetic radiation spectrum.

## SEXUAL BEHAVIOR

Fairly detailed observations of sexual behavior in *Aphytis* were recorded by Rao and DeBach (1969a), Khasimuddin and DeBach (1975) and Gordh and DeBach (1976, 1978). The following account is based on their papers.

There is no evidence of a premating period, or of any biological clocks controlling mating in *Aphytis*. Both females and males are sexually receptive immediately after emergence, and virgin females will mate with conspecific males at virtually any time of the day or night.

Upon approaching a conspecific female, the male appears excited, beating his wings and displaying some rapid and erratic movements. A virgin female will usually respond by ceasing all movement and remaining stationary. The male then mounts her dorsum from the rear, orients his body parallel to hers, and proceeds with precoital courtship. The fore tarsi of the male are placed either on the female's compound eyes or fore tibiae, his middle legs are not in contact with her body, and his hind legs are on the anterior margins of her wings. The male rapidly palpitates the tips of the female's antennae with his own (Figure 54 A, E), while his wings continue to vibrate. This precoital phase lasts 2–4 seconds. A sexually receptive female reacts by raising and slightly extending the abdomen (Figure 54 B, F), which presumably serves to dilate the gonopore, located near the base of the ovipositor. The male then quickly moves to the rear, lowers his abdomen below that of the female, and copulates with her. He holds on to the female by the fore- and middle legs, while the hind legs and wings touch the substrate and provide some additional support (Figure 54 C, G). The coital phase lasts 3–4 seconds, after which the male returns to his previous position on the dorsum of the female. He rests for a short while, then proceeds to vibrate the antennae, palpitating the tips of the female's antennae while his mandibles may touch her antennal pedicels (Figure 54 D, H). The male's middle legs are moved semaphorally, whereas the wings

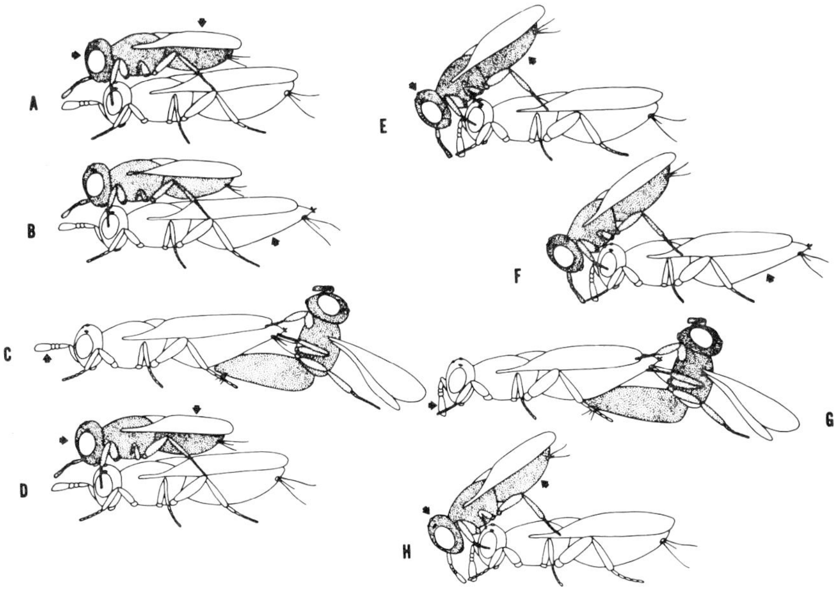


Figure 54. Courtship and mating positions (semidiagrammatic). A–D. *Aphytis lingnanensis* Compere. E–H. *Aphytis melinus* DeBach. (From Gordh and DeBach, 1978.)

are fanned rapidly and returned to a position over the gaster. Postcoital courtship lasts about 1–2 minutes, after which the male either dismounts or is dislodged by the female.

Sexual behavior in *Aphytis* is evidently controlled by species-specific sex pheromones. The male is strongly attracted to a pheromone released by a conspecific virgin female, apparently from the thoracic region (Rao and DeBach, 1969a), whereas a male sex pheromone serves to quiet the virgin female and render her sexually receptive (Khasimuddin and DeBach, 1975). Males and females do not normally respond to members of the opposite sex belonging to other, even closely related, species.

Visual stimuli do not seem to be essential for successful insemination in *Aphytis*, as this is normally accomplished even in total darkness. Amputation of the wings does not affect sexual behavior, whereas amputation of the antennae prohibits mating altogether (Rao and DeBach, 1969a).

The *Aphytis* female is essentially monogamous. She normally mates only once in her lifetime (usually soon after emergence), and the sperm stored in the spermatheca is presumably sufficient to fertilize all her eggs. An inseminated female will refuse subsequent matings. The male, on the other hand, is definitely polygamous, and is capable of mating with several females in rapid succession. In laboratory experiments, individual males of *A. lingnanensis* copulated with 27 females during their lifetime, and these matings resulted in an average of 155 female progeny per male (Gordh and DeBach, 1975).

## HOST SELECTION AND OVIPOSITION

The process of host selection by parasites is commonly considered to consist of four distinct steps: 1) habitat selection; 2) host finding; 3) host acceptance; 4) host suitability (Doutt, 1964).

Very little is known about habitat selection in *Aphytis*. Because of the aggregative distribution of armored scale insects, *Aphytis* adults often emerge in close proximity to their hosts. Nevertheless, long-range dispersal does occur, and ability to select a suitable habitat would therefore be of definite advantage to the parasites, especially at low host densities.

Smith (1957) demonstrated a differential response of two *Aphytis* species to different host plants. When released in equal numbers in a greenhouse containing *Yucca filipendula* and *Cycas revoluta* plants infested with the California red scale, *Aonidiella aurantii* (Maskell), *A. lingnanensis* outnumbered *A. chrysomphali* 3.1:1 in parasitized scales on the *Yucca*, 81.0:1 on the *Cycas*. However, it is an open question whether that result was actually due to differential attraction to the host plants themselves, or to the possibly different attractiveness or suitability of host scales growing on these plants. On the other hand, Quednau and Hübsch (1964) offered *A. coheni* females a choice between oleander scales on citrus leaves and on artificial paper leaves, and concluded that "when *A. coheni* searches for its hosts citrus leaves do not exercise an attractive influence upon the parasite." However, as pointed out by Baker (1976), their conclusion is questionable in view of the very small size of their experimental chamber.

Finding the host itself is accomplished by a random search of the surface of the host plant. The female parasite moves rapidly, palpitating the surface with the antennae, occasionally resorting to short flights. Quednau and Hübsch (1964) stated that "*Aphytis* is able to recognize its hosts only when in direct contact with their surfaces," but DeBach and Sundby (1963) reported that the female parasites seemed to perceive the scale insects, and went directly to them, from a distance of about 1 cm.

More recently, Sternlicht (1973) reported that *Aphytis* adults (sex unspecified) were caught in sticky traps baited with virgin females of the California red scale. He further presented some data indicating that adults of *A. melinus* and *A. coheni* were attracted to virgin hosts in olfactometer tests, which he interpreted as attraction to the sex pheromone produced by the scale insects. It should, however, be pointed out that Sternlicht's assertion was entirely based on circumstantial evidence and was not supported by his experimental data. If indeed *Aphytis* females are attracted to their hosts from a considerable distance (and this is questionable), the factors involved in such attraction are at present unknown. It is, of course, possible that some component of the host, or some unknown substance produced by it, does act as a kairomone, but so far no hard evidence has been presented to link the presumed attractant with the specific sex pheromone produced by virgin scale insect females. In California, pheromone traps designed to catch California red scale males do commonly trap large numbers of *Aphytis melinus*. However, it appears that the yellow color of the sticky-board trap has more to do with the catch of *Aphytis* than does the pheromone, because far fewer *A. melinus* are trapped by white sticky-board pheromone traps (D. Moreno, personal communication, 1976).

Determination of host acceptability is the most important step in the complex selection process. Once an *Aphytis* female has encountered a potential host, she proceeds to examine it with the antennae, mouthparts, tarsi and tip of the abdomen, as well as with the ovipositor. Several authors have described the examination and oviposition sequence in some detail, including Fisher (1952) for *A. lingnanensis*, DeBach and Landi (1961) for *A. lepidosaphes*, Quednau and Hübsch (1964) for *A. coheni*, and Abdelrahman (1974c) for *A. melinus*. The following account is based mainly on their observations.

The *Aphytis* female moves rapidly over the surface of the host plant, palpitating scales and other objects with the antennae, perhaps hesitating momentarily over several different scales or stopping abruptly between scales, to remain motionless for a while before continuing her search. Upon selecting a particular scale for more thorough investigation, she proceeds to make several exploratory examinations over and around the periphery of the scale. *A. melinus* performs this in an orderly fashion: she stands on the host (the California red scale, which is nearly circular) with her head toward the periphery, then moves quickly from the center to the periphery and back, with antennae and mouthparts tapping the host; she then turns through about 30°, repeats the steps, and continues turning and stepping until the entire surface of the host has been explored. This exploratory phase, which also involves feeling the scale cover with the tip of the abdomen, takes from 1 to 4 minutes in *A. lepidosaphes*, only 9 seconds in *A. coheni*. If the *Aphytis* female has not abandoned the host at the end of this phase, she then selects a site for penetrating the hard scale cover with the ovipositor, usually near the margin.

The drilling procedure in *A. lepidosaphes* consists of rapid thrusts of the twisting, turning ovipositor at an angle of contact of approximately 45° with the scale surface. The rapid vertical movements of the stylets are accompanied by a slower rotary movement in both a clockwise and counterclockwise direction, approaching 180° in each direction. The tenacity with which the parasite seems to grip the surface of the scale with the tarsi and pull downward gives one the impression that the operation requires a very strenuous effort. The actual time required for penetrating the scale cover varies in *A. lepidosaphes* from slightly over 30 seconds to nearly 4 minutes, and it is not uncommon for a parasite to cease drilling altogether for perhaps a minute or two, and on occasion to leave the scale momentarily and then return and resume drilling.

Just prior to complete penetration of the scale cover, the drilling tempo slows down considerably. After the cover has been pierced, the ovipositor explores the scale body rather slowly with the tips of the stylets, exhibiting amazing flexibility. This is followed by slow, deliberate thrusts of the ovipositor into the scale body itself and extensive internal probing of the posterior portion of the scale body. This may be repeated several times.

Just seconds before actual oviposition, the rapid pulsation of the stylets ceases, the ovipositor is withdrawn from the scale body, and its tip contacts the body wall. In *A. lingnanensis* the ovipositor is seen moving gently, almost imperceptibly, back and forth, exuding a viscous droplet from an orifice located a short distance from its tip. This substance precedes the egg on its way down the ovipositor and forms a tiny adhesive pad on the body of the scale insect host. Next, the egg balloons out of the ovipositor and is affixed on its side to the adhesive pad. The ovipositor may then be withdrawn or, in a gregarious species such as *A. lepidosaphes*, additional probing may be followed by the deposition of a second egg, adjacent to the first.

The elaborate sequence of exploration and probing prior to oviposition apparently serves to provide the *Aphytis* female with the stimuli required for determination of the acceptability of the host, as well as to modify the host and ensure its continuing suitability for the development of parasite progeny.

Quednau and Hübsch (1964) have demonstrated that whereas physical stimuli, such as the shape of the scale cover, do not play a significant role in the process of host recognition by *Aphytis*, chemical stimuli are of great importance. Their experiments with *A. coheni* have indicated the presence of a water-soluble "stimulating" substance in scale covers, that is destroyed or deactivated by exposure to high temperature. More recently Baker (1976), working with *A. mytilaspidis*, showed that characteristics of the scale body as well as those of the scale cover are perceived by the ovipositing *Aphytis* female. The parasites were capable of discriminating between the bodies of preferred and unpreferred host species, even when these were offered to them under various experimental combinations of covers and bodies.

Avoidance of superparasitism is an important aspect of host selection. Although specific marking of hosts has not been reported in *Aphytis*, Abdelrahman (1974c) has shown that females of *A. melinus* can recognize a recently parasitized host by antennal examination, presumably due to a "spoor effect," i.e., an odor left by the wasps as they move around, that lasts for several hours. Even after this wears off, examination and probing with the ovipositor can still determine whether or not the host is parasitized.

Recognition of host stage is as important to *Aphytis* as is the recognition of host species. *Aphytis* females oviposit only in hosts whose body is free beneath the scale cover. Molt stages are not accepted as hosts, apparently because their body becomes attached to the cover, and also because their integument becomes heavily sclerotized and is too hard to be pierced by the ovipositor. Similarly, mature, mated females of *Aonidiella* species, which become heavily sclerotized and have their body attached to the scale cover as if in preparation for another molt, are not accepted as hosts by *Aphytis* (DeBach and White, 1960; DeBach 1969a; Baker, 1976). First-instar larvae of scale insects are also usually not parasitized, presumably because of their small size.

The ovipositional probe, performed by the *Aphytis* female inside the body of the host prior to oviposition, apparently serves not only to determine the acceptability of the host, but also to prevent it from reaching an unsuitable developmental stage. The deep probe apparently causes partial paralysis of the host (Fisher, 1952; DeBach and Landi, 1961). Its main effect is to prevent the host from becoming sclerotized and molting (Abdelrahman, 1974c; Baker, 1976). Whether this is done by physical mutilation with the ovipositor or by the injection of venom is for the time being unknown but Beard's work (1971) and our observations would favor the venom hypothesis.

Not much is known about the effects of host unsuitability on *Aphytis*. Fisher (1952) reported that eggs of *A. lingnanensis* that were transferred to healthy, nonparasitized hosts failed to develop on them, which indicates that the ovipositional probe is indeed essential for successful development of *Aphytis* progeny. DeBach and Landi (1961) noted that freshly laid eggs and early larval stages of *A. lepidosaphes*, a monophagous species, failed to complete their development when transferred to unpreferred hosts. However, this may have again been due to the unsuitability of healthy, "unprobed" hosts, rather than to host specificity.

*Aphytis* females may lay their eggs on either the dorsal or the ventral aspect of the host's body. Some species do not seem to have any preference in this respect, whereas others, such as *A. lepidosaphes* or the Australian *A. chrysomphali*, definitely prefer the ventral position (DeBach and Landi, 1961; Abdelrahman, 1974c). In *A. melinus*, young mated females reportedly deposit their female eggs on the dorsal aspect and their male eggs on the ventral aspect of the host, whereas unmated and very old mated females, which deposit male eggs only, place them in both positions (Abdelrahman, 1974c).

Some species of *Aphytis* are essentially solitary. In *A. maculicornis*, *A. paramaculicornis* and *A. hispanicus*, for instance, rarely are two eggs found on a single host, and usually only one of them is able to complete development (Huffaker, Kennett and Finney, 1962; Gerson, 1968). Other species, such as *A. lepidosaphes*, are normally gregarious (DeBach and Landi, 1961). Figure 38 (p. 33) shows 5 pupae of *A. lepidosaphes* that have developed on a single host. Most species of *Aphytis* are probably facultatively gregarious. In such species the number of parasites per host appears to be correlated with the size of the latter, as well as with the relative density of host and parasite populations.

The *Aphytis* female is synovigenic, eggs developing and maturing in the ovaries continuously through much of the life span. Rössler and DeBach (1973) have shown that each of the two ovaries of *A. mytilaspidis* is composed of 3 ovarioles, each contain-

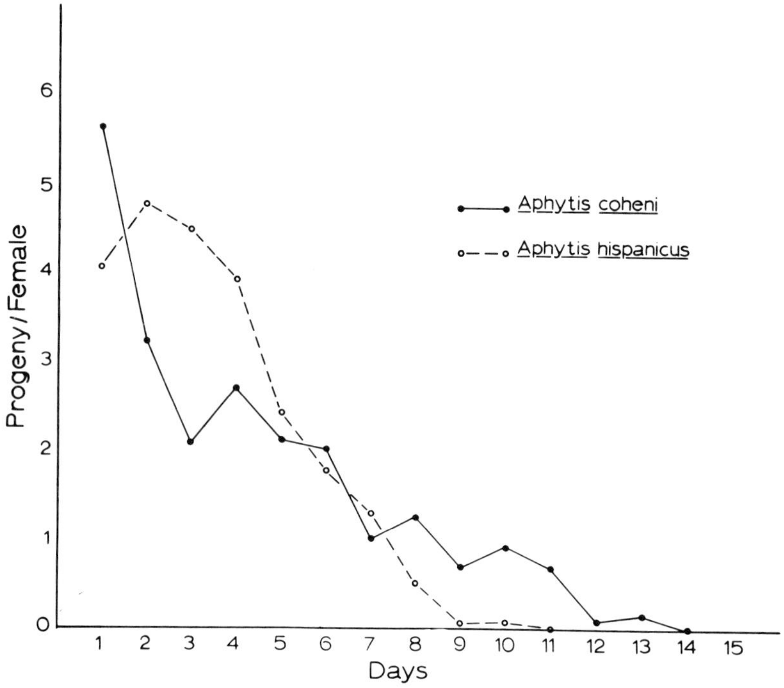


Figure 55. Average daily oviposition in *Aphytis*.
(From Gerson, 1968, and Avidov, Balshin and Gerson, 1970.)

ing 4 oocytes in various stages of yolk accumulation. Hence, at any given time, 24 eggs are in various advanced stages of maturation. Chumakova (1961) indicated a similar situation in *A. proclia*.

Total average fecundity is of the order of 30–60 progeny per female. The preoviposition period is very short; in *A. lingnanensis*, for instance, it does not exceed 4–12 hours (Fisher, 1952). Most of the eggs are invariably laid during the first few days after emergence (Figure 55). If the females are kept for several days in the absence of hosts, their total fecundity decreases drastically, apparently due to ovarial ovisorption.

## HOST-FEEDING

Adult *Aphytis* wasps derive their carbohydrate nutrition in nature mainly from nectar and honeydew, although the latter may vary considerably in nutritional value according to its source (e.g., Avidov, Balshin and Gerson, 1970). The wasps are capable of feeding on a dry film of sugar (Bartlett, 1962), and can be maintained successfully in laboratory cultures with honey as food. However, like other synovigenic insects, the females require proteinaceous nutrition for continuous ovigenesis. This they obtain by predatory host-feeding.

Marchal (1909) was apparently the first to notice the host-feeding habit in *Aphytis*.

Similar observations were recorded soon thereafter by Quayle (1910) and Imms (1916), as well as by several subsequent authors. Fisher (1952) described host-feeding by *A. lingnanensis* in some detail. After drilling through the scale cover, the female thrusts the ovipositor into the body of the host, moving it around in several directions and pulling it back and forth, sometimes for as long as 2 minutes. The ovipositor is then used to pull the scale insect's body close to the cover, after which it starts rotating while moving slowly up and down and simultaneously exuding a highly viscous substance that is built up around the shaft. When the ovipositor is at last carefully withdrawn, a tiny strawlike feeding tube remains. The female then turns around, applies the mouthparts to the tube and feeds on the body fluids of the host. From time to time she may reopen the feeding tube with the ovipositor (see Gulmahamad and DeBach, 1978).

Unlike a host on which oviposition by *Aphytis* has taken place, a host-fed scale insect soon develops large brown necrotic spots and dies within several hours to a few days.

Certain host stages that are not acceptable for oviposition by *Aphytis*, such as first-instar larvae and first molt stages, are readily utilized (in fact, may even be preferred) for host-feeding. However, second molt stages, in which the integument is heavily sclerotized, are not host-fed upon, and the same is true for mature, mated females of *Aonidiella* spp. (Abdelrahman, 1974c; Gulmahamad and DeBach, 1978).

Females of *A. lepidosaphes* apparently do not have to host-feed prior to oviposition (DeBach and Landi, 1961). On the other hand, newly emerged females of *A. aonidiae* exhibit marked preference for host-feeding over oviposition in the first host they encounter (Gulmahamad and DeBach, 1978). In *A. lingnanensis*, males have also been occasionally observed to feed on recently vacated feeding punctures (Fisher, 1952).

# ECOLOGY

Ecology is commonly divided arbitrarily into four levels, that of the individual, that of the population, that of the community and that of the ecosystem. This discussion will emphasize aspects of field population ecology of *Aphytis*, i.e., intrinsic parasite-host population interactions in relation to host population regulation and the modifying effect of extrinsic environmental factors, both biotic and physical, on the regulatory ability of *Aphytis*. The latter factors are obviously also components of the community and ecosystem. Ecology of individuals, of course, cannot be excluded from such a discussion but this is treated more fully in other sections, including Reproductive Biology (p. 41) and Adult Behavior (p. 42).

## HOST POPULATION REGULATION

The *fact* of host population regulation at low levels by *Aphytis* acting alone, or essentially so, has been proven for several species by the use of experimental check methods (see DeBach and Huffaker, 1971; DeBach, Huffaker and MacPhee, 1976). These include the control of the California red scale, *Aonidiella aurantii* (Maskell), by *Aphytis chrysomphali* (DeBach, Fleschner and Dietrick, 1948; DeBach, 1958a) as well as by *A. lingnanensis* (DeBach et al., 1950), *A. melinus* (DeBach and Huffaker, 1971), and *A. coheni* (Ben-Dov and Rosen, 1969); of the dictyospermum scale, *Chrysomphalus dictyospermi* (Morgan), by *A. melinus* (DeBach and Argyriou, 1967); of the purple scale, *Cornuaspis* [= *Lepidosaphes*] *beckii* (Newman) by *A. lepidosaphes* (DeBach, 1969a; DeBach and Argyriou, 1967); of the olive scale, *Parlatoria oleae* (Colvée), by *A. paramaculicornis* (recorded at the time as the "Persian strain" of *A. maculicornis*) (Huffaker, Kennett and Finney, 1962) and of the San Jose scale, *Quadraspidiotus perniciosus* (Comstock), by *A. aonidiae* (Gulmahamad and DeBach, 1978).

Such proof has only been developed for evaluation of the effectiveness of natural enemies in a few other cases involving other genera. This lack is one reason that claims for biological control are not always readily accepted. However, careful observation of before-and-after changes in applied biological control projects can be equated with experimental comparisons and also considered as adequate proof (DeBach and Huffaker, 1971). Most of the preceding species have been included in such large biological control projects. In addition, the Florida red scale, *Chrysomphalus aonidum* (L.), has been excellently controlled in several countries by *A. holoxanthus* (DeBach, Rosen and Kennett, 1971), and the rufous scale, *Selenaspidus articulatus* (Morgan), has been reduced to low levels by *A. roseni* in Peru (DeBach and Rosen, 1976a).

In addition to the preceding, our field observations in many countries over the years, often bolstered by natural "experimental checks" including ant-inhibition of parasites ("the biological check method") and "micro-exclusion cages" composed of spider webbing or debris that protects scale insects from parasites, strongly indicate that various other species of *Aphytis* are similarly effective and that *Aphytis* species are the predominant group in the population regulation of armored scale insects. Often *Aphytis* constitutes the only parasite species present in agro-ecosystems and sometimes in natural ecosystems, so its true regulatory ability acting alone can be evaluated.

Life-table data are commonly considered to provide the information necessary for a determination of key mortality factors in a natural population. However, with *Aphytis* species they have been inadequate to provide satisfactory proof that the parasites are indeed regulatory. In fact, life-table data probably never can provide *proof* of regulation. As DeBach and Huffaker (1971) put it, "the important point . . . is . . . that neither precise laboratory studies nor quantitative field studies of parasitization furnish adequate means of evaluating the effectiveness of a parasite in host population regulation and control."

Typically, with *Aphytis* species that have been proven to be regulatory by appropriate experimental check methods, only a low average percent parasitization, often less than 25%, occurs at any given moment in time with populations in equilibrium at low levels. However, parasitism is not the only cause of mortality by *Aphytis*. *Aphytis* species commonly kill as many or more scale insects by predatory host-feeding as by parasitism (DeBach, Fleschner and Dietrick, 1953; DeBach and White, 1960; DeBach and Sundby, 1963). This mortality, however, is cryptic and field counts (life-table data) reveal only a small portion of it. The reason is that most scale insects killed by host-feeding have to be recorded merely as dead from unknown causes, because they die

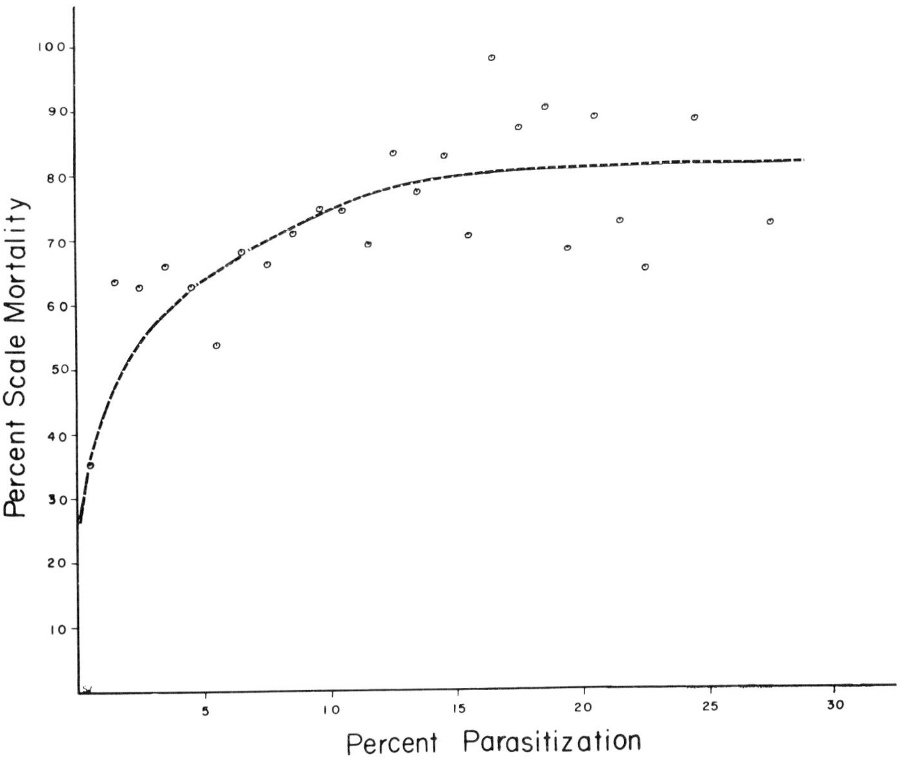


Figure 56. Relation between percent parasitization by *Aphytis melinus* and field mortality of the California red scale. (From DeBach, 1969a.)

and dry up within a day or so and then the cause of death cannot be ascertained. (For a description of host-feeding in *Aphytis*, see p. 48.)

A further complication is that there are usually two to three generations of *Aphytis* to one of the host, but the generations of both generally show overlap. Thus, even though we know intuitively that there is a high cumulative summation of total generation mortality of the scale insect host caused by both parasitism and host-feeding, there is no practical way of accurately quantifying this summation (DeBach, Fleschner and Dietrick, 1948; DeBach, 1969a). We know, therefore, that a relatively low degree of parasitization by *Aphytis* will be associated with a much greater total *Aphytis*-caused mortality, as shown in Figure 56.

For example, the averaged data upon which Figure 56 is based show that 1% parasitization is associated with about 35% total scale mortality, 5% parasitization with some 63% total scale mortality, 10% parasitization with 75% scale mortality and 25% parasitization with 84% mortaily (see DeBach, 1969a, pp. 806–810).

The *mechanisms* that make *Aphytis* species so effective as compared with other parasite genera are not readily evident. Obviously, many *Aphytis* species are highly density-dependent, i.e., they respond rapidly to any tendency for the host population to increase. The fact that *Aphytis* species usually have two to three generations to one of the host is very important for the potential rate of parasite increase. For example, a "typical" *Aphytis* species in nature may produce 40 progeny per female parent per generation, whereas a typical diaspidid scale insect host will approximate perhaps 200 progeny per female parent per generation. This might appear to favor the scale insect but, assuming a 50:50 sex ratio for each, the parasite can potentially produce from 400 to 4000 progeny during the period of one host generation, depending on whether it has two or three generations to one of the host. The intrinsic rate of natural increase ($r_m$) of *Aphytis chrysomphali* and *A. melinus* in laboratory experiments was 3.1 to 5.0 times that of *Aonidiella aurantii*, depending on parasite species and temperature (Abdelrahman, 1974b).

*Aphytis* species evidently have good searching ability, because they commonly do maintain scale insect populations at very low levels. The precise how and why of this remains unknown. Observations show that adult females move rapidly and vigorously over vegetation in search of hosts. Long-distance dispersal in *A. melinus*, for instance, is known to be exceptional: when first colonized on dictyospermum and California red scale in Greece it moved some 75–100 kilometers per year (DeBach and Argyriou, 1967; see Figure 66, p. 81). On the other hand, *A. lingnanensis*, which can also regulate scale populations at low levels, was found to disperse only about 10 tree rows per year when first colonized in California citrus groves (DeBach, 1954). Rapid long-distance dispersal, therefore, is not critical to effective host population regulation in established systems.

Another factor favoring the parasite is that most *Aphytis* species attack hosts that exhibit continuous chronological availability on perennial host plants. However, the same conditions apply to other armored scale insect parasites in the genera *Aspidiotiphagus*, *Comperiella*, *Habrolepis*, *Physcus* and *Prospaltella*, and none of these groups even begins to compare favorably with *Aphytis* in host population regulation. DeBach and Rosen (1976a, Table 20), for instance, have shown that for 33 of the most important armored scale insect pests of the world, *Aphytis* species are listed 46 times as major natural enemies, whereas all other parasite species in all other genera are listed only 34 times and all predators just 4 times. Thus, although we can demonstrate that *Aphytis* species are often highly effective in host population regulation, at present we can only roughly approximate some of the basic capabilities that underlie such effectiveness. The preceding discussion has emphasized that *Aphytis* species are far above average in regulatory efficiency and often achieve regulation at low host levels. However, as will be shown in the following sections, the regulatory potential of *Aphytis* may be negated at times or in places by certain adverse factors.

## DISTRIBUTION AND ABUNDANCE

The geographical distribution of any given species of *Aphytis* is, of course, ultimately limited by climatic factors. What we are more concerned with here are the factors that further limit distribution and affect abundance or efficiency in host population regulation within tolerable climatic zones. Only a few species have been studied intensively in this regard, principally those attacking the California red scale, *Aonidiella aurantii*. *Aphytis lingnanensis* will be emphasized as a model for other species in this discussion because it has been studied intensively, both in the field and in the laboratory. Although *A. lingnanensis* has now been displaced by competition with *A. melinus* in certain California areas included in this phase of the discussion, this fact has no bearing on "The Effect of Physical Factors." The phenomenon of competitive displacement will be discussed under "The Effect of Biotic Factors" (pp. 69–79).

## THE EFFECT OF PHYSICAL FACTORS

**Temperature extremes** have been shown to be by far the most important natural cause of mortality (as well as of other adverse consequences) of *Aphytis lingnanensis*, both in the laboratory and in the field. Some of the first field studies in California (before the introduction of *A. melinus*) showed that *A. lingnanensis* and *A. chrysomphali* populations were more or less greatly reduced by low winter temperatures. The effect varied in different years and was relatively less in coastal (mild winter) areas than in inland valley (colder winter) areas. Close to 100% pupal mortality was recorded at times in the field in inland areas. The *Aphytis* pupal mortality index also increased decidedly during the hot dry months of July, August and September in the harsher inland climatic areas. These results were considered to explain why natural control of the California red scale usually occurred in untreated experimental groves in coastal areas and only rarely in such groves in inland areas. The indications were that subfreezing or average minimum temperatures were not the best index of mortality. Daily winter maximum temperatures (above 15.6°C) had a favorable modifying effect, depending on frequency and intensity (DeBach, Fisher and Landi, 1955). Similar effects of winter cold on *Aphytis* were reported by Lord and MacPhee (1953). They showed that natural control of the oystershell scale, *Lepidosaphes ulmi* (L.), by *Aphytis mytilaspidis* usually occurs in the mild-wintered Annapolis Valley of Nova Scotia but that the severe winters (−28.9°C or below) of central New Brunswick limit the success of the parasite.

Similar effects of weather have been reported for *A. paramaculicornis*. This imported parasite greatly reduced the olive scale, *Parlatoria oleae* (Colvée), in California but completely satisfactory effectiveness was prevented because the parasite was intolerant of the hot, dry summers of the Central Valley. Also winter cold hindered normal reproductive activity of the parasite, so that the only highly successful period of activity was spring (Huffaker, Kennett and Finney, 1962; DeBach, Rosen and Kennett, 1971).

Laboratory studies of *A. lingnanensis* were carried out to help interpret the field results previously indicated for California. It was found that survival of all stages was adversely affected by low constant temperatures of from −1.1°C to 10.0°C, and in some cases even 15.6°C, depending on the duration of exposure. The adult stage was the most susceptible, with 100% mortality occurring after less than 24 hours' exposure to −1.1°C. The pupal stage was the most resistant. Effects on egg and larval stages were intermediate, i.e., 100% mortality occurred after 2–4 days' exposure to −1.1°C. Eggs did not hatch up to 12.8°C constant, and little or no development occurred at 15.6°C constant. High temperatures (32.2°C) greatly lowered the life expectancy of adults and the survival of immature stages (DeBach, Fisher and Landi, 1955).

A subsequent study (DeBach and Sisojević, 1960), involving comparisons between *A. lingnanensis* and *A. chrysomphali*, showed further that the effective progeny production per lifetime of mated *A. lingnanensis* females was greatly reduced by only moderately low constant temperatures, being 27.4 progeny per parent female at 30.6°C but only 10.5 progeny at 19.4°C. On the other hand, within this temperature range the trend was reversed for *A. chrysomphali*, being 5.0 progeny per female at 19.4°C and only 1.9 progeny at 30.6°C.

The fact that progeny production of *A. lingnanensis* was significantly reduced by cool temperatures (as distinct from cold ones) led to further research involving the effect of short-term (24 hours) suboptimum constant temperatures (15.6°C to 20.6°C range) on sex ratio and progeny production by *A. lingnanensis* and *A. melinus*. Preliminary tests had shown the threshold of adult activity to be about 14.4°C to 15.6°C. Virgin females and males were brought together and held at 15.6°C, 17.2°, 18.9°, 20.6° and 26.7°C (check) for only 24 hours, and the females were then removed to an optimum 26.7°C and allowed to oviposit for 17 days—essentially their entire lifetime. Progeny production and sex ratio were then measured. The sex ratio was adversely affected below 26.7°C; the lower the temperature, the greater the effect. After the short one-day exposure to 15.6°C, for example, 73.9% of *A. lingnanensis* progeny were males, as compared with only 33.5% males at 26.7°C; with *A. melinus*, 90.9% of the progeny were males at 15.6°C, as against 48.9% at 26.7°C.

Total progeny production (both males and females) was also severely reduced. This was unexpected for such a short exposure to suboptimum but not really low temperatures. Following the 15.6°C exposure, *A. lingnanensis* produced only 18.4 total progeny per test female, as compared with 32.2 total progeny for those held at 26.7°C. This was a reduction of 42.9%. With *A. melinus*, only 11.0 total progeny per female were produced following the 15.6°C exposure, as compared with 28.2 produced at 26.7°C, or a reduction of 61%. Even following the relatively mild exposure of 20.6°C, the decrease in total progeny production was significant.

When the combined effects of poor sex ratio and reduced progeny production were calculated, the adverse effects were even more striking. This was measured by *effective* progeny production, i.e., by the number of female progeny per female parent. For instance, with *A. melinus*, after 24 hours at 15.6°C effective progeny production was only 1.0 female progeny per female parent, as compared with 14.4 in the control—a striking reduction of 93.1%; with *A. lingnanensis* it was 4.8, as against 21.4—a reduction of 77.6% (DeBach and Argyriou, 1966).

The aforementioned results could be ascribed to the effect of cool temperatures on two or more of several biological or physiological processes, including successful copulation, successful insemination, survival or vitality of sperm, or survival or vitality of ova; however, these various possibilities were not analyzed individually in these tests.

The effect on sperm survival of subjecting adults to sublethal cold was determined later in other studies. When female *Aphytis lingnanensis*, mated at 26.7°C, were then subjected to −1.1°C for 6 hours and the survivors (about 50%) removed and allowed to oviposit at an optimal temperature, the progeny were virtually all males (1779 males: 1 female). Inasmuch as the normal insectary sex ratio is 70% females or better, this means that the sperm in the spermathecae of the mated females were nearly all killed. In another test, males were subjected to the same temperature, −1.1°C, for 6 hours, and then mated to young virgin females reared and held under optimal conditions. Their progeny were 97% males, showing that males were likewise rendered essentially sterile by this exposure. Obviously, the adverse effects of such exposures are greater on sperm than on adults (DeBach, Fisher and Landi, 1955).

From the standpoint of adverse effects on host population regulation, sterility of *A. lingnanensis* adults resulting from exposure to −1.1°C for 6 hours becomes a major consideration, because such exposures occur quite commonly during the winter in southern California. Obviously, if such exposures result in the production of practically all male progeny, populations of *A. lingnanensis* would decline rapidly.

Cold sterilization of sperm in *A. lingnanensis* adults was confirmed by additional studies (DeBach and Rao, 1968), using an exposure of −1.1°C for 8 hours. Most adults survived, but nearly 100% sperm mortality occurred in the spermathecae of mated females and in the testes of males. The males were apparently permanently sterilized. In addition, it was found that cold-sterilized mated females would not mate again when normal males were provided under optimal conditions, nor would previously cold-treated virgin females mate with normal males under optimal conditions. Thus this sublethal effect of cold on adults is actually the equivalent of nearly 100% direct mortality.

**The effect of temperature on the duration of development** of *Aphytis* species is an important aspect influencing their regulatory ability. Different species, of course, have different rates of development, although many are quite similar. According to Rao and DeBach (1969a), *A. lingnanensis*, *coheni*, *africanus*, *melinus*, *fisheri* and *holoxanthus* require either 12 or 13 days at 26.7°C constant to complete their development from egg to adult. The other species tested, *lepidosaphes*, was significantly different, requiring 17 days. *A. chrysomphali* was indicated to have a shorter life cycle at all constant temperatures tested (30.6°, 25.0° and 19.4°C) (DeBach and Sisojević, 1960), but in other comparative tests *chrysomphali* had a substantial advantage only at 15.6° C (DeBach and White, 1960). The threshold of development has been determined for various species as follows: *chilensis*: 14.1°C (Rosen and Eliraz, 1978), *hispanicus*: 12.9°C (Gerson, 1968), *coheni*: 12.0°C (Avidov, Balshin and Gerson, 1970), *melinus*: 11.0°C and *chrysomphali*: 8.5°C (Abdelrahman, 1974b). The above results indicate, as is to be expected, that certain species have evolved to become adapted to rather different climatic regimes.

An illustration of how the duration of development relates to different constant temperatures is given for *A. chilensis* by Rosen and Eliraz (1978). This relationship can be represented by an equilateral hyperbola, which indicates that the developmental threshold is 14.1°C. At 19°C the mean developmental period is 48.8 days, at 24°C it is 25.5 days, at 28°C it is 18.6 days, and at 32°C total mortality occurred. Thus a drop of 9°C below the optimal constant temperature of 28°C results in a prolongation of development amounting to 30.2 additional days, or more than 2.5 times the optimal period, and an increase of 4°C above the optimum results in 100% mortality. Aside from a marked reduction in the relative rate of increase of the parasite, the fact that generation time is prolonged by cold would result in nature in all immature stages being subjected to cold for longer periods, and so would compound the adverse effects.

It might seem that the longer *Aphytis* adults survive in nature (or in the laboratory), the more effective they would be as natural enemies, but the indications are that this is not necessarily so. When the longevity of adult *A. lingnanensis* was determined over a wide range of constant temperatures, it became evident that the optimal temperature for adult survival was on the cool side—about 15.6°C—although 21.1°C was nearly as favorable. Above and below these temperatures survival decreased rapidly and markedly, so that at −1.1°C and above 32.2°C mortality occurred very rapidly (DeBach, Fisher and Landi, 1955; DeBach and Sisojević, 1960).

It will be recalled that the optimal temperature for adult survival (about 15.6°C) coincides with the temperature which adversely affected ova and sperm in adults, so that both progeny production and sex ratio were seriously upset. Thus, although adults may live longer at cool temperatures, their effective progeny production (female progeny per female parent) may be seriously reduced as compared with that occurring at higher temperatures that are more optimal for most other phases of development, survival of ova and sperm, and reproduction.

A good deal of work has been done in Australia on *Aphytis* species that parasitize the California red scale, *Aonidiella aurantii*. From an extensive series of comparative tests on *A. melinus* and *A. chrysomphali*, conducted under various constant temperature regimes, Abdelrahman (1974b) found that 1) pupal mortality was inversely related to temperature within the range 20–30°C; 2) *A. melinus* adults lived longer than *A. chrysomphali* at all temperatures; 3) *A. melinus* produced significantly more female progeny at all temperatures (20°, 25° and 30°C), and especially at 30°C, than *A. chrysomphali*; 4) the duration of development was longer for *A. chrysomphali* than for *A. melinus* at 30°C, but shorter at 20° and 25°C; 5) *A. chrysomphali* had a higher $r_m$ at 20° and 25°C than *A. melinus*, but a much lower one at 30°C. He concluded that *A. chrysomphali* is more adapted to cold and less to heat than *A. melinus*. Despite these many differences, he states surprisingly: "it follows that *A. chrysomphali* and *A. melinus*, since they coexist [in semiarid inland southern Australia], must be equally well adapted." Obviously they are not equally well adapted to temperature extremes, and in any event the axiom of inequality holds that no two species are exactly equal in their requisites or responses. The key word in the citation is "coexist" and, as we discuss later under interspecific competition, true coexistence certainly does not occur (see DeBach, 1966, p. 197). In our opinion, competitive displacement of *A. chrysomphali* by *A. melinus* is probably gradually taking place in inland southern Australia and the so-called "coexistence" will continue for only a few more years.

**The effect of humidity** on *Aphytis* has been found to be much less pronounced than that of temperature. Insofar as has been demonstrated, it is only the desiccating effect of low humidity that is detrimental. Inasmuch as the immature stages are situated under the protective scale covering of their host and against a plant surface, they usually exist in a highly humid microenvironment and are less exposed than adults to adverse humidity conditions.

Laboratory tests on adult *A. lingnanensis* were conducted at three relative humidities, low (20%), medium (50%) and high (80%), and at five temperatures ranging from 10.0° to 32.2°C. The results showed that adults are very seriously affected by low humidities. At 20% RH they lived on the average less than $\frac{1}{3}$ as long as at 80% RH. The extreme effect, however, occurred at the high temperatures combined with low humidity. At 26.7° and 32.2°C and 20% RH, adults lived only one tenth to one twentieth as long as at 15.6° to 21.1°C and 80% RH. In the field in southern California, counts made before and after a "north" wind with ambient humidities ranging from 10 to 20% RH showed virtual disappearance of adults after three days (DeBach, Fisher and Landi, 1955).

**The effect of light** on *Aphytis* species in the field has not been studied to our knowledge but it undoubtedly plays a role in flight initiation and in searching. It is well known that *Aphytis melinus* adults are trapped in large numbers on yellow sticky-board California red scale pheromone traps, as compared with white ones. Thus, wavelength may have an effect. In the laboratory, *A. lingnanensis* females are positively phototactic; the higher the light intensity, the more attractive it seems to be (DeBach and White, 1960).

Oviposition by *A. lingnanensis* will occur in total darkness, but less so than under normal light (2:3.3). However, there are indications that oviposition may be enhanced under subdued light intensity (4 ft candles), as compared with intense light (120 ft candles), especially for short periods. In the insectary, oviposition under natural window light during the nine daylight hours between 0800 and 1700 hr was over four times as great as during the 15 hours between 1700 and 0800 hr (DeBach and White, 1960). This may indicate a further factor involved in reduced progeny production in the field during the winter, when daylight hours are shorter. Direct sunlight itself (insolation), however, can be conceived of as mitigating the adverse effects of low winter temperatures on adult activity and perhaps on the survival of immatures. Those individuals in direct sunlight will attain substantially higher body temperatures and will be able to oviposit, etc., even though the ambient temperature may be below the activity threshold.

The question naturally arises as to how indicative laboratory findings are of what actually occurs in the field. The answer is that they are often of real help in interpretation, but they are sometimes misleading. For example, results of tests previously reported (DeBach, Fisher and Landi, 1955) showed that low constant temperatures (−1.1° to 10.0°C) were always detrimental to *Aphytis*, the ultimate effect depending on duration and intensity. Field data, however, indicated that periods of high temperatures following low temperatures during the winter may alleviate the adverse effects (DeBach, Fisher and Landi, 1955). Accordingly, laboratory tests were run to explore this.

In these tests newly laid *A. lingnanensis* eggs were exposed to a constant temperature of 4.4°C for 12 hours and then transferred to a constant temperature of 26.7°C for 12 hours—a rough approximation of a mild winter diurnal cycle. This was repeated until the life cycle was completed. Surprisingly, very little mortality occurred even though the individuals were subjected to a total of 16.5 days at 4.4°C during their life cycle. By comparison, eggs subjected to 4.4°C constant for 8 days suffered 100% mortality. The developmental period was slightly more than doubled, as would be expected, but the actual time spent at 26.7°C (16.5 days) was only a little longer than the checks at 26.7°C constant (15 days). In effect the diurnal reversal from low to optimal temperature practically negated or neutralized the otherwise adverse effect of the low temperature. This emphasizes that constant temperature data considered in isolation can be quite misleading. A further illustration from the same test series was obtained when eggs were subjected to 15.6°C constant until the life cycle was completed. The latter temperature represents the mean of 4.4°C for 12 hours and 26.7°C for 12 hours, yet the life cycle was much longer at the constant cool temperature of 15.6°C and survival was also lower (DeBach, Fisher and Landi, 1955).

Although such results are not unexpected, they stress that it is difficult to duplicate in the laboratory the complex interactions that occur in the field, even when controlled variable-temperature cabinets are used. For one thing, we rarely know the actual conditions a tiny *Aphytis* adult or immature stage experiences in its microhabitat on a plant surface which may or may not be periodically subjected to sunlight or protected from wind, low humidity, etc. Nor do laboratory tests show what happens to all individual stages of the parasite throughout the year under variable temperature and humidity conditions.

Abdelrahman (1974a) questioned the validity of laboratory tests utilizing extreme constant temperatures, and so instead he utilized a programmed diurnal cycle that represented cold winter days or hot summer days for testing *A. chrysomphali* and *A. melinus*. He concluded that all stages of *A. chrysomphali* are more tolerant of extreme cold than the respective stages of *A. melinus*. Pupae of both species were the most tolerant stage in both cold and heat tests. Adults were least tolerant of heat, larvae least tolerant of cold. Eggs and adults showed intermediate tolerance of cold and eggs and larvae showed intermediate tolerance of heat. He also states that fertility of *A. melinus* is not affected by low temperatures but he gives no data for fertility, stating only that the sex ratios were normal.

Abdelrahman recognizes that the difference between these latter results and those of DeBach and coauthors may be interspecific, but he suggests that their (DeBach, et al.) results are more probably due to the shock of the parasites being abruptly transferred from 28° to −1°C, saying that: "It is therefore doubtful that such an effect would occur in nature." We would merely point out here that the transfer of *A. lingnanensis* from 26.7° to 20.6°C constant resulted in a significant decrease in fecundity and this is a rather small temperature change. More important, the following discussion will show that field tests of adult *A. lingnanensis* under conditions of actual winter cold resulted in drastic reductions in fecundity. We find it difficult to believe that the fecundity of *A. melinus* (which is taxonomically and biologically very similar to *A. lingnanensis*) would not be reduced by extreme winter cold.

Abdelrahman (op. cit.) concurrently tested the effects of heat and cold on the host, *Aonidiella aurantii*, and from these laboratory results suggested that winter would be less critical for both *Aphytis* species than for the host. In California, at least, the fact is that *A. chrysomphali*, *A. melinus* and *A. lingnanensis* are severely decimated over the winter with respect to the host, as shown by field counts. He suggests further that "summer would therefore be a more critical period for [*A. chrysomphali* and *A. melinus*] than winter, which causes more mortality to red scale and less to its natural enemies." Again, we have to disagree (see Figure 57 for summer-winter comparisons for *A. lingnanensis*).

Questions like those above led to tests designed to show the effect of actual field

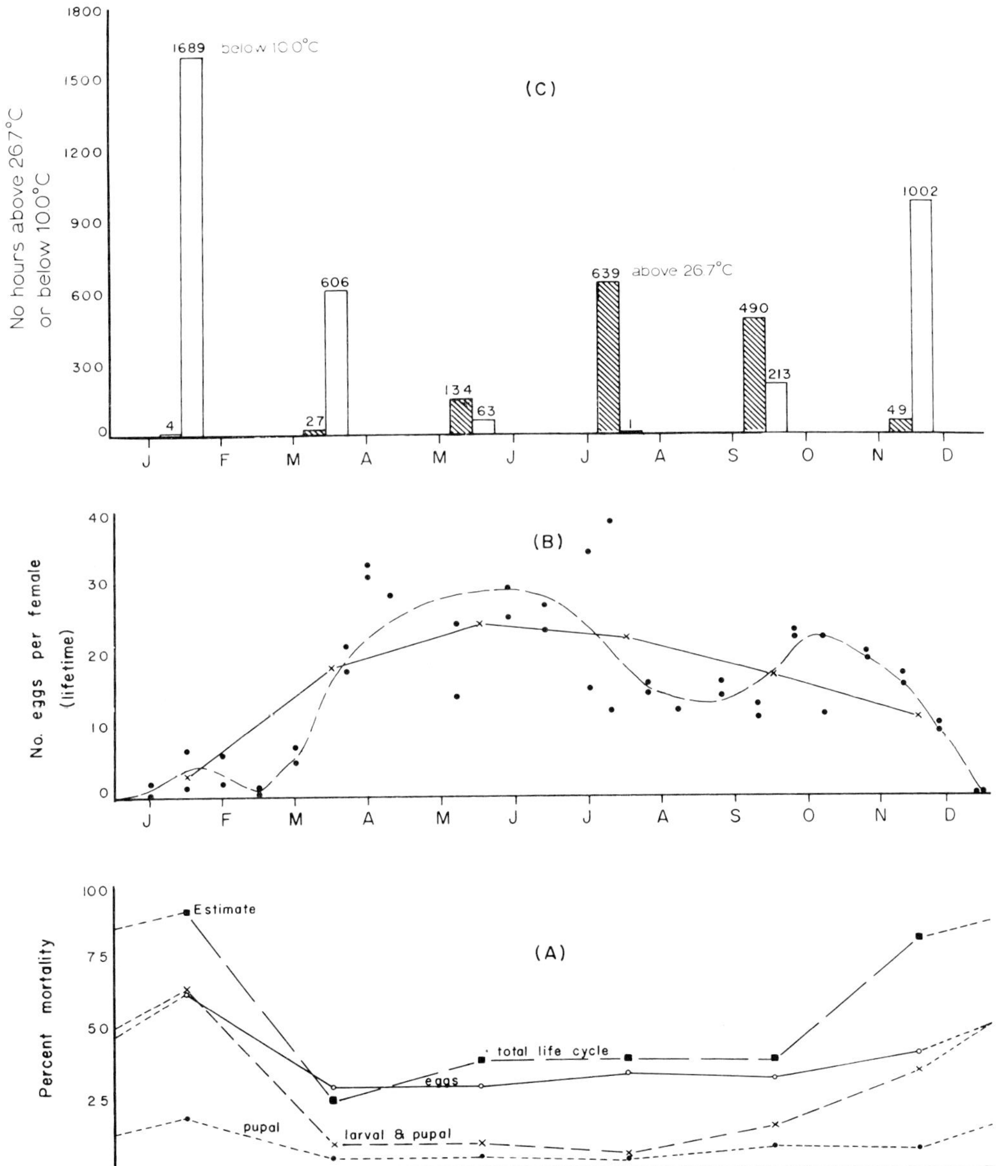


Figure 57. The effect of weather during successive periods of the year on mortality of all stages individually and on the entire developmental cycle of *Aphytis lingnanensis*.
(From DeBach, 1965a.)

weather, beginning at successive two-week periods during the year, on mortality of *A. lingnanensis* eggs, larvae and pupae alone, or on the entire life cycle, and on oviposition (fecundity) of adults (DeBach, 1965a). At the beginning of each two-week interval, laboratory-infested lemons with host scale insects bearing either parasite eggs, larvae or pupae, as well as adults confined with scale-infested lemons for oviposition, were placed in suitable containers, shaded from the direct sun in the field, and kept there until development to the next stage was indicated or, in the case of life-cycle studies, until adult emergence occurred. Mortality counts were then made, or total oviposition was recorded in the case of adult females. Checks were run in the laboratory under optimum conditions (26.7 ± 1.1°C, 60 ± 5% RH).

Figure 57A shows the general effect of weather on average mortalities for each developmental stage and on the complete life cycle during the year, and Figure 57B shows the effect on oviposition by females which, in part at least, reflects adult mortality. As would be expected, the mortality occurring during the total life cycle is greater than that of any one stage and reaches an average high level of 90% or above during the four winter months of November, December, January and February. A slight average increase is indicated in midsummer. Mortality during the egg stage is seen to constitute by far the greatest component of total life-cycle mortality during the summer, but is little or no greater than larval-pupal mortality (larvae kept in the field until emergence as adults) during the coldest months, when high average mortalities surpassed 60% for these stages. The only noticeable increase in pupal mortality during the year was in the January–February period, when it averaged 19%. Obviously the winter that year (1954–1955) was not nearly as severe as in 1951–1952, when DeBach, Fisher and Landi (1955) reported a peak of over 90% pupal mortality in the field. It more closely approximated the winter of 1952–1953, when they recorded average pupal mortalities of about 35%. It can be seen from the temperature summation (Figure 57C) that the highest mortalities are associated with the maximum number of hours below 10.0°C.

Oviposition or fecundity of female parasites is graphed in Figure 57B by two-week intervals (original data) and by bimonthly averages. The bimonthly averages (solid line) show a generalized trend from a low of 2.5 eggs per female in January–February to a high production of 24 eggs per female in May–June. This represents a winter decrease in fecundity of about 90% as compared with the early-summer maximum, and this alone accounts for a great reduction of the parasites' potential during the winter. When the actual results occurring from the various two-week exposures of adults in the field are considered (dots and dashed curve), three exposures (dots) are seen during December and January, when no oviposition whatever occurred. There is also a strong indication that fecundity is depressed by as much as 50% during the hottest and perhaps driest periods of the summer (July–September).

The use of bimonthly average mortalities in Figure 57 A masks the more extreme mortalities that occurred from time to time in response to short-term weather conditions. The most critical exposure periods occurred in late December and early January, when 100% mortality befell the four larval-pupal exposures and three of the four egg exposures, with 96% mortality occurring to the fourth egg exposure. It will also be recalled that there was no oviposition during the same periods in three exposures and only 1.9 eggs per female occurred in the fourth. Only two life-cycle exposures were initiated during December and January and no parasite completed its development. Thus, during this period nearly everything in the field except pupae was wiped out. Strangely enough, the only appreciable pupal mortality of the year (about 35% and 30%) occurred during the next two periods, in late January and February, when egg and larval mortalities were slightly decreased. This may be due to a cumulative effect of cold on preceding stages as well as on pupae, being finally expressed by pupal mortality.

The only significant mortality occurring during the summer period was to eggs, the highest being 90% for one exposure period, and this was associated with the year's highest temperature of 38.3°C.

One other adverse effect of weather on *A. lingnanensis* was on the length of the life cycle. This is depicted in Figure 58, where it is evident that during the coldest part of the year the life cycle is about three times longer than during the summer. Obviously, the longer the life cycle, the longer the period of potential exposure to adverse conditions. Furthermore, the potential rate of increase of the parasite population is greatly slowed, and when we recall that high mortalities along with greatly reduced fecundity

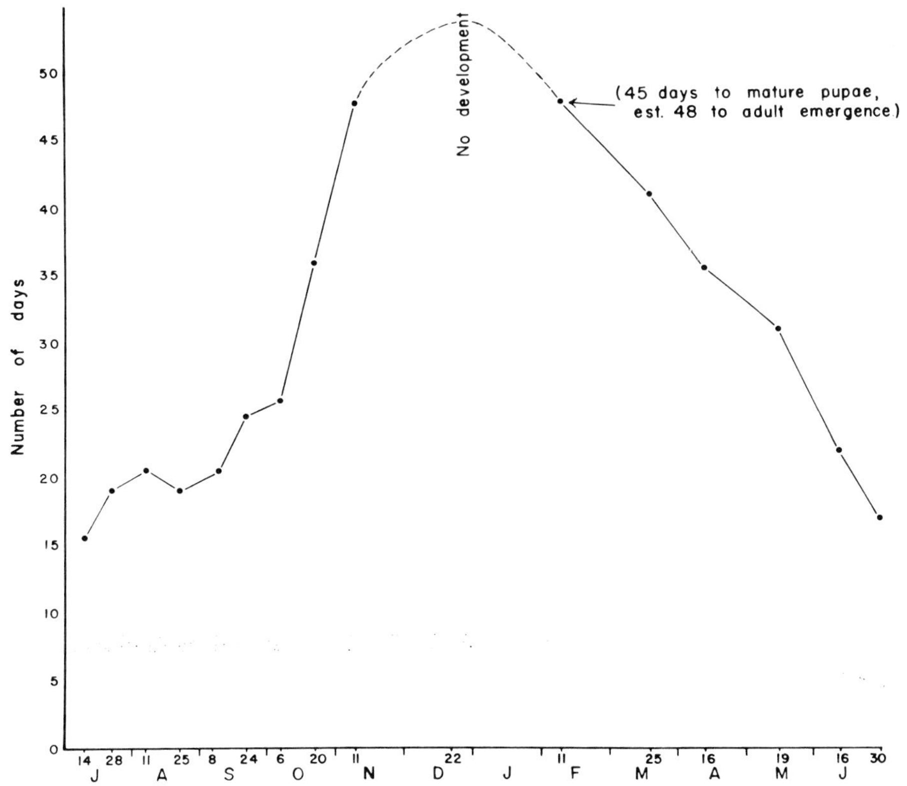


Figure 58. The effect of weather during successive periods of the year on the length of the average life cycle of *Aphytis lingnanensis* beginning on the date indicated. (From DeBach, 1965a.)

accompany the much lengthened life cycle, it might seem fairly obvious that parasite populations should decline to very low levels over the winter. However, even from such data as these it is difficult to calculate very precisely what the total effect would be on a year-round basis; accordingly, subsequent tests were conducted in experimental field population cages to measure directly the cumulative total effect of summer and winter weather on established, dynamic *A. lingnanensis* populations that had an inherent potential for quite considerable increase.

Three cloth-covered population cages were set up in southern California, one at a mild-weather coastal site, one at a severe-weather inland site and the third, a control, in the insectary at Riverside under optimal conditions. They were shaded from direct sun but otherwise exposed to ambient weather. The initial populations consisted of 1000 individuals, including equal numbers of each stage—eggs, larvae, pupae and adults. A surplus of hosts was always made available so that rapid initial growth was potentially possible and a high population density could be attained and maintained indefinitely. Weather was the only variable that could restrict population growth in these tests. One series of tests representing a winter period was run from 15 December to 15 April and another representing a summer period was run from 1 May to 9 October (DeBach, 1965a).

The results showed striking effects of adverse winter and summer weather conditions as compared with the control. A total cumulative production of over 300,000 to nearly 500,000 individuals occurred in the insectary control cages. By contrast, unusually high ambient temperatures during the summer exposure at the inland area site (45.0°C maximum during the week of 3–10 June) resulted in extermination of the population. Again, by way of contrast, the summer population at the mild coastal site increased 28-fold, nearly half as much as that attained in the insectary control. At that site the average summer maximum of 28.7°C and other parameters were highly favorable, whereas conditions at the inland site were much more extreme.

Populations at both the coastal and inland sites showed drastic decreases during the winter exposure period, even though this was not an unusually severe winter. At the inland site only 32 live parasites were left at the end of the winter period as against an initial 1000 (about a 97% decrease), and at the coastal site only 220 parasites were left (about a 78% decrease). Even though both populations were very low, there were seven times as many survivors at the coastal site as at the inland one. When corrections were made for sex ratios, the final effective (i.e., female) population at the coastal site was eleven times that of the inland site. These decreases contrast with the *potential* total production of both field population cages, which was some 300-fold, and the potential for increase (initial population to final population) which was 60-fold, as shown by the insectary control.

The small differences in temperatures between the coastal and inland sites may not seem impressive, but it must be stressed that small differences can be critical for *Aphytis*. The coastal site had only a 1.7°C higher average minimum and a 2.2°C higher average maximum during the winter period, although the extremes at the inland site were greater. It is emphasized that these population decreases occurred even though weekly minimum temperatures never fell below freezing, except on two occasions at the inland site when −1.1°C was reached (see DeBach, 1969a, p. 810).

Various conclusions can be drawn from all these tests. In certain California districts, ambient weather conditions become so unfavorable at times during winter and summer that extinction might be expected. However, although marked population reductions do occur, extinction, except perhaps in "spots", does not. Evidently with *A. lingnanensis* protective microshelters and some genetically resistant individuals permit limited survival of a population under otherwise periodically intolerable conditions. The host, the California red scale, on the other hand, is much less affected by weather extremes and is known to be able to achieve and maintain very high average population densities in all citrus areas of California, provided parasites or pesticides do not control it (see DeBach, 1958c, p. 759).

This can be considered a case of parasite population regulation by climate. When climate regulates an otherwise effective parasite, it naturally permits the host population to exist at a much higher average density than it otherwise would. This, briefly, is because the *rate* at which host mortality increases with host density determines the equilibrium position (see DeBach, 1965a, fig. 5). If a potentially effective parasite suffers excessive mortality from weather, obviously its rate of increase is reduced and hence the host population may be permitted to attain a higher density than if the parasite were able to exert its maximum effect.

The adverse effect of extreme winter and summer weather on parasites may be likened to the effects of periodic pesticide treatments for insect control. If applied regularly, perhaps semiannually as winter and summer weather extremes are, then "control" of the parasite population may go on indefinitely. Evidently, the number of protected microhabitats plus some genetically resistant individuals set an average minimum population density below which *A. lingnanensis* does not drop in southern California. This minimum may vary from year to year. The remainder of the year is spent rebuilding the population toward the previous peak, but the disruption in effectiveness is never overcome in severe climatic areas.

On the other hand, in more favorable climatic areas, regulation of the host population at low densities by *A. lingnanensis* can and does occur in spite of a certain amount of weather-caused mortality, as was just discussed for the mild coastal-site population cages. In the most favorable areas, where the host population is maintained at very low levels, it follows that the parasite population also must exist at very low levels. Thus *abundance* of parasites is not a criterion for judging their success or efficiency. A parasite may be rare for a number of reasons: because it is a poor searcher, because of adverse effects of the physical or biotic environment, or because, as mentioned above, it is highly effective in host population regulation.

Conversely, a parasite may be abundant and not necessarily effective in host population regulation. In fact, parasite abundance reflects the fact that the host is or was abundant and this can mean that the parasite is intrinsically relatively ineffective in host population regulation. This takes us back to the earlier discussion in this section that showed that experimental check methods are the most reliable means of determining whether or not a parasite is responsible for the host population density observed in nature.

Certain other physical factors, besides temperature extremes and, to a lesser extent, low humidity, may act to disrupt the potential of *Aphytis* species in host population regulation. These are principally chemical pesticides and airborne dust.

**The adverse impact of chemical pesticides** upon *Aphytis* species is extensive in relation to numerous toxicants, is often explosively drastic in the field and has long-term effects that are often overlooked. Only a brief glimpse of some of these adverse effects can be given here. The chlorinated hydrocarbons, organophosphorus compounds and carbamates, including DDT, endrin, dieldrin, parathion and carbaryl among others, have been shown in experimental field tests to cause severe upsets of various species important in biological control, including *A. chrysomphali*, *A. lingnanensis*, *A. lepidosaphes*, *A. melinus*, *A. holoxanthus*, and *A. paramaculicornis* (DeBach, Dietrick, Fleschner and Fisher, 1950; DeBach, Rosen and Kennett, 1971; Huffaker, Kennett and Finney, 1962; DeBach, 1958a).

The adverse effect of insecticides can be strikingly different in different climatic zones. DDT, toxaphene, endrin and dieldrin were simultaneously tested in plots in citrus groves having California red scale under complete biological control in a coastal area and in an inland area. All insecticides eventually caused host population increases of from 40- to 60-fold in the inland area, as compared with untreated checks. In the coastal area only DDT caused impressive increases but these were over 200-fold. The *Aphytis* species responsible for biological control (*A. melinus* inland and *A. lingnanensis* in the coastal area) were different in the two climatic zones but the results are not thought to be due as much to differential effects on the two parasite species as to interactions among weather, the insecticides and the parasites (DeBach, 1965a, Table II).

Laboratory studies have indicated a large spectrum of chemical pesticides as having a high toxicity for *Aphytis* species. Bartlett (1963) tested 61 pesticides against adults of *A. lingnanensis* and found that 30 of them had high toxicity whereas an average of only 13 compounds were highly toxic for six coccinellid species tested. Twenty-one pesticide formulations were bioassayed by Rosen (1967a) against *A. holoxanthus*; about half of these caused significant reductions in fecundity as well as reduced longevity of adults.

Laboratory results may be taken only as indications or guidelines. Adverse effects in the field may be greater or smaller than bioassay tests indicate, depending on dosage, type of formulation, mode of application, time of application, etc.

A typical host population explosion, resulting from the chemical inhibition of an *Aphytis lepidosaphes* population by repeated applications of a toxic insecticide that had little effect on the host purple scale population, is shown in Figure 59A. An increase of about 100-fold from the use of endrin, as compared with the undisturbed (control) host population, is evident. We have duplicated this phenomenon time and again in southern California, particularly with DDT and the California red scale and its parasites, *A. chrysomphali*, *A. lingnanensis* or *A. melinus*, as is shown, for instance, in Figure 59B which involved *A. melinus*. Recently, in a residential area of San Diego where the Japanese beetle, *Popillia japonica* Newman, was under an eradication program using carbaryl (Sevin®) during two summer seasons, purple scale on citrus trees was caused to increase over 4000-fold as compared with untreated trees that remained under complete biological control. The purple scale parasite, *Aphytis lepidosaphes*, was virtually eliminated in the eradication zone (DeBach and Rose, 1977).

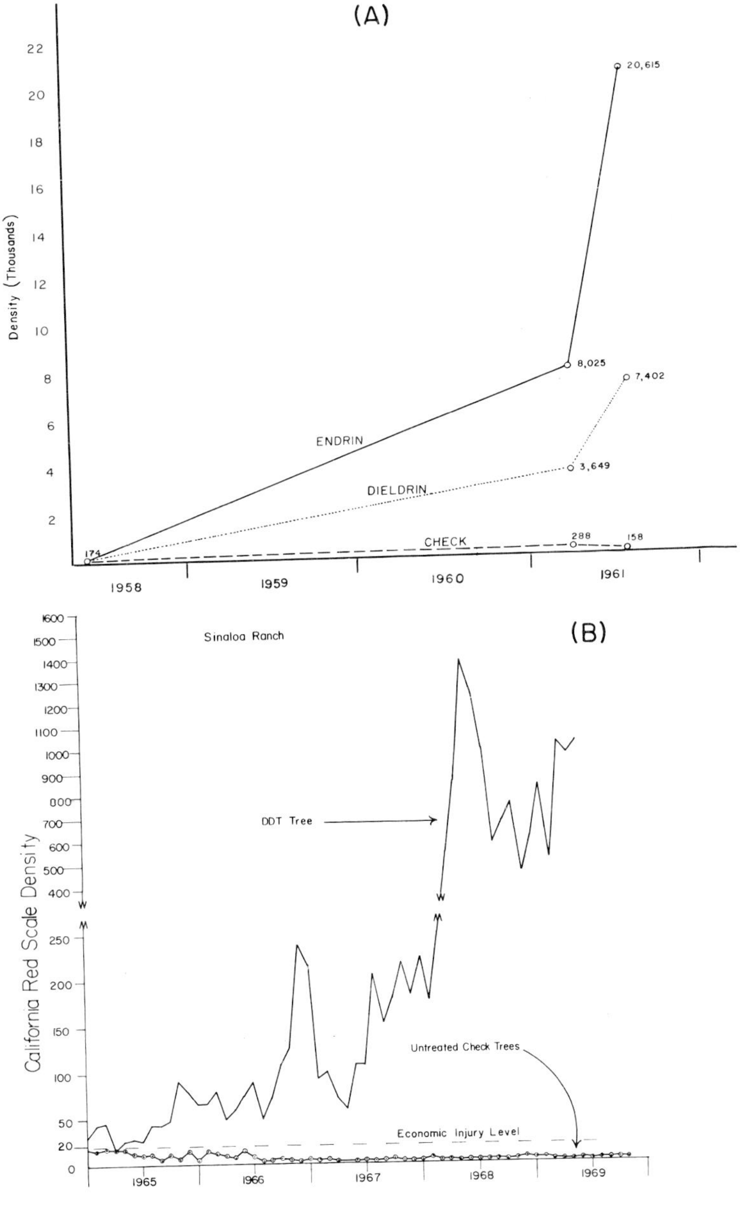


Figure 59. Increase in scale insect densities on citrus resulting from the use of insecticides which inhibit parasite activity but not the scale insects:

A. Effect of endrin and dieldrin on *Aphytis lepidosaphes* resulting in increases of *Cornuaspis* [= *Lepidosaphes*] *beckii*. (From DeBach, 1969a.)

B. Effect of DDT on *Aphytis melinus* resulting in increase of *Aonidiella aurantii*. (From DeBach, Rosen and Kennett, 1971.)

When such massive upsets in natural balance occur, satisfactory biological control can take an inordinately long time to be reattained. One cannot just quit spraying and expect even a very effective parasite to reduce a badly upset host population by the next year (DeBach, 1969a). For example, the upset of the purple scale caused by endrin, depicted in Figure 59 A, took more than three years to be reduced by *A. lepidosaphes* to the original level, even though reservoirs of the parasite were present on untreated trees surrounding the treated trees (Figure 60 A). Similarly, it took three years for a California red scale population, upset as a result of DDT decimating *A. melinus*, to be reduced to the original level (Figure 60 B).

The broad ecological explanation for a normally effective host–parasite population interaction being disrupted by long-term toxic pesticide residues is very similar to that given for upsets caused by adverse weather. More or less regular periodic depression of the parasite population reduces the parasites' effective rate of population increase relative to that of the host, with the result that the host rises to a new and often much higher equilibrium position. The simple fact is that a potentially effective parasite is turned into an ineffective one. The parasite may, but need not necessarily, be greatly reduced; in fact, percent parasitization can be surprisingly high and parasites quite abundant under such outbreak or upset conditions. The degree of increase of the host population is a relative matter depending on how severely and how frequently the parasite population (i.e., its rate of increase) is depressed.

Ecological methods of integrated control or pest management have been investigated in order to conserve *Aphytis* populations in commercial orchards. Strip-treatment of alternate pairs of tree rows with an annual oil spray in southern California citrus orchards left refuges for the purple scale parasite, *A. lepidosaphes*, while reducing pesticide usage 50% and still maintaining commercial control (DeBach and Landi, 1961).

According to Rosen (1967a), of all the materials tested in Israel for scale insect control, only mineral oil proved to be safe for *Aphytis*. He found that the malathion-bait spray used against the Mediterranean fruit fly, *Ceratitis capitata* (Wiedemann), was highly toxic in laboratory tests, but considered that it might be ecologically selective to *A. holoxanthus* in the field and thus not cause upsets of the Florida red scale. On the other hand, in Australia Abdelrahman (1973) found malathion to be relatively more toxic for *A. melinus* than for its host, the California red scale, and concluded that an integrated control program for citrus utilizing full-coverage malathion sprays was not possible. In South Africa, Bedford (1975, 1976) has recommended various integrated control measures aimed to a considerable extent at conserving *Aphytis* populations, and a significant degree of success has been achieved.

Miticides often appear to be less upsetting to parasite populations than insecticides. Over a three-year period in experimental plots, the use of the miticide chlorobenzilate in combination with tedion, kelthane or trithion did not upset perfect biological control of the California red scale, *Aonidiella aurantii*, by *Aphytis chrysomphali* and *A. lingnanensis*. Chlorobenzilate with delnav resulted in an increase of some 2.5-fold of the scale population but this represents only a minor upset (DeBach, Landi and Jeppson, 1959).

**Airborne dust** is an additional factor adversely affecting *Aphytis* populations both in

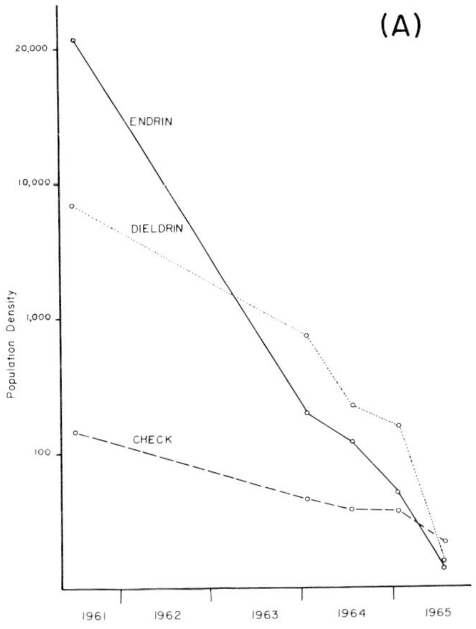


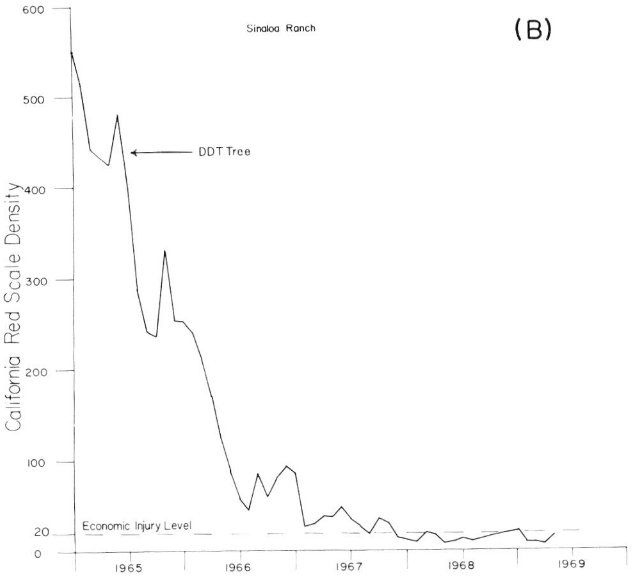


Figure 60. Lengthy period required to reattain biological control of scale insects on citrus after insecticide-induced outbreaks:
A. Case of *Aphytis lepidosaphes* and its host, *Cornuaspis* [= *Lepidosaphes*] *beckii*
(From DeBach, 1969a.)
B. Case of *Aphytis melinus* and its host, *Aonidiella aurantii*.
(From DeBach, Rosen and Kennett, 1971.)

nature and on crops. Its effect is frequently akin to that of highly toxic pesticides, especially when the dust is fine and is deposited regularly. It is commonly noticeable alongside dusty roads. The mode of action of dust particles has not been demonstrated, but they are thought to abrade the integument of intersegmental membranes, causing death by desiccation (Bartlett, 1951). *Aphytis* adults that have been exposed to dust can be observed spending considerable time vigorously cleaning themselves. In nature, armored scale insect populations are commonly found to be noticeably heavier on chronically dusty plants, and from the few experiments that have been conducted this can well be ascribed to adverse effects on *Aphytis* and perhaps on other parasites.

Tests using artificial applications of road dust to citrus trees having the California red scale under biological control by *Aphytis* showed marked increases in the scale population within seven months, as shown in Figure 61. When very lightly dusty trees, in a test plot which was under satisfactory biological control by *A. melinus*, were rinsed with a light water spray at regular weekly intervals, the California red scale populations decreased even further and averaged about half the density on the unwashed trees over a two-year period (DeBach, 1969a). Thus even very light, barely noticeable, dust deposits can interfere with *Aphytis*. When *Aphytis* adults are placed on dusty fruit they are killed so rapidly that oviposition is markedly reduced. For example, one test involved exposing 40 laboratory scale-infested fruits to normal fallout of field dust for thirty days in the Riverside area of California. The fruits were then returned to the laboratory and 20 of them were cleaned by rinsing with water. Ten female *A. lingnanensis*, placed on each fruit for a 24-hour oviposition period, laid only about half as many eggs per female on the dusty as compared with the clean fruit (DeBach, 1969a).

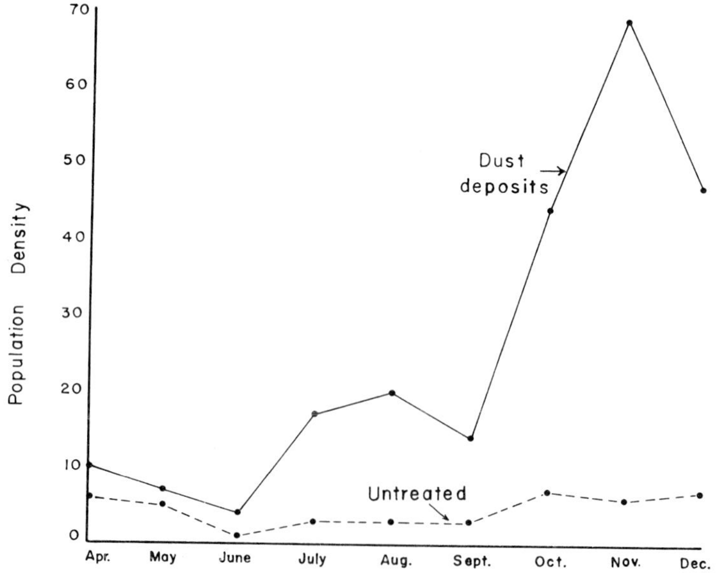


Figure 61. Population increase of *Aonidiella aurantii*, induced by the adverse effect of artificially applied airborne dust on the population of its parasite *Aphytis lingnanensis*. (From DeBach, 1965a.)

## THE EFFECT OF BIOTIC FACTORS

Various biotic factors can have a more or less important influence on the distribution and abundance of *Aphytis*. Principal influences include the host scale insect, the host plant, food for adults, natural enemies of *Aphytis*, intraspecific competition and interspecific competition.

**The host scale insect** affects the fecundity of *Aphytis* in several ways. A given *Aphytis* species will usually have several potential host species, one or two of which will be most "preferred" (Rosen and DeBach, 1977b). This is another way of saying that *Aphytis* species have a rather high degree of host specificity and can discriminate between more suitable and less suitable hosts (Baker, 1976). The more "preferred" the host, the higher the fecundity of *Aphytis* as compared with a less preferred host, and this is a key factor in their efficiency in population regulation. In two series of comparative laboratory tests, *A. mytilaspidis* produced 75 to 80 eggs on the cactus scale, *Diaspis echinocacti* (Bouché), as compared with 25 eggs on the California red scale, *Aonidiella aurantii* (Maskell), and 30 eggs on the oleander scale, *Aspidiotus nerii* Bouché [= *A. hederae* (Vallot)] (Baker, 1976). *A. lingnanensis* produced over twice as many female progeny on oleander scale as on California red scale in comparable tests (DeBach and White, 1960). Using two semispecies of *A. mytilaspidis*, Khasimuddin and DeBach (1976c) showed in comparative laboratory tests that one produced an average of about 206 progeny on cactus scale, but produced none on 5 other scale insect species; the other semispecies produced an average of about 179 progeny on cactus scale, as well as an average of about 32 on the latania scale, *Hemiberlesia lataniae* (Signoret), but produced none on 4 other scale insect species. In all likelihood the semispecies of *A. mytilaspidis* that attacked *H. lataniae*, but had a very low fecundity on it, would not be able to limit populations of this host satisfactorily in the field.

Sex ratio is also affected by the host scale insect species; the poorer the sex ratio, the poorer the relative rate of increase of the parasite population. The sex ratio of *A. lingnanensis*, reared on three scale insect species on the same host plant in the laboratory, was 55% females on *Aonidiella aurantii*, 67% females on *Quadraspidiotus perniciosus* (Comstock), and 80% females on *Aspidiotus nerii* (DeBach and White, 1960). In subsequent comparative tests the sex ratio of *A. lingnanensis* was 35% females on *Aonidiella aurantii* and 63% females on *Aspidiotus nerii* (DeBach and Sundby, 1963).

The developmental stage of the scale insect host is highly important with some host species but of less consequence with others. About 90% of the total duration of all stages of *Aspidiotus nerii* that are large enough to support development of *Aphytis* are successfully attacked, including the egg-laying adult. On the other hand, *Aonidiella aurantii* has more heavily sclerotized adult egg-laying stages that are attached to the dorsal and ventral scale coverings. Such stages are not parasitized by *Aphytis* and because of this only about 20% of the life cycle of *Aonidiella aurantii* is susceptible to parasitism (DeBach, 1969a). This makes satisfactory biological control of such a species more difficult than of one that can be parasitized during most of its life cycle. Baker (1976) concluded that the molt stages were unsuitable because of their hard integument. The stage (or size) of the scale also affects the sex ratio. *Aphytis* progeny tends to be predominantly male on smaller scales, and the females produced are small and have a low reproductive potential.

From an ecological viewpoint it can be very important to an *Aphytis* species to have alternate host species in the same ecosystem, especially if the primarily preferred host is one such as the California red scale, which may at times and in places consist predominantly of the adult female stage that cannot be parasitized. Currently, we are placing artificial infestations of oleander scale in citrus groves in attempts to maintain *Aphytis melinus* populations during periods when the California red scale largely consists of adult females that are unsuitable for parasitism. Because of periodic asynchrony of *Aphytis* with the host, evolutionary strategy which is certainly toward specialization (i.e., toward high host specificity) in this group may generally have stopped short of monophagy.

**The host plant** may have an affect on *Aphytis* fecundity and efficiency in more than one way. As a microhabitat it can provide either good or poor conditions for escape from, or moderation of, a harsh physical environment. It has been observed many times in various countries that *Aphytis* species can be effective against the California red scale on large trees but ineffective on young, small trees. We have observed similar effects in the field between adjacent different host plant species. Of course, part of this effect may be indirect, operating through an effect of the host plant on the host scale insect. The size of scale insects (and presumably nutrition for parasites) is affected by the host plant (Smith, 1957). The same author indicates that given a choice, *A. lingnanensis* parasitized more California red scale on sago palm than on yucca. The same scale insect species, present on potato tubers, was not parasitized at all. This was largely ascribed to the fact that the potato cuticle grows over the scale covering and protects it from the parasite.

*Aphytis chrysomphali*, when given a simultaneous choice in laboratory cages of *Aonidiella aurantii*-infested citron melons, grapefruit, oranges and potato tubers, preferred to oviposit in the scale insects on the host plants in the order named. Further tests, comparing oviposition on *A. aurantii* on grapefruit with that on potato tubers, showed that *A. chrysomphali* laid over three times as many eggs on the former as on the latter where an equal choice was given (Bartlett and Fisher, 1950). Thus it seems logical that different degrees of biological control of the same scale insect species in the field can result from the same *Aphytis* species, when the scale insect host occurs on different host plant species.

**Food** for adult *Aphytis* is a requisite that is continuously necessary. We learned years ago that *Aphytis* species cannot be cultured unless honey, sugar-water or a similar carbohydrate food is always provided in the cage. Most adults die within about 24 hours without such food (DeBach and White, 1960). Obviously in nature a similar food must be available and apparently it generally is, judging from the frequent success of *Aphytis* in biological control.

The main sources of carbohydrate food in nature are evidently honeydew from homopterous insects and plant nectars. Citrus nectar appears to be an excellent food for *Aphytis* adults, as good as honey or even better, whereas the quality of honeydew differs markedly with its insect source (see Gerson, 1968; Avidov, Balshin and Gerson, 1970). Attempts have been made in the USSR to improve the efficiency of *A. proclia* in the biological control of the San Jose scale, *Quadraspidiotus perniciosus*, by planting a *Phacelia* cover crop to provide nectar and/or honeydew. Three crops of *Phacelia* increased the proportion of parasitized scales from 5% to 76% (see van den Bosch and Telford, 1964). Some similar attempts have been made in commercial citrus orchards in California. The technique offers interesting possibilities to manipulate the ecology of *Aphytis*.

Proteinaceous food for adults is necessary for *Aphytis* to maintain adequate fecundity (DeBach and White, 1960). This is acquired by predatory host-feeding, which is discussed in the section on "Adult Behavior" (p. 48). Host-feeding has also already been mentioned in the preceding discussion of "Host Population Regulation" as often being the predominant cause of host mortality in the field. As far as has been observed, this phenomenon does not interfere (at least in a practical way) with the host population regulatory ability of *Aphytis*. In fact, it may provide a mechanism for *Aphytis* species to reduce more rapidly host populations that have been upset for one reason or another.

**Natural enemies** of *Aphytis* are of little consequence from a population dynamics standpoint, with the exception of the adverse effect of ants on adults. Traboulsi (1968) listed several predators and hyperparasites that are apparently more or less incidental. In South Africa, *Aphytis holoxanthus* was subject to attack by *Marietta exitiosa* Compere, but no harmful influence of this secondary parasite on the primary *Aphytis* was shown (Cilliers, 1971). In California on occasion, but more particularly with *A. lepidosaphes*, we have reared *Marietta mexicana* (Howard) as a hyperparasite of *Aphytis*. It is so rare as to be inconsequential.

**Honeydew-seeking ants** probably rate next in importance to weather extremes and pesticides as a detrimental influence on the ability of *Aphytis* species to regulate host population densities at low levels. The ants affect *Aphytis* mainly by interference with adults and to a certain extent by direct predation on them. However, *Aphytis* adults usually escape being captured by ants, because at the first contact they tend to jump and fly. This results principally in *Aphytis* adults being constantly disturbed and interfered with if ants are at all abundant. The effect is to greatly reduce the effective progeny production, or *R* (see DeBach, 1966) and therefore the relative rate of increase of the parasite population with respect to that of the host. As a result, host populations will exist at a much higher level in the presence of honeydew-seeking ants, often as much as 50- to 100-fold or more. In California we have seen striking increases in various armored scale insect populations induced by ant interference with *Aphytis*, including *Aonidiella aurantii*, *Cornuaspis* [= *Lepidosaphes*] *beckii*, *Aspidiotus nerii* and *Quadraspidiotus perniciosus*. Since armored scale insects produce no honeydew, ants do not tend them and hence have no direct effect on them, as they might have on honeydew-producing coccids, yet their indirect effect may be just as pronounced. (See DeBach, Dietrick and Fleschner, 1951; DeBach, 1958a; Steyn, 1958.)

Interference with *Aphytis* by ants can be so drastic that certain ant species have been purposefully utilized as a "biological check method" to evaluate the effectiveness of *Aphytis* in biological control. Differences between host population densities on ant-free trees where parasites are undisturbed, as compared with ant-infested trees where parasite efficiency is greatly reduced, demonstrate the degree of biological control being exerted (DeBach, Fleschner and Dietrick, 1951; DeBach and Huffaker, 1971). We have observed that the effect of ants depends to some extent on their mobility and aggressiveness, which varies with the species.

**Competition** in *Aphytis* can be conveniently divided into intraspecific and interspecific categories. A voluminous literature exists concerning intraspecific competition among parasites in general, under such headings as superparasitism, cannibalism, interference, discrimination, etc. All these subjects represent inherent characteristics of individuals that, when operating in a population, act as feedback mechanisms to prevent parasites from increasing or decreasing excessively, so that neither host nor parasite becomes exterminated. Rather, a characteristic state of balance results. Thus, intraspecific competition such as superparasitism should not be considered as detrimental, except under artificial conditions when optimization of insectary production is sought.

Only fragmentary knowledge of the intrinsic mechanisms involved in maintaining balance exists for *Aphytis*. *Aphytis* species are able to discriminate between parasitized and unparasitized scale insects and tend to avoid oviposition in parasitized hosts (J. C. van Lenteren, personal communication), thus avoiding superparasitization unless the pressure to oviposit is too great. Competition between supernumerary larvae of the same species results in somewhat decreased survival to the pupal stages, but there are differences in degree between species. When there are three larvae per scale, 40% of them die with *A. lingnanensis*, 34% with *A. fisheri* and only 6% with *A. melinus*. This compares with 100% survival of each when only one larva is present per scale. The cause of mortality was not ascertained but is thought to be starvation and perhaps occasional accidental injury. Interspecific competition between larvae produces somewhat greater effects than intraspecific competition (DeBach and Sundby, 1963).

The density of adult females has marked effects on fecundity. In comparative tests, four adult females per host-infested lemon fruit gave 31.8 progeny per female for *A. lingnanensis*, 25.9 for *A. fisheri* and 24.1 for *A. melinus*, whereas with 40 adult females per fruit (a 10-fold increase in density), progeny per female was 7.0 for *A. lingnanensis*, 7.1 for *A. fisheri* and 6.5 for *A. melinus* (DeBach and Sundby, 1963). Previously, DeBach and Sisojević (1960) had shown that increasing density of adult females from 2 to 20 per infested fruit had striking effects on progeny production with *A. lingnanensis* (86% reduction) but rather moderate effects with *A. chrysomphali* (45% reduction). Any increase in the adult density of *A. lingnanensis* above the minimum (2) caused reductions in progeny production, whereas with *A. chrysomphali* there was an optimum density (5) both below and above which progeny production decreased.

Perhaps the most important long-term result of intraspecific competition in *Aphytis* is evolution leading to speciation. Selection acting against the just discussed mechanisms involved in intraspecific as well as in interspecific competition should operate to result in host population regulation at lower and lower levels, because this will eliminate competition from other natural enemy species having lesser searching capacities. This means that such a "successful" *Aphytis* species will be scarce, even rare. However, from an evolutionary standpoint the success of a species is measured not in terms of numbers but rather in its ability to persist and successfully adapt to changing environmental conditions. The commonness of infraspecific forms (semispecies, etc.), as in the *lingnanensis, mytilaspidis* or *maculicornis* complexes, indicates marked ability of *Aphytis* to adapt and evolve rapidly.

Research on interspecific competition with *Aphytis* has resulted in significant contributions to basic ecological theory and to the theory and practice of biological control. Two distinct phases are involved, and the results are completely different in many respects. Competition between two or more ecological homologues, which are usually closely related species of *Aphytis*, results in competitive displacement (eradication) or exclusion of one species by another. As will be discussed later, this is not detrimental to biological control or host population regulation. On the other hand, competition between nonhomologues, which usually involves a species of *Aphytis* competing with one or more endoparasitic species for the same host, does not result in displacement but rather in complementary action that again tends to increase the total degree of biological control, i.e., the average host population density is decreased.

Nearly all of the studies on competitive displacement in nature of one species of *Aphytis* by another, and of the theory and consequences involved, have been conducted in California and reported upon in a series of papers that, inter alia, depict the chronological history of development of the field phenomena from their first vague comprehension, through occasional contradictory indications, up to their more or less complete understanding. Supportive laboratory studies were also conducted in an effort to explain the mechanisms involved. (See DeBach, 1950; DeBach, 1954; DeBach and Sisojević, 1960; DeBach and White, 1960; DeBach and Sundby, 1963; DeBach, 1965a, 1965b, 1966, 1969a; DeBach, Rosen and Kennett, 1971; DeBach and Rosen, 1976a.) There is too much involved in these studies to attempt to cover them in much detail, and so a brief summary follows.

Originally only one species, *A. chrysomphali*, parasitized the California red scale, *Aonidiella aurantii*, in southern California. The species became accidentally imported and established probably somewhat prior to 1900. It was generally distributed throughout the citrus districts wherever *Aonidiella aurantii* occurred by the 1920's, was sometimes locally abundant and occasionally—but not generally—thought to be important in biological control of that pest. It remained the only species established in California up to 1948, existing at that time in all coastal, intermediate and interior citrus districts (an area of some 4000 square miles), but was scarce in the interior districts due to adverse effects of climate already discussed.

*Aphytis lingnanensis* was purposely introduced from southern China in 1947–1948 and became established at once. It has exactly the same essential needs and habits as *A. chrysomphali*, i.e., they are ecological homologues. *A. lingnanensis* rapidly displaced *A. chrysomphali* in the interior and intermediate areas but more slowly in the coastal areas, so that by 1958 *A. chrysomphali* became virtually restricted to strictly coastal areas (see Figure 62). The next year, 1959 (see Figure 63), showed a further significant reduction of *A. chrysomphali*. Subsequently it was gradually further displaced, until today it exists in only a few small pockets immediately adjacent to the ocean (Figure 65). The mild microclimates of these pockets are apparently fairly optimal to *chrysomphali* and give it a relative advantage so that it is able to exclude *lingnanensis*.

Meanwhile, beginning in 1957, *Aphytis melinus*, an ecological homologue of both *A. lingnanensis* and *A. chrysomphali*, was newly imported from India and Pakistan

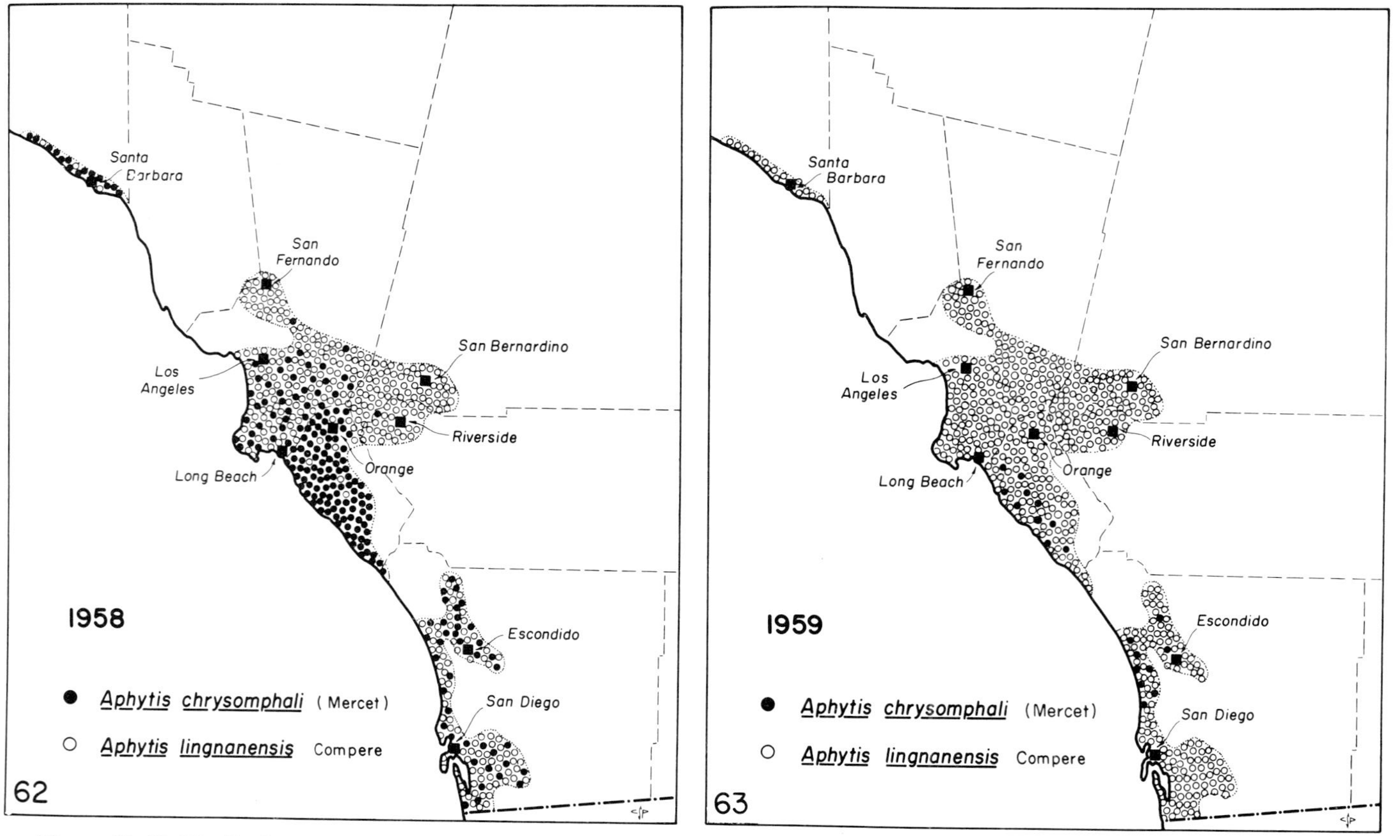


Figures 62, 63. Distribution and relative abundance of *Aphytis chrysomphali* and *Aphytis lingnanensis* in southern California. Semidiagrammatic. (From DeBach and Sundby, 1963.)
Figure 62 shows the amount of competitive displacement of *A. chrysomphali* by *A. lingnanensis* during the decade 1948–1958. Figure 63 shows the almost complete additional displacement of *A. chrysomphali* by *A. lingnanensis* by 1959.

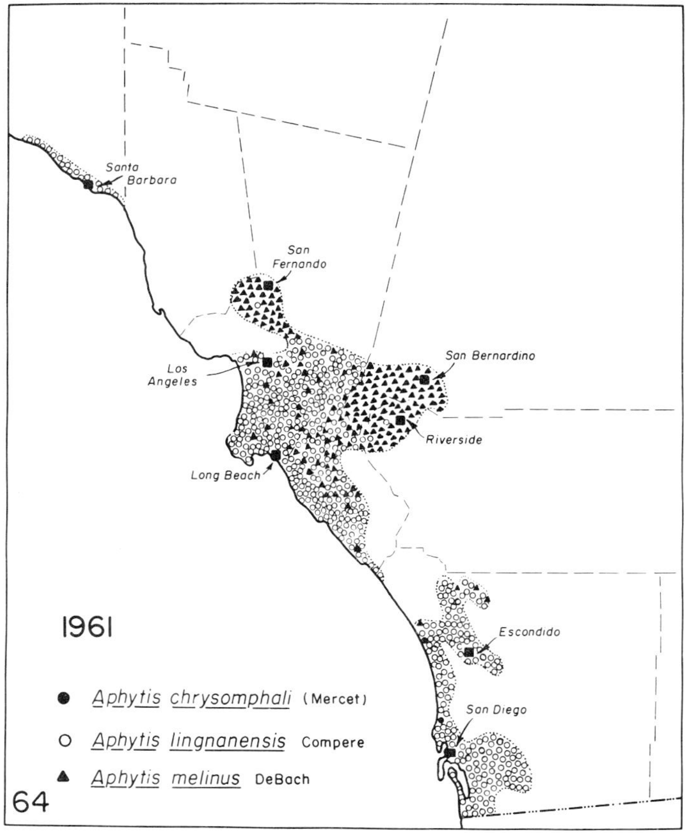


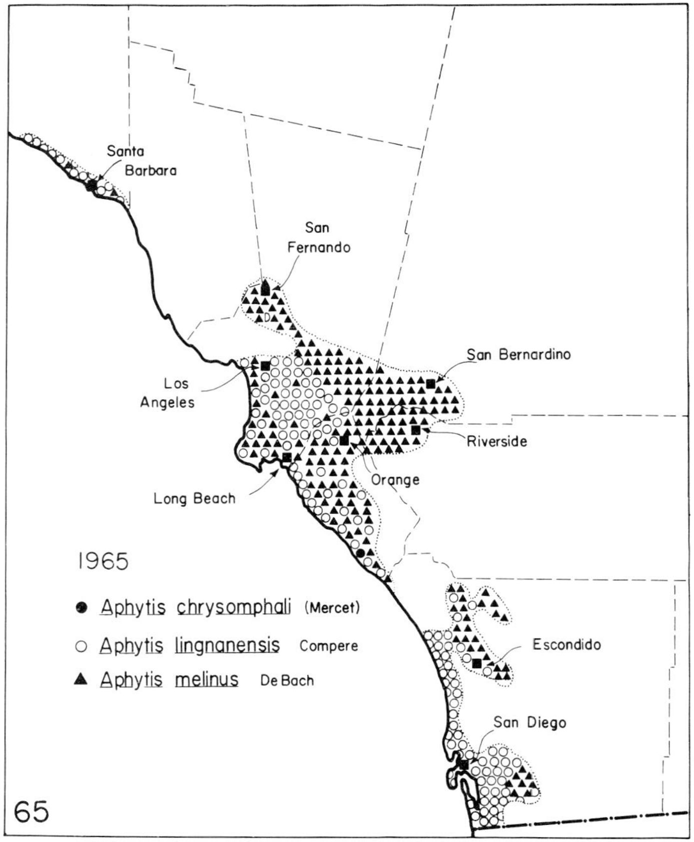


Figures 64, 65. Distribution and relative abundance of *Aphytis chrysomphali*, *Aphytis lingnanensis* and *Aphytis melinus* in southern California. Semidiagrammatic.
Figure 64 illustrates the virtual extinction of *A. chrysomphali* and the early progress of *A. melinus* in interior areas. (From DeBach and Sundby, 1963.) Figure 65 shows the further spread of *A. melinus* and gradual displacement of *A. lingnanensis* by 1965 in interior areas. (From DeBach, Rosen and Kennett, 1971.)

and colonized extensively in all citrus areas. It became rapidly and rather generally established and by 1960–1961 it had spread well through interior areas and largely replaced *A. lingnanensis*, which had earlier replaced *chrysomphali* in such areas (see Figure 64). *A. lingnanensis* meanwhile remained dominant in the intermediate and most of the coastal areas and precluded the establishment of *A. melinus* in the latter. By 1965 *A. melinus* had virtually completed the displacement of *A. lingnanensis* from all interior areas and had further decimated and replaced *lingnanensis* in the intermediate areas (see Figure 65). Since then the situation has remained more or less static: at present (1977) *A. melinus* is the only *Aphytis* found on *Aonidiella aurantii* in interior areas, and *A. lingnanensis* the only one found in coastal areas with the exception of the few isolated pockets immediately bordering on the ocean, where *A. chrysomphali* is still found. Both *A. melinus* and *A. lingnanensis* exist in varying proportions in intermediate areas, as would be expected in a zone of distributional overlap.

These results show that ecological homologues, that is, two or more species having truly identical ecological niches (i.e., attacking the same host stages, requiring the same adult food, utilizing the same alternate hosts, etc.), cannot coexist in a particular defined habitat—but they may exist close together in adjacent areas having slightly different physical conditions, because the winner is determined by its effective progeny production, *R* (DeBach, 1966). However, it should be noted that relative *R* values can be reversed in different environments, thus reversing the winner (DeBach, 1966). It is significant that competitive displacement occurred in places in spite of food (the host scale insect *Aonidiella aurantii*) being abundant. Thus actual food scarcity—and food (i.e., hosts) is the requisite being competed for—is not necessary for competitive displacement to occur. Neither is active aggression of any sort. This is contrary to the opinion of many ecologists (DeBach and Sundby, 1963). Very briefly, interaction between intrinsic fecundity and environmental factors produces a certain *R* value (female progeny (that reproduce) per female parent) and the species with the greater *R* value will be the winner among competing ecological homologues. The old idea of parasites winning only because they are "intrinsically" or "extrinsically" superior is thus only part of the picture.

In spite of all this intense and continuing competition between parasites, in spite of what is termed multiple parasitism, the extent of biological control has gradually increased until today California red scale populations are much lower in general than they have ever been in southern California. Thus competition between parasite species is not harmful to the degree of host population regulation exerted; rather it is beneficial even though one parasite species may be eliminated. It seems evident that the replacement of one *Aphytis* species by another means that the second is more effective in the given habitat and hence produces better biological control.

From careful analysis of all the detailed data obtained, both in the laboratory and in the field, we see no possibility that the observed results could have been predicted with any accuracy beforehand. Laboratory studies on the biological characteristics of *A. chrysomphali* as compared with *A. lingnanensis*, and of the latter as compared with *A. melinus*, including data on longevity, fecundity, temperature responses, etc., perhaps give some indications but still do not permit us to explain with any degree of certainty why, for example, *A. melinus* is the better adapted parasite (i.e., the winner) in interior climatic areas whereas *A. lingnanensis* remains so in coastal areas. Most laboratory data would seem to indicate that *A. lingnanensis* should have been the winner in all areas (DeBach and Sundby, 1963), except perhaps over *chrysomphali* in immediate coastal areas (DeBach and Sisojević, 1960).

These results have powerful implications for ecological theory and, of course, for biological control which is a phase of applied ecology. Traditional importation policy in biological control, which has been questioned by some, seems to be clarified. Very briefly it means that we can reduce promising natural enemy candidates for importation to a relative few, based on host specificity, evidence of field effectiveness abroad, etc., but we cannot predict with certainty which single species, or which combination, will be the best even in one new habitat, much less in the variety of habitats usually inhabited by a major pest. Thus, multiple importations of as many natural enemies from abroad as show any promise are indicated, since competition between parasite species is not detrimental to host population regulation (see DeBach, 1971a).

In one instance, somewhat contradictory conclusions have been advanced which indicate that *A. chrysomphali* and *A. melinus* can coexist as parasites of *Aonidiella aurantii* in the same habitat. Laboratory experiments on these two species of *Aphytis* that attack *Aonidiella aurantii* in South Australia by Abdelrahman (1974a, 1974b) led him to the conclusion that *A. chrysomphali* is more adapted to cold and less to heat than *A. melinus*, and that this adequately explains why the two species *coexist* (our italics) and fluctuate in relative abundance seasonally and annually in extreme inland environments, *A. chrysomphali* holding the advantage in winter and *A. melinus* in summer. He apparently did not conceive of the possibility of an ongoing process of competitive displacement of *A. chrysomphali* (accidentally established for many years) by *A. melinus* (introduced from California in 1961) being in operation. This is odd in light of the fact that since about 1970 we have been apprised by field entomologists of Victoria, Australia, that *A. melinus* appears to be replacing *A. chrysomphali* here and there. We do not consider that two ecological homologues such as *A. chrysomphali* and *A. melinus* can coexist indefinitely in the same habitat and act in a complementary manner in the control of the California red scale, as Abdelrahman alleges. We predict that the "coexistence" will last for a matter of only a few years. On the basis of his own results and his conclusion that *A. chrysomphali* and *A. melinus* "coexist" in extreme inland environments in South Australia, Abdelrahman (1974b, p. 228) appears puzzled by the "fact" that in California *A. chrysomphali* fails to exist in extreme inland environments. Even though he cites the appropriate literature (DeBach and Sundby, 1963), he has overlooked the fact that *A. chrysomphali* did exist in such interior area environments in California for some 50 years following its original establishment, but was eliminated by competitive displacement from them within about 10 years following the importation of its ecological homologue, *A. lingnanensis*, in 1948.

On the other hand, additional cases of competitive displacement between species of *Aphytis* have been reported. In the Peloponnesus, Greece, *A. melinus* was introduced in 1962, and has since largely displaced *A. chrysomphali* from populations of *Aonidiella aurantii* and *Chrysomphalus dictyospermi* (Morgan) on citrus (DeBach and Argyriou, 1967; Argyriou, 1974). *A. melinus* appears to have displaced *A. chrysomphali* from certain important citrus-growing areas in Israel. However, the latter has remained dominant in certain coastal areas, whereas *A. coheni* remains dominant in the hot inland valleys (DeBach and Rosen, 1976a).

Competition between parasite species that are not ecological homologues (i.e., they have different ecological niches) does not necessarily or usually lead to elimination of one by the other, but rather the two may complement each other so that better host population regulation occurs. Niche difference between parasite species can be subtle. If, for example, one attacks eggs, another larvae and a third pupae, all have different ecological niches and can theoretically coexist. This occurred in the studies previously reported for California. Aside from the displacement between the *Aphytis* species, two species of internal parasites of *Aonidiella aurantii* coexisted with one or the other species of *Aphytis* during and following the displacement process. *Prospaltella perniciosi* Tower and *Comperiella bifasciata* Howard attack slightly different stages of the host, including molt stages and ovipositing females, and hence are not strict ecological homologues of *Aphytis* species that are restricted to second and third instars. Neither of these two endoparasites is as effective a parasite of *Aonidiella aurantii* as are the *Aphytis* species, but *P. perniciosi* complements the work of *Aphytis* in coastal areas and *C. bifasciata* performs a similar role in inland areas. This gave rise in inland areas to the interesting ecological anomaly of the most efficient parasite, *A. melinus*, completely displacing the second most effective parasite, *A. lingnanensis*, but permitting the least efficient, *C. bifasciata*, to persist (DeBach and Sundby, 1963; DeBach, 1966; DeBach, Rosen and Kennett, 1971).

The best documented case concerning complementary action between parasite species in enhancing host population regulation concerns *A. paramaculicornis* (recorded at the time as the "Persian strain" of *A. maculicornis*) and *Coccophagoides utilis* Doutt (Huffaker and Kennett, 1966; DeBach, Rosen and Kennett, 1971).

*Aphytis paramaculicornis*, the "Persian *Aphytis*," was imported into California from Iran and Iraq in 1951 and colonized in the field for control of the olive scale, *Parlatoria oleae* (Colvée). Early results were very promising. *A. paramaculicornis* became easily established and attained levels of parasitization of 90% or more. Scale populations were generally reduced by over 90% and in certain favorable localities *A. paramaculicornis* produced good control. However, in general, even this striking reduction was economically unsatisfactory in many cases, because if as much as a single scale occurs on an olive fruit it may be culled. This indicated that additional suppressive natural enemies were necessary (Huffaker, Kennett and Finney, 1962).

This led in 1957 to the importation of *Coccophagoides utilis*, an endoparasite, from Pakistan and its colonization in California. By 1961 it was well established at two release sites, after which it was mass-reared, widely colonized and generally established. *C. utilis* is not an ecological homologue of *A. paramaculicornis*, because it attacks earlier stages; hence the two can coexist. It proved to be an effective addition, and its introduction into California resulted in complete control of the olive scale.

*Coccophagoides utilis* attains a moderately high average parasitization of about 40% on each of the two annual generations of the host. It apparently would not have controlled the olive scale alone, but it substantially adds to host mortality during the summer (when *A. paramaculicornis* is depressed by high temperatures) and contributes additional mortality to the fall scale generation. Thus *C. utilis* acts as a complementary factor in the regulation of olive scale population density.

Although a considerable degree of competition (i.e., multiple parasitism) may occur, the two parasites perform in a compensatory manner and the total parasitization resulting from their combined action is higher on an annual basis than that produced by *A. paramaculicornis* alone, and considerably higher than that produced by *C. utilis* alone. However, to a large extent, parasitism by *C. utilis* merely replaces mortality of olive scale which would have been caused by *A. paramaculicornis* acting alone. Thus, during the spring generation, for example, although *C. utilis* may parasitize some 38% of the scales and *A. paramaculicornis* 50% (for a total of 88%) when acting together, *A. paramaculicornis* would produce about 83% if acting alone. This means that the so-called indispensable contribution to olive scale mortality by *C. utilis* would be only about 5% rather than the observed 38%. It is emphasized, nonetheless, that this "indispensable parasitism" by *C. utilis*, although accounting for only a small proportion of total mortality, actually represents an addition which assures reliable, permanent biological control of *Parlatoria oleae* at extremely low population levels (Huffaker and Kennett, 1966; DeBach, Rosen and Kennett, 1971).

These results confirm our conclusions that, virtually without exception, interspecific competition, either between ecological homologues or nonhomologues, is not disadvantageous but rather beneficial as far as host population regulation is concerned. This is true even though one or more of the individual parasite species involved suffer from competition ranging from a reduction in their effective rate of increase to total elimination or displacement. As an ecological basis for biological control endeavors, this means that multiple species importations of new natural enemies should always be the objective in importation projects (DeBach, 1971c).

# ECONOMIC IMPORTANCE AND UTILIZATION

The species of *Aphytis* are now generally recognized as the most effective natural enemies of armored scale insects. However, from the early days of economic entomology until about the last quarter-century, their potential value was generally unappreciated.

Le Baron (1870) was apparently the first to draw attention to the great economic importance of what is now known as a species of *Aphytis*. Writing about *A. mytilaspidis* as a natural enemy of the oystershell scale, *Lepidosaphes ulmi* (L.), on apple in Illinois, he stated that "more than twice as many lice are destroyed by the Chalcis . . . than by all other causes combined." Observing that the oystershell scale was "loosening its hold upon the apple tree," he concluded: ". . . many, no doubt, have prided themselves upon the successful application of some infallible wash, or patent nostrum; but underneath all this goodly show, busily intent upon the accomplishment of her own curious economy, and heedless of the momentous results she is effecting in human interests, works unseen our infinitesimal friend, the Apple-tree Bark-louse parasite . . ."

Le Baron was also the first to attempt the utilization of *Aphytis* in biological control.

In 1871–1872 he transferred *A. mytilaspidis* from Geneva, Illinois, where it was abundant, to Galena, Illinois, where it was not supposed to occur, and "at the end of a year evidence was obtained to show that the parasites had certainly become domiciled in their new quarters" (Howard, 1881).

Howard (1895) acknowledged the general importance of the Aphelinidae as natural enemies of armored scale insects. However, with special reference to *A. mytilaspidis* he cautioned that "it is plain that the good work of these parasites has been overestimated." In England, Imms (1916), too, concluded that *A. mytilaspidis* was of little economic value.

As early as 1906, George Compere recognized the outstanding economic value of a species of *Aphytis* (presumably *A. lingnanensis*) as a natural enemy of the California red scale, *Aonidiella aurantii* (Maskell) on citrus in China, and enthusiastically advocated its importation into California. Unfortunately, his advice was ignored due to Howard's misidentification of that species as one already occurring in California (see Compere, 1961).

Various subsequent workers have expressed a variety of views. It is only since the late 1940's that the great economic importance of *Aphytis* has become more or less generally accepted.

As discussed in the preceding section, the species of *Aphytis* have some of the most important attributes of effective natural enemies. Their life cycle is considerably shorter than that of their hosts, they are usually capable of inflicting high mortality in host populations both through oviposition and through predatory host-feeding, they are rather host-specific, and are relatively free of attack by hyperparasites. Most significantly, they appear to have an excellent searching ability. Although the various species obviously differ one from the other in some of these attributes, and in spite of their sensitivity to certain environmental conditions, the species of *Aphytis* are generally superior in effectiveness to the predators and endoparasites of armored scale insects.

Flanders (1971) contended that the ectoparasitic species of *Aphytis* are invariably "intrinsically superior" to their endoparasitic competitors. In other words, whenever actual competition on an individual host scale occurs, *Aphytis* supposedly wins by destroying the host along with the developing endoparasite. Hence his conclusion that "the intrinsically inferior endoparasite in order to be competitive must be extrinsically superior, that is, able to find its host (to subsist) at lower host densities . . ." Which is to say that if endoparasites can survive competition with *Aphytis*, they "must" be better searchers, capable of controlling host populations at lower levels than do the species of *Aphytis*. Thus, he reasoned, endoparasites and not *Aphytis* are the more effective natural enemies of armored scale insects.

Unfortunately, this rather circuitous reasoning is not supported by hard evidence. For one thing, Flanders' thesis overlooks the fact that, after a certain developmental stage, an endoparasitized scale is usually rejected by the ovipositing *Aphytis* female (DeBach, Rosen and Kennett, 1971). Thus the endoparasite becomes, in a sense, intrinsically superior to *Aphytis*. Then, of course, the coexistence of *Aphytis* and endoparasites can usually be explained by the fact that they are not strict ecological homologues (DeBach and Sundby, 1963): endoparasites may attack first-instar larvae and molt stages of their hosts, which are not available to *Aphytis* (see p. 47), and often have different alternative hosts. Most important, if endoparasites were indeed capable of maintaining host populations at densities lower than the equilibrium established by *Aphytis*, they would eventually displace the species of *Aphytis* simply by reducing the host populations to levels at which *Aphytis* could not exist.

The bulk of available records, including our own personal experience in various countries, indicates that more often than not the opposite is the case. Species of *Aphytis* are usually the most abundant, and often appear to be the most effective, natural enemies of their respective hosts in various agricultural and natural habitats. Numerous successful biological control projects (see below) also attest to the effectiveness of *Aphytis* species as natural enemies. Rarely have endoparasites effected complete biological control of armored scale insect pests by themselves, although in certain cases they have been very useful in complementing the work of *Aphytis* (see DeBach, Rosen and Kennett, 1971; Rosen and DeBach, 1978). Whereas a detailed evaluation of the potential importance of endoparasites or predators is outside the scope of this monograph, there is little doubt that the species of *Aphytis*, as a group, are by far the most important natural enemies of armored scale insects, and should be emphasized in all biological control projects directed against those serious pests.

Naturally occurring species of *Aphytis* have been instrumental in the natural control of many potentially injurious pests in various ecosystems, without deliberate intervention by man. *A. aonidiae* on the San Jose scale, *Quadraspidiotus perniciosus* (Comstock);

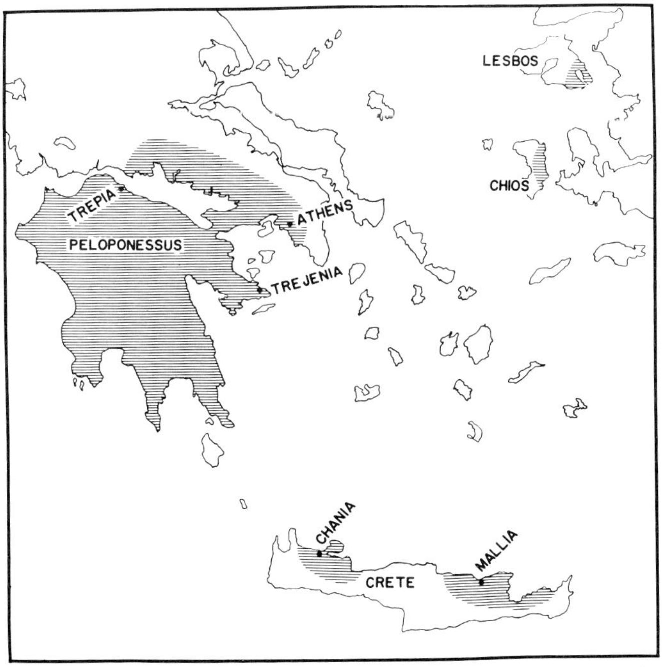


Figure 66. Spread of *Aphytis melinus* DeBach in southern Greece. This species was colonized at Athens, Trejenia, Chania and Mallia during October–November 1962. The shaded areas show its known approximate distribution in December 1966. (From DeBach and Argyriou, 1967.)

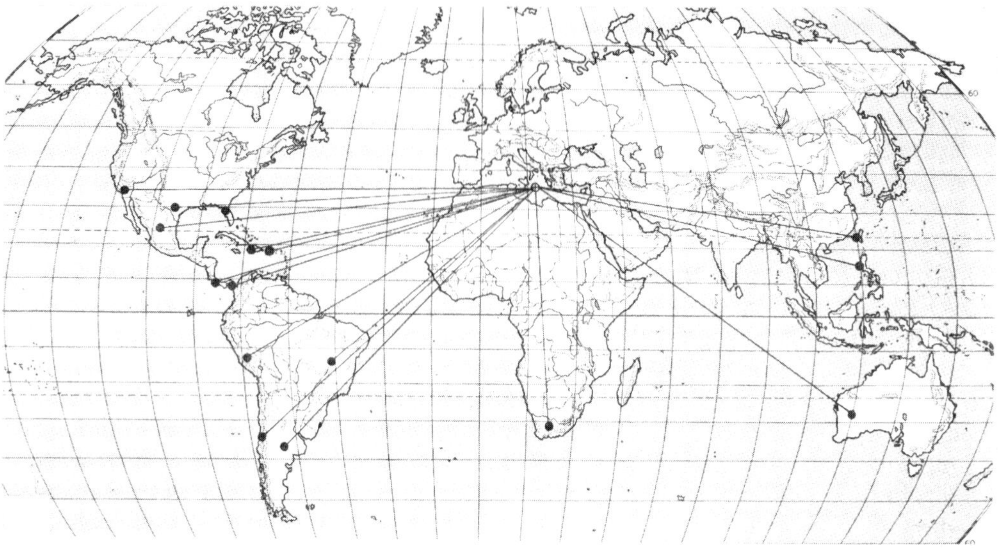

Figure 67. Ecesis of *Aphytis chrysomphali* (Mercet) from a hypothetical center of origin in the Mediterranean Basin. (From DeBach, 1971b.)

*A. hispanicus* on the chaff scale, *Parlatoria pergandii* Comstock; and *A. mytilaspidis* on the oystershell scale, *Lepidosaphes ulmi* (L.), are but a few examples (see Lord, 1947; Dean, 1965; Gulmahamad and DeBach, 1978; and others).

Imms (1916), noting that adult wasps of *A. mytilaspidis* seldom resort to flight, concluded that "the migratory powers of the insect are, therefore, extremely limited." This is certainly not the case with other members of the genus. As mentioned earlier (p. 42), *Aphytis* adults often fly on sunny days. When introduced into new areas, certain species of *Aphytis* have been known to disperse very rapidly over long distances from the original points of release. (See, for instance, Figure 66, showing the rapid dispersal of *A. melinus* in Greece.)

Other species have dispersed around the world of their own accord, or have been accidentally transported with their hosts by man. DeBach (1971b) documented numerous cases of fortuitous biological control that have resulted from accidental ecesis (migratory establishment) of *Aphytis* spp. *A. lepidosaphes*, for instance, has successfully colonized as many, or more, countries as have been purposely colonized by man, and has effected substantial to complete biological control of the purple scale, *Cornuaspis* [= *Lepidosaphes*] *beckii* (Newman), in nearly all of them. Figure 67 depicts the ecesis of another species, *A. chrysomphali*, now nearly cosmopolitan, from a hypothetical center of origin in the Mediterranean Basin.

## UTILIZATION IN BIOLOGICAL CONTROL

Several species of *Aphytis* have played a major role in eminently successful biological control projects directed against serious armored scale insect pests. Extensive reviews of these projects have recently been presented by Rosen and DeBach (1978) and DeBach and Rosen (1976a). Following is a brief account of some outstanding cases. Additional details are discussed under the various species concerned.

The introduction of *A. lingnanensis* into California from China in 1947 against the California red scale, *Aonidiella aurantii* (Maskell), marked a significant turning point in the long campaign against this serious pest of citrus, as well as in the appreciation of *Aphytis* spp. as effective biological control agents. Substantial biological control was achieved, especially in the coastal citrus areas of southern California. *A. lingnanensis* was also utilized successfully in an experimental augumentation program of periodic releases against the California red scale in certain districts of California, and was transferred and established on this pest in several other countries. It is currently being established in Florida, with promising initial results, against the citrus snow scale, *Unaspis citri* (Comstock).

*A. melinus*, introduced into California in 1956–1957 from India and Pakistan, soon proved to be the most effective known natural enemy of the California red scale. Displacing *A. lingnanensis* (except in coastal zones) as it spread rapidly in southern California (see p. 73), it has become the dominant parasite of this pest in most of the citrus-growing areas, effecting a marked improvement in biological control. Since 1959, *A. melinus* has been transferred from California to numerous other countries, where it has become an important parasite of the California red scale. In addition, it has proved to be an effective natural enemy of the yellow scale, *Aonidiella citrina* (Coquillett), the oleander scale, *Aspidiotus nerii* Bouché, and the dictyospermum scale, *Chrysomphalus dictyospermi* (Morgan), in various areas.

*A. lepidosaphes* was introduced into California from China in 1948–1949 against the purple scale, *Cornuaspis* [= *Lepidosaphes*] *beckii* (Newman), another serious pest of citrus. Even though it was initially only partly successful in controlling the purple scale, its establishment in California was shown experimentally to have made an effective integrated control program feasible. *A. lepidosaphes* has since become established, either by purposeful introduction or by ecesis, in 24 additional countries and has effected substantial to complete biological control of purple scale in all of them.

*A. paramaculicornis* (recorded at the time as the "Persian strain" of *A. maculicornis*) was introduced into California from Iran in 1951 against the olive scale, *Parlatoria oleae* (Colvée), a serious pest of olive, deciduous fruit trees and ornamentals. It readily became established, and soon proved to be a very effective natural enemy. Together with an introduced endoparasite, *A. paramaculicornis* has been the main factor responsible for the complete biological control of olive scale throughout California.

*A. holoxanthus* was introduced into Israel in 1956–1957 from Hong Kong against the Florida red scale, *Chrysomphalus aonidum* (L.), which was then the most important pest of citrus in that country. Complete biological control was achieved within two or three years, and has been maintained in Israel ever since. *A. holoxanthus* has subsequently been introduced into several other countries, and has effected complete biological control of the Florida red scale in all of them.

More recently, *A. roseni* was introduced into Peru from Uganda in 1970 against the rufous scale, *Selenaspidus articulatus* (Morgan), yet another serious pest of citrus. By 1975, it had effected complete biological control.

Needless to say, these and other projects have resulted in substantial economic savings to agriculture in the various countries involved.

## ROLE OF *APHYTIS* IN THE DEVELOPMENT OF BIOLOGICAL CONTROL

Research on species of *Aphytis* has not only resulted in outstanding biological control successes, but has also helped establish, elucidate or illustrate some of the most important basic principles, concepts and practices of biological control (see DeBach, 1969a, 1976). These include 1) the importance of systematics and biology in biological control; 2) the role of indigenous natural enemies in biological control; 3) the ecological basis for importation policy; 4) methods of augmenting natural-enemy populations; 5) the effects of adverse factors on natural-enemy populations and some methods for their mitigation; 6) methods for evaluating the effectiveness of natural enemies.

**Systematics, Biology and Biological Control.** As pointed out by Rosen and DeBach (1973), species of *Aphytis* have furnished some good examples of the great importance of correct identification of natural enemies for the success of biological control projects. In fact, potential success in the biological control of the California red scale in California was delayed for more than half a century, because various species of *Aphytis* attacking this pest in the Orient were misidentified as *A. chrysomphali*, a relatively inefficient parasite already present in California. For several decades, repeated attempts were made to discover and import endoparasites of the pest, whereas species of *Aphytis* were largely ignored or, ironically, actually eliminated from cultures as undesirable contaminants. The two most effective known enemies of the California red scale, *A. lingnanensis* and *A. melinus*, were not recognized as distinct species until 1948 and 1956, respectively, and were then successfully introduced from the Orient into California, and subsequently also into other countries (see also DeBach, Rosen and Kennett, 1971; Rosen and DeBach, 1978).

Another species of *Aphytis* may serve as a further example. *A. holoxanthus*, currently known as the most effective natural enemy of the Florida red scale, was apparently first reared from this host near the turn of the century, but was then ignored and forgotten due to its misidentification as another species. Its true identity was only revealed as late as 1960, several years after its successful introduction into Israel. Meanwhile it was thought to be *A. lingnanensis*. Needless to say, complete biological control of the Florida red scale could have been achieved decades ago, had this important parasite been recognized much earlier as a distinct species (DeBach, 1960; DeBach, Rosen and Kennett, 1971; Rosen and DeBach, 1973).

From the standpoint of practical biological control, we are vitally interested in whether natural enemies differ from one another, regardless of our ability to tell them apart morphologically. All grades of specific or subspecific genetically based difference may be important (DeBach, 1969b). The importance of this biosystematic approach was demonstrated by the successful biological control of the olive scale in California. Four so-called "biological strains" of *Aphytis*, all believed at the time to be morphologically indistinguishable but possessing distinct biological attributes, were introduced into California from various countries in the early 1950's. Fortunately, although they were all referred to *A. maculicornis*, a species already recorded as established in California, the various allopatric stocks were cultured and colonized separately. It soon became evident that only one of them, the biparental "Persian strain," was showing promise of being an effective natural enemy. Consequently, efforts were concentrated on that "strain"—now recognized here as a distinct species, *A. paramaculicornis*—which has indeed become the major factor in the complete biological control of the olive scale in California (Huffaker, Kennett and Finney, 1962; DeBach, Rosen and Kennett, 1971).

Crossing tests with known pure species cultures of *Aphytis* have served to confirm the distinct identity of such important natural enemies as *A. melinus*, *A. holoxanthus* and others, thus providing a further illustration of the great importance of biosystematic studies for biological control (see DeBach, 1959, 1960; Rao and DeBach, 1969a; also p. 206).

Detailed knowledge of the biology and ecology of both the host and its natural enemies may be an essential prerequisite for the success of biological control projects, especially when methods of augmentation are considered. The development of sophisticated mass-rearing and periodic-colonization programs, utilizing *Aphytis* species (see p. 86), was based on such knowledge. Likewise, studies on the effects of low temperature on *Aphytis*, discussed in some detail in the Ecology section (p. 53), have had a direct effect on accepted procedures and practices of biological control. When it was realized that the low temperatures commonly used for storing parasites in biological control projects may result in the release of completely sterile individuals, this common practice had to be abandoned (DeBach, 1969a).

**Naturally Occurring Biological Control.** The role of indigenous natural enemies in the "natural" biological control of potentially injurious pests has often been overlooked by economic entomologists. Yet it is a well-known fact that numerous species of arthropods are potential pests of agricultural crops, but never attain economic injury levels because their populations are suppressed by locally established natural enemies, without any deliberate human intervention. Unfortunately, only a few such cases have been adequately documented. Among those few are several species of *Aphytis*.

Even before the introduction of more efficient parasites, *A. chrysomphali* was shown to be capable, at times and in places, of effecting satisfactory biological control of the California red scale in California (DeBach, Fleschner and Dietrick, 1948). Further examples of this important aspect of biological control are provided by *A. mytilaspidis* on the oystershell scale, *Lepidosaphes ulmi* (L.), in Canada (Lord, 1947); *A. hispanicus* on the chaff scale, *Parlatoria pergandii* Comstock, in Texas (Dean, 1965); *A. aonidiae* on the San Jose scale, *Quadraspidiotus perniciosus* (Comstock), in California (Gulmahamad and DeBach, 1978); and others.

**Importation of Natural Enemies.** Of the three main approaches to applied biological control—importation, augmentation and conservation—the importation of exotic natural enemies is by far the most promising. This has been demonstrated repeatedly by successful biological control projects, in which various imported species of *Aphytis* have played a major role (see p. 82).

The usual importation policy, accepted by the great majority of practicing biological control experts, is to introduce as many promising natural enemies as may be found throughout the range of distribution of the pest. This policy, which is based on both theoretical and practical considerations (see DeBach, 1969a, 1971b, 1971c, 1976; Rosen, 1973), has been the subject of a rather fierce controversy. Certain authors (e.g., Turnbull and Chant, 1961; Watt, 1965) have criticized multiple importations as unscientific and as possibly detrimental to biological control, their main argument being that competition among natural-enemy species following multiple importations may induce pest populations to increase. As discussed under Ecology (pp. 78–79), ecological studies with several species of *Aphytis*—especially in connection with the biological control of the California red scale and the olive scale in California—have helped resolve this controversy and confirm the sound ecological basis for multiple importations. These studies have clearly shown that different natural enemies may exist in different parts of the range of a pest, that such enemies may exhibit preferential adaptation to different habitats, and that competition among them following their release in a given habitat, resulting either in competitive displacement or in complementary action, can only lead to enhanced biological control of the pest.

The phenomenon of ecesis (migratory establishment), which has been documented for several species of *Aphytis* (see p. 82), also emphasizes the advisability of extending the search for natural enemies throughout the range of the pest.

**Augmentation of Natural Enemies.** Augmentation of parasites or predators to increase their effectiveness involves their direct manipulation either by mass production and periodic colonization or by some type of planned genetic improvement (DeBach and Hagen, 1964). Both approaches have been investigated extensively with certain species of *Aphytis*.

*Periodic colonization* programs have unfortunately often been undertaken on an arbitrary basis, without sufficient scientific justification. However, periodic colonization is indeed indicated as a possible approach when natural enemies are seriously decimated at intervals, as we have shown that *Aphytis* is by weather extremes in certain zones (see p. 62). This approach to biological control was therefore tried in California in relation to the California red scale, first with *A. chrysomphali* and later with *A. lingnanensis* (DeBach, Dietrick, Fleschner and Fisher, 1950; DeBach and White, 1960).

Periodic colonization requires an adequate mass-production system. An efficient, economical mass-culture system was developed for *A. lingnanensis*, involving a factitious host, the oleander scale, which was found to yield a higher parasite production than the natural host, the California red scale. A factitious host plant, banana squash, was found to be superior to other host plants both for scale insect and for parasite production. Optimum densities of both host and parasite were determined experimentally for maximum parasite production. Finally, a new concept in mass production was advanced which, instead of using a so-called "sting stock" of females which were permitted to lay all their eggs and die, utilized the entire daily production of parasites to lay eggs, but only left them with hosts long enough for the oviposition of about one egg per female. The females were then collected for field colonization. The resultant daily insectary production will remain constant with this method; each female produced replaces itself with just one daughter and is then colonized in the field. This system made it possible to rear and colonize 400,000 females per acre for the equivalent cost of an insecticidal spray (DeBach, 1969a, 1976).

The method gave good results in intermediate climatic zones, but was not consistent enough in interior areas to be practical. The techniques developed have now been applied to the more effective *A. melinus*, and several commercial insectaries are producing this parasite in California for mass release against the California red scale.

*Genetic manipulation* is possible either through artificial selection or through hybridization. Again, both approaches have been attempted with *Aphytis*.

Selection of *A. lingnanensis* for resistance to climatic extremes over a period of five years (more than 100 generations) resulted in a population markedly more tolerant of both high and low temperatures. Although no practical field results were obtained (see p. 537), this demonstrated the principle to be sound. Similar efforts are now being made with *A. melinus* (White, DeBach and Garber, 1970; DeBach, 1976).

Interspecific hybridization could be of great importance in biological control, inasmuch as the resultant hybrids might differ from either parental species in their host preferences or in other biological attributes. Laboratory experiments with *Aphytis* species have demonstrated that such hybridization is indeed possible. These experiments have produced at least three interspecific hybrids which were completely reproductively isolated from one parent and partially from the other. Although initially these hybrid cultures were only poorly viable, their fecundity improved rapidly with succeeding generations in the laboratory, until it soon equaled or surpassed that of either parental species (Rao and DeBach, 1969a; DeBach, 1969a).

**Conservation and Integrated Control.** Conservation of established natural-enemy species involves manipulation of the environment to favor such species, either by eliminating or mitigating adverse factors or by providing lacking requisites (see van den Bosch and Telford, 1964; DeBach, 1974).

The adverse effects on *Aphytis* of various environmental factors, such as meteorological conditions, dust, honeydew-seeking ants and chemical pesticides, have been studied in considerable detail and are discussed in the Ecology section (pp. 53–72). Realization of such adverse effects has led to further importation efforts and to the development of various augmentation techniques, as well as to some attempts at environmental modification.

Little can be done about mitigating the adverse effects of meteorological conditions on *Aphytis*. Perhaps the best solution to this problem lies in searching for additional exotic species or strains that are better adapted to climatic extremes. Developing such strains in the laboratory is another possibility. However, the use of permanent cover-crops and overhead sprinkler-type irrigation have been suggested as possible means of modifying humidity conditions in citrus groves, and the same methods have also been suggested for dust control (DeBach, 1951). Chemical control of ants is also definitely recommended for the conservation of *Aphytis* species wherever aggressive honeydew-seeking ants, such as the Argentine ant, *Iridomyrmex humilis* (Mayr), are present (DeBach, 1951, 1958b).

Attempts to add lacking requisites for *Aphytis* have included planting a *Phacelia* cover-crop in order to provide nectar and honeydew for *A. proclia* (see p. 70), and infesting an English ivy cover-crop with oleander scale in order to provide alternative hosts for the species of *Aphytis* parasitic on the California red scale (see p. 70 and DeBach, 1958b).

Species of *Aphytis* have played a major role in several successful integrated control programs. In southern California, the integrated control of citrus pests by strip treatments (see p. 66 and DeBach and Landi, 1961) was made potentially possible by the partial success of *A. lepidosaphes* in controlling the purple scale. In Israel, the successful integrated control of citrus pests has been largely based on conservation of *A. holoxanthus* following its spectacular success in controlling the Florida red scale (Rosen, 1967b, 1974; Harpaz and Rosen, 1971). Similarly, introduced species of *Aphytis* have been the key to integrated control of citrus pests in South Africa (Bedford, 1975, 1976).

**Evaluation of the Effectiveness of Natural Enemies.** Evaluating the actual effectiveness of native or imported natural enemies in the field is one of the most problematic, yet undoubtedly one of the most crucial, aspects of biological control. Studies with *Aphytis* have contributed overwhelmingly to the development of scientific evaluation methods (see DeBach and Bartlett, 1964; DeBach and Huffaker, 1971; DeBach, Huffaker and MacPhee, 1976).

Attempts to assess the impact of *Aphytis* species on host populations by life-table studies have been unsuccessful. Such studies, which are of course only as reliable as the sampling techniques on which they are based, cannot ascertain cause-and-effect in the observed host and parasite population changes. DeBach and Bartlett (1964) have summarized the results of this approach as follows: "After more than ten years of study of a single species of parasite (*Aphytis lingnanensis* Comp.) of the California red scale we have been unable to develop a practical sample which will accurately depict *total* parasite-caused mortality. The many complexities involved, such as host-feeding and multivoltinism of the parasite, render impractical anything other than a sample which gives parasite and host population trends." (For a more detailed discussion see p. 50.)

On the other hand, experimental methods have proved to be most suitable for demonstrating and evaluating the effectiveness of natural enemies in host population regulation. Several experimental "check methods" have now been developed, largely through research with *Aphytis* species. All these methods employ the same basic principle; that is, they exclude or inhibit natural enemies in certain field units, and permit comparison with other units (leaves, branches, or entire trees) having unimpeded natural-enemy activity. If the pest population increases where the natural enemies are excluded or seriously interfered with, as compared with the area of normal natural-enemy activity, then the natural enemies are proven to be responsible for biological control at the population level observed in the normal undisturbed area. Exclusion of natural enemies may be accomplished by a variety of mechanical, chemical or biological methods.

Mechanical exclusion involves the comparison of host population densities in closed (enemy-free) and open cages. Chemical exclusion, involving light applications of a relatively selective pesticide (i.e., one that is considerably more detrimental to *Aphytis* than to the scale insect host), is the easiest to execute and may yield dramatic results (see p. 64 and Figure 59). A combination of the two techniques, using paired, open sleeve cages, one of each pair being impregnated with a selective pesticide, is probably the best method. Finally, the "biological check method" employs aggressive honeydew-seeking ants to interfere with the activity of *Aphytis* and other natural enemies; naturally occurring ants are eliminated from certain trees, but are left undisturbed on others.

# SYSTEMATICS

## FAMILY APHELINIDAE

The Aphelinidae constitute a moderate-sized family of chalcidoid Hymenoptera, comprising 41 known genera and some 700 described species. A few genera are hyperparasitic and two comprise egg parasites, but the bulk of the family consists of primary parasites of sternorrhynchous Homoptera—coccids, aphids and aleyrodids.

Morphologically, the Aphelinidae show affinity to both the Eulophidae and the Encyrtidae. They resemble the former in the structure of the mesonotum (parapsidal grooves complete, axillae advanced into the base of the parapsides) and in the small number of antennal segments, the latter in the shape of the mesopleura, in the saltatorial mid-tibial spur, and in the presence of a well-developed strigil on the foreleg. Biologically, their host spectrum resembles that of the Encyrtidae.

Opinions as to the systematic position of this interesting group have differed greatly among specialists. Dalman (1820) apparently regarded his genus *Aphelinus* as a subdivision of his larger genus *Entedon*, which has been subsequently referred to the Eulophidae. Walker (1839) regarded *Aphelinus* as "nearly allied" to *Encyrtus*. The group was first assigned family rank by Foerster (1856), who established the family Myinoidae (based on the genus *Myina* Nees, a synonym of *Aphelinus* Dalman). Thomson (1875) referred his tribes Aphelinina and Eulophina to two separate sections of the family Pteromalidae, and placed the Aphelinina next to the Encyrtina. Howard (1895), too, in the first modern revision of the group, regarded the subfamily Aphelininae as closely allied to the Encyrtinae and Eupelminae. On the other hand, Ashmead (1904) insisted that the Aphelininae have "no affinity whatever" with the Encyrtidae, and included them in the Eulophidae. Ashmead's classification was accepted by Schmiedeknecht (1909), whereas Girault (1913, 1915) again argued that the Aphelininae ought to be classified as a subfamily of the Encyrtidae. Mercet (1929) treated the Aphelinidae as a separate family, but pointed out that, morphologically as well as biologically, they form a homogenous group with the Encyrtidae and Signiphoridae.

Most subsequent authors have treated the Aphelinidae as a distinct family. However, Peck (1951) and Burks (1967) again reduced them to a subfamily of the Eulophidae.

Compere and Annecke (1961) reiterated the close ties between the Aphelinidae and the Encyrtidae, whereas Peck, Bouček and Hoffer (1964) regarded the Aphelinidae as transitional between the Eulophidae and the Trichogrammatidae. Recently, Riek (1970) has treated the Aphelininae as a subfamily of the Encyrtidae.

We agree with Compere and Annecke (1961). In our view, the Aphelinidae should be considered a distinct family, which is more closely related to the Encyrtidae and Signiphoridae than to any other group of Chalcidoidea.

Various concepts of the higher organization of this family are currently recognized. Early workers (e.g., Ashmead, 1904; Howard, 1907) divided the Aphelinidae (then regarded as the subfamily Aphelininae) into two tribes—the Aphelinini with 5-segmented tarsi and the Pteroptricini with 4-segmented tarsi. The first modern classification was offered by De Santis (1946, 1948, 1967), who divided the bulk of the family into two main subfamilies, the Aphelininae, with a distinct speculum in the forewing and 3- to 6-segmented antennae, and the Coccophaginae, with an aspeculate forewing and usually 7- to 9-segmented antennae. De Santis also recognized a third subfamily, the Calesinae, in which he included a single aberrant genus, *Cales* Howard.

Ferrière (1965) retained the Aphelininae and Coccophaginae, but grouped all the genera with 4-segmented tarsi in a separate subfamily, the Pteroptricinae, in which he placed, among others, the aphelinine genus *Eretmocerus* Haldeman as well as the calesine genus *Cales*. Ferrière also recognized the Eriaporinae as a fourth subfamily of the Aphelinidae.

Nikol'skaya and Yasnosh (1966), on the other hand, retained the Aphelininae, Calesinae and Coccophaginae, but divided the last further to establish the subfamilies Azotinae and Prospaltellinae. They did not recognize the Pteroptricinae as a distinct group, but placed most pteroptricine genera in the Prospaltellinae and returned *Eretmocerus* to the Aphelininae. Also, they considered the Eriaporinae as an ancient group related to both the Elasmidae and Aphelinidae, rather than a bona fide member of the latter family.

The various authors are in fair agreement regarding the characterization and composition of the subfamily Aphelininae. All three schemes include the genus *Aphytis* in this subfamily.

Recently, Yasnosh (1976) offered an elaborate classification of the Aphelinidae into 7 subfamilies—Aphelininae, Aphytinae, Azotinae, Calesinae, Coccophaginae, Physcinae and Prospaltellinae. The subfamily Aphytinae was further split into the tribes Aphytini and Centrodorini. We cannot accept some of the characters Yasnosh ascribed to the Aphytini (e.g., shape of the pronotum and of the male genitalia) and are not satisfied that so much splitting is justified for a family the size of the Aphelinidae at the present state of our knowledge. Although none of the existing classifications appears to be entirely acceptable, the various subfamilies not necessarily representing natural groups (see Gordh and Rosen, 1973), the broader concepts of De Santis (op. cit.) seem to be most satisfactory for the time being.

# GENUS *APHYTIS* HOWARD

*Aphytis* Howard, 1900, Canad. Entomol., **32**: 168.
Type: *Aphytis chilensis* Howard. Monobasic.
*Prospaphelinus* De Gregorio, 1914, N. Ann. Agric. Sicil., **3** (4):227 (as a subgenus of *Aphelinus*).
Type: *Aphelinus* (*Prospaphelinus*) *silvestrii* De Gregorio. Monobasic.
*Paraphytis* Compere, 1925, Trans. Amer. Entomol. Soc., **51**: 129–130.
Type: *Paraphytis vittata* Compere. Monobasic.

**Distinguishing Characters.** *Body* robust, about 1 mm long, usually yellowish or grayish, sometimes mottled. *Head* transverse, as wide as the thorax; frontovertex wide, ocelli in an equilateral or obtuse triangle, the posterior pair closer to the orbits than to the occipital margin; eyes large, usually finely setose. *Mandibles* usually well developed, with 2 denticles and a truncation; maxillary palpi usually 2-segmented, labial palpi 1-segmented. *Antennae* inserted close to oral margin, usually 6-segmented: scape cylindrical, with a short radicle; pedicel subconical or pyriform, longer than wide; funicle usually 3-segmented, the first two segments short, the third considerably longer; club undivided, longer and wider than the penultimate segment. *Thorax* somewhat longer than wide; pronotum short, usually composed of two triangular sclerites; mesoscutum trapezoidal, bearing several pairs of setae; axillae advanced into parapsides, with about half their length produced beyond the scuto-scutellar line; scutellum transverse, bearing 2 pairs of setae; metanotum bandlike, with an anteromedian apodeme; endophragma elongate, broadly rounded at apex. *Propodeum* long, usually with a sculptured posteromedian salient, posterior margin bearing scalelike crenulae. *Gaster* subsessile, somewhat longer than the thorax; ninth and tenth abdominal tergites fused into an apical syntergum; seventh abdominal sternite not reaching apex of gaster. *Ovipositor* weakly exserted, with the basal part enclosed by the sterna and the apical part exposed; outer ovipositor plates with a longitudinal ridge from base to about three fourths length. *Legs* with 5-segmented tarsi; mid-tibial spur usually somewhat shorter than the corresponding basitarsus. *Forewings* well developed, about 2.5-3.0 times as long as wide, hyaline or clouded, sometimes mottled; disk rather densely setose except at base, with a well-defined oblique speculum; delta-shaped area basad of speculum bearing several rows of setae; marginal fringe short, usually not exceeding one third width of disk; marginal vein longer than the submarginal, stigmal vein short, postmarginal vein absent. *Hind wings* lanceolate, narrowly rounded at apex; marginal fringe not exceeding width of disk. *Male* usually not exhibiting marked sexual dimorphism, differing from the female mainly in the somewhat paler coloration, the more hyaline forewings, and in the smaller number of sensilla on the antennal club, sometimes also in reduced funicular segments. *Genitalia* with well-developed digital sclerites bearing a single claw; parameres absent.

Some of the main morphological characteristics of the adult *Aphytis* female are shown in Figure 68.

Several species deviate markedly from the above scheme in one character or another, yet resemble *Aphytis* in general habitus and have been included in the genus on this

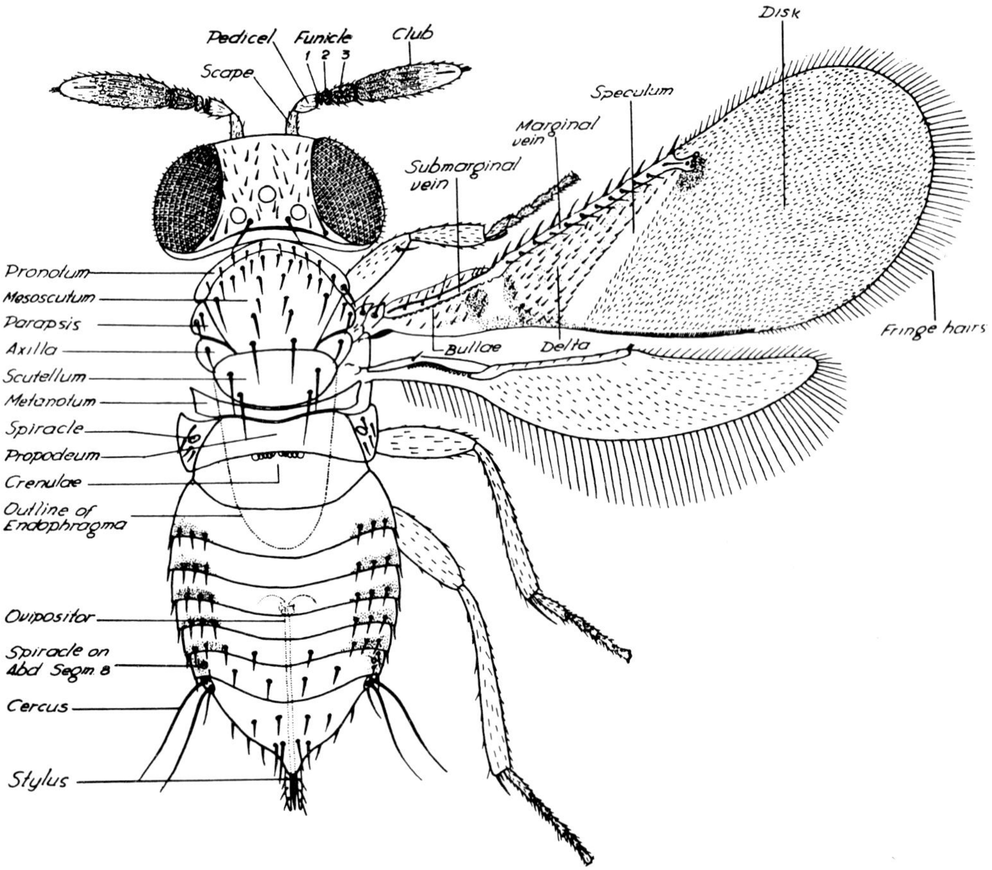


Figure 68. *Aphytis chilensis* Howard, ♀: semidiagrammatic representation of general morphology (from DeBach, 1964; the same terminology is used in this monograph, except that "stylus" = ovipositor sheaths).

basis. Certain species have reduced mouthparts, others have 4- or 5-segmented antennae or an acrenulate propodeum. Erecting separate genera for such aberrant species would tend to obscure their natural relationships and would not contribute to a better understanding of the group. As pointed out by Mayr (1942, p. 22), "even though a certain character is widespread in a certain taxonomic group, this does not prove that all forms that lack this character do not belong to this group. . . . This is the reason why it is so important that all classifications be based on the greatest possible number of different characters."

**Allied Genera.** Early workers included the species of *Aphytis* in the older genus *Aphelinus* Dalman, which *Aphytis* resembles superficially (see Historical Review, p. 6). However, the two genera are not very closely related. They differ markedly in their biology, all the known species of *Aphytis* being ectoparasites of armored scale insects whereas all the species of *Aphelinus* develop endoparasitically in aphids. Timberlake (1924) was apparently the first to notice that this difference in host relations is correlated with significant differences in the structure and mode of action of the ovipositor. Following is a quotation from Compere (1955), who discussed these differences in some detail (see also Figure 69):

"*Aphelinus* and *Aphytis* differ in the manner of oviposition, and this difference is correlated with fundamental differences in the structural features of the abdomen and ovipositor.

"When *Aphelinus* oviposits, the entire ovipositor everts; the whole apparatus swings outward. From the everted base the shaft swings in an arc of approximately 90° and extends horizontally backward, and at the same time the free tenth tergum and the freely articulated styli tilt upward and stand semierect. In freshly killed specimens all the components of the ovipositor can be withdrawn from the abdomen by manipulation, without injury to other structures. In museum specimens on slides, species of *Aphelinus* may be distinguished easily from those of *Aphytis* by the separate tenth tergum (the free apical dorsal sclerite) and the sterna which enclose the retracted ovipositor to the apex of the abdomen.

"When *Aphytis* oviposits, only the shaft swings downward; the other components of the ovipositor are not everted. The modified dorsal remnant of the ninth segment, or ninth tergum, is immovably fused to the dorsal sclerite of the tenth segment, or tenth tergum, to form a rigid suspensorium, a syntergum. In freshly killed specimens, only

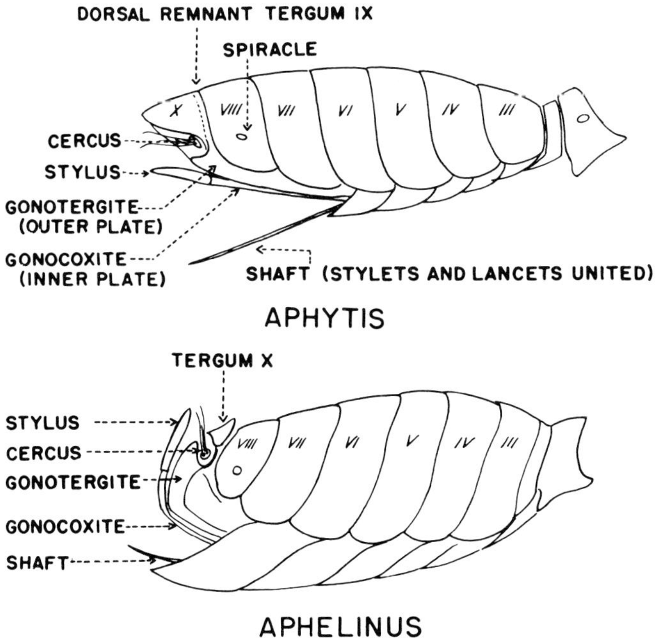


Figure 69. Abdomen and ovipositor of *Aphytis* and *Aphelinus*, lateral view (diagrammatic). (From Compere, 1955.)

the shaft can be protracted by manipulation without damage to a specimen. In museum specimens on slides, the species of *Aphytis* may be easily distinguished from those of *Aphelinus* by the syntergum and the sterna which enclose only the basal part of the retracted ovipositor."

*Aphelinus* also differs from *Aphytis* in the shape of the propodeum, which is usually shorter and does not bear marginal crenulae.

Although the distinction between *Aphytis* and *Aphelinus* has been unanimously accepted for quite a while, four Australian species, described some years ago, had remained hidden in *Aphelinus* until recently and have been transferred to *Aphytis* only during the course of the present revision. These species are *columbi* (Girault), *ovidii* (Girault), *ruskini* (Girault) and *stellaris* (Girault).

The aphid-attacking genera of Aphelinidae have recently been revised by Mackauer (1972). Besides *Aphelinus*, this group at present includes the genera *Mesidia* Foerster, *Mesidiopsis* Nowicki, and *Protaphelinus* Mackauer. All these genera are similar to *Aphelinus* in the structure of the abdomen and mode of oviposition, and hence cannot be considered closely allied with *Aphytis*. We have some reservations as to the validity of the last-named genus, but further discussion of this group is outside the scope of the present study.

Four aphelinid genera may be regarded as most closely related to *Aphytis*. These are *Marlattiella* Howard, *Marietta* Motschulsky, *Centrodora* Foerster and *Bestiola* Nikolskaja, all of which share with *Aphytis* the basic structure of the abdomen and ovipositor. *Debachiella* Gordh and Rosen and *Tumidiscapus* Girault also belong to this group. In previous classifications (e.g., Compere, 1955; Ferrière, 1965; Nikol'skaya and Yasnosh, 1966), some of these genera were separated from *Aphytis* on the basis of unreliable characters. Thus, *Marlattiella* was considered distinct mainly because of its peculiar, 4-segmented antennae; *Marietta* was separated from *Aphytis* mainly by its mottled wings, body and appendages, *Centrodora* mainly by its elongate body, ovipositor, legs and wings. *Aphytis* was considered to have 6-segmented antennae, unmottled wings, and a predominantly yellow, rather robust body, with the ovipositor, legs and wings not unusually elongate.

A detailed comparative study proved some of these characters to be unreliable for generic classification. Bona fide species of *Aphytis* may have pictured wings, a heavily maculate body, and 4 to 6 segments in the antennae. The main diagnostic characters separating *Aphytis* from closely related genera are 1) the well-developed, relatively long propodeum with a distinct, sculptured median salient, usually bearing posteromarginal crenulae; 2) the chaetotaxis and venation of the forewings (delta-shaped area basad of the speculum with several rows of setae, marginal vein longer than the submarginal); 3) the absence of parameres in the male genitalia. Following are some further notes on the genera most closely related to *Aphytis*.

The genus *Marlattiella* at present includes the single species *M. prima* Howard (Figures 70–75, p. 96), an Oriental parasite of *Lopholeucaspis japonica* (Cockerell). The 4-segmented antenna of the female (Figure 71) has commonly been regarded as the main distinguishing character of *Marlattiella*. Compere (1936) wrote: "This genus is most closely related to *Aphytis* Howard; the principal difference is in the antennae, which are four-jointed in *Marlattiella* . . ." Again, in 1955, Compere stated: "In my opinion *Marlattiella* and *Aphytis* are closely related although the antennae are very different." Tachikawa (1962) described the male of *M. prima*, which is characterized by a 3-segmented antenna with an enormously enlarged club. He noted that "*Marlattiella* is closely related not only to *Aphytis*, but also to *Eretmocerus* Haldeman."

In our opinion, *Marlattiella*, as represented by *M. prima*, is certainly distinct from *Aphytis*, and not only because of antennal differences (Rosen and DeBach, 1970). The propodeum of *M. prima* (Figures 72, 73) is short, not much longer than the metanotum, more or less uniformly sculptured, and lacks the marginal crenulae; it is entirely different from the propodeum of *Aphytis*. Also, the head of *M. prima* (Figure 70) is rather elongate, the eyes are relatively small, and the ocellar triangle is acute; the mesoscutum (Figure 72) bears 3 pairs of setae; and the longitudinal ridge on the outer ovipositor plates (Figure 74) starts at about the anterior third of the outer margin. The forewings of *M. prima* (Figure 75) also differ from those of *Aphytis* in the absence of a row of setae along the posterior margin, below the delta.

Compere (1936) described the African species *secunda*, which he assigned to *Marlattiella* mainly on the basis of its 4-segmented antennae. We previously regarded *secunda* as a dubious species of *Aphytis*, and suggested that it may possibly constitute a "connecting link" between the two genera (Rosen and DeBach, 1970). However, further study led us to the conclusion that this species is a bona fide member of *Aphytis*, quite closely related to the *funicularis* complex (see p. 672), and apparently not at all related to *Marlattiella*.

The genus *Marietta* (Figures 76–82, pp. 97, 98) comprises secondary parasites of various coccoids. In the past, this genus was considered to differ from *Aphytis* mainly in its generally mottled habitus (see Figures 76, 77, 81), and the two genera were thought to intergrade—so much so that at one time Girault (1915, p. 45) even considered *Perissopterus* Howard, a synonym of *Marietta*, to be synonymous with *Aphelinus*. Compere (1936) presented the commonly accepted concept of *Marietta* as follows: "Species of *Marietta* are usually easy to recognize because of spots or maculations on the wings and body, and by the diamond-shaped sculpture on the metanotum just behind the apex of the scutellum. This genus is closely related to *Aphytis* Howard, in which the species are generally yellow without definite spots or maculations on wings or body, and the metanotum is smooth across the center. There is scarcely any gap separating some of the faintly mottled species of *Marietta* from those of *Aphytis*." In 1955, Compere added: "*Marietta* and *Aphytis* are not clearly distinguishable, owing to the existence of many intergrading species. The genus to which an intergrade is assigned depends on personal opinion as to the genotype the intergrade resembles most closely."

We do not accept this view. On the one hand, general pigmentation and wing pattern cannot be regarded as reliable generic criteria for separating *Marietta* from *Aphytis*. Quite a few species, recognized here as bona fide members of *Aphytis*, are heavily pigmented and have distinctly mottled forewings, whereas at least one bona fide species of *Marietta*, *M. connecta* Compere (Figure 82), has unmottled, uniformly infumated forewings. On the other hand, the species of *Marietta* and *Aphytis* can be easily separated by consistent structural characters that have hitherto been largely ignored.

The most important diagnostic characters distinguishing *Aphytis* from *Marietta* are

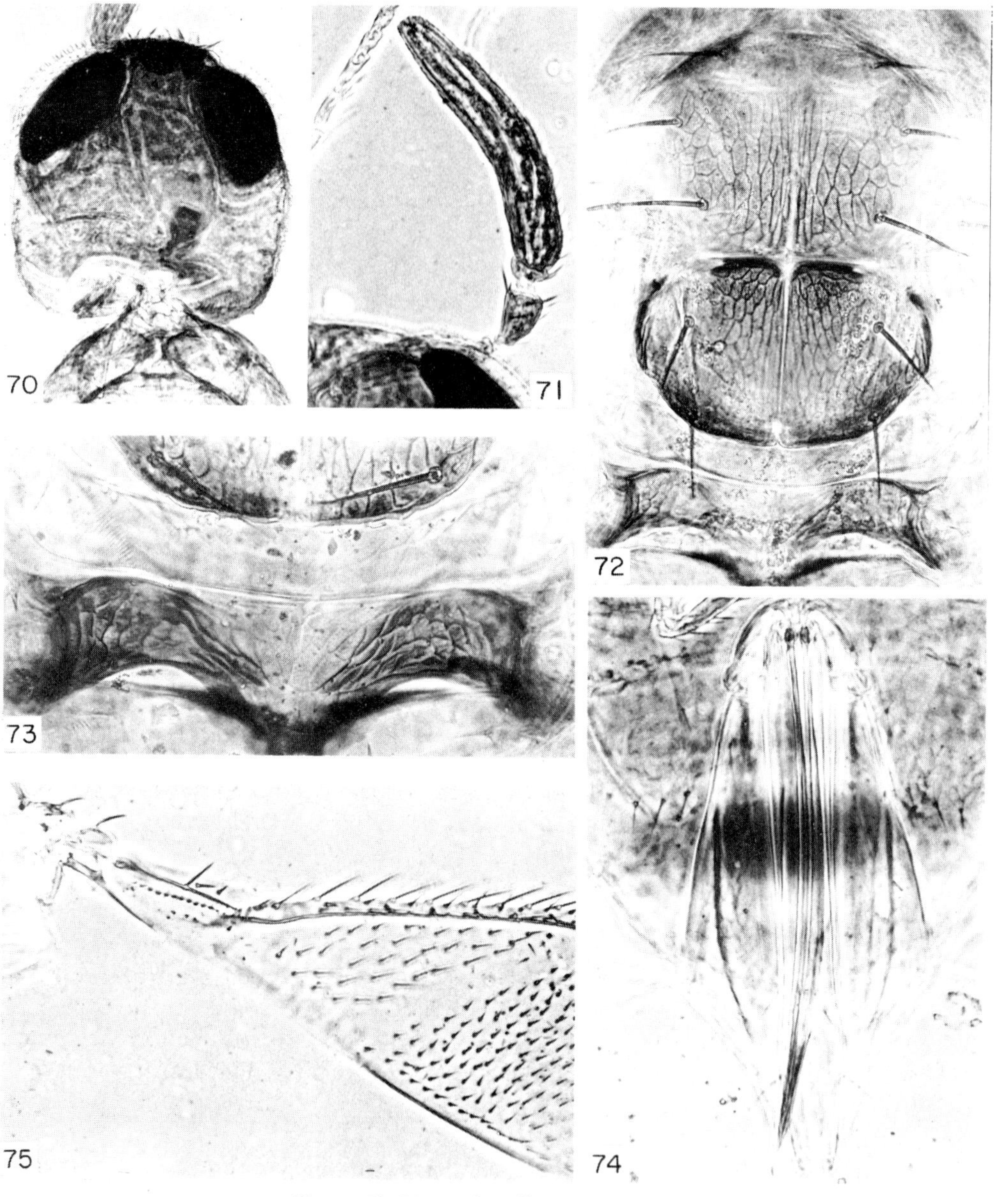


Figures 70–75. *Marlattiella prima* Howard, ♀.
70. Head. 71. Antenna. 72. Thorax and propodeum. 73. Metanotum and propodeum. 74. Ovipositor and plates. 75. Proximal half of forewing.
(From Rosen and DeBach, 1970.)

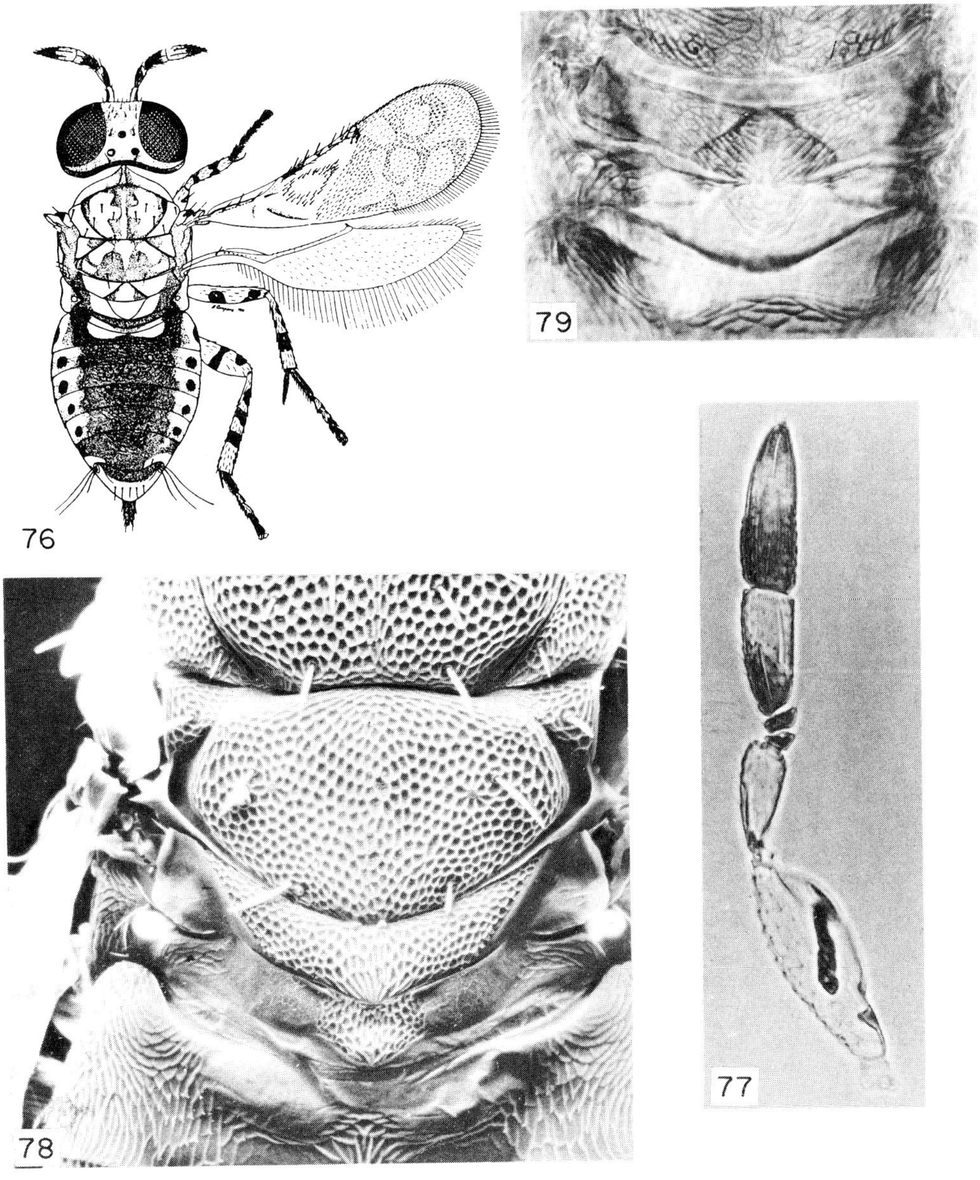


Figures 76–79. *Marietta* spp.
76. *M. carnesi* (Howard), ♀: imago (from Compere, 1936). 77. *M. mexicana* (Howard), ♀: antenna. 78. *M. exitiosa* Compere, ♀: thorax, propodeum and base of gaster (SEM, from Kfir and Rosen, in preparation). 79. *Marietta* sp., ♂: metanotum, propodeum and base of gaster.

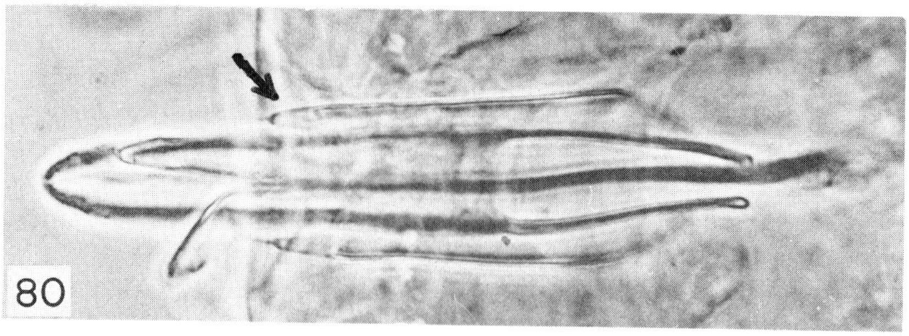


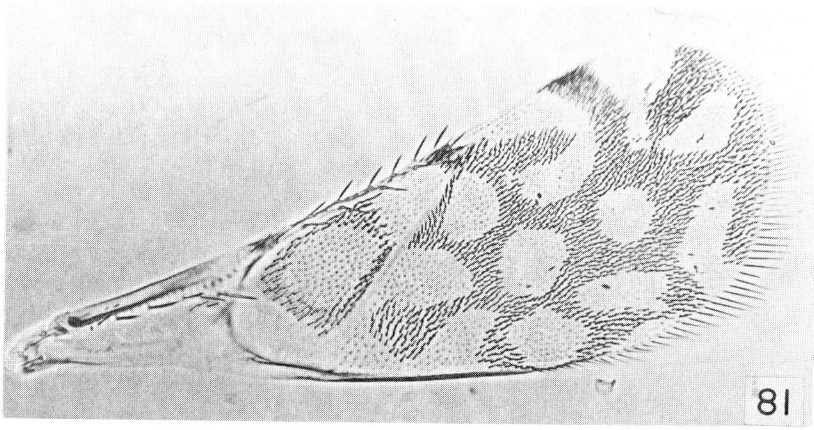


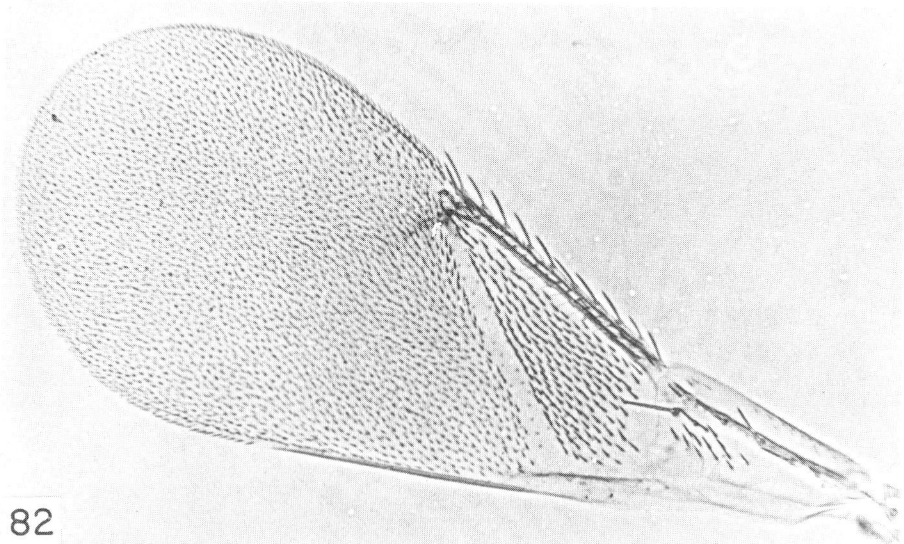


Figures 80–82. *Marietta* spp.

80. *Marietta* sp., ♂: genitalia; paramere indicated by arrow. 81. *M. exitiosa* Compere, ♀: forewing (from Kfir and Rosen, in preparation). 82. *M. connecta* Compere, paratype ♀: forewing.

the structure, shape and relative proportions of the metanotum and propodeum. In *Aphytis* the metanotum is relatively short, weakly sculptured, with a rather robust anteromedian apodeme, and the propodeum is quite long, strongly sculptured medially, and bears marginal crenulae. In *Marietta* the metanotum is relatively long, usually bears the median "diamond-shaped" sculpture described by Compere (Figures 78, 79), and the anteromedian apodeme is very slender, hardly visible in many species; the propodeum is short, subequal to the metanotum, weakly sculptured, and lacks the marginal crenulae. The shape of the male genitalia is also different: all the species of *Marietta* known to us have distinct parameres (Figure 80), which are always absent in *Aphytis*. Also, in the species of *Aphytis* that have mottled wings, the mottled effect is usually produced by contrasting coarse, dark setae and fine, hyaline setae, whereas in *Marietta* there is also integumentary infuscation.

In certain species of *Marietta* the antennal club appears to be separated from the third funicular segment only by a suture, not by a constriction (see Compere, 1936; Annecke and Insley, 1972; also Figure 76). This, however, cannot be regarded as a reliable generic character, inasmuch as a normal constriction is evident in other species of *Marietta* (e.g., Figure 77), whereas it is absent in certain bona fide species of *Aphytis*, such as *dealbatus* and *melanostictus*.

The following 19 species have been transferred from *Marietta* to *Aphytis* on the basis of these morphological criteria: *angeloni* (Girault), *capillatus* (Howard), *ciliatus* (Dodd), *costalimai* (Gomes), *cowperi* (Girault), *distonota* (Girault), *emersoni* (Girault), *haywardi* (Blanchard), *lessingi* (Girault), *maculatipennis* (Dozier), *maculatipes* (Girault), *nigripes* (Compere), *noumeaensis* (Howard), *novicapillata* (Girault), *peculiaris* (Girault), *perissoptroides* (Girault), *punctaticorpus* (Girault), *romae* (Girault), and *vittata* (Compere). Nine of these are valid species, while the rest are at present regarded as unrecognizable or synonyms. All the other species listed by Compere (1936) and Gomes (1942) in *Marietta* are bona fide members of that genus.

Our morphological criteria for separating *Aphytis* from *Marietta* are corroborated by biological evidence. As far as is currently known, all the species transferred here from *Marietta* to *Aphytis* develop as primary ectoparasites of armored scale insects. Needless to say, all these species are now recognized as being available for biological control.

The genus *Centrodora* is very closely related to *Aphytis*. Girault (1913, p. 180) synonymized *Paraphelinus* Perkins, a synonym of *Centrodora*, with *Aphelinus*, claiming that forms of the two genera intergrade. Waterston (1917) disagreed, but added that "while *Paraphelinus* is distinct from *Aphelinus*, it is closely approached by the small yellow-bodied group (at present placed in *Aphelinus*) which attack Coccids." [i.e., *Aphytis*]. Compere (1955) also suspected that the species of *Aphytis* and *Centrodora* may possibly intergrade.

Biologically, the two genera are very distinct. Whereas the species of *Aphytis* are ectoparasitic upon armored scale insects, those of *Centrodora* usually develop as primary parasites in the eggs of Orthoptera or auchenorrhynchous Homoptera. Certain species are known to diverge from this general habit: *C. speciosissima* (Girault) has been reared also from puparia of the Hessian fly, *Mayetiola destructor* (Say) (Diptera: Cecidomyidae) (McConnell, 1916; Gahan, 1933), whereas *C. xiphidii* (Perkins) has been recorded also as a facultative hyperparasite, having been reared from the cocoons of *Haplogonatopus* (Hymenoptera: Dryinidae) (Waterston, 1917), but none are known to attack armored scale insects.

Morphologically, the species of *Centrodora* usually differ from those of *Aphytis* in their generally elongate habitus: the body and wings are elongate and relatively narrow, the ovipositor is long and strongly exserted, and the antennae are long, with elongate funicular segments and a beaklike club, slightly curved at the apex. The propodeum of *Centrodora* is rather weakly sculptured, sometimes relatively short. However, certain species such as *C. penthimiae* Annecke (Figures 83–87) are rather similar to *Aphytis* in their general habitus and antennal proportions, and have a relatively long propodeum (Annecke, 1965; see also Hayat, 1973). An undetermined species of *Centrodora* in the collection of the Division of Biological Control, University of California, Riverside, has a well-developed propodeum bearing minute, distinct crenulae. Nevertheless, such species of *Centrodora*, which are often also generally yellow, can still be distinguished from *Aphytis* by several consistent morphological characters (see Annecke, 1965). In *Centrodora*, the entire venation of the forewing (Figure 87) is relatively short, not exceeding one half the greatest wing length; the marginal vein is shorter than, or subequal to, the submarginal vein; the male genitalia (Figure 86) have distinct parameres, and each of the digital sclerites bears two claws; in addition, the mesoscutum and scutellum are distinctly grooved along the midline. In *Aphytis*, on the other hand, the venation is considerably longer than one half the greatest wing length, the marginal vein is considerably longer than the submarginal vein, the male genitalia lack distinct parameres, each digital sclerite bears a single claw, and the mesonotal groove is much less pronounced. Slight as these differences are, when combined with biological information they can be relied upon for separating the species of these two genera.

The genera *Tumidiscapus* and *Debachiella* are most closely related to *Centrodora*. The species of *Tumidiscapus* develop as egg parasites, but *T. flavus* Girault has been reared also from Hessian fly puparia (Gahan, 1919, 1933). They differ from the species *Centrodora* in that the antennal scape of the male is greatly enlarged and the second funicular segment is longer than, or subequal to, the third segment in both sexes. As pointed out by Gahan (1919), "it is likely that the study of additional species will make it necessary to sink *Tumidiscapus* as a synonym of *Centrodora*." *Debachiella* differs from *Centrodora* in having aspeculate forewings, and is presumably parasitic upon armored scale insects (Gordh and Rosen, 1973).

The genus *Bestiola* is based on the single species *B. mira* Nikolskaja (Figures 88–95, p. 102), a parasite of *Odonaspis secreta* (Cockerell) on bamboo in the Caucasus (Nikol'skaya, 1963). It resembles *Aphytis* in most diagnostic characters, including the robust metanotal apodeme (Figure 92) and the well-developed propodeum which, although rather weakly sclerotized and sculptured, bears indistinct, elongate crenulae (Figures 93, 94). *Bestiola* differs from *Aphytis* mainly in the chaetotaxis of the forewings (Figures 88, 95), with a tuft of long, coarse setae in place of the delta-shaped area basad of the speculum. The antennal scape of the female (Figure 90) bears a minute tuberculous sensillum (in *Aphytis* such sensilla are present only in the males of certain species); the pronotum (Figure 91) is composed of two broadly joined plates and the pronotal

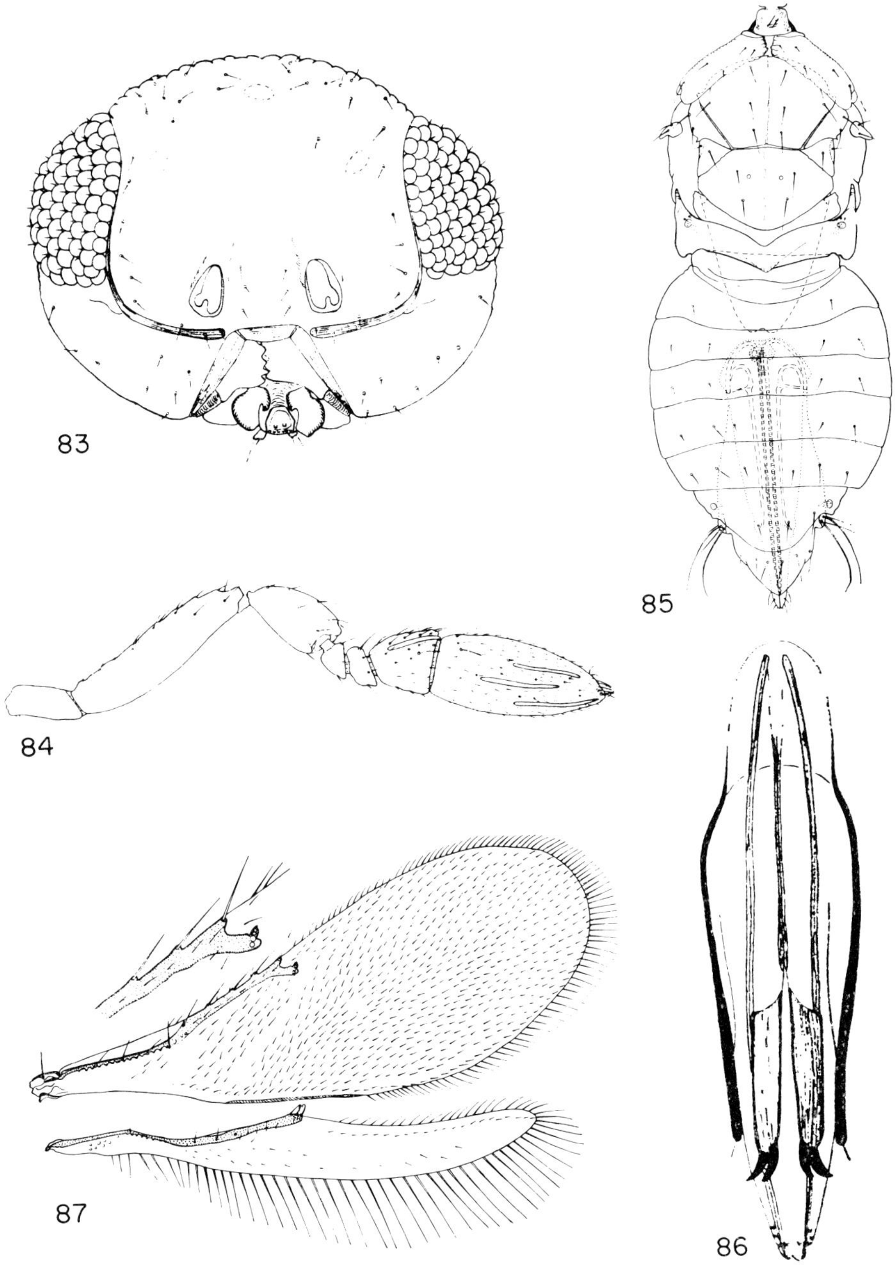


Figures 83–87. *Centrodora penthimiae* Annecke
83. ♀ head, frontal view. 84. ♀ antenna. 85. ♀ thorax and abdomen.
86. ♂ genitalia. 87. ♀ wings, with venation of forewing enlarged.
(From Annecke, 1965.)

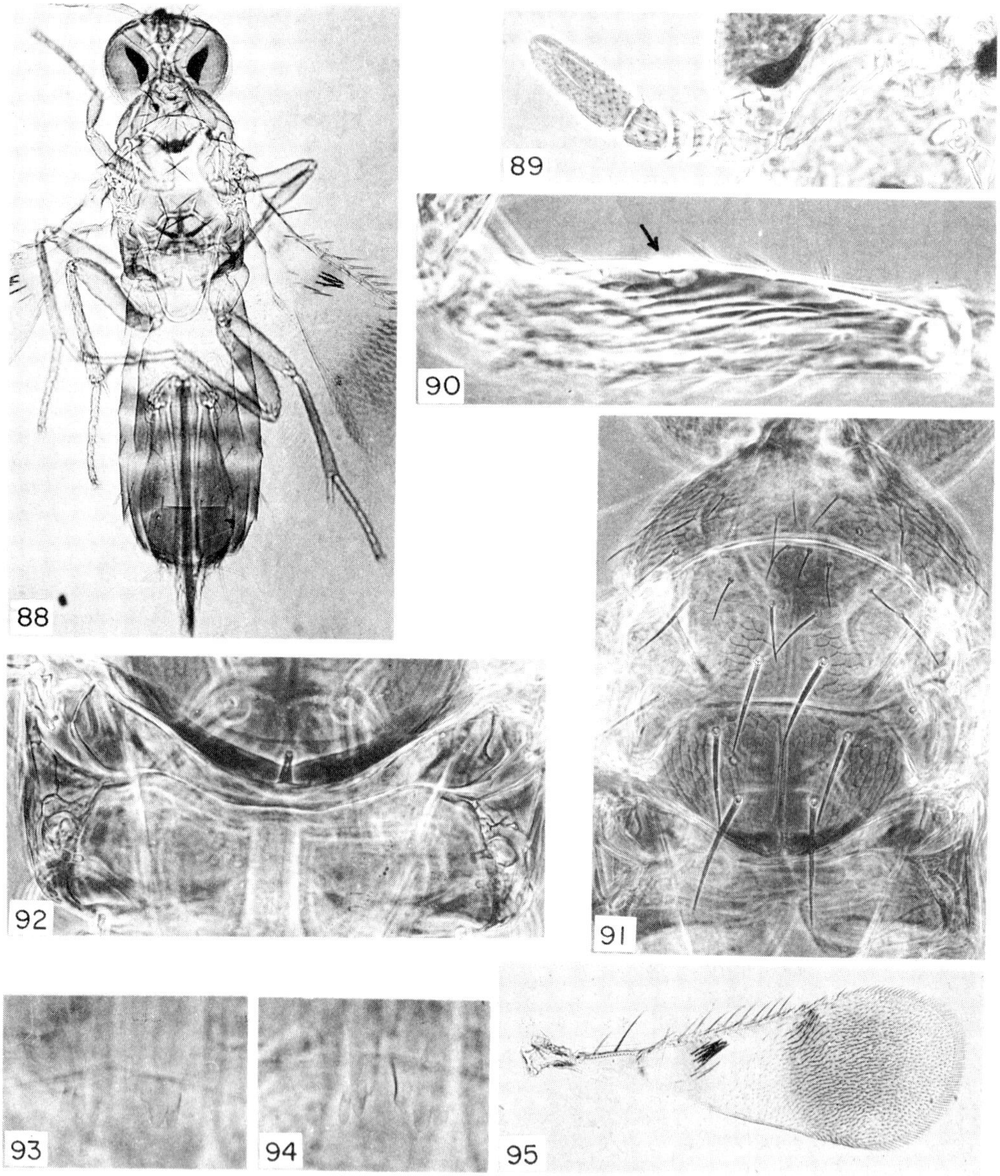


Figures 88–95. *Bestiola mira* Nikolskaja, ♀
88. Imago. 89. Antenna. 90. Antennal scape, showing sense organ. 91. Thorax and propodeum. 92. Metanotum and propodeum; note metanotal apodeme. 93, 94. Propodeal crenulae. 95. Forewing.

spiracle is rather conspicuous (in *Aphytis* these plates are usually—but not always—triangular and the spiracle is usually hidden); the gaster (Figure 88) is cylindrical, much more elongate than in *Aphytis*, whereas the endophragma is relatively short. The male of *Bestiola* is unknown.

Hayat (1972) erected the genus *Eriaphytis* to include the species *E. orientalis* Hayat, an Indian parasite of *Cerococcus* sp. (Hymenoptera: Asterolecaniidae) which he regarded as closely related to *Aphytis* due to the presence of an abdominal syntergum. In our opinion *Eriaphytis*, with an elongate scutellum, a short propodeum, a short marginal vein and 7-segmented antennae (1132), is rather remotely related to *Aphytis* and should not be included in the same group of genera.

The genera *Signiphora* Ashmead and *Aphytis* have some apparent similarities. According to Quezada, DeBach and Rosen (1973), "the peculiar antennae, with reduced funicular segments and a long, undivided club; the structure of the metanotum, with two oblique sutures on the sides and a small anteromedian apodeme; the relatively long propodeum, with a cut-off, triangular median salient . . . all these typical characters of the Signiphoridae have similar counterparts in *Aphytis*." Obviously, *Signiphora* and *Aphytis* are markedly different in many respects. Whether this similarity represents a true phylogenetic relationship or merely the convergence of certain derived characters is for the moment an open question.

The following key provides a shortcut to the identification of those aphelinine genera that are considered here as most closely related to *Aphytis*, although like most dichotomous keys it does not include all the various diagnostic characters discussed in the preceding paragraphs.

## KEY TO *APHYTIS* AND CLOSELY RELATED GENERA

1. Venation of forewings short, not exceeding half wing length; marginal vein shorter than submarginal vein . . . . . . . . . . . . . . . . . . . . . . . . . . . . . . . . . . . 2
— Venation of forewings considerably longer than half wing length; marginal vein considerably longer than submarginal vein . . . . . . . . . . . . . . . . . . . . . . . . . . . 4

2(1). Forewings with a distinct speculum . . . . . . . . . . . . . . . . . . . . . . . . . . . . . . . . . 3
— Forewings aspeculate . . . . . . . . . . . . . . . . . . . . . . . . . . . . . . . . . . *Debachiella*

3(2). Second funicular segment at least as long as the third segment; male antennal scape greatly enlarged . . . . . . . . . . . . . . . . . . . . . . . . . . . . . . . . *Tumidiscapus*
— Second funicular segment considerably shorter than the third; scape of male not unusually enlarged . . . . . . . . . . . . . . . . . . . . . . . . . . . . . . . . . . . . *Centrodora*

4(1). Propodeum short, subequal to the metanotum, acrenulate . . . . . . . . . . . . . . . . 5
— Propodeum long, considerably longer than the metanotum, bearing marginal crenulae . . . . . . . . . . . . . . . . . . . . . . . . . . . . . . . . . . . . . . . . . . . . . . . . . . 6

5(4). Antennae 6-segmented (often 5-segmented in the male); body and appendages maculate; forewings usually mottled . . . . . . . . . . . . . . . . . . . . . . . . . . *Marietta*
— Antennae 4-segmented (3-segmented in the male); body and appendages not conspicuously maculate; forewings hyaline . . . . . . . . . . . . . . . . . . . . *Marlattiella*

6(4). Forewings with a delta-shaped area basad of speculum bearing several rows of setae . . . . . . . . . . . . . . . . . . . . . . . . . . . . . . . . . . . . . . . . . . . . . *Aphytis*
— Forewings with a tuft of long setae basad of speculum . . . . . . . . . . . . . *Bestiola*

Available information about the biology of these genera leads to an interesting question: Can this be regarded as a monophyletic group? Can *Aphytis*, comprising primary parasites of armored scale insects, be regarded as really closely related to *Marietta*, a genus of hyperparasites, and to *Centrodora*, a genus of primary egg parasites? In other words, does our classification reflect close natural relationships, or is it more or less artificial?

Closer examination readily reveals that the actual mode of development of *Aphytis* and *Marietta* is not as different as the terms "primary" and "secondary" imply. Basically, both develop in a very similar ecological situation, as ectoparasites on a soft-bodied host enclosed in an empty "shell." *Aphytis* develops externally on the soft body of an armored scale insect beneath the covering scale, whereas *Marietta* develops externally on the soft body of an encyrtid or aphelinid parasite within the empty, "mummified" scale insect host. We have no data on the actual mode of development of *Centrodora* and *Tumidiscapus* as egg parasites, but we suspect that they, too, will eventually be found to have a similar ecological development, i.e., to develop externally on the full-grown embryo within the egg shell. The fact that species of these genera have been reared also from dipterous puparia appears to lend support to this assumption. We assume that they develop externally on the soft-bodied pupa enclosed in the puparium. No information is at present available on the exact mode of development of *Marlattiella* and *Bestiola* in armored scale insects, but we shall not be surprised if, like *Aphytis*, they too eventually prove to be ectoparasitic.

Thus, although great gaps still exist in our knowledge, these genera do appear to be closely related biologically as well as morphologically, and may indeed have evolved from a common ancestor.

# LIST OF *APHYTIS* SPECIES

Following is an alphabetical list of all the names that have been applied to the species of *Aphytis*, along with notes indicating their current systematic position. Names of valid species are printed in boldface; synonyms, homonyms and unrecognizable species are in italics. Misidentifications recorded in the entomological literature have not been included in the list.

| | |
|---|---|
| *abnormis* (Howard 1881, *Aphelinus*) | = Unrecognizable |
| **acrenulatus** DeBach and Rosen, 1976 | |
| **acustapidis** n.sp. | |
| **africanus** Quednau, 1964 | |
| *alami* Agarwal, 1964 | = Unrecognizable |
| *albidus* (Westwood, 1837, *Agonioneurus*) | = **mytilaspidis** (Le Baron) |
| **amazonensis** n.sp. | |
| *angeloni* (Girault, 1932, *Marietta*) | = Unrecognizable |
| **angustus** Compere, 1955 | |
| **anneckei** DeBach and Rosen, 1976 | |
| **anomalus** Compere, 1955 | |
| **antennalis** n.sp. | |
| **aonidiae** (Mercet, 1911, *Aphelinus*) | |
| **argenticorpus** n.sp. | |
| *argentinus* (Brèthes, 1916, *Aphelinus*) | = **hispanicus** (Mercet) |
| **australiensis** DeBach and Rosen, 1976 | |
| **bedfordi** n.sp. | |
| *bovelli* (Malenotti, 1918, *Aphelinus*) | = **hispanicus** (Mercet) |
| **capensis** DeBach and Rosen, 1976 | |
| **capillatus** (Howard, 1907, *Perissopterus*) | |
| *capitis* (Rust, 1915, *Aphelinus*) | = **chilensis** Howard |
| *caucasicus* Tshumakova, 1964 | = Unrecognizable |
| **cercinus** Compere, 1955 | |
| **chilensis** Howard, 1900 | |
| **chrysomphali** (Mercet, 1912, *Aphelinus*) | |
| **ciliatus** (Dodd, 1917, *Aphelinus*) | |
| *ciliatus* Quednau, 1964 | = **setosus** DeBach and Rosen |
| *citrinus* Compere, 1955 | = **aonidiae** (Mercet) |
| **cochereaui** DeBach and Rosen, 1976 | |
| **coheni** DeBach, 1960 | |
| **columbi** (Girault, 1932, *Aphelinus*) | |
| **comperei** DeBach and Rosen, 1976 | |
| **confusus** DeBach and Rosen, 1976 | |
| **costalimai** (Gomes, 1942, *Marietta*) | |
| *cowperi* (Girault, 1923, *Perissopterus*) | = **capillatus** (Howard) |
| **cylindratus** Compere, 1955 | |

**dealbatus** Compere, 1955
**debachi** Azim, 1963
**desantisi** DeBach and Rosen, 1976
*diaspidioti* Tshumakova, 1957 = Unrecognizable
**diaspidis** (Howard, 1881, *Aphelinus*)
*distonotus* (Girault, 1932, *Marietta distonota*) = **columbi** (Girault)
*dubius* De Santis, 1948 = **aonidiae** (Mercet)
*dubius* var. *intermedia* De Santis, 1948 = **aonidiae** (Mercet)
*emersoni* (Girault, 1923, *Perissopterus*) = **capillatus** (Howard)
**equatorialis** n.sp.
**erythraeus** (Silvestri, 1915, *Aphelinus*)
**fabresi** DeBach and Rosen, 1976
**faurei** Annecke, 1963
**fisheri** DeBach, 1959
*flavus* (Ashmead, 1880, *Trichogramma*) = Unrecognizable
*flavus* (Nees, 1834, *Eulophus*) = Unrecognizable
*flavus* Quednau, 1964 = **rolaspidis** DeBach and Rosen
**funicularis** Compere, 1955
*fuscipennis* (Howard, 1881, *Aphelinus*) = **diaspidis** (Howard)
**gordoni** DeBach and Rosen, 1976
**griseus** Quednau, 1964
*grotiusi* (Girault, 1917, *Aphelinus*) = Unrecognizable
**haywardi** (Blanchard, 1948, *Marietta*)
**hispanicus** (Mercet, 1912, *Aphelinus maculicornis* var. *hispanica*)
**holoxanthus** DeBach, 1960
**hyalinipennis** n.sp.
**ignotus** Compere, 1955
**immaculatus** Compere, 1955
**japonicus** DeBach and Azim, 1962
**lepidosaphes** Compere, 1955
*lessingi* (Girault, 1932, *Marietta*) = **capillatus** (Howard)
**libanicus** Traboulsi, 1969
*limonus* (Rust, 1915, *Aphelinus*) = Unrecognizable
**lingnanensis** Compere, 1955
**longicaudus** n.sp.
*longiclavae* (Mercet, 1911, *Aphelinus*) = **chilensis** Howard
**luteus** (Ratzeburg, 1852, *Coccobius*)
**maculatipennis** (Dozier, 1933, *Marietta*)
*maculatipes* (Girault, 1917, *Marietta*) = Unrecognizable
**maculicornis** (Masi, 1911, *Aphelinus*)
*madagascariensis* (Risbec, 1952, *Prospaphelinus*) = **diaspidis** (Howard)
**malayensis** n.sp.
**mandalayensis** n.sp.
**margaretae** DeBach and Rosen, 1976
**mazalae** DeBach and Rosen, 1976

| | |
|---|---|
| **melanostictus** Compere, 1955 | |
| **melinus** DeBach, 1959 | |
| **merceti** Compere, 1955 | |
| **mimosae** DeBach and Rosen, 1976 | |
| *minutissimus* (Girault, 1913, *Aphelinus*) | = Unrecognizable |
| **moldavicus** Jasnosh, 1966 | |
| **mytilaspidis** (Le Baron, 1870, *Aphelinus*) | |
| **neuter** Jasnosh and Myartseva, 1971 | |
| *newtoni* (Girault, 1913, *Aphelinus*) | = Unrecognizable |
| **nigripes** (Compere, 1936, *Marietta*) | |
| **notialis** De Santis, 1965 | |
| **noumeaensis** (Howard, 1907, *Perissopterus*) | |
| *novicapillatus* (Girault, 1917, *Marietta novicapillata*) | = **capillatus** (Howard) |
| **obscurus** DeBach and Rosen, 1976 | |
| **opuntiae** (Mercet, 1912, *Aphelinus*) | |
| *opuntiae* Risbec, 1952 | = **diaspidis** (Howard) |
| *ovidii* (Girault, 1919, *Aphelinus*) | = **diaspidis** (Howard) |
| *pallidus* (Ratzeburg, 1852, *Coccobius*) | = Unrecognizable |
| **paramaculicornis** DeBach and Rosen, 1976 | |
| **peculiaris** (Girault, 1932, *Marietta*) | |
| *perissoptroides* (Girault, 1915, *Aphelinus*) | = Unrecognizable |
| **perplexus** n.sp. | |
| **philippinensis** DeBach and Rosen, 1976 | |
| **phoenicis** DeBach and Rosen, 1976 | |
| **pilosus** DeBach and Rosen, 1976 | |
| **pinnaspidis** n.sp. | |
| **proclia** (Walker, 1839, *Aphelinus*) | |
| *punctaticorpus* (Girault, 1917, *Perissopterus*) | = Unrecognizable |
| *quaylei* (Rust, 1915, *Aphelinus*) | = **chrysomphali** (Mercet) |
| *riadi* Delucchi, 1964 | = **chilensis** Howard |
| *risbeci* Annecke and Insley, 1971 | = **diaspidis** (Howard) |
| **rolaspidis** DeBach and Rosen, 1976 | |
| *romae* (Girault, 1932, *Marietta*) | = **capillatus** (Howard) |
| **roseni** DeBach and Gordh, 1974 | |
| *ruskini* (Girault, 1915, *Aphelinus*) | = Unrecognizable |
| **salvadorensis** n.sp. | |
| **secundus** (Compere, 1936, *Marlattiella secunda*) | |
| **sensorius** DeBach and Rosen, 1976 | |
| **setosus** DeBach and Rosen, 1976 | |
| *signiphoroides* (Brèthes, 1913, *Trichogrammatoidea*) | = **chilensis** Howard |
| *silvestrii* (De Gregorio, 1914, *Aphelinus*) | = ? **chrysomphali** (Mercet) |
| *simplex* (Zehntner, 1897, *Aphelinus*) | = Unrecognizable |
| *spiniferus* Compere and Annecke, 1961 | = **costalimai** (Gomes) |

| | |
|---|---|
| *stellaris* (Girault, 1933, *Aphelinus*) | = **columbi** (Girault) |
| *sugonjaevi* Jasnosh, 1972 | = **proclia** (Walker) |
| **taylori** Quednau, 1964 | |
| **testaceus** Tshumakova, 1961 | |
| **theae** (Cameron, 1891, *Aphelinus*) | |
| **tucumani** n.sp. | |
| **ulianovi** Girault, 1932 | |
| **vandenboschi** DeBach and Rosen, 1976 | |
| *variolosum* Alam, 1956 | = Unrecognizable |
| **vittatus** (Compere, 1925, *Paraphytis vittata*) | |
| **wallumbillae** (Girault, 1924, *Aphelinus*) | |
| **yasumatsui** Azim, 1963 | |
| *zonatus* Alam, 1956 | = **proclia** (Walker) |

Altogether, 137 names have been applied to species of *Aphytis*. Of these, 90 represent valid species, 18 are at present unrecognizable, and the rest are synonyms or homonyms.

## SUBGENERIC CLASSIFICATION

Mercet (1932) proposed the division of *Aphytis* into two subgenera, and resurrected De Gregorio's (1914, 1915) subgeneric name *Prospaphelinus* for one of them. Following is a translation of Mercet's key to his subgenera:

Female antennae 6-segmented but appear to be 5-segmented; first funicular segment smaller than the second; ovipositor exserted. Male antennae 4-segmented . . . . . . . . . . . . . . . . . . . . . . . . *Aphytis* Howard, s.str.
Both female and male antennae 6-segmented; first two funicular segments subequal; ovipositor concealed . . . . . . . . . . . . . . . . . . . . . . . . . . . . . . . . . . . . . . . . . . . . . . . . . . . . . . . . . *Prospaphelinus* De Gregorio

Mercet included only the generotype, *A. chilensis*, and one other species—*A. longiclavae*, a synonym of *chilensis*—in the nominate subgenus, and referred all other European species to the subgenus *Prospaphelinus*. His subgeneric classification was accepted by De Santis (1946, 1948). Compere (1955), on the other hand, did not follow Mercet's lead, and stated his opinion that "this needlessly complicates the classification without serving any useful purpose." We agree with Compere's conclusion.

De Gregorio (1914, 1915) originally proposed *Prospaphelinus* as a subgenus of *Aphelinus*. His classification was based on erroneous diagnostic characters. He characterized *Prospaphelinus* mainly by the presence of minute setae between the ommatidia of the compound eyes, which he assumed were absent in "typical" *Aphelinus*, and by the presence of "two bifid setae on the posterior part of the abdomen"—presumably the cerci. This was obviously wrong, inasmuch as all the known species of *Aphelinus* and *Aphytis* possess inter-ommatidial setae as well as well-developed cerci. Mercet (1930) was therefore correct in synonymizing *Prospaphelinus* with *Aphytis*.

The diagnostic criteria proposed by Mercet (1932) cannot be relied upon for subgeneric classification of *Aphytis*. His reference to the ovipositor should be disregarded, since the ovipositor of *chilensis* does not differ in any significant way from that of other members of the genus. In the female antennae of *chilensis*, the first funicular segment is indeed greatly reduced, considerably smaller than the second segment; and is triangular in shape. (It will be recalled that this segment was overlooked by Howard, who erroneously assumed that the antennae of *chilensis* were only 5-segmented.) The male antennae do appear 4-segmented in *chilensis*, with the first two funicular segments reduced to minute rings. However, acceptance of antennal configuration as the basis for subgeneric classification would result in the formation of unnatural groups. Thus, for instance, *A. capillatus*, which is quite similar to *chilensis* in this respect but differs markedly in many other important characters, would have to be included in the same subgenus, whereas *A. merceti*, which is rather closely related to *chilensis* in most characters including the male antennae but has the first two funicular segments subequal in the female, would have to be placed elsewhere.

The subgenera as defined by Mercet are not mutually exclusive. In several species of *Aphytis*, such as *costalimai*, *cochereaui* and *fabresi*, the first funicular segment of the female is triangular and considerably smaller than the second segment, whereas the male antennae are clearly 6-segmented, although the first two funicular segments are sometimes quite reduced. Such species would not fit into any of Mercet's subgenera, and a third subgenus would probably have to be established to accommodate them. Also, species with the first funicular segment of the female considerably smaller than the second segment appear to intergrade with species in which these segments are subequal, and there is no clear demarcation line separating the two groups. On the other hand, sexual dimorphism of the antennae, as expressed by reduced segmentation or by the presence of long setae on the male antennae, is evident in such unrelated species as *vittatus*, *chilensis*, *cercinus*, *costalimai*, *pilosus* and *roseni*, differing markedly in many other respects including the antennal configuration of the female.

Then, too, for the sake of consistency, additional subgenera would have to be erected for such species as *faurei* and *secundus*, with 4-segmented antennae in both sexes; for *funicularis* and allied species with 5-segmented antennae in both sexes; for *vittatus* and *cercinus* with 6-segmented female antennae and 5-segmented male antennae (it will be recalled that in 1925 Compere established the genus *Paraphytis* on this basis); and so forth. All this would result in a highly artificial classification, exaggerating the importance of a relatively minor character. We consider overall similarity, as expressed by the general pigmentation and chaetotaxis of the body, the forewing pattern, the relative length of the propodeum, the shape of the crenulae and many other characters to be at least as important as antennal characters in the classification of *Aphytis*.

We therefore prefer to use informal, more flexible "species groups" (Compere, 1955; Rosen and DeBach, 1976, 1977a) instead of Mercet's subgenera. Such groups are more convenient in the framework of our present knowledge, because they are less rigid and do not have any formal systematic status.

Compere (1955) mentioned three groups of species in the genus *Aphytis*—the **proclia**, **mytilaspidis** and **chrysomphali** groups. With 90 valid species in the genus, we now recognize seven more or less distinct groups—the **vittatus**, **chilensis**, **proclia**, **mytilaspidis**, **lingnanensis**, **chrysomphali** and **funicularis** groups. These may be briefly characterized as follows:

The **vittatus** group includes the "*Marietta*-like" species of *Aphytis*, with distinctly mottled forewings and heavily, extensively pigmented body and appendages. Structural characters are rather variable in this large, heterogeneous group, which may be further subdivided into several complexes.

No other groups possess distinctly mottled wings.

The **chilensis** group includes species with marked sexual dimorphism in the structure of the antennae, the male antennae exhibiting a reduced number of segments and an enlarged, elongate club. Usually rather strongly pigmented, although paler than in the preceding group.

No other groups exhibit marked sexual dimorphism.

The **proclia** group includes generally grayish or dusky species, with a typical pattern of black cephalic pigmentation. This group appears to be much more uniform than the preceding groups.

All remaining groups are essentially yellow in general coloration.

The **mytilaspidis** group includes species with a relatively short propodeum, small crenulae, and coarse, dark thoracic setae.

The **lingnanensis** group includes species with a relatively long propodeum and large, overlapping crenulae. Thoracic setae are variable in this group.

The **chrysomphali** group includes species with a very long propodeum, small, non-overlapping crenulae, and slender, pale thoracic setae.

The **funicularis** group includes small species with reduced mouthparts and 5-segmented antennae in both sexes. The propodeum is rather short, but the shape of the propodeal crenulae and other morphological characters vary considerably among the members of this small, aberrant group.

Several species cannot be assigned at present to any of these groups. Some appear to be intermediate, others are aberrant or present unique characteristics. Group designations may change with the accumulation of further knowledge, and additional groups may eventually be formed with some of the unassigned species.

## SOME ZOOGEOGRAPHICAL CONSIDERATIONS

Available information about the distribution of the species of *Aphytis* in the main zoogeographical regions of the world (as defined by Wallace, 1876) is tabulated in Appendix 1 (p. 747).

The parasite fauna of vast areas in the world is still very poorly known, and our knowledge of the geographical distribution of the species of *Aphytis* is therefore woefully incomplete. Moreover, since many species were undoubtedly transferred inadvertently by man from one region to another with their rather cryptic scale-insect hosts on imported plant material, their present distribution does not necessarily yield any clue to their origin (see also Rosen, 1969; DeBach, 1971b; also p. 82 here). Nevertheless, the large amount of material at hand appears to permit some preliminary conclusions regarding the pattern of geographical distribution of the main *Aphytis* groups.

Thus, the **vittatus** group appears to be centered in the South Pacific, ranging from Australia to South America, with a few representatives in southeastern Asia. Bona fide members of this group have not been recorded from Africa or from the Holarctic region, but two related species occur in California. The **chilensis** group seems to be centered in southern Africa, with one Australian and one nearly cosmopolitan representative. The **proclia** group appears to be mainly Palearctic and Oriental, with some South American and African elements; only widely distributed or cosmopolitan members of this group have been recorded from Australia or North America. Several species related to the **proclia** group occur in the Neotropical and Ethiopian regions. The **mytilaspidis** group appears to be overwhelmingly Palearctic (or Holarctic), but several species related to it occur in the Ethiopian region, and one in the Neotropical region. The **lingnanensis** and **chrysomphali** groups are centered in the Far East (i.e., in the Oriental and eastern Palearctic region), with a few representatives in Africa and in South America. The small **funicularis** group is apparently widely distributed, with two Oriental members, one Ethiopian and one Australian member; it has not yet been recorded from the Holarctic or Neotropical regions. Finally, most of the "unassigned species" of *Aphytis* occur in the Ethiopian region.

Altogether, 18 species of *Aphytis* have been recorded from the Australian region, 23 from the Neotropical region, 23 from the Ethiopian region, 27 from the Oriental region, 23 from the Palearctic region, and only 12 from the Nearctic region. (These totals do not include purposefully introduced or unrecognizable species, but they include the occurrence of a species in two or more regions.)

The Australian region is mainly represented by the **vittatus** and **chilensis** groups. The **mytilaspidis** group has not been recorded from this region, the **lingnanensis** group is represented by one species, and only cosmopolitan members of the **proclia** and **chrysomphali** groups occur there.

The Neotropical region is represented mainly by the **vittatus** and **proclia** groups.

The Ethiopian region presents an interesting *Aphytis* fauna. It is the center of distribution of the **chilensis** group, shows a strong representation of the **proclia** group, and has a few endemic members of the **lingnanensis** and **chrysomphali** groups. The **vittatus** and **mytilaspidis** groups are not represented in this region, but several species related to the **mytilaspidis** group occur here. Many "unassigned" species of *Aphytis* occur in the Ethiopian region, and various aberrant forms—e.g., reduced mouthparts, reduced antennal segmentation, hairy antennae, etc.—are more abundant here than in any other zoogeographical region.

The Oriental region shows a strong representation of the **proclia**, **lingnanensis** and **chrysomphali** groups, as well as some elements of the **vittatus** group. The **chilensis** and **mytilaspidis** groups have not been recorded from that region, while the small **funicularis** group is represented by two members.

The Palearctic region does not have any representatives of the **vittatus** group and only one cosmopolitan member of the **chilensis** group, but it appears to be the center of distribution of the **mytilaspidis** group and shows a strong representation of the **proclia** and **chrysomphali** groups. The eastern Palearctic region appears to have a lot in common with the Oriental region.

The Nearctic region has the poorest *Aphytis* fauna. The **vittatus** group is not represented in this region, except for two related species. The **lingnanensis** group is represented by a single Neotropical species that occurs also in Baja California, while the **chilensis** and **chrysomphali** groups are represented only by cosmopolitan species. The **proclia** and **mytilaspidis** groups are the only ones showing a relatively strong representation in the Nearctic region, but they, too, are represented by widely distributed species that are apparently not endemic to this region.

Although no area should be neglected in the search for additional species of *Aphytis*, the Australian, Ethiopian, Neotropical and Oriental regions are probably the most promising.

# PHYLOGENY

The species of *Aphytis* present a wide variety of morphological characters. They range in general habitus from heavily pigmented, coarsely setose species with mottled wings, resembling the species of *Marietta*, to immaculate, finely setose species with perfectly hyaline wings. In the absence of any fossil record, and with the 90 known species presumably representing but a small fraction of all extant members of the genus (see p. 213), our ideas concerning the phylogeny of *Aphytis* can at best be regarded as preliminary working hypotheses.

General pigmentation and wing pattern appear to offer the main key to the phylogenetic status of the various species groups of *Aphytis*. On the basis of a comparative study of *Aphytis* and closely related genera, we assume that both *Aphytis* and *Marietta* have evolved from a dark, *Marietta*-like ancestor. Dark pigmentation and mottled wings are therefore considered to represent primitive or ancestral traits, as opposed to yellow coloration and hyaline wings which are presumably advanced or recent characteristics. The **vittatus** group would thus appear to be the most primitive species group of *Aphytis*, whereas the **chrysomphali** group would be the most advanced, with the **proclia** group occupying an intermediate position.

Several other characters appear to be associated with dark pigmentation, and are therefore regarded as primitive. These include coarse inter-ommatidial setae in the compound eye; numerous, coarse or dark thoracic setae; numerous coarse setae along the submarginal vein and in the delta area of the forewing; setae of unequal length along the marginal vein; specialized sense organs or sensory areas on the antennal scape or club of the male; a relatively long female ovipositor; a broad, v-shaped syntergum; cercal plates bearing 3 long setae in the female; and a slender metanotal apodeme. Some of these characters are common in related genera but appear in only a few species of *Aphytis*. On the other hand, we regard fewer, finer or paler setae, absence of specialized sense organs, a shorter ovipositor and a robust metanotal apodeme as advanced characteristics.

Interpretation of propodeal characters in this context is not easy. A well-developed propodeum bearing marginal crenulae is the main diagnostic character separating *Aphytis* from other aphelinid genera (see p. 94). We regard lengthening of the propodeum in relation to the metanotum as an advanced trait in *Aphytis*. However, it does not necessarily follow that small propodeal crenulae are more primitive than large, overlapping ones. Indeed, the opposite seems to be true. Large, conspicuous crenulae are quite common in primitive, heavily pigmented species of *Aphytis*, and there may have been a trend toward their reduction in some of the highly evolved groups.

In the developmental stages, black or dark pupal pigmentation appears to be more primitive than pale or yellow coloration, but there is considerable variation in this character.

On the basis of all these characters, we consider the **vittatus** group to represent the most primitive element of *Aphytis*. The **chilensis** group is also rather primitive, occupying an intermediate position between the **vittatus** and **proclia** groups, or—more likely—representing an offshoot of the former. The **proclia** group apparently evolved from ancestral forms related to the **vittatus** group, whereas both the **mytilaspidis** and **lingnanensis** groups have evolved independently from the **proclia** group. Although generally more advanced than the **proclia** group, these two groups present interesting combinations of relatively primitive and advanced characteristics. Finally, the **chrysomphali** group is probably the most highly evolved component of *Aphytis*.

Some of these groups may be polyphyletic in origin. Certain members of the **chrysomphali** group, for instance, apparently evolved from the **lingnanensis** group, whereas others may have evolved from the **mytilaspidis** group. Our current knowledge does not provide any clues to the phylogenetic status of the small, aberrant **funicularis** group or the various unassigned species of *Aphytis*.

Thus, on the basis of present knowledge, we assume that the genus *Aphytis* originated from a *Marietta*-like ancestor in what is now the South Pacific region. It gradually extended its distribution into the Neotropical, Oriental and Ethiopian regions, parts of which were included in the common southern continent, Gondwanaland, up to the Jurassic Period (see Gressitt, 1974), and subsequently into the Palearctic and Nearctic regions, and evolved into less strongly pigmented, paler, and finally yellow forms. It would therefore appear that the primitive forms of *Aphytis*, represented by the **vittatus** group, evolved in the Jurassic or early Cretaceous, before continental drift shaped the continents into their present form. When eventually reconstructed, the phylogenetic tree of *Aphytis* may somewhat resemble the cladogram proposed in Figure 96.

Interpretation of the phylogenetic status of individual species is very difficult. Consider, for instance, the following pair of closely related species: *Aphytis chilensis*, a uniparental, nearly cosmopolitan species, and *A. columbi*, a biparental species known only from Australia. They are virtually identical in the male sex, but the females of *columbi* are considerably paler than those of *chilensis*. As pointed out by Mayr (1963, p. 411), "though superficially appearing a 'more primitive' type of reproduction, parthenogenesis in recent animals is evidently in all cases secondarily derived from sexual reproduction." Uniparental species, abundant as they may be in certain groups, can therefore be assumed to have originated from biparental ancestors. On the other hand, we have already stated that pale coloration is considered to be an advanced (i.e., recent) characteristic in *Aphytis*. Can we, therefore, assume that the biparental *columbi* represents the ancestral species from which the uniparental *chilensis* evolved? Can we further assume that *columbi*, being biparental, subsequently evolved into its present paler form, while the uniparental *chilensis*, with presumably less opportunity

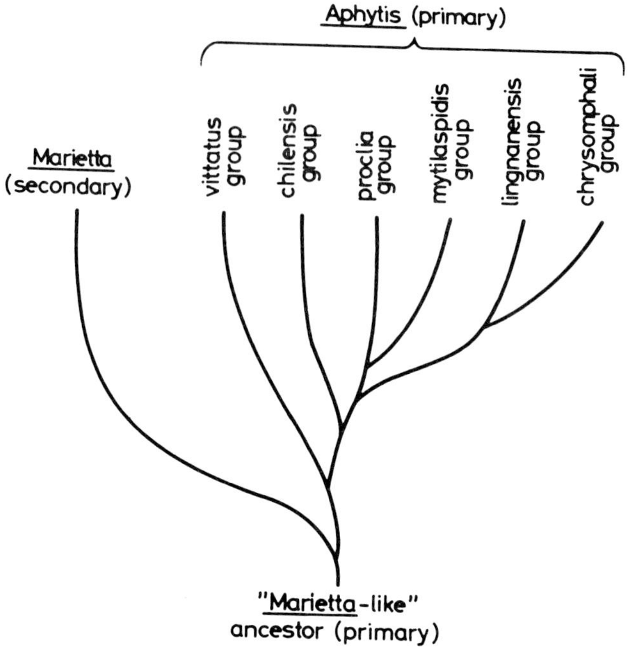


Figure 96. A proposed cladogram for *Marietta* and the main species groups of *Aphytis* (the **funicularis** group is not included). (From Rosen and DeBach, 1976.)

for natural selection, has retained the darker ancestral form? Superficially, these hypotheses appear quite plausible. However, our experience with the uniparental species of *Aphytis* indicates that they are about as variable, and therefore presumably also as amenable to natural selection, as the biparental species. Rössler and DeBach (1972b), for instance, in a comparative study of a uniparental and a biparental "form" of *A. mytilaspidis*, found no evidence for increased homogeneity in the uniparental form. In fact, size-linked (or host-affected) variation in *A. chilensis* is so great that a small "variant" of that species, reared from an unusual host, was erroneously described as a distinct species (see Gerson, 1967). A considerable amount of additional genetic information will have to be accumulated before the phylogenetic status of individual species can be intelligently interpreted.

The preceding broad outline of the phylogeny of *Aphytis* is corroborated by consideration of the phylogeny of the hosts of *Aphytis*, the armored scale insects. Available information about the known hosts of the various species of *Aphytis* is listed and tabulated in Appendix 2 (p. 752). We have followed the classification proposed by Borkhsenius (1966), who sees the Diaspididae and Phoenicococcidae as two closely related families and divides the Diaspididae into five subfamilies: Xanthophthalminae, Diaspidinae, Leucaspidinae (= Parlatoriinae auctt.), Odonaspidinae and Aspidiotinae. Current ideas on the phylogeny of the Diaspididae (Brown and McKenzie, 1962; Borkhsenius, 1965) can be summarized as follows: The family is believed to have originated from the Asterolecaniidae. The monobasic subfamily Xanthophthalminae is the most primitive, probably occupying an intermediate position between the more primitive Phoenicococcidae and the true Diaspididae. The subfamily Odonaspidinae is also very primitive, presumably representing an offshoot of the Diaspidinae. The main evolutionary pathway of the family was from the Diaspidinae through the Leucaspidinae to the most highly evolved Aspidiotinae.

There are no authenticated records of any species of *Aphytis* parasitizing the Asterolecaniidae, Phoenicococcidae, Xanthophthalminae or Odonaspidinae. (Interestingly, the closely related aphelinid genus *Bestiola* is parasitic on *Odonaspis*.) On the other hand, species of *Aphytis* have been abundantly recorded from the Diaspidinae, Leucaspidinae and Aspidiotinae, with the exception of certain tribes or subtribes that are either the most primitive in their respective groups (e.g., the Antakaspidini which are the most primitive Diaspidinae, the Coccomytilina which are the most primitive Lepidosaphidini, the Sclopetaspidina which are the most primitive Chionaspidini, the Rugaspidiotina and Howardiina which are the most primitive Diaspidini, and the Comstockiellini which are the most primitive Aspidiotinae, probably related to the Odonaspidinae), or include aberrant pupillarial forms (e.g., the Ancepaspidini and the Gymnaspidina) (see Table 5 in Appendix 2).

We realize, of course, that our data are far from complete. Some of the groups for which there is no record of parasitism by *Aphytis* are relatively small, comprising mostly rare species that are not often collected even by coccidologists, and do not include any important pest species. The chances of rearing *Aphytis* from them would therefore be minimal. However, although further extensive collecting may yield exceptions, the broad outline does appear to have some evolutionary significance. Parasitism by *Aphytis* appears to be restricted to the mainstream of the Diaspididae, hence the genus in its present form must have evolved only after the separation of such groups as the Phoenicococcidae, Xanthophthalminae and Odonaspidinae.

The association of the main species groups of *Aphytis* with the five subfamilies of the Diaspididae is summarized in Tables 6 and 7 (Appendix 2). Available information indicates that the **proclia** and **mytilaspidis** groups are by far the dominant parasites of the Diaspidinae and Leucaspidinae, whereas the Aspidiotinae are heavily attacked also by the **lingnanensis** and **chrysomphali** groups. Moreover, it appears as if the more primitive groups of *Aphytis* are more evenly distributed among the three susceptible subfamilies of the Diaspididae, whereas the more recent groups, especially the **lingnanensis** and **chrysomphali** groups, are overwhelmingly associated with the more recent Aspidiotinae (see also Rosen and DeBach, 1977b). The fact that the **funicularis** group has so far been recorded only from the Diaspidinae may be a clue to the relationships of this aberrant group with the more primitive elements of *Aphytis*. Again, although economically important genera of scale insects, such as *Aspidiotus*, *Chrysomphalus*, *Aonidiella* or *Quadraspidiotus* (all of them in the Aspidiotinae), have undoubtedly received more research emphasis and therefore have many species of *Aphytis* recorded from them, the broad outline appears to indicate parallel evolution of the Diaspididae and the main species groups of *Aphytis*.

# MORPHOLOGY OF ADULTS

Snodgrass' (1935) classical treatise, "Principles of Insect Morphology," is still regarded as the basic general text. The morphology of adult Hymenoptera was recently discussed in considerable detail by Richards (1956) and Riek (1970), and Eady (1968) has clarified the terminology of microsculpure in Hymenoptera. For accounts of the morphology of the Aphelinidae see also De Santis (1948) and Nikol'skaya and Yasnosh (1966). Imms (1916) treated the internal anatomy of *Aphytis* in some detail. Following is a discussion of some of the more salient aspects of the external morphology of adult *Aphytis*, with notes on interspecific variation.

As in most other parasitic and aculeate Hymenoptera, the body of *Aphytis* is divided into three well-defined regions: 1) the **head**, bearing the eyes, mouthparts and antennae; 2) the **mesosoma**, comprising the three thoracic segments, bearing the legs and wings, to which the first segment of the abdomen (*propodeum*) is firmly attached; 3) the **gaster**, comprising all subsequent abdominal segments, separated from the propodeum by a well-marked constriction and bearing the ovipositor and genitalia.

## THE HEAD AND ITS APPENDAGES (Figures 97–144)

**The head** of *Aphytis* is hypognathous, with its long axis vertical and the mouthparts situated ventrally. In lateral view (Figure 97) the head is more or less triangular, with the large *compound eyes* occupying more than half its lateral surface. The cheeks, or *genae*, are longer than the diameter of the eye, undivided, reticulate-striated and sparsely setose. In dorsal view (Figures 98–100) the head is considerably wider than long, as wide as the thorax or wider. The *frontovertex* is nearly parallel-sided, somewhat wider than long, occupying a little less than half the width of the head; it is reticulate and usually bears two pairs of long setae along the acute occipital margin (only one pair in *melanostictus*), in addition to numerous shorter setae.

**The ocelli** (Figures 98–102) are lenticular, arranged in an equilateral or obtuse triangle. The posterior ocelli are situated at about their own diameter (long axis) from the inner orbits, about one-and-a-half to two diameters from the occipital margin, and about three diameters from one another.

**The compound eyes** (Figures 97–100, 103–107) are large, globular, multifaceted. They are invariably furnished with numerous *inter-ommatidial setae*, which are coarse and dark in some members of the primitive **vittatus** and **chilensis** groups (e.g., Figures 103, 104), more slender and pale in the more advanced species of *Aphytis* (Figures 105–107). When viewed under a low-power ($\times$ 30) dissecting microscope against a white background, coarsely setose eyes appear distinctly hairy, whereas finely setose eyes may appear glabrous.

In frontal view (Figure 115) the head is nearly rectangular, considerably wider than high. The *frons* is reticulate except on the antennal scrobes, sparsely setose along the inner orbits and between the antennal sockets. The *antennal scrobes* are smooth,

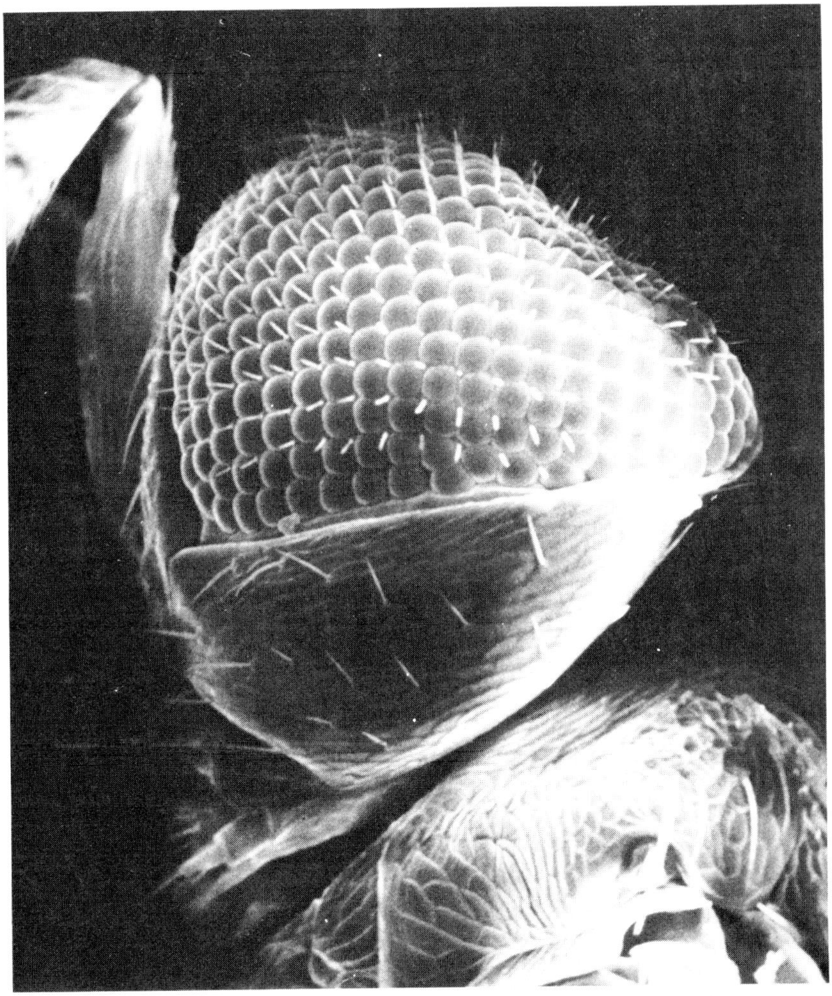

Figure 97. *Aphytis chilensis* Howard, ♀: head and pronotum, lateral view. (From Rosen and DeBach, 1976.)

elongate, somewhat convergent submedian grooves accepting the antennal scapes at rest. They are usually broadly concave, shallow and rather vaguely defined. However, in *funicularis* (Figure 1130, p. 662) the scrobes are narrow, deeply impressed and sharply outlined. The antennal sockets, or *toruli* (Figures 108, 115, 117), are situated at the level of the lower margins of the compound eyes, just above the clypeus; they are oval or subtriangular, and are surrounded by a narrow, ringlike *antennal sclerite*. The *clypeus* (Figures 108, *Cl*; 109, 111, 115) is a small, smooth, transverse sclerite, distinctly emarginate ventrally; the *epistomal suture*, defining it dorsally and on the sides, is obliterated centrally in certain species (e.g., Figure 108). The clypeus bears a

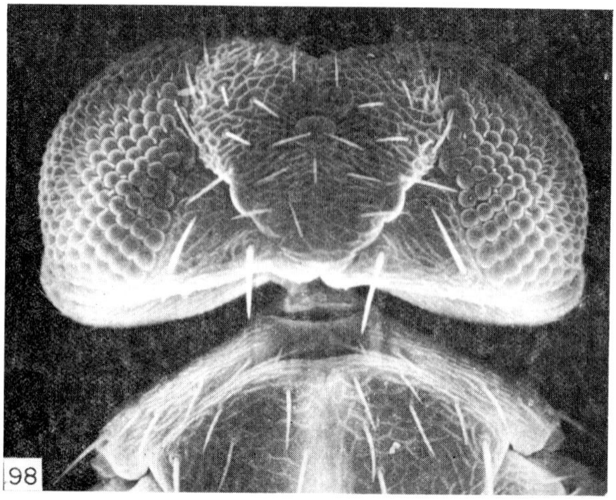

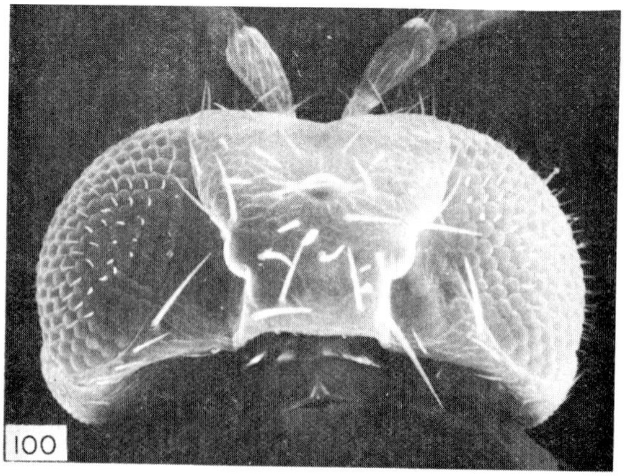


Figures 98–100. Head, dorsal view
98. *Aphytis vandenboschi* DeBach and Rosen, ♀; note also pronotum.
99. *Aphytis chilensis* Howard, ♀. (From Rosen and DeBach, 1976.)
100. *Aphytis melinus* DeBach, ♀.

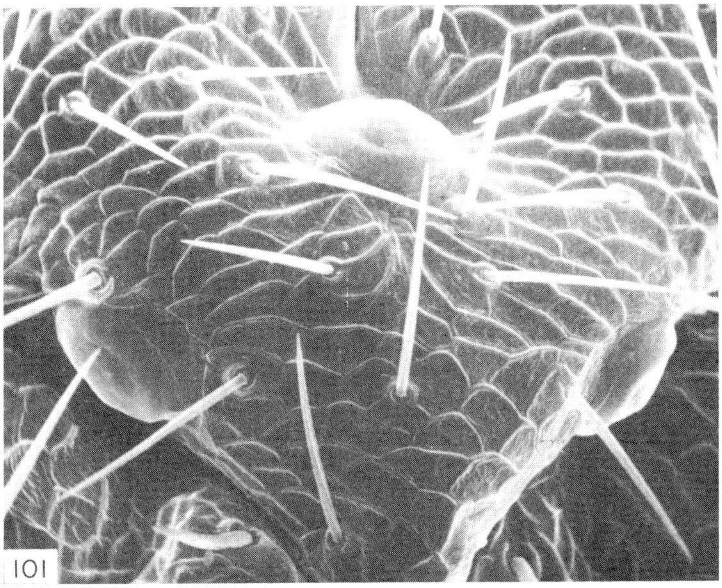


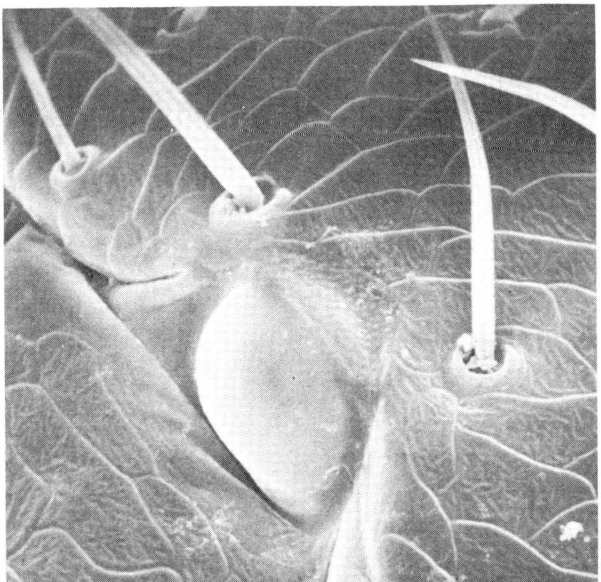


Figures 101, 102. Ocelli
101. *Aphytis chilensis* Howard, ♀. 102. *Aphytis melinus* DeBach, ♀: lateral ocellus.

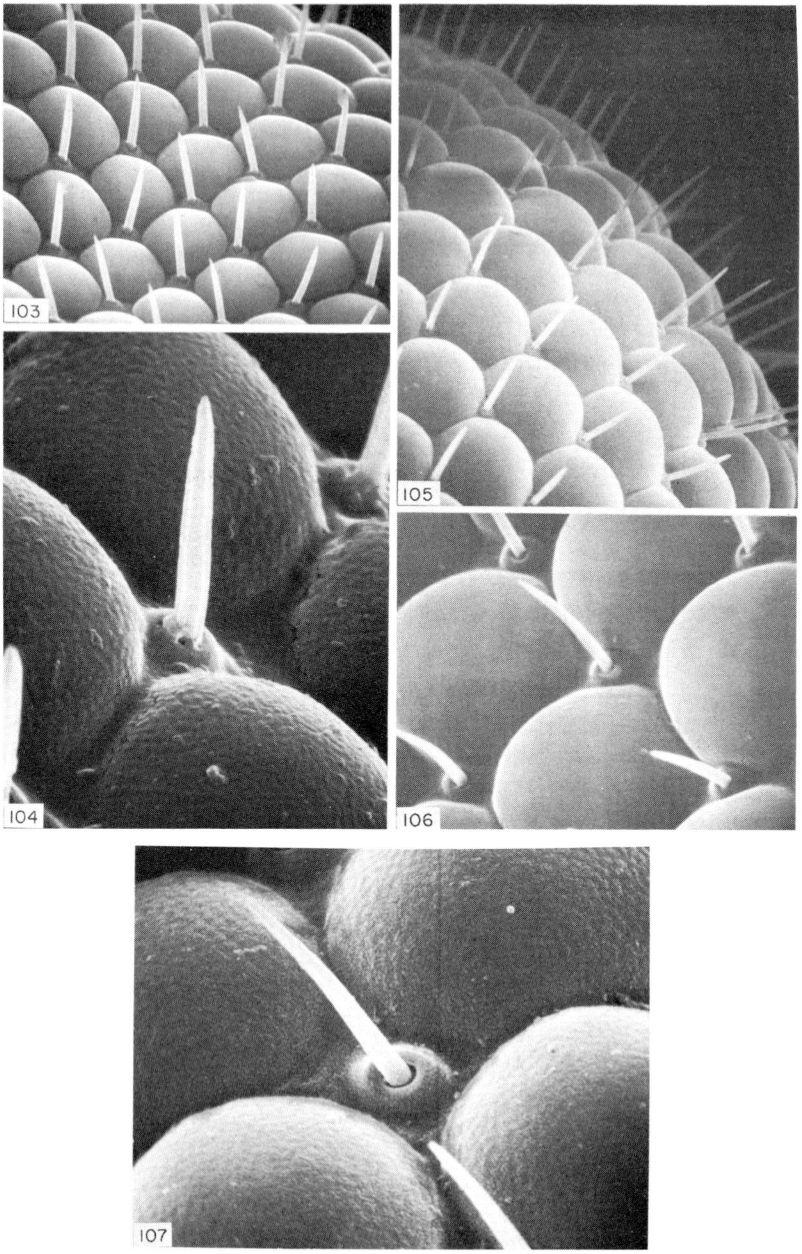


Figures 103–107. Compound eyes
103, 104. *Aphytis chilensis* Howard, ♀; note coarse inter-ommatidial setae.
105, 106. *Aphytis melinus* DeBach, ♀; note fine inter-ommatidial setae.
107. *Aphytis vandenboschi* DeBach and Rosen, ♀.

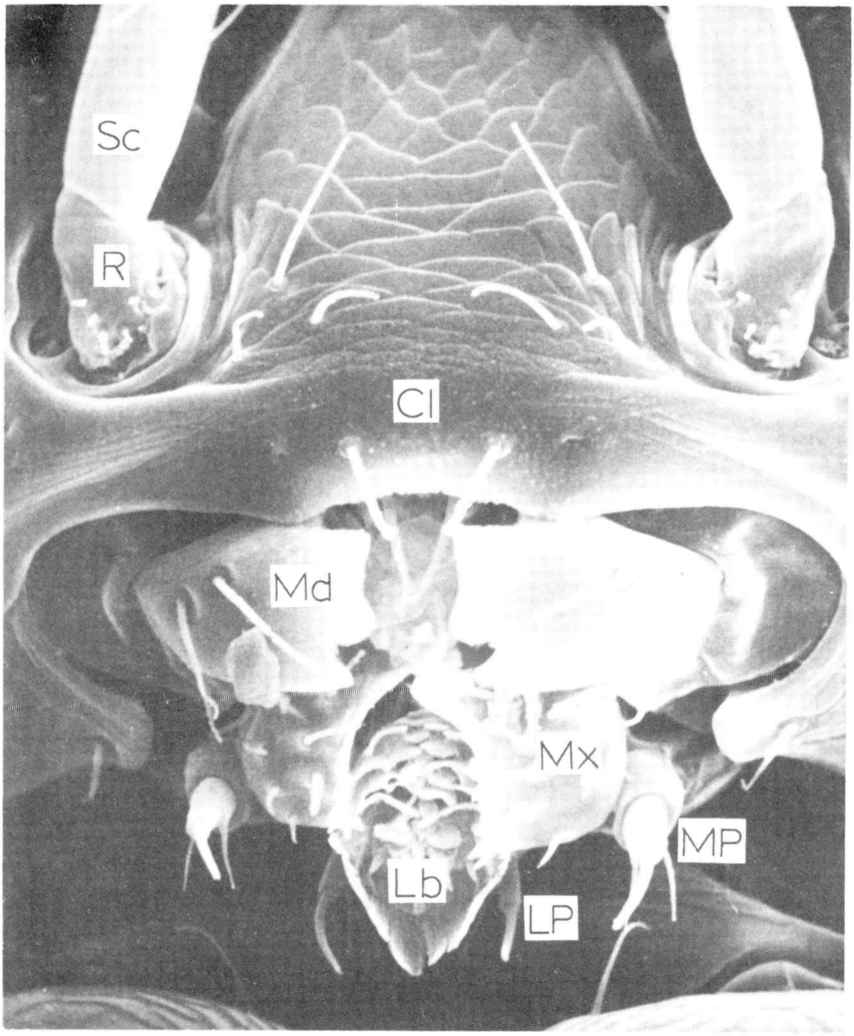


Figure 108. *Aphytis lingnanensis* Compere, ♀: mouthparts and bases of antennae, frontal view *Cl* = clypeus; *Lb* = labium; *LP* = labial palpus; *Md* = mandible; *MP* = maxillary palpus; *Mx* = maxilla; *R* = radicle; *Sc* = scape.

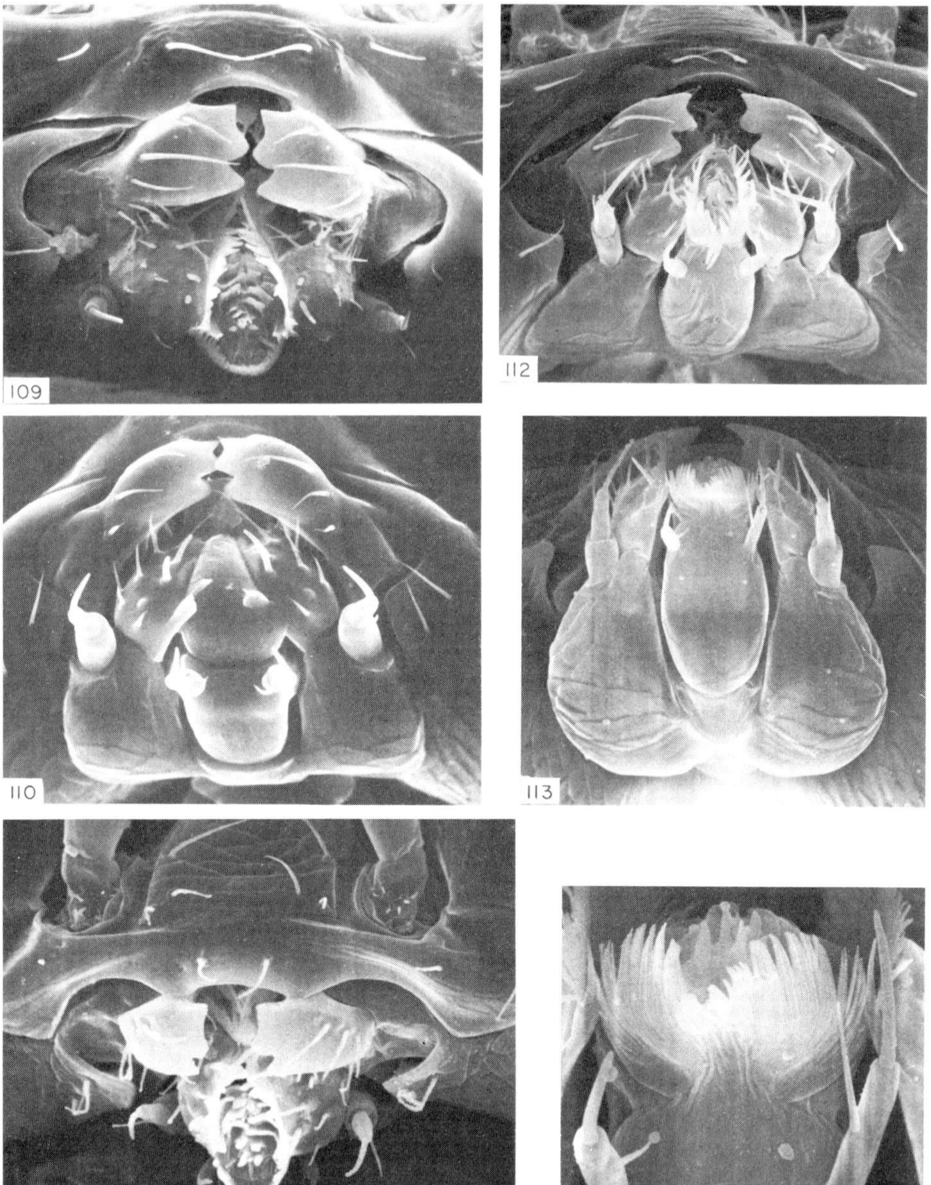


Figures 109–114. Mouthparts
109, 110. *Aphytis chilensis* Howard, ♀: 109. Frontal view. 110. Fronto-ventral view.
111. *Aphytis melinus* DeBach, ♀, frontal view. 112. *Aphytis mytilaspidis* (Le Baron), ♂, fronto-ventral view.
113–114. *Aphytis vandenboschi* DeBach and Rosen, ♀, ventral view: 113. Maxillae and labium. 114. Labium.
(Figure 110 from Rosen and DeBach, 1976.)

pair of submedian setae and a pair of minute apertures, the *anterior tentorial pits*, situated on each side of the setae. The subocular or *genal sutures* run from the lower corners of the compound eyes to the epistomal suture. They are usually moderately curved (Figure 115), but in *salvadorensis* (Figures 1298, 1307, pp. 723–724) they curve at an acute angle and then run vertically parallel to each other. In many members of the **proclia** group, the genal sutures are heavily sclerotized and pigmented (e.g., Figures 565 (p. 429) and 586 (p. 434)).

**The mouthparts** (Figures 108–114) usually consist of a pair of well-developed mandibles and a labio-maxillary complex. The *labrum* is concealed beneath the clypeus. The *mandibles* (Figure 108, *Md*) are basically bidentate, with a distinct ventral denticle and a central denticle merging into a broad dorsal truncation. In the Chalcidoidea, each mandibular denticle is usually marked by a separate gland leading to it. Only two such glands are evident in the mandibles of *Aphytis* (e.g., Figures 329 (p. 312) and 880 (p. 568)), and therefore the dorsal truncation cannot be regarded as homologous to a third denticle. The truncation is only very slightly developed in *cercinus* (Figure 502, p. 366), whereas in *equatorialis* it looks more like a third denticle. Each mandible bears 2–3 setae on its outer aspect.

Reduction of the mandibles is evident in several species of *Aphytis* belonging to different species groups. They are short, broad, nonoverlapping and nearly truncate in *merceti* (Figure 475, p. 360), *faurei* (Figure 496, p. 364), *anneckei* (Figure 1032, p. 626), *roseni* (Figure 1210, p. 706), *setosus* (Figure 1259, p. 716) and *ignotus* (Figure 1316, p. 726); small and tridentate in *debachi* (Figure 1048, p. 629); small and bidentate in *erythraeus* (Figures 1193 and 1198, p. 703); minute and pointed in *funicularis* (Figure 1129, p. 662); and vestigial, virtually invisible in *ulianovi*, *gordoni* (Figure 1163, p. 667), *theae* (Figure 1172, p. 669) and *secundus*.

The *maxillae* (Figure 108, *Mx*) and *labium* (Figure 108, *Lb*) can be readily discerned as separate units when viewed from the rear (Figure 113). The *maxillary palpi* (Figure 108, *MP*) are usually 2-segmented (Figures 108–113). However, they are only 1-segmented in *merceti*, *faurei*, all four members of the **funicularis** group and *secundus* (see Figures 496 (p. 364) and 1146 (p. 665)). The *labial palpi* (Figure 108, *LP*) are always 1-segmented. Both the maxillary and labial palpi bear a subapical seta and a stouter apical spine (Figures 113, 114).

**The antennae** (Figures 115, 116) are elbowed, or *geniculate*, the pedicel forming the pivot between the elongate scape and the flagellum. In most species they are 6-segmented in both sexes, comprising the scape, pedicel, a 3-segmented funicle and an undivided club (1131), the first two funicular segments being considerably smaller than the third. However, quite a few species exhibit a reduced number of funicular segments in either one or both sexes, so that their antennae are 5-segmented (1121), or even only 4-segmented (1111).

The *scape* (Figures 115, 116) is usually slender, cylindrical or nearly so, several times as long as wide, and as long as the club or somewhat longer. In many species the scape of the male is somewhat shorter and thicker than that of the female, sometimes considerably so (e.g., *maculatipennis* (Figure 292, p. 305) and *acrenulatus* (Figures 712, 713, p. 457)). The short *radicle* at the base of the scape (Figures 108, *R*; 115, 117) is usually longer than wide; it is furnished with several minute, fingerlike sensilla (Figures

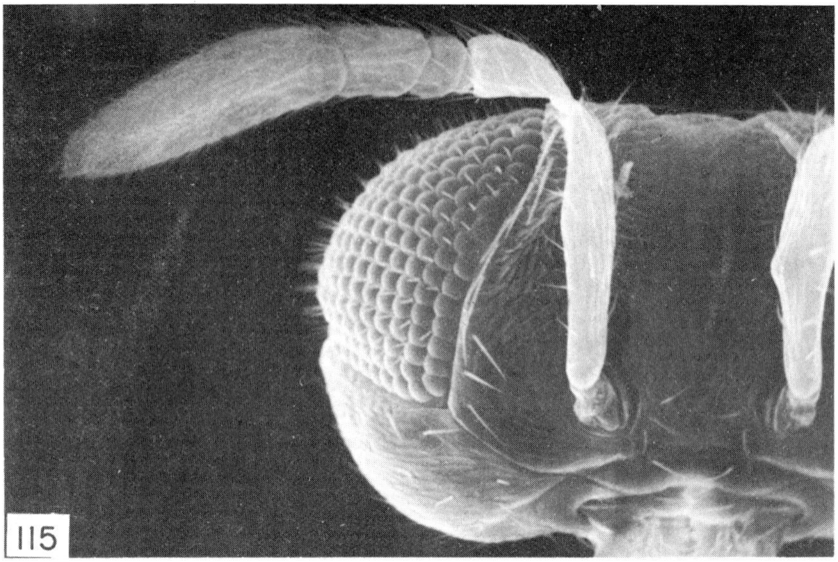


Figure 115. *Aphytis chilensis* Howard, ♀: head (frontal view) and antenna.

Figure 116. *Aphytis vandenboschi* DeBach and Rosen, ♀: antenna.

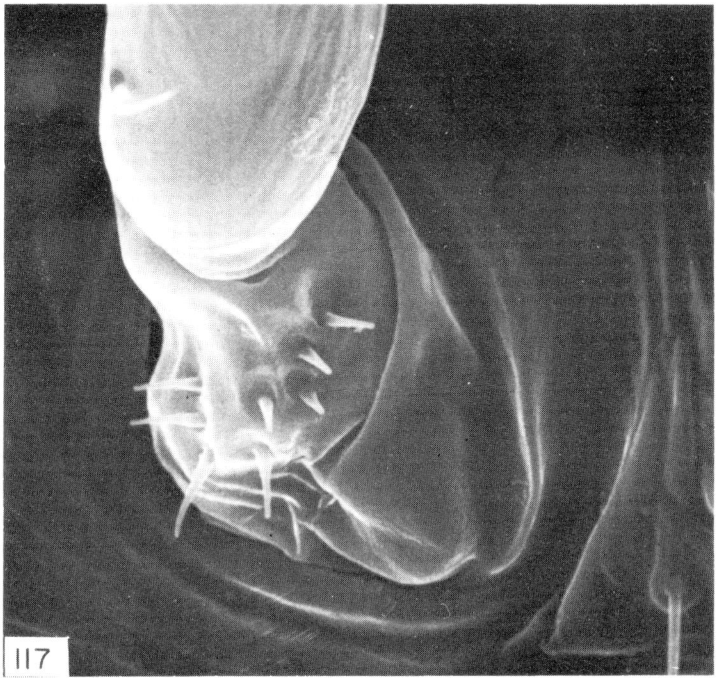

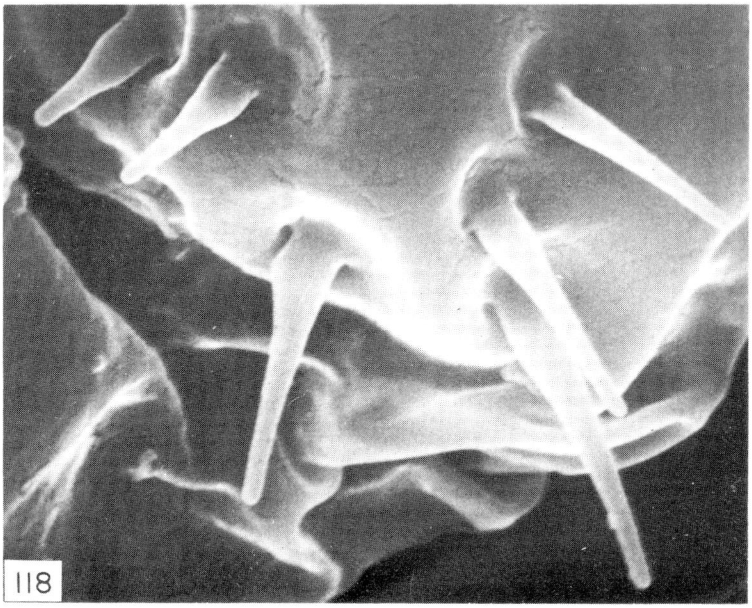

Figures 117, 118. *Aphytis chilensis* Howard, ♀
117. Base of antenna; note sensilla on radicle. (From Rosen and DeBach, 1976.)
118. Sensilla on antennal radicle.

117, 118)—probably modified *sensilla trichodea* or elongate *sensilla basiconica*—which presumably touch the antennal sclerite when the radicle moves, thus providing information on the position of the antenna.

The scape is sparsely setose. Other than that, no sense organs are evident on it in the female sex, whereas in the males of quite a few species the scape bears various specialized sensilla on its ventral aspect. These vary between species, in shape as well as in number: an elongate plate, bearing several rows of discoid sensilla in *maculatipennis* (Figure 292, p. 305) or one row of minute elongate sensilla in *costalimai* (Figure 305, p. 308); a similar plate bearing one row of tuberculous sensilla in *funicularis* (Figures 1140, 1141, p. 663); an oval plate bearing 2 or 3 dentate sensilla in *merceti* (Figures 483–486, p. 362) and in *faurei* (Figure 498, p. 364); an irregular plate bearing two discoid sensilla in *melanostictus* (Figures 461, 462, p. 337); one or two discoid or oval plates bearing a pair of tuberculous sensilla in *amazonensis* (Figures 725, 726, p. 459), *margaretae* (Figures 937–939, p. 578), *gordoni* (Figure 1165, p. 668), *theae* (Figures 1181, 1182, pp. 670–671), *longicaudus* (Figures 1240, 1241, p. 712) and *salvadorensis* (Figures 1308, 1309, p. 724); or a small plate bearing a single tuberculous sensillum, in *fabresi* (Figure 352, p. 315) and in *capillatus* (Figures 368, 369, p. 318). An indistinct ridge is evident on the ventral aspect of the scape in the males of the **mytilaspidis** group (Figures 119, 120; see also Figure 759, p. 506).

The *pedicel* (Figure 121) is conical or pyriform, widening toward its distal end. It is sparsely setose, about $1\frac{1}{2}$ times to a little over twice as long as wide, and usually longer than the third segment of the funicle. Along its distal rim are several *sensilla campaniformia* (Figures 123, 124).

The *funicle* is usually 3-segmented (Figures 115, 116), but reduction of this number has occurred in several species. Thus, the antennae are only 4-segmented, with a single funicular segment (1111), in both sexes of *faurei* (Figures 491, 497, pp. 363–364) and *secundus* (Figures 1188–1190, p. 702), 5-segmented (1121) in the females of *malayensis* (Figure 466, p. 338), *mandalayensis* (Figure 469, p. 339) and *antennalis* (Figure 547, p. 374) (males are not known in these species), and in both sexes of the **funicularis** group (e.g., Figures 1131 and 1180, pp. 662 and 670). In *funicularis* a rudimentary, incomplete suture on the club marks the site of a missing third funicular segment (Figure 1133). In the male sex, either one or both of the first two funicular segments have become greatly reduced, or are entirely absent, in *vittatus* (Figure 269, p. 300), *peculiaris* (Figure 278, p. 303), *cochereaui* (Figure 338, p. 313), *capillatus* (Figures 367, 370, 371, pp. 318–319) and *hyalinipennis* (Figure 443, p. 332), and in the males of the **chilensis** group (Figures 483, 487, 509, 523, 538, pp. 362–372), whereas the females of these species have the usual 3-segmented funicle.

The first segment of the funicle is triangular and considerably smaller than the second segment in *chilensis* (Figures 121, 122; also Figure 515, p. 368) and in most members of the **vittatus** group (e.g., *costalimai* (Figure 296, p. 306), *cochereaui* (Figure 330, p. 312), *capillatus* (Figure 359, p. 317), *anomalus* (Figure 400, p. 324) and related species). In several species, the first segment is triangular or obliquely truncate but not much smaller than the second segment (e.g., *vittatus* (Figure 264, p. 300) and *melanostictus* (Figure 452, p. 335)). In most other species of *Aphytis*, the first funicular segment is more or less quadrate or globular, about as long as wide or somewhat longer,

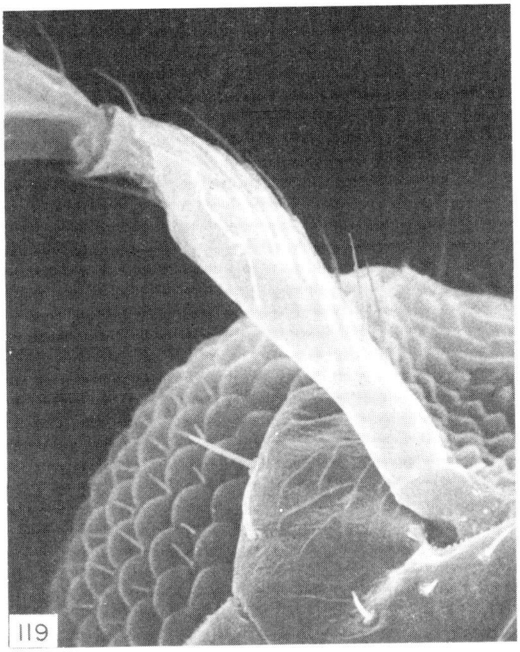


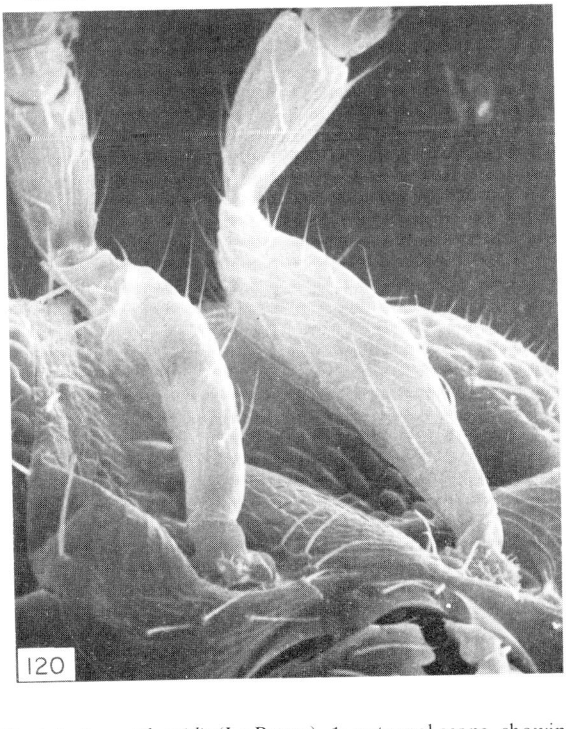


Figures 119, 120. *Aphytis mytilaspidis* (Le Baron), ♂: antennal scape, showing ventral ridge.

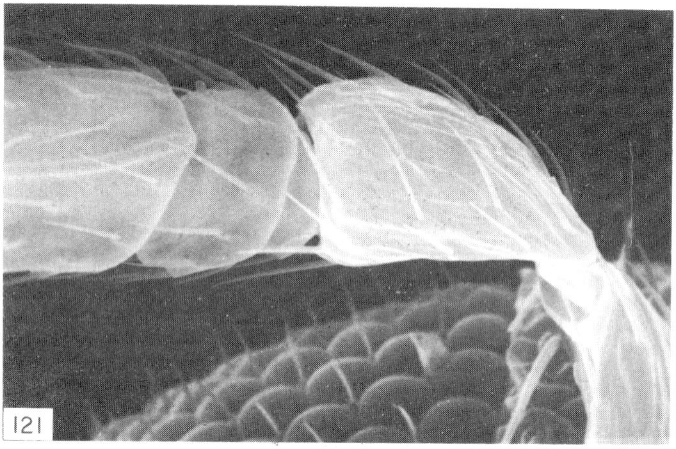


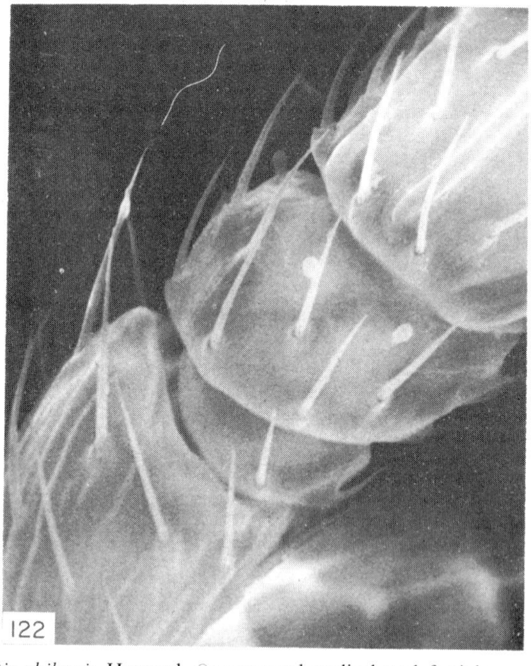


Figures 121, 122. *Aphytis chilensis* Howard, ♀: antennal pedicel and funicle; note reduced first funicular segment and sensilla on second segment.

and subequal to the second segment (Figures 123, 125). (For a discussion of this character see under Subgeneric Classification, p. 108.) The second segment of the funicle is usually subrectangular, wider than long. In most species it is somewhat shorter and wider than the preceding segment.

Each of the first two funicular segments usually bears a single modified *sensillum basiconicum* in a dorso-distal position (Figures 123, 125–127); in *chilensis* the second segment bears several such sensilla (Figure 122). The shape of these minute sense organs may be clavate as in *chilensis* (Figures 122, 128) or more or less curved as in *melinus* (Figure 126) or *vandenboschi* (Figure 127). They are conspicuously enlarged, spindle-shaped, and occupy a ventral position in the male of *sensorius* (Figures 1125, 1126, p. 645). The first funicular segment also bears a funnel-shaped sense pore—probably a *sensillum campaniformium*—on its proximal aspect, facing the pedicel (Figures 123, 124).

The third segment of the funicle (Figure 129) is usually more or less symmetrical, about as long as wide or longer. However, it is trapezoidal and distinctly wider than long in *cochereaui* (Figures 330, 338, pp. 312–313), *capillatus* (Figure 359, p. 317) and several other species, very short and transverse in the males of *roseni* (Figure 1222, p. 708) and *longicaudus* (Figure 1240, p. 712), strongly asymmetrical and obliquely truncate in the males of *mazalae* (Figures 1104, 1105, pp. 640–641) and *sensorius* (Figure 1124, p. 645). In addition to setae, sense pores and *sensilla basiconica* similar to those present on the first two funicular segments (Figures 129–132), the third segment bears several *longitudinal sensilla* (Figures 129, 130)—narrow, elongate, tapering ridges of thin cuticle, which appear transparent in slide mounts. These sense organs—presumably modified *sensilla placodea*—were termed *rhinaria* by Richards (1956), *glumes* by Riek (1970). In the female of *pilosus*, the third funicular segment bears a single unusually long seta in addition to numerous short setae (Figures 1246, 1247, p. 713).

The *club* is elongate, tapering, usually about 2 to 4 times as long as wide and several times longer than the preceding segment, but somewhat shorter than the scape. It is enormously enlarged—and considerably longer than the scape—in *faurei* (Figures 491, 497, pp. 363–364), *antennalis* (Figure 547, p. 374) and in the males of the **chilensis** group (Figures 484, 509, 523, 538, pp. 362–372). In most species, the club is clearly separated from the penultimate segment by a distinct constriction. However, in *dealbatus* (Figures 449, 450, p. 334) and *melanostictus* (Figures 452, 460, pp. 335, 337) the two segments are separated only by a suture, so that the antennae of these species appear to have a 2-segmented funicle and a 2-segmented club.

Like the third segment of the funicle, the club bears numerous short setae—presumably *sensilla trichodea*—as well as several *sensilla basiconica*, sense pores and *longitudinal sensilla* (Figures 133–135). The latter are very numerous in species with an enlarged club. The apex of the club is furnished with several elongate, fingerlike sensilla. They are somewhat curved in the female, and the one at the very tip of the club is differentiated and appears to have an apical pore (Figures 134–140), whereas in the male they are straight and uniform in shape (Figures 141, 142). The apical sensillum is enlarged and rather conspicuous in the females of *dealbatus* (Figures 449, 450, p. 334) and *melanostictus* (Figure 452, p. 335).

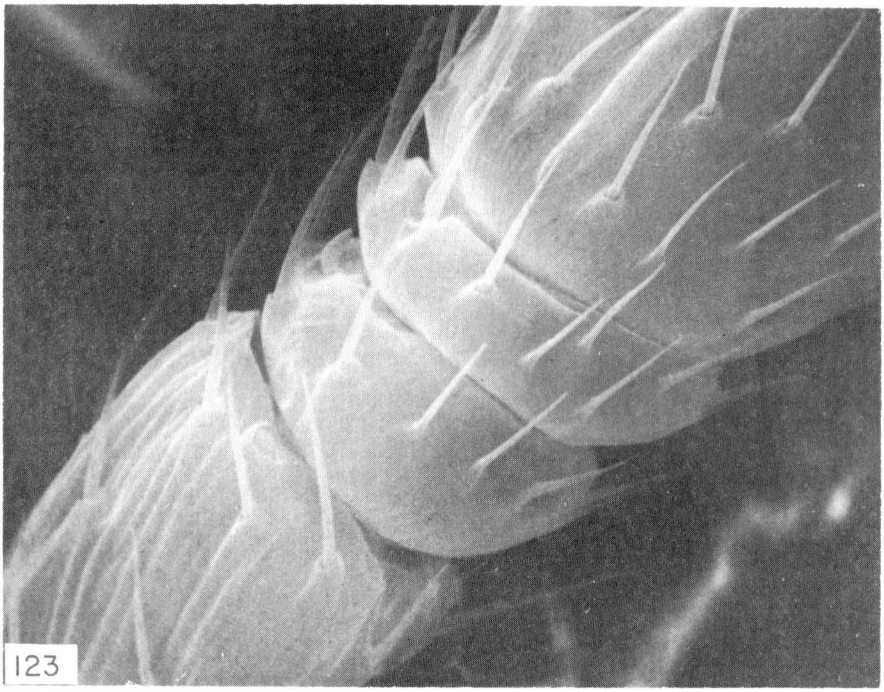

Figures 123, 124. *Aphytis vandenboschi* DeBach and Rosen, ♀
123. Antennal pedicel and funicle. 124. Sensilla on pedicel (left) and on first funicular segment (right).

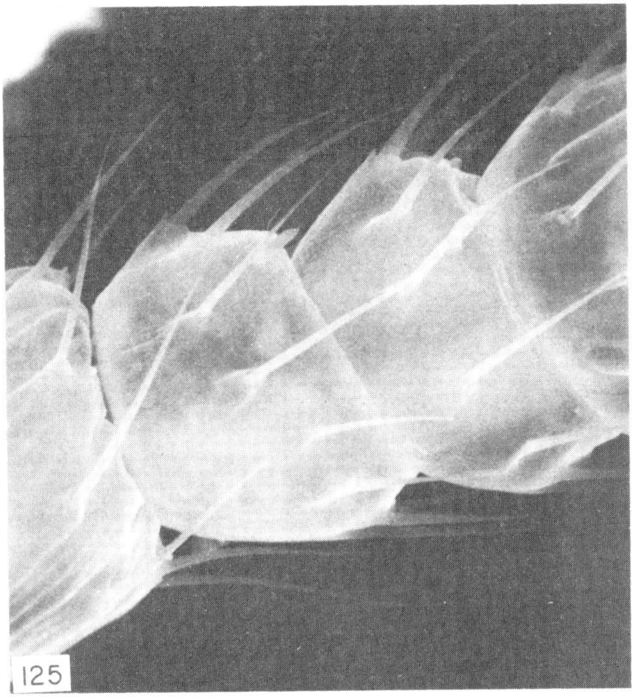

125

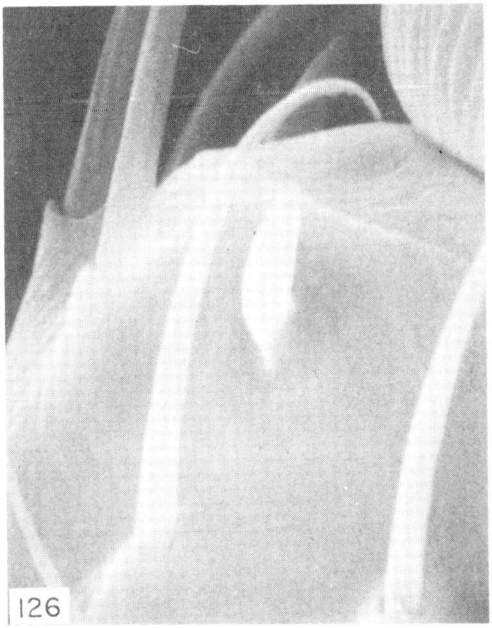

126

Figures 125, 126. *Aphytis melinus* DeBach, ♀
125. First two funicular segments. 126. Sensillum on dorsal aspect of first funicular segment.

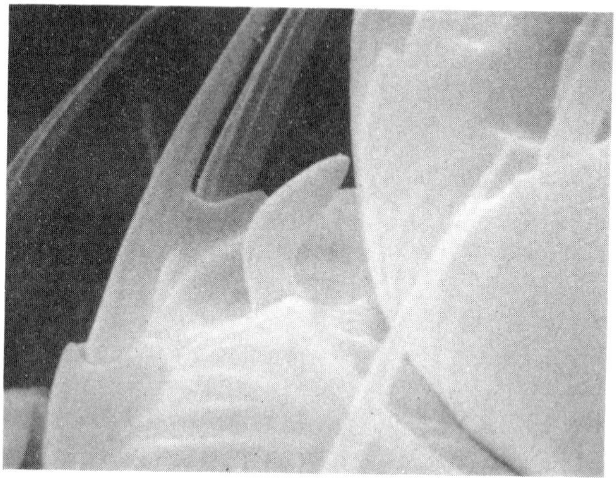

Figure 127. *Aphytis vandenboschi* DeBach and Rosen, ♀: sensillum on dorsal aspect of first funicular segment.

Figure 128. *Aphytis chilensis* Howard, ♀: sensillum on dorsal aspect of second funicular segment.

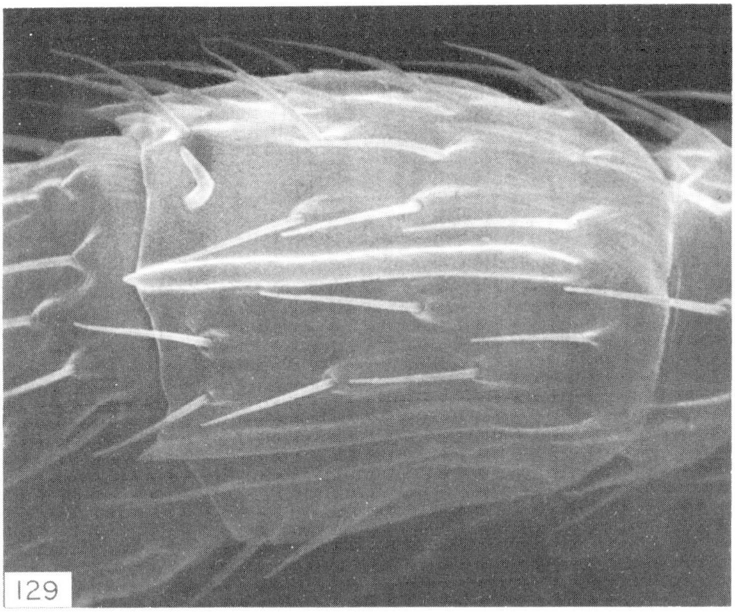

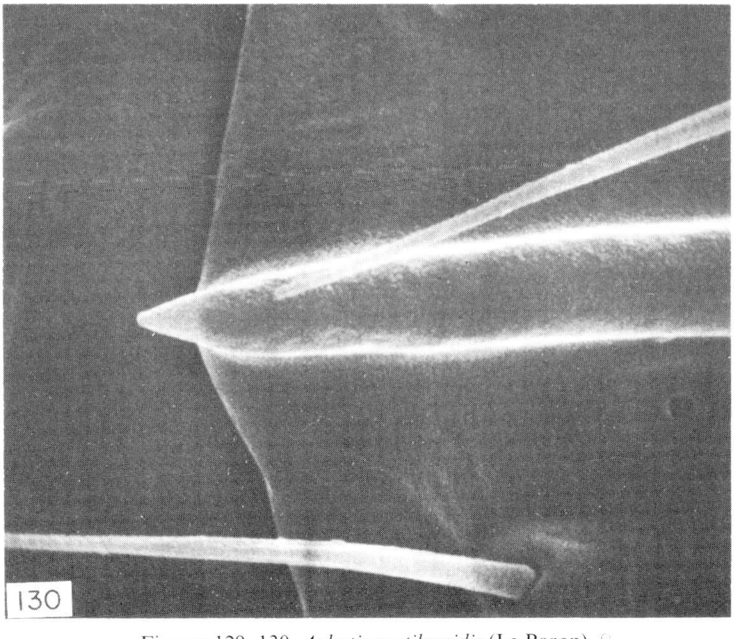

Figures 129, 130. *Aphytis mytilaspidis* (Le Baron), ♀
129. Third funicular segment, showing various sensilla. 130. Longitudinal sensillum and sense pore on third funicular segment.

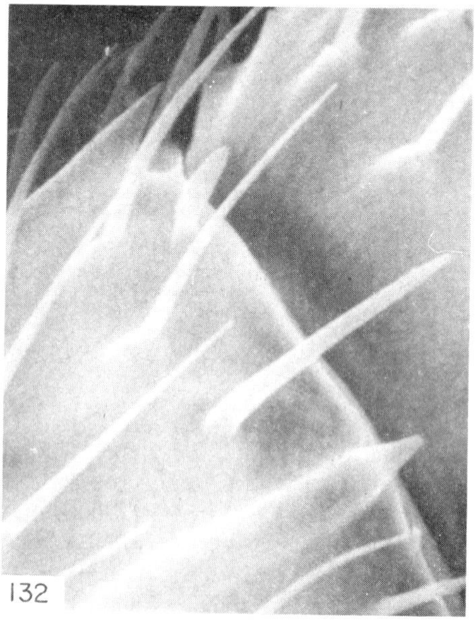


Figures 131, 132. Sensilla on dorsal aspect of third funicular segment 131. *Aphytis chilensis* Howard, ♀. 132. *Aphytis melinus* DeBach, ♀.

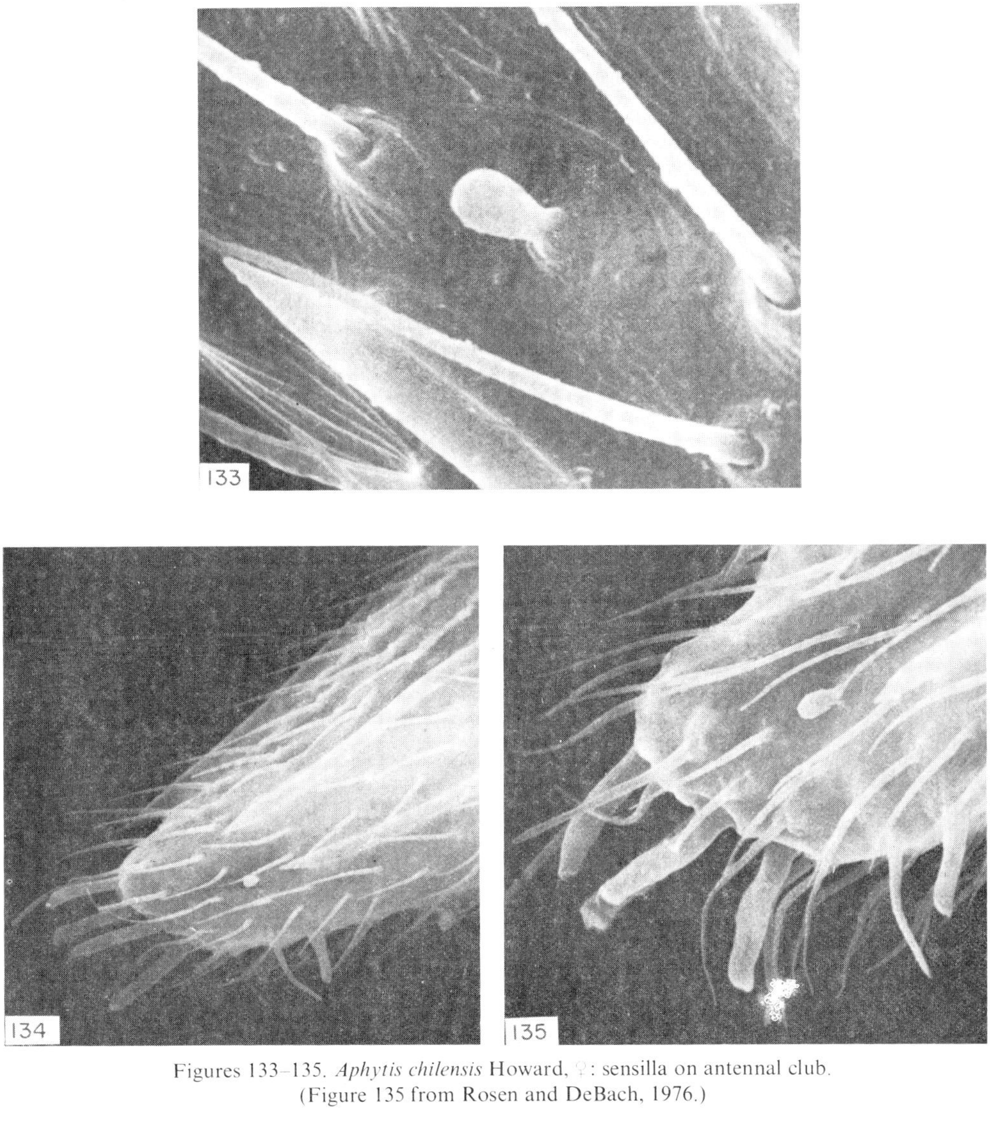


Figures 133–135. *Aphytis chilensis* Howard, ♀: sensilla on antennal club.
(Figure 135 from Rosen and DeBach, 1976.)

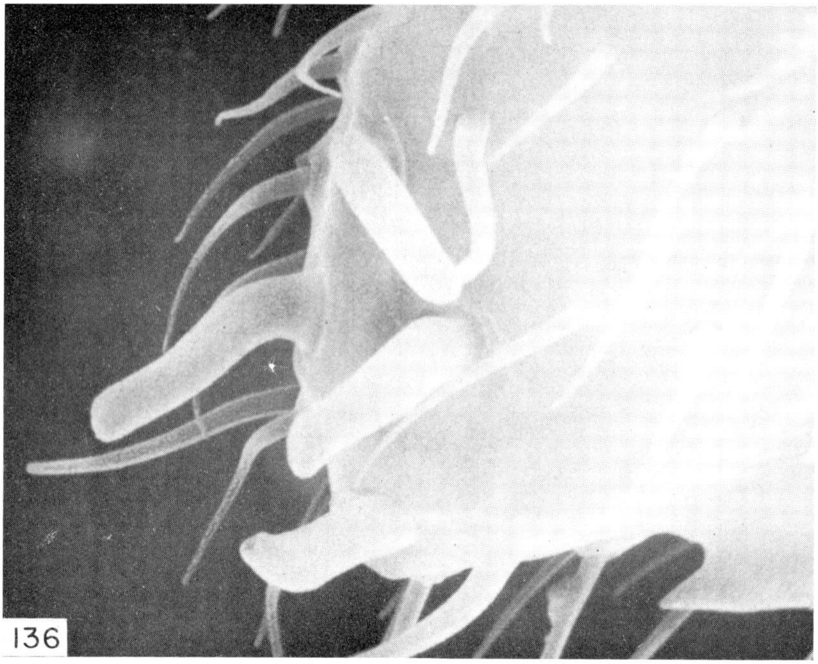


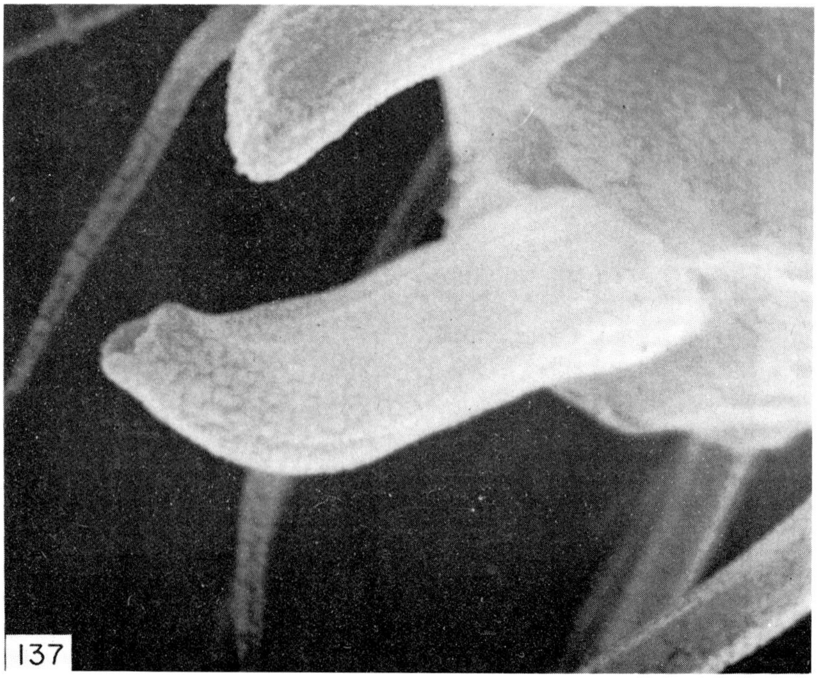


Figures 136, 137. *Aphytis chilensis* Howard, ♀: sensilla on tip of antennal club.

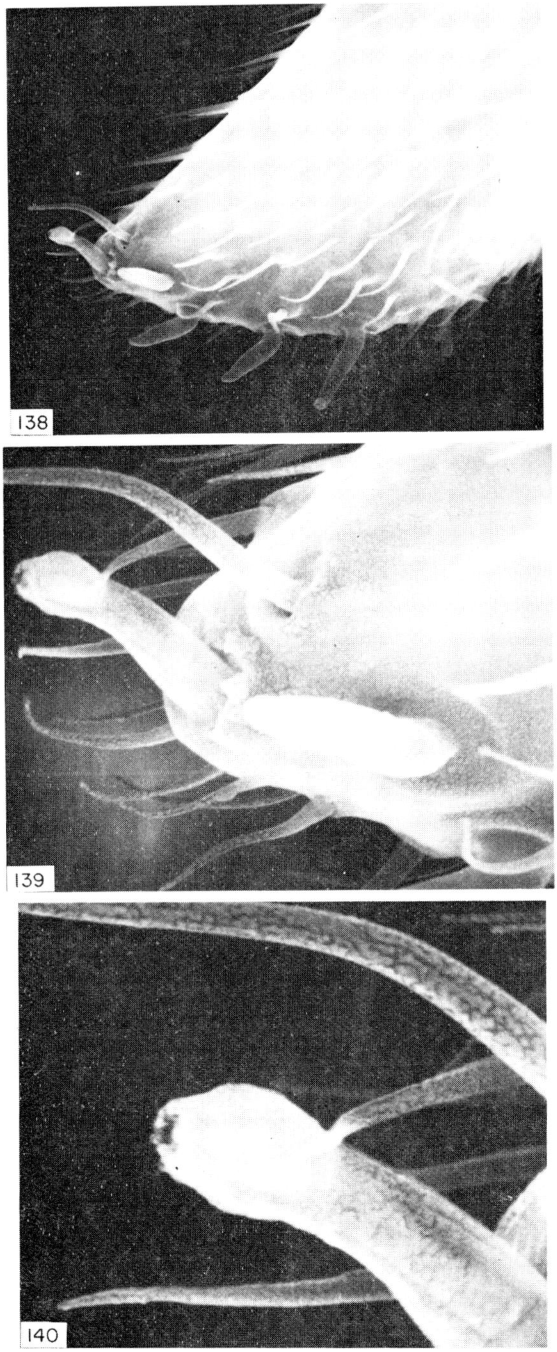


Figures 138–140. *Aphytis melinus* DeBach, ♀: sensilla on tip of antennal club.

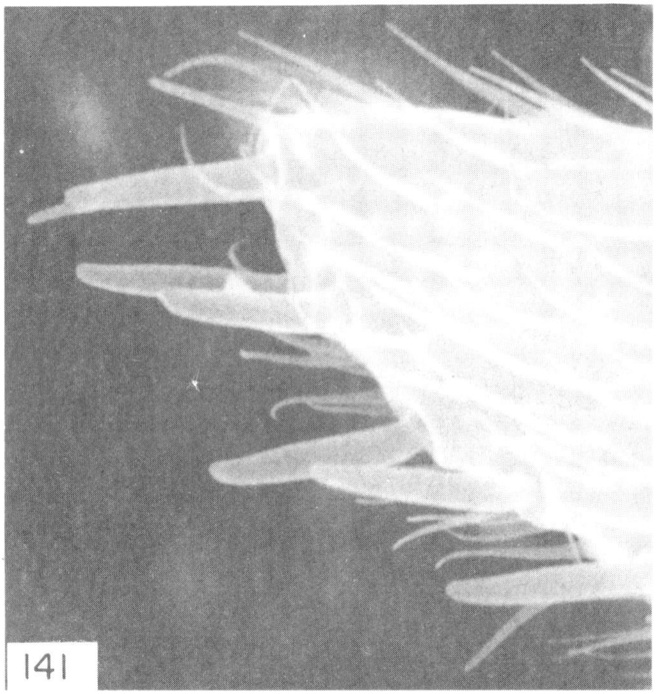

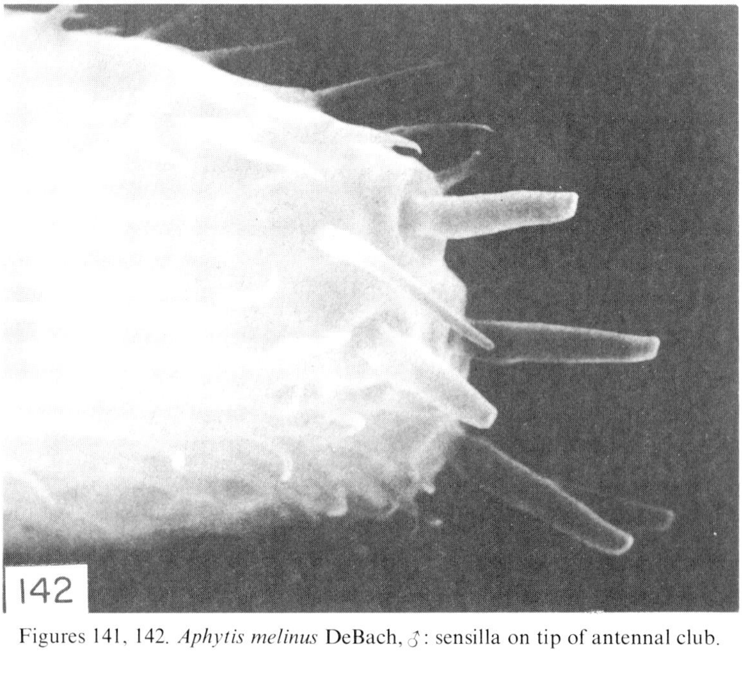

Figures 141, 142. *Aphytis melinus* DeBach, ♂: sensilla on tip of antennal club.

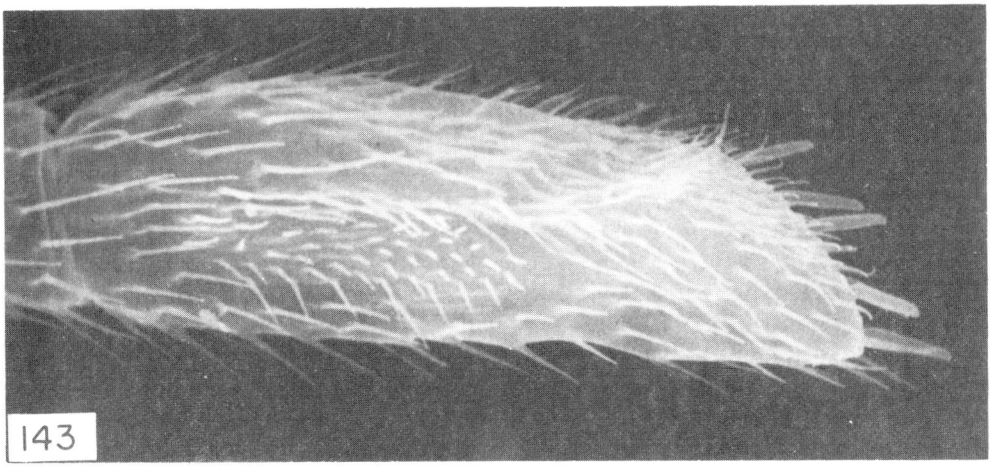

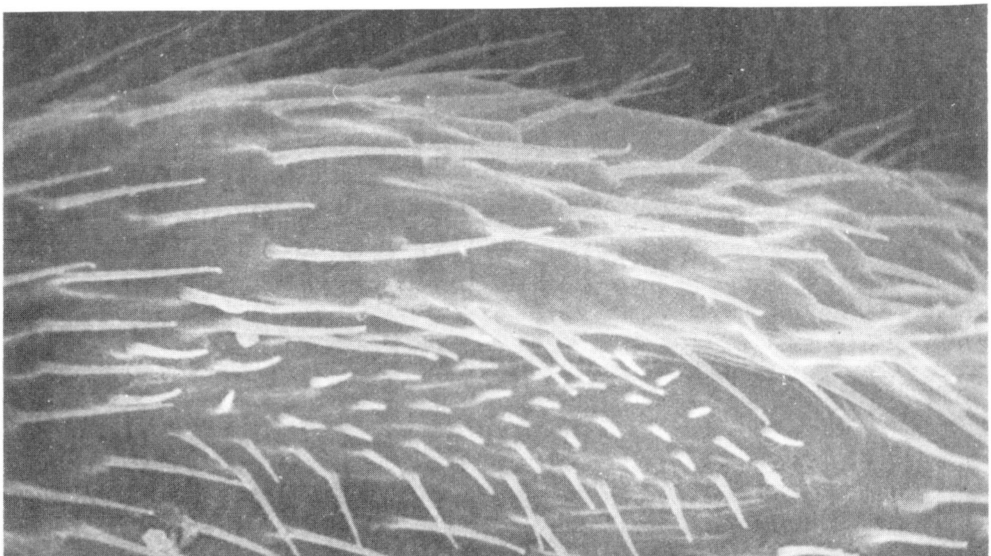

Figures 143, 144. *Aphytis melinus* DeBach, ♂: antennal club, showing specialized sensory area on ventral aspect, partially cut off from the surrounding areas by an indistinct groove.

A specialized sensory area is evident on the ventral aspect of the club in the males of many species. This usually takes the shape of a small oval area or plate, bearing numerous minute setae. It may be bordered by one or several rows of short setae with large sockets, as in *cochereaui* (Figure 339, p. 313) and *fabresi* (Figure 353, p. 316), or by a distinct formation, superficially resembling a longitudinal sensillum, which appears to cut it off from the surrounding areas (e.g., *capillatus* (Figure 367, p. 318), the **proclia** group (Figures 583 and 654, pp. 432 and 445) and *pilosus* (Figure 1255, p. 715)). In other species, such as the **lingnanensis** group, it appears to be only partially cut off by a less distinct groove (Figures 143, 144, also Figure 893, p. 570), whereas in others it is not visibly bordered (e.g., *chilensis* (Figure 525, p. 370), *columbi* (Figure 539, p. 372) and *roseni* (Figure 1223, p. 708)). The sensory area is absent in the **mytilaspidis** and **funicularis** groups and in most members of the **chrysomphali** group.

The setae on the antennae are usually rather short. However, in the males of several species the third funicular segment and the club bear long setae, so that the antennae appear unusually hairy. Such aberrant antennae are present in the males of *maculatipennis* (Figure 292, p. 305), *costalimai* (Figure 304, p. 308), *perplexus* (Figure 323, p. 311), *erythraeus* (Figure 1199, p. 704), *roseni* (Figure 1222, p. 708), *longicaudus* (Figure 1239, p. 711), *pilosus* (Figure 1253, p. 714) and *setosus* (Figure 1270, p. 717), diverse species that are otherwise not closely related to each other.

## THE MESOSOMA (Figures 145–207)

**The neck**, or *cervix* (Figure 145), is an intersegmental membrane, attached to the occipital foramen and connecting the head to the thorax.

**The pronotum** (Figures 148, *Pn* ; 151, 152) is very short in dorsal view, several times as wide as long. It is reticulate and bears several short setae, with a row of longer setae along its posterior margin, the lateral pair being considerably longer and coarser than the others. In most species of *Aphytis*, the pronotum is composed of two triangular plates which are narrowly joined medially (Figures 884 and 894, pp. 568 and 570). This was listed as a generic character by Alam (1956a). However, the pronotal plates are broadly joined in *acrenulatus* (Figure 707, p. 455), *roseni* (Figure 1213, p. 706) and *setosus* (Figure 1263, p. 716).

The anterior pair of spiracles (usually termed the "prothoracic" spiracles) is concealed beneath the postero-lateral corners of the pronotum (Figures 146, 147). This is the only pair of spiracles in the thorax of *Aphytis*.

**The mesonotum** (Figures 148–152) is composed of several sclerites. The *mesoscutum* (Figures 148, 149, *Ms*) is trapezoidal, somewhat wider than long, distinctly grooved along the longitudinal midline. It is bordered laterally by a pair of complete, convergent *parapsidal sutures*, which separate an elongate, triangular *parapsis* (*Pa*) on each side of the mesoscutum (see Compere and Rosen (1970) for a discussion of the mesoscutum and parapsides in the Hymenoptera). The *axillae* (*Ax*) are small, triangular, wider than long. They are slightly advanced into the parapsides, with only about half their length, at the most, produced beyond the scuto-scutellar line. The *scutellum* (*Scl*) is oval or nearly pentagonal, considerably wider than long, shorter than the mesoscutum and similarly grooved along the midline. The *tegulae* are small sclerites laterad of the parapsides, at the bases of the forewings (see Figure 151).

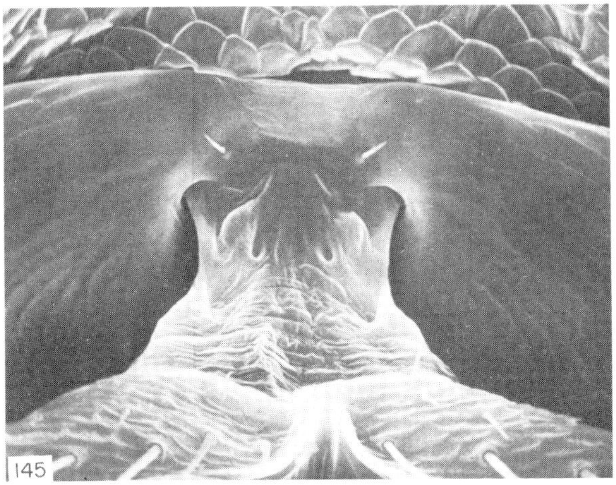


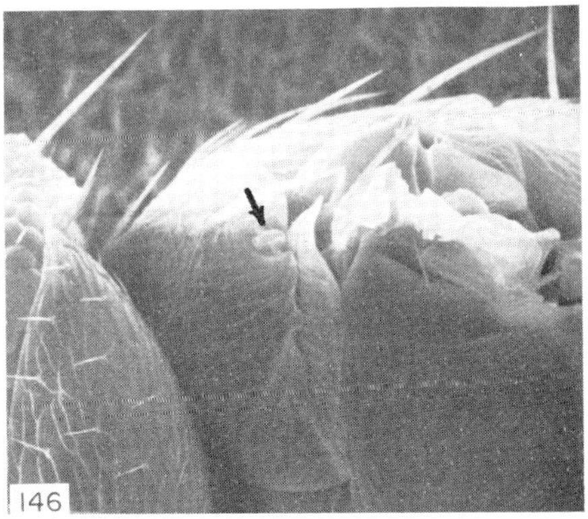


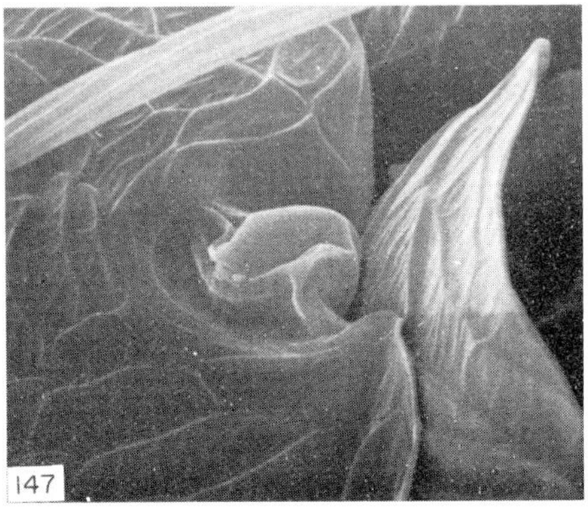


Figures 145–147. Cervix and pronotum
145. *Aphytis chilensis* Howard, ♀: occipital margin, cervix and pronotum, dorsal view.
146, 147. *Aphytis mytilaspidis* (Le Baron), ♀: 146. Pronotum, lateral view; the spiracle is indicated by an arrow. 147. Pronotal spiracle.

The mesothoracic postnotum is represented by a large, invaginated *endophragma* (Figure 150, *En*), extending backward into the base of the gaster. It is roundly triangular, about as long as wide, and is readily visible in cleared, slide-mounted specimens (e.g., Figures 307 (p. 308), 556 (p. 427), 767 (p. 508) and 1122 (p. 644)).

The mesonotal sclerites are rather coarsely reticulate, with a delicate infrasculpture within the cells (Figures 153–160). The mesoscutum bears several pairs of long, distinctly costate setae. Their number is subject to size-linked individual variation (see p. 193), but also varies among species. They tend to be numerous—up to 50 or more—in the more primitive species of *Aphytis* (e.g., *costalimai* (Figure 298, p. 306) and *cochereaui* (Figure 331, p. 312)), fewer—only 3 to 6 pairs—in the more advanced species. The posterior pair and one seta at each antero-lateral corner of the mesoscutum are usually considerably longer and coarser than the others. Each parapsis usually bears 2–3 (rarely 1 or 4) short setae. Each axilla bears a single seta (occasionally 2), about as long as the shortest setae on the mesoscutum. However, the axillae of several species are asetose (e.g., *haywardi* (Figure 409, p. 325), *obscurus* (Figure 418, p. 327), *melanostictus* (Figure 454, p. 335), *tucumani* (Figure 692, p. 453) and *salvadorensis* (Figures 1300 and 1310, pp. 723 and 725)). The scutellum bears 2 pairs of long setae which are least as long and as coarse as the longest mesoscutal setae.

The mesonotal setae are very coarse in members of the **vittatus**, **chilensis** and **proclia**

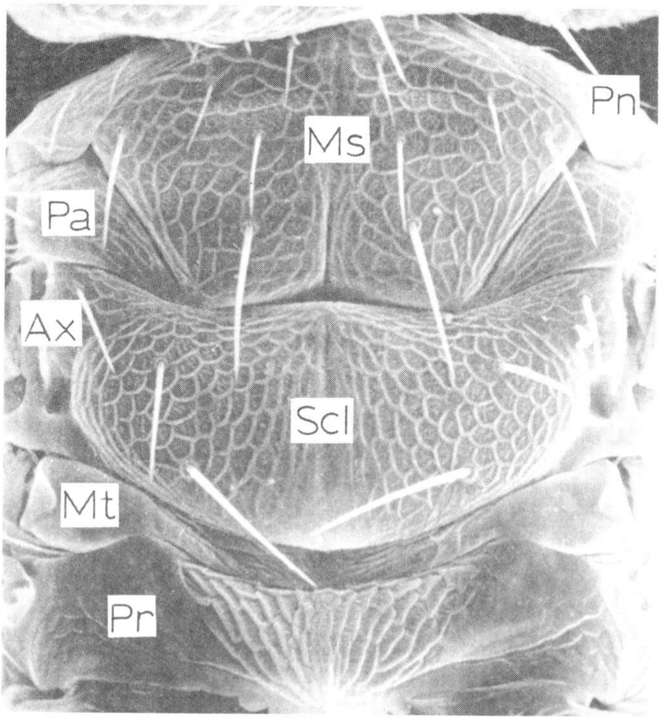


Figure 148. *Aphytis paramaculicornis* DeBach and Rosen, ♀: mesosoma, dorsal view. *Ax* = axilla; *Ms* = mesoscutum; *Mt* = metanotum; *Pa* = parapsis; *Pn* = pronotum; *Pr* = propodeum; *Scl* = scutellum.

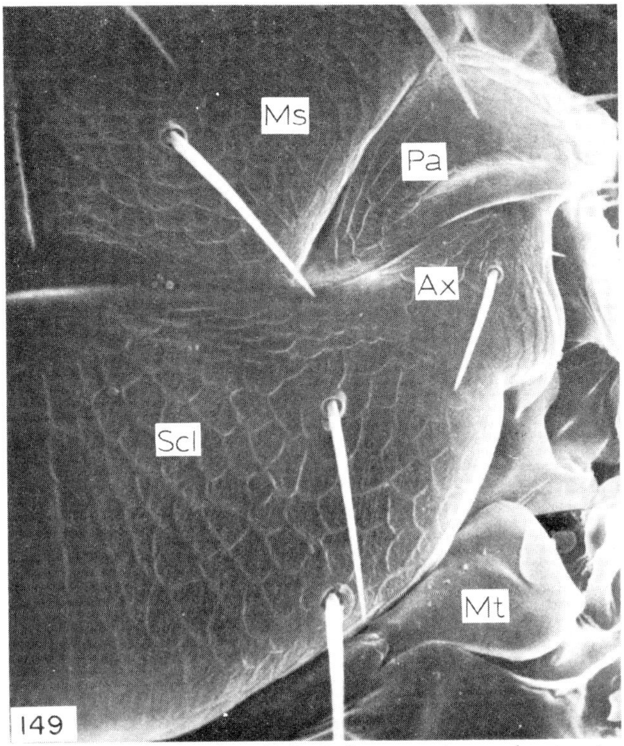


Figure 149. *Aphytis melinus* DeBach, ♀: mesonotum and metanotum. *Ax* = axilla; *Ms* = mesoscutum; *Mt* = metanotum; *Pa* = parapsis; *Scl* = scutellum.

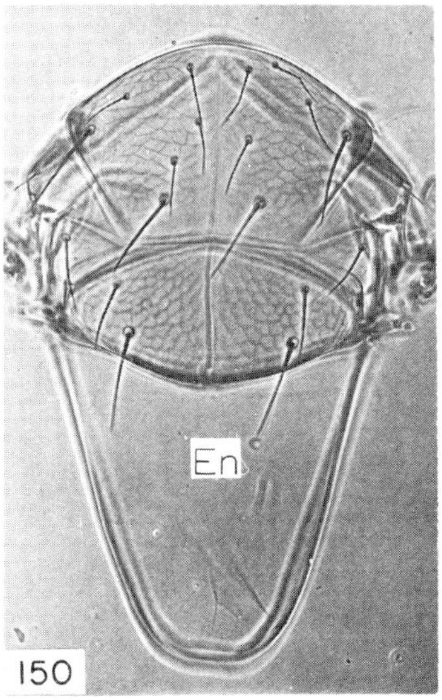


Figure 150. *Aphytis lingnanensis* Compere, ♀: mesonotum. *En* = endophragma.

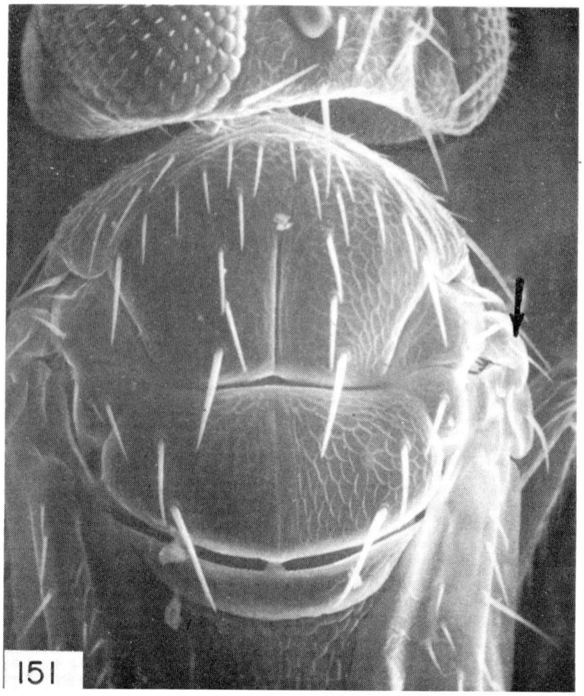


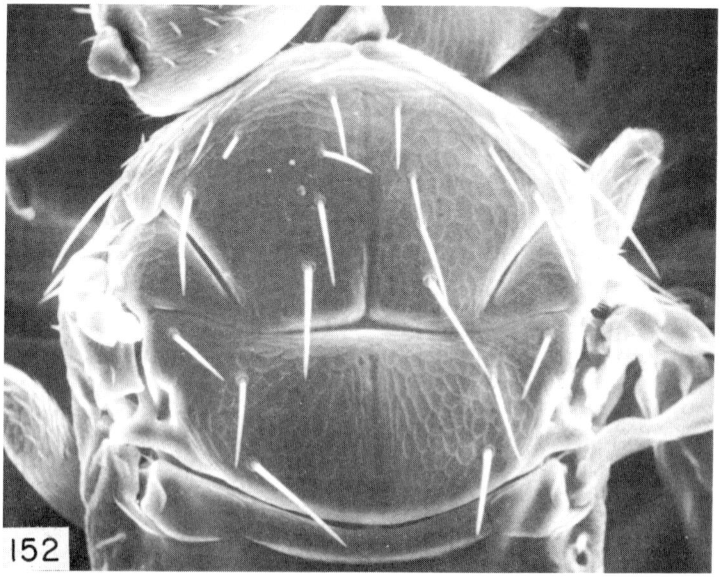


Figures 151, 152. Thorax, dorsal view
151. *Aphytis chilensis* Howard, ♀; tegula indicated by an arrow; note also anteromedian apodeme of metanotum.
152. *Aphytis lepidosaphes* Compere, ♀.

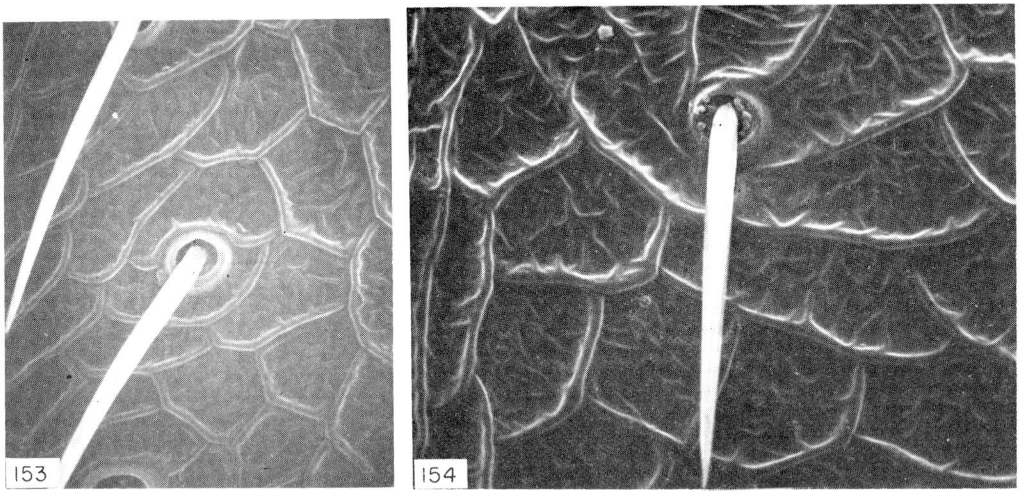

Figures 153, 154. Sculpture and setae on mesoscutum
153. *Aphytis chilensis* Howard, ♀. 154. *Aphytis mytilaspidis* (Le Baron), ♀.

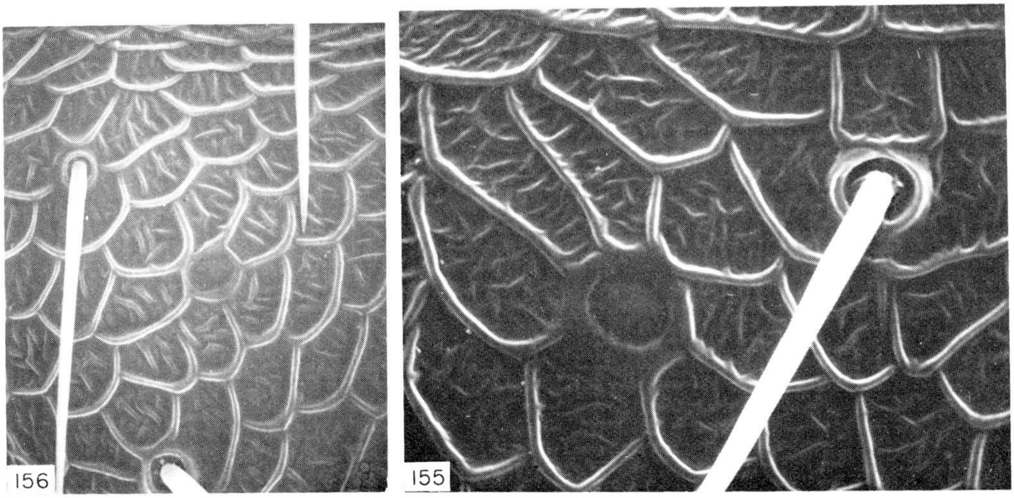

Figures 155, 156. Sculpture, setae and discoid sensillum on scutellum
155. *Aphytis chilensis* Howard, ♀. 156. *Aphytis paramaculicornis* DeBach and Rosen, ♀.

groups; when viewed under a phase-contrast microscope they may appear as if composed of distinct filaments (e.g., Figure 332, p. 312). They are dark and coarse in the **mytilaspidis** group, pale and slender in the **chrysomphali** group and in some members of the **lingnanensis** group.

The scutellum bears a pair of minute discoid *sensilla placodea* (Figures 155–160). They are usually about equidistant from the anterior and posterior pairs of scutellar setae, or somewhat closer to the anterior pair. However, in certain species they may be situated very close to the anterior pair of setae (e.g., *malayensis* (Figure 467, p. 338), *pilosus* (Figure 1249, p. 713) and *salvadorensis* (Figure 1310, p. 725)), whereas in others, such as *capillatus* (Figure 373, p. 319) they are considerably closer to the posterior pair. These sensilla have a very thin cuticle and appear transparent in cleared specimens, thus resembling empty hair sockets. In fact, they occasionally give rise to setae (see, for instance, Figures 574 (p. 430) and 750 (p. 505)). Other rare freaks include supernumerary sensilla (Figures 361 (p. 317), 656 (p. 445) and 812 (p. 518)) or supernumerary scutellar setae (Figure 895, p. 570).

On the sides of the parapsides and axillae, just below the tegulae, there are several minute, fingerlike sensilla (Figure 161; see also Figures 151, 152). These sensilla touch the bases of the forewings, presumably serving to register their movements.

**The metanotum** (Figures 148, 149, *Mt*; 152, 169) is short, bandlike, straight or moderately arcuate, sometimes distinctly v-shaped (e.g., *vittatus*, Figures 265, 266, p. 300). It is asetose, transversely reticulate or reticulate-strigose centrally, widening into a smooth, unsculptured lobe on each side. The metanotum bears a small *anteromedian apodeme* which fits beneath the posterior margin of the scutellum (see Figure 151). This apodeme is usually quite robust in *Aphytis*, shorter than the median length of the metanotum or about as long (e.g., Figures 332 (p. 312) and 817 (p. 519)). However, in a few species it is very slender (e.g., Figures 299 (p. 306) and 1301 (p. 723)). In dissected specimens, a pair of ligaments is seen arising from the apex of the metanotal apodeme (Figure 332, p. 312).

The lateral lobes of the metanotum often appear to be distinctly folded (Figures 149, 162, 163). Immediately beneath the folds, or a short distance laterad, a few fingerlike sensilla are present (Figures 162–165) which appear to touch the bases of the hind wings (Figure 162) and presumably serve to register their movements. Similar sensilla are present nearby, on the sides of the propodeum (Figure 163). Each of the metanotal side-lobes themselves bears a pair of minute *sensilla basiconica* (Figures 164, 166–168).

**The propodeum** (Figures 148, *Pr*; 169) is well developed in *Aphytis*, considerably longer than the metanotum (ranging from nearly 3 times as long in the **mytilaspidis** group up to 9 times in certain members of the **chrysomphali** group). It is distinctly shorter than the scutellum in most species, nearly as long as the scutellum in some. The propodeum bears the first pair of abdominal spiracles, with a more or less distinct longitudinal carina immediately mesad of the spiracle on each side. It is reticulate laterally and on a broad central area, sometimes also transversely reticulate or striated along the posterior margin, but largely smooth submedially, between the sculptured central and lateral areas.

The sculptured area on the central portion of the propodeum is directly continuous with the sculptured area on the metanotum, the two together forming a trapezoidal or

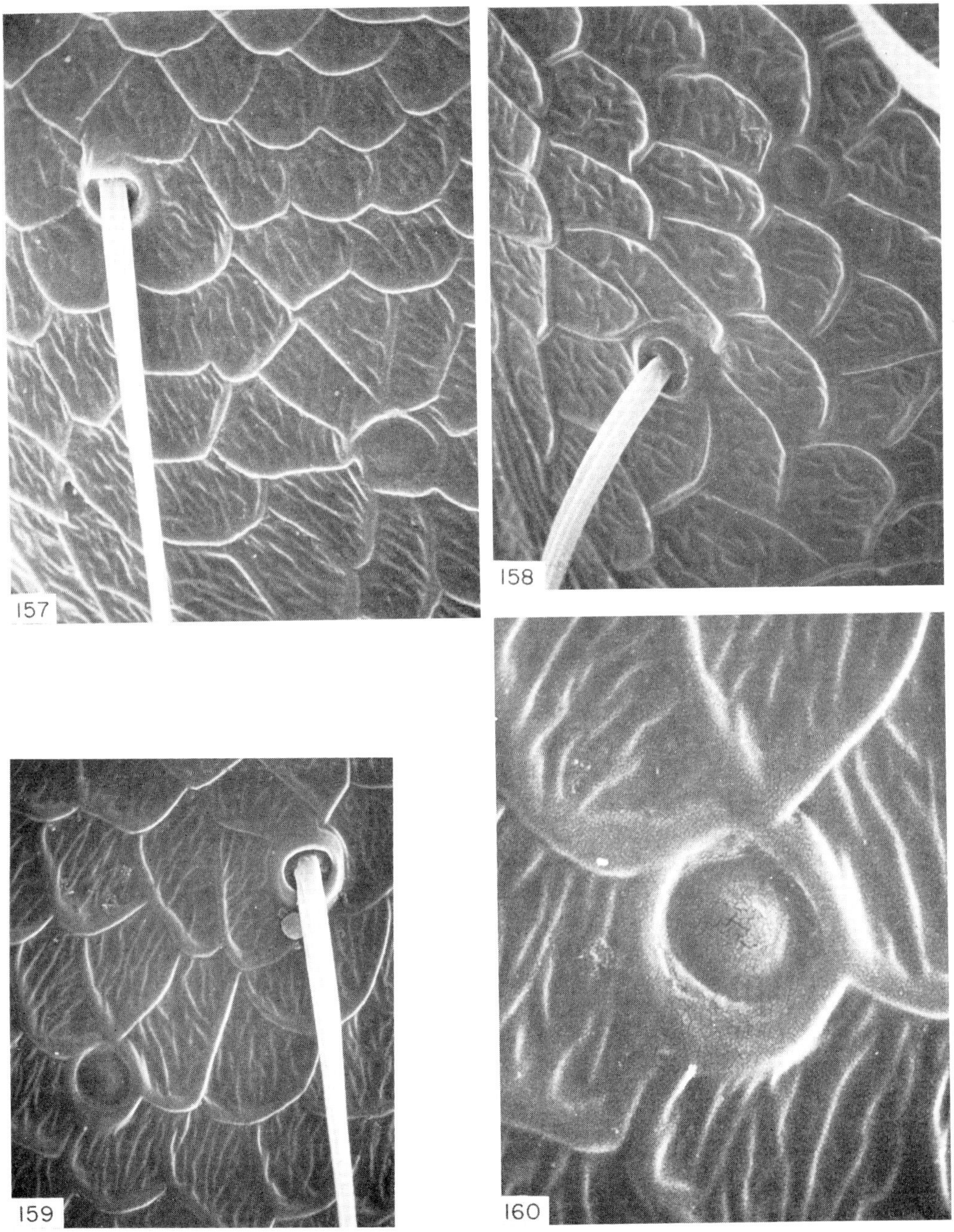


Figures 157–160. Sculpture, setae and discoid sensillum on scutellum
157. *Aphytis lingnanensis* Compere, ♀. 158. *Aphytis mytilaspidis* (Le Baron), ♀.
159, 160. *Aphytis melinus* DeBach, ♀.

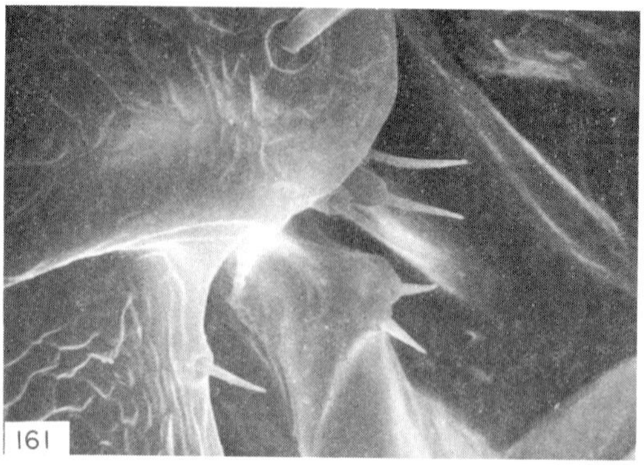


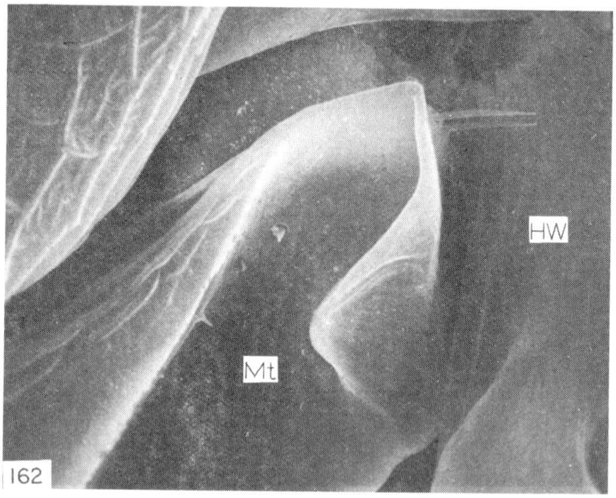


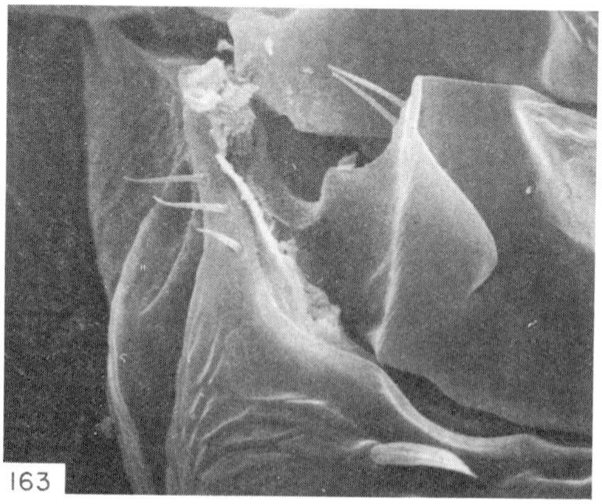


Figure 161. *Aphytis vandenboschi* DeBach and Rosen, ♀: sides of parapsis (above) and axilla (below), showing sensilla.

Figures 162, 163. *Aphytis mytilaspidis* (Le Baron), ♀: 162. Sensilla on side of metanotum. *HW* = hind wing; *Mt* = metanotum. 163. Sensilla on sides of metanotum and propodeum.

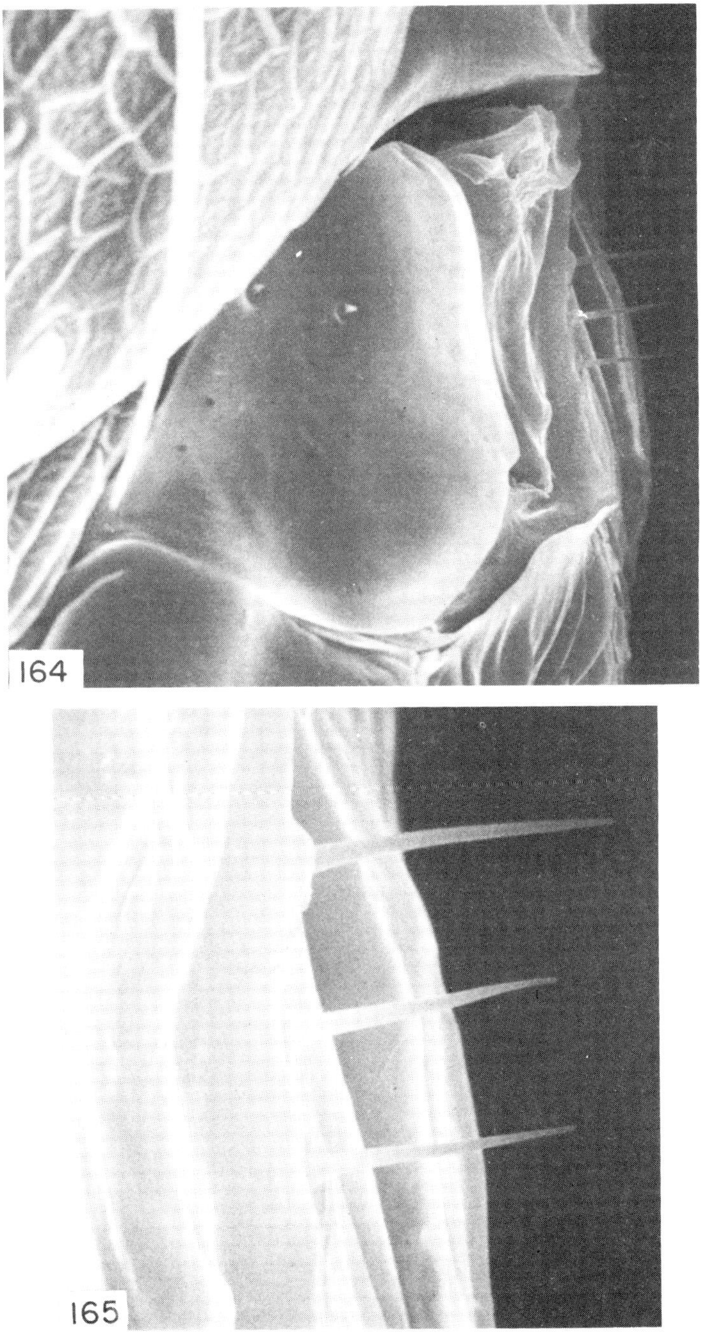


Figures 164, 165. *Aphytis chilensis* Howard, ♀
164. Sensilla on lateral lobe of metanotum and on metapleuron. 165. Sensilla on metapleuron.

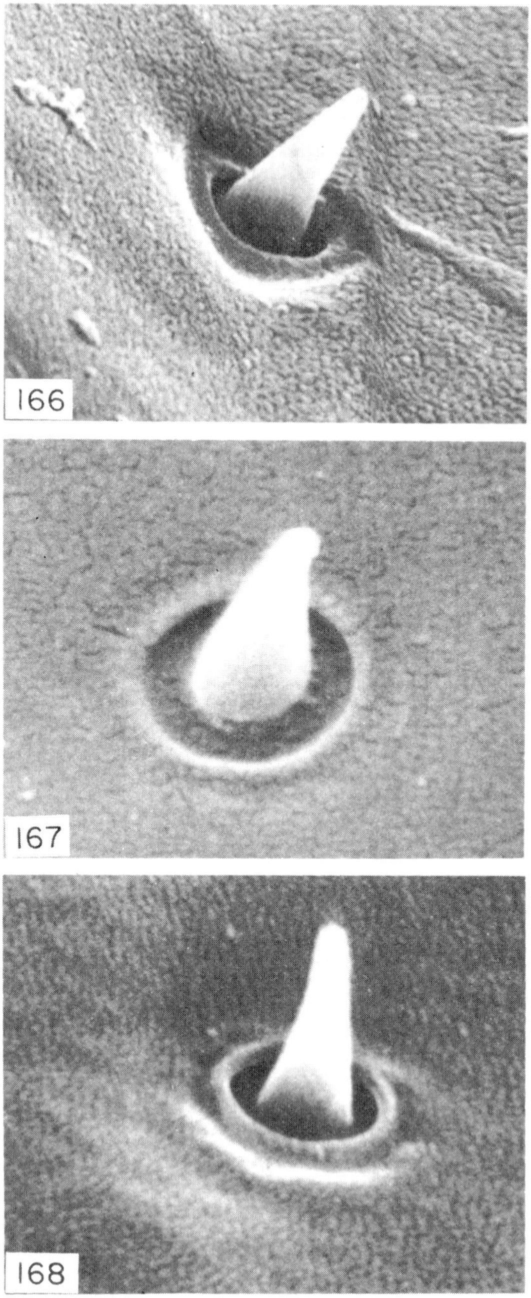


Figures 166–168. Sensillum on lateral lobe of metanotum
166. *Aphytis chilensis* Howard, ♀. 167. *Aphytis lingnanensis* Compere, ♀.
168. *Aphytis melinus* DeBach, ♀.

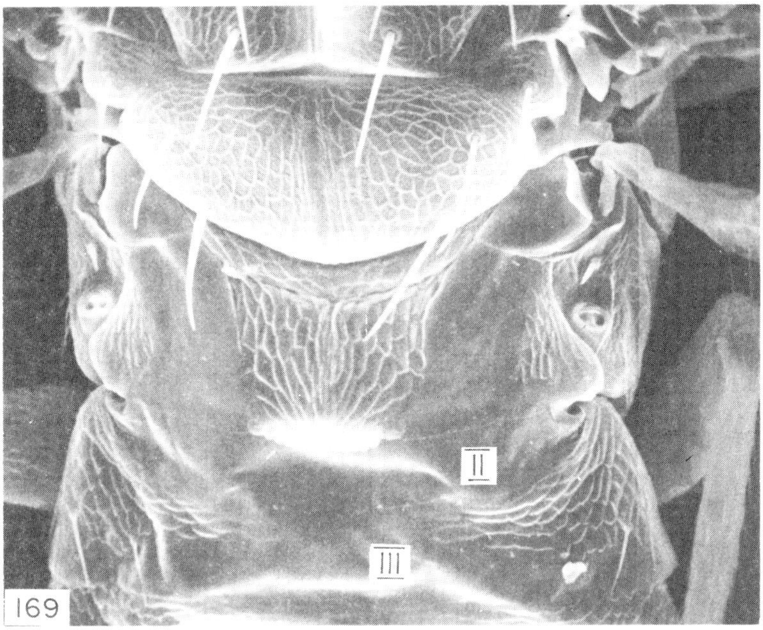


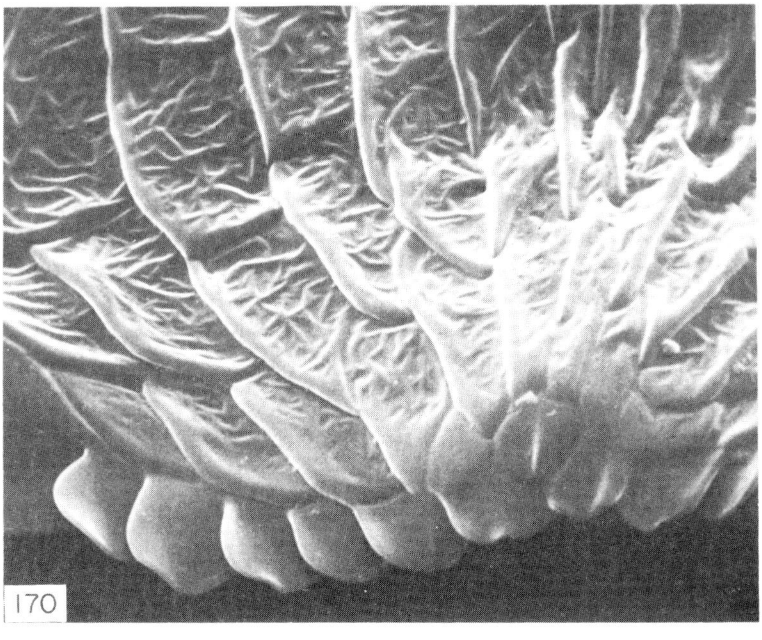


Figures 169, 170. *Aphytis chilensis* Howard, ♀
169. Scutellum, metanotum, propodeum and base of gaster. *II*, *III* = second and third abdominal tergites.
170. Propodeal sculpture and marginal crenulae. (From Rosen and DeBach, 1976.)

triangular area. Such sculpture is not common in the Aphelinidae, but is somewhat similar to the peculiar "propodeal triangle" of the Signiphoridae (see Quezada, DeBach and Rosen, 1973).

The posterior margin of the propodeum (Figures 170–182) is sometimes produced into a rounded or bilobed median salient. Its central portion overhangs the base of the gaster in rooflike fashion (see Figures 171, 172, 176, 178) and bears several small, scalelike *crenulae*. The latter may be arranged in a continuous row (Figure 170) or in two distinct groups (Figure 179), and vary in shape, size and number among species, being elongate (Figure 173) or nearly rounded (Figure 174), overlapping (Figures 170, 179–182) or nonoverlapping (Figures 173–178). They are lacking altogether in *faurei* (Figure 492, p. 363) and in *acrenulatus* (Figures 709, 710, 715, pp. 456–457). Their function is at present unknown.

The propodeum is asetose, except for a few fine setae laterad of the spiracles. Additionally, a small stout spine is present anterad of each propodeal spiracle (Figures 169, 183–194). It is usually situated quite close to the spiracle, and may sometimes overlap it (e.g., Figures 187, 188). The spine varies in shape, being subconical or lamellate, entire or dentate, but this cannot be regarded as a specific character, as it is subject to considerable individual variation (see Figures 189, 190 and 191, 192). In fact, the shape of the propodeal spine may differ even on the two sides of a single specimen (see Figures 193, 194).

On the ventral aspect of the mesosoma, the **propleura** (Figure 195, *Pp*) are a pair of subtriangular, tapering plates converging into the occipital foramen of the head. The fore coxae are hinged to the bases of the propleura (see Figures 522 (p. 369), 1231 (p. 709), 1280 (p. 719) and 1290 (p. 721)). Between the propleura lies the small, pentagonal **prosternum** (Figures 195, *Ps*; 196, 199). An interesting feature of the prosternum is the presence of two conspicuous orifices near its posterior corners (Figures 196–199).

Immediately behind the prosternum, on the membrane separating it from the mesosternum and just above the fore coxae, is a pair of *epicoxal pads*—small, membranous pads bearing numerous minute spines (Figures 196–201). The function of these epicoxal pads is at present unknown, but their peculiar shape and location suggest that they may be stridulatory organs, on which the fore coxae scrape to produce sound that is then amplified by the cavities in the prosternum acting as echo chambers.

**The prepectus** (i.e., the differentiated anterior part of the mesepisternum) is not continuous in *Aphytis*, but is composed of two convergent but widely separated triangular plates, located on each side of the epicoxal pads (see Annecke and Insley (1970) and Figures 708 (p. 455) and 962 (p. 582)).

**The mesosternum** is composed of two large plates, with a Y-shaped suture, the *median sternal groove*, between them indicating the endosternal *furca*. An orifice at the base of the stem (Figures 202–205) indicates the point of invagination of the furca. This orifice has been suggested by Gordh and DeBach (1978) as the presumed site of release of sexual pheromones in *Aphytis*. A minute sensory seta (presumably a *sensillum trichodeum*) is present near the tip of each arm of the suture (Figures 199, 206, 207). Between the two arms, immediately behind the epicoxal pads, there is a minute sense organ or apodeme, indicated as a mere pointed bump on the surface (Figures 196, 199)

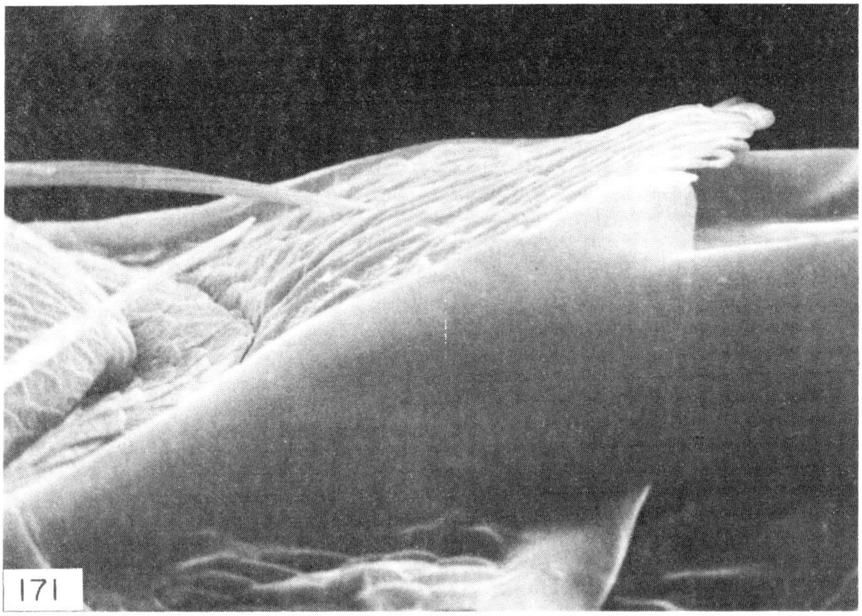


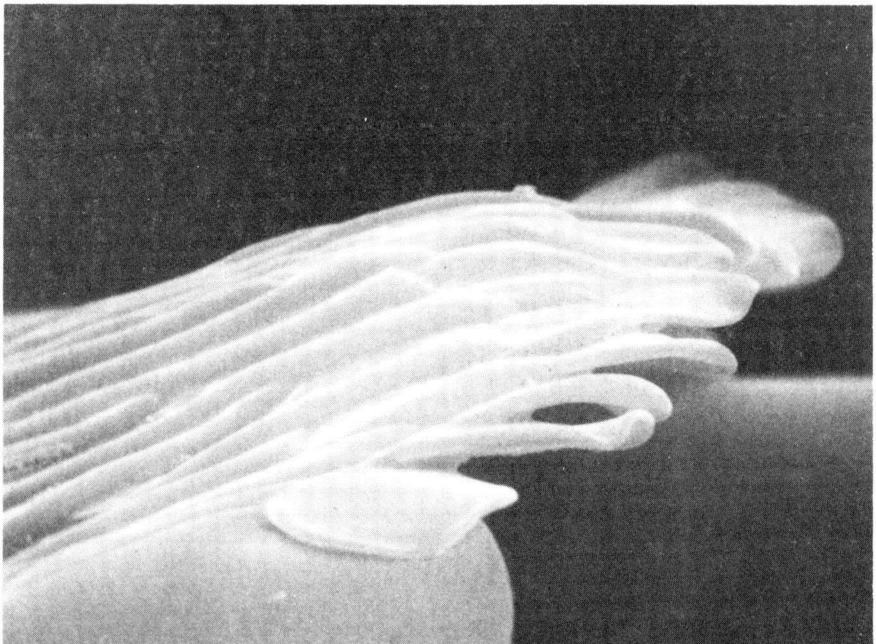


Figures 171, 172. *Aphytis chilensis* Howard, ♀: scutellum, metanotum and propodeum, lateral view.

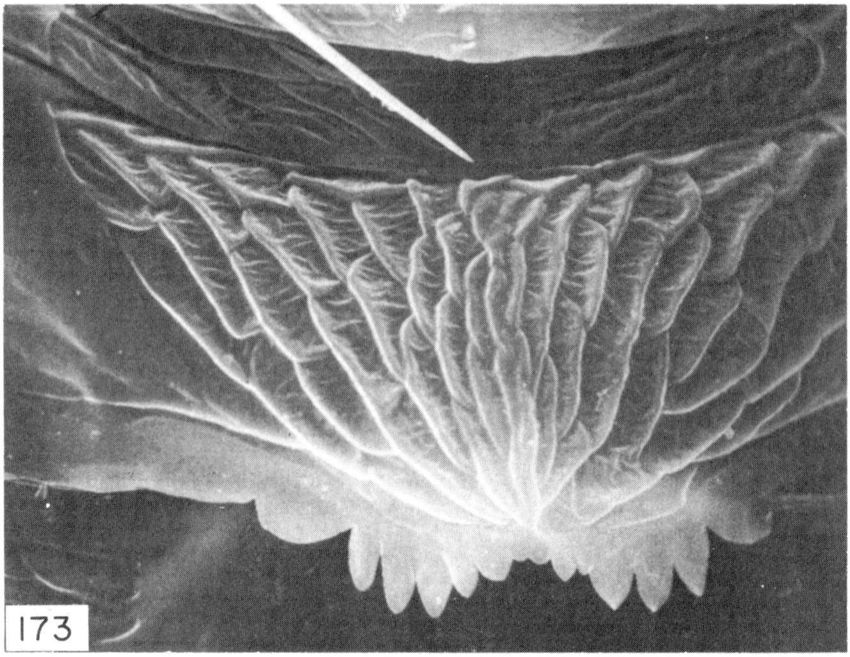

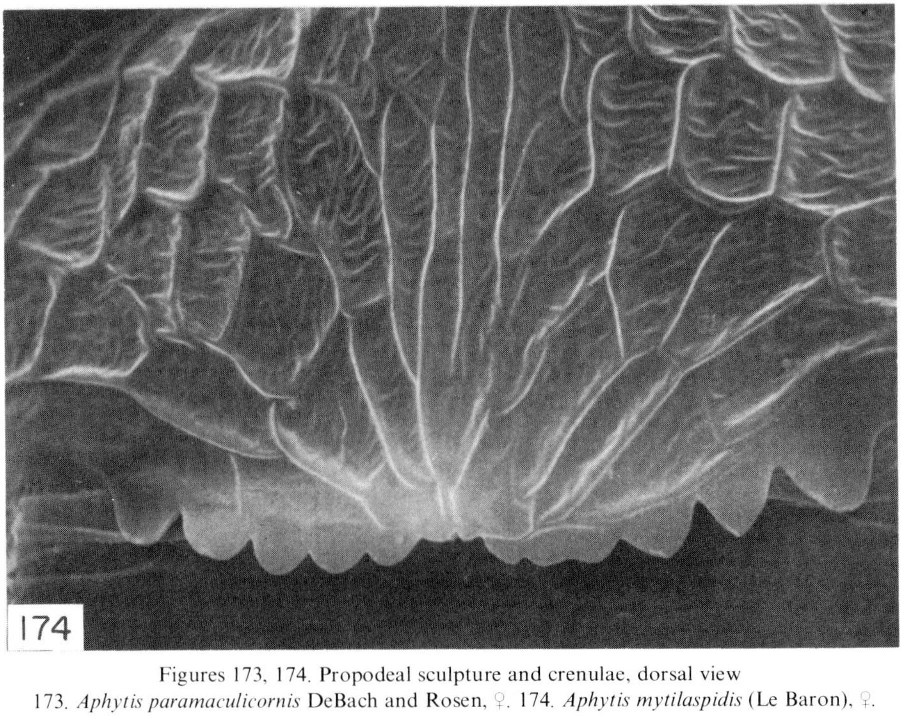

Figures 173, 174. Propodeal sculpture and crenulae, dorsal view
173. *Aphytis paramaculicornis* DeBach and Rosen, ♀. 174. *Aphytis mytilaspidis* (Le Baron), ♀.

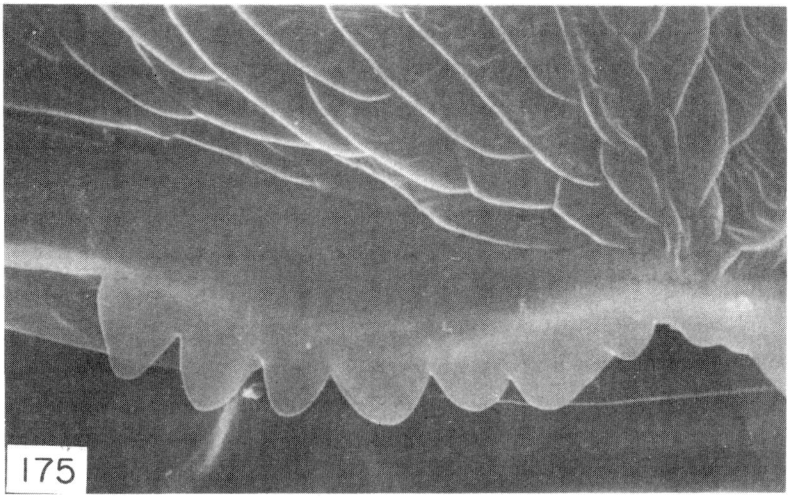

Figures 175, 176. *Aphytis vandenboschi* DeBach and Rosen, ♀: propodeal sculpture and crenulae
175. Dorsal view. 176. Lateral view.

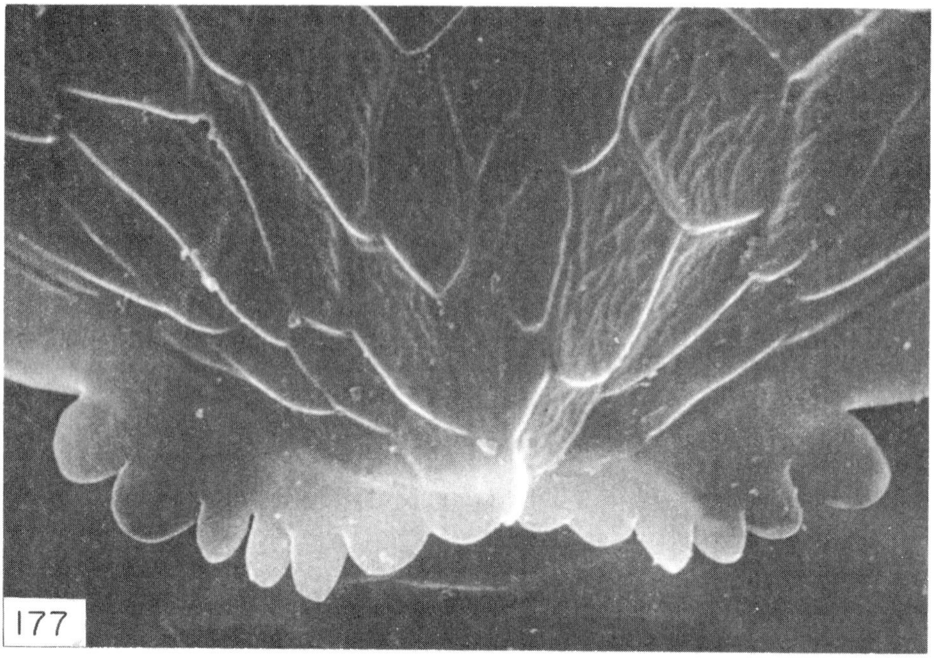


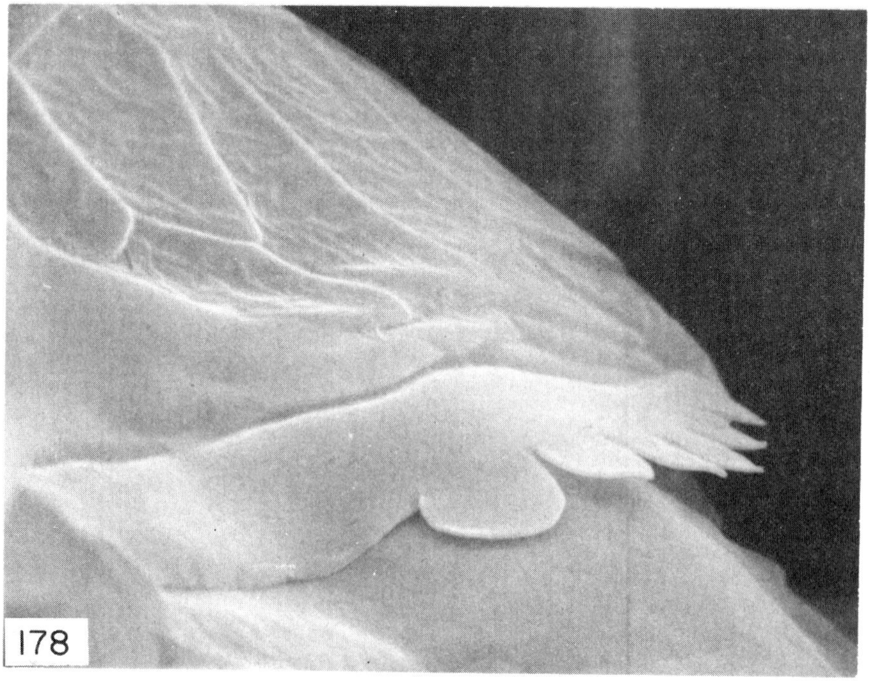


Figures 177, 178. *Aphytis lepidosaphes* Compere, ♀: propodeal sculpture and crenulae
177. Dorsal view. 178. Lateral view.

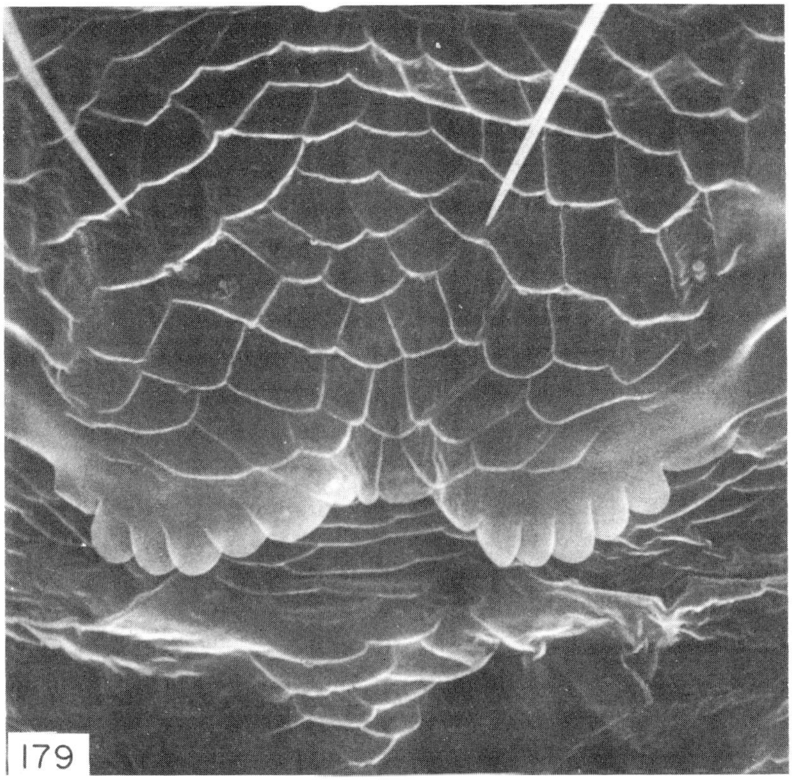

179

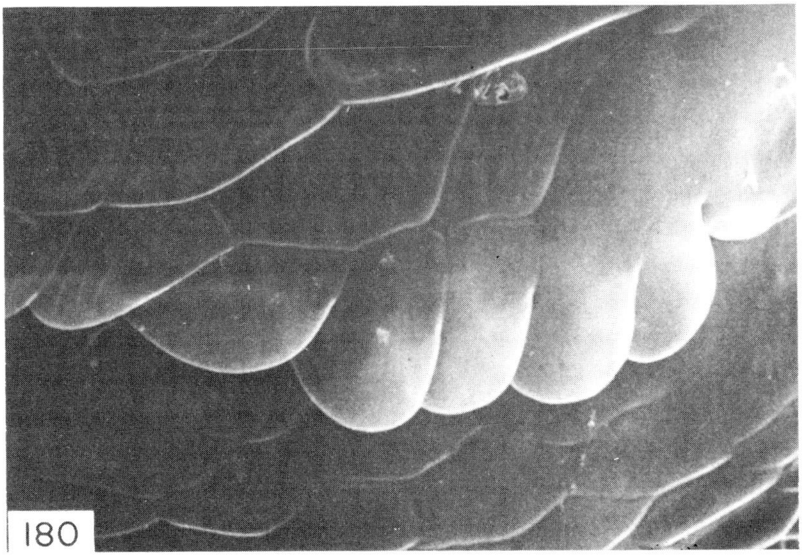

180

Figures 179, 180. *Aphytis melinus* DeBach, ♀: propodeal sculpture and crenulae; note also sculpture of second abdominal tergite.

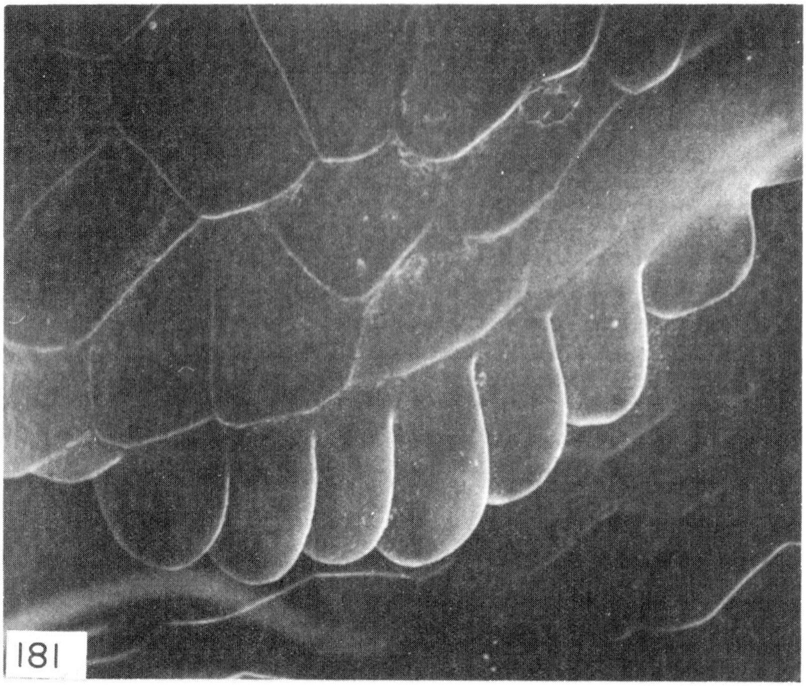


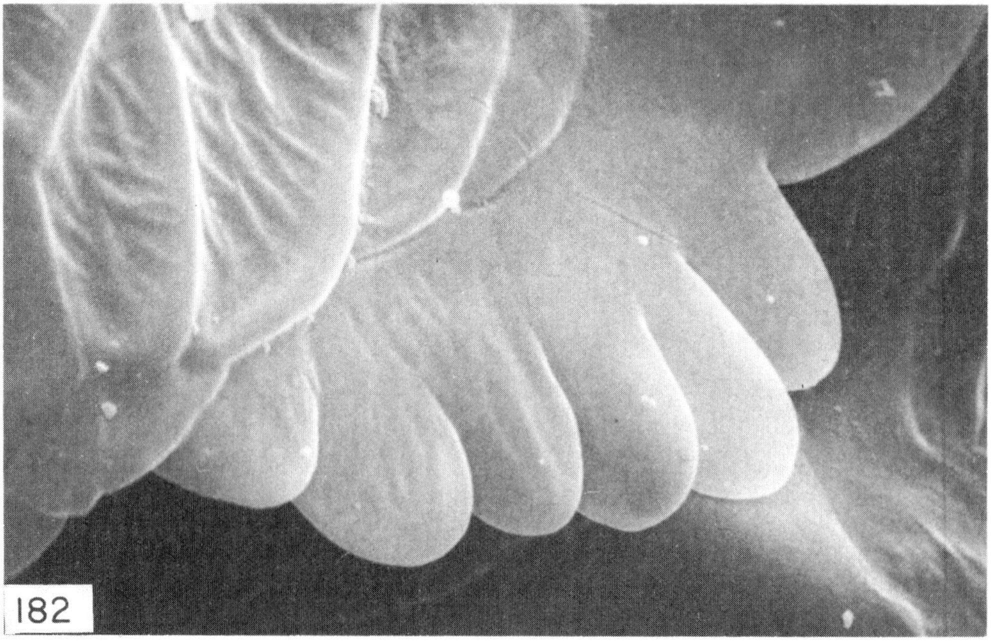


Figures 181, 182. Propodeal sculpture and crenulae
181. *Aphytis lingnanensis* Compere, ♀. 182. *Aphytis fisheri* DeBach, ♂.

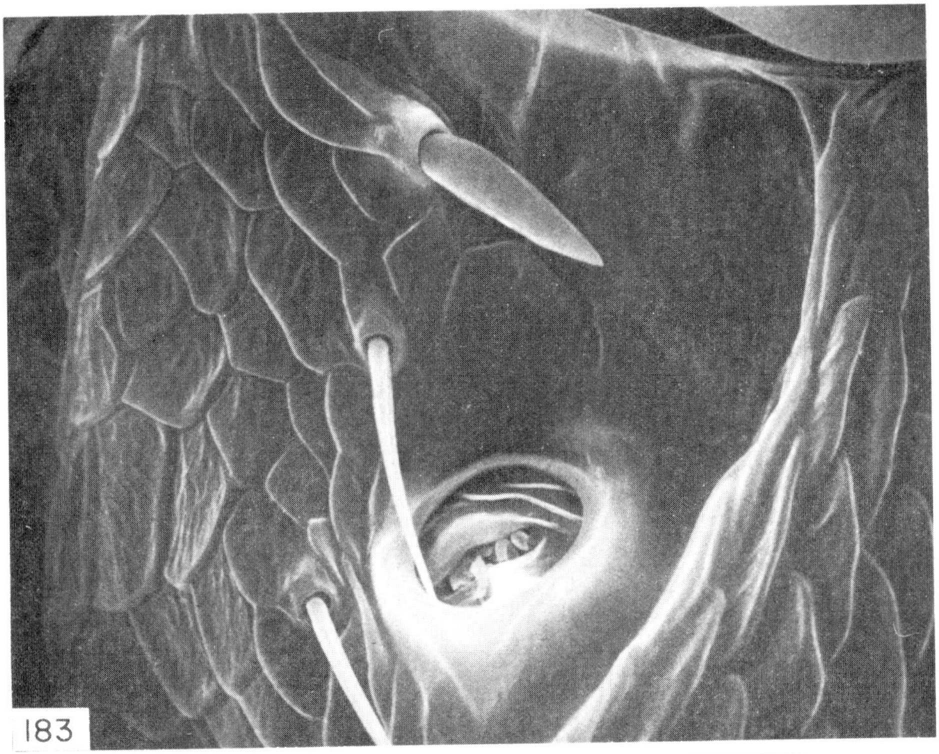


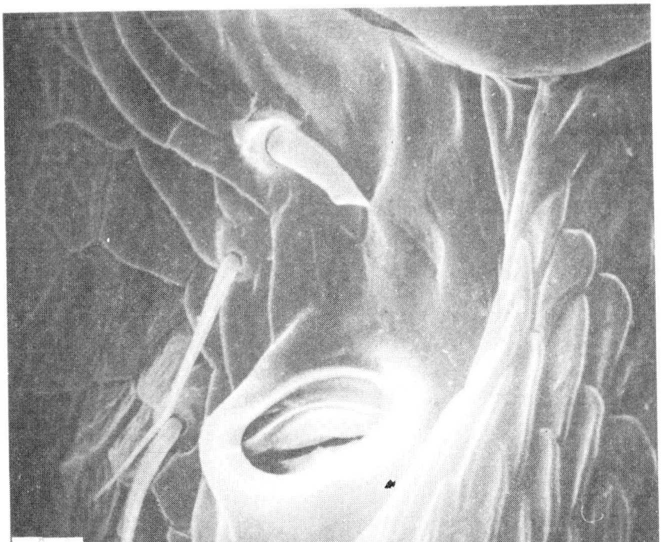


Figures 183, 184. *Aphytis chilensis* Howard, ♀: propodeal spiracle and spine (Figure 183 from Rosen and DeBach, 1976.)

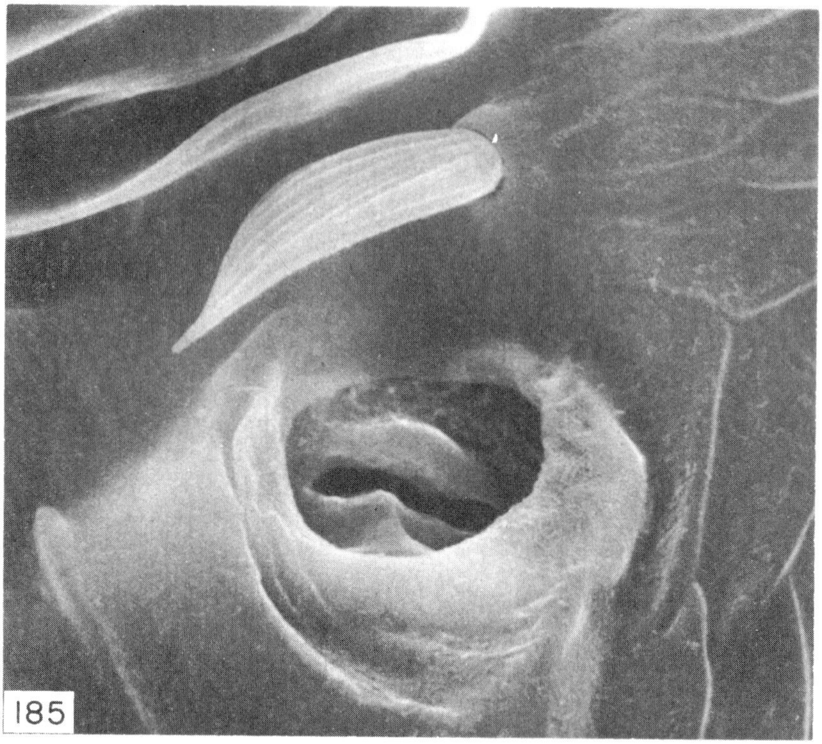

185

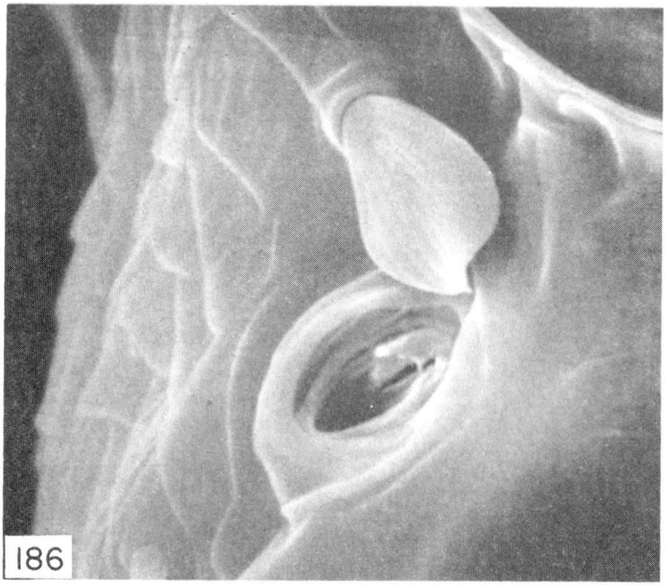

186

Figures 185, 186. Propodeal spiracle and spine
185. *Aphytis mytilaspidis* (Le Baron), ♀. 186. *Aphytis melinus* DeBach, ♀.

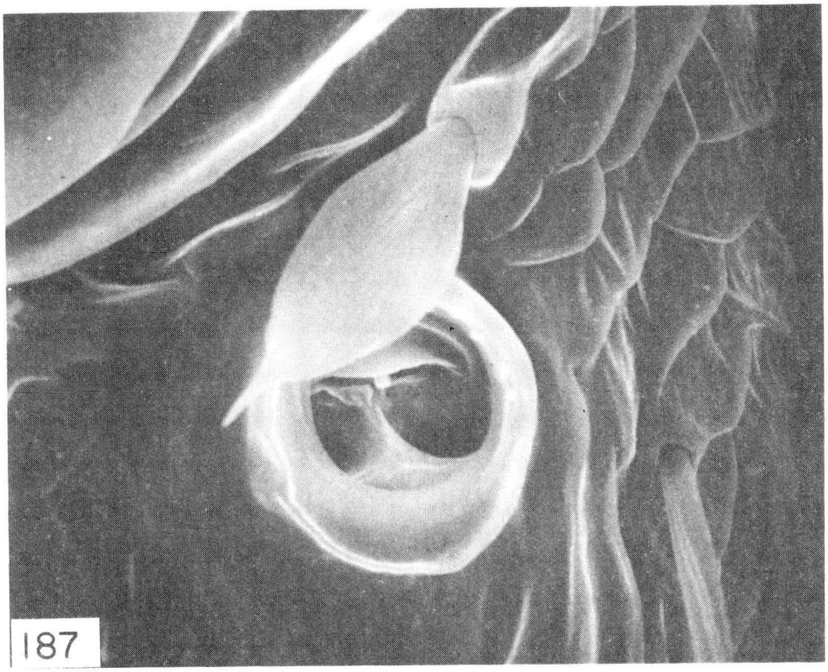


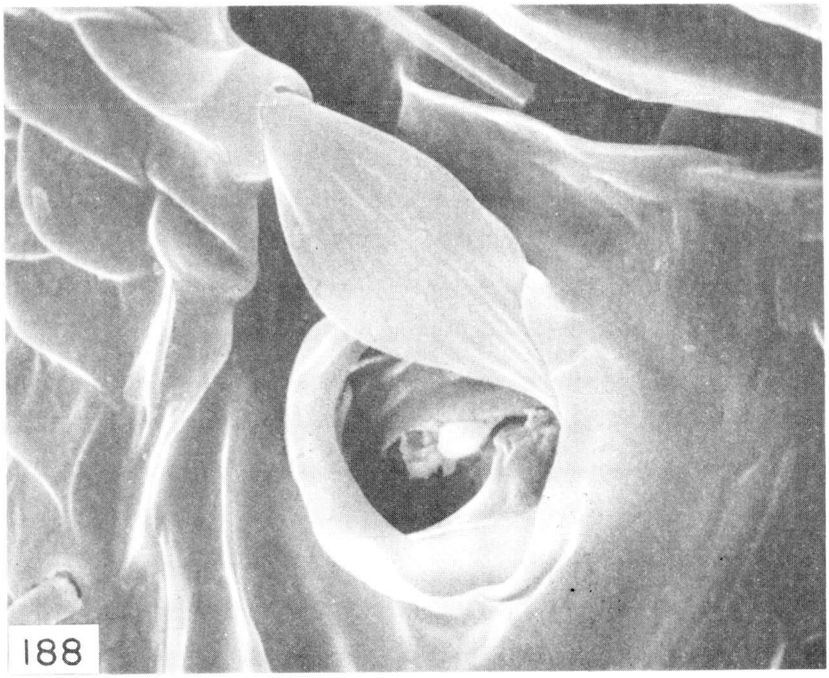


Figures 187, 188. *Aphytis vandenboschi* DeBach and Rosen, ♀: propodeal spiracle and spine.

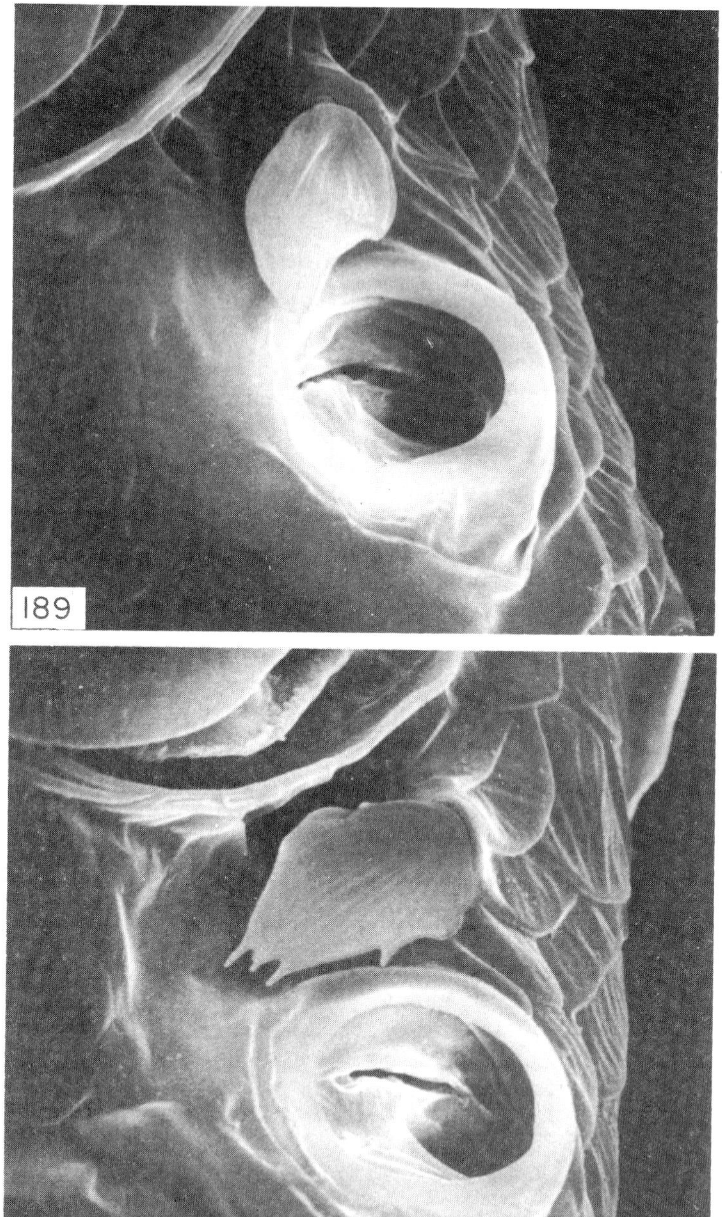


Figures 189, 190. *Aphytis lingnanensis* Compere, ♀: propodeal spiracle and spine; note individual variation.

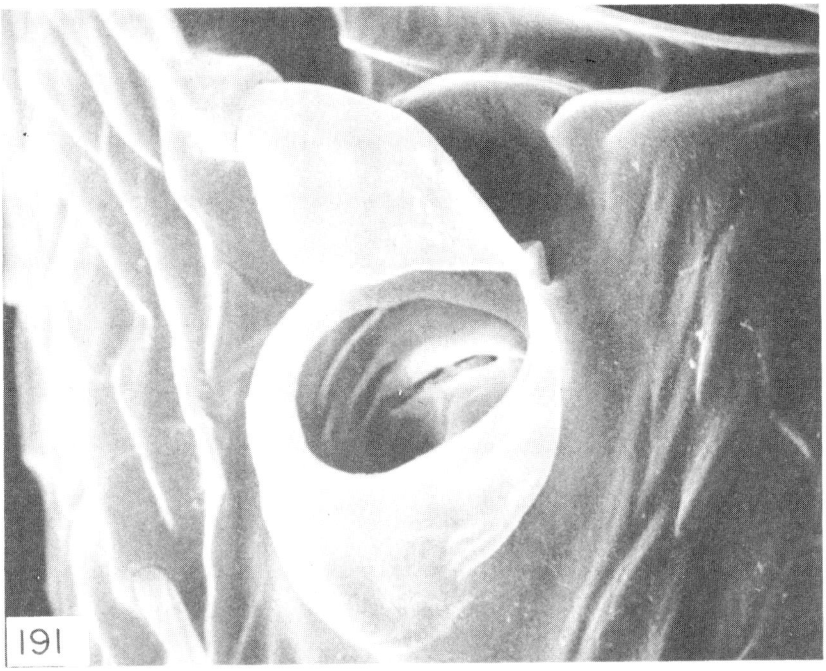

Figures 191, 192. *Aphytis paramaculicornis* DeBach and Rosen, ♀: propodeal spiracle and spine; note individual variation.

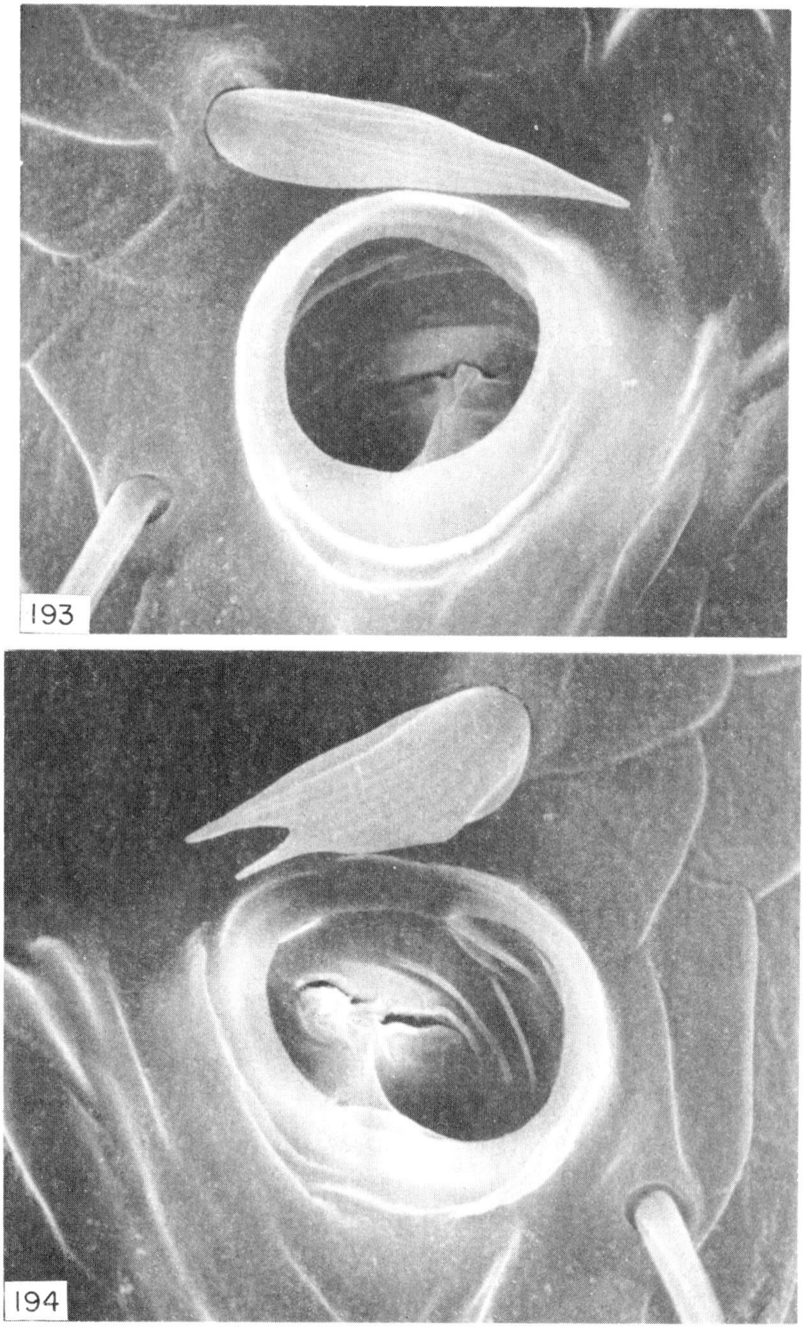


Figures 193, 194. *Aphytis lepidosaphes* Compere, ♀: propodeal spiracle and spine; note individual variation (taken from both sides of the same specimen).

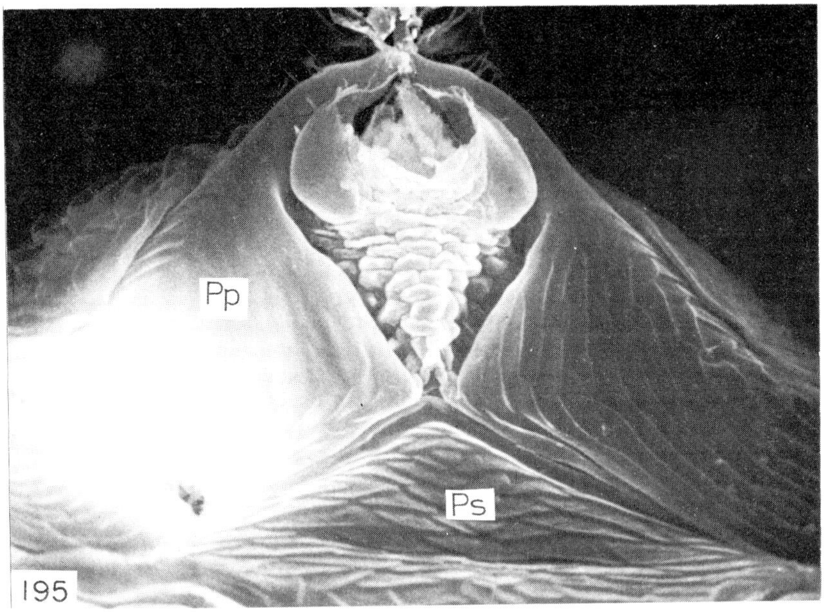


Figure 195. *Aphytis vandenboschi* DeBach and Rosen, ♀: cervix, propleura and prosternum, ventral view. *Pp* = propleuron; *Ps* = prosternum.

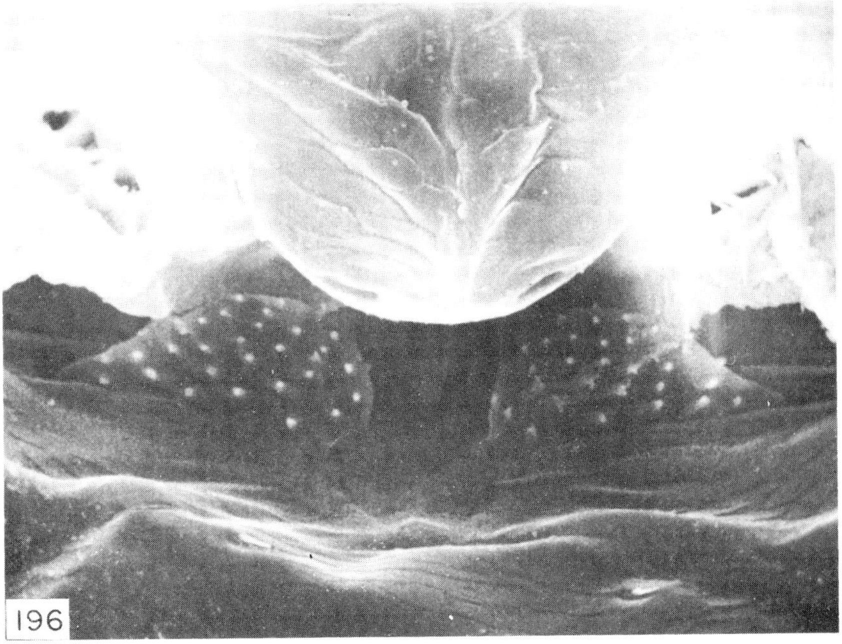


Figure 196. *Aphytis paramaculicornis* DeBach and Rosen, ♂: prosternum and epicoxal pads; note orifices in prosternum (the fore coxae have been removed).

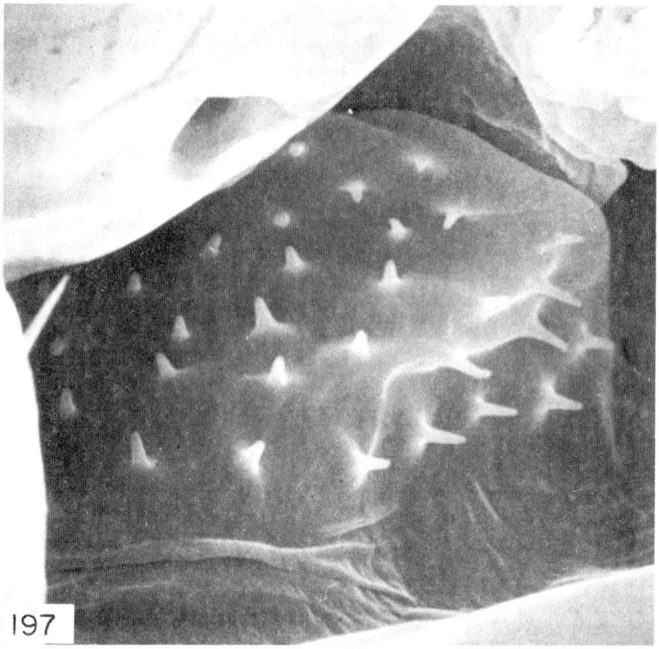

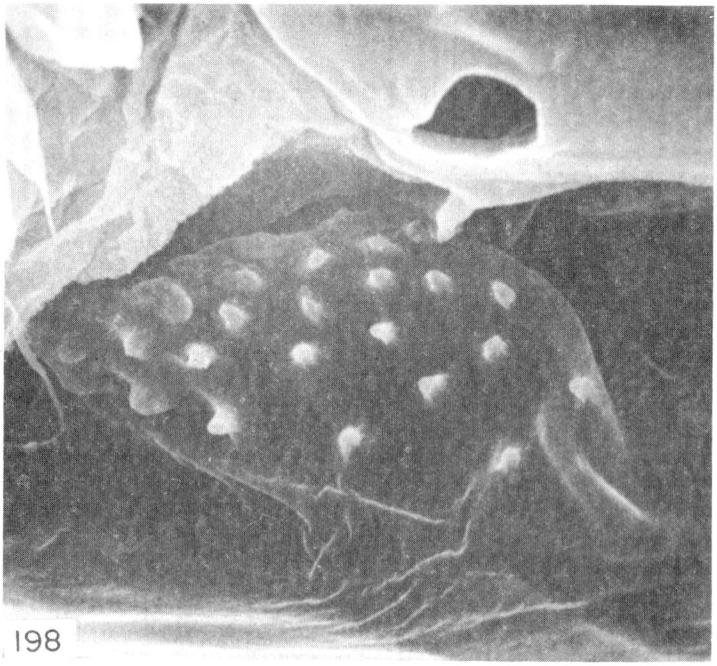

Figures 197, 198. Epicoxal pad and prosternal orifice
197. *Aphytis lingnanensis* Compere, ♀. 198. *Aphytis chilensis* Howard, ♀.

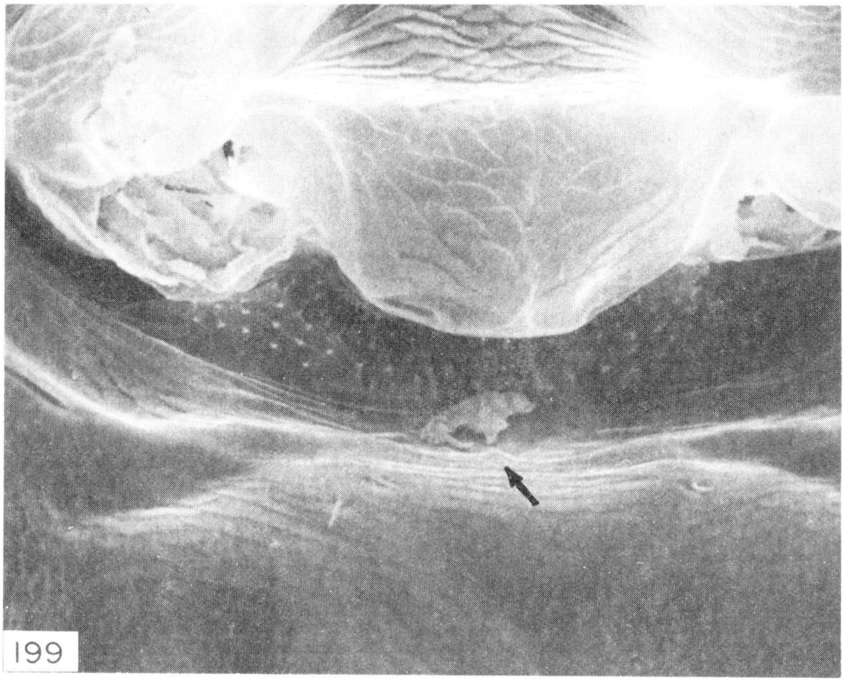


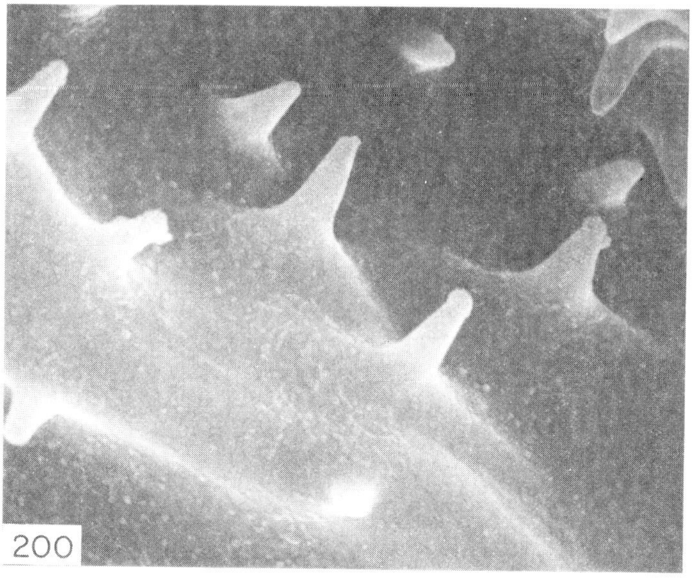


Figures 199, 200. *Aphytis vandenboschi* DeBach and Rosen, ♀

199. Prosternum, epicoxal pads and mesosternum; arrow indicates a sensillum on the mesosternum; note also Y-shaped outline of mesosternal furca (the fore coxae have been removed).

200. Close-up of epicoxal pad.

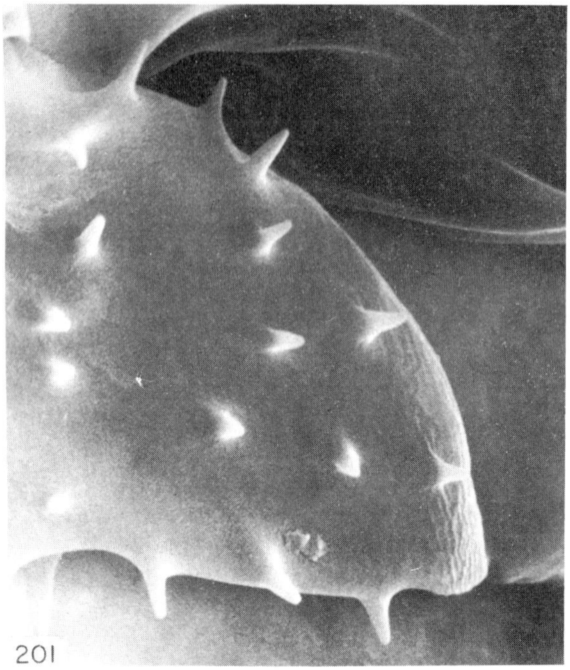

Figure 201. *Aphytis melinus* DeBach, ♀: epicoxal pad.

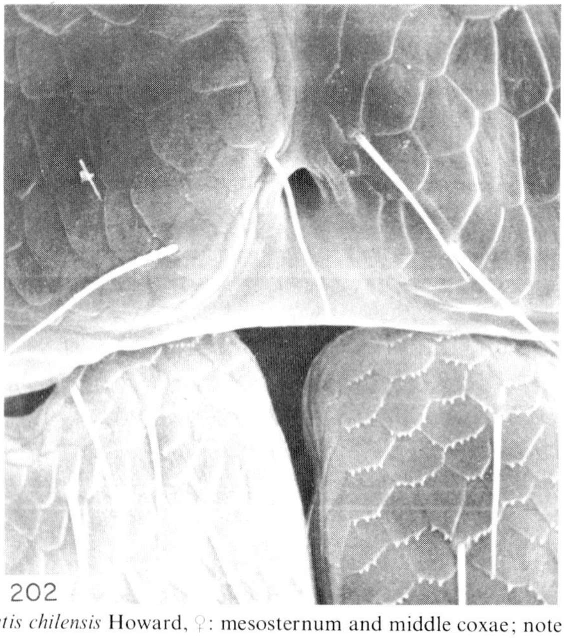

Figure 202. *Aphytis chilensis* Howard, ♀: mesosternum and middle coxae; note orifice at base of mesosternal furca.

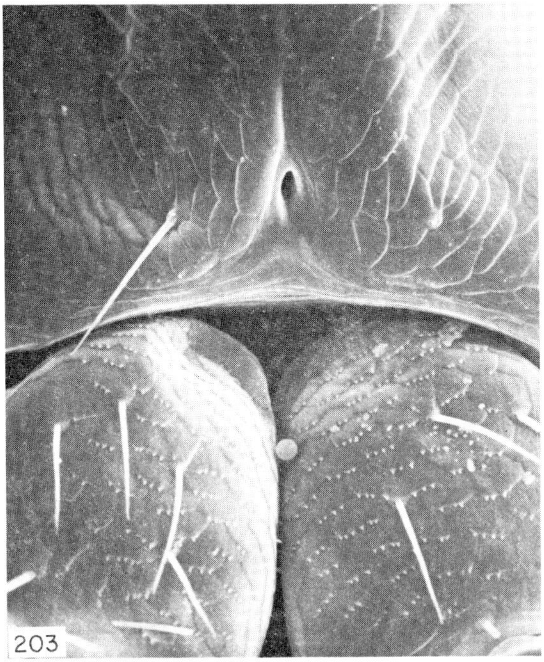


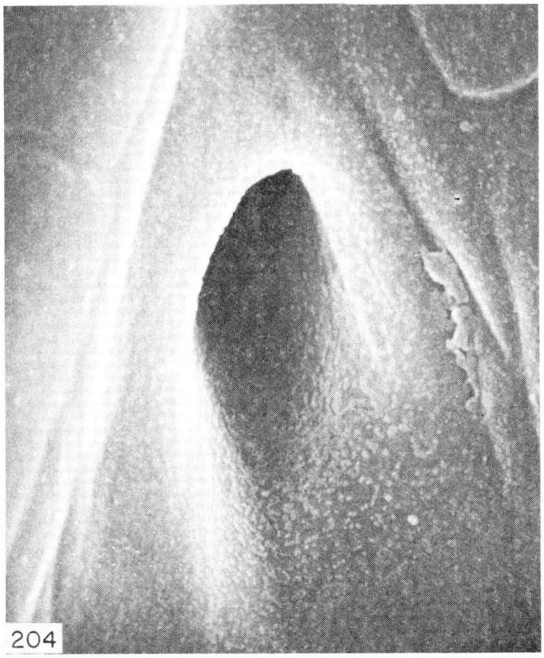


Figures 203, 204. *Aphytis lingnanensis* Compere, ♀
203. Mesosternum and middle coxae. 204. Close-up of orifice at base of mesosternal furca.

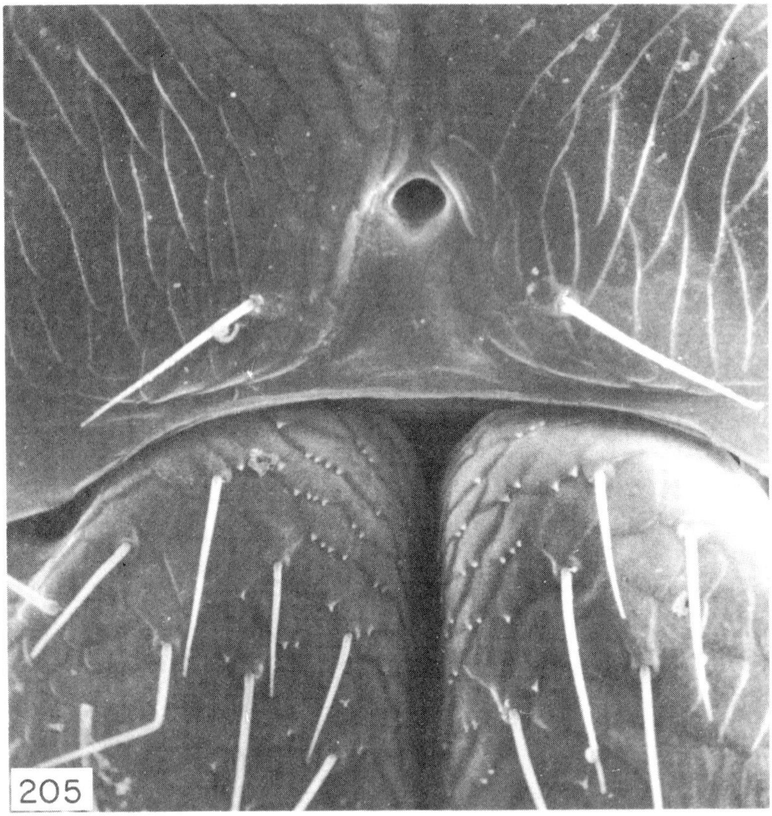


Figure 205. *Aphytis mytilaspidis* (Le Baron), ♀: mesosternum and middle coxae.

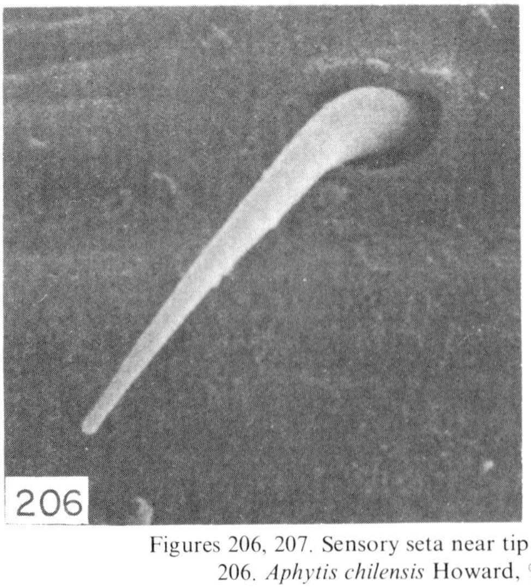


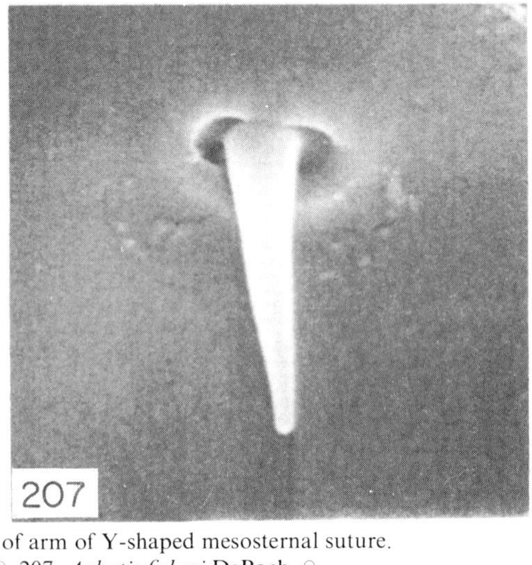


Figures 206, 207. Sensory seta near tip of arm of Y-shaped mesosternal suture.
206. *Aphytis chilensis* Howard, ♀. 207. *Aphytis fisheri* DeBach, ♀.

but readily evident as a tiny rod in cleared specimens (e.g., Figures 356 (p. 316), 522 (p. 369), 1039 (p. 627), 1047 (p. 628), etc.).

**The metasternum** is composed of two short, transverse plates (see Figure 1040, p. 627). The sternum of the first abdominal segment is not visible as a distinct sclerite.

## THE GASTER (Figures 169, 208–220)

The gaster of *Aphytis* comprises segments II–X of the abdomen. (It will be recalled that the propodeum, which has become part of the mesosoma, is the first abdominal segment.) The gaster is subsessile, lacking a distinct petiole but separated from the propodeum by a deep constriction on each side (Figure 169).

The small, transverse second abdominal tergite (Figure 169, *II*) can easily be overlooked. The third abdominal tergite is the largest, extending anteriorly to outflank the second tergite on each side. Tergites IV–VII (Figure 208) are shorter, transverse, with straight anterior and posterior margins. The eighth tergite is subequal in length to the seventh (Figure 208) or considerably longer (Figure 209). It bears the second (posterior) pair of abdominal spiracles (Figures 208–211). Two lateral pieces of the eighth tergite—*laterotergites*—are present on the ventral aspect of the gaster (Figure 213, *VIII LT*). Remnants of the ninth and tenth tergites are fused to form the apical *syntergum* (see p. 000), which bears a pair of small setiferous plates, the *cerci* or *pygostyles* (Figures 208–211). In the **vittatus** group and in a few other primitive species, the syntergum is short and broad, sometimes arcuate or v-shaped, with the cerci situated close to the posterior spiracles (e.g., *maculatipennis* (Figure 290, p. 305), *costalimai* (Figure 302, p. 307), *anomalus* (Figure 405, p. 324), *obscurus* (Figure 421, p. 328), *hyalinipennis* (Figure 441, p. 332), *mandalayensis* (Figure 473, p. 339), *cercinus* (Figure 507, p. 366) and *salvadorensis* (Figure 1304, p. 724)). It is triangular, with a relatively long cauda, in *merceti* (Figure 481, p. 361) and *chilensis* (Figure 208; see also Figure 520, p. 369). In the more advanced species of *Aphytis*, the syntergum is triangular and more elongate, with the cerci about equidistant from the posterior spiracles and from the tip of the gaster (e.g., *lepidosaphes* (Figure 209); see also *proclia* (Figure 562, p. 428), *diaspidis* (Figure 650, p. 444), *mytilaspidis* (Figure 743, p. 504), *lingnanensis* (Figure 888, p. 569), *chrysomphali* (Figure 1022, p. 624), *roseni* (Figure 1220, p. 707), etc.). The cauda, representing the remnant of the tenth abdominal tergite, is deeply incised in *funicularis* (Figure 1137, p. 663); it is exceptionally elongate in *longicaudus* (Figures 1236, 1237, p. 711). Modified lateral parts of the ninth abdominal tergite form the *outer ovipositor plates*, which are partly exposed on the ventral aspect of the gaster (Figure 213, *IX L*).

The second abdominal tergite is asetose, entirely smooth or only faintly striated laterally, sometimes reticulate on a small central area (e.g., Figures 266 (p. 300), 710 (p. 456) and 887 (p. 569)). Tergites III–VII are reticulate laterally, with a few fine setae in one or two short, transverse rows on each reticulate area (Figures 169, 208). The third tergite is usually transversely reticulate also mesad of the lateral sculptured areas, sometimes narrowly reticulate across center. Tergites IV–VI are usually smooth or only faintly striated centrally (Figure 208), but in a few species they may be more extensively sculptured, transversely striated anteriorly and longitudinally striated posteriorly (e.g., *merceti* (Figure 480, p. 361), *faurei* (Figure 494, p. 363) and *debachi*

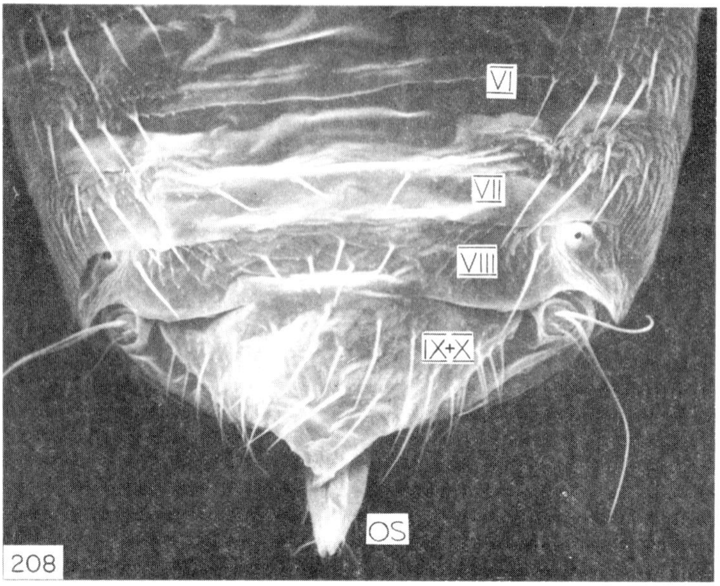


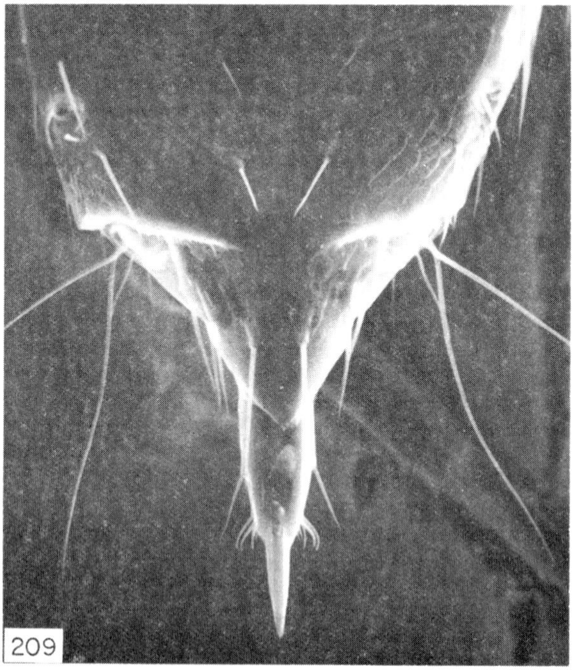


Figures 208, 209. Posterior part of gaster, dorsal view
208. *Aphytis chilensis* Howard, ♀. *VI, VII, VIII* = sixth, seventh and eighth abdominal tergites; *IX* + *X* = syntergum; *OS* = ovipositor sheaths.
209. *Aphytis lepidosaphes* Compere, ♀.

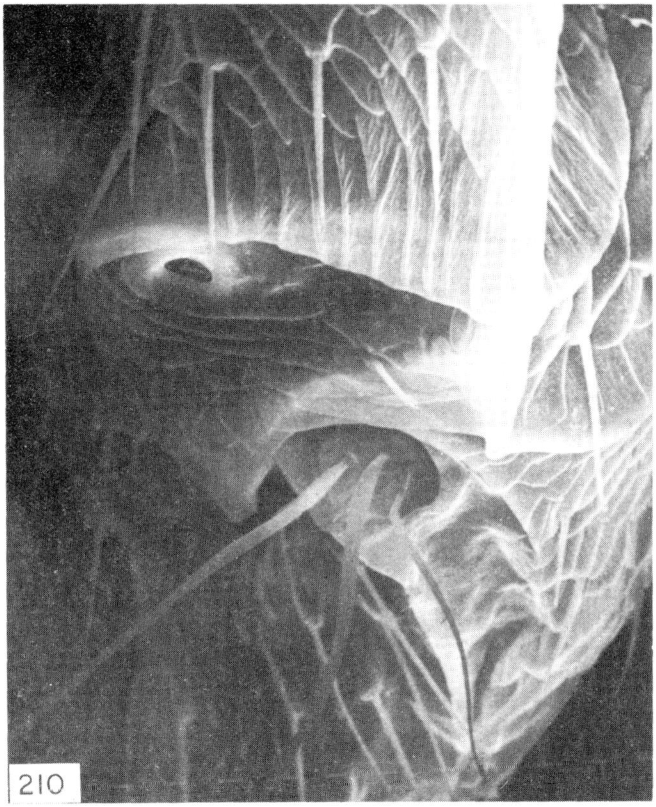


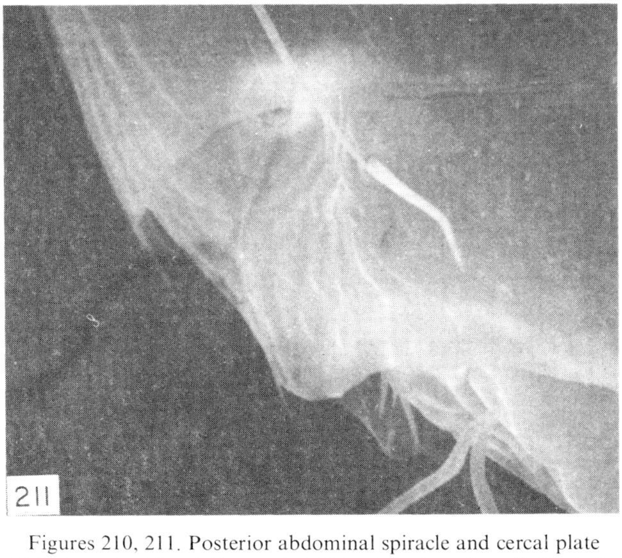


Figures 210, 211. Posterior abdominal spiracle and cercal plate
210. *Aphytis chilensis* Howard, ♀. 211. *Aphytis melinus* DeBach, ♀.

(Figure 1054, p. 630)). The seventh tergite is usually somewhat more distinctly sculptured centrally than the preceding sclerites. Tergites III–VI are asetose centrally, whereas the seventh tergite bears several fine setae in a transverse row between the two lateral setiferous areas (Figure 208). A single pair of submedian setae occurs on the seventh tergite in the more advanced species of *Aphytis* (Figure 209). The eighth tergite is usually delicately reticulate. It bears several setae in a transverse row between the spiracles in the primitive species (Figure 208), 4 setae in the more advanced species (Figure 209). The syntergum is faintly reticulate or reticulate-punctate. It bears several transverse rows of setae in the primitive species (Figure 208), one arcuate row in the more advanced species (Figure 209).

**The cerci** bear three long setae and one short seta in the females of some members of the **vittatus** group (e.g., *vittatus* (Figure 267, p. 300), *maculatipennis* (Figure 290, p. 305), *costalimai* (Figure 302, p. 307)) and in *cercinus* (Figure 507, p. 366), two long setae and one short seta in the females of most other species (Figures 208–211) and in the males of all species.

In the female, six sternites—representing abdominal segments II–VII—are present on the ventral aspect of the gaster. The second abdominal sternite is relatively small. The third, fourth, fifth and sixth sternites are unmodified, transverse or arcuate plates. The last visible (i.e., seventh) sternite—also called the *subgenital plate*—is the largest, with a bilobed or deeply arcuate posterior margin, enclosing the basal part of the ovipositor (Figure 213). Sternites II–III are asetose, while each of sternites IV–VI usually bears a transverse row of fine setae. The sternites are delicately reticulate, usually also faintly stippled centrally. In the female of *ignotus* (Figure 1322, p. 727), the posterior abdominal sternites are extensively stippled.

**The ovipositor** is usually only weakly exserted. The *shaft*, or *terebra*, is exposed for most of its length on the ventral aspect of the gaster, only its basal quarter or third being enclosed by the sternites (Figure 213; see also p. 93). The shaft varies in length, being only a little longer than the middle tibia in some species (e.g., *haywardi griseus*), considerably over twice as long in others (e.g., *merceti*); however, this character is sometimes also subject to considerable individual variation (see Figure 659, p. 446). As in other parasitic Hymenoptera, the ovipositor shaft is composed of three parts—a pair of serrated stylets (the *first valvulae*) and an unpaired, smooth component representing the united *second valvulae* (Figures 214, 215). The elongate *ovipositor sheaths* (*gonostyli* or *third valvulae*) are furnished with several setae, especially near their apex (Figures 208, *OS*; 209, 212–214). The sheaths vary in length, from nearly one fifth to about three fourths length of the middle tibia. The *outer ovipositor plates* have a distinct longitudinal ridge, starting at their base and fading out at about three fourths their length (see Figures 396 (p. 323) and 636 (p. 441)).

In the male, eight sternites are present, representing abdominal segments II–IX. The ninth sternite is bilobed or deeply arcuate, sometimes appearing as if composed of two narrowly joined, triangular plates. Sternites IV–IX usually bear transverse rows of fine setae (e.g., Figures 542 (p. 373) and 570 (p. 429)). The posterior sternites are often stippled centrally, sometimes very conspicuously so—especially in males of the **lingnanensis** and **funicularis** groups, but in several members of other groups as well (e.g., *merceti* (Figure 489, p. 362), *lingnanensis* (Figure 896, p. 570), *coheni* (Figure 911,

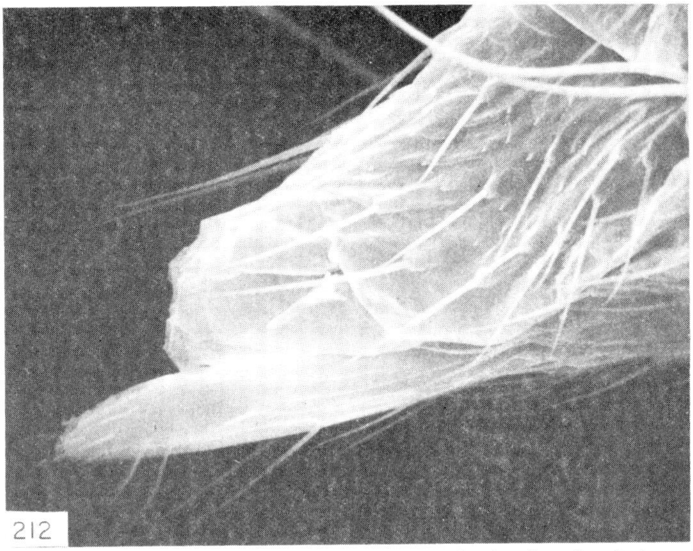


Figure 212. *Aphytis chilensis* Howard, ♀: tip of gaster and ovipositor sheath, lateral view.

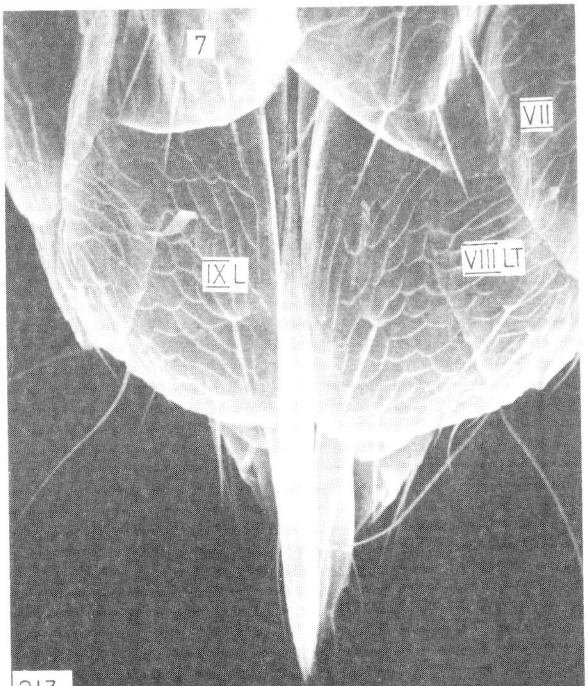


Figure 213. *Aphytis mytilaspidis* (Le Baron), ♀: posterior part of gaster and ovipositor, ventral view. 7 = seventh abdominal sternite, or subgenital plate; *VII* = seventh abdominal tergite; *VIII LT* = laterotergite, or lateral piece of eighth abdominal tergite; *IX L* = modified lateral part of ninth abdominal tergite.

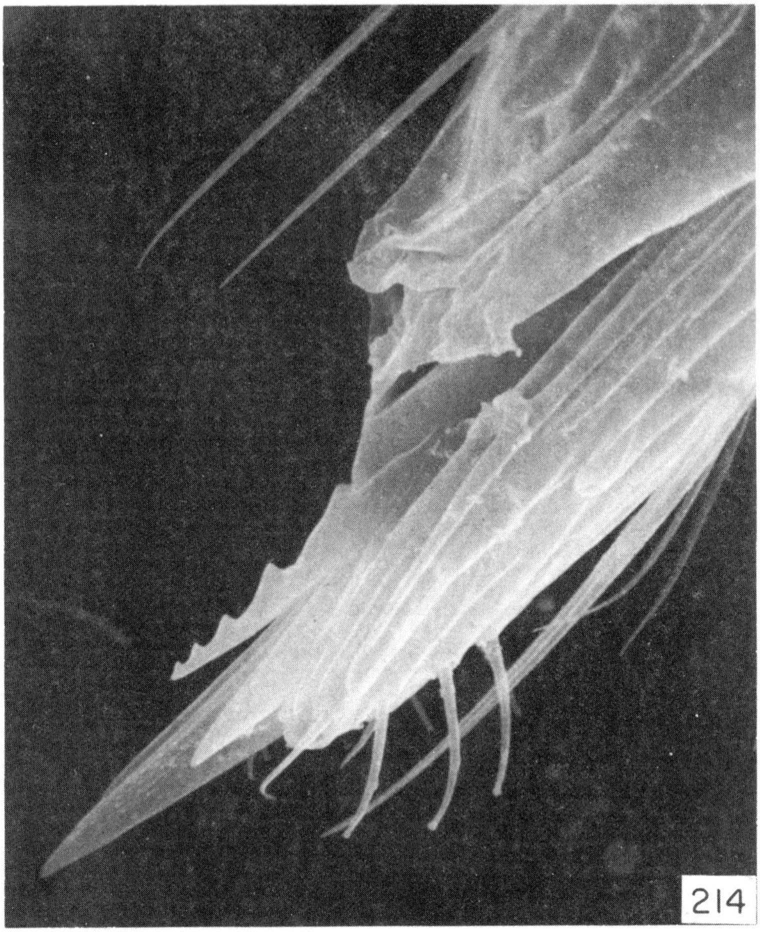

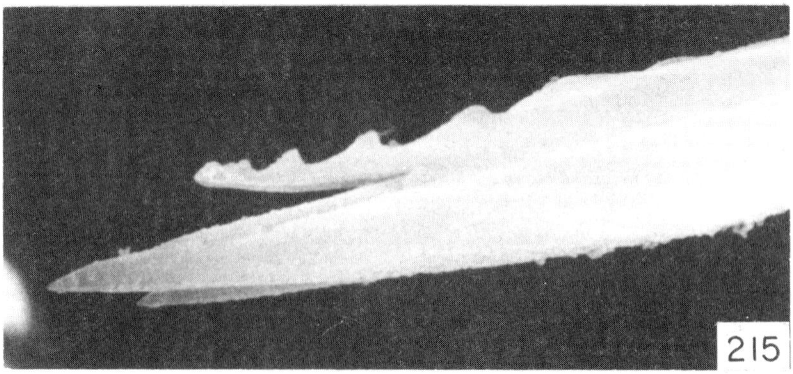

Figures 214, 215. Tip of ovipositor, showing the 3 components of the shaft
214. *Aphytis melinus* DeBach, ♀. 215. *Aphytis paramaculicornis* DeBach and Rosen, ♀.

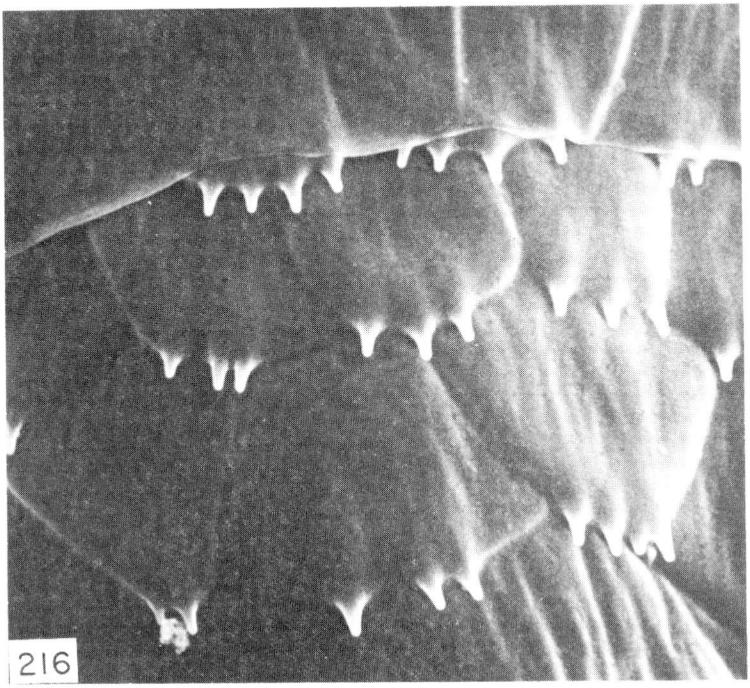


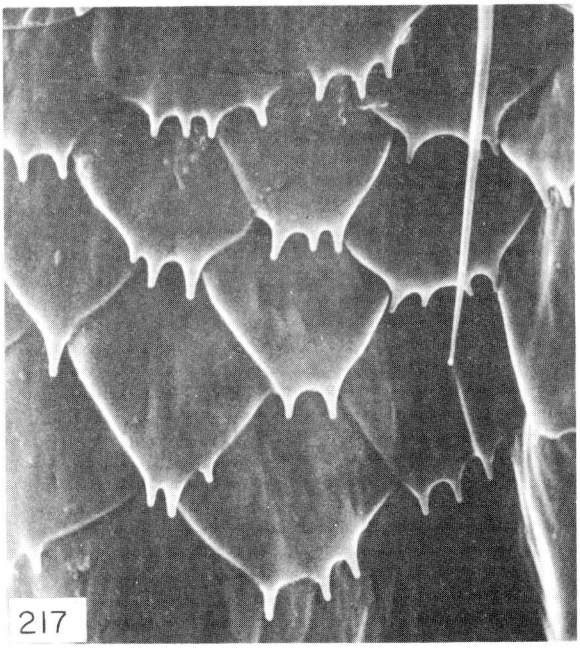


Figures 216, 217. Stippling of posterior abdominal sternites
216. *Aphytis lingnanensis* Compere, ♂. 217. *Aphytis melinus* DeBach, ♂.

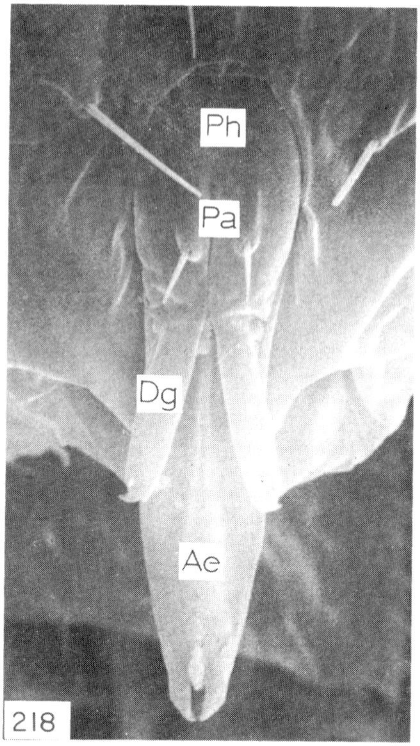


Figure 218. *Aphytis mytilaspidis* (Le Baron), ♂: genitalia
*Ae* = aedeagus; *Dg* = digital sclerite; *Pa* = papillae; *Ph* = phallobase.

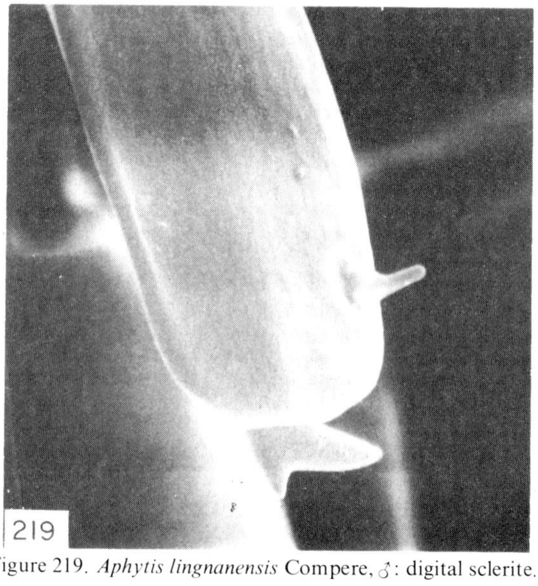


Figure 219. *Aphytis lingnanensis* Compere, ♂: digital sclerite.

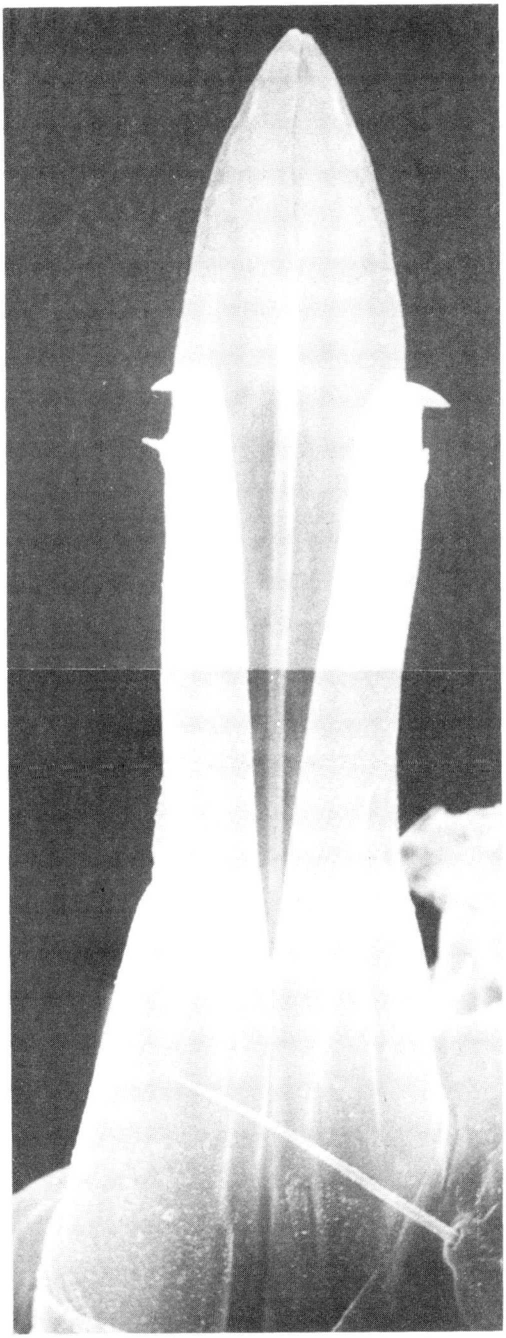

Figure 220. *Aphytis paramaculicornis* DeBach and Rosen, ♂: genitalia (a composite scanning electron micrograph).

p. 573), *melinus* (Figure 975, p. 585), *fisheri* (Figure 1001, p. 589), *funicularis* (Figure 1144, p. 664), *ulianovi* (Figure 1158, p. 666), *gordoni* (Figure 1168, p. 668) and *ignotus* (Figure 1327, p. 728)). The stippled effect is produced by the presence of minute stout spines on the posterior margins of the sternal reticulae (Figures 216, 217).

**The male genitalia** of *Aphytis* (Figures 218–220) are characterized by the complete absence of *parameres*. (The terminology of Snodgrass (1941) has been followed here, but see also Michener (1956) and Richards (1956).) The genitalia are relatively large, ranging from about two thirds length to full length of the middle tibia. They usually consist of a basal *caulis* or *phallobase* (Figure 218, *Ph*) and a narrower distal *aedeagus* (*Ae*). Only the basal (anterior) part of the phallobase is concealed by the abdominal sternites. The phallobase bears a pair of elongate, movable lobes, the *digital sclerites* (Figures 218, *Dg*; 219), each furnished with a short subapical spine and a single apical claw. Anterad of the digital sclerites, the phallobase bears a pair of small submedian *papillae* (Figure 218, *Pa*), each bearing a single seta. In cleared specimens, a pair of internal *apodemes* can be seen running from the base of the aedeagus and into the phallobase (see Figure 355, p. 316). In certain members of the **vittatus** and **proclia** groups, a median longitudinal rod is present in the phallobase, between the aedeagal apodemes (e.g., *maculatipennis* (Figure 294, p. 305), *perplexus* (Figure 326, p. 311) and *proclia* (Figure 570, p. 429)). Such a longitudinal rod is absent in members of the more advanced species groups of *Aphytis*.

There are a few exceptions to this general pattern. In the genitalia of *obscurus* (Figure 425, p. 328) and *angustus* (Figure 436, p. 330), for instance, the phallobase and aedeagus are not clearly differentiated and the digital sclerites, papillae and apodemes are absent altogether. The genitalia of *tucumani* (Figure 701, p. 454) are similar but possess distinct papillae. In *salvadorensis* (Figure 1313, p. 725), the undifferentiated phallobase is truncate basally, bears well-developed digital sclerites but lacks distinct apodemes. The relative proportions of these parts may also vary. Thus, in *merceti* (Figure 489, p. 362) the aedeagus is relatively long and the digital sclerites are short and slender. In *cochereaui* (Figure 341, p. 313) the phallobase is about as narrow as the aedeagus, and the digital sclerites are relatively short and slender. Relatively short but robust digital sclerites are present in *chilensis* (Figure 526, p. 370), *columbi* (Figure 543, p. 373), *confusus* (Figure 640, p. 442), *roseni* (Figure 1230, p. 709) and *setosus* (Figure 1277, p. 719). The papillae are unusually elongate in *funicularis* (Figure 1144, p. 664) and *erythraeus* (Figure 1204, p. 704).

## THE LEGS (Figures 221–227)

The forelegs of *Aphytis* are separated from the other two pairs. The fore coxae are attached to the propleura on each side of the prosternum, whereas the middle coxae are attached submedially to the posterior edge of the mesosternum (see Figures 202, 203, 205), and the hind coxae are attached to the sides of the metasternum, laterad and only slightly posterad of the middle coxae (see Figure 1231, p. 709).

The coxae exhibit a reticulate sculpture similar to that of the sternal sclerites (Figure 221). The hind coxae are considerably larger than the other two pairs. Minute finger-

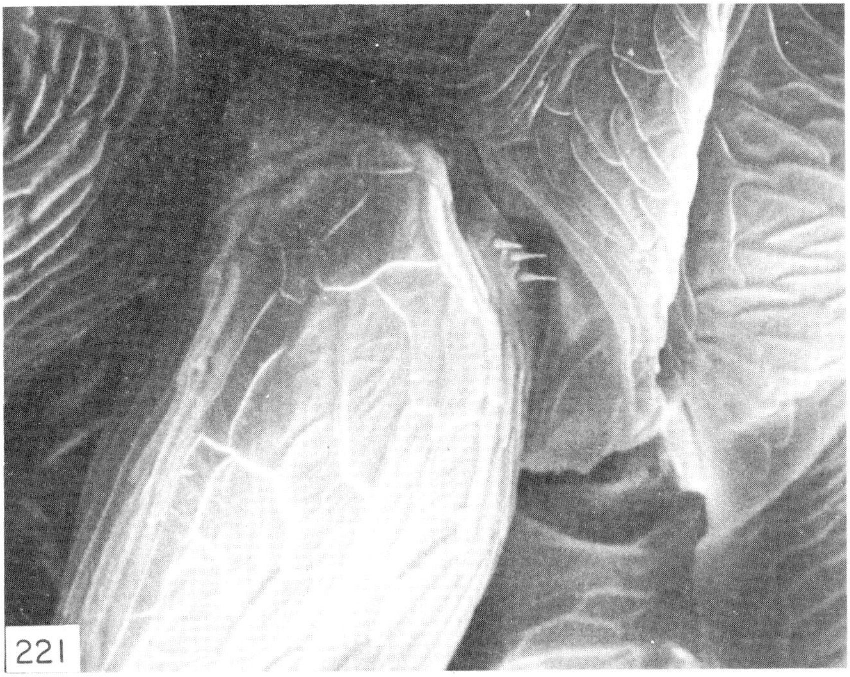

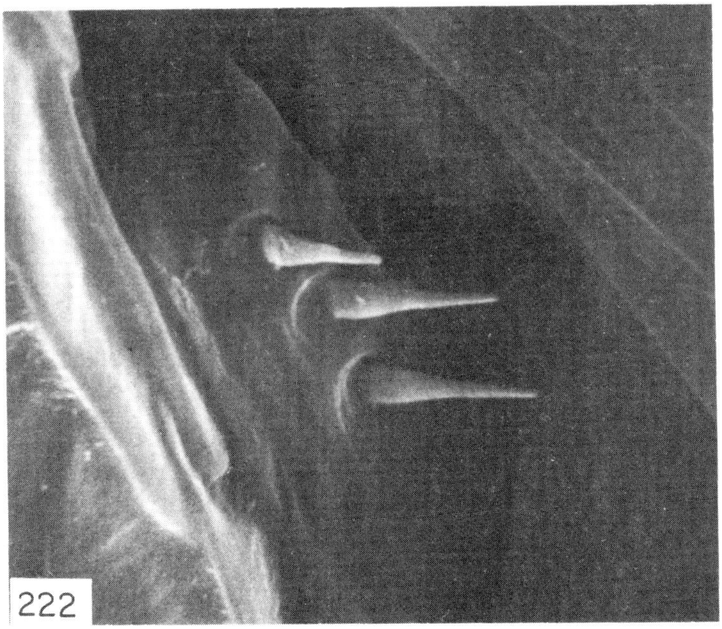

Figures 221, 222. *Aphytis vandenboschi* DeBach and Rosen, ♀
221. Hind coxa, lateral view; note 3 sensilla at base. 222. Close-up of the sensilla.

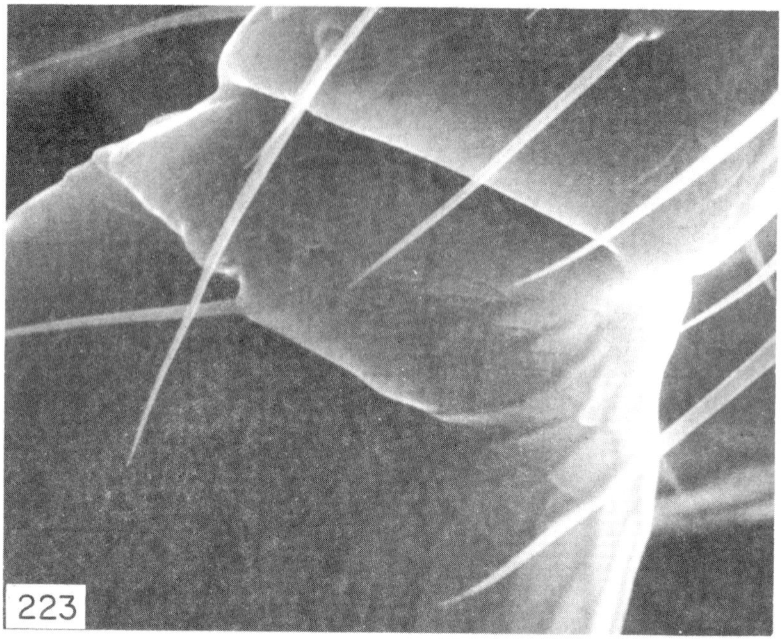


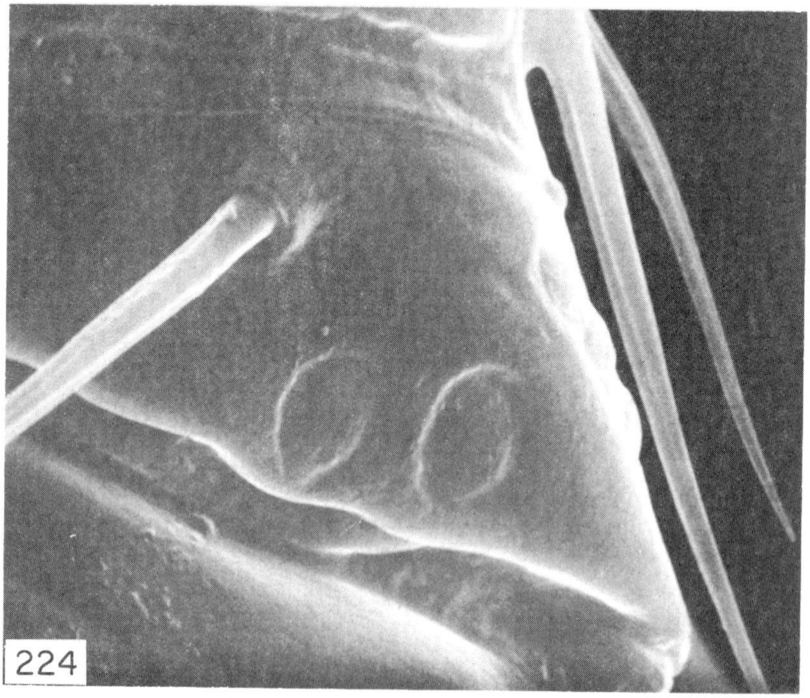


Figures 223, 224. Middle trochanter, showing discoid sensilla
223. *Aphytis chilensis* Howard, ♀. 224. *Aphytis melinus* DeBach, ♀.

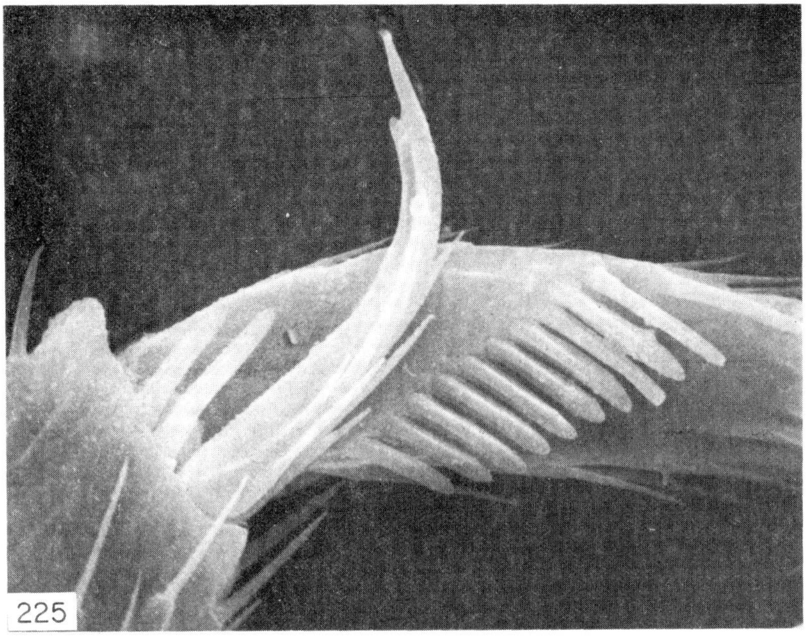


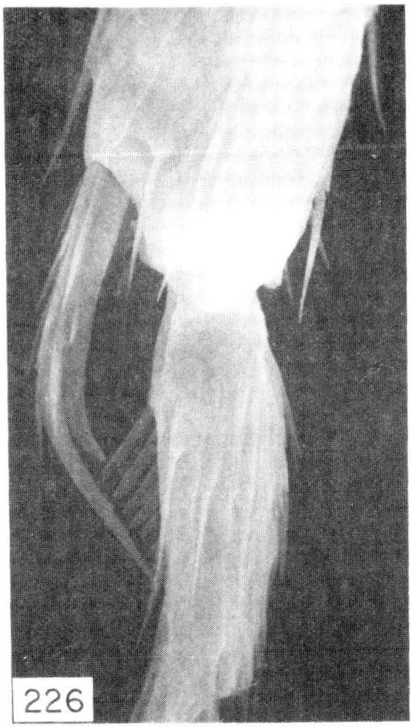


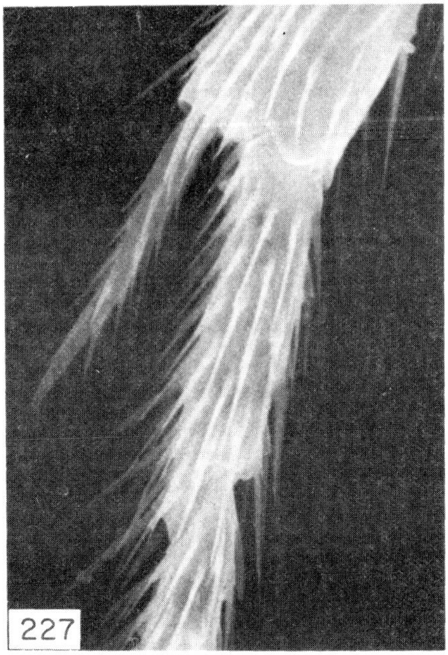


Figure 225. *Aphytis chilensis* Howard, ♀: strigil of foreleg, inner view. (From Rosen and DeBach, 1976.)
Figure 226. *Aphytis vandenboschi* DeBach and Rosen, ♀: strigil of foreleg, outer view.
Figure 227. *Aphytis chilensis* Howard, ♀: mid-tibial spur and basitarsus.

like sensilla at their base (Figures 221, 222) presumably serve to register the movements and attitudes of the hind legs.

All trochanters bear a few discoid *sensilla placodea* (Figures 223, 224). The base of the femur in all legs is slightly constricted and bears similar sensilla. These sensilla appear to have a very thin cuticle, and are visible as small pale disks in cleared specimens (see Figure 744, p. 504). A small glandlike structure can be seen in all femora of cleared specimens, with a long duct leading from it to the distal end of the femur (Figure 744). The function of this peculiar organ is unknown.

The forelegs bear a well-developed comblike antenna-cleaner; or *strigil* (Figures 225, 226), consisting of an arcuate, bifid tibial spur and an oblique row of spines on the inner aspect of the basitarsus. The saltatorial *mid-tibial spur* (Figure 227) is large, straight and stout, usually somewhat shorter than the corresponding basitarsus.

The tarsi of all legs are 5-segmented.

## THE WINGS (Figures 228–241)

**The forewings** of *Aphytis* are well developed, rounded apically, about twice to three times as long as wide. The *marginal fringe* is short, usually not exceeding one third width of the disk but often one tenth width of this or shorter. At rest, the forewings overlap one another on the back (Figure 228), extending beyond the tip of the gaster.

As in other chalcidoids, the venation of the forewings in *Aphytis* is greatly reduced, consisting of a short *submarginal vein* and a longer *marginal vein*, from the distal end of which branches a short, subsessile *stigmal vein*. A *postmarginal vein* is entirely absent.

The submarginal vein bears several long, coarse setae along its dorsal aspect (Figure 229). In the more advanced species, usually only a pair of such setae is present, the distal one being considerably longer than the proximal, with another long seta near the junction of the submarginal and marginal veins. A series of small rounded or oval tubercles, or *bullae*, is present along the ventral aspect of the submarginal vein (Figures 233–237). Their function is unknown.

The marginal vein bears a row of long, prominent setae, arising from tuberculous bases, along its anterior edge (Figure 230). They are usually subequal in length, but in a few species they decrease considerably in length toward the distal end of the vein (e.g., *costalimai* (Figures 303 and 308, pp. 307 and 308) and *salvadorensis* (Figure 1306, p. 724). Another row of shorter setae is borne dorsally along the center of the vein (Figure 230). Two minute disks—presumably *sensilla placodea*—are present at the base of the marginal vein. Four discoid sensilla are arranged in two pairs on the stigmal vein (Figures 231, 232; see also Figure 826, p. 521).

The dorsal aspect of the disk is densely covered with short setae. A conspicuous landmark on this aspect is the *speculum*, an oblique bare streak extending backward from the stigmal vein to the posterior margin of the wing (Figure 230). The speculum is interrupted by a single row of setae running along the posterior margin, composed of short setae distally and a few longer setae proximally. Proximad of the speculum, below the marginal vein, there is a triangular area known as the *delta*, bearing several oblique rows of setae. The setae in the delta are usually noticeably longer, coarser and sparser than the discal setae distad of the speculum. Their number is subject to con-

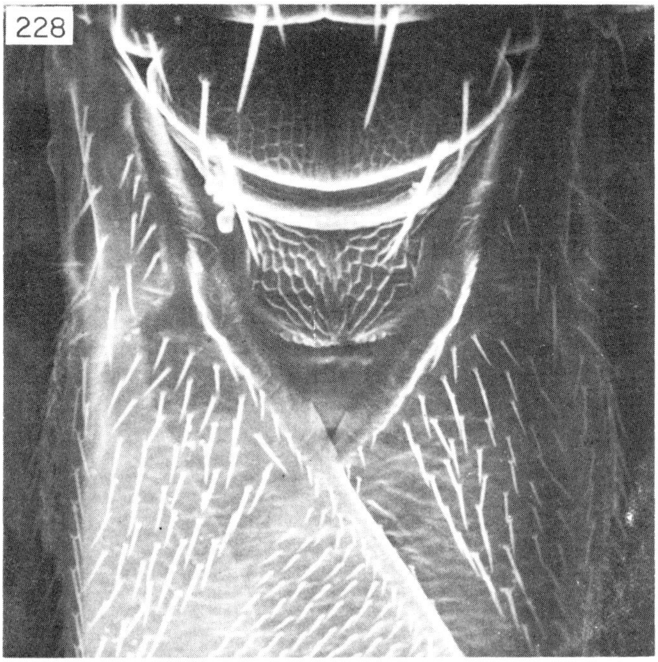


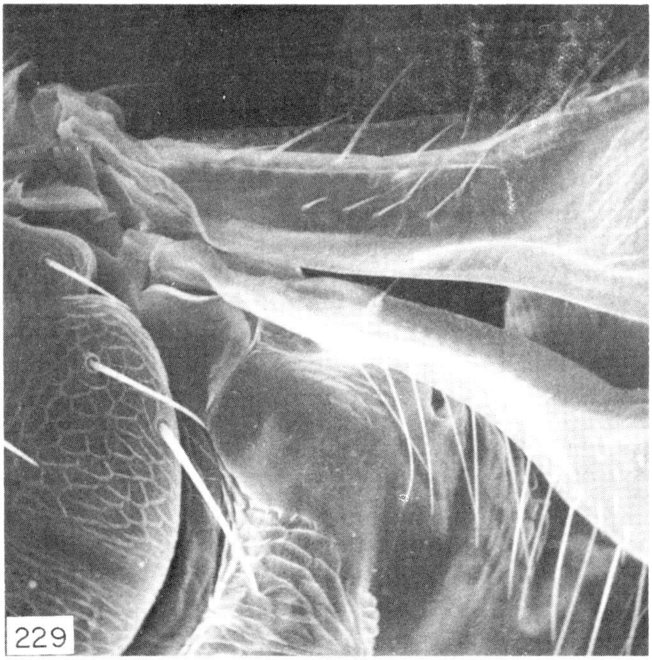


Figures 228, 229. *Aphytis chilensis* Howard, ♀: wings
228. Forewings at rest, dorsal view. 229. Bases of fore- and hind wings, dorsal view.

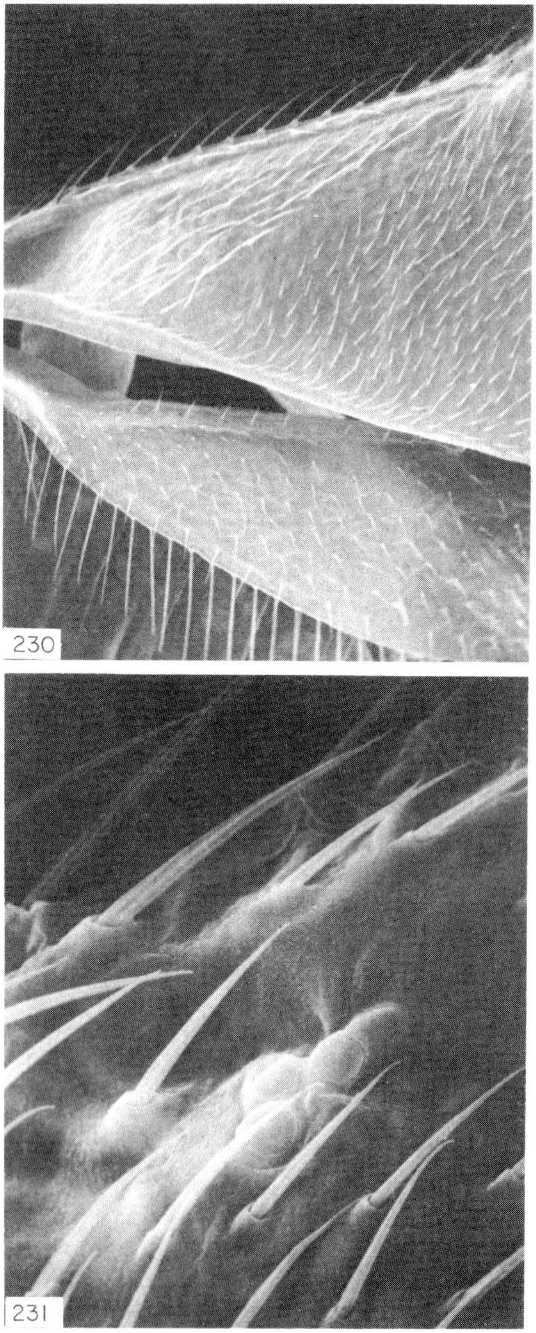


Figures 230, 231. *Aphytis chilensis* Howard, ♀: wings
230. Fore- and hind wings, dorsal view. 231. Stigmal vein of forewing, dorsal view.

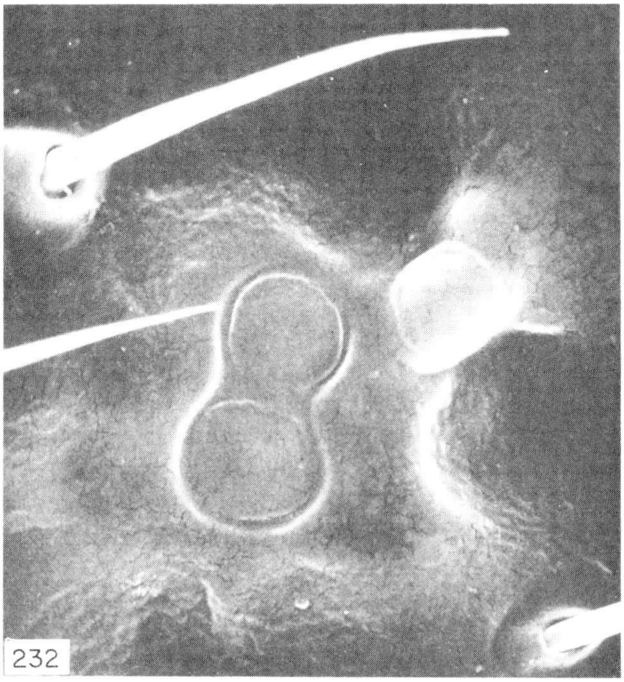


Figure 232. *Aphytis melinus* DeBach, ♀: stigmal vein of forewing, dorsal view.

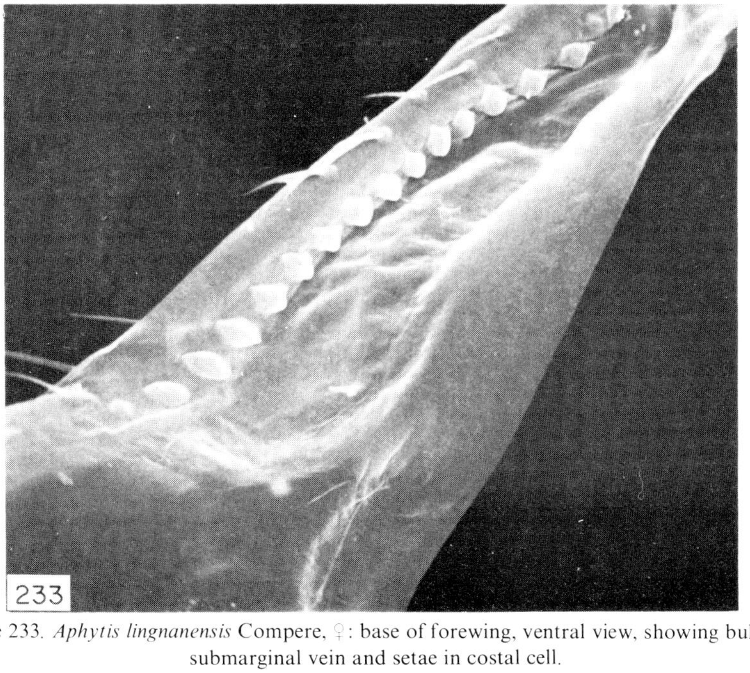


Figure 233. *Aphytis lingnanensis* Compere, ♀: base of forewing, ventral view, showing bullae on submarginal vein and setae in costal cell.

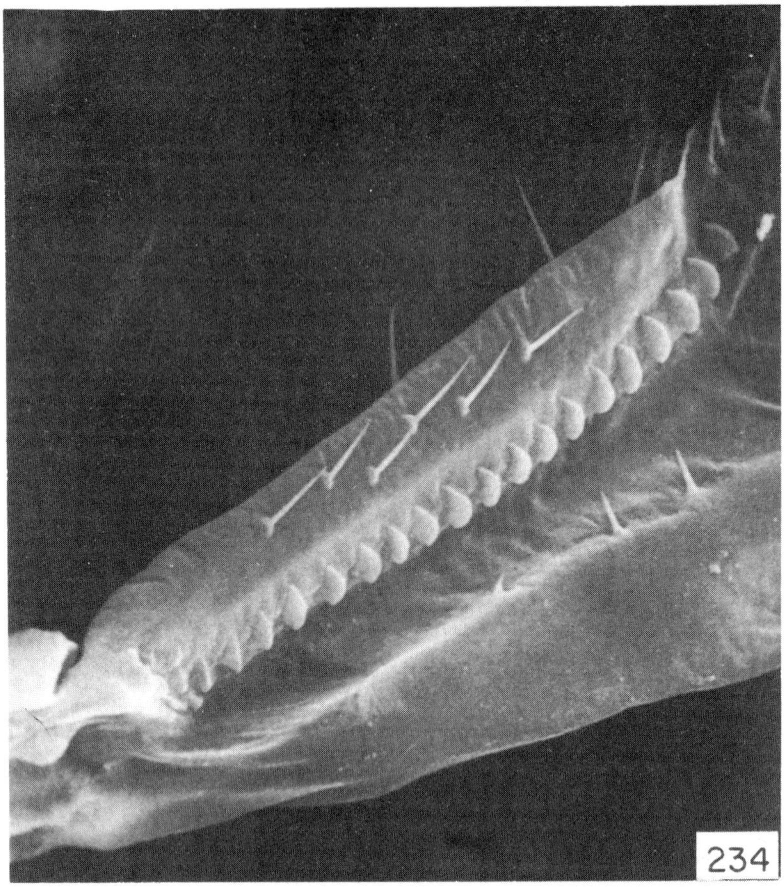


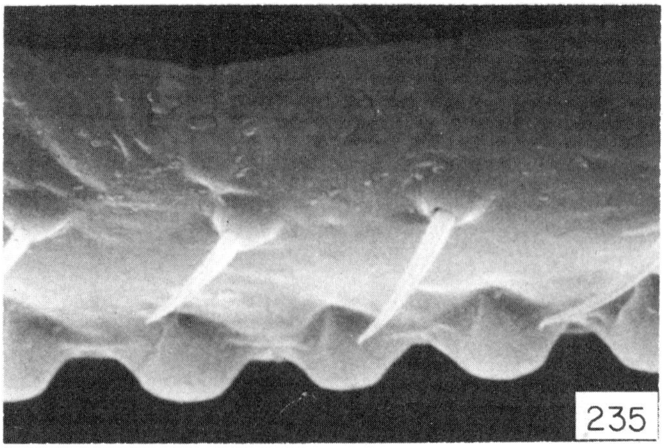


Figures 234, 235. *Aphytis chilensis* Howard, ♀: base of forewing, ventral view, showing bullae on submarginal vein and setae in costal cell.

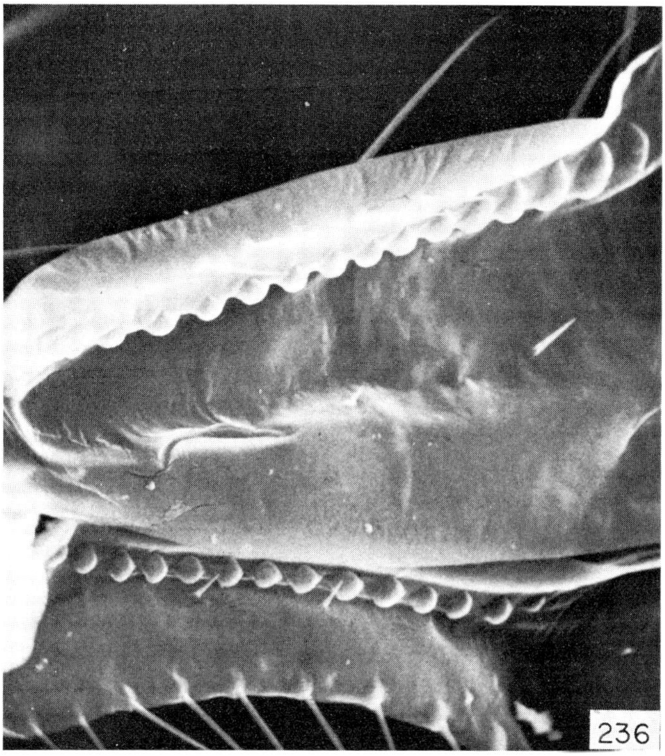

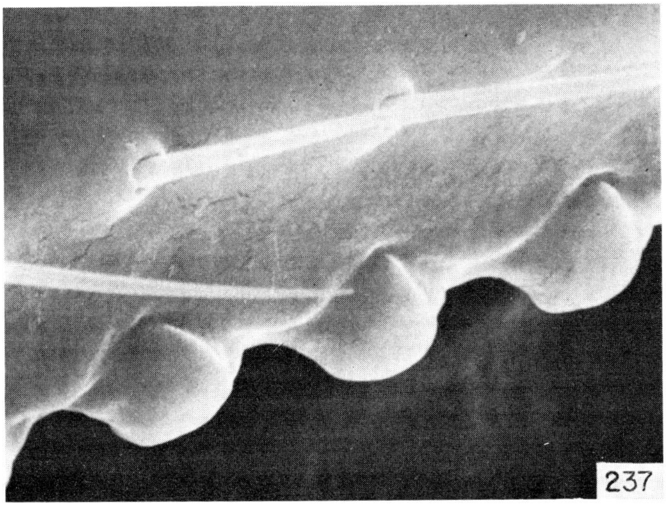

Figures 236, 237. *Aphytis melinus* DeBach, ♀: wings, ventral view
236. Bases of fore- and hind wings, showing bullae on submarginal veins.
237. Close-up of bullae on submarginal vein of forewing.

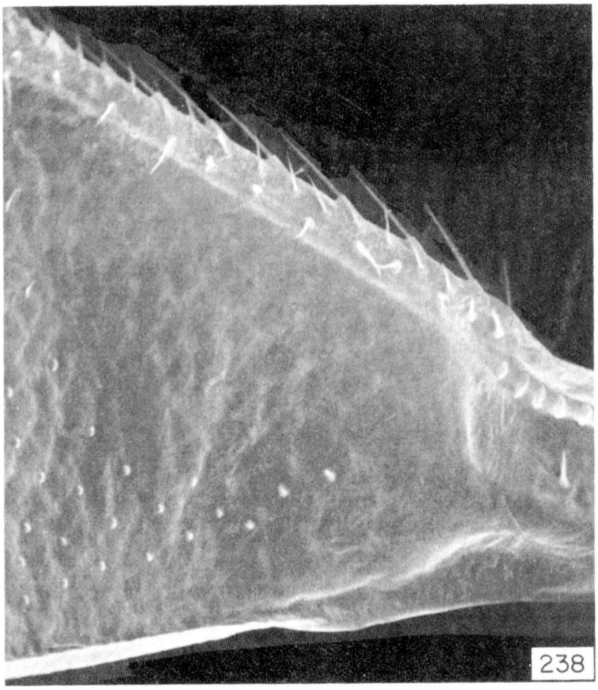


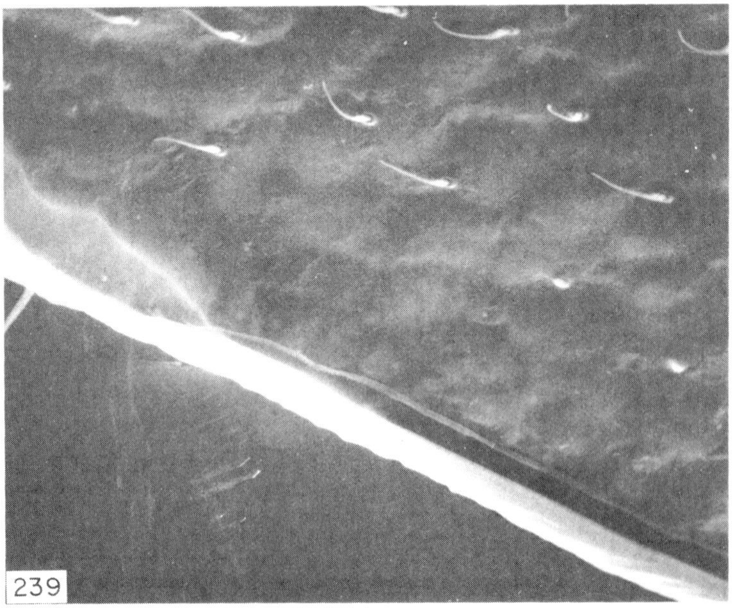


Figures 238, 239. Forewing, ventral view
238. *Aphytis mytilaspidis* (Le Baron), ♂; note tubercles and frenal fold.
239. *Aphytis melinus* DeBach, ♀: note tubercles, setae and frenal fold.

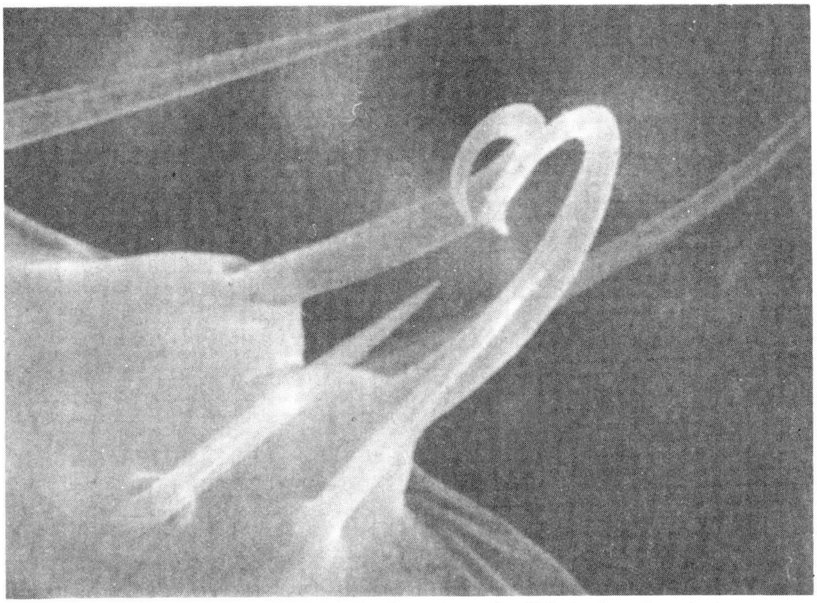

Figure 240. *Aphytis melinus* DeBach, ♀: hamuli of hind wing, dorsal view.

Figure 241. *Aphytis lepidosaphes* Compere, ♀: hamuli of hind wing hooked to frenal fold of forewing, ventral view.

siderable size-linked individual variation (see p. 193), but varies also among species; they are much more numerous—up to 150 or more—in the **vittatus**, **chilensis** and **proclia** groups than in the more advanced species groups. The delta is usually quite clearly separated from the row of setae along the posterior margin of the wing, widely separated from a small group of setae below the submarginal vein.

In most species of *Aphytis*, the discal setae are rather uniform, except for the differences between the delta and the area distad of the speculum. However, in the females of the **vittatus** group the forewings bear patches of coarse, dark setae against a background of fine, pale setae (e.g., Figures 268, 291, 303, 337, 366, 406, 414, 422, etc., pp. 300–328). Many species also exhibit some integumental infuscation or cloud on the forewings, especially below the stigmal vein (e.g., Figures 652, 653, p. 444). In others, the forewings are perfectly hyaline.

The narrow *costal cell*, delimited by the submarginal vein along the anterior margin of the wing, usually bears a few fine setae ventrally in a longitudinal row (Figures 233–237), often also a few coarser setae dorsally near the distal end of the cell.

The ventral aspect of the disk bears sparse, fine setae and minute tubercles (Figures 238, 239). The forewing is folded ventrally along the central part of the posterior margin, forming the *frenal fold* (Figures 238, 239), an important component of the wing-coupling mechanism.

**The hind wings** are shorter and considerably narrower than the forewings, roundly pointed at the apex. The *marginal fringe* is longer, usually exceeding half the width of the disk. The *submarginal vein* bears a series of *bullae* ventrally (Figure 236). The *marginal vein* bears a row of setae dorsally along its center (Figure 230). At the apex of the marginal vein there is a small group of hooked setae, or *hamuli* (Figure 240), which can be hooked onto the frenal fold of the forewing (Figure 241), thus coupling the fore- and hind wings in flight.

The hind wings are usually hyaline.

## SEXUAL DIMORPHISM

Sexual dimorphism is usually weakly expressed in *Aphytis*. In most species, the males are essentially similar to the females in structure, chaetotaxis, sculpture and general coloration. Except for primary sexual differences in the shape of the genitalia and abdominal sternites, sexual dimorphism is expressed mainly in antennal characters and in wing pattern.

Males are usually smaller than their conspecific females, and have fewer setae on the mesoscutum, along the marginal vein and in the delta area of the forewings. The males of the **chilensis** group exhibit a reduced number of funicular segments in the antennae, as do several members of the **vittatus** group (see p. 126). In other species the third funicular segment and the antennal club of the male bear unusually long setae (see p. 140). In quite a few species, the antennal scape or club of the male are furnished with specialized sense organs that are absent in the female. On the other hand, the antennal club of the male usually bears fewer longitudinal sensilla than in the female (except in species with a greatly elongate club in the male sex).

The forewings of the male are usually somewhat shorter, broader and paler than those of the female. In the **vittatus** group, the male forewings usually lack the mottled pattern of the female. When the cerci of the female bear three long setae, those of the male bear only two.

Males are usually somewhat paler than their conspecific females. However, in *mazalae* (Figure 1108, p. 641) and *erythraeus* (Figure 1205, p. 705) the males differ from the females in that the middle tibiae are conspicuously marked with black at the apex.

# INTRASPECIFIC VARIATION AND EVALUATION OF TAXONOMIC CRITERIA

As is the case with most other animal groups, the taxonomy of *Aphytis* has been based mainly on the morphology of dead adult specimens. Some of the characters used to advantage in the past are to be found in papers by Mercet (1912b), Malenotti (1918a, 1918b), De Santis (1948) and Compere (1955). Characters commonly used, particularly with reference to females, include 1) number of antennal segments; 2) ratio of length to width of the antennal club; 3) ratio between the lengths of the various antennal segments; 4) number, size and relative melanization of the setae on the mesonotal sclerites; 5) shape of the crenulae on the posterior margin of the propodeum; 6) number of setae in the abdominal cerci; 7) wing shape; 8) ratio of the length of the longest setae in the marginal fringe of the forewing to the greatest width of the disk; 9) number of setae along the marginal vein; 10) number of bullae along the submarginal vein; 11) number of setae in the delta area of the forewing; 12) relative visibility of the interommatidial setae on the compound eye; 13) sculpture on the thorax; 14) general color of live adults; 15) degree of duskiness (melanization) of the integument and its location on the body, head, antennae, legs and wings, especially in cleared specimens; 16) pigmentation of the pupa. (See DeBach, 1964.)

Some of these characters are rather variable. Before the actual revision of individual species could be commenced, the validity of morphological characters for species determination had to be evaluated. Intraspecific variation was therefore investigated in several species of *Aphytis*. Clones derived from single females of thelytokous (uniparental) species were particularly useful in this respect. Inasmuch as little or no genetic variation should be involved, individuals of such clones were reared under different environmental conditions, on different hosts, etc., in order to assess the effect of environmental factors on phenetic variation.

Variation proved to be great in many of the taxonomic characters which had been commonly used with *Aphytis* species. Most variation between individuals of the same species is due to differences in size. Size depends on larval nutrition, that is, on whether the larvae develop on a small or on a large scale-insect host, or whether two or more larvae develop on a single host. Very small individuals are difficult to relate to their normal-sized sisters. The length and number of setae present on a given sclerite or organ are probably the most variable characters. The larger the individual of a given species, the greater will usually be the number of setae on the body, especially on the mesoscutum and pronotum but also on the head and abdomen. In *chilensis*, for instance, the number of mesoscutal setae may vary from 10 in small specimens to as many as 22 in large ones (e.g., Figures 517 and 518, p. 368; for illustrations of variation in other species, see Figures 739–740, 788–789, 815–816 and 901–904). However, the extent of this size-linked variation may differ with the species, being, for instance, much more pronounced in *diaspidis* than in the closely related *proclia* (see Figure 657 and p. 407). The number of setae on the posterior abdominal tergites is quite variable in the more primitive species of *Aphytis*, much less so in the more advanced species.

Wing proportions, and to some extent also antennal and other proportions, vary considerably with size. Small specimens usually have proportionately narrower forewings than large specimens of the same species. Smaller wings bear fewer setae in the delta area than larger wings. In a uniparental clone of *mytilaspidis* from Crete, for instance, the number of setae in the delta was found to vary from about 50 in small specimens to over 90 in large ones. The same applies to the number of setae along the marginal vein and to the number of bullae along the submarginal vein. On the other hand, the length of the marginal fringe is inversely correlated with the width of the wing: the narrower the wing, the longer will usually be the hairs of the marginal fringe, both in the absolute and in the relative sense. In the same uniparental clone of *mytilaspidis*, for example, the length of the marginal fringe varied from about one eleventh the width of the disk in large specimens to about one fifth in small ones, and the fringe hairs of the latter were actually up to one-and-a-half times longer than those of the former. The longer fringe may partly compensate the smaller individual for the reduced area of its narrower wings.

The proportions of the antennal segments (i.e., the ratio of length to width of a given segment—especially the club—and the ratio between the lengths of the various segments) are usually quite reliable diagnostic characters (e.g., Figure 584, p. 433). However, minute specimens may have shorter antennal segments than large specimens of the same species (compare, for instance, the proportions of the third funicular segment in Figures 523 and 524, p. 369).

The relative length of the ovipositor usually tends to be inversely correlated with the size of specimens. Again, the extent of this size-linked variation may differ markedly with the species, being much more pronounced in *diaspidis* than in *proclia* (see Figure 659 and p. 407). In certain cases this character may aid in specific identification (e.g., Figure 585, p. 433).

The number of propodeal crenulae is quite variable with size, but may nevertheless serve as an auxiliary diagnostic character in certain cases (e.g., Figure 658, p. 446). On the other hand, the shape of the crenulae is one of the most stable characters in *Aphytis*, although in certain species even this may be subject to individual variation (e.g., *melanostictus*, Figures 456 and 457, p. 336).

The relative length of the propodeum is a quite reliable, useful character. Other stable characters are the shape of the mouthparts (especially the number of segments in the maxillary palpi), the presence or absence of specialized sensilla on the antennae, the shape of the abdominal syntergum, the number of setae in the cerci, the shape and proportions of the male genitalia, etc.

The degree of coarseness and melanization of the body setae is usually quite stable and useful for species identification. So is also the wing pattern, although in some of the species with maculated wings this, too, may be subject to considerable variation (e.g., *capillatus*, Figures 364–366, p. 318).

The pattern of pigmentation of the body and appendages—a character often disregarded by systematists—is probably the least variable character in *Aphytis*. Unlike the *intensity* of melanization (i.e., the amount of melanin deposited in the exoskeleton), which may sometimes vary with size, the *pattern* (i.e., the location and extent of melanized areas) is surprisingly constant. In particular, the relation between clear and melanized areas in the antennae, the presence of absence of melanization on the head, on the thoracic sterna, on the posterior margin of the scutellum, on the propodeum and crenulae, on the gastral tergites, at the base of the ovipositor and on the legs are excellent diagnostic characters with very limited intraspecific variability. However, it should be remembered that different mounting procedures and microscopic techniques may affect apparent melanization in *Aphytis*. Phase-contrast microscopy, for instance, tends to enhance the duskiness of the integument, causing clear areas to appear melanized (compare Figures 297 and 298 (p. 306), 453 and 454 (p. 335), or 768 and 769 (p. 508)), whereas treatment in KOH and mounting in balsam tend to eliminate melanin (as well as to distort the specimens). For most reliable results, the pattern of pigmentation should be studied only in cleared specimens (acetic acid—lactophenol method, see p. 15) mounted in Hoyer's medium, under a binocular stereoscopic microscope, with a white background. On the other hand, structural and sculptural characters are seen to advantage under a high-power phase-contrast microscope.

The position of the specimens on the slide may sometimes affect the apparent shape or proportions of various parts or organs (see, for instance, the effect of position on the shape of the antennal segments in Figures 737 (p. 503), 851 (p. 526) and 1173 (p. 669)). Excessive pressure of the cover slip often causes unnatural flattening of specimens. For positive identification, large series of properly mounted specimens should be studied whenever possible.

Finally, the emphasis just placed on certain diagnostic characters should not lead to the presumption that the other characters mentioned in this section are of little value for the identification of *Aphytis* species. Some are perfectly good for a particular species, but they should be used with care. Usually it will be found that a combination of several—perhaps many—characters will be necessary in order to separate one species of *Aphytis* from another.

## NOTES ON MEASUREMENTS

The position of a specimen on the slide sometimes makes accurate measurement of certain parts or organs virtually impossible, especially when the parts in question are not perfectly horizontal but assume an oblique position on the slide. This is often true for the setae along the wing veins, but also for antennal segments, legs, etc.

Throughout the specific descriptions, the "length" of a sclerite always refers to the longitudinal axis of the body, whereas "width"—and the terms "broad" or "narrow"—refer to the transverse axis. The same applies, of course, to the antennae and wings.

Whenever two parts of the body overlapped (e.g., head and thorax, metanotum and propodeum, etc.), they were measured separately. In measuring the length of a specimen, care was taken to deduct "gaps" caused by mounting.

Measurements of antennal segments were taken, whenever possible, from antennae in profile. Their length and width were always measured along their longest aspect; however, the length of the pedicel was usually measured along its ventral aspect. When longitudinal sensilla happened to be situated on the edge of the club or third funicular segment, they were probably sometimes missed when the sensilla were counted.

When the components of the ovipositor shaft became separated, the length of the shaft was measured from the two minute sclerites at its base to the tip of the unpaired component (united second valvulae). The ovipositor sheaths were usually measured along their outer aspect.

In counting the bullae along the submarginal vein of the forewing, those on the base of the marginal vein were also included. In counting the "prominent" setae along the marginal vein, only those arising from elevated bases along the very edge of the vein were included. The length of setae was usually measured from their base, including the part concealed within the socket.

# BIOSYSTEMATIC STUDIES

"Classical" systematics, based on the morphological characteristics of dead adult specimens, cannot provide satisfactory solutions to all the intricate taxonomic problems presented by the species of *Aphytis*. The identification and separation of *Aphytis* species are very difficult, owing mainly to the minute size of the species and to the relative scarcity of reliable diagnostic characters. The various species groups contain numerous complexes of cryptic or sibling species, as well as various biological races and other infraspecific entities exhibiting differing degrees of morphological and biological distinctness. In fact, practically every one of the originally described, more common species of *Aphytis* is now known to include a complex of cryptic forms. The recognition of such forms is, of course, of supreme practical importance for biological control, since they may attack different host species or may otherwise have different inherent capabilities, affecting their efficacy as natural enemies.

"Sibling species" are commonly defined as morphologically indistinguishable (see Mayr, 1963). Such species can be recognized as distinct only through biosystematic research, biological differences often providing the first clue to their distinctness. However, their "sibling" status is relative, reflecting the depth of our knowledge of a given group. In some instances, further study may eventually reveal some cryptic morphological differences; in others, the species remain indistinguishable. Members of the **lingnanensis** group of *Aphytis* may serve to illustrate this point. *A. lingnanensis*, for instance, would have been considered a sibling of *A. chrysomphali* in the past, and was indeed confused with that species for many years, but since the diagnostic value of the shape of the propodeal crenulae was discovered, separation of these two species has become relatively easy and they can no longer be regarded as siblings by present-day criteria. Further, both *A. melinus* and *A. holoxanthus* would have been considered siblings of *lingnanensis*, until the validity of sternal pigmentation as a diagnostic character was realized. *A. holoxanthus* may still be regarded as a sibling of *melinus*, although the two species can usually be separated by slight differences in the antennal proportions. Finally, *A. fisheri* is still considered a sibling of *melinus* in the adult stage, but further study may yield new characters that may eventually facilitate their separation on morphological grounds, thus canceling their "sibling" status. All these species differ markedly in salient biological attributes, as well as in their performance as biological control agents.

Ever since the advent of the so-called "new systematics" (Huxley, 1940), most practicing taxonomists have accepted the biological definition of the species concept, and have been more or less aware of the importance of biosystematic research. Unfortunately, however, this knowledge has often remained theoretical, inasmuch as very few systematists have at their disposal the time and facilities required for biosystematic investigations. In our present revision of *Aphytis*, we have endeavored to correlate the results of comparative morphological research with those of various biosystematic studies. These have included investigations of biological, developmental, cytological, reproductive, ethological and biochemical characteristics.

## DEVELOPMENTAL CHARACTERISTICS

**Morphology of Eggs and Larvae.** Fisher (1952) compared the larvae of several species of *Aphytis*, and noted some slight differences in the shape of their mandibles, spiracles and spiracular tubercles. Benassy (1955) compared the larvae of *A. mytilaspidis* and *A. proclia* and described some differences in the shape and dimensions of their mandibles, but subsequent studies by Traboulsi (1969) and by Eliraz and Rosen (1978) failed to confirm his findings.

During the present revision, Eliraz and Rosen (1978) made detailed comparisons of the eggs and larvae of six species—*chilensis, hispanicus, mytilaspidis, coheni, melinus* and *chrysomphali*—representing five of the seven species groups of *Aphytis*. Their results were rather disappointing: The eggs and larvae of all six species were found to be practically indistinguishable in all significant morphological characters, except for the number and position of the minute cuticular tubercles and pores in the cephalic area of the third-instar larva (Figures 242–246). However, although these obscure structures were observed to differ in the representatives of the five species groups, they proved to be identical in the larvae of *coheni* and *melinus*, closely related species belonging to the same group. At best, this is a very difficult diagnostic character of limited practical value.

**Pupal Pigmentation.** Taylor (1935) was apparently the first to notice differences in pupal pigmentation between closely related species of *Aphytis*. He reported that the "form" of *A. chrysomphali* parasitic upon the coconut scale, *Temnaspidiotus destructor* (Signoret), in Fiji was characterized by pale, lemon-yellow pupae, whereas another "form" of this species, parasitic upon the Florida red scale, *Chrysomphalus aonidum* (L.), in Java, was characterized by dark pigmentation on the ventral aspect of the pupae. Since the two forms were at that time considered to be conspecific, this was not regarded

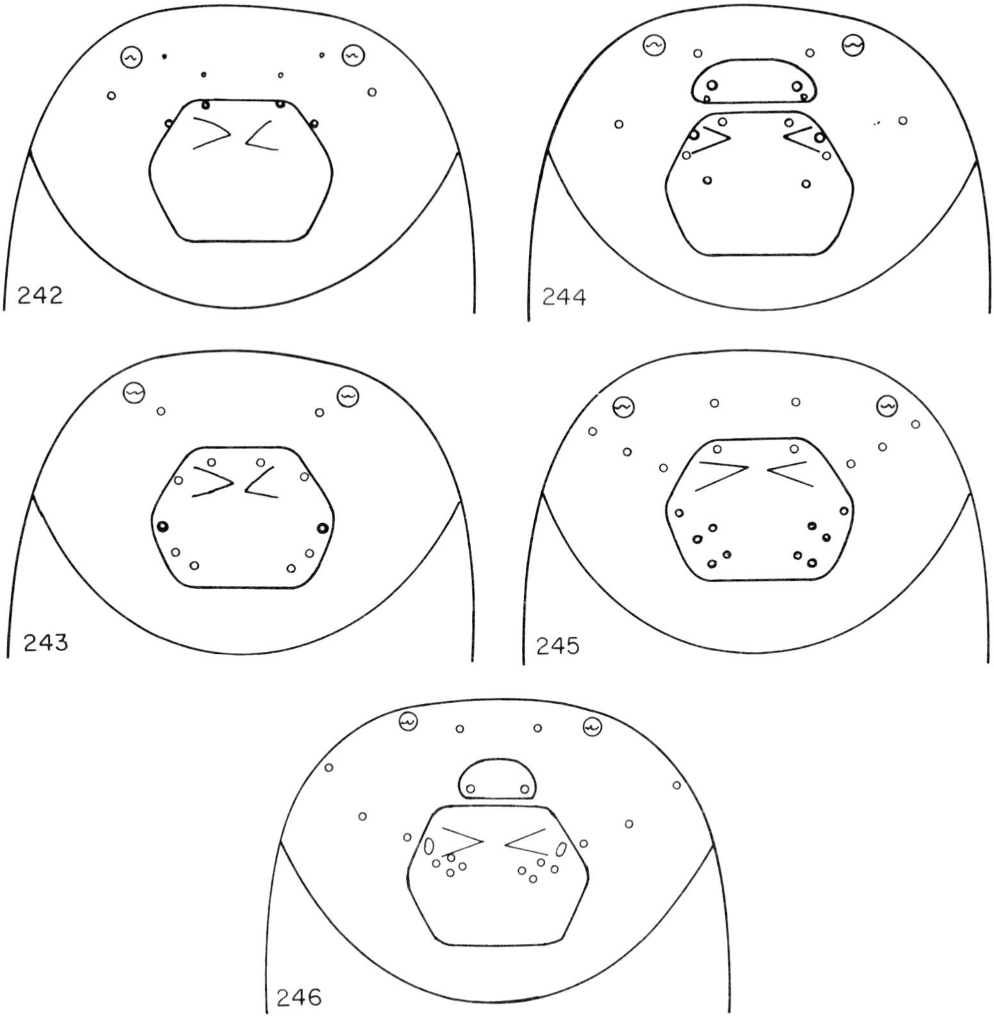


Figures 242–246. Arrangement of cuticular tubercles in cephalic region of third-instar larvae (diagrammatic; from Eliraz and Rosen, 1978).
242. *Aphytis chilensis* Howard. 243. *Aphytis mytilaspidis* (Le Baron). 244. *Aphytis melinus* DeBach and *Aphytis coheni* DeBach. 245. *Aphytis hispanicus* (Mercet). 246. *Aphytis chrysomphali* (Mercet).

as a reliable diagnostic character. However, we now believe that the latter "form" was in fact a distinct species, most probably *A. holoxanthus*.

In some instances, pupal pigmentation actually provided the first clue to the distinctness of cryptic *Aphytis* species. *A. lingnanensis*, for instance, was first recognized as distinct from *A. chrysomphali* on the basis of this character (Compere, 1955). More recently, DeBach (1959) used it to separate several closely related species of *Aphytis* parasitic upon the California red scale, *Aonidiella aurantii* (Maskell), in the Orient.

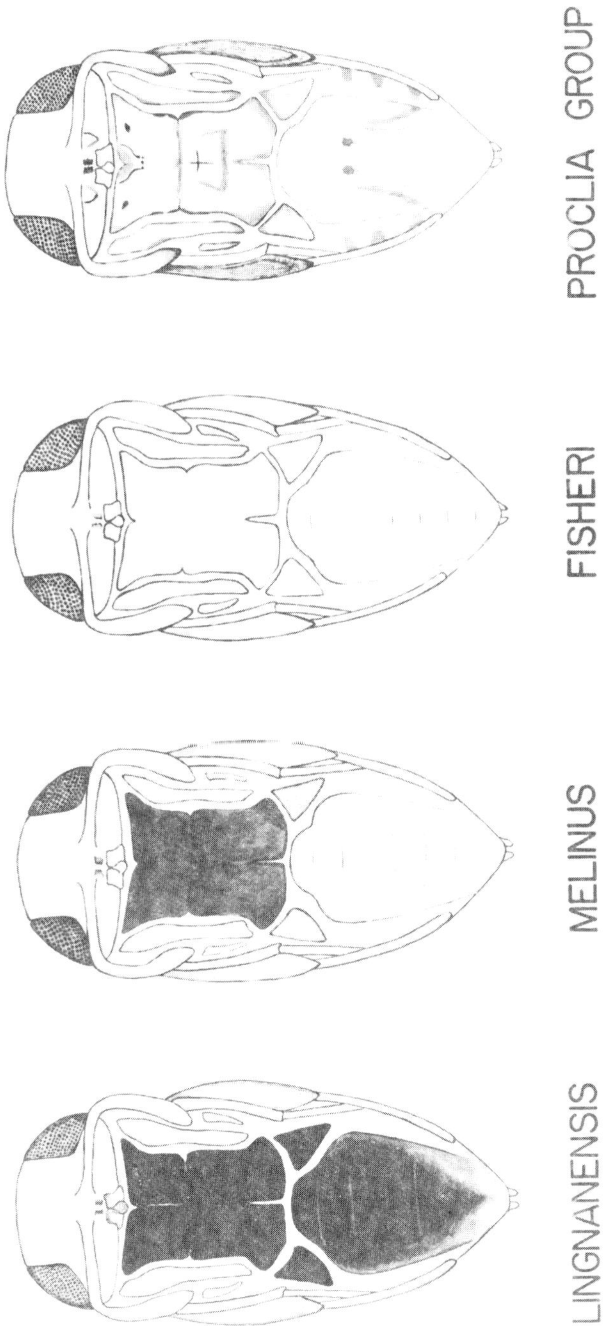


Figure 247. *Aphytis* pupae, ventral view, showing four types of pigmentation. (From DeBach, 1959.)

Although pupal pigmentation is somewhat variable at times, it certainly may be regarded as an important supplementary diagnostic character, and may even serve as a convenient shortcut to the separation of certain closely related species of *Aphytis*. As mentioned earlier in this section, *A. holoxanthus*, *A. melinus*, and *A. fisheri*, for example, are sibling or near-sibling species that are almost inseparable in the adult stage. However, they do differ markedly in pupal pigmentation. The pupae of *holoxanthus* exhibit black pigmentation on both the thoracic and the abdominal sterna, those of *melinus* are pigmented only on the thoracic sterna, whereas the pupae of *fisheri* are entirely yellow. (Care should be taken to compare only mature, green-eyed pupae, because the pigmentation of younger pupae may be incompletely developed.)

Several characteristic pigmentation patterns of *Aphytis* pupae are presented in Figure 247.

**Meconia.** The larvae of all known species of *Aphytis* produce distinctive meconial pellets prior to pupation. Protected by the covering scale of the parasitized host, the meconia remain there, unchanged, long after the parasite has emerged—lasting evidence of the former presence of *Aphytis*. They are readily obtainable by simply turning the covering scale over, with the aid of a dissecting needle or fine forceps, under a low-power dissecting microscope. If consistent specific differences could be shown to exist between them, the meconia would conceivably provide a convenient adjunct in the identification of *Aphytis* species. However, most students of *Aphytis* have either ignored the meconia altogether, or else considered them to be highly variable (e.g., Traboulsi, 1969). The first published attempt to utilize some characteristics of the meconia in the systematics of *Aphytis* was recently made by Yasnosh (1972), who referred mainly to their number, color, general shape and arrangement.

The detailed morphology of *Aphytis* meconia was first explored by DeBach, Rose and Rosen (1978) during the present revision. A comparative scanning electron microscope investigation of the meconial pellets of 12 species and strains, representing six of the seven species groups, revealed specific differences between them in their size and general shape, as well as in the surface patterns and striations on them (Figures 248–261). Although more should be known about intraspecific variation in this respect, meconial characters do appear to provide an additional diagnostic tool for the identification of certain species of *Aphytis* (e.g., *chilensis*, Figures 249, 250), and even for the separation of certain closely related entities, especially when used in combination with other characters such as the pigmentation of pupae or exuviae.

## OTHER BIOLOGICAL CHARACTERISTICS

**Duration of Life Cycle.** Species of *Aphytis* may differ markedly in the duration of their development from egg to adult under similar conditions (see p. 55). It appears that, as a rule, the life cycle is longer in members of the more primitive species groups than in the more advanced groups of *Aphytis*. Quednau (1965) included differences in the length of the life cycle in his proposed procedure for identifying mixed populations of *Aphytis* species reared from the California red scale in South Africa. However, extensive rearing tests under various constant temperatures have

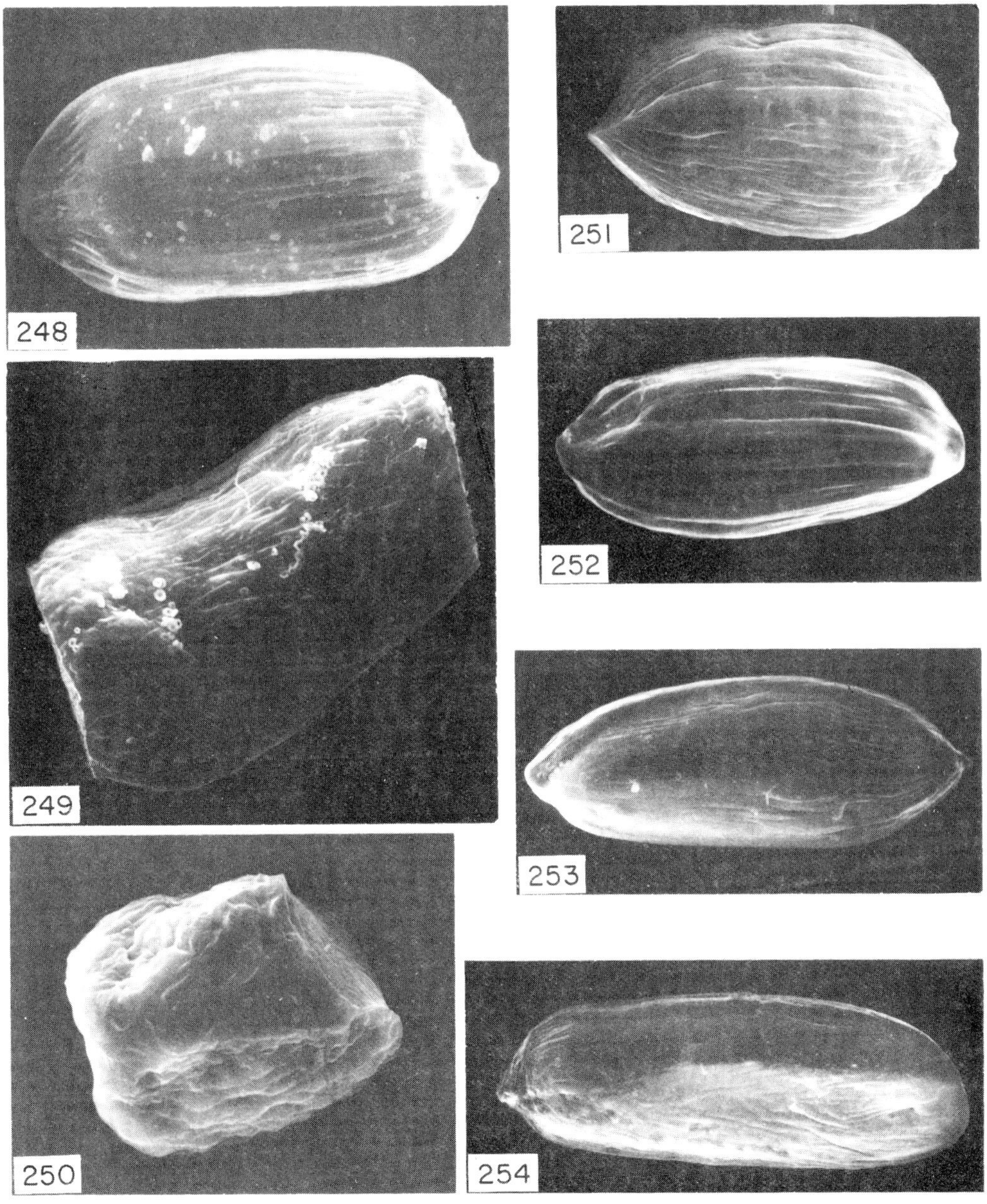


Figures 248–254. Meconial pellets
248. *Aphytis costalimai* (Gomes). 249, 250. *Aphytis chilensis* Howard. 251. *Aphytis paramaculicornis* DeBach and Rosen. 252. *Aphytis aonidiae* (Mercet). 253. *Aphytis lingnanensis* Compere, red-scale strain. 254. *Aphytis lingnanensis* Compere, snow-scale strain.
(SEM micrographs; from DeBach, Rose and Rosen, 1978.)

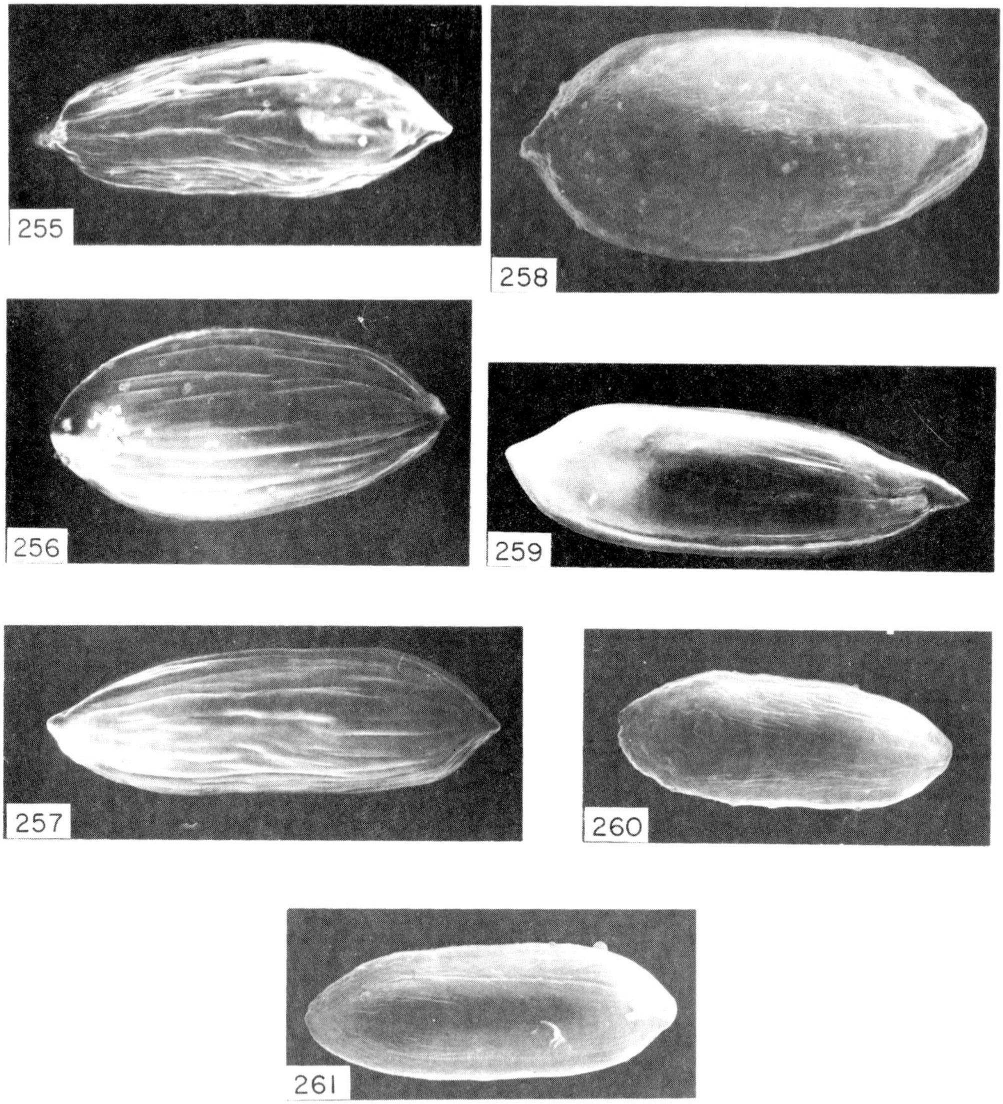


Figures 255–261. Meconial pellets
255. *Aphytis holoxanthus* DeBach. 256. *Aphytis melinus* DeBach. 257. *Aphytis yasumatsui* Azim. 258. *Aphytis fisheri* DeBach. 259. *Aphytis africanus* Quednau. 260, 261. *Aphytis chrysomphali* (Mercet). (From DeBach, Rose and Rosen, 1978.)

shown that the duration of development of closely related species of *Aphytis* is remarkably similar (e.g., Rao and DeBach, 1969a), and that normal biological variability even under strictly standardized conditions usually results in overlapping ranges for such species. Although some species exhibit high pupal mortality at certain high temperatures (P. DeBach, unpublished data), this, too, is an impractical biosystematic criterion for separating closely related species of *Aphytis*.

**Longevity of Adult Wasps.** Extensive comparative tests were carried out at Riverside to determine the longevity of adult *Aphytis* wasps under various constant temperatures. All tests were run under 50% relative humidity and in the absence of hosts, with honey provided as food for the parasites. Some representative results are presented in Table 1.

TABLE 1. Number of days required to reach 50% mortality of adult *Aphytis* wasps under constant temperatures (from Rao and DeBach, 1969a and unpublished data)

| | 4.4°C | 10.0°C | 15.6°C | 21.1°C | 26.7°C |
|---|---|---|---|---|---|
| *A. chilensis* | 23 | 67 | 45 | 17 | 14 |
| *A. lingnanensis* | 3 | 5 | 70 | 35 | 15 |
| *A.* "2002" | — | 43 | 78 | 40 | 21 |
| *A. coheni* | 2 | 8 | 40 | 32 | 14 |
| *A.* "khunti" | 4 | 8 | 48 | 38 | 12 |
| *A. africanus* | — | 77 | 57 | 41 | 24 |
| *A. melinus* | 3 | 84 | 104 | 53 | 20 |
| *A. fisheri* | 10 | 98 | 103 | 47 | 24 |
| *A. holoxanthus* | 3 | 94 | 89 | 46 | 24 |

Such differences in mean longevity may be used to a certain extent as a diagnostic criterion for the separation of species. *A. chilensis*, for instance, representing the **chilensis** group, was significantly distinct at 21.1°C from all the other species tested, which are all members of the **lingnanensis** group. On the other hand, *A. coheni* and *A.* "khunti", which yielded almost identical results, are indeed very closely related and cannot be regarded as distinct species (see p. 542). But these two were quite distinct from the mean longevity obtained for nearly all the others at 10.0°C, and from most of them at 15.6°C. In fact, with few exceptions, combinations of results from several temperatures would probably serve to separate any given species of *Aphytis* from all the others. However, *A. melinus* and *A. fisheri* yielded very similar results, and both would be very difficult to separate by mean longevity from the closely related *A. holoxanthus*, whereas on the other hand *A.* "2002", which is almost conspecific with *A. lingnanensis* (see pp. 208 and 539), seems quite distinct from this species at 10.0°C.

Species of *Aphytis* may also differ in their susceptibility to extreme high or low temperatures. Thus, comparative longevity tests at 36.1°C and 1.7°C have indicated that heat tolerance by *A. coheni* and *A.* "khunti" is definitely inferior to that by *A. lingnanensis*, *A. africanus* and *A. fisheri*, whereas tolerance for cold was outstandingly superior in *A. chilensis* but not very different among the other species tested (P. DeBach, unpublished data).

For comparable results, experimental procedures employed in longevity tests should be strictly standardized, and handling of the parasites reduced to a minimum. Even slightly different conditions may result in significant differences in longevity. Thus, in tests carried out by Quednau (1964a) with some of the species listed in our Table 1, at 26.7°C but under 75% RH and in the presence of hosts (and with daily transfers to fresh hosts), the longevity of *Aphytis* females was 19.8 ± 2.2 days for *A. lingnanensis*, 9.9 ± 1.2 days for *A. coheni*, 17.5 ± 2.1 days for *A. africanus*, 18.1 ± 1.9 days for *A. melinus*, and 14.8 ± 0.1 days for *A. holoxanthus*.

Thus, although longevity tests may demonstrate the existence of significant biological differences between species of *Aphytis* and may contribute important information to our knowledge of these species, at present it is hard to visualize them as a practical biosystematic tool.

**Fecundity.** Like the duration of the life cycle and adult longevity, potential fecundity—i.e., average total progeny production per female on a preferred host under optimal conditions—is undoubtedly characteristic of the species. Unfortunately, however, this is also one of the most variable biological parameters. The fecundity of *Aphytis* females may be drastically affected by such factors as the ambient temperature and humidity, the amount of light, the host scale insect and host plant, the number of parasites that have developed on an individual host, etc. (see Ecology section, pp. 53–72). The amount and exact method of handling of the parasites in laboratory experiments may also have a direct effect on their fecundity. Fecundity tests carried out with the same species by different researchers, or even by the same researcher at different times, may therefore yield different results. Table 2, presenting results of fecundity tests performed by Quednau (1964a) and by Rao and DeBach (1969a) with the same 5 species of *Aphytis*, illustrates this point.

TABLE 2. Average total progeny production per *Aphytis* female: comparison of results obtained under different experimental procedures

| Species | Fecundity according to Quednau (1964a)* | Fecundity according to Rao and DeBach (1969a)** |
|---|---|---|
| *A. lingnanensis* | 76.3 ± 3.0 | 36.6 ± 1.60 |
| *A. coheni* | 28.8 ± 3.4 | 28.7 ± 1.00 |
| *A. africanus* | 19.8 ± 4.0 | 19.0 ± 1.03 |
| *A. melinus* | 51.5 ± 8.9 | 36.1 ± 1.74 |
| *A. holoxanthus* | 41.3 ± 7.1 | 32.3 ± 1.63 |

* Host: California red scale artificially pasted on citrus leaves; daily transfer of parasites; 26.7°C; 75% RH.

** Host: California red scale for *A. africanus*, oleander scale for the others; no transfer of parasites; 26.7°C; 50% RH.

Although the results obtained by these researchers with *A. coheni* and *A. africanus* were very similar, those obtained with the other species—especially with *A. lingnanensis*—were not. Standardization of conditions and experimental procedures, plus minimization of handling, are therefore of even more crucial importance in comparative fecundity tests than in any other biological experiments with *Aphytis* species.

Assessing potential fecundity as an absolute value characteristic of the species is impractical not only for the reasons discussed above, but also because the most suitable host for a particular species of *Aphytis* may not be available in laboratory culture, or may not even be known. On the other hand, a comparative study of the fecundity of several species on a given host, under uniform conditions, may serve to indicate biological differences between them and may even aid in their separation.

**Host Preference Patterns.** Differential fecundity of a parasite on different host species is, of course, an important manifestation of its host specificity. More than other biological parameters, patterns of host preference may serve as a very useful biosystematic tool for the separation of *Aphytis* species.

The hosts of quite a few species of *Aphytis* are still unknown. Host records in the entomological literature are largely unreliable, due to the frequent misidentification of both host and parasite species that occurred almost as a rule until recently. Nevertheless, during the present revision we have accumulated authenticated host records for 76 of the 90 valid species of *Aphytis* (see Appendix 2). Of these, 37 species (or 48.7%) have so far been reared from a single host species, 10 have been obtained from two hosts, and 12 from three hosts. In other words, 59 species of *Aphytis* (or 77.6% of all species for which reliable host records are available to us) have been reared from only 3 or fewer host species. Only 9 species (or 11.8%) have been reared from 9 or more host species. Significantly, the latter include some that are already known to be complex "superspecies", comprising an array of closely related forms, such as *A. lingnanensis* or *A. mytilaspidis*.

We naturally realize that our data are far from complete. It is expected that many of the species now recorded from a single host may eventually be obtained from several additional host species. However, most of the species of *Aphytis* do appear to be rather narrowly oligophagous. Although the actual range may vary from one to about 25 hosts species per *Aphytis* species, we assume that 5 accepted hosts may represent a "typical" figure, with one scale-insect species being *the* preferred host and the other accepted host species being much less preferred (Rosen and DeBach, 1977b).

The *actual* host range of a parasite in a given habitat may be distinctly narrower than its *potential* host range, due to processes of competitive displacement. In other words, a species of *Aphytis* may be perfectly capable of developing and reproducing on a given host, but may never do so in the field because it is being continually displaced from that particular ecological niche by another, more suitable species (see pp. 73–77). This should be taken into consideration when compiling information on the host range of a given parasite species. Comparative laboratory tests are therefore required if data on host preference are to be used in biosystematics.

A few species of *Aphytis* are strictly monophagous. *A. lepidosaphes*, for instance, has so far been reared only from the purple scale, *Cornuaspis* [= *Lepidosaphes*] *beckii* (Newman), and would not accept any other host in laboratory experiments (DeBach and Landi, 1961). Since none of the close relatives of *A. lepidosaphes* is known to develop on the purple scale, the host record may provide an easy shortcut to the identification of this species.

*A. cylindratus* is another highly specific parasite. Although not monophagous, it appears to be restricted to a single genus of scale insects. Extensive surveys in Japan, Brazil and Trinidad, in which numerous samples of various potential hosts were collected, have indicated that this species attacks only species of the genus *Pseudaonidia* Cockerell. It has never been obtained from any other host in the same habitats (Rosen and DeBach, 1977b).

Other species may exhibit rather distinctive patterns of host preference. *A. africanus*, for example, will not develop on the oleander scale, *Aspidiotus nerii* Bouché [= *A. hederae* (Vallot)], which is commonly used as a good laboratory host for many of its close relatives. This sort of "negative specificity" may also serve as a convenient biosystematic criterion (Quednau, 1965; Rosen and DeBach, 1977b).

Acceptance or rejection of a given host species may indicate the possible distinctness of closely related forms of *Aphytis*. Thus, for instance a population of *A. paramaculicornis* from Iran develops readily on the oleander scale, which is unacceptable as a host to other allopatric populations referable to this species (Khasimuddin and DeBach, 1976b). Similarly, two sympatric populations of *A. mytilaspidis* from Greece differ in that one accepts the latania scale, *Hemiberlesia lataniae* (Signoret), as a host while the other does not (Khasimuddin and DeBach, 1976c).

Even when the host spectra of several *Aphytis* species completely overlap, as they often do, their differential fecundity on each host species is likely to be significantly different. An index of host suitability may therefore be obtained by presenting various scale-insect species, individually or in multiple-choice experiments, to each species of *Aphytis*. Theoretically, such indexes may serve to "fingerprint" a given parasite species and separate it from all others (P. DeBach, unpublished data). In practice, however, when the differences are quantitative rather than qualitative, it is again very difficult to employ such data for the separation of closely related species, due to problems of inherent variability and standardization of experimental procedures.

## CYTOLOGICAL CHARACTERISTICS

The chromosomes of several species and strains of *Aphytis* were studied in some detail by Rössler and DeBach (1973 and unpublished data). In *A. mytilaspidis*, both arrhenotokous and thelytokous forms have a diploid number of chromosomes, with one metacentric and four subtelocentric pairs. Individual chromosomes at the second meiotic metaphase range in size from 3.5 to 6 microns, the metacentric chromosome being the largest. Although preliminary comparative studies have so far not yielded any significant differences in the number, shape or size of the chromosomes of *Aphytis* species, this approach should certainly be pursued much further before any definite conclusions may be drawn about its possible applicability.

## REPRODUCTIVE CHARACTERISTICS

**Reproductive Isolation.** The biological criterion of reproductive isolation is the key to the modern "biological" species concept. Mayr, Linsley and Usinger (1953) define species as "groups of actually (or potentially) interbreeding natural populations which are reproductively isolated from other such groups" (see also Mayr, 1963). Hybridization tests may therefore be considered as providing the ultimate proof of the specific status of closely related populations, at least with bisexual organisms. Indeed, experimental hybridization has served as a powerful biosystematic tool in the study of *Aphytis* populations. The distinctness of quite a few sibling or near-sibling species of *Aphytis* was confirmed by extensive reciprocal crossing tests that proved them to be reproductively isolated from one another (see DeBach, 1959, 1960).

Arrhenotokous species are especially amenable to studies of reproductive isolation. Inasmuch as all fertilized eggs of such species develop into females, the degree of success or failure of a heterogamic cross is indicated by the proportion of female progeny resulting in the $F_1$ generation. Thus, an all-male $F_1$ progeny indicates lack of fertilization and may be taken as evidence for complete reproductive isolation, whereas a normal sex ratio in the $F_1$ generation indicates conspecificity.

A more accurate measure of reproductive isolation in arrhenotokous organisms was proposed by Rao and DeBach (1969a:545). Their "*P–D isolation index*" takes into account both the percentage and the relative number of female (= hybrid) progeny per parental female in a heterogamic cross, as compared with standard values obtained in a homogamic cross, as well as the degree of fertility or sterility of the $F_1$ hybrids. This index is calculated as follows:

$$P\text{–}D \textit{ isolation index} = A \times B \times C \qquad \text{where}$$

$$A = \frac{\text{Percent female progeny in heterogamic cross}}{\text{Standard percent female progeny in homogamic cross}}$$

$$B = \frac{\text{Average number of female progeny per parental female in heterogamic cross}}{\text{Standard average number of female progeny per parental female in homogamic cross}}$$

$C$ = the "fertility factor", being unity when the $F_1$ hybrid progeny are all fertile and zero when they are all sterile.

The standard values used in the denominator of $A$ and $B$ are those of the female parent in the heterogamic cross of the numerator.

When the index is unity, no reproductive isolation is indicated and the populations under study are considered to be conspecific. On the other hand, when the index is zero, reproductive isolation is complete and the populations under study are considered to represent "good" species in relation to each other. However, reciprocal crossing tests with closely related allopatric populations quite often indicate the existence of *partial* reproductive isolation between them (in other words, gene flow between them is possible but is more or less restricted). Such populations have been designated "*semispecies*" (Mayr, 1963), a very useful concept stressing the often-forgotten fact that evolution is a continuous process. Obviously, this term covers a wide range of

| ♂ \ ♀ | *lingnanensis* | "2002" | "khunti" | *coheni* | "R-65-23" | *africanus* | *lepidosaphes* | *fisheri* | *holoxanthus* | *melinus* |
|---|---|---|---|---|---|---|---|---|---|---|
| *lingnanensis* | **F 66·1** | F 13·9 | N | St 1·2 | St 16·3 | N | N | N | N | N |
| "2002" | F 25·3 | **F 64·0** | N | F 1·7 | F 26·0 | N | N | N | N | N |
| "khunti" | F 6·7 | F 1·14 | **F 73·1** | F 51·2 | N | N | N | N | N | N |
| *coheni* | St 8·6 | F 8·7 | F 49·7 | **F 74·7** | N | N | N | N | N | N |
| "R-65-23" | N | F 46·3 | N | F 6·5 | **F 60·0** | — | — | N | N | — |
| *africanus* | N | N | N | N | — | **F 51·0** | N | N | N | N |
| *lepidosaphes* | N | N | N | N | — | N | **F 52·0** | N | N | N |
| *fisheri* | N | N | N | N | N | N | N | **F 57·6** | N | N |
| *holoxanthus* | N | N | N | N | N | N | N | N | **F 63·6** | F 12.2 |
| *melinus* | N | N | N | N | — | N | N | St 0·54 | N | **F 64·1** |

Figure 262. Hybridization experiments: outcome of reciprocal crosses between various members of the **lingnanensis** group. Dark-bordered squares represent homogamic crosses. Numbers represent percentage of female progeny. F = fertile hybrids; St = sterile hybrids; N = no hybrids; — = cross not attempted. (From Rao and DeBach, 1969a.)

different degrees of reproductive isolation, and semispecies are therefore considered as being "weak" or "strong" accordingly. For a discussion of some of the nomenclatural problems involved with semispecies, the reader is referred to p. 560.

An extensive study by Rao and DeBach (1969a) of reproductive isolation among various members of the **lingnanensis** group of *Aphytis* may serve to illustrate how complicated this can get. In an attempt to surmount the ethological barriers of sexual isolation and induce heterogamic mating, the females used in crossing tests were temporarily anesthetized with chloroform, while the presence of crushed homogamic females was used to excite the males. Some results of these studies are depicted in Figures 262 and 263.

Consider, for instance, the following complex of three allopatric, sibling forms: *A. lingnanensis* (ex the California red scale, *Aonidiella aurantii* (Maskell) from southern China), *A.* "2002" (ex the coconut scale, *Temnaspidiotus destructor* (Signoret), from Puerto Rico), and *A.* "R-65-23" (ex the citrus snow scale, *Unaspis citri* (Comstock), from Florida). Morphologically they are all indistinguishable. However, reciprocal crossing tests have shown that, whereas *lingnanensis* and "R-65-23" are only partially isolated reproductively from "2002", they are completely isolated from each other. In other words, *lingnanensis* and "R-65-23" may be considered as good species in re-

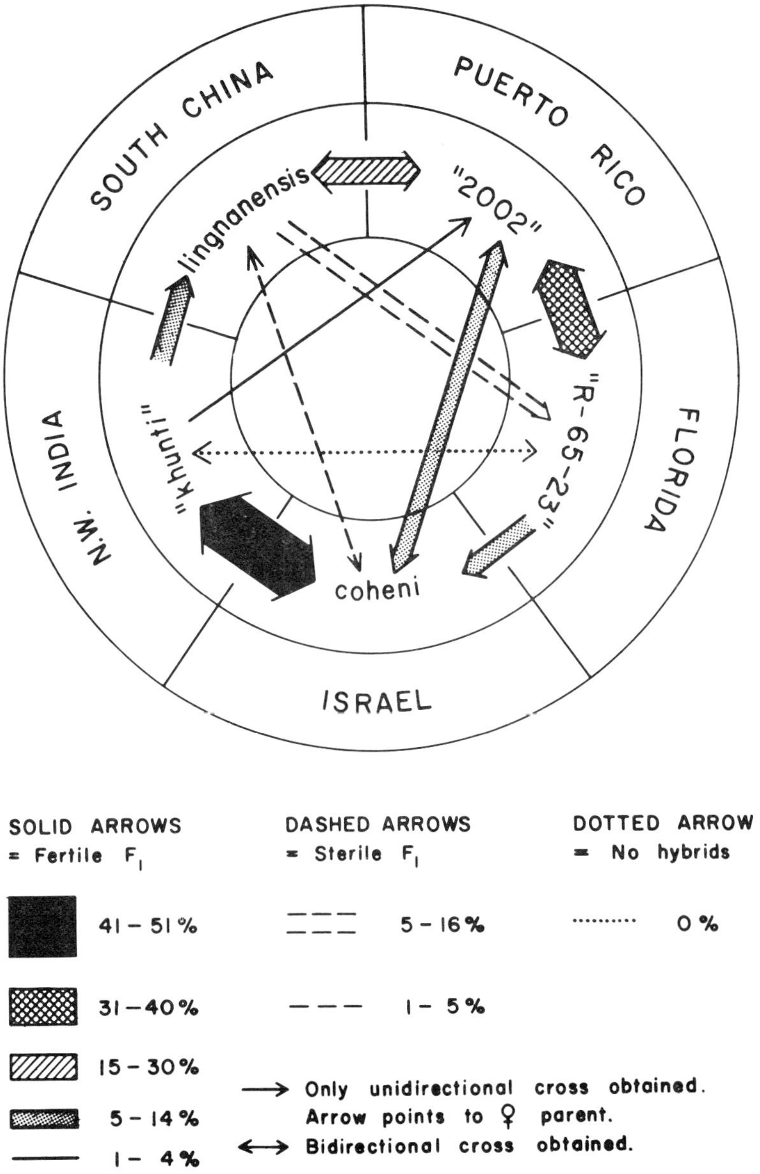


Figure 263. Crossing relations within the *lingnanensis–coheni* complex, showing percentages of $F_1$ progeny that are female and the nature of the $F_1$ hybrids. For example, *lingnanensis* crossed with "2002" produces between 15 and 30 percent fertile hybrid female progeny, whereas crossed with *coheni* it produces from 1 to 5 percent sterile hybrid female progeny. Normal intraspecific $F_1$ progeny production consists of 60 to 75 percent females.

(From Rao and DeBach, 1969a.)

lation to each other, but both are semispecies in relation to "2002". Gene flow between *lingnanensis* and "R-65-23" is potentially possible via "2002", but this does not occur in nature because their populations are allopatric.

For a further discussion of these studies see pp. 532, 539 and 542. Rao and DeBach (1969c) carried their hybridization studies one step further when they attempted to back-cross some of their hybrids with the parental species. Some of the interspecific hybrids between *Aphytis* species proved to be at least partially isolated reproductively from either one or both parent species.

Further studies of reproductive isolation among various other species of *Aphytis* were reported by Khasimuddin and DeBach (1976a, 1976b, 1976c) and by Gordh and DeBach (1978). In some of these subsequent tests, crushed homogamic males were used instead of chloroform anesthetization to induce receptive behavior in the test females. These studies demonstrated that similar complexes of sibling species and semispecies are rather common in some of the main species groups of *Aphytis*. Thus, in spite of the various limitations and complications discussed herein, the study of reproductive isolation appears to provide one of the best available biosystematic criteria for the separation of closely related species of *Aphytis*.

**Sexual Isolation.** When experimental hybridization indicates the existence of reproductive isolation between two populations of *Aphytis*, this outcome may be due to various isolating mechanisms, such as inviability of "alien" sperm in the spermathecae of heterogamic females, incompatibility of heterogamic sperm and egg cells (gametic isolation), inviability of hybrid zygotes or subsequent developmental stages, or sterility of hybrid adults. When crossing tests are performed without any artificial "mating inducers," their outcome may to a large extent also reflect the existence of sexual (ethological) isolating mechanisms. In other words, the populations under study may not interbreed simply because the males fail to inseminate heterogamic females. Such sexual isolation, ascribable to the existence of various behavioral barriers, is in itself a valuable–and relatively easy—biosystematic criterion.

Rao and DeBach (1969b) investigated sexual isolation among several members of the **lingnanensis** group. In their reciprocal multiple-choice experiments, males of a given species or strain were confined with homogamic and heterogamic females together for limited periods, after which the females were dissected and their spermathecae examined for the presence of sperm. The females were marked by being fed with colored honey, which showed readily through their abdominal integument. The results of such experiments reflect both the choice made by the males between homogamic and heterogamic females and the degree to which these females accept or repel them.

The data obtained from multiple-choice experiments were used to calculate the *coefficient of joint isolation*, developed by Levene (1949). To determine the degree of sexual isolation between two populations (designated "strain 1" and "strain 2"), first the coefficients of isolation ($K_{1,2}$ for the experiment using strain 1 males and $K_{2,1}$ for the reciprocal experiment using strain 2 males) are calculated as follows:

$$K_{1,2} = \frac{\log q_{1,1} - \log q_{1,2}}{\log q_{1,1} + \log q_{1,2}}$$

$$K_{2,1} = \frac{\log q_{2,2} - \log q_{2,1}}{\log q_{2,2} + \log q_{2,1}}$$

$$\text{where } q_{1,1} = \frac{\text{number of strain 1 females not inseminated when placed with strain 1 males and strain 2 females}}{\text{number of strain 1 females placed with strain 1 males}}$$

$$\text{and } q_{1,2} = \frac{\text{number of strain 2 females not inseminated when placed with strain 1 males and strain 1 females}}{\text{number of strain 2 females placed with strain 1 males}}$$

$q_{2,1}$ and $q_{2,2}$ are similarly defined when strain 2 males are used.

Then, the coefficient of joint isolation ($K_{1\text{and}2}$) is calculated from the following formula:

$$K_{1\text{and}2} = \frac{K_{1,2} + K_{2,1}}{2}$$

A value of 1.00 for the coefficient of joint isolation indicates complete sexual isolation between the populations under study, whereas zero indicates random mating and hence absence of sexual isolation. Intermediate values are indicative of partial sexual isolation.

Another index proposed by Levene (1949), the *coefficient of excess insemination* ($M_{1,2}$), is calculated from the following formula:

$$M_{1,2} = \frac{K_{1,2} - K_{2,1}}{2}$$

and is especially useful in cases where gene flow occurs in one direction only.

Rao and DeBach's (1969b) studies of sexual isolation in *Aphytis* largely corroborated their studies of reproductive isolation. Thus, certain closely related species such as *A. melinus* and *A. holoxanthus*, or *A. melinus* and *A. fisheri*, exhibited complete sexual isolation, whereas others such as *A. coheni* and *A.* "khunti", or *A. lingnanensis* and *A.* "2002", which were considered as semispecies in relation to each other on the basis of reproductive isolation studies, indeed proved to be only partially isolated sexually.

These studies of reproductive and sexual isolation led Rao and DeBach (1969b) to propose the following quantitative criteria for delimiting the species and semispecies categories, based on the coefficients of joint isolation:

| *Coefficient of Joint Isolation* | *Category* |
|---|---|
| 0–0.05 | Same species |
| 0.06–0.95 | Semispecies |
| 0.96–1.00 | Different species |

Combined with information on hybrid viability and hybrid fertility, these admittedly arbitrary limits provide a quantitative expression of phylogenetic relationship between closely related populations, and indicate the extent of divergence of a semispecies.

**Thelytoky.** Uniparental (thelytokous) species, in which each individual is reproductively isolated from all the others, are commonly recognized as an exception to the "biological" species concept (e.g., Mayr, 1963). In the case of *Aphytis* this is a rather large exception, comprising probably at least one fourth of all known species. Of the 90 species of *Aphytis* regarded as valid in the present revision, at least 15 are known, or presumed, to be uniparental, including such widely distributed and well-known species as *chilensis*, *diaspidis*, *mytilaspidis* and *chrysomphali*. In addition, males have not yet been recorded in 16 other species, quite a few of which may eventually also prove to be uniparental, but these species are at present known only from small samples and there is no definite information about their mode of reproduction.

When a uniparental taxon is homogeneous and clearly distinguishable morphologically from all its known congeners, and the morphological gap separating it from other taxa is as great as that commonly observed between biparental species in the same group, such a taxon can be regarded as a distinct species for all practical purposes (Mayr, 1963, 1969; Enghoff, 1976). We have followed this accepted taxonomic practice in our studies of *Aphytis*. However, when a uniparental taxon appears to be closely related to a biparental species, complex taxonomic and nomenclatural problems may arise.

Should a uniparental "form" be regarded as a distinct species in relation to a morphologically indistinguishable biparental "form" on the premise that the two are reproductively isolated from each other? Recent studies by Rössler and DeBach (1972a) have demonstrated that even this is not always true. They found that a uniparental form of *A. mytilaspidis*, differing in host preference from a sibling biparental form, is capable of mating with it and producing viable hybrids (see also p. 41). So these particular uniparental and biparental forms are only partially reproductively isolated and cannot be regarded as more than semispecies in relation to each other. In other cases they are completely isolated and no introgression occurs, as was found with *A. maculicornis* and *A. paramaculicornis*, sibling taxa recognized here as distinct species (see p. 387). Thus, although a different mode of reproduction may serve as an obvious distinguishing character and may be indicative of the existence of other biological differences, no simple rule can be set forth and it appears that each case should be decided on its own merits.

Finally, uniparental taxa that are morphologically indistinguishable from each other present perhaps the most perplexing taxonomic problems of all. It may be argued that "any group of reproductively isolated organisms which behaves like a biological species (that is, plays a significantly different ethological or ecological role in the habitat) should be considered to be a valid species" (DeBach, 1969b). Differences in biological, ecological or ethological characteristics, in host preference or habitat specificity etc., may all serve to indicate the possible distinctness of such taxa. Whatever their formal systematic status, such sibling uniparental entities ("ethological species") may possess distinct attributes as natural enemies and should therefore be recognized for possible utilization in biological control projects.

## ETHOLOGICAL CHARACTERISTICS

Behavioral characteristics have not been used much in the systematics of parasitic Hymenoptera. The degree of sexual isolation between closely related populations is, of course, a measure of ethological barriers preventing successful mating between the individuals of such populations. It is easily demonstrated with *Aphytis* that females usually reject heterogamic males. Direct observation of the mating preferences of males and females may therefore provide a simple means of distinguishing between species when live cultures are available.

Rao and DeBach (1969a) demonstrated the existence of a female sex pheromone that attracts conspecific males and mediates copulation in *Aphytis*, whereas Khasimuddin and DeBach (1975) showed that a male sex pheromone serves to render the female sexually receptive to conspecific males. The high species specificity of these pheromones, especially of the male pheromone, obviously makes this a good biosystematic tool.

Specific behavioral patterns may differ even in closely related species of *Aphytis*. Thus, for instance, a detailed study by Gordh and DeBach (1978) recently showed that courtship and mating behavior are both qualitatively and quantitatively different in such closely related species as *A. lingnanensis* and *A. melinus*. Parameters such as the position of the female's antennae or the angle of the male's body during the various phases of courtship and mating are consistently different in these species (see Figure 54, p. 43) and can be used to distinguish between them. This interesting approach should certainly be developed much further.

## BIOCHEMICAL CHARACTERISTICS

Both paper chromatography and slab-gel electrophoresis have been investigated in preliminary studies as diagnostic tools for the separation of *Aphytis* species. Very few—or even single—individuals of *Aphytis* are usually sufficient for meaningful results. For the most part there seem to be no differences between males and females of the same species. Preliminary tests have indicated that significant differences may indeed be detected between sibling species and even between certain semispecies of *Aphytis*. These studies will be continued.

## CONCLUSION

Several biosystematic criteria are now available as an adjunct to morphological characters for the separation of *Aphytis* species. Of these, sexual and reproductive isolation, host preference patterns and pupal pigmentation at present appear to be the most useful, while ethological and biochemical studies are promising approaches that deserve further research emphasis.

We probably know more, from a biosystematic viewpoint, about *Aphytis* than is known about any other genus of parasitic Hymenoptera. Yet this appears to be a mere beginning. The 90 species of *Aphytis* included in the present revision were obtained from about 90 different species of hosts (see Appendix 2). Thus, a ratio of roughly 1:1 between host and parasite species is indicated on a worldwide basis. With some 2000 described species of armored scale insects now recorded in the world (see Borkhsenius, 1966), and so many more that have not yet been described, it is evident that most of the *Aphytis* fauna of the world still awaits discovery.

# KEY TO SPECIES

The following key should serve mainly as a guide to the detailed descriptions and figures presented in the following sections, which ought to be consulted for more reliable identification.

Large series of both sexes are often required for identification of *Aphytis* species. However, females generally provide the best diagnostic characters, and the key is therefore based mainly on female characteristics. References to males have been included only when these differ markedly from the females and contribute significantly to species identification.

A separate key to species with hairy antennae in the male sex is presented on p. 690.

1. Both female and male antennae 4-segmented . . . . . . . . . . . . . . . . . . . . . . . . . 2
— Female antennae 5- or 6- segmented . . . . . . . . . . . . . . . . . . . . . . . . . . . . . . . . 3

2(1). Funicular segment small, trapezoidal, somewhat wider than long, shorter than the pedicel and not bearing longitudinal sensilla; antennal club less than $3\frac{1}{2}$ times as long as wide, about as long as the scape; coloration entirely yellow; wings hyaline . . . . . . . . . . . . . . . . . . . . . . . . . . . . . . . . . . *secundus*, p. 672
— Funicular segment distinctly longer than wide, considerably longer than the pedicel and bearing numerous longitudinal sensilla; antennal club enormously elongate, more than 5 times as long as wide, twice longer than the scape; propodeal crenulae absent; head, body and appendages extensively marked with fuscous; gastral tergites very coarsely sculptured; forewings with a broad, rather faint fuscous crossband . . . . . . . . . . . . . . . . . . . . . . . . *faurei*, p. 344

3(1). Female antennae 5-segmented . . . . . . . . . . . . . . . . . . . . . . . . . . . . . . . 4
— Female antennae 6-segmented . . . . . . . . . . . . . . . . . . . . . . . . . . . . . . . 10

4(3). Both funicular segments small, considerably shorter than the pedicel, wider than long and not bearing longitudinal sensilla. . . . . . . . . . . . . . . . . . . . . . . . . 5
— First funicular segment minute, the second segment large, considerably longer than the pedicel and longer than wide, bearing numerous longitudinal sensilla; antennal club enormously enlarged, $4\frac{1}{2}$ times as long as wide; general coloration yellow, with some fuscous markings; wings hyaline . . . . . . . *antennalis*, p. 357

5(4). Head, body and appendages extensively marked with blackish; mandibles well developed, maxillary palpi 2-segmented . . . . . . . . . . . . . . . . . . . . . . . . . 6
— General coloration yellow; mandibles minute or absent, maxillary palpi 1-segmented . . . **the funicularis group** . . . . . . . . . . . . . . . . . . . . . . . . . . . 7

6(5). Antennal scape slender; club long, tapering, about $3\frac{3}{4}$ times as long as wide, pale basally and black on distal half; mesonotum with 4 dark brown longitudinal stripes; forewings hyaline, with a small infuscated cloud below stigma . . . . . . . . *malayensis*, p. 294

— Antennal scape robust; club robust, less than 3 times as long as wide, blackish basally and pale on distal third; mesonotum largely fuscous, without distinct longitudinal dark stripes; forewings infuscated, with a subapical hyaline crossband . . . . . . . . *mandalayensis*, p. 296

7(5). Thoracic sterna infuscated or dusky . . . . . . . . 8

— Thoracic sterna immaculate, concolorous with body; propodeal crenulae minute, triangular, nonoverlapping; ovipositor short; male antennal scape bearing a ventral plate with 2 tuberculous sensilla . . . . . . . . *theae*, p. 656

8(7). Thoracic sterna faintly dusky in the female, lightly infuscated in the male . . . . 9

— Thoracic sterna strongly infuscated in both sexes; mesoscutum with numerous (15 or more) setae, each parapsis with 3; propodeal crenulae minute, triangular, nonoverlapping; ovipositor long; forewings broad, delta area densely setose; male antennal scape bearing a ventral plate with several tuberculous sensilla . . . . . . . . *funicularis*, p. 647

9(8). Propodeal crenulae minute, rounded, nonoverlapping; mesoscutum with few (8–12) setae, each parapsis with 1 seta; forewings narrow, delta area sparsely setose; male antennal scape bearing 2 distinct sensilla . . . . . . . . *gordoni*, p. 653

— Propodeal crenulae larger, rounded, overlapping; mesoscutum with numerous (usually 15 or more) setae, each parapsis with 3; forewings broad, delta area densely setose; male antennal scape not bearing any specialized sensilla. . . . . . . . *ulianovi*, p. 650

10(3). Antennal club separated from the penultimate segment only by a suture, the two together semifusiform and terminating in a prominent projecting sensillum; general coloration yellow, rather extensively marked with blackish or fuscous and faintly mottled with pale spots . . . . . . . . 11

— Antennal club separated from the penultimate segment by a distinct constriction . . . . . . . . 12

11(10). Forewings with 2 infumated crossbands bearing somewhat coarser setae, alternating with 2 hyaline areas bearing extremely fine, transparent setae . . . . . . . . *dealbatus*, p. 290

— Forewings without hyaline areas bearing transparent setae, nearly uniformly setose but clouded below the stigma and below the junction of the submarginal and marginal veins, the clouded areas bearing somewhat coarser setae; in the male, apex of middle tibiae swollen and contrastingly marked with black . . . . . . . . *melanostictus*, p. 291

12(10). Female forewings mottled or banded, with a contrasting pattern of areas bearing coarse, dark setae and areas bearing fine, hyaline setae (in the males of most species, the forewings are hyaline or infumated, uniformly setose or nearly so) . . . **the vittatus group** . . . . . . . . . . . . . . . . . . . . . . . . . . . . . . . . 13

— Female forewings hyaline or clouded but rather uniformly setose . . . . . . . . . 30

13(12). Female forewings mottled, with several patches or clouds of dark setae . . . . . 14

— Female forewings largely infuscated or with a fuscous crossband . . . . . . . . 27

14(13). Cerci bearing 3 long setae and 1 short seta in the female, 2 long setae and 1 short seta in the male . . . . . . . . . . . . . . . . . . . . . . . . . . . . . . . . . . . . . . . . 15

— Cerci bearing 2 long setae and 1 short seta in both sexes . . . . . . . . . . . . . . . . . 20

15(14). Female with 4 distinct, longitudinal dark stripes on mesonotum; forewings extensively mottled, with a pattern of anastomosing areas bearing coarse, dark setae; male antennae 5-segmented . . . . . . . . . . . . . . . . . . *vittatus*, p. 236

— Female without 4 complete longitudinal stripes on mesonotum; forewings with several well-defined patches of coarse, dark setae . . . . . . . . . . . . . . . . . . 16

Female forewings largely hyaline, with a patch of dark setae in the delta and only 2 patches distad of speculum, below the stigma; dorsum of gaster with several dark crossbands; male antennae 5-segmented . . . . . *peculiaris*, p. 238

— Female forewings with a patch of dark setae in the delta and 4 or 5 patches distad of speculum; dorsum of gaster with 2 broad crossbands or with a large central blotch; antennae of known males 6-segmented, the club and third funicular segment bearing long setae . . . . . . . . . . . . . . . . . . . . . . . . . . . . . . 17

17(16). Female with a pair of submedian dark stripes on both mesoscutum and scutellum; gaster with 2 dark crossbands; patches on forewings relatively small; setae along anterior margin of marginal vein subequal in length; ovipositor less than twice length of middle tibia; male with 3 rows of sensilla on antennal scape, and gaster dark basally and distally, pale centrally . . . . *maculatipennis*, p. 241

— Female with submedian dark stripes present only on scutellum; patches on forewings more extensive; setae along anterior margin of marginal vein de-decreasing in length toward apex of vein; ovipositor longer, at least twice length of middle tibia; known males with 1 row of sensilla on antennal scape, and gaster entirely dark brown . . . . . . . . . . . . . . . . . . . . . . . . . . . . . . . 18

18(17). Propodeal crenulae rounded, overlapping; gaster of female dark brown basally and distally, pale centrally and on sides; forewings with a complete v-shaped crossband below stigma . . . . . . . . . . . . . . . . . . . . . . . . . . *costalimai*, p. 244

— Propodeal crenulae elongate, nonoverlapping; gaster of female with a continuous dark blotch centrally; arms of v-shaped crossband widely separated at apex . . . . . . . . . . . . . . . . . . . . . . . . . . . . . . . . . . . . . . . . . . . . . . . 19

19(18). Propodeal crenulae of female distinct, tapering; mesoscutum and scutellum lightly suffused with brownish; abdominal tergites III–VI with short transverse dark stripes on sides; mesoscutum with 20–35 setae . . . . . . . *acutaspidis*, p. 248

— Propodeal crenulae of female very indistinct, obscured by coarse reticulation of second abdominal tergite, more roundly elongate; mesoscutum and scutellum more extensively suffused with brownish; sides of abdominal tergites III–VI pale; mesoscutum with 40–55 setae . . . . . . . . . . . . . . . . . . *perplexus*, p. 249

20(14). Female forewings distad of speculum with 3 extensive patches and a subapical crossband of dark setae . . . . . . . . . . . . . . . . . . . . . . . . . . . . . . . . . . . . . 21

— Female forewings distad of speculum with 8 smaller patches and an apical or subapical cloud of dark setae . . . . . . . . . . . . . . . . . . . . . . . . . . . . . . . . . 23

21(20). Female extensively marked with blackish, gastral tergites with complete crossbands and a continuous central blotch; subapical crossband on forewings narrow, well defined . . . . . . . . . . . . . . . . . . . . . . . . . . . . . . . . . . . . . . . . 22

— Female pale, gastral tergites with fuscous lateral spots; subapical crossband on forewings broad, less distinct; male antennae 6-segmented, the first 2 funicular segments minute, ringlike, the scape bearing a single sensillum . . . . . . . . . . . . . . . . . . . . . . . . . . . . . . . . . . . . . . . . . . . . . . . . . . . . *fabresi*, p. 259

22(21). Mesonotal sclerites of female largely whitish, margins and sutures lined with black, bases of setae surrounded by blackish spots; metanotum white, propodeum with a black anterocentral triangle followed by a white v-shaped stripe, outer ovipositor plates black laterad of the longitudinal ridge, the latter extending almost to end of plate; male antennae 6-segmented, the first 2 funicular segments minute, ringlike, the scape not bearing a sensillum . . . . . . . . . . . . . . . . . . . . . . . . . . . . . . . . . . . . . . . . . . . . . . . . . . . . . . . . . . . . . . . . . *cochereaui*, p. 251

— Thorax and propodeum of female more uniformly suffused with brownish, without a contrasting pattern of black and white areas; outer ovipositor plates immaculate, the longitudinal ridge fading out at about one half length of plate . . . . . . . . . . . . . . . . . . . . . . . . . . . . . . . . . . . . . . . *noumeaensis*, p. 256

23(20). Female antennae short, stout, the club less than $2\frac{1}{2}$ times as long as wide; forewings less than 3 times as long as wide; propodeal crenulae large, elongate . . . . . . . . . . . . . . . . . . . . . . . . . . . . . . . . . . . . . . . . . . . . . . . . . . 24

— Female antennae more elongate, the club nearly 3 times as long as wide; forewings narrow, fully 3 times as long as wide, with a v-shaped subapical cloud; abdominal tergites III–VII with 5 complete blackish crossbands . . . . . . . . . . . . . . . . . . . . . . . . . . . . . . . . . . . . . . . . . . . . . . . . *argenticorpus*, p. 257

24(23). Eyes of female coarsely setose; genal sutures marked with blackish; abdominal tergites III–VI with 4 complete blackish crossbands . . . . . . . . . . . . . . . . . . . 25

— Eyes of female finely setose; genal sutures immaculate; crossbands on abdominal tergites largely obliterated centrally . . . . . . . . . . . . . . . . . . . . . . . . . . . 26

25(24). Female forewings with a relatively small, distinct apical cloud; abdominal crossbands broad; seventh abdominal tergite with short, transverse lateral stripes and faintly infuscated across center; mesoscutum with 22 or fewer setae; propodeal crenulae numerous (up to 11 + 11 in large specimens); male antennae essentially 4-segmented, the first 2 funicular segments reduced to indistinct rudiments, the scape bearing a tuberculous sensillum . . . . . . . . . . . . . . . . . . . . . . . . . . . . . . . . . . . . . . . . . . . . . . . . . . . . . . . . *capillatus*, p. 262

— Apical cloud on forewings more extensive, merging with the subapical clouds; abdominal crossbands narrow; seventh tergite pale between the lateral stripes; mesoscutum with 24–32 setae; propodeum with fewer (7 + 7) crenulae . . . . . . . . . . . . . . . . . . . . . . . . . . . . . . . . . . . . . . . . . . . . . . . . . . . . . . . . *nigripes*, p. 268

26(40). Relatively large, more than 1 mm long; propodeum nearly 3 times as long as the metanotum, more than $\frac{4}{5}$ length of scutellum, with a prominent median salient; mesoscutum with 13–16 setae, each parapsis with 3; forewings about $2\frac{1}{2}$ times as long as wide, the marginal vein bearing 14–17 setae along anterior margin . . . . . . . . . . . . . . . . . . . . . . . . . . . . . . . . . . . . . . . . . . . . . . . . *australiensis*, p. 272

— Smaller, less than 1 mm long; propodeum shorter, about twice as long as the metanotum, less than $\frac{3}{5}$ length of scutellum, the median salient less prominent; mesoscutum with 8–9 setae, each parapsis with 1 seta; forewings broad, about $2\frac{1}{5}$ times as long as wide, the marginal vein bearing 9 setae along anterior margin . . . . . . . . . . . . . . . . . . . . . . . . . . . . . . . . . . . . . . . . . *ciliatus*, p. 270

27(13). Forewings fuscous at base and with a fuscous crossband below stigma; propodeum shorter than the scutellum . . . . . . . . . . . . . . . . . . . . . . . . . . . . . . 28

— Forewings largely fuscous except apex; propodeum distinctly longer than the scutellum; endophragma and syntergum elongate; eyes finely setose; antennae short and thick, the club only about $2\frac{1}{4}$ times as long as wide; mesoscutum with 10–11 setae; head, body and appendages extensively marked with fuscous . . . . . . . . . . . . . . . . . . . . . . . . . . . . . . . . . . *angustus*, p. 282

28(27). Thorax and abdomen extensively fuscous or blackish; female cerci bearing 2 long setae and 1 short seta; eyes finely setose; antennae short, the club less than $2\frac{1}{2}$ times as long as wide, pale apically; each parapsis with 2 setae, axillae asetose; ovipositor short, not more than $1\frac{1}{2}$ times length of middle tibia . . . . 29

Thorax largely pale, abdomen dark brown centrally, pale on sides and distally; female cerci bearing 3 long setae and 1 short seta; eyes coarsely setose; antennae slender, the club $3\frac{1}{2}$ times as long as wide, black on distal half; mesoscutum with 10–19 setae, each parapsis with 3, each axilla with 1 seta; propodeal crenulae small, nearly rectangular; ovipositor nearly twice length of middle tibia . . . . . . . . . . . . . . . . . . . . . . . . . . . . . . . . . . . . . . . . . *anomalus*, p. 275

29(28). Mesonotal sclerites fuscous, with blackish spots around bases of setae; antennal pedicel and third funicular segment blackish; mesoscutum with 20–22 setae; ovipositor very short, not exceeding $1\frac{1}{5}$ times length of middle tibia, setae along anterior margin of marginal vein subequal in length . . . . . . . . . . . . . *haywardi*, p. 277

— Mesonotal sclerites more deeply blackish, without distinct spots around bases of setae; antennal pedicel and third funicular segment pale; mesoscutum with 10–14 setae; ovipositor somewhat longer, up to $1\frac{1}{2}$ times length of middle tibia; setae along anterior margin of marginal vein decreasing in length toward apex of vein; male forewings banded as in the female, the genitalia lacking digital sclerites . . . . . . . . . . . . . . . . . . . . . . . . . . . . . . . . . . . *obscurus*, p. 279

30(12). Dorsum of gaster conspicuously banded, with 4 or 5 black crossbands; antennae short and stout, the first funicular segment triangular, the club less than $2\frac{1}{2}$ times as long as wide; propodeum short, only twice as long as the metanotum, with large, nonoverlapping crenulae . . . . . . . . . . . . . . . . . . . . . . . . . . . . . . 31

— Without the above combination of characters; coloration yellow or infuscated but without conspicuous black crossbands on dorsum of gaster . . . . . . . . . . 32

31(30). Extensively marked with black, with 5 complete crossbands on abdominal tergites III–VII; eyes coarsely setose; mesoscutum with 22–23 setae; propodeal crenulae elongate; male antennae essentially 4-segmented, the first 2 funicular segments reduced to minute but distinct annelli . . . . . . . *hyalinipennis*, p. 285

— Paler, with 4 partly obliterated crossbands on abdominal tergites III–VI; eyes finely setose; mesoscutum with 15–18 setae; propodeal crenulae more rounded . . . . . . . . . . . . . . . . . . . . . . . . . . . . . . . . . . *wallumbillae*, p. 287

32(30). Both female and male antennae clearly 6-segmented . . . . . . . . . . . . . . . . . . . 36

— Female antennae 6-segmented, with an elongate club (usually more than $3\frac{1}{2}$ times as long as wide); male antennae 4- or 5-segmented, with the first 2 funicular segments greatly reduced or absent and an enormously enlarged club bearing numerous longitudinal sensilla; setae on head, thorax and abdomen coarse, dark . . . **the chilensis group** . . . . . . . . . . . . . . . . . . . . . . . . . . . . . . . . 33

33(32). Cerci bearing 2 long setae and 1 short seta in both sexes; propodeum relatively long, the crenulae rounded, overlapping; ovipositor relatively long; male antennae essentially 4-segmented, the first 2 funicular segments reduced to indistinct annelli . . . . . . . . . . . . . . . . . . . . . . . . . . . . . . . . . . . . . . . . . 34

— Cerci bearing 3 long setae and 1 short seta in the female, 2 long setae and 1 short seta in the male; propodeum short, emarginate, not exceeding $\frac{1}{2}$ length of scutellum, the crenulae elongate, nonoverlapping; mandibles well developed, bidentate, maxillary palpi 2-segmented; female antennae with the first 2 funicular segments subequal; male antennae 5-segmented, with a minute but distinct annellus . . . . . . . . . . . . . . . . . . . . . . . . . . . . . *cercinus*, p. 346

34(33). Extensively marked with fuscous; eyes finely setose; mandibles reduced, short and broad, maxillary palpi 1-segmented; gastral tergites very coarsely sculptured; female antennae with the first 2 funicular segments subequal; ovipositor very long, considerably more than twice length of middle tibia; in the male, antennal scape widened, bearing 2–3 dentate sensilla, gastral sternites densely stippled, genitalia with long, slender digital sclerites lacking an apical claw . . . . . . . . . . . . . . . . . . . . . . . . . . . . . . . . . . . . . . . . . . . . . . . . *merceti*, p. 341

— Paler; eyes coarsely setose; mandibles well developed, maxillary palpi 2-segmented; gastral sculpture delicate; female antennae with the first funicular segment triangular, considerably smaller than the second; ovipositor shorter, less than twice length of middle tibia; in the male, antennal scape slender, not bearing specialized sensilla, gastral sternites not stippled, genitalia with short, stout digital sclerites bearing an apical claw . . . . . . . . . . . . . . . . . . . . 35

35(34). Head, thorax and propodeum rather extensively marked with fuscous; abdominal tergites III–VII with short, transverse fuscous stripes on sides; thoracic sterna strongly infuscated, the mesosternal furca black; antennal club of female fuscous, the apical part distinctly paler; males very rare . . . . . . . . . . . . . . . . . . . . . . . . . . . . . . . . . . . . . . . . . . . . . . . . . . . . . *chilensis*, p. 349

— Head, thorax and propodeum paler; abdominal tergites III–VII immaculate or with faint dusky spots on sides (transverse stripes absent but dark setae produce a similar effect); thoracic sterna faintly dusky; antennal club of female uniformly, faintly dusky, the apical part not contrastingly pale; males abundant . . . . . . . . . . . . . . . . . . . . . . . . . . . . . . . . . . . . . . . . *columbi*, p. 354

36(32). Head without distinct black bars and margins . . . . . . . . . . . . . . . . . . . . . . . 47

— A narrow, well-defined transverse black bar on occiput, on each side of foramen; genal sutures distinctly marked with black; general coloration more or less dusky, with fuscous markings; forewings distinctly clouded . . . **the proclia group** . . . . . . . . . . . . . . . . . . . . . . . . . . . . . . . . . . . . . . . . . . . . . . . 37

37(36). Female with dorsum of gaster pale centrally, with short, transverse fuscous stripes on sides, at most with a complete crossband on third and seventh abdominal tergites . . . . . . . . . . . . . . . . . . . . . . . . . . . . . . . . . . . . . . . . 38

— Female with dorsum of gaster rather uniformly fuscous or dusky . . . . . . . . . 45

38(37). Third and seventh abdominal tergites with a complete fuscous crossband . . . 39

— Either one or both crossbands obliterated centrally, represented by short lateral stripes . . . . . . . . . . . . . . . . . . . . . . . . . . . . . . . . . . . . . . . . . . . . . . 43

39(38). Basal part of antennal club paler than the funicle, apex of club blackish; propodeal crenulae elongate . . . . . . . . . . . . . . . . . . . . . . . . . . . . . . . . . . . . 40

— Antennal pedicel, funicular segments and base of club uniformly, rather strongly infuscated, apex of club blackish; propodeal crenulae distinctly rounded . . . . . . . . . . . . . . . . . . . . . . . . . . . . . . . . . . . . . . . . . . . . . . . . . 42

40(39). Funicular segments of female strongly infuscated, pedicel and basal two thirds of club pallid in distinct contrast; ovipositor relatively long, $1\frac{2}{3}$ times to nearly twice length of middle tibia, with a pair of conspicuous dark spots at base of shaft; forewings narrow, usually about 3 times as long as wide; antennal club of male distinctly paler than the funicle . . . . . . . . . . . . . . . . . . . . . . . 41

— Pedicel and funicular segments of female fuscous, basal part of club somewhat paler; ovipositor usually about $1\frac{1}{2}$ times length of middle tibia; forewings usually about $2\frac{2}{3}$ times as long as wide; mesoscutum with 9–15 setae; antennal pedicel, funicle and club of male uniformly infuscated . . . . . . . . *proclia*, p. 377

41(40). Uniparental; antennae slender, club of female usually more than 3 times as long as wide; in the rare males, the antennal club is uniformly pale, more than $3\frac{1}{2}$ times as long as wide . . . . . . . . . . . . . . . . . . . . . . . . . . . *maculicornis*, p. 383

— Biparental; antennae somewhat less slender, club of female usually about 3 times as long as wide; in the abundant males, the antennal club is uniformly dusky, not more than $3\frac{1}{2}$ times as long as wide . . . . . . *paramaculicornis*, p. 387

42(39). Uniparental; antennal club about 3 times as long as wide; mesoscutum usually with 10–13 setae; seventh abdominal tergite bearing 3–5 fine setae in a transverse row between the lateral setiferous areas, eighth tergite with 6–10 setae in a transverse row between the spiracles; propodeum short, not more than 3 times as long as the metanotum, not exceeding $\frac{2}{3}$ length of scutellum; relative length of ovipositor inversely correlated with size, from $1\frac{1}{3}$ times to twice length of middle tibia . . . . . . . . . . . . . . . . . . . . . . . . . . . . . . . . . . . . *vandenboschi*, p. 400

— Biparental; antennal club short, less than 3 times as long as wide; mesoscutum usually with 15 setae; seventh abdominal tergite bearing 2 fine submedian setae, eighth tergite with 4 setae in a transverse row between the spiracles; propodeum longer, 3–4 times as long as the metanotum, $\frac{3}{4}$ to $\frac{4}{5}$ length of scutellum; ovipositor about $1\frac{3}{5}$ to $1\frac{2}{3}$ length of middle tibia . . . *philippinensis*, p. 397

43(38). Uniparental; third abdominal tergite with a complete fuscous crossband; antennal club rather thick, not tapering to a pointed apex; propodeal crenulae elongate . . . . . . . . . . . . . . . . . . . . . . . . . . . . . . . . . . . . . . . . . . . . . . . . 44

— Biparental; third abdominal tergite with transverse fuscous stripes on sides, pale centrally; antennal club strongly tapering to a pointed apex; propodeal crenulae rounded, closely appressed, conspicuously blackish; in the male, abdominal tergites with more or less complete, rather faintly fuscous crossbands; genitalia relatively short, with short digital sclerites . . . . . . . . . *confusus*, p. 402

44(43). Genal sutures heavily sclerotized, strongly infuscated from oral margin to about $\frac{3}{4}$ distance to eye; antennal club more than $2\frac{1}{2}$ times as long as wide; pedicel, funicle and basal part of club uniformly, strongly infuscated, apical third of club blackish; propodeal crenulae elongate, faintly infuscated; ovipositor not more than $1\frac{3}{4}$ times length of middle tibia; forewings relatively narrow . . . . . . . . . . . . . . . . . . . . . . . . . . . . . . . . . . . . . . . . *hispanicus*, p. 390

— Genal sutures less heavily sclerotized, rather faintly infuscated; antennae shorter, thicker, the club only about $2\frac{1}{3}$ times as long as wide; pedicel, funicle and club paler, tip of club with a conspicuous black spot; propodeal crenulae fewer, somewhat wider, distinctly blackish; ovipositor longer, up to twice length of middle tibia; forewings somewhat broader . . . . . . . *comperei*, p. 394

45(37). Gastral tergites faintly infuscated; antennal pedicel, funicle and club uniformly dusky . . . . . . . . . . . . . . . . . . . . . . . . . . . . . . . . . . . . . . . . . . 46

— Gastral tergites strongly infuscated, with distinct transverse crossbands and a continuous central blotch; antennal pedicel, funicle and club fuscous, apex of club usually blackish; mesonotal sclerites extensively and strongly infuscated, mesoscutum with numerous (usually 15 or more) setae; propodeal crenulae rounded; relative length of ovipositor inversely correlated with size, $1\frac{1}{3}$ to $1\frac{3}{4}$ times length of middle tibia; males usually rare, with complete fuscous crossbands on gastral tergites, antennal flagellum uniformly fuscous . . . . . . . . . . . . . . . . . . . . . . . . . . . . . . . . . . . . . . . . . . . . . . . . . . . . . *diaspidis*, p. 405

46(45). Propodeum long, more than 6 times as long as the metanotum and at least as long as the scutellum, the crenulae minute, few, sometimes entirely absent; eighth abdominal tergite with 4 setae in a transverse row between the spiracles; ovipositor short, not more than $1\frac{1}{3}$ times length of middle tibia, the sheaths about $\frac{1}{3}$ length of middle tibia; forewings at least $2\frac{1}{2}$ times as long as wide; in the male, antennal scape widened, genitalia with short digital sclerites . . . . . . . . . . . . . . . . . . . . . . . . . . . . . . . . . . . . . . . . . . . . . . . . . . *pinnaspidis*, p. 409

— Propodeum short, not more than 3 times as long as the metanotum or $\frac{3}{4}$ length of scutellum, the crenulae rounded, rather wide and low, nearly forming a continuous ledge; eighth abdominal tergite with 6–8 setae in a transverse row between the spiracles; ovipositor long, about twice length of middle tibia, the sheaths about $\frac{1}{2}$ length of middle tibia; forewings broader, clearly less than $2\frac{1}{2}$ times as long as wide . . . . . . . . . . . . . . . . . . . . . . . . . *testaceus*, p. 411

47(36). Posterior margin of propodeum arcuate or straight, not deeply emarginate . . 48

— Propodeum short, deeply emarginate and distinctly bilobed; crenulae elongate, irregular, nonoverlapping; general coloration pale, faintly suffused with dusky; propodeal lobes, crenulae and thoracic sterna conspicuously fuscous; forewings nearly hyaline . . . . . . . . . . . . . . . . . . . . . . . . . . . . . . . . . . . *desantisi*, p. 424

48(47). General coloration distinctly and rather uniformly grayish, thoracic sterna blackish; forewings hyaline or faintly infumated . . . . . . . . . . . . . . . . . . . . . . 49

— General coloration yellow or pale . . . . . . . . . . . . . . . . . . . . . . . . . . . . . . . . 52

49(48). Propodeum with well-developed crenulae; pronotal plates triangular; setae on head and thorax dark . . . . . . . . . . . . . . . . . . . . . . . . . . . . . . . . . . . . . . . . . 50

— Propodeum with a continuous ledge on posterior margin, the crenulae absent; pronotal plates broadly joined; setae on head and thorax slender, pale; female antennae stout; ovipositor relatively long; forewings narrow; male considerably darker, the antennal scape widened, the club short; genitalia robust, about as long as the middle tibia, with short digital sclerites . . . . . *acrenulatus*, 419

50(49). Propodeal crenulae large, rounded, overlapping; ovipositor short; cerci about equidistant from posterior spiracles and tip of gaster; axillae bearing one seta; male genitalia with well-developed apodemes and digital sclerites . . . . . . . . . 51
— Propodeal crenulae small, rather irregular, usually nonoverlapping; ovipositor long, about twice length of middle tibia; cerci situated near posterior spiracles; axillae asetose; male genitalia lacking apodemes or digital sclerites, the aedeagus and phallobase forming a single unit . . . . . . . . . . . . . . . . . . . *tucumani*, p. 416

51(50). Antennal club $2\frac{1}{2}$–3 times as long as wide; ovipositor very short, not exceeding $1\frac{1}{3}$ times length of middle tibia, the sheaths less than $\frac{1}{3}$ length of middle tibia; in the male, antennal scape not widened, not bearing specialized sensilla . . . . . . . . . . . . . . . . . . . . . . . . . . . . . . . . . . . . . . . . . . . . . . *griseus*, p. 414
— Antennae relatively small, the club less than $2\frac{1}{2}$ times as long as wide; ovipositor somewhat longer, $1\frac{2}{5}$ times length of middle tibia, the sheaths more than $\frac{1}{3}$ length of middle tibia; in the male, antennal scape somewhat widened and bearing specialized sensilla . . . . . . . . . . . . . . . . . . . . . . . *amazonensis*, p. 422

52(48). Setae along anterior margin of marginal vein subequal in length; genal sutures moderately curved, divergent . . . . . . . . . . . . . . . . . . . . . . . . . . . . . . . . . 53
— Setae along anterior margin of marginal vein decreasing considerably in length toward apex of vein; genal sutures curved at an acute angle, then running vertically parallel to each other; mesoscutum with 10 coarse, dark setae, axillae asetose; propodeal crenulae moderately large, rounded and overlapping; in the male, antennal scape bearing minute sensilla; genitalia truncate basally, without distinct apodemes; middle tibiae tipped with black . . . . . . . . . . . . . . . . . . . . . . . . . . . . . . . . . . . . . . . . . . . . . . . . . . . . . . . . . . . . . . . . . . *salvadorensis*, p. 696

53(52). Female syntergum triangular or distinctly caudate, but the cauda not unusually long . . . . . . . . . . . . . . . . . . . . . . . . . . . . . . . . . . . . . . . . . . . . . . . . . . . . 54
— Female syntergum produced into a very long, densely setose cauda; antennal club very long; mesoscutum with numerous dark setae; propodeal crenulae elongate, overlapping; ovipositor very long, over twice length of middle tibia; in the male, cauda absent; antennal scape thick, bearing an oval plate with 2 minute sensilla, the club and third funicular segment bearing long setae . . . . . . . . . . . . . . . . . . . . . . . . . . . . . . . . . . . . . . . . . . . . . . . . *longicaudus*, p. 681

54(53). Propodeal crenulae large and overlapping . . . **the lingnanensis group** . . . . . . 55
— Propodeal crenulae either large but nonoverlapping, or small . . . . . . . . . . . 63

55(54). Thoracic sterna dusky . . . . . . . . . . . . . . . . . . . . . . . . . . . . . . . . . . . . . . 56
— Thoracic sterna immaculate . . . . . . . . . . . . . . . . . . . . . . . . . . . . . . . . . . 59

56(55). Propodeal crenulae smooth, concolorous with body and strongly overlapping; antennal scape of the male not bearing specialized sensilla, the club with a ventral sensory area bearing short setae; fully developed (green-eyed) pupae distinctly marked with black on both thoracic and abdominal sterna . . . . . . . 57

— Propodeal crenulae striated longitudinally, conspicuously blackish, closely appressed and only slightly overlapping; antennal scape of male bearing an oval plate with 2 minute sensilla, the club not bearing a specialized sensory area; pupae entirely yellow, with dusky wing-pads . . . . . . . . . . . . *margaretae*, p. 545

57(56). Propodeal crenulae elongate; in the female, antennal club about 3 times as long as wide, ovipositor sheaths up to $\frac{1}{2}$ length of middle tibia; in the male, posterior abdominal sternites densely, conspicuously stippled . . . . . . . . . . . 58

— Propodeal crenulae somewhat smaller, less elongate and more oblique; in the female, antennal club somewhat shorter, less than 3 times as long as wide; ovipositor sheaths somewhat shorter, about $\frac{2}{5}$ length of middle tibia; in the male, posterior abdominal sternites very weakly stippled . . . . . . . . *africanus*, p. 542

58(57). Body setae relatively slender and pale, those on sides of gaster usually invisible or barely discernible at × 120 magnification (under a stereoscopic microscope, on a white background); usually 10–12 setae on mesoscutum, 30–50 setae in delta area of forewings; thoracic sterna rather faintly dusky, mesosternum often with only the longitudinal stem of the furca fuscous; gaster immaculate . . . . . . . . . . . . . . . . . . . . . . . . . . . . . . . . . . . . . . . . . . . *lingnanensis*, p. 533

— Body setae somewhat coarser and darker, those on sides of gaster usually distinct at × 120 magnification; usually 12–14 setae on mesoscutum, 50–70 setae in delta area of forewings; thoracic sterna more strongly infuscated, including a triangle between the arms of the mesosternal furca; basal segments of gaster faintly dusky . . . . . . . . . . . . . . . . . . . . . . . . . . . . . . . . . *coheni*, p. 539

59(55). Propodeal crenulae elongate; female antennal club long, distinctly more than 3 times as long as wide; ovipositor at least $1\frac{2}{3}$ times length of middle tibia, the sheaths up to $\frac{1}{2}$ length of middle tibia; in the male, posterior abdominal sternites densely, conspicuously stippled . . . . . . . . . . . . . . . . . . . . . . . . . . . . . . . . . . . 60

— Propodeal crenulae wider, less elongate; female antennal club less than 3 times as long as wide; ovipositor short, $1\frac{1}{3}$ to $1\frac{1}{2}$ times length of middle tibia, the sheaths not more than $\frac{2}{5}$ length of middle tibia; in the male, posterior abdominal sternites very weakly stippled . . . . . . . . . . . . . . . . . . . . . . . *equatorialis*, p. 564

60(59). Female antennae rather strongly infumated, the club robust, usually not more than $3\frac{1}{2}$ times as long as wide; thoracic setae dark and coarse; ovipositor relatively long, $1\frac{4}{5}$ times to twice length of middle tibia; forewings broad, not more than $2\frac{1}{2}$ times as long as wide, with about 30–45 setae in the delta; fully developed (green-eyed) pupae with black or blackish pigmentation on both thoracic and abdominal sterna . . . . . . . . . . . . . . . . . . . . . . . . . . . . . . . . . . 61

— Female antennae faintly infumated, long and slender, the club $3\frac{2}{5}$ to fully 4 times as long as wide; thoracic setae paler and more slender; ovipositor somewhat shorter; forewings somewhat narrower, with about 40–60 setae in the delta; pupae pigmented only on thoracic sterna, or entirely yellow . . . . . . . . 62

61(60). Fully developed (green-eyed) pupae with well-defined black pigmentation on both thoracic and abdominal sterna; male genitalia with the digital sclerites usually a little over $\frac{1}{4}$ the combined length of aedeagus and apodemes . . . . . . . . . . . . . . . . . . . . . . . . . . . . . . . . . . . . . . . . . . *holoxanthus*, p. 548
— Pupal abdominal sterna with paler, more diffuse, less clearly defined pigmentation; male genitalia with somewhat shorter digital sclerites, usually a little less than $\frac{1}{4}$ the combined length of aedeagus and apodemes; thoracic setae somewhat paler . . . . . . . . . . . . . . . . . . . . . . . . . . . . . *yasumatsui*, p. 558

62(60). Hind wings narrow, width of disk $1\frac{1}{5}$ to $1\frac{1}{2}$ times length of longest hairs in marginal fringe; fully developed (green-eyed) pupae with black pigmentation on thoracic sterna, the abdominal sterna yellow . . . . . . . . . . . . *melinus*, p. 552
— Hind wings somewhat broader, width of disk $1\frac{3}{5}$ to $1\frac{4}{5}$ times length of longest hairs in marginal fringe; pupae entirely yellow . . . . . . . . . . . . . . . *fisheri*, p. 561

63(54). Propodeal crenulae large and conspicuous, rounded, nonoverlapping, forming a continuous row; thoracic sterna dusky; female antennae short and robust . . 64
— Propodeal crenulae small . . . . . . . . . . . . . . . . . . . . . . . . . . . . . . . . . . . . . . . 65

64(63). Antennal club of female about $2\frac{2}{5}$ to $2\frac{3}{4}$ times as long as wide, the third funicular segment bearing an unusually long seta, longer than width of segment; forewings narrow, $2\frac{1}{2}$ to 3 times as long as wide; in the male, antennal club and third funicular segment elongate, bearing long setae . . . . . . . . . *pilosus*, p. 684
— Antennal club of female shorter, about twice as long as wide, the third funicular segment not bearing an unusually long seta; forewings broad, $2\frac{1}{3}$ times as long as wide; male antennae short and thick, not bearing long setae. . . . . . . . . . . . . . . . . . . . . . . . . . . . . . . . . . . . . . . . . . . . . . . . . . . . *mimosae*, p. 694

65(63). Propodeal crenulae rather low and wide, forming a wavy ledge on posterior margin of propodeum . . . . . . . . . . . . . . . . . . . . . . . . . . . . . . . . . . . . . . . 66
— Propodeal crenulae small or minute but usually discrete, not forming a wavy ledge on posterior margin of propodeum . . . . . . . . . . . . . . . . . . . . . . . . . . 67

66(65). Thoracic sterna dusky; mesoscutum with 9–11 setae; antennal club $2\frac{2}{3}$ to 3 times as long as wide; ovipositor about twice length of middle tibia . . . *taylori*, p. 494
— Thoracic sterna immaculate; mesoscutum with 15–16 setae; antennal club thicker, not more than $2\frac{1}{2}$ times as long as wide; ovipositor a little less than twice length of middle tibia . . . . . . . . . . . . . . . . . . . . . . . . . . *bedfordi*, p. 692

67(65). Mesoscutum with about 15–20 setae . . . . . . . . . . . . . . . . . . . . . . . . . . . . 68
— Mesoscutum with 8–16 setae (8–12 in most species, sometimes fewer than 8, rarely as many as 15 or 16) . . . . . . . . . . . . . . . . . . . . . . . . . . . . . . . . . . . 70

68(67). Antennal club of female long, more than 3 times as long as wide; propodeal crenulae nonoverlapping; thoracic sterna immaculate in both sexes; in the male, antennae not bearing long setae, genital papillae minute, middle tibiae concolorous . . . . . . . . . . . . . . . . . . . . . . . . . . . . . . . . . . . . . . . . 69

— Antennal club of female shorter, robust, less than 3 times as long as wide; propodeal crenulae crowded, slightly overlapping; thoracic sterna immaculate in the female, distinctly infuscated in the male; mandibles reduced, with 2 minute denticles; forewings very broad; in the male, antennal club and third funicular segment bearing long setae, genital papillae elongate, middle tibiae tipped with black . . . . . . . . . . . . . . . . . . . . . . . . . . . . . . *erythraeus*, p. 674

69(68). Abdominal sternites very extensively stippled in both sexes; mandibles reduced; propodeal crenulae oval; ovipositor long, about twice length of middle tibia, the sheaths up to $\frac{3}{5}$ length of middle tibia . . . . . . . . . . . . . . . . . . *ignotus*, p. 699

— Abdominal sternites not stippled; mandibles well developed; propodeal crenulae rounded; ovipositor shorter, not more than $1\frac{3}{5}$ times length of middle tibia, the sheaths $\frac{2}{5}$ length of middle tibia . . . . . . . . . . . . . . . . . *notialis*, p. 492

70(67). Thoracic setae dark, coarse; antennal club short, not more than 3 times as long as wide; propodeum short, usually less than 4 times as long as the metanotum; male antennae with an indistinct ridge on the scape and without a specialized sensory area on the club; pupae predominantly black . . . **the mytilaspidis group** . . . . . . . . . . . . . . . . . . . . . . . . . . . . . . . . . . . . . . . . . . . . . 71

— Without the above combination of characters; either thoracic setae paler, or antennal club longer, or propodeum longer . . . . . . . . . . . . . . . . . . . . . . . . . 77

71(70). Thoracic sterna dusky; propodeal crenulae rounded; mesoscutum usually with 8–12 setae . . . . . . . . . . . . . . . . . . . . . . . . . . . . . . . . . . . . . . . . . 72

— Thoracic sterna immaculate . . . . . . . . . . . . . . . . . . . . . . . . . . . . . . . . . . . . 74

72(71). Ovipositor not more than $1\frac{3}{4}$ times length of middle tibia . . . . . . . . . . . . . . . . 73

— Ovipositor somewhat longer, $1\frac{2}{3}$ times to fully twice length of middle tibia . . . . . . . . . . . . . . . . . . . . . . . . . . . . . . . . . . . . . . . . . . . . . . . . . . *moldavicus*, p. 475

73(72). Pale; dorsum of thorax and abdomen generally immaculate (but posterior margin of scutellum lined with blackish, posterior margin of propodeum centrally and the crenulae distinctly dusky) . . . . . . . . . . . *mytilaspidis*, p. 464

— Considerably darker; pronotum, 2 conspicuous blotches on mesoscutum, and a small triangle at center of second abdominal tergite, fuscous; gastral tergites lightly but distinctly infuscated (margins of scutellum and propodeum and the crenulae infuscated as in *mytilaspidis*, but more strongly so) . . . . *opuntiae*, p. 473

74(71). Uniparental; propodeal crenulae triangular, pale; mesoscutum usually with 6–12 setae . . . . . . . . . . . . . . . . . . . . . . . . . . . . . . . . . . . . . . . . . . . . . 75

— A large, biparental species parasitic upon *Leucaspis* spp. on pine; propodeal crenulae rounded, dusky; mesoscutum usually with 10–16 setae . . . . . . . . . . . . . . . . . . . . . . . . . . . . . . . . . . . . . . . . . . . . . . . . . . . . . . . . . . *luteus*, p. 486

75(74). Propodeum short, usually 3–4 times as long as the metanotum and $\frac{1}{2}$ to $\frac{2}{3}$ length of scutellum; ovipositor not more than $1\frac{3}{4}$ times length of middle tibia, the sheaths $\frac{1}{3}$ to $\frac{2}{5}$ length of middle tibia . . . . . . . . . . . . . . . . . . . . . . . . . . . . . . 76

— Propodeum somewhat longer, usually $4$–$4\frac{1}{2}$ times as long as the metanotum but not exceeding $\frac{3}{5}$ length of scutellum, the crenulae somewhat wider, sometimes irregular; ovipositor somewhat longer, $1\frac{2}{3}$ times to twice length of middle tibia, the sheaths $\frac{2}{5}$ to $\frac{1}{2}$ length of middle tibia; mesoscutum with 6–8 setae; parasitic upon the parlatoria date scale . . . . . . . . . . . . . . . . . . *phoenicis*, p. 485

76(75). Pale yellow; posterior margin of scutellum narrowly lined with blackish or fuscous; thoracic setae very dark . . . . . . . . . . . . . . . . . . . . . . *aonidiae*, p. 476

— Entirely pale yellow, posterior margin of scutellum concolorous; thoracic setae somewhat paler . . . . . . . . . . . . . . . . . . . . . . . . . . . . . *libanicus*, p. 483

77(70). Propodeum relatively long, more than 4 times length of the metanotum or $\frac{3}{5}$ length of scutellum . . . . . . . . . . . . . . . . . . . . . . . . . . . . . . . . . . 78

— Propodeum short, less than 3 times as long as the metanotum, $\frac{1}{2}$ length of scutellum; antennal club elongate, tapering, considerably more than 3 times as long as wide; thoracic sterna immaculate, thoracic seate pale; mesoscutum with 13 setae; forewings broad . . . . . . . . . . . . . . . . . . . . . . . . . *neuter*, p. 490

78(77). Thoracic setae nearly as dark as in the **mytilaspidis** group; mesoscutum with 12–16 setae, each parapsis with 3 or 4 . . . . . . . . . . . . . . . . . . . . . . . . . . . . 79

— Thoracic setae paler; mesoscutum usually with 10–12 setae, rarely with as many as 15, each parapsis with 2 . . . . . . . . . . . . . . . . . . . . . . . . . . . . . . . . . . . 80

79(78). Thoracic sterna immaculate; propodeum 5–6 times as long as the metanotum, the crenulae minute, irregular, very indistinct; antennal club short; ovipositor relatively long, $1\frac{3}{4}$ to $1\frac{4}{5}$ times length of middle tibia; forewings with about 30–45 setae in the delta . . . . . . . . . . . . . . . . . . . . . . . . . . *rolaspidis*, p. 497

— Thoracic sterna dusky; propodeum 5–6 times as long as the metanotum, the crenulae minute but distinct, rounded or somewhat elongate; ovipositor very short, $1\frac{1}{4}$ to $1\frac{1}{3}$ times length of middle tibia; forewings with about 45–65 setae in the delta . . . . . . . . . . . . . . . . . . . . . . . . . . . . . . . *capensis*, p. 499

80(78). Male antennae bearing long setae on the club and third funicular segment; mandibles reduced; pronotal plates broadly joined; thoracic sterna dusky . . . 81

— Male antennae not bearing long setae; pronotal plates triangular, narrowly joined; pupae generally yellow . . . **the chrysomphali group** . . . . . . . . . . . . . . 82

81(80). Female antennae with an elongate club, distinctly more than $3\frac{1}{2}$ times as long as wide; third funicular segment of female not bearing an unusually long seta; mesoscutum with 10–13 pale setae; propodeum very long, distinctly longer than the scutellum, the crenulae strongly overlapping; ovipositor long, at least twice length of middle tibia; in the male, antennal club more than 4 times as long

as wide, third funicular segment wider than long . . . . . . . . . . . . . *roseni*, p. 678
— Female antennae with a shorter club, 3–3½ times as long as wide; third funicular segment of female bearing a very long seta; mesoscutum with 10 darker setae; propodeum very long but shorter than the scutellum, the crenulae nonoverlapping; ovipositor less than twice length of middle tibia; in the male, antennal club less than 4 times as long as wide, third funicular segment elongate . . . . . . . . . . . . . . . . . . . . . . . . . . . . . . . . . . . . . . . . . . . . . . . . . . *setosus*, p. 687

82(80). Thoracic sterna dusky . . . . . . . . . . . . . . . . . . . . . . . . . . . . . . . . . . . . . . . 83
— Thoracic sterna immaculate; antennal club short, not more than 3 times as long as wide . . . . . . . . . . . . . . . . . . . . . . . . . . . . . . . . . . . . . . . . . . . . . . . . . 87

83(82). Mesoscutum usually with 10 setae . . . . . . . . . . . . . . . . . . . . . . . . . . . . . . . . 84
— Mesoscutum with 10–15 (usually 12) setae, including 4 submedian pairs; antennal club about 3 times as long as wide or somewhat longer; propodeum 4–4½ times as long as the metanotum, ¾ length of scutellum, the crenulae small, rounded, nonoverlapping; ovipositor relatively long, 1¾ times to twice length of middle tibia; male antennal club lacking a specialized sensory area; a specific parasite of the purple scale . . . . . . . . . . . . . . . . . *lepidosaphes*, p. 607

84(83). Antennal club long, considerably more than 3 (often 4) times as long as wide; ovipositor less than twice length of middle tibia, the sheaths not more than ½ length of middle tibia . . . . . . . . . . . . . . . . . . . . . . . . . . . . . . . . . . . . . . . . 85
— Antennal club short, 2½ to 3 times as long as wide; ovipositor longer, about twice length of middle tibia, the sheaths nearly ⅗ length of middle tibia; mandibles reduced, tridentate; propodeum 5–7 times as long as the metanotum, about as long as the scutellum, the crenulae rounded and somewhat elongate, nonoverlapping, the 2 sets widely separated . . . . . . . . . . . . . . . . . . . *debachi*, p. 601

85(84). Propodeum very long, at least as long as the scutellum . . . . . . . . . . . . . . . . . 86
— Propodeum usually 4–5 times as long as the metanotum and not more than ¾ length of scutellum, the crenulae closely appressed, usually somewhat overlapping; antennal club long and thick, cylindrical, 4 times as long as wide; forewings broad; in the male, antennal club 3–4 times as long as wide, with a partly cut-off sensory area on ventral aspect . . . . . . . . . . . . *cylindratus*, p. 604

86(85). Mandibles well developed; propodeum 6–8 times as long as the metanotum, the crenulae usually nonoverlapping; ovipositor not more than 1⅔ times length of middle tibia; forewings relatively narrow, 2½ to 2⅔ times as long as wide, with 25–40 setae in the delta; thoracic sterna faintly dusky, with a conspicuous longitudinal black line on the stem of the mesosternal furca; a uniparental species, the rare males lacking a specialized sensory area on the ventral aspect of the antennal club . . . . . . . . . . . . . . . . . . . . . . . . . . . . *chrysomphali*, p. 593
— Mandibles reduced, lacking distinct denticles; propodeum about 6 times as long as the metanotum, the crenulae strongly overlapping; ovipositor somewhat

longer, a little over $1\frac{4}{5}$ times length of middle tibia; forewings broad, not more than $2\frac{2}{5}$ times as long as wide, with 40–60 setae in the delta; thoracic sterna more strongly infuscated; male antennal club with an elongate ventral plate bearing short setae . . . . . . . . . . . . . . . . . . . . . . . . . . . . . . . . . *anneckei*, p. 598

87(82). Mesoscutum with 8–10 setae . . . . . . . . . . . . . . . . . . . . . . . . . . . . . . . . . . . . . 88

— Mesoscutum with 12–15 setae, including 4 submedian pairs; propodeum 5–7 times as long as the metanotum, the crenulae triangular, closely spaced, sometimes slightly overlapping at base . . . . . . . . . . . . . . . . . . . *immaculatus*, p. 610

88(87). Middle tibiae pale, concolorous with body; antennal flagellum of female uniformly dusky; thoracic setae pale; forewings with 25–40 setae in the delta; propodeal crenulae elongate . . . . . . . . . . . . . . . . . . . . . . . . . . . . . . . . . 89

— Middle tibiae with a brownish spot at apex in the female, conspicuously tipped with black in the male; antennal flagellum of female strongly infuscated, tip of club blackish; forewings with 30–60 setae in the delta; propodeal crenulae rounded; female antennae with the third funicular segment asymmetrical, ventral aspect distinctly longer than the dorsal; in the male, antennal club and third funicular segment markedly asymmetrical . . . . . . . . . . . *mazalae*, p. 616

89(88). Forewings very narrow, a little over 3 times as long as wide; endophragma elongate, about $1\frac{1}{2}$ times as long as wide; generally yellow, tibiae and tarsi of all legs rather strongly, contrastingly dusky; male antennae with the first 2 funicular segments bearing unusually elongate, spindle-shaped sensilla, the third funicular segment markedly asymmetrical . . . . . . . . . . . . *sensorius*, p. 620

— Forewings not more than $2\frac{4}{5}$ times as long as wide; endophragma about as long as wide at base; generally yellow, legs concolorous, the tibiae and tarsi darker yellow; in the male, sensilla on first 2 funicular segments not unusually elongate, third funicular segment only slightly asymmetrical . . . . . . . . . . . . . . . . . . . . . . . . . . . . . . . . . . . . . . . . . . . . . . . . . . . . . . . . . . .

# PART II

# SPECIFIC DESCRIPTIONS

# I. THE VITTATUS GROUP

This large, heterogeneous group includes the "*Marietta*-like" species of *Aphytis*, characterized by mottled forewings in the female sex, with a pattern of areas bearing coarse, dark setae against a background of fine, hyaline setae. The body and appendages are usually heavily pigmented in this group.

Although superficially resembling *Marietta*, the members of the **vittatus** group differ markedly from the species of this genus in possessing a well-developed, crenulate propodeum and in having the antennal club clearly separated from the penultimate segment by a distinct constriction. Unlike *Marietta*, in most members of the **vittatus** group the forewing pattern is produced mainly by contrasting dark and pale setae, not by integumentary infuscation. Males in this group further differ from *Marietta* in having essentially unmottled wings and lacking genital parameres. Like all other species of *Aphytis*, all known members of the **vittatus** group are primary parasites of armored scale insects. This is, of course, in contrast to the *Marietta* species, which are all hyperparasites.

The forewing pattern is basically similar in most members of this group, consisting of a ring or patch of dark setae in the delta, a submedian patch distad of the speculum, an oblong patch below the stigmal vein, another one opposite it against the posterior margin of the wing, and several subapical patches, all of which may be broken up or united in various combinations, completely or partly obliterated, etc. (e.g., Figures 268, 291, 303, 337, 346 and 365).

Other morphological characters are rather variable between species. The eyes are coarsely setose in most forms, finely setose in a few. The mouthparts are well developed in all known species, with strong mandibles and 2-segmented maxillary palpi. Antennal segmentation is reduced in the males of several members, but not in others. The syntergum is v-shaped or triangular, considerably wider than long, with the cercal plates situated close to the posterior spiracles, bearing 3 long setae and 1 short seta in the females of some species, 2 long setae and 1 short seta in others and in all males. The shape and proportions of the antennal segments, the number of mesoscutal setae, the shape of the propodeum and crenulae, the length of the ovipositor, the shape of the male genitalia, etc., vary considerably among the members of this group.

The **vittatus** group at present includes 18 bona fide members, ranging in distribution from the Australian to the Neotropical region, with a few representatives in the Oriental region. Six additional species, resembling members of the group in several important characters and in general habitus, but differing markedly in having hyaline wings or

5-segmented antennae in the female sex, or in having the antennal club separated from the penultimate segment only by a suture, have not been included in the group but are obviously closely related. These include two Australian, two Oriental and two Nearctic species. They are discussed under "Related Species" at the end of this section (see p. 284).

Several complexes of closely related species are evident within the **vittatus** group.

1. The *vittatus* complex includes at least 3 Oriental species. *A. vittatus* (Compere) is the most distinctly pigmented species of *Aphytis*, with a conspicuous pattern of 4 longitudinal dark stripes on the thorax. It also has the most intricate forewing pattern, with anastomosing dark rays against a hyaline background. *A. peculiaris* (Girault) is similar to *vittatus* in body coloration, but has a greatly reduced forewing pattern. Both these species have 5-segmented antennae in the male sex. *A. malayensis* n.sp. is also similar in body coloration, but differs markedly in the 5-segmented antennae and hyaline wings of the female. It is likely that additional related species will eventually be discovered in the Orient.

2. The *costalimai* complex, including 4 Neotropical species—*costalimai* (Gomes), *acutaspidis* n.sp., *perplexus* n.sp. and *maculatipennis* (Dozier)—is characterized by a few well-defined dark blotches on the forewing of the female and by hairy male antennae. The representatives of this complex can be separated from one another by differences in the shape of the propodeal crenulae and in details of coloration and wing pattern. *A. maculatipennis* is rather similar to *vittatus* in general coloration, but has been included in the *costalimai* complex because of its wing pattern and the hairy male antennae, as well as its geographical distribution.

In the peculiar shape of the male antennae, the members of the *costalimai* complex are quite similar to a heterogeneous assemblage of species including *erythraeus*, *roseni*, *longicaudus*, *pilosus* and *setosus* (see under "Unassigned Species," p. 672), which are otherwise not related to the **vittatus** group.

3. The *cochereaui* complex, including *cochereaui* DeBach and Rosen, *noumeaensis* (Howard), *argenticorpus* n.sp. and *fabresi* DeBach and Rosen, ranges in distribution from New Caledonia to Australia. The New Caledonian members of this complex are characterized by the subapical patches of dark setae on the forewing of the female forming a continuous subapical crossband, and by the peculiar 6-segmented antennae of the male, with 2 small ring segments and an enlarged club bearing a specialized sensory area; in both sexes the cerci bear 2 long setae and 1 short seta. *A. fabresi* is the palest, presumably most advanced member of this complex. *A. argenticorpus*, an Australian species, is included because it resembles *cochereaui* in general habitus, although it differs in having the forewing pattern broken into more numerous separate areas, essentially similar to the pattern of the *capillatus* complex.

The peculiar male antennae of the *cochereaui* complex are rather similar to those characteristic of the **chilensis** group.

4. The *capillatus* complex includes 4 closely related Australian species—*capillatus* (Howard), *nigripes* (Compere), *ciliatus* (Dodd) and *australiensis* DeBach and Rosen—characterized by an extensive forewing pattern in the female sex, with 3 submedian patches of dark setae distad of the speculum and several subapical patches. The female antennae are short and stout in the members of this complex, the abdomen is more or

less banded, and the cerci bear 2 long setae and 1 short seta. The male, known only in *capillatus*, has essentially 4-segmented antennae, with a minute ring segment and the rudiment of another, and a robust club bearing a specialized sensory area.

The *capillatus* complex is obviously very closely related to the *cochereaui* complex; like the latter, it shows some affinity to the **chilensis** group. The four members of this complex can be separated mainly by the shape of the propodeum, by the coarseness of the inter-ommatidial setae in the eyes, and by various details of coloration and wing pattern.

Several other Australian species are here considered synonymous with *capillatus*. Two additional Australian species, *hyalinipennis* n.sp. and *wallumbillae* (Girault), resemble *capillatus* in important diagnostic characters but differ markedly in having hyaline wings.

5. The *anomalus* complex includes 3 Neotropical species, *anomalus* Compere, *haywardi* (Blanchard) and *obscurus* DeBach and Rosen, which differ from all other members of the **vittatus** group in having a single crossband on the forewing, with both an infuscated integument and coarse, dark setae. *A. angustus* Compere, an Oriental species, is similar in that it has a largely infuscated forewing, with the apical portion pale and bearing fine, hyaline setae. Another Oriental species, *A. mandalayensis* n.sp., is similar in wing pattern but differs markedly in having 5-segmented antennae in the female sex.

Two Nearctic species, *dealbatus* Compere and *melanostictus* Compere, resemble the species of *Marietta* in that the antennal club is separated from the third funicular segment by a suture only, not by a constriction as in all other species of *Aphytis*, but are otherwise quite similar in habitus to the **vittatus** group. These species may represent another possible link between the genera *Aphytis* and *Marietta*. They are the only known Nearctic species that are related to the **vittatus** group.

Synonyms in the **vittatus** group include *cowperi* (Girault), *emersoni* (Girault), *lessingi* (Girault), *novicapillatus* (Girault), *romae* (Girault) and *spiniferus* Compere and Annecke.

Several species of *Aphytis*, regarded here as unrecognizable (see below, p. 729), are apparently closely related to the **vittatus** group. These include *angeloni* (Girault), *maculatipes* (Girault), *perissoptroides* (Girault) and *punctaticorpus* (Girault).

This heterogeneous, probably polyphyletic, group is presumably the most primitive species group in the genus *Aphytis*. Primitive characters abound in members of this group. Heavy pigmentation of the body and appendages ; an extensive forewing pattern in the female sex; numerous coarse setae on the submarginal vein; unequal setae along the marginal vein, decreasing in length toward the apex of the vein; cerci bearing 3 long setae and 1 short seta in the female sex; a long ovipositor; and specialized sense organs on the antennal scape of the male—all these are regarded as representing primitive traits in *Aphytis*.

Quite a few characteristics of the more advanced species groups apparently originated in the **vittatus** group. As pointed out above, sexual dimorphism in the shape of the antennae, characteristic of the **chilensis** group, is shared by members of the *cochereaui* and *capillatus* complexes. Likewise, the small, triangular first funicular segment of female *chilensis*, the cephalic pigmentation of the **proclia** group, the cut-off sensory area on the ventral aspect of the antennal club of the males of several species groups, even overlapping propodeal crenulae, are all present in various members of the **vittatus** group. As additional species are accumulated, it is hoped that the affinities between members of this and other groups will become clearer.

### 1. **Aphytis vittatus** (Compere)

(Figures 264–269)

*Paraphytis vittata* Compere, 1925, Trans. Amer. Entomol. Soc., **51**: 129–133.
*Marietta vittata*: Compere, 1936, Univ. Calif. Publ. Entomol., **6**: 311.
*Aphytis vittatus*: DeBach and Rosen, 1976, Ann. Entomol. Soc. Amer., **69**: 541.

This Oriental species can be readily recognized by the distinctive pigmentation of the female, with 4 longitudinal dark brown stripes on the mesonotum; female forewings extensively mottled, with a pattern of anastomosing fuscous areas against a hyaline background; eyes coarsely setose; antennae 6-segmented; propodeal crenulae small, nonoverlapping; ovipositor long; cerci with 3 long setae and 1 short seta. Male considerably paler, lacking the longitudinal stripes, with hyaline wings and 5-segmented antennae; cerci with 2 long setae and 1 short seta.

*Female*. Eyes coarsely setose. Mandibles well developed, essentially bidentate, with a very weak dorsal truncation; maxillary palpi 2-segmented, labial palpi 1-segmented. Antennal scape (Figure 264) slender, about $6\frac{1}{3}$ times as long as wide, about $1\frac{1}{5}$ times longer than the club; pedicel relatively short, about $1\frac{1}{3}$ times as long as wide, a trifle longer than the third segment of the funicle; first funicular segment trapezoidal, ventral margin longer than the dorsal, about $1\frac{3}{5}$ to $1\frac{2}{3}$ times as wide as long; second segment trapezoidal, dorsal margin longer than the ventral, somewhat shorter and wider than the first segment, about twice to $2\frac{1}{2}$ times as wide as long; third funicular segment about as long as wide or slightly wider than long, bearing 1–2 longitudinal sensilla, with the dorsal margin somewhat longer than the ventral; club separated from third funicular segment by a distinct constriction, tapering, about $2\frac{1}{2}$ times as long as wide, about $2\frac{2}{3}$ times longer and somewhat wider than the preceding segment, bearing 8 longitudinal sensilla.

Setae on head and thorax dark, rather coarse and distinctly costate, the largest ones apparently composed of distinct filaments; they are readily visible in cleared specimens under low magnification ($\times 30$), except when obscured by the pigmentation of the body. Vertex with 2 pairs of long setae along the occipital margin, in addition to several slightly smaller setae; mesoscutum with 11–14 setae (only their bases are shown in Figure 265): 4–5 in a longitudinal row in each of the two submedian dark stripes, 1 or 2 on each of the two lateral stripes, the posterior pair considerably coarser than the others; each parapsis with 3 setae, each axilla with 1 seta, scutellum with 4 setae on the inner borders of the lateral dark stripes; the pair of discoid sensilla on the scutellum is usually much closer to the anterior pair of setae than to the posterior pair. Frontovertex, pronotum, and mesonotal sclerites reticulate, heavily so on pigmented areas. Scutellum elongate, trapezoidal, only slightly shorter than the mesoscutum. Metanotum v-shaped, reticulate except on the sides, heavily so on the central pigmented area; anteromedian apodeme slender, distinctly shorter than median length of metanotum.

Propodeum (Figure 266) $2\frac{3}{4}$ to 4 times as long as the metanotum, a little over $\frac{2}{5}$ length of the elongate scutellum, reticulate on a central salient and just mesad of the spiracles, bearing 4 + 4 to 4 + 5 small, rounded, nonoverlapping crenulae on the posterior margin.

Second abdominal tergite (Figure 266) coarsely reticulate on the central pigmented area just below the crenulae; tergites III–VII reticulate on the sides, bearing a few setae on each reticulate area; seventh tergite with 6–7 setae on each side; eighth tergite delicately reticulate across, with a transverse row of 9–14 setae between the spiracles; syntergum broad, v-shaped, bearing 13–20 setae in 1–2 transverse rows. Cerci (Figure 267) situated very close to the posterior spiracles, with 3 long setae and 1 short seta. Ovipositor shaft well over twice as long as the middle tibia (2.09–2.22), ovipositor sheaths about $\frac{1}{2}$ length of middle tibia (0.45–0.51).

Mid-tibial spur rather slender, about $\frac{4}{5}$ length of the corresponding basitarsus (but about as long as the basitarsus in a smaller specimen).

Forewing (Figure 268) about $2\frac{3}{5}$ to $2\frac{3}{4}$ times as long as wide; marginal fringe not exceeding $\frac{1}{5}$ width of disk. Delta area of disk about as densely setose as area distad of speculum, with 141–219 setae in the specimens at hand, not separated from setose area below distal half of submarginal vein, slightly separated from row of fine setae along posterior margin of wing; costal cell with a longitudinal row of fine setae, and 1 to 3 coarse setae near the apex; submarginal vein bearing 4–5 setae and 17–21 bullae; marginal vein bearing 8–13 prominent setae along anterior margin, these subequal or slightly decreasing in length toward apex of vein, somewhat longer than setae in a row along center of vein.

Forewing with an intricate pattern of lightly infuscated areas bearing dark, coarse setae alternating with hyaline areas bearing transparent, fine setae: a large infuscated ring in delta area below marginal vein, its center and surrounding areas hyaline; distad of speculum, anastomosing infuscated rays against a hyaline background. Hind wing entirely hyaline, the discal setae uniform, very fine, transparent.

General coloration light yellow, frontovertex with a trace of orange; dark brown markings as follows: oral margin suffused; pronotum with a pair of submedian spots and a spot on each side; mesoscutum and scutellum (Figure 265) with a pair of submedian longitudinal stripes and a stripe on each side, which on the scutellum merge into the submedian stripes on the posterior margin; anterior margin of mesoscutum, parapsidal sutures and sides of parapsides; central part of metanotum; a v-shaped infuscation along posterior margin of propodeum, starting mesad of the spiracles and converging on the median salient; crenulae; a median blotch on second abdominal tergite; posterior margin of third tergite; transverse stripes on each side of all subsequent tergites, including the syntergum; thoracic sterna and margins of endophragma strongly infuscated. Mandibular denticles black. Antennal scape and first 2 funicular segments dusky, pedicel and third funicular segment pallid, club dark brown. All femora with a fuscous blotch near the apex; all tibiae with 2 broad, distinct rings; all tarsi fuscous, the middle basitarsus darker. Veins of forewing brownish.

Length 0.83–1.24 mm.

*Male.* Very similar in structure to the female, differing mainly in antennal characters, wing pattern, and coloration.

Antennae (Figure 269) 5-segmented (1121); scape slender, $6\frac{1}{3}$ times as long as wide, about $1\frac{1}{3}$ times longer than the club; pedicel $1\frac{1}{2}$ times as long as wide, about $1\frac{1}{4}$ times longer than second segment of funicle; first funicular segment trapezoidal, only slightly wider than long; second funicular segment considerably larger, symmetrical, somewhat longer than wide; club oval, about $2\frac{1}{3}$ times as long as wide, about $2\frac{1}{2}$ times longer and somewhat wider than the preceding segment; 1–2 longitudinal sensilla on second funicular segment, 2–3 on club.

Mesoscutum with 13–14 setae. Propodeum about 4 times as long as the metanotum, about $\frac{2}{5}$ length of scutellum; crenulae 3 + 6. Cerci with 2 long setae and 1 short seta. Genitalia about $\frac{2}{3}$ length of middle tibia (0.67); digital sclerites long, nearly $\frac{2}{5}$ the combined length of aedeagus and apodemes (0.37).

Forewing $2\frac{3}{5}$ times as long as wide, entirely hyaline, with a faint cloud below the junction of the submarginal and marginal veins, uniformly setose without alternating areas bearing fine and coarse setae; delta area with 61–84 setae in 6–9 rows, these somewhat longer and sparser than the setae distad of speculum.

General coloration light yellow, without the conspicuous dark stripes on the mesonotum; metanotum entirely yellow; pronotum, parapsidal sutures, propodeum, second and third abdominal tergites as in the female, subsequent abdominal tergites yellow, except for a pair of dark transverse blotches on seventh tergite; antennae entirely pale; legs suffused with fuscous, without distinct rings on the tibiae; middle basitarsus contrastingly dark brown.

Length about 1 mm.

Redescribed from 2♀♀ and 3♂♂ (♂ allotype and paratypes), reared by F. Silvestri from *Paralepidosaphes tubulorum* (Ferris), Amoy, China, January 24, 1925; 3♀♀, apparently of the same lot, from Silvestri's collection; 1♀ from a shipment made by J. L. Gressitt from China, June 23 1950. The ♀ holotype and one additional ♀ paratype, kept at the U.S. National Museum, Washington, D.C., were also seen.

**Notes.** This interesting species unquestionably belongs in *Aphytis*. The genus *Paraphytis,* established by Compere (1925) for its reception and subsequently declared as a synonym of *Marietta* (Compere, 1936), therefore falls into synonymy with *Aphytis*.

DeBach and Rosen (1976b) have changed the name *vittata* to *vittatus*, to conform with the gender of *Aphytis*.

## 2. **Aphytis peculiaris** (Girault) n. comb.

(Figures 270–283)

*Marietta peculiaris* Girault, 1932, New Lower Hymenoptera from Australia and India. Priv. Publ., Brisbane, 6pp., page 2.
*Marietta peculiaris*: Hayat, 1974, J. Nat. Hist., **8**: 345–347.

This peculiar Indian species may be recognized in the female sex by the incomplete submedian stripes on the mesonotum, by the presence of several more or less complete crossbands on the abdomen, and by the nearly hyaline forewings, with 2 patches of dark, coarse setae; eyes coarsely setose; antennae slender, contrastingly pigmented; mesonotal setae numerous; propodeal crenulae numerous, crowded; cerci with 3 long setae and 1 short seta. The hitherto undescribed male is considerably paler, with nearly hyaline wings and 5-segmented antennae; cerci with 2 long setae and 1 short seta.

*Female.* Eyes coarsely setose. Mandibles well developed, essentially bidentate, with a short, steep dorsal truncation; maxillary palpi 2-segmented, labial palpi 1-segmented. Antennae (Figures 270, 271) slender; scape about 6 to $7\frac{1}{3}$ times as long as wide, nearly $1\frac{1}{2}$ times longer than the club; pedicel not widening appreciably toward apex, $2\frac{1}{7}$ to $2\frac{1}{2}$ times as long as wide; first 2 funicular segments nearly rectangular, somewhat wider than long, the second segment slightly shorter and wider than the first; third funicular segment about as long as the pedicel, about $1\frac{3}{5}$ times as long as wide, bearing 3–5 longitudinal sensilla; club $3\frac{1}{7}$ to $3\frac{1}{4}$ times as long as wide, about $2\frac{1}{6}$ to nearly $2\frac{1}{2}$ times longer and somewhat wider than the preceding segment, bearing 10–11 longitudinal sensilla.

Setae on head and thorax coarse, dark, easily visible under $\times$ 30 magnification; those on sides of abdomen finer. Vertex with a pair of long setae along the occipital margin, in addition to numerous shorter setae. Mesoscutum (Figures 272–274) with 21 setae, the posterior pair considerably longer than the others, each parapsis with 4 setae, each axilla with 1 seta; scutellum with 4 setae, the discoid sensilla situated at about the level of the anterior pair of setae. Frontovertex, pronotum and mesonotal sclerites reticulate. Scutellum roundly trapezoidal, about $\frac{9}{10}$ median length of mesoscutum or nearly so. Metanotum arcuate, reticulate except on the sides; anteromedian apodeme short, slender, about $\frac{1}{2}$ or $\frac{3}{5}$ median length of metanotum.

Propodeum (Figures 273–276) about $3\frac{1}{5}$ to $3\frac{2}{5}$ times as long as the metanotum, nearly $\frac{1}{2}$ length of scutellum, reticulate on the sides and on a central trapezoidal area; crenulae 10 + 10 to 11 + 14, small, distinct, rounded, crowded, sometimes overlapping or arranged in 2 rows (Figures 275, 276).

Second abdominal tergite (Figures 273–275) reticulate on a small central area just below the propodeal crenulae; tergites III–VII reticulate on the sides, bearing a few setae on each reticulate area; third tergite distinctly reticulate across, tergites IV–VII more obscurely so on dark crossbands; eighth tergite delicately reticulate, with a transverse row of 11 setae between the spiracles; syntergum broad, v-shaped, reticulate, bearing about 20 setae in a transverse row. Cerci situated very close to the posterior spiracles, with 3 long setae and 1 short seta. Ovipositor shaft about twice as long as the middle tibia (1.85–2.02), ovipositor sheaths about $\frac{1}{2}$ length of middle tibia (0.44–0.50).

Mid-tibial spur robust, nearly as long as the corresponding basitarsus (0.79–0.87).

Forewing (Figure 277) relatively broad, about $2\frac{1}{4}$ to $2\frac{1}{3}$ times as long as wide; marginal fringe short, not exceeding $\frac{1}{9}$ width of disk. Delta area with about 200 setae, about as densely and uniformly setose as the area distad of speculum, not separated from the extensive setose area below the submarginal vein or from the row of setae along the posterior margin of the wing; costal cell with a row of fine setae, the 3 distal ones considerably coarser and longer than the others. Submarginal vein bearing 6 coarse setae and 19–20 bullae. Marginal vein bearing 10–11 prominent, subequal setae along anterior margin, these decreasing only slightly in length toward apex of vein, only slightly longer than the setae in a row along center of vein.

Forewing hyaline, with two patches of dark setae: a triangular patch in delta just proximad of speculum, another patch extending below stigmal vein to a point about midway between anterior and posterior margins of wing, widening posteriorly; a fainter, much less conspicuous patch some distance below the latter. At the two conspicuous patches, the integument of the wing is faintly infuscated and the setae are considerably longer and coarser than the surrounding discal setae. A narrow fuscous streak along posterior margin of wing, distad of speculum.

General coloration (Figures 272, 273) yellow, with dark brown markings as follows: 2 longitudinal, submedian stripes on pronotum, mesoscutum and scutellum, those on the scutellum curving out at both ends, continuing along posterior margin; parapsidal sutures; scuto-scutellar suture; posterior margin of metanotum centrally; sides of propodeum mesad of spiracles, posterior margin of propodeum and the crenulae; the reticulate area on second abdominal tergite just below the crenulae; more or less complete crossbands on tergites III–VII and on syntergum, sometimes obliterated centrally (this character is variable in the specimens at hand); sides of syntergum around cerci; thoracic sterna, internal apodemes, and lateral margins of endophragma. Head with some faint markings around oral margin; mandibular denticles dark brown. First and second funicular segments and antennal club dark brown, other antennal segments contrastingly pallid. Legs yellow, femora tinted with brownish ventrally, especially near apex, all tibiae with 2 dark brown rings or blotches; first tarsal segment of all legs dark brown; mid-tibial spur varying from pale to dark brown. Wing veins lined with brownish.

Length 1.15–1.31 mm.

*Male*. Similar to the female in structure, chaetotaxis and sculpture, differing mainly in antennal characters, wing pattern, and coloration.

Antennae (Figure 278) 5-segmented (1121); scape slender, $5\frac{2}{3}$ times as long as wide, about $1\frac{2}{5}$ times longer than the club; pedicel about as long as the second segment of the funicle; first funicular segment small, oblong, $1\frac{1}{4}$ times as long as wide; second funicular segment considerably larger, elongate, nearly twice as long as the first, apparently bearing 2–3 longitudinal sensilla; club elongate, twice longer than the preceding segment, apparently bearing a few longitudinal sensilla.

Mesoscutum (Figure 279) with 22 setae, each parapsis with 4 setae, each axilla with 1 seta. Propodeum (Figure 280) about 3 times as long as the metanotum, $\frac{2}{5}$ length of scutellum; crenulae 10 + 11, as in the female. Cerci with 2 long setae and 1 short seta. Genitalia (Figure 281) robust, about $\frac{2}{3}$ length of middle tibia (0.69); digital sclerites long, nearly $\frac{1}{2}$ the combined length of aedeagus and apodemes (0.46).

Forewing (Figure 283) somewhat broader and more sparsely setose than in the female, $2\frac{1}{9}$ times as long as wide; delta area with 90 setae in 9 rows. Forewing nearly entirely hyaline, with just a faint trace of the pattern of the female; a few setae in delta (on proximal border of speculum) and below stigmal vein somewhat coarser and darker than the others.

General coloration yellow, marked with dark brown or fuscous as follows: pronotum submedially; parapsidal and scuto-scutellar sutures faintly; a streak at base of forewing, below the dusky tegula; propodeum on sides mesad of spiracles, posterior margin faintly at center; second abdominal tergite centrally; a crossband on third and seventh abdominal tergites; a pair of sublateral blotches on eighth abdominal tergite. Antennae and legs entirely pale, except for mid-tibial spur and first 3 segments of middle tarsus, which are contrastingly dark brown (Figure 282).

Length 1.11 mm.

Redescribed from 1♀ (holotype), reared by L. J. Newman "from red scale, India," no additional data; also 2♀♀, 1♂ and the head of another ♀, reared by G. W. Angalet from *Aonidiella ?orientalis* (Newstead) on rose, Khunti, India, August 31, 1956. The male specimen is hereby designated allotype.

Holotype partly cleared and mounted in balsam, in the collection of the Department of Agriculture, South Perth, Western Australia. The other specimens are cleared, mounted in balsam, in the collection of the Division of Biological Control, University of California, Riverside.

**Notes.** Girault (1932) originally described this species from a single female specimen, and assigned it to the genus *Marietta*. Hayat (1974) retained it in *Marietta*, but made the following comments: "*Marietta peculiaris* belongs to a group of species which are intermediate between typical *Marietta* and typical *Aphytis*. Girault's species comes close to *M. vittata* (Compere), but is distinctive in several respects, specially antennal dimensions, maculations on abdomen and wing infuscation." Hayat's brief redescription and figures of the holotype female are quite accurate, except that the cerci bear 3 long setae and 1 short seta, not 2 + 1 as shown by him.

Like the preceding species, *peculiaris* is unquestionably a bona fide member of *Aphytis*. The female is rather similar to *vittatus* and *maculatipennis* in the coloration of the thorax and antennae. It differs from both these species in that the submedian stripes on the mesonotum are less distinct, in the presence of crossbands on most abdominal tergites, in the reduced wing pattern, elongate antennae and more numerous mesoscutal setae and propodeal crenulae. The male, described here for the first time, resembles the male of *vittatus* in the 5-segmented antennae, but may be readily distinguished from the latter by the more numerous mesoscutal setae and propodeal crenulae, by the presence of a dark crossband on the seventh abdominal tergite and a pair of dark blotches on the eighth tergite, and by the distinctive pigmentation of the mid-tibial spur.

**Potential Value for Biological Control.** This little-known Oriental species should be taken into account as a potential natural enemy of *Aonidiella* spp.

### 3. **Aphytis maculatipennis** (Dozier) n. comb.

(Figures 284–295)

*Marietta maculatipennis* Dozier, 1933, Proc. Entomol. Soc. Wash., **35**:88–89.
*Marietta maculatipennis*: Compere, 1936, Univ. Calif. Publ. Entomol., **6**:312.

This Caribbean species can be recognized by the following combination of characters: Eyes coarsely setose; propodeum with an emarginate median salient, bearing irregular crenulae. Female with 2 submedian dark stripes on mesonotum and 2 dark crossbands on abdomen, forewing with a pattern of several small patches of dark, coarse setae, cerci with 3 long setae and 1 short seta. Male antenna 6-segmented, the scape bearing 3 rows of sensilla, the club and penultimate segment bearing relatively long setae; forewing hyaline, with a fuscous spot below the stigma.

*Female.* Eyes (Figure 284) coarsely setose. Mandibles well developed; maxillary palpi 2-segmented, labial palpi 1-segmented. Antennal scape (Figure 285) about $3\frac{4}{5}$ to 5 times as long as wide, about as long as or slightly longer than the club; pedicel about $1\frac{1}{3}$ to $1\frac{3}{5}$ times as long as wide, about as long as the third segment of the funicle; first funicular segment triangular, about $1\frac{1}{3}$ to $1\frac{3}{5}$ times as wide as long; second segment symmetrical, considerably wider and shorter than the first, $2\frac{2}{3}$ to $3\frac{1}{2}$ times as wide as long; third funicular segment somewhat longer than wide, bearing 4–7 longitudinal sensilla; club separated from preceding segment by a distinct constriction, $2\frac{1}{3}$ to $2\frac{3}{4}$ times as long as wide, $2\frac{1}{3}$ to $2\frac{3}{5}$ times longer and somewhat wider than preceding segment, bearing 10–14 longitudinal sensilla.

Setae on head, thorax and sides of abdomen coarse, dark, readily visible under × 30 magnification. Vertex (Figure 284) with 2 pairs of long setae along the occipital margin in addition to several smaller setae; mesoscutum (Figure 286) with 11–16 setae, 8 of which are arranged in 4 submedian pairs, the posterior pair considerably longer and coarser than the others; each parapsis with 3 setae, each axilla with 1 seta; scutellum with 4 setae, the discoid sensilla considerably closer to the anterior than to the posterior pair. Face aciculate between scrobes; frontovertex, pronotum and mesonotal sclerites reticulate. Scutellum oval, about $\frac{9}{10}$ median length of mesoscutum. Metanotum relatively short medially, reticulate except on the sides; anteromedian apodeme slender, distinctly longer than median length of metanotum.

Propodeum (Figures 287–289) rather short, with a distinct, often emarginate median salient, about $3\frac{1}{5}$ to $4\frac{3}{4}$ times as long as the metanotum, $\frac{1}{2}$ length of scutellum, reticulate-strigose on the sides and on a median trapezoidal area; crenulae 2 + 2 to 4 + 4, small, irregular, nonoverlapping. Shape of median salient and crenulae somewhat variable, but 2 distinct lobes usually present.

Second abdominal tergite (Figures 287, 288), faintly sculptured on the center; tergites III–VII reticulate on the sides, bearing 2–3 strong setae on or just mesad of the reticulate areas, third and seventh tergites transversely striated between the reticulate areas; eighth tergite reticulate, with a transverse row of 8–12 setae between the spiracles; syntergum (Figure 290) moderately v-shaped, short and broad, reticulate, bearing a transverse row of 11–14 setae. Cerci situated close to and somewhat laterad of the posterior spiracles, with 3 long setae and 1 short seta. Ovipositor shaft about $1\frac{2}{3}$ to $1\frac{4}{5}$ times as long as the middle tibia (1.64–1.84), ovipositor sheaths about $\frac{1}{3}$ to $\frac{2}{5}$ length of middle tibia (0.30–0.40).

Mid-tibial spur as long as, or a trifle shorter than, the corresponding basitarsus.

Forewing (Figure 291) $2\frac{1}{4}$ to nearly $2\frac{1}{2}$ times as long as wide; marginal fringe not exceeding $\frac{1}{6}$ width of disk. Delta area nearly as densely setose as area distad of speculum, with 94–141 setae in 8–10 indistinct rows, clearly separated by narrow, bare stripes from 2–3 rows of setae along posterior margin of wing and from 3–4 rows of setae below distal half of submarginal vein; costal cell with a row of 4–5 fine setae along proximal half, and 1–2 prominent setae near apex. Submarginal vein bearing 5 coarse setae and 15–19 bullae. Marginal vein bearing 8–11 (usually 10) prominent setae along anterior margin, these subequal or somewhat decreasing in length toward apex of vein, considerably longer than the setae in a row along center of vein.

Forewing moderately mottled, with patches of coarse, dark setae against a background of fine, hyaline setae: a triangular dark patch in delta, just proximad of speculum; a circular patch somewhat distad of speculum; an oblong blotch below stigma; a fainter one opposite it, against posterior margin of wing; a pair of circular patches distad of stigma, about halfway between the oblong blotches and apex of wing; the integument of the wing is very faintly infumated at the patches, the mottled effect being produced mainly by the contrasting setae themselves; a slightly infumated, arcuate band below junction of submarginal and marginal veins; a narrow fuscous streak along posterior margin of wing distad of speculum.

General coloration yellow, with dark brown markings as follows: a small, faint spot on vertex posterior to ocelli; 2 spots on occiput near foramen; oral margin; center and posterior margin of pronotum; 2 longitudinal, submedian stripes on mesoscutum and scutellum, being arcuate on the scutellum and curving out on the posterior margin (Figure 286); parapsidal sutures; lateral margins of axillae; anterior margin of metanotum centrally; propodeum mesad of spiracles and along anterior and posterior margins (posterior part of median salient and crenulae pale); margins and large, central v-shaped infuscation on second abdominal tergite; broad crossbands on third and seventh abdominal tergites, short transverse stripes on sides of fourth and fifth tergites, faint spots on sides of sixth and on center of eighth tergite. Thoracic sterna infuscated. Antennae (Figure 285) pale, first 2 funicular segments infuscated, club dark brown. All femora with a conspicuous spot or ring near the apex, fore tibiae with one ring, middle and hind tibiae with 2 rings (Figure 290). Wing venation brownish.

Length 0.88–0.96 mm.

*Male.* Differs from the female mainly in the structure and chaetotaxis of the antennae, wing pattern and general coloration.

Antennal scape (Figure 292) rather short, widened ventrally, somewhat less than 3 times as long as wide, bearing a ventral, elongate plate with 3 longitudinal rows of roughly discoid sensilla; pedicel and first 2 funicular segments more or less as in the female; third funicular segment $1\frac{2}{5}$ times as long as wide, bearing 2–3 longitudinal sensilla and numerous long setae; club elongate, nearly $3\frac{1}{2}$ times as long as wide, distinctly longer than the scape, nearly 3 times longer and considerably wider than the preceding segment, bearing 5–7 longitudinal sensilla and numerous long setae.

Mesoscutum (Figure 293) with 8–17 setae; mid-tibial spur about as long as the corresponding basitarsus; cerci with 2 long setae and 1 short seta.

Genitalia (Figure 294) about $\frac{4}{5}$ length of middle tibia (0.77–0.86), with a distinct elongate rod between the apodemes; digital sclerites about $\frac{1}{4}$ the combined length of aedeagus and apodemes (0.24–0.29).

Forewing (Figure 295) hyaline, with an infuscated spot below stigma, without a pattern of contrasting coarse and fine setae; delta with 44–62 setae; marginal vein with 6–9 prominent, subequal setae along anterior margin.

General coloration yellow; the distinct longitudinal stripes on mesoscutum and scutellum missing; dark brown as follows: center of pronotum; parapsidal sutures; a pair of faint spots on mesoscutum, another pair near anterior margin of scutellum; posterior margin of scutellum; propodeum, second and third abdominal tergites as in the female; an almost continuous, transverse blotch on sixth, seventh and eighth tergites. Antennae uniformly infumated. Legs as in the female.

Length 0.57–0.72 mm.

Redescribed from 5♀♀ and 8♂♂ (paratypes), reared by H. L. Dozier from "*Diaspidiotus* sp. nov." on mahogany, Côte Plage, Haiti, June 28, 1931, remounted from one balsam slide to several Hoyer's slides, after clearing.

Holotype in U.S. National Museum, Washington, D.C.

**Notes.** *A. maculatipennis* is rather similar to *vittatus* in the general coloration of the thorax and antennae. It may, however, be readily distinguished from this species by the absence of lateral longitudinal stripes on the mesonotum, by the less extensive pattern of the forewing, as well as by various other details of structure and coloration. The male of *maculatipennis* differs markedly from the male of *vittatus* in that the antennae are 6-segmented, the scape bears sensilla and the club and third funicular segment bear long setae.

The wing pattern of *maculatipennis* closely resembles that of *costalimai*, but is much less extensive than in the latter species. *A. maculatipennis* also differs markedly from *costalimai* in general coloration, being considerably more extensively pigmented on the mesonotum and much less so on the abdominal tergites, in having the setae along the anterior margin of the marginal vein essentially subequal in length, and in the shape of the sensilla on the antennal scape of the male. On the other hand, the peculiar long setae on the antennal flagellum of the male are similar to those of *costalimai* and *perplexus*, as well as to those of *erythraeus*, *roseni*, *longicaudus*, *pilosus* and *setosus*. For the time being, *maculatipennis* is considered a member of the *costalimai* complex.

## 4. **Aphytis costalimai** (Gomes)

(Figures 296–309)

*Marietta costa-limai* Gomes, 1942, Bol. Esc. Nac. Agron., Rio de Janeiro, **2** (1941):23–25, est. 1.
*Marietta costa-limai* Gomes, 1942, Rodriguesia, Rio de Janeiro, **4**:54–56.
*Aphytis spiniferus*: Compere and Annecke, 1961, J. Entomol. Soc. S. Afr., **24**:27–28.
*Aphytis costa-limai*: DeBach, 1963, Rev. Brasil. Biol., **23**:35–38.

This distinctive South American species can be recognized by the following combination of characters: Eyes coarsely setose; mesoscutum with numerous setae; propodeal crenulae rounded, overlapping; setae along marginal vein of forewing decreasing sharply in length toward apex. Female forewing with an extensive pattern of well-defined dark patches; abdomen dark brown basally and distally, pale centrally and on the sides; ovipositor long; cerci with 3 long setae and 1 short seta. Male forewing nearly hyaline; antennal scape bearing 1 row of sensilla, the club and penultimate segment bearing long setae; abdomen entirely dark brown; genitalia with rather short digital sclerites.

*Female.* Eyes coarsely setose. Mandibles well developed, heavily sclerotized; maxillary palpi 2-segmented, labial palpi 1-segmented. Antennal scape (Figure 296) $4\frac{1}{3}$ to $5\frac{1}{6}$ times as long as wide, about as long as or slightly longer than the club; pedicel $1\frac{3}{5}$ to $1\frac{3}{4}$ times as long as wide; first funicular segment triangular, about $1\frac{2}{3}$ to $1\frac{4}{5}$ times as wide as long; second segment symmetrical, considerably larger than the first segment, about $1\frac{3}{4}$ to $2\frac{1}{4}$ times as wide as long; third funicular segment $1\frac{1}{5}$ to $1\frac{1}{2}$ times as long as wide, subequal to or somewhat longer than the pedicel, bearing 2–4 longitudinal sensilla; club $2\frac{1}{2}$ to $3\frac{1}{4}$ times (but usually less than 3 times) as long as wide, about $2\frac{1}{4}$ to $2\frac{3}{5}$ times longer and somewhat wider than the preceding segment, bearing 7–14 longitudinal sensilla.

Setae on head, thorax and sides of abdomen coarse, dark, readily visible under $\times 30$ magnification. Vertex with 2 pairs of long setae along occipital margin, in addition to numerous shorter setae; mesoscutum (Figures 297, 298) with numerous setae (18–49), the posterior pair considerably longer and coarser than the others; each parapsis with 3 setae, each axilla with 1 seta; scutellum with 4 setae, the discoid sensilla considerably closer to the anterior than to the posterior pair. Frontovertex, pronotum and mesonotal sclerites reticulate. Scutellum trapezoidal, $\frac{4}{5}$ to nearly full median length of mesoscutum. Metanotum (Figures 297–299) v-shaped, obliquely reticulate-strigose except on sides; anteromedian apodeme slender, considerably longer than median length of metanotum.

Propodeum (Figures 299–301) 3 to $3\frac{1}{2}$ times as long as metanotum, $\frac{3}{5}$ to $\frac{2}{3}$ length of scutellum, reticulate-strigose on a median trapezoidal salient and just mesad of spiracles; median salient usually overlapping central part of second abdominal tergite; crenulae 3 + 3 to 6 + 6, distinct, rounded, overlapping, often obscured by the sculpture and pigmentation of the second tergite.

Second abdominal tergite (Figure 301) coarsely reticulate centrally, just below median salient of propodeum; tergites III–VII reticulate on the sides, bearing 2–5 strong setae on each reticulate area; third and fourth tergites transversely striated between the lateral reticulate areas, fifth tergite faintly so in large specimens, seventh tergite reticulate-striated across; eighth tergite transversely reticulate, bearing 13–24 setae in 1–2 arcuate, transverse rows between the spiracles (Figure 302); syntergum deeply v-shaped, reticulate, bearing numerous setae in 2–3 arcuate rows. Cerci (Figure 302) situated very close to and laterad of posterior spiracles, with 3 long setae and 1 short seta. Ovipositor shaft long, $2\frac{1}{5}$ to $2\frac{3}{4}$ times as long as the middle tibia (2.14–2.75), ovipositor sheaths about $\frac{1}{2}$ length of middle tibia (0.43–0.54).

Mid-tibial spur $\frac{3}{4}$ to nearly entire length of the corresponding basitarsus.

Forewing (Figure 303) $2\frac{1}{2}$ to nearly 3 times as long as wide; marginal fringe not exceeding $\frac{1}{6}$ width of disk (usually shorter). Delta area as densely setose as area distad of speculum; 2–3 irregular rows of setae below distal two thirds of submarginal vein, and 2 rows along posterior margin of wing below delta, are narrowly separated from the delta; costal cell with a row of fine setae along the proximal half and 1–2 coarse setae near the apex. Submarginal vein bearing 5–7 coarse setae and 14–23 bullae. Marginal vein bearing 7–12 prominent setae along anterior margin, these decreasing considerably in length toward apex of vein, the first (proximal) seta over twice as long as the last (distal) one, followed by a few small, "regular" setae on distal portion of vein; a row of subequal setae along center of vein.

Forewing mottled, with a distinctive pattern of conspicuous patches of coarse, dark setae against a background of fine, hyaline setae, similar to that of *A. maculatipennis*: a small patch of coarse setae in the delta, medially against the speculum; a v-shaped crossband below stigma, pointing toward apex of wing and extending across the wing (the arms of the v are sometimes narrowly separated at the apex), and 3 circular patches: 1 proximad of the crossband, between the arms of the v and sometimes almost merging with the latter, 2 distad of the crossband; the integument of the wing is very faintly infumated at the patches, the mottled effect being produced mainly by the contrasting setae; an infumated, arcuate band below junction of submarginal and marginal veins.

General coloration orange-yellow to whitish, with dark brown markings. Dorsum of head orange-yellow, face white, lower parts of face and cheek surrounding oral margin blackish. Pronotum dark brown centrally; mesoscutum light orange-yellow, suffused with brownish submedially, the anterior margin and parapsidal sutures narrowly dark brown; scutellum pale yellowish, with a pair of longitudinal, submedian brownish stripes, turning outward and continuing along posterior margin; scutello-axillar sutures dark brown; metanotum pale, central portion of anterior margin narrowly dark brown; propodeum pale; central portion of anterior margin, sides, and posterior margin broadly dark brown; crenulae dark brown. Dorsum of abdomen whitish, with dark brown markings as follows: triangular area on center of second tergite, third and fourth tergites except on the sides, sometimes narrowly on center of fifth tergite, seventh and eighth tergites except on the sides, and posterior margin of syntergum broadly; ovipositor sheaths dark brown. Thoracic and abdominal sterna fuscous. Internal apodemes and margins of endophragma dark brown. Antennal scape pale, ventral margin narrowly dark brown on basal half; pedicel slightly suffused with blackish; first 2 funicular segments and basal half of club dark brown, third funicular segment entirely pale, apical half of club orange-yellow. Legs whitish, with dark brown markings as follows: ventral margin of middle coxa, fore femur suffused ventrally, middle femur narrowly margined, a spot near apex of hind femur, an incomplete ring on fore tibia, ventral margin of middle tibia narrowly along basal half, 2 distinct rings on hind tibia; all tarsi yellow, pretarsi dark brown. Wing veins light brownish to colorless.

Length 0.81–1.18 mm.

*Male*. Similar to the female in general structure, sculpture and chaetotaxis, differing mainly in the structure of the antennae, wing pattern, and general coloration.

Antennal scape (Figures 304, 305) rather thick, $3\frac{1}{4}$ to a little over 4 times as long as wide, somewhat shorter than the club, bearing a ventral, elongate plate with 1 row of minute, elongate sensilla (Figure 305); pedicel about $1\frac{1}{2}$ to $1\frac{3}{5}$ times as long as wide, distinctly shorter than the third segment of the funicle; first 2 funicular segments about as in the female, third segment about $1\frac{2}{5}$ to $1\frac{3}{5}$ times as long as wide, bearing 1–2 longitudinal sensilla and numerous long setae; club about 3 times as long as wide, $2\frac{1}{4}$ to $2\frac{1}{2}$

times longer and somewhat wider than preceding segment, bearing 4–7 longitudinal sensilla and numerous long setae.

Propodeum (Figure 306) as in the female.

Cerci (Figure 307) with 2 long setae and 1 short seta. Genitalia about $\frac{2}{3}$ length of middle tibia (0.64–0.66), with a distinct, spatulate rod between the apodemes; digital sclerites short, about $\frac{1}{5}$ the combined length of aedeagus and apodemes (0.20–0.22).

Forewing (Figure 308) nearly hyaline; delta area slightly infumated, more strongly so below junction of submarginal and marginal veins; an elongate, infumated cloud below stigma, another one opposite it against posterior margin of wing, the two forming an almost continuous, lightly infumated crossband. Delta area less densely setose than the area distad of speculum, with about 60 somewhat coarser setae in 7 rows; setae immediately below stigma somewhat coarser and darker than the surrounding discal setae.

Dorsum of head orange-yellow, lower parts of face and cheeks more extensively blackish than in the female. Mesoscutum and scutellum entirely yellow, except for a pair of brownish spots near anterior margin of scutellum; mesonotal sutures lined with dark brown; propodeum brownish centrally and on posterior margin. Abdomen (Figure 307) entirely dark brown. Antennal scape pale, with a fuscous stripe along ventral margin; pedicel pale, funicle and club fuscous; hind femur suffused with dark brown dorsally; hind tibia dark brown centrally, ends pale; hind basitarsus fuscous.

Length 0.68–0.86 mm.

Redescribed from numerous specimens, including 1♀ paratype, reared by J. G. Gomes from the Florida red scale, *Chrysomphalus aonidum* (L.), on citrus, Rio de Janeiro State (km 34 on Rio de Janeiro-São Paulo Highway), Brazil, January 15, 1940; several ♀ paratypes of *Aphytis spiniferus*, collected by S. E. Flanders from citrus trees infested with *C. aonidum*, Rio de Janeiro and Limeira, Brazil, March and April 1958; and abundant ♀♂ material, reared by P. DeBach from the same host on citrus and from *Abgrallapsis cyanophylli* (Signoret) on ornamentals, as well as from *Pseudaonidia trilobitiformis* (Green) on citrus, at various localities in the states of Rio de Janeiro and Minas Gerais, Brazil, March to May, 1962.

Type series in the collections of the Escola Nacional de Agronomia and the Divisão de Defesa Sanitária Vegetal, Rio de Janeiro.

**Additional Material.** 5♀♀, reared by E. J. Hembleton from *Chrysomphalus aonidum* (L.), Campinas, Brazil, March 14, 1938; 14♀♀ reared by R. R. Toledo, same host and locality, March 1940; poorly preserved in balsam, U.S. National Museum; important diagnostic characters very difficult to discern due to inadequate method of mounting.

**Notes.** *A. costalimai* is rather closely related to *maculatipennis*. The wing pattern of the female is very similar to that of *maculatipennis*, the dark patches being more extensive in *costalimai*, the 2 oblong patches of *maculatipennis* uniting to form the v-shaped crossband in *costalimai*. This species also differs from *maculatipennis* in details of coloration, in the well-developed, overlapping crenulae, more numerous mesoscutal setae and sharply decreasing setae along the marginal vein of both sexes, in the longer ovipositor of the female, and in the shape of the sensilla on the antennal scape and shorter digital sclerites of the male.

Although some variation is evident in the pigmentation of the abdominal tergites of the female, the general pattern is the same, with the sides and center of the abdomen pale. *A. acutaspidis* and *A. perplexus*, described here as new species, differ from *costalimai* in having a continuous dark blotch on the dorsum of the abdomen, as well as in the shape of the propodeal crenulae and in certain other details of structure and pigmentation.

**Biology and Potential Value for Biological Control.** *A. costalimai* is a primary ectoparasite of the Florida red scale, *Chrysomphalus aonidum* (L.), in Brazil. The pupae (Figure 309) are shining black. This little-known species was recorded by DeBach (1963) as a rather common and partially effective natural enemy of this injurious pest on citrus in Brazil. No attempt has so far been made to transfer it to other countries.

## 5. **Aphytis acutaspidis** n.sp.

(Figures 310–316)

This South American species is very closely related to *A. costalimai*, differing from the latter mainly in the more extensive pigmentation of the thorax and abdomen, in the elongate, nonoverlapping crenulae on the propodeum, and in some details of wing pattern. It is described below relative to *costalimai*.

*Female.* Eyes coarsely setose. Mandibles well developed; maxillary palpi 2-segmented, labial palpi 1-segmented. Antennal scape (Figure 310) about 5 to $5\frac{3}{4}$ times as long as wide, somewhat longer than the club; pedicel rather slender, about twice as long as wide or nearly so, about as long as or slightly shorter than the third segment of the funicle; first funicular segment triangular, about $1\frac{3}{5}$ to $1\frac{4}{5}$ times as wide as long; second segment symmetrical, considerably larger than the first, $1\frac{2}{3}$ times to nearly twice as wide as long; third funicular segment about $1\frac{3}{5}$ times as long as wide, bearing 3–4 longitudinal sensilla; club tapering, a little less than 3 times as long as wide, about $2\frac{1}{4}$ to $2\frac{1}{2}$ times longer and a little wider than the preceding segment, bearing 10–11 longitudinal sensilla.

Setae on head, thorax and abdomen as in *costalimai*; mesoscutum (Figures 311, 312) with 42–54 setae. Sculpture of head and thorax as in *costalimai*. Other thoracic characters as in *costalimai*.

Propodeum (Figures 311, 313, 314) 3 to $3\frac{3}{5}$ times as long as the metanotum, about $\frac{1}{2}$ to $\frac{3}{5}$ length of scutellum, reticulate-strigose on a trapezoidal median salient and just mesad of spiracles; the median salient is nearly triangular, narrower posteriorly than in *costalimai*, bearing $3+4$ to $5+5$ elongate, tapering, nonoverlapping, somewhat irregular crenulae.

Abdominal sculpture and chaetotaxis (Figures 311, 313–315) as in *costalimai*, but all the tergites are coarsely, transversely striated centrally, between the lateral reticulate areas. Syntergum (Figure 315) v-shaped, as in *costalimai*. Cerci situated very close to and laterad of posterior spiracles, with 3 long setae and 1 short seta. Ovipositor shaft distinctly somewhat over twice as long as the middle tibia (2.12–2.20), ovipositor sheaths about $\frac{1}{2}$ length of middle tibia (0.48–0.51).

Mid-tibial spur about $\frac{4}{5}$ length of the corresponding basitarsus.

Forewing (Figure 316) about $2\frac{2}{5}$ to $2\frac{1}{2}$ times as long as wide, marginal fringe not exceeding $\frac{1}{7}$ width of disk. Delta area as densely setose as the area distad of speculum; 3–4 rows of setae below distal half of submarginal vein, 3 along posterior margin of wing below delta; costal cell with a row of fine setae along proximal half and 1–3 prominent setae near apex. Submarginal vein bearing 6–7 coarse setae and 14–20 bullae. Marginal vein bearing 12–14 prominent setae along anterior margin, these decreasing considerably in length toward apex of vein, as in *costalimai*.

Forewing pattern as in *costalimai*, except that the v-shaped blotch of *costalimai* is rather widely separated by a clear area at the apex in *acutaspidis*.

Coloration very similar to *costalimai*; mesoscutum and scutellum more uniformly suffused with brownish; dorsum of abdomen (Figure 315) with a continuous dark brown blotch centrally on tergites III–VIII, and short, dark transverse stripes on sides of tergites III–VI. Antennal scape with a rather conspicuous longitudinal dark brown stripe; pedicel and third funicular segment lined with brownish ventrally.

Length 1.18–1.28 mm.

*Male*. Unknown.

Described from 4♀♀ (holotype and paratypes), reared by P. DeBach from *Acutaspis albopicta* (Cockerell) on coconut palm, Rural University, km 47 on the old Rio de Janeiro-São Paulo highway, Estado do Rio de Janeiro, Brazil, July 14, 1962; mounted in Hoyer's medium on 3 slides; 2 paratypes headless.

Type series in the collection of the Division of Biological Control, University of California, Riverside.

**Notes:** *A. acutaspidis* is very closely related to *costalimai*, but may be readily separated from this species by several valid characters. The pigmentation of the thorax and abdomen is considerably more extensive in *acutaspidis*; the forewing pattern is different from that of *costalimai* in that the arms of the v-shaped crossband are separated at the apex, whereas these arms are consistently united in *costalimai*; the elongate, non-overlapping propodeal crenulae are probably the most reliable character separating *acutaspidis* from *costalimai*.

The scale insect host was found on the basal part of the frond, near the trunk of coconut palms. *A. acutaspidis* is definitely a primary ectoparasite. This species has black pupae and cast skins and irregular meconial pellets.

### 6. **Aphytis perplexus** n.sp.

(Figures 317–327)

This South American species is very closely related to *costalimai* and *acutaspidis*, differing from both species in the very indistinct crenulae which are non- or only slightly overlapping, as well as in certain details of coloration and wing pattern. It is described below relative to *costalimai*.

*Female*. Eyes coarsely setose. Mandibles well developed; maxillary palpi 2-segmented, labial palpi 1-segmented. Antennal scape (Figure 317) $4\frac{3}{4}$ to $5\frac{3}{4}$ times as long as wide, a trifle longer than the club; pedicel about $1\frac{1}{2}$ times to nearly twice as long as wide, as long as or somewhat longer than the third segment of the funicle; first funicular segment triangular, $1\frac{3}{4}$ times to fully twice as wide as long; second segment symmetrical, considerably larger than the first, $1\frac{3}{4}$ to $2\frac{1}{4}$ times as wide as long; third funicular segment about $1\frac{1}{7}$ to $1\frac{1}{3}$ times as long as wide, bearing 1–3 longitudinal sensilla; club $2\frac{3}{4}$ to fully 3 times as long as wide, $2\frac{3}{5}$ to 3 times longer and somewhat wider than the preceding segment, bearing 9–11 longitudinal sensilla.

Setae on head, thorax and abdomen as in *costalimai*; mesoscutum (Figure 318) with 21–35 setae. Sculpture of head, thorax and abdomen as in *costalimai*, central abdominal tergites vaguely transversely striated between the lateral reticulate areas. Anteromedian apodeme of metanotum $1\frac{2}{3}$ times to nearly twice median length of metanotum.

Propodeum (Figures 319–321) $3\frac{3}{4}$ to 4 times as long as the metanotum, $\frac{3}{5}$ to $\frac{3}{4}$ length of scutellum, reticulate-strigose on a median trapezoidal salient and just mesad of spiracles, the median salient considerably narrower posteriorly than in *costalimai*; crenulae (Figures 319, 320) 2 + 3 to 3 + 4, very indistinct, obscured by the coarse reticulation on the second abdominal tergite, roundly elongate, not or only slightly overlapping.

Abdominal tergites (Figure 321) as in *costalimai*. Cerci with 3 long setae and 1 short seta. Ovipositor shaft about twice as long as the middle tibia (1.92–2.13), ovipositor sheaths about $\frac{2}{5}$ length of middle tibia (0.39–0.45).

Mid-tibial spur $\frac{4}{5}$ to $\frac{9}{10}$ length of the corresponding basitarsus.

Forewing (Figure 322) $2\frac{1}{3}$ to nearly $2\frac{1}{2}$ times as long as wide, marginal fringe not exceeding $\frac{1}{5}$ width of disk in the smallest specimen. Discal setae as in *costalimai*. Submarginal vein with 6 coarse setae and 14–15 bullae; setae on marginal vein as in *costalimai*.

Forewing pattern very similar to that of *costalimai*, but the arms of the v-shaped crossband are rather widely separated at the apex as in *acutaspidis*; the upper (anterior) arm merges with the corresponding circular patch in all but the smallest specimen.

Coloration very similar to *costalimai*; mesoscutum and scutellum (Figure 318) more uniformly suffused with brownish, more extensively so than in *acutaspidis*; dorsum of abdomen (Figure 321) with a continuous dark brown blotch centrally on tergites III–VIII, the sides pale. Antennal scape with a conspicuous longitudinal dark brown stripe, pedicel narrowly lined with brown ventrally, basal two thirds of club dark brown, darker than in the preceding two species.

Length 0.79–1.17 mm.

*Male*. Very similar to *costalimai*, differing mainly in the structure of the propodeum and in details of coloration.

Antennal scape (Figure 323) 4 times as long as wide, somewhat shorter than the club, bearing 1 row of sensilla as in *costalimai*; pedicel about $1\frac{1}{3}$ times as long as wide; third funicular segment somewhat over $1\frac{2}{3}$ times as long as wide, about $1\frac{1}{3}$ times longer than the pedicel, bearing 2–3 longitudinal sensilla and numerous long setae; club $3\frac{2}{3}$ times as long as wide, $2\frac{1}{2}$ times longer and somewhat wider than the preceding segment, bearing 5–7 longitudinal sensilla and numerous long setae.

Median salient of propodeum (Figure 324) considerably narrower posteriorly than in *costalimai*, posterior margin emarginate, bearing 3 + 3 to 3 + 4 nonoverlapping crenulae.

Cerci (Figure 325) with 2 long setae and 1 short seta. Genitalia (Figure 326) $\frac{2}{3}$ to $\frac{3}{4}$ length of middle tibia, with a conspicuous longitudinal rod between the apodemes; digital sclerites short, about $\frac{1}{5}$ length of middle tibia (0.18–0.19).

Forewing (Figure 327) as in *costalimai*, the infumated crossband below stigma somewhat more distinct, continuous, not composed of two separate clouds.

All mesonotal sclerites rather uniformly suffused with brownish, most extensively so on mesoscutum; abdomen (Figure 325) uniformly dark brown, apparently somewhat paler on sides of tergites.

Length 0.85 mm.

Described from 4♀♀ and 2♂♂ (♀ holotype, ♂ allotype, and paratypes; the ♂ paratype headless), reared by P. DeBach from the latania scale, *Hemiberlesia lataniae* (Signoret), on an ornamental palm, Rural University, Rio de Janeiro State, Brazil, April 25, 1962.

Type series in the collection of the Division of Biological Control, University of California, Riverside.

**Notes.** *A. perplexus* is obviously very closely related to both *costalimai* and *acutaspidis*. It is nevertheless considered here to be a distinct species, and may be separated by several valid characters. The propodeal crenulae, which in the female can be discerned only with the aid of a high-power, oil-immersion phase-contrast objective, are nonoverlapping, more elongate than in *costalimai* but broader and more rounded than in *acutaspidis*. The thorax and antennal club are more extensively pigmented than in both *costalimai* and *acutaspidis*; abdominal pigmentation differs from *costalimai* in the presence of a continuous dark brown blotch, not interrupted by pale color centrally, from *acutaspidis* in the absence of short transverse stripes on the sides of tergites III–VI. The forewing pattern differs from *costalimai* in that the arms of the v-shaped crossband are widely separated at the apex and from *acutaspidis* in that the anterior arm tends to merge with the anterior circular patch.

**Potential Value for Biological Control.** This little-known species may prove to be of some value in the biological control of the latania scale, which is an important pest of avocado in California and elsewhere.

### 7. **Aphytis cochereaui** DeBach and Rosen

(Figures 328–342)

*Aphytis cochereaui* DeBach and Rosen, 1976, Ann. Entomol. Soc. Amer., **69**: 541–542.

This striking South Pacific species can be readily recognized by the following characters: Head and thorax largely pale, abdomen dark brown; mesoscutal setae numerous;

propodeal crenulae mostly triangular, nonoverlapping, cerci with 2 long setae and 1 short seta in both sexes. Female forewing with an extensive pattern of patches of dark, coarse setae, but less contrasting than in the preceding species, the setae along the marginal vein subequal in length. Male antennae 6-segmented, with 2 minute ring segments and a greatly enlarged club bearing a peculiar sensory area on the ventral surface; forewing uniformly infumated from base to stigma.

*Female.* Eyes (Figure 328) coarsely setose; ocelli in a nearly equilateral triangle, the posterior pair about their own diameter from the orbits. Mandibles (Figure 329) well developed, with a ventral denticle and a dorsal truncation; maxillary palpi 2-segmented, labial palpi 1-segmented. Antennal scape (Figure 330) somewhat widened ventrally, $3\frac{1}{2}$ to 5 times as long as wide (usually about 4 times), about as long as the club; pedicel $1\frac{1}{3}$ to $1\frac{2}{3}$ times as long as wide, distinctly longer than the third segment of the funicle; first funicular segment triangular, nearly twice as wide as long; second segment considerably larger than the first, trapezoidal, with the dorsal aspect longer than the ventral, fully twice as wide as long or somewhat wider; third funicular segment nearly symmetrical, dorsal aspect slightly longer than the ventral, somewhat to about $1\frac{1}{4}$ times as wide as long, bearing 3–5 longitudinal sensilla; club moderately tapering, $2\frac{3}{5}$ to a little over 3 times as long as wide, $3\frac{1}{4}$ to $3\frac{4}{5}$ times longer and about as wide as the preceding segment, bearing 15–24 longitudinal sensilla.

Seate on head and thorax coarse, dark, readily visible under $\times$ 30 magnification, apparently composed of distinct filaments (see Figure 332). Vertex with 2 pairs of long setae along the occipital margin, in addition to numerous short setae. Mesoscutum (Figure 331) with numerous setae (23–46), the posterior pair considerably longer and coarser than the others; each parapsis with 3 setae, each axilla with 1 seta; scutellum with 4 setae, the discoid sensilla somewhat closer to the posterior than to the anterior pair. Frontovertex, pronotum and mesonotal sclerites densely reticulate. Scutellum ellipsoidal, somewhat over $\frac{2}{3}$ median length of mesoscutum. Metanotum (Figures 332, 333) relatively long, of uniform length, arcuate, coarsely reticulate except on the sides; anteromedian apodeme (Figure 332) robust, about $\frac{2}{3}$ median length of metanotum.

Propodeum (Figures 333–335) relatively long, only 2 to $2\frac{3}{5}$ times longer than the long metanotum but usually about $\frac{5}{6}$ length of scutellum, coarsely reticulate on a triangular median salient (in direct continuation of the reticulation on the metanotum; anterior margin of propodeum, normally overlapped by metanotum, smooth); crenulae 6+7 to 9 + 10, mostly nonoverlapping, elongate and tapering on center of posterior margin, hardly distinguishable from sculpture near tip of median salient, becoming larger and more rounded toward the sides.

Second abdominal tergite smooth; tergites III–VII reticulate on the sides, bearing several long, dark, rather slender setae in a transverse row on each reticulate area; no distinct sculpture on the center of these tergites, but the transverse plates are very clearly distinguishable; eighth tergite lightly aciculate-reticulate across the center, more strongly reticulate on the sides, around the spiracles; syntergum (Figure 336) triangular, considerably wider than long, faintly reticulate, the anterior margin somewhat arcuate; a transverse row of fine setae across seventh tergite, 1–2 rows across eighth tergite, 3 rows across syntergum. Cerci situated very close to posterior spiracles, with 2 long setae and 1 short seta. Ovipositor shaft $1\frac{1}{2}$ to $1\frac{3}{4}$ times as long as the middle tibia (1.51–1.77), ovipositor sheaths about $\frac{2}{5}$ length of middle tibia (0.36–0.42). Ridge on outer ovipositor plates reaching almost to end of plate (Figure 336).

Mid-tibial spur robust, usually about $\frac{4}{5}$ length of the corresponding basitarsus; strong peglike spines on the latter and near apex of middle tibia.

Forewing (Figure 337) usually $2\frac{1}{2}$ to $2\frac{3}{4}$, rarely 3 times as long as wide; marginal fringe usually $\frac{1}{10}$ width of disk, about $\frac{1}{6}$ width in smallest specimens. Delta area as densely setose as the area distad of speculum, not clearly separated from an equally densely setose area below submarginal vein and from a short row of coarse setae along posterior margin of wing; costal cell with a longitudinal row of fine setae and 2–3 coarse setae near the apex. Submarginal vein bearing 6–7 coarse setae and 16–24 bullae. Marginal vein bearing 11–16 prominent setae along anterior margin, these subequal or slightly decreasing in length toward apex of vein, somewhat longer than the setae in a row along center of vein.

Forewing moderately mottled, with patches of coarse, dark setae against a background of areas bearing fine, transparent setae; the integument of the wing is not pigmented, the mottled effect being produced by the contrasting setae alone. Coarse, dark setae below submarginal vein; a large dark ring in delta, as well as along proximal margin of speculum and at tip of delta; a large patch just distad of speculum; an oblong patch below stigma, another one opposite it against posterior margin of wing; a crossband about halfway between the oblong patches and apex of wing, widest at center. Basal half of wing generally slightly infumated.

General coloration orange-yellow with white areas and extensive black or blackish markings. Vertex orange-yellow, paler on occipital margin; face whitish; occiput blackish on both sides of foramen, a blackish streak along fronto-genal and fronto-clypeal sutures. Propleura blackish distally; mesoscutum whitish, more or less suffused with pale orange-yellow, all margins narrowly black, with irregular blackish markings, mainly around bases of setae; other mesonotal sclerites dirty whitish; scutellum suffused with blackish submedially, with 2 blackish spots near anterior margin, posterior margin narrowly black; metanotum white, anterior margin narrowly black; propodeum (Figure 333) dirty whitish, grading into blackish on the sides; median salient (sculptured area) with a black triangle on the anterior part, followed by a white v-shaped stripe posteriorly; crenulae dusky. Second abdominal tergite blackish; tergites III–VIII (Figure 336) with alternating black and white transverse crossbands which are most conspicuous on the sides and are obscured medially by a continuous blackish blotch; the prominent transverse ridge on the posterior margin of the eighth tergite, just above the cerci, black; syntergum white; ovipositor sheaths and lateral portions of outer ovipositor plates, laterad of ridge, black. Sides and underparts of thorax whitish, sterna extensively infuscated; internal apodemes and margins of endophragma black. Antenna (Figure 336) with alternating black and white transverse crossbands which are most and third funicular segment dorsally, and tip of club; in cleared specimens, the transparent longitudinal sensilla and bases of the setae stand out in contrast to the black background of the club. Legs white, the following parts suffused with black or blackish: fore femur faintly, mainly dorsally, fore tibia on proximal two thirds, with a spot on the very tip, fore tarsus uniformly suffused; middle coxa ventrally, middle femur with a broad stripe dorsally, middle tibia with a narrow streak ventrally and a spot on the tip; hind coxa ventrally, hind femur ventrally on distal half, hind tibia ventrally on proximal half and broadly on apex; middle tibia sometimes turning to orange-yellow toward apex. Wing veins dark brown.

Length 0.89–1.35 mm.

*Male.* Similar to the female in structure, chaetotaxis and sculpture, differing mainly in the shape of the antennae, wing pattern and details of coloration.

Eyes coarsely setose. Antennal scape (Figure 338) 4–5 times as long as wide; pedicel about $1\frac{1}{4}$ times as long as wide, usually distinctly shorter than the third segment of the funicle; first 2 funicular segments minute, subequal, very flattened rings, the first triangular, the second more nearly symmetrical; third funicular segment trapezoidal, dorsal aspect longer than the ventral, about as long as wide, bearing numerous longitudinal sensilla; club robust, about $2\frac{1}{2}$ to 3 times as long as wide, considerably (1.10–1.30 times) longer than the scape, about $2\frac{2}{3}$ to $3\frac{2}{5}$ times longer and somewhat wider than the preceding segment, bearing numerous, relatively short longitudinal sensilla; ventral surface of club bearing a peculiar sensory area (Figure 339) consisting of an oval area with delicate, minute setae, bordered by 3–4 rows of short, dense setae on one side and one row of sparser small setae on the other side; this sensory area is best seen in flattened antennae.

Mesoscutum with 15–43 setae. Propodeum (Figure 340) somewhat shorter than in the female, only about $1\frac{2}{3}$ to 2 times as long as the metanotum, about $\frac{3}{5}$ length of scutellum; crenulae 4 + 4 to 6 + 8, usually somewhat more rounded, more readily distinguishable than in the female. Genitalia (Figure 341) about $\frac{3}{5}$–$\frac{2}{3}$ length of middle tibia (0.57–0.71), with a conspicuous, spatulate rod between the apodemes; digital sclerites slender, about $\frac{1}{5}$ to $\frac{1}{4}$ the combined length of aedeagus and apodemes (0.19–0.25).

Forewing (Figure 342) usually $2\frac{1}{4}$ to $2\frac{1}{2}$ (rarely $2\frac{3}{5}$) times as long as wide, marginal fringe not exceeding $\frac{1}{6}$ width of disk. Delta area somewhat less densely setose than the area distad of the speculum, with about 5–10 rows of somewhat coarser setae; no pattern of contrasting coarse and fine setae on disk. Submarginal vein bearing 5–7 coarse setae and 13–24 bullae. Marginal vein bearing 8–14 prominent, subequal setae along anterior margin.

Forewing not mottled, slightly, uniformly infumated from base to somewhat beyond stigmal vein, with a distinct, darker spot below the stigma.

General coloration of head and thorax orange-yellow, abdomen dark brown. Vertex deep orange-yellow, face whitish; pronotum, mesonotal sclerites and metanotum deep orange-yellow, with black or blackish markings as in the female; propodeum orange-yellow, with a black triangle on anterior part of median salient. Abdomen entirely dark brown. Sides and underparts of thorax whitish, sterna infuscated. Antennal scape pale, with a dark spot at the apex; other parts of antenna uniformly infumated. Coloration of legs very similar to that of the female.

Length 0.59–1.08 mm.

Described from the following material: 17♀♀ and 14♂♂ (♂ allotype and paratypes), reared by P. Cochereau from a mixture of armored scale insects on *Citrus nobilis*,

Noumea, New Caledonia, October 1967; 1♂ (paratype), reared by P. Cochereau from an undetermined host on *Casuarina equisetifolia*, Noumea, October 2, 1967; 1♂ (paratype), reared by P. DeBach from an undetermined host on *Citrus aurantium*, Noumea, April 21, 1968; 5♀♀, 3♂♂ (♀ holotype and paratypes), reared by P. DeBach from the purple scale, *Cornuaspis* [= *Lepidosaphes*] *beckii* (Newman), on *Murrayea exotica*, Sarramea, New Caledonia, April 24, 1968; 13♀♀ and 7♂♂ (paratypes), reared by P. Cochereau from *L. beckii* on same plant, Noumea, May 16, 1968; 1♂ (paratype), reared by G. Fabres from *C. beckii* on an undetermined plant, Noumea, March 20, 1969; all mounted in Hoyer's medium. Several ♀♂ specimens, reared at the insectary of the University of California, Riverside, on the cactus scale, *Diaspis echinocacti* (Bouché), during the latter half of 1968, have also been studied but are not included in the type series. Some of the latter specimens have been dissected on slides or mounted on points.

Type series in the collection of the Division of Biological Control, University of California, Riverside.

**Notes.** Fabres (1973, 1974, 1975) used our manuscript name of *cochereaui*, obtained from personal communication with the authors, but did not provide any description or distinguishing characters. As pointed out by DeBach and Rosen (1976b), his treatment of "*Aphytis cochereaui* Rosen" should be regarded as a *nomen nudum*, whereas our description (DeBach and Rosen, loc. cit.) constitutes the original description of this species.

According to Fabres (op. cit.), *A. cochereaui* is a biparental, solitary, primary ectoparasite of the purple scale on citrus in New Caledonia. It attacks ovipositing females of this host, laying its egg in the egg-mass of the scale insect beneath the covering scale, and the young larva feeds exclusively on host eggs. The adult female may host-feed on male scales. This species has not yet been obtained from any other scale insect in nature. However, since the purple scale presumably is an introduced species in New Caledonia, this endemic parasite must have evolved on another host. In fact, it may be reared in the laboratory on the cactus scale, and seems to do better on that host than on the purple scale.

Fabres (1977) reared *A. cochereaui* in the laboratory under an average temperature of 22°C and 79% RH, and reported that the longevity of adults averaged 8 days under these conditions. Fecundity was very low, averaging a total of 5 eggs per female (maximum 9). The sex ratio was 1:1. The life cycle was long, averaging 35 days under these conditions.

The pupa and cast skin of *cochereaui* are entirely black.

*A. cochereaui* resembles *chilensis* in the small, triangular first funicular segment of the female. Likewise, the peculiar male antenna of *cochereaui*, with 2 ringlike segments and an enlarged club, resembles the male antennae of *chilensis* and other members of the **chilensis** group. It is quite possible that *cochereaui* represents a link between the **vittatus** group and the **chilensis** group.

This species was named in honor of Dr. Paul Cochereau, formerly of ORSTOM, Noumea, New Caledonia and now at ORSTOM, Bouake , Ivory Coast.

**Potential Value for Biological Control.** According to Fabres (1974, 1975), *A. cochereaui* is a rather ineffective parasite of purple scale in New Caledonia, and does not seem to compare well with *A. lepidosaphes*. However, it does have the advantage of being more resistant to dry climatic conditions. It should be considered as a possible candidate for introduction into areas where *lepidosaphes* does not effect complete control of this serious pest of citrus.

*A. cochereaui* is very closely related to *A. noumeaensis* (Howard), differing from this species only in certain details of coloration. Since *noumeaensis* is known only from a single, partly cleared female specimen, it is redescribed below relative to *cochereaui*.

### 8. **Aphytis noumeaensis** (Howard)

(Figures 343, 344)

*Perissopterus noumeaensis* Howard, 1907, U.S. Dept. Agr. Bur. Entomol. Tech. Ser. No. **12**, Part 4:87.
*Marietta noumeaensis*: Compere, 1936, Univ. Calif. Publ. Entomol., **6**:312.
*Aphytis noumeaensis*: DeBach and Rosen, 1976, Ann. Entomol. Soc. Amer., **69**:541.

This little-known South Pacific species is undoubtedly an *Aphytis*. It is very closely related to *cochereaui*, from which it differs mainly in details of coloration. It is described here relative to *cochereaui*.

*Female*. Eyes as coarsely setose as in *cochereaui*. Antennae very similar to those of *cochereaui*, but the first 2 funicular segments appear more flattened, distinctly over twice as wide as long; third funicular segment about as long as wide, more distinctly trapezoidal; club about $2\frac{8}{9}$ times as long as the preceding segment (however, all these apparent differences may have been affected by differences in the mounting procedure).

Structure, chaetotaxis and sculpture of head, thorax and propodeum essentially as in *cochereaui*. In the single specimen available for study, the mesoscutum (Figure 343) bears about 30 setae; one parapsis with 3 setae, the other with 2; the discoid sensilla on the scutellum not clearly seen; propodeal crenulae present but not clearly seen (Figure 344), apparently similar to those of *cochereaui*.

Abdominal tergites apparently as in *cochereaui*. Cerci not clearly seen. Ovipositor and sheaths as in *cochereaui*; ridge on outer ovipositor plates fading at about $\frac{1}{2}$ length of plate.

Mid-tibial spur as in *cochereaui*.

Forewing fully $2\frac{3}{4}$ times as long as wide; marginal fringe about $\frac{1}{7}$ width of disk; row of coarse setae along posterior margin of wing, below delta, longer than in *cochereaui*; otherwise as in *cochereaui*; bullae on submarginal vein not clearly seen; marginal vein bearing 14–16 prominent, subequal setae along anterior margin. Pattern of forewing as in *cochereaui*.

General coloration very similar to *cochereaui*, differing as follows: Mesoscutum, scutellum, metanotum and propodeum (Figures 343, 344) uniformly suffused with brownish, midline of mesoscutum and scutellum and small circular areas surrounding the anterior pair of setae on scutellum paler; no darker spots surrounding bases of setae on mesoscutum; no contrast in coloration between metanotum and propodeum; posterior parts of propodeum, surrounding median salient, indistinctly paler, without conspicuous v-shaped white stripe. Transverse ridge above cerci pale; outer ovipositor plates immaculate; ovipositor sheaths faintly infuscated. Middle tibia infuscated on both ends, without the narrow ventral streak; hind femur faintly infuscated dorsally on distal half.

Length 0.83 mm.

*Male*. Unknown.

Redescribed from 1♀ (holotype, single specimen known), reared by A. Koebele from an *Aspidiotus* sp. ("not *destructor*") on coconut palm, Noumea, New Caledonia, October 1899.

Holotype in U. S. National Museum (Cat. No.10314).

**Notes.** *A. noumeaensis* differs from *cochereaui* mainly in the more uniformly fuscous coloration of the mesonotal sclerites, metanotum and propodeum, without the contrasting white and dark pattern characteristic of *cochereaui*, and in the paler ovipositor plates and sheaths. Antennal differences may not be significant, but the length of the ridge on the outer ovipositor plates may be regarded as a reliable diagnostic character.

A comparison between *noumeaensis* and *cochereaui* raises an interesting question. The two species, although very similar, appear to be morphologically distinct. However, both were collected at the same locality, almost 70 years apart. Are they indeed two discrete, closely related, sympatric species, or is it possible that "*noumeaensis*" has evolved into "*cochereaui*" during the last 7 decades? Only the recovery of *noumeaensis* from New Caledonia will be considered sufficient proof of the distinctness of these two species.

### 9. **Aphytis argenticorpus** n.sp.

(Figures 345, 346)

This Australian species appears to be closely related to *cochereaui*, differing from the latter species mainly in rather conspicuous details of the generally paler coloration and more extensive wing pattern. The single specimen available for study is not sufficiently cleared, and certain characters—especially details of sculpture—cannot be seen. It is described below relative to *cochereaui*.

*Female*. Eyes (Figure 345) coarsely setose; ocellar triangle nearly right-angled, apparently somewhat more obtuse than in *cochereaui*. Antennal scape about 5 times as long as wide, a little longer than the club; pedicel (Figure 345) about $1\frac{2}{3}$ times as long as wide, nearly $1\frac{1}{2}$ times longer than the third segment of the funicle; first funicular segment triangular, about $1\frac{1}{4}$ times as wide as long; second segment more symmetrical, somewhat trapezoidal, a trifle shorter and considerably wider than the first segment, about $1\frac{3}{4}$ times as wide as long; third funicular segment somewhat trapezoidal, somewhat wider than long, bearing 4 longitudinal sensilla; club nearly 3 times as long as wide, about $3\frac{1}{2}$ times longer and slightly wider than the preceding segment, bearing about 13 longitudinal sensilla.

Setae on head and thorax coarse, dark, apparently composed of distinct filaments, probably a little more slender than in *cochereaui*. Vertex with 2 pairs of long setae along occipital margin, in addition to numerous short setae. Mesoscutum apparently with 16 setae, each parapsis with 3, each axilla with 1 seta; scutellum with 4 setae, the discoid sensilla closer to the posterior than to the anterior pair. Frontovertex, pronotum, and mesonotal sclerites apparently reticulate. Metanotum relatively long, anteromedian apodeme and sculpture unseen.

Propodeum, as far as can be seen in the uncleared specimen, apparently long, $2\frac{1}{4}$ times as long as the metanotum but not much shorter than the scutellum (36:40); sculpture and crenulae not seen.

Abdominal tergites III–VII reticulate on the sides, bearing several long, dark, rather slender setae in a transverse row on each reticulate area, the transverse tergal plates apparently clearly distinguishable; seventh tergite with a transverse row of 4 fine setae across center; eighth tergite apparently aciculate-reticulate across center, with a transverse row of 8 fine setae between spiracles; syntergum bearing about 18 setae in 2 transverse rows. Cerci with 2 long setae and 1 short seta. Ovipositor shaft nearly $1\frac{3}{5}$ times as long as the middle tibia (1.55), ovipositor sheaths $\frac{2}{5}$ length of middle tibia (0.40).

Mid-tibial spur apparently a little less than $\frac{3}{4}$ length of the corresponding basitarsus.

Forewing (Figure 346) relatively narrow, fully 3 times as long as wide; marginal fringe not exceeding $\frac{1}{8}$ width of disk. Delta area as densely setose as the area distad of the speculum, clearly separated from an equally densely setose area below the submarginal vein, but not from the row of setae along the posterior margin of the wing; costal cell with a row of 5 fine setae along proximal three fifths, and 5–6 coarse setae near apex. Submarginal vein bearing 5–6 coarse setae and 24–25 bullae. Marginal vein bearing 12 prominent, subequal setae along anterior margin, these somewhat longer than the setae in a row along center of vein.

Forewing extensively mottled, with patches of coarse, dark setae against a background of areas bearing fine, transparent setae; the integument of the wing is not pigmented, the mottled effect being produced by the contrasting setae. Dark patches as follows: below distal part of submarginal vein; in delta as in *cochereaui*: a large ring, as well as proximal margin of speculum and a patch at tip of delta; a v-shaped patch, with one arm projecting disto-posterad from stigma, the other arm almost fused with the tip of a subapical, v-shaped crossband; an oval patch against anterior margin of wing, about halfway between the stigma and the anterior arm of the v-shaped crossband; six distinct patches in posterior half of wing, distad of speculum.

General coloration orange-yellow, extensively marked with silvery on head, thorax and abdomen. A pair of transverse black stripes on occiput, on each side of foramen. Parapsidal sutures black, posterior margins of mesoscutum and scutellum narrowly lined with blackish; fuscous spots surrounding bases of mesonotal setae; propodeum fuscous antero-medially, the sutures mesad of spiracles and the lateral parts of posterior margin black. Second abdominal tergite fuscous; blackish crossbands along posterior margins of tergites III–VII; transverse ridge on posterior margin of eighth tergite lined with blackish; ovipositor sheaths blackish, margins of ovipositor plates and ridge on outer plate lined with blackish. Thoracic sterna strongly infuscated; lateral margins of endophragma blackish. Antenna pale, club rather faintly infuscated. All femora with blackish markings ventrally and on sides; all tibiae with black tips and 2 distinct, incomplete blackish rings; apical segment of all tarsi blackish. Wing veins colorless.

Length 1.14 mm.

*Male*. Unknown.

Redescribed from 1♀ (holotype) mounted in balsam, labeled "*Perrissopterus argenticorpus* Gir., Type ♀, Queensland", with no additional data. The slide carried, under separate cover slips, the type of *Aphytis angeloni* (Girault) (see p. 730) and specimens of *Ufens pretiosa* (Girault) (Hymenoptera: Trichogrammatidae). The specimen of *argenticorpus* is marked as type No. 3762, Queensland Museum.

**Notes.** Girault apparently never published the description of *argenticorpus*. The species is, therefore, described here as new, with Girault's manuscript name.

*A. argenticorpus* differs markedly from *cochereaui* in coloration. It is generally lighter than *cochereaui*, lacking the black markings on the face, the large black blotch on the abdomen, and the black markings on the antennae; the tibiae are distinctly spotted or banded. In the forewing, the delta is clearly separated from the setose area below the submarginal vein, and the dark patches are broken into more separate areas than in *cochereaui*, the resulting pattern being conspicuously different. The apparently smaller number of mesoscutal setae may be another character separating this Australian species from *cochereaui*.

## 10. **Aphytis fabresi** DeBach and Rosen

(Figures 347–357)

*Aphytis fabresi* DeBach and Rosen, 1976, Ann. Entomol. Soc. Amer., **69**:542.

This interesting South Pacific species is rather similar to *cochereaui* in general structure and wing pattern, differing mainly in coloration: Female pale, with infuscated antennae and spots on sides of abdominal tergites; forewing rather faintly but extensively mottled. Male almost entirely pale, with infuscated antennae; forewing hyaline with a dark spot below the stigma; antennal scape bearing a single sensillum, the club with a specialized sensory area ventrally.

*Female*. Eyes coarsely setose. Mandibles well developed, as in *cochereaui*; maxillary palpi 2-segmented, labial palpi 1-segmented. Antennal scape (Figure 347) slender, $4\frac{2}{3}$ to 6 times as long as wide, about as long as the club; pedicel about $1\frac{2}{5}$ to $1\frac{1}{2}$ times as long as wide, distinctly longer than the third segment of the funicle; first funicular segment triangular, $1\frac{4}{5}$ to 2 times as wide as long; second segment symmetrical, considerably larger than the first, about $2\frac{1}{5}$ to $2\frac{3}{5}$ times as wide as long; third funicular segment about as long as wide, bearing 3–5 longitudinal sensilla; club tapering, $2\frac{4}{5}$ to $3\frac{1}{4}$ times as long as wide, $3\frac{1}{4}$ to $3\frac{1}{2}$ times longer and somewhat wider than the preceding segment, bearing 12–16 longitudinal sensilla.

Setae on head and thorax coarse, dark, readily visible under × 30 magnification; those on sides of abdomen more slender, visible under × 60 magnification; the largest setae appear as if composed of distinct filaments. Vertex with 2 pairs of long setae along occipital margin, in addition to numerous shorter setae. Mesoscutum (Figure 348) with 16–24 setae, the posterior pair and 1 seta at each antero-lateral corner considerably longer and coarser than the others; each parapsis with 2 setae, each axilla with 1 seta (2 in one specimen); scutellum with 4 setae, the discoid sensilla about equidistant from the two pairs. Frontovertex, pronotum and mesonotal sclerites reticulate. Scutellum ellipsoidal, about $\frac{3}{4}$ median length of mesoscutum. Metanotum arcuate, reticulate except on the sides; anteromedian apodeme robust, distinctly shorter than median length of metanotum.

Propodeum (Figure 349) about $2\frac{1}{7}$ to $2\frac{2}{3}$ times as long as the metanotum, about $\frac{3}{4}$ length of scutellum, reticulate on a wide trapezoidal central area, weakly striated mesad of spiracles; median salient only weakly protruding posteriorly; crenulae 7 + 7 to 10 + 10, small, rounded, sometimes somewhat overlapping.

Second abdominal tergite smooth, tergites III–VII reticulate on the sides, bearing 7–8 setae on each reticulate area; 2–4 fine setae in a transverse row across seventh tergite; eighth tergite very faintly reticulate, with 7–10 fine setae in 1–2 transverse rows; syntergum triangular, considerably wider than long, very faintly reticulate, bearing 2 rows of setae. Cerci situated very close to posterior spiracles, with 2 long setae and 1 short seta. Ovipositor shaft about $1\frac{2}{3}$ to $1\frac{3}{4}$ times as long as the middle tibia (1.64–1.73), ovipositor sheaths about $\frac{2}{5}$ length of middle tibia (0.37–0.44).

Mid-tibial spur about $\frac{7}{8}$ length of the corresponding basitarsus.

Forewing (Figure 350) clearly over $2\frac{1}{2}$ times as long as wide (2.54–2.70); marginal fringe not exceeding $\frac{1}{8}$ width of disk. Delta area about as densely setose as the area distad of the speculum, well separated from 1–3 rows of setae below distal $\frac{5}{8}$ of submarginal vein and from row of setae along posterior margin of wing; costal cell with a longitudinal row of fine setae and 1–3 coarse setae near the apex. Submarginal vein bearing 4–6 coarse setae and 16–22 bullae. Marginal vein bearing 10–14 prominent setae along anterior margin, these decreasing somewhat in length toward apex of vein, the proximal setae at the most $1\frac{1}{3}$ times as long as the distal ones; a row of subequal setae along center of vein, a little shorter than the proximal setae along the anterior margin.

Forewing rather extensively mottled, with patches of coarse, dark setae against a background of fine, hyaline setae, this pattern being sometimes rather faint; dark setae forming a ring in the delta, a circular patch distad of speculum, an oblong blotch below stigma, another one opposite to it against posterior margin, and a large transverse blotch near apex of wing; a few dark setae below junction of submarginal and marginal veins and along proximal margin of speculum.

General coloration yellow, with fuscous markings as follows: 2 faint streaks on each side of oral margin, 2 faint spots near anterior margin of scutellum, posterior margin of scutellum narrowly, anterior part of propodeum centrally, spots on sides of abdominal tergites III–VII, ovipositor sheaths, anterior parts of lateral margins of endophragma. Thoracic sterna immaculate. Mandibular denticles dark brown. Antennal scape pale, with a dark brown spot at the tip; pedicel and funicular segments infuscated, club more strongly so on basal two thirds. Legs pale, with a small dark brown spot on the tip of each tibia. Wing veins pale brownish to colorless.

Length 0.85–1.07 mm.

*Male*. Rather similar to the female, differing mainly in the structure of the antennae, in wing pattern and in details of coloration.

Eyes finely setose. Antennae (Figure 351) 6-segmented; scape (Figure 352) about $4\frac{1}{4}$ to $5\frac{3}{5}$ times as long as wide, bearing a single tuberculous sense organ on the ventral aspect; pedicel $1\frac{1}{4}$ to $1\frac{1}{2}$ times as long as wide, about $1\frac{1}{4}$ to $1\frac{1}{3}$ times longer than the third segment of the funicle; first 2 funicular segments small, ringlike, the first triangular, $2\frac{1}{4}$ to $2\frac{1}{2}$ times as wide as long; second segment more symmetrical, slightly shorter and wider than the first, about $2\frac{3}{4}$ to $3\frac{1}{3}$ times as wide as long; third funicular segment trapezoidal, dorsal aspect the longest, about as long as wide, bearing 1–3 longitudinal sensilla; club $2\frac{1}{2}$ to nearly 3 times as long as wide, $3\frac{3}{5}$ to $3\frac{4}{5}$ times longer and distinctly wider than the preceding segment, bearing 6–11 longitudinal sensilla; a specialized sensory area, bearing minute setae and surrounded by one row of short setae on ventral aspect of club (Figure 353).

Structure, chaetotaxis and sculpture of head and thorax as in the female. Propodeum (Figure 354) shorter, $1\frac{3}{4}$ to 2 times as long as the metanotum, $\frac{3}{5}$ length of scutellum; crenulae 4 + 5 to 7 + 7.

Genitalia (Figure 355) about $\frac{3}{4}$ to $\frac{4}{5}$ length of middle tibia; phallobase narrowing anteriorly, forming a distinct bottleneck; no longitudinal rod between apodemes; digital sclerites slender, short, about $\frac{1}{5}$ the combined length of aedeagus and apodemes (0.20–0.23).

Forewing (Figure 357) somewhat wider than in the female, $2\frac{1}{3}$ to nearly $2\frac{1}{2}$ times as long as wide; marginal fringe not exceeding $\frac{1}{5}$ width of disk in smallest specimens. Forewing uniformly setose, without a pattern of contrasting coarse and fine setae; delta with 4–7 rows of setae, these considerably longer and sparser than the setae distad of speculum. Marginal vein bearing 7–12 prominent setae along anterior margin.

Forewing hyaline; a distinct fuscous spot below stigma; an infuscated stripe along posterior margin of wing below submarginal vein and delta; a fuscous streak along posterior margin of wing distad of speculum.

General coloration yellow, with fuscous markings as in the female but spots on sides of abdominal tergites absent; antennal pedicel, funicle and club rather uniformly fuscous.

Length 0.59–0.92 mm.

Described from the following material: 1♀ (paratype), reared by P. Cochereau from an undetermined armored scale insect on coconut, Anse Vata, Noumea, New Caledonia, October 1967; 3♀♀ (paratypes), remounted in Hoyer's medium from a balsam slide labeled "New Caledonia" in H. Compere's collection, with no additional data; 1♀, 23♂♂ (paratypes and ♂ allotype), captured by G. Fabres in a sticky trap, Noumea,

New Caledonia, March 20, 1969; 6♀♀ (♀ holotype and paratypes), captured by G. Fabres in a sticky trap, Noumea, New Caledonia, no date.

Type series in the collection of the Division of Biological Control, University of California, Riverside.

**Notes.** *A. fabresi* is structurally quite closely related to *cochereaui*, but is more similar is coloration to *capillatus* and *ciliatus*. In the shape of the female antennae, propodeum and forewing pattern, *fabresi* is very much like *cochereaui*, differing from the latter species mainly in the generally paler coloration, in the fewer mesoscutal setae, and in the broader, less distinct subapical crossband on the forewing of the female. The male antenna of *fabresi* is 6-segmented as in *cochereaui*, with 2 ringlike segments and a rather enlarged club, but bears a tuberculous sense organ on the scape as in *capillatus*; the club differs from that of male *cochereaui* in bearing normal longitudinal sensilla; the sensory area on the venter of the club is different from that of both *cochereaui* and *capillatus*. *A. fabresi* also differs from *cochereaui* in the absence of a longitudinal rod in the male genitalia. It further differs from *capillatus* in the paler coloration, the more elongate antennae of the female, the shape of the crenulae and the structure of the male antennae. This peculiar species, presenting an odd mixture of primitive and advanced characters, may represent a connecting link between the *Aphytis* fauna of New Caledonia and that of the Australian continent.

This species was named in honor of Dr. Gerard Fabres, ORSTOM, Noumea, New Caledonia.

## 11. **Aphytis capillatus** (Howard)

(Figures 358–383)

*Perissopterus capillatus* Howard, 1907, U.S. Dept. Agr. Bur. Entomol. Tech. Serv., **12**(4):87.
*Perissopterus capillatus*: Girault, 1913, Mem. Queensl. Mus., **2**:184–185.
*Marietta novicapillata* Girault, 1917, Insecut. Inscit. Menstr., **5**:32; n. syn.
*Perissopterus cowperi* Girault, 1923, Insecut. Inscit. Menstr., **11**:144; n.syn.
*Perissopterus emersoni* Girault, 1923, Insecut. Inscit. Menstr., **11**:144; n.syn.
*Marietta lessingi* Girault, 1932, New Lower Hymenoptera from Australia and India. Priv. Publ., Brisbane, 6 pp., page 1; n.syn.
*Marietta romae* Girault, 1932, New Lower Hymenoptera from Australia and India. Priv. Publ., Brisbane, 6 pp., page 2; n.syn.
*Marietta capillata*: Compere, 1936, Univ. Calif. Publ. Entomol., **6**:307, 311.
*Marietta novicapillata*: Compere, 1936, Univ. Calif. Publ. Entomol., **6**:315.
*Marietta cowperi*: Compere, 1936, Univ. Calif. Publ. Entomol., **6**:315.
*Marietta emersoni*: Compere, 1936, Univ. Calif. Publ. Entomol., **6**:315.
*Aphytis* sp. (♂): Rosen and DeBach, 1970, Mushi, **43**:42–43.
*Aphytis capillatus*: DeBach and Rosen, 1976, Ann. Entomol. Soc. Amer., **69**:542.

This Australian species can be readily recognized by the following combination of characters: Female generally yellow, with 4 conspicuous dark crossbands on abdominal tergites; forewing extensively mottled; eyes coarsely setose; antennae short, stout; propodeal crenulae large, nonoverlapping; cerci with 2 long setae and 1 short seta. Male yellow, abdomen extensively marked with fuscous crossbands and blotches, wings hyaline; antenna essentially 4-segmented, with a minute ring segment and an indistinct rudiment of another; scape bearing one tuberculous sense organ, penultimate segment small, not bearing longitudinal sensilla, club with a cut-off sensory area.

*Female.* Eyes coarsely setose, the setae visible under $\times 60$ magnification even in uncleared specimens. Mandibles well developed, with 2 denticles and a truncation; maxillary palpi 2-segmented, labial palpi 1-segmented. Antennae (Figures 358, 359) rather short, stout; scape about $3\frac{3}{5}$ to 4 times as long as wide in cleared specimens, somewhat longer than the club; pedicel about $1\frac{1}{3}$ to $1\frac{3}{5}$ times as long as wide, about $1\frac{1}{3}$ to $1\frac{3}{5}$ times longer than the third segment of the funicle; first funicular segment triangular, about $1\frac{3}{4}$ times as wide as long; second segment nearly symmetrical, considerably larger than the first, $1\frac{3}{4}$ to nearly 3 times as wide as long; third funicular segment short, trapezoidal, dorsal aspect longer than the ventral, about $1\frac{1}{5}$ to $1\frac{2}{5}$ times as wide as long, bearing 1–2 longitudinal sensilla; club short and stout, usually only about 2 to $2\frac{1}{4}$ times as long as wide, about 3 to $3\frac{3}{4}$ times longer and distinctly wider than the preceding segment.

Setae on head, thorax and sides of abdomen coarse, dark, readily visible under $\times 30$ magnification, the larger ones apparently composed of distinct filaments (see Figure 362). Vertex with 2 pairs of long setae along occipital margin, in addition to numerous shorter setae. Mesoscutum (Figure 360) with 9–22 setae, the posterior pair and 1 seta at each antero-lateral corner considerably longer and coarser than the others; each parapsis with 1–3 setae, each axilla with 1 seta; scutellum (Figures 360, 361) with 4 setae, the posterior pair considerably longer and coarser than the anterior pair, the discoid sensilla closer to the posterior pair. Frontovertex, pronotum and mesonotal sclerites reticulate, with a distinct infrasculpture within the cells. Scutellum ellipsoidal, $\frac{2}{3}$ to $\frac{4}{5}$ median length of mesoscutum. Metanotum (Figures 361, 362) arcuate, relatively long, reticulate except on the sides; anteromedian apodeme robust, considerably shorter than median length of metanotum.

Propodeum (Figures 360, 362) about $2\frac{1}{4}$ to $3\frac{1}{4}$ times as long as the metanotum, about $\frac{2}{3}$ to $\frac{4}{5}$ length of scutellum, heavily reticulate on a triangular median salient which is continuous with the reticulate area on the metanotum; crenulae 6 + 7 to 11 + 11, large, conspicuous, roundly elongate, nonoverlapping.

Second abdominal tergite smooth; tergites III–VII reticulate on the sides, bearing a few rather strong setae in a transverse row on each reticulate area; tergites III–VI essentially smooth, very faintly transversely striated between the lateral reticulate areas; seventh tergite delicately reticulate across, faintly longitudinally striated posteriorly, bearing 2–4 fine setae in a transverse row between the lateral setiferous areas; eighth tergite delicately reticulate, with a transverse row of 4–7 setae between spiracles; syntergum (Figure 363) triangular, short, considerably wider than long, faintly reticulate, with up to 20 setae in an arcuate row. Cerci situated close to the posterior spiracles, with 2 long setae and 1 short seta. Ovipositor shaft usually $1\frac{3}{5}$ to $1\frac{4}{5}$ times as long as the middle tibia (1.63–1.78), ovipositor sheaths about $\frac{1}{3}$ to $\frac{2}{5}$ length of middle tibia (0.34–0.42).

Forewing (Figures 364–366) usually about $2\frac{3}{5}$ to $2\frac{2}{3}$ times as long as wide (about $2\frac{1}{4}$ times in one specimen); marginal fringe not exceeding $\frac{1}{6}$ width of disk, usually $\frac{1}{8}$ or shorter. Delta area about as densely setose as the area distad of speculum, not clearly separated from row of setae along posterior margin of wing, widely separated from 2–3 rows of setae below distal three fifths of submarginal vein; costal cell with a few fine setae in a row along proximal half or three fifths, and 2–4 coarse setae near the apex. Submarginal vein bearing 4–7 coarse setae (usually 5) and 15–23 bullae. Marginal vein bearing 9–13 prominent, subequal setae along anterior margin, these subequal in length to the setae in a row along center of vein.

Forewing extensively mottled, with patches of coarse, dark setae against a background of fine, transparent setae; the integument of the wing is entirely hyaline, the mottled effect being produced by the contrasting setae alone. Patches of coarse setae as follows: below submarginal vein; a large incomplete ring in delta, continued by a narrow streak along proximal border of speculum and a small group at apex of delta; 3 more or less fused patches in a triangle just distad of speculum in lower (posterior) part of disk; an oblong patch extending postero-distad from stigma, widening distally; a circular patch a short distance below it; 2 smaller, fainter patches below the latter, against posterior margin of wing, sometimes more or less fused with the latter patch; a circular patch against anterior margin of wing, about halfway between stigma and apex of wing; a large apical cloud, varying in size and shape, distinct in some specimens, more extensive and merging with the subapical clouds in others; apical margin of wing hyaline.

General coloration dirty yellow with fuscous and black markings. Occiput with a pair of transverse black bars on each side of foramen; genal suture lined with black from oral margin to about $\frac{1}{3}$ distance to eyes. Pronotum fuscous centrally; all margins of mesoscutum and posterior margin of scutellum lined with black; scutellum infuscated submedially and along lateral margin; fuscous spots surrounding bases of mesoscutal and scutellar setae; propodeum infuscated antero-medially, the sutures mesad of spiracles and the posterior margin lined with black. Second abdominal tergite strongly infuscated; tergites III–VI with complete, rather broad, blackish crossbands; seventh tergite with short transverse blackish stripes on sides, faintly infuscated across center; transverse ridge on posterior margin of eighth tergite narrowly lined with fuscous. Lateral margins of endophragma lined with black, fading into fuscous posteriorly; margins of ovipositor plates lined with fuscous, ovipositor sheaths fuscous on distal two thirds or so. Propleura marked with blackish along ventral margin; thoracic sterna strongly infuscated. Antennal scape with a longitudinal black stripe along ventral margin; pedicel, funicular segments and club uniformly, rather strongly infuscated, the club sometimes darker. All femora marked with blackish; all tibiae with 3 incomplete, blackish rings; apical segment of all tarsi blackish. Wing veins pale.

Length 0.60–1.03 mm.

*Male.* Differs from the female mainly in the structure of the antennae, wing pattern, and details of coloration.

Eyes finely setose. Antennae (Figures 367–372) essentially 4-segmented, with one minute ring segment and the rudiment of another; scape (Figures 367–369) about 4 to $4\frac{1}{2}$ times as long as wide, usually distinctly shorter than the club, bearing a small plate with one discoid, tuberculous sensillum ventrally at about midlength; pedicel $1\frac{1}{4}$ to $1\frac{2}{5}$ times as long as wide; first funicular segment (Figures 370, 371) reduced to a minute but distinct, triangular "ring," about 3 times as wide as long; second segment represented by an indistinct rudiment (Figures 370, 371), bearing 1 dorsal seta; subsequent funicular segment much larger, usually considerably wider than long, only $\frac{1}{2}$ to $\frac{2}{3}$ length of the pedicel, not bearing longitudinal sensilla; club (Figure 372) large, $2\frac{1}{2}$ to $3\frac{1}{2}$ times as long as wide, about 4–6 times longer and considerably wider than the preceding segment, bearing 6–12 longitudinal sensilla, with an oval, distinctly bordered, cut-off area bearing short setae on the ventral surface.

Mesoscutum (Figure 373) with 10–20 setae. Propodeum (Figure 374) about $2\frac{1}{4}$ to $2\frac{2}{3}$ times as long as the metanotum, $\frac{2}{3}$ to $\frac{3}{4}$ length of scutellum; crenulae $5 + 5$ to $9 + 9$. Seventh abdominal tergite more coarsely reticulate than in the female, bearing 2–5 setae in a transverse row between the lateral setiferous areas.

Genitalia (Figure 375) about $\frac{3}{4}$ to $\frac{4}{5}$ length of middle tibia (0.73–0.85); no longitudinal rod between apodemes; digital sclerites rather robust, about $\frac{1}{4}$ the combined length of aedeagus and apodemes (0.24–0.29).

Forewing (Figure 376) about $2\frac{1}{5}$ to $2\frac{2}{5}$ times as long as wide, entirely hyaline, rather uniformly setose, without a contrasting pattern of coarse and fine setae. About 4–7 rows of setae in delta, these considerably longer, coarser and sparser than the setae distad of speculum.

General coloration (Figure 377) very similar to that of the female, except as follows: sides of abdominal tergites pale; tergites III–VII with conspicuous black crossbands on posterior margin, between the lateral setiferous areas; sixth tergite with a pair of circular, sublateral black blotches in front of crossband; seventh tergite with a pair of large, transverse black blotches in front of crossband, merging medially; eighth tergite with a short, transverse black stripe on center of posterior margin.

Length 0.64–0.97 mm.

Redescribed from the following material: 14♀♀, 1♂ (including 11♀♀, syntypes), reared by A. Koebele from *Lepidosaphes pallens* (Maskell) on *Xanthorrhoea*, Sydney, New South Wales, Australia, December 20, 1899, all mounted in balsam on one slide, together with at least 4 additional species of Chalcidoidea; 24♀♀, 26♂♂, reared by G. Compere from *Neomorgania eucalypti* (Maskell) on *Eucalyptus*, Perth, Western Australia, no date, remounted in Hoyer's medium from one balsam slide; 5♀♀, 12♂♂, collected by G. Compere "on Laurestine and Pittosporum," Perth, Western Australia, no date, remounted in Hoyer's medium from one balsam slide; 1♂, reared by S. E. Flanders from *Aspidiotus nerii* Bouché [= *A. hederae* (Vallot)], same locality and date, remounted in Hoyer's medium; 3♂♂, reared by E. F. Riek from an undetermined armored scale insect on mistletoe on ironwood (*Aeacla* sp.), 30 miles west of Cobar, New South Wales, Australia, November 24, 1949, remounted in Hoyer's medium from points in the Australian National Insect Collection; 1♂, reared by M. M. H. Wallace from the California red scale, *Aonidiella aurantii* (Maskell), on citrus, Herne Hill, Swan River Valley, Western Australia, March 1967, mounted in Hoyer's medium.

**Notes.** Howard (1907) listed 11♀♀, designated syntypes, in his original description of *capillatus*. The original slide bearing the type series, kept in the U.S. National Museum, also includes 2 headless female specimens, one brownish female, considerably darker than the rest due to coagulation of the body contents, and—under another cover glass—a headless male specimen. Since all these specimens appear to be conspecific with the others, they were included in the material used for the present redescription. One female specimen, marked with a circle of ink, is designated lectotype. The slide is labeled "Type No. 10313, U.S.N.M."

The type specimens (Figures 358, 364) are not cleared, and requests to clear and remount them in Hoyer's medium have been turned down by the U.S. National Museum. Several important characters, such as the shape of the propodeal crenulae and details of sculpture, cannot be seen clearly. As far as can be determined, the type series is conspecific with the other series listed above, the only apparent difference being in the number of mesoscutal setae: 14–22 in females of the type series, 9–14 in females collected by G. Compere. Parts of the above redescription were, of necessity, based entirely on the remounted series. The female antennae were usually rather flattened in the remounted specimens.

In the type series, the apical cloud of the forewing is rather small, well separated from the subapical clouds (Figure 364). In the other series, the shape of this cloud is rather variable (Figures 365, 366), being similar to that of the type series in certain specimens and much more extensive, more or less merging with the subapical clouds, or arcuate, in others. This character evidently cannot be relied upon for the separation of *capillatus* from other species. On the other hand, the pattern of body pigmentation appears to be remarkably consistent in all the specimens available for study.

Five Australian species, described by A. A. Girault, are recognized here as synonyms of *capillatus*. These species—*novicapillata*, *cowperi*, *emersoni*, *lessingi* and *romae*—are briefly discussed in the following paragraphs.

*Marietta novicapillata* was described by Girault (1917a) from 5♀♀, reared by G. Compere from *Neomorgania eucalypti* (Maskell) on mistletoe growing on *Eucalyptus*, Perth, Western Australia. The specimens are mounted in balsam on a single slide (No. 20689, U.S. National Museum) and, although uncleared, are in very good shape, with some details of sculpture and even the propodeal crenulae quite clearly visible.

Girault separated *novicapillata* from *capillatus* only by the wing pattern, the apical cloud on the forewing being more extensive. However, the forewing pattern of *novicapillata* appears to be well within the range of variation of *capillatus*, as understood here. Since no other differences can be detected, *novicapillata* is considered to be a junior synonym of *capillatus*.

The transfer of *novicapillata* from *Marietta* into *Aphytis* necessitates a change of gender, to *novicapillatus*.

*Perissopterus cowperi* was apparently described from a single female specimen, captured in forest, Tingalpa (Brisbane), Queensland, Australia. Girault (1923) compared it with *ciliatus*, another Australian species that is closely related to *capillatus* (see p. 270). His original description was very brief: "As *ciliatus* Dodd but abdomen with four distinct cross-stripes from base, marginal spot in place of a fifth, fringes twice longer (one-fifth wing width), wing pattern slightly different."

The holotype ♀ of *cowperi* (No. 3759, Queensland Museum) is mounted in balsam and, although only partly cleared, is in fairly good shape. The specimen is small, 0.83 mm long, apparently with 8 mesoscutal setae. It is rather pale, but this may be due to the method of preparation, and the 4 abdominal crossbands are clearly evident. The forewing pattern (Figure 378) is quite extensive, the apical cloud merging with the subapical ones. Although the propodeal crenulae cannot be seen, we do not consider these slight differences to be of any diagnostic value, and regard the holotype of *cowperi* as a small specimen of *capillatus*.

Another female specimen, preserved on a slide in the Queensland Museum, was determined as *cowperi* in Girault's handwriting, but there is at present no way of telling whether it was indeed part of the type series. This specimen is somewhat larger than the holotype, 0.90 mm long, and does not differ at all from *capillatus* as interpreted here.

*Perissopterus emersoni* was described by Girault (1923), on the same page as *cowperi*, from a single female specimen obtained by "sweeping flowering *Leptospermum*, Wynnum [Queensland, Australia], September 23, 1921." Girault described it as follows: "As *cowperi* but a sixth cross-stripe on abdomen beyond the indicated fifth indicated by transverse marginal spots; a patch of discal cilia at base, three longitudinal lines deep from venation; fringes one-tenth wing width; 12–13 setae upon marginal vein; a narrow dusky line on each side mesad of thorax. Lateral margin of propodeum black."

The holotype of *emersoni*, a female specimen (No. 3760, Queensland Museum), is in rather poor shape. It is mounted in balsam without clearing, the dorsum of the thorax, propodeum and base of gaster is split along the midline, and the antennae are folded and cannot be seen clearly. This is a relatively large specimen, 1.07 mm long, exhibiting characteristics of large specimens such as numerous setae on the mesoscutum (apparently 16 or more) and on the marginal vein, a short marginal fringe on the forewing, etc. There are 4 conspicuous crossbands on the abdomen and a fifth is indicated by lateral spots, but there is no sixth crossband as described, except for the dusky ridge on the posterior margin of the eighth abdominal tergite. This specimen does not differ from *capillatus* in any significant character. The forewing pattern (Figure 379) is similar to that of the type series of *capillatus*, with a small, triangular apical cloud (this appears to be a characteristic of large specimens in *capillatus*). We regard *emersoni* as another junior synonym of *capillatus*.

*Marietta lessingi* was described from a single female specimen, captured on a window, Indooroopilly (Brisbane), Queensland, Australia, July 1932. Girault (1932) compared it with *literata* (Girault), which from the description appears to be a species of *Marietta* (the type of *literata* was not available for study). The type specimen (No. 7483, Queensland Museum) (Figures 380–382) was remounted in Hoyer's medium from Girault's original balsam slide, and was found to be identical to *capillatus*. The apical cloud of the forewing (Figure 380) is arcuate, as described by Girault, but *lessingi* is similar in this respect to several specimens in the *capillatus* series (see, for instance, Figure 366). The pigmentation is somewhat stronger than in *capillatus*, but the pattern is identical and the slight difference is likely due to mounting procedures. For the time being at least, *lessingi* is considered to be yet another junior synonym of *capillatus*.

*Marietta romae* was described by Girault (1932) from 3♀♀ and 3♂♂, reared by D. O. Atherton and A. A. Girault from *Chionaspis* sp. on *Atalanta*, Roma, Queensland, Australia, January 17, 1931. All the specimens were mounted in balsam on one slide (No. 4945, Queensland Museum); 1♀ and 1♂ have been cleared and remounted in Hoyer's medium. This species, too, was found to be identical to *capillatus*. The specimens are small, and the forewing pattern of the female is rather faint, with an extensive apical cloud (Figure 383), similar to that of several specimens in the *capillatus* series.

The *capillatus* complex also includes *A. nigripes* (Compere), *A. ciliatus* (Dodd) and *A. australiensis* DeBach and Rosen, redescribed here as distinct species. *A. maculatipes* (Girault) and *A. perissoptroides* (Girault), listed here as unrecognizable species (see p. 729), appear to be closely related to this complex. *A. hyalinipennis* n.sp. and *A. wallumbillae* (Girault) are also closely related in structure and coloration, but differ markedly from *capillatus* in having hyaline, nonmottled wings. Additional biological information is required for a better understanding of this interesting complex of little-known species. It is, of course, possible that some of the series considered here as conspecific will eventually prove to be biologically distinct. On the other hand, additional information on intraspecific variation may eventually lead to synonymizing *nigripes*, or even *ciliatus*, with *capillatus*. Above all, the type series of *capillatus* should be properly cleared and remounted, in order to clarify once and for all the specific identity and affinites of *capillatus* and closely related species.

## 12. **Aphytis nigripes** (Compere)

(Figures 384–388)

*Marietta nigripes* Compere, 1936, Univ. Calif. Publ. Entomol., **6**: 312.
*Aphytis nigripes*: DeBach and Rosen, 1976, Ann. Entomol. Soc. Amer., **69**: 542.

This problematic, large Australian species is very closely related to *capillatus*, differing from the latter species mainly in the larger number of mesoscutal setae and the smaller number of propodeal crenulae, as well as in certain minor details of coloration and wing pattern.

*Female.* Eyes moderately coarsely setose, the setae readily visible under ×60 magnification. Mandibles well developed; maxillary palpi 2-segmented, labial palpi 1-segmented. Antennal scape (Figure 384) about 4 to $4\frac{1}{2}$ times as long as wide, a little longer than the club; pedicel about $1\frac{2}{3}$ times as long as wide, about $1\frac{3}{5}$ times longer than the third segment of the funicle; first funicular segment trapezoidal, $1\frac{1}{3}$ times as wide as long; second segment more symmetrical, slightly trapezoidal, larger, about as long as and considerably wider than the first segment, $1\frac{2}{5}$ to $1\frac{2}{3}$ times as wide as long; third funicular segment trapezoidal, dorsal aspect considerably longer than the ventral, about as long as wide or somewhat wider than long, bearing 2–3 longitudinal sensilla; club rather thick, truncate, $2\frac{1}{3}$ to nearly $2\frac{1}{2}$ times as long as wide, $3\frac{1}{5}$ to $3\frac{1}{2}$ times longer and considerably wider than the preceding segment, bearing 13–14 longitudinal sensilla; scape, pedicel and funicle furnished with strong, relatively long setae, club with minute setae; flagellar segments attached at an angle to each other.

Setae on head, thorax and sides of abdomen dark, coarse, readily visible under ×30 magnification; the longest setae appear to be composed of distinct filaments. Vertex with several pairs of long setae, in addition to numerous short setae. Mesoscutum (Figure 385) with 24–32 relatively long setae, each parapsis with 2–3, each axilla with 1–2; scutellum (Figure 386) with 4 setae, the discoid sensilla closer to the posterior than to the anterior pair. Scutellum nearly $\frac{3}{4}$ median length of mesoscutum. Frontovertex, pronotum, and mesonotal sclerites reticulate. Metanotum (Figure 386) arcuate, rather long, reticulate except on the sides; anteromedian apodeme robust, clavate, about $\frac{3}{4}$ median length of metanotum.

Propodeum (Figure 387) short, about twice as long as the metanotum, $\frac{3}{5}$ to $\frac{2}{3}$ length of scutellum, reticulate on a triangular, rather prominent median salient; crenulae 7+7, relatively large, distinct, elongate, nonoverlapping.

Second abdominal tergite short, faintly, transversely striated centrally; tergites III–VII reticulate on the sides, with faint indications of transverse striation across the center, bearing several strong setae in a short transverse row on each reticulate area; seventh tergite with faint longitudinal striation and bearing a transverse row of 5 fine setae between the lateral setiferous areas; eighth tergite reticulate-striated across, bearing a long seta mesad of each spiracle, and a transverse row of 4 shorter setae across the center; syntergum triangular, considerably wider than long, reticulate, bearing 18 setae in 2–3 transverse rows, the anterior margin arcuate. Cerci situated very close to posterior spiracles, with 2 long setae and 1 short seta. Ovipositor shaft about $1\frac{3}{5}$ times as long as the middle tibia (1.55–1.65), ovipositor sheaths about $\frac{1}{3}$ length of middle tibia (0.33–0.35).

Mid-tibial spur rather slender, about $\frac{4}{5}$ length of the corresponding basitarsus.

Forewing (Figure 388) $2\frac{1}{2}$ times as long as wide or nearly so; marginal fringe short, not exceeding $\frac{1}{11}$ width of disk. Delta area about as densely setose as the area distad of speculum, not clearly separated from row of setae along posterior margin of wing, widely separated from 3–4 rows of setae below distal three quarters of submarginal vein; costal cell with a longitudinal row of fine setae, these separated by a short bare space from 3–5 coarse setae near apex. Submarginal vein bearing 4–6 coarse setae and 22–24 bullae. Marginal vein bearing 10–15 prominent, subequal setae along the anterior margin, these subequal in length to the setae in a row along center of vein.

Forewing mottled, with patches of coarse, dark setae against a background of fine, transparent setae; the integument of the wing is entirely hyaline except for a very small fuscous cloud below the stigma, the mottled effect being produced by the contrast between the dark and pale setae. Pattern similar to *capillatus* but apical cloud very extensive, merging with the subapical clouds, apical margin of wing narrowly hyaline.

General coloration dirty yellow, dorsum of thorax and abdomen faintly mottled with white or silvery; fuscous or black markings as follows: a pair of transverse, rather faint stripes on the occiput, on each side of the foramen; genal sutures, faintly from oral margin to about $\frac{1}{3}$ length; anterior and posterior margins of mesoscutum, parapsidal sutures, posterior margin of scutellum, faint spots surrounding bases of thoracic setae; sutures mesad of propodeal spiracles, posterior margin of propodeum laterally; lateral margins of endophragma anteriorly; anterior and posterior margin of second abdominal tergite laterally; narrow transverse crossbands on abdominal tergites III–VI, short transverse stripes on sides of seventh tergite. Second abdominal tergite faintly infuscated; seventh tergite not infuscated between the lateral stripes; transverse ridge on posterior margin of eighth tergite fuscous; ovipositor sheaths infuscated toward apex, margins of ovipositor plates fuscous. Thoracic sterna dusky. Antennal scape with a longitudinal blackish stripe on the ventral margin; pedicel, funicular segments and club uniformly dusky. All femora suffused with fuscous ventrally and toward apex, all tibiae with fuscous rings, these markings best seen in a cleared specimen; apical segment of all tarsi contrastingly dark brown or blackish. Wing veins yellowish.

Length 1.39–1.52 mm.

*Male*. Unknown.

Redescribed from 2♀♀ (paratypes), reared by H. Compere from an undetermined coccid, Cape Banks, Botany Bay, Sydney, New South Wales, Australia, September 29 and October 16, 1927; mounted in balsam on 2 slides, one of them cleared in KOH. The holotype ♀ and another paratype ♀, kept at the U. S. National Museum, were also examined (U. S. N. M. Type No. 101627, mounted in balsam on one slide, same data, October 16, 1927).

**Notes.** *A. nigripes* is evidently very closely related to *capillatus*, and may eventually be recognized as a junior synonym of this species. It differs from *capillatus* mainly in its larger size, in the more extensive apical cloud on the forewing, in the considerably narrower crossbands on the abdomen and absence of any infuscation across the seventh tergite, in the more numerous mesoscutal setae (24–32, not 14–18 as described by Compere, 1936), and in the fewer, less conspicuous propodeal crenulae. The difference in wing pattern alone would not be considered sufficient for separating *nigripes* from *capillatus*, as this character appears to be rather variable in the latter species. The apparent difference in the shape of the abdominal crossbands (narrow in *nigripes*, broad and extensive in *capillatus*) may have been affected by different mounting procedures, although this is rather unlikely. The larger number of mesoscutal setae, as well as slight differences in the proportions of the antennal segments, may be related to the considerably larger size of *nigripes* (numerous specimens of *capillatus*, reared from various hosts, usually do not exceed 1 mm in length). The smaller number of propodeal crenulae appears to be significant in this context: if indeed *nigripes* merely represented large specimens of *capillatus*, one would expect the crenulae to be much more numerous in such specimens.

In view of this combination of differences, it is deemed advisable to retain *nigripes* as a distinct species, at least until more morphological and biological information becomes available on the various forms comprising the *capillatus* complex.

## 13. **Aphytis ciliatus** (Dodd)

(Figures 389–392)

*Aphelinus ciliatus* Dodd, 1917, Trans. Roy. Soc. S. Austral., **41**:353.
*Marietta ciliata*: Compere, 1936, Univ. Calif. Publ. Entomol., **6**:307, 311.
*Aphytis ciliatus*: DeBach and Rosen, 1976, Ann. Entomol. Soc. Amer., **69**:542.

This small Australian species is closely related to *capillatus*, but may be readily distinguished from this species by the strikingly different abdominal pigmentation, with the crossbands mostly obliterated; thoracic sterna faintly infuscated; eyes finely setose;

mesonotal setae few; propodeum short, without a prominent median salient, crenulae numerous. *A. ciliatus* is redescribed below relative to *capillatus*.

*Female.* Eyes finely setose, the setae invisible even under × 120 magnification. Mouthparts as in *capillatus*. Antennae (Figure 389) short, similar to those of *capillatus*; scape 4 times as long as wide or nearly so, a little longer than the club; pedicel $1\frac{1}{3}$ to nearly $1\frac{1}{2}$ times as long as wide, about $1\frac{1}{8}$ to $1\frac{1}{3}$ times longer than the third segment of the funicle; first funicular segment triangular, $1\frac{1}{2}$ to $1\frac{3}{5}$ times as wide as long; second segment much more symmetrical, slightly trapezoidal, somewhat shorter and considerably wider than the first segment, fully $2\frac{1}{2}$ times as wide as long; third funicular segment trapezoidal, dorsal aspect longer than the ventral, nearly $1\frac{1}{3}$ times as wide as long, bearing 3 longitudinal sensilla; club about $2\frac{2}{5}$ times as long as wide, $2\frac{3}{4}$ to $3\frac{1}{4}$ times longer than the preceding segment, bearing 12 longitudinal sensilla.

Setae on head, thorax and sides of abdomen dark, coarse. Mesoscutum (Figure 390) with 8–9 rather short, stout setae, each parapsis with 1 seta, each axilla with 1 seta; scutellum with 4 setae, the posterior pair considerably longer and coarser than any other setae on the body, apparently composed of distinct filaments, the discoid sensilla a little closer to the posterior than to the anterior pair. Scutellum oval, nearly $\frac{4}{5}$ median length of mesoscutum. Otherwise structure and sculpture of head and thorax as in *capillatus*.

Propodeum (Figure 391) short, about twice as long as the metanotum, a little less than $\frac{3}{5}$ length of scutellum; median salient less prominent than in *capillatus*, central reticulate area trapezoidal; crenulae 8 + 10 or 9 + 9, occupying a wide area on posterior margin of propodeum, large, mostly distinct and nonoverlapping except centrally, where they are more crowded and occasionally overlap.

Structure and sculpture of abdomen essentially as in *capillatus*; second tergite apparently smooth; seventh tergite with 1–2 fine setae between the reticulate areas; eighth tergite with a transverse row of 4–5 fine setae between spiracles; syntergum bearing 8–9 setae in 2 transverse rows. Cerci as in *capillatus*. Ovipositor shaft about $1\frac{3}{5}$ times as long as the middle tibia (1.59–1.64), ovipositor sheaths short, a little more than $\frac{1}{4}$ length of middle tibia (0.27).

Mid-tibial spur a little shorter than the corresponding basitarsus.

Forewing (Figure 392) relatively wide, about $2\frac{1}{5}$ times as long as wide, essentially similar to that of *capillatus*. Delta area rather clearly separated from sparse row of setae along posterior margin of wing, widely separated from 2–3 rows of setae below distal two thirds of submarginal vein; costal cell with a row of fine setae along proximal three fifths and 2 coarse setae near apex. Submarginal vein bearing 4 coarse setae and 16–22 bullae. Marginal vein bearing 9 prominent, subequal setae along anterior margin, these subequal in length to the setae in a row along center of vein.

Forewing pattern essentially as in *capillatus*.

General coloration (see Figure 390) similar to that of *capillatus*, but differing markedly as follows: fuscous markings on genal sutures apparently absent; second abdominal tergite rather strongly infuscated; fuscous crossband on third abdominal tergite obliterated centrally; tergites IV–VII pale, with short, transverse fuscous stripes on the sides decreasing in size toward seventh tergite; ovipositor plates pale, the arcs dusky;

thoracic sterna very faintly infuscated; antennae and legs essentially as in *capillatus*.
Length 0.82–0.85 mm.

*Male*. Unknown.

Redescribed from 2♀♀ (syntypes), reared by G. F. Hill from *Aspidiotus* sp. on *Eucalyptus miniatus*, Darwin, Northern Territory, Australia, June 7, 1915; remounted from balsam in Hoyer's medium on two slides. One of the specimens, with 9 mesoscutal setae, designated lectotype.

Type series in the Queensland Museum (#3958).

**Notes.** In the original description, Dodd (1917) described *ciliatus* as "pale lemon-yellow, the legs and antennae concolorous." This statement misled Compere (1936), in his key to the species of *Marietta*, to refer to *ciliatus* as having immaculate legs, in contrast to the spotted or banded legs of other species. In fact, the legs of *ciliatus* are distinctly marked with fuscous or blackish, being very similar in that respect to those of *capillatus*.

*A. ciliatus* is known only from the two type specimens. Some of the diagnostic characters ascribed to this species may be related to the small size of these specimens. However, the distinctive coloration leaves no doubt but that this little-known species is indeed distinct from *capillatus*. As a matter of fact, whereas *ciliatus* resembles *capillatus* in the structure of the antennae, etc., it is much more similar to *fabresi* in abdominal pigmentation.

As pointed out by DeBach and Rosen (1976b), since *ciliatus* Dodd, 1917, is recognized as a bona fide species of *Aphytis*, the name *Aphytis ciliatus* Quednau, 1964, necessarily becomes a junior secondary homonym.

## 14. **Aphytis australiensis** DeBach and Rosen

(Figures 393–397)

*Aphytis australiensis* DeBach and Rosen, 1976, Ann. Entomol. Soc. Amer., **69**:542.

This Australian species appears to be very closely related to *ciliatus*, and may eventually be recognized as a synonym of the latter. However, it differs markedly from *ciliatus* in the relatively long propodeum, with a conspicuously prominent median salient, and in the shorter ovipositor. To avoid unnecessary repetition, the following description occasionally refers to *ciliatus* or *capillatus*.

*Female*. Eyes finely setose, the setae invisible even under × 120 magnification. Mandibles well developed; maxillary palpi 2-segmented, labial palpi 1-segmented. Antennal scape (Figure 393) $4\frac{1}{4}$ to $5\frac{1}{4}$ times as long as wide, a little longer than the club; pedicel $1\frac{2}{5}$ to $1\frac{3}{5}$ times as long as wide, $1\frac{1}{4}$ to $1\frac{1}{2}$ times longer than the third segment of the funicle; first funicular segment triangular, $1\frac{1}{2}$ to a little over $1\frac{4}{5}$ times as wide as long; second segment nearly symmetrical, a little longer and considerably wider than the first, usually over twice as wide as long ($1\frac{2}{3}$ to $2\frac{1}{2}$ times); third funicular segment $1\frac{1}{8}$ to $1\frac{1}{4}$ times as wide as long, bearing 3–4 longitudinal sensilla; club $2\frac{1}{6}$ to nearly $2\frac{1}{2}$ times as long as wide, $3\frac{1}{5}$ to a little over $3\frac{1}{2}$ times longer than the preceding segment, bearing 12–16 longitudinal sensilla.

Setae on head, thorax and sides of abdomen dark, coarse, readily visible under ×60 magnification. Vertex with one pair of long setae along occipital margin, in addition to numerous shorter setae. Mesoscutum with 13–16 setae, longer and somewhat less stout than in *ciliatus*; each parapsis with 3 setae, each axilla with 1 seta; scutellum with 4 setae, the posterior pair considerably longer and coarser than any other setae, apparently composed of distinct filaments, the discoid sensilla closer to the posterior than to the anterior pair. Scutellum oval, about $\frac{2}{3}$ median length of mesoscutum. Sculpture of head and thorax as in *capillatus* and *ciliatus*. Metanotal apodeme about $\frac{2}{3}$ median length of metanotum.

Propodeum (Figures 394, 395) relatively long, $2\frac{2}{3}$ to nearly 3 times as long as the metanotum, distinctly over $\frac{4}{5}$ length of scutellum (0.82–0.90); median salient very prominent, central reticulate area triangular; crenulae numerous, 8+9 to 12+13, large, elongate, mostly nonoverlapping but with a few irregular crenulae in a second row, which occasionally overlap on the "regular" crenulae.

Structure and sculpture of abdomen (Figure 396) essentially as in *capillatus* and *ciliatus*; second tergite short, smooth; seventh tergite with 4–7 fine setae in a transverse row between the reticulate areas; eighth tergite with a transverse row of 5–8 fine setae between spiracles, the lateral pair considerably longer than the others; syntergum triangular, wider than long, bearing 18–26 setae in 2–3 rows. Cerci as in *capillatus* and *ciliatus*. Ovipositor shaft relatively short, usually a little less than $1\frac{1}{2}$ times as long as the middle tibia (1.44–1.53), ovipositor sheaths a little over $\frac{1}{3}$ length of middle tibia (0.35–0.36).

Mid-tibial spur about $\frac{4}{5}$ length of the corresponding basitarsus.

Forewing (Figure 397) narrower than in *ciliatus*, $2\frac{2}{5}$ to $2\frac{2}{3}$ times as long as wide; marginal fringe not exceeding $\frac{1}{7}$ width of disk. Delta area widely separated from 2–4 rows of setae below distal three fifths of submarginal vein; row of setae along posterior margin of wing usually starting at disto-posterior corner of delta; costal cell with a row of fine setae along proximal half or three fifths, and 2–4 coarse setae near apex. Submarginal vein bearing 4–6 coarse setae and 17–21 bullae. Marginal vein bearing 14–17 prominent, subequal setae along anterior margin, these subequal in length to the setae in a row along center of vein.

Forewing pattern essentially as in *capillatus*; apical cloud more extensive than in *ciliatus*, but with a pair of rather distinct subapical clouds.

General coloration very similar to that of *ciliatus*; fuscous markings on occiput very faint, often missing; no markings on genal sutures; propodeum distinctly infuscated antero-medially; second abdominal tergite strongly infuscated; fuscous crossband on third abdominal tergite obliterated centrally (Figure 396), sometimes nearly complete; crossband on fourth tergite broadly obliterated centrally; tergites V–VII with short transverse fuscous stripes on the sides; eighth tergite with two sublateral fuscous blotches; otherwise as in *ciliatus*.

Length 1.15 to 1.38 mm.

*Male*. Unknown.

Described from 7♀♀ (holotype and paratypes), reared by E. M. Ehrhorn from *Chionaspis* sp. on mistletoe on *Eucalyptus*, Waroongus (or Waroonga), New South Wales, Australia, August 26, 1923; 1♀ mounted in balsam, the others remounted in Hoyer from points that stood in the collection of P. H. Timberlake.

Type series in the collection of the Division of Biological Control, University of California, Riverside.

**Notes.** *A. australiensis* is undoubtedly specifically distinct from *capillatus*. Since both species are represented by specimens of about the same size, the finely setose eyes of *australiensis* may be regarded as an additional valid character separating this species from *capillatus*. On the other hand, *australiensis* appears to be very similar to *ciliatus* in coloration, wing pattern, and most structural characters. The apparent differences in the number of setae on the mesoscutum and along the marginal vein, as well as the relatively narrower forewing, relatively shorter ovipositor and somewhat more slender antennae may all be related to the considerably larger size of *australiensis*. However, the strikingly different shape of the propodeum, with a very prominent median salient, appears to constitute a valid diagnostic character. For the time being, we regard the differences between *australiensis* and *ciliatus* to be of sufficient magnitude and consistency for the separation of *australiensis* as a distinct species. However, further studies and additional biological information may eventually show that *australiensis* is actually a large "form" of *ciliatus*.

One specimen of *australiensis* has one abnormal antenna, with the club fused to the third funicular segment, only partially separated by a vestigial suture.

**Additional Material.** 1♀ (Figures 398, 399), reared by P. DeBach from *Chrysomphalus aonidum* (L.) or *Cornuaspis beckii* (Newman) on orange, Ayr, Queensland, Australia, August 7, 1971, agrees with the type series of *australiensis* in most structural characters, including the prominent propodeal salient (Figure 398), but differs in details of coloration. The abdominal crossbands are entirely absent and the third abdominal tergite is entirely pale; tergites IV–VII with short stripes or spots on the sides. This specimen also differs from the type series in having the eyes moderately coarsely setose; it has 17 mesoscutal setae, about 8 + 9 propodeal crenulae and 11 setae along the marginal vein, and the ovipositor shaft is about $1\frac{2}{3}$ times as long as the middle tibia. Forewing pattern (Figure 399) similar to the type series, the apical cloud less extensive. This single specimen may represent a distinct, undescribed species. However, until more information is available on the biology and intraspecific variation of this interesting but little-known complex, it is for the time being referred to *australiensis*.

## 15. **Aphytis anomalus** Compere

(Figures 400–406)

*Aphytis anomalus* Compere, 1955, Univ. Calif. Publ. Entomol., **10**:286.

This South American species may be readily recognized by the few, small, nonoverlapping, nearly rectangular propodeal crenulae, the rather slender antennae, the large number of bullae on the submarginal vein of the forewing, and the distinctive coloration: distal half of antennal club black, remainder of antenna clear; thorax essentially pale, abdomen dark brown centrally, pale on the sides and at the apex. Forewing with an obscure infuscated crossband below the stigma, bearing coarse setae. Cerci with 3 long setae and 1 short seta.

*Female*. Eyes coarsely setose. Mandibles well developed, nearly bidentate, with a strong ventral tooth and a dorsal truncation; maxillary palpi 2-segmented, labial palpi 1-segmented. Antennal scape (Figure 400) slender, nearly 7 times as long as wide, about $1\frac{1}{5}$ times longer than the club; pedicel slender, somewhat over $2\frac{1}{4}$ times as long as wide, about $1\frac{1}{6}$ times longer than the third segment of the funicle; first funicular segment triangular, about $1\frac{1}{2}$ times as wide as long; second segment symmetrical, somewhat shorter and considerably wider than the first, nearly twice as wide as long; third funicular segment $1\frac{1}{2}$ times as long as wide, bearing 5–6 longitudinal sensilla; club elongate, tapering, about $3\frac{1}{2}$ times as long as wide, about $2\frac{3}{4}$ times longer and somewhat wider than the preceding segment, bearing 14 longitudinal sensilla.

Setae on vertex, thorax and sides of abdomen coarse, dark, easily visible under ×30 magnification. Mesoscutum (Figure 401) with 19 setae, the posterior pair considerably longer and coarser than the others; each parapsis with 3 setae, each axilla with 1 seta; scutellum with 4 setae, the discoid sensilla considerably closer to the anterior than to the posterior pair. Frontovertex, pronotum and mesonotal sclerites reticulate. Scutellum trapezoidal, a little over $\frac{4}{5}$ median length of mesoscutum. Metanotum faintly reticulate centrally; anteromedian apodeme slender, twice the median length of metanotum.

Propodeum (Figures 404, 405) about $3\frac{1}{5}$ times as long as the metanotum, about $\frac{1}{2}$ length of scutellum, faintly reticulate centrally and just mesad of spiracles; crenulae 3 + 3, small, rectangular, nonoverlapping but closely appressed.

Second abdominal tergite (Figures 404, 405) reticulate centrally; tergites III–VII (Figure 405) delicately reticulate on the sides, more coarsely so on central pigmented areas, bearing 2–3 prominent setae on each side; eighth tergite transversely reticulate, with 13 setae in 2 transverse rows between spiracles; syntergum deeply v-shaped, reticulate, bearing 19 setae. Cerci situated very close to and laterad of posterior spiracles, with 3 long setae and 1 short seta. Ovipositor shaft nearly twice as long as the middle tibia (1.95), ovipositor sheaths about $\frac{1}{2}$ length of middle tibia (0.49).

Mid-tibial spur about $\frac{4}{5}$ length of the corresponding basitarsus.

Forewing (Figure 406) about $2\frac{1}{2}$ times as long as wide; marginal fringe short, not exceeding $\frac{1}{9}$ width of disk. Delta area with 10 rows of setae, less densely setose than the area distad of speculum, narrowly separated from row of setae along posterior margin of wing and from 3–4 rows of setae below submarginal vein; costal cell with a row of short, fine setae along proximal half, and 2 coarse setae near apex. Submarginal vein bearing 5 coarse setae and 28–30 bullae. Marginal vein bearing 1 or 2 very long setae at base and a row of 11–12 prominent setae along anterior margin, these decreasing considerably in length toward apex of vein, the first few setae about $1\frac{3}{4}$ times as long as those near apex; a row of subequal setae along center of vein.

Forewing with a broad infuscated crossband below the stigmal vein (see Figure 406), fading posteriorly, bearing dark, coarse, almost peglike setae; area between speculum and crossband infumated, distad of crossband hyaline; an arcuate band below junction of submarginal and marginal veins; a small distinct cloud bearing a patch of coarse, dark setae in delta just proximad of speculum. Hind wing with an infuscated spot below base of marginal vein.

General coloration pale yellow with dark brown markings, dorsum of head possibly orange-yellow. Two brownish spots on occiput, cheeks suffused with brownish. Center of pronotum brown; a pair of diffuse brownish spots on anterior part of mesoscutum, another pair near anterior margin of scutellum; parapsidal sutures dark brown; posterior margin of propodeum between spiracles, and crenulae, dark brown, a longitudinal pale stripe at center of propodeal salient (Figure 404). Second abdominal tergite dark brown; subsequent abdominal tergites with a continuous dark brown pattern (Figure 405), covering most of third, seventh and eighth tergites, narrowing on tergites IV–VI; sides of abdomen and the syntergum pale. Thoracic sterna strongly infuscated. Mandibular denticles dark brown. Antennae pale, first 2 funicular segments dusky, distal half of club contrastingly black. Legs pale yellow, hind tibiae slightly infuscated centrally. Wing veins brownish.

Length 1.38 mm.

*Male*. Unknown.

Redescribed from the ♀ holotype, captured by H. Compere on *Cassia imperialis*, Bom Retiro, Estado Rio de Janeiro, Brazil, September 8, 1934; remounted in Hoyer's medium from the original balsam slide.

Holotype in the collection of the Division of Biological Control, University of California, Riverside.

**Additional Material.** 7♀♀ (Figures 402, 403, 406), reared by P. DeBach from *Chrysomphalus dictyospermi* (Morgan) on rose, Belo Horizonte, Minas Gerais, Brazil, April 30, 1962. These specimens agree with the holotype in all significant diagnostic characters, and are unquestionably conspecific; they are, however, smaller, ranging from 0.81 to 1.20 mm in length, and differ accordingly from the holotype in the number of mesoscutal setae (10–19, see Figure 402), in the number of prominent setae along the anterior margin of the marginal vein (7–10), in the relative length of the marginal fringe of the forewing (up to over $\frac{1}{5}$ width of disk in the smallest specimen), and in that the infumated areas on the forewing are fainter, with the small cloud in the delta rather indistinct (Figure 406); they have 46–102 setae in the delta. This series may serve to illustrate the range of intraspecific variation in *anomalus*.

## 16. **Aphytis haywardi** (Blanchard)

(Figures 407–414)

*Marietta haywardi* Blanchard, 1948, in: De Santis, Rev. Mus. La Plata (N.S.), **5**: 146–149.
*Aphytis haywardi*: DeBach and Rosen, 1976, Ann. Entomol. Soc. Amer., **69**: 541.

This little-known, dark South American species may be recognized by the extensive fuscous pigmentation of the thorax, abdomen, antennae and legs, by the asetose axillae, and by the distinctive forewing pattern, with a broad fuscous cloud bearing dark, coarse setae. The finely setose eyes, the short ovipositor, the large, irregular propodeal crenulae, the rather numerous mesoscutal setae, the cerci bearing 2 long setae and 1 short seta, and the dark crossband on the hind wing may serve as additional distinguishing characters.

*Female*. Eyes finely setose. Mandibles well developed, with a strong ventral tooth, the central and dorsal denticles forming a truncation; maxillary palpi 2-segmented, labial palpi 1-segmented. Antennae (Figures 407, 408) rather short; scape slender, $4\frac{2}{5}$ to a little over 5 times as long as wide, about $1\frac{1}{5}$ times as long as the club; pedicel about $1\frac{2}{5}$ to $1\frac{1}{2}$ times as long as wide, somewhat longer than the third segment of the funicle; first funicular segment trapezoidal, ventral aspect longer than the dorsal, about $1\frac{1}{2}$ to $1\frac{2}{3}$ times as wide as long; second segment symmetrical, somewhat shorter and wider than the first segment, about $1\frac{2}{3}$ to $1\frac{4}{5}$ times as wide as long; third funicular segment trapezoidal, dorsal aspect longer than the ventral, about as long as wide, bearing 3 longitudinal sensilla; club rather short, $2\frac{1}{5}$ to $2\frac{1}{3}$ times as long as wide, about $2\frac{1}{5}$ to $2\frac{3}{5}$ times longer and somewhat wider than the preceding segment, bearing 7 longitudinal sensilla.

Setae on head and thorax short, coarse, rather dark; those on the thorax visible only under ×120 magnification against the dark background of body coloration, those on the head visible under ×60 magnification; setae on sides and apex of abdomen more slender, visible under ×120 magnification. Vertex with 2 pairs of relatively long setae along occipital margin, in addition to numerous short setae. Mesoscutum (Figure 409) with 20–22 setae (17–18 according to the original description), the posterior pair and 1 seta at each antero-lateral corner considerably longer and coarser than the others; each parapsis with 2 setae, axillae asetose; scutellum with 4 setae, the discoid sensilla somewhat closer to the anterior than to the posterior pair. Frontovertex, pronotum and mesonotal sclerites conspicuously reticulate, with a distinct infrastructure within the cells. Scutellum oval, $\frac{4}{5}$ median length of mesoscutum or longer. Metanotum (Figures 410, 411) relatively long, moderately arcuate, conspicuously reticulate except on the sides; anteromedian apodeme slender, bifid, $\frac{3}{4}$ to nearly full median length of metanotum.

Propodeum (Figures 410, 411) about 3 to $3\frac{1}{2}$ times as long as the metanotum, about $\frac{4}{5}$ length of scutellum, longitudinally reticulate mesad of spiracles, reticulate-strigose on a triangular central area; 2 + 2 to 4 + 4, large, irregular, somewhat overlapping crenulae on posterior margin, on each side of the central reticulate area, sometimes with a few smaller, irregular crenulae between the 2 sets; the large crenulae are conspicuous in cleared specimens.

Second abdominal tergite transversely striated anteriorly on both sides, reticulate centrally; tergites III–VII reticulate on the sides, bearing a few setae in a short transverse row on each reticulate area; third and seventh tergites transversely reticulate across center; tergites IV–VI with some transverse striation across center; seventh tergite bearing 2 submedian setae between the lateral setiferous areas; eighth tergite reticulate, with a transverse, arcuate row of 8–9 setae along posterior margin between spiracles; syntergum (Figure 412) short, broad, slightly arcuate, reticulate, bearing 10–13 setae in a transverse row. Cerci situated very close to the posterior spiracles, with 2 long setae and 1 short seta. Ovipositor short, the shaft only $1\frac{1}{7}$ to $1\frac{1}{5}$ times as long as the middle tibia (1.14–1.20), the sheaths about $\frac{1}{5}$ length of middle tibia (0.19–0.20).

Mid-tibial spur about $\frac{4}{5}$ length of the corresponding basitarsus.

Forewing (Figures 413, 414) $2\frac{3}{5}$ to $2\frac{4}{5}$ times as long as wide; marginal fringe not exceeding $\frac{1}{7}$ width of disk. Delta area with 74–87 setae in 7 rows, these sparser and somewhat longer than the setae distad of speculum, well separated from row of setae along posterior margin of wing and from a small group of setae in 1–3 rows below distal three fifths of submarginal vein; costal cell with 1–5 fine setae in a row along proximal half or so, and 1–2 coarse setae near apex. Submarginal vein bearing 4–5 coarse setae and 17–18 bullae. Marginal vein bearing 7–8 prominent, subequal, rather short setae along anterior margin, about $1\frac{1}{5}$ to $1\frac{2}{5}$ times as long as the setae in a row along center of vein.

Forewing strongly infuscated on basal portion up to base of delta and on a broad crossband below stigmal vein, darkest below junction of submarginal and marginal vein and immediately below stigma, the setae on the infuscated areas considerably coarser and darker than those on the hyaline parts of the wing; setae near apex of wing darker than those on the area just distad of the fuscous crossband, but the integument of the wing is not infuscated near the apex; a fuscous streak along central portion of posterior margin of wing, distad of speculum.

Hind wing hyaline, infumated at base, with a very faintly dusky crossband, bearing darker setae, below apex of venation.

General coloration yellow, extensively marked with fuscous and blackish. Occiput with a faint, broad transverse fuscous stripe beyond ocelli; lower part of head blackish, from foramen to oral margin. Pronotum fuscous, center and margins black; mesonotal sclerites fuscous, with a pale median line; borders of median line, anterior margin of mesoscutum and all margins of scutellum blackish; blackish spots around bases of mesonotal setae; tegulae fuscous; a black streak at base of forewing, below tegula; metanotum and propodeum blackish on central reticulate areas, margins lined with black, a black line mesad of spiracle on each side of propodeum, crenulae distinctly blackish. Second abdominal tergite blackish on antero-lateral and central reticulate areas, anterior margin lined with black on either side; third, sixth and seventh tergite with complete blackish crossbands along anterior margin; fourth and fifth tergite with short transverse blackish bands centrally, pale on the sides; eighth tergite and syntergum entirely blackish; ovipositor sheaths blackish. Thoracic sterna strongly infuscated. Internal apodemes, lateral margins of endophragma, most of outer ovipositor plates and outer margin of inner plates black. Antennal scape pale, with a central fuscous blotch; pedicel and third funicular segment entirely blackish, first and second funicular segments contrastingly pale; basal two thirds of club black, apical third pale. All femora extensively marked with fuscous, fore tibia largely blackish, middle and hind tibiae with 2 broad blackish bands, apical tarsal segment of all legs blackish. Submarginal vein of forewing lined with black, marginal vein pale.

Length 0.78–0.98 mm.

*Male*. Unknown.

Redescribed from 3♀♀, apparently paratypes but not labeled as such, reared by K. Hayward from "? *Lecaniodiaspis*" sp. (see notes below) on *Baccharis*, Concordia, Entre Rios, Argentina, October 23 and November 12, 1934. Remounted in Hoyer's medium from the uncleared balsam slides.

Types at the Instituto de Patologia Vegetal, I.N.T.A., Castelar, Buenos Aires, Argentina.

**Notes.** Blanchard (in litt.) classified this species as an *Aphytis*, and it was listed as such by Lizer y Trelles (1939), Hayward (1941) and De Santis (1941). However, the original description was published only in De Santis' (1948) monograph on the Aphelinidae of Argentina, where *haywardi* was classified under *Marietta*. The structure of the metanotum, propodeum, antennae, etc. leaves no doubt that this is a bona fide member of *Aphytis*, though one of the most extensively infuscated species in this genus. The species was originally described from 15♀♀ and appears to be uniparental.

The host record on the labels of the specimens at hand ("?*Lecaniodiaspis*") is evidently erroneous. Both Lizer y Trelles (1939) and Hayward (1941), as well as De Santis (1941, 1948) and De Santis and Esquivel (1966), recorded *haywardi* as a parasite of *Lepidosaphes espinosai* Porter in Argentina.

**Additional Material.** 1♀, reared by H. L. Parker from an unidentified scale insect on *Baccharis*, Montevideo, Uruguay; mounted on a slide in the U.S. National Museum.

## 17. **Aphytis obscurus** DeBach and Rosen

(Figures 415–427)

*Aphytis obscurus* DeBach and Rosen, 1976, Ann. Entomol. Soc. Amer., **69**: 541.

This very dark, biparental South American species is closely related to *haywardi*, differing from the latter species mainly in the pale antennal pedicel, longer ovipositor, fewer mesoscutal setae, smaller, paler, inconspicuous propodeal crenulae, unequal setae along marginal vein of forewing, and in certain details of coloration and wing pattern. The male genitalia lack digital sclerites. *A. obscurus* is described below relative to *haywardi*.

*Female*. Eyes and mouthparts as in *haywardi* (mandibles as in Figure 415). Antennae (Figures 416, 417) similar; scape about $4\frac{1}{4}$ to a little over 5 times as long as wide, as long as the third segment of the funicle or a little longer; first funicular segment trapezoidal, about $1\frac{1}{2}$ to $1\frac{3}{4}$ times as wide as long; second segment symmetrical, somewhat shorter and wider than the first, $1\frac{3}{4}$ times to a little over twice as wide as long; third funicular segment trapezoidal, as long as wide or a little longer than wide, bearing 3–4 longitudinal sensilla; club somewhat longer than in *haywardi*, $2\frac{2}{5}$ to $2\frac{4}{5}$ times as long as wide, $2\frac{2}{3}$ to $2\frac{4}{5}$ times longer and somewhat wider than the preceding segment, bearing 8–9 longitudinal sensilla.

Structure, chaetotaxis and sculpture of head and thorax essentially as in *haywardi*; mesoscutum (Figure 418) with 10–14 setae, axillae asetose. Scutellum more trapezoidal. Epicoxal pads apparently smaller than in other species.

Propodeum (Figures 418–420) $2\frac{2}{3}$ to $3\frac{1}{2}$ times as long as the metanotum, $\frac{3}{4}$ to $\frac{9}{10}$ length of scutellum, sculpture essentially as in *haywardi*; crenulae less conspicuous, apparently smaller and fewer than in *haywardi*, nonoverlapping, situated only at apex of triangular reticulate area (Figure 420), largely obscured by reticulation of second abdominal tergite.

Structure, chaetotaxis and sculpture of abdominal tergites similar to *haywardi*; reticulate area at center of second tergite larger; third tergite nearly smooth centrally, with only faint indications of transverse striation; reticulation on seventh and eighth tergites and on syntergum apparently more delicate than in *haywardi*. Syntergum (Figure 421) short, broad, arcuate. Cerci situated very close to posterior spiracles, with 2 long setae and 1 short seta. Ovipositor longer than in *haywardi*, the shaft about $1\frac{1}{3}$ to $1\frac{1}{2}$ times as long as the middle tibia (1.36–1.53), the sheaths nearly $\frac{1}{3}$ length of middle tibia (0.29–0.32).

Mid-tibial spur $\frac{3}{4}$ to $\frac{4}{5}$ length of the corresponding basitarsus.

Forewing (Figure 422) somewhat narrower than in *haywardi*, about 3 times as long as wide, marginal fringe longer, not exceeding $\frac{1}{3}$ width of disk (usually about $\frac{1}{4}$ to $\frac{1}{5}$). Delta with 48–81 setae in 6–8 rows, these sparser and considerably longer than the setae distad of speculum, usually rather clearly separated from setae in 2–3 rows along posterior margin of wing, widely separated from a rather large group of setae in 2–3 rows below distal two thirds to three quarters of submarginal vein; costal cell with a row of fine setae along proximal half or so, and 1–2 coarse setae near apex. Submarginal vein bearing 4–5 coarse setae and 15–22 bullae. Marginal vein bearing 6–9 prominent setae along anterior margin, these decreasing in length toward apex of vein, the proximal setae $1\frac{1}{5}$ to $1\frac{1}{2}$ times as long as the distal ones, considerably (about $1\frac{1}{2}$ to 2 times) longer than the setae in a row along center of vein.

Forewing pattern essentially similar to that in *haywardi*; infuscated area at base of wing more extensive, covering base of delta; discal setae near apex of wing more contrastingly darker and coarser than in *haywardi*; long setae along proximal part of anterior margin of marginal vein hyaline, sometimes hyaline at base and dark apically, the shorter setae along distal part of vein dark (see Figure 426).

Hind wing pattern more contrasting than in *haywardi*, the basal part and central crossband more distinctly infumated (see Figure 427), discal setae near tip of wing contrastingly darker than the setae just beyond crossband, marginal fringe dark at dark parts of wing, contrastingly pale at hyaline parts.

General coloration essentially as in *haywardi* but apparently darker, although this may be due to different clearing and mounting procedures. Head as in *haywardi*.

Mesoscutum and scutellum (Figure 418) uniformly, deeply blackish, except for a contrastingly hyaline median line and some hyaline along antero-lateral margins; borders of reticulation cells lined with black; no black spots around bases of mesonotal setae; pale median line clearly evident also on propodeum (Figures 418–420); crenulae pale or faintly fuscous. Abdomen, thoracic sterna and internal structures as in *haywardi*. Antennal scape (Figure 416) strongly infuscated on proximal half or so, pale distally; pedicel and first 2 funicular segments entirely white; third funicular segment usually entirely white, sometimes faintly dusky distally; club black on proximal two thirds or so, yellow apically as in *haywardi*. Legs rather extensively infuscated, middle and hind tibiae not as clearly banded as in *haywardi*. Wing veins as in *haywardi*.

Length 0.76–1.11 mm.

*Male*. Very similar to the female in structure, chaetotaxis, sculpture and coloration, differing slightly in the shape of the antennae.

Antennal scape (Figure 423) wider than in the female, about $3\frac{3}{5}$ times as long as wide, a little longer than the club; pedicel $1\frac{1}{4}$ to $1\frac{2}{5}$ times as long as wide, somewhat longer than the third segment of the funicle; first 2 funicular segments as in the female, third segment up to $1\frac{1}{4}$ times as long as wide, bearing 2–3 longitudinal sensilla; club somewhat shorter than in the female, about $2\frac{1}{5}$ to $2\frac{1}{3}$ times as long as wide, $2\frac{2}{3}$ to over $2\frac{4}{5}$ times longer than the preceding segment, bearing 4–5 longitudinal sensilla.

Mesoscutum with 10–12 setae. Propodeum (Figure 424) about $2\frac{1}{2}$ to 3 times as long as the metanotum, $\frac{3}{4}$ to $\frac{4}{5}$ length of scutellum; crenulae as in the female. Genitalia (Figure 425) with a central longitudinal rod but lacking any visible digital sclerites, apodemes or papillae, about $\frac{2}{3}$ length of middle tibia (0.63–0.70).

Forewing (Figure 426) as in the female, a little more extensively infuscated on basal part; delta with 45–57 setae; submarginal vein bearing 16–18 bullae; marginal vein bearing 6–7 prominent setae along anterior margin, the proximal setae about $1\frac{1}{5}$ to $1\frac{1}{3}$ times as long as the distal ones. Hind wing (Figure 427) as in the female.

General coloration as in the female; third funicular segment usually rather faintly dusky.

Length 0.73–1.03 mm.

Described from 17♀♀ and 6♂♂ (♀ holotype, ♂ allotype, and paratypes), reared by H. Zimmermann from the cactus scale, *Diaspis echinocacti* (Bouché), Martin Garcia Island, Argentina, August 1971 (received for study from Dr. D. P. Annecke, Pretoria, South Africa; his code No. T3893).

Holotype in the collection of the Division of Biological Control, University of California, Riverside.

**Notes.** Although *obscurus* is very closely related to *haywardi*, it may be recognized at a glance by the entirely pale antennal pedicel (the pedicel is uniformly blackish in *haywardi*). We also regard the longer ovipositor, the fewer mesoscutal setae, the paler, inconspicuous crenulae and the narrower, more extensively infuscated forewing, with more numerous setae in the basal area and unequal setae along the marginal vein, as valid diagnostic characters separating *obscurus* from *haywardi*. The last character, as well as the generally heavier pigmentation and the biparental mode of reproduction, seem to indicate that *obscurus* is a more primitive species than *haywardi*.

Although the intensity of pigmentation may have been influenced by clearing and mounting procedures, *obscurus* appears to be considerably darker than *haywardi*. In fact, this is probably the darkest known species of *Aphytis*.

The holotype ♀ has one supernumerary seta on the scutellum.

## 18. **Aphytis angustus** Compere

(Figures 428–433)

*Aphytis angustus* Compere, 1955, Univ. Calif. Publ. Entomol., **10**:286.

This little-known Oriental species appears to be rather closely related to *obscurus*, but may be readily separated from the latter species by the relatively long endophragma, propodeum, ovipositor and syntergum; the forewing of *angustus* is narrower at the base, less densely setose, and largely infuscated except the apex; general coloration extensively marked with fuscous but less so than in *obscurus*; antennae mostly dusky.

*Female*. Eyes finely setose. Mandibles well developed; maxillary palpi 2-segmented, labial palpi 1-segmented. Antennae (Figures 428, 429) short; scape slender, $5\frac{1}{3}$ to 6 times as long as wide, about $1\frac{1}{4}$ times longer than the club; pedicel $1\frac{1}{4}$ to $1\frac{1}{3}$ times as long as wide, about as long as, or slightly longer than, the third segment of the funicle; first funicular segment short, trapezoidal or nearly triangular, about twice as wide as long; second segment nearly symmetrical, about as long as but distinctly wider than the first segment, 3 to $3\frac{1}{2}$ times as wide as long; third funicular segment short, trapezoidal, dorsal aspect longer than the ventral, about $1\frac{1}{6}$ to $1\frac{1}{4}$ times as wide as long, bearing 3–4 longitudinal sensilla; club short, thick, about 2 to $2\frac{1}{4}$ times as long as wide, $2\frac{2}{3}$ to nearly 3 times longer and somewhat wider than the preceding segment, bearing 8–9 longitudinal sensilla.

Setae on head and thorax dark, coarse, visible under ×30 magnification; those on abdomen more slender but visible under ×60 magnification. Vertex with 2 pairs of long setae along occipital margin, in addition to numerous shorter setae. Mesoscutum with 10–11 setae, the posterior pair and 1 seta at each antero-lateral corner considerably longer and coarser than the others; each parapsis with 1 long and 1 short seta, each axilla with 1 seta; scutellum with 4, the discoid sensilla about equidistant from the two pairs. Frontovertex, pronotum and mesonotal sclerites reticulate. Scutellum about $\frac{3}{4}$ median length of mesoscutum. Metanotum (Figure 430) relatively long, rather strongly arcuate, reticulate except on the sides; anteromedian apodeme stout, long, $\frac{4}{5}$ to nearly full median length of metanotum. Endophragma (Figure 432) long, about $1\frac{7}{10}$ as long as wide at base.

Propodeum (Figures 430, 431) long, $3\frac{2}{5}$ to $3\frac{3}{5}$ times as long as the metanotum, distinctly (up to $1\frac{1}{5}$ times) longer than the scutellum, faintly longitudinally striated mesad of spiracles, reticulate-strigose on a triangular median salient; crenulae not clearly visible in the uncleared types, apparently few, small, rounded, probably somewhat overlapping.

Second abdominal tergite transversely reticulate across, the sculpture fading somewhat centrally; tergites III–VII with a few fine setae on each side, the sculpture on the sides not visible in the uncleared types; third and seventh tergites transversely reticulate across center, tergites IV–VI transversely striated; seventh tergite longitudinally striated anterad of the reticulate area, bearing 2–3 fine setae in a transverse row along posterior margin, between the lateral setiferous areas; eighth tergite longitudinally striated anteriorly, transversely reticulate posteriorly, with 10–12 setae in 1–2 transverse rows between spiracles; syntergum (Figure 432) long, arcuate, conspicuous, bearing 33–34 setae in several transverse rows. Cerci situated very close to posterior spiracles, with 2 long setae and 1 short seta. Ovipositor shaft about $1\frac{3}{4}$ to $1\frac{4}{5}$ times as long as the middle tibia (1.72–1.84), ovipositor sheaths long, nearly $\frac{1}{2}$ length of middle tibia (0.45).

Mid-tibial spur a little shorter than the corresponding basitarsus.

Forewing (Figure 433) long, narrow, fully 3 times as long as wide or somewhat longer; marginal fringe not exceeding $\frac{1}{4}$ width of disk. Delta area with 60–67 setae in 8–9 rows, these considerably longer, coarser and somewhat sparser than the setae distad of speculum, not clearly separated from 2–3 rows of setae along posterior margin of wing, widely separated from 3 setae in a row below anterior half of submarginal vein; disk rather uniformly setose distad of speculum, but the setae on the hyaline apical part considerably finer and paler than those on the infumated part of the disk; costal cell with a row of 4–5 fine setae along proximal two thirds and 1 coarse seta near apex. Submarginal vein bearing 3 coarse setae and 1 fine seta and 17–21 bullae. Marginal vein bearing 10–12 prominent, unequal setae along anterior margin decreasing noticeably in length toward apex, the proximal setae $1\frac{3}{5}$ to $1\frac{4}{5}$ times as long as the distal ones and up to twice longer than the setae in a row along center of vein.

Forewing distinctly infuscated, more strongly so at base of delta and below stigmal vein, with dark setae as well as an integumentary infuscation, the apical part contrastingly hyaline and bearing pale setae; a fuscous streak along central portion of posterior margin, distad of speculum.

General coloration yellow, extensively marked with brown and blackish. Head with narrow black stripes on oral margin and genal sutures, broader stripes on occiput on each side of foramen. Pronotum fuscous centrally and on lateral margins, propleura lined with blackish; mesoscutum with 2 large fuscous blotches; scutellum fuscous anteriorly, posterior margin blackish; metanotum and propodeum largely fuscous, margins lined with blackish. Abdominal tergites strongly infuscated centrally, the sides pale (originally described by Compere, from fresh material, as "translucent white"), posterior margin of eighth tergite lined with blackish; ovipositor plates lined with blackish, sheaths blackish. Endophragma and thoracic sterna strongly infuscated. Antennal scape (Figure 429) pale, fuscous apically; rest of antenna dusky, basal half of club more distinctly so, distal half of club pale. Forelegs pale, the tarsi dusky; middle femur and tibia and hind tibia marked with blackish, hind basitarsus contrastingly dusky. Wing veins dusky.

Length 0.91 mm.

Redescribed from 2♀♀ (holotype and paratype), obtained from a shipment from China (host unknown), February 28, 1949; material collected by J. L. Gressitt; "mounted before shrinkage in gum mar."

Type series in the U.S. National Museum of Natural History, Washington, D.C.

*Male.* A single male specimen, reared from *Lepidosaphes* material on *Agalma,* Taiwan, December 3, 1952 ("S&R No. 1031, Maalin No. 238"), is rather similar to the female types of *angustus* in the most important characters and is tentatively referred to this species. This presumed male of *angustus* (Figures 434–436), mounted in Hoyer's medium and kept in the collection of the Division of Biological Control, University of California, Riverside, is described below relative to the female.

Antennal scape wider than in the female, $3\frac{1}{2}$ times as long as wide, bearing 2 peglike sense organs on antero-ventral aspect (Figure 434); pedicel about $1\frac{2}{5}$ times as long as wide, about as long as the third segment of the funicle; first funicular segment a little over twice as wide as long, second segment $2\frac{2}{3}$ times as wide as long, third segment symmetrical, a trifle longer than wide, bearing 4 longitudinal sensilla; club a little less than twice as long as wide, about twice longer than the preceding segment, bearing 7 longitudinal sensilla.

Mesoscutum with 12 setae. Propodeum (Figure 435) shorter than in the female, $2\frac{2}{3}$ times as long as the metanotum, but still fully as long as the scutellum; crenulae 4+5, small, rounded, slightly overlapping. Syntergum long, bearing 21 setae.

Genitalia (Figure 436) similar to those of *obscurus*, long, narrow, $\frac{3}{4}$ length of the middle tibia, with a central longitudinal rod but no visible digital sclerites, apodemes or papillae.

Forewing wider than in the female, $2\frac{2}{3}$ times as long as wide; delta with about 50 setae; marginal vein bearing 9 prominent setae along anterior margin, the proximal setae about $1\frac{1}{2}$ times longer than the distal ones; disk rather uniformly setose distad of speculum.

Forewing nearly hyaline, faintly dusky at base of delta and below stigmal vein.

Otherwise structure, chaetotaxis, sculpture and coloration as in the female.

Length 0.96 mm.

**Notes.** *A. angustus* is possibly not a bona fide member of the **vittatus** group as defined here, but may be regarded as closely related to this group because of the heavy pigmentation of the body and infuscated wings. It is rather closely related to *obscurus*, but may be separated from this species by the shape of the propodeum and syntergum, the length of the ovipositor and other readily obvious characters of structure and coloration. The presumed male of *angustus* is also similar to the male of *obscurus* in the shape of the genitalia. However, additional reared material is required for positive determination of the specific identity of the presumed male.

## RELATED SPECIES

The following 6 species are obviously related to the **vittatus** group, although for various reasons they cannot be considered bona fide members of this group.

## 19. **Aphytis hyalinipennis** n.sp.

(Figures 437–445)

This interesting South Pacific species is very similar to *capillatus* in the conspicuously banded abdomen, short antennae, large, nonoverlapping propodeal crenulae, and other details of coloration and structure, but differs markedly in the perfectly hyaline, nonmottled forewing of the female. The male differs from *capillatus* in the presence of two readily visible "ring" segments in the antenna and in the absence of a specialized sense organ on the scape.

*Female.* Eyes moderately coarsely setose, the setae clearly visible under $\times 60$ magnification. Mandibles well developed; maxillary palpi 2-segmented, labial palpi 1-segmented. Antennae (Figure 437) short, rather stout; scape about 5 times as long as wide, distinctly longer than the club; pedicel $1\frac{3}{5}$ to $1\frac{2}{3}$ times as long as wide, about $1\frac{2}{5}$ to $1\frac{3}{5}$ times longer than the third segment of the funicle; first funicular segment triangular, $1\frac{1}{3}$ times as wide as long; second segment trapezoidal, dorsal aspect longer than the ventral, about as long as but considerably wider than the first segment, $1\frac{4}{5}$ times as wide as long; third funicular segment trapezoidal, dorsal aspect longer than the ventral, somewhat wider than long, bearing 1 longitudinal sensillum; club thick, $2\frac{1}{7}$ to $2\frac{2}{5}$ times as long as wide, about 3 to $3\frac{3}{5}$ times longer and considerably wider than the preceding segment, bearing 11–12 longitudinal sensilla; scape, pedicel and funicle furnished with strong, relatively long setae, club with shorter setae.

Setae on head, thorax and sides of abdomen dark, coarse, readily visible under $\times 60$ magnification; the longest setae appear as if composed of distinct filaments. Vertex with a pair of long setae along occipital margin, in addition to numerous shorter setae. Mesoscutum (Figures 438, 439) with 22–23 setae, the posterior pair considerably longer and coarser than the others; each parapsis with 3 rather short setae, each axilla with 1 long seta; scutellum with 4 setae, the discoid sensilla considerably closer to the posterior than to the anterior pair. Frontovertex, pronotum, and mesonotal sclerites reticulate. Scutellum oval, nearly $\frac{3}{4}$ median length of mesoscutum. Metanotum rather long, reticulate except on the sides, anteromedian apodeme robust, about $\frac{3}{5}$ median length of metanotum.

Propodeum (Figures 438, 440) only about twice as long as the metanotum, $\frac{3}{4}$ length of scutellum, with a rather prominent median salient, reticulate on a median triangular area, smooth on the sides; crenulae about 9+9, large, elongate, nonoverlapping.

Second abdominal tergite smooth; tergites III–VII (Figure 438) reticulate on the sides, bearing rather numerous strong setae in 2 short, transverse rows on each reticulate area; tergites III–VI faintly transversely striated across; seventh tergite more distinctly so, bearing 5–7 fine setae in a transverse row between the lateral setiferous areas; eighth tergite reticulate, with a transverse row of 6–8 fine setae between spiracles and 1–2 longer setae laterad of spiracles; syntergum (Figure 441) short, triangular, considerably wider than long, reticulate, bearing numerous strong setae in 2 transverse rows. Cerci situated close to posterior spiracles, with 2 long setae and 1 short seta.

Ovipositor shaft about $1\frac{1}{2}$ times as long as the middle tibia (1.49–1.53), ovipositor sheaths nearly $\frac{1}{3}$ length of middle tibia (0.30–0.31).

Mid-tibial spur stout, a little shorter than the corresponding basitarsus.

Forewing (Figure 442) about $2\frac{3}{5}$ times as long as wide; marginal fringe not exceeding $\frac{1}{10}$ width of disk. Disk uniformly setose, without contrasting coarse and fine setae; delta area rather densely setose, bearing 112–162 setae in 9–11 rows, these longer and somewhat sparser than the setae distad of speculum, narrowly separated from row of setae along posterior margin of wing, widely separated from 3–4 rows of setae below distal two thirds of submarginal vein; costal cell with a row of fine setae separated by a short bare space from 2–4 coarse setae near apex. Submarginal vein bearing 5–6 coarse setae and 19–22 bullae. Marginal vein bearing 11–14 prominent, subequal setae along anterior margin, these about as long as the setae in a row along center of vein.

Forewing entirely hyaline, very faintly dusky below junction of submarginal and marginal veins; posterior margin of wing narrowly lined with fuscous distad of speculum.

General coloration yellow, with extensive black and fuscous markings as follows (Figure 438): Oral margin and lower portion of genal sutures lined with fuscous; a pair of conspicuous transverse black bars on occiput, on each side of foramen. Pronotum narrowly fuscous centrally and on sides; mesoscutum pale, with faint fuscous blotches anteriorly and faint spots around bases of posterior setae, anterior and posterior margins and parapsidal sutures black; scutellum with a pair of antero-median fuscous blotches, extending posteriorly in two faint submedian stripes, with fuscous blotches around bases of setae, posterior margin black; metanotum narrowly lined with black on sides; propodeum lined with blackish on lateral and posterior margins, fuscous on anterior part of median sculptured area. Second abdominal tergite fuscous, lateral portions of anterior and posterior margins lined with black; tergites III–VII with broad, complete fuscous crossbands along posterior margins, third tergite also with a pair of antero-lateral fuscous blotches, tergites IV–VII with transverse antero-median blotches anterad of crossbands; eighth tergite fuscous, posterior margin blackish; syntergum faintly fuscous; ovipositor plates lined with black, sheaths entirely fuscous. Thoracic sterna strongly infuscated; endophragma margined with black, fading posteriorly. Mandibles dark brown on distal half, the denticles black. Antennal scape extensively infuscated ventrally; pedicel, funicle and club uniformly fuscous. All femora with blackish markings, all tibiae with 2 indistinct, incomplete blackish rings, the 2 distal segments of all tarsi black. Wing veins pale, narrowly lined with blackish.

Length 1.19–1.34 mm.

*Male.* Essentially similar to the female, differing mainly in the structure of the antennae and in the more extensive pigmentation.

Eyes moderately coarsely setose. Antennae (Figure 443) essentially 4-segmented, but with 2 distinct, readily visible "ring" segments; scape 4 times as long as wide, lacking any specialized sense organs; pedicel about $1\frac{3}{10}$ times as long as wide, about $1\frac{2}{5}$ times longer than the third segment of the funicle; first funicular segment reduced to a minute, triangular annellus which is twice as wide as long; second segment rudimentary but distinct, considerably shorter than but about as wide as the first annellus, $4\frac{1}{2}$ times as wide as long; third funicular segment trapezoidal, dorsal aspect longer than the ventral,

$1\frac{1}{3}$ times as wide as long, not bearing any longitudinal sensilla; club large, thick, $2\frac{1}{2}$ times as long as wide, considerably longer than the scape, $4\frac{1}{2}$ times longer and considerably wider than the preceding segment, bearing about 11 longitudinal sensilla, apparently with a cut-off sensory area on the ventral aspect.

Sculpture and chaetotaxis about as in the female; mesoscutum (Figure 444) with 26 setae. Scutellum $\frac{2}{3}$ median length of mesoscutum. Propodeum (Figure 445) about $2\frac{1}{4}$ times as long as the metanotum, $\frac{2}{3}$ length of scutellum; median salient prominent, crenulae 7+7.

Genitalia not seen clearly due to position of single male specimen on slide; apparently no distinct rod between apodemes.

Forewing essentially as in the female.

Coloration as in the female, except as follows: pronotum more extensively blackish; fuscous patches on mesoscutum larger, more distinct; submedian fuscous stripes on scutellum more pronounced; abdominal tergites III–VII extensively infuscated centrally anterad of crossbands, merging into a continuous, large central blotch.

Length 0.97 mm.

Described from 2♀♀ and 1♂ (♀ holotype, ♂ allotype and ♀ paratype), reared by G. Fabres from an undetermined armored scale insect on *Casuarina equisetifolia*, Noumea, New Caledonia, August 1971; all mounted in Hoyer's medium on one slide, together with 1♂ of *Metaphycus* sp. The holotype ♀ is the lowest specimen in the row.

Type series in the collection of the Division of Biological Control, University of California, Riverside.

**Notes.** *A. hyalinipennis* appears to be closely related to the *capillatus* complex. It closely resembles *capillatus* in coloration, as well as in other important diagnostic characters such as the structure of the antennae and the shape of the propodeal crenulae; it is rather similar to *australiensis* in the presence of a prominent propodeal salient. However, it differs markedly from all members of the *capillatus* complex in the perfectly hyaline, nonmottled forewing of the female.

Although *hyalinipennis* cannot be considered a bona fide member of the **vittatus** group because of the nonmottled wings, it is nevertheless obviously closely related to this group of primitive, heavily pigmented species.

## 20. Aphytis wallumbillae (Girault)

(Figures 446–448)

*Aphelinus wallumbillae* Girault, 1924, *Homo perniciosus* and New Hymenoptera, Priv. Publ., Brisbane, 4 pp., page 4.
*Aphytis wallumbillae*: Mercet, 1932, Eos, **8**:355.

This enigmatic Australian species appears to be very closely related to *hyalinipennis*, differing from the latter species mainly in the somewhat less extensive pigmentation, in the finely setose eyes, and (possibly) in the more rounded crenulae.

*Female.* Eyes finely setose. Mandibles well developed; maxillary palpi 2-segmented, labial palpi 1-segmented. Antennal scape (Figure 446) about $4\frac{2}{5}$ times as long as wide, a little longer than the club; pedicel $1\frac{3}{10}$ times as long as wide, about $1\frac{1}{3}$ times longer than the third segment of the funicle; first funicular segment triangular, nearly twice as wide as long; second segment nearly symmetrical, attached to the oblique aspect of the preceding segment, about as long as and considerably wider than the first segment, $2\frac{2}{5}$ times as wide as long; third funicular segment trapezoidal, about $1\frac{1}{3}$ times as wide as long, bearing 2–3 longitudinal sensilla; club about twice as long as wide, nearly $3\frac{1}{2}$ times longer and distinctly wider than the preceding segment, bearing 13–14 longitudinal sensilla.

Setae on head, thorax, and sides of abdomen dark, coarse, readily visible under ×60 magnification; longest setae on thorax appear as if composed of distinct filaments and are clearly visible under ×30 magnification. Vertex with 2 pairs of long setae along occipital margin, in addition to numerous short setae. Mesoscutum with 15–18 setae; each parapsis with 2–3 setae, each axilla with 1 seta; scutellum with 4 setae, the discoid sensilla closer to the posterior than to the anterior pair. Frontovertex, pronotum, and mesonotal sclerites reticulate. Scutellum oval, nearly $\frac{3}{4}$ median length of mesoscutum. Metanotum arcuate, reticulate except on the sides, anteromedian apodeme rather robust, about $\frac{2}{3}$ median length of metanotum.

Propodeum (Figure 447) relatively short, about twice as long as the metanotum, only about $\frac{1}{2}$ length of scutellum, reticulate on the sides and on a trapezoidal median salient, sculpture changing to strigose posteriorly; crenulae 7+8, large, rounded, distinct, nonoverlapping.

Second abdominal tergite short, smooth; tergites III–VII reticulate on the sides, bearing several setae on each reticulate area; tergites III–VI with faint indications of transverse striation across center; seventh tergite delicately reticulate and longitudinally striated across center, bearing a transverse row of 4–6 fine setae between the lateral groups of coarser setae; eighth tergite faintly reticulate, with a transverse row of 7–8 setae between spiracles; syntergum delicately reticulate, bearing 14 setae. Cerci with 2 long setae and 1 short seta. Ovipositor shaft $1\frac{1}{2}$ to $1\frac{3}{5}$ times as long as the middle tibia (1.50–1.58), ovipositor sheaths about $\frac{1}{4}$ to $\frac{1}{3}$ length of middle tibia (0.27–0.32).

Mid-tibial spur as long as or a little shorter than the corresponding basitarsus.

Forewing (Figure 448) about $2\frac{1}{2}$ times as long as wide; marginal fringe short, not exceeding $\frac{1}{11}$ width of disk. Delta area with about 62 setae in 5–7 rows, these sparser and considerably longer than the setae distad of speculum, not clearly separated from sparse row of setae along posterior margin of wing, widely separated from a few setae below distal two thirds of submarginal vein; costal cell with a row of fine setae along proximal half or three fifths, and 2–3 coarse setae near apex. Submarginal vein bearing 5–6 coarse setae and 19 bullae. Marginal vein bearing 9–11 prominent, subequal setae along anterior margin, these about as long as the coarse setae in a row along center of vein.

Forewing entirely hyaline.

General coloration yellow, with silvery markings on pronotum, mesoscutum, scutellum and metanotum, and blackish markings as follows: parapsidal sutures, posterior margin of mesoscutum and scutellum, spots surrounding bases of thoracic setae, sutures mesad of propodeal spiracles, lateral parts of posterior margin of propodeum, second abdominal tergite anteriorly, crossbands on tergites III–VI, partly obliterated centrally, lateral margins of endophragma anteriorly; ovipositor sheaths fuscous distally, margins of ovipositor plates fuscous. Thoracic sterna dusky. Two faint, fuscous, transverse stripes on occiput, on each side of foramen. Antennal scape with a longitudinal blackish stripe ventrally; pedicel, funicular segments and club faintly, uniformly dusky. Legs pale, hind femora dusky distally; all tibiae with 2 incomplete fuscous rings and a black spot at apex, all tarsi with the distal segment blackish. Wing veins pale.

Length 0.89 mm.

*Male*. Unknown.

Redescribed from the type series: 4♀♀ (3 bodies, 4 heads), "from a lecaniid, Brigalow, Wallumbilla [Queensland, Australia], October 17, 1923"; had all been mounted without clearing in balsam, on one slide (#3775, Queensland Museum); 2♀♀, in very poor shape, and 3 heads were cleared and remounted in Hoyer's medium. The remaining ♀, uncleared, on the original balsam slide, is designated lectotype.

**Notes.** *A. wallumbillae* is evidently very closely related to *hyalinipennis* in structure as well as in coloration. It appears to differ from this species mainly in details of pigmentation: the crossbands on abdominal tergites III–VI are partly obliterated centrally, whereas the blotches anterad of the crossbands, and the crossband on the seventh tergite, are absent, and the eighth tergite and syntergum are pale in *wallumbillae*. Also, the eyes are finely setose, the propodeal crenulae appear to be more rounded, less elongate than in *hyalinipennis*, and there are fewer setae on the mesoscutum and in the delta area of the forewing. Some of these apparent differences may be related to the smaller size of *wallumbillae*, or even to the poor shape of the material at hand. However, the differences in coloration alone would seem sufficient for retaining *wallumbillae* and *hyalinipennis* as two distinct species.

Like *hyalinipennis*, *wallumbillae* may be regarded as related to the **vittatus** group, although not a bona fide member of that group.

Girault's original slide includes a single male specimen. As far as can now be determined, this poorly preserved specimen is identical to *capillatus*, as understood here. Girault's original description does not refer to the sex of his type specimens, but evidently pertains to the females only. Since his slide is labeled "*Aphelinus wallumbillae* Gir. Types, ♀" in Girault's own handwriting, it can be safely assumed that this male specimen was not intended as part of the type series. Until more information is available on this little-known species, the male of *wallumbillae* remains unknown.

The host record given by Girault—"from lecaniid"—was apparently incorrect. We have no reason to believe that this species is indeed different from all known species of *Aphytis*, which are all primary ectoparasites of armored scale insects.

*A. ruskini* (Girault), an unrecognizable Australian species of *Aphytis*, appears to be closely related to *wallumbillae* and may in fact be a senior synonym (see p. 740).

## 21. **Aphytis dealbatus** Compere

(Figures 449–451)

*Aphytis dealbatus* Compere, 1955, Univ. Calif. Publ. Entomol., **10**:286–287.

This little-known North American species differs from all known species of *Aphytis*, except *melanostictus*, in that the antennal club of the female is separated from the penultimate segment by a suture only, without a distinct constriction, and terminates in a prominent projecting sensillum. It may be further recognized by the alternating infumated and hyaline bands on the forewing.

*Female*. Eyes finely setose. Mandibles well developed, with 2 denticles and a dorsal truncation; maxillary palpi 2-segmented, labial palpi 1-segmented. Antennal scape (Figure 449) slender, 5–6 times as long as wide, about $1\frac{1}{2}$ times longer than the club; pedicel (Figures 449, 450) slender, about twice as long as wide, somewhat longer than the third segment of the funicle; first funicular segment trapezoidal, about $1\frac{7}{10}$ times as wide as long; second segment more nearly symmetrical, slightly shorter and wider than the first segment, about $2\frac{1}{3}$ times as wide as long; third funicular segment and the club separated only by a suture, without a constriction, the two together semifusiform, widest at the suture; third funicular segment distinctly (about $1\frac{1}{5}$ times) longer than wide, bearing 2–3 longitudinal sensilla; club short, a little over twice as long as wide, only about $1\frac{4}{5}$ times as long as the preceding segment, bearing 6–7 longitudinal sensilla, terminating in a prominent projecting sensillum.

Setae on thorax coarse, dark; mesoscutum with 10–12 setae, the posterior pair and 1 seta at each antero-lateral corner considerably longer and coarser than the others. Vertex, pronotum and mesonotum apparently reticulate; metanotum striated centrally.

Propodeum about 3 times as long as the metanotum, nearly $\frac{1}{2}$ length of scutellum, apparently reticulate centrally and just mesad of the spiracles; crenulae unseen. Abdominal tergites apparently reticulate on the sides. Syntergum triangular; cerci situated considerably closer to posterior spiracles than to tip of abdomen, with 2 long setae and 1 short seta. Ovipositor about $1\frac{4}{5}$ times as long as the middle tibia; ovipositor sheaths $\frac{2}{5}$ length of middle tibia.

Mid-tibial spur only about $\frac{2}{3}$ length of the corresponding basitarsus.

Forewing (Figure 451) about $2\frac{2}{5}$ times as long as wide; marginal fringe short, not exceeding $\frac{1}{8}$ width of disk. Delta area with 112–154 setae in about 10 rows, only somewhat less densely setose than the area distad of speculum, rather clearly separated from row of setae along posterior margin of wing and from a triangular patch of setae below submarginal vein; costal cell with a row of 2–4 short, very fine setae along proximal half or so, and 1 coarse seta near apex. Submarginal vein bearing 5 coarse setae and 18 bullae. Marginal vein bearing 8–12 prominent, subequal setae along anterior margin, these considerably longer than the setae in a row along center of vein (up to $1\frac{1}{2}$ times as long in the holotype, even more so in the smaller paratype).

Forewing with 2 infumated, rather obscure crossbands, bearing somewhat coarser setae, alternating with 2 hyaline areas bearing exceedingly fine setae; one crossband below junction of submarginal and marginal veins, at base of delta, another below the stigmal vein, with a darker spot just below the vein; one hyaline area between the crossbands, another just distad of the second band; the apical part of the wing, distad of the second hyaline area, is very slightly infumated; a fuscous streak along posterior margin of wing, distad of speculum.

"General color yellow to white, with light brown to blackish as follows: greater part of antennae slightly suffused, the ends of scape pale; parapsidal sutures, margins of propodeum mesad of the spiracles; sides of abdomen; pro-, meso-, and metasternum; all femora and tibiae, partly suffused or spotted" (Compere 1955). Wing veins brownish.

Length 0.91–1.13 mm.

*Male*. Unknown.

Redescribed from 2♀♀ (holotype and paratype), reared by T. W. Fisher from *Lepidosaphes ulmi* (L.) on black willow, Placerville, California, August 26, 1952, mounted without clearing in gum mar.

Types in the collection of the Division of Biological Control, University of California, Riverside.

**Notes.** As pointed out by Compere (1955), *A. dealbatus* is "suggestive of *Marietta* in the conformation of antennae, the rudimentary hyaline spots on front wings, and the faint white spots on the mesoscutum and scutellum." The two type specimens are uncleared, and important characters such as the structure of the propodeum, the presence of crenulae, the anteromedian apodeme of the metanotum, chaetotaxis, and details of sculpture are not clearly visible. No additional material is available for study. There is, however, little doubt that this species has been correctly placed in the genus *Aphytis*. It is obviously closely related to the **vittatus** group.

We have made several attempts to obtain additional material of *dealbatus* from collections of the type host on the type host plant in the type locality, but have been unsuccessful. The material reared was all *A. mytilaspidis*.

## 22. **Aphytis melanostictus** Compere

(Figures 452–465)

*Aphytis melanostictus* Compere, 1955, Univ. Calif. Publ. Entomol., **10**:287.

This peculiar North American species is similar to *dealbatus* in the structure of the antennae, the club being separated from the penultimate segment by a suture only, without a constriction, and bearing a prominent projecting sensillum at apex. It may be further recognized by the rather heavy thoracic pigmentation, the uniformly setose, nearly hyaline forewings with a distinct fuscous cloud below the stigma, the asetose axillae and the triangular, nonoverlapping, sometimes irregular propodeal crenulae. The male is characterized by apically swollen, black-tipped middle tibiae, and by specialized sense organs on the antennal scape.

*Female.* Eyes finely setose. Mandibles well developed, essentially bidentate, with 2 strong denticles and a short, oblique dorsal truncation; maxillary palpi 2-segmented, labial palpi 1-segmented. Antennal scape (Figure 452) slender, about $4\frac{1}{2}$ to over 6 times as long as wide, considerably longer than the club ($1\frac{1}{3}$ to $1\frac{1}{2}$ times); pedicel about $1\frac{1}{2}$ to $1\frac{7}{10}$ times as long as wide, about as long as the third segment of the funicle or nearly so; first funicular segment triangular or trapezoidal, about $1\frac{2}{5}$ times as wide as long; second segment nearly symmetrical, somewhat shorter and wider than the first, about twice as wide as long; third funicular segment and the club separated only by a suture, without a constriction, the two together semifusiform; third funicular segment a little longer than wide, bearing 2–5 longitudinal sensilla; club about $2\frac{1}{3}$ to $2\frac{1}{2}$ times as long as wide, about twice longer than the preceding segment, bearing 6–8 longitudinal sensilla and terminating in a prominent projecting sensillum.

Setae on head and thorax mostly short, coarse, dark, the larger ones easily seen in cleared specimens under $\times 30$ magnification; those on abdomen paler, more slender, invisible even under $\times 120$ magnification. Vertex with one pair of larger setae beyond posterior ocelli, anterad of occipital margin, in addition to numerous short setae. Mesoscutum (Figures 453, 454) with 8–14 setae, the posterior pair considerably longer and coarser than the others; each parapsis with 2–3 setae (sometimes 2 coarse setae and 1 minute sensory seta); axillae asetose (rarely with 1 seta); scutellum with 4 setae, the discoid sensilla considerably closer to the anterior than to the posterior pair. Frontovertex, pronotum and mesonotal sclerites reticulate, with a distinct infrasculpture within the cells. Scutellum a little over $\frac{4}{5}$ median length of mesoscutum. Metanotum (Figures 455, 456) reticulate-strigose except on the sides; anteromedian apodeme slender, considerably shorter than median length of metanotum.

Propodeum (Figures 455–457) about 3 to $3\frac{1}{2}$ times as long as the metanotum, about $\frac{2}{3}$ to $\frac{3}{4}$ length of scutellum, reticulate-strigose on a central trapezoidal salient, longitudinally striated just mesad of spiracles; crenulae $4 + 4$ to $7 + 8$, elongate, triangular, nonoverlapping, sometimes quite irregular.

Second abdominal tergite (Figures 455–457) smooth; tergites III–VII reticulate on the sides, bearing 3–7 setae on each reticulate area; third tergite faintly transversely striated across center; seventh tergite faintly longitudinally striated posteriorly between the lateral reticulate areas, bearing 2–4 fine setae across center; eighth tergite faintly reticulate-punctate, with a transverse arcuate row of 6–13 setae between the spiracles; syntergum v-shaped, relatively short and broad, bearing 12–14 setae in a transverse, arcuate row. Cerci situated very close to the posterior spiracles, with 2 long setae and 1 short seta. Ovipositor shaft about $1\frac{1}{2}$ to $1\frac{2}{3}$ times as long as the middle tibia (1.51–1.67); ovipositor sheaths about $\frac{1}{3}$ to $\frac{2}{5}$ length of middle tibia (0.32–0.43).

Mid-tibial spur (Figure 458) about $\frac{2}{3}$ to $\frac{4}{5}$ length of the corresponding basitarsus.

Forewing (Figure 459) about $2\frac{2}{5}$ to $2\frac{1}{2}$ times as long as wide; marginal fringe usually not exceeding $\frac{1}{7}$ width of disk. Disk nearly uniformly setose, the setae on clouded areas only slightly coarser than the others; delta area with 75–264 setae in 7–14 rows, these somewhat longer and somewhat sparser than the setae distad of speculum, rather clearly separated from row of setae along posterior margin of wing and from several short, longitudinal rows of setae below distal half of submarginal vein; costal cell with a row of very fine setae along proximal half or so, and 2–4 coarse setae near apex.

Submarginal vein bearing 3–5 coarse setae and 14–19 bullae. Marginal vein bearing 7–11 prominent, subequal setae along anterior margin, up to about $1\frac{2}{5}$ times longer than the setae in a row along center of vein.

Forewing hyaline, with a narrow, curved infumated crossband below the junction of the submarginal and marginal veins, and a small, rather strongly infuscated cloud below the stigma; a faint cloud against posterior margin of wing, opposite the cloud below stigma; a fuscous streak along posterior margin of wing, distad of speculum.

"General color yellow to orange, with white maculations on dorsum of head, thorax, propodeum, sides of abdomen" (Compere 1955). Head immaculate. Margins of pronotum, mesonotal sclerites and metanotum lined with black or blackish; mesoscutum and scutellum with submedian fuscous blotches; tegulae fuscous; a short black streak at base of forewing, below tegula; propodeum fuscous anteriorly, blackish on sides. Anterior margins of second and third abdominal tergites lined with black on sides; second tergite fuscous centrally; tergites III–VII with conspicuous blackish spots on lateral reticulate areas, immaculate centrally; eighth tergite immaculate; anterior margin of syntergum lined with fuscous. Thoracic sterna dusky, internal apodemes and ovipositor plates lined with black or blackish. Antennal scape pale, dusky anteroventrally; rest of antenna uniformly dusky, tip of club fuscous. All femora largely pale, tipped with fuscous apically; all tibiae with 2 broad, rather diffuse fuscous bands. Wing veins lined with brownish.

Length 0.67–1.14 mm.

*Male.* Essentially similar to the female in structure, chaetotaxis and sculpture, differing mainly in details of antennal structure and coloration.

Antennal scape (Figures 460–462) somewhat thicker than in the female, about $3\frac{4}{5}$ to $4\frac{3}{5}$ times as long as wide, with a distinct, well-defined, somewhat irregular plate on ventral surface bearing 2 discoid sensilla (rarely 3 sensilla; sometimes 2 adjoining plates, with one sensillum on each); pedicel about $1\frac{1}{4}$ times as long as the third segment of the funicle; first funicular segment about $1\frac{1}{5}$ to $1\frac{1}{3}$ times as wide as long, second segment about $1\frac{3}{4}$ times as wide as long; third funicular segment and the club separated only by a suture, as in the female; third funicular segment bearing 2 longitudinal sensilla; club about $2\frac{2}{3}$ times as long as wide, about $2\frac{1}{2}$ times longer than the preceding segment, bearing 3–6 longitudinal sensilla.

Mesoscutum (Figure 463) with 7–9 setae. Propodeum (Figure 463) $2\frac{2}{3}$ to $3\frac{2}{3}$ times as long as the metanotum, about $\frac{3}{5}$ length of scutellum; crenulae 3+4 to 6+6, as in the female. Middle tibia (Figure 464) swollen apically, contrastingly marked with black at apex.

Genitalia (Figure 465) robust, about $\frac{2}{3}$ to $\frac{3}{4}$ length of middle tibia (0.64–0.74), with a distinct rod between the apodemes; digital sclerites about $\frac{1}{3}$ the combined length of aedeagus and apodemes (0.30–0.38).

Forewing with 38–104 setae in delta, arranged in 5–9 rows; pattern as in the female.

General coloration of head and thorax as in the female; abdominal tergites III–VII without blackish spots on sides. Antennal club uniformly dusky. Middle tibia as above.

Length 0.55–1.10 mm.

Redescribed from the type series (numerous ♀♂ specimens, reared by S. E. Flanders from the walnut scale, *Quadraspidiotus juglansregiae* (Comstock), on walnut, Mira Loma, California, March 10–15, 1936; uncleared, mounted in balsam); additional material, from same host on same host plant: 5♀♀, 3♂♂, reared by D. Gerling, Loma Linda, California, October 11, 1962; 1♀, 1♂, reared by J. Hall, Grand Terrace, California, September 21, 1964; also 2♀♀, reared by P. DeBach from *Clavaspis ?subsimilis* (Cockerell) on *Bursera microphylla*, Arroyo de las Parras, Loreto, Baja California Sur, Mexico, April 6, 1975; 11♀♀, 11♂♂, reared by P. DeBach from the same host on "Picante de Cimaron," same locality, April 7, 1975 (also 7 pupae); 9♀♀, 12♂♂, same data, June 27, 1975; all cleared and mounted in Hoyer's medium. The specimens from Baja California appear decidedly darker than those from the United States.

Type series in the collection of the Division of Biological Control, University of California, Riverside.

**Notes.** The pupa of *melanostictus* is entirely black.

The "white maculations" described by Compere (1955) can barely be seen in the uncleared, slide-mounted type series, and are invisible in properly cleared material. On the other hand, the various fuscous markings, as well as most structural characters, are best seen in cleared specimens.

*A. melanostictus* appears to be rather closely related to *dealbatus*, sharing with this species the peculiar, "*Marietta*-like" structure of the antennae. In the female sex, it differs from *dealbatus* mainly in that the forewing is almost uniformly setose, without alternating areas bearing coarse and fine setae, and with a relatively small fuscous cloud, rather than a crossband, below the stigma. (Other differences cannot be ascertained due to the poor condition of the *dealbatus* types.) In the male sex, *melanostictus* may be distinguished from all other species of *Aphytis* by the swollen, black-tipped middle tibiae, similar to those of *mazalae* and *erythraeus*, in combination with the peculiar antennae (sensilla on scape, no constriction between club and penultimate segment) and the pattern of the forewing.

Like *dealbatus*, *A. melanostictus* may be regarded as an aberrant species of *Aphytis*, probably related to the **vittatus** group.

### 23. **Aphytis malayensis** n.sp.

(Figures 466–468)

This aberrant Southeast Asian species superficially resembles *vittatus* in the distinctive coloration of the female, with 4 longitudinal dark brown stripes on the mesonotum. It is readily recognizable by the peculiar, 5-segmented antennae and by the hyaline wings, as well as by the shorter ovipositor and various details of chaetotaxis, sculpture and coloration.

*Female.* Eyes coarsely setose. Mandibles well developed, essentially bidentate, with some indication of a dorsal truncation; maxillary palpi 2-segmented, labial palpi 1-segmented. Antennae (Figure 466) 5-segmented (1121), with 2 small funicular segments and an elongate club; scape rather slender, nearly 5 times as long as wide; pedicel about $1\frac{2}{5}$ times as long as wide; first funicular segment trapezoidal, almost triangular, ventral aspect longer than the dorsal, $1\frac{1}{3}$ times as wide as long; second segment symmetrical, distinctly shorter and wider than the first, twice as wide as long; club long, tapering, about $3\frac{3}{4}$ times as long as wide, somewhat over $1\frac{1}{4}$ times longer than the scape, bearing about 12 longitudinal sensilla.

Setae on head, thorax, and sides of abdomen dark, coarse, readily visible under ×30 magnification except when obscured by the pigmentation of the body. Vertex with two pairs of long setae along occipital margin, in addition to numerous shorter setae. Mesoscutum (Figure 467) with 13 setae, the posterior pair considerably longer and coarser than the others; each parapsis with 4 setae, each axilla with 1 seta; scutellum with 4 subequal setae, the discoid sensilla very close to the anterior pair. Frontovertex, pronotum, and mesonotal sclerites heavily reticulate. Scutellum trapezoidal, somewhat over $\frac{4}{5}$ median length of mesoscutum. Metanotum arcuate, obliquely reticulate except on the sides; anteromedian apodeme about $\frac{3}{4}$ median length of metanotum.

Propodeum (Figure 468) rather short, with a prominent median salient, about $3\frac{1}{2}$ times as long as the metanotum, nearly $\frac{1}{2}$ length of scutellum, reticulate on the sides, reticulate-strigose on a trapezoidal central salient; crenulae 5+5, small, rounded, closely spaced but nonoverlapping.

Second abdominal tergite (Figure 468) coarsely reticulate on a small central area just below the crenulae; tergites III–VII reticulate on the sides, with 3–5 setae in a transverse row on each reticulate area and 1 seta a short distance mesad; third and seventh tergites delicately, transversely reticulate across; eighth tergite delicately reticulate, with a transverse row of 6 setae between spiracles; syntergum short, broadly v-shaped, delicately reticulate, bearing 12 setae in a transverse row. Cerci situated very close to posterior spiracles, with 3 long setae and 1 short seta. Ovipositor shaft about $1\frac{4}{5}$ times as long as the middle tibia (1.81), ovipositor sheaths somewhat over $\frac{1}{3}$ length of middle tibia (0.37).

Mid-tibial spur somewhat shorter than the corresponding basitarsus.

Forewing $2\frac{3}{5}$ times as long as wide; marginal fringe not exceeding $\frac{1}{6}$ width of disk. Discal setae rather uniform, without contrasting groups of coarse and fine setae; delta area with 115–117 setae, about as densely setose as the area distad of speculum, the setae somewhat coarser, not clearly separated from setose area below submarginal vein or from row of setae along posterior margin of wing; costal cell with 4 fine setae in a row along proximal half and 1 coarse seta near apex. Submarginal vein bearing 4 coarse setae and 18–19 bullae. Marginal vein bearing 10 prominent, subequal setae along anterior margin, these about $1\frac{2}{5}$ times as long as the setae in a row along center of vein.

Forewing hyaline, with a small infuscated cloud below stigma and a narrow fuscous streak along posterior margin of wing distad of speculum.

General coloration pale, with a distinctive pattern of dark brown markings as follows: oral margin and lower third of genal sutures; pronotum centrally and along posterior margin; mesoscutum and scutellum (Figure 467) with a pair of submedian longitudinal stripes and a stripe on each side, which on the scutellum merge into the submedian stripes along the posterior margin; anterior margin of mesoscutum narrowly, parapsidal sutures, antero-lateral margin of parapsis narrowly; anterior, posterior and lateral margins of metanotum; transverse stripe along posterior margin of propodeum on each side; v-shaped blotch centrally on second abdominal tergite, surrounding the small central reticulate area and covering posterior half of same; short transverse stripes on sides of tergites III–VII. A small, faintly fuscous blotch on vertex beyond posterior ocelli; anterior part of median salient of propodeum suffused with fuscous, posterior part and the crenulae pale; a faintly fuscous band across third abdominal tergite; thoracic sterna and margins of endophragma strongly infuscated. Mandibles brown on distal half, the denticles black. Antennal scape (Figure 466) and proximal half of club pale, pedicel faintly dusky, funicular segments fuscous, distal half of club black. All femora faintly suffused with dusky distally; fore tibia with a broad, distinct fuscous ring; middle and hind tibiae with 2 broad rings; basitarsi of all legs dark brown, subsequent tarsal segments pale, except second tarsal segment of foreleg which is dusky; strigil of foreleg and mid-tibial spur fuscous. Veins of forewing brownish.

Length 0.85 mm.

*Male*. Unknown.

Described from 1♀ (holotype), reared by Chua Tock Hing from a sample of *Saissetia nigra* (Nietner) collected in Malaysia (undoubtedly from an undetermined armored scale contaminating the sample).

Holotype in the collection of the Division of Biological Control, University of California, Riverside.

**Notes.** In spite of the peculiar structure of the female antenna, this interesting species undoubtedly belongs in *Aphytis*. The shape of the propodeum, crenulae and forewing and the general habitus certainly resemble *Aphytis* more than any other genus, and there appears to be no justification for the establishment of a separate genus to accommodate *malayensis*. It is assumed that this species, like all known species in the genus, is a primary parasite of armored scale insects. This, however, should be verified by rearing additional material from carefully determined hosts.

*A. malayensis* does not seem to be closely related to the **funicularis** group, which currently includes 4 minute, yellow species possessing 5-segmented antennae and reduced mouthparts. On the other hand, in view of its aberrant antennae and hyaline wings, *malayensis* cannot be considered a bona fide member of the **vittatus** group, in spite of its overall similarity to *vittatus*. For the time being, it is regarded as an aberrant species closely related to the **vittatus** group.

## 24. **Aphytis mandalayensis** n.sp.

(Figures 469–474)

This aberrant Southeast Asian species resembles *malayensis* in having peculiar, 5-segmented antennae in the female sex. It may be further recognized by the robust antennal scape, relatively long propodeum with elongate crenulae, and long ovipositor; general coloration extensively marked with fuscous; forewing fuscous with a subapical hyaline crossband.

*Female.* Eyes finely setose. Mandibles well developed, with 2 denticles and a short truncation; maxillary palpi 2-segmented, labial palpi 1-segmented. Antennae (Figures 469, 470) 5-segmented (1121), with 2 small funicular segments, the club with an incomplete transverse suture at about the proximal third, which if complete would have separated a normal third funicular segment; scape rather robust, and about $3\frac{1}{3}$ to nearly 4 times as long as wide, distinctly widening disto-ventrally; pedicel short, a little over $1\frac{1}{4}$ times as long as wide; first funicular segment nearly triangular, about $2\frac{1}{2}$ to nearly 3 times as wide as long; second segment symmetrical, somewhat longer and considerably wider than the first, about $2\frac{2}{3}$ to a little over 3 times as wide as long; club robust, about $2\frac{2}{5}$ to $2\frac{3}{4}$ times as long as wide, a little longer than the scape, bearing 14–15 longitudinal sensilla.

Setae on head, thorax and sides of abdomen rather coarse, dark, readily visible under $\times 30$ magnification. Vertex with 2 pairs of relatively long setae along occipital margin, in addition to numerous somewhat shorter setae. Mesoscutum (Figure 471) with 16 setae, the posterior pair and 1 seta at each antero-lateral corner considerably longer and coarser than the others; each parapsis with 2 setae, each axilla with 1 seta; scutellum with 4 setae, the discoid sensilla about equidistant from the two pairs. Frontovertex, pronotum and mesonotal sclerites reticulate, with a distinct infrastructure within the cells. Scutellum oval, about $\frac{3}{4}$ median length of mesoscutum. Metanotum (Figures 471, 472) reticulate except on the sides, anteromedian apodeme robust, about $1\frac{1}{6}$ to $1\frac{1}{4}$ times longer than median length of metanotum.

Propodeum (Figures 471–473) rather long, about $3\frac{1}{4}$ to $3\frac{2}{3}$ times as long as the well-developed metanotum, a trifle longer than the scutellum, reticulate on the sides, reticulate-strigose on a trapezoidal median salient; crenulae 7 + 7, elongate, rounded apically, closely appressed or slightly overlapping.

Second abdominal tergite (Figure 473) smooth; tergites III–VII reticulate on the sides, with 4–6 setae in a short transverse row on each reticulate area, smooth centrally except for some transverse striation; seventh tergite with 2–3 fine setae between the lateral setiferous areas; eighth tergite reticulate across, with a transverse row of 12–13 setae between spiracles; syntergum (Figure 473) broad, v-shaped, faintly reticulate-punctate, with 21–23 setae in about 3 irregular transverse rows. Cerci situated very close to posterior spiracles, with 2 long setae and 1 short seta. Ovipositor relatively long, the shaft nearly twice as long as the middle tibia (1.89–1.96), the sheaths about $\frac{1}{2}$ length of middle tibia (0.51–0.54).

Mid-tibial spur somewhat shorter than the corresponding basitarsus.

Forewing (Figure 474) about $2\frac{1}{2}$ to $2\frac{3}{5}$ times as long as wide; marginal fringe not exceeding $\frac{1}{5}$ width of disk. Delta area with 86–99 setae in 9–11 rows, these considerably longer and somewhat sparser than the setae distad of speculum, not clearly separated from 2–3 irregular rows of setae along posterior margin of wing, somewhat more clearly separated from 2–3 irregular rows of setae below entire length of submarginal vein; the setae bordering the speculum are considerably longer and coarser than the other setae in the delta; setae distad of speculum not uniform: those below stigma are noticeably coarser than the others, whereas those on the hyaline band are exceedingly fine, pale; costal cell with a row of fine setae along proximal three fifths or so, and 1–2 coarse setae near apex. Submarginal vein bearing 2 coarse, long setae, the proximal one being only $\frac{1}{2}$ to $\frac{2}{3}$ length of the distal, and 19–22 bullae. Marginal vein bearing 9–10 prominent setae along anterior margin, these decreasing noticeably in length toward apex, the proximal ones about $1\frac{1}{3}$ to $1\frac{3}{5}$ times as long as the distal and up to $1\frac{2}{3}$ times longer than the longest setae in an uneven row along the center of the vein.

Forewing rather uniformly infuscated from base to stigmal vein, with a broad, arcuate hyaline crossband, followed by a lightly infumated apical area; a fuscous streak along posterior margin of wing.

Hind wing (Figure 474) lightly infumated, with a broad subapical hyaline crossband bearing extremely fine, pale setae.

General coloration yellow, extensively marked with brown and blackish. Lower part of head broadly fuscous, with narrow black stripes on oral margin and genal sutures. Pronotum fuscous centrally, propleura lined with black; mesoscutum and scutellum largely fuscous, pale along midline and on antero-lateral corners of both sclerites; metanotum pale, lined with blackish and fuscous along margins, sometimes also fuscous centrally; propodeum largely fuscous, margins lined with blackish, with a pale v-shaped mark on median salient, outer crenulae blackish. Second abdominal tergite fuscous; tergites III–VIII strongly infuscated centrally, each tergite with an anterior blackish crossband and a posterior paler fuscous crossband, pale on the sides, with variable small fuscous spots on sides of tergites III–VI; posterior margin of eighth tergite narrowly pale; syntergum pale, narrowly lined with blackish along anterior margin. Ovipositor plates lined with blackish, sheaths broadly lined with blackish along outer margin; endophragma and thoracic sterna strongly infuscated. Antennal scape (Figure 470) pale, broadly marked with a longitudinal black stripe; pedicel dusky, funicular segments and basal two thirds or so of club blackish, distal third of club pale. Legs largely pale; fore and middle femora sometimes tipped with dusky, fore tarsus fuscous; middle tibia dusky basally, blackish apically; hind femur with a fuscous blotch subapically on outer surface; hind tibia marked with blackish on both ends; basitarsi of both middle and hind legs pale, subsequent tarsal segments blackish. Wing veins dusky.

Length 0.96 mm.

*Male*. Unknown.

Described from 2♀♀ (holotype and paratype), reared by P. DeBach from ?*Aonidiella aurantii* (Maskell) on rose, Mandalay, Burma, February 7, 1957. Both specimens are cleared and mounted in Hoyer's medium, holotype with head separate, paratype dissected.

Type series in the collection of the Division of Biological Control, University of California, Riverside.

**Notes.** Like *malayensis*, this peculiar species undoubtedly belongs in *Aphytis* and may be regarded as an aberrant species closely related to, but not a bona fide member of, the **vittatus** group. It resembles *malayensis* in the aberrant, 5-segmented antennae, but is otherwise rather similar in general habitus, forewing pattern and various structural characters to *angustus*, an Oriental member (or close relative) of the **vittatus** group.

Like *malayensis*, *mandalayensis* cannot be regarded as closely related to the members of the **funicularis** group, which also have 5-segmented antennae; its affinities with the **vittatus** group are readily apparent. Reduction in the number of antennal segments must have evolved independently in various groups of *Aphytis*.

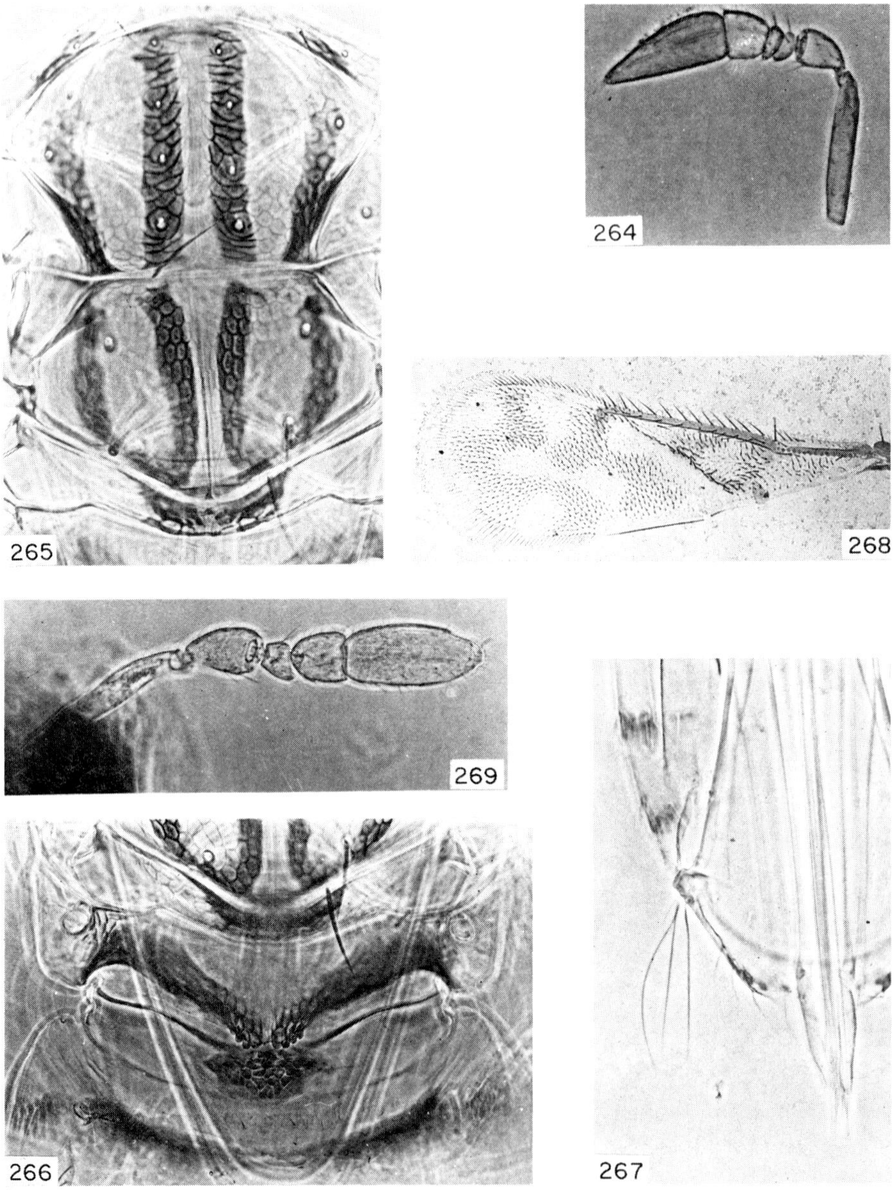


Figures 264–269. *Aphytis vittatus* (Gompere)
264. ♀: Antenna (paratype). 265. ♀: Thorax (paratype). 266. ♀: Metanotum, propodeum and base of gaster (paratype). 267. ♀: Cercal setae (paratype). 268. ♀: Forewing. 269. ♂: Antenna (paratype).

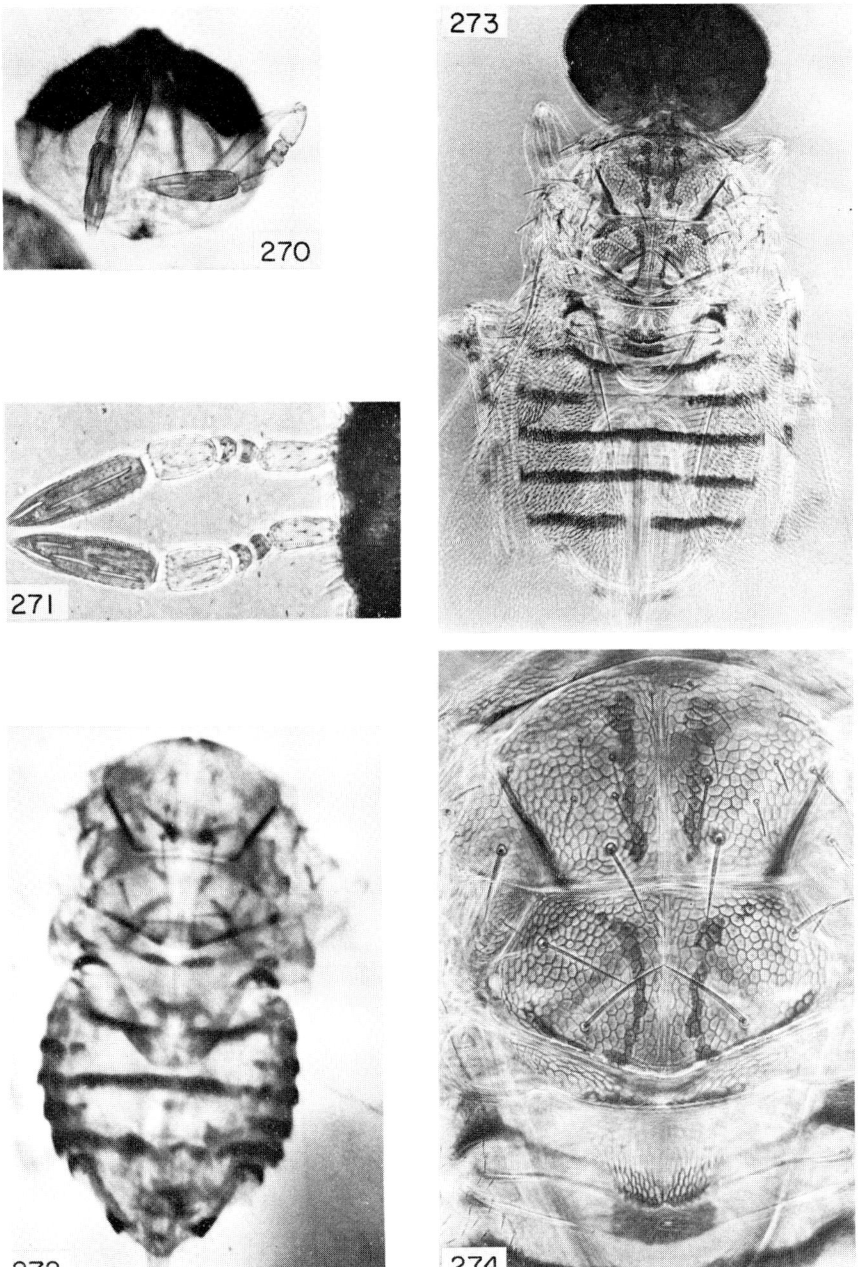


Figures 270–274. *Aphytis peculiaris* (Girault), ♀
270. Head and antennae (holotype). 271. Antennae. 272. Body, showing pigmentation (holotype; photo: Max Badgley). 273. Body; note more complete abdominal crossbands (Angalet's material). 274. Thorax and propodeum.

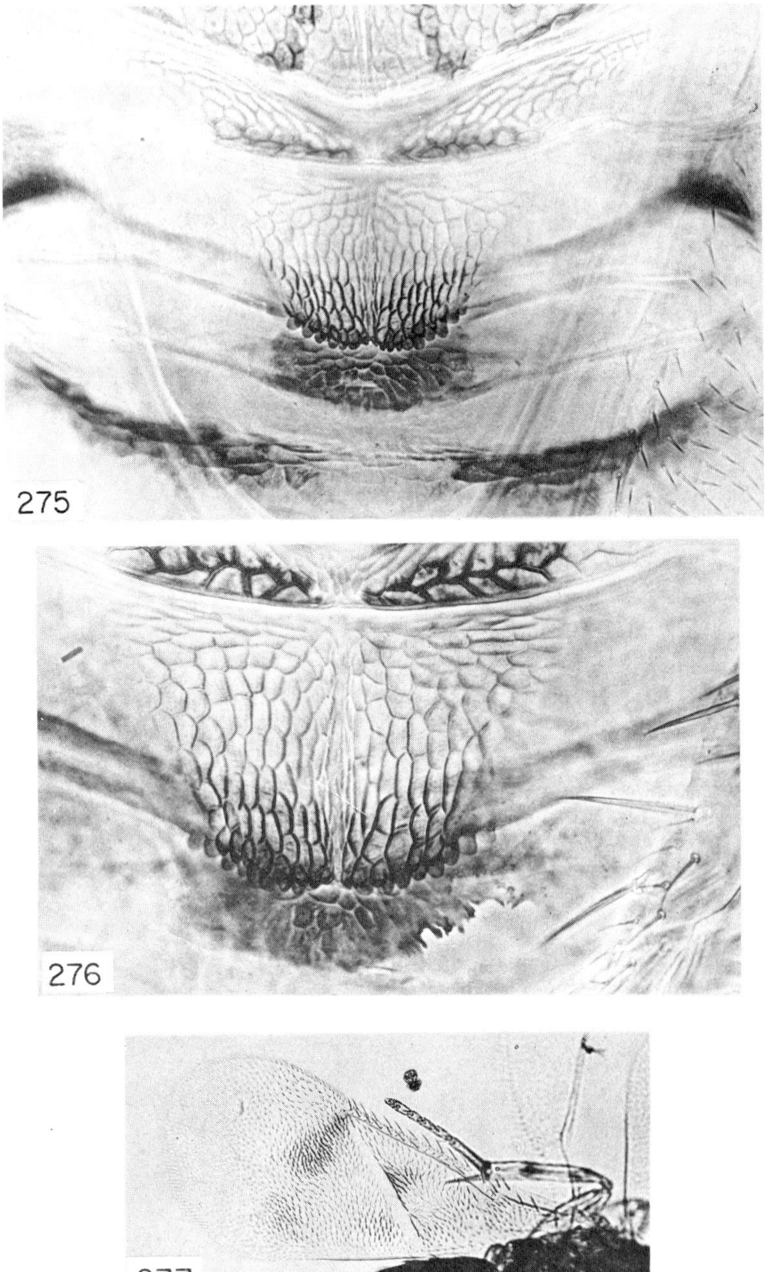


Figures 275–277. *Aphytis peculiaris* (Girault), ♀
275. Metanotum, propodeum and base of gaster. 276. Median salient of propodeum and crenulae.
277. Forewing and middle leg (holotype).

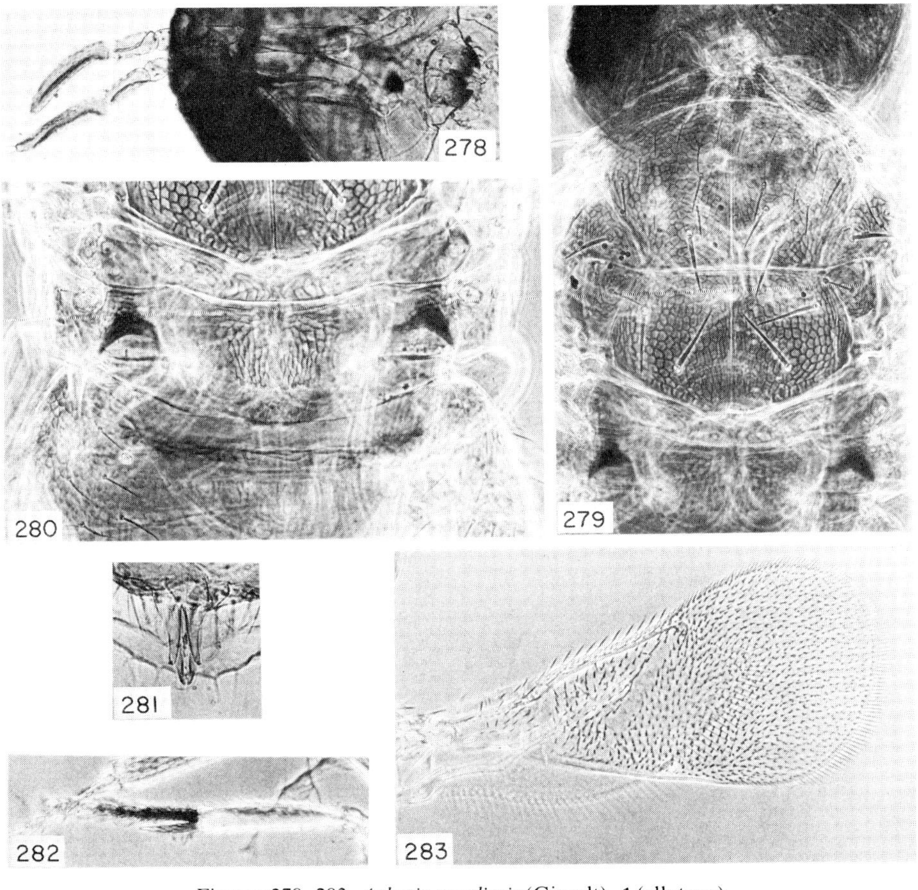


Figures 278–283. *Aphytis peculiaris* (Girault), ♂ (allotype)
278. Antennae; note also mandibles. 279. Thorax. 280. Metanotum, propodeum and base of gaster. 281. Genitalia. 282. Middle tibia and tarsus. 283. Forewing (overlapped by hind wing).

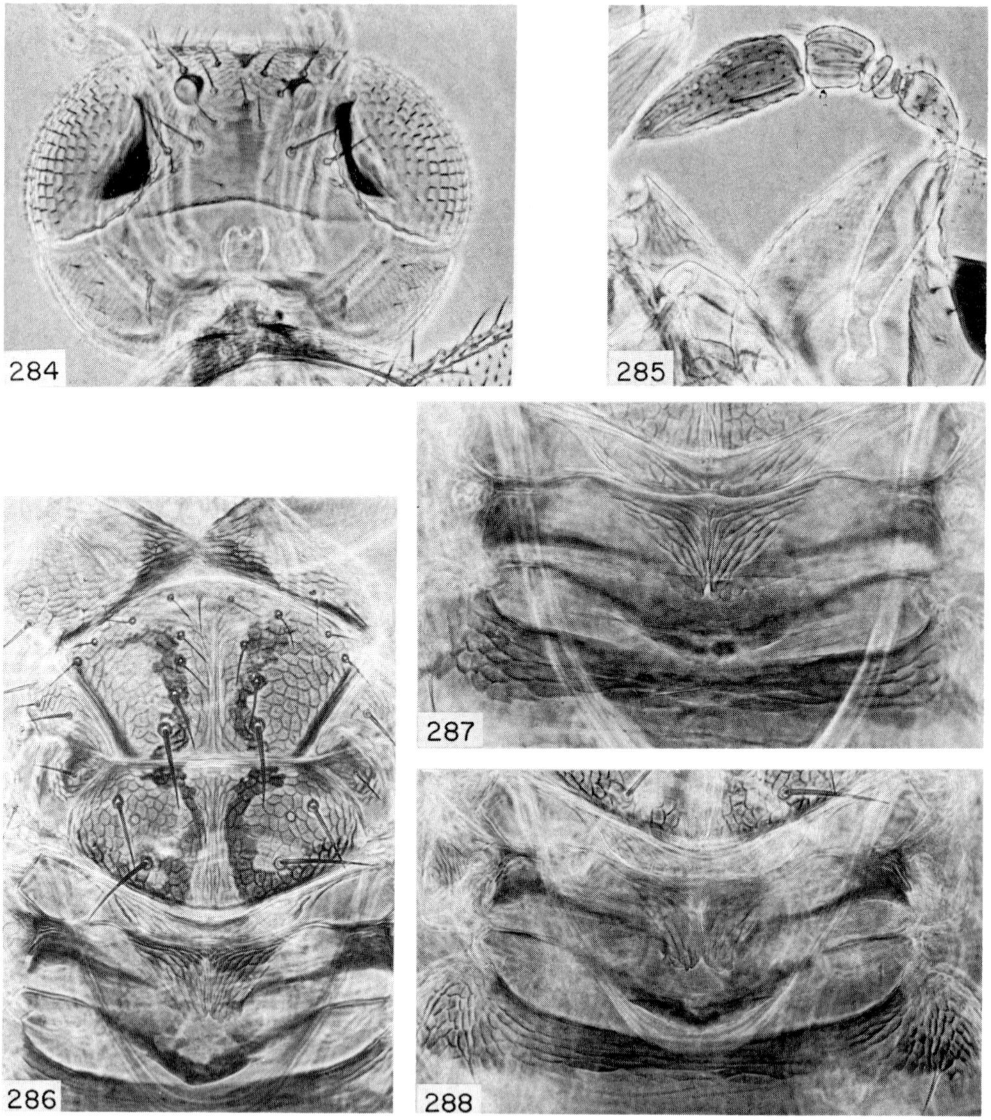


Figures 284–288. *Aphytis maculatipennis* (Dozier), ♀ (paratypes)
284. Head, dorsal view; note coarse inter-ommatidial setae in compound eyes. 285. Antenna. 286. Thorax, propodeum and base of gaster. 287, 288. Metanotum, propodeum and base of gaster.

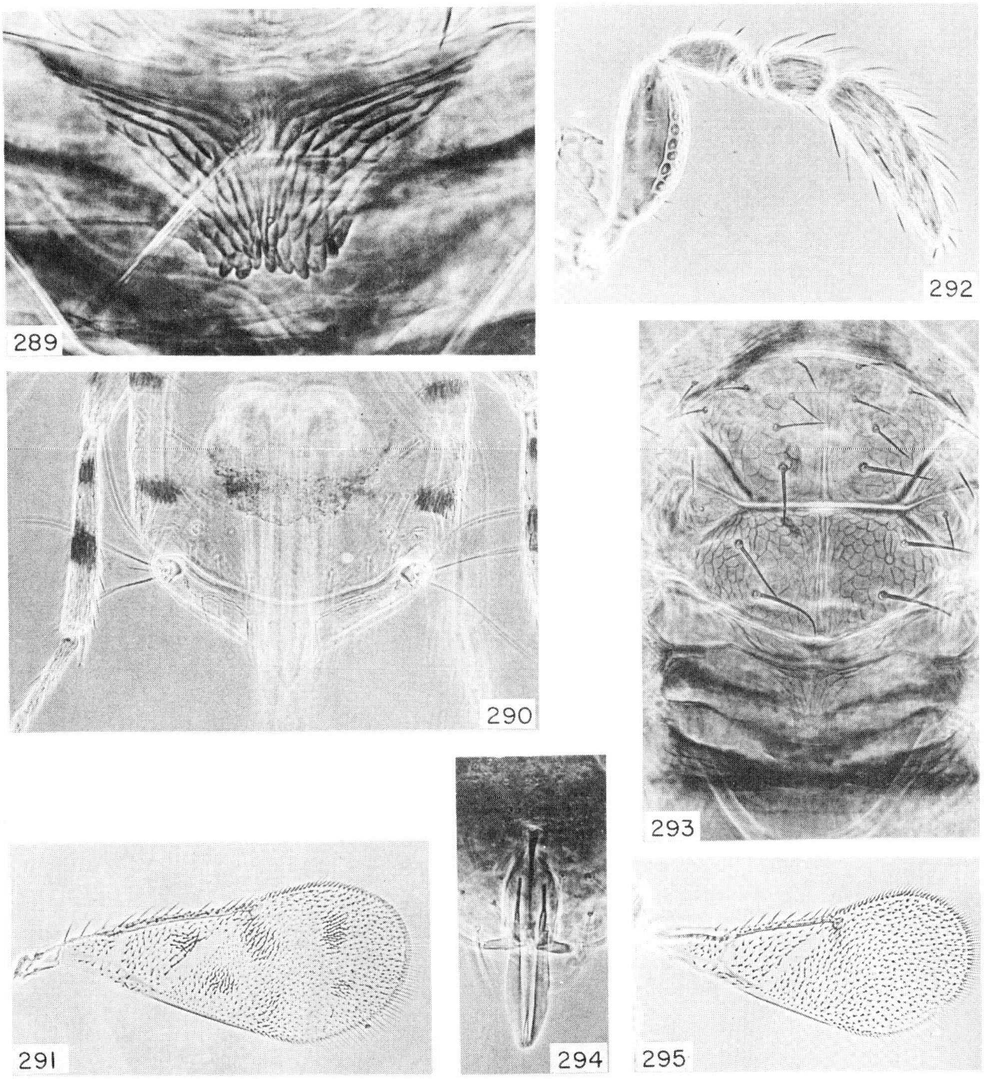


Figures 289–295. *Aphytis maculatipennis* (Dozier) (paratypes)
289. ♀: Median salient of propodeum and crenulae. 290. ♀: Posterior abdominal tergites, showing syntergum and cerci; note also pigmentation of middle tibia. 291. ♀: Forewing. 292. ♂: Antenna. 293. ♂: Thorax, propodeum and base of gaster. 294. ♂: Genitalia. 295. ♂: Forewing.

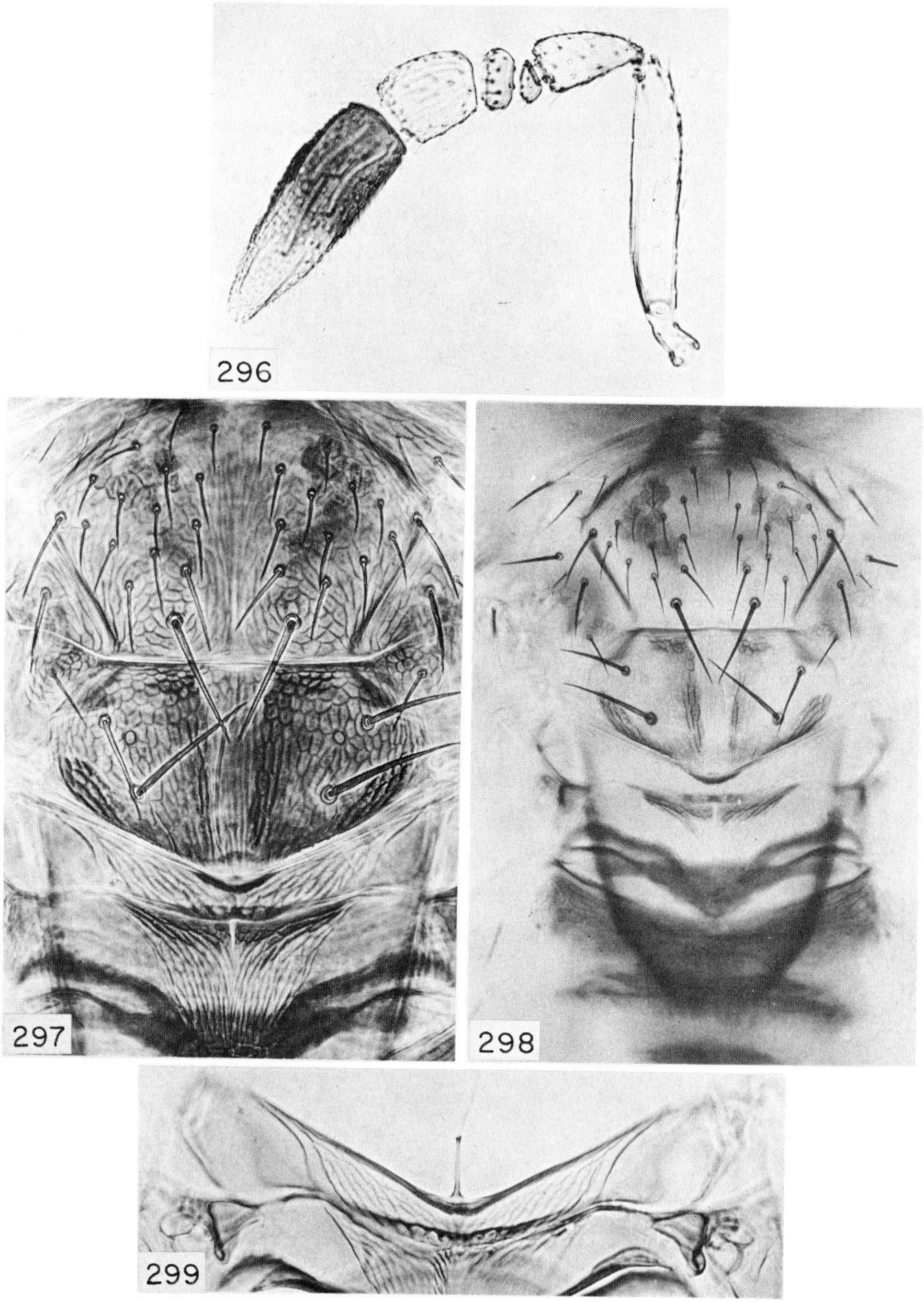


Figures 296–299. *Aphytis costalimai* (Gomes), ♀
296. Antenna. 297. Thorax and propodeum; phase-contrast photograph stressing chaetotaxis and sculpture. 298. Thorax and propodeum; same specimen as in Figure 297, bright-field photograph stressing pigmentation. 299. Metanotum.

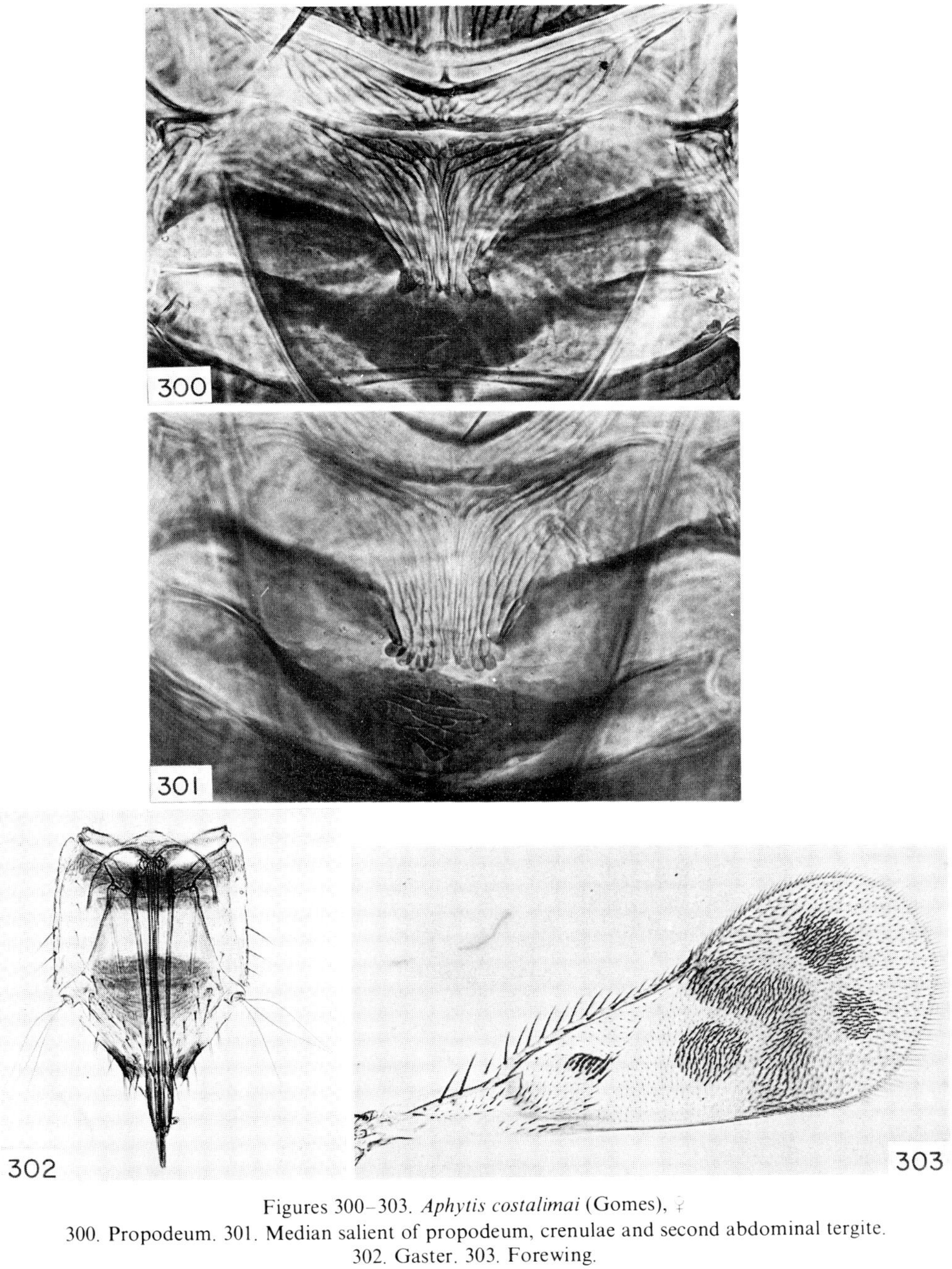


Figures 300–303. *Aphytis costalimai* (Gomes), ♀
300. Propodeum. 301. Median salient of propodeum, crenulae and second abdominal tergite. 302. Gaster. 303. Forewing.

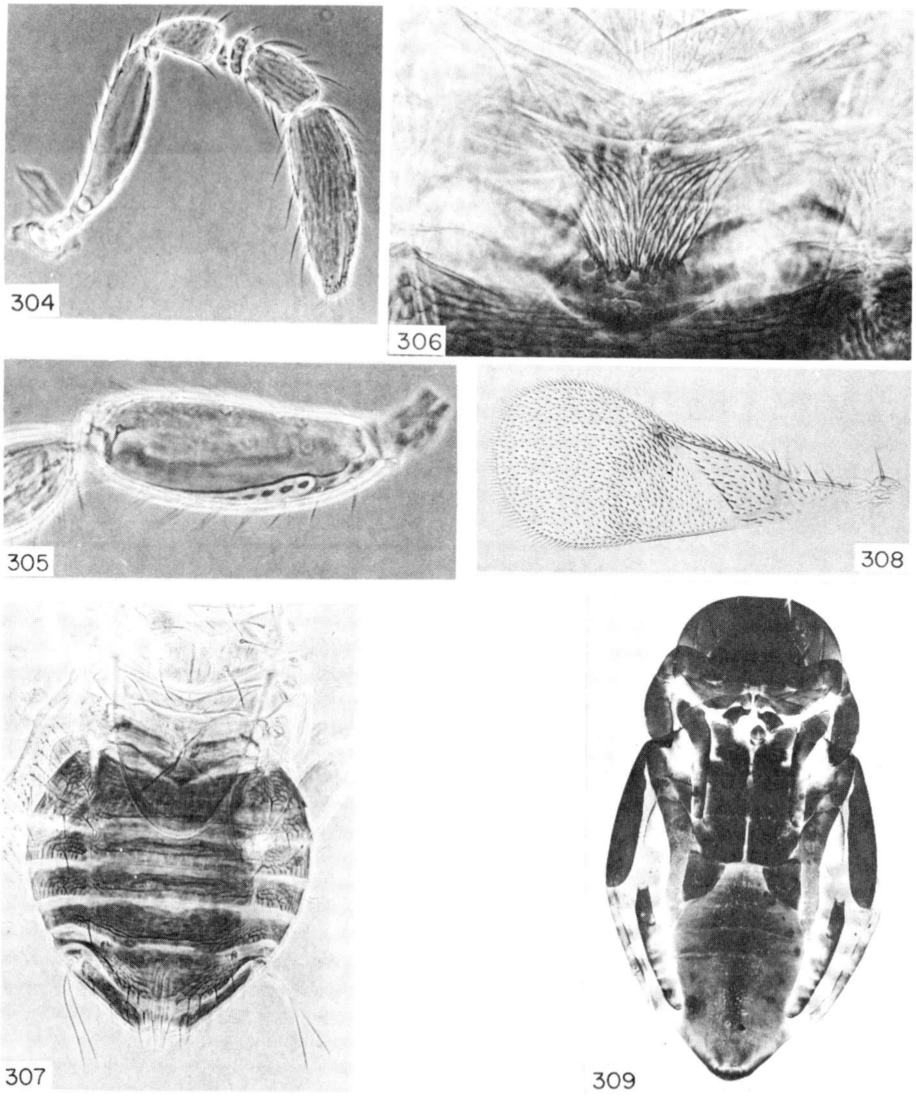


Figures 304–308. *Aphytis costalimai* (Gomes), ♂
304. Antenna. 305. Antennal scape, showing sense organs. 306. Propodeum. 307. Abdomen, dorsal view. 308. Forewing.
Figure 309. *Aphytis costalimai*: ♀ pupa, ventral view.

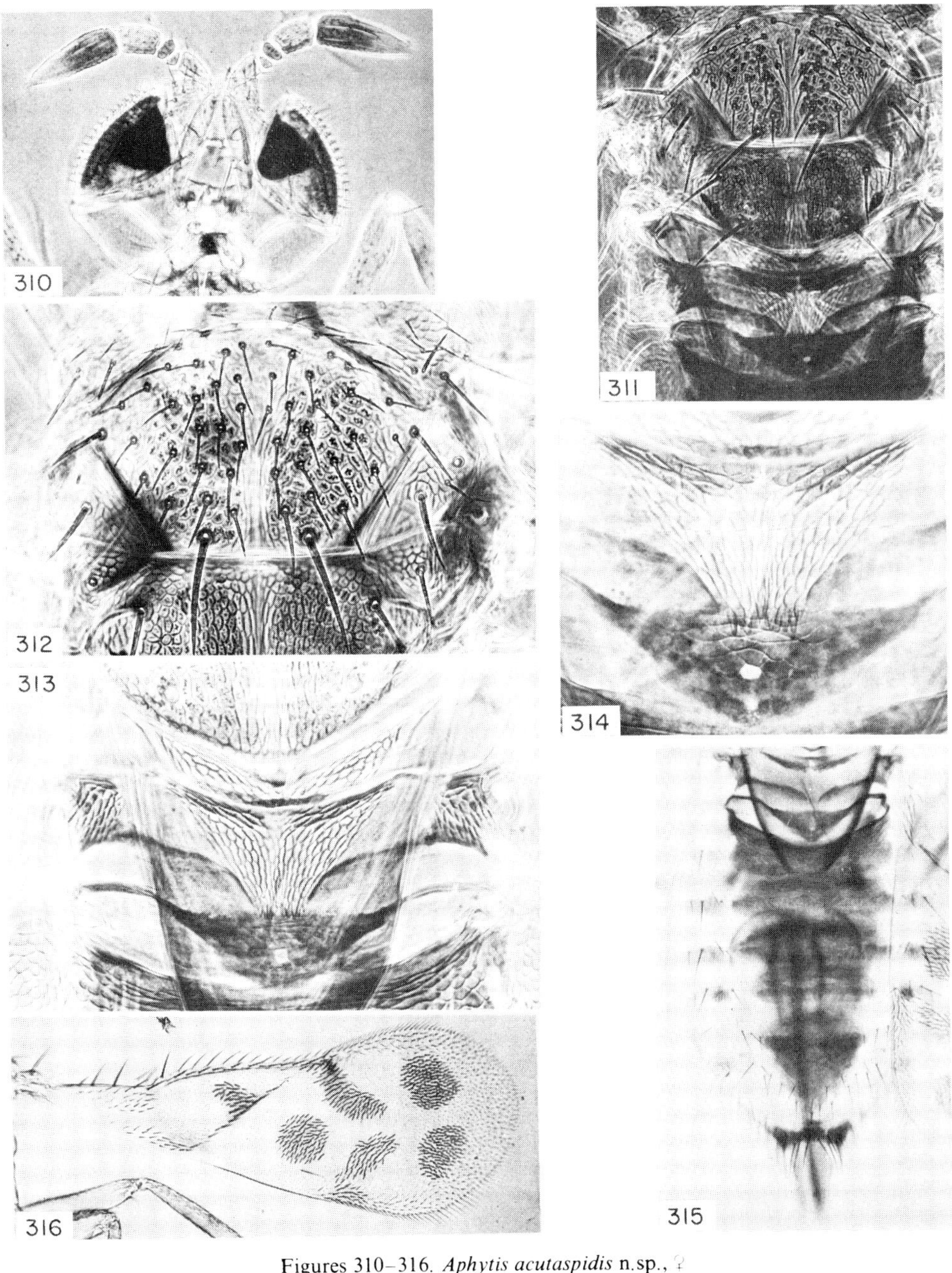


Figures 310–316. *Aphytis acutaspidis* n.sp., ♀
310. Head and antennae (holotype). 311. Thorax, propodeum and base of gaster (holotype). 312. Mesoscutum (paratype). 313. Metanotum, propodeum and base of gaster (paratype). 314. Median salient of propodeum, crenulae and base of gaster (holotype). 315. Abdomen, showing pattern of pigmentation (paratype). 316. Forewing (holotype).

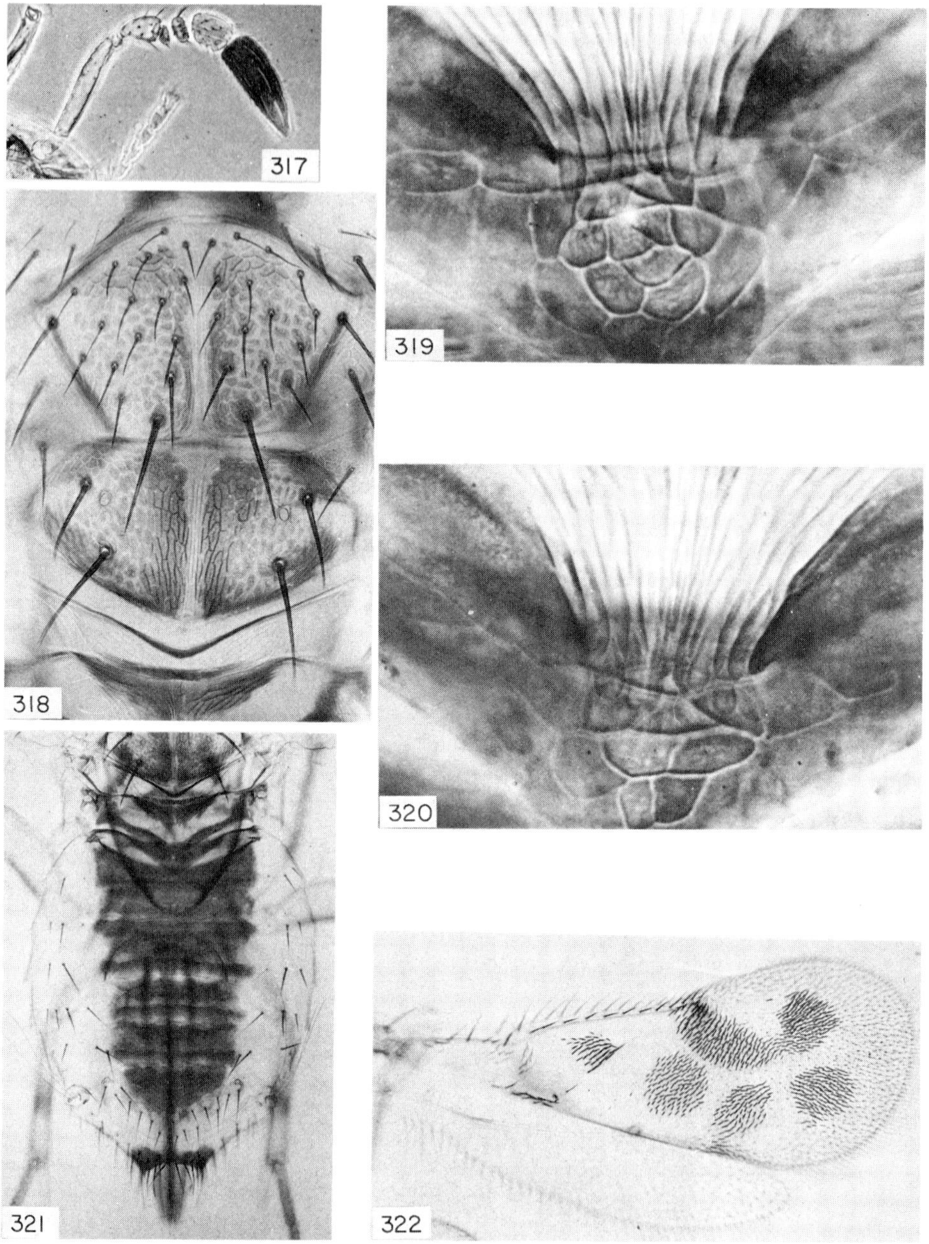

Figures 317–322. *Aphytis perplexus* n.sp., ♀
317. Antenna (paratype). 318. Thorax (paratype). 319. Propodeal crenulae and second abdominal tergite (holotype). 320. Propodeal crenulae and second abdominal tergite (paratype). 321. Abdomen (holotype). 322. Wings (holotype).

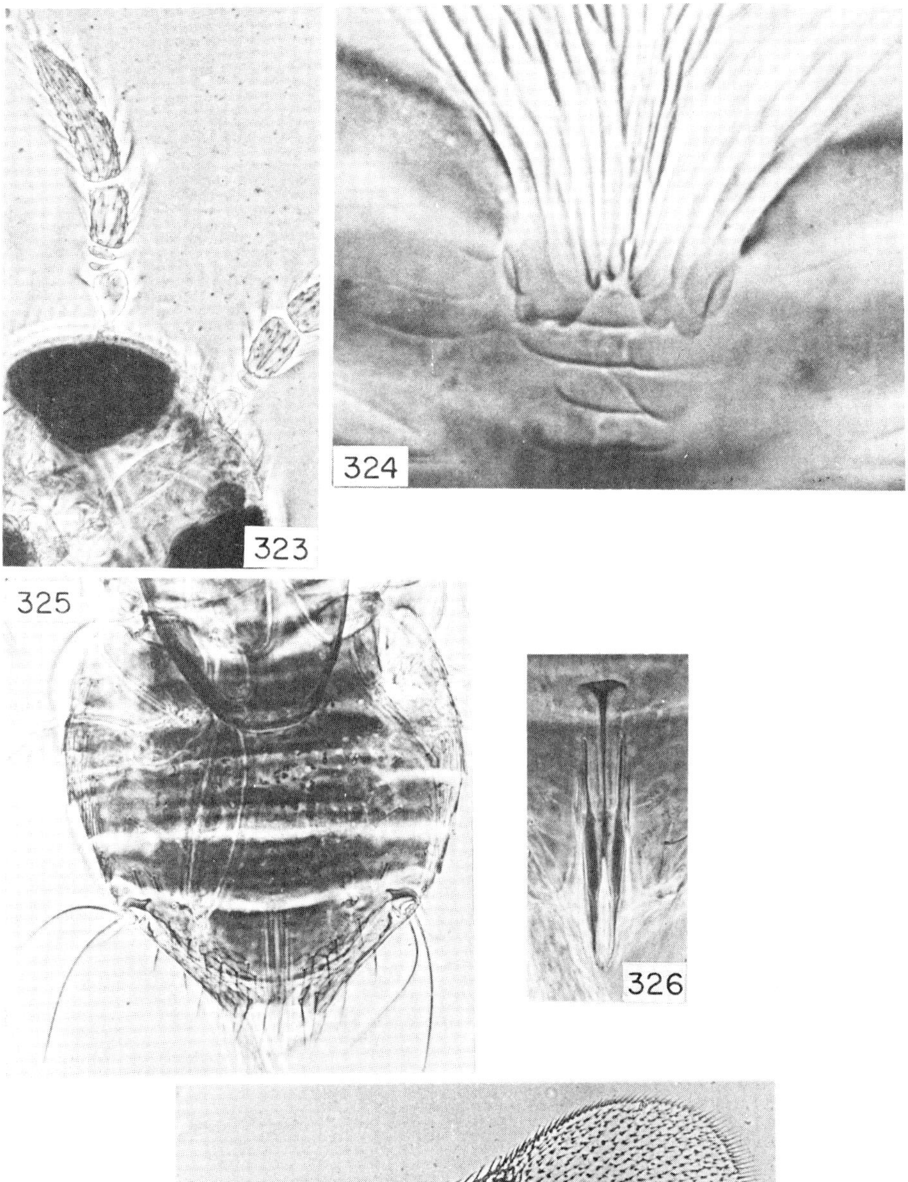


Figures 323–327. *Aphytis perplexus* n.sp., ♂
323. Antennae (allotype). 324. Propodeal crenulae (allotype). 325. Gaster (paratype). 326. Genitalia (paratype). 327. Forewing (paratype).

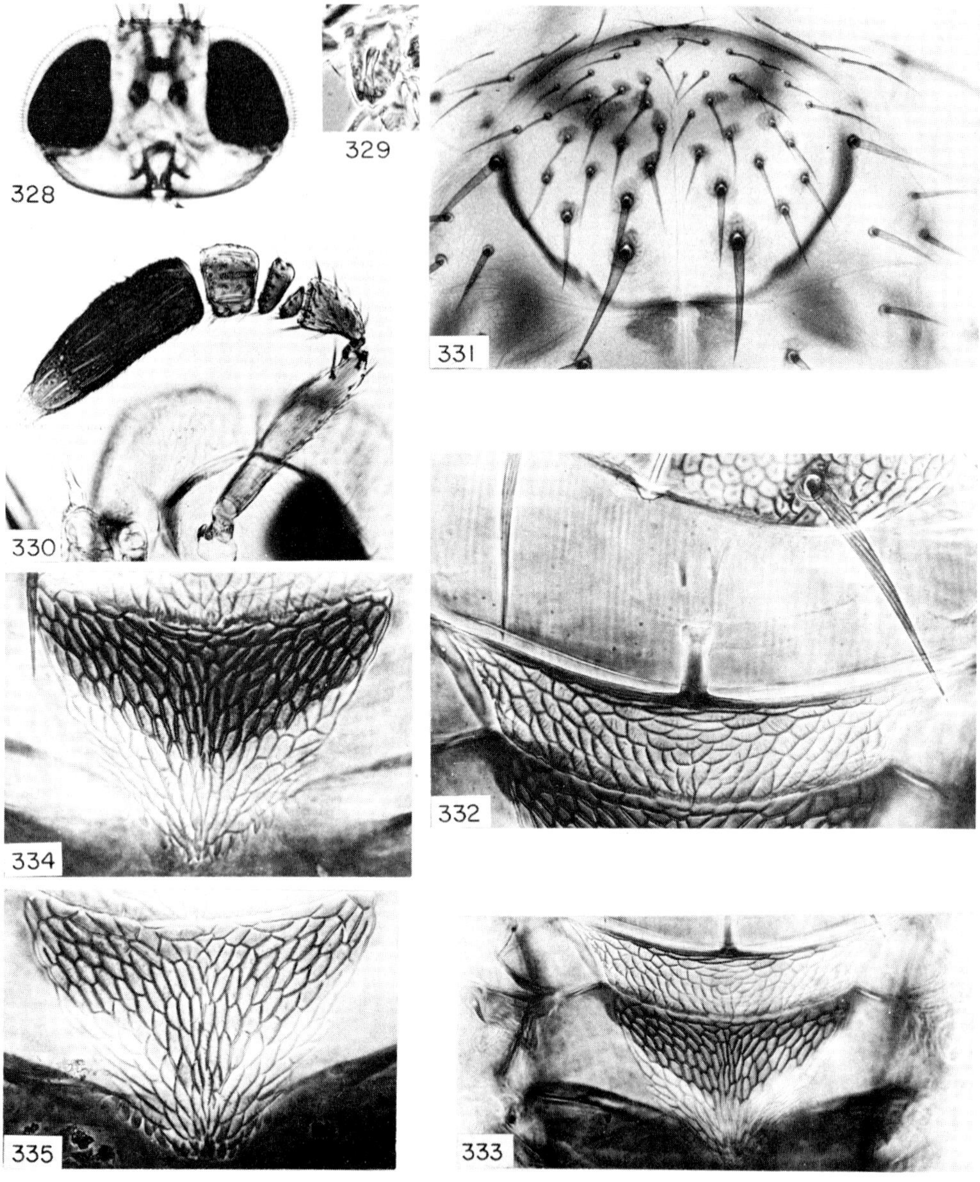


Figures 328–335. *Aphytis cochereaui* DeBach and Rosen, ♀
328. Head, showing eyes and ocelli (holotype). 329. Mandible. 330. Antenna (paratype). 331. Mesoscutum (holotype). 332. Metanotum, showing anteromedian apodeme; note coarse scutellar seta, composed of distinct filaments (paratype). 333. Metanotum and propodeum (paratype). 334. Median salient of propodeum and crenulae (holotype). 335. Median salient of propodeum and crenulae (paratype).

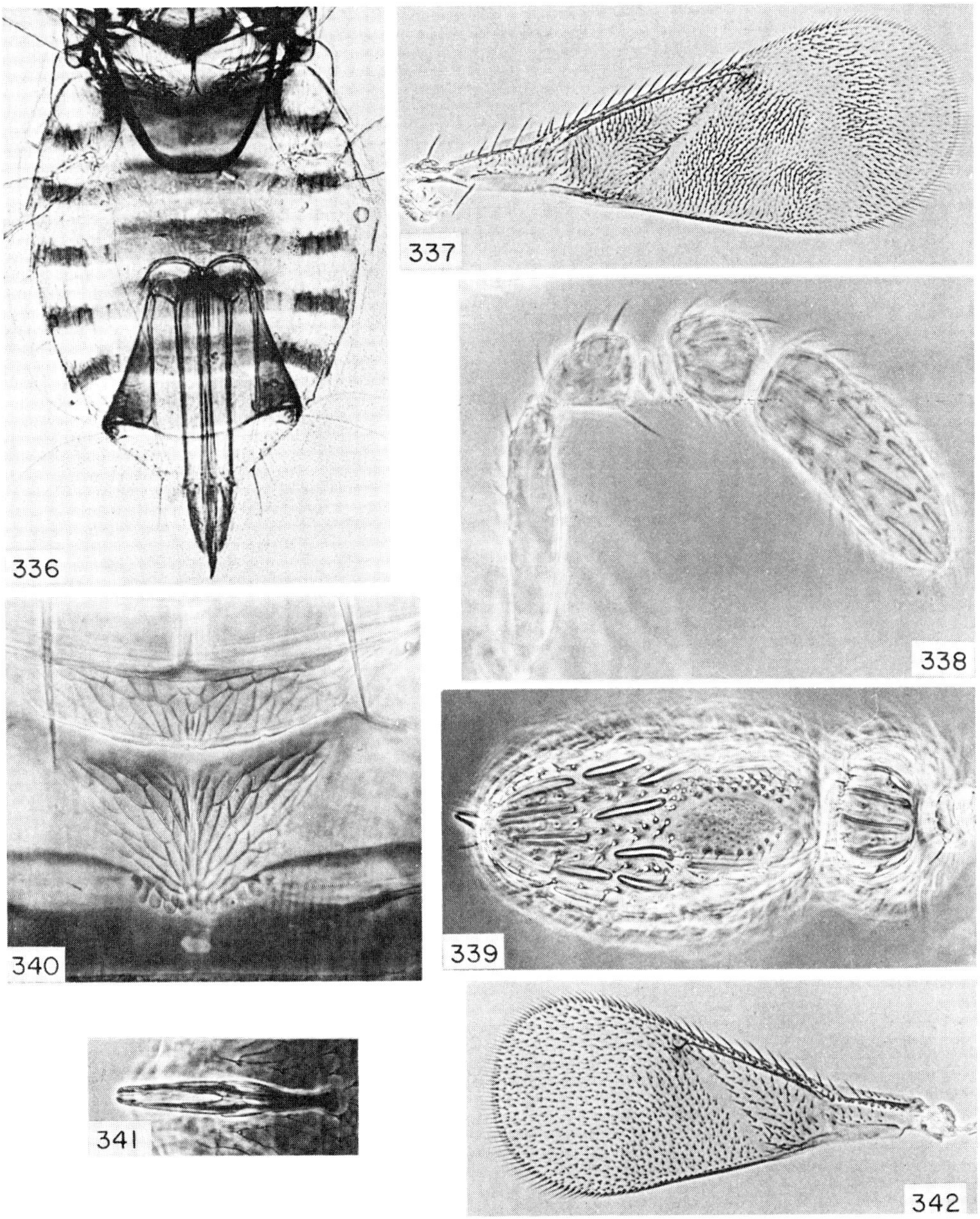


Figures 336–342. *Aphytis cochereaui* DeBach and Rosen
336. ♀: Abdomen (holotype). 337. ♀: Forewing. 338. ♂: Antenna (paratype). 339. ♂: Antennal club, flattened to show ventral sensory area. 340. ♂: Metanotum and propodeum (paratype). 341. ♂: Genitalia (paratype). 342. ♂: Forewing.

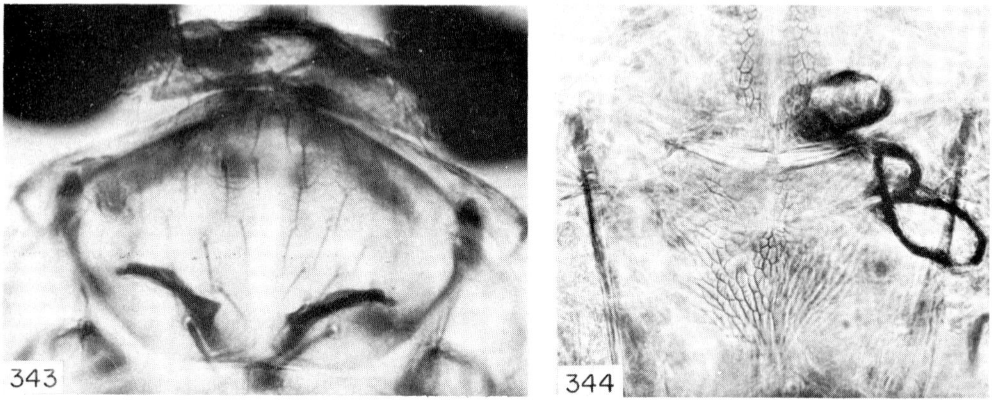


Figures 343–344. *Aphytis noumeaensis* (Howard), ♀ (holotype)
343. Mesoscutum. 344. Metanotum and propodeum.

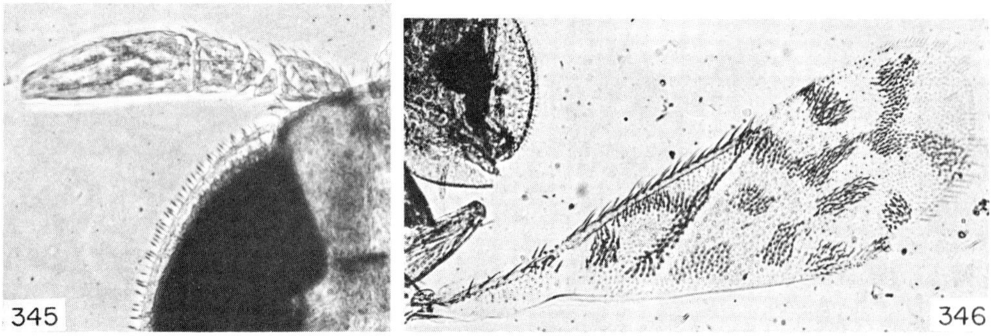


Figures 345–346. *Aphytis argenticorpus* n.sp., ♀ (holotype)
345. Antenna; also note coarse setae on compound eye. 346. Forewing.

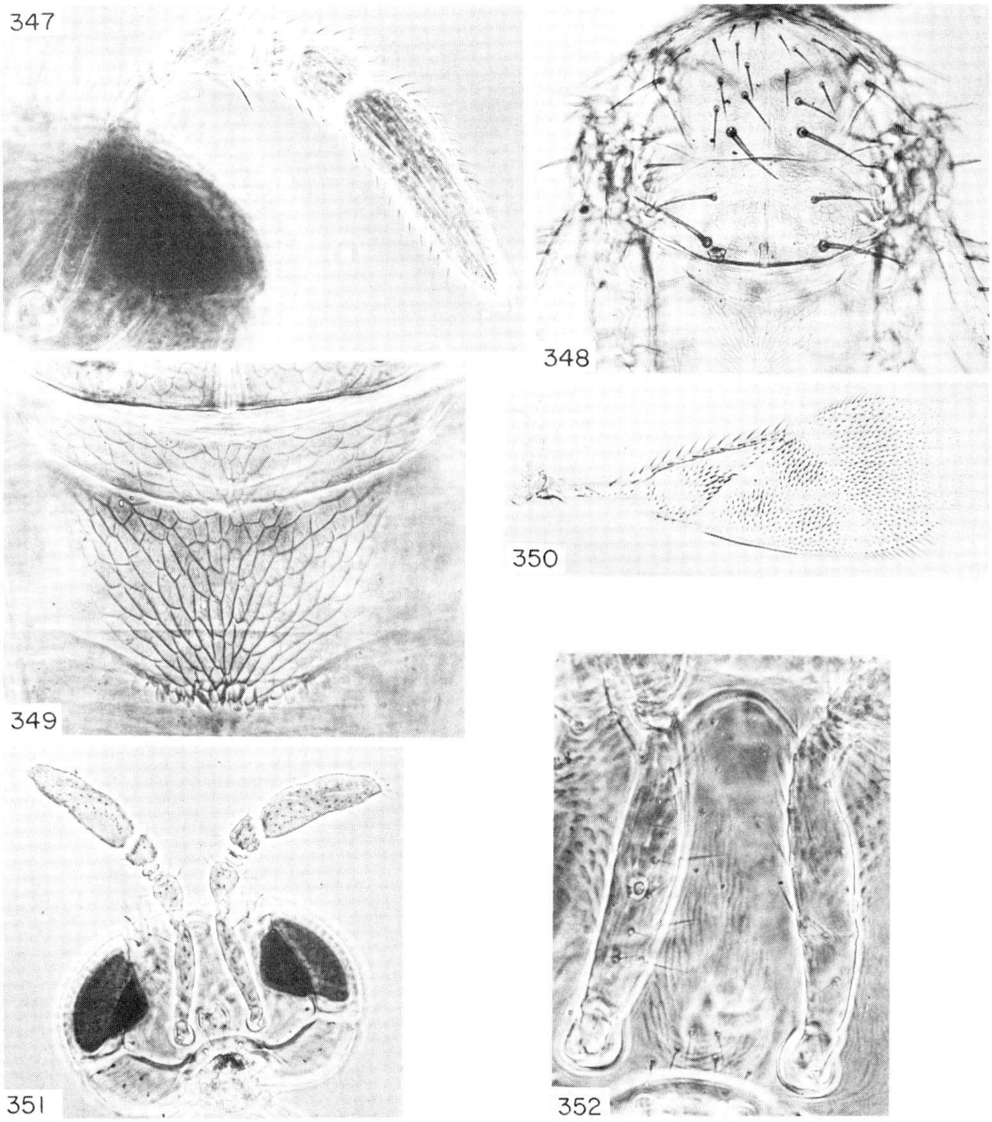


Figures 347–352. *Aphytis fabresi* DeBach and Rosen
347. ♀: Antenna (holotype). 348. ♀: Thorax and propodeum (holotype). 349. ♀: Metanotum and propodeum (holotype). 350. ♀: Forewing (paratype). 351. ♂: Head and antennae (allotype). 352. ♂: Antennal scape, showing tuberculous sense organ in frontal and lateral view (paratype).

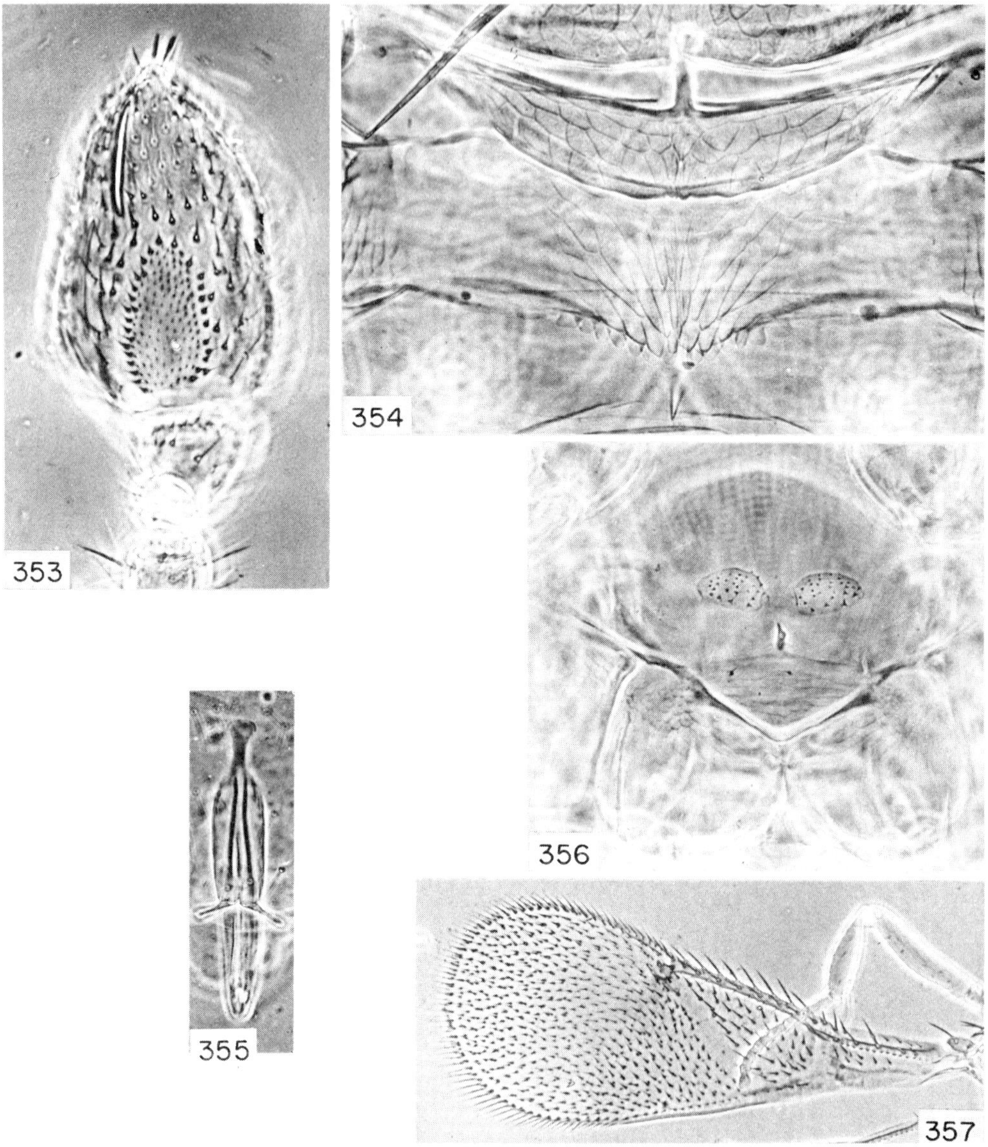


Figures 353–357. *Aphytis fabresi* DeBach and Rosen, ♂ (paratypes)
353. Antennal club, flattened to show ventral sensory area. 354. Metanotum and propodeum. 355. Genitalia. 356. Epicoxal pads between prosternum and mesosternum. 357. Forewing.

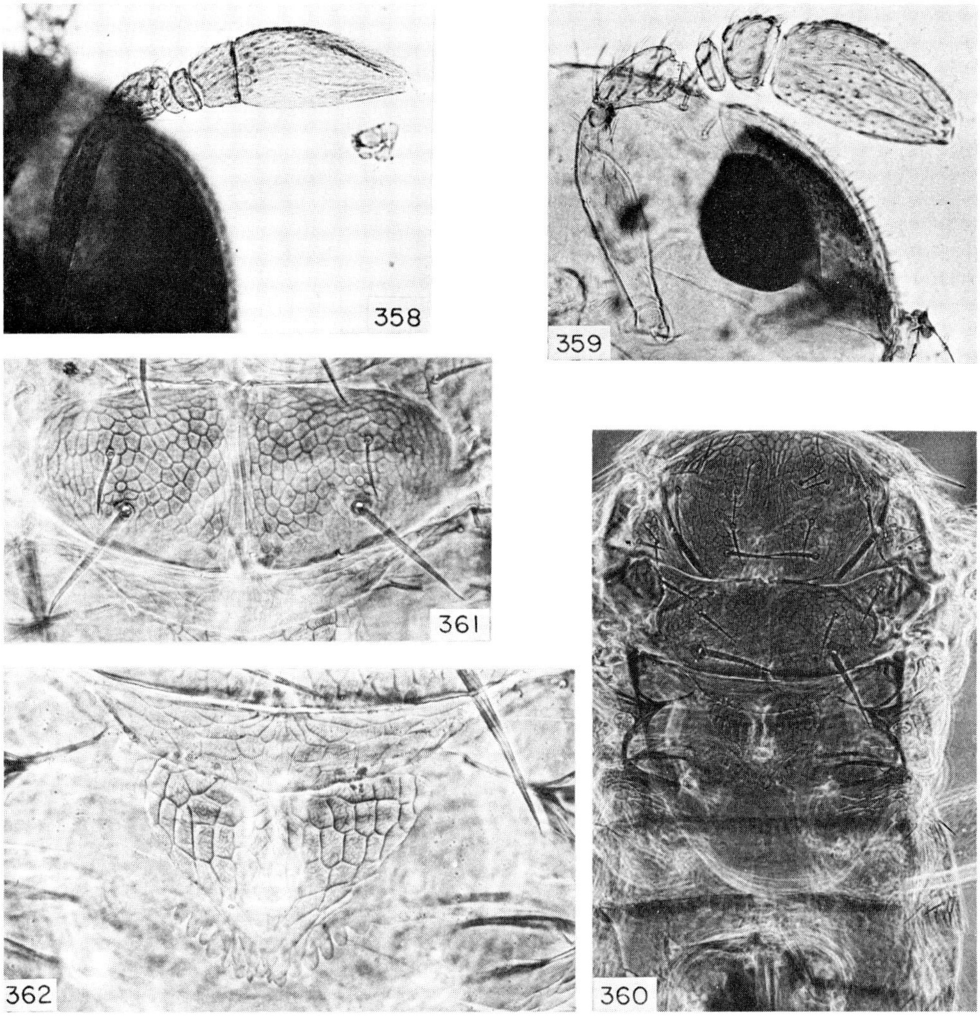


Figures 358–362. *Aphytis capillatus* (Howard), ♀
358. Antenna (paralectotype). 359. Antenna. 360. Thorax, propodeum and base of gaster. 361. Abnormal scutellum with 3 discoid sensilla. 362. Metanotum, propodeum and crenulae.

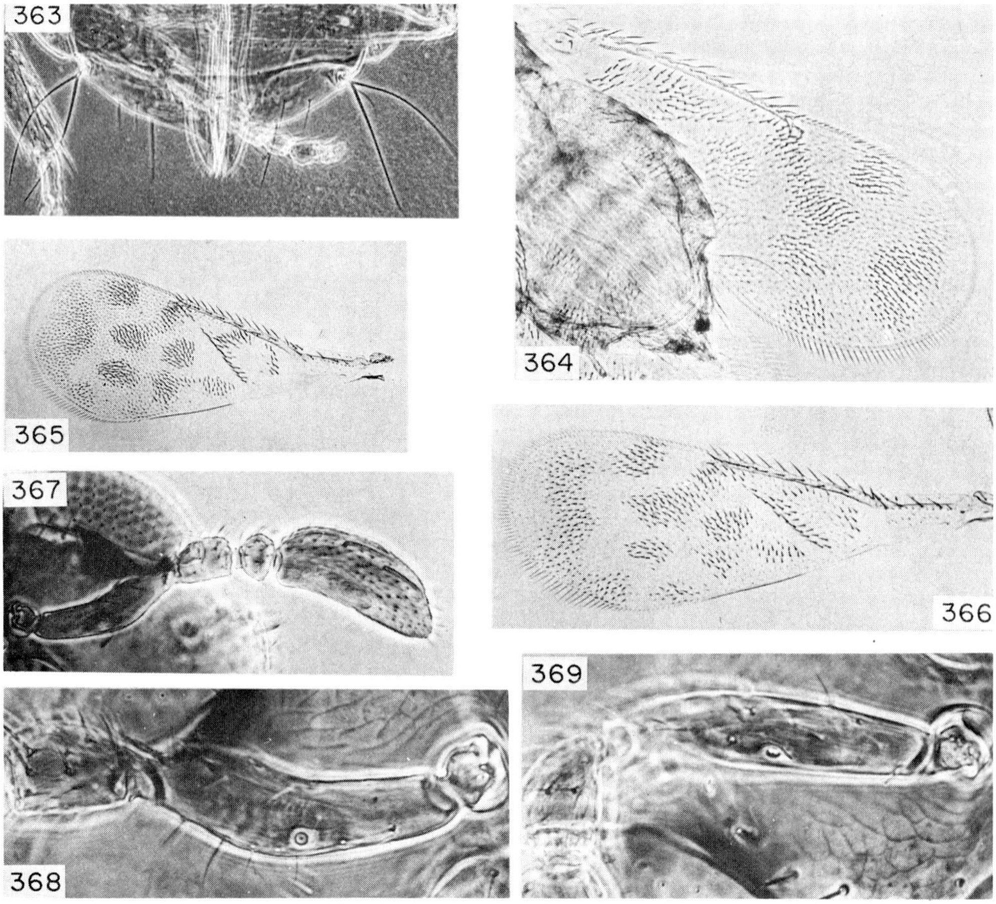


Figures 363–369. *Aphytis capillatus* (Howard)
363. ♀: Syntergum and cerci. 364. ♀: Forewing,(paralectotype). 365. ♀: Forewing. 366. ♀: Forewing, showing "*lessingi*-like" pattern. 367. ♂: Antenna. 368. ♂: Antennal scape, showing sense organ in frontal view. 369. ♂: Antennal scape, showing sense organ in lateral view.

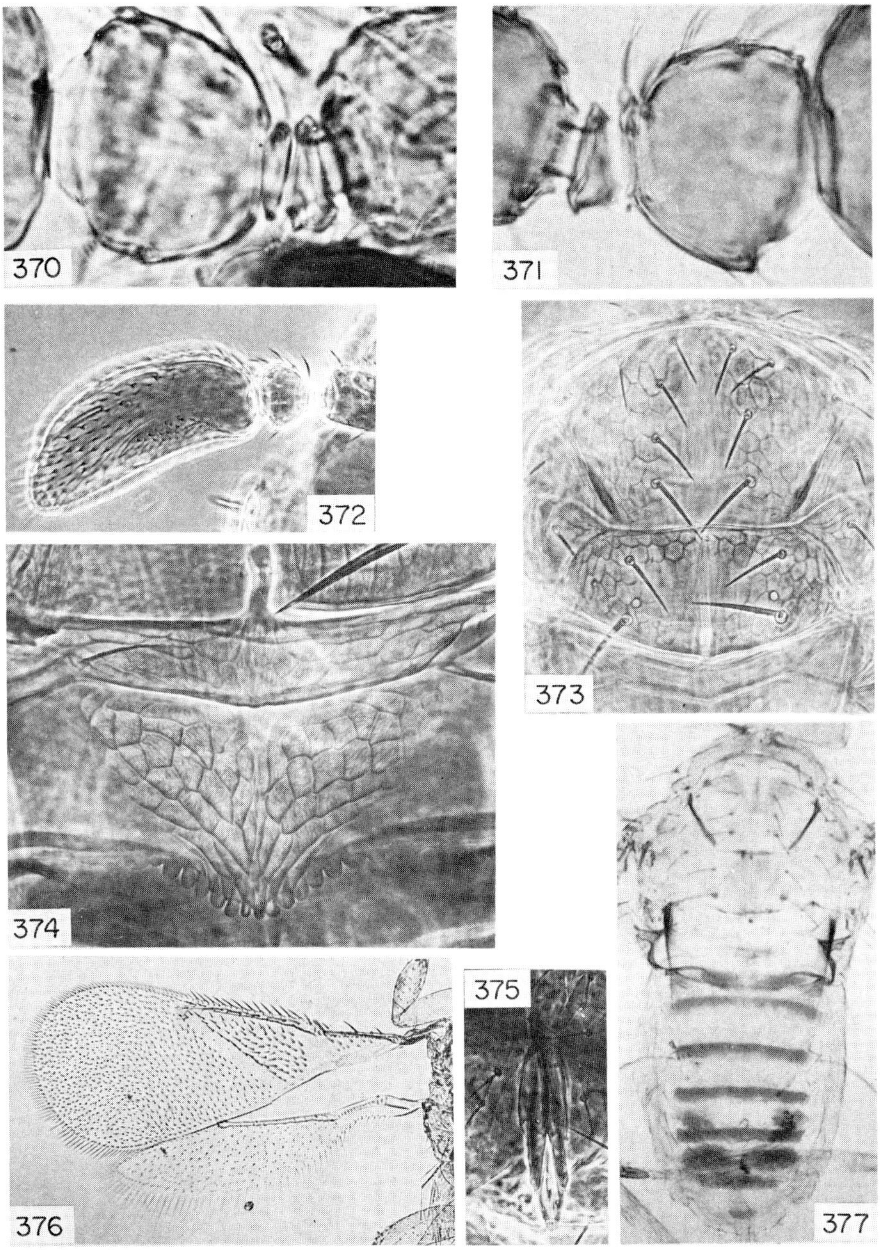


Figures 370–377. *Aphytis capillatus* (Howard), ♂
370, 371. Antennal funicle, showing the two "ring segments." 372. Antenna, showing cut-off sensory area on club. 373. Thorax. 374. Metanotum, propodeum and crenulae. 375. Genitalia. 376. Wings. 377. Thorax and abdomen, showing general coloration.

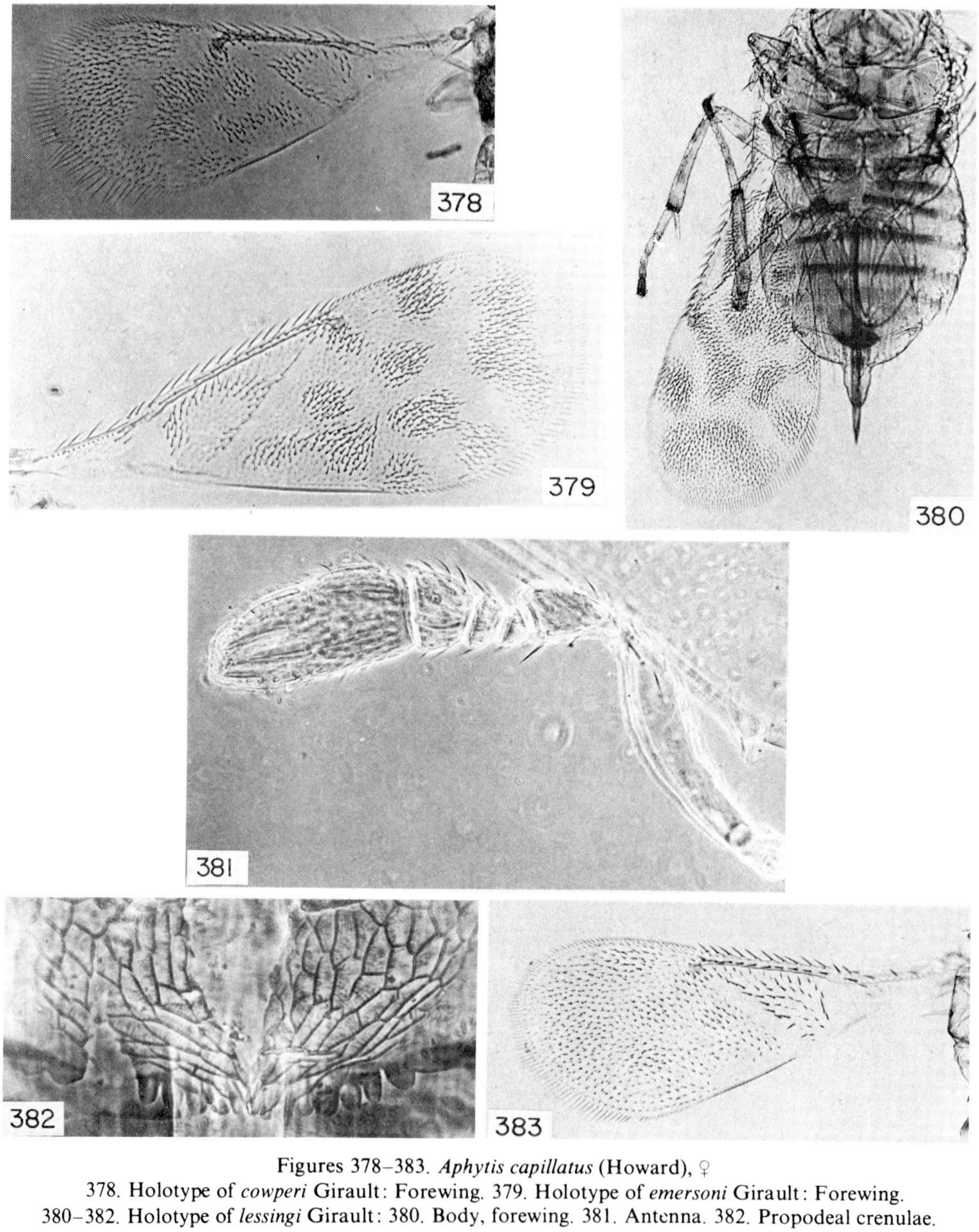


Figures 378–383. *Aphytis capillatus* (Howard), ♀
378. Holotype of *cowperi* Girault: Forewing. 379. Holotype of *emersoni* Girault: Forewing.
380–382. Holotype of *lessingi* Girault: 380. Body, forewing. 381. Antenna. 382. Propodeal crenulae.
383. Type of *romae* Girault: Forewing.

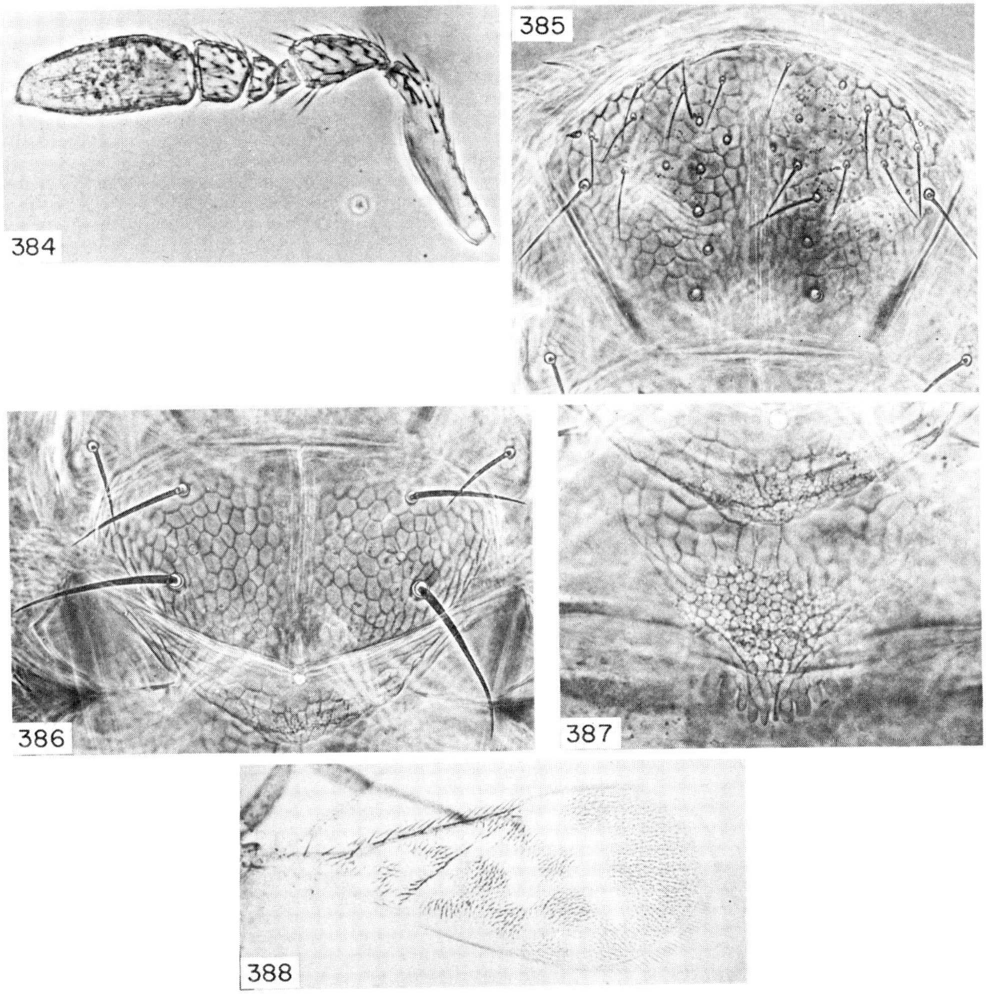


Figures 384–388. *Aphytis nigripes* (Compere), ♀ (paratypes)
384. Antenna. 385. Mesoscutum; note numerous setae. 386. Scutellum and metanotum. 387. Median salient of propodeum and crenulae. 388. Forewing.

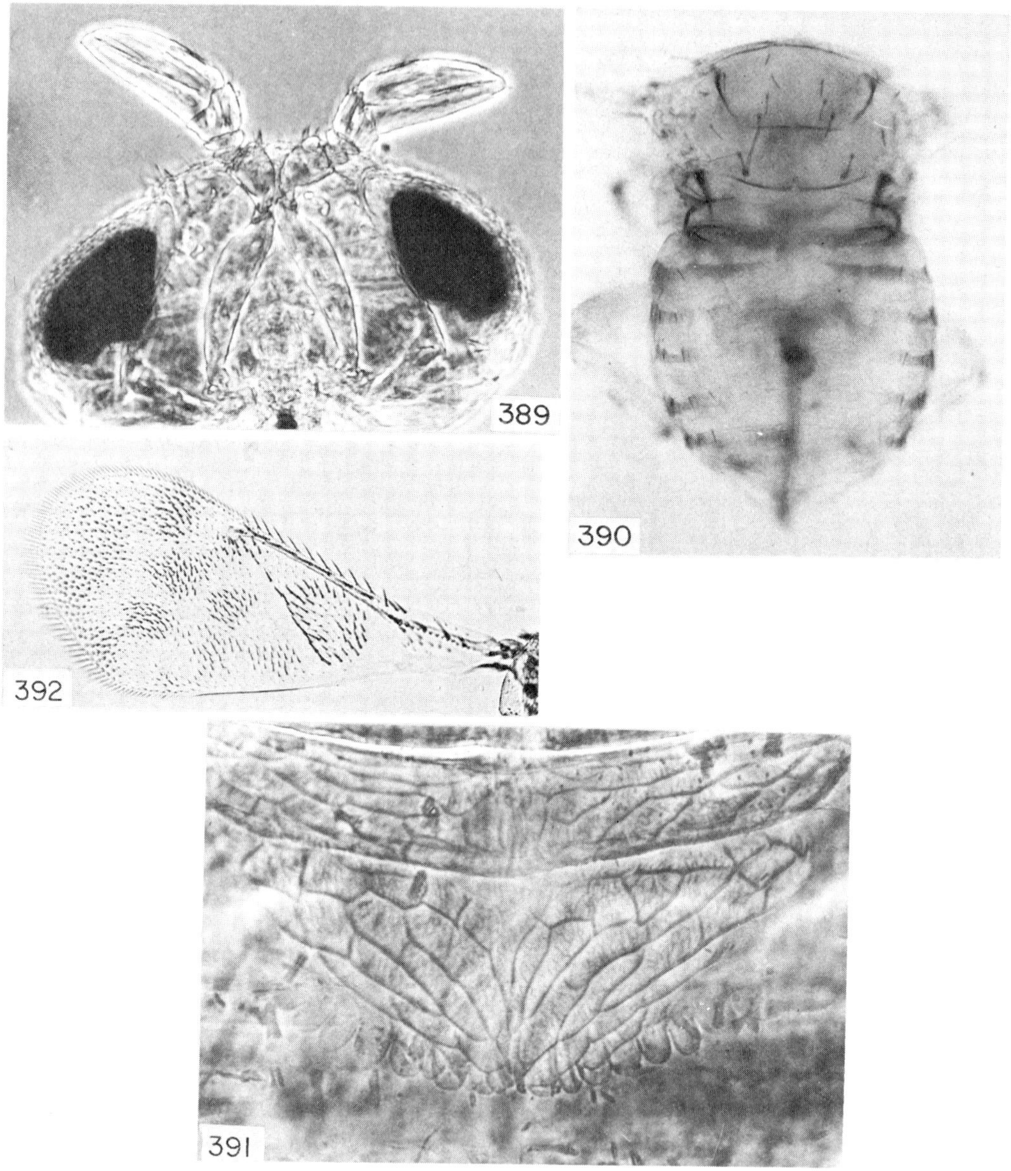


Figures 389–392. *Aphytis ciliatus* (Dodd), ♀
389. Antennae (lectotype). 390. Thorax and abdomen (lectotype). 391. Median part of metanotum, propodeum and crenulae (paralectotype). 392. Forewing (lectotype).

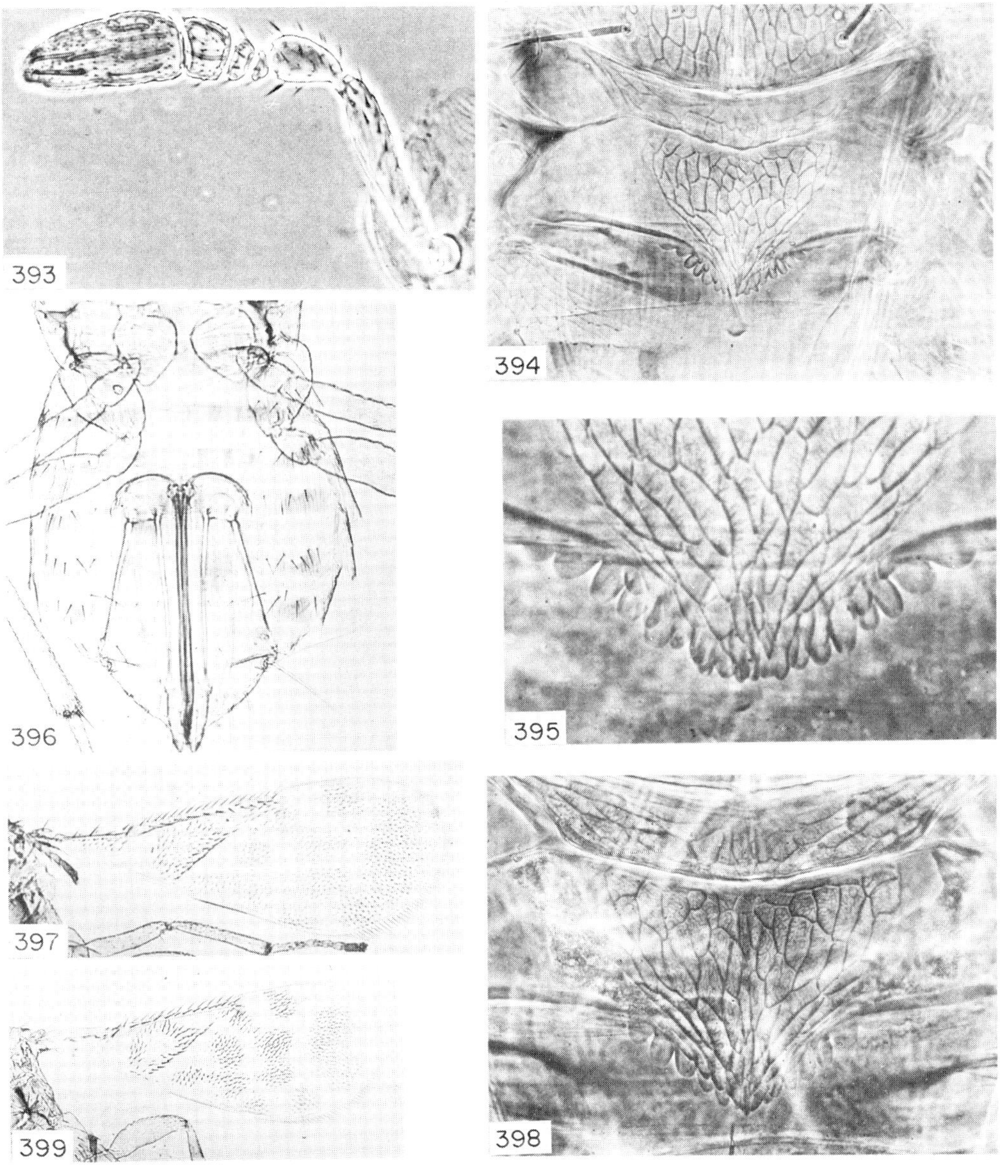


Figures 393–397. *Aphytis australiensis* DeBach and Rosen, ♀
393. Antenna (paratype). 394. Metanotum and propodeum (paratype). 395. Propodeal crenulae (paratype).
396. Abdomen (holotype). 397. Forewing (holotype).
Figures 398–399. *Aphytis ?australiensis* (1971 ♀ specimen)
398. Median salient of propodeum and crenulae. 399. Forewing.

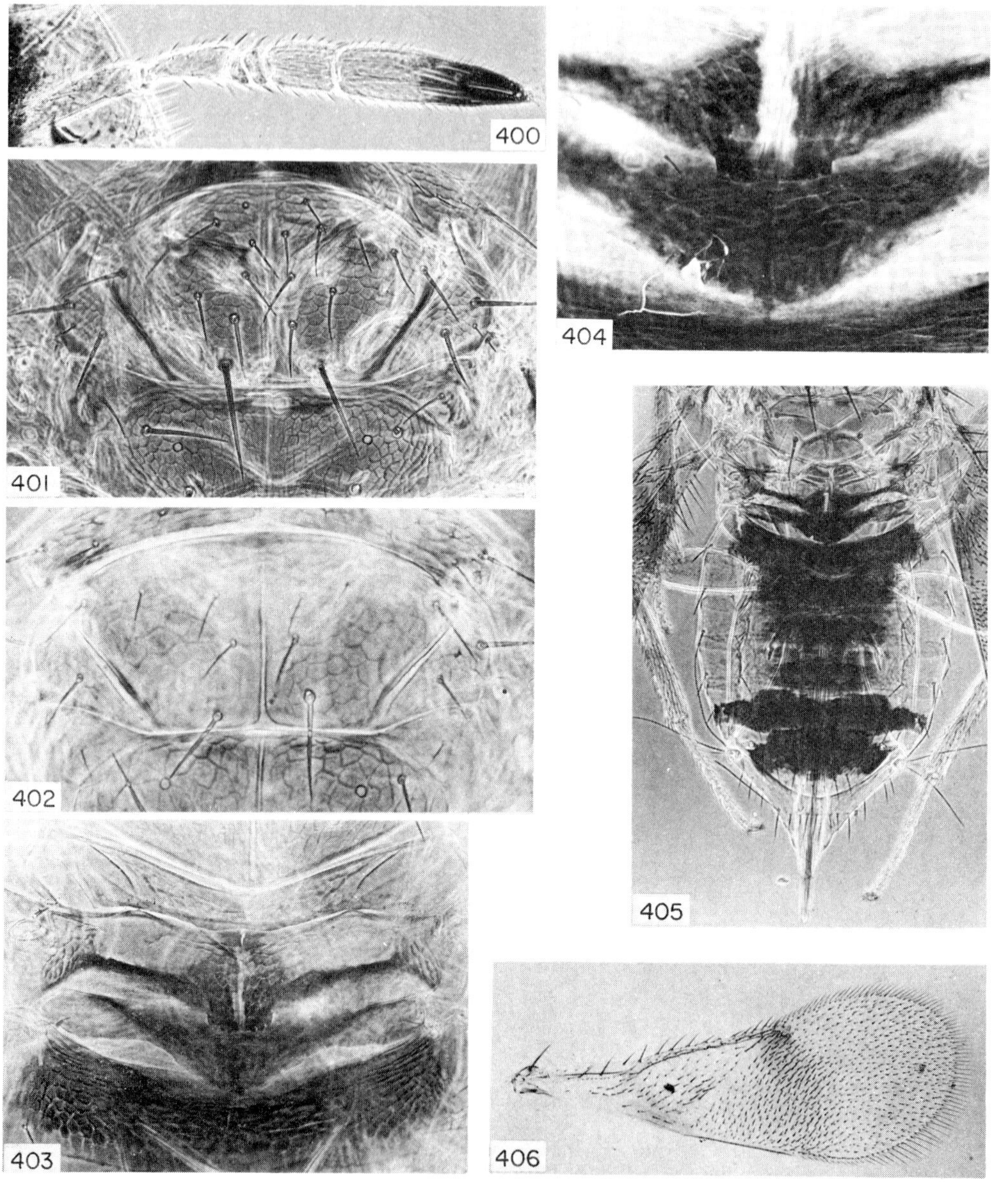


Figures 400–406. *Aphytis anomalus* Compere, ♀
400. Antenna (holotype). 401. Mesoscutum (holotype). 402. Mesoscutum (smallest specimen). 403. Metanotum, propodeum and base of gaster. 404. Median salient of propodeum, crenulae and second abdominal tergite (holotype). 405. Abdomen (holotype). 406. Forewing.

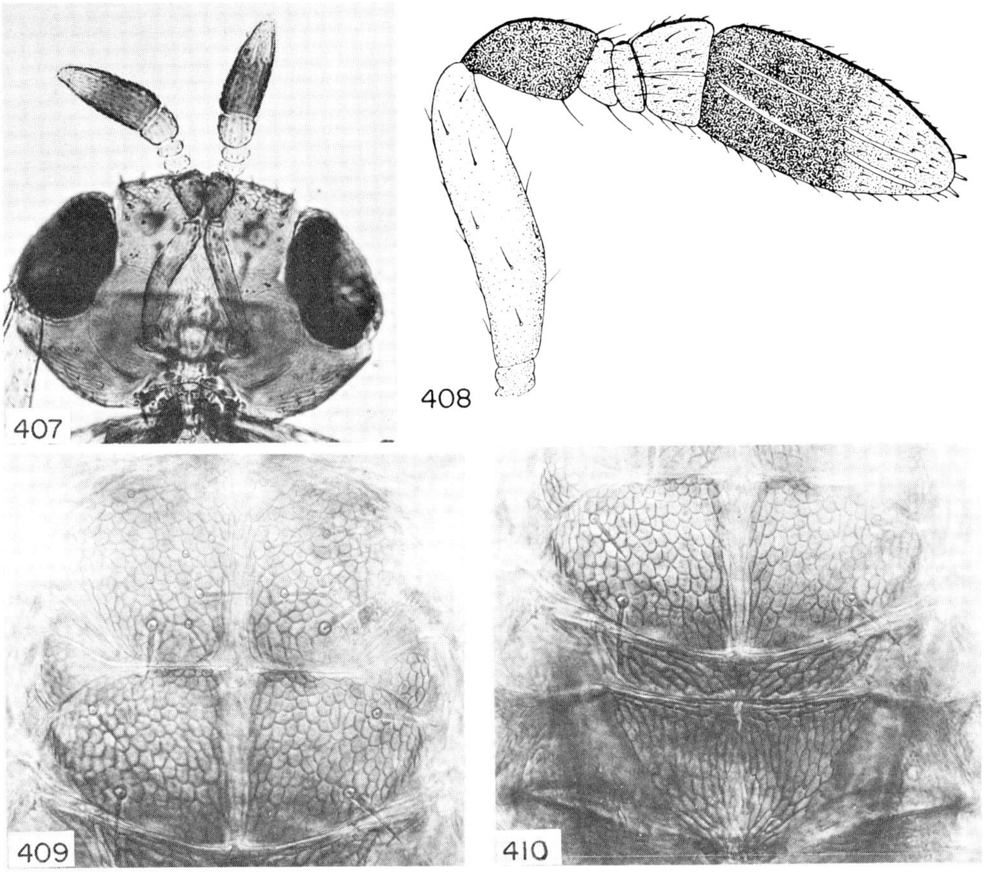

Figures 407–410. *Aphytis haywardi* (Blanchard), ♀
407. Head and antennae (paratype). 408. Antenna (from De Santis, 1948). 409. Mesonotum (paratype).
410. Scutellum, metanotum and propodeum (paratype).

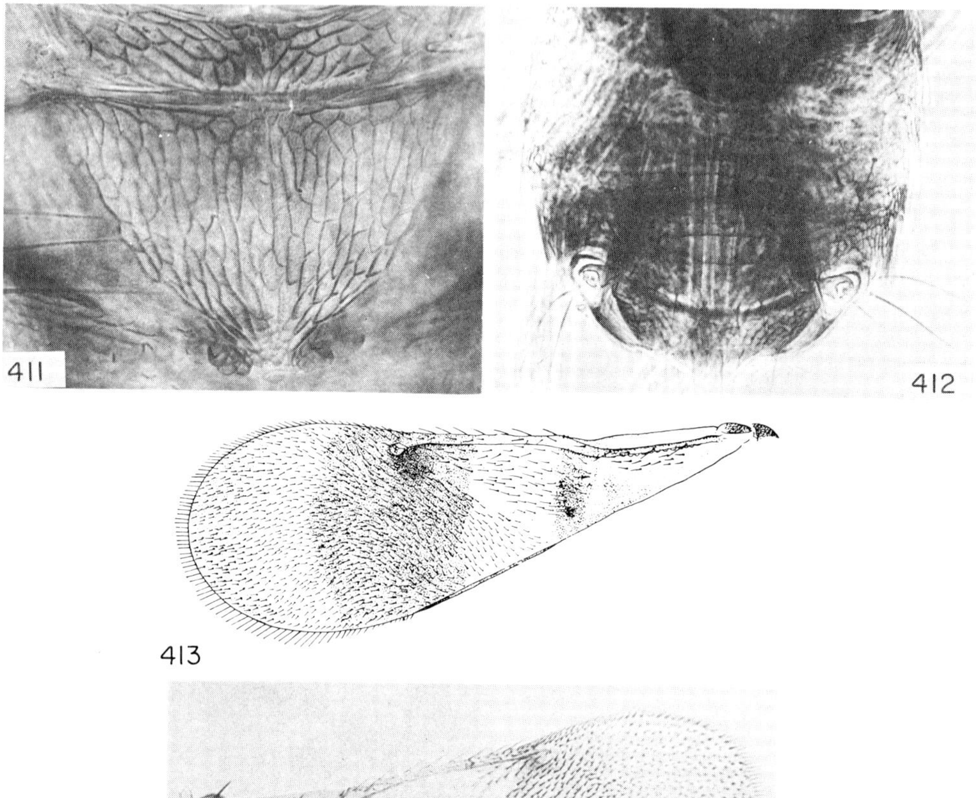


Figures 411–414. *Aphytis haywardi* (Blanchard), ♀
411. Metanotum and propodeum (paratype). 412. Gaster, showing syntergum and cerci (paratype). 413. Forewing (from De Santis, 1948). 414. Forewing (paratype).

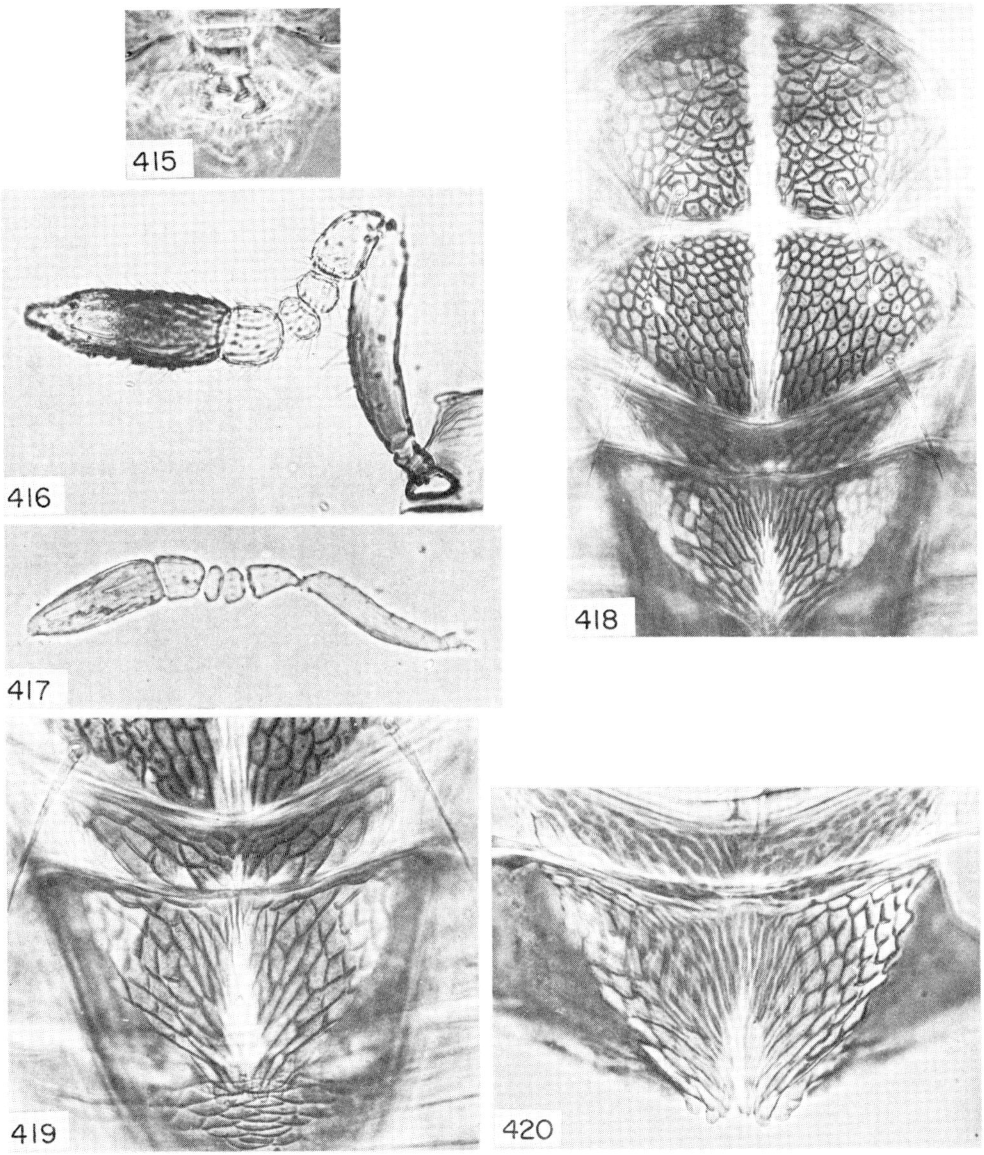

Figures 415–420. *Aphytis obscurus* DeBach and Rosen, ♀ (paratypes)
415. Mandibles. 416, 417. Antennae. 418. Thorax and propodeum. 419. Metanotum, propodeum and second abdominal tergite. 420. Metanotum and propodeum, dissected to show crenulae.

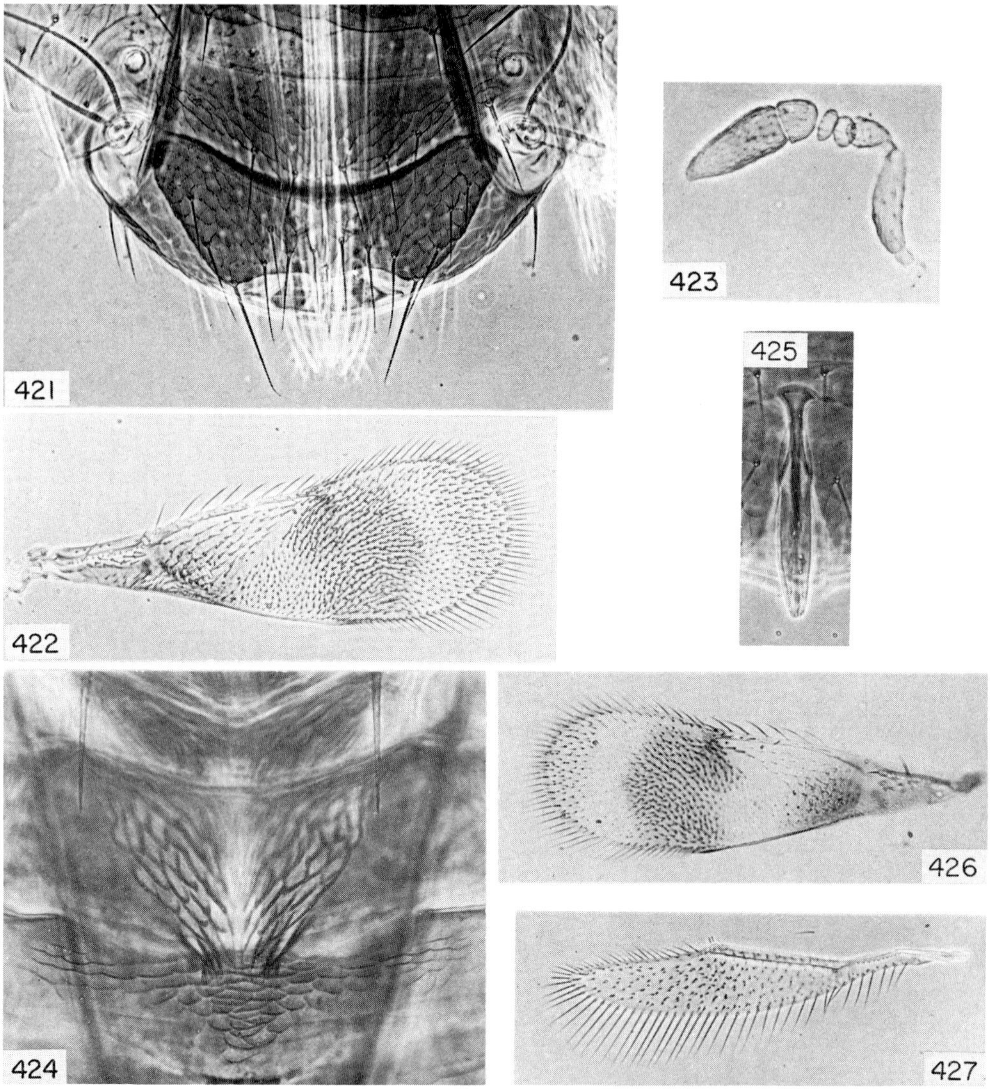


Figures 421–427. *Aphytis obscurus* DeBach and Rosen
421. ♀: Syntergum and cerci; note also spiracles on eighth abdominal tergite (paratype). 422. ♀: Forewing (paratype). 423. ♂: Antenna (paratype). 424. ♂: Metanotum, propodeum and second abdominal tergite (allotype). 425. ♂: Genitalia (paratype). 426. ♂: Forewing (paratype). 427. ♂: Hind wing (paratype).

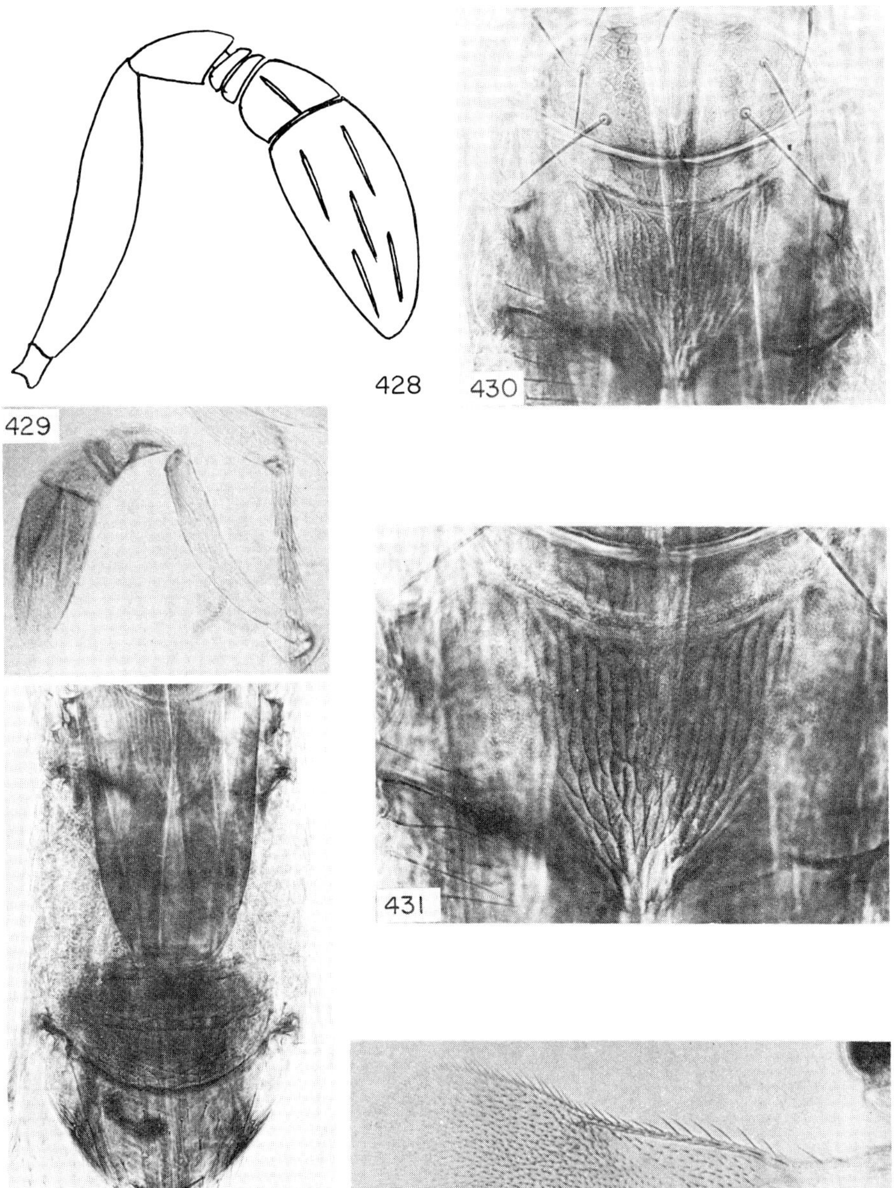


Figures 428–433. *Aphytis angustus* Compere, ♀
428. Antenna (paratype; from Compere, 1955). 429. Antenna (paratype). 430. Scutellum, metanotum and propodeum (paratype). 431. Propodeum (paratype). 432. Abdomen; note elongate endophragma and syntergum (paratype). 433. Forewing (holotype).

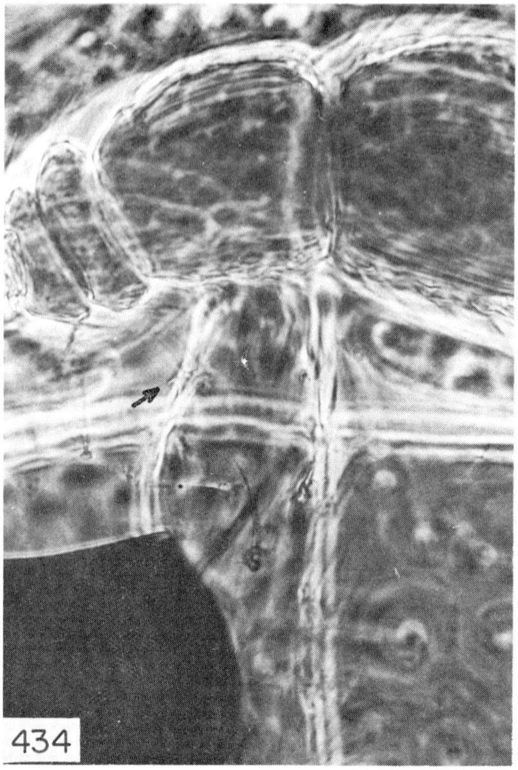


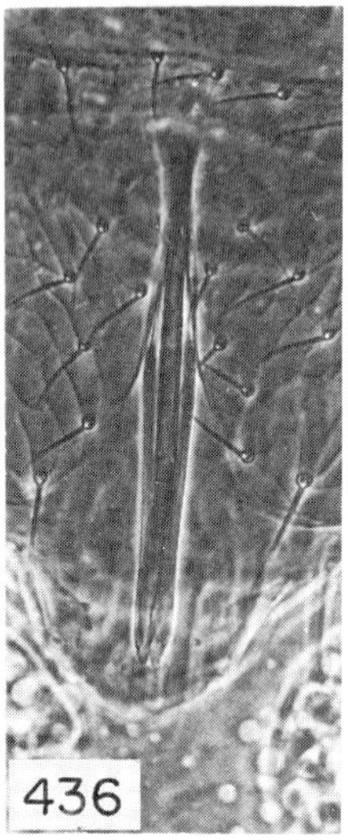


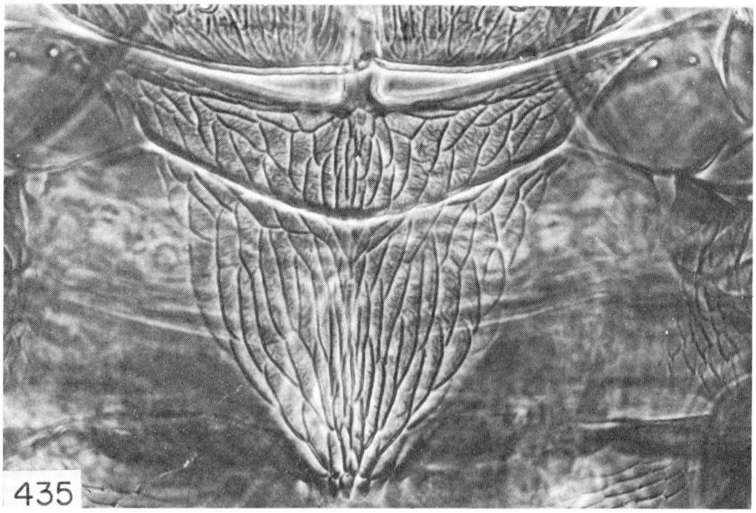


Figures 434–436. Presumed ♂ of *Aphytis angustus* Compere
434. Antennal scape and funicle; arrow points at sense organ on scape. 435. Metanotum and propodeum. 436. Genitalia.

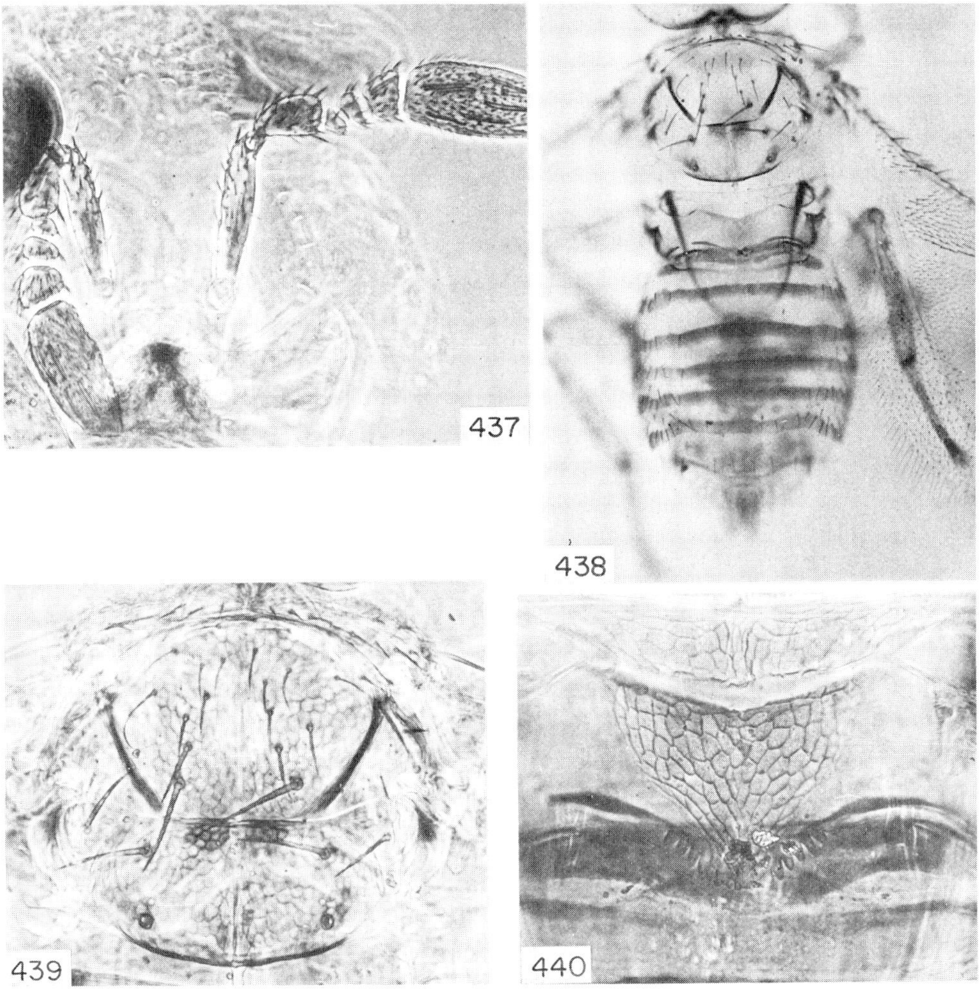


Figures 437–440. *Aphytis hyalinipennis* n.sp., ♀
437. Antennae (paratype). 438. Part of head, thorax and abdomen, showing chaetotaxis and general coloration (holotype). 439. Mesonotum (holotype). 440. Metanotum, propodeum and base of gaster (holotype).

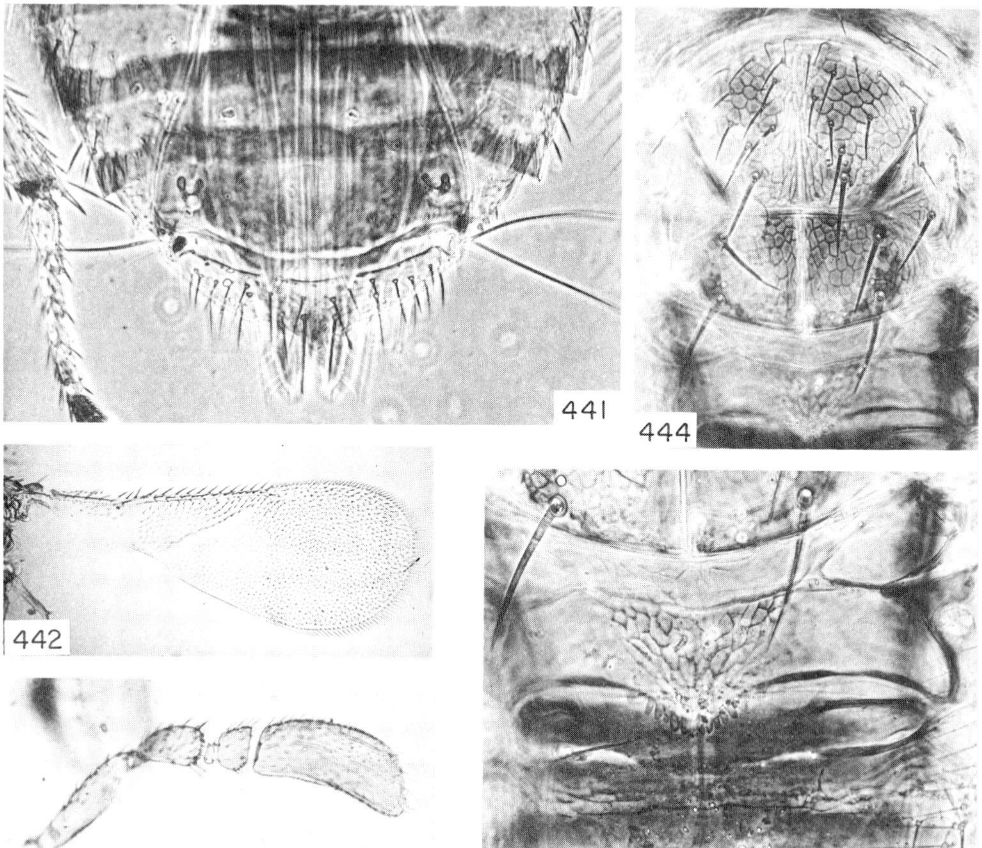


Figures 441–445. *Aphytis hyalinipennis* n.sp.
441. ♀: Posterior abdominal tergites (holotype). 442. ♀: Forewing (paratype). 443. ♂: Antenna (allotype). 444. ♂: Thorax (allotype). 445. ♂: Metanotum, propodeum and base of gaster (allotype).

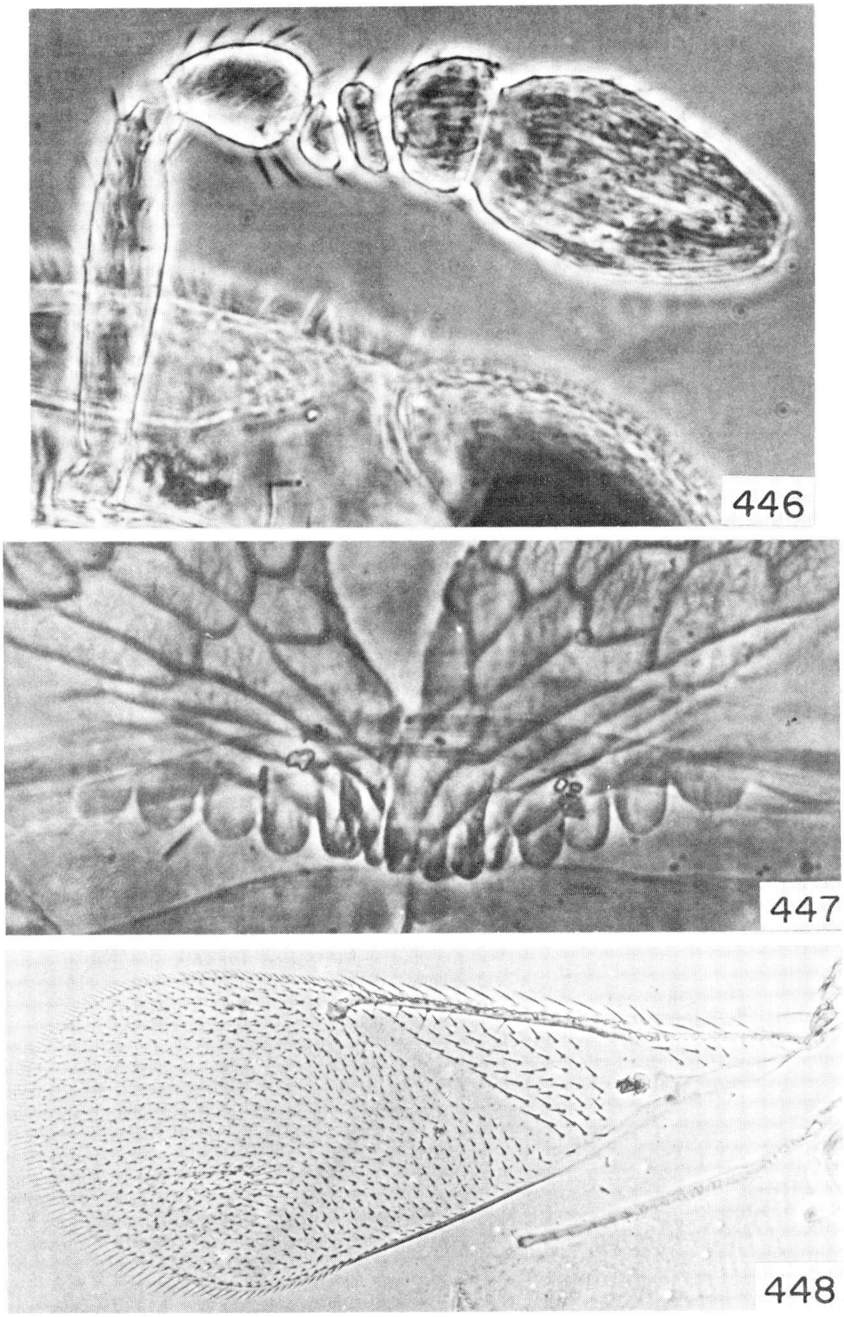


Figures 446–448. *Aphytis wallumbillae* (Girault), ♀
446. Antenna (paralectotype). 447. Propodeal crenulae (paralectotype). 448. Forewing (lectotype).

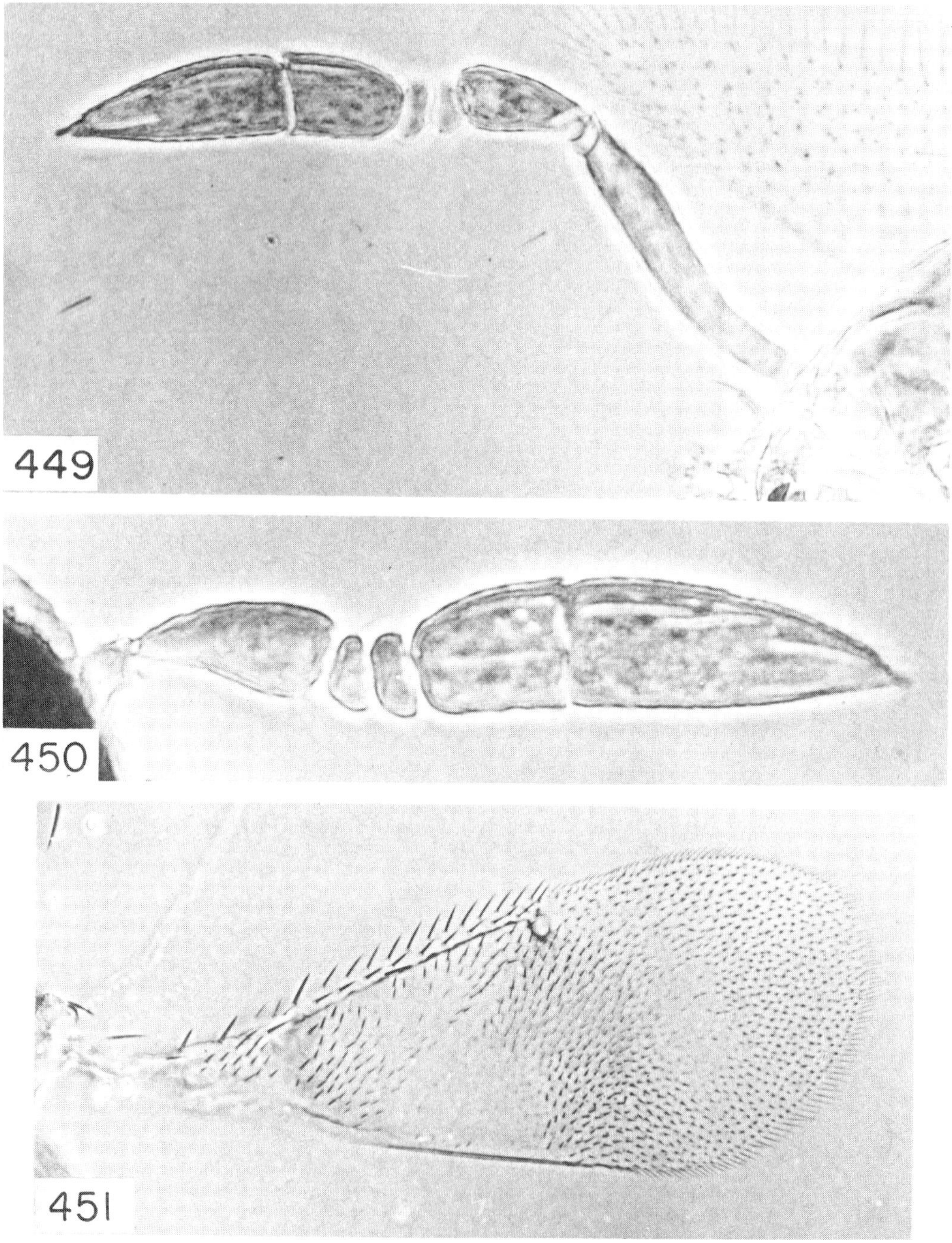


Figures 449–451. *Aphytis dealbatus* Compere, ♀
449. Antenna (holotype). 450. Antennal pedicel, funicle and club (paratype). 451. Forewing (holotype).

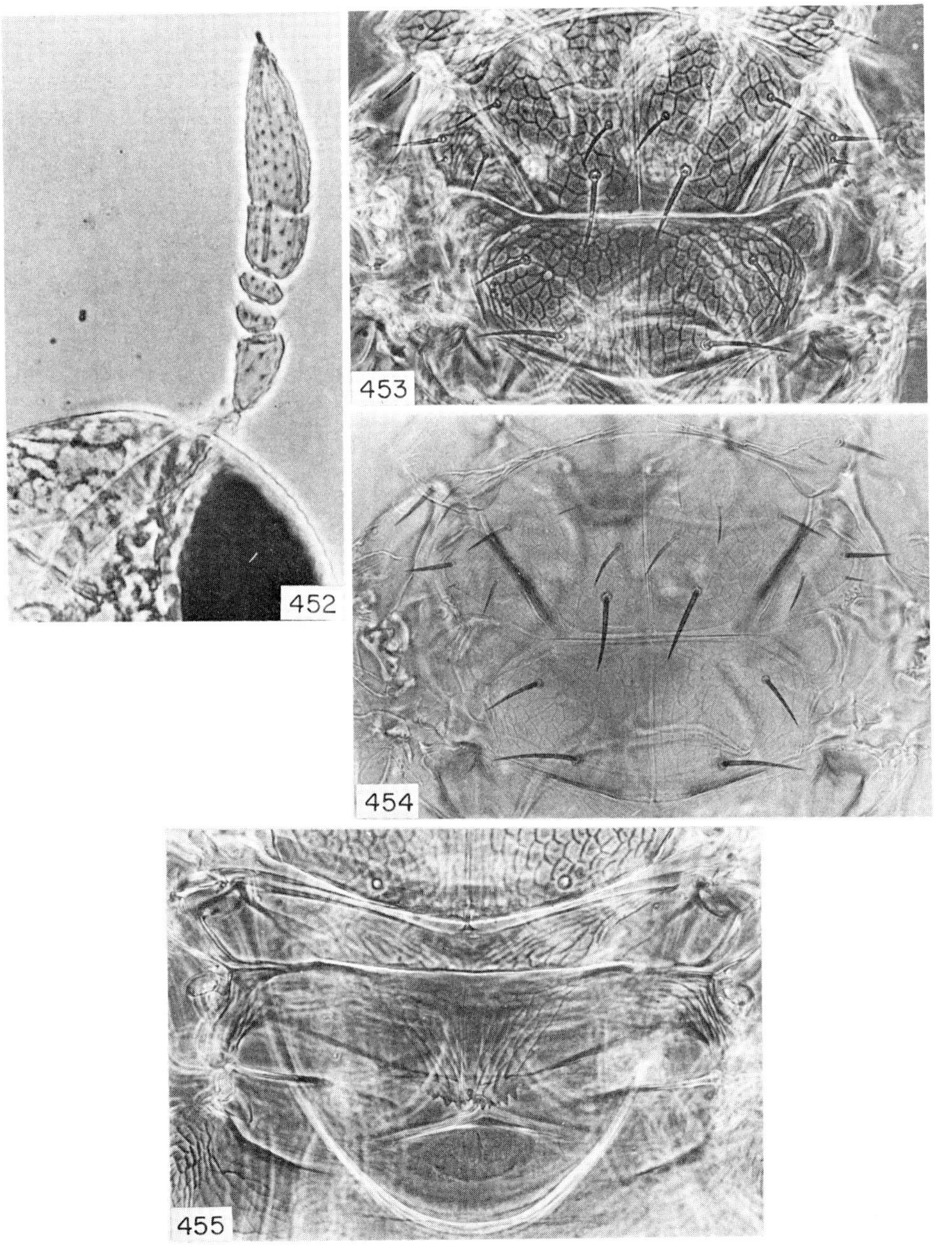


Figures 452–455. *Aphytis melanostictus* Compere, ♀
452. Antenna; note suture separating club from third funicular segment. 453. Thorax, with emphasis on structure and sculpture. 454. Thorax, same specimen as in Figure 453, with emphasis on chaetotaxis and pigmentation. 455. Metanotum, propodeum and base of gaster; note irregular crenulae.

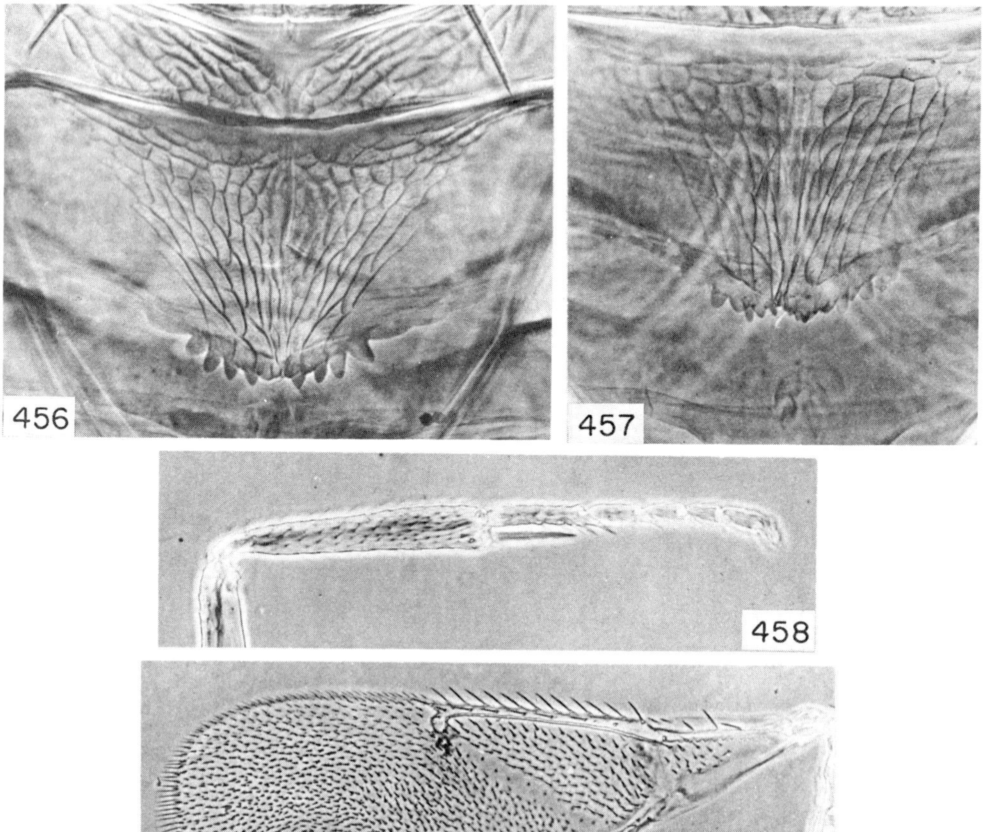


Figures 456–459. *Aphytis melanostictus* Compere, ♀
456, 457. Propodeal crenulae. 458. Middle tibia and tarsus. 459. Forewing.

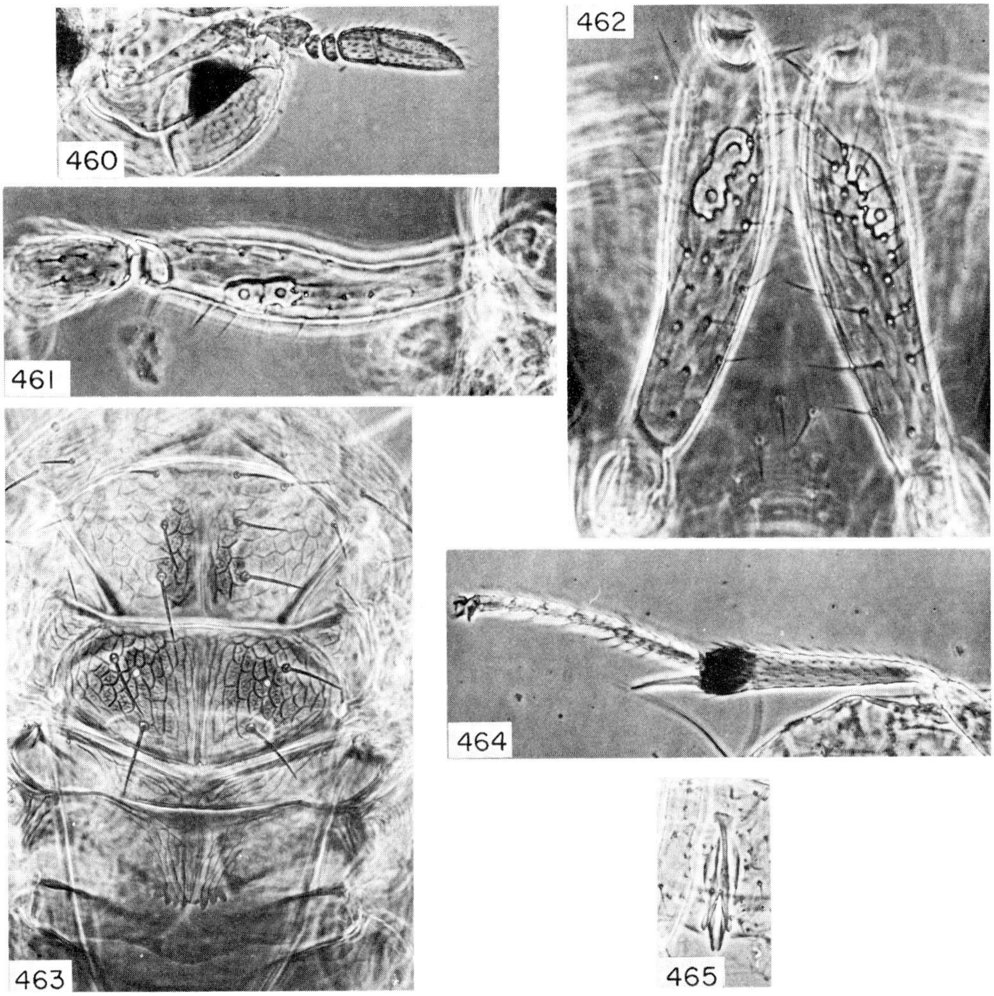


Figures 460–465. *Aphytis melanostictus* Compere, ♂
460. Antenna. 461, 462. Antennal scape, showing specialized sensilla. 463. Thorax, propodeum and base of gaster. 464. Middle tibia and tarsus. 465. Genitalia.

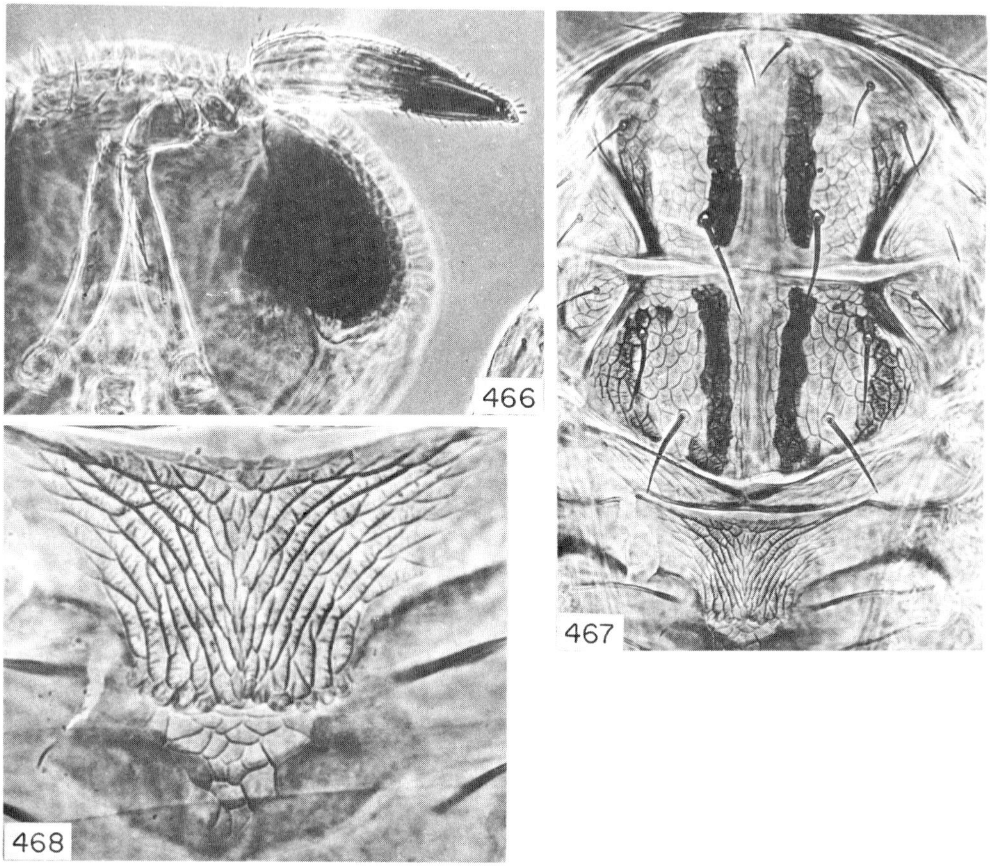


Figures 466–468. *Aphytis malayensis* n.sp., ♀ (holotype)
466. Antenna. 467. Thorax and propodeum. 468. Median salient of propodeum and crenulae; note also sculpture of second abdominal tergite.

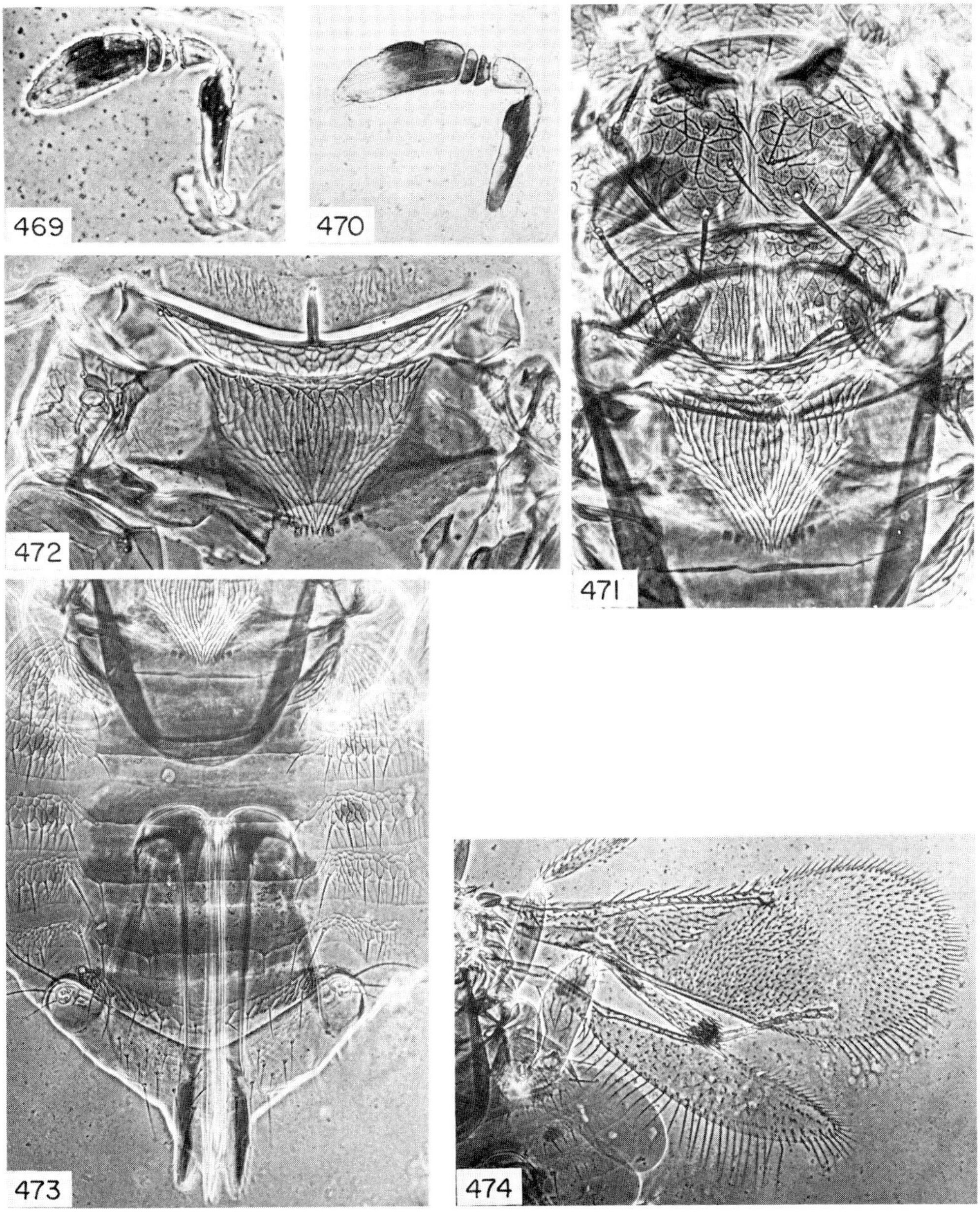


Figures 469–474. *Aphytis mandalayensis* n.sp., ♀

469, 470. Antennae (paratype). 471. Thorax and propodeum (holotype). 472. Metanotum and propodeum (paratype). 473. Propodeum and gastral tergites; note shape of syntergum and position of cerci; gastral tergites flattened (holotype). 474. Wings (holotype).

# II. THE CHILENSIS GROUP

This small, rather heterogenous group at present includes *cercinus* Compere, *chilensis* Howard, *columbi* (Girault), *faurei* Annecke, and *merceti* Compere. Of these members, three are Ethiopian, one is Australian, and one is nearly cosmopolitan in distribution. One Oriental species, *antennalis* n.sp., is provisionally referred to this group.

The **chilensis** group is characterized by pronounced sexual dimorphism, expressed mainly in the peculiar structure of the male antennae: the first 2 funicular segments are greatly reduced or absent, so that the antenna appears to be 4-segmented, and the club is very strongly developed, elongate, resembling that of the male of *Marlattiella* or *Eretmocerus*. Like other species groups in *Aphytis*, the **chilensis** group as understood here is not an exclusive, clearly delimited taxon. Thus, members of the *cochereaui* and *capillatus* complexes, at present included in the **vittatus** group (see above), have rather similar (although considerably shorter) male antennae and are probably related to the **chilensis** group. Discovery of additional species may eventually help clarify the relationships between these groups. In fact, it seems quite possible that the **chilensis** group evolved from ancestral forms related to the more primitive **vittatus** group.

The peculiar aberrant structure of the male antennae does not feature in any of the other species groups of *Aphytis*. Sexual dimorphism is evident also in members of the *costalimai* complex (in the **vittatus** group) and of the *erythraeus* complex (see under "Unassigned Species", pp. 674–692), in which the male antennae bear unusually long setae, but these complexes do not appear to be closely related to the **chilensis** group.

*A. faurei* and *A. antennalis* have similar, "aberrant" antennae in the female sex. The male of *faurei* is very closely related to the male of *merceti*, and that species is therefore regarded as a bona fide member of the **chilensis** group. The male of *antennalis* being unknown, this species is for the time being regarded as associated with this group.

Another characteristic common to all members of the **chilensis** group is the long antennal club of the female. Other characters are rather variable; pigmentation ranges from strongly to rather faintly infuscated; the wings are uniformly setose, dusky but not mottled. The body setae are usually dark and heavy in this group.

In pigmentation, wing pattern and general habitus, the **chilensis** group appears to occupy an intermediate position between the **vittatus** and **proclia** groups. *A. merceti*, with its coarse sculpture, heavy pigmentation and peculiar sense organs on the male antennal scape, is probably the most primitive member of the group. On the other hand, *chilensis* and *columbi* are very similar in general appearance to the **proclia** group and, but for the structure of the male antennae, could have been included in this group.

The following are considered to be synonyms of bona fide members of the **chilensis** group: *capitis* (Rust), *distonotus* (Girault), *longiclavae* (Mercet), *riadi* Delucchi, *signiphoroides* (Brèthes), and *stellaris* (Girault).

The members of the **chilensis** group may be easily separated from each other by obvious characters such as the shape of the propodeum and crenulae, the male genitalia, the structure of the female or male antennae, or details of coloration.

## 25. **Aphytis merceti** Compere

(Figures 475–489)

*Aphytis merceti* Compere, 1955, Univ. Calif. Publ. Entomol., **10**:299.
*Aphytis merceti*: Quednau, 1964, J. Entomol. Soc. S. Afr., **27**:96–98.

This large South African species may be easily recognized by the following combination of peculiar characters: Mandibles reduced, maxillary palpi 1-segmented; propodeum long, crenulae overlapping; abdominal sculpture coarse; ovipositor long, with enlarged triangular plates and long sheaths; male antenna essentially 4-segmented, the first 2 funicular segments reduced to minute "rings", the scape bearing specialized sensilla, the club elongate and bearing numerous sensilla; male genitalia long, with slender digital sclerites.

*Female*. Eyes finely setose. Mandibles (Figure 475) reduced, short and broad, rather weakly sclerotized, the denticles minute; maxillary palpi 1-segmented, labial palpi 1-segmented. Antennal scape (Figure 476) about $4\frac{1}{2}$ to over 5 times as long as wide; pedicel $1\frac{1}{4}$ to $1\frac{1}{2}$ times as long as wide; first 2 funicular segments symmetrical, subequal, about twice as wide as long, the second segment somewhat wider than the first; third funicular segment $1\frac{1}{4}$ to $1\frac{1}{2}$ times as long as wide, distinctly longer than the pedicel, bearing 3–5 longitudinal sensilla; club long, tapering gradually, usually $3\frac{3}{5}$ to 4 times as long as wide, distinctly longer than the scape, usually over 3 times longer and somewhat wider than the preceding segment, bearing 15–25 longitudinal sensilla.

Setae on head and thorax coarse, dark, readily visible under $\times 30$ magnification. Vertex with a pair of long setae just beyond the posterior ocelli, another pair near the posterior corners of the eyes, in addition to several somewhat shorter setae. Mesoscutum (Figures 477, 478) with 16–26 setae, the posterior pair and 1 seta at each anterolateral corner considerably longer and coarser than the others; each parapsis with 2 small, inconspicuous setae; each axilla with 1 seta; scutellum with 4 setae, the discoid sensilla a little closer to the anterior than to the posterior pair. Frontovertex, pronotum, and mesonotal sclerites reticulate; pronotal plates apparently more broadly joined than in other species of *Aphytis*; scutellum oval, about $\frac{3}{4}$ to $\frac{4}{5}$ median length of mesoscutum, reticulate anteriorly and on the sides, reticulate-strigose centrally. Metanotum (Figure 479) reticulate except on the sides; anteromedian apodeme moderately robust, about $\frac{3}{4}$ median length of metanotum.

Propodeum (Figure 479) relatively long, $3\frac{1}{3}$ to 4 times as long as the metanotum, about $\frac{4}{5}$ length of scutellum, reticulate-strigose on the sides and on a relatively wide median area; crenulae 6 + 7 to 10 + 10, large, overlapping, the 2 sets widely separated, sometimes with indications of a few small, nonoverlapping crenulae between them.

Second abdominal tergite (Figures 479, 480) with a small, coarsely reticulate area medially, just below median salient of propodeum; abdominal tergites III–VII reticulate on the sides, bearing 2–6 setae on each reticulate area; seventh tergite with 2–4 fine setae between the lateral setiferous areas; third tergite (Figure 480) coarsely reticulate across anterior half, longitudinally striated on posterior half; tergites IV–VII transversely striated or vaguely reticulate across anterior half, longitudinally striated on posterior half; eighth tergite delicately reticulate, with a transverse row of 4–6 setae between the spiracles; syntergum (Figure 481) triangular, nearly as long as wide, with an elongate cauda, bearing a transverse row of setae. Base of gaster densely stippled ventrally, with transverse rows of dense punctations. Cerci situated considerably closer to posterior spiracles than to tip of abdomen, with 2 long setae and 1 short seta. Ovipositor shaft long, usually $2\frac{1}{5}$ to $2\frac{3}{5}$ times as long as the middle tibia (2.19–2.63); triangular plates and basal arcs conspicuously large; ovipositor sheaths exceptionally long, about $\frac{2}{3}$ to $\frac{4}{5}$ length of middle tibia (0.69–0.80).

Mid-tibial spur rather slender, somewhat shorter than the corresponding basitarsus.

Forewing (Figure 482) fully $2\frac{1}{2}$ times as long as wide; marginal fringe not exceeding $\frac{1}{7}$ width of disk, usually considerably shorter. Delta area nearly as densely setose as the area distad of speculum, with 57–195 somewhat longer setae in 9–12 rows; the delta reaches the posterior margin of the wing, but is clearly separated by a bare area from a group of up to 4 rows of setae below the distal half of the submarginal vein; costal cell with 1, rarely 2 rows of fine setae along proximal half, and 1 coarse seta near the apex. Submarginal vein bearing 2 coarse, unequal setae, the proximal seta about $\frac{1}{2}$ to $\frac{2}{3}$ length of the distal, and 15–23 bullae; another long seta near junction of submarginal and marginal veins, on or just below posterior margin of submarginal vein. Marginal vein bearing 10–13 prominent, subequal setae along anterior margin, these decreasing only slightly in length toward apex of vein, the proximal seta no more than $1\frac{1}{3}$ times longer than the distal, only somewhat longer than the setae in a row along center of vein.

Forewing lightly infumated below submarginal vein; an arcuate, fuscous crossband below junction of marginal and submarginal veins; a distinct fuscous cloud below stigma, fading into an obscure, broad, rather faint crossband; a narrow fuscous line along central portion of posterior margin of wing; apical part of wing and most of the area surrounding speculum distinctly more hyaline than rest of wing.

General coloration yellow to white, extensively marked with brownish or blackish as follows: transverse stripes on occiput on each side of foramen; narrow lines on genal sutures from oral margin to about halfway to eye; concealed part of pronotum medially; all margins of mesoscutum; posterior margin of scutellum, mesoscutum and scutellum faintly suffused; propodeum just mesad of spiracles, faintly suffused centrally; anterior margin of second abdominal tergite; broad transverse stripes across all subsequent tergites, most pronounced on third tergite and on syntergum; thoracic sterna; internal apodemes including margins of endophragma. Antenna (Figure 476) uniformly blackish, scape and tip of club paler; all legs yellowish, tibiae rather uniformly fuscous. Wing veins yellowish.

Length 1.02–1.68 mm.

*Male.* Essentially similar to the female in structure, chaetotaxis, sculpture and coloration, differing mainly in the structure of the antennae.

Antennal scape (Figures 483–486) reticulate, widened ventrally, usually only slightly over 3 times as long as wide, ventral aspect bearing an oval plate with 2–3 short, rather flattened, denticlelike sensilla; pedicel about $1\frac{1}{7}$ to $1\frac{1}{4}$ times as long as wide, $\frac{3}{5}$ to $\frac{3}{4}$ times as long as the third segment of the funicle; first 2 funicular segments reduced to minute, vestigial annelli (Figure 487), the antenna appearing 4-segmented; third funicular segment distinctly longer than wide, bearing 6–8 longitudinal sensilla; club elongate, about 5 times as long as wide or longer, $1\frac{2}{3}$ times as long as the scape, about 4 times longer than and as wide as the preceding segment, bearing 24–30 rather long longitudinal sensilla.

Chaetotaxis and sculpture of head and thorax as in the female; mesoscutum with 12–16 setae. Propodeum (Figure 488) and abdomen essentially as in the female, the longitudinal striation on the abdominal tergites sometimes obscure; abdominal sterna (Figure 489) densely stippled.

Genitalia (Figure 489) long, nearly as long as the middle tibia or even somewhat longer (0.87–1.08), with a short, narrow phallobase and an elongate aedeagus; digital sclerites long, slender, about $\frac{1}{4}$ to $\frac{1}{3}$ the combined length of aedeagus and apodemes (0.26–0.34), not bearing an apical claw.

Forewing slightly wider than in the female, about $2\frac{2}{5}$ to $2\frac{1}{2}$ times as long as wide; delta with 34–106 setae; submarginal vein with 10–18 bullae; marginal vein with 6–11 prominent, subequal setae along anterior margin.

Wing pattern and general coloration as in the female; dorsum of abdomen more uniformly fuscous.

Length 0.76–1.16 mm.

Redescribed from the following material, all from South Africa: 2♀♀, 3♂♂ (♂ allotype and paratypes), reared by E. W. Rust from *Lindingaspis rossi* (Maskell), Newlands, Cape Province, February 10–15, 1923; 3♀♀ (♀holotype and paratypes), reared by E. W. Rust from a sample of white wax scale, *Gascardia destructor* (Newstead),* Roseband, Cape Province, September 18–20, 1924; 3♀♀, 3♂♂ (paratypes), reared by E. W. Rust from *Melanaspis corticosa* (Brain), Groote Schuur, Cape Town, January 21–February 3, 1925; 5♀♀ labeled "E. W. Rust, S. Africa"; 2♂♂ labeled "Import #A98"; 2♀♀, 1♂ reared by B. R. Bartlett from *Lindingaspis rossi* (Maskell) on scotch broom, Stellenbosch, April 11, 1958; 1♀ labeled S&R 1847, from a shipment of black scale, *Saissetia oleae* (Olivier),* on oleander twigs, collected by B. R. Bartlett, Aliwal North, May 29, 1958; 1♀, reared by C. J. Cilliers from *Chrysomphalus aonidum* (L.) on citrus, Durban, Natal, March 1964; 5♀♀, 1♂, reared by Y. B. Whitehead from an undetermined host

* Probably from armored scale insects contaminating the sample.

on Australian bottlebrush, Stellenbosch, Cape Province, November 1965; 11♀♀, captured by V. C. Moran on a sticky yellow screen, Grahamstown, Cape Province, August–October 1966; 1♀, 5♂♂, reared by H. P. Insley from *Chrysomphalus aonidum* (L.) on *Acacia melanoxylon*, Darling, Cape Province, October 1968; 13♀♀, 14♂♂, reared by F. Honiball from *Lindingaspis rossi* (Maskell) on *Hedera helix*, Citrusdal, Cape Province, February 26, 1969.

Type series in the collection of the Division of Biological Control, University of California, Riverside.

**Notes.** *A. merceti* is one of the largest, most coarsely sculptured species of *Aphytis*. It differs from most other species in having reduced mouthparts, and may be readily recognized by the distinctive sculpture, long ovipositor, and peculiar male antennae and genitalia.

This interesting species is undoubtedly a parasite of armored scale insects. The records of soft scales as hosts are apparently erroneous.

In the ♀ holotype specimen, a freak of chaetotaxis is noted: on the mesoscutum (Figure 478), 2 setae arise from one base.

## 26. **Aphytis faurei** Annecke

(Figures 490–501)

*Aphytis faurei* Annecke, 1963, Ann. Mag. Nat. Hist., Ser. 13, **6**:343–345.

This little-known, aberrant South African species resembles *merceti* in coloration, chaetotaxis and distinctive sculpture, as well as in the reduced mouthparts. It can be easily distinguished from all other species of *Aphytis* by the peculiar, 4-segmented antennae of the female, with an elongate funicular segment and a very long club bearing numerous sensilla, and by the absence of distinct propodeal crenulae. The hitherto undescribed male is very similar to the female, differing from the male of *merceti* mainly in the absence of distinct crenulae and in details of antennal structure. *A. faurei* is redescribed below relative to *merceti*.

*Female.* Eyes finely setose; ocelli somewhat larger than in *merceti*. Mandibles short and broad, as in *merceti*; both maxillary and labial palpi 1-segmented. Antennae (Figures 490, 491) rather similar to those of ♂ *merceti*, essentially 4-segmented (1111), with 1 vestigial ring segment and a slight rudiment of another; scape slender, about $4\frac{1}{2}$ to over 5 times as long as wide; pedicel as long as wide or somewhat longer than wide; funicular segment about $1\frac{1}{4}$ to $1\frac{1}{2}$ times as long as wide, $1\frac{3}{5}$ times to nearly twice longer than the pedicel, bearing 14–15 longitudinal sensilla; club enormously elongate, about $5\frac{1}{2}$ to over 7 times as long as wide, about twice as long as the scape, $4\frac{1}{4}$ to $4\frac{1}{2}$ times longer and somewhat wider than the preceding segment, bearing about 60 longitudinal sensilla.

Chaetotaxis of head as in *merceti*. Thorax and abdomen as in *merceti*, the thoracic setae somewhat longer; mesoscutum with 18–29 setae in the 4 specimens at hand.

Propodeum (Figure 492) rather similar to that of *merceti*, $3\frac{1}{3}$ to $3\frac{3}{4}$ times as long as the metanotum, $\frac{3}{4}$ to $\frac{4}{5}$ length of scutellum, with a uniform, smooth, slightly serrate ledge along posterior margin but usually without any distinct crenulae; 1♀ at hand with a few very small, rounded, nonoverlapping crenulae on one side (Figure 493). Syntergum triangular, apparently somewhat longer than in *merceti*; cerci and ovipositor as in *merceti*; triangular ovipositor plates somewhat smaller than in *merceti*. Chaetotaxis and sculpture as in *merceti* (see Figure 494).

Forewing (Figure 495) somewhat wider than in *merceti*, $2\frac{1}{4}$ to $2\frac{1}{2}$ times as long as wide; marginal fringe not exceeding $\frac{1}{9}$ width of disk. Setae in delta (58–123 in 7–10 rows) somewhat sparser than in *merceti*; only a few setae below distal half of submarginal vein. In the 4 specimens at hand, submarginal vein bearing 17–23 bullae, marginal vein bearing 11–16 prominent, subequal setae along anterior margin. Otherwise as in *merceti*.

Coloration and wing pattern as in *merceti*.

Length 1.13–1.62 mm.

*Male*. Very similar to the female in structure, chaetotaxis, sculpture and coloration, differing mainly in the shape of the antennal scape.

Eyes finely setose. Mandibles (Figure 496) reduced, as in the female; maxillary and labial palpi 1-segmented. Antennae (Figures 497, 498) essentially 4-segmented (1111), as in ♂ *merceti* but the ring segments are more reduced, vestigial, as in the female; scape delicately reticulate, widened ventrally, about $3\frac{2}{5}$ times as long as wide, ventral aspect bearing an oval plate with 2 short, broad, denticlelike sensilla (Figure 498); pedicel about $1\frac{1}{5}$ times as long as wide; funicular segment about $1\frac{3}{5}$ times as long as wide, fully twice longer than the pedicel, bearing 21–23 longitudinal sensilla; club about $5\frac{1}{3}$ times as long as wide, $2\frac{1}{4}$ times longer than the scape, $3\frac{2}{3}$ times longer and slightly wider than the preceding segment, bearing 75–85 longitudinal sensilla.

In the single male at hand, the mesoscutum bears 17 setae; scutellum about $\frac{3}{4}$ median length of mesoscutum; anteromedian apodeme $\frac{7}{10}$ median length of metanotum. Propodeum (Figures 499, 500) a little less than 3 times as long as the metanotum, $\frac{3}{5}$ length of scutellum, with a uniform ledge along posterior margin, one minute crenula on one side. Abdominal sterna stippled.

Genitalia (Figure 501) long, a little longer than the middle tibia; digital sclerites long, slender, about $\frac{3}{10}$ the combined length of aedeagus and apodemes, not bearing an apical claw.

Forewing somewhat wider than in ♂ *merceti*, about $2\frac{1}{4}$ times as long as wide; delta with 45–47 setae; submarginal vein with 18 bullae; marginal vein bearing 11 prominent, subequal setae along anterior margin.

Wing pattern and general coloration essentially as in ♂ *merceti*.

Length 0.97 mm.

Redescribed from 1♀ (paratype), collected by D. P. Annecke with *Africaspis chionaspiformis* (Newstead) and *Melanaspis corticosa* (Brain) on a leafless tree, Ngwavuma, Natal, South Africa, September 1961; 2♀♀ caught by V. C. Moran on sticky yellow trap, Grahamstown, Cape Province, South Africa, August 1966; 1♀, 1♂ collected by G. L. Prinsloo, Jozini Dam, Zululand, South Africa, March 1973.

**Notes.** *A. faurei* is obviously very closely related to *merceti*. The female antennae, curiously, resemble those of male *merceti*, differing from the latter mainly in the shape of the scape. Unlike other members of the **chilensis** group, *faurei* exhibits rather little sexual dimorphism, the antennae of the male being very similar to those of the female.

Annecke (1963) described *A. faurei* with "crenulae on median caudal margin of propodeum very small, only slightly overlapping." His figure (p. 344, fig. 18) does not show any distinct crenulae, except for a very slight serration on the marginal ledge of the propodeum. In the material at hand, 3♀♀ (including one paratype) do not have any distinct crenulae (see Figure 492), whereas the female and male from Zululand have minute crenulae on one side of the propodeum (Figures 493, 499), which may be considered as no more than an irregular serration of the propodeal ledge.

Annecke (op. cit.) correctly placed *faurei* in *Aphytis*. As pointed out on p. 94, the number of antennal segments cannot be relied upon for a natural generic classification of *Aphytis* and related taxa. In the structure of the propodeum, as well as in other characteristics, *faurei* is much more closely related to *Aphytis* than to *Marlattiella*, the absence of distinct crenulae apparently being a secondary development. *A. faurei* is therefore regarded as an aberrant species of *Aphytis* and a bona fide member of the **chilensis** group.

## 27. **Aphytis cercinus** Compere

(Figures 502–514)

*Aphytis cercinus* Compere, 1955, Univ. Calif. Publ. Entomol., **10**:302.
*Aphytis cercinus*: Quednau, 1964, J. Entomol. Soc. S. Afr., **27**:93–94.

This interesting South African species can be recognized by the short, emarginate propodeum with elongate, nearly fingerlike, somewhat irregular crenulae, by the rather extensive pigmentation and slightly infumated wings. Female antennae rather slender; cerci with 3 long setae and 1 short seta. The hitherto undescribed male has 5-segmented antennae, with one distinct annellus and a greatly enlarged club; male genitalia normal for *Aphytis*.

*Female*. Eyes moderately coarsely setose, the setae visible under $\times 60$ magnification. Mandibles (Figure 502) well developed, essentially bidentate or with a slight dorsal truncation; maxillary palpi 2-segmented, labial palpi 1-segmented. Antennal scape (Figure 503) slender, about 5 to $6\frac{1}{2}$ times as long as wide; pedicel about $1\frac{2}{3}$ times to fully twice as long as wide, about as long as or somewhat longer than the third segment of the funicle; first funicular segment slightly trapezoidal, about $1\frac{1}{5}$ to $1\frac{3}{5}$ times as wide as long; second segment symmetrical, slightly shorter and wider than the first, $1\frac{1}{2}$ to 2 times as wide as long; third funicular segment symmetrical, elongate, about $1\frac{1}{2}$ times to nearly twice as long as wide (usually $1\frac{2}{3}$ times), bearing 2–4 longitudinal sensilla; club long, tapering, about $3\frac{4}{5}$ times as long as wide, about as long as the scape, $2\frac{1}{2}$ to $2\frac{4}{5}$ times longer and somewhat wider than the preceding segment, bearing 9–13 sensilla.

Setae on head, thorax and sides of abdomen coarse, dark, readily visible under × 30 magnification. Vertex with a pair of long setae along occipital margin, in addition to numerous smaller setae. Mesoscutum (Figure 504) with 10–14 setae, the posterior pair considerably longer and coarser than the others; each parapsis with 3 (rarely 4) setae, each axilla with 0–1 seta; scutellum with 4 setae, the discoid sensilla usually considerably closer to the anterior than to the posterior pair. Frontovertex, pronotum and mesonotal sclerites reticulate. Scutellum somewhat trapezoidal, $\frac{3}{4}$ to about full median length of mesoscutum. Metanotum (Figures 504, 505) v-shaped, rather short medially, rather weakly reticulate-aciculate except on the sides; anteromedian apodeme rather slender, about as long as or somewhat shorter than median length of metanotum.

Propodeum (Figures 504–506) short, about 3–4 times as long as the short metanotum but only about $\frac{2}{5}$–$\frac{1}{2}$ length of scutellum, rather weakly reticulate-aciculate on a narrow, prominent, bilobed median salient; crenulae 3 + 4 to 6 + 6, distinct, elongate, non-overlapping, somewhat irregular.

Second abdominal tergite (Figures 505, 506) smooth or obscurely sculptured centrally, just below median salient of propodeum; tergites III–VII strongly reticulate on the sides, bearing several strong setae in 1–2 short, transverse rows on each reticulate area, with 1 seta usually just mesad of same; third tergite (Figure 505) with a narrow, coarsely sculptured transverse band, tergites IV–VII smooth or very faintly transversely striated across center; eighth tergite (Figure 507) faintly reticulate, with a transverse row of 5–8 setae between spiracles; syntergum (Figure 507) short, v-shaped, transversely striated anteriorly, bearing about 9–14 setae in a transverse row. Cerci situated close to posterior spiracles, with 3 long setae and 1 short seta. Ovipositor shaft about $1\frac{1}{3}$ to $1\frac{1}{2}$ times as long as the middle tibia (1.35–1.49), ovipositor sheaths about $\frac{1}{3}$ length of middle tibia (0.32–0.38).

Mid-tibial spur somewhat shorter than the corresponding basitarsus (about as long in small specimens).

Forewing (Figure 508) relatively wide, about $2\frac{1}{4}$ to nearly $2\frac{1}{2}$ times as long as wide (usually about $2\frac{1}{3}$ times); marginal fringe not exceeding $\frac{1}{5}$ width of disk. Delta area with 41–101 setae in about 5–10 rows, these longer and somewhat sparser than the setae distad of speculum, well separated from row of setae along posterior margin of wing and from 2–4 rows of setae below distal half of submarginal vein; costal cell with a row of 3–6 fine setae along proximal half, and 1–4 coarse setae near apex. Submarginal vein bearing 3–4 coarse setae, an additional one just below posterior margin of vein distally, and 15–22 bullae. Marginal vein bearing 7–14 prominent, subequal setae along anterior margin, these somewhat longer than the setae in a row along center of vein.

Forewing hyaline, with a small, distinct, fuscous cloud below stigma, fading into an obscure, faint crossband; faintly infumated below junction of submarginal and marginal veins; a narrow fuscous streak along posterior margin of wing, distad of speculum.

General coloration yellow, with rather extensive black or fuscous markings as follows: Pronotum blackish submedially and on posterior margin; parapsidal sutures pronouncedly black; mesoscutum with faint fuscous spots submedially; scutellum with a pair of broad longitudinal blackish stripes submedially, these curving out on anterior and posterior margins, often forming complete circles, both anterior and posterior margins of scutellum lined with black; metanotum fuscous medially, anterior and posterior margins lined with blackish; posterior margin of propodeum broadly black between spiracles and median salient, the latter faintly fuscous. Second abdominal tergite with a conspicuous, central blackish spot below median salient of propodeum; third tergite with a conspicuous, narrow blackish crossband anteriorly on sculptured area; tergites III–VII with short, transverse fuscous stripes on side; center and apex of abdomen and ovipositor sheaths pale. Prosternum and metasternum fuscous, mesosternum rather faintly so. Antenna faintly, rather uniformly dusky, tip of club fuscous. Legs pale. Wing veins lined with brownish.

Length 0.93–1.22 mm.

*Male*. Essentially similar to the female in structure, chaetotaxis, sculpture and coloration, differing mainly in the structure of the antennae and in details of abdominal coloration.

Eyes rather finely setose. Antennae (Figure 509) 5-segmented, with one minute annellus and a greatly enlarged club; scape slender, about 5 times as long as wide, not bearing any specialized sense organs; pedicel short, about $1\frac{1}{5}$ to $1\frac{1}{3}$ times as long as wide; the single annellus distinct, triangular, about twice as wide as long or somewhat wider; funicular segment long, symmetrical, about $1\frac{4}{5}$ times as long as wide, about twice longer than the pedicel, bearing 10–14 longitudinal sensilla, the latter about as long as the entire segment and distributed evenly around it; club long, about $4\frac{1}{2}$ to $4\frac{4}{5}$ times as long as wide, $1\frac{2}{5}$ to $1\frac{3}{5}$ times longer than the scape, about $2\frac{1}{3}$ to $2\frac{3}{5}$ times longer than and as wide as the preceding segment, bearing about 20–30 longitudinal sensilla.

Mesoscutum (Figure 510) with 12–16 setae, each parapsis with 2–3 setae. Propodeum (Figure 510) somewhat shorter than in the female, about $2\frac{4}{5}$ to $3\frac{1}{2}$ times as long as the metanotum, $\frac{2}{5}$ length of scutellum; crenulae (Figures 511, 512) $3+5$ to $6+7$, as in the female. Cerci with 2 long setae and 1 short seta.

Genitalia (Figure 513) about $\frac{2}{3}$ to $\frac{3}{4}$ length of middle tibia (0.67–0.77), rather robust, with a distinct rod between the apodemes; digital sclerites robust, about $\frac{1}{3}$ the combined length of aedeagus and apodemes (0.31–0.36).

Forewing (Figure 514) as in the female but even somewhat broader, not more than $2\frac{1}{4}$ times as long as wide; marginal fringe not exceeding $\frac{1}{6}$ width of disk.

Forewing hyaline, with a small fuscous spot below stigma, faintly infumated below junction of submarginal and marginal veins.

Coloration of head and thorax very similar to that of the female; the pair of submedian stripes on scutellum very faint or absent, the fuscous blotches near anterior and posterior margins present (Figure 510); propodeum, second and third abdominal tergites as in the female; tergites IV–V pale, sixth tergite faintly infuscated centrally, pale on the sides; seventh tergite with a transverse fuscous blotch, sides pale; eighth tergite with 2 faint, sublateral fuscous blotches mesad of spiracles. Antenna faintly dusky, funicle and club appearing darker due to the presence of numerous longitudinal sensilla.

Length 0.79–1.19 mm.

Redescribed from 6♀♀ (holotype and paratypes), reared by E. W. Rust from *Aspidiotus* sp. on mistletoe, Durban, Natal, South Africa, December 18, 1925; 3♀♀, 1♂, reared by J. Munting from *Africaspis terminaliae* Munting on *Terminalia sericea*, Warmbaths, Transvaal, South Africa, November 5, 1964; 41♀♀, 62♂♂, reared by H. P. Insley from *Tecaspis visci* (Brain) on *Loranthus* sp., Heuningsvlei, N.C.P., South Africa, September 1968; all mounted in Hoyer's medium on numerous slides.

Type series in the collection of the Division of Biological Control, University of California, Riverside.

**Notes.** The type series is in rather poor shape, the specimens somewhat crushed, the antennal proportions differing somewhat from those of the abundant, well-preserved recent specimens from South Africa listed above. Also, the ovipositor appears somewhat longer, and the forewing pattern is a little more pronounced in the type series.

Compere (1955) described *cercinus* with "dorsum of head, thorax and abdomen yellow, grading to white on the sides and underparts." This coloration pattern was probably observed in uncleared specimens before mounting, and is not readily evident in the cleared, slide-mounted material.

The male of *cercinus* had not been previously described.

*A. cercinus* presents a peculiar combination of primitive and advanced characters: the short propodeum, the cerci with 3 long setae, and the rather extensive pigmentation may be considered primitive, whereas the nearly hyaline wings, the subequal first 2 funicular segments of the female, and the lack of specialized sense organs on the antennal scape of the male are probably advanced characters. The male genitalia of *cercinus* differ from those of other members of the **chilensis** group in being of the normal *Aphytis* type.

## 28. **Aphytis chilensis** Howard

(Figures 97, 99, 101, 103, 104, 109, 110, 115, 117, 118, 121, 122, 128, 131, 133–137, 145, 151, 153, 155, 164–166, 169–172, 183, 184, 198, 202, 206, 208, 210, 212, 223, 225, 227–231, 234, 235; and 515–530)

*Aphytis chilensis* Howard, 1900, Canad. Entomol., **32**: 168; Howard, 1902, Rev. Chilena Hist. Nat, **6**: 172; Timberlake, 1924, Proc. Hawaii. Entomol. Soc., **5**: 411–412; Compere, 1955, Univ. Calif. Publ. Entomol., **10**: 283–285; Quednau, 1964, J. Entomol. Soc. S. Afr., **27**: 94–96; Gerson, 1967, Ann. Entomol. Soc. Amer., **60**: 1116–1118.

*Aphelinus longiclavae* Mercet, 1911, Assoc. Espan. Prog. Cienc. Congr., Valencia 1910, **5**: 128; 1912, Trab. Mus. Cienc. Nat., Madrid, **10**: 76–80.

*Trichogrammatoidea signiphoroides* Brèthes, 1913, An. Mus. Nac. Nat. Buenos Aires, **24**: 99–100.

*Aphelinus capitis* Rust, 1915, Entomol. News, **26**: 73–74; Gahan, 1924, Proc. U.S. Nat. Mus., **64**: 11.

*Aphelinus signiphoroides*: Brèthes, 1916, An. Mus. Nac. Hist. Nat. Buenos Aires, **27**: 429.

*Aphelinus chilensis*: Mercet, 1921, R. Soc. Espan. Hist. Nat., 50th Anniv. Vol., pp. 299–300.

*Aphytis (Aphytis) longiclavae*: Mercet, 1932, Eos **8**: 362–363; De Santis, 1948, Rev. Mus. La Plata (n.s.) Zool., **5**: 116–120.

*Aphytis riadi* Delucchi, 1964, Rev. Path. Veg. Entomol. Agr., Fr., **43**: 136–139; Gerson, 1967, Ann. Entomol. Soc. Amer., **60**: 1116–1118.

For a more extensive bibliography, see De Santis (1948) and Peck (1963).

This cosmopolitan, uniparental species may be separated from other members of the **chilensis** group by the long propodeum with overlapping crenulae, and by the reduced first funicular segment of the female. In the rare males, the antennal scape does not bear specialized sense organs, the first 2 funicular segments are greatly reduced but distinct, the long club bears a small sensory area, and the digital sclerites of the genitalia are short.

*Female*. Eyes (Figures 97, 99, 103, 104) moderately coarsely setose, the setae about as coarse as in *cercinus* or *anomalus*, less so than in *cochereaui* (larger specimens usually have coarser setae than smaller specimens). Mandibles (Figures 109, 110) well developed, with 2 denticles and a dorsal truncation; maxillary palpi (Figure 110) 2-segmented, labial palpi 1-segmented. Antennal scape (Figures 115, 515) $4\frac{1}{2}$ to $5\frac{1}{2}$ times as long as wide; pedicel about $1\frac{3}{5}$ to $1\frac{3}{4}$ times as long as wide, as long as or somewhat longer than the third segment of the funicle; first funicular segment (Figures 121, 122, 515) small, triangular, about $1\frac{2}{3}$ to $1\frac{4}{5}$ times as wide as long; second segment symmetrical, considerably larger than the first segment ($1\frac{3}{5}$ to $1\frac{3}{4}$ times longer, $1\frac{1}{2}$ times wider), about $1\frac{1}{2}$ to $1\frac{3}{5}$ times as wide as long; third funicular segment symmetrical, distinctly wider than the preceding segment, $1\frac{1}{6}$ to nearly $1\frac{1}{2}$ times as long as wide, bearing 2–4 longitudinal sensilla; club long, tapering, $3\frac{1}{4}$ to over 4 times as long as wide, considerably longer than the scape, 3 to $3\frac{1}{2}$ times longer and somewhat wider than the preceding segment, bearing 9–16 longitudinal sensilla.

Setae on head, thorax, and sides of abdomen dark, coarse, readily visible under $\times 30$ magnification. Vertex with 2 pairs of long setae along occipital margin, in addition to numerous shorter setae. Mesoscutum (Figures 151, 516–518) with 10–22 setae (usually 16–18), the posterior pair and 1 seta at each antero-lateral corner longer and coarser than the others; each parapsis with 2 relatively short setae, each axilla with 1 seta; scutellum with 4 setae, the discoid sensilla about equidistant from the two pairs. Frontovertex, pronotum and mesonotal sclerites reticulate. Scutellum (Figures 151, 169, 516) oval, relatively short, $\frac{2}{3}$ to $\frac{3}{4}$ median length of mesoscutum, appearing slightly strigose medially. Metanotum reticulate except on the sides; anteromedian apodeme robust, distinctly bifid at apex, about as long as or slightly shorter than median length of metanotum.

Propodeum (Figures 169–172, 516, 519) long, $2\frac{1}{2}$ to $3\frac{1}{2}$ times as long as the metanotum, usually about $\frac{4}{5}$ length of scutellum (0.75–0.88), reticulate-strigose just mesad of the spiracles and on a relatively wide central area; crenulae 4 + 5 to 8 + 8, relatively large, rounded, overlapping, the 2 sets forming a continuous row.

Second abdominal tergite smooth; tergites III–VII reticulate on the sides, with several setae in a short transverse row on, and just mesad of, each reticulate area; seventh tergite bearing 2–4 setae in a transverse row between the reticulate areas; eighth tergite delicately reticulate, with a transverse row of 6–8 setae between spiracles; syntergum (Figures 208, 520) triangular bearing 2–4 irregular rows of setae. Cerci (Figures 208, 210, 520) situated considerably closer to posterior spiracles than to tip of abdomen, with 2 long setae and 1 short seta. Ovipositor shaft about $1\frac{3}{5}$ to $1\frac{4}{5}$ times as long as the middle tibia (1.61–1.86), ovipositor sheaths about $\frac{1}{2}$ length of middle tibia (0.44–0.53).

Mid-tibial spur (Figure 227) about $\frac{4}{5}$ length of the corresponding basitarsus.

Forewing (Figures 228–231, 234, 235, 521) fully $2\frac{1}{2}$ times as long as wide or slightly longer; marginal fringe not exceeding $\frac{1}{4}$ width of disk (usually $\frac{1}{6}$ or shorter). Delta area with 30–99 setae in 5–8 rows, these considerably longer and sparser than the setae distad of speculum, not clearly separated from row of setae along posterior margin of wing, very well separated from 1–2 rows of setae below distal two thirds of submarginal vein; costal cell with a row of fine setae, separated by a short space from 2–3 coarse setae near apex. Submarginal vein bearing 4–7 coarse setae and 13–20 bullae. Marginal vein bearing 6–16 prominent, subequal setae along anterior margin (usually 11–13), these decreasing only slightly in length toward apex of vein, about $1\frac{1}{2}$ times as long as the setae in a row along center of vein.

Forewing with a distinct dark spot below the stigma which fades posteriorly into a broad, obscure, faintly fuscous crossband; a fuscous cloud below apex of submarginal vein; an arcuate, fuscous crossband at base of delta, the bare area below junction of submarginal and marginal veins paler; a narrow blackish streak along posterior margin of wing, distad of speculum; apical margin of wing distinctly paler than rest of disk.

General coloration pale yellowish, with fuscous markings. A narrow, transverse, strongly pigmented black bar on occiput, on each side of foramen; sometimes also faintly fuscous lines around oral margin. Central part of pronotum fuscous; anterior margin of mesoscutum, parapsidal sutures, posterior margin of scutellum, and posterior margin of metanotum centrally, lined with blackish; mesoscutum and scutellum suffused with fuscous, median line pale (scutellum usually with a pair of submedian spots anteriorly); propodeum faintly suffused with fuscous on median reticulate area, narrowly blackish along posterior margin, including the posterior margin of the median salient and the crenulae. Second abdominal tergite, sometimes also third tergite, suffused with fuscous; tergites III–VII with dark transverse stripes on sides; center and apex of abdomen immaculate; ovipositor sheaths usually immaculate, rarely tipped with blackish (as in some specimens from California and all specimens from South Africa). Thoracic sterna (Figure 522) strongly infuscated, especially the mesosternal furca ("Y") and triangular plate. Lateral margins of endophragma lined with blackish anteriorly; ovipositor plates immaculate. Antennal scape pale, subsequent segments dusky; club more strongly infuscated, apical part distinctly paler. All tibiae and fore tarsi slightly infuscated. Wing veins lined with brownish.

Length 0.50–1.08 mm.

*Male.* Essentially similar to the female, differing mainly in the structure of the antennae and in some details of abdominal structure, sculpture and coloration.

Antennae (Figures 523–525) appearing 4-segmented, with the first 2 funicular segments vestigial. Scape slender, 4 to $4\frac{1}{2}$ times as long as wide; pedicel short, $1\frac{1}{3}$ to $1\frac{1}{2}$ times as long as wide, somewhat wider at apex than maximum width of scape; first 2 funicular segments reduced to nearly invisible ringlike rudiments; third funicular segment usually distinctly longer than wide (in some slides appearing as long as wide, compare Figures 523 and 524), considerably longer than the pedicel, bearing numerous longitudinal sensilla (about 8–15 in some specimens); club enormously enlarged, $3\frac{1}{3}$ to $4\frac{1}{2}$ times as long as wide, considerably longer than the scape, 3–4 times longer than and about as wide as the preceding segment, bearing numerous long, slender longitudinal sensilla; a small, specialized sensory area, bearing minute setae, on ventral aspect of club (Figure 525).

Structure, chaetotaxis and sculpture of head and thorax as in the female. Propodeum somewhat shorter than in the female, about $\frac{3}{4}$ length of scutellum (0.69–0.83), $2\frac{1}{2}$ to $3\frac{1}{3}$ times as long as the metanotum; crenulae 4 + 4 to 7 + 7, as in the female. Abdominal tergites III–VII reticulate on the sides as in the female, delicately transversely striated between reticulate areas, more coarsely so on seventh tergite. Chaetotaxis of abdomen as in the female. Genitalia (Figure 526) usually nearly as long as the middle tibia; aedeagal apodemes long; digital sclerites short, stout, rather strongly sclerotized, about $\frac{1}{6}$ the combined length of aedeagus and apodemes (0.12–0.18).

Mid-tibial spur nearly as long as the corresponding basitarsus.

Forewing somewhat wider than in the female, $2\frac{1}{4}$ to $2\frac{1}{2}$ times as long as wide. Delta area with 22–60 setae in 4–7 rows (Figure 527 shows the wings of a small specimen); marginal vein bearing 7–12 prominent setae along anterior margin; otherwise as in the female. Wing pattern as in the female.

Coloration similar to that of the female, but generally more extensively fuscous. Head with black lines on genal sutures in addition to those on occiput. Abdominal tergites II–VIII fuscous, without stripes on the sides; seventh tergite with a conspicuous, transverse black blotch centrally. Antennal scape pale, pedicel faintly dusky, third funicular segment and club uniformly fuscous. Tip of middle tibia conspicuously black.

Length 0.54–0.93 mm.

Redescribed from hundreds of specimens. *A. chilensis* is a nearly cosmopolitan, thelytokous species. It is known mainly as a parasite of the oleander scale, *Aspidiotus nerii* Bouché [= *A. hederae* (Vallot)]. At hand are numerous specimens, reared from this host in California, Mexico, Europe, the Middle East and South Africa, as well as specimens from *Aspidiotus hedericola* Leonardi from Spain, *Chrysomphalus dictyospermi* (Morgan) from Argentina, *Hemiberlesia lataniae* (Signoret) from California and Greece, *Hemiberlesia rapax* (Comstock) from California and Mexico, *Aonidiella aurantii* (Maskell) from California, *Aonidia lauri* (Bouché) from Greece, *Carulaspis caruelii* (Targioni-Tozzetti) [= *C. minima* (Targioni-Tozzetti)] from California, *Diaspis echinocacti* (Bouché) from California, *Furchadiaspis zamiae* (Morgan) from California, *Parlatoria pergandii* Comstock from Lebanon, *Parlatoria oleae* (Colvée) from Italy and Greece, *Parlatoria pittospori* Maskell from New Zealand, *Phenacaspis pinifoliae* (Fitch) from California, and *Rolaspis* sp. nr. *incisa* Munting from South Africa; also specimens reared from unidentified scale insect hosts in South Africa and Australia. The latter include 11♀♀, collected by G. Compere "on Laurestine and Pittosporum", Perth, Western Australia, no date (together with *capillatus*; remounted from one balsam slide in the U.S. National Museum; #805 in Compere's original ledger).

**Notes.** The pupa of *chilensis* (Figure 530) is uniformly black, except for the venter of the abdomen, which is partly grayish yellow.

Compere (1955) related the interesting history of this cosmopolitan species (see also the historical review in the present monograph, p. 7). Howard (1900) described *chilensis* from a single female specimen, reared from *Aspidiotus nerii* Bouché on ivy (presumably English ivy, *Hedera helix* L.), collected at Rancagua, Chile. He erroneously considered this specimen to have only 5 segments in the antennae (1121) and established the new genus *Aphytis* to accommodate it. We have studied the holotype specimen, mounted on a slide in the U.S. National Museum. Although the specimen is not cleared and the antennae are folded beneath the head, with modern microscopy it is not difficult to distinguish the minute first funicular segment overlooked by Howard. The antenna of *chilensis* is indeed 6-segmented, as assumed by Timberlake (1924) and Compere (1955). The propodeal crenulae and other distinguishing characters are also visible in the holotype, and there remains no doubt whatever that this unique specimen is identical to the species now commonly recognized by that name.

Specimens of *longiclavae* (Mercet) (including 1♀ from Madrid, determined by Mercet) and *capitis* (Rust) (holotype and paratype) in the U.S. National Museum are undoubtedly conspecific with *chilensis*. We have not examined the type of *signiphoroides* (Brèthes), and have followed Compere (1955) in regarding that species as yet another synonym of *chilensis*.

*A. riadi* was described by Delucchi (1964) from material reared from chaff scale, *Parlatoria pergandii* Comstock, at Fanar, Lebanon. We have studied a paratype specimen (Figures 528, 529) and agree with Gerson (1967) that *riadi* is indeed conspecific with *chilensis*.

As demonstrated by Gerson (1967) and in the foregoing description, *A. chilensis* exhibits a wide range of morphological variation, especially in chaetotaxis, which is mainly related to the size of specimens (compare, for instance, Figures 517 and 518, showing variation in mesoscutal chaetotaxis). Thus, the number of setae on the pronotum may vary from 4 in small specimens to 13 in large specimens of the same clone (all descendants of a single female), and the number of setae in the forewing delta may vary from 17 to 86. On the other hand, basic color patterns, such as the pigmentation of the head or thoracic sclerites, appear to be much more stable in *chilensis* and other species of *Aphytis* than are certain structural characters.

*A. chilensis* resembles several species in the **vittatus** group, such as *costalimai*, *cochereaui* and *capillatus*, in the reduced, triangular first funicular segment of the female. As mentioned earlier, we do not consider this character to be of subgeneric value, as suggested by various previous authors. Nevertheless, this peculiar similarity may be an indication that *chilensis* is indeed related to these species. Inasmuch as the male antennae of *chilensis* also resemble those of *cochereaui* and *capillatus*, it is quite conceivable that *chilensis* evolved from an ancestral form resembling those members of the **vittatus** group. On the other hand, *chilensis* is obviously very closely related to the **proclia** group in general coloration and habitus, differing from the members of this group mainly in the distinctive structure of the male antennae. Thus, this interesting species may represent a connecting link between the **vittatus** and **proclia** groups.

**Biology.** The developmental history of *A. chilensis* was recently described in detail by Rosen and Eliraz (1978) (see also p. 17 here). The development of this species is relatively slow; the life cycle from oviposition to adult emergence is 45–61 days (average 48.8) at 19°C, 24–30 days (average 25.5) at 24°C, and 17–21 days (average 18.6) at 28°C; total mortality occurred at a constant temperature of 32°C. The theoretical threshold of development was calculated as 14.1°C; a thermal constant of 270.2 days-degrees is required for the completion of development. At 28°C, egg incubation takes 2–3 days, larval development (including the prepupal stage) 10–12 days, and pupal development 6–7 days. The species is thelytokous, males being very rare.

In laboratory studies in California, *A. chilensis* females showed marked preference for the oleander scale, accepted the cactus scale and (rarely) the latania scale as hosts, but would not oviposit in the California red scale or in the yellow scale. Mean progeny on oleander scale at 26.7°C was 26 per female.

**Potential Value for Biological Control.** Although other species of *Aphytis* appear to be more effective natural enemies of the oleander scale, *A. chilensis* should not be ignored as a potential candidate for introduction against that pest in areas where this parasite is not yet present.

## 29. **Aphytis columbi** (Girault)

(Figures 531–546)

*Aphelinus columbi* Girault, 1932, New Lower Hymenoptera from Australia and India. Priv. Publ., Brisbane, Australia, 6 pp., page 1.
*Marietta distonota* Girault, 1932, ibid: page 1; n. syn.
*Aphelinus stellaris* Girault, 1933, Some Beauties Inhabitant Not of Commercial Boudoirs But of Nature's Bosom, Notably New Insects. Priv. Publ., Brisbane, Australia, 5 pp., page 4; n. syn.
*Aphytis columbi*: DeBach and Rosen, 1976, Ann. Entomol. Soc. Amer., **69**: 542.

This Australian species is very closely related to *chilensis*, differing from the latter species mainly in being apparently arrhenotokous, with rather abundant males, in the generally larger size, and in the paler coloration of the female. The male is virtually indistinguishable from *chilensis*. To avoid unnecessary repetition, *columbi* is redescribed below relative to *chilensis*.

*Female.* Eyes about as coarsely setose as in *chilensis*. Mouthparts as in *chilensis*. Antennae (Figure 531) essentially as in *chilensis*; pedicel about $1\frac{1}{2}$ times as long as wide or somewhat longer, as long as or a little shorter than the third segment of the funicle; first funicular segment triangular, considerably smaller than the second; third funicular segment nearly $1\frac{1}{2}$ times as long as wide, bearing 2–4 longitudinal sensilla; club $3\frac{1}{2}$ to 4 times as long as wide, 3 to $3\frac{1}{3}$ times longer and a little wider than the preceding segment, bearing 10–15 longitudinal sensilla.

Structure, chaetotaxis and sculpture of head and thorax as in *chilensis*. Mesoscutum (Figure 532) with 14–28 setae. Propodeum (Figures 532–535) somewhat longer than in *chilensis*, $3\frac{1}{3}$ to 4 times as long as the metanotum, usually nearly as long as the scutellum (0.91–1.07); sculpture as in *chilensis*; crenulae 6 + 6 to 8 + 8, overlapping, as in *chilensis*. Abdominal tergites, cerci and ovipositor essentially as in *chilensis*.

Mid-tibial spur as in *chilensis*.

Forewing (Figure 536) a little longer than in *chilensis*, $2\frac{2}{3}$ to $2\frac{3}{4}$ times as long as wide, otherwise as in *chilensis*; submarginal vein bearing 4–6 coarse setae and 17–23 bullae; marginal vein bearing 11–15 prominent, subequal setae along anterior margin. Forewing pattern similar to that of *chilensis*, but somewhat paler.

General coloration considerably paler than in *chilensis*, less extensively marked with fuscous. The black bar on the occiput is sometimes absent or partly obliterated. Dorsum of thorax as in *chilensis*, but the mesoscutum and scutellum paler, less extensively suffused with fuscous; propodeum narrowly marked with black only on posterior margin of median salient, including crenulae (Figure 534). Second abdominal tergite very faintly infuscated; sides of tergites III–VII immaculate, or with a faint dusky spot, without the distinct transverse stripes of *chilensis*; however, the dark setae on the sides of the tergites sometimes produce a similar effect; tips of ovipositor sheaths blackish. Thoracic sterna (Figure 537) faintly dusky, without the distinct, black mesosternal furca of *chilensis*. Endophragma as in *chilensis*. Antenna as in *chilensis*, but the club is uniformly, faintly dusky, the apical part not constrastingly pale. Legs as in *chilensis*.

Length 0.96–1.23 mm.

*Male*. Virtually indistinguishable from *chilensis* in structure, chaetotaxis, sculpture and coloration. Antennae (Figures 538–540) as in *chilensis*, the specialized sensory area on the ventral aspect of the club somewhat more distinct. Mesoscutum with 15–24 setae, apparently somewhat more slender than in *chilensis*. Propodeum apparently a little longer than in *chilensis*, about $\frac{4}{5}$ length of scutellum (0.80–0.86), 3 to $3\frac{3}{4}$ times as long as the metanotum. Abdomen (Figures 541, 542) and genitalia (Figures 542, 543) essentially as in *chilensis*. Forewing (Figure 544) as in *chilensis*; marginal vein bearing 11–15 prominent, subequal setae along anterior margin.

Length 0.86–0.95 mm.

Redescribed from the holotype of *Aphelinus columbi* Girault (♀, Indooroopilly, Australia, March 15, 1931; Queensland Museum type #4944); the type series of *Marietta distonota* Girault (7♂♂, no data, Queensland Museum type # 3756; 1♀, "reared from a diaspine scale on gum leaves", Indooroopilly, Australia, March 27, 1931; Queensland Museum type #4965); additional specimens determined by A. A. Girault as *M. distonota* (♂, "ex *Diaspis rosae*" [= *Aulacaspis rosae* (Bouché)], Brisbane, Australia, May 18, 1932; 17♀♀, 5♂♂, Booroy, Australia, February 1933; 1♀, Brisbane, March; 1♀, 1♂, Brisbane, June 1933); the type of *Aphelinus stellaris* Girault (♂, no data); and numerous recent specimens, all from Australia, as follows: 2♀♀ collected by S. Neser ("with soft scale") on *Hakea sericea*, Picnic Point, Wilson's Promontory, Victoria, July 1964; 1♀, 1♂, reared by W. Thompson from the purple scale, *Cornuaspis* [= *Lepidosaphes*] *beckii* (Newman), on citrus, Sydney, New South Wales, June 14, 1966; 1♀, same data, July 12, 1966; 61♀♀, 5♂♂, reared by G. J. Snowball from *C. beckii* on citrus, Oxford Falls, near Sydney, October 9, 1966; 33♀♀, 2♂♂, same data, September–October 1966; 12♀♀, 6♂♂, reared by W. Thompson, same host and locality, October 1966; 5♀♀, 4♂♂, reared by G. J. Snowball, same host and locality, April 4, 1967; 1♂, reared by G. J. Snowball, same host, Lindfield, New South Wales, January 25, 1968; 1♀, 4♂♂, reared by G. J. Snowball, same host, Narara, New South Wales, September 4, 1968; 1♀, 1♂, reared by W. Thompson from the Florida red scale, *Chrysomphalus aonidum* (L.), on grapefruit, Sydney, June 1966; 4♀♀, 2♂♂, reared by D. P. Sands from the same host on *Beilschmiedia elliptica*, Botanical Gardens, Sydney, June 26–July 1, 1968.

**Notes.** The pupa of *columbi* is entirely black.

We at first considered this Australian parasite of the purple scale to be an undescribed species. However, a study of material loaned by the Queensland Museum revealed that this species was described *three* times by Girault, under two different genera. *Aphelinus columbi* and *Marietta distonota* were described on the same page of a privately published paper (Girault, 1932). *A. columbi* was described from a single female specimen; the description of *M. distonota* was apparently based on male specimens (Figure 545), although the type material includes some female specimens as well (Figure 546). *Aphelinus stellaris* was described by Girault (1933) from a single male specimen. Comparison of the types leaves no doubt that these 3 species are conspecific, as is also the abundant recent material from Australia listed above. Since the description of *A. columbi* preceded that of *M. distonota* by a few lines on the page, priority should be given to the name *columbi*.

The name *distonota* is hereby changed to *distonotus*, to conform with the gender of *Aphytis*.

Most of the structural differences separating *columbi* from *chilensis*, such as the longer propodeum, longer forewing and somewhat more numerous setae, may be correlated with the generally larger size of *columbi*. Actually, most of these characters appear to fall within the range of variation of *chilensis*. The paler pigmentation of the female and the uniformly dusky antennal club appear to be the only valid, consistent morphological characters separating *columbi* from *chilensis*. The abundance of males in the samples of *columbi* seems to indicate a biological difference between the two species, as does the distinctive host preference: at present, *columbi* is known mainly from purple scale, *chilensis* mainly from oleander scale, and never to our knowledge from purple scale even where both purple scale and *chilensis* are numerous and sympatric.

We assume that in *Aphytis*, as in most other parasitic Hymenoptera, arrhenotoky is the original mode of reproduction while thelytoky represents a more recent evolutionary development. Therefore, if *columbi* is indeed arrhenotokous, as indicated by the relative abundance of males in the material at hand, it may be regarded as representing the ancestral species from which the thelytokous, cosmopolitan *chilensis* apparently evolved. On the other hand, the relatively pale pigmentation of the female of *columbi* is considered an evolutionally "advanced" character, which was probably acquired only after the speciation of *chilensis* from the ancestral stock. This hypothesis implies that the thelytokous *chilensis*, with presumably less genetic variation, has evolved at a slower rate and hence has retained the ancestral pigmentation lost by the isolated, biparental, faster evolving *columbi*. The fact that numerous samples of *chilensis*, obtained from diverse hosts and habitats in various zoogeographical regions, are all remarkably uniform in coloration appears to corroborate this hypothesis (see also p. 113).

Since both the purple scale and the Florida red scale are presumably not native to Australia, *A. columbi* must have originally evolved on another, presumably endemic, host on this continent.

**Potential Value for Biological Control.** *A. columbi* should be considered as a potentially valuable addition to the parasite complex of purple scale, wherever this serious pest of citrus has not been brought under complete biological control.

## RELATED SPECIES

The following species is considered to be related to, but not a bona fide member of, the **chilensis** group.

### 30. **Aphytis antennalis** n. sp.

(Figures 547–551)

This aberrant Oriental species resembles *faurei* in the peculiar structure of the antennae, with a conspicuously elongate funicular segment and club. However, it has one distinct annellus, so that the antenna is clearly 5-segmented, and differs markedly from *faurei* and other members of the **chilensis** group in all other significant characters. The relatively few mesoscutal setae, the long propodeum with a distinct median salient and small, nonoverlapping crenulae, the long ovipositor with narrow plates, the broad hyaline wings with a sparsely setose delta area, and the fuscous crossbands on the abdominal tergites are additional distinguishing characters.

*Female*. Eyes finely setose. Mandibles well developed; maxillary palpi 2-segmented, labial palpi 1-segmented. Antennae (Figure 547) 5-segmented (1121), with one minute annellus and an enormous club; scape short, 4 times as long as wide, a little less than $\frac{1}{2}$ length of club; pedicel relatively small, about $1\frac{3}{10}$ times as long as wide; first funicular segment a minute, triangular annellus, about twice as wide as long; second funicular segment very large, much larger than the first, about $2\frac{1}{4}$ times as long as wide, $2\frac{3}{5}$ times as long as the pedicel, bearing numerous longitudinal sensilla; club enormously enlarged, $4\frac{1}{2}$ times as long as wide, $2\frac{3}{8}$ times as long as the preceding segment, bearing numerous longitudinal sensilla.

Setae on head, thorax, and sides of abdomen moderately coarse, dark, visible under $\times 30$ magnification. Vertex with 2 pairs of long setae along occipital margin, in addition to numerous shorter setae. Mesoscutum (Figure 548) with 10 subequal, rather long setae, each parapsis with 2 short setae, each axilla with 1 seta; scutellum with 4 setae, the discoid sensilla about equidistant from the two pairs. Frontovertex and pronotum reticulate, mesoscutum and scutellum rather uniformly so; parapsides and axillae more delicately reticulate, the cells more elongate than on the mesoscutum. Scutellum nearly $\frac{4}{5}$ median length of mesoscutum. Metanotum (Figure 549) relatively short, reticulate except on the sides; anteromedian apodeme rather slender, somewhat longer than median length of metanotum.

Propodeum (Figure 549) long, 5 times as long as the metanotum, about $\frac{7}{10}$ length of scutellum, reticulate on a triangular, prominent median salient and on the sides; crenulae 5 + 7, small, distinct, rounded, nonoverlapping, the 2 sets distinctly separated.

Second abdominal tergite transversely striated anteriorly, on both sides of median salient of propodeum; tergites III–VII reticulate on the sides, bearing a few (3–8) setae on each reticulate area; third tergite delicately transversely striated across; tergites IV–VII not sculptured between reticulate areas, but tergal plates distinct; seventh tergite bearing 2 setae between the reticulate areas; eighth tergite delicately reticulate across, with a group of small setae below each spiracle and 4 longer setae in a transverse row between the spiracles; syntergum triangular, faintly reticulate-aciculate, bearing an arcuate row of 6 setae. Cerci situated somewhat closer to posterior spiracles than to tip of abdomen, with 2 long setae and 1 short seta. Ovipositor and sheaths long, plates relatively narrow (Figure 550); shaft about twice as long as the middle tibia (2.08), sheaths nearly $\frac{2}{3}$ length of middle tibia (0.63).

Mid-tibial spur about as long as the corresponding basitarsus.

Forewing (Figure 551) relatively broad, about $2\frac{1}{4}$ times as long as wide; marginal fringe not exceeding $\frac{1}{8}$ width of disk. Delta area rather sparsely setose, bearing 34 setae in 4 rows, these well separated from a distinct row of setae along posterior margin of wing, widely separated from a few scattered setae below submarginal vein; costal cell with a row of fine setae along proximal half, and 1 coarse seta near apex. Submarginal vein bearing 2 coarse setae, the proximal seta a little less than $\frac{3}{5}$ length of the distal, 1 much shorter seta distad, 1 long seta at junction of submarginal and marginal veins, and 21–22 bullae. Marginal vein bearing 8–9 prominent, subequal setae along anterior margin, these about $1\frac{2}{3}$ times as long as the setae in a row along center of vein.

Forewing hyaline, slightly infumated below junction of submarginal and marginal veins and just below stigma; a narrow fuscous streak along posterior margin of wing distad of speculum.

General coloration yellow, with some fuscous markings. Occiput dark brown on each side of foramen; posterior margin of scutellum dark brown; lateral and posterior margins of propodeum and crenulae fuscous, as is the central part of second abdominal tergite below median salient of propodeum; faintly infuscated crossbands on abdominal tergites III–VII; thoracic sterna fuscous. Antenna yellow, faintly dusky, second funicular segment more strongly so. Legs and ovipositor sheaths yellow. Wing veins pale, narrowly lined with brownish.

Length 1.40 mm.

*Male*. Unknown.

Described from a single ♀ specimen (holotype), collected by S. E. Flanders on citrus, Taipo, Hong Kong, March 26, 1954 (S&R #1227–3). Host unknown.

Type in the collection of the Division of Biological Control, University of California, Riverside.

**Notes.** Although *antennalis* resembles the type species of *Marlattiella* in the structure of the antennae, in the mesoscutal chaetotaxis and in the narrow ovipositor plates, the presence of a well-developed propodeum with distinct marginal crenulae, as well as the general habitus, unmistakably place this aberrant species in the genus *Aphytis*. Since the peculiar, 5-segmented antennae superficially resemble those of female *faurei* and are very similar to those of male *cercinus*, *antennalis* is provisionally listed here as associated with the rather heterogenous **chilensis** group.

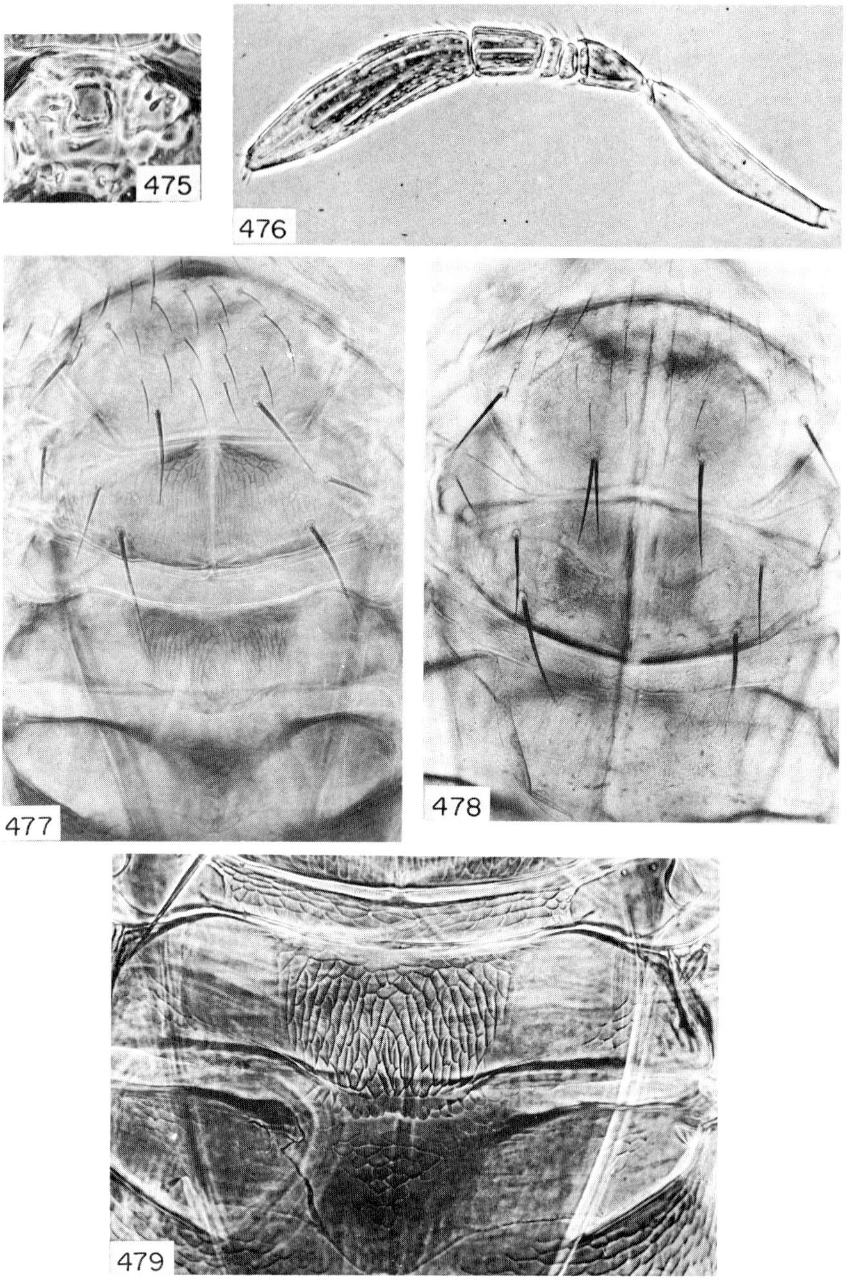


Figures 475–479. *Aphytis merceti* Compere, ♀
475. Mandibles. 476. Antenna. 477. Thorax, propodeum and base of gaster. 478. Thorax, showing freak of chaetotaxis on mesoscutum (holotype). 479. Metanotum, propodeum and base of gaster.

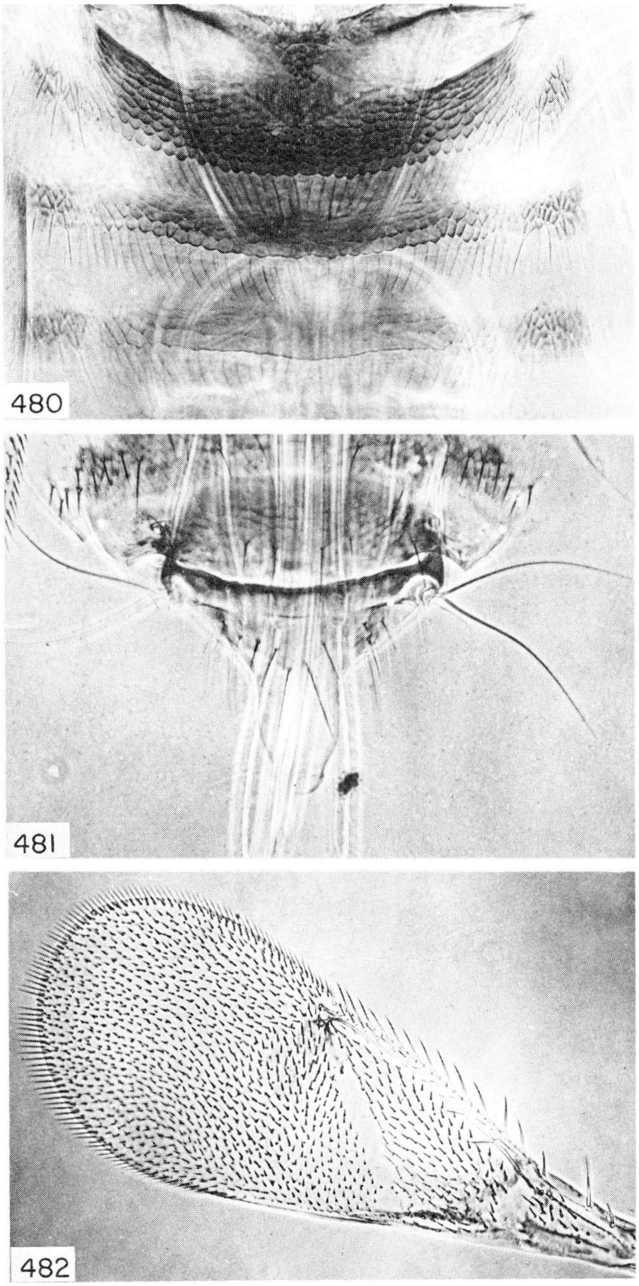

Figures 480–482. *Aphytis merceti* Compere, ♀
480. Abdominal tergites II–V, showing distinctive sculpture. 481. Posterior abdominal tergites, showing spiracles, syntergum and cerci. 482. Forewing.

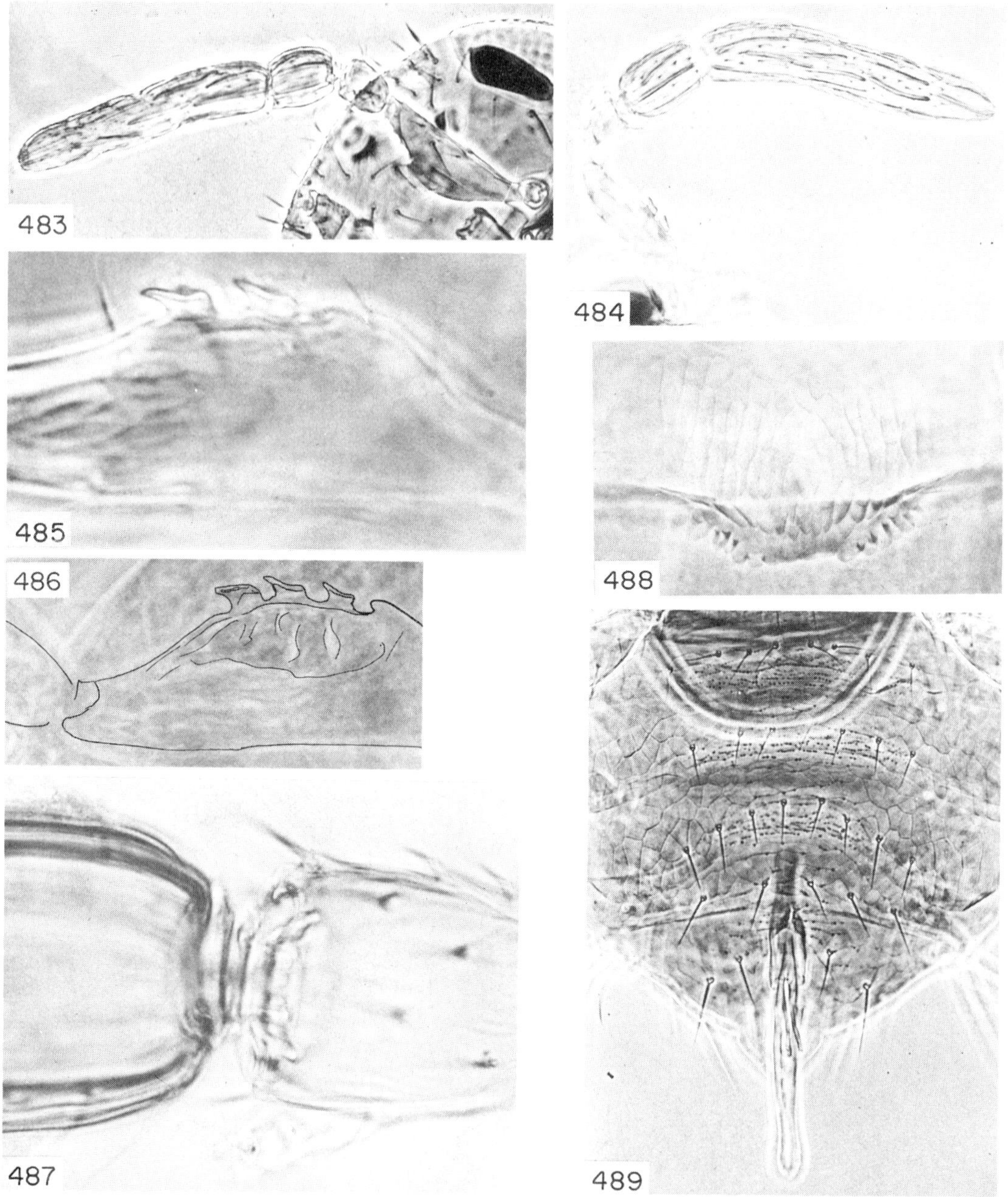

Figures 483–489. *Aphytis merceti* Compere, ♂
483, 484. Antennae. 485, 486. Antennal scape, showing sense organs (Figure 486 is retouched). 487. Vestigial annelli in antennal funicle. 488. Propodeal crenulae. 489. Venter of abdomen and genitalia.

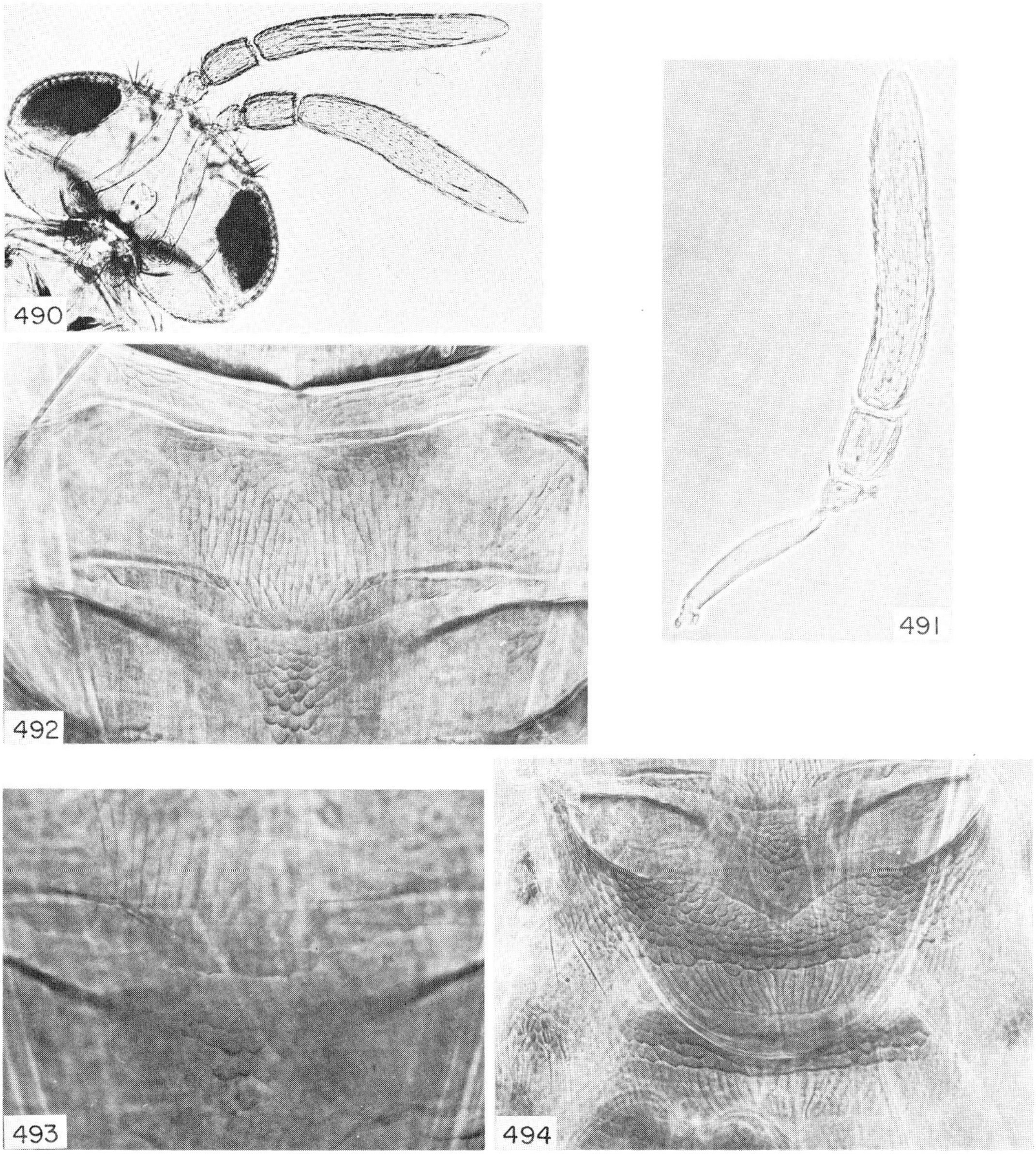


Figures 490–494. *Aphytis faurei* Annecke, ♀
490. Head and antennae. 491. Antenna (paratype). 492. Metanotum, propodeum and second abdominal tergite. 493. Propodeum; note minute crenulae on one side. 494. Abdominal tergites II–VI, showing distinctive sculpture.

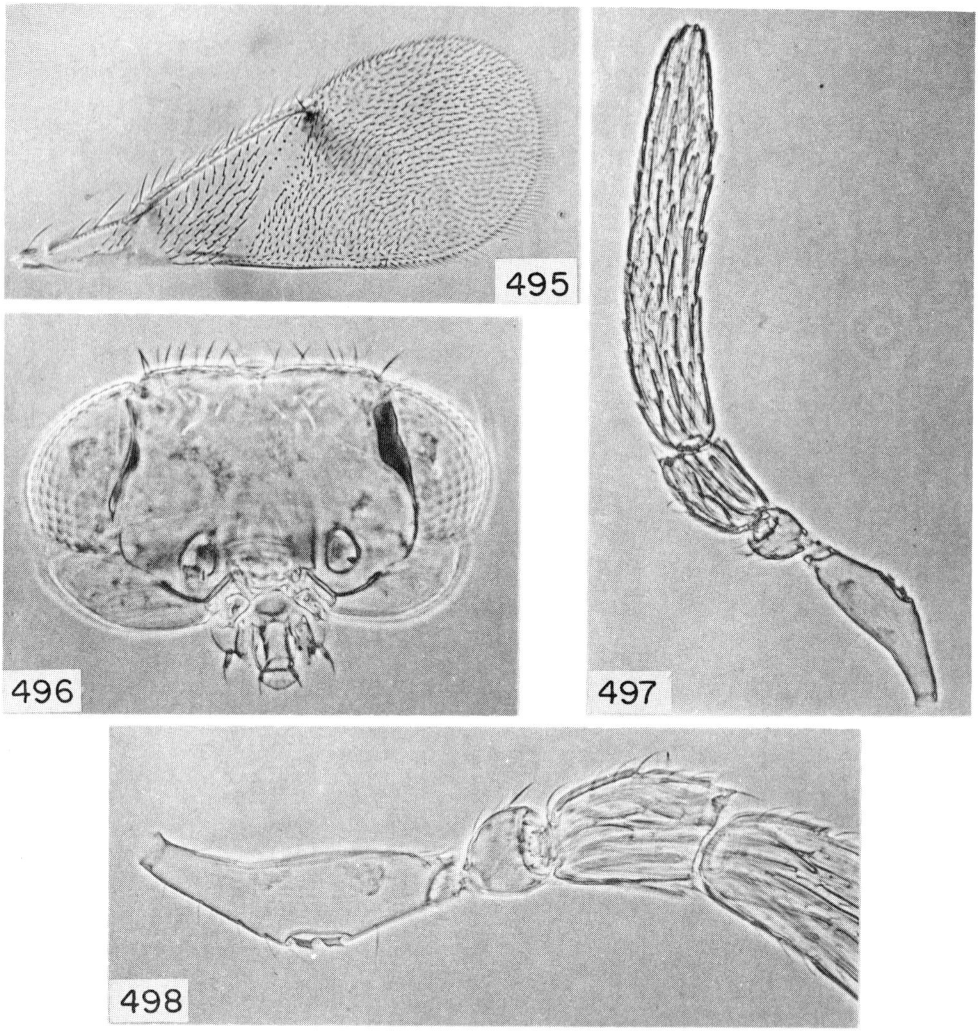


Figures 495–498. *Aphytis faurei* Annecke
495. ♀: Forewing (paratype). 496. ♂: Head and mouthparts; note reduced mandibles, 1-segmented palpi. 497. ♂: Antenna. 498. ♂: Basal part of antenna, showing sense organs on scape.

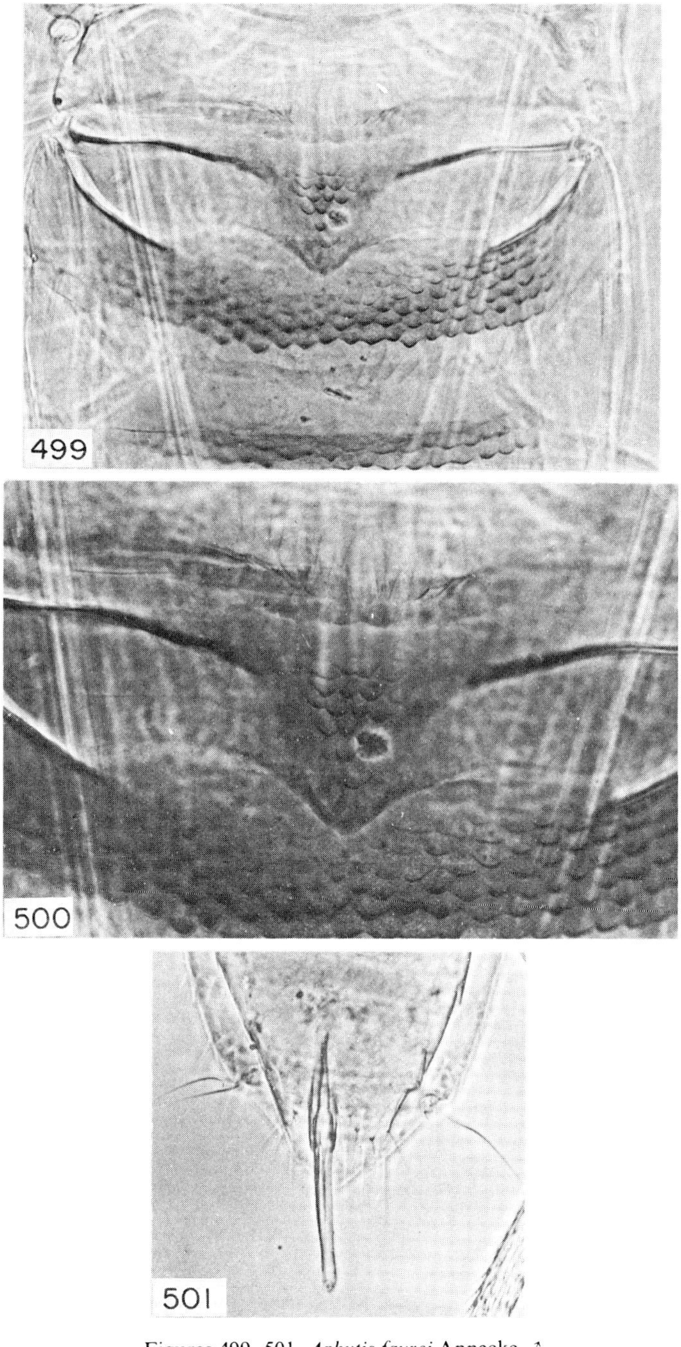


Figures 499–501. *Aphytis faurei* Annecke, ♂
499, 500. Propodeum and base of gaster; note single crenula on one side. 501. Genitalia.

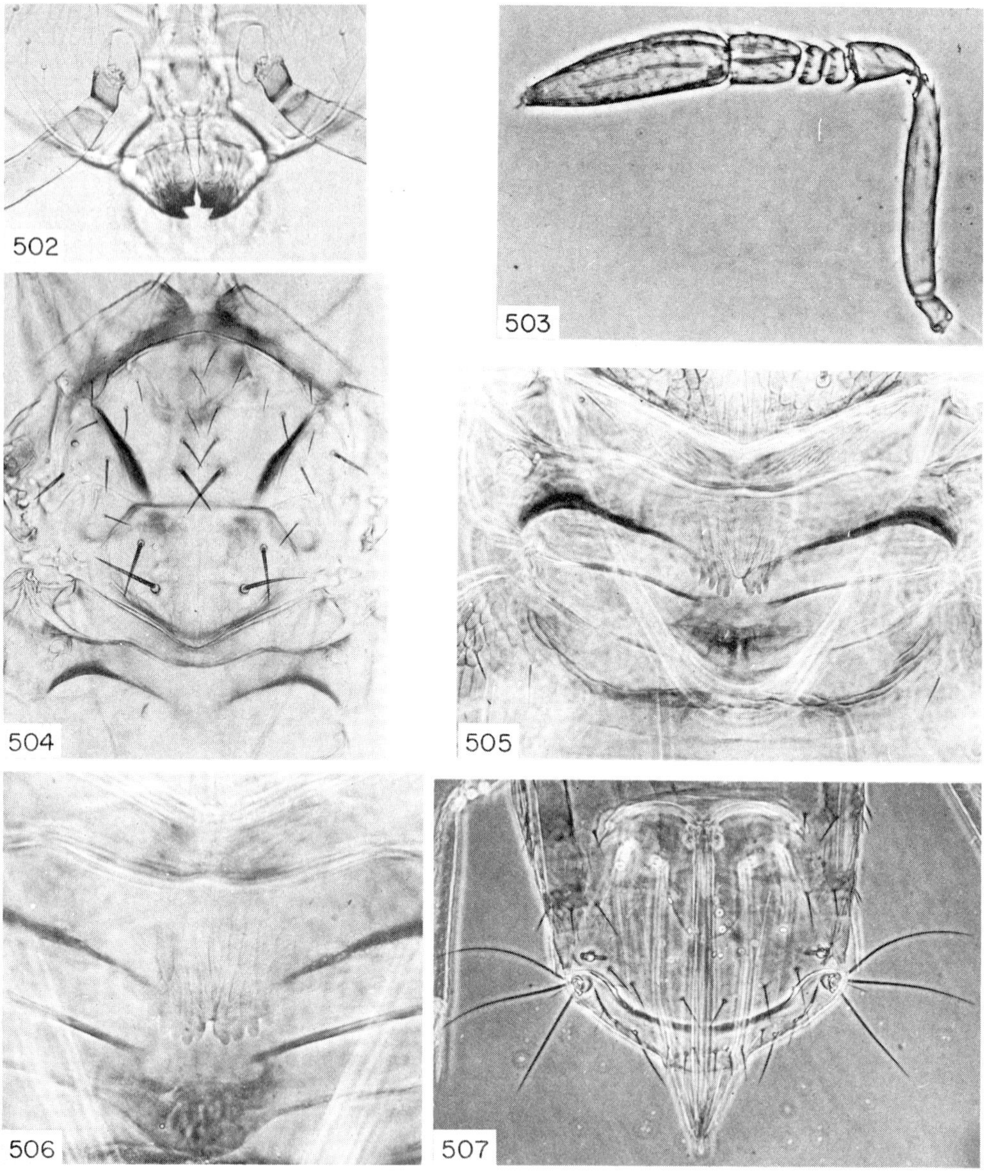


Figures 502–507. *Aphytis cercinus* Compere, ♀
502. Mandibles. 503. Antenna. 504. Thorax and propodeum. 505. Metanotum, propodeum and base of gaster (paratype). 506. Median salient of propodeum and crenulae. 507. Posterior abdominal tergites, showing spiracles, syntergum and cerci.

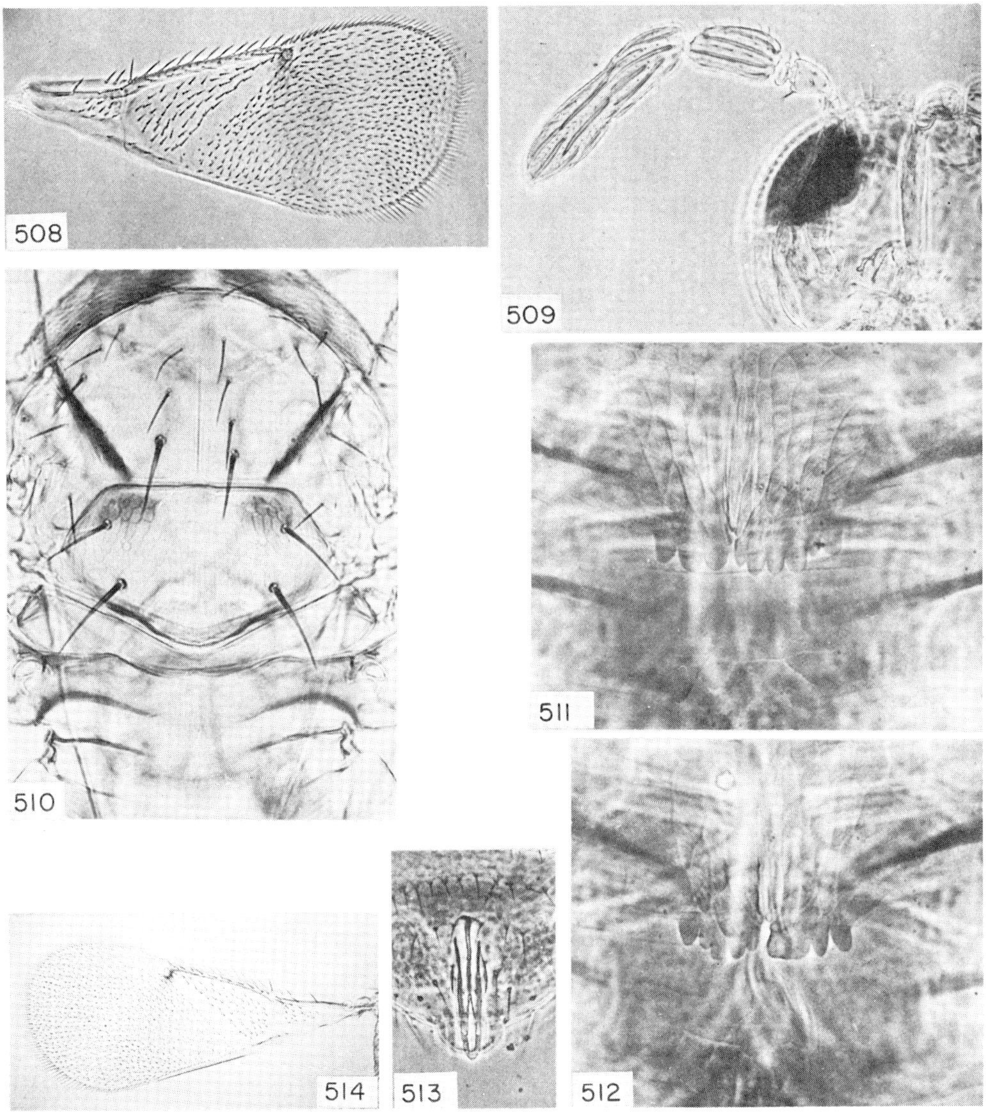


Figures 508–514. *Aphytis cercinus* Compere
508. ♀: Forewing (holotype). 509. ♂: Antenna. 510. ♂: Thorax, propodeum and base of gaster. 511, 512. ♂: Propodeal crenulae. 513. ♂: Genitalia. 514. ♂: Forewing.

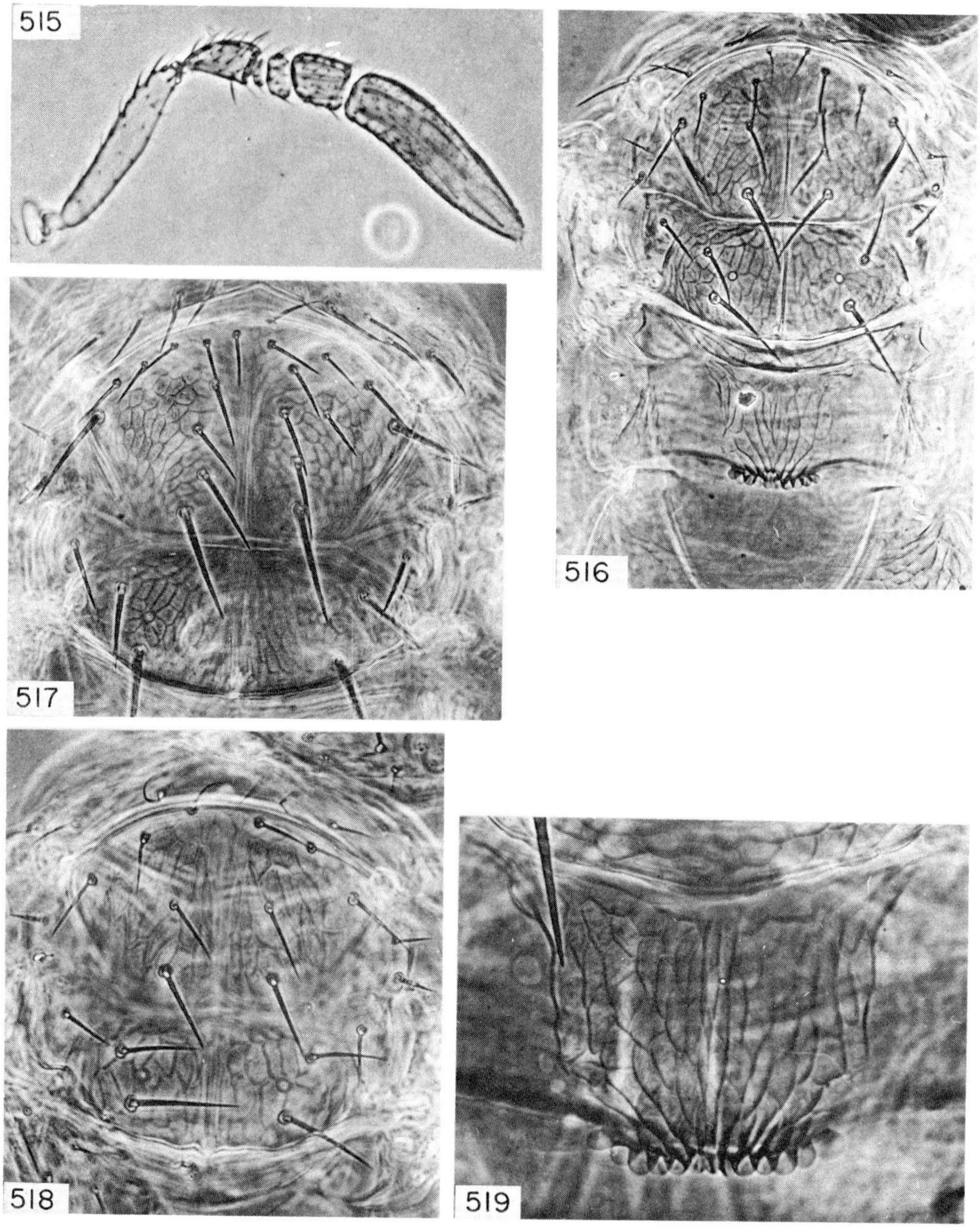


Figures 515–519. *Aphytis chilensis* Howard, ♀
515. Antenna. 516. Thorax, propodeum and base of gaster. 517, 518. Mesonotum, showing size-linked variation in the number of mesoscutal setae. 519. Median salient of propodeum and crenulae.

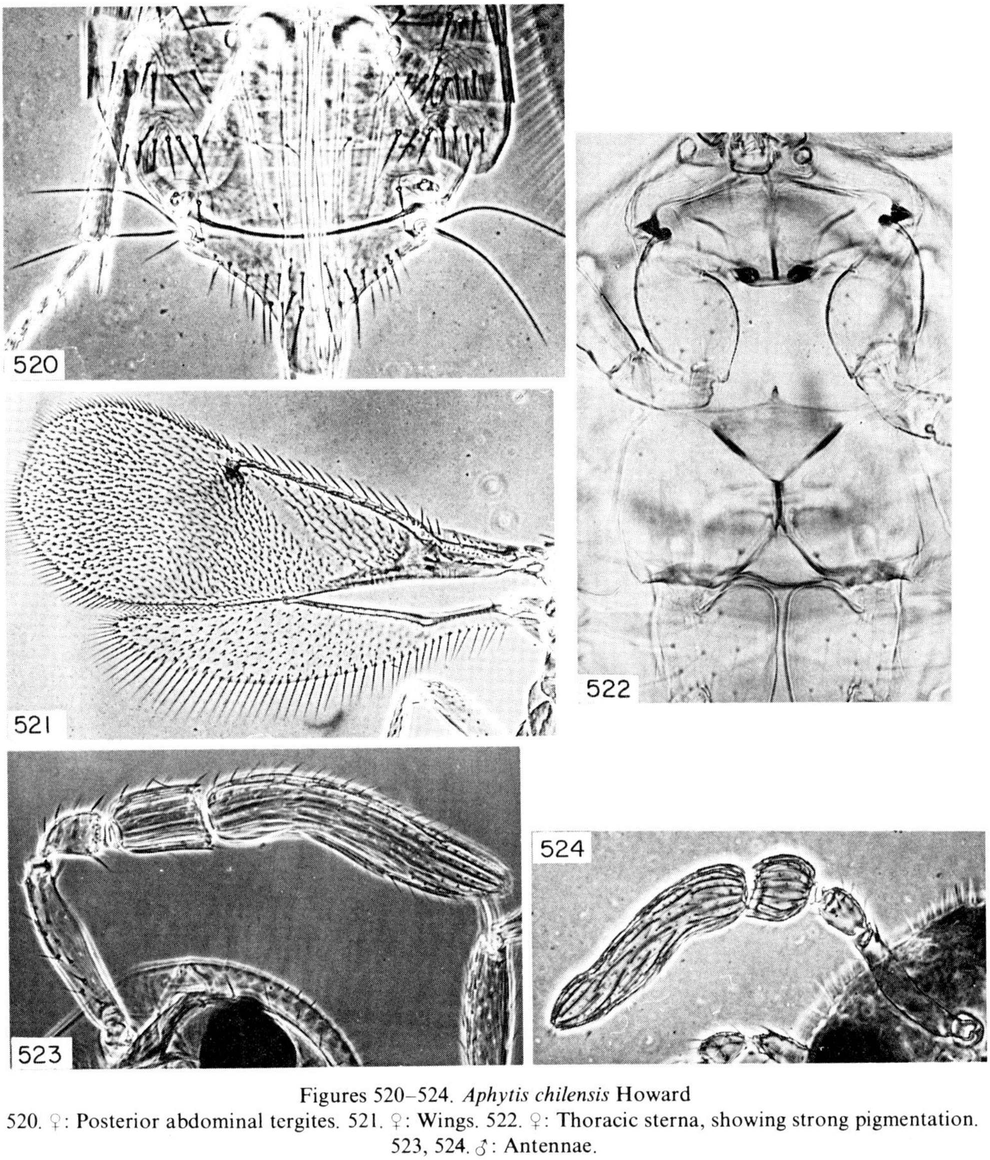


Figures 520–524. *Aphytis chilensis* Howard
520. ♀: Posterior abdominal tergites. 521. ♀: Wings. 522. ♀: Thoracic sterna, showing strong pigmentation. 523, 524. ♂: Antennae.

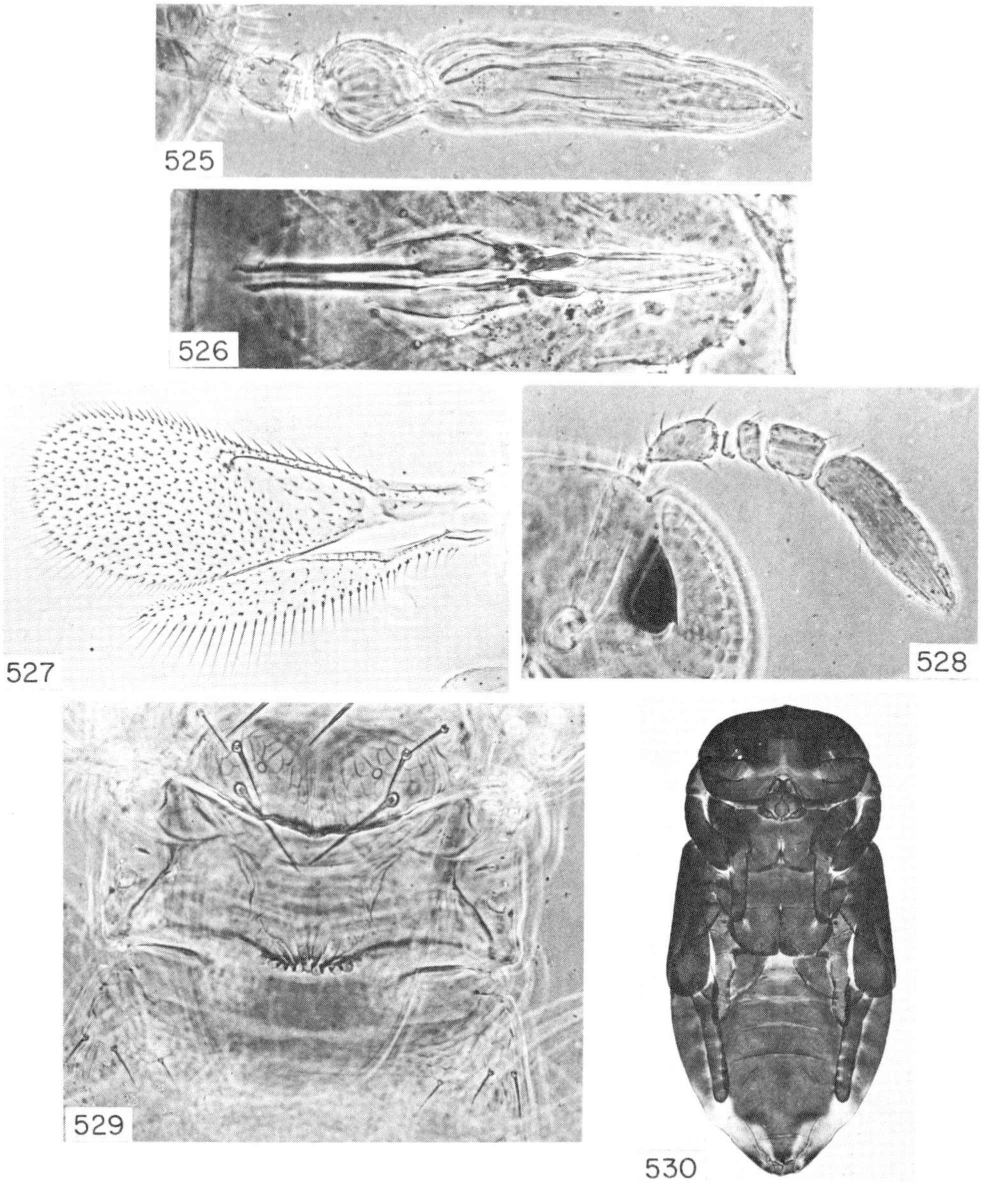


Figures 525–530. *Aphytis chilensis* Howard
525. ♂: Antennal club, showing specialized sensory area on ventral aspect. 526. ♂: Genitalia. 527. ♂: Wings (small specimen). 528, 529. Paratype ♀ of *riadi* Delucchi: 528. Antenna. 529. Thorax, propodeum and base of gaster. 530. ♀ pupa, ventral view.

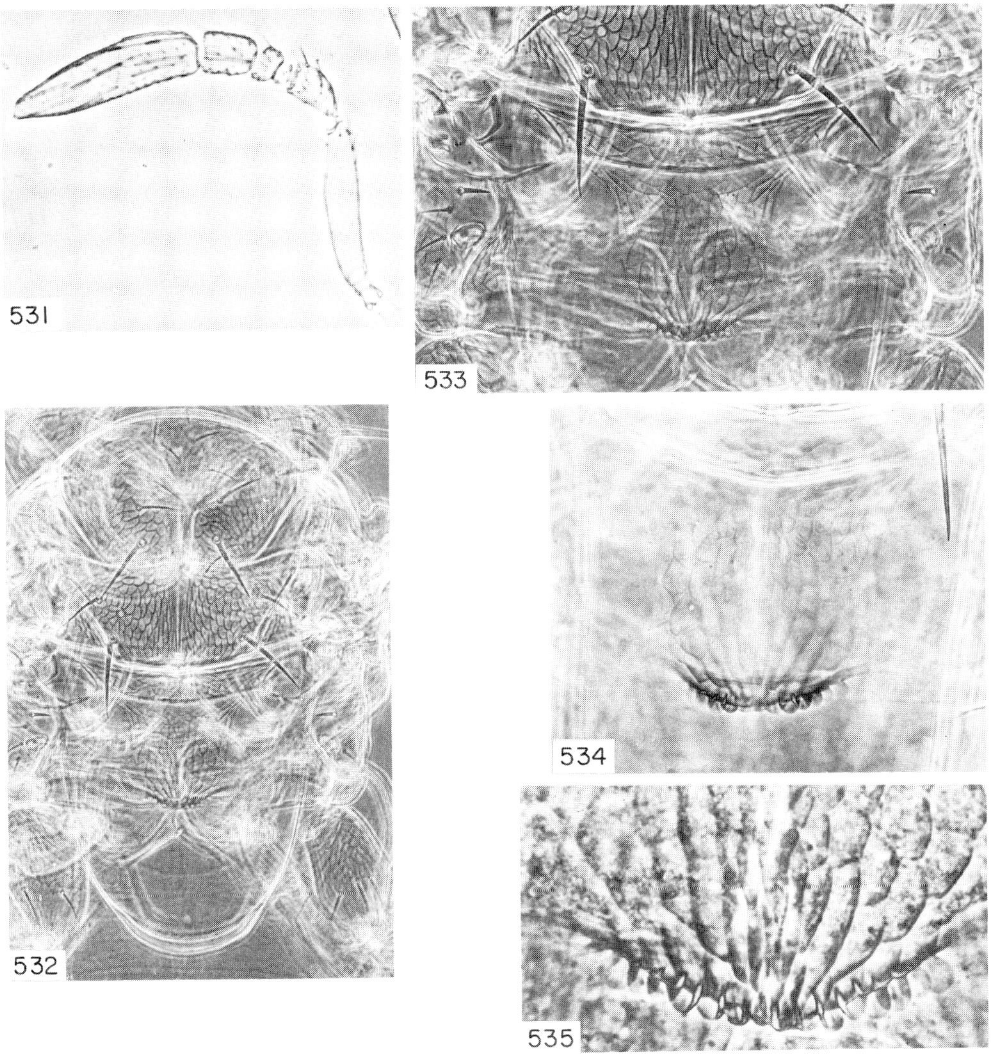


Figures 531–535. *Aphytis columbi* (Girault), ♀
531. Antenna. 532. Thorax and propodeum. 533. Metanotum and propodeum. 534. Propodeum, showing coloration of crenulae. 535. Propodeal crenulae.

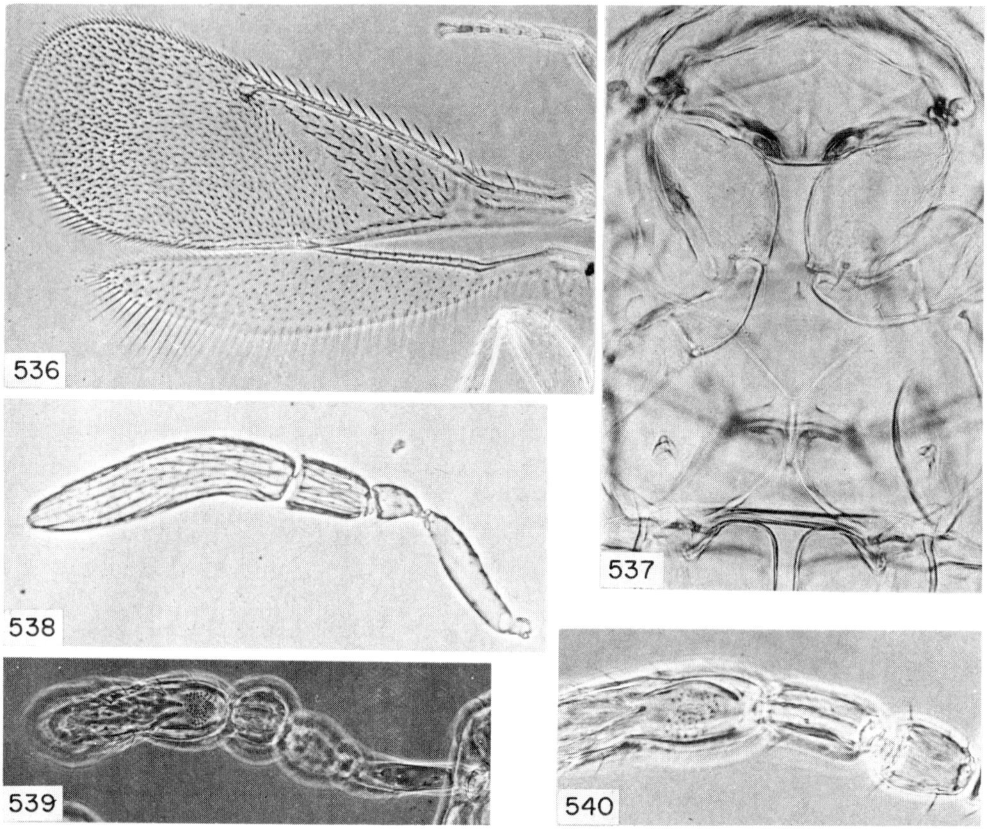


Figures 536–540. *Aphytis columbi* (Girault)
536. ♀: Wings. 537. ♀: Thoracic sterna, showing light pigmentation. 538. ♂: Antenna. 539, 540. ♂: Antenna, showing specialized sensory area on ventral aspect of club.

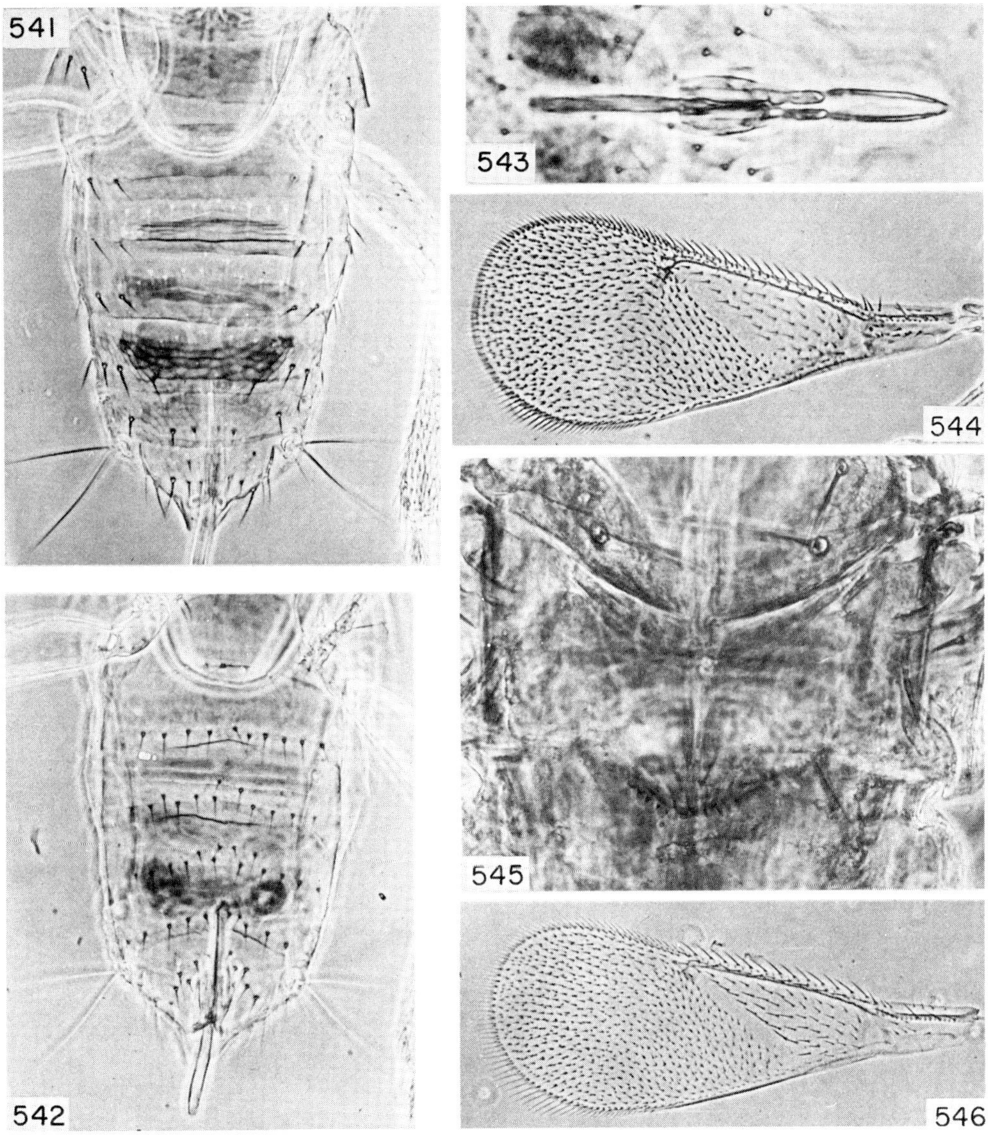


Figures 541–546. *Aphytis columbi* (Girault)
541. ♂: Abdomen, dorsal view. 542. ♂: Abdomen, ventral view. 543. ♂: Genitalia. 544. ♂: Forewing. 545. Type ♂ of *distonotus* (Girault): Propodeum. 546. Type ♀ of *distonotus* (Girault): Forewing.

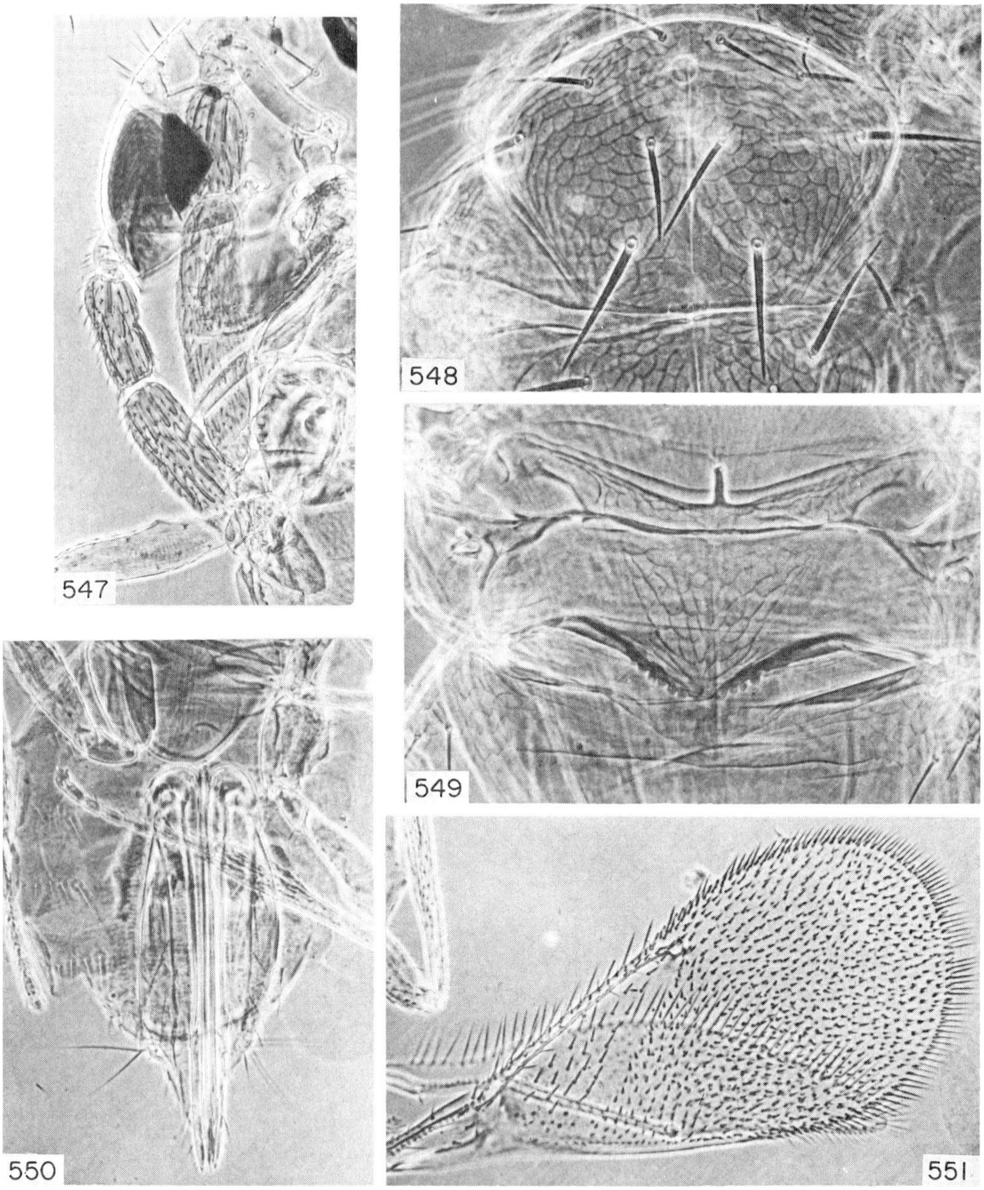


Figures 547–551. *Aphytis antennalis* n. sp., ♀ (holotype)
547. Antennae. 548. Mesoscutum. 549. Metanotum, propodeum and base of gaster. 550. Venter of abdomen, showing ovipositor and plates. 551. Forewing (partly overlapped by hind wing).

# III. THE PROCLIA GROUP

This large, remarkably uniform group is characterized by a distinctive pattern of cephalic pigmentation, with rather conspicuous black bars on the occiput and genal sutures. General coloration more or less dusky, with fuscous markings on the body and appendages; thoracic sterna and internal apodemes strongly pigmented; forewings uniformly setose but distinctly clouded below the stigma and below the junction of the submarginal and marginal veins; antennae 6-segmented in both sexes.

The pupa in this group is generally mottled with fuscous; wing pads and appendages appear fuscous to black; dark areas or furcae are visible on both dorsal and ventral surfaces of thorax and abdomen; abdominal segments with dark patches dorso-laterally. However, it should be noted that the pupal skin itself is usually not melanized at all. It is straw-yellow, and the melanization seen is that of the developing adult. Coarse, black spines are also usually evident through the integument of the fully developed pupa (see DeBach, 1959, and Figure 247). This is in contrast to other species, such as *chilensis* or the members of the **mytilaspidis** group, in which the pupal skin itself is melanized.

The **proclia** group appears to occupy an intermediate position between the **vittatus** group on the one hand and the paler **mytilaspidis** and **lingnanensis** groups on the other. The characteristic cephalic pigmentation obviously originated in ancestral forms related to the primitive **vittatus** group, from which the **proclia** group must have evolved. The distinctive pattern is found in members of both the **vittatus** and **chilensis** groups, but not in any of the other species groups of *Aphytis*.

As conceived here, the **proclia** group at present includes the following 11 members: *comperei* DeBach and Rosen, *confusus* DeBach and Rosen, *diaspidis* (Howard), *hispanicus* (Mercet), *maculicornis* (Masi), *paramaculicornis* DeBach and Rosen, *philippinensis* DeBach and Rosen, *pinnaspidis* n.sp., *proclia* (Walker), *testaceus* Tshumakova, and *vandenboschi* DeBach and Rosen. Quite a few additional species were referred to the **proclia** group by various previous authors, but are recognized here as synonyms. These include *argentinus* (Brèthes), *bovelli* (Malenotti), *fuscipennis* (Howard), *madagascariensis* (Risbec), *opuntiae* Risbec, *ovidii* (Girault), *risbeci* Annecke and Insley, *sugonjaevi* Jasnosh, and *zonatus* Alam.

The **proclia** group is widely distributed. At the present status of our knowledge it seems to be more abundantly represented in the Palearctic and Oriental regions, less so in the Neotropical, Nearctic and Ethiopian regions. Of the bona fide members of this group, only *diaspidis* has so far been recorded from the Australian region (New

Zealand); however, *minutissimus* (Girault), listed here as an unrecognizable member of the **proclia** group (see p. 736), was described from Australia. Several members of this group are nearly cosmopolitan in distribution.

Five species—*acrenulatus* DeBach and Rosen, *amazonensis* n.sp., *desantisi* DeBach and Rosen, *griseus* Quednau and *tucumani* n.sp.—resemble the **proclia** group in their generally dusky coloration but lack the characteristic cephalic pigmentation and are therefore regarded as related to, but not bona fide members of, this group. Three of these species are Neotropical, one is Ethiopian and one is possibly Oriental.

The three oldest species in the **proclia** group, namely *proclia* (Walker), *diaspidis* (Howard) and *fuscipennis* (Howard), were the cause of considerable confusion in the past. Mercet (1932) synonymized both of Howard's species with *proclia*, and his synonymy was generally accepted by subsequent workers (see De Santis, 1948; Nikol'skaya, 1952; Peck, 1963). Compere (1955) agreed that Mercet's conclusion may be correct, but questioned the advisability of suppressing "the well-known name *diaspidis*, the application of which is fairly certain, for the little-known name *proclia*, the application of which is less certain." Nikol'skaya and Yasnosh (1966) compared specimens of *diaspidis* from California with specimens of *proclia* from the USSR and concluded that they represented two different species. Examination of the types led us to the conclusion that *diaspidis* is indeed distinct from *proclia*, while *fuscipennis* is a junior synonym of *diaspidis*.

*A. diaspidis* can be readily separated from *proclia* by the different pattern of abdominal pigmentation: in *proclia* there are fuscous crossbands on tergites III and VII, short lateral stripes on tergites IV–VI, and the center of the abdomen is immaculate; in *diaspidis*, on the other hand, there is a distinct fuscous crossband on each of tergites III–VII, often forming a continuous fuscous blotch on the center of the abdomen. These species also differ in the pattern of antennal pigmentation, in the number of mesoscutal setae, in the shape and number of propodeal crenulae, and in the relative length of the ovipositor (see p. 407). These diagnostic characters in various combinations, as well as other details of pigmentation and chaetotaxis, also serve to separate the other members of the **proclia** group.

Although we have gone a long way toward separating the various members of this complex group, this appears to be a mere beginning. Several species in the **proclia** group evidently contain morphologically indistinguishable, but biologically distinct "forms," which may eventually be recognized as sibling species. Thus, for instance, an Oriental "strain" of *proclia* parasitic on the California red scale may be distinct from the typical Holarctic form. Likewise, biparental "strains" of *diaspidis* may or may not be conspecific with the common uniparental form. Several uniparental and biparental "strains" of *maculicornis* have been reported by biological control workers (see Hafez and Doutt, 1954; Huffaker, Kennett and Finney, 1962). We have taken the formal step of separating the biparental *paramaculicornis* from the uniparental *maculicornis* (DeBach and Rosen, 1976b), but there is some evidence indicating that each of these taxa may still consist of several sibling species (Khasimuddin and DeBach, 1976a, 1976b). Such information may, of course, be crucial for the success of biological control.

The material available is not sufficient for drawing definite conclusions regarding possible evolutionary trends within the **proclia** group. We assume that paler forms have evolved from darker ones, and that uniparental forms must have evolved from biparental ancestral stocks. Judging by pigmentation, chaetotaxis and general habitus, *diaspidis* would appear to be the most primitive member of the group. Discovery of additional species and further biosystematic studies may help clarify the complex affinities between the members of this group.

## 31. **Aphytis proclia** (Walker)

(Figures 552–574)

*Aphelinus proclia* Walker, 1839, Monogr. Chalc., **1**:9.
*Aphelinus proclia*: Mercet, 1912, Trab. Mus. Cienc. Nat. Madrid, **10**:104–105.
*Aphytis (Prospaphelinus) proclia*: Mercet, 1932, Eos, **8**:363–364.
*Aphytis zonatus* Alam, 1956, Trans. R. Entomol. Soc. London, **108**:370–372.
*Aphytis diaspidis*, Silwood material: Alam (nec Howard), 1956, Trans. R. Entomol. Soc. London, **108**:372–373; misident.
*Aphytis proclia*: Ferrière, 1965, Hymenoptera Aphelinidae d'Europe et du Bassin Méditerranéen, pp. 93–95; Nikol'skaya and Yasnosh, 1966, Aphelinids of the European Part of the USSR and the Caucasus, pp. 197–199 (in Russian); Graham, 1976, Syst. Entomol., **1**:134.
*Aphytis sugonjaevi* Jasnosh, 1972, Entomol. Obozr., **51**:247–248 (in Russian); n.syn.

This widespread, biparental Holarctic species can be recognized by the cephalic pigmentation characteristic of the group, with conspicuous black markings on the occiput and genal sutures. It may be separated from other species in the **proclia** group by the following combination of details of structure and pigmentation: Antennal club relatively short; mesoscutal setae not numerous (9–15); propodeal crenulae fairly numerous, narrow, elongate, nonoverlapping; ovipositor relatively short; mesoscutum mostly pale; center of abdomen immaculate, short fuscous stripes on sides of tergites, seventh tergite with a crossband; antennal flagellum fuscous, base of club paler fuscous, tip of club blackish.

Since *proclia* is the oldest species in the **proclia** group and has given the group its name, it is redescribed here in considerable detail. Other species in the group are subsequently described relative to *proclia*.

*Female*. Eyes finely setose. Mandibles (Figures 552, 565) well developed, with 2 denticles and a dorsal truncation; maxillary palpi 2-segmented, labial palpi 1-segmented. Antennal scape (Figure 552) slender, $4\frac{1}{2}$ to 6 times as long as wide, considerably longer than the club; pedicel relatively slender, usually $1\frac{2}{3}$ to $1\frac{4}{5}$ times as long as wide, somewhat longer than the third segment of the funicle; first funicular segment (Figures 552, 553, 571) trapezoidal, about $1\frac{1}{4}$ times as wide as long; second segment more symmetrical, somewhat shorter and wider than the first, over $1\frac{1}{2}$ times to twice as wide as long; third funicular segment slightly longer than wide, bearing 1–3 longitudinal sensilla; club $2\frac{1}{2}$ to somewhat over 3 times as long as wide (see also Figure 584), $2\frac{1}{2}$ to nearly 3 times longer and somewhat wider than the preceding segment, bearing 5–7 longitudinal sensilla.

Setae on head, thorax and sides of abdomen dark, coarse, readily visible under × 30 magnification; the longest setae appear as if composed of fused, indistinct filaments. Vertex with a pair of long setae along occipital margin, in addition to numerous shorter setae. Mesoscutum (Figures 554–556) with 9–15 setae, this number being more or less directly correlated with the size of the specimens, the posterior pair and 1 seta at each antero-lateral corner considerably longer and coarser than the others; each parapsis usually with 2 short setae, rarely with 3; each axilla with 1 seta; scutellum with 4 setae, the posterior pair somewhat longer than the anterior pair, the discoid sensilla usually somewhat closer to the anterior than to the posterior pair. Frontovertex, pronotum and mesonotal sclerites reticulate, with a distinct infrasculpture within the cells. Scutellum oval, usually $\frac{3}{4}$ to over $\frac{4}{5}$ median length of mesoscutum. Metanotum (Figure 557) somewhat curved, transversely reticulate except on the sides, the posterior margin usually overlapping the anterior margin of the propodeum; anteromedian apodeme stout, bifid at apex, about as long as or slightly shorter than median length of metanotum.

Propodeum (Figures 556–559, 572) usually over $3\frac{1}{2}$ times as long as the metanotum ($2\frac{3}{4}$ to $4\frac{1}{4}$), about $\frac{2}{3}$ length of scutellum ($\frac{3}{5}$–$\frac{3}{4}$), delicately reticulate on the sides, reticulate-strigose on a wide central area, very narrowly transversely striated along posterior margin; crenulae 5 + 5 to 10 + 10, elongate, relatively narrow, nonoverlapping.

Second abdominal tergite (Figures 557, 558) slightly transversely reticulate anteriorly on both sides, smooth centrally; tergites III–VII (Figures 560, 561) reticulate on the sides, with several setae in 1 or 2 short transverse rows on each reticulate area; third tergite reticulate also mesad of the lateral setiferous areas, transversely striated centrally; tergites IV–VI with some faint indications of transverse striation centrally; seventh tergite (Figures 561, 562) delicately reticulate across, bearing 3–5 fine setae in a transverse row across center; eighth tergite delicately reticulate, with a transverse row of 6–10 setae between the spiracles; syntergum (Figures 561, 562) triangular, wider than long, delicately striate-punctate, bearing 8–18 setae (usually 10–15) in 1 or 2 irregular transverse rows. Cerci situated closer to posterior spiracles than to tip of abdomen but not very close to spiracles, with 2 long setae and 1 short seta. Ovipositor shaft usually about $1\frac{1}{2}$ times as long as the middle tibia (1.41–1.61) (see also Figure 585), ovipositor sheaths $\frac{1}{3}$ to $\frac{2}{5}$ length of middle tibia (0.33–0.41).

Mid-tibial spur usually slightly shorter than the corresponding basitarsus.

Forewing (Figure 563) usually about $2\frac{2}{3}$ times as long as wide (always over $2\frac{1}{2}$ times; nearly 3 times as long as wide in smallest specimens); marginal fringe not exceeding $\frac{1}{4}$ width of disk (usually about $\frac{1}{6}$). Delta area with 51–163 setae in 8–13 rows, these considerably longer and somewhat sparser than the setae distad of speculum, well separated from row of setae along posterior margin of wing and from a small group of setae in 1–3 rows below distal half of submarginal vein; costal cell with a row of fine setae along proximal three fifths or so, and 1–2 coarse setae near apex. Submarginal vein bearing 2 coarse setae, the proximal seta about $\frac{2}{3}$ length of the distal, and 15–23 bullae. Marginal vein bearing 7–13 prominent, subequal setae along anterior margin (usually 10–12), these about $1\frac{1}{2}$ times as long as the setae in a row along center of vein.

Forewing with a distinct dark spot below stigma, merging into a paler, broader, quite well-defined fuscous crossband, which stands out rather clearly in contrast with the hyaline background; an arcuate, fuscous crossband at base of delta; a fainter fuscous cloud on setiferous area below apex of submarginal vein; the bare area below junction of submarginal and marginal veins hyaline; a narrow fuscous streak along posterior margin of wing, distad of speculum.

General coloration pale yellowish with fuscous markings. A transverse, strongly pigmented black bar on occiput, on each side of foramen (Figure 564); two faintly fuscous patches on occiput, between black bar and ocelli; genal sutures sclerotized, marked with black from oral margin to somewhat over halfway to eye (Figure 565); oral margin broadly infuscated. Central part of pronotum and apical parts of propleura fuscous; mesoscutum usually pale or rather faintly infuscated, the midline and lateral parts paler, sometimes with a pair of faint fuscous blotches, the anterior margin, parapsidal sutures, and central part of posterior margin lined with blackish; scutellum more strongly infuscated, with fuscous markings anteriorly, submedially and on the sides, the pale longitudinal midline more conspicuous, the posterior pair of setae arising from pale spots, the posterior margin lined with blackish; metanotum fuscous antero-medially and on the sides; propodeum infuscated on central reticulate area, the sutures mesad of spiracles and posterior margin lined with blackish; crenulae rather faintly infuscated. Second abdominal tergite infuscated centrally, the anterior, lateral, and latero-posterior margins lined with blackish; center of abdomen immaculate; tergites III–VI with short, transverse fuscous stripes on lateral reticulate areas, third tergite also with a narrow fuscous crossband anteriorly; seventh tergite with a fuscous crossband; the transverse ridge along posterior margin of eighth tergite black; syntergum suffused with fuscous; ovipositor sheaths immaculate. Thoracic sterna strongly infuscated. Internal apodemes, lateral margins of endophragma, the ridge and lateral parts of outer ovipositor plates blackish. Antennal scape pale, with a longitudinal fuscous stripe; pedicel and funicular segments fuscous, basal part of club usually somewhat paler, apex of club blackish. All femora with some fuscous markings, all tibiae entirely fuscous. Wing veins lined with brownish.

Length 0.82–1.20 mm (usually over 1 mm).

*Male*. Very similar to the female in structure, chaetotaxis, sculpture and coloration.

Antennae (Figures 566, 567) differing as follows: third funicular segment usually with a single longitudinal sensillum, club with 3–4 sensilla; a ventral, partly cut-off sensory area on the club, bearing short setae.

Mesoscutum (Figure 568) with 11–15 setae. Propodeum (Figure 569) about 3 times as long as the metanotum, a little less than $\frac{2}{3}$ length of scutellum; crenulae $5+6$ to $7+8$.

Genitalia (Figure 570) about $\frac{4}{5}$ length of middle tibia or a little longer (0.80–0.89), with a longitudinal rod arising from the junction of the digital sclerites, parallel to and between the aedeagal apodemes; digital sclerites rather long, about $\frac{1}{3}$ the combined length of aedeagus and apodemes (0.31–0.34).

Forewing about $2\frac{1}{2}$ to $2\frac{2}{3}$ times as long as wide, essentially as in the female; delta with 49–92 setae; marginal vein bearing 7–12 prominent, subequal setae along anterior margin; pattern as in the female, the crossband below stigma fading posteriorly.

General coloration somewhat paler than in the female. Third abdominal tergite pale on sides, tergites IV–VI entirely pale; seventh and eighth tergites and syntergum with broad, rather obscure, faint fuscous crossbands. Antennal club more or less uniformly infuscated, not blackish at apex. Wing pattern fainter than in the female.

Length 0.82–1.06 mm.

Redescribed from the lectotype (♀, removed from point and remounted in Hoyer's medium by P. DeBach, 1963; the slide, kept at the British Museum (Natural History), is labeled "*Aphelinus proclia* Walker—This is the type—M. de V. G.—Type. 5. 1727—Britain"; it was designated lectotype by Graham (1976)); the holotype of *Aphytis zonatus* Alam (♀, reared by S. M. Alam from *Quadraspidiotus zonatus* (Frauenfeld) on *Quercus robur* L., Silwood Park, England, June 26, 1953; cleared and remounted in Hoyer's medium by P. DeBach, 1963; slide in the British Museum (Natural History) labeled "Type. 5. 1677"); 1♀ of Alam's "*Aphytis diaspidis*, Silwood material," reared by S. M. Alam from *Chionaspis salicis* (L.) on *Tilia* sp., Silwood Park, England, August 12, 1953 (slide in British Museum (Natural History) remounted in Hoyer's medium by P. DeBach, 1963); 33♀♀, 11♂♂, reared by P. S. Messenger from *Chionaspis salicis* (L.) on *Tilia* sp., Silwood Park, England, March–June 1965; also 5♀♀, 77♂♂ from a laboratory culture of the same stock, Riverside, 1965–1966.

**Additional Material.** Since *A. proclia* was frequently misidentified in the past, a complete list of the material determined by us as belonging to this species is given herewith: 6♀♀, reared by F. Silvestri from the white peach scale, *Pseudaulacaspis pentagona* (Targioni-Tozzetti), Portici, Italy, May 8, 1916; 9♀♀, same data, May 24, 1917; 4♀♀, same data, May 18, 1927; 9♀♀, same host and collector, other data uncertain, Italy; 1♀, reared November 18, 1955, from the same host collected by V. Delucchi, Pescia (Toscana), Italy; 7♀♀, reared by E. S. Grinfeld from the San Jose scale, *Quadraspidiotus perniciosus* (Comstock), Maikop, N. Caucasus, USSR, June 1932; 2♀♀, reared by E. S. Sugonyayev, same host, Krasnodar Territory (Black Sea Coast of the Caucasus), USSR, October 10, 1950; 2♀♀, reared by V. Yasnosh, same host, Tskaltuvo, Caucasus, USSR, September 5, 1962; 3♀♀, 1♂, same host, Kishinev, Moldavia, USSR, June 18, 1961; 30♀♀, 12♂♂, reared by E. Dietrick, same host, New Haven, Connecticut, USA, August 1953; 6♀♀, 4♂♂ from a laboratory culture in Riverside, originally from the San Jose scale collected by P. Garman, Connecticut, October 1954; 28♀♀, 15♂♂, reared by C. Benassy, same host, Vallée de la Durance, France, September 1960; 1♀, 3♂♂, reared by M. Kosztarab from *Quadraspidiotus forbesi* (Johnson) on *Pyrus americana*, Dawes Arboretum, Lake Co., Ohio, USA, September 17, 1961; 2♀♀, reared by E. S. Sugonyayev from *Chionaspis salicis* (L.) on *Salix* sp., Pushkin, Leningrad Region, USSR, July 9, 1965; 2♀♀, reared by M. Kosztarab from *Chionaspis salicis-nigrae* (Walsh) on *Salix* sp., Conkles Hollow, Ohio, May 19, 1960; 3♀♀, 1♂, reared by M. Kosztarab from *Chionaspis americana* Johnson on *Ulmus americana*, Putnam Co., Indiana, USA, March 25, 1961; 1♀, reared by M. Kosztarab from *Lepidosaphes ulmi* (L.) on *Syringa vulgaris*, Columbus, Ohio, September 2, 1961; 2♀♀, reared by G. P. Georghiou from an undetermined host on apple, Pera, Cyprus, February 6, 1968; 1♀,

1♂, reared by J. Erdös from *Diaspidiotus* sp. on pear, Tompa, Hungary, October 1955; 1♀, reared by P. DeBach from the dictyospermum scale, *Chrysomphalus dictyospermi* (Morgan), on orange, Jalupa, Vera Cruz, Mexico, June 21, 1954; 26♀♀, reared by J. R. Quezada, same host, on rose, San Salvador, El Salvador, February 26, 1970; and 44♀♀, 22♂♂ from a laboratory culture in Riverside, originally from California red scale material, *Aonidiella aurantii* (Maskell), collected by P. DeBach in Maymyo, Burma, February 1957, propagated in Riverside on the latania scale, *Hemiberlesia lataniae* (Signoret) and California red scale. The latter stock was at first recorded by DeBach (1959) as the "Burma *hispanicus*."

Another stock of *proclia*, obtained from Connecticut in 1958 and reared in Riverside on latania scale and oleander scale, *Aspidiotus nerii* Bouché, was apparently contaminated in the insectary and eventually taken over by *diaspidis*. Samples taken from that culture in November and December 1958 contained only 2♀♀ and 1♂ of *proclia* and numerous ♀♀ of *diaspidis*.

**Notes.** The type of *A. zonatus* Alam (Figures 571, 572), as well as the specimens recorded by Alam (1956) as "*A. diaspidis*, Silwood material" (Figure 555), are undoubtedly conspecific with *proclia*.

Yasnosh (1972) described *A. sugonjaevi* as a sibling species of *proclia*, parasitic on *Chionaspis salicis* (L.) in the Leningrad Region, USSR, differing slightly from *proclia* in the chaetotaxis of the forewing: marginal fringe somewhat shorter ($\frac{1}{9}$–$\frac{1}{10}$ width of disk in *sugonjaevi*, as against $\frac{1}{7}$–$\frac{1}{8}$ in *proclia*), setae in delta somewhat more numerous (10–12 rows in *sugonjaevi*, 9–11 in *proclia*), setae along marginal vein somewhat more numerous (10–12 in *sugonjaevi*, 9–11 in *proclia*). She also noted that the meconial pellets of *sugonjaevi* are smaller and more numerous, and regarded the host record as significant, considering *proclia* to be a parasite of the San Jose scale.

We cannot accept *sugonjaevi* as a distinct species on the basis of the evidence presented so far. The morphological differences listed above are not significant, the data given for *sugonjaevi* clearly falling within the range of variation of *proclia* as understood here. Moreover, we have based our redescription of *proclia* on material reared from *C. salicis* in England, and do not regard the host record as significant in this case. We have examined 2♀♀ of *sugonjaevi* (reared by E. S. Sugonyayev from *C. salicis*, Pushkin, Leningrad Region, USSR, August 15, 1955, determined by V. A. Yasnosh), and find them identical to *proclia*. We therefore consider *sugonjaevi* to be a junior synonym of *proclia*.

In general coloration, *proclia* is rather similar to *chilensis*. The pattern of pigmentation is virtually identical in the various specimens listed above. In some of them, however, the dark areas appear faded, considerably paler than in other specimens. These paler specimens include the lectotype, the type of *zonatus*, and the specimen of Alam's "Silwood Park *diaspidis*," which have all been remounted from points, as well as the specimens from the USSR, which had been preserved in alcohol and could not be satisfactorily cleared. It is very probable that preservation on points or in alcohol caused the fading of pigments in these specimens. The actual intensity of pigmentation in slide-mounted specimens cannot be relied upon as a diagnostic character, especially when different methods of preserving or mounting are employed. Point-mounted specimens are even poorer for determining the degree of pigmentation. On the other hand, the *pattern* of pigmentation appears to be consistent, and may be regarded as a valid diagnostic character for the separation of species in the **proclia** group. Therefore, all the above-mentioned series are considered conspecific, regardless of the apparent differences in the intensity of pigmentation on infuscated areas.

Several specimens from Ohio and England show an aberration of scutellar chaetotaxis: 3 pairs of setae are present on the scutellum, and the discoid sensilla are absent, having evidently been replaced by the extra pair of setae (Figure 574).

Two series of specimens, obtained from *Lepidosaphes* spp. in China and South Africa, are very similar to typical material of *proclia* in general morphology and coloration, but appear to differ somewhat in having a longer propodeum and narrower wings. This material includes 3♀♀ and 2 nearly emerged ♀♀ in their pupal cases, reared at Riverside on February 18–19, 1925, from *Lepidosaphes* sp. (?*tubulorum* Ferris) on *Magnolia pumila*, sent by F. Silvestri from Canton, China; 1♀ from *Lepidosaphes* sp. on *Pandanus*, sent by F. Silvestri from China (no additional data); 28♀♀, reared by E. W. Rust from *Lepidosaphes pinnaeformis* (Bouché) [= ?*Cornuaspis beckii* (Newman)], Durban, Natal, South Africa, August 20, 1925 and February 2–12, 1926. No males are present in the series at hand.

The propodeum in these specimens (Figure 573) appears to be relatively longer than in typical material, $\frac{3}{4}$ to over $\frac{4}{5}$ length of scutellum. The forewing is usually at least fully 3 times, often $3\frac{1}{4}$ to $3\frac{1}{3}$ times, as long as wide; marginal fringe $\frac{1}{3}$–$\frac{1}{4}$ width of disk in the South African series, $\frac{1}{3}$–$\frac{1}{6}$ in the Chinese specimens. Mesoscutum with 10 setae in the South African series, 10–12 in the Chinese specimens. Otherwise very similar to typical material. Length 0.53–0.91 mm.

Compere (1955) discussed these series under *A. fuscipennis* (Howard), a species considered here to be a synonym of *diaspidis* (see p. 407). For the time being, these parasites of *Lepidosaphes* spp. are provisionally referred to *proclia*. However, it is quite possible that additional biological information will eventually indicate that they represent a distinct species.

Ferrière (1965) recorded *A. proclia* from *Quadraspidiotus perniciosus* (Comstock) in Switzerland, France, Germany, Italy, Yugoslavia and Algeria; from *Pseudaulacaspis pentagona* (Targioni-Tozzetti) in France and Switzerland; from *Chionaspis salicis* (L.) in Germany and Austria; and from *Quadraspidiotus ostreaeformis* (Curtis), *Hemiberlesia lataniae* (Signoret) and *H. rapax* (Comstock) in Morocco. Yasnosh (1972) recorded this species also from European USSR, Crimea, Caucasus, Soviet Central Asia and the Maritime Provinces (Soviet Far East), as a parasite of *Q. perniciosus*, *Q. pyri* (Lichtenstein) and *Epidiaspis leperii* (Signoret). Other authors have listed numerous additional hosts (see Peck, 1963; Nikol'skaya and Yasnosh, 1966), but these data may pertain to other members of the **proclia** group and need confirmation.

De Santis (1948) recorded *A. proclia* from Argentina and followed Mercet (1932) in synonymizing *A. diaspidis* (Howard) and *A. fuscipennis* (Howard) with *A. proclia*. We disagree with that synonymy. At hand are 2♀♀ and 1♂ from Argentina, determined by De Santis as *A. proclia*, which in our opinion are *diaspidis*. We have not seen any South American specimens of *A. proclia*, although we know that the species does occur in the Neotropical region, as evidenced by material from Mexico and El Salvador listed above.

**Biology.** Alam's (1956b) account of the biology of "*Aphytis diaspidis* (Silwood material)" apparently refers to *proclia* as understood here. He recorded this species as a biparental, solitary ectoparasite of second-instar larvae and young adult females of *Chionaspis salicis* (L.) on *Tilia* in England, and described the anatomy of the larva in some detail. According to him, the species is univoltine in England, overwintering as larvae.

The material available from England, which is considered to represent the typical form of *proclia*, appears to be biparental. The same is apparently true for the Connecticut material. In the USSR, *A. proclia* is reported to be biparental under the continental climate of the northern Caucasus, uniparental on the subtropical coast of the Black Sea (Chumakova, 1964). These records would benefit from experimental corroboration. We do not have first-hand evidence of a similar change in the mode of reproduction, affected by climatic conditions, in any species of *Aphytis*.

It is rather unlikely that the "strain" of *proclia* obtained from the California red scale in Burma, although morphologically indistinguishable from typical material, is indeed conspecific with the Holarctic form. A biosystematic study of this complex would certainly prove worthwhile. The "form" attacking *Lepidosaphes* spp. should also be investigated biologically.

**Utilization in Biological Control.** *A. proclia* from Burma was successfully propagated on California red scale and extensively colonized in southern California during 1957–1958, but failed to become established (Rosen and DeBach, 1978). Efforts to obtain this promising natural enemy again will be continued. Likewise, the typical form of *proclia* should be included in all biological control projects directed against the San Jose scale, white peach scale, and other important pests that serve as hosts to this parasite.

## 32. **Aphytis maculicornis** (Masi)

(Figures 575–579)

*Aphelinus maculicornis* Masi, 1911, Bol. Lab. Zool. Gen. Agr. Portici, **5**:158–159.
*Aphytis maculicornis* (part): Hafez and Doutt, 1954, Canad. Entomol., **86**:90–96; Compere, 1955, Univ. Calif. Publ. Entomol., **10**:295–296; Ferrière, 1965, Hymenoptera Aphelinidae d'Europe et du Bassin Méditerranéen, pp. 95–96; Nikol'skaya and Yasnosh, 1966, Aphelinids of the European Part of the USSR and the Caucasus, pp. 200–201 (in Russian).

This uniparental, Palearctic species can be recognized by the distinctive coloration of the antennae: female antennae with the pedicel pale and the basal part of the club entirely hyaline, in strong contrast to the blackish apex and fuscous funicle; male club uniformly pale. *A. maculicornis* is very closely related to *proclia*, differing from the latter species mainly in antennal structure and coloration, as well as in certain other details of structure and chaetotaxis. It is redescribed below relative to *proclia*.

*Female.* Eyes and mouthparts as in *proclia*. Antennae (Figures 575, 576) a little more slender than in *proclia*; pedicel usually twice as long as wide or nearly so (rarely

$1\frac{2}{3}$ times as long as wide), about $1\frac{1}{4}$ to $1\frac{1}{3}$ times longer than the third segment of the funicle; third funicular segment $1\frac{1}{4}$ to $1\frac{1}{2}$ times as long as wide; club usually over 3 times as long as wide, often considerably longer ($2\frac{3}{4}$ to $3\frac{1}{2}$, but rarely less than 3 times as long as wide).

Structure, chaetotaxis and sculpture of head and thorax as in *proclia*. Setae moderately coarse, a little more slender than in *proclia*; mesoscutum (Figure 577) usually with 8–10 setae, rarely with 6–7 or 11–12; discoid sensilla on scutellum usually about equidistant from the anterior and posterior pairs of setae.

Propodeum (Figures 578, 579) essentially as in *proclia*, about $\frac{2}{3}$ to $\frac{3}{4}$ median length of scutellum; crenulae 4 + 4 to 8 + 8, usually elongate, nonoverlapping, but often appearing somewhat broader and a little more rounded than in *proclia*.

Sculpture and chaetotaxis of abdominal tergites as in *proclia*. Cerci as in *proclia*. Ovipositor shaft relatively longer, about $1\frac{2}{3}$ times to nearly twice as long as the middle tibia (1.66–1.96), ovipositor sheaths $\frac{2}{5}$ to $\frac{1}{2}$ length of middle tibia (0.41–0.53).

Forewing somewhat narrower than in *proclia*, usually $2\frac{3}{4}$ to over 3 times as long as wide (rarely $2\frac{2}{3}$); marginal fringe often relatively long, $\frac{1}{3}$ to $\frac{1}{6}$ width of disk. Delta area sometimes not clearly separated from row of setae along posterior margin of wing. Marginal vein usually bearing 7–10 prominent setae along anterior margin (as few as 4–6 in minute specimens). Otherwise essentially as in *proclia*. Forewing pattern as in *proclia*.

General coloration essentially as in *proclia* but usually somewhat paler. Mesoscutum usually almost entirely pale or with a pair of faint, antero-submedian fuscous blotches, sometimes suffused with faint fuscous; scutellum more strongly infuscated, mainly submedially, but less extensively so than in *proclia*, without the distinct pale spots around the bases of the posterior setae. Sides of third and fourth abdominal tergites sometimes entirely pale; syntergum immaculate. Two dark spots at base of ovipositor shaft, often much more conspicuous than in *proclia*. Otherwise head, thorax, abdomen and internal apodemes as in *proclia*. Antennal scape (Figure 576) pale, with a longitudinal fuscous stripe; pedicel pale or faintly infuscated, funicular segments strongly infuscated; basal two thirds of club entirely hyaline, in strong contrast to the funicle; apical portion of club blackish. Legs as in *proclia*, but middle and hind tibiae rather faintly infuscated below middle, darker toward both ends.

Length 0.66–1.16 mm.

*Male*. The rare male is very similar to the male of *proclia*, differing from the latter mainly in details of antennal coloration. Structure of antenna and genitalia essentially as in *proclia*; the antennal club is somewhat over $3\frac{1}{2}$ times as long as wide.

Antennal scape pale, with a longitudinal fuscous stripe; pedicel rather faintly infuscated, funicular segments more strongly so; club uniformly pale or faintly infuscated, considerably paler than the funicle.

Redescribed from numerous thelytokous specimens, reared from the olive scale, *Parlatoria oleae* (Colvée), in various Mediterranean countries.

**Distinguishing Characters.** The distinctive antennal coloration provides the best diagnostic character separating *maculicornis* from *proclia*. In the female of *proclia*, the basal portion of the club is infuscated, though usually somewhat paler than the funicle; the difference in pigmentation is slight, and is evident only when the antenna is mounted in perfect lateral position. In the female of *maculicornis*, the basal portion of the club is immaculate, in striking contrast to the strongly infuscated funicle. Likewise, the club and funicle are concolorous in the male of *proclia*, whereas in the male of *maculicornis* the club is distinctly paler than the funicle.

Additional characters are the somewhat more slender antennae (see Figure 584), the somewhat fewer mesoscutal setae and crenulae, the longer ovipositor shaft (see Figure 585) and sheaths, and the somewhat narrower forewing of *maculicornis*. Two dark spots at the base of the ovipositor are more conspicuous in *maculicornis*. In general, specimens of *maculicornis* appear somewhat paler and more slender than specimens of *proclia*.

**Notes and Records.** *A. maculicornis* was originally described from 3♀♀, reared from an unidentified scale insect common on *Ligustrum vulgare* in Rome, Italy. The types were not available for study and are apparently lost. One of us (P.D.) corresponded extensively and personally searched for them in the most likely museums of Italy.

Compere (1955) applied the name *maculicornis* to the common parasite of the olive scale, *Parlatoria oleae* (Colvée), from Italy. Since 1939, efforts have been made to introduce parasites of *P. oleae* into California from all over this pest's range of distribution. In the course of extensive studies that followed, several stocks, all morphologically indistinguishable from *maculicornis*, were obtained from that and other hosts in various countries. They were found to be uniparental around the Mediterranean Basin, biparental in Iraq, Iran and India (see Hafez and Doutt, 1954; Huffaker, Kennett and Finney, 1962). We have restricted the concept of *maculicornis* to apply to the uniparental stocks only (see below).

*Parlatoria oleae* very likely infests *Ligustrum vulgare* in Rome (Compere, 1955). Circumstantial evidence indicates that Masi's original description was based on the uniparental, Mediterranean parasite of *P. oleae*. Since no good series from Italy was available for study, the above redescription is based on uniparental series reared from *P. oleae* in Greece, Israel, etc.

As pointed out by Compere (1955), "The species identified as *maculicornis* is not fully in agreement with Masi's description. In some respects the description is vague. For example, in the figure of the wing, the stigma is omitted, and proportions of the antennal segments are not described. Masi compared the antenna to that of *mytilaspidis*. The description leaves much to the imagination. . . ."

Nikol'skaya and Yasnosh (1966) noted that Masi's species apparently had the marginal fringe of the forewing considerably longer than in the species discussed by Compere. However, the relative length of the marginal fringe is a very variable character, dependent to a large extent on the size of specimens.

Hafez and Doutt (1954) have shown that uniparental stocks of *maculicornis*, obtained from *P. oleae* in Spain and in California (the latter apparently introduced from Egypt; see Huffaker, Kennett and Finney, 1962), differ in certain biological characteristics, such as fecundity and duration of development. They suggested that two sibling species were involved. However, inasmuch as these and other uniparental stocks of *maculicornis* are morphologically indistinguishable and are apparently identical in their host preferences, we at present consider them to represent a single species.

A possible exception may be a thelytokous stock obtained from *Parlatoreopsis longispinus* (Newstead) in Egypt. These relatively small specimens differ from "typical" *maculicornis* in having 10–15 mesoscutal setae (usually 13) and sometimes a rather short ovipositor (1.45 to 1.71 times as long as the middle tibia). They differ in these respects also from specimens reared from *Parlatoria oleae* in Egypt, which are inseparable from other uniparental series of *maculicornis*. For the time being, this series is considered to be an aberrant form of *maculicornis*.

Uniparental series of *maculicornis* in the collection of the Division of Biological Control, University of California, Riverside, were obtained from *Parlatoria oleae* in Spain, Italy, Greece, Israel, Egypt, Soviet Central Asia (Tadzhikistan), the Caucasus (Armenia), and California; from *Aspidiotus nerii* Bouché [= *A. hederae* (Vallot)] in Greece and Mexico; from *Hemiberlesia lataniae* (Signoret) in Greece; from *Epidiaspis leperii* (Signoret) in the Caucasus (Georgia); and from *Quadraspidiotus juglansregiae* (Comstock) in California. Since *maculicornis* was not known from North America prior to its introduction into California from Egypt in 1948–1949, it may be assumed that the series from California, and possibly also from Mexico, derive from the Egyptian importation (see Huffaker et al., 1962).

Ferrière (1965) recorded *A. maculicornis* from *P. oleae* in Algeria. Compere (1955) pointed out that Mercet's (1930, 1932) records of *A. maculicornis* should be referred to *hispanicus*. Likewise, Quednau's (1964b) record of *A. maculicornis* from South Africa is apparently erroneous. Material determined by Quednau as *maculicornis* was found to include *confusus* DeBach and Rosen and *comperei* DeBach and Rosen, but no authentic specimens of *maculicornis*. Other authors may have confused *maculicornis* with *proclia*.

Specimens of the various biparental series are virtually indistinguishable from those of the uniparental *maculicornis*, except for a slight tendency of the antennal club to be somewhat less slender, less elongate than in the uniparental series, being more similar in this respect to the club of *proclia* (see Figure 584). However, our laboratory studies have indicated that the biparental stocks are completely isolated reproductively from the uniparental *maculicornis*. Unlike other similar complexes (e.g., *mytilaspidis*), where uniparental females are known to mate readily with biparental males and produce hybrid offspring (Rössler and DeBach, 1972a), uniparental females of *maculicornis* apparently never copulate with males of the biparental stocks, and hybridization never takes place. In view of the different mode of reproduction, the complete reproductive isolation, and the presence of slight morphological differences, the biparental series are recognized as representing a biologically distinct sibling species and are described here as *paramaculicornis* DeBach and Rosen.

**Biology and Ecology.** According to Hafez and Doutt (1954), development of the uniparental *maculicornis* is slower than that of the biparental form (recognized here as *paramaculicornis*). In their laboratory tests, the uniparental cultures required 20–23 days for peak emergence, as compared with 17 days for the biparental cultures.

In Georgia, USSR, uniparental *A. maculicornis* has 6 annual generations on 2 generations of the olive scale. Overwintering larvae pupate in February, and many perish if the weather becomes cold again. High temperatures, combined with low relative humidities in August, also cause high mortality. Effectiveness is also reduced because of asynchronization of host and parasite development in June (Yasnosh and Mindiashvili, 1971; Yasnosh, 1972).

### 33. **Aphytis paramaculicornis** DeBach and Rosen

(Figures 148, 156, 173, 191, 192, 196, 215, 220; and 580–583)

*Aphytis maculicornis* (part): Hafez and Doutt, 1954, Canad. Entomol., **86**:90–96; Compere, 1955, Univ. Calif. Publ. Entomol., **10**:295–296; Huffaker, Kennett and Finney, 1962, Hilgardia, **32**:569–582.
*Aphytis paramaculicornis* DeBach and Rosen, 1976, Ann. Entomol. Soc. Amer., **69**:542–543.

This biparental, Southwest Asian species is virtually indistinguishable from *maculicornis* in structure, chaetotaxis, sculpture and coloration. It differs from *maculicornis* in the arrhenotokous mode of reproduction and in the abundance of males, and is completely reproductively isolated from that species. Since a formal, detailed description would be superfluous, a few comparative notes follow.

*Female.* The antennal club (Figure 580) tends to be somewhat less slender, less elongate than in *maculicornis*, $2\frac{1}{2}$ to $3\frac{1}{3}$ times as long as wide (usually about 3 times, see Figure 584), more similar in this respect to the club of *proclia*. Some specimens with 11–13 mesoscutal setae. Propodeum (Figures 148, 173, 581) essentially as in *maculicornis*, the crenulae tending to be somewhat more elongate, narrower, more resembling those of *proclia*. Ovipositor as in *maculicornis*, somewhat longer than in *proclia* (see Figure 585). Forewing (Figure 582) essentially as in *maculicornis*. Distinctive antennal coloration as in *maculicornis*.

*Male.* The antennal club (Figure 583) tends to be somewhat less elongate than in *maculicornis*, usually $3\frac{1}{3}$ to $3\frac{1}{2}$ times as long as wide, uniformly dusky but paler than the funicle. Otherwise similar to the rare males of *maculicornis*. Genitalia (Figure 220) as in *maculicornis* and *proclia*.

Described from numerous specimens (♀ holotype, ♂ allotype and paratypes), reared in the insectary at Riverside on the cactus scale, *Diaspis echinocacti* (Bouché), originally from the olive scale, *Parlatoria oleae* (Colvée) collected in Iran in 1951 by A. M. Boyce (= "Persian *Aphytis*"); laboratory stock received from Albany, California, from material recovered from olive scale in the field in the San Joaquin Valley.

Type series in the collection of the Division of Biological Control, University of California, Riverside.

**Notes.** Hafez and Doutt (1954) considered the biparental stocks obtained from *Parlatoria oleae* in Iraq, Iran and India to be conspecific, whereas Huffaker, Kennett and Finney (1962) presented data indicating that the "Persian" and "Indian" stocks may be reproductively isolated. Recent studies by Khasimuddin and DeBach (1976a, 1976b) have indicated that three morphologically indistinguishable strains: "Persian *Aphytis*," originally from *P. oleae* on olive in Iran; a stock obtained from the chaff scale, *Parlatoria pergandii* Comstock, on citrus in southern California; and another stock obtained from the San Jose scale, *Quadraspidiotus perniciosus* (Comstock), on apple in Pakistan, show a high degree of reproductive isolation as well as some differences in host specificity, and may be regarded as sibling species or strong semispecies. Although these populations can produce occasional viable and fertile hybrids in the laboratory, this is apparently effectively prevented in nature by means of ethological isolating mechanisms, in which pheromones appear to play a major role. Since no information is at present available regarding the possible existence of intergrading populations, we have treated these and other biparental series as conspecific and have used them for the description and figures of *paramaculicornis*. However, should distinct species be described from this complex in the future, the name *paramaculicornis* should be applied to the form commonly known as "Persian *Aphytis*."

*A. paramaculicornis* can at present be considered as a sibling of *maculicornis*. However, in certain characters, such as antennal structure and possibly the shape of the propodeal crenulae, *paramaculicornis* appears to occupy an intermediate position between *proclia* and *maculicornis*, and the differences between it and *proclia* are very slight. Even antennal coloration cannot always be relied upon for separating *paramaculicornis* from *proclia*. In certain series from Southeast Asia, the basal portion of the female club is faintly infuscated, certainly not immaculate, but considerably paler than in typical *proclia*. Although such series are at present referred to *proclia*, their position may change when more is known about their biology.

Biparental series of *A. paramaculicornis* in the collection of the Division of Biological Control, University of California, Riverside, were obtained from *Parlatoria oleae* in Iraq, Iran, India, Pakistan, and California; from *Parlatoria pergandii* Comstock in California, from *Parlatoria blanchardi* (Targioni-Tozzetti) in Saudi Arabia; from *Quadraspidiotus perniciosus* (Comstock), *Lepidosaphes* sp. and undetermined hosts in Pakistan; and from (?) *Aonidiella aurantii* (Maskell) in the San Joaquin Valley, California. This species was introduced into California from Iraq, Iran and India.

**Biology.** "Persian *Aphytis*" is a solitary ectoparasite of the olive scale; gregarious parasitism (i.e., the development of two parasites on a single host) is a rare occurrence with this species. Nearly mature, third-stage female scales are the preferred host stage, but immature males may also be parasitized. Host-feeding occurs, but is not an appreciable mortality factor.

Unlike other "strains" of *paramaculicornis*, "Persian *Aphytis*" readily parasitizes the oleander scale, *Aspidiotus nerii* Bouché [= *A. hederae* (Vallot)] in laboratory experiments. It may be reared in the insectary on the cactus scale, *Diaspis echinocacti* (Bouché), and latania scale, *Hemiberlesia lataniae* (Signoret).

At 28°C and 65% RH, the eggs of this species have an incubation period of 3 days; larval development takes 7 to 9 days and the pupal stage 6 to 9 days. The mean time from oviposition to peak emergence is 17 days under these conditions. The preoviposition period is less than 24 hours, and the peak of oviposition is attained by females 3 to 4 days old. A generation may therefore be completed in about 21 days. At 27°C and 55% RH, an average 30 eggs are laid by the female during an average life period of 20 days.

In California, reproduction of *A. paramaculicornis* virtually ceases in midwinter, and the species overwinters in the immature stages. Reproduction in midsummer is also greatly restricted due to heavy mortality of the adults. Thus, about 4 to 6 effective generations of the parasite may develop annually in California in fall and spring (Huffaker, Kennett and Finney, 1962; Rosen and DeBach, 1978; Khasimuddin and DeBach, 1976b).

**Utilization in Biological Control.** Accounts of the successful biological control of the olive scale, *Parlatoria oleae*, by the "Persian *Aphytis*" on olive and deciduous fruit trees in California (see Doutt, 1954; Huffaker, Kennett and Finney, 1962; Huffaker and Kennett, 1966; DeBach, Rosen and Kennett, 1971; Rosen and DeBach, 1978) now obviously refer to *paramaculicornis* rather than to *maculicornis*.

The olive scale became a major pest of olive in California following its establishment near Fresno about 1934. It also became a major pest of many deciduous fruit crops and of ornamental shrubs and trees. In 1951, great efforts were made to obtain natural enemies of olive scale throughout its range of distribution, including its supposed native home in northern India, West Pakistan and the Middle East. Several species of parasites were obtained and released in California, including four so-called "biological strains" of *maculicornis* and, as we now know, *paramaculicornis*. It early became evident that *A. paramaculicornis*, obtained from Iran and Iraq and first colonized during 1952, was the only natural enemy giving some degree of control in field plots. Consequently, insectary production, colonization and field study were concentrated on this parasite. Over 27 million *A. paramaculicornis* were colonized during 1952–1960 at several hundred sites in 24 counties in California.

Early results were very promising. *A. paramaculicornis* became readily established, and proved capable of attaining levels of parasitization of 90 percent or more even after host densities had been reduced to low levels. Scale populations were generally lowered by more than 90 percent. However, this drastic reduction proved to be economically unsatisfactory in many cases, because if even one scale occurs on an olive fruit it may be culled. Greater effectiveness was prevented because the parasite proved intolerant of the hot, dry summers and cold winters of the Central Valley.

In certain favorable locations, *A. paramaculicornis* alone produced good control. However, full economic control was not achieved in California until a second parasite, *Coccophagoides utilis* Doutt, was introduced in 1957 from Pakistan and established. Complementary action of the ectoparasitic *A. paramaculicornis* and the endoparasitic *C. utilis* eventually brought about complete biological control of the olive scale throughout California.

The origin of the "strain" of *A. paramaculicornis* (or sibling species?) parasitizing the chaff scale, *Parlatoria pergandii* Comstock, on citrus in southern California is a mystery. It was first discovered in 1963 near Escondido, where *A. hispanicus* had been colonized against chaff scale one year earlier, and was at first mistaken for this species.

After it was correctly identified, it was assumed to be identical to the "Persian" *paramaculicornis* parasitizing olive scale in California. However, recent studies have indicated that these two forms are biologically distinct. This is the only "form" of *paramaculicornis* known to parasitize chaff scale, and it appears to be quite an effective natural enemy, having reduced the scale's populations near Escondido to very low levels. It should be considered for importation against chaff scale in other countries.

*A. paramaculicornis* was introduced in 1970 from California into Soviet Georgia (Yasnosh, 1972).

## 34. **Aphytis hispanicus** (Mercet)

(Figures 586–603)

*Aphelinus maculicornis* var. *hispanica* Mercet, 1912, Trab. Mus. Cienc. Nat. Madrid, Ser. Zool., **10**:81–82.
*Aphelinus argentinus* Brèthes, 1916, An. Mus. Nac. Hist. Nat. Buenos Aires, **27**:428–429; n. syn.
*Aphelinus bovelli* Malenotti, 1918, Redia, **13**:78–81; n. syn.
*Aphytis maculicornis*: Mercet (nec Masi), 1930, Rev. Biol. Forest. Limnol., Ano **2**, Ser. B, No. 2:51, 53–54.
*Aphytis (Prospaphelinus) maculicornis*: Mercet (nec Masi), 1932, Eos, **8**:363.
*Aphytis argentinus*: De Santis, 1938, Anuar. Rural Prov. Buenos Aires, **6**:268; De Santis, 1941, Bol. Soc. Brasil. Agron., **4**:14.
*Aphytis (Prospaphelinus) maculicornis* var. *argentinus*: De Santis, 1948, Rev. Mus. La Plata, N.S., **5** (Zool.): 131–133.
*Aphytis hispanicus*: Compere, 1955, Univ. Calif. Publ. Entomol., **10**:294–295; Ferrière, 1965, Hymenoptera Aphelinidae d'Europe et du Bassin Méditerranéen, pp. 92–93; Nikol'skaya and Yasnosh, 1966, Aphelinids of the European part of the USSR and the Caucasus, pp. 195–196 (in Russian).
*Aphytis bovelli*: Compere, 1955, Univ. Calif. Publ. Entomol., **10**:296–297.
*Aphytis maculicornis* var. *argentinus*: Compere, 1955, Univ. Calif. Publ. Entomol., **10**:297.
*Aphytis hispanica*: Nikol'skaya and Yasnosh, 1968, Trud. Vsesoyuz. Entomol. Obshchest., **52**:11.

This uniparental, South Palearctic species is very closely related to *proclia*, differing from the latter species in the mode of reproduction and in certain details of structure and coloration. It can be recognized by the following combination of characters: Antennae short, rather stout, club less than 3 times as long as wide; pedicel, funicle and most of club uniformly fuscous, apical third of club blackish; genal sutures sclerotized and black to about $\frac{3}{4}$ distance from oral margin to eye; forewing narrow; abdominal coloration paler than in *proclia*, the crossband on seventh tergite obliterated centrally in the female; male antennae uniformly fuscous.

*Female*. Eyes and mouthparts as in *proclia*. Antennae (Figures 586–588) somewhat shorter, more compressed; scape and pedicel essentially as in *proclia*; first funicular segment $1\frac{1}{3}$ to over $1\frac{1}{2}$ times as wide as long, second segment twice as wide as long or wider; third funicular segment somewhat longer than wide, bearing 2 longitudinal sensilla (rarely 1); club usually over $2\frac{1}{2}$ but distinctly less than 3 times as long as wide, about $2\frac{1}{5}$ to $2\frac{1}{2}$ times longer than the preceding segment, bearing 6 longitudinal sensilla (rarely 5).

Structure, chaetotaxis and sculpture of head and thorax essentially as in *proclia*. Setae more slender than in *proclia* but readily visible under $\times 60$ magnification; vertex with 2 pairs of long setae along occipital margin, in addition to numerous shorter setae. Mesoscutum (Figures 589, 590) with 10–13 setae (often 11–12); discoid sensilla on scutellum about equidistant from the anterior and posterior pairs of setae.

Propodeum (Figures 591, 592) very similar to that of *proclia*, $3\frac{1}{2}$ to $4\frac{1}{2}$ times as long as the metanotum, $\frac{2}{3}$ to $\frac{3}{4}$ length of scutellum; crenulae 6 + 6 to 8 + 8, elongate, rather narrow, nonoverlapping, as in *proclia*.

Sculpture and chaetotaxis of abdominal tergites (Figure 593) essentially as in *proclia*. Second tergite a little more distinctly striated anteriorly on the sides (Figure 591), often faintly reticulate centrally; seventh tergite bearing 2–6 fine setae in a transverse row across center; eighth tergite with a transverse row of 5–9 setae between spiracles; syntergum bearing 7–12 setae. Cerci as in *proclia*. Ovipositor shaft about $1\frac{1}{2}$ to $1\frac{3}{4}$ times as long as the middle tibia (1.46–1.71), ovipositor sheaths about $\frac{2}{5}$ length of middle tibia (0.37–0.41).

Forewing (Figure 594) relatively narrow, $2\frac{3}{4}$ to nearly 3 times as long as wide; marginal fringe $\frac{1}{3}$ to $\frac{1}{6}$ width of disk; otherwise essentially as in *proclia*. Delta area with 59–161 setae in 9–12 rows, these sparser and considerably longer than the setae distad of speculum, not clearly separated from row of setae along posterior margin of wing, widely separated from 1–3 rows of setae below distal half of submarginal vein; costal cell with a row of fine setae along proximal half or three fifths, and usually 2 coarse setae near apex. Submarginal vein bearing 2 coarse setae, the proximal one about $\frac{2}{3}$ to $\frac{3}{4}$ length of the distal, and 18–22 bullae. Marginal vein bearing 8–12 prominent, subequal setae along anterior margin.

Forewing pattern similar to that of *proclia*, but the fuscous cloud below the stigma is weak, rather obscure, less distinctly defined than in *proclia*.

General coloration similar to that of *proclia*, but paler. Head without the pair of faintly infuscated patches on the occiput, otherwise as in *proclia*; genal sutures (Figure 586) heavily sclerotized, strongly infuscated from oral margin to about $\frac{3}{4}$ distance to eyes. Mesoscutum and scutellum pale, each with a pair of faint fuscous blotches anterosubmedially, scutellum sometimes very faintly infuscated along both sides of midline; otherwise thoracic sclerites and sutures as in *proclia*; propodeum pale, the sutures mesad of spiracles and the posterior margin lined with blackish; crenulae faintly infuscated. Abdominal tergites as in *proclia*, but the fuscous crossband on the seventh tergite is obliterated centrally, leaving a pair of transverse lateral stripes, the latter somewhat wider than on the preceding tergites; transverse ridge on posterior margin of eighth tergite faintly infuscated; syntergum and ovipositor sheaths immaculate. Thoracic sterna strongly infuscated. Internal apodemes and outer margins of endophragma blackish; outer parts of outer ovipositor plates fuscous. Antennal scape pale, with a longitudinal fuscous stripe; pedicel, funicular segments and club uniformly, strongly infuscated; apical third of club blackish. Legs as in *proclia*.

Length 0.73–1.00 mm.

Redescribed from 3♀♀, reared by R. G. Mercet from the chaff scale, *Parlatoria pergandii* Comstock, on orange and sour orange, Valencia, Spain, November 27, 1911 (holotype), November 29 and December 5, 1911; 8♀♀, reared by P. DeBach from *Parlatoria pergandii* on sour orange, Valencia, Spain, July 28–29, 1963; 10♀♀, reared by P. DeBach from *Chrysomphalus dictyospermi* (Morgan) on Valencia orange, Valencia, Spain, July 30, 1963.

Holotype at the Museo de Ciencias Naturales, Madrid, Spain.

**Notes.** Specimens of *A. hispanicus* in the collection of the Division of Biological Control, University of California, Riverside, were reared from *Parlatoria pergandii* Comstock in Israel, Turkey, Trinidad, Brazil, Mexico, Texas and California; from *Parlatoria cinerea* Doane and Hadden in Trinidad, from *Chrysomphalus dictyospermi* (Morgan) and *Acutaspis scutiformis* (Cockerell) in Brazil; from California red scale material (probably chaff scale) in Morocco; and from undetermined hosts in Italy, Jamaica, Taiwan and Florida. Yasnosh (1972) recorded *A. hispanicus* also as a parasite of *P. pergandii*, *P. oleae* (Colvée), *Aspidiotus nerii* Bouché, *Lopholeucaspis japonica* (Cockerell), *Mytilaspis conchiformis* (Gmelin) [= *Lepidosaphes ficus* (Signoret)] and *Insulaspis pallida* (Green) in the Caucasus.

*Aphytis hispanicus* is a thelytokous species. Males are not present in the Spanish material at hand, on which the above redescription was based, and are very rare elsewhere. Rosen (1965), for instance, recorded only about 2% males among several hundred specimens examined in Israel. In fact, the male of *hispanicus* has hitherto remained undescribed. The following description is based on rare male specimens reared from chaff scale in Israel and Texas.

*Male.* Essentially similar to the female in structure, chaetotaxis, sculpture and coloration. Very similar to the male of *proclia*, differing mainly in the somewhat shorter antennae, somewhat narrower wings, and in details of coloration.

Antennae (Figure 595) rather stout; third funicular segment somewhat longer than wide, bearing 1–2 longitudinal sensilla; club rather short, $2\frac{1}{3}$ to nearly 3 times as long as wide, usually about $2\frac{1}{2}$ times longer than the preceding segment, bearing 3–4 longitudinal sensilla with a partly cut-off sensory area on ventral aspect.

Mesoscutum with 10–12 setae. Propodeum (Figure 596) about 3 to $3\frac{2}{3}$ times as long as the metanotum, about $\frac{3}{5}$ to $\frac{2}{3}$ length of scutellum; crenulae 5 + 5 to 6 + 6, elongate, as in the female.

Genitalia about $\frac{3}{4}$ to $\frac{9}{10}$ length of middle tibia (0.76–0.91), with a distinct rod between the apodemes; digital sclerites rather long, stout, about $\frac{1}{3}$ the combined length of aedeagus and apodemes (0.30–0.36).

Forewing (Figure 597) relatively narrow, $2\frac{3}{4}$ to about 3 times as long as wide; marginal fringe about $\frac{1}{4}$ to $\frac{1}{6}$ width of disk. Submarginal vein bearing 15–22 bullae. Marginal vein bearing 6–10 prominent, subequal setae along anterior margin. Forewing pattern as in the female.

General coloration somewhat paler than in the male of *proclia*. Abdominal tergites pale or very faintly infumated; seventh tergite with a pair of small fuscous blotches on the sides, eighth tergite with a pair of obscure submedian blotches, sometimes with faint indications of fuscous crossbands on those tergites. Antennal pedicel, funicle and club uniformly infuscated, as in *proclia*.

Length 0.62–1.02 mm.

In the female sex, *A. hispanicus* may be separated from *proclia* by the somewhat shorter, thicker antennae, the somewhat narrower forewing, the paler general coloration, and by details of antennal coloration. In previous publications, the relative length of the marginal fringe of the forewing was often used as one of the main distinguishing characters of *hispanicus*. This, unfortunately, is an entirely unreliable character, being affected to a large extent by the size of specimens.

Mercet's specimens of *hispanicus*, including the holotype, are in excellent shape but appear to be somewhat flattened on the slides by the pressure of the cover-glass. This has affected their shape, especially the apparent proportions of the antennal segments. Thus, the antennal club appears to be only $2\frac{1}{3}$ to $2\frac{1}{2}$ times as long as wide in these specimens, as compared with $2\frac{1}{2}$ (rarely) to nearly 3 times as long as wide in the other, better-mounted specimens from Spain.

*A. argentinus* (Figures 598–600) was described from a single female specimen, reared by A. de Llamas from *Parlatoria pergandii* Comstock, Misiones, Argentina, June 1, 1915. The type specimen was received on loan from Dr. Manuel Jose Viana of the Museo Argentino de Ciencias Naturales "Bernadino Rivadavia", Buenos Aires. It is cleared and mounted on a slide, and although the mounting medium is deteriorating, the specimen is still in very good shape. As suspected by Compere (1955), *argentinus* appears to be identical to *hispanicus*. As noted by Brèthes (1916), the mid-tibial spur of *argentinus* (Figure 600) is long, about as long as the corresponding basitarsus. This is common in small specimens of *hispanicus*, and should not be regarded as a specific, or "varietal", character. In view of the host record and similarity, *argentinus* is considered a junior synonym of *hispanicus*.

*A. bovelli* (Figures 601–603) was described from 5♀♀, reared by John H. Bovell from *Chrysomphalus dictyospermi* (Morgan), Bridgetown, Barbados. A female specimen, received on loan from the U.S. National Museum, Washington, was reared by Bovell from this host in that locality, October 28, 1917 and is apparently part of Malenotti's type series. (This appears to be the specimen mentioned by Compere (1955).) The specimen is mounted on a slide and, although brown with deposits, is in fairly good shape. It appears to be identical to *hispanicus*. Malenotti (1918b) noted that the pedicel, first two funicular segments, and apical half of the club of *bovelli* are more or less infuscated. This was misinterpreted by Compere (1955) to mean that "the third segment of the funicle and basal part of club [are] pallid in contrast to the pedicel, first two segments of the funicle, and apical one-half of the club which are fuscous"—but this description is not corroborated by the specimen at hand. *A. bovelli* is considered to be another junior synonym of *hispanicus*.

**Biology.** *A. hispanicus* is a thelytokous, solitary ectoparasite of the chaff scale, *Parlatoria pergandii* Comstock. Second-instar larvae, male pupae and adult females of this host may be parasitized, but the latter stage is clearly preferred. The egg may be deposited on either the dorsum or the venter of the body of the scale insect host, or among its eggs. The parasite can be reared in the laboratory on the latania scale, *Hemiberlesia lataniae* (Signoret). Development on this host averaged 34.3 days at 20°C, 16 days at 28°C; the theoretical threshold of development was calculated as 12.9°C, and the thermal constant as 240 days-degrees. Average fecundity was reported to be 23.8

progeny per female. At 28°C and 55% RH, with honey as food, adult longevity (to 75% mortality) was 24 days (Gerson, 1968).

**Related Forms.** Agarwal (1964:319) described an "Aligarh form" of *A. hispanicus*, reared from *Mytilaspis* sp. on *Jasminum* sp. at Aligarh, India. Unfortunately, his material was not available for study, while his description lacks important diagnostic characters such as the shape of the propodeum and crenulae. The "Aligarh form" is apparently similar to *hispanicus* in general coloration, except for the antennal club and third funicular segment of the female, which were described as uniformly light dusky. We are unable to determine the identity of this so-called "form" on the basis of available evidence. Agarwal's reference to his specimens as "endoparasites" is presumably incorrect.

A closely related, near sibling species, *A. comperei* DeBach and Rosen, has been found by us mixed with the populations of *A. hispanicus* in several countries. Records of *hispanicus* should therefore be regarded with caution unless adequately cleared slide-mounted material is available for confirmation.

## 35. **Aphytis comperei** DeBach and Rosen

(Figures 604–611)

*Aphytis maculicornis*: Quednau (nec Masi), 1964, J. Entomol. Soc. S. Afr. **27**:99 (part, misident.)
*Aphytis comperei* DeBach and Rosen, 1976, Ann. Entomol. Soc. Amer., **69**:543.

This widely distributed uniparental species, probably of Oriental origin, is very closely related to *hispanicus* and was apparently always confused with it until its recent description as a new species. It may be distinguished from *hispanicus* by the shorter, thicker antennae, with a definite black spot on the tip of the club, by the fewer, distinctly pigmented propodeal crenulae, by the somewhat broader wings, as well as by certain other details of coloration, chaetotaxis and wing pattern. To avoid unnecessary repetition, *comperei* is described below relative to *proclia* and *hispanicus*.

*Female.* Eyes and mouthparts as in *proclia* and *hispanicus*. Antennae (Figures 604, 605) distinctly shorter and thicker; scape as in *proclia* and *hispanicus*, $4\frac{1}{2}$ to nearly 6 times as long as wide, about $1\frac{1}{5}$ times as long as the club; pedicel shorter, $1\frac{2}{5}$ to $1\frac{2}{3}$ times as long as wide, about as long as or somewhat longer than the third segment of the funicle; first funicular segment trapezoidal, $1\frac{3}{5}$ times to fully twice as wide as long; second segment more symmetrical, sometimes saucer-shaped, as long as or slightly shorter than the first segment but considerably wider, usually 2 to $2\frac{1}{2}$ times as wide as long; third funicular segment about as long as wide, bearing 2–3 longitudinal sensilla; club about $2\frac{1}{3}$ times as long as wide ($2\frac{1}{5}$ to $2\frac{2}{5}$), usually about $2\frac{1}{3}$ times longer (2.07–2.62) and as wide as or slightly wider than the preceding segment, bearing 6–7 longitudinal sensilla.

Structure, chaetotaxis and sculpture of head and thorax similar to that of *hispanicus*. Vertex with 2 pairs of long setae along occipital margin, in addition to numerous shorter setae. Mesoscutum (Figure 606) with 10–11 (usually 10) setae; discoid sensilla on scutellum about equidistant from the anterior and posterior pairs of setae. Metanotum (Figure 607) somewhat shorter than in *hispanicus*; anteromedian apodeme usually somewhat more triangular, broader at base, more distinctly bifid at apex than in *hispanicus*, often somewhat longer than median length of metanotum.

Propodeum (Figures 606–608) shorter than in *hispanicus*, $3\frac{1}{3}$ to $4\frac{1}{5}$ times as long as the metanotum, somewhat less than $\frac{3}{5}$ to $\frac{2}{3}$ length of scutellum; central reticulate-strigose area broader than in *hispanicus*, transverse striation along posterior margin more distinct; crenulae fewer, 3 + 4 to 5 + 5, usually somewhat wider, more rounded than in *hispanicus*.

Sculpture of abdominal tergites similar to that of *proclia* and *hispanicus*. Second tergite faintly reticulate centrally; seventh tergite bearing 2 fine setae between the reticulate, setiferous lateral areas; eighth tergite with a transverse row of 4 setae between spiracles; syntergum (Figure 609) broad, more distinctly reticulate-striated, bearing 6–9 setae. Cerci situated closer to posterior spiracles than to tip of abdomen, as in *proclia* and *hispanicus*. Ovipositor shaft longer, $1\frac{2}{3}$ times to fully twice as long as the middle tibia (1.69–2.00), ovipositor sheaths about $\frac{2}{5}$ length of middle tibia (0.38–0.40).

Mid-tibial spur somewhat shorter than the corresponding basitarsus.

Forewing (Figures 610, 611) broader than in *hispanicus*, $2\frac{1}{2}$ to nearly $2\frac{2}{3}$ times as long as wide; marginal fringe usually short, not exceeding $\frac{1}{5}$ width of disk. Delta area less densely setose than in *hispanicus*, with 51–96 setae in 7–9 rows, these considerably longer and sparser than the setae distad of speculum, not clearly separated from row of setae along posterior margin of wing, widely separated from a few setae in 1–3 rows below distal half of submarginal vein; costal cell with a row of fine setae along proximal half or three fifths, and 1 coarse seta near apex. Submarginal vein bearing 2 coarse setae, the proximal one about $\frac{2}{3}$ to $\frac{4}{5}$ length of the distal, and 15–19 bullae. Marginal vein bearing 7–11 prominent, subequal setae along anterior margin.

Forewing pattern similar to that of *proclia*, but the fuscous crossband below the stigma is interrupted by hyaline centrally, forming 2 distinct fuscous clouds—one below the stigma, the other opposite it against the posterior margin of the wing.

General coloration similar to that of *hispanicus*. Genal sutures less heavily sclerotized, rather faintly infuscated. Mesoscutum faintly infuscated submedially just above posterior margin; propodeal crenulae blackish, concolorous with posterior margin of propodeum (Figures 606–608). Antennal pedicel, funicular segments and club more or less uniformly infuscated, paler than in *hispanicus*; tip of club black, the black pigmentation being confined to a small distinct area at the very tip (Figures 604, 605). Otherwise as in *hispanicus*.

Length 0.77–1.07 mm (usually under 1 mm).

*Male*. Unknown.

Described from the following material, all included in the type series: 3♀♀, reared by H. Compere from undetermined coccids on orange and pummelo, Fanling, New Territories, Hong Kong, November 18–24, 1932; 2♀♀, reared by J. L. Gressitt from

"*Aonidiella aurantii* (Maskell) and *Chrysomphalus aonidum* (L.) material" on citrus, near Canton, China, June and December 1948; 1♀ reared by D. W. Clancy from purple scale material (*Cornuaspis beckii* (Newman)) on citrus, Lake Alfred, Florida, September 22, 1958; 2♀♀, reared by A. G. Selhime from the chaff scale, *Parlatoria pergandii* Comstock, on Valencia orange, Goldenrod, Florida, January 3, 1961; 10♀♀, reared by H. Maltby together with *A. hispanicus* from *Parlatoria pergandii* on citrus, Ciudad Victoria, Mexico, July 18, 1961; 6♀♀, reared by H. Suarez together with *A. hispanicus* from *Parlatoria pergandii* on citrus, Linares, Nuevo Leon, Mexico, October 1961; 5♀♀, reared by P. DeBach together with *A. hispanicus* from *Parlatoria pergandii* on sour orange, Los Mochis, Sinaloa, Mexico, January 17, 1967; 4♀♀, collected by D. P. Annecke with the Florida red scale, *Chrysomphalus aonidum* (L.), "on naartjie," Durban, Natal, South Africa, May 1961; 37♀♀, reared by H. A. Dean together with *A. hispanicus* from *Parlatoria pergandii* on citrus, Weslaco, Texas, May 12, 1965; 1♀ (holotype), reared by W. Hart from "*Aonidiella aurantii* material" on citrus, McAllen, Texas, August 1968; 41♀♀, reared by L. W. van Whervin together with *A. hispanicus* from an undetermined host (recorded as probably purple scale) on citrus, Joalmi, Clarendon, Jamaica, West Indies, February 28, 1968; 36♀♀, same data, Denbign Kraal, Jamaica.

Holotype in the collection of the Division of Biological Control, University of California, Riverside.

**Additional Material.** Several specimens of *comperei* were found in the U.S. National Museum, mounted on a slide labeled as follows: "*Mytilaspis gloverii* on orange, Hong Kong, China—a few, always badly parasitized, *Parlatoria pergandii*?, Koebele" (together with several other species of *Aphytis, Marietta* and *Prospaltella*). It appears that, like *A. holoxanthus* (see p. 550), *A. comperei* could have been available for utilization in biological control ever since the early days of exploration, had its identity been recognized at the time.

**Notes.** *A. comperei* appears to be mixed with *hispanicus* on citrus in several countries. The peculiar coloration of the antennae, with a black spot at the tip of the club, the relatively pale genal sutures and dark propodeal crenulae may provide a shortcut to its separation from *hispanicus*. Both species are thelytokous, and crossing tests are of course impossible; nevertheless, the consistent morphological differences between them are considered sufficient for the description of *comperei* as a distinct species.

Although some of the host records listed above are undoubtedly confused, circumstantial evidence indicates that *comperei* is principally a parasite of the chaff scale, *Parlatoria pergandii* Comstock. Chaff scale occurs on citrus wherever *comperei* has been reared, and is often mixed with populations of California red scale, Florida red scale or purple scale, where it may go unnoticed due to its rather cryptic coloration. We assume that *comperei* is of Oriental origin and has spread to other zoogeographical regions together with its host on citrus. It should be investigated as a potential candidate for introduction against chaff scale wherever this serious pest causes economic injury to citrus.

We have named this species in honor of our late colleague and friend, Mr. Harold Compere, Specialist (Emeritus), Division of Biological Control, University of California, Riverside, in recognition of his everlasting contribution to the study of parasitic Hymenoptera. His pioneering work on the species of *Aphytis*, which culminated in his excellent 1955 paper, laid the solid foundation upon which the present study has been based.

## 36. **Aphytis philippinensis** DeBach and Rosen

(Figures 612–624)

"*Aphytis hispanicus* (Mercet): The Assam form": DeBach, 1959, Ann. Entomol. Soc. Amer., **52**: 358–360.
*Aphytis philippinensis* DeBach and Rosen, 1976, Ann. Entomol. Soc. Amer., **69**: 543.

This biparental, Southeast Asian species is rather closely related to *proclia* and *hispanicus*. It can be separated from both species by the more numerous setae on the mesoscutum, by the fewer, rounded propodeal crenulae, by the fewer setae on the posterior abdominal tergites, and by the absence of a distinct cut-off sensory area on the antennal club of the male. Abdominal coloration as in *proclia*, antennal coloration as in *hispanicus*.

*Female*. Eyes finely setose. Mandibles well developed, with 2 denticles and a dorsal truncation; maxillary palpi 2-segmented, labial palpi 1-segmented. Antennal scape (Figures 612, 613) 4 to $5\frac{2}{3}$ times as long as wide, about $1\frac{1}{4}$ times longer than the club; pedicel usually $1\frac{2}{3}$ times as long as wide or slightly longer, somewhat longer than the third segment of the funicle; first funicular segment trapezoidal, $1\frac{1}{3}$ to $1\frac{2}{3}$ times as wide as long; second segment more symmetrical, shorter and somewhat wider than the first, $1\frac{4}{5}$ to $2\frac{1}{2}$ times (usually over twice) as wide as long; third funicular segment somewhat longer than wide, bearing 2–3 longitudinal sensilla; club rather short, $2\frac{2}{5}$ to over $2\frac{4}{5}$ times as long as wide, $2\frac{1}{4}$ to nearly $2\frac{2}{3}$ times longer and somewhat wider than the preceding segment, bearing 6–8 longitudinal sensilla.

Setae on head, thorax, and sides of abdomen dark, moderately coarse, readily visible under $\times 60$ magnification. Vertex with 2 pairs of long setae along occipital margin, in addition to numerous short setae. Mesoscutum (Figures 614, 615) with 11–19 setae (average 15); each parapsis with 2 short setae, each axilla with 1 seta; scutellum with 4 setae, the discoid sensilla usually equidistant from the two pairs. Frontovertex, pronotum, and mesonotal sclerites reticulate. Scutellum (Figure 614) oval, usually about $\frac{3}{4}$ median length of mesoscutum. Metanotum (Figures 614, 616) reticulate except on the sides; anteromedian apodeme about $\frac{2}{3}$ to $\frac{4}{5}$ median length of metanotum.

Propodeum (Figures 616–618) 3–4 times as long as the metanotum, $\frac{3}{4}$ to $\frac{4}{5}$ length of scutellum, delicately reticulate on the sides, reticulate-strigose on a relatively wide central area, narrowly transversely striated along posterior margin; crenulae 4 + 5 to 6 + 6, rounded, sometimes slightly overlapping.

Second abdominal tergite (Figures 616–618) faintly transversely striated anteriorly on the sides, smooth centrally; tergites III–VI reticulate on the sides, bearing 2–5 setae in a transverse row on each reticulate area; third tergite (Figure 616) reticulate on a wider area, clearly transversely striated across center; tergites IV–VI with indications of transverse striation across center; seventh tergite (Figure 619) reticulate on the sides, bearing 4–7 setae on each reticulate area, delicately reticulate-striated and bearing 2 fine setae across center; eighth tergite delicately reticulate, with a transverse row of 4 setae between spiracles; syntergum (Figure 619) broad, triangular, faintly striate-punctate, bearing a transverse row of 12–17 setae. Cerci situated closer to posterior spiracles than to tip of abdomen, with 2 long setae and 1 short seta. Ovipositor shaft $1\frac{3}{5}$ to somewhat over $1\frac{2}{3}$ times as long as the middle tibia (1.59–1.71), ovipositor sheaths $\frac{1}{3}$ to $\frac{2}{5}$ length of middle tibia (0.33–0.39).

Mid-tibial spur somewhat shorter than the corresponding basitarsus.

Forewing (Figures 620, 621) about $2\frac{1}{2}$ to $2\frac{2}{3}$ times as long as wide; marginal fringe short, not exceeding $\frac{1}{6}$ width of disk. Delta area densely setose, with 87–158 setae in 11–14 rows, these considerably longer but only slightly sparser than the setae distad of speculum, not clearly separated from row of setae along posterior margin of wing, widely separated from 1–4 rows of setae below distal two thirds of submarginal vein; costal cell with a row of fine setae along proximal three fifths or two thirds, and 1–2 coarse setae near apex. Submarginal vein bearing 2 coarse setae and 16–20 bullae. Marginal vein bearing 8–12 prominent, subequal setae along anterior margin.

Forewing with a dark spot below stigma, merging into a broad, obscure, faintly infuscated cloud; another obscure cloud opposite it against posterior margin of wing; a darker, arcuate crossband at base of delta, separated by a paler area from a fuscous cloud on setiferous area below submarginal vein; a narrow fuscous streak along posterior margin of wing, distad of speculum.

General coloration similar to that of *proclia*, yellowish with fuscous and blackish markings. Occiput with a transverse black bar on each side of foramen; genal sutures marked with black from oral margin to about halfway to eyes; lower part of face and oral margin broadly suffused with blackish. Central part of pronotum and apical parts of propleura blackish; mesoscutum suffused with fuscous, with a pair of darker blotches anteriorly, the midline paler, the anterior and posterior margins lined with blackish, parapsidal sutures fuscous; scutellum more strongly infuscated, mainly on both sides of the pale midline, with a pair of antero-submedian blackish blotches, the posterior margin lined with blackish; metanotum strongly infuscated centrally along anterior margin and on the sides; propodeum pale, the sides of anterior margin, sutures mesad of spiracles, and posterior margin lined with blackish; crenulae pale, sometimes not clearly visible. Second abdominal tergite strongly infuscated centrally, the anterior, lateral, and latero-posterior margins lined with blackish; tergites III–VII with short, transverse blackish stripes on lateral reticulate areas, third tergite also with a narrow, transverse blackish crossband anteriorly, seventh tergite with a paler but distinct fuscous crossband, center of abdomen pale; eighth tergite faintly infuscated posteriorly, transverse ridge on posterior margin blackish; syntergum and ovipositor sheaths immaculate. Thoracic sterna strongly infuscated or blackish. Internal apodemes, lateral margins of endophragma, ridge and lateral parts of outer ovipositor plates blackish; triangular plates and second (posterior) rami of ovipositor arms fuscous. Antenna

(Figure 612) relatively dark; scape pale with a broad longitudinal fuscous stripe; pedicel, funicular segments and most of club uniformly blackish, apical portion of club black. All femora with fuscous markings, all tibiae entirely fuscous. Wing veins lined with brownish.

Length about 1 mm (0.94–1.18 mm).

*Male.* Similar to the female in structure, chaetotaxis and sculpture.

Antennal club (Figure 622) bearing 3–4 longitudinal sensilla, without a well-defined cut-off sensory area on the ventral aspect as in *proclia*. Mesoscutum with 12–16 setae. Propodeum with 3 + 4 to 5 + 5 crenulae. Forewing (Figure 624) somewhat wider than in the female, $2\frac{2}{5}$ to $2\frac{1}{2}$ times as long as wide.

Genitalia (Figure 623) shorter than in *proclia*, about $\frac{3}{5}$ to $\frac{3}{4}$ length of middle tibia (0.61–0.77), the rod between apodemes absent or vestigial; digital sclerites about $\frac{1}{3}$ the combined length of aedeagus and apodemes (0.31–0.33).

General coloration considerably paler than in the female. Black bars on occiput often less conspicuous. Mesonotal sclerites paler, but the antero-submedian fuscous blotches present on mesoscutum and scutellum. Abdominal tergites without the transverse blackish stripes on lateral sculptured areas, tergites IV–VIII with faintly infuscated crossbands across center (sometimes missing from fourth and fifth tergites, rather broad on seventh and eighth tergites). Antennal pedicel, funicular segments and club uniformly fuscous, paler than in the female, apex of club not blackish. Otherwise as in the female.

Length 0.76–0.90 mm.

Described from 13♀♀ and 6♂♂, reared by P. DeBach from the California red scale, *Aonidiella aurantii* (Maskell), on pummelo and lemon, Gauhati, Assam, India, March 26, 1957; 43♀♀, 8♂♂ (♀ holotype, ♂ allotype, and paratypes), reared by V. G. Ortega from the Florida red scale, *Chrysomphalus aonidum* (L.), on coconut palm, Mandaue, Cebu, Philippines, June 14, 1965; 2♂♂, reared by V. P. Rao from *Aonidiella aurantii* on rose, Hebbal, Bangalore, India, August 1966.

Type series in the collection of the Division of Biological Control, University of California, Riverside.

**Notes.** The Assam and Philippines series are considered to be conspecific. The slight differences between these stocks are apparently attributable to different methods of preparation of the specimens. The Philippines specimens appear to be somewhat flattened on the slides, with a resulting effect on the proportions of the antennal segments, etc. To avoid possible confusion, the type series is restricted to the Philippines specimens.

*A. philippinensis* is rather closely related to *proclia* and *hispanicus*. It resembles *proclia* in general coloration, *hispanicus* in the proportions and coloration of the antennal segments. It differs markedly from both species in the fewer, more rounded propodeal crenulae, as well as in the larger number of mesoscutal setae and the fewer setae on the abdominal tergites; the constant, small number of setae on the seventh and eighth abdominal tergites appears to be a valid diagnostic character in this case.

The male of *philippinensis* differs markedly from that of *proclia* and of *hispanicus* in the absence of a well-defined cut-off sensory area on the antennal club, as well as in the shorter genitalia, lacking a distinct rod between the apodemes. The female also differs from *proclia* in antennal coloration.

*A. philippinensis* is suggestive of *diaspidis* in the number of mesoscutal setae and in the shape and number of propodeal crenulae. It differs markedly from *diaspidis* in abdominal coloration, the center of the abdomen being pale as in *proclia*, in the fewer number of setae on the abdominal tergites and, apparently, also in the biparental mode of reproduction.

**Utilization in Biological Control.** *A. philippinensis* was found to be the dominant parasite of California red scale in Assam. Shipments were made to California in March 1957, but the parasites did not survive the journey (Rosen and DeBach, 1978). This interesting species should receive more attention as a potentially important natural enemy of California red scale and Florida red scale, both of which are well-known, serious pests of citrus, other fruit trees, and ornamentals.

## 37. **Aphytis vandenboschi** DeBach and Rosen

(Figures 98, 107, 113, 114, 116, 123, 124, 127, 161, 175, 176, 187, 188, 195, 199, 200, 221, 222, 226; and 625–630)

*Aphytis diaspidis*: Azim (nec Howard), 1963, J. Fac. Agric. Kyushu Univ., **12**:276–278; misident.
*Aphytis vandenboschi* DeBach and Rosen, 1976, Ann. Entomol. Soc. Amer., **69**:543.

This uniparental Japanese species may be separated from other members of the **proclia** group by a combination of characters. It resembles *proclia* in the coloration and chaetotaxis of the abdominal tergites, but is more similar to *diaspidis* in having somewhat fewer, more rounded propodeal crenulae, in the relative length of the ovipositor, and in that the basal part of the antennal club is about as strongly infuscated as the funicle. In the number of mesoscutal setae, as well as in general habitus, *vandenboschi* may probably be considered as occupying an intermediate position between *proclia* and *diaspidis*. It is described below relative to *proclia*.

*Female.* Eyes (Figure 98) finely setose. Mouthparts (Figure 113) as in *proclia*. Antennae (Figures 116, 123, 625) essentially as in *proclia*; third funicular segment about as long as wide or somewhat longer than wide, bearing 1–3 longitudinal sensilla; club about 3 times as long as wide ($2\frac{4}{5}$ to $3\frac{1}{5}$), $2\frac{1}{2}$ to $2\frac{4}{5}$ times longer than, and about as wide as or slightly wider than, the preceding segment, bearing 5–7 longitudinal sensilla.

Chaetotaxis and sculpture of head and thorax essentially as in *proclia*; vertex with 2 pairs of long setae along occipital margin, in addition to numerous shorter setae. Mesoscutum (Figure 626) usually with 10–13 setae (6 in the smallest specimen, up to

15 in large ones), each parapsis with 2, each axilla with 1 seta; scutellum with 4 setae, the discoid sensilla closer to the anterior than to the posterior pair. Scutellum about $\frac{3}{4}$ median length of mesoscutum. Metanotum (Figure 627) as in *proclia*; anteromedian apodeme shorter than median length of metanotum.

Propodeum (Figures 175, 176, 627, 628) relatively short, about $2\frac{2}{3}$ to 3 times as long as the metanotum, usually $\frac{3}{5}$ to $\frac{2}{3}$ length of scutellum, sculptured as in *proclia*; crenulae usually fewer than in *proclia*, 3 + 4 to 8 + 8, rather rounded, not slender and elongate as in *proclia*.

Sculpture and chaetotaxis of abdominal tergites (Figure 629) as in *proclia*; seventh tergite bearing 3–5 fine setae in a transverse row between the lateral setiferous areas; eighth tergite with 6–10 setae in a transverse row between spiracles; syntergum triangular, with 10–14 setae in 1–2 rows. Cerci as in *proclia*. Ovipositor shaft showing marked variation in relative length, inversely correlated with size of body: $1\frac{1}{3}$ times as long as the middle tibia in large specimens, to $1\frac{3}{5}$ times in small specimens, up to twice as long as the middle tibia in the smallest specimen (see also Figure 659); ovipositor sheaths about $\frac{2}{5}$ length of middle tibia (0.36–0.40).

Forewing (Figure 630) essentially as in *proclia*; marginal vein bearing 7–10 prominent, subequal setae along anterior margin; pattern as in *proclia*.

General coloration of head, thorax and abdomen essentially as in *proclia*, probably somewhat paler. The faint fuscous patches on the occiput are not evident; abdominal tergites III–VI with short, transverse fuscous stripes on lateral reticulate areas, third tergite also with a narrow fuscous crossband anteriorly; seventh tergite with a complete fuscous crossband; center of abdomen pale. Antennal scape pale, with a longitudinal fuscous stripe ventrally; pedicel, funicular segments and basal part of club uniformly, rather strongly infuscated, tip of club blackish.

Length 0.57–1.14 mm.

*Male*. Unknown.

Described from numerous ♀ specimens (holotype and paratypes), reared in the insectary at Riverside on the oleander scale, *Aspidiotus nerii* Bouché [= *A. hederae* (Vallot)], originally obtained by R. van den Bosch from the San Jose scale, *Quadraspidiotus perniciosus* (Comstock), on pear, Kashiwabaru, Fukuoka, Japan, May 20, 1964.

Type series in the collection of the Division of Biological Control, University of California, Riverside.

**Notes.** *A. vandenboschi* exhibits slight but consistent differences from all other species in the **proclia** group. It differs from *proclia* in the somewhat fewer, more rounded crenulae, in antennal coloration and in the pronounced variation in the relative length of the ovipositor; from *hispanicus* in the shape of the crenulae; from *philippinensis* in the more numerous setae on the posterior abdominal tergites; from *maculicornis*, *paramaculicornis* and *comperei* in details of antennal pigmentation; and from *diaspidis* in many details of coloration, especially in the absence of a central blotch on the abdomen.

**Biology.** *A. vandenboschi* is apparently identical to the species recorded by Azim (1963b, 1963c) from Japan as *A. diaspidis*. He reported it as a common, thelytokous parasite of the San Jose scale, *Quadraspidiotus perniciosus* (Comstock), attacking second-instar and young adult females but not ovipositing females or male prepupae. Although up to 6 eggs may be laid on one scale, only one adult parasite usually emerges from each parasitized host. Fecundity was recorded as 22 eggs or more per female. The life cycle in the laboratory in Japan totaled 16–19 days: egg incubation 3–4 days, larval development 6–7 days, the prepupal stage 1 day, pupal development 6–7 days. Adult longevity averaged 19 days, with honey as food. Host-feeding was observed.

Azim (1963c) recorded this species also from the bifasciculate scale, *Chrysomphalus bifasciculatus* Ferris, the white peach scale, *Pseudaulacaspis pentagona* (Targioni-Tozzetti) and *Lepidosaphes* sp., and observed it in the laboratory to oviposit also in the Asiatic red scale, *Aonidiella taxus* Leonardi. At the Riverside insectary, *vandenboschi* accepted about equally well the oleander scale, *Aspidiotus nerii* Bouché [= *A. hederae* (Vallot)], the cactus scale, *Diaspis echinocacti* (Bouché) and the latania scale, *Hemiberlesia lataniae* (Signoret), but fecundity on these hosts was rather low. The life cycle on oleander scale takes 15–19 days at 26.7°C, with peak emergence occurring on the 18th day.

**Potential Value for Biological Control.** Azim (1963c) recorded *vandenboschi* as a quite effective parasite of the San Jose scale in Japan. This species should be regarded as an important candidate for introduction against this serious pest of deciduous trees.

We have named this species in honor of Dr. Robert van den Bosch, Division of Biological Control, University of California, Berkeley, an assiduous foreign explorer and outstanding proponent of biological control, who collected the type material in Japan.

## 38. **Aphytis confusus** DeBach and Rosen

(Figures 631–641)

*Aphytis maculicornis*: Quednau (nec Masi), 1964, J. Entomol. Soc. S. Afr., **27**:99 (part, misident.)
*Aphytis confusus* DeBach and Rosen, 1976, Ann. Entomol. Soc. Amer., **69**:543.

This biparental South African species may be readily separated from other members of the **proclia** group by the peculiar shape of the antennal club of the female, which tapers strongly to a pointed apex; by the rounded, closely appressed propodeal crenulae, which are rather conspicuously blackish; and by the short digital sclerites of the male genitalia, in combination with details of chaetotaxis and coloration.

*Female.* Eyes finely setose. Mandibles (Figure 631) well developed, with 2 denticles and a dorsal truncation; maxillary palpi 2-segmented, labial palpi 1-segmented. Antennal scape (Figures 631, 632) slender, $4\frac{1}{2}$ to $5\frac{1}{4}$ times as long as wide, considerably longer than the club; pedicel usually rather slender, a little over $1\frac{1}{2}$ times to nearly twice as long as wide, as long as or somewhat longer than the third segment of the funicle; first funicular segment trapezoidal, a little over $1\frac{1}{2}$ times to nearly twice as wide as long; second segment symmetrical, compressed, considerably shorter and wider than the first, $2\frac{1}{3}$ to nearly 3 times as wide as long; third funicular segment a little longer than wide, bearing 2 longitudinal sensilla; club tapering strongly toward a rather pointed apex, $2\frac{1}{2}$ to 3 times as long as wide, $2\frac{2}{5}$ to $2\frac{3}{4}$ times longer than the preceding segment and about as wide, bearing 6 longitudinal sensilla.

Setae on head, thorax and abdomen generally slender, except for the lateral setae on the pronotum, the antero-lateral and posterior pairs on the mesoscutum, and the posterior pair on the scutellum, which are visible under $\times 60$ magnification; setae on abdomen invisible even under $\times 120$ magnification. Vertex with 2 pairs of coarse, medium-long setae along occipital margin, in addition to numerous fine, short setae. Mesoscutum with 9–11 setae, slender except for the posterior pair and 1 seta at each antero-lateral corner; each parapsis with 2 short setae, each axilla with 1 seta; scutellum (Figure 633) with 4 setae, the discoid sensilla about equidistant from the two pairs. Frontovertex, pronotum and mesonotal sclerites reticulate. Scutellum oval, about $\frac{4}{5}$ median length of mesoscutum. Metanotum (Figures 633, 634) reticulate except on the sides, tending toward transversely strigose along anterior margin; anteromedian apodeme slender, distinctly shorter than median length of metanotum.

Propodeum (Figures 633–635) about 3–4 times as long as the metanotum, $\frac{2}{3}$ to $\frac{3}{4}$ median length of scutellum, faintly reticulate on the sides, reticulate-strigose on a relatively wide central area; crenulae 5 + 5 to 6 + 7, conspicuous, rounded, closely appressed, sometimes slightly overlapping.

Second abdominal tergite (Figure 635) transversely reticulate-striated anteriorly on the sides, smooth centrally; tergites III–VII (Figure 636) reticulate on the sides, bearing 2–4 setae on each reticulate area, third and seventh tergites reticulate on a wider area, with indications of transverse striation centrally; seventh tergite bearing 2 fine setae submedially; eighth tergite with a transverse row of 4–6 setae between spiracles; syntergum triangular, broad, with a transverse row of 7–8 setae. Cerci situated closer to posterior spiracles than to tip of abdomen, with 2 long setae and 1 short seta. Ovipositor shaft about $1\frac{1}{2}$ times as long as the middle tibia (1.44–1.52), ovipositor sheaths $\frac{1}{3}$ to $\frac{2}{5}$ length of middle tibia (0.32–0.39).

Mid-tibial spur over $\frac{4}{5}$ length of the corresponding basitarsus.

Forewing (Figure 637) a little over $2\frac{1}{2}$ to $2\frac{3}{4}$ times as long as wide; marginal fringe not exceeding $\frac{1}{5}$ width of disk. Delta area with 77–97 setae in 8–10 rows, these considerably longer and sparser than the setae distad of speculum, clearly separated from row of setae along posterior margin of wing, widely separated from a few setae in 1–2 rows below distal half of submarginal vein; costal cell with a row of fine setae along proximal two thirds, and 1 coarse seta near apex. Submarginal vein bearing 2 coarse setae, the proximal one about $\frac{3}{5}$ length of the distal, and 15–23 bullae. Marginal vein bearing 9–11 subequal, relatively short, prominent setae along anterior margin, these about $1\frac{1}{2}$ to $1\frac{3}{4}$ times longer than the setae in a row along center of vein.

Forewing with a distinct dark spot below the stigma which fades into a paler, but distinct, triangular fuscous cloud; a small fuscous cloud opposite it against posterior margin of wing; a fuscous cloud below distal part of submarginal vein, separated by a paler area from an arcuate, narrow fuscous crossband below base of marginal vein; a narrow fuscous streak along posterior margin of wing, distad of speculum.

General coloration similar to that of *proclia*, yellowish with fuscous markings. Occiput with a transverse black bar on each side of foramen; lower part of face and oral margin suffused with blackish; genal sutures blackish from oral margin to about $\frac{2}{3}$ distance to eye. Pronotum fuscous to blackish centrally; mesoscutum with 2 fuscous blotches antero-submedially, parapsidal sutures lined with blackish; scutellum more extensively fuscous, somewhat paler on midline and on sides, posterior margin lined with blackish; metanotum fuscous on reticulate area antero-medially, paler posteriorly and on sides; propodeum with the sutures mesad of the spiracles and the posterior margin lined with blackish; crenulae blackish, distinct. Second abdominal tergite fuscous centrally, lateral parts of anterior and posterior margins lined with blackish; tergites III–VII with short transverse blackish stripes on lateral reticulate areas, pale centrally; eighth tergite with 2 sublateral blotches, connected centrally by a faintly fuscous stripe. Thoracic sterna black or blackish. Internal apodemes blackish; endophragma (Figures 633, 636) broadly lined with blackish on lateral margins; outer ovipositor plates infuscated along lateral margins. Antennal scape mostly fuscous; pedicel fuscous, paler than the flagellum; funicular segments and club strongly, rather uniformly fuscous, club somewhat more blackish toward apex. All femora pale, marked with fuscous toward apex; all tibiae rather strongly infuscated. Wing veins lined with brownish.

Length about 1 mm (0.94–1.07 mm).

*Male*. Very similar to the female in structure, chaetotaxis, sculpture and coloration.

Antennal pedicel (Figure 638) $1\frac{1}{2}$ to $1\frac{3}{4}$ times as long as wide; third funicular segment bearing 1–2 longitudinal sensilla; club $2\frac{1}{3}$ to over 3 times as long as wide, bearing 2–3 longitudinal sensilla, with a ventral, partly cut-off sensory area bearing minute setae as in *proclia*.

Mesoscutum with 9–10 setae. Propodeal crenulae (Figure 639) as in the female, 3 + 3 to 6 + 6. Marginal vein of forewing (Figure 641) with 6–10 prominent, subequal setae along anterior margin.

Genitalia (Figure 640) $\frac{2}{3}$ to $\frac{3}{4}$ length of middle tibia (0.67–0.75), with a distinct rod between the apodemes; digital sclerites relatively short, about $\frac{1}{5}$ the combined length of aedeagus and apodemes (0.20–0.22).

General coloration and wing pattern as in the female; abdominal tergites with more or less complete, rather faintly fuscous, transverse crossbands. Antennal flagellum uniformly fuscous, club not blackish toward apex; pedicel paler.

Length 0.75–0.85 mm.

Described from 11♀♀ and 3♂♂ (♀holotype, ♂ allotype, and paratypes), reared by D. P. Annecke from *Ledaspis distincta* (Leonardi) on *Protea caffra*, Fountains, Pretoria, South Africa, August 1961; 1♀, 1♂ (paratypes), reared by D. P. Annecke from *Africaspis chionaspiformis* (Newstead) or *Melanaspis corticosa* (Brain), Ngwavuma, Natal, South Africa, September 1961.

Type series in the collection of the Division of Biological Control, University of California, Riverside.

**Notes.** Quednau (1964b) did not recognize *confusus* as a new species, but misidentified it as *maculicornis*. This species differs from *maculicornis* in the unusually tapering antennal club of the female, in the short digital sclerites in the male genitalia, and in the shape of the crenulae, which are dark, conspicuous, rounded and closely appressed in both sexes (in *maculicornis* the crenulae are pale, inconspicuous, rather elongate, and well separated). The coloration of the antennal flagellum is a good distinguishing character: the flagellum is uniformly infuscated in both sexes of *confusus*; in the female of *maculicornis*, the basal portion of the club is hyaline and stands in strong contrast to the infuscated funicle; in the male of *maculicornis* the entire club is considerably paler than the funicle.

## 39. **Aphytis diaspidis** (Howard)

(Figures 642–656)

*Aphelinus diaspidis* Howard, 1881, U.S. Commiss. Agr. Ann. Rept. for 1880, p. 355.
*Aphelinus fuscipennis* Howard, 1881, U.S. Commiss. Agr. Ann. Rept. for 1880, p. 356; n. syn.
*Aphelinus ovidii* Girault, 1919, Hymenoptera Chalcidoidea nova Australiensis. Priv. Publ., Brisbane, 3 pp., page 1; n. syn.
*Aphytis (Prospaphelinus) proclia*: De Santis (nec Walker), 1948, Rev. Mus. La Plata (N.S.), **5**, Zool., 133–138.
*Aphytis opuntiae* Risbec (nec Mercet), 1952, Mem. Inst. Sci. Madagascar, Ser. E., **2**:150–152; n. syn.
*Prospaphelinus madagascariensis* Risbec, 1952, Mem. Inst. Sci. Madagascar, Ser. E., **2**:156–158; n. syn.
*Aphytis diaspidis*: Compere, 1955, Univ. Calif. Publ. Entomol., **10**:289–294; Quednau, 1964, J. Entomol. Soc. S. Afr., **27**:98–99.
*Aphytis madagascariensis*: Annecke and Insley, 1971, S. Afr. Dept. Agr. Tech. Serv. Entomol. Mem., **23**:28.
*Aphytis risbeci* Annecke and Insley, 1971, S. Afr. Dept. Agr. Tech. Serv. Entomol. Mem., **23**:29 (new name for *A. opuntiae* Risbec); n. syn.

Extensive bibliographies of *diaspidis* are given by De Santis (1948) and Peck (1963) under *A. proclia*. Some of the references to *diaspidis* in the literature may actually pertain to other species of *Aphytis*. For example, Compere (1955) pointed out that "from 1894 to 1915, *diaspidis* was applied incorrectly to the common red-scale-inhabiting species now known as *Aphytis chrysomphali* (Mercet)."

This cosmopolitan species is the most extensively pigmented member of the **proclia** group. It is quite closely related to *proclia* and was frequently confused with this species in the past, but may be readily separated from it by several obvious characters. The general pattern of pigmentation is much more extensive in *diaspidis* than in *proclia*, with a large fuscous blotch on the center of the abdomen; antennal pedicel, funicle and base of club uniformly fuscous; mesoscutal setae more numerous than in *proclia*, propodeal crenulae fewer and more rounded. Male with complete fuscous crossbands on abdominal tergites.

*Female.* Head, mouthparts and antennae (Figure 642) essentially as in *proclia*. Structure of setae as in *proclia*; mesoscutum (Figures 643–645) with 9–21 setae (average in material at hand 14.5), this number being in direct correlation with the size of

specimens. Other details of thoracic structure, chaetotaxis and sculpture as in *proclia*.
Propodeum (Figures 643, 646–649) essentially as in *proclia*; crenulae usually wider, more rounded, fewer (3 + 3 to 6 + 7), somewhat less distinct than in *proclia*. Abdominal tergites (Figure 650) essentially as in *proclia*, the transverse striation on the center of tergites IV–VI somewhat more pronounced. The relative length of the ovipositor shaft is in inverse correlation with the size of specimens, varying from about $1\frac{1}{3}$ times length of the middle tibia in larger specimens to $1\frac{2}{3}$–$1\frac{3}{4}$ length of the middle tibia in the smallest specimens (see Figure 659); ovipositor sheaths as in *proclia*.
Legs and wings (Figure 652) essentially as in *proclia*; submarginal vein sometimes bearing a few supernumerary coarse setae in large specimens; marginal vein bearing 6–12 prominent, subequal setae along anterior margin.
Forewing pattern (Figure 652) very similar to that of *proclia*, but the general background is slightly more infumated and the fuscous crossband below the stigma is broader, somewhat less prominent. Figure 653 shows details of integumental pigmentation below the stigma.
General pattern of coloration similar to that of *proclia* but considerably darker. Lower part of head usually more extensively blackish. Mesoscutum rather extensively, usually strongly, infuscated except for a distinct pale longitudinal midline, the pigmentation fading gradually toward the sides; posterior margin entirely lined with fuscous; scutellum more extensively infuscated than in *proclia*, midline and sublateral areas pale; propodeum lined with fuscous also along anterior margin. Otherwise, head, thorax and propodeum as in *proclia*. Second abdominal tergite as in *proclia*; subsequent tergites with distinct, transverse fuscous crossbands, darker on lateral reticulate areas but continuous across center, sometimes forming a continuous fuscous blotch on center of abdomen. Thoracic sterna, internal apodemes, ovipositor plates and legs as in *proclia*. Antennal scape pale, with a longitudinal fuscous stripe; pedicel, funicular segments and most of club uniformly fuscous, usually more strongly so than in *proclia*; tip of club usually blackish, but in a few small specimens the club appears uniformly infuscated.
Length 0.61–1.26 mm.

*Male*. Very similar to the male of *proclia*, differing mainly in details of coloration.
Antennae (Figures 654, 655) as in *proclia*, with a similar, partly cut-off area on ventral aspect of club (Figure 655). Mesoscutum with 8–17 (usually 11–15) setae. Propodeal crenulae as in the female, fewer than in *proclia*, 2 + 3 to 5 + 7. Genitalia essentially as in *proclia*, $\frac{2}{3}$ to $\frac{4}{5}$ length of middle tibia (0.67–0.79); digital sclerites about $\frac{1}{3}$ the combined length of aedeagus and apodemes (usually 0.29–0.33; 0.38 in one specimen).
Forewing pattern similar to the female but somewhat fainter.
Coloration of head, thorax and propodeum similar to that of the female but somewhat paler and less extensive. Abdominal tergites III–VIII with complete fuscous crossbands. Antennal scape as in the female, flagellum uniformly infuscated.
Length 0.61–1.11 mm.

Redescribed from numerous specimens reared in California from the greedy scale, *Hemiberlesia rapax* (Comstock), the latania scale, *H. lataniae* (Signoret), the oleander scale, *Aspidiotus nerii* Bouché [= *A. hederae* (Vallot)], the walnut scale, *Quadraspidiotus juglansregiae* (Comstock), the San Jose scale, *Q. perniciosus* (Comstock), the California red scale, *Aonidiella aurantii* (Maskell), the cactus scale, *Diaspis echinocacti* Bouché), and the pittosporum scale, *Parlatoria pittospori* Maskell. This common California form agrees in all morphological characters with the type series and is uniparental (thelytokous); males occur occasionally but are apparently not essential for reproduction.

The type series of *diaspidis* is in the U.S. National Museum of Natural History (#2596).

**Notes.** The distinctive pattern of pigmentation provides the easiest diagnostic characters by which *diaspidis* may be separated from *proclia*: the center of the abdomen is infuscated in *diaspidis*, immaculate in *proclia*; the mesonotum is more extensively infuscated in *diaspidis*; the antennal funicle and base of club are uniformly infuscated in *diaspidis*, whereas in *proclia* the basal part of the club appears paler than the funicular segments.

In most morphological characters there is considerable overlap between *diaspidis* and *proclia*. However, when plotted against the size (length) of specimens, certain characters show distinctly different trends in the two species. The number of mesoscutal setae, for instance, is positively correlated with the size of specimens in both species, but the trend is much steeper in *diaspidis*, and the number of setae in larger specimens (over 1 mm long) is larger in this species than in specimens of *proclia* of comparable size (see Figure 657). On the other hand, the number of propodeal crenulae is considerably larger in *proclia*, and the trend is markedly steeper in this species (see Figure 658); the crenulae of *diaspidis* are also more rounded, less elongate than in *proclia*. In *diaspidis*, the relative length of the ovipositor is in marked inverse correlation with the size of specimens; such correlation is evident also in *vandenboschi*, but is not apparent in *proclia* (see Figure 659).

Howard (1881) based his original description of *diaspidis* on 9♀♀ and 2♂♂, reared from the rose scale, *Aulacaspis rosae* (Bouché), in Florida, February 20, 1880, and applied that name also to 2♀♀ reared from the same host in California. Compere (1955) suggested that the Florida type series may have represented a biparental (arrhenotokous) form.

*Aphelinus fuscipennis* was described by Howard (1881) from 9♀♀, reared from *Mytilaspis* [= *Lepidosaphes*] sp. in California, Florida and the District of Columbia. Two female specimens of the type series of *fuscipennis* (U.S. National Museum of Natural History, #2598) (Figures 645, 648) were removed from the tags on which they had been mounted, and were then cleared and mounted in Hoyer's medium; they were found to be conspecific with *diaspidis*. Since *diaspidis* has page precedence, we consider *fuscipennis* as a junior synonym.

The female type of *Aphelinus ovidii* Girault was received on loan from the Queensland Museum and was also studied, along with 3♀♀ determined by A. A. Girault as *ovidii* (Figure 649), all from Australia. They were also found to be conspecific with *diaspidis*.

*Aphytis opuntiae* was described by Risbec (1952) from a single (dry) female specimen, reared by R. Paulian from *Pseudaulacaspis pentagona* (Targioni-Tozzetti) on *Solanum auriculatum*, Tsimbazaza, Madagascar, February 24, 1950. Since the name *opuntiae* is preoccupied in *Aphytis*, Annecke and Insely (1971) renamed this species *risbeci*. The type specimen was received on loan from the Muséum National d'Histoire Naturelle, Paris, and was cleared and mounted in Hoyer's medium. This is undoubtedly another synonym of *diaspidis*. Risbec's figure of the antenna of *opuntiae*, showing an elongate second funicular segment and a bi-segmented club, is erroneous.

*Prospaphelinus madagascariensis* was described by Risbec (1952) from 3♀♀ (syntypes), reared from *Pseudaulacaspis pentagona* (Targioni-Tozzetti) on *Solanum auriculatum*, Tsimbazaza, Madagascar, February 27, 1950. The specimens, cleared but rather poorly mounted on one slide, were received for study from the Muséum National d'Histoire Naturelle, Paris, and were found to be conspecific with *diaspidis*. Risbec's figure of *madagascariensis*, showing an encyrtid-type structure of the thorax, is mostly erroneous.

Two series at hand apparently represent biparental "forms" of *diaspidis*: I) 13♀♀, 13♂♂, relatively small, reared by W. White from *Quadraspidiotus juglansregiae* on English walnut, Riverside, California, January 20, 1967; II) 9♀♀, 23♂♂, relatively large, reared by O. Beingolea from *Diaspis echinocacti* (Bouché) on cactus, Lima, Peru, June 8, 1961. The females of these series do not differ in any structural characters, nor in coloration, from the common uniparental form. The males are also very similar to those of the uniparental form, except for a slight difference in the appearance of the genitalia. In the occasional males of the common uniparental form, the proximal part of the phallobase is sometimes not distinct, and the genitalia therefore appear somewhat shorter than in *proclia*, $\frac{2}{3}$ to $\frac{4}{5}$ length of the middle tibia; in the males of the biparental series, the phallobase is clearly delimited and the genitalia appear identical to those of *proclia*, and usually somewhat longer than in the uniparental form (0.70–0.87 length of middle tibia). For lack of better distinguishing characters or any conclusive biological information, these series are for the time being considered to be conspecific with *diaspidis*. However, if further studies eventually prove that they are indeed reproductively isolated from the uniparental form, these biparental series will have to be regarded as representing a distinct species. Figures 654 and 656 were taken from males of those series.

**Additional Material.** *A. diaspidis* appears to be a cosmopolitan, polyphagous species. Specimens which are here referred to *diaspidis* were reared from *Hemiberlesia* sp. in Greece and Crete; from the latania scale, *H. lataniae* (Signoret), in Greece, Crete, Cyprus, Lebanon and Brazil; from the oleander scale, *Aspidiotus nerii* Bouché [= *A. hederae* (Vallot)], in Greece, Crete, Israel, South Africa, Chile and Mexico; from the San Jose scale, *Quadraspidiotus perniciosus* (Comstock), in California, Ohio, Georgia, Argentina, Chile, Pakistan and New Zealand; from the cactus scale, *Diaspis echinocacti* (Bouché), on Santa Cruz Island (California) and in Mexico, Peru and Greece; from the pineapple scale, *D. bromeliae* (Kerner), in Hawaii; from the white peach scale, *Pseudaulacaspis pentagona* (Targioni-Tozzetti) in South Africa and Argentina; from the rose scale, *Aulacaspis rosae* (Bouché), in Hawaii and Ruanda; from the Florida red scale, *Chrysomphalus aonidum* (L.), in South Africa; from the dictyospermum scale, *C.*

*dictyospermi* (Morgan), in Brazil; from *Acutaspis agavis* (Townsend and Cockerell) in Bermuda; from the rufous scale, *Selenaspidus articulatus* (Morgan) in Peru; from the oystershell scale, *Lepidosaphes ulmi* (L.), in Greece and Canada; from (?) the Glover scale, *Insulaspis* [= *Lepidosaphes*] *gloverii* (Packard) in Iran; from *Chionaspis* sp. in Sri Lanka; from the California red scale, *Aonidiella aurantii* (Maskell) in South Africa; and from undetermined hosts in Iran, Cyprus, Israel, Italy, France, Australia and Eritrea (the latter stock was propagated in Riverside on *H. lataniae*). Whenever a sufficient number of specimens is available, these series appear to represent uniparental populations.

The stock of "so-called *diaspidis*" from Connecticut, recorded by Compere (1955), has been discussed in the present revision under *A. proclia* (see p. 381).

A slight aberration was noticed in one male specimen from Peru (biparental series): scutellum (Figure 656) with 3 discoid sensilla and 4 setae.

## 40. **Aphytis pinnaspidis** n. sp.

(Figures 660–670)

This distinctive South American species may be easily separated from all other species in the **proclia** group by the long propodeum, minute crenulae, short ovipositor and digital sclerites, as well as by details of sculpture, chaetotaxis and coloration.

*Female.* Eyes finely setose. Mandibles well developed, with 2 denticles and a dorsal truncation; maxillary palpi 2-segmented, labial palpi 1-segmented. Antennal scape (Figure 660) about 4–5 times as long as wide, as long as or just slightly longer than the club; pedicel usually $1\frac{2}{3}$ to $1\frac{4}{5}$ times as long as wide, as long as or slightly longer than the third segment of the funicle; first funicular segment slightly trapezoidal, about $1\frac{1}{3}$ to $1\frac{1}{2}$ times as wide as long; second segment more symmetrical, subequal to the first segment or slightly shorter and wider, about $1\frac{1}{2}$ to 2 times as wide as long; third funicular segment about $1\frac{1}{4}$ times as long as wide, bearing 2–3 longitudinal sensilla; club at least fully 3 times as long as wide (3 to $3\frac{1}{3}$ times), about $2\frac{2}{3}$ times longer and slightly wider than the preceding segment, bearing 6–8 longitudinal sensilla.

Setae on head, thorax, and sides of abdomen dark, coarse, rather readily visible under $\times 30$ magnification, but less so than in *proclia*. Vertex with 2 pairs of long setae along occipital margin, in addition to numerous shorter setae. Mesoscutum (Figure 661) with 10–12 setae, rarely with 9; each parapsis with 2 short setae, each axilla with 1 seta; scutellum with 4 setae, the discoid sensilla somewhat closer to the posterior than to the anterior pair. Frontovertex, pronotum, and mesonotal sclerites reticulate. Scutellum oval, usually about $\frac{3}{4}$ median length of mesoscutum. Metanotum (Figures 662, 663) short, transversely reticulate along anterior margin, the sculpture fading posteriorly, smooth on the sides; the posterior margin straight, not overlapping the anterior margin of the propodeum; anteromedian apodeme rather slender, somewhat shorter to somewhat longer than median length of metanotum.

Propodeum (Figures 661–664) relatively long, over 6 times as long as the metanotum, fully as long as or somewhat longer than the scutellum, reticulate on the sides and on a very wide central area, the cells wider, more rounded than in *proclia*, not strigose; crenulae, when present, minute, nonoverlapping, 3 + 4 or less, sometimes barely recognizable, sometimes absent altogether.

Second abdominal tergite (Figure 664) transversely reticulate on the sides, smooth centrally; tergites III–VII (Figure 665) reticulate on the sides, with a few setae in short, transverse rows on reticulate areas; third tergite transversely striated centrally; tergites IV–VI with some very faint indications of transverse striation centrally; seventh tergite delicately reticulate across, bearing 2 setae across center; eighth tergite faintly reticulate, with a transverse row of 4 setae between spiracles; syntergum triangular, broad, usually bearing 6 setae in a transverse row. Cerci situated only somewhat closer to posterior spiracles than to tip of abdomen, with 2 long setae and 1 short seta. Ovipositor shaft short, $1\frac{1}{4}$ to $1\frac{1}{3}$ times as long as the middle tibia (1.23–1.34); ovipositor sheaths about $\frac{1}{3}$ length of middle tibia (0.30–0.35).

Mid-tibial spur somewhat shorter than the corresponding basitarsus.

Forewing (Figure 666) usually smaller than in specimens of *proclia* of comparable size, at least fully $2\frac{1}{2}$ times as long as wide (nearly 3 times in small specimens); marginal fringe not exceeding $\frac{1}{5}$ width of disk (usually about $\frac{1}{6}$). Delta area with 24–60 setae in 4–6 rows, these considerably longer and sparser than the setae distad of speculum, not clearly separated from row of setae along posterior margin of wing, widely separated from a few setae in 1–2 rows below distal half of submarginal vein; costal cell with a row of fine setae along central portion, and 1 coarse seta near apex. Submarginal vein bearing 2 coarse setae, the proximal one about $\frac{2}{3}$ length of the distal, and 16–21 bullae. Marginal vein bearing 7–11 prominent, subequal setae along anterior margin, these over $1\frac{1}{2}$ times to twice as long as the setae in a row along center of vein.

Forewing more or less uniformly infumated on proximal three quarters to somewhat beyond stigma, more strongly so below apex of submarginal vein and base of marginal vein; apical portion of wing paler; no distinct crossband below stigmal vein; a narrow fuscous streak along posterior margin of wing, distad of speculum.

General coloration gray, the fuscous pigmentation more extensive but generally paler than in *proclia*. A transverse black bar on occiput, on each side of foramen (sometimes rather diffuse); lower part of head fuscous, genal sutures not strongly infuscated. Pronotum and propleura uniformly fuscous; mesoscutum rather uniformly fuscous except paler midline and triangular antero-lateral portions, parapsidal sutures blackish; scutellum somewhat more strongly infuscated, the midline, anterior margin and spots surrounding anterior pair of setae paler, posterior margin blackish; metanotum fuscous on anterior sculptured part and on sides, pale along posterior margin; propodeum fuscous on broad (sculptured) median portion, more faintly so on sides, the oblique suture mesad of spiracles and lateral parts of posterior margin lined with fuscous. Abdominal tergites uniformly, rather faintly infuscated. Thoracic sterna blackish. Internal apodemes and endophragma not pigmented; ridge and lateral parts of outer ovipositor plates faintly infuscated. Antennal scape pale, with a longitudinal fuscous stripe; pedicel, funicle and club uniformly, rather faintly infuscated. All femora marked faintly with fuscous, all tibiae with a broad, longitudinal, ventral fuscous stripe. Wing veins lined with brownish.

Length 0.63–1.00 mm.

*Male.* Very similar to the female in structure, chaetotaxis, sculpture and coloration.

Antennal scape (Figure 667) somewhat widened, 3–4 times as long as wide; third funicular segment with 1–2 longitudinal sensilla, club with 2–3; sensory area on ventral aspect of club bearing short setae, but not cut off as in *proclia*. Thorax and propodeum (Figure 668) as in the female.

Genitalia (Figure 669) shorter than in *proclia*, about $\frac{2}{3}$ length of middle tibia (0.60–0.71); no distinct rod between apodemes; digital sclerites relatively short, about $\frac{1}{5}$ to $\frac{1}{4}$ the combined length of aedeagus and apodemes (0.18–0.25).

Wings (Figure 670) as in the female.

Length 0.54–0.80 mm.

Described from 21♀♀ and 12♂♂ (paratypes), reared by P. DeBach from the lesser snow scale, *Pinnaspis strachani* (Cooley), on *Solanum juribeba*, Vitoria, Pernambuco, Brazil, April 11, 1962; 20♀♀, 17♂♂ (♀ holotype, ♂ allotype, and paratypes), reared by P. DeBach from the same host on *Hibiscus* sp., Rural University, Rio de Janeiro State, Brazil, June 11, 1962.

Type series in the collection of the Division of Biological Control, University of California, Riverside.

**Additional Material.** 3♀♀, reared by E. I. Schlinger from *P. strachani*, Villa Union, Sinaloa, Mexico, January 29, 1964; 21♀♀, 30♂♂, reared by P. DeBach from *Pinnaspis* nr. *strachani* on date palm, Buena Vista, Baja California Sur, June 22, 1966; 3♀♀, 3♂♂, reared by P. DeBach from the California red scale, *Aonidiella aurantii* (Maskell), on lime, La Paz, Baja California Sur, Mexico, January 15, 1967; 1♀, reared by J. Quezada from an unknown host (possibly the Florida red scale) on citrus, Cuyulitan, El Salvador, June 27, 1972.

**Notes.** The long, broadly reticulate propodeum, reduced crenulae, short ovipositor, short male genitalia with short digital sclerites, reduced number of setae in delta of forewing, and pale gray coloration may serve as distinguishing characters for *A. pinnaspidis*. The wings are relatively short, about 0.60 mm in female specimens 0.90–1.00 mm long.

*A. pinnaspidis* is quite closely related to *A. griseus* Quednau, a species considered here as related to, but not a member of, the **proclia** group. *A. pinnaspidis* clearly belongs to the **proclia** group on account of the characteristic cephalic pigmentation.

## 41. **Aphytis testaceus** Tshumakova

(Figures 671–678)

*Aphytis testaceus* Tshumakova, 1961, Entomol. Obozr., **40**: 321 (in Russian).

*Aphytis testaceus*: Nikol'skaya and Yasnosh, 1966, Aphelinids of the European Part of the USSR and the Caucasus, pp. 196–197 (in Russian).

*Aphytis testacea*: Nikol'skaya and Yasnosh, 1968, Trud. Vsesoyuz. Entomol. Obshchest., **52**: 12 (in Russian).

This little-known Palearctic species may be easily recognized by the short propodeum, wide low crenulae, relatively long ovipositor, and faintly fuscous general coloration. It is tentatively placed in the **proclia** group on account of the characteristic cephalic pigmentation.

*Female.* Eyes finely setose. Mandibles well developed; maxillary palpi 2-segmented, labial palpi 1-segmented. Antennae (Figures 671, 672) quite distinctly cylindrical; scape about 4 to $4\frac{1}{2}$ times as long as wide, only slightly longer than the club; pedicel $1\frac{3}{5}$ to $1\frac{4}{5}$ times as long as wide, about as long as or slightly longer than the third segment of the funicle; first funicular segment trapezoidal, about $1\frac{2}{5}$ to $1\frac{3}{5}$ times as wide as long; second segment more symmetrical, considerably shorter and somewhat wider than the first segment, about twice to $2\frac{1}{4}$ times as wide as long; third funicular segment about $1\frac{1}{5}$ to $1\frac{1}{2}$ times as long as wide, bearing 2–3 longitudinal sensilla; club usually distinctly cylindrical, about $2\frac{3}{4}$ to 3 times as long as wide, $2\frac{1}{3}$ to $2\frac{1}{2}$ times longer and somewhat wider than the preceding segment, bearing 8–9 longitudinal sensilla.

Setae on head and thorax rather fine, relatively pale, barely visible under $\times 60$ magnification; setae on abdomen even less conspicuous, mostly invisible under $\times 120$ magnification. Vertex with 2 pairs of medium-long setae along occipital margin, in addition to numerous short setae. Mesoscutum (Figures 673, 674) with 10–14 setae, each parapsis with 2–3 short setae, each axilla with 1 seta; scutellum with 4, the discoid sensilla about equidistant from the two pairs. Frontovertex, pronotum and mesonotal sclerites reticulate. Scutellum about $\frac{2}{3}$ to $\frac{4}{5}$ median length of mesoscutum. Metanotum (Figures 674–676) reticulate except on the sides; anteromedian apodeme rather slender, shorter than median length of metanotum.

Propodeum (Figures 675–677) short, about $2\frac{1}{2}$ to 3 times as long as the metanotum, $\frac{3}{5}$ to $\frac{3}{4}$ length of scutellum, rather broadly reticulate on a relatively wide central area and on the sides; crenulae 6 + 7 to 7 + 8, rounded, rather wide and low, closely appressed, nearly forming a continuous ledge.

Second abdominal tergite (Figures 675–677) transversely striated anteriorly on the sides, smooth or faintly reticulate centrally; tergites III–VI reticulate on the sides, with 2–4 setae on each reticulate area, and clearly transversely striated centrally; seventh tergite reticulate on the sides, bearing 3–5 setae on each reticulate area, reticulate-striated centrally with 2–3 setae across center; eighth tergite delicately reticulate across, with a transverse row of 6–8 setae between spiracles (including 1 short seta immediately below each spiracle); syntergum triangular, finely punctate, bearing a transverse row of 8–12 setae. Cerci situated only somewhat closer to posterior spiracles than to tip of abdomen, with 2 long setae and 1 short seta. Ovipositor shaft relatively long, about twice as long as the middle tibia (1.89–2.13); ovipositor sheaths $\frac{1}{2}$ length of middle tibia or nearly so (0.45–0.52).

Mid-tibial spur about $\frac{3}{4}$ to $\frac{4}{5}$ length of the corresponding basitarsus.

Forewing (Figure 678) relatively wide, $2\frac{1}{3}$ to $2\frac{2}{5}$ times as long as wide; marginal fringe short, not exceeding $\frac{1}{6}$ width of disk. Delta area with 94–108 setae in 6–9 rows, these sparser and considerably longer than the setae distad of speculum, not clearly separated from row of setae along posterior margin of wing, widely separated from a few setae in 1–3 rows below distal half of submarginal vein; costal cell with a few fine setae in a row along proximal half or three fifths, and 1 prominent seta near apex. Submarginal vein bearing 2 coarse setae and 18–22 bullae. Marginal vein bearing 6–9 prominent, subequal setae along anterior margin, these about $1\frac{1}{2}$ to $1\frac{2}{3}$ times longer than the setae in a row along center of vein.

Forewing slightly infuscated below stigmal vein; a rather faintly infuscated cloud below submarginal vein, separated by a paler area from an arcuate crossband below base of marginal vein; a narrow fuscous streak along posterior margin of wing, distad of speculum.

General coloration yellowish, with rather faint fuscous markings. Occiput with a transverse blackish bar on each side of foramen; lower part of face and oral margin suffused with blackish; genal sutures blackish from oral margin to about halfway to eyes. Mesoscutum and scutellum very faintly suffused with fuscous; posterior margin of scutellum lined with blackish; propodeum with the lateral parts of the anterior margin, the sutures mesad of the spiracles, the posterior margin and the crenulae lined with blackish. Second abdominal tergite fuscous centrally, lateral parts of anterior and posterior margins lined with blackish; subsequent abdominal tergites with very faint, transverse fuscous crossbands; posterior margin of eighth tergite lined with fuscous. Thoracic sterna blackish, much darker than rest of body. Triangular plates at base of ovipositor infuscated. Antennae uniformly dusky. All tibiae faintly infuscated. Wing veins colorless or faintly infuscated.

Length 0.99–1.23 mm.

*Male*. Unknown.

Redescribed from 1♀, reared by V. Talizky from the "typical" host, the European fruit scale, *Quadraspidiotus ostreaeformis* (Curtis), Kishinev, Moldavia, USSR, June 10, 1958 (determined by V. A. Yasnosh); 2♀♀, reared by Ye. S. Sugonyayev from the same host on *Caragona arborescens*, same locality, May 15–16, 1959 (determined by V. A. Yasnosh); and 2♀♀, reared by M. Kosztarab from the same host on *Corylus avellana*, Budaörs: Kamaraerdö, Hungary, 1955 (misidentified as *A. bovelli* by J. Erdös).

Holotype in the collection of the Zoological Institute, USSR Academy of Sciences, Leningrad.

**Notes.** The above redescription, based on these specimens, differs in certain respects from Chumakova's (1961) original description and from the redescription by Nikol'skaya and Yasnosh (1966).

*A. testaceus* has also been recorded in the USSR as a parasite of *Epidiaspis leperii* (Signoret) and *Lepidosaphes ulmi* (L.) (Nikol'skaya and Yasnosh, 1966).

In the absence of known males, *A. testaceus* is tentatively placed in the **proclia** group. If males are ever discovered, this species might prove to be more closely related to the **chilensis** group. In general coloration, propodeal sculpture, etc., *testaceus* is also somewhat similar to *pinnaspidis* and *griseus*. It differs markedly from both species in the relatively long ovipositor, the short propodeum, and the peculiar shape of the crenulae.

**Biology.** *A. testaceus* is a rather specific parasite of the European fruit scale, *Quadraspidiotus ostreaeformis* (Curtis), on apple in Kabardino-Balkaria (USSR), and its life cycle is closely synchronized with that of its host. The species is univoltine, entering diapause at the end of July and overwintering as fully grown larvae. Reproduction is reportedly thelytokous (Chumakova, 1961; Yasnosh, 1972).

**Potential Value for Biological Control.** *A. testaceus* is a potentially important natural enemy of the European fruit scale and should be included in all biological control projects directed against this widespread pest of deciduous fruit trees and ornamentals.

## RELATED SPECIES

The following 5 species, although lacking some important characteristics of the **proclia** group, appear to be quite closely related to that group.

### 42. **Aphytis griseus** Quednau

(Figures 679–690)

*Aphytis griseus* Quednau, 1964, J. Entomol. Soc. S. Afr., **27**: 100–102, fig. 5a–e.

This South African species may be recognized by the distinctive, rather uniform grayish coloration of the body and antennae, with contrastingly blackish thoracic sterna, and by the relatively long propodeum with rather large, rounded, overlapping crenulae. It is apparently related to the **proclia** group, but lacks the characteristic cephalic pigmentation and wing pattern, and cannot be considered a bona fide member of that group.

*Female.* Eyes finely setose. Mandibles well developed, usually with 2 denticles and a dorsal truncation; maxillary palpi 2-segmented, labial palpi 1-segmented. Antennal scape (Figures 679, 680) about $4\frac{1}{2}$ to $5\frac{1}{2}$ times as long as wide, usually considerably longer than the club; pedicel slender, nearly twice as long as wide, considerably (1.30–1.40 times) longer than the third segment of the funicle; first funicular segment trapezoidal, $1\frac{1}{4}$ to $1\frac{1}{2}$ times as wide as long; second segment somewhat more symmetrical, slightly shorter and wider than the first segment, usually $1\frac{1}{2}$ to 2 times as wide as long; third funicular segment usually about $1\frac{1}{4}$ times as long as wide, bearing 1–2 longitudinal sensilla; club $2\frac{1}{2}$ to somewhat over 3 times as long as wide, $2\frac{1}{2}$ to $2\frac{3}{4}$ times longer and somewhat wider than the preceding segment, bearing 4–8 longitudinal sensilla.

Setae on head and thorax rather coarse, dark, fairly visible under × 60 magnification, but less so than in *pinnaspidis*; setae on sides of abdomen more slender, less readily visible. Vertex with 2 pairs of long setae along occipital margin, in addition to several shorter setae. Mesoscutum (Figure 681) with 9–13 setae (usually 12), the posterior pair and 1 seta at each antero-lateral corner considerably longer than the others; each parapsis with 2 short setae, each axilla with 1 seta; scutellum with 4, the discoid sensilla usually closer to the anterior than to the posterior pair. Frontovertex, pronotum, and mesonotal sclerites reticulate. Scutellum about $\frac{2}{3}$ to $\frac{3}{4}$ median length of mesoscutum. Metanotum (Figure 681) somewhat curved, reticulate except on the sides, slightly overlapping on the anterior margin of the propodeum in the largest specimens; anteromedian apodeme relatively small, slender, as long as or shorter than median length of metanotum.

Propodeum (Figures 681–683) relatively long, about $3\frac{1}{2}$ to nearly 5 times as long as the metanotum, fully $\frac{4}{5}$ length of scutellum or slightly longer, rather broadly reticulate on a wide central area and on the sides; crenulae 4 + 4 to 7 + 8, rather large, rounded, mostly overlapping.

Second abdominal tergite finely reticulate centrally, transversely striated along anterior margin; tergites III–VII reticulate on the sides, with a few setae on each reticulate area; third tergite transversely striated centrally, tergites IV–VI with faint indications of transverse striation; seventh tergite delicately reticulate across, bearing 2 setae across center; eighth tergite delicately reticulate across, with a transverse row of 4 setae between spiracles; syntergum (Figure 684) broad, with a distinct triangular cauda, finely punctate, bearing 6–8 setae in an irregular transverse row. Cerci about equidistant from posterior spiracles and from tip of abdomen, with 2 long setae and 1 short seta. Ovipositor shaft short, about $1\frac{1}{6}$ to $1\frac{1}{3}$ times as long as the middle tibia (1.17–1.34); ovipositor sheaths somewhat over $\frac{1}{4}$ length of middle tibia (0.27–0.30).

Mid-tibial spur about $\frac{3}{4}$ to $\frac{4}{5}$ length of the corresponding basitarsus.

Forewing (Figure 685) usually about $2\frac{1}{2}$ times as long as wide ($2\frac{1}{3}$ to $2\frac{3}{5}$); marginal fringe not exceeding $\frac{1}{3}$ width of disk (usually about $\frac{1}{5}$). Delta area with 37–97 setae in 6–9 irregular rows, less densely setose than in *proclia* but more so than in *pinnaspidis*, the setae sparser and considerably longer than those distad of speculum; delta clearly separated from row of setae along posterior margin of wing, widely separated from a few setae below distal half of submarginal vein; costal cell with a few fine setae in a row along proximal half and 1 prominent seta near apex. Submarginal vein bearing 2 coarse setae, the proximal one about $\frac{1}{2}$ to $\frac{3}{4}$ length of the distal, and 16–21 bullae. Marginal vein bearing 9–10 prominent, subequal setae along anterior margin; the setae in a row along center of vein decreasing considerably toward apex of vein, the proximal ones coarse and about $\frac{4}{5}$ length of the setae along anterior margin, the distal ones considerably shorter.

Forewing hyaline, very faintly dusky below submarginal vein.

General coloration grayish yellow, all parts of body more or less uniformly tinged with light fuscous. Occiput without black bars; oral margin and genal sutures faintly infuscated. Posterior margin of scutellum blackish. Thoracic sterna (Figure 686) blackish, much darker than rest of body. Antennae uniformly infuscated. All tibiae faintly infuscated. Wing veins colorless.

Length 0.67–1.17 mm.

*Male*. Very similar to the female in structure, chaetotaxis, sculpture and coloration.

Antennal scape $3\frac{1}{2}$ to 5 times as long as wide; third funicular segment with 1 longitudinal sensillum, club with 2–4 sensilla; specialized sensory area on ventral aspect of club bearing minute setae but not cut off as in *proclia* (Figure 687).

Mesoscutum with 8–13 setae (usually 10). Propodeum (Figure 688) as in the female. Gaster elongate.

Genitalia (Figures 689, 690) about $\frac{2}{3}$ to $\frac{3}{4}$ length of middle tibia (0.66–0.74); no longitudinal rod between apodemes; digital sclerites about $\frac{1}{3}$ the combined length of aedeagus and apodemes (0.28–0.33).

Marginal vein of forewing with 7–10 prominent, subequal setae along anterior margin. Length 0.72–1.03 mm.

Redescribed from 19♀♀ and 14♂♂ (syntypes), reared by F. W. Quednau from *Nelaspis exalbida* (Cockerell) on *Alöe arborescens*, Pretoria, South Africa, April–May, 1961; 4♀♀, 4♂♂, reared by M. Naude from an unknown host on *Alöe* sp., Kaapsche Hoop Pad, South Africa, January 1965; 14♀♀, 19♂♂, reared by M. Kerny from *Nelaspis exalbida* on *Alöe* sp., Baberton and Louis Trichardt, Transvaal, South Africa, September 1965.

A female specimen (Figure 679) on bottom right of Quednau's ♀ syntype slide deposited in the collection of the Division of Biological Control, University of California, Riverside, is designated lectotype. Other specimens in the syntype series are considered paralectotypes.

**Additional Material.** 16♀♀, 13♂♂, reared by E. W. Rust from "*Chionaspis* sp." (probably the typical host?) on *Alöe*, Tse-Tse Fly Laboratory, Zululand, Natal, December 1925 to June 1926; numerous ♀♂ specimens, reared from *Nelaspis exalbida* and an undetermined host on *Alöe*, Cape Province and Transvaal, South Africa, 1965–1969; also 6♀♀, 3♂♂, reared by H. P. Insley from *Rolaspis lounsburyi* (Cooley) on *Alöe*, Mosselbaai, Cape Province, South Africa, March 1970.

**Notes.** *A. griseus* is quite closely related to *pinnaspidis*. It resembles *pinnaspidis* in the short ovipositor, in the structure of the male antennae, in the broad sculpture on the propodeum, and in the generally gray coloration. It differs markedly from *pinnaspidis* in the shorter propodeum bearing relatively large, overlapping crenulae, in the larger wings, in the longer digital sclerites of the male, and in lacking the cephalic pigmentation and wing pattern characteristic of the **proclia** group. *A. griseus* may be considered to represent a hypothetical connecting link between the **proclia** group and paler groups of *Aphytis*, possibly the **lingnanensis** group.

## 43. **Aphytis tucumani** n.sp.

(Figures 691–702)

This interesting South American species resembles *griseus* in the rather extensively grayish coloration and in the absence of "*proclia*-type" cephalic pigmentation. It may be further recognized by the following characters: forewing infuscated at base, without a distinct cloud below stigma; axillae asetose; propodeum trapezoidal, with a few small, irregular crenulae; male genitalia lacking digital sclerites and apodemes. Like *griseus*, this species is regarded here as related to the **proclia** group.

*Female.* Eyes finely setose. Mandibles well developed, with a strong ventral spine, a ventral denticle and a dorsal truncation; maxillary palpi 2-segmented, labial palpi 1-segmented. Antennal scape (Figure 691) about $5\frac{2}{5}$ to $5\frac{4}{5}$ times as long as wide, somewhat longer than the club; pedicel about $1\frac{2}{3}$ to $1\frac{4}{5}$ times as long as wide, $1\frac{1}{4}$ to $1\frac{1}{3}$ times longer than the third segment of the funicle; first 2 funicular segments subglobular, subequal, up to $1\frac{1}{3}$ times as wide as long, the first slightly trapezoidal; third funicular segment about $1\frac{1}{4}$ times as long as wide, bearing 2 longitudinal sensilla; club about $3\frac{1}{3}$ times as long as wide, slightly over 3 times longer and somewhat wider than the preceding segment, bearing 6–8 longitudinal sensilla.

Setae on head and thorax moderately coarse and dusky, visible under $\times 60$ magnification; those on sides of abdomen more slender and paler, barely visible under $\times 120$ magnification. Vertex with 2 pairs of long setae along occipital margin, in addition to numerous shorter setae. Mesoscutum (Figure 692) with 9–10 setae, the posterior pair and 1 seta at each antero-lateral corner considerably longer and coarser than the others; each parapsis with 1 or 2 setae, axillae asetose; scutellum with 4 setae, the discoid sensilla considerably closer to the anterior than to the posterior pair. Frontovertex, pronotum and mesonotal sclerites reticulate. Scutellum oval, nearly as long as the mesoscutum. Metanotum (Figures 692–694) transversely reticulate except on the sides; anteromedian apodeme slender, considerably shorter than median length of metanotum.

Propodeum (Figures 692–694) usually $3\frac{1}{2}$ to $4\frac{1}{3}$ times as long as the metanotum ($2\frac{4}{5}$ in a small specimen), about $\frac{1}{2}$ to $\frac{3}{4}$ length of scutellum, trapezoidal, with a distinct, broad median salient, rather faintly reticulate centrally and on the sides; crenulae $2 + 3$ to $4 + 6$, small, rounded, rather irregular, sometimes overlapping.

Second abdominal tergite (Figures 693, 694) reticulate on a small triangular area centrally, transversely reticulate-striated anteriorly, faintly transversely striated posteriorly; tergites III–VII (Figure 695) reticulate on the sides, with 2–3 setae in a transverse row on each reticulate area, distinctly transversely striated across center; seventh tergite asetose between the lateral setiferous areas; eighth tergite delicately reticulate-punctate, with 4–5 setae in a transverse, arcuate row between spiracles; syntergum (Figure 695) v-shaped, delicately reticulate-punctate, bearing 8–9 setae in a transverse row. Cerci situated near posterior spiracles, with 2 long setae and 1 short seta. Ovipositor shaft twice as long as the middle tibia or nearly so (1.85–2.08), ovipositor sheaths not clearly separated from inner plates, about $\frac{2}{5}$ to $\frac{1}{2}$ length of middle tibia (0.37–0.50).

Mid-tibial spur somewhat shorter than the corresponding basitarsus.

Forewing (Figure 696) about $2\frac{1}{2}$ to $2\frac{2}{3}$ times as long as wide; marginal fringe not exceeding $\frac{1}{5}$ width of disk. Delta area with 57–94 setae in 6–9 rows, these considerably longer and somewhat sparser than the setae distad of speculum, rather clearly separated from row of setae along posterior margin of wing and from a few setae below distal half of submarginal vein; costal cell with 1–3 fine setae along proximal half or so, and 1–2 coarser setae near apex. Submarginal vein bearing 2 coarse, subequal setae, the proximal one only somewhat shorter than the distal, and 17–23 bullae. Marginal vein bearing 6–9 prominent, subequal setae along anterior margin, about $1\frac{1}{3}$ times longer than the setae in a row along center of vein.

Forewing faintly infumated on proximal half or so, much more distinctly infuscated at base of delta and below apex of submarginal vein; a narrow fuscous streak along posterior margin of wing, distad of speculum.

General coloration grayish, rather extensively infumated. Lower half of face fuscous; genal sutures lined with blackish; occiput without black bars, with a faint fuscous blotch surrounding the foramen. Mesoscutum and scutellum with large, rather faint but distinct fuscous blotches, lateral margins and a wide median stripe hyaline; posterior margin of scutellum blackish; anterior and posterior margins of metanotum fuscous; posterior margin of propodeum and anterior margin of second abdominal tergite blackish. All subsequent abdominal tergites with more or less distinct fuscous crossbands. Thoracic sterna blackish. Antennal scape lined with fuscous ventrally, rest of antenna uniformly dusky. All legs uniformly dusky. Wing veins lined with brownish.

Length 0.95–1.14 mm.

*Male*. Very similar to the female in structure, chaetotaxis, sculpture and coloration.

Antennal scape (Figure 697) somewhat thicker than in the female, about 4 to $4\frac{3}{5}$ times as long as wide; third funicular segment $1\frac{1}{4}$ to $1\frac{2}{5}$ times as long as wide, bearing 2 (rarely 3) longitudinal sensilla; club $2\frac{4}{5}$ to $3\frac{1}{5}$ times as long as wide, $2\frac{1}{2}$ to $2\frac{3}{4}$ times longer than the preceding segment, bearing 3–5 longitudinal sensilla but no specialized sensory area.

Mesoscutum (Figure 698) with 8–12 (usually 10) setae; axillae asetose. Propodeum (Figures 698, 699) 3 to $4\frac{1}{3}$ times as long as the metanotum; crenulae 2 + 2 to 4 + 5, as in the female.

Genitalia (Figure 701) about $\frac{3}{4}$ to $\frac{4}{5}$ length of middle tibia (0.72–0.80), the aedeagus and phallobase forming a single undifferentiated unit, with a distinct longitudinal rod and distinct papillae but lacking digital sclerites or apodemes.

Forewing (Figure 702) essentially as in the female; delta with 57–70 setae in 6–7 rows; submarginal vein with 15–21 bullae; marginal vein bearing 6–8 prominent, subequal setae along anterior margin.

General coloration as in the female; antennal club tipped with blackish apically.

Length 0.85–1.03 mm.

Described from 5♀♀, 8♂♂ (♀ holotype, ♂ allotype, and paratypes), reared by P. Fidalgo from *Diaspis ?echinocacti* (Bouché) on *Cereus* sp., Tucumán, S.M. de Tucumán, Argentina, June 13, 1972; 9♂♂ (paratypes), reared by P. Fidalgo from the same host on the same host plant, Chaco, 20 km west of Rio Saenz Peña, Argentina, December 9, 1972.

Type series in the collection of the Division of Biological Control, University of California, Riverside.

**Notes.** At the present status of our knowledge it is very difficult to determine the affinities of *A. tucumani*. Like *griseus* and the following 3 species it appears to resemble the **proclia** group in general coloration, but lacks the characteristic cephalic pigmentation and therefore cannot be considered a bona fide member of that group. It differs from most known species of *Aphytis* in the asetose axillae and in the peculiar genitalia of the male, lacking both the digital sclerites and the apodemes but possessing the papillae.

### 44. **Aphytis acrenulatus** DeBach and Rosen

(Figures 703–717)

*Aphytis acrenulatus* DeBach and Rosen, 1976, Ann. Entomol. Soc. Amer., **69**:543.

This peculiar species from Mauritius may be readily recognized by the absence of propodeal crenulae, as well as by the stout antennae, broadly joined pronotal plates, relatively long ovipositor and male genitalia, narrow wings, specialized sensilla on the antennal scape of the male, and uniformly grayish coloration (considerably darker in the male). Like *griseus* and *tucumani*, it lacks the characteristic cephalic pigmentation and may be regarded as associated with, but not a member of, the **proclia** group.

*Female*. Eyes finely setose. Mandibles (Figure 703) well developed, with 2 denticles and a dorsal truncation; maxillary palpi 2-segmented, labial palpi 1-segmented. Antennae (Figures 704, 705) rather stout; scape only about 4 to $4\frac{1}{2}$ times as long as wide, about as long as the club; pedicel $1\frac{2}{5}$ to $1\frac{2}{3}$ times as long as wide, up to $1\frac{2}{5}$ times longer than the third segment of the funicle; first funicular segment trapezoidal, $1\frac{2}{3}$ times to nearly twice as wide as long; second segment symmetrical, somewhat shorter and wider than the first segment, about 2 to $2\frac{2}{5}$ times as wide as long; third funicular segment trapezoidal, about as long as wide, bearing 2–3 longitudinal sensilla; club stout, not tapering markedly, about $2\frac{1}{3}$ to $2\frac{3}{4}$ times as long as wide, $2\frac{3}{5}$ to 3 times longer and distinctly wider than the preceding segment, bearing 7–8 longitudinal sensilla.

Setae on head, thorax and abdomen slender, pale, invisible even under $\times 120$ magnification (on a white background under a stereoscopic microscope). Vertex with 1 pair of long setae on sides of occipital margin, in addition to numerous shorter setae. Mesoscutum (Figure 706) with 9–14 (usually 10–11) setae, the posterior pair somewhat longer than the others; each parapsis with 2 setae (rarely with 1), each axilla with 1; scutellum with 4, the discoid sensilla about equidistant from the two pairs. Frontovertex, pronotum and mesonotal sclerites rather faintly reticulate. Pronotal plates (Figure 707) more broadly joined than in most other species of *Aphytis*. Epicoxal pads (Figure 708) large, closely spaced, almost contiguous. Scutellum oval, nearly as long as the mesoscutum (Figure 706). Metanotum (Figures 706, 709, 710) transversely reticulate anteriorly, smooth posteriorly and on the sides; anteromedian apodeme rather slender, usually somewhat longer than median length of metanotum.

Propodeum (Figures 706, 709, 710) about 4 to $4\frac{1}{2}$ times as long as the metanotum, $\frac{3}{5}$ to $\frac{3}{4}$ length of scutellum, broadly reticulate on a wide central area; crenulae absent, posterior margin of propodeum with a distinct, continuous ledge, notched medially.

Second abdominal tergite (Figure 710) rather coarsely reticulate on a small central area, with short transverse striation anteriorly on each side; tergites III–VII reticulate on the sides, bearing 2–6 fine setae in a short transverse row on each reticulate area;

third tergite transversely striated across center, tergites IV–VI with faint indications of transverse striation; seventh tergite faintly reticulate mesad of lateral setiferous areas, faintly striated centrally, bearing a pair of submedian setae which are closer to each other than to the lateral setae; eighth tergite rather long, faintly reticulate, with 6–10 setae in an irregular transverse row between spiracles; syntergum rather narrow, apparently smooth, with a distinct elongate cauda, bearing 7–8 setae in an irregular transverse row. Cerci about equidistant from posterior spiracles and tip of abdomen, with 2 long setae and 1 short seta. Ovipositor relatively long, the shaft $1\frac{3}{4}$ times to a little over twice as long as the middle tibia (1.73–2.07), the sheaths about $\frac{1}{2}$ length of middle tibia (0.43–0.55).

Mid-tibial spur nearly as long as the corresponding basitarsus.

Forewing (Figure 711) rather narrow, about 3 times as long as wide or nearly so; marginal fringe relatively long, not exceeding $\frac{1}{3}$ width of disk. Delta area with 31–50 setae in 4–5 rows, these considerably sparser and longer than the setae distad of speculum, well separated from row of setae along posterior margin of wing, widely separated from 2–3 setae below distal part of submarginal vein; costal cell with a row of fine setae along proximal half or three fifths, devoid of setae near apex but with 1 coarse seta on the vein. Submarginal vein bearing 2 coarse setae, the proximal one nearly $\frac{3}{4}$ length of the distal, and 14–20 bullae. Marginal vein bearing 7–9 prominent, subequal setae along anterior margin, considerably longer than the setae in a row along center of vein ($1\frac{3}{5}$ times longer in the holotype).

Forewing nearly hyaline, very faintly infumated on proximal half, somewhat more distinctly so below junction of submarginal and marginal vein and at base of delta; a faint fuscous streak along posterior margin of wing distad of speculum.

General coloration uniformly rather faintly dusky; occiput suffused with blackish; pronotum somewhat more distinctly fuscous than rest of thorax, anterior margin of mesoscutum lined with blackish; abdominal tergites with broad, faint fuscous bands. Thoracic sterna about as dusky as body, prosternum somewhat more distinctly so. Antenna uniformly fuscous. Legs concolorous with body. Wing veins brownish.

Length 0.86–1.25 mm (the largest specimens may have been distorted by mounting, the abdomen extended, and may therefore appear larger than they really are; in such specimens, the cerci appear closer to the tip of the abdomen than to the posterior spiracles).

*Male*. Essentially similar to the female, differing mainly in the shape of the antennae and in the darker general coloration.

Antennal scape (Figures 712, 713) considerably wider than in the female, about $2\frac{3}{4}$ times as long as wide, about $1\frac{1}{4}$ times longer than the club, bearing 3 minute, discoid, slightly depressed sensilla on antero-ventral surface (these are usually very hard to see when the antenna is flattened on the slide); pedicel $1\frac{1}{3}$ to $1\frac{2}{5}$ times as long as wide, $1\frac{1}{3}$ times longer than the third segment of the funicle; first 2 funicular segments as in the female, third segment about as long as wide, bearing 1–2 longitudinal sensilla; club a little over twice as long as wide, $2\frac{2}{3}$ times longer and distinctly wider than the preceding segment, bearing 1–2 longitudinal sensilla.

Mesoscutum (Figure 714) with 8–10 setae, somewhat darker than in the female;

scutellum about as long as the mesoscutum; metanotal apodeme up to $1\frac{2}{5}$ times as long as median length of metanotum. Propodeum (Figures 714, 715) as in the female.

Genitalia (Figure 716) large, robust, about as long as the middle tibia (0.96–1.03); digital sclerites about $\frac{1}{4}$ the combined length of aedeagus and apodemes (0.24–0.28).

Forewing (Figure 717) essentially as in the female; marginal fringe nearly $\frac{1}{2}$ width of disk; delta with 19–24 setae in 3–4 rows; submarginal vein bearing 12–16 bullae; marginal vein bearing 5–7 prominent setae along anterior margin, up to about twice longer than the setae in a row along center of vein. Forewing pattern as in the female.

General coloration rather uniformly dusky, essentially as in the female but considerably darker; bands on abdominal tergites and pigmentation of thorax and legs more distinct than in the female; antennal scape uniformly blackish or dark brown, rest of antenna uniformly, strongly infuscated.

Length 0.62–0.74 mm.

Described from 23♀♀ and 4♂♂ (holotype ♀, allotype ♂ and paratypes), reared by J. R. Williams from *Aspidiella zingiberi* Mamet, Montagne Longue, Mauritius, March 1971; (first mounted on 3 Hoyer slides, subsequently remounted on separate slides, several specimens dissected; most photographs taken before remounting).

Type series in the collection of the Division of Biological Control, University of California, Riverside.

**Notes.** Despite the absence of propodeal crenulae, there is no doubt whatsoever that *acrenulatus* is a bona fide member of the genus *Aphytis*, the absence of crenulae and formation of a continuous ledge apparently representing secondary, aberrant development, as in *faurei* (see p. 346). The affinities of this peculiar species are not clear at present. Like *griseus*, it is regarded for the time being as associated with the **proclia** group on account of the generally dusky coloration, although it differs from members of this group in obvious structural characters. It may be somewhat related to *pinnaspidis*, a South American representative of the **proclia** group characterized by minute crenulae. The presence of sensilla on the antennal scape of the male may be regarded as a primitive character. Discovery of additional species of *Aphytis* may eventually shed some light on the affinities of *acrenulatus* and related species.

The relative length of the ovipositor and male genitalia, as compared with the middle tibia, a valid diagnostic character of *acrenulatus*, may be partly due to the presence of relatively short legs; the ovipositor itself does not appear very long in relation to the length of the abdomen.

The holotype and 1♂ paratype specimen have a supernumerary seta on the scutellum (Figure 714).

Since its only known scale-insect host occurs both in Mauritius and in Thailand (Borkhsenius, 1966), it is possible that *acrenulatus* represents an Oriental, rather than an Ethiopian, element on the island of Mauritius.

## 45. **Aphytis amazonensis** n. sp.

(Figures 718–728)

This unique South American species resembles the **proclia** group in general coloration. However, like *griseus* it lacks the characteristic cephalic pigmentation and wing pattern, and therefore cannot be considered a bona fide member of that group. It may be further distinguished by the few large, overlapping propodeal crenulae, by the sparsely setose delta of the forewing, by the short antennal club, and by the specialized sense organs on the antennal scape of the male.

*Female*. Eyes finely setose. Mandibles well developed; maxillary palpi 2-segmented, labial palpi 1-segmented. Antennae (Figure 718) relatively small; scape cylindrical, about $5\frac{1}{3}$ times as long as wide, about $1\frac{1}{4}$ times longer than the club; pedicel about $1\frac{3}{5}$ times as long as wide, $1\frac{3}{5}$ times longer than the third segment of the funicle; first funicular segment fairly symmetrical, $1\frac{1}{2}$ times as wide as long; second segment saucer-shaped, considerably shorter and wider than the first segment, fully $2\frac{1}{2}$ times as wide as long; third funicular segment nearly $1\frac{1}{3}$ times as wide as long, bearing 1 longitudinal sensillum; club a little over twice as long as wide, $3\frac{1}{4}$ times longer and somewhat wider than the preceding segment, bearing 3 longitudinal sensilla.

Setae on head, thorax and abdomen dark, rather slender, readily visible only under $\times 120$ magnification; only the setae on the scutellum and the longest setae on the mesoscutum are visible also under $\times 60$ magnification. Vertex with 2 pairs of long setae along occipital margin, in addition to numerous short setae. Mesoscutum (Figure 719) with 10 setae, the posterior pair and 1 seta at each antero-lateral corner considerably longer and coarser than the others; each parapsis with 2 short setae, each axilla with 1 seta; scutellum with 4 long setae, the discoid sensilla about equidistant from the two pairs. Frontovertex, pronotum, and mesonotal sclerites reticulate; reticulation on mesoscutum and scutellum broad, with a delicate infrasculpture within the large cells. Scutellum about $\frac{3}{4}$ median length of mesoscutum. Metanotum (Figures 719–721) reticulate-striated except on the sides; anteromedian apodeme (Figure 720) long, slender, about $1\frac{3}{5}$ times as long as the metanotum.

Propodeum (Figures 719, 721, 722) relatively long, over $4\frac{1}{2}$ times as long as the metanotum, $\frac{4}{5}$ median length of scutellum, reticulate on the sides, rather widely reticulate-strigose on a distinctly trapezoidal central area; crenulae (Figure 722) $4 + 4$, large, conspicuous, rounded, overlapping, the inner pair considerably smaller than the others.

Second abdominal tergite transversely striated on the sides, smooth centrally; tergites III–VII reticulate on the sides, bearing 3–4 setae in a transverse row on each reticulate area; third tergite transversely striated across center; seventh tergite faintly reticulate across, with 2 setae mesad of the lateral transverse rows; eighth tergite delicately reticulate across, bearing 4 setae in a transverse row between spiracles; syntergum triangular, faintly striate-punctate, bearing 6 setae. Cerci situated somewhat closer to posterior spiracles than to tip of abdomen, with 2 long setae and 1 short seta. Ovipositor shaft $1\frac{2}{5}$ times as long as the middle tibia (1.40), ovipositor sheaths a little over $\frac{1}{3}$ length of middle tibia (0.37).

Mid-tibial spur long, about as long as the corresponding basitarsus.

Forewing (Figure 723) fully $2\frac{1}{2}$ times as long as wide; marginal fringe about $\frac{1}{4}$ width of disk. Delta area sparsely setose, with 29–30 setae in 4–5 rows, these considerably longer and sparser than the setae distad of speculum, clearly separated from row of setae along posterior margin of wing, widely separated from a group of 4–5 setae below distal two fifths of submarginal vein; costal cell with a row of fine setae along proximal two thirds and 1–2 coarse setae near apex. Submarginal vein bearing 2 coarse setae, the proximal one a little over $\frac{2}{3}$ length of the distal, and 14 bullae. Marginal vein bearing 7 long, prominent, subequal setae along anterior margin, these fully twice as long as the setae in a row along center of vein.

Forewing uniformly dusky on proximal half, the cloud extending a little beyond venation; a paler area below junction of submarginal and marginal veins; a blackish streak along posterior margin, distad of speculum.

General coloration uniform gray. Head yellowish, occiput suffused with blackish around foramen, but without distinct black bars. Pronotum entirely gray; mesoscutum and scutellum uniformly gray with a narrow, pale longitudinal midline, the anterior margin of mesoscutum, parapsidal sutures, and posterior margin of scutellum lined with black; metanotum gray centrally and on the sides; propodeum gray on sculptured areas on sides and center, pale on smooth sublateral areas; crenulae blackish, darker than central area; sutures mesad of spiracles blackish. All abdominal tergites uniformly gray. Thoracic sterna blackish, darker than dorsum of body. Antennae uniformly gray. All femora marked with gray, all tibiae uniformly gray. Wing veins lined with grayish.

Length about 0.70 mm.

*Male*. Similar to the female in structure, chaetotaxis, sculpture and coloration.

Antennal scape (Figures 724–726) somewhat widened, about 3 to $3\frac{1}{2}$ times as long as wide, bearing 2 concave plates on ventral aspect, each with a minute, truncate, tuberculous sensillum; pedicel about $1\frac{1}{2}$ times longer than the third segment of the funicle; the latter about as long as wide or slightly wider than long; club $2\frac{1}{4}$ to $2\frac{2}{5}$ times as long as wide, $2\frac{2}{3}$ to $2\frac{4}{5}$ times longer than the preceding segment, bearing 2–3 longitudinal sensilla.

Mesoscutum (Figure 727) with 13 setae. Scutellum about $\frac{4}{5}$ median length of mesoscutum. Metanotal apodeme about $1\frac{2}{5}$ times as long as the metanotum. Propodeum (Figures 727, 728) over $3\frac{1}{2}$ times as long as the metanotum, about $\frac{3}{4}$ median length of scutellum.

Genitalia about $\frac{3}{4}$ to $\frac{4}{5}$ length of middle tibia (0.73–0.83), without a longitudinal rod between apodemes; digital sclerites apparently sclerotized, rather short, a little over $\frac{1}{4}$ the combined length of aedeagus and apodemes (0.27–0.28).

Mid-tibial spur somewhat longer than the corresponding basitarsus. Forewing $2\frac{1}{2}$ to $2\frac{2}{3}$ times as long as wide; delta area with 22–27 setae; submarginal vein with 13–15 bullae; marginal vein bearing 6–7 prominent setae along anterior margin.

Length about 0.67–0.75 mm.

Described from 1♀ (holotype) and 2♂♂ (allotype and paratype), reared by P. DeBach from an undetermined armored scale insect on a wild jungle plant, Acapa Terr., Brazil, July 25, 1962.

Type series in the collection of the Division of Biological Control, University of California, Riverside.

**Notes.** *A. amazonensis* presents some characteristics of several species groups, and may prove to be a link between them. It may be considered as related to the **proclia** group in general coloration; to the **lingnanensis** group in the long propodeum and large, overlapping crenulae; to the **mytilaspidis** group in the short antennal club. The digital sclerites in the male genitalia of *amazonensis* resemble those of *chilensis* and *columbi* of the **chilensis** group; and the peculiar sense organs on the antennal scape of the male are apparently similar to those of several species in the **vittatus** group. For the time being, *amazonensis* may be classified as an aberrant species, somewhat related to the **proclia** group. In general coloration, this species is rather similar to *pinnaspidis*, *testaceus* and *griseus*.

## 46. **Aphytis desantisi** DeBach and Rosen

(Figures 729–736)

*Aphytis desantisi* DeBach and Rosen, 1976, Ann. Entomol. Soc. Amer., **69**:543.

This distinctive South American species is somewhat similar to *griseus* in the faintly dusky general coloration and infuscated thoracic sterna. It may be readily recognized by the short, distinctly bilobed propodeum with irregular, nonoverlapping crenulae, as well as by the paler coloration and longer ovipositor. Like *griseus*, it lacks the characteristic cephalic pigmentation and may be regarded as only remotely related to the **proclia** group.

*Female.* Eyes finely setose. Mandibles (Figure 729) well developed, with 2 denticles and a dorsal truncation; maxillary palpi 2-segmented, labial palpi 1-segmented. Antennal scape (Figures 730, 731) slender, $4\frac{1}{4}$ to a little over 5 times as long as wide, a little longer than the club; pedicel rather long, $1\frac{1}{2}$ to $1\frac{3}{4}$ times as long as wide, $1\frac{1}{2}$ to $1\frac{3}{5}$ times longer than the third segment of the funicle; first funicular segment subglobular, somewhat trapezoidal, ventral aspect longer than the dorsal, $1\frac{1}{4}$ to $1\frac{1}{2}$ times as wide as long; second segment nearly symmetrical, considerably shorter and somewhat wider than the first segment, $1\frac{2}{3}$ to $2\frac{1}{5}$ times as wide as long; third funicular segment short, trapezoidal, dorsal aspect longer than the ventral, usually a little wider than long (up to $1\frac{1}{4}$ times), bearing 1 longitudinal sensillum; club $2\frac{3}{5}$ to somewhat over 3 times as long as wide (usually less than 3 times), $3\frac{1}{4}$ to $3\frac{1}{2}$ times longer and somewhat wider than the preceding segment, bearing 4–5 longitudinal sensilla.

Setae on thorax dark, coarse, visible under $\times 30$ magnification; those on head and abdomen paler, slender, indistinct even under $\times 60$ magnification. Vertex with 2 pairs of long setae along occipital margin, in addition to numerous shorter setae. Mesoscutum (Figure 732) with 8–10 setae, the posterior pair and 1 seta at each anterolateral corner considerably longer and coarser than the others; each parapsis with 2 setae, each axilla with 1 seta; scutellum with 4 setae, the discoid sensilla usually closer to the anterior than to the posterior pair. Frontovertex, pronotum and mesonotal sclerites reticulate. Scutellum oval, $\frac{3}{4}$ to $\frac{4}{5}$ median length of mesoscutum. Metanotum somewhat curved, reticulate except on the sides; anteromedian apodeme slender, usually distinctly longer (up to $1\frac{1}{3}$ times) than median length of metanotum.

Propodeum (Figures 732–734) short, distinctly bilobed, often with a deep cleft between the lobes (Figure 733), about $2\frac{2}{5}$ to 3 times as long as the metanotum, about $\frac{1}{2}$ length of scutellum, delicately longitudinally striated on the sides, rather broadly reticulate on a wide central area above and including the two lobes; crenulae 3 + 3 to 6 + 8, usually elongate, irregular, nonoverlapping.

Second abdominal tergite (Figures 732, 734) distinctly delimited, transversely striated centrally and on the sides; tergites III–VII reticulate on the sides, with 2–5 setae in a transverse row on each reticulate area; third and seventh tergites reticulate also mesad of the lateral setiferous areas, transversely striated centrally; tergites IV–VI faintly transversely striated centrally; seventh tergite bearing 2 submedian setae, mesad of the lateral setiferous areas; eighth tergite delicately reticulate-punctate, with a transverse row of 4 setae between the spiracles; syntergum (Figure 735) triangular, very faintly sculptured, slightly punctate, bearing 5–6 setae in a transverse row. Cerci situated somewhat closer to posterior spiracles than to tip of abdomen, with 2 long setae and 1 short seta. Ovipositor shaft about $1\frac{2}{3}$ to $1\frac{4}{5}$ times as long as the middle tibia (1.69–1.85); ovipositor sheaths $\frac{2}{5}$–$\frac{1}{2}$ length of middle tibia (0.43–0.51).

Mid-tibial spur long, usually a little longer than the corresponding basitarsus.

Forewing (Figure 736) about $2\frac{2}{5}$ to $2\frac{7}{10}$ times as long as wide (usually over $2\frac{1}{2}$ times); marginal fringe rather long but not exceeding $\frac{1}{3}$ width of disk. Delta area with 32–50 setae in 4–6 rows, these considerably longer and sparser than the setae distad of speculum, not clearly separated from row of setae along posterior margin of wing, widely separated from 2–4 setae below distal third of submarginal vein; costal cell with a row of fine setae along proximal three fifths or so, no setae near apex. Submarginal vein bearing 2 coarse setae, the proximal one about $\frac{3}{5}$ to $\frac{3}{4}$ length of the distal, and 11–16 bullae. Marginal vein bearing 6–10 (usually 8) prominent, subequal setae along anterior margin, about $1\frac{1}{2}$ to $1\frac{3}{4}$ times as long as the setae in a row along center of vein.

Forewing hyaline, very faintly infumated on proximal half up to stigma, more distinctly so below distal part of submarginal vein and at base of delta; a narrow fuscous streak along posterior margin of wing, distad of speculum.

General coloration pale, faintly suffused with dusky. Head immaculate. Posterior margin of scutellum blackish; a short blackish streak at base of forewing, below tegula; metanotum faintly dusky centrally, posterior margin faintly lined with fuscous; propodeal lobes and crenulae rather conspicuously fuscous. Abdominal tergites faintly suffused with dusky, more noticeably so at base and sides of gaster, posterior margin of eighth tergite faintly lined with fuscous; ovipositor sheaths very faintly dusky. Thoracic sterna strongly infuscated. Triangular plates and arcs at base of ovipositor fuscous. Antennal scape faintly infuscated distally, rest of antenna uniformly dusky. Wing veins hyaline, lined with brownish.

Length 0.65–0.87 mm.

*Male*. Unknown.

Described from 12♀♀ (holotype and paratypes), reared by A. Teran from the California red scale, *Aonidiella aurantii* (Maskell), on orange, El Siambon, Tucumán, Argentina, June 6, 1969.

Type series in the collection of the Division of Biological Control, University of California, Riverside.

**Notes.** *A. desantisi* differs from almost all other species of *Aphytis* in the distinctive shape of the propodeum, which is rather similar to that of *cercinus* (see p. 346). We regard it as somewhat related to the **proclia** group, on account of the very faintly dusky general coloration and fuscous thoracic sterna. However, it is decidedly paler than any other species related to that group, and may not belong there at all. In fact, it may be more closely related to the **mytilaspidis** group, or even to the **lingnanensis** group.

**Potential Value for Biological Control.** As a parasite of the California red scale, *Aonidiella aurantii* (Maskell), this species should be considered for importation into countries where that pest is a serious problem.

This species is named in honor of Professor Luis De Santis, Universidad Nacional de La Plata, Argentina.

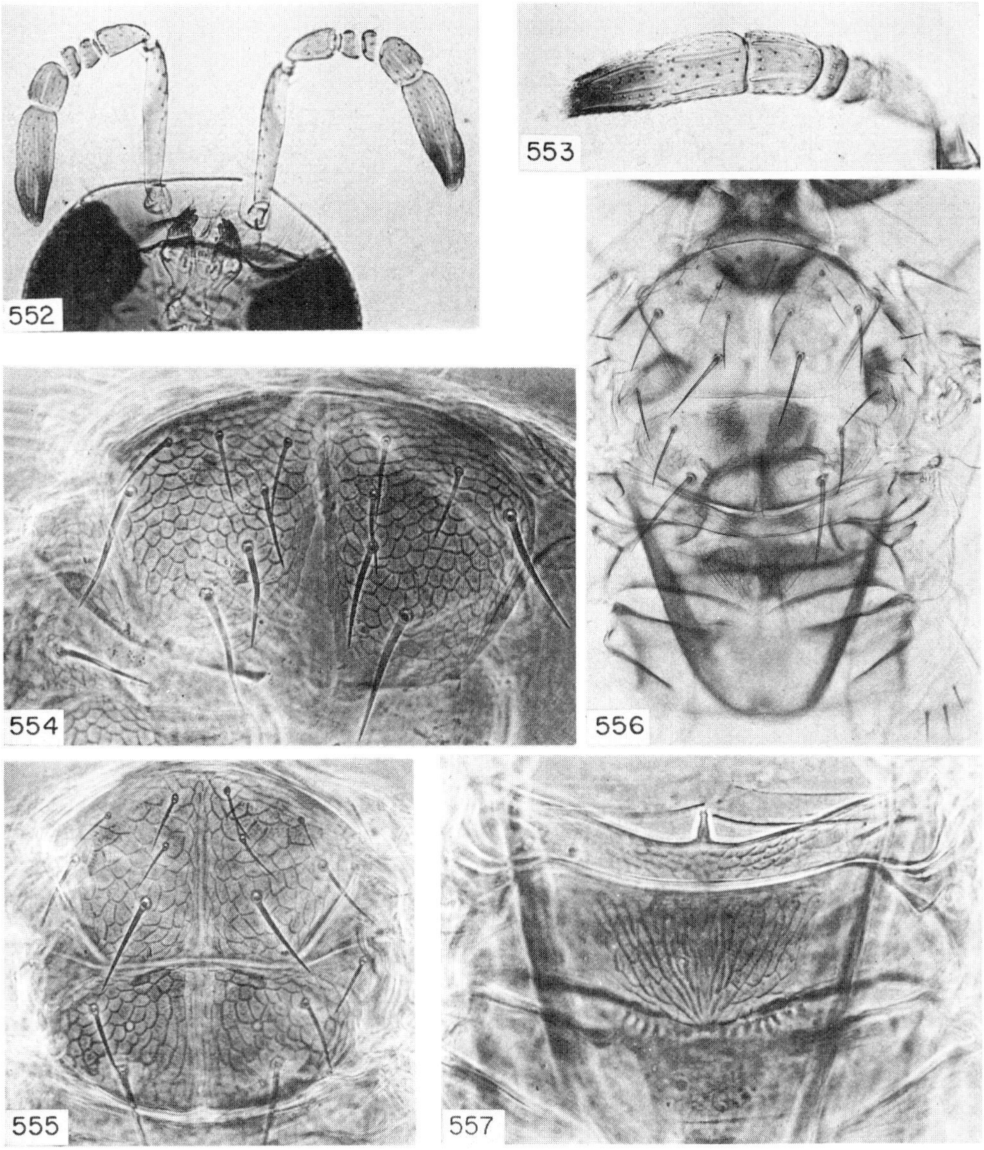


Figures 552–557. *Aphytis proclia* (Walker), ♀
552. Antennae and mandibles. 553. Antennal pedicel and flagellum. 554. Mesoscutum (holotype).
555. Mesonotum (type of Alam's "*diaspidis*, Silwood material"). 556. Thorax and propodeum.
557. Metanotum and propodeum (Ohio material).

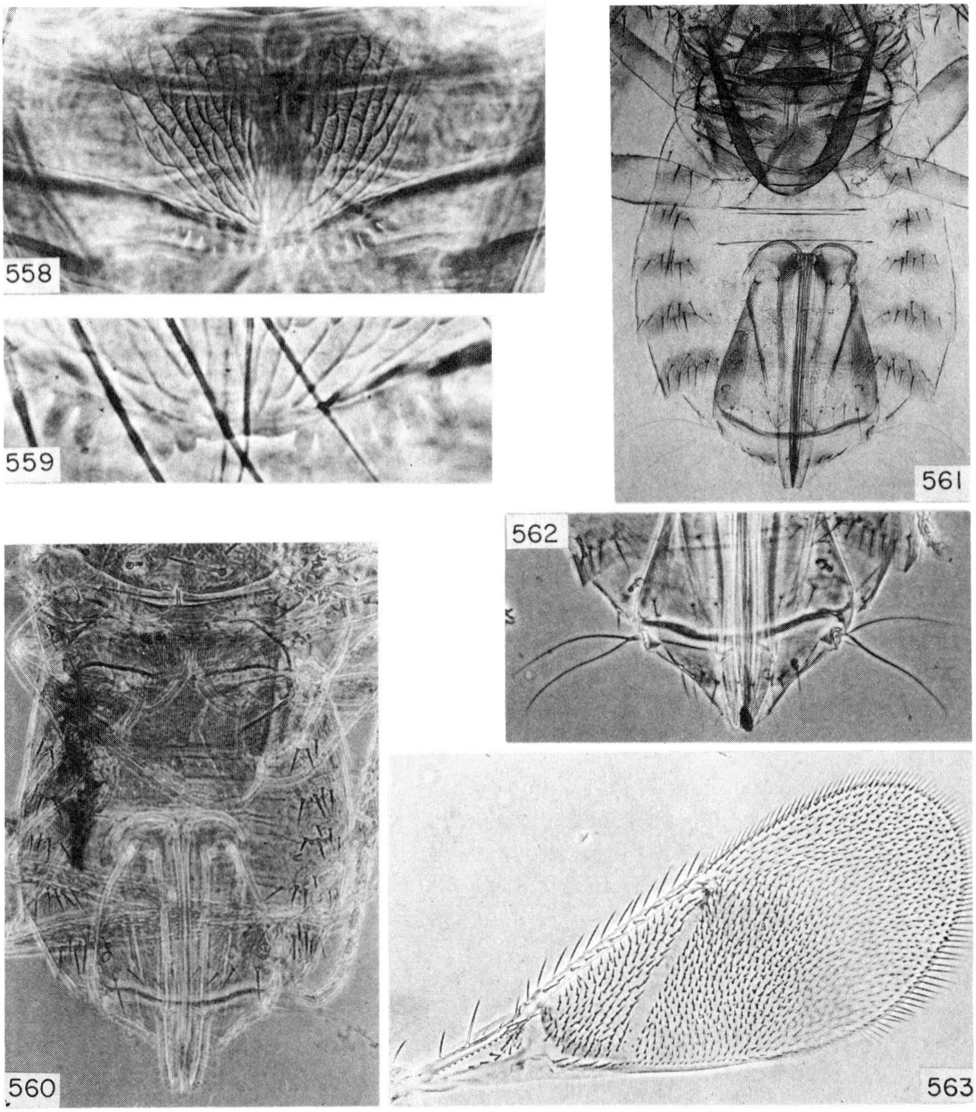


Figures 558–563. *Aphytis proclia* (Walker), ♀
558. Propodeum and crenulae. 559. Propodeal crenulae; the 2 sets are widely separated in this specimen (overlapped by hind wing; USSR material). 560 Abdomen, dorsal view (holotype). 561. Abdomen, showing chaetotaxis and pigmentation of tergites; note also endophragma. 562. Posterior abdominal tergites; note cerci and posterior spiracles. 563. Forewing.

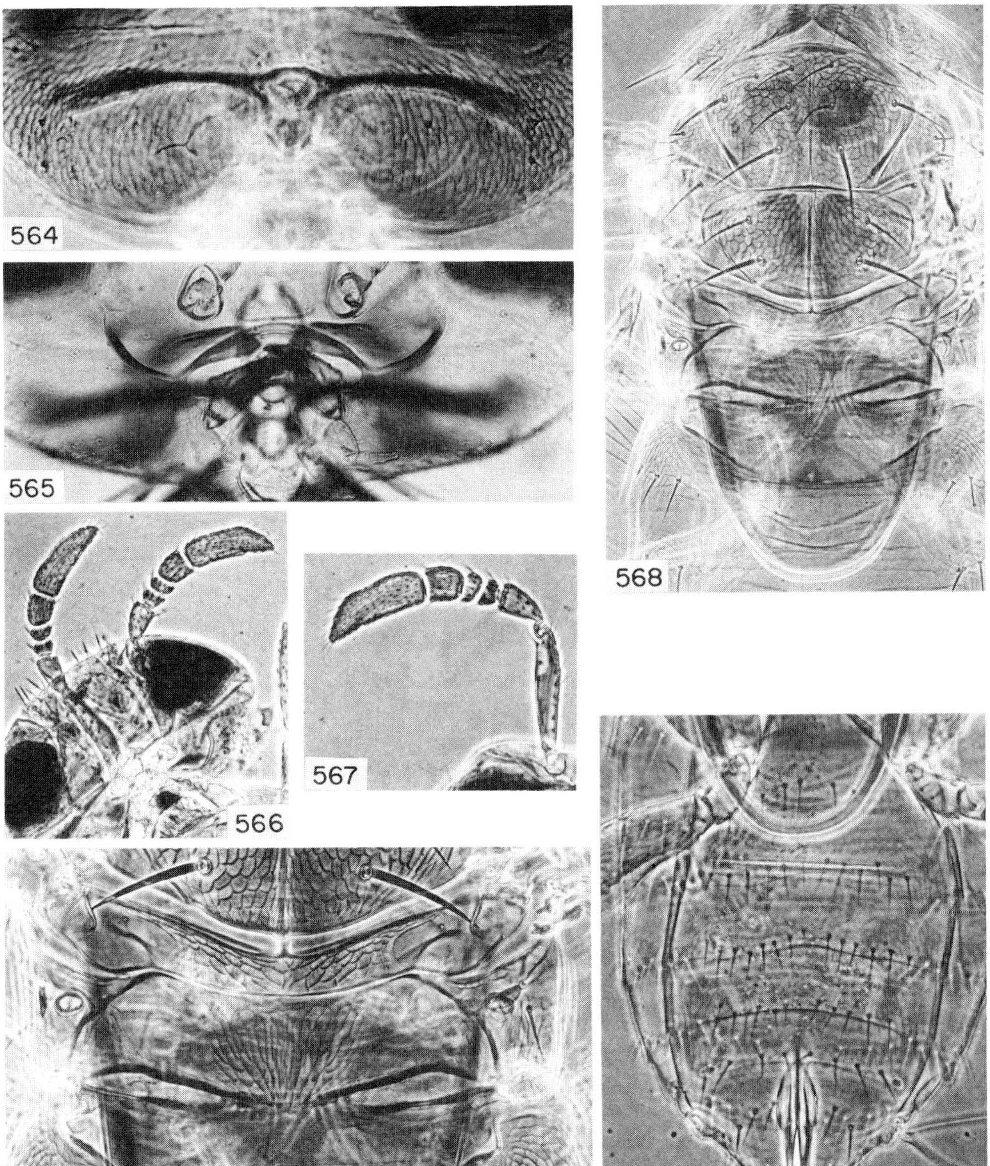


Figures 564–570. *Aphytis proclia* (Walker)

564. ♀: Occiput, showing distinctive pigmentation (holotype). 565. ♀: Genal sutures, showing distinctive pigmentation. 566, 567. ♂: Antennae; note partly cut-off area on ventral aspect of club. 568. ♂: Thorax, propodeum and base of gaster. 569. ♂: Metanotum, propodeum and base of gaster. 570. ♂: Abdominal sternites and genitalia.

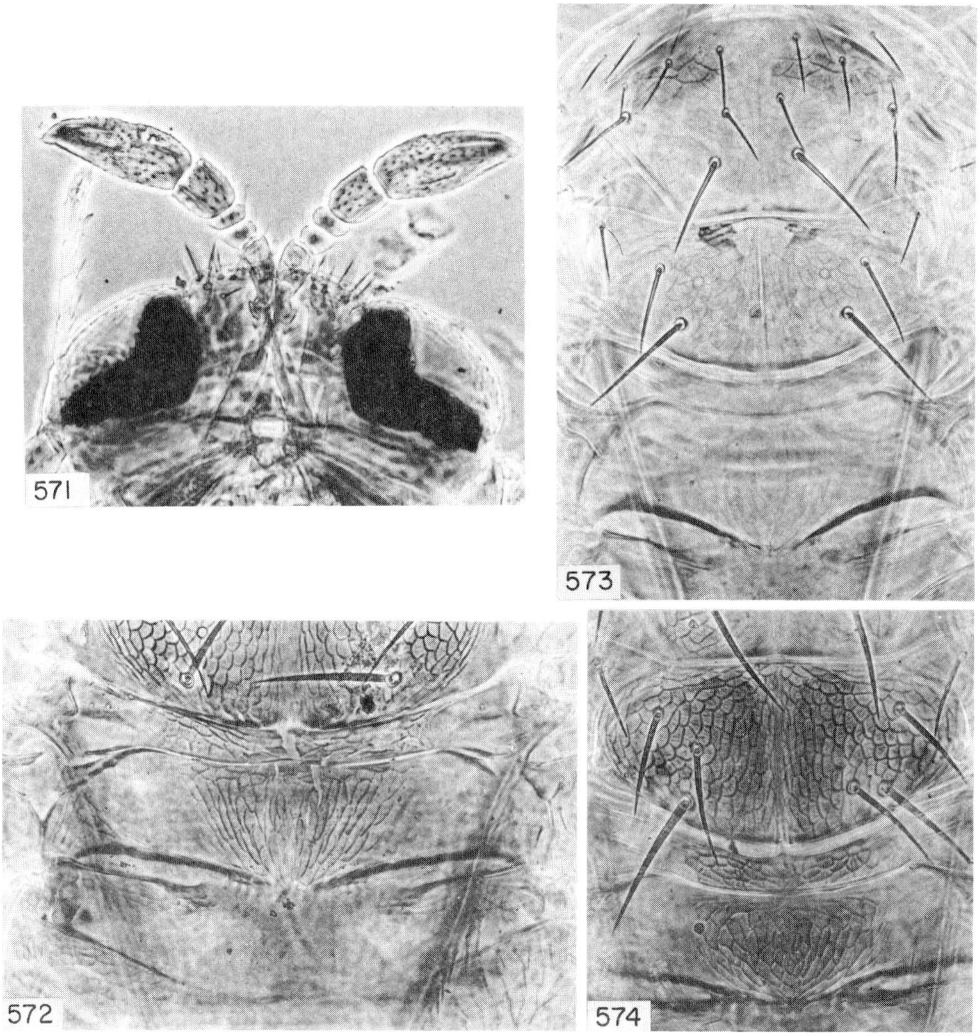


Figures 571–574. *Aphytis proclia* (Walker), ♀
571, 572. Holotype of *zonatus* Alam: 571. Head and antennae. 572. Metanotum, propodeum and base of gaster.
573. Thorax and propodeum (ex *Lepidosaphes* sp., China).
574. Scutellum, showing aberrant chaetotaxis with 3 pairs of setae, discoid sensilla absent (Ohio material).

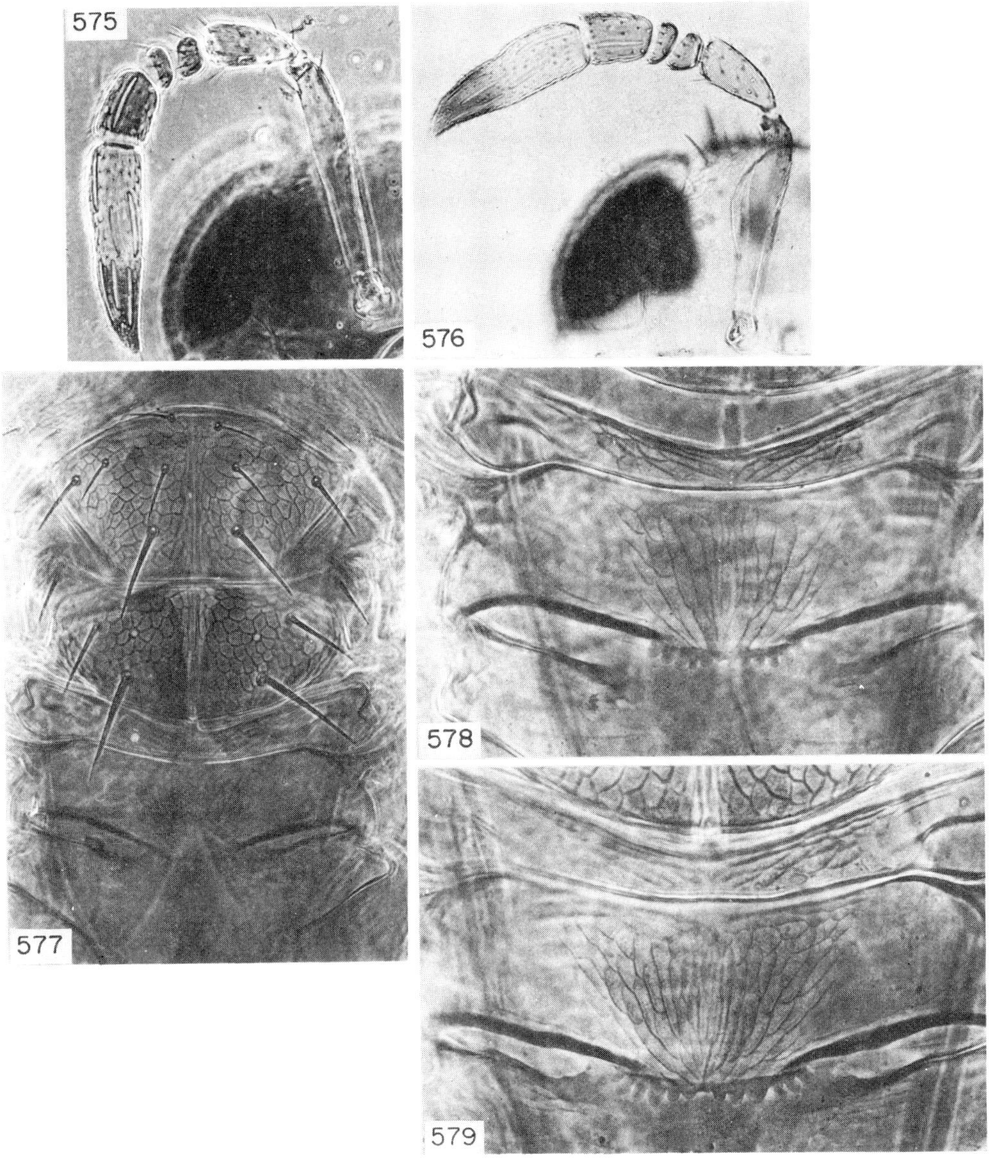


Figures 575–579. *Aphytis maculicornis* (Masi), ♀
575. Antenna; note elongate club. 576. Antenna, showing distinctive coloration. 577. Thorax and propodeum. 578, 579. Metanotum, propodeum and crenulae.

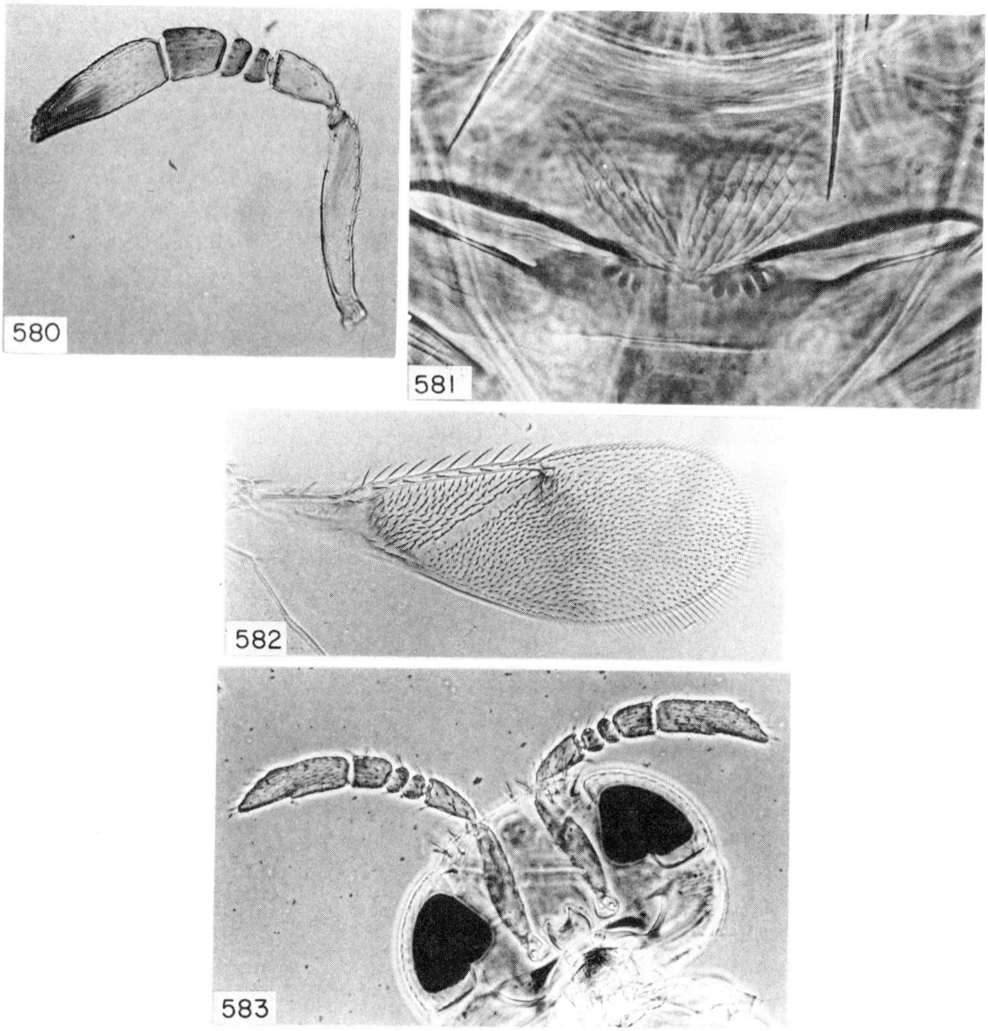


Figures 580–583. *Aphytis paramaculicornis* DeBach and Rosen
580. ♀: Antenna (Pakistan material). 581. ♀: Propodeum and crenulae (Indian material). 582. ♀: Forewing (Pakistan material). 583. ♂: Head and antennae; note sclerotized genal suture and cut-off sensory area on ventral aspect of club (Persian material).

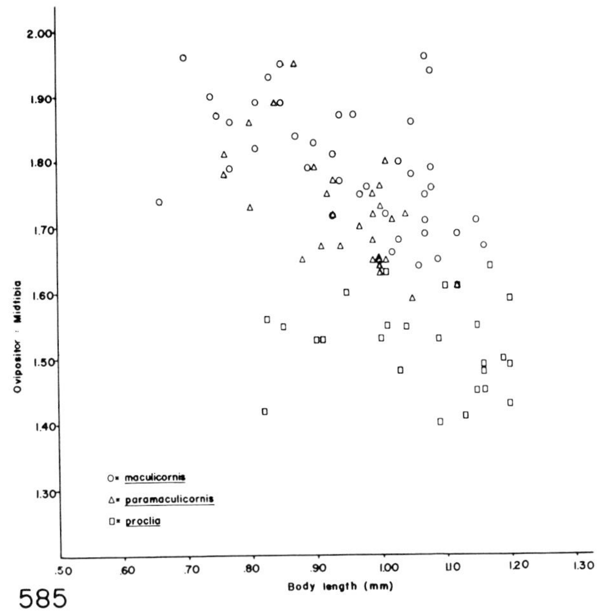


585

Figure 585. Scatter diagram, showing the relative length of the ovipositor in relation to body length in females of *A. proclia* (Walker), *A. maculicornis* (Masi) and *A. paramaculicornis* DeBach and Rosen.

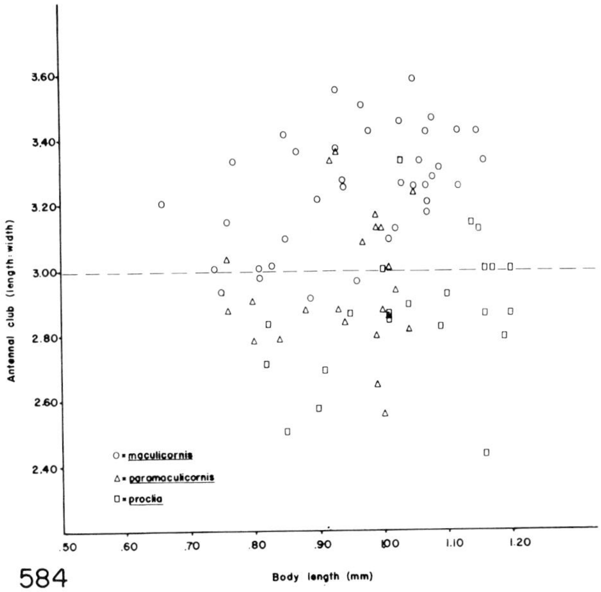


584

Figure 584. Scatter diagram, showing proportions of antennal club in relation to body length in females of *A. proclia* (Walker), *A. maculicornis* (Masi) and *A. paramaculicornis* DeBach and Rosen.

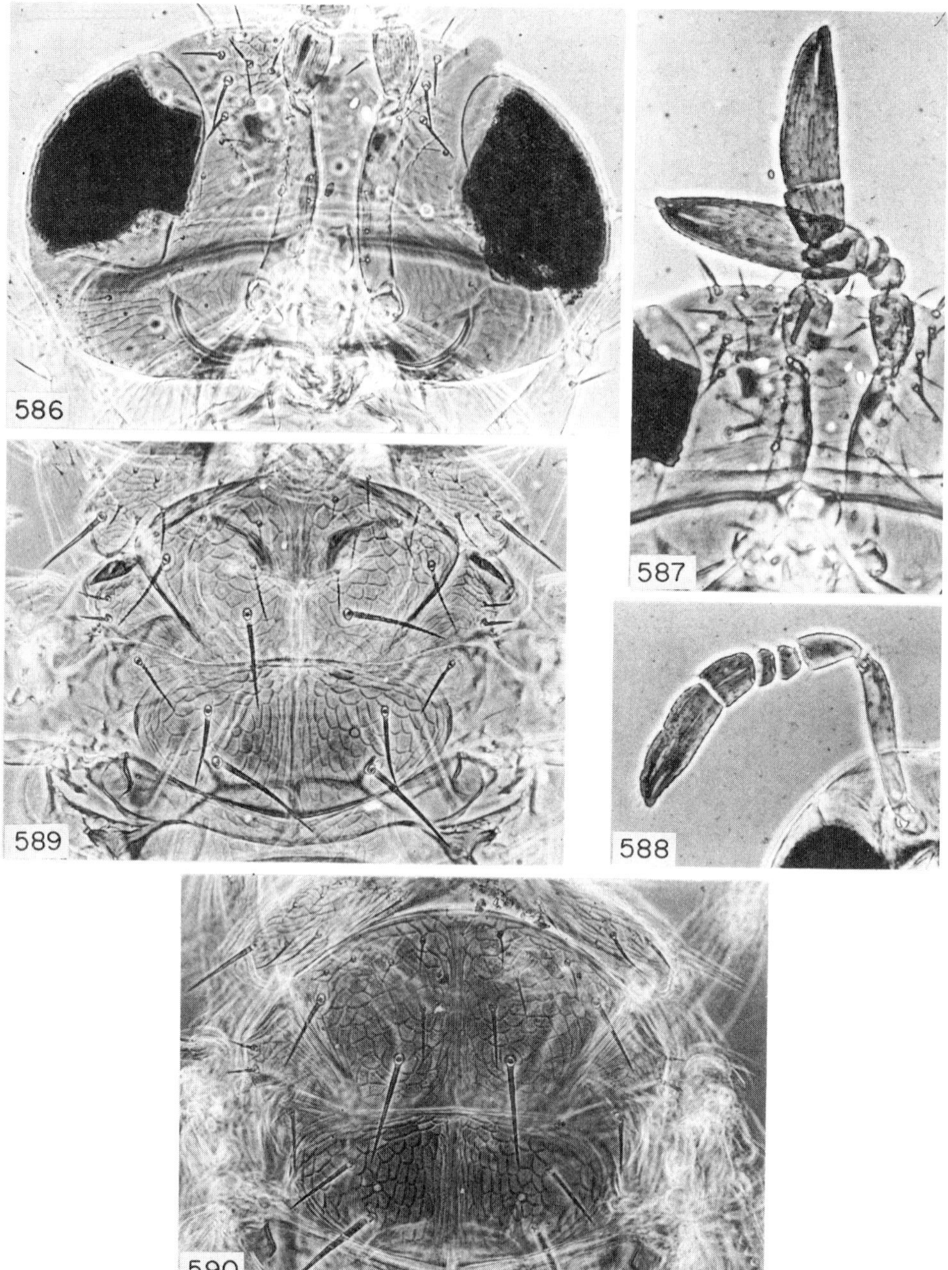


Figures 586–590. *Aphytis hispanicus* (Mercet), ♀
586. Head, showing genal sutures and antennal scape (holotype). 587. Antennae (holotype). 588. Antenna. 589. Thorax (holotype). 590. Thorax.

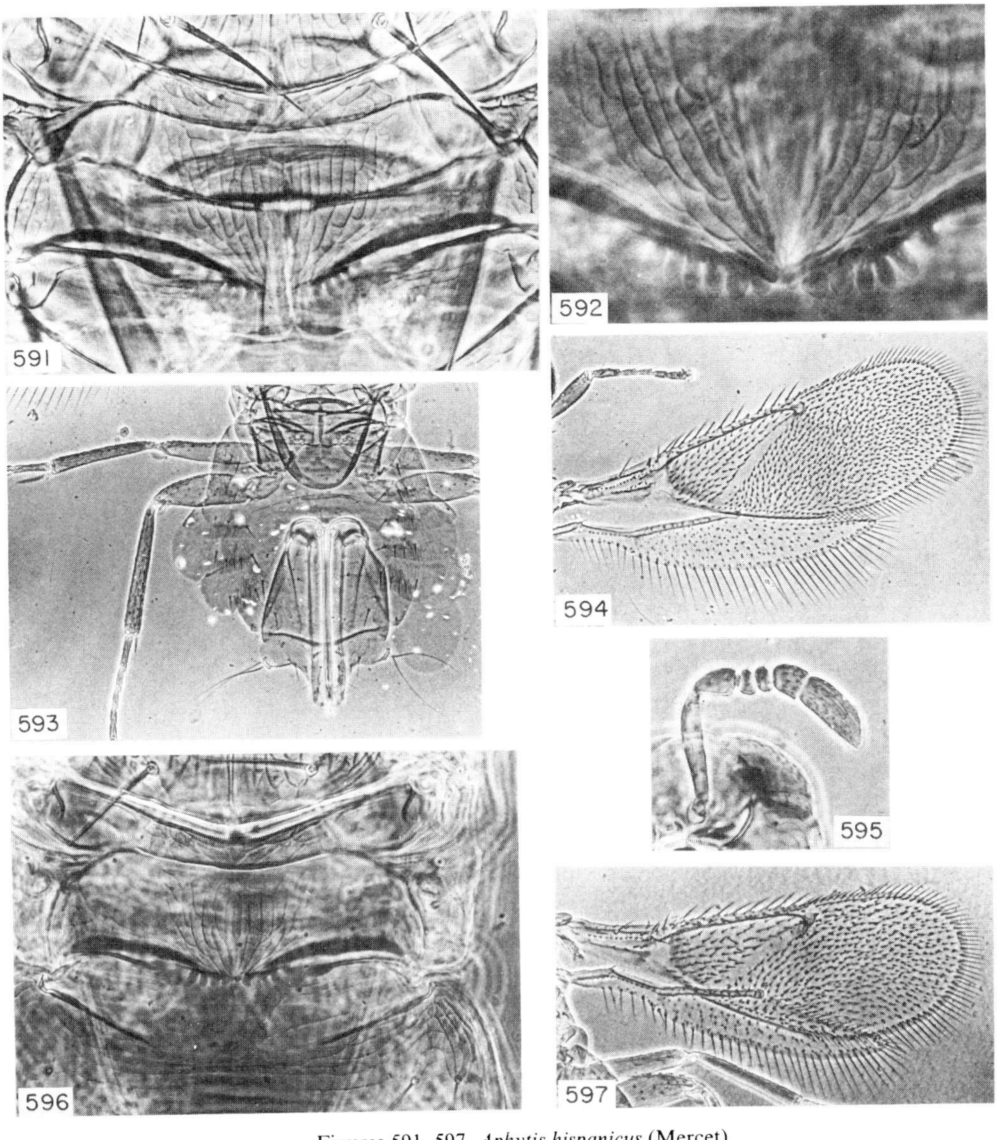


Figures 591–597. *Aphytis hispanicus* (Mercet)
591. ♀: Metanotum, propodeum and crenulae (holotype). 592. ♀: Propodeal crenulae. 593. ♀: Abdomen and legs (holotype). 594. ♀: Wings (holotype). 595. ♂: Antenna. 596. ♂: Metanotum, propodeum and base of gaster. 597. ♂: Wings.

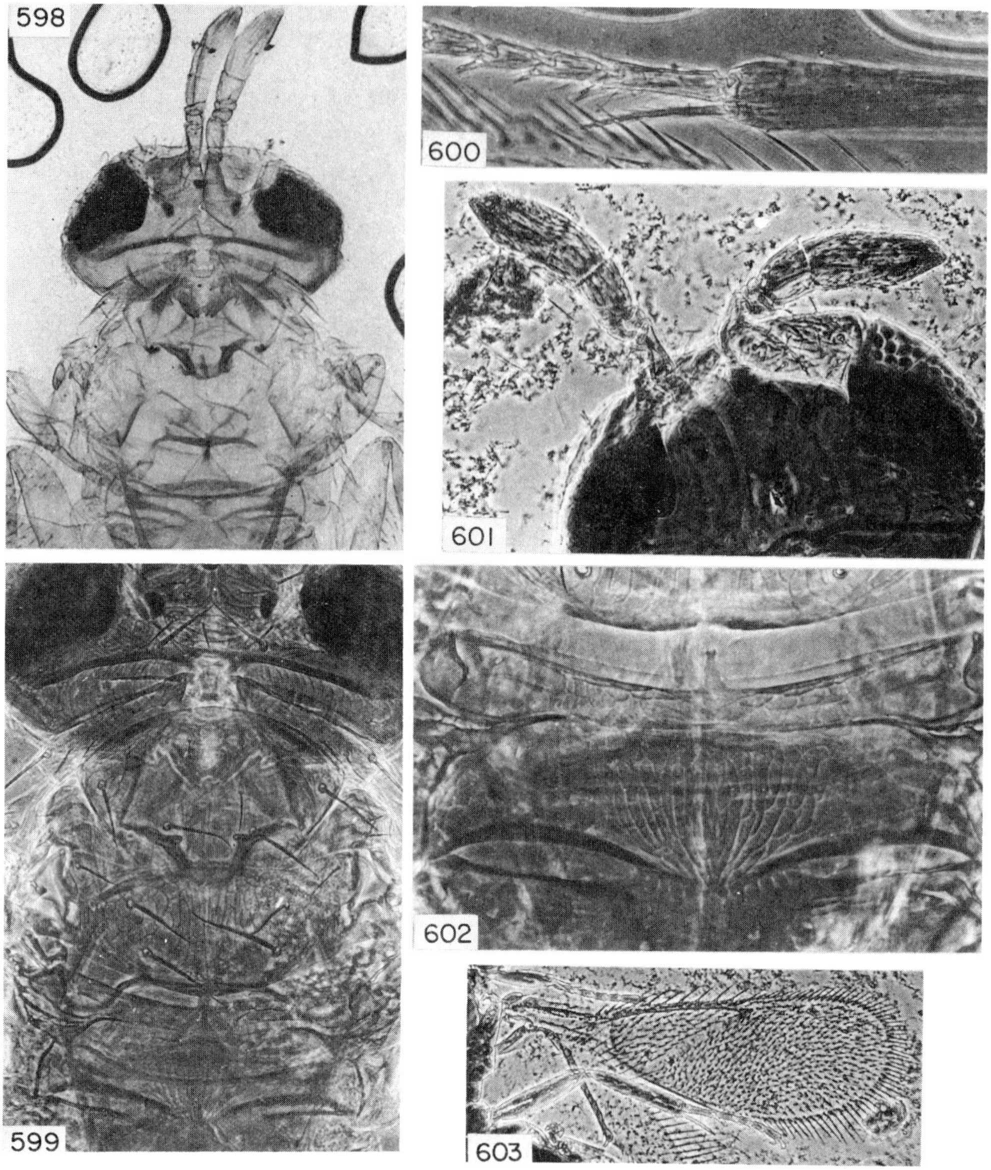


Figures 598–603. *Aphytis hispanicus* (Mercet), ♀
598–600. Holotype of *argentinus* (Brèthes): 598. Head, antennae, part of body. 599. Vertex, thorax and propodeum. 600. Mid-tibial spur and basitarsus.
601–603. Type of *bovelli* (Malenotti): 601. Antennae. 602. Metanotum, propodeum and crenulae. 603. Forewing.

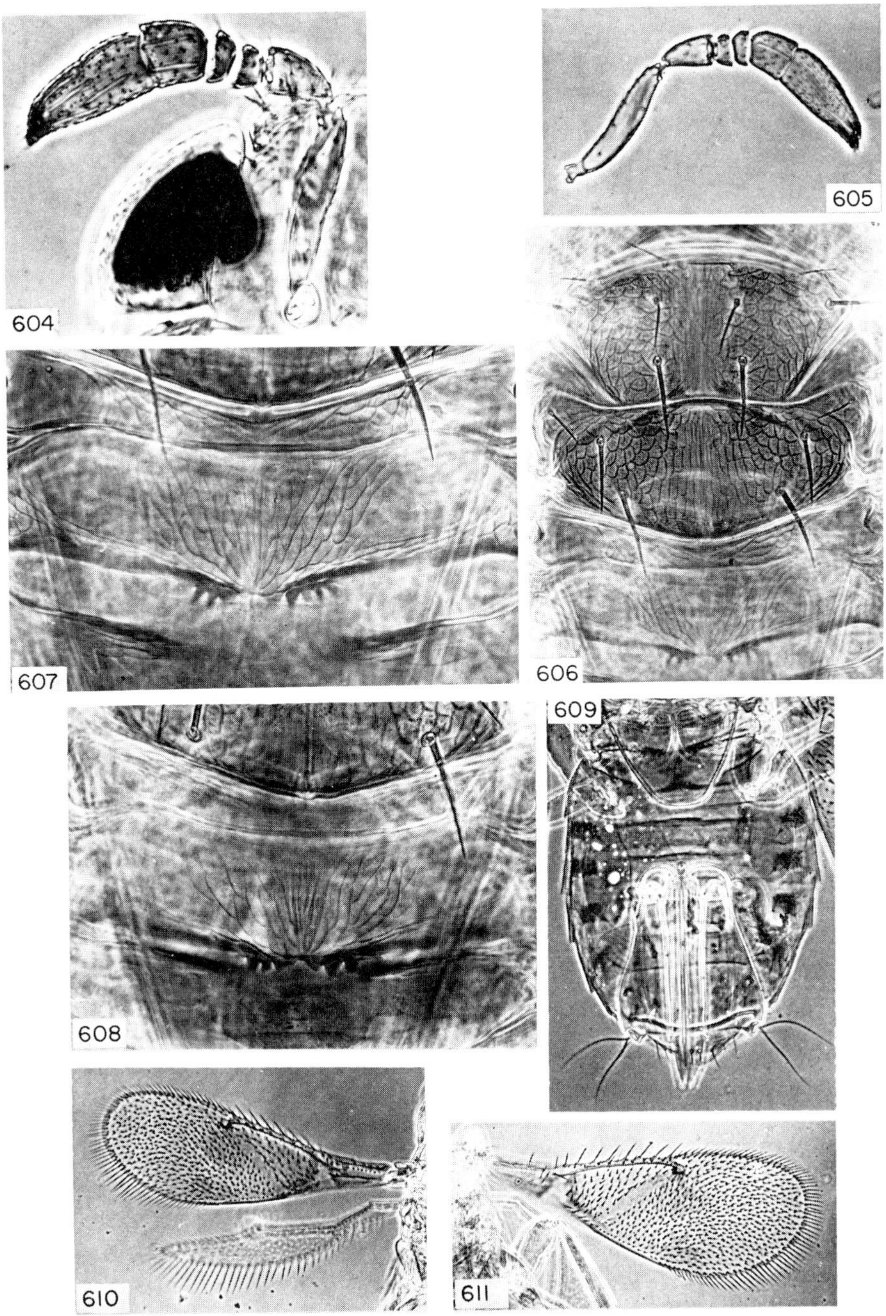


Figures 604–611. *Aphytis comperei* DeBach and Rosen, ♀ (paratypes)
604, 605. Antenna. 606. Thorax and propodeum. 607, 608. Metanotum, propodeum and crenulae.
609. Abdomen. 610. Wings. 611. Forewing.

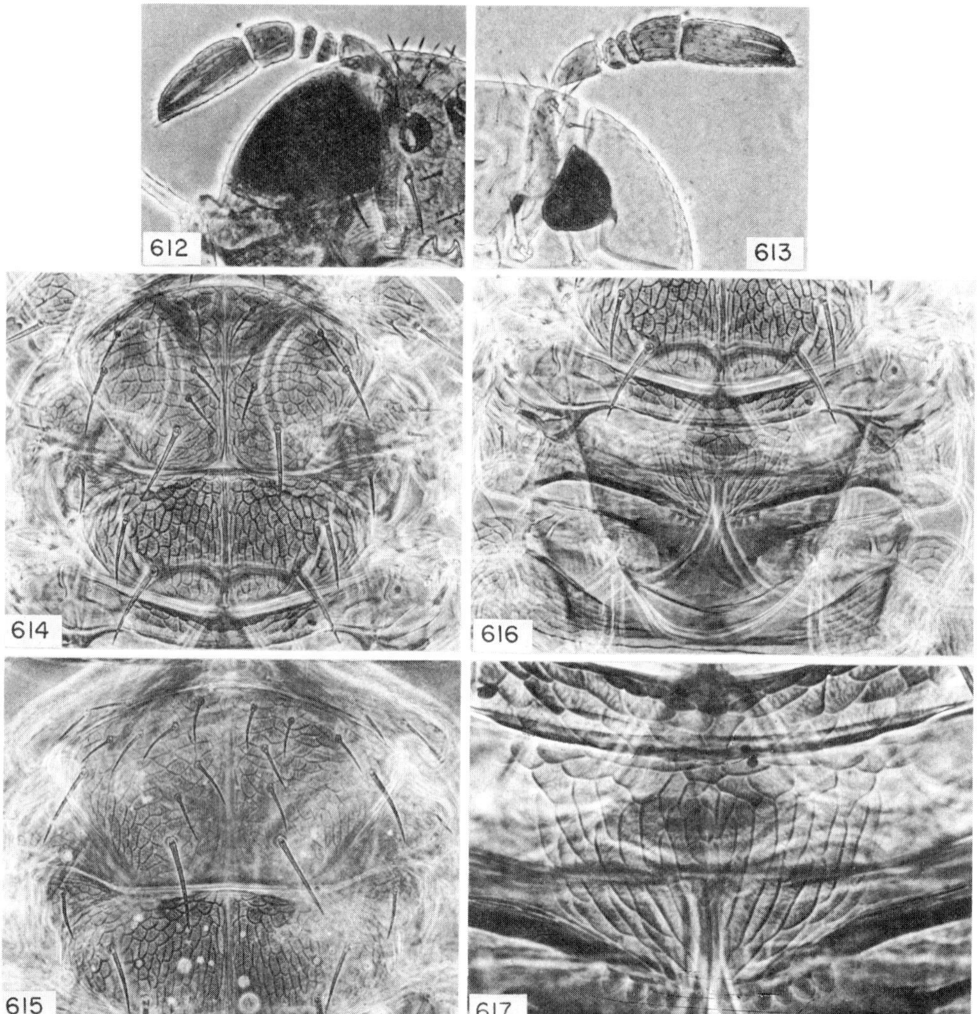


Figures 612–617. *Aphytis philippinensis* DeBach and Rosen, ♀
612. Antenna (holotype). 613. Antenna (paratype). 614. Thorax (holotype). 615. Mesonotum (Assam material). 616. Metanotum, propodeum and base of gaster (holotype). 617. Metanotum, propodeum and crenulae (holotype).

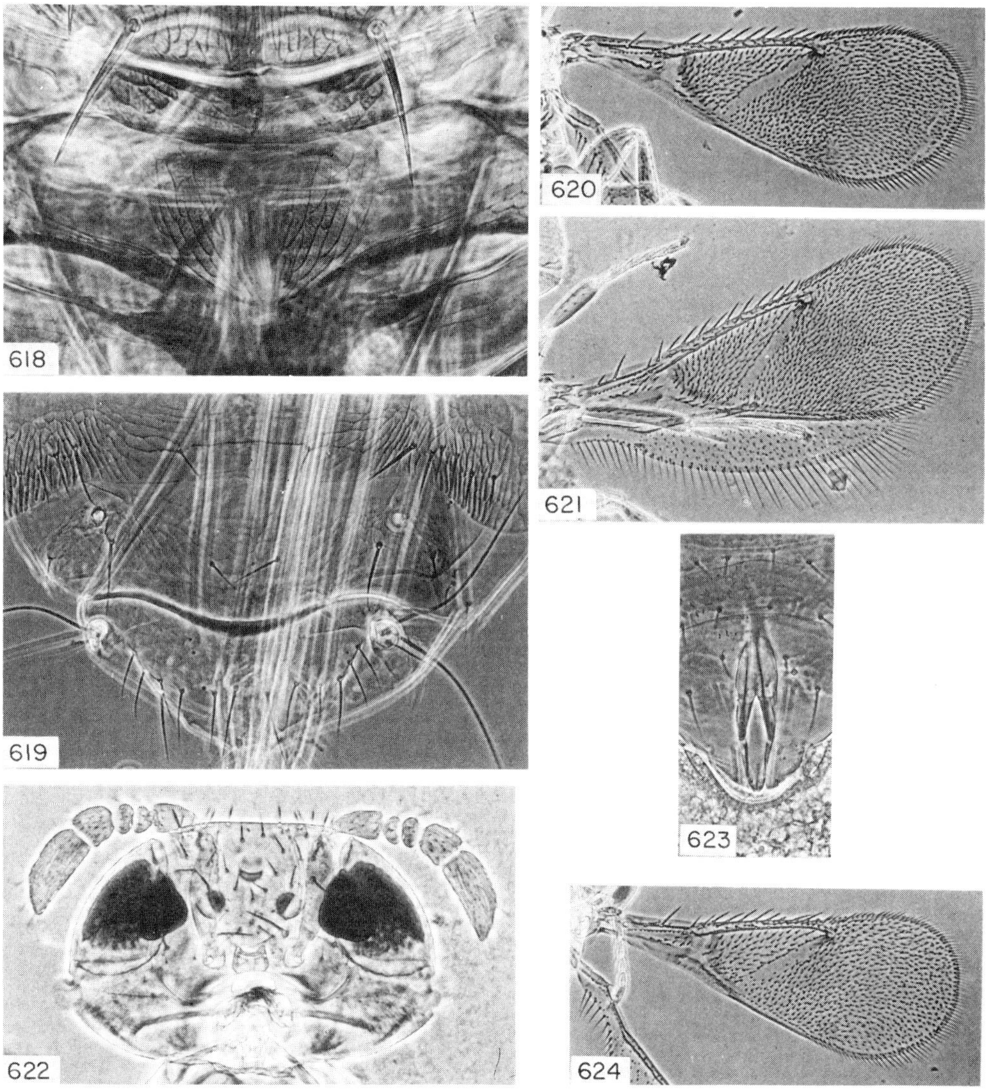


Figures 618–624. *Aphytis philippinensis* DeBach and Rosen
618. ♀: Metanotum, propodeum and crenulae (paratype). 619. ♀: Posterior abdominal tergites (VII, VIII and syntergum), showing sculpture and chaetotaxis (holotype). 620. ♀: Forewing (holotype). 621. ♀: Wings (paratype). 622. ♂: Head and antennae (paratype). 623. ♂: Genitalia (paratype). 624. ♂: Forewing (allotype).

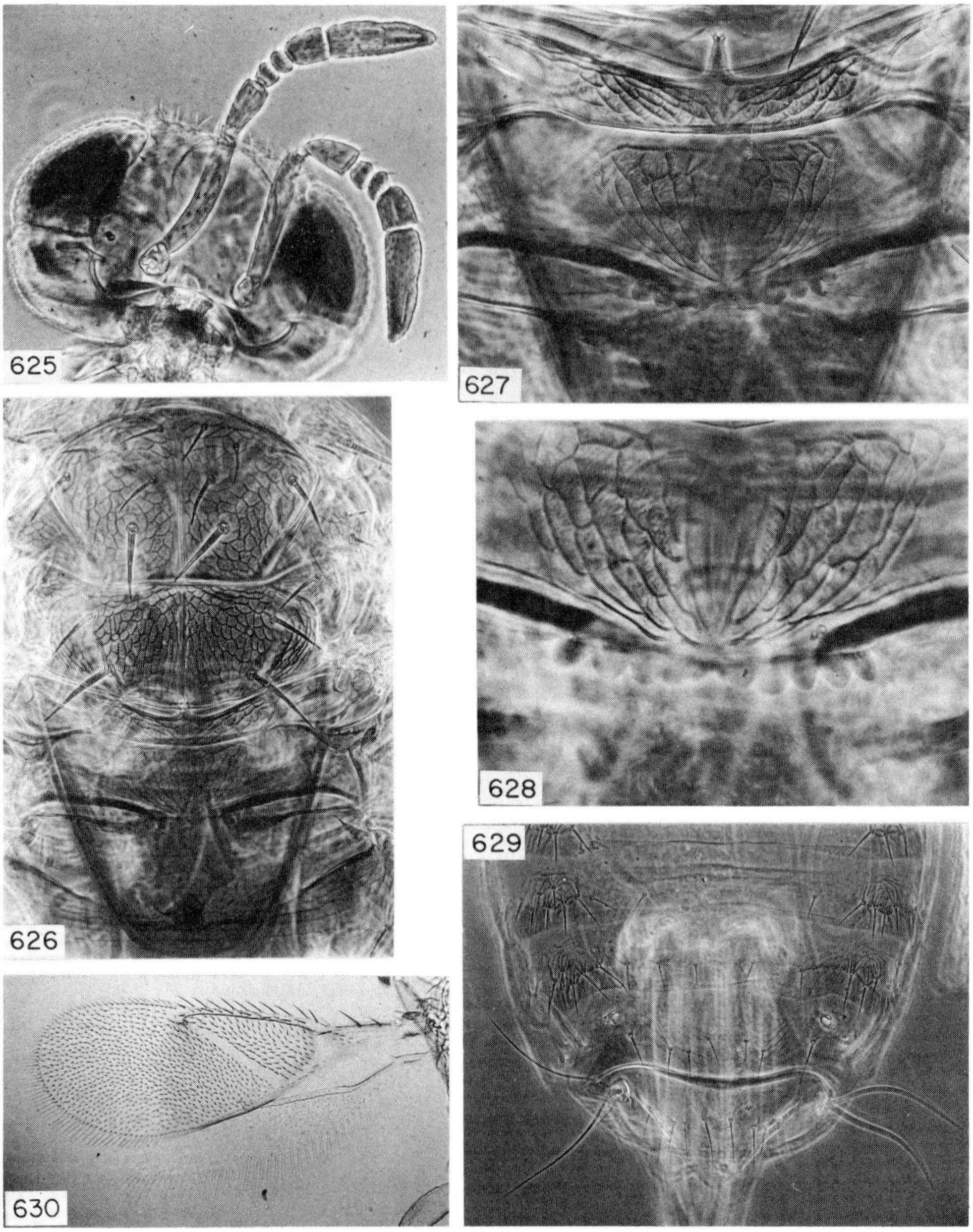


Figures 625–630. *Aphytis vandenboschi* DeBach and Rosen, ♀ (paratypes)
625. Head and antennae. 626. Thorax, propodeum and base of gaster. 627. Metanotum, propodeum and crenulae. 628. Propodeal crenulae. 629. Abdominal tergites V–VIII and syntergum, showing sculpture and chaetotaxis. 630. Wings.

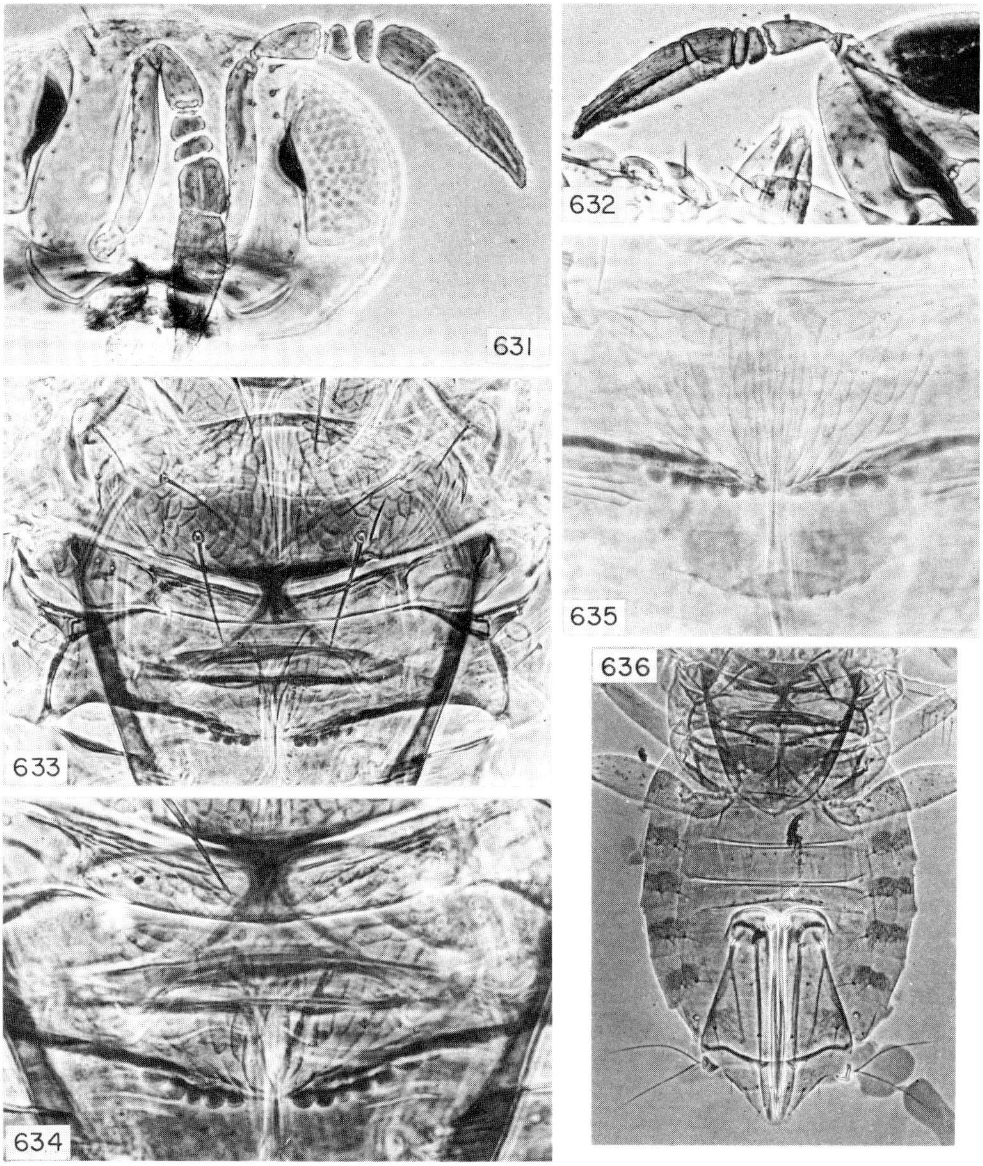


Figures 631–636. *Aphytis confusus* DeBach and Rosen, ♀
631. Head and antennae (paratype). 632. Antenna (paratype). 633. Scutellum, metanotum and propodeum (paratype). 634. Metanotum, propodeum and crenulae (paratype). 635. Propodeum, crenulae and second abdominal tergite (paratype). 636. Abdomen; note also endophragma (holotype).

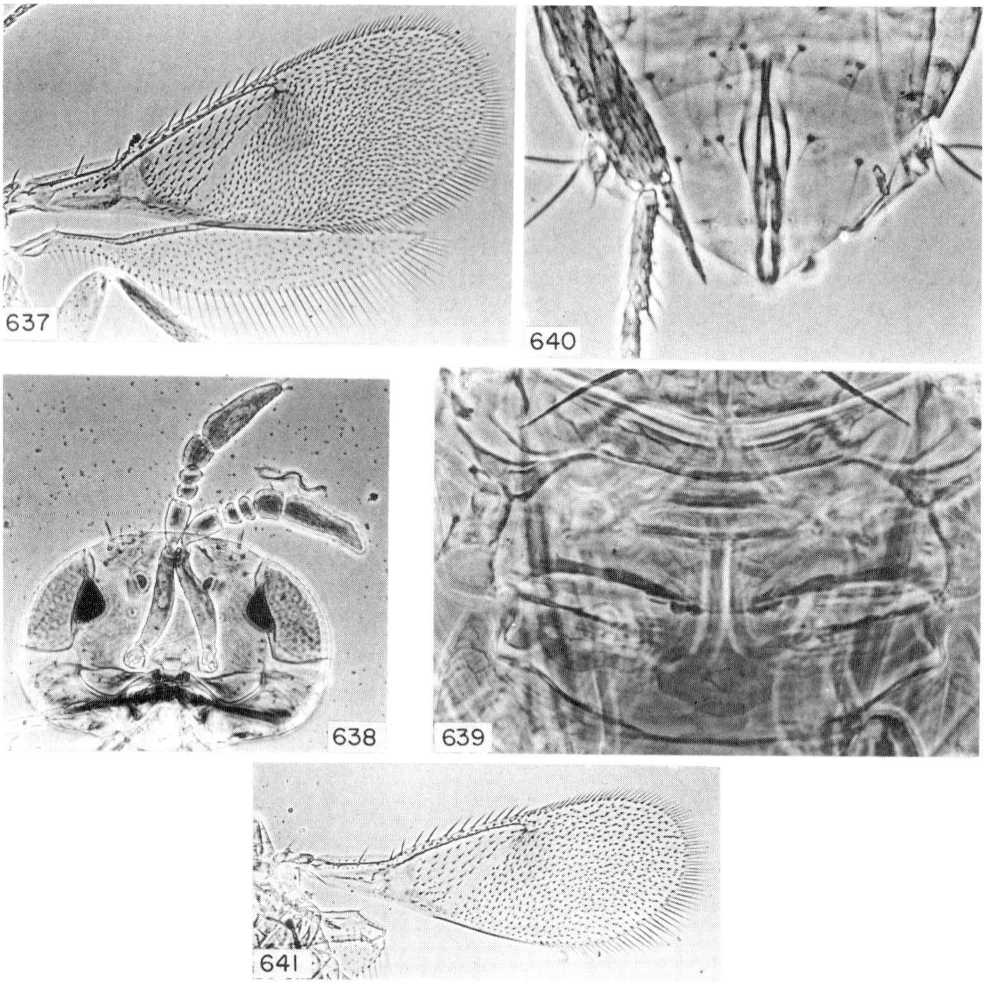


Figures 637–641. *Aphytis confusus* DeBach and Rosen
637. ♀: Wings (paratype). 638. ♂: Head and antennae (paratype). 639. ♂: Metanotum, propodeum and base of gaster (paratype). 640. ♂: Genitalia; note also cerci and mid-tibial spur (allotype). 641. ♂: Forewing (paratype).

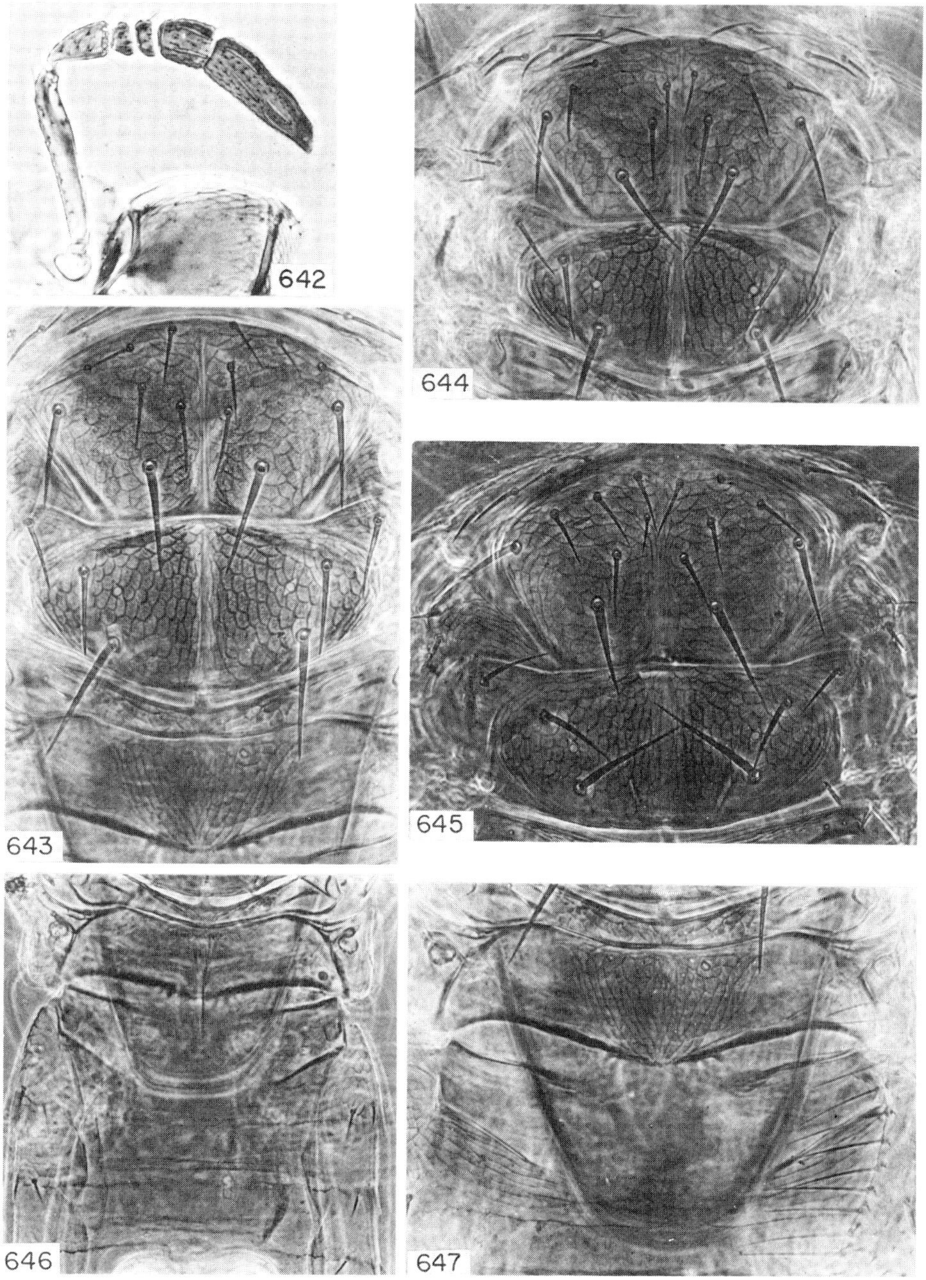


Figures 642–647. *Aphytis diaspidis* (Howard), ♀
642. Antenna. 643. Thorax and propodeum. 644. Thorax. 645. Mesonotum (syntype of *fuscipennis* (Howard)). 646. Metanotum, propodeum and abdominal tergites II–V (minute specimen). 647. Metanotum, propodeum and base of gaster.

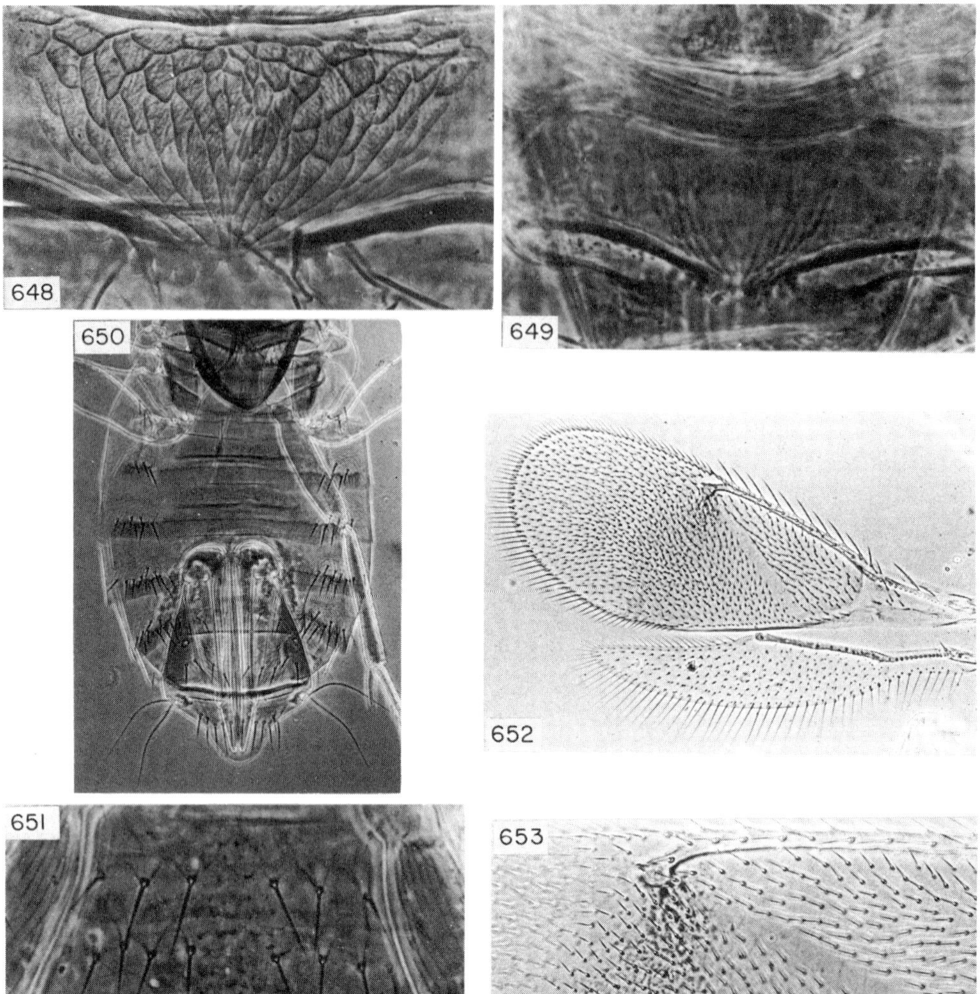


Figures 648–653. *Aphytis diaspidis* (Howard), ♀
648. Propodeum and crenulae (syntype of *fuscipennis* (Howard)). 649. Propodeum and crenulae (*ovidii* (Girault), det. Girault). 650. Abdomen dorsal view. 651. Venter of abdomen, showing chaetotaxis and sculpture. 652. Wings. 653. Integumental pigmentation below stigmal vein of forewing.

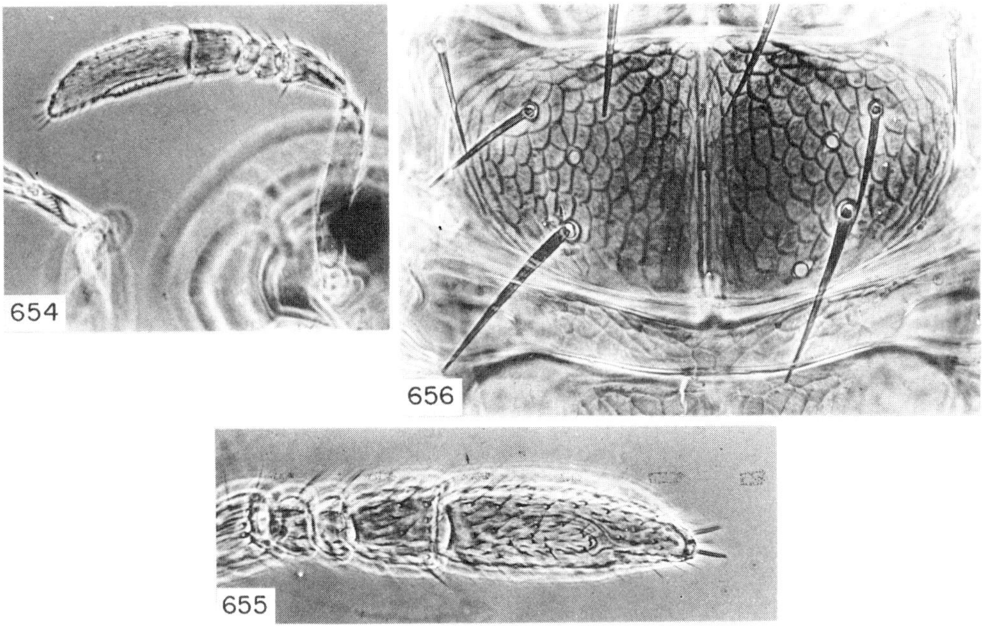


Figures 654–656. *Aphytis diaspidis* (Howard), ♂
654. Antenna; note cut-off sensory area on ventral aspect of club. 655. Antennal funicle and club, ventral view, showing cut-off sensory area. 656. Aberrant scutellum, with 3 discoid sensilla and 4 setae.

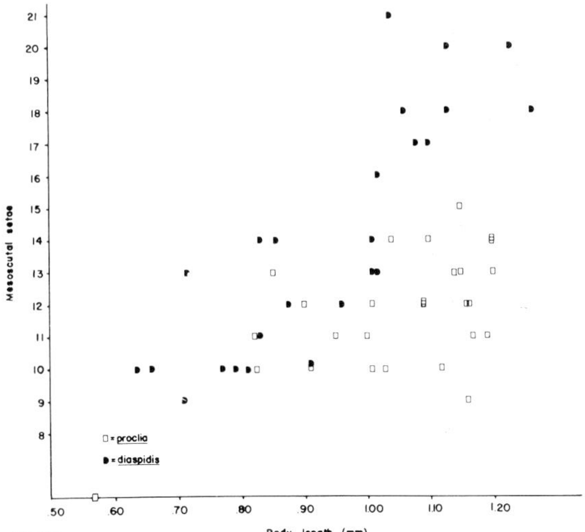


657

Figure 657. Scatter diagram, showing number of mesoscutal setae in relation to body length in females of *A. proclia* (Walker) and *A. diaspidis* (Howard).

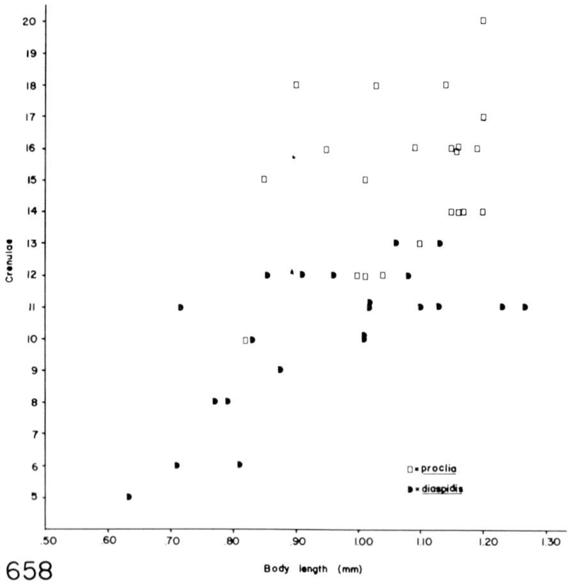


658

Figure 658. Scatter diagram, showing number of propodeal crenulae in relation to body length in females of *A. proclia* (Walker) and *A. diaspidis* (Howard).

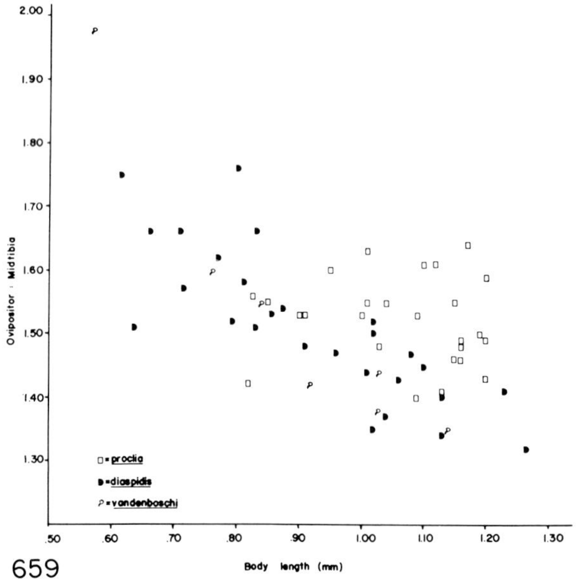


659

Figure 659. Scatter diagram, showing the relative length of the ovipositor in relation to body length in females of *A. proclia* (Walker), *A. diaspidis* (Howard) and *A. vandenboschi* DeBach and Rosen.

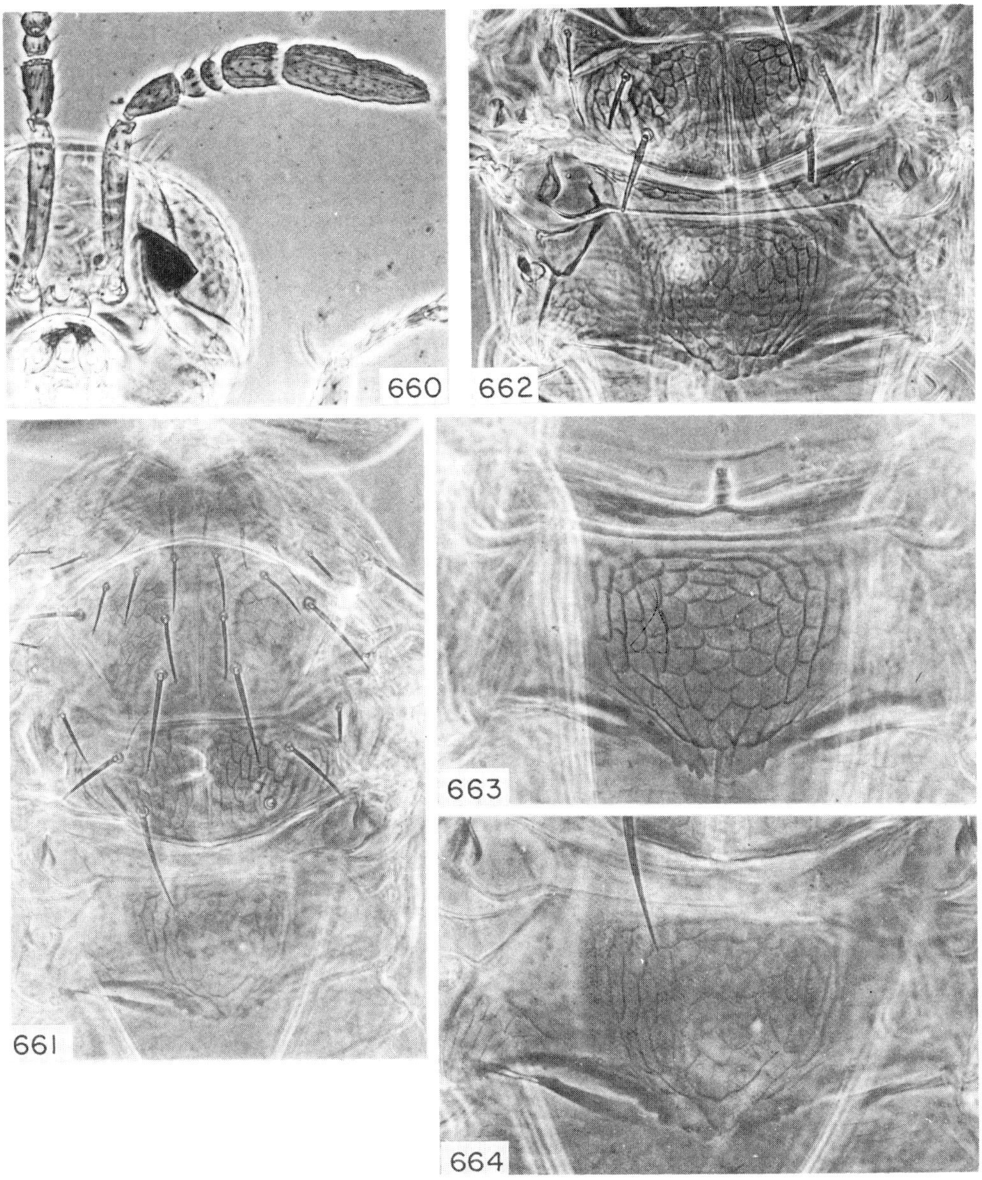


Figures 660–664. *Aphytis pinnaspidis* n.sp., ♀

660. Antenna (holotype). 661. Thorax and propodeum (paratype). 662. Scutellum, metanotum and propodeum (paratype). 663. Metanotum, propodeum and crenulae. 664. Propodeum and crenulae (paratype).

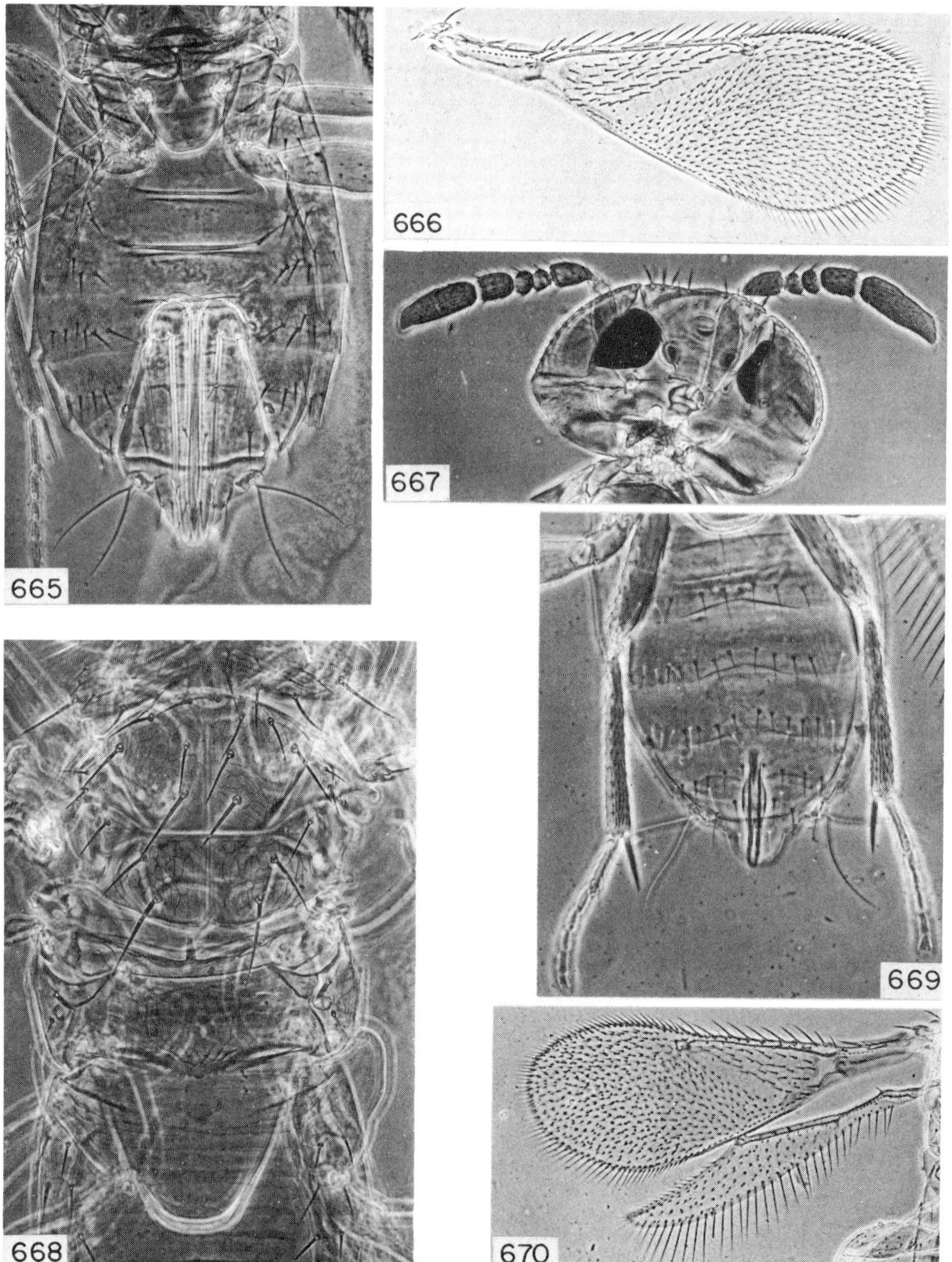


Figures 665–670. *Aphytis pinnaspidis* n.sp. (paratypes)
665. ♀: Abdomen. 666. ♀: Forewing. 667. ♂: Head and antennae. 668. ♂: Thorax, propodeum and base of gaster. 669. ♂: Venter of abdomen and genitalia, showing also middle legs. 670. ♂: Wings.

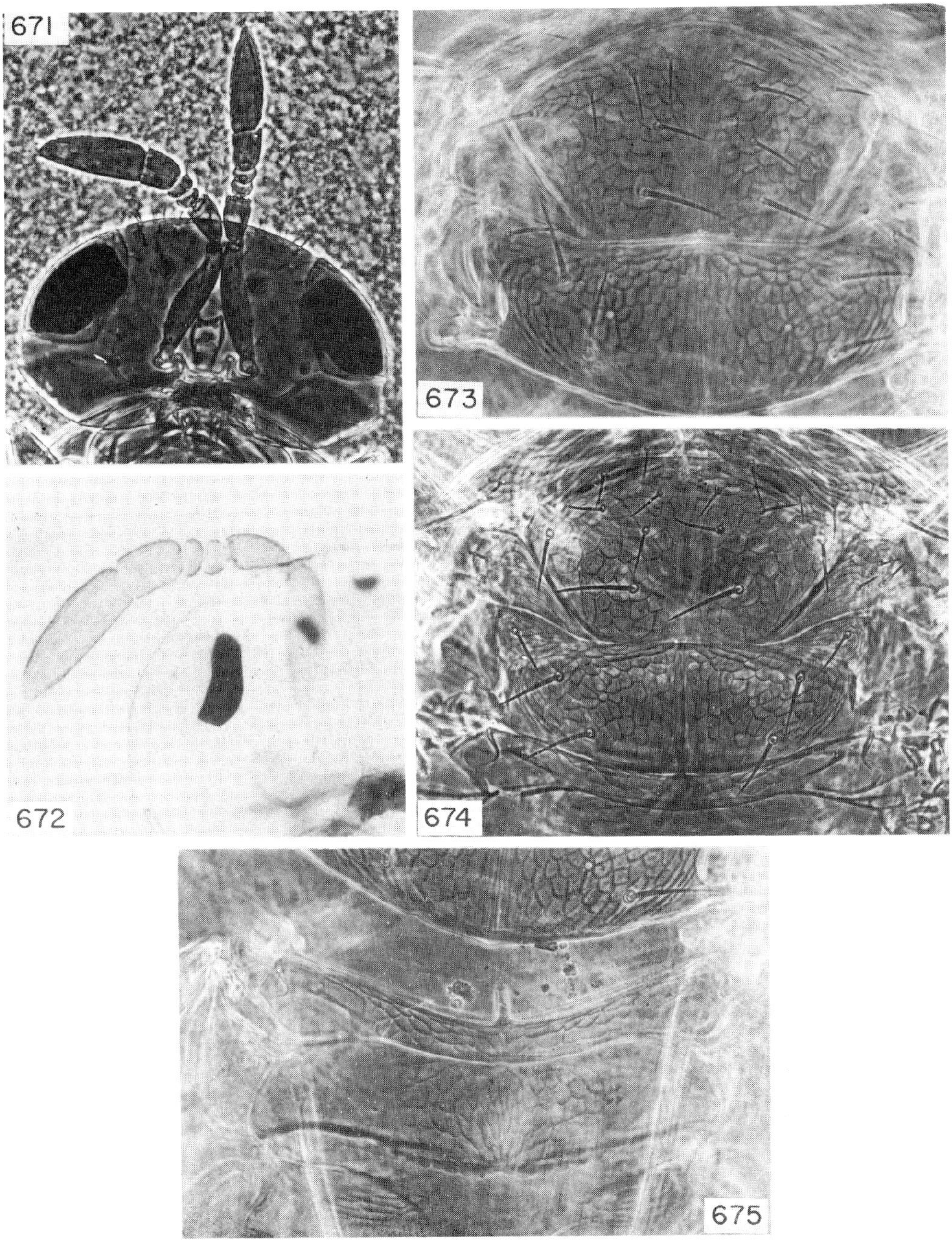


Figures 671–675. *Aphytis testaceus* Tshumakova, ♀
671. Head and antennae. 672. Antenna. 673. Mesonotum. 674. Thorax. 675. Metanotum, propodeum and crenulae.

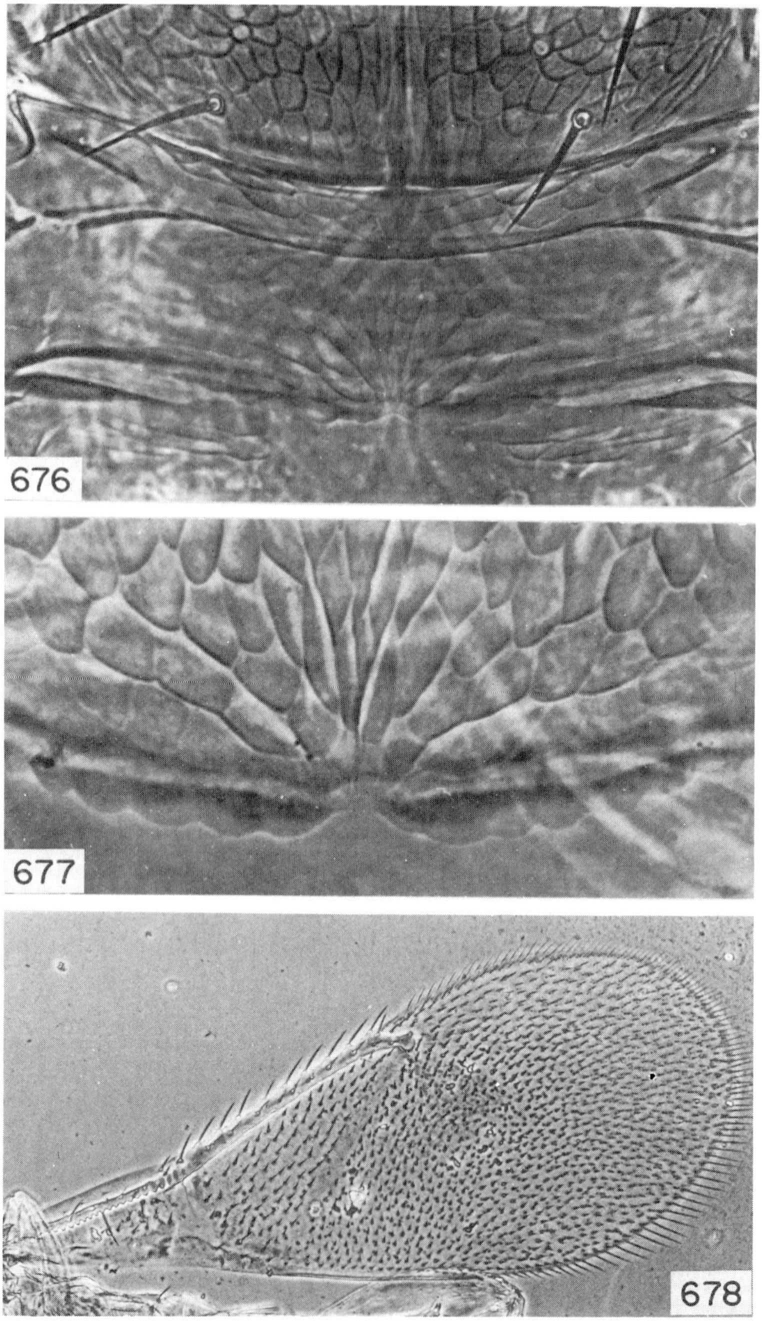


Figures 676–678. *Aphytis testaceus* Tshumakova, ♀
676. Metanotum, propodeum and crenulae. 677. Propodeal crenulae. 678. Forewing.

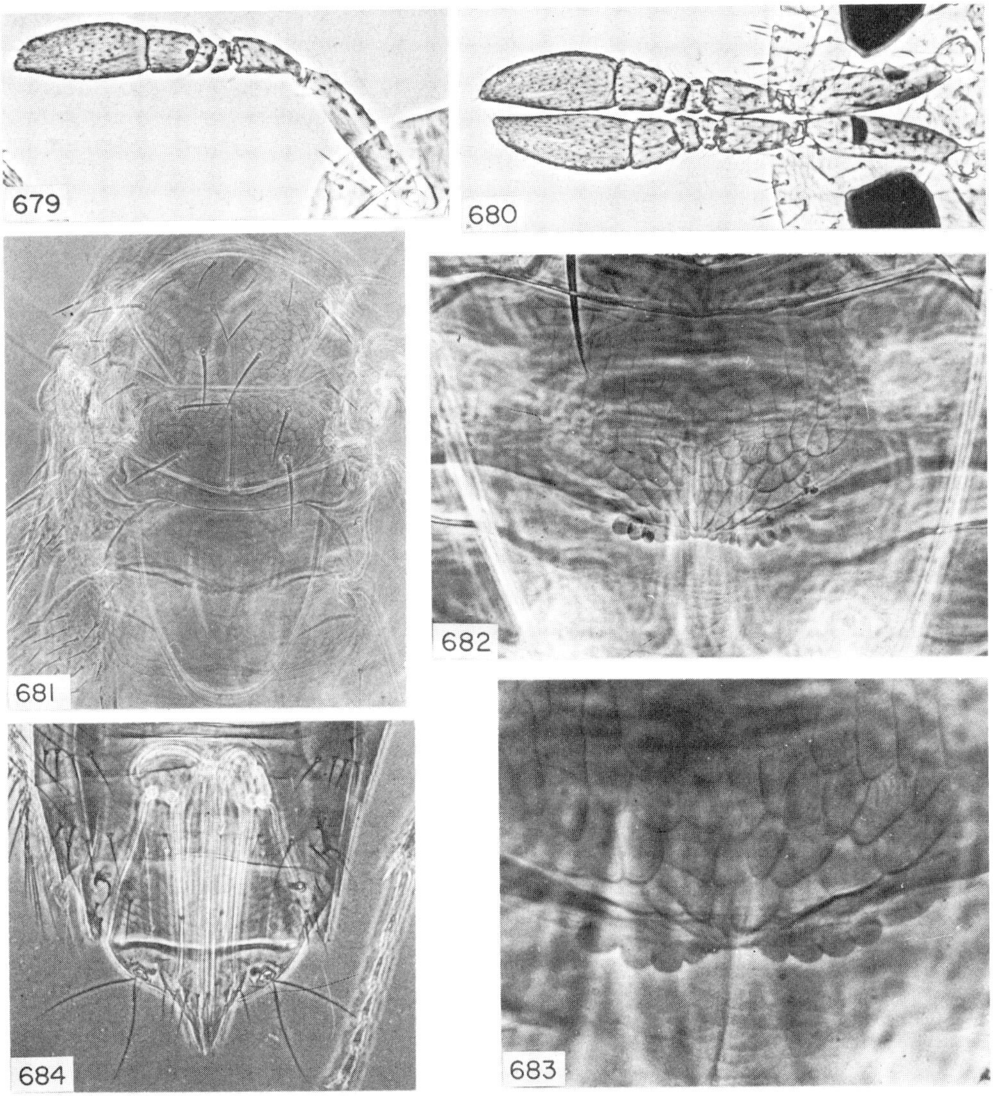


Figures 679–684. *Aphytis griseus* Quednau, ♀

679. Antenna (lectotype). 680. Antennae (paralectotype). 681. Thorax, propodeum and base of gaster. 682. Propodeum and crenulae (paralectotype). 683. Propodeal crenulae (paralectotype). 684. Posterior abdominal tergites, showing cerci and spiracles.

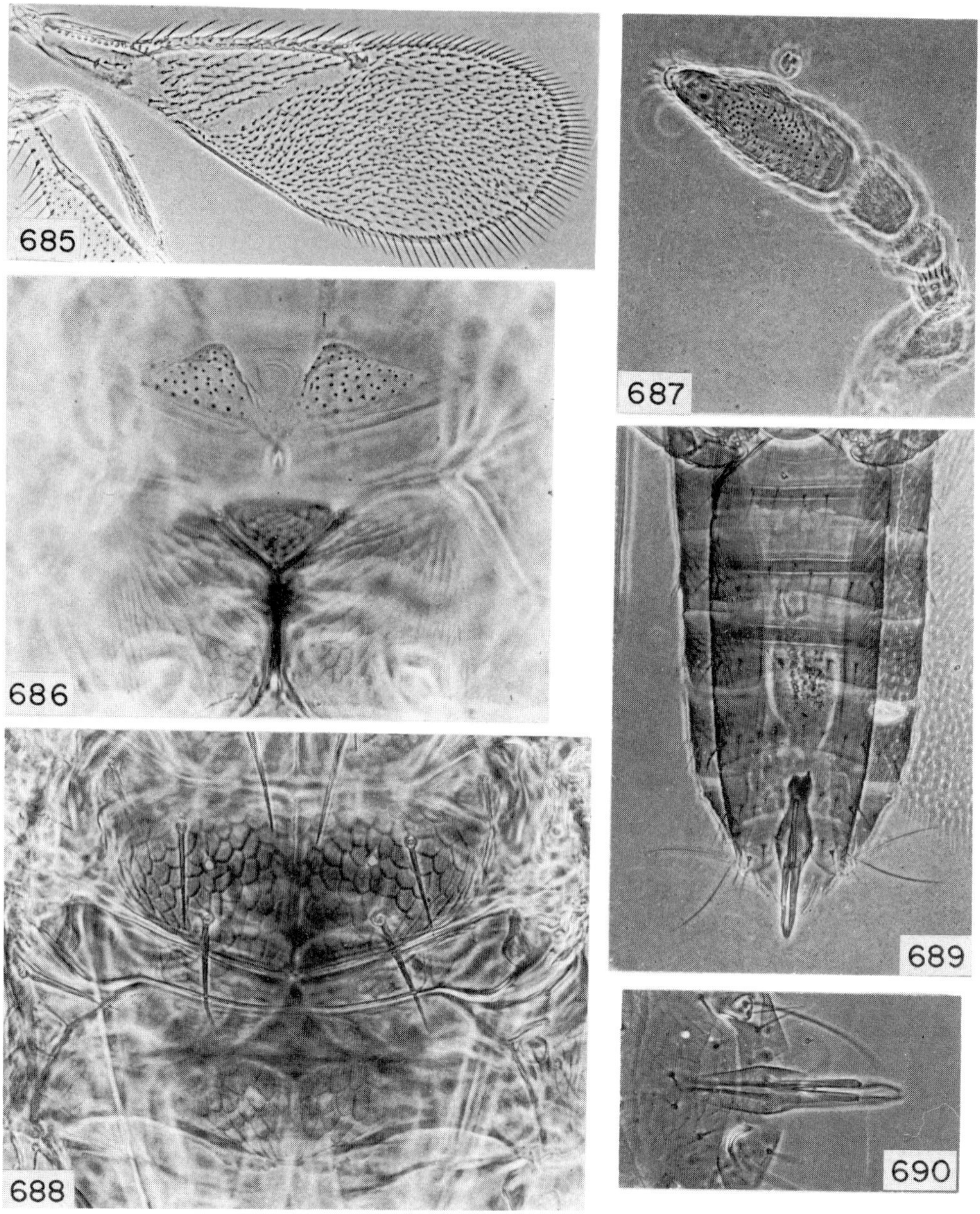


Figures 685–690. *Aphytis griseus* Quednau
685. ♀: Forewing. 686. ♀: Mesosternum, showing pigmentation; note also epicoxal pads anterad of mesosternal furca. 687. ♂: Antennal club, showing specialized sensory area on ventral aspect (paralectotype). 688. ♂: Scutellum, metanotum and propodeum (paralectotype). 689. ♂: Venter of abdomen and genitalia. 690. ♂: Genitalia (paralectotype).

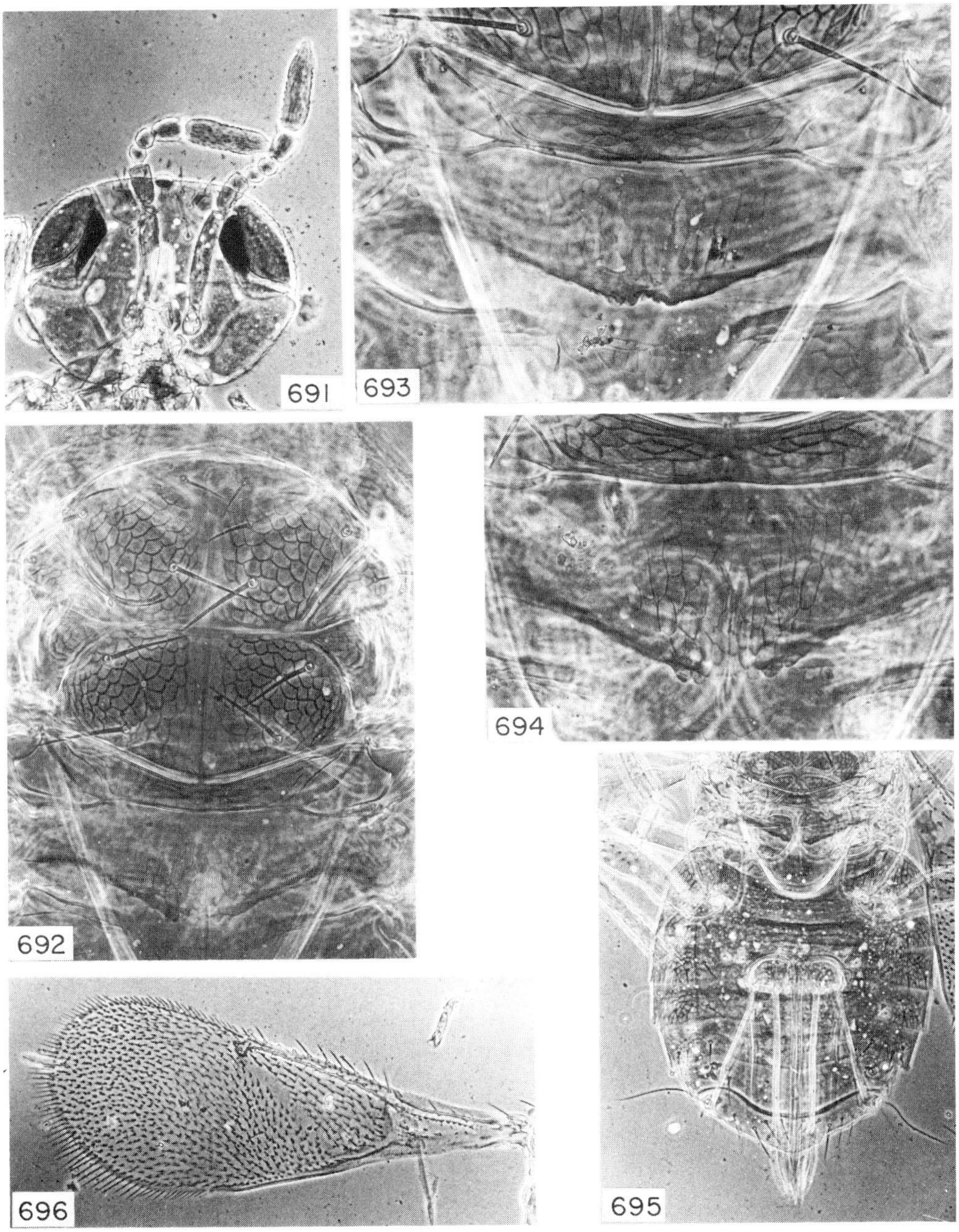


Figures 691–696. *Aphytis tucumani* n.sp., ♀
691. Head and antennae (holotype). 692. Thorax and propodeum (paratype). 693. Metanotum, propodeum and base of gaster (paratype). 694. Metanotum, propodeum and crenulae (holotype). 695. Abdomen (cercal setae missing; paratype). 696. Forewing (holotype).

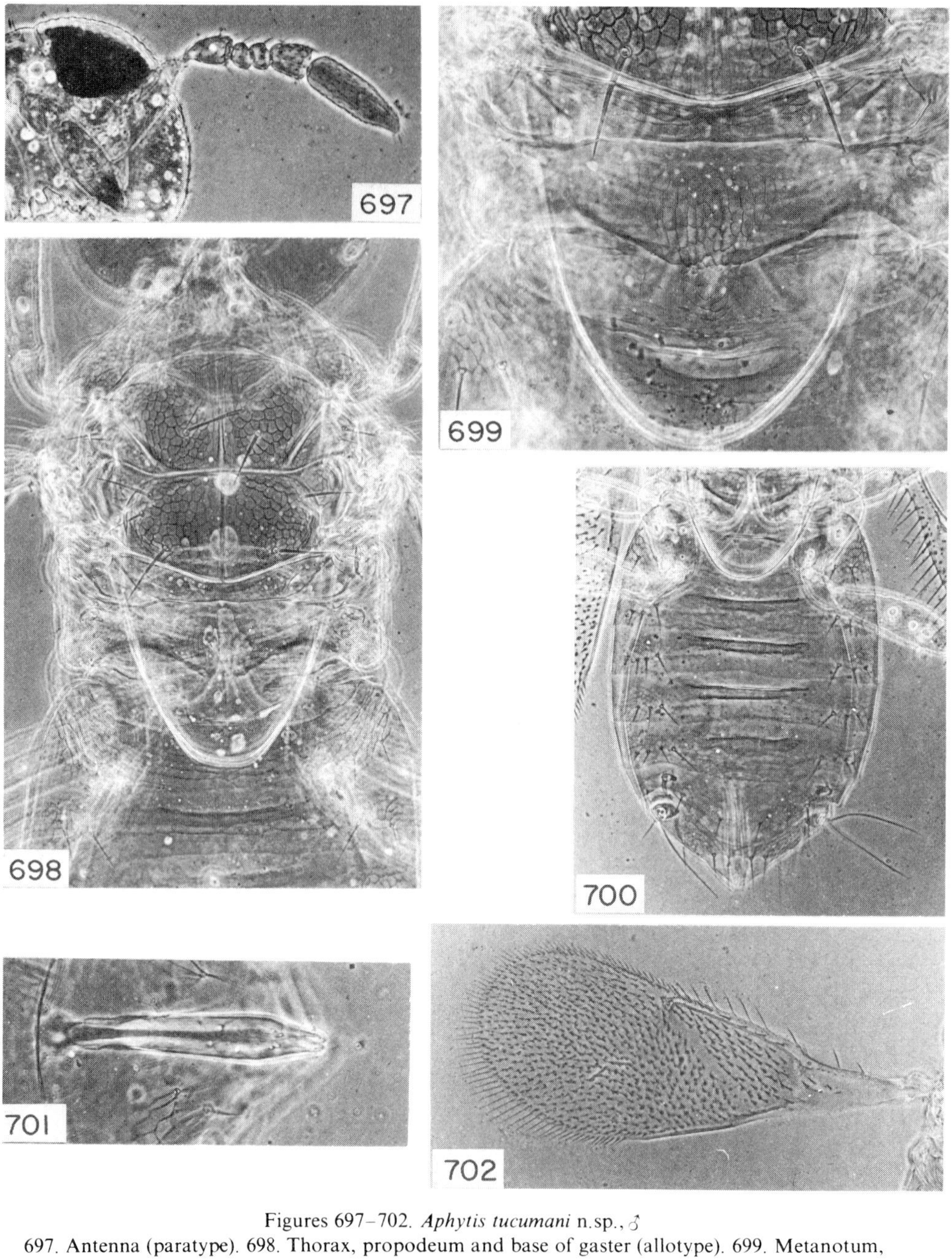


Figures 697–702. *Aphytis tucumani* n.sp., ♂
697. Antenna (paratype). 698. Thorax, propodeum and base of gaster (allotype). 699. Metanotum, propodeum and base of gaster (paratype). 700. Abdomen, dorsal view (paratype). 701. Genitalia; note absence of digital sclerites and apodemes (paratype). 702. Forewing (allotype).

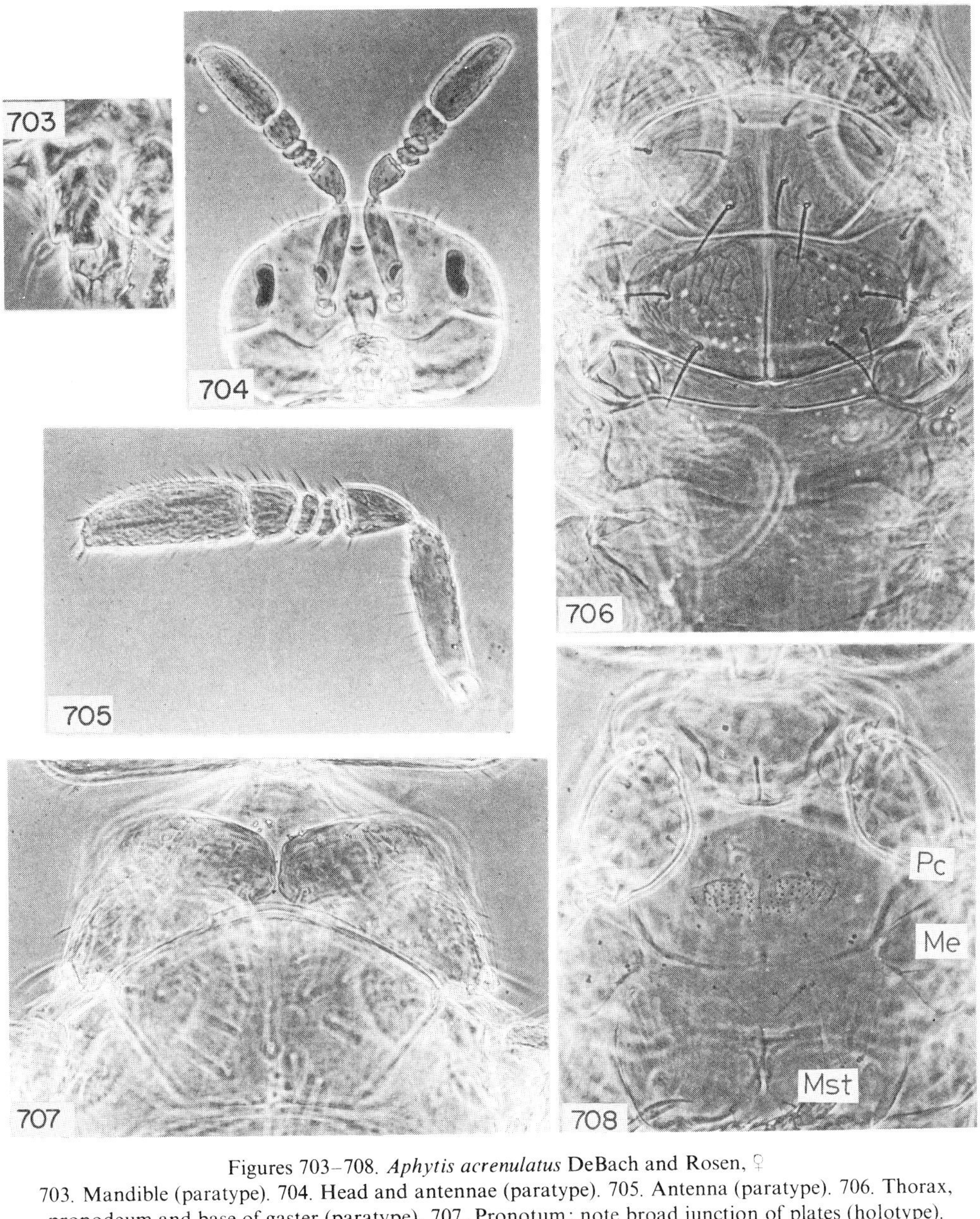


Figures 703–708. *Aphytis acrenulatus* DeBach and Rosen, ♀
703. Mandible (paratype). 704. Head and antennae (paratype). 705. Antenna (paratype). 706. Thorax, propodeum and base of gaster (paratype). 707. Pronotum; note broad junction of plates (holotype). 708. Venter of thorax, showing epicoxal pads (paratype).
*Me* = mesepisternum; *Mst* = mesosternum; *Pc* = prepectus.

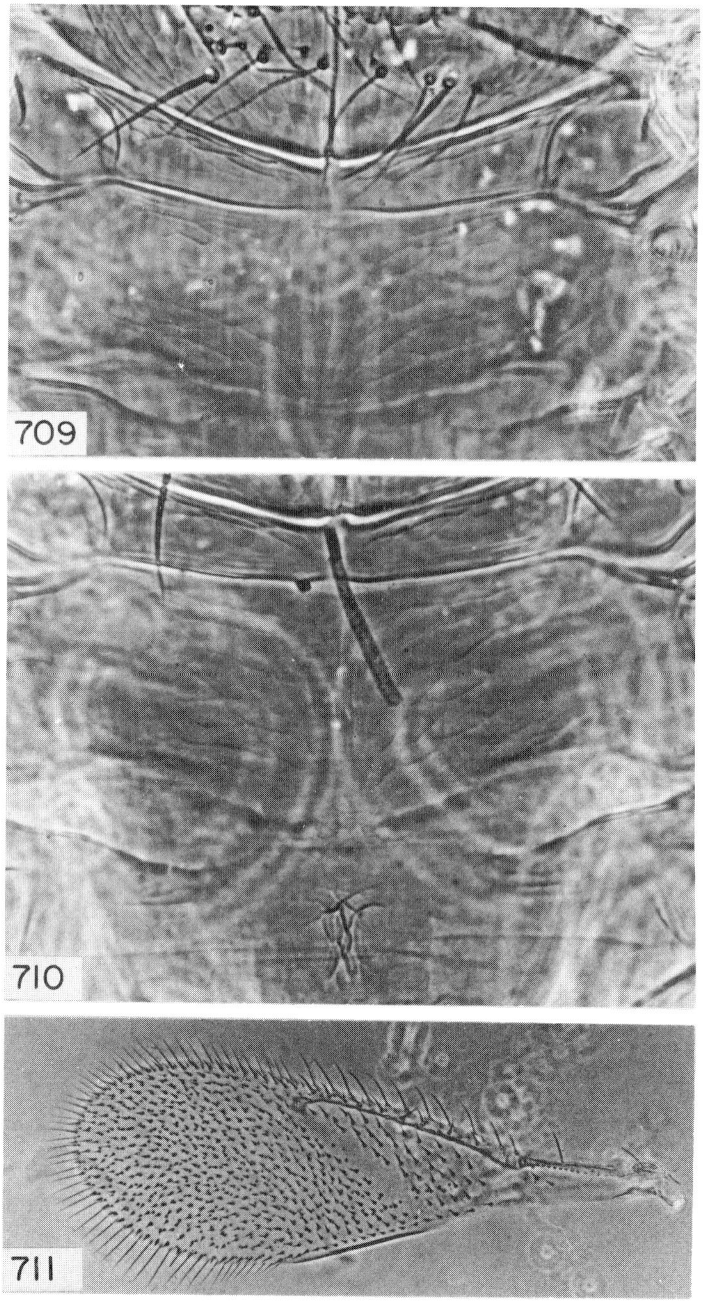


Figures 709–711. *Aphytis acrenulatus* DeBach and Rosen, ♀ (paratypes)
709. Metanotum and propodeum. 710. Metanotum, propodeum and second abdominal tergite; note median notch in propodeal ledge and sculpture at center of second tergite. 711. Forewing.

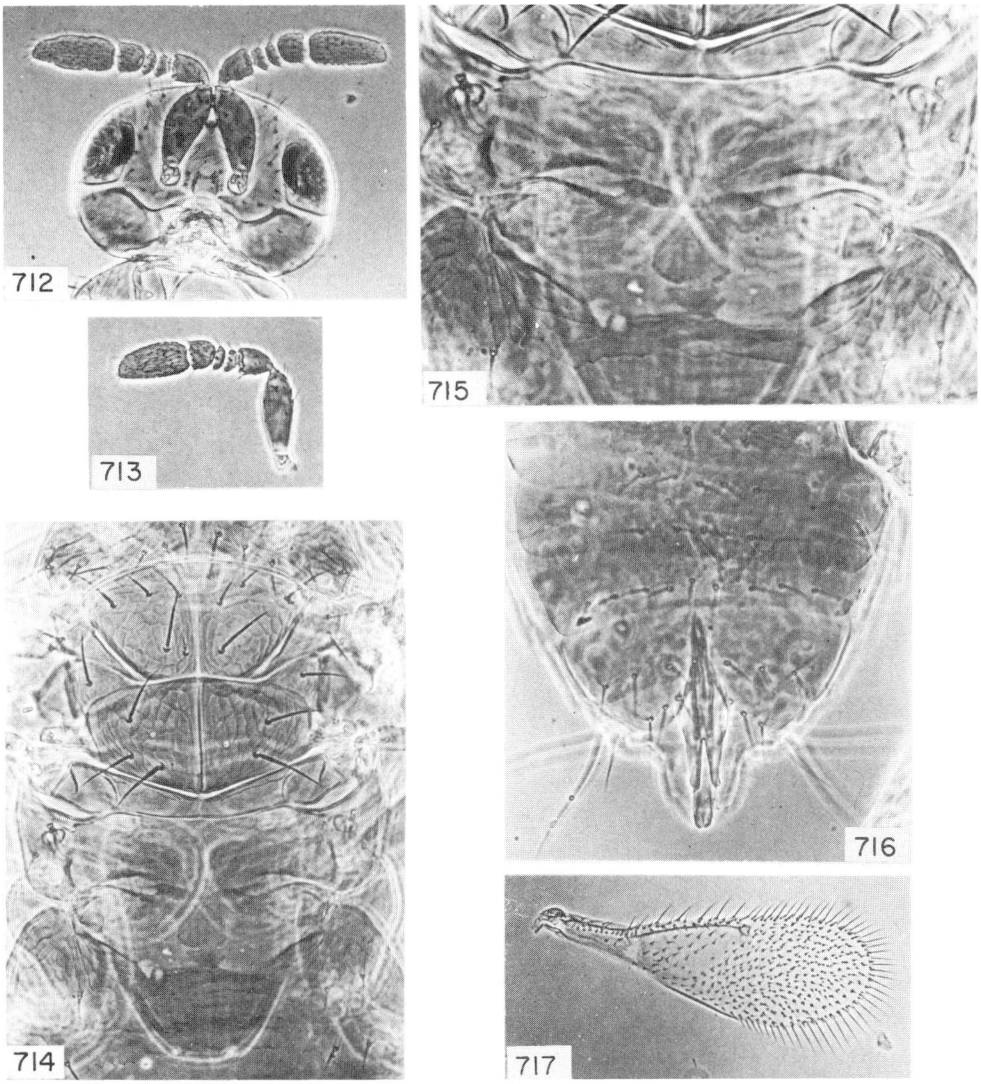


Figures 712–717. *Aphytis acrenulatus* DeBach and Rosen, ♂ (paratypes)
712. Head and antennae. 713. Antenna. 714. Thorax, propodeum and base of gaster. 715. Metanotum, propodeum and base of gaster. 716. Genitalia. 717. Forewing.

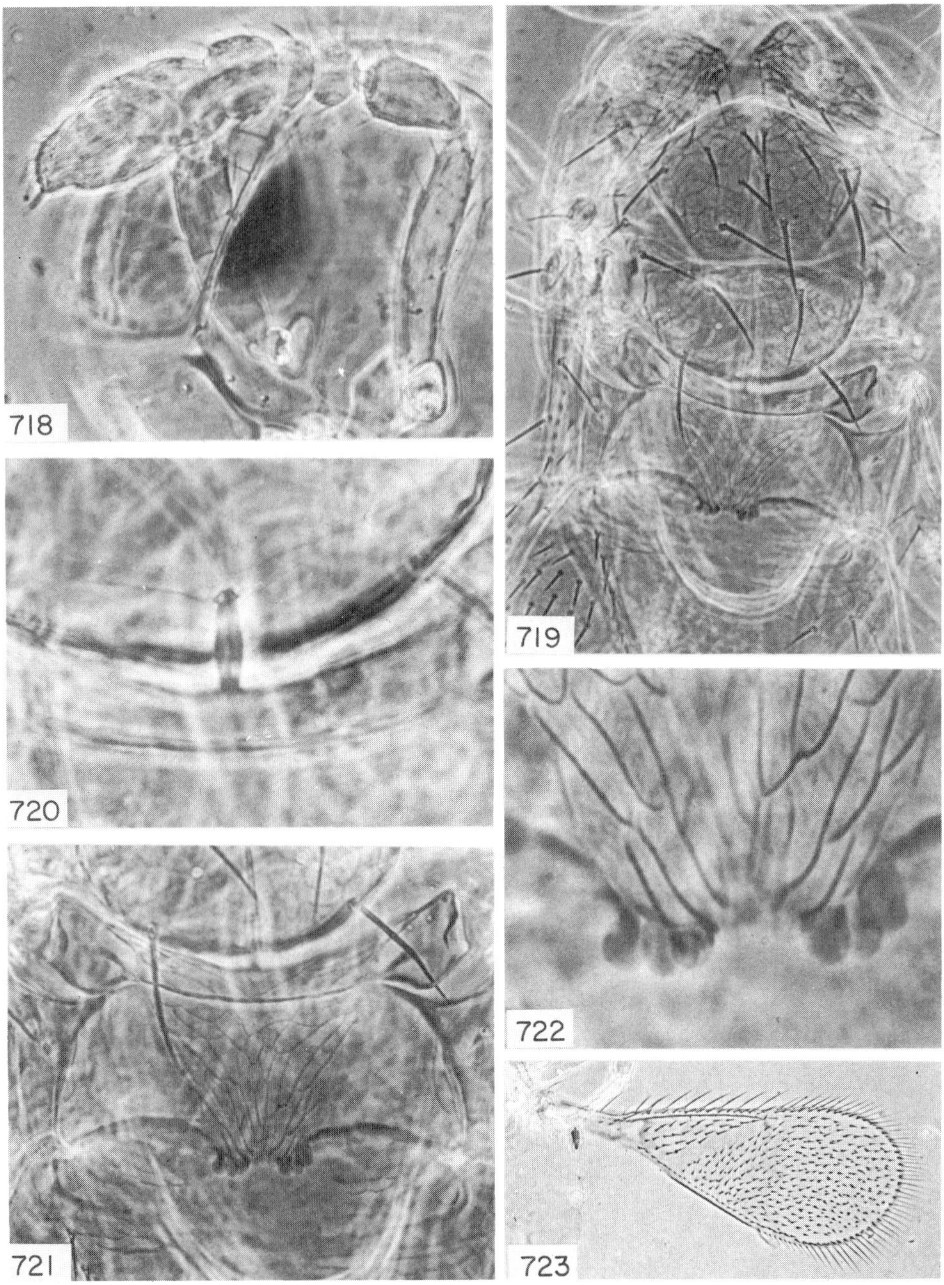


Figures 718–723. *Aphytis amazonensis* n.sp., ♀ (holotype)
718. Antenna. 719. Thorax, propodeum and base of gaster. 720. Metanotum and anteromedian apodeme. 721. Metanotum, propodeum and crenulae. 722. Propodeal crenulae. 723. Forewing.

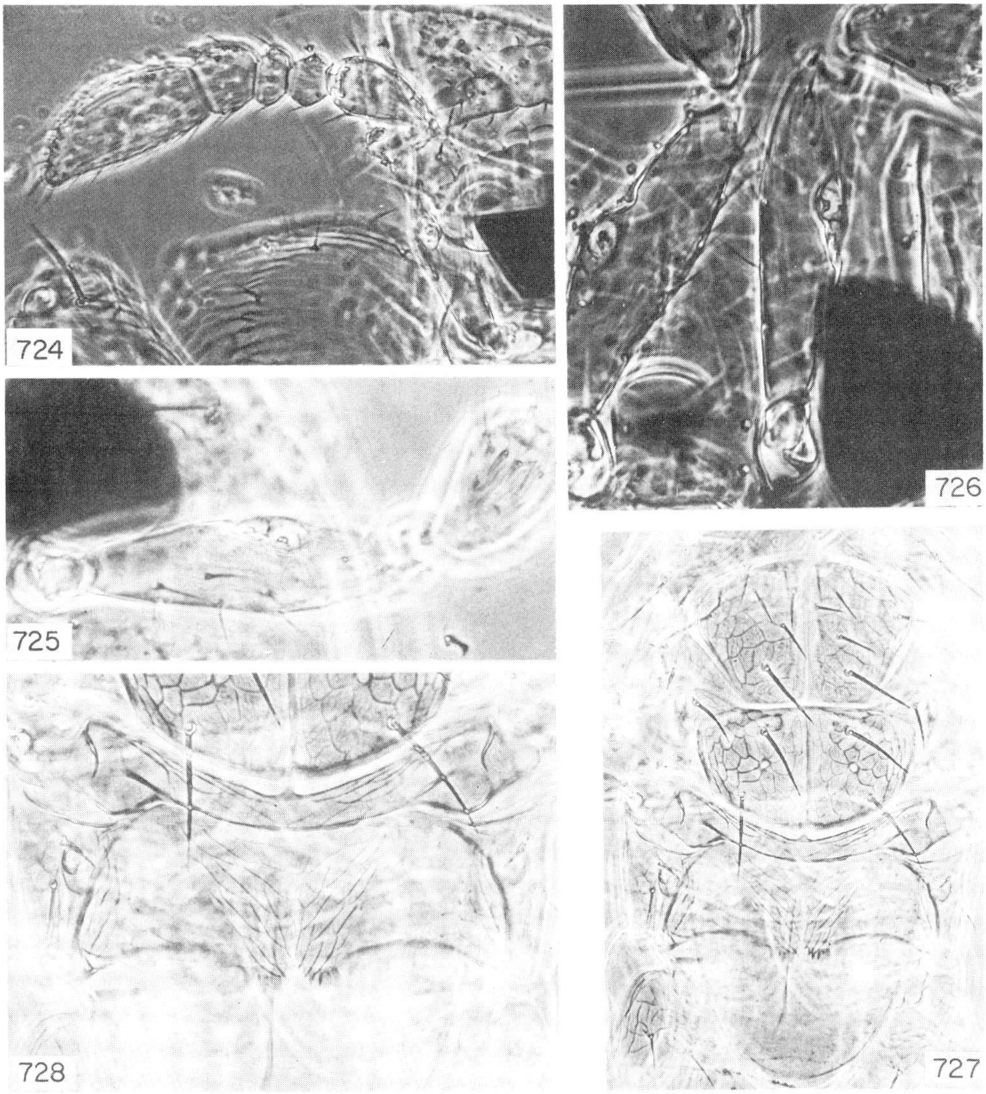


Figures 724–728. *Aphytis amazonensis* n.sp., ♂
724. Antenna (allotype). 725. Sense organs on antennal scape (allotype). 726. Sense organs on antennal scape (paratype). 727. Thorax, propodeum and base of gaster (paratype). 728. Metanotum, propodeum and crenulae (paratype).

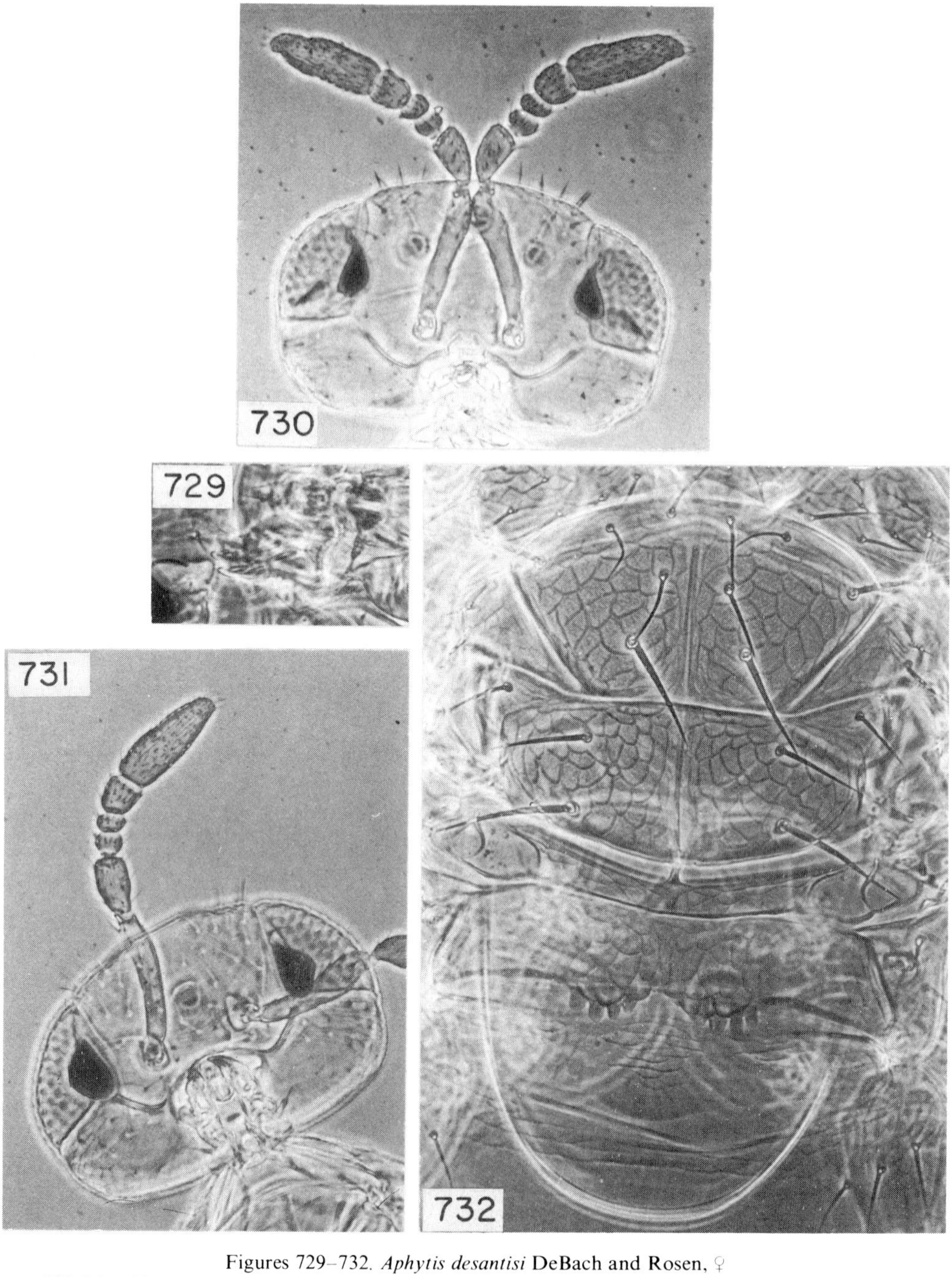


Figures 729–732. *Aphytis desantisi* DeBach and Rosen, ♀
729. Mandible (paratype). 730. Head and antennae (holotype). 731. Head and antenna (paratype). 732. Thorax, propodeum and base of gaster (holotype).

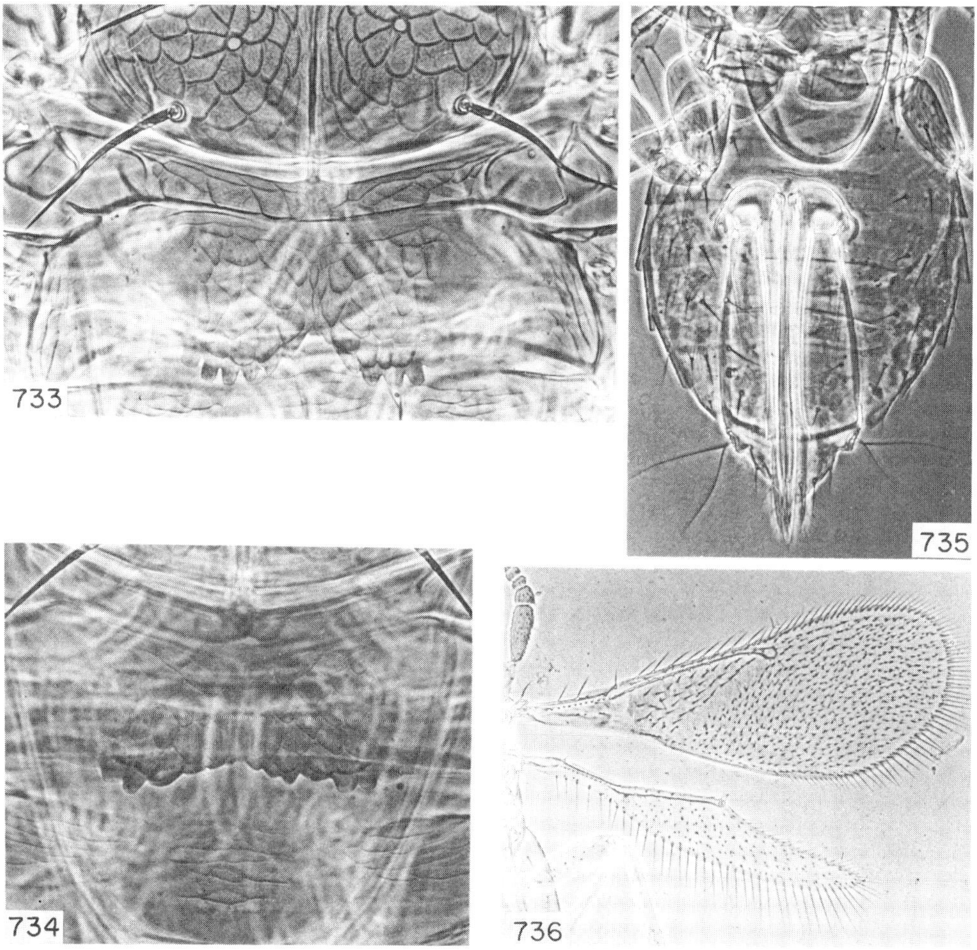


Figures 733–736. *Aphytis desantisi* DeBach and Rosen, ♀ (paratypes)
733. Metanotum and propodeum; note deep cleft between the two propodeal lobes. 734. Propodeum and second abdominal tergite. 735. Abdominal tergites; note posterior spiracles and cerci. 736. Wings.

# IV. THE MYTILASPIDIS GROUP

This highly uniform, Holarctic group can be recognized by the following combination of characters: Coloration predominantly yellow; thoracic setae coarse, black; wings uniformly setose, hyaline or nearly so; antennal club relatively short; propodeum relatively short; crenulae small, nonoverlapping; male antennae lacking a specialized sensory area on ventral aspect of club; pupa predominantly black.

The **mytilaspidis** group is undoubtedly one of the more highly evolved groups of species in the genus *Aphytis*. Although the coarse, dark thoracic setae and black pupae may be regarded as primitive characters, in most other morphological characters as well as in its Holarctic distribution this group appears to be rather far removed from the presumed ancestral type of *Aphytis*, represented by the **vittatus** group. To the taxonomist, this group has probably caused more confusion and frustration than any other group of species in *Aphytis*.

The lack of easy distinguishing characters had in the past led to overemphasis of minor—and often unreliable—morphological characters in the **mytilaspidis** group. The standards commonly used for separating species in this group were different from those applied in other groups, where easy "key" characters are more abundant. Thus, characters such as the number of setae along the marginal vein or on the mesoscutum, or the relative length of the marginal fringe of the forewing, which are now recognized as representing size-linked intraspecific variation, were commonly used as the main basis for the separation of several species in the **mytilaspidis** group. On the other hand, valid, reliable characters such as the pigmentation of the thoracic sterna, or the shape of the propodeal crenulae, were largely ignored or misinterpreted in this group.

The two "key" species in the group are *mytilaspidis* (Le Baron), with pigmented thoracic sterna and dusky, rounded crenulae, and *aonidiae* (Mercet), with immaculate thoracic sterna and pale, triangular crenulae. Unfortunately, *mytilaspidis* was originally described from large specimens, with relatively numerous setae on the body and wings and a short marginal fringe on the forewing, whereas *aonidiae* was originally described from minute specimens, with relatively few setae and a long marginal fringe, and these characters were commonly used for the recognition and separation of these species. As we now know, large specimens as compared with small specimens of the same species will show the same extent of variation as attributed to *mytilaspidis* and *aonidiae*. To complicate matters even further, the types of *mytilaspidis* were lost. Thus, for some ninety years, *mytilaspidis* was interpreted as an inherently large species, with all the characteristics of a large species, and was described as entirely yellow, with no reference to the pigmented sterna (see Compere, 1955; Quednau, 1964b; and others), whereas since its description in 1912, *aonidiae* was interpreted as a small species, with all the characteristics of a small species. Several species, such as *dubius* De Santis and *citrinus* Compere, were actually described as intermediate, in the length of the marginal fringe, etc., between *mytilaspidis* and *aonidiae*. Large specimens, regardless of sternal pigmentation, were commonly referred to *mytilaspidis*, whereas small specimens were assigned to *aonidiae* or to one of the "intermediate" species. As we now understand them, both *mytilaspidis* and *aonidiae* are represented by large as well as small specimens, exhibiting a rather wide range of intraspecific variation in the number of setae, relative length of the marginal fringe, etc., similar to the range common in other species of *Aphytis*. As for *dubius* and *citrinus*, they appear to be medium-sized or large specimens of *aonidiae*, and at present cannot be separated from this species by any valid morphological criteria.

We have made vigorous attempts to resolve these questions by conducting interspecific hybridization tests, but thus far have not been successful in obtaining the cultures needed for such tests and maintaining them in the laboratory.

As understood here, the **mytilaspidis** group at present includes the following 7 species: *aonidiae* (Mercet), *libanicus* Traboulsi, *luteus* (Ratzeburg), *moldavicus* Jasnosh, *mytilaspidis* (Le Baron), *opuntiae* (Mercet) and *phoenicis* DeBach and Rosen. Synonyms include *albidus* (Westwood), *citrinus* Compere, *dubius* De Santis and *dubius* var. *intermedia* De Santis. Recent biosystematic studies have shown that some of these species include uniparental as well as different biparental "forms". It was shown that a uniparental form (having rare males) may cross with its sibling biparental species, so that restricted but nonetheless significant introgression occurs in the laboratory. Thus these should be considered conspecific until more evidence proves otherwise. However, other uniparental (thelytokous) forms, especially ones lacking sibling biparental species, may be considered to have specific status. The sibling biparental forms exhibit various degrees of reproductive isolation, from none (conspecific strains) through partial (strong semispecies) to complete (distinct species) (see Rössler and DeBach, 1972a,b; Khasimuddin and DeBach, 1967c). We wish to acquire more evidence before describing any of these sibling forms, showing complete reproductive isolation, as new species.

Three Ethiopian species: *capensis* DeBach and Rosen, *rolaspidis* DeBach and Rosen, and *taylori* Quednau, one Neotropical species: *notialis* De Santis, and one Palearctic species: *neuter* Jasnosh and Myartseva, are regarded as related to the **mytilaspidis** group, probably representing a link with the **chrysomphali** group, to which they show some affinity. Certain other species of *Aphytis*, previously referred to the **mytilaspidis** group, such as *japonicus* DeBach and Azim, with a relatively long propodeum, pale thoracic setae and yellow pupae, are at present considered as members of the **chrysomphali** group.

At least four species, regarded here as unrecognizable, appear to be related to the **mytilaspidis** group. These include *abnormis* (Howard), *alami* Agarwal, *diaspidioti* Tshumakova, and *variolosum* Alam.

## 47. **Aphytis mytilaspidis** (Le Baron)

(Figures 112, 119, 120, 129, 130, 146, 147, 154, 158, 162, 163, 174, 185, 205, 213, 218, 238; and 737–764)

*Agonioneurus albidus* Westwood, 1837, Phil. Mag., Ser. 3, **10**:442; n.syn.
*Aphelinus mytilaspidis* Le Baron, 1870, Amer. Entomol. Bot., **2**:360–362.
*Aphelinus mytilaspidis*: Howard, 1881, U.S. Commiss. Agr. Ann. Rept. for 1880, pp. 354–355; Howard, 1895, U.S. Dept. Agr. Div. Entomol. Tech. Ser., **1**:25–26; Masi, 1911, Boll. Lab. Zool. Gen. Agr. Portici, **5**:156–158; Mercet, 1912, Trab. Mus. Cienc. Nat. Madrid, **10**:82–84; Imms, 1916, Quart. J. Microsc. Sci., N.S., **61**:217–274.
*Aphytis mytilaspidis*: Mercet, 1930, Rev. Biol. Forest. Limnol., Ser. B, **2**:54; Mercet, 1932, Eos, **8**:360; Compere, 1955, Univ. Calif. Publ. Entomol., **10**:309–310; Ferrière, 1965, Hymenoptera Aphelinidae d'Europe et du Bassin Méditerranéen, pp. 90–91; Nikol'skaya and Yasnosh, 1966, Aphelinids of the European part of the USSR and the Caucasus, pp. 203–204; Traboulsi, 1969, Ann. Soc. Entomol. Fr. (N.S.), **5**:59–66.
*Aphytis (Prospaphelinus) albidus*: Novitzky, 1961, Entomol. Mon. Mag., **97**:195.
*Aphytis albidus*: Graham, 1976, Syst. Entomol., **1**:133–134.

For additional references see Peck (1963) and Nikol'skaya and Yasnosh (1966).

*A. mytilaspidis*, a Holarctic species, may be distinguished from other members of the **mytilaspidis** group by the dusky thoracic sterna, the generally yellow coloration, and the distinct, more or less rounded, pigmented crenulae.

The pigmented thoracic sterna are the most important recognition character of *mytilaspidis*. Because this character has not been noted or utilized significantly by any of the authors listed above, their descriptions may have referred to more than one species. Host records listed for *mytilaspidis* in the literature should therefore be regarded as questionable.

Le Baron's (1870) original description of *mytilaspidis* is very poor by modern standards. We were unable to locate the types, if they ever existed. The species was described from specimens reared from the oystershell scale, *Lepidosaphes ulmi* (L.), on apple in Kane and DuPage Counties, Illinois. Since no such material was available to us, we attempted to duplicate it as closely as possible in order to obtain neotype specimens. At our request, C. E. White kindly reared a series of specimens in 1965 from *L. ulmi* on privet, Urbana, Illinois, and sent them to us. The following redescription is based on this neotype series.

*Female*. Eyes finely setose. Mandibles (see Figure 112, p. 122) well developed, with 2 denticles and a dorsal truncation; maxillary palpi 2-segmented, labial palpi 1-segmented. Antennal scape (Figures 737, 738) slender, 5 to $6\frac{1}{4}$ times as long as wide, about $1\frac{1}{4}$ times as long as the club; pedicel usually $1\frac{2}{3}$ to $1\frac{4}{5}$ times as long as wide, somewhat longer than the third segment of the funicle; first funicular segment trapezoidal, usually about $1\frac{1}{4}$ to $1\frac{1}{2}$ times as wide as long; second segment nearly symmetrical, considerably shorter and often somewhat wider than the first segment, $1\frac{1}{2}$ times to about twice as wide as long; third funicular segment about $1\frac{1}{4}$ to $1\frac{2}{5}$ times as long as wide, bearing 2–3 longitudinal sensilla; club a little over $2\frac{1}{2}$ to nearly 3 times as long as wide (usually distinctly less than 3 times), $2\frac{1}{4}$ to $2\frac{1}{2}$ times longer and distinctly wider than the preceding segment, bearing 5–6 longitudinal sensilla.

Setae on head and thorax dark, coarse, readily visible under $\times 30$ magnification (Figures 739, 740; see also Figures 154, 158), those on abdomen slender, paler, invisible even under $\times 120$ magnification (in cleared slide-mounted specimens under a stereoscopic microscope with a white background). Vertex with 2 pairs of long setae along occipital margin, in addition to numerous short setae. Mesoscutum (Figures 739, 740) with 8–13 setae, the posterior pair and 1 seta at each antero-lateral corner noticeably longer and coarser than the others; each parapsis with 1–3 (usually 2) short setae; each axilla with 1 seta; scutellum with 4, the discoid sensilla about equidistant from the two pairs. Frontovertex, pronotum, and mesonotal sclerites reticulate, with a distinct infrasculpture within the cells, especially on mesoscutum and scutellum. Scutellum about $\frac{3}{4}$ to a little over $\frac{4}{5}$ median length of mesoscutum. Metanotum (Figures 739, 741, 742) somewhat curved, transversely reticulate except on the sides, usually slightly overlapping the anterior margin of the propodeum; anteromedian apodeme about as long as or slightly shorter than median length of metanotum.

Propodeum (Figures 739, 741, 742) rather short, about $2\frac{3}{4}$ to 3 times as long as the metanotum, about $\frac{3}{5}$ length of scutellum (0.54–0.64), rather faintly reticulate on the sides, broadly reticulate centrally and with a distinct infrasculpture within the cells; crenulae 5+5 to 7+8, distinct, rounded, nonoverlapping (see also Figure 174).

Second abdominal tergite slightly, transversely reticulate anteriorly on both sides, smooth centrally; tergites III–VII (Figure 743) reticulate on the sides, with several setae in a short transverse row on each reticulate area; third and seventh tergites transversely striated mesad of reticulate areas, the striation fading centrally; tergites IV–VI smooth centrally, sometimes with very faint indications of transverse striation; seventh tergite bearing 2 setae submedially, between the lateral setiferous areas; eighth tergite delicately reticulate-punctate across, this nearly fading centrally, with a transverse row of 4–6 setae between spiracles (usually with an additional, shorter seta immediately mesad of each spiracle, obviously part of a lateral group of short setae); syntergum (Figure 743) triangular, delicately punctate, bearing 8–11 setae in a transverse row. Cerci nearly equidistant from posterior spiracles and tip of abdomen or somewhat closer to the former, with 2 long setae and 1 short seta. Ovipositor shaft varying from less than $1\frac{1}{2}$ times as long as the middle tibia in large specimens to nearly $1\frac{3}{4}$ times in small specimens (1.44–1.73); ovipositor sheaths about $\frac{1}{3}$ length of middle tibia (0.28–0.37).

Mid-tibial spur considerably shorter than the corresponding basitarsus (0.74–0.83).

Forewing (Figure 745) usually broad, $2\frac{1}{3}$ to $2\frac{2}{5}$ times as long as wide (considerably narrower, $2\frac{3}{4}$ times as long as wide, in minute specimens); marginal fringe varying from about $\frac{1}{10}$ width of disk in large specimens to over $\frac{1}{4}$ width of disk in small specimens. Delta area with 31–72 setae in 5–8 rows, these considerably longer and sparser than the setae distad of speculum, quite distinct from row of setae along posterior margin of wing, widely separated from a few setae in a small group below distal two fifths or half of submarginal vein; costal cell usually with 4–5 fine setae in a row along proximal three fifths and 1 coarse seta near apex. Submarginal vein bearing 2 coarse setae, the proximal one about $\frac{3}{5}$ to $\frac{3}{4}$ length of the distal, and 13–21 bullae. Marginal vein bearing 6–12 prominent, subequal setae along anterior margin (this number being more or less in correlation with the size of specimens), these somewhat longer than the setae in a row along center of vein (up to $1\frac{2}{5}$ times longer in smaller specimens).

Forewing generally hyaline, very faintly infumated on proximal half, more distinctly so below submarginal vein and at base of delta; a fuscous streak along posterior margin of wing, distad of speculum.

General coloration pale yellow; thoracic sterna distinctly, rather strongly infuscated (see Figure 763); dorsum of thorax and abdomen generally immaculate; posterior margin of scutellum narrowly lined with blackish; a short black streak at base of forewing, below tegula; posterior margin of propodeum centrally and the crenulae distinctly dusky; plates around base of ovipositor faintly dusky, other internal apodemes and endophragma immaculate. Antennal scape pale, except for a fuscous spot on the ventral margin; rest of antenna uniformly, rather strongly dusky. Tibiae and tarsi of all legs uniformly, faintly dusky, appearing darker yellow in slide mounts. Wing veins pale.

Length 0.75–1.26 mm.

*Male*. Very similar to the female in structure, chaetotaxis, sculpture and coloration.

Antennal scape (Figure 746) strongly infuscated on distal half or so, with what looks like an indistinct ridge or elevated area on the ventral aspect (see also Figures 119, 120 and 759); pedicel about $1\frac{1}{3}$ times as long as the third segment of the funicle; club appearing more truncate than in the female in lateral view, with 3–4 longitudinal sensilla, lacking any specialized, cut-off sensory area on the ventral surface.

Mesoscutum (Figure 747) with 10 setae. Propodeum (Figure 748) a little over 3 times as long as the metanotum, about $\frac{3}{5}$ length of scutellum; crenulae as in the female.

Genitalia (Figure 218) about $\frac{2}{3}$ length of middle tibia (0.63–0.72); digital sclerites about $\frac{1}{4}$ the combined length of aedeagus and apodemes.

Forewing (Figure 749) about $2\frac{1}{3}$ times as long as wide.

Length 1.14–1.27 mm.

Redescribed from 39♀♀ and 2♂♂ (neotype series), reared from the oystershell scale, *Lepidosaphes ulmi* (L.), collected by C. E. White on privet, *Ligustrum* sp., Urbana, Illinois, February 16, 1965. This series obviously represents a uniparental population. One female of this series (slide No. II A12:6, 4th row from the right, 7th specimen from top) is designated neotype. The slide, marked accordingly, is deposited in the collection of the Division of Biological Control, University of California, Riverside. Additional slides of the neotype series have been deposited in the U.S. National Museum of Natural History, Washington, D.C., in the British Museum (Natural History), London, in the collection of the Zoological Institute, USSR Academy of Sciences, Leningrad, in the collection of the Plant Protection Research Institute, Pretoria, and in the collection of the Department of Entomology, the Hebrew University, Rehovot.

**Notes.** The pupa of *mytilaspidis* is entirely black, as shown in Figure 764.

*Agonioneurus albidus* is a senior synonym of *mytilaspidis*. Westwood's (1837) original description of *albidus* was very brief:

"Totus pallidè flavescenti-albidus; oculis obscuris; alis immaculatis albidis, longè ciliatis, callositate stigmaticali magìs conspicuâ quam in *Ag. basalii*, angulum parvum cum margine alarum formanti, antennarum articulo 5to praecedenti multò majori.—Long. corp. $\frac{1}{3}$ lin. Expans. alar. $\frac{5}{6}$ lin.

"Habitat—? In mus. nostr."

Novitzky (1961) correctly placed *albidus* in the genus *Aphytis* but made the following comments:

"*A. albidus* Westw. is a different species from *A. mytilaspidis* Le Baron; I cannot indicate with certainty any synonyms, except for a species named by Förster '*Myina citrina* Först. i.l.' and reared from '*Coccus crataegi*', apparently at Aachen, Western Germany. I took this species myself on spruce in Central Poland (Skierniewice) in 1927. The Central European species of *Aphytis* are badly in need of revision."

Graham (1976) reported that *albidus* was represented in Westwood's collection at Oxford by three syntype specimens, one male (the specimen examined by Novitzky) and two females, all mounted on cards. The two females were labeled in Westwood's handwriting: "on apple trees (end of June) infested by *Coccus conchiformis*, on which I saw one stationed so that I have no doubt it is its parasite." Graham pointed out that Westwood's *"Coccus conchiformis"* is the species now known as *Lepidosaphes ulmi* (L.), and considered *albidus* to belong to the **mytilaspidis** group of *Aphytis*.

We received the two female syntypes on loan from the Hope Department of Entomology, Oxford (courtesy of Dr. M. W. R. de V. Graham). Permission was granted to remove them from the rectangular cardboard tag onto which they had been both pasted, and the specimens were then cleared and remounted in Hoyer's medium on two separate slides. Both are in good shape. One of them is designated lectotype; the other (with part of one wing under a separate cover slip) is a paralectotype.

Examination of these type specimens leaves no doubt whatsoever that they are conspecific with *mytilaspidis*. The name *albidus* is therefore a senior synonym of *mytilaspidis*. However, according to the International Code of Zoological Nomenclature (ICZN, 1961), Article 23(b), "A name that has remained unused as a senior synonym in the primary zoological literature for more than fifty years is to be considered a forgotten name (nomen oblitum)." Inasmuch as *albidus* has remained unrecognizable for 140 years, its suppression by the International Commission on Zoological Nomenclature is hereby recommended for the sake of nomenclatural stability, and will be formally requested in a forthcoming paper. We shall, therefore, continue to apply the name *mytilaspidis*.

The species originally described as *A. mytilaspidis* was apparently uniparental. Following his description, Le Baron (1870) noted: "all the specimens that I have examined have proved to be females." He also reported the parasite to be bivoltine, overwintering as fully grown larvae. Imms (1916), in a detailed study, confirmed these observations for what was undoubtedly the same species in England.

DeBach (1964) reported a "form" of *mytilaspidis* parasitizing *L. ulmi* in Greece, which he temporarily designated "*mytilaspidis*-form 3a". Like *mytilaspidis* from Illinois and England, "3a" is uniparental and exhibits winter diapause in the field. The diapause, however, is apparently facultative and may be induced by the univoltine host; when reared on the cactus scale, *Diaspis echinocacti* (Bouché), in the insectary of the Division of Biological Control, University of California, Riverside, "3a" developed normally throughout winter, and no diapause occurred.

The specimens of "3a" are essentially identical to the neotypes of *mytilaspidis*, differing only in minor details of coloration. The dusky areas generally appear somewhat darker in "3a", the fuscous spot on the antennal scape appears to be more distinct, and there is a small, faintly infuscated triangle or half-circle at the center of the second abdominal tergite (Figures 750, 751) (in the neotypes, the latter area sometimes appears very faintly infumated under phase contrast but is entirely hyaline under a stereoscopic microscope). We do not consider these slight differences sufficient to justify the separation of "3a", and this form is regarded here as conspecific with the neotypes of *mytilaspidis*.

Two other uniparental forms from Greece, designated "3b" and "3c", were obtained from *Aspidiotus hedericola* Leonardi in Attica and *Hemiberlesia lataniae* (Signoret) in Crete, respectively (DeBach, 1964). Preliminary studies indicated that these forms were incapable of developing on cactus scale, whereas "3a" accepted this scale insect as a host in the laboratory. However, subsequent rearings have not confirmed these early observations, and both forms proved to develop readily on cactus scale in the Riverside insectary.

"Form 3b" (Figures 752, 753) is identical in coloration to "3a". "Form 3c" (Figures (754–759) was originally described as being somewhat darker than other forms, "including distinct duskiness of the prothorax and traces of duskiness on the abdomen" (DeBach, 1964). Most of the original specimens of "3c" from Greece do appear to be slightly darker, as described. Also, specimens reared from *Hemiberlesia lataniae* in Israel in 1964 appear to have an identical color pattern. However, this color variation is not consistent in the original Greek material, and is entirely absent in specimens of an insectary stock supposedly derived from that material. Also, slight duskiness of the prothorax occasionally occurs in other forms of *mytilaspidis* (such as in some original specimens of the biparental "form 2", discussed below). We regard the faint duskiness of "3c" as falling within the range of variation of *mytilaspidis*.

In both "3b" and some specimens of "3c", the ovipositor sheaths appear to be slightly longer than in the neotype series, being about $\frac{2}{5}$ the length of the middle tibia (0.37–0.44 in "3b" and in the insectary stock of "3c"); the ovipositor shaft is about $1\frac{3}{5}$ to $1\frac{4}{5}$ as long as the middle tibia, and the propodeum appears to be a trifle longer than in the neotype series, being 3 to $3\frac{4}{5}$ times as long as the metanotum. These variations are not shared by some of the original specimens of "3c". Considering the usual range of variation in these characters, these slight differences are not considered significant, and the two forms are considered conspecific with the neotype series of *mytilaspidis*.

The male of *mytilaspidis* was first described by Howard (1881), who recorded "many male and female specimens." Most of the numerous specimens labeled as *mytilaspidis* in the collection of the U.S. National Museum are mounted in balsam without clearing, and their identity cannot be ascertained without careful remounting. It is therefore difficult to confirm now whether Howard (1881, 1895) was actually referring to a biparental form of *mytilaspidis*. However, such biparental forms certainly do exist, and will be discussed below.

DeBach (1964) recorded a biparental form of *mytilaspidis*, which he temporarily designated "*mytilaspidis*-form 2", parasitizing the cactus scale, *Diaspis echinocacti* (Bouché), on *Opuntia* sp. in Greece. But for the abundance of males, this arrhenotokous form (Figures 760–762) cannot be distinguished from the neotype series of *mytilaspidis*. (In some of the original Greek specimens, the antennal scape of the female appears to be somewhat more extensively infuscated, the small dusky spot is present on the second abdominal tergite, and the general coloration is somewhat darker. These specimens, however, are rather poorly preserved and not perfectly cleared; specimens obtained in 1969 from an insectary stock of "form 2" are identical to the neotypes in practically all details of structure and coloration.) "Form 2" does not develop on the latania scale, *Hemiberlesia lataniae* (Signoret), which is the original host of the uniparental "form 3c". However, males of "form 2" are capable of mating, though at a reduced incidence, with the uniparental females of "3c", and hybrids are readily produced. On the other hand, mating does not occur with the uniparental females of forms "3a" and "3b". To complicate matters still further, the occasional males of "3b" do mate readily with the biparental females of "form 2".

Rössler and DeBach (1972a, 1972b) recently reported the results of an in-depth study of the biosystematic relations between the uniparental "form 3c" and the biparental "form 2" of *mytilaspidis*. The two forms show only partial sexual isolation, and a thelytokous female, once inseminated by an arrhenotokous male, will utilize the sperm as efficiently as an arrhenotokous female and will produce viable and fertile hybrids. Other biological differences between the two forms were found, aside from the important difference in host preference. The uniparental form appears adapted to warmer conditions, its overall fecundity is lower, but the production of female progeny per parent female is higher than in the biparental form. Apart from a difference in the number of setae in the delta area of similar-sized specimens, no significant morphological differences were found between the two forms. Arguments pro and con could be advanced as to whether these two forms are conspecific. We think they should be regarded as conspecific at present and that more evidence is needed before considering the formal designation of the biparental form as a distinct species.

We assume that *mytilaspidis* first evolved as a biparental species, giving rise to uniparental forms at a later evolutionary stage. Some of these sibling forms have developed different host preferences, adapting themselves to new host species which have enabled them to avoid competition with sympatric biparental populations. In some of these forms there still is a small percentage of males, and the females are capable of mating and laying fertilized eggs; thus in certain cases they are not completely reproductively isolated from sympatric biparental populations. In other forms, divergence has apparently progressed further, and they are no longer capable of exchanging genetic material with the biparental forms. The partially isolated "forms" may be regarded as incipient species. Completely isolated forms, especially ones with different biologies and behavior, may be regarded as valid species. As more information is gathered, additional "forms" of *mytilaspidis* can be expected to be discovered, exhibiting various degrees of reproductive isolation and different host preferences. Until much more is known about them, all these forms will be regarded by us as the components of one complex species, which is evidently undergoing rigorous processes of speciation at the present time.

DeBach (1964) obtained a biparental *Aphytis* from *Carulaspis visci* (Schrank) on *Cupressus sempervirens* in Greece. It was of the general "*aonidiae* type", that is, rather small, with relatively few setae on the forewing and a rather long marginal fringe, and was designated "*aonidiae*-form 2". However, DeBach (op. cit.) made the following observation: "However, the possibility exists that *aonidiae*-form 2 might represent small specimens of *mytilaspidis*-form 2 inasmuch as both have sternal pigmentation and possess males. Specimens of *aonidiae*-form 2 are nearly all small because their host, *Carulaspis visci*, usually is small. It is known that small specimens of any given *mytilaspidis*-form tend to approach the *aonidiae*-form in their characteristics. Actually, further experimental tests are necessary in order to determine whether *aonidiae*-form 2 will develop on *Diaspis echinocacti* and still retain its characteristics or whether it will turn out to be *mytilaspidis*-form 2 when reared on *Diaspis*."

Examination of all the *Aphytis* material collected in Greece in 1962–1963, as well as additional shipments received in 1969, shows that specimens obtained from *Carulaspis visci* range all the way from the "*aonidiae* type" to typical "*mytilaspidis* type", with several specimens occupying an intermediate position in the length of the marginal fringe, etc.

Khasimuddin and DeBach (1976c) recently reported on a biosystematic study of two sympatric arrhenotokous forms of *mytilaspidis* obtained from Greece: one from the cactus scale (apparently the "*mytilaspidis*-form 2" discussed above), the other from *Carulaspis visci* (apparently the "*aonidiae*-form 2"). They found the two forms to be partially reproductively isolated, with gene flow possible only in one direction, i.e., from males of the *C. visci* form to females of the cactus scale form. They also noted that the two forms differed in their host preferences—the *C. visci* form developed on latania scale in the laboratory, whereas the cactus scale form would not—as well as in adult survival rate under different temperatures, and concluded that they could be regarded as strong semispecies.

A series of 21♀♀ and 7♂♂, reared by P. DeBach from *Lepidosaphes ulmi* on black willow near Placerville, California, October 1, 1966, apparently represents another biparental form of *mytilaspidis*. Traboulsi (1969) also recorded a biparental form of *mytilaspidis* parasitizing *Lepidosaphes ulmi* in Lebanon.

An interesting uniparental form was obtained by DeBach in 1963 from *Aonidia lauri* (Bouché) on *Laurus nobilis* in Spain. Like the small specimens from *Carulaspis*, many of the specimens in this series resemble the so-called "*aonidiae*-type", with a rather long marginal fringe on the forewing, relatively few setae on the mesoscutum and marginal vein, sometimes with relatively few propodeal crenulae. However, these characters exhibit a wide range of variation in this series, clearly correlated with the size of the specimens. All the specimens of this "form" possess the sternal pigmentation and other characteristics of *mytilaspidis*; the larger specimens are indistinguishable from small specimens in the neotype series. For the time being, at least, we consider this form to fall within the range of morphological variation of *mytilaspidis*.

*Aphytis diaspidioti* Tshumakova was listed by Ferrière (1965) and by Nikol'skaya and Yasnosh (1966) as a synonym of *mytilaspidis*. However, *diaspidioti* was described as an entirely yellow species, and no reference was made to the pigmentation of the thoracic sterna. Until the types of *diaspidioti* are carefully re-examined, we have to regard this species as an unidentifiable member of the **mytilaspidis** group (see p. 731).

Ferrière (1965) also listed *Aphytis variolosum* Alam as a synonym of *mytilaspidis*. We cannot accept this synonymy; *A. variolosum* is an unidentifiable species, apparently with pale thoracic sterna (see p. 741).

Traboulsi (1969) recorded four morphologically indistinguishable "forms" of *mytilaspidis* from Lebanon, and named them as follows: "*Aphytis mytilaspidis* forme *ulmi*", a biparental parasite of *Lepidosaphes ulmi*; "*Aphytis mytilaspidis* forme *echinocacti*", a uniparental parasite of *Diaspis echinocacti*; "*Aphytis mytilaspidis* forme *ficus*", a uniparental parasite of *Lepidosaphes ficus* (Signoret) [=*Mytilaspis conchiformis* (Gmelin)] and "*Aphytis mytilaspidis* forme *hedericola*", a uniparental parasite of *Aspidiotus hedericola* Leonardi. According to the International Code of Zoological Nomenclature (1961, Art. 10b, 15, 45e), a name proposed, after 1960, explicitly as the name of a "variety" or "form" is to be regarded as of infrasubspecific rank, and is not available. If the taxon in question is subsequently elevated to a rank of the species group (subspecies or species), its name becomes available, but takes the authorship and date of its elevation. The "form" names proposed by Traboulsi are therefore unavailable and have at present no nomenclatural value.

**Material Examined.** Specimens in the collection of the Division of Biological Control, University of California, Riverside, recognized as *A. mytilaspidis*, were reared from *Lepidosaphes ulmi* (L.) on privet, *Ligustrum* sp., in Illinois, on *Tibes odoratum* and *Syringa vulgaris* in Ohio, on an undetermined plant in Quebec, on *Salix nigra* and *Salix* sp. in California, on *Populus* sp. in Yugoslavia, on *Populus alba* in Hungary, on *Ligustrum japonicum*, *Ligustrum vulgare*, *Celastrum scaudens* and rose in Greece, on rose in Crete, on carob in Cyprus, on broom, *Genista* sp., in Lebanon and on *Crataegus* sp. in Moldavia (USSR); from *Lepidosaphes* sp. on *Caragona* sp. in Kabardino-Balkaria (USSR); from *Mytilaspis conchiformis* (Gmelin) [=*Lepidosaphes ficus* (Signoret)] on fig in Italy and on an undetermined plant in Georgia (USSR); from *Aonidia lauri* (Bouché) on *Laurus nobilis* in Spain; from *Aonidiella aurantii* (Maskell) on rose and on *Heteromeles* sp. in California; from *Aonidiella orientalis* (Newstead) on citrus in Iran; from *Aspidiotus hedericola* Leonardi on *Hedera helix* in Greece; from *Aspidiotus nerii* Bouché [=*A. hederae* (Vallot)] on an undetermined ornamental shrub in Yugoslavia; from *Aulacaspis rosae* (Bouché) on rose in Crete; from *Carulaspis caruelii* (Targioni-Tozzetti) [=*C. minima* (Targioni-Tozzetti)] on *Cupressus* sp. in California; from *Carulaspis visci* (Schrank) on *Juniperus chinensis* and *Juniperus* sp. in Ohio, on *Cupressus sempervirens* and *Thuja horizontalis* in Greece and on *Cupressus* sp. in

Morocco; from *Chionaspis* sp. on *Phragmites communis* in Israel; from *Chrysomphalus dictyospermi* (Morgan) on orange in Morocco and on an unspecified plant in Georgia (USSR); from *Diaspis echinocacti* (Bouché) on *Opuntia littoralis* on Santa Cruz Island, California, on *Opuntia zacuapanensis* in Spain and on *Opuntia* sp. in Greece and California; from *Epidiaspis leperii* (Signoret) on prune in California and on an undetermined plant in Georgia (USSR); from *Hemiberlesia lataniae* (Signoret) on *Ficus* sp. and on *Ampelopsis* sp. in Crete; from *Insulaspis* juniperi (Lindinger) on an undetermined plant in Georgia (USSR); from *Leucaspis riccae* Targioni-Tozzetti on olive in Israel; from *Parlatoria oleae* (Colvée) on peach in Greece; from *Phenacaspis pinifoliae* (Fitch) on *Pinus sylvestris* in Ohio and on *Pinus* sp. in California and on Santa Cruz Island; from *Quadraspidiotus juglansregiae* (Comstock) on English walnut in California; from *Quadraspidiotus perniciosus* (Comstock) on peach in California; and from *Quadraspidiotus zonatus* (Frauenfeld) on fig in Greece. Also from unidentified hosts on *Plantanus orientalis* in Greece, on *Pistacia terebinthus* in Cyprus, on oleander in Italy, and on citrus in Spain.

Some of the records listed above may be erroneous, due to misidentification of scale insects. Others represent occasional rearings of a few individuals, and may not reflect the normal host range of *mytilaspidis*. All these specimens have the distinguishing characters of *mytilaspidis* and essentially agree with the description of the neotype series. Slight differences in the extent of infuscation of the female antennal scape, occasional faint duskiness of the pronotum and traces of faint duskiness on the abdomen, and slight differences in the relative length of the ovipositor sheaths, are not considered significant but as representing the range of intraspecific variation in *mytilaspidis*. Until much more is known about possible biological differences, all this material is referred to *mytilaspidis*.

Ferrière (1965) and Nikol'skaya and Yasnosh (1966) recorded *A. mytilaspidis* also from the Oriental region (India, Sri Lanka) and from Japan, and listed numerous additional host species, but since they made no reference to sternal pigmentation it is doubtful that all their records refer to the species interpreted here as *mytilaspidis*. The same can be said of the records made by Mercet (1912b) from Sri Lanka and by Nakayama (1921) from Japan.

**Biology.** Griswold (1925) reported that *A. mytilaspidis* developed 3, possibly 4 generations at Ithaca, New York, and recorded it as both an ectoparasite and an egg predator of the oystershell scale, *Lepidosaphes ulmi* (L.). She commonly observed the females to host-feed on this scale insect. Imms (1916) reported that this species is bivoltine in England, overwintering as fully grown larvae, but doubted that it could develop on host eggs. Lord and MacPhee (1953) recorded 3 generations in Canada and reported that *A. mytilaspidis* is severely limited by low winter temperatures. DeBach (unpublished) observed two generations at Athens, Greece, with the second dispausing as fully grown larvae among the eggs of *L. ulmi*.

The developmental period of a thelytokous form of *mytilaspidis* ("3c") on cactus scale in the laboratory was about $34\frac{1}{2}$ days at 21°C, 21 days at 27°, $18\frac{1}{2}$ days at 32°.

Adult mortality reached 50% after 22–23 days at 27°, after 10–11 days at 32°. Data for the arrhenotokous "form 2" were somewhat different (Rössler and DeBach, 1972b).

**Utilization in Biological Control.** *A. mytilaspidis* is recorded as an important natural enemy of the oystershell scale, *Lepidosaphes ulmi* (L.), on apple in Nova Scotia, Canada, quite capable of controlling heavy outbreaks of this serious pest. Certain fungicides were found to be destructive to this parasite, and their replacement with selective materials resulted in successful integrated control (Lord, 1947).

*A. mytilaspidis* was first introduced from Italy into California against the fig scale, *Mytilaspis conchiformis* (Gmelin) [=*Lepidosaphes ficus* (Signoret)] in 1939,* but was not released because it was believed to be established on both fig scale and oystershell scale in California. In 1947 a "strain" of *mytilaspidis*, designated "*Aphytis* C", was introduced into California from France and was reared in the laboratory on latania scale. When released against the fig scale, this introduced strain proved much more efficient than the locally established form. It soon became established and proceeded to spread. Parasitism rose to 67–100 %, and fig-scale populations were markedly reduced (see Rosen and DeBach, 1978). This emphasizes that the various "strains" and semispecies of *mytilaspidis* should receive more attention as potentially effective natural enemies of several injurious pests.

The following two species are very closely related to *mytilaspidis*, and may eventually prove to be junior synonyms of this species.

### 48. **Aphytis opuntiae** (Mercet)

(Figures 765–772)

*Aphelinus opuntiae* Mercet, 1912, Trab. Mus. Cienc. Nat. Madrid, **10**:84–87.

*Aphytis opuntiae*: Mercet, 1930, Rev. Biol. Forest. Limnol., Ser. B, **2**:54; Mercet, 1932, Eos, **8**:360; Compere, 1955, Univ. Calif. Publ. Entomol., **10**:305–306; Ferrière, 1965, Hymenoptera Aphelinidae d'Europe et du Bassin Méditerranéen, p. 89; Nikol'skaya and Yasnosh, 1966, Aphelinids of the European Part of the USSR and the Caucasus, p. 192 (key; in Russian).

If correctly interpreted here, this uniparental Palearctic species is very closely related to *A. mytilaspidis*, differing only in the considerably darker coloration: mesoscutum with a pair of fuscous blotches, abdominal tergites distinctly infuscated. To avoid unnecessary repetition, *opuntiae* is redescribed below relative to *mytilaspidis*.

*Female*. Structure, chaetotaxis and sculpture essentially as in *mytilaspidis*. Antennal club (Figures 765, 766) sometimes fully 3 times as long as wide, bearing 5–7 longitudinal sensilla. Mesoscutum (Figures 767–769) with 8–11 setae, each parapsis with 2 (rarely 1); metanotal apodeme nearly as long as, or distinctly longer than, median length of metanotum.

Propodeum (Figure 770) usually 3 to $3\frac{1}{2}$ times as long as the metanotum (only $2\frac{1}{4}$ times in a minute specimen); crenulae 3+4 to 7+9, distinct, nonoverlapping, some-

* Some of these records may refer to the species recognized here as *opuntiae* (see below).

what more triangular in shape than those of *mytilaspidis*. Syntergum with 6–8 setae in a transverse row. Ovipositor sheaths 0.35–0.40 length of middle tibia.

Wings as in *mytilaspidis*. Marginal fringe of forewing (Figure 771) varying from $\frac{1}{9}$ to $\frac{1}{7}$ width of disk in large specimens to nearly $\frac{1}{3}$ width of disk in small specimens; marginal vein bearing 5–13 prominent, subequal setae along anterior margin, this number being in correlation with the size of specimens. Forewing pattern as in *mytilaspidis*.

General coloration (Figure 767) dusky yellowish, considerably darker than in any other member of the **mytilaspidis** group. Thoracic sterna strongly infuscated; posterior margin of scutellum lined with black; a short black streak at base of forewing, below tegula; posterior margin of propodeum and the crenulae blackish; pronotum, two conspicuous blotches on mesoscutum, and a small triangle on center of second abdominal tergite, fuscous. Abdominal tergites lightly, but usually very distinctly, infuscated; plates around base of ovipositor dusky. Antennal scape (Figure 765) noticeably, rather extensively dusky in parts, rest of antenna uniformly, very strongly infuscated. Tibiae and tarsi of all legs distinctly dusky. Wing veins lined with brownish.

Length 0.70–1.20 mm.

*Male*. Structure, chaetotaxis and sculpture, including antenna and genitalia, essentially as in *mytilaspidis*. As in the female, two fuscous blotches are present on the mesoscutum, and the pronotum and abdominal tergites are distinctly dusky.

Length 0.84–0.98 mm.

Redescribed from a series of 613♀♀ and 5♂♂, reared by P. DeBach from the cactus scale, *Diaspis echinocacti* (Bouché), on *Opuntia* sp., Almuñcar, Spain, August 3, 1963; all mounted in Hoyer's medium, with the dissected parts of several additional females.

**Notes.** A female pupa of *opuntiae* is shown in Figure 772.

Mercet (1912b) originally described *opuntiae* as a parasite of an unidentified scale insect on *Opuntia vulgaris*, collected in Palma de Mallorca, Balearic Islands. He subsequently (1930) recorded the host as *Diaspis echinocacti*. Mercet (1912b) regarded *opuntiae* as closely related to *mytilaspidis*, but described it with 18 setae on the mesoscutum, arranged in 5 transverse rows. The first row, along the anterior margin of the mesoscutum, was described as comprising 8 setae. Holotype in the Museo de Ciencias Naturales, Madrid.

Compere (1955) discussed *opuntiae* after studying Mercet's type in 1932. He listed the following as distinguishing characters: "Wings and antennae suggestive of *mytilaspidis*, except that the antennae are possibly less slender. Mesoscutum with at least 16 setae, and these seemingly larger than in *mytilaspidis* and other species of that group." He then noted: "At one time I regarded *opuntiae* as a questionable synonym of *mytilaspidis*. I now regard it as a distinct species that can be easily separated from *mytilaspidis* by the greater number and large size of the setae on the mesoscutum." Compere presented a figure, drawn from the type, showing the large number of setae on the mesoscutum.

We have not seen the type of *opuntiae*. However, we consider the number of mesoscutal setae mentioned by Mercet and Compere to be a mistake. Mercet apparently described *opuntiae* from a single specimen. From the original description, and especially from Compere's figure, it is evident that the type specimen must have been distorted in preparation, the pronotum having been pushed back over the mesoscutum, so that the pronotal setae appeared to be on the mesoscutum and were counted as the anterior row of mesoscutal setae. This happens quite commonly in slide-mounted specimens of *Aphytis*.

We consider the above-described specimens reared from cactus scale on *Opuntia* in Spain to be identical to the species obtained by Mercet on the same host plant in the Balearic Islands. If our interpretation is correct, it appears that *opuntiae* is very similar to *mytilaspidis* in all morphological characters, including the number of mesoscutal setae. The principal difference between the two species seems to be in the darker coloration of *opuntiae*. This is indicated in Mercet's (1912b) original description, which had *opuntiae* as "dirty yellow, with duskiness," whereas *mytilaspidis* was described by him as "uniformly lemon yellow." If this character is indeed consistent as seems evident from our large series, this difference should justify the separation of *opuntiae* as a distinct species. Since the name *opuntiae* exists, and the taxon is morphologically recognizable, we prefer to retain it as distinct, at least until evidence to the contrary is presented. However, *opuntiae* may eventually turn out to be simply an extreme color variation of *mytilaspidis*, probably related to the faintly dusky specimens of "form 3c". This could only be determined by hybridization tests, which are impractical in view of the scarcity of males in both forms.

*A. opuntiae* is a uniparental species, males comprising less than one percent of the population sampled.

A series of 15♀♀, obtained from a shipment of the fig scale, *Mytilaspis conchiformis* (Gmelin) [ = *Lepidosaphes ficus* (Signoret)], from Agropoli, Italy (1939?), and 4♀♀ from the same host on fig, Portici, Italy, 1948, have the same color pattern of *opuntiae*, and are tentatively referred to this species. So is also a single ♀ specimen reared by P. DeBach from mixed scale insects on walnut, Roviais, Euboea Island, Greece, October 1962.

## 49. **Aphytis moldavicus** Jasnosh

(Figures 773–776)

*Aphytis moldavicus* Jasnosh, 1966, in: Nikol'skaya and Yasnosh, Aphelinids of the European Part of the USSR and the Caucasus, p. 207 (in Russian).
*Aphytis moldavica*: Nikol'skaya and Yasnosh, 1968, Trud. Vsesoyuz. Entomol. Obshchest., **52**: 15.

The female of this Palearctic species was described as differing from *mytilaspidis* in the following characters: antennal scape shorter, only 4 times as long as wide; mesoscutum with 8 setae in 3 rows; forewing with 5–6 rows of setae in the delta; from *aonidiae* in the shorter antennal scape and in the shorter marginal fringe of the forewing ($\frac{1}{5}$–$\frac{1}{6}$ width of disk). The rare male was described with a more strongly setose antenna.

At hand is a series of 9♀♀ (paratypes), reared from *Lepidosaphes ulmi* (L.) and *Lepidosaphes* sp., Kishinev and Dubossary, Moldavia, USSR, 1958–1961; also 3♀♀, reared from *L. ulmi*, Kuliab, Soviet Central Asia, July 1958, determined by V. A. Yasnosh. The specimens are rather poorly preserved, having been kept in alcohol prior to mounting in Hoyer's medium. Nevertheless, they appear to have the dusky thoracic sterna and other characteristics of *mytilaspidis*, whereas the distinguishing characters given by Yasnosh are size-related and hence not very reliable. Thus, the antennal scape (Figure 773) in the specimens at hand varies from 4 to $6\frac{1}{4}$ times as long as wide; the mesoscutum (Figure 774) is furnished with 7–10 setae; and the marginal fringe of the forewing (Figure 776) varies from $\frac{1}{6}$ to $\frac{1}{3}$ the width of the disk. The propodeum and crenulae are shown in Figures 774, 775. In all significant diagnostic characters, these specimens seem to fall within the range of morphological variation of *mytilaspidis*, except that the ovipositor tends to be longer: the shaft varies from $1\frac{2}{3}$ to fully twice as long as the middle tibia (1.68–2.05), the sheaths about $\frac{2}{5}$ the length of the middle tibia (0.35–0.42). Length 0.67–1.02 mm.

Judging from the specimens at hand, we would consider *moldavicus* to be a synonym of *mytilaspidis*, but for the somewhat longer ovipositor. Better preserved material, including male specimens, should be examined before a final decision can be made. For the time being, *moldavicus* is retained as a questionable species, differing from *mytilaspidis* in the longer ovipositor of the female.

Nikol'skaya and Yasnosh (1966) mentioned 3♂♂ among the type series of *moldavicus*, and listed *Epidiaspis leperii* (Signoret) and *Quadraspidiotus pyri* (Lichtenstein) as additional hosts. However, Yasnosh (1972) subsequently stated that the male of *moldavicus* was unknown, and listed only *L. ulmi* as a host of this species. She suggested that *moldavicus* may be identical to "*mytilaspidis*-form 3a" from Greece.

## 50. **Aphytis aonidiae** (Mercet)

(Figures 777–798)

*Aphelinus aonidiae* Mercet, 1911, Bol. R. Soc. Españ. Hist. Nat., **11**:511–514.
*Aphelinus aonidiae*: Mercet, 1912, Trab. Mus. Cienc. Nat. Madrid, **10**:63–67.
*Aphytis aonidiae*: Mercet, 1930, Rev. Biol. Forest. Limmol., Ser. B, **2**:51; Mercet, 1932, Eos, **8**:360; Compere, 1955, Univ. Calif. Publ. Entomol., **10**:310–312; Ferrière, 1965, Hymenoptera Aphelinidae d'Europe et du Bassin Méditerranéen, pp. 88–89; Nikol'skaya and Yasnosh, 1966, Aphelinids of the European Part of the USSR and the Caucasus, pp. 205–206 (in Russian).
*Aphytis (Prospaphelinus) dubius* De Santis, 1948, Rev. Mus. La Plata, Zool. (N.S.), **5**:128–129; n. syn.
*Aphytis dubius*: Compere, 1955, Univ. Calif. Publ. Entomol., **10**:310.
*Aphytis (Prospaphelinus) dubius* var. *intermedia* De Santis, 1948, Rev. Mus. La Plata, Zool. (N.S.), **5**:129–130; n. syn.
*Aphytis dubius* var. *intermedius*: Compere, 1955, Univ. Calif. Publ. Entomol., **10**:312.
*Aphytis citrinus* Compere, 1955, Univ. Calif. Publ. Entomol., **10**:312–313; n. syn.
*Aphytis citrinus*: Peck, 1963, Canad. Entomol. Suppl. **30**:249.

*A. aonidiae* can be readily distinguished from *mytilaspidis* by the generally paler coloration, especially the immaculate thoracic sterna; the propodeal crenulae are somewhat more triangular, less rounded, and paler than in *mytilaspidis*; the setae on the head are considerably paler than in *mytilaspidis*.

This Holarctic species has been greatly misinterpreted in the taxonomic literature. In order to clarify some of the confusion, we shall start with a redescription of Mercet's holotype specimen (Figures 777–779). It should be noted that this is a small and probably atypical specimen.

*Female*. Eyes finely setose. Mandibles well developed, with 2 denticles and a dorsal truncation; maxillary palpi 2-segmented, labial palpi 1-segmented. Antennae (Figure 777) relatively short and thick; scape moderately slender, about $4\frac{1}{5}$ times as long as wide, $1\frac{1}{4}$ times as long as the club; pedicel about $1\frac{2}{3}$ times as long as wide, $1\frac{1}{5}$ times longer than the third segment of the funicle; first funicular segment trapezoidal, about $1\frac{1}{3}$ times as wide as long; second segment nearly symmetrical, as wide as but considerably shorter than the first segment, nearly twice as wide as long; third funicular segment about $1\frac{1}{4}$ times as long as wide, bearing a single longitudinal sensillum; club about $2\frac{1}{4}$ times as long as wide, somewhat over twice longer and distinctly wider than the preceding segment, bearing 6 longitudinal sensilla.

Setae on thorax (Figure 778) dark, coarse, readily visible under $\times 30$ magnification; setae on head considerably paler, barely visible under $\times 60$ magnification; those on abdomen pale, slender, invisible even under $\times 120$ magnification. Vertex with 2 pairs of long setae along occipital margin, in addition to numerous short setae. Mesoscutum with 6 setae, each parapsis with 1 short seta, each axilla with 1 seta; scutellum with 4, the discoid sensilla about equidistant from the two pairs or somewhat closer to the anterior pair. Frontovertex, pronotum and mesonotal sclerites reticulate, infrasculpture evident within the cells on mesoscutum and scutellum. Scutellum (Figures 778, 779) about $\frac{4}{5}$ median length of mesoscutum. Metanotum (Figure 779) reticulate except on the sides; anteromedian apodeme considerably longer than median length of metanotum (in the holotype specimen the metanotum is somewhat crushed centrally, and thus appears to be exceedingly short; this is considered to be an artifact).

Propodeum (Figure 779) rather short, about $\frac{3}{5}$ median length of scutellum (in the holotype specimen the propodeum appears about 4 times as long as the distorted metanotum), faintly reticulate on the sides, broadly reticulate centrally; posterior margin slightly bilobed; crenulae 5 + 5, distinct, triangular, nonoverlapping.

Second abdominal tergite distinctly, transversely striated anteriorly on both sides, smooth centrally; tergites III–VII reticulate on the sides, with 2–3 setae in a short, transverse row on each reticulate area; tergites III and VII transversely striated mesad of reticulate areas, the striation fading centrally, sixth tergite with faint indications of transverse striation, fourth and fifth tergites smooth centrally; seventh tergite bearing 2 setae submedially, between the lateral setiferous areas; eighth tergite faintly reticulate-punctate across, this fading centrally, with a transverse row of 4 setae between spiracles; syntergum triangular, faintly punctate, bearing 5 setae in a transverse row. Cerci about equidistant from posterior spiracles and from tip of abdomen, with 2 long setae and 1 short seta. Ovipositor shaft a little over $1\frac{3}{5}$ times as long as the middle tibia (1.63); ovipositor sheaths a little over $\frac{1}{3}$ length of middle tibia (0.35).

Mid-tibial spur somewhat shorter than the corresponding basitarsus.

Forewing about $2\frac{1}{2}$ times as long as wide; marginal fringe about $\frac{1}{3}$ width of disk. Delta area with 44–50 setae in 5–6 rows (two wings of same specimen), these considerably longer and sparser than the setae distad of speculum, quite distinct from row of setae along posterior margin of wing, widely separated from a small group of 3 setae below distal two fifths of submarginal vein; costal cell with 2 fine setae somewhat proximad of middle and 1 coarse seta near apex. Submarginal vein bearing 2 coarse setae, the proximal one about $\frac{1}{2}$–$\frac{3}{5}$ length of the distal, and 14 bullae. Marginal vein bearing 8 prominent, subequal setae along anterior margin, about $1\frac{3}{5}$ times longer than the setae in a row along center of vein.

Forewing generally hyaline, distinctly infumated below junction of marginal and submarginal veins and at base of delta, the two areas separated by hyaline; a fuscous streak along posterior margin of wing, distad of speculum.

General coloration pale yellow; thoracic sterna immaculate, concolorous with body; dorsum of thorax and abdomen, including propodeum and crenulae, immaculate; posterior margin of scutellum narrowly lined with blackish; tegulae faintly dusky, a short black streak below them, at base of forewing. Plates around base of ovipositor faintly dusky. Antennal scape pale, very faintly infuscated ventrally; rest of antenna uniformly, rather strongly dusky. Tibiae of all legs rather faintly dusky, apical segment of all tarsi distinctly dusky, obviously darker than the preceding segments. Wing veins pale.

Length of holotype difficult to determine, as the specimen appears somewhat flattened on the slide, with the head separated and the abdomen probably distended. Mercet (1912b) recorded the length of *aonidiae* as 0.60–0.63 mm.

*Male*. Unknown.

Redescribed from 1♀ (holotype), reared from *Aonidia lauri* (Bouché) [probably on *Laurus nobilis*], Toledo, Spain; cleared and mounted on a slide, identified as the holotype by Ch. Ferrière.

Holotype at the Museo de Ciencias Naturales, Madrid.

**Notes.** Since *A. aonidiae* was described from minute specimens reared from *Aonidia lauri*, it has been narrowly interpreted in the literature as a minute species, differing from *mytilaspidis* in the longer fringe of the forewing, the fewer setae on the thorax and forewing, etc. We consider these characters to be associated with the minute size of the holotype and other specimens reared from *Aonidia lauri*. Thus in our interpretation the only valid differences between *aonidiae* and *mytilaspidis* are those listed as distinguishing characters on p. 477, namely the paler coloration, the immaculate thoracic sterna, the paler cephalic setae, and the somewhat more triangular shape of the propodeal crenulae. The following discussion of various series assigned by us to *aonidiae* will demonstrate, and hopefully somewhat clarify, the range of morphological variation of this species as understood here.

A series of ♀ specimens, reared from *Aonidia lauri* on *Laurus nobilis* in Spain (1♀, Madrid, det. Mercet (Figure 780); 23♀♀, Valencia, P. DeBach, July 29, 1963), are undoubtedly conspecific with the holotype of *aonidiae*. The specimens are relatively small, 0.53–0.71 mm in length. They agree with the holotype in coloration and in all morphological characters, but exhibit the following range of variation:

Antennal scape about $4\frac{1}{2}$ to $5\frac{1}{2}$ times as long as wide; pedicel $1\frac{1}{2}$ to 2 times as long as wide; first funicular segment $1\frac{1}{3}$ to $1\frac{1}{2}$ times as wide as long, second segment $1\frac{4}{5}$ to $2\frac{1}{3}$ times as wide as long; third funicular segment usually about $1\frac{1}{4}$ times as long as wide, bearing 1–2 longitudinal sensilla; club $2\frac{1}{4}$ to $2\frac{3}{4}$ times as long as wide, 2 to nearly 3 times longer than the preceding segment, bearing 3–6 longitudinal sensilla.

Mesoscutum with 4–8 setae, each parapsis with 1–2. Metanotal apodeme $1\frac{1}{5}$ to $1\frac{2}{5}$ times median length of metanotum. Propodeum a little over 3 to $4\frac{1}{6}$ times as long as the metanotum, $\frac{1}{2}$ to $\frac{2}{3}$ length of scutellum; crenulae 2+3 to 5+6. Syntergum with 5–7 setae in a transverse row. Ovipositor shaft $1\frac{2}{5}$ to $1\frac{3}{4}$ times as long as the middle tibia (1.61–1.77), ovipositor sheaths $\frac{1}{3}$ to $\frac{2}{5}$ length of middle tibia (0.33–0.42).

Forewing very faintly, almost imperceptibly infumated on basal half, in addition to the pattern at base, $2\frac{2}{5}$ to $2\frac{1}{2}$ as long as wide; marginal fringe varying from over $\frac{1}{3}$ to less than $\frac{1}{5}$ width of disk. Delta with 21–44 setae; submarginal vein bearing 10–17 bullae. Marginal vein bearing 5–8 prominent setae along anterior margin, $1\frac{1}{3}$ to $1\frac{4}{5}$ times longer than the setae along center of vein.

A paratype specimen of *A. dubius* was examined (1♀, reared by M. Griot from the San Jose scale, *Quadraspidiotus perniciosus* (Comstock), Angel Gallardo, Santa Fe, Argentina, August 8, 1938; remounted in Hoyer's medium). The specimen is relatively large, 1.11 mm in length. In coloration and most morphological characters it falls within the range of variation of *aonidiae*. Antennal club $2\frac{4}{5}$ times as long as wide; mesoscutum with 8 setae; crenulae 6+7; forewing (Figure 781) faintly infumated on basal half, marginal fringe $\frac{1}{4}$ to $\frac{1}{5}$ width of disk; delta with 58–60 setae; marginal vein with 8 prominent setae along anterior margin. The slight differences are very probably associated with the larger size of this specimen. Thus we cannot find sufficient justification for retaining *dubius* as a distinct species; it should be regarded as a junior synonym of *aonidiae*.

The single known specimen of *A. dubius* var. *intermedia* was also remounted and studied (1♀, holotype, reared by L. De Santis from the oleander scale, *Aspidiotus nerii* Bouché [= *A. hederae* (Vallot)] on olive, Patagones, Buenos Aires, Argentina, July 1938). This specimen (Figures 782–784) is a little smaller than the type of *dubius*, but larger than the type of *aonidiae*: length 0.94 mm. It agrees with *aonidiae* in general coloration and structural characters, but the posterior margin of the propodeum is faintly infuscated (in *aonidiae*, this area appears distinctly dusky under phase contrast, but not under the dissecting microscope). Mesoscutum with 9 setae; propodeum (Figure 783) $4\frac{1}{6}$ times as long as the metanotum, crenulae 5+6. Forewing (Figure 784) faintly infumated on basal half, a little less than $2\frac{2}{5}$ times as long as wide; marginal fringe between $\frac{1}{3}$ and $\frac{1}{4}$ width of disk (0.29); delta with 47–50 setae; marginal vein with 8–9 prominent, subequal setae along anterior margin. Again, these slight differences do not appear to be significant, and "*dubius* var. *intermedia*" is regarded here as another synonym of *aonidiae*.

*A. citrinus* was characterized by Compere (1955) as follows: "Difficult or perhaps impossible to distinguish from *aonidiae* (Mercet), *dubius* De Santis, and *dubius* var. *intermedius* De Santis on a morphological basis. If my sketches are correct, *citrinus* can be separated from *aonidiae* by the more numerous hairs on the front wings and the shorter antennal club. The characters used here to separate *citrinus* from *dubius* are just about the same as those used by De Santis to separate var. *intermedius* from *dubius*, namely, the longer marginal fringe of the front wings. The most reliable clue to the identity of *citrinus* is the host relationship. So far as known, it is parasitic only on yellow scale, *Aonidiella citrina* (Coq.)."

Morphologically, *citrinus* indeed seems inseparable from *aonidiae*. At hand is an adequate series of cleared, remounted syntype specimens (14♀♀, reared by O. Hemphill from yellow scale on orange, Visalia, California, December 20, 1948, part of Compere's large "cotype" series, originally mounted in balsam). They are intermediate in size, 0.70–0.96 mm in length, and essentially agree with *aonidiae* in general coloration, structure, chaetotaxis and sculpture (Figures 785–794). The antennal club (Figures 786, 787) tends to appear slightly more slender, $2\frac{1}{2}$ to $2\frac{4}{5}$ times as long as wide; mesoscutum (Figures 788–790) with 6–9 setae, each parapsis with 2; propodeum (Figures 788–789, 791–793) $3\frac{1}{3}$ to $4\frac{1}{5}$ times as long as the metanotum, about $\frac{2}{3}$ length of scutellum; crenulae 4+4 to 6+6. Forewing (Figure 794) faintly infumated on basal half, somewhat broader, $2\frac{1}{3}$ to nearly $2\frac{1}{2}$ times as long as wide; marginal fringe $\frac{1}{5}$ to over $\frac{1}{4}$ width of disk; delta with 40–69 setae; submarginal vein with 14–19 bullae; marginal vein bearing 7–10 prominent, subequal setae along anterior margin. In some specimens, the posterior margin of the propodeum appears faintly dusky.

A female specimen of that series (slide No. IIA2:9 in the Riverside *Aphytis* collection) is designated lectotype of *citrinus* (Figures 786, 791). The other specimens in Compere's syntype series are paralectotypes.

On the basis of morphological criteria, *citrinus* is regarded here as yet another junior synonym of *aonidiae*. When dealing with uniparental forms, there is not much that can be done by way of hybridization tests in order to ascertain specific identity. However, the alleged specificity of *citrinus* to yellow scale should be further investigated. Compere's (loc. cit.) record was at least incomplete: "*citrinus*" was reared on occasion in California from the California red scale, *Aonidiella aurantii* (Maskell), as well as from yellow scale (P. DeBach, unpublished data). While the yellow-scale "form" has not yet been tested on other scale-insect hosts, recent studies carried out in our laboratory have indicated that another presumed "form" of *aonidiae*, obtained from the San Jose scale in southern California (see below), is capable of developing also in yellow scale. For the time being, therefore, *citrinus* is considered to represent part of the range of morphological and biological variation of *aonidiae*.

The following quotation from Compere (1955) is of historical interest: "In all probability the species that currently parasitizes yellow scale in California is the same as that which commonly parasitized yellow scale in the San Gabriel Valley in the early days of citriculture, and to which Alexander Craw applied the name "golden chalcid." Circumstantial evidence indicates that the species was probably accidentally introduced into California from Japan on yellow scale some time before 1892." However, evidence now indicates that *aonidiae* is a Palearctic species and probably adopted the yellow scale after its ecesis to North America.

A uniparental stock obtained from the San Jose scale, *Quadraspidiotus perniciosus* (Comstock), on almond in Ventura County, California, apparently represents large specimens of *aonidiae*. Specimens from an insectary culture of this stock, reared on the latania scale, *Hemiberlesia lataniae* (Signoret), on potato, are relatively large: 0.72–1.13 mm in length. They agree with *aonidiae* in general coloration and structure, the larger specimens differing slightly—mainly in characters that are probably correlated with their larger size: Antennae (Figure 795) somewhat more slender than in other series of *aonidiae*; scape $5\frac{1}{4}$ to $6\frac{2}{3}$ times as long as wide, pedicel $1\frac{2}{3}$ to $2\frac{1}{4}$ times, club about 3 to $3\frac{1}{4}$ times as long as wide, bearing 5–7 longitudinal sensilla; mesoscutum (Figures 796, 797) with 8–12 setae, each parapsis with 2; anteromedian apodeme nearly as long as, or distinctly longer than, median length of metanotum; propodeum (Figure 798) often distinctly bilobed, about $3\frac{1}{2}$ to $4\frac{1}{3}$ times as long as the metanotum, crenulae 4 + 5 to 7 + 8, sometimes less distinctly triangular than in other series; syntergum with 6–10 setae in a transverse row; marginal fringe of forewing varying from a little less than $\frac{1}{5}$ width of disk in a small specimen to about $\frac{1}{8}$ in large specimens; delta with 36–64 setae; submarginal vein with 16–21 bullae; marginal vein bearing 8–12 prominent, subequal setae along anterior margin. The specimens at hand appear somewhat paler than other series: antennal scape entirely pale; posterior margin of scutellum only faintly dusky (but clearly so under phase contrast); apical tarsal segment concolorous with the preceding segments, not contrastingly dusky.

In spite of these morphological differences, this parasite of the San Jose scale is regarded here—at least for the time being—as conspecific with *aonidiae*.

Because of its large size, the short marginal fringe of the forewing, and the relatively numerous setae, this stock was at first provisionally designated in our laboratory "*mytilaspidis*-form with clear sterna". Since the differences between *mytilaspidis* and *aonidiae*, as understood here, were not recognized by previous workers, the records referring to *mytilaspidis* in the literature may have included similar large specimens of *aonidiae*. A case in point is the "form" recorded by DeBach (1964) from Greece as "*mytilaspidis*-form 1", a uniparental parasite of *Hemiberlesia lataniae* with unpigmented thoracic sterna. We now consider this "form" to be conspecific with *aonidiae*.

An interesting series of specimens (205♀♀, 2♂♂, reared by S. C. Warner from the tan oak scale, *Aspidaspis densiflorae* (Bremner), on *Quercus* sp., Oak Glen, California, December 3, 1963) illustrates the range of variation of *aonidiae*. The specimens range from minute to large, 0.53–1.03 mm in length, and several characters vary accordingly. Mesoscutum with 4–11 setae (but each parapsis always with 2); anteromedian apodeme varying from slightly shorter than, to fully $1\frac{1}{2}$ times as long as, median length of metanotum; propodeum a little less than 3 to fully $4\frac{1}{2}$ times as long as the metanotum, crenulae 3 + 4 to 6 + 7. Forewing $2\frac{1}{3}$ to $2\frac{3}{5}$ times as long as wide; marginal fringe varying from about $\frac{1}{8}$ to over $\frac{1}{3}$ width of disk; delta with 21–78 setae in 3 to 7 rows; submarginal vein with 12–22 bullae; marginal vein bearing 5–12 prominent, subequal setae along anterior margin. Posterior margin of scutellum distinctly lined with black in some specimens, very faintly dusky in others, with many specimens occupying an intermediate position; antennal scape ranging from entirely pale to distinctly infuscated; apical tarsal segment distinctly dusky in some specimens, concolorous with preceding segments in others. Forewing pattern varying in intensity, from barely noticeable in small specimens to very distinct in large specimens.

*Male.* The 2 ♂♂ included in this series are similar to the females in structure, chaetotaxis, sculpture and coloration. As in the male of *mytilaspidis*, the antennal scape is not so slender, a little over 4 times as long as wide, and with an indistinct, slightly infuscated ridge or elevated area on the ventral aspect; club truncate in lateral view, with 3 longitudinal sensilla and lacking a specialized sensory area on the ventral surface.

Genitalia about $\frac{2}{3}$ to $\frac{3}{4}$ length of middle tibia (0.67–0.74); digital sclerites a little over $\frac{1}{4}$ the combined length of aedeagus and apodemes (0.26–0.28).

Length 0.70–0.72 mm.

Pupal pigmentation is overall black in all the series referred here to *aonidiae*.

**Conclusion.** *A. aonidiae* appears to be a widespread, uniparental species, parasitizing a variety of armored scale insects in various parts of the world, and exhibiting a considerable degree of size-linked variation. Biparental forms are unknown. Minute specimens may have few (4–7) mesoscutal setae, a somewhat shorter antennal club, narrower wings and a relatively long marginal fringe (up to $\frac{1}{3}$ width of disk, or more), whereas large specimens may have more numerous setae (10–12 on the mesoscutum), a somewhat more slender antennal club, broader wings, and a short marginal fringe ($\frac{1}{8}$ width of disk). All the series discussed here are essentially similar in structure and coloration and are, at least for the time being, regarded as representing populations of one species. The range of variation of *aonidiae*, as understood here, parallels that of *mytilaspidis*: both species are represented by large as well as minute specimens, and differ mainly in coloration (the easiest distinguishing character being the pigmentation of the thoracic sterna).

Some populations of *aonidiae* may have distinctive host preferences or other biological differences. Even if this proves to be the case, we as yet have insufficient justification for separating such populations as distinct species. Much more will have to be known about them first. For the researcher in applied biological control, however, such biological differences between closely related populations may prove to be of supreme importance and should not be overlooked.

**Material Examined.** Specimens in the collection of the Division of Biological Control, University of California, Riverside, recognized here as *A. aonidiae*, were reared from *Aonidia lauri* (Bouché) on *Laurus nobilis* in Spain and Greece; from *Aonidiella aurantii* (Maskell) on citrus in California; from *Aonidiella citrina* (Coquillett) on citrus in California; from *Aonidiella orientalis* (Newstead) on citrus in Iran; from *Aspidaspis densiflorae* (Bremner) on oak in California; from *Aulacaspis rosae* (Bouché) on rose in Crete; from *Chrysomphalus dictyospermi* (Morgan) on *Pistacia terebinthus* in Greece and on orange in Spain; from *Dynaspidiotus britannicus* (Newstead) on *Rhamnus aluternus* in Greece; from *Hemiberlesia lataniae* (Signoret) on *Boxus sempervirens*, on *Crataegus pyracantha* and on *Laurus nobilis* in Greece, on *Ficus* sp. in Crete, and on Bignoniaceae in Israel; from *Insulaspis juniperi* (Lindinger) on an undetermined plant in Georgia

(USSR); from *Lepidosaphes* sp. on fig in Greece; from "*Parlatoria pergandii* Comstock material on sour orange" in Spain; from *Quadraspidiotus juglansregiae* (Comstock) on English walnut in California; from *Quadraspidiotus ostreaeformis* (Curtis) on poplar in Argentina; from *Quadraspidiotus perniciosus* (Comstock) on peach, on prune, on almond, on flowering quince, on rose and on *Pyracantha hybrida* in California, on *Chaenomeles japonica* in Ohio; and on an undetermined plant in Armenia (USSR). Also from unidentified hosts on *Ampelopsis* sp. in Greece and on *Laurus* in Cyprus.

*A. aonidiae* also occurs on *Aspidiotus nerii* Bouché [= *A. hederae* (Vallot)] in Israel.

**Biology.** In laboratory tests, females of the "form" of *aonidiae* obtained from the San Jose scale in California (see above) lived for an average of 35.4 ± 6.9 days when fed honey at 25.6 ± 1.1°C and 50 ± 5 percent RH. Their fecundity on San Jose scale (on citron melon) averaged 85.1 eggs per female under these conditions. The parasites readily host-fed on San Jose scale, a female destroying an average of 62.0 first-instar larvae, or 75.3 second-instar larvae and prepupal males, or 37.1 third-instar female scales during her lifetime. Reproduction was thelytokous, with rare males occurring occasionally in the population (Gulmahamad and DeBach, 1978).

**Potential Value for Biological Control.** Field studies and parasite exclusion tests have recently demonstrated that *A. aonidiae* is an important natural enemy of the San Jose scale, *Quadraspidiotus perniciosus* (Comstock), in southern California and is effecting substantial biological control of this serious pest (Gulmahamad and DeBach, 1978). More attention should be paid to this parasite in the extensive projects of biological control directed against the San Jose scale in Europe and elsewhere, which have so far greatly overemphasized endoparasites (see Rosen and DeBach, 1978).

The following three species, *A. libanicus*, *A. phoenicis* and *A. luteus*, are very closely related to *aonidiae*, and may eventually be recognized as synonyms of this species.

## 51. **Aphytis libanicus** Traboulsi

(Figures 799–805)

*Aphytis libanicus* Traboulsi, 1969, Ann. Soc. Entomol. Fr. (N.S.), **5**:66–67.

This Mediterranean species appears to be very closely related to *aonidiae*, and may eventually prove to be a synonym of the latter. If our interpretation is correct, *libanicus* differs slightly from *aonidiae* in being uniformly yellow, without any fuscous markings whatsoever, and with somewhat paler thoracic setae.

In order to avoid unnecessary repetition, *libanicus* is discussed below relative to *aonidiae*.

We have not seen the types of *libanicus*, which were reared from the holly scale, *Dynaspidiotus britannicus* (Newstead), on olive and carob in Lebanon. The species was described as entirely yellow, with immaculate thoracic sterna and pale thoracic setae, sometimes with the posterior margin of the scutellum very faintly infuscated.

At hand are specimens, reared by D. Rosen from *Dynaspidiotus britannicus* (Newstead) and *Leucaspis riccae* Targioni-Tozzetti on olive in the northern part of Israel. The specimens (147♀♀, 3♂♂) were collected in 1961 and were apparently not perfectly cleared; in several slides, the Hoyer's medium is deteriorating and the specimens appear in rather poor shape. They are very similar to specimens of *aonidiae* (syntypes of *citrinus*, etc.) in all morphological characters, but appear to be entirely yellow, without even a trace of infuscation on the posterior margin of the scutellum, and the thoracic setae appear somewhat paler than in the series referred to *aonidiae*.

*Female.* Eyes and mouthparts as in *aonidiae*. Antennal club (Figures 799, 800) $2\frac{3}{5}$ to $3\frac{1}{4}$ times as long as wide (usually less than 3 times). Mesoscutum (Figures 801, 802) with 7–13 setae, each parapsis with 2 (rarely 3), each axilla with 1 seta. Propodeum (Figures 801–803) $3\frac{1}{10}$ to $4\frac{1}{4}$ times as long as the metanotum (usually less than 4 times), crenulae 4+5 to 7+7, triangular, nonoverlapping, as in *aonidiae*. Ovipositor shaft about $1\frac{1}{2}$ to $1\frac{3}{4}$ times as long as the middle tibia (1.51–1.73), ovipositor sheaths usually $\frac{1}{3}$ to $\frac{2}{5}$ length of middle tibia (0.33–0.40).

Forewing (Figure 804) about $2\frac{2}{5}$ to $2\frac{3}{5}$ times as long as wide; marginal fringe varying from over $\frac{1}{4}$ to about $\frac{1}{9}$ width of disk. Delta area with 29–72 setae in 4–7 rows. Submarginal vein with 13–22 bullae; marginal vein bearing 7–13 prominent, subequal setae along anterior margin. Forewing pattern as in *aonidiae*.

General coloration entirely yellow, thoracic sterna and posterior margin of scutellum concolorous; a short black streak present at base of forewing, below tegula. Legs nearly concolorous with body, not distinctly dusky. Antennal scape pale, rest of antenna uniformly, very faintly dusky.

Length 0.70–1.14 mm.

*Male.* The rare males are very similar to the rare males of *aonidiae*, except for the absence of infuscation on the posterior margin of the scutellum.

Antennal scape apparently with an indistinct, slightly infuscated ridge; club about $2\frac{1}{2}$ times as long as wide, without a cut-off sensory area on the ventral aspect. Genitalia 0.69–0.71 length of middle tibia; digital sclerites 0.24–0.30 the combined length of aedeagus and apodemes. Forewing about $2\frac{1}{3}$ times as long as wide. Coloration and wing pattern as in the female.

Length 0.61 mm.

Redescribed from the following specimens, all reared by D. Rosen in Israel: From *Leucaspis riccae* on olive: 2♀♀, Kinnereth, March 7, 1961; 32♀♀, 2♂♂, western shore of the Sea of Galilee, August 27, 1961; 69♀♀, 1♂, same locality, September 27, 1961; 18♀♀, same locality, October 25, 1961. From *Dynaspidiotus britannicus* on olive, Beit She'an Valley: 13♀♀, March 22, 1961; 2♀♀, August 27, 1961; 11♀♀, October 5, 1961.

According to Traboulsi (1969), the holotype and numerous paratypes of *libanicus* are kept in his personal collection, whereas other paratypes have been deposited at the Identification Center of the O.I.L.B., Muséum d'Histoire Naturelle, Geneva, Switzerland.

**Notes.** A female pupa of *libanicus* is shown in Figure 805.

The differences between *libanicus*, as interpreted here, and *aonidiae* are very slight. This is especially true with regard to certain series referred to *aonidiae*, such as the specimens obtained from the San Jose scale in California and Ohio. In these specimens, the posterior margin of the scutellum is very faintly infuscated. Careful biological studies are required in order to help clarify the specific identity of these series relative to *libanicus*.

## 52. **Aphytis phoenicis** DeBach and Rosen

(Figures 806–813)

*Aphytis phoenicis* DeBach and Rosen, 1976, Ann. Entomol. Soc. Amer., **69**: 543–544.

This uniparental Middle Eastern species is very closely related to *aonidiae*, and may eventually prove to be another synonym of the latter. It appears to differ from *aonidiae* mainly in the somewhat longer ovipositor and propodeum and in the somewhat wider, sometimes irregular, propodeal crenulae.

In order to avoid unnecessary repetition, *phoenicis* is described below relative to *aonidiae*.

*Female*. Eyes and mouthparts as in *aonidiae*. Antenna (Figure 806) essentially as in *aonidiae*; scape slender, $4\frac{1}{3}$ to $5\frac{3}{4}$ times as long as wide; pedicel a little less than $1\frac{1}{2}$ times to nearly twice as long as wide; first funicular segment about $1\frac{3}{10}$ to $1\frac{1}{2}$ times as wide as long, second segment $1\frac{4}{5}$ to over twice as wide as long; third funicular segment $1\frac{1}{10}$ to $1\frac{1}{3}$ times as long as wide, bearing 1–2 longitudinal sensilla; club $2\frac{2}{3}$ to $2\frac{4}{5}$ times as long as wide, $2\frac{1}{4}$ to $2\frac{3}{4}$ times longer than the preceding segment, bearing 5–6 longitudinal sensilla.

Structure and chaetotaxis of head, thorax and abdomen essentially as in *aonidiae*, sculpture similar but somewhat fainter. Mesoscutum (Figures 807, 808) with 6–8 setae, each parapsis with 2, each axilla with 1 seta. Scutellum nearly as long as the mesoscutum (an abnormal scutellum with one supernumerary sensillum is shown in Figure 812). Anteromedian apodeme $1\frac{1}{5}$ to $1\frac{2}{5}$ times median length of metanotum.

Propodeum (Figures 807–811) relatively long, $3\frac{1}{2}$ to $5\frac{1}{5}$ times as long as the metanotum (usually 4 to $4\frac{1}{2}$ times), about $\frac{3}{5}$ length of scutellum; crenulae 3+3 to 4+5, rather variable: generally triangular (Figure 809), nonoverlapping, but somewhat wider and lower than in *aonidiae*; sometimes irregular, very wide or forming a continuous ledge (Figures 810, 811).

Ovipositor relatively long; shaft about $1\frac{2}{3}$ to nearly twice as long as the middle tibia (1.69–1.96, usually over $1\frac{4}{5}$); sheaths $\frac{2}{5}$ to $\frac{1}{2}$ length of middle tibia (0.40–0.49).

Mid-tibial spur usually somewhat shorter, sometimes a little longer, than the corresponding basitarsus.

Forewing (Figure 813) moderately narrow, nearly $2\frac{1}{2}$ to $2\frac{3}{5}$ times as long as wide; marginal fringe varying from $\frac{1}{6}$ to well over $\frac{1}{3}$ width of disk. Chaetotaxis essentially as in *aonidiae*; delta with 18–50 setae in 3–6 rows; submarginal vein bearing 14–18 bullae; marginal vein bearing 5–9 prominent, subequal setae along anterio· margin, these $1\frac{3}{10}$ to $1\frac{4}{5}$ times longer than setae in a row along center of vein.

Forewing generally hyaline, infumated below submarginal vein and at base of delta; under phase contrast, the basal half of the wing appears very faintly infumated; a fuscous streak along posterior margin of wing, distad of speculum.

General coloration similar to *aonidiae*: entirely pale yellow, thoracic sterna immaculate; posterior margin of scutellum narrowly lined with blackish; a short black streak at base of forewing, below tegula; antennal scape pale, rest of antenna uniformly, rather faintly dusky; tibiae and tarsi of all legs dusky, apical tarsal segment not contrastingly dark; wing veins colorless.

Length 0.55–0.88 mm.

*Male*. Unknown.

Described from numerous specimens, reared by H. E. Martin from the parlatoria date scale, *Parlatoria blanchardi* (Targioni-Tozzetti), on date palm, *Phoenix dactylifera*, in Saudi Arabia (1♀, holotype, Aunaiza, March 7, 1967; 5♀♀, paratypes, Riyadh-Erka, March 30, 1967; 5♀♀, paratypes, Riyadh-Erka, April 30, 1967; 24♀♀, paratypes, Riyadh-Erka, May 24, 1967; 72♀♀, paratypes, Riyadh-Erka, June 11, 1967); also 13♀♀, reared by M. Kehat from same host on same host plant, Eilat, Israel, August 11, 1961.

Type series in the collection of the Division of Biological Control, University of California, Riverside.

**Notes.** *A. phoenicis* is known only from the parlatoria date scale in desert oases in the Middle East. The differences between *phoenicis* and *aonidiae*, although slight, appear to be consistent enough to facilitate the identification of *phoenicis*. Although the ranges of variation of *phoenicis* and *aonidiae* may overlap somewhat, the relative length of the ovipositor and sheaths appears to be a reliable character for their separation. Until biological studies indicate otherwise, we feel *phoenicis* should be recognized as a distinct species.

**Potential Value for Biological Control.** *A. phoenicis* was recorded by Kehat et al. (1974) as a widely distributed, highly effective parasite of the parlatoria date scale in Israel. Its successful re-establishment, together with other natural enemies, in groves from which it had been eliminated by indiscriminate application of nonselective pesticides, has resulted in effective biological control of the parlatoria date scale. *A. phoenicis* should be regarded as a promising candidate for introduction against this serious pest of date palms.

## 53. **Aphytis luteus** (Ratzeburg)

(Figures 814–829)

*Coccobius luteus* Ratzeburg, 1852, Die Ichneumonen der Forstinsecten, **3**: 196.
*Aphelinus luteus*: Dalla Torre, 1898, Cat. Hym., **5**:221; Schmiedeknecht, 1909, Gen. Insect. **97**:452.
*Ablerus* (?) *luteus*: Mercet, 1912, Trab. Mus. Cienc. Nat. Madrid, **10**:136.
*Aphytis* "undescribed species Z": DeBach, 1964, Ann. Inst. Phytopathol. Benaki (N.S.), **7**: 12, 14, 15.
*Coccobius luteus*: Bouček, 1964, Beitr. Entomol., **14**:667.
*Aphytis luteus*: DeBach and Rosen, 1976, Ann. Entomol. Soc. Amer., **69**:544.
*Aphytis luteus*: Graham, 1976, Syst. Entomol., **1**: 133–134 (as possible syn. of *A. albidus* (Westwood)).

A large, biparental European parasite of *Leucaspis* spp. on pine, *A. luteus* is similar to *aonidiae* in the immaculate thoracic sterna and in most morphological characters, but may be recognized by the rounded, dusky propodeal crenulae and somewhat shorter ovipositor, as well as by the somewhat more numerous mesoscutal setae and deeper yellow coloration. The males are large, approximately equal to the females in size.

*Female*. Eyes finely setose. Mandibles well developed, with 2 denticles and a dorsal truncation; maxillary palpi 2-segmented, labial palpi 1-segmented. Antennal scape (Figure 814) slender, 5 to $5\frac{4}{5}$ times as long as wide, $1\frac{1}{5}$ to $1\frac{1}{3}$ times longer than the club; pedicel relatively slender, $1\frac{3}{4}$ times to fully twice as long as wide, a little longer than the third segment of the funicle; first funicular segment trapezoidal, only $1\frac{1}{12}$ to $1\frac{3}{10}$ times as wide as long; second segment nearly symmetrical, considerably shorter and about as wide as the first segment, $1\frac{1}{3}$ times to fully twice as wide as long; third funicular segment relatively long, about $1\frac{1}{3}$ to $1\frac{2}{3}$ times as long as wide, bearing 2–3 longitudinal sensilla; club $2\frac{3}{5}$ to a little over 3 times as long as wide, only 2 to $2\frac{2}{5}$ times longer and somewhat wider than the preceding segment, bearing 6–8 longitudinal sensilla.

Setae on thorax coarse, dark, readily visible under ×30 magnification; setae on head somewhat paler, but readily visible under ×60 magnification; those on abdomen paler, more slender, barely visible under ×60 magnification. Vertex with 2 pairs of long setae along occipital margin, in addition to numerous shorter setae. Mesoscutum (Figures 815, 816) with 10–16 (usually 14) setae, the posterior pair and 1 seta at each antero-lateral corner distinctly longer and coarser than the others; each parapsis with 2–3 (usually 3) shorter setae, each axilla with 1 seta; scutellum with 4, the discoid sensilla about equidistant from the two pairs or somewhat closer to the anterior pair. Frontovertex, pronotum, and mesonotal sclerites reticulate. Scutellum about $\frac{7}{10}$ to $\frac{4}{5}$ median length of mesoscutum. Metanotum (Figure 817) distinctly curved, v-shaped, reticulate except on the sides, usually slightly overlapping the anterior margin of the propodeum; anteromedian apodeme $\frac{3}{4}$ length to fully as long as metanotum.

Propodeum (Figures 817, 818) about $2\frac{1}{2}$ to $3\frac{1}{4}$ times as long as the metanotum (usually less than 3 times), about $\frac{3}{5}$ length of scutellum (0.56–0.64), faintly striated on the sides and along posterior margin, rather broadly reticulate centrally; crenulae 5+6 to 8+8, distinct, rounded, nonoverlapping.

Second abdominal tergite transversely reticulate anteriorly on both sides, less distinctly also antero-centrally; tergites III–VII reticulate on the sides, with several setae in a short, transverse row on each reticulate area; third tergite reticulate also mesad of the lateral setiferous areas, transversely striated across center; tergites IV–V with very faint indications of transverse striation; tergites VI–VII faintly reticulate mesad of the lateral setiferous areas, the reticulation fading toward center; seventh tergite bearing 2–4 setae across center; eighth tergite delicately reticulate, this almost fading centrally, with a transverse row of 4–8 setae between spiracles, usually with an additional, shorter seta immediately mesad of each spiracle; syntergum triangular, faintly reticulate-punctate, bearing 8–13 setae in 1–2 transverse rows. Cerci about equidistant from posterior spiracles and from tip of abdomen, with 2 long setae and 1 short seta.

Ovipositor shaft a little less than $1\frac{1}{2}$ to $1\frac{3}{4}$ times as long as the middle tibia (1.45–1.75); ovipositor sheaths about $\frac{1}{4}$ to $\frac{1}{3}$ length of middle tibia (0.27–0.34).

Forewing (Figure 819) relatively broad, about $2\frac{1}{5}$ to $2\frac{2}{5}$ times as long as wide (2.22–2.43); marginal fringe short, not exceeding $\frac{1}{6}$ width of disk (usually $\frac{1}{10}$ or shorter). Delta area with 34–86 setae in 4–7 rows, these considerably longer and sparser than the setae distad of speculum, clearly distinct from row of setae along posterior margin of wing, widely separated from a few setae in a small group below distal portion of submarginal vein; costal cell with 2–5 setae in a row along central portion, widely separated from a single coarse seta near apex. Submarginal vein bearing 2 coarse setae, the proximal one about $\frac{3}{5}$ to $\frac{3}{4}$ length of the distal, and 15–26 bullae. Marginal vein bearing 9–16 prominent, subequal setae along anterior margin, these somewhat longer than the setae in a row along center of vein (up to $1\frac{1}{3}$ times longer in smallest specimen).

Forewing generally perfectly hyaline, infumated only below junction of submarginal and marginal veins and at base of delta; a fuscous streak along posterior margin of wing, distad of speculum.

General coloration deep yellow with a slight orange tint; thoracic sterna concolorous, immaculate; posterior margin of scutellum narrowly lined with blackish; central portion of posterior margin of propodeum and crenulae noticeably dusky; tegula dusky; a short black streak near base of forewing, below tegula; plates around base of ovipositor faintly dusky. Antennal scape fuscous ventrally, rest of antenna rather uniformly, quite strongly dusky. Tibiae of all legs faintly dusky; apical tarsal segment distinctly blackish (sometimes also one or more of the preceding segments). Wing veins colorless, lined with brownish.

Length 0.87–1.50 mm.

*Male.* Essentially similar to the female in structure, chaetotaxis, sculpture and coloration.

Antennal scape (Figure 820) somewhat wider than in the female, 4 to $4\frac{4}{5}$ times as long as wide, apparently with an indistinct ridge or elevated area ventrally, similar to that of *mytilaspidis*; pedicel up to $1\frac{1}{4}$ times as long as the third segment of the funicle; third funicular segment bearing 1–2 longitudinal sensilla; club truncate in lateral view, $2\frac{1}{7}$ to $2\frac{2}{3}$ times longer than the preceding segment, bearing 3–4 longitudinal sensilla, lacking a cut-off sensory area on the ventral surface. Thorax and propodeum (Figures 821–824) as in the female; propodeal crenulae (Figures 823, 824) 4+4 to 7+7.

Genitalia (Figure 825) $\frac{2}{3}$ to nearly $\frac{4}{5}$ length of middle tibia (0.66–0.78); digital sclerites a little over $\frac{1}{4}$ the combined length of aedeagus and apodemes (0.26–0.30).

Stigmal vein of forewing shown in Figure 826.

Length 0.91–1.25 mm.

Redescribed from a series of 20♀♀ and 18♂♂, reared by C. Benassy from *Leucaspis signoreti* Targioni-Tozzetti on pine, Corsica, 1967.

Graham (1976) reported that the remains of Ratzeburg's collection in Eberswalde, East Germany, included three card-pointed female specimens, which he regarded as syntypes of *Coccobius luteus*. One of them, labeled "Coccus Pini" [= ?*Leucaspis pini* (Hartig)] and "luteus Rtz" in Ratzeburg's handwriting, was designated as lectotype by

Bouček (1964). We received the lectotype and one of the female paralectotypes on loan from the Ratzeburg collection and had them cleared and remounted in Hoyer's medium. Although somewhat smaller and in rather poor shape, these specimens (Figures 827–829) agree in all important characters with the series from Corsica described above and are undoubtedly conspecific. The only differences are evidently related to the smaller size of the types, which have the mesoscutum with 7–8 setae, each parapsis with 1–2, and general coloration seemingly paler, but this may be due to aging.

Howard (1895) suspected that *luteus* belonged to the genus *Ablerus* Howard, and his assumption was adopted by Mercet (1912b). Both authors had apparently not seen the types, and based their conclusions on Ratzeburg's brief and inadequate description. A study of the types leaves no doubt that *luteus* is a valid member of *Aphytis*.

DeBach's (1964) "*Aphytis* Z" is apparently conspecific with *luteus*. At hand are several specimens, reared by P. DeBach from *Leucaspis pini* (Hartig) on *Pinus halpensis*, Athens, Greece (1♀, September 20, 1962; 6♀♀, 4♂♂, October 17, 1962; 9♀♀, 8♂♂, February 3, 1963), mostly damaged and in poor condition. They appear to be conspecific with the type series, although in some of the females the propodeum appears 4 times as long as the metanotum, or even longer (this is not true in other females, and may be an artifact of mounting).

Graham (1976) considered *luteus* as a possible synonym of *A. albidus* (Westwood). We regard *albidus* as synonym of *mytilaspidis* (see p. 467).

**Additional Material.** 1♀, reared from *Leucaspis signoreti* on *Pinus nigra*, Corsica, November 12, 1963; 1♂, reared by M. Kosztarab from *Leucaspis* sp. on *Pinus nigra*, Budapest, Hungary, July 15, 1955; 1♀, reared by G. P. Georghiou from an undetermined host on pine, Atnalassa, Cyprus, February 14, 1968.

**Notes.** The dusky, rounded propodeal crenulae—similar to those of *mytilaspidis*—appear to be a valid, reliable diagnostic character separating *luteus* from *aonidiae*. In *aonidiae*, the crenulae are paler, and distinctly triangular in shape. The slight morphological differences, the biparental mode of reproduction, and the unique habitat, *Leucaspis* spp. on pine, leave little doubt that *luteus* is a distinct, valid species. The presence of unusually large males may be a significant character. The males of *luteus* are usually about as large as the females, in contrast to most other species of *Aphytis* where they are distinctly smaller. Males of *luteus* are as large as or larger than females of many other species.

Mercet (1931, 1932) recorded under the name *A. abnormis* (Howard) a species of *Aphytis*, parasitizing *Leucaspis signoreti* on *Pinus laricio* in Corsica. This identification was questioned by Ferrière (1965), who referred similar material from Corsica to *A. mytilaspidis*. The species recorded by Mercet as *abnormis* was very probably the same as the one recognized here as *luteus*. However, since we have not seen the actual specimens discussed by Mercet, we have not listed his designation in our synonymy. *A. abnormis*, which was obtained from *Lepidosaphes ulmi* (L.) on *Salix* in the United States, is unidentifiable at present (see p. 729).

# RELATED SPECIES

The following 5 species are considered to be closely related to, but not bona fide members of, the **mytilaspidis** group. In certain morphological characteristics they may be regarded as occupying an intermediate position between the **mytilaspidis** and **chrysomphali** groups.

## 54. **Aphytis neuter** Jasnosh and Myartseva

(Figures 830–833)

*Aphytis neuter* Jasnosh and Myartseva, 1971, Izv. Akad. Nauk Turkmen. SSR, Ser. Biol. Nauk, **6**:36–37.

This little-known species from Soviet Central Asia may be recognized by the following combination of characters: Coloration entirely yellow, thoracic sterna immaculate; forewing broad, hyaline; antennal club elongate, over 3 times as long as wide, tapering to a pointed apex; thoracic setae short, pale; propodeum short, crenulae rounded, nonoverlapping. The following redescription is based on a single paratype specimen.

*Female.* Eyes finely setose. Mandibles well developed, apparently with 2 denticles and a dorsal truncation; maxillary and labial palpi not seen. Antennal scape (Figure 830) $4\frac{3}{4}$ times as long as wide, a little longer than the club; pedicel about $1\frac{4}{5}$ times as long as wide, about $1\frac{1}{5}$ times longer than the third segment of the funicle; first funicular segment trapezoidal, $1\frac{1}{2}$ times as wide as long; second segment more nearly symmetrical, somewhat shorter and wider than the first segment, $1\frac{4}{5}$ times as wide as long; third funicular segment nearly $1\frac{1}{2}$ times as long as wide, bearing 2 longitudinal sensilla; club elongate, $3\frac{2}{5}$ times as long as wide, tapering to a pointed apex, about $2\frac{2}{3}$ times longer and somewhat wider than the preceding segment, bearing 6–7 longitudinal sensilla.

Setae on thorax rather short, pale, the largest ones seemingly composed of distinct filaments, barely visible under $\times 60$ magnification; setae on head barely visible under $\times 120$ magnification; those on abdomen invisible even under $\times 120$ magnification. Vertex with 2 pairs of medium-long setae along occipital margin, in addition to numerous shorter setae. Mesoscutum (Figure 831) with 13 setae, the posterior pair and 1 seta at each antero-lateral corner considerably longer and coarser than the others; each parapsis with 2 setae, each axilla with 1 seta; scutellum with 4 setae, the discoid sensilla about equidistant from the two pairs. Frontovertex, pronotum and mesonotal sclerites reticulate. Scutellum about $\frac{3}{4}$ median length of mesoscutum. Metanotum (Figures 831, 832) mildly arcuate, reticulate except on the sides; anteromedian apodeme stout, about as long as median length of metanotum.

Propodeum (Figures 831, 832) short, a little less than 3 times as long as the metanotum, about $\frac{1}{2}$ length of scutellum, without a distinct median salient, reticulate on the sides, rather broadly reticulate on a wide central area; posterior margin distinctly notched medially; crenulae 4+5, rounded, nonoverlapping.

Second abdominal tergite transversely reticulate-striated on the sides, smooth centrally; tergites III–VII reticulate on the sides, bearing a few fine setae in a short transverse row on each reticulate area, with faint indications of transverse striation across center, the latter considerably more pronounced on third tergite; seventh tergite bearing 2 submedian setae; eighth tergite delicately reticulate-punctate, with a transverse row of 4 setae between spiracles and 1 shorter seta immediately mesad of each spiracle; syntergum triangular, faintly reticulate-punctate, bearing 8 setae in a transverse row. Cerci closer to posterior spiracles than to tip of abdomen, with 2 long setae and 1 short seta. Ovipositor shaft about $1\frac{3}{5}$ times as long as the middle tibia (1.62); ovipositor sheaths about $\frac{2}{5}$ length of middle tibia (0.39).

Mid-tibial spur a little shorter than the corresponding basitarsus.

Forewing (Figure 833) broad, $2\frac{1}{3}$ times as long as wide; marginal fringe not exceeding $\frac{1}{6}$ width of disk. Delta area with 60–65 setae in 6–7 rows, these considerably longer and sparser than the setae distad of speculum, rather distinct from row of setae along posterior margin of wing, widely separated from a few dorsal setae below distal half of submarginal vein; setae in costal cell not seen. Submarginal vein bearing 2 coarse setae, the proximal one $\frac{7}{10}$ length of the distal, with a third, shorter seta anterad, and 21–23 bullae. Marginal vein bearing 7 prominent, subequal setae along anterior margin, these about $1\frac{1}{3}$ times longer than the setae in a row along center of vein.

Forewing perfectly hyaline; a faintly fuscous streak along posterior margin, distad of speculum.

General coloration pale yellow, without any dark markings except for a short black streak at base of forewing, below tegula; thoracic sterna immaculate; triangular ovipositor plates faintly dusky. Antennal scape pale, rest of antenna uniformly, faintly dusky. Legs concolorous with body. Wing veins hyaline.

Length 0.95 mm.

*Male*. Unknown.

Redescribed from 1♀ (paratype), reared by E. A. Borovkov from *Quadraspidiotus slavonicus* (Green), Dushanbe, Tadjikistan (Soviet Central Asia), USSR, June 2, 1950; sent to us by Dr. V. A. Yasnosh, Leningrad.

Type series in the collection of the Zoological Institute, USSR Academy of Sciences, Leningrad.

**Notes.** Yasnosh and Myartseva (1971) described *neuter* from a series of 217♀♀, all reared from *Quadraspidiotus slavonicus* (Green) in Soviet Central Asia (Uzbekistan, Tadjikistan, Turkmenia), 1937–1970. They correctly placed this species in an intermediate position between the **mytilaspidis** and **chrysomphali** groups: it has the short propodeum of the former group, the pale setae and elongate antennal club of the latter group. The peculiar, tapering shape of the club appears to set *neuter* apart from all known members of these two groups.

Yasnosh (1972) reported that the pupa and exuvia of *neuter* are entirely black, as in *mytilaspidis* and *moldavicus*. This species therefore appears to be rather closely related to the **mytilaspidis** group. The presence of numerous females but no males in the type series strongly indicates that *neuter* is a uniparental, thelytokous species.

Although the single specimen available to us for study was labeled "*Aphytis slavonicus*

Jasnosh et Myartseva," we have no doubt that it is indeed a paratype of *A. neuter*; "*slavonicus*" must have been an early manuscript name for this species. In fact, that single specimen appears to be quite representative of the species, as can be seen from a comparison with the original description. Yasnosh and Myartseva (1971) described *neuter* as 0.8–1.0 mm long, with 10–13 setae on the mesoscutum, propodeum $\frac{2}{3}$ length of scutellum, forewing $2\frac{1}{2}$ times as long as wide, with 10–12 setae along the marginal vein.

## 55. **Aphytis notialis** De Santis

(Figures 834–842)

*Aphytis notialis* De Santis, 1965, Rev. Soc. Entomol. Argent., **27**:75–77 (1964).

This biparental South American species may be recognized by the following combination of characters: General coloration yellow, thoracic sterna immaculate; forewing large, broad, hyaline; antennal club rather long, thick; mesoscutum bearing numerous (15–19) long, rather coarse setae; propodeum about 4 times as long as the metanotum, crenulae small, rounded; ovipositor sheaths rather long; male antennae lacking a specialized sensory area on the ventral aspect; pupa predominantly black.

*Female.* Eyes finely setose. Mandibles well developed, with 2 denticles and a dorsal truncation; maxillary palpi 2-segmented, labial palpi 1-segmented. Antennal scape (Figures 834, 835) slender, about 5–6 times as long as wide, only slightly longer than the club; pedicel elongate, $1\frac{4}{5}$ to 2 times as long as wide, considerably (up to $1\frac{3}{10}$ times) longer than the third segment of the funicle; first funicular segment somewhat trapezoidal, about $1\frac{1}{5}$ to $1\frac{1}{2}$ times as wide as long; second segment symmetrical, usually shorter and wider than the first segment, $1\frac{1}{2}$ times to fully twice as wide as long; third funicular segment about $1\frac{1}{4}$ to $1\frac{1}{2}$ times as long as wide, bearing 2–3 longitudinal sensilla; club relatively long and thick, about 3 to $3\frac{2}{3}$ times as long as wide (usually over 3 times), $2\frac{1}{2}$ to 3 times longer and somewhat wider than the preceding segment, bearing 5–8 longitudinal sensilla.

Setae on thorax dark, rather coarse but longer than in *mytilaspidis*, readily visible under $\times 30$ magnification; setae on head paler, visible only under $\times 120$ magnification; those on abdomen slender, invisible even under $\times 120$ magnification. Vertex with 2 pairs of long setae along occipital margin, in addition to numerous shorter setae. Mesoscutum (Figures 836, 837) with 15–21 setae, the posterior pair and 1 seta at each antero-lateral corner considerably longer and coarser than the others; each parapsis usually with 3 (rarely 2) short setae, each axilla with 1 longer seta; scutellum with 4, the discoid sensilla considerably closer to the anterior than to the posterior pair. Frontovertex, pronotum and mesonotal sclerites reticulate. Scutellum oval, $\frac{3}{4}$ to over $\frac{4}{5}$ median length of mesoscutum. Metanotum (Figures 836–838) slightly arcuate, reticulate except on the sides; anteromedian apodeme $\frac{2}{3}$ to full length of metanotum.

Propodeum (Figures 837, 838) $3\frac{3}{4}$ to $4\frac{2}{5}$ times as long as the metanotum, about $\frac{2}{3}$ to $\frac{4}{5}$ length of scutellum, reticulate on the sides and on a distinct median salient, faintly transversely striated along posterior margin; crenulae 5+5 to 10+10, small, distinct, rounded, nonoverlapping.

Second abdominal tergite transversely striated anteriorly on the sides, smooth centrally; tergites III–VII reticulate on the sides, bearing a few fine setae in an irregular transverse row on each reticulate area; third tergite reticulate also mesad of the lateral setiferous areas, smooth centrally; tergites IV–VI with faint indications of transverse striation marking posterior borders; seventh tergite transversely reticulate-striated across center, bearing a pair of submedian setae; eighth tergite delicately reticulate, with a transverse row of 4–6 setae between spiracles and 1 seta immediately below each spiracle; syntergum triangular, faintly reticulate anteriorly, punctate, bearing 6–7 setae in a transverse row. Cerci about equidistant from posterior spiracles and from tip of abdomen, with 2 long setae and 1 short seta. Ovipositor shaft about $1\frac{1}{2}$ to $1\frac{3}{5}$ times as long as the middle tibia (1.46–1.61); ovipositor sheaths about $\frac{2}{5}$ length of middle tibia (0.36–0.42).

Mid-tibial spur long, $\frac{4}{5}$ to nearly full length of the corresponding basitarsus.

Forewing (Figure 839) large, broad, about $2\frac{1}{4}$ to $2\frac{1}{3}$ times as long as wide; marginal fringe not exceeding $\frac{1}{6}$ width of disk. Delta area rather sparsely setose, with 32–59 setae in 4–5 rows, these considerably longer and sparser than the setae distad of speculum, not clearly separated from row of setae along posterior margin of wing, widely separated from 4–5 setae below distal portion of submarginal vein; costal cell with a row of short setae along proximal two thirds or so, and 1 coarse seta near apex. Submarginal vein bearing 2 coarse setae, the proximal one about $\frac{3}{5}$ to $\frac{4}{5}$ length of the distal (sometimes with 3 setae), and 16–23 bullae. Marginal vein bearing 8–14 (usually 10 or more) prominent, rather short, subequal setae along anterior margin, these only slightly longer than the setae in a row along center of vein.

Forewing nearly perfectly hyaline, very faintly infumated below junction of submarginal and marginal veins and at base of delta; a fuscous streak along posterior margin of wing, distad of speculum.

General coloration entirely yellow, legs concolorous, thoracic sterna immaculate; posterior margin of scutellum faintly lined with fuscous; a short, blackish streak at base of forewing, below tegula; plates around base of ovipositor faintly dusky. Antennal pedicel, funicle and club faintly, uniformly dusky. Wing veins pale.

Length 0.81–1.22 mm.

*Male*. Essentially similar to the female in structure, chaetotaxis, sculpture and coloration.

Antennae (Figure 840) as in the female; club bearing 3 longitudinal sensilla, apparently without a cut-off sensory area on the ventral surface.

Mesoscutum with 14–19 setae. Propodeum (Figures 841, 842) $3\frac{3}{5}$ to a little over 4 times as long as the metanotum; crenulae 5 + 5 to 9 + 10, as in the female.

Genitalia about $\frac{2}{3}$ length of middle tibia (0.68–0.70); digital sclerites about $\frac{3}{10}$ the combined length of aedeagus and apodemes (0.29–0.32).

Length 0.66–0.91 mm.

Redescribed from 3♀♀ and 4♂♂ (♀ holotype, and paratypes), reared by I. S. de Crouzel from the oleander scale, *Aspidiotus nerii* Bouché [= *A. hederae* (Vallot)], on *Aucuba japonica*, Llavallol, Castelar (Prov. de Buenos Aires), Argentina, December 1963; 4♀♀, 1♂, same data (apparently part of type series but not labeled as such); 1♂ (allotype), reared by G. Kuschel from an unrecorded host, Santiago, Chile, February 17, 1953; 4♀♀, reared by S. Rojas from *A. nerii* on ivy, La Cruz, Argentina, July 22, 1966; 3♀♀, 1♂, reared from *A. nerii*, Pocochay and Olmué, Chile, August–November 1969.

**Notes:** The type series of *notialis* was received for study from Prof. L. De Santis, of the Museo de La Plata, Argentina; most of the types, except 1♀ and 1♂ paratypes, are not sufficiently cleared. In his original description, De Santis (1965) listed a ♂ specimen from Argentina as allotype, whereas the specimen from Chile was listed as paratype. However, the slide bearing the specimen from Chile is actually labeled allotype.

De Santis (op. cit.) considered *notialis* to be closely related to *opuntiae*. This was based on the common misinterpretation of *opuntiae* as a species characterized by numerous mesoscutal setae (see p. 474). Of the species of the **mytilaspidis** group, *notialis* appears to be most closely related to *luteus*, with which species it shares the immaculate sterna, rounded crenulae and broad wings. It may be separated from *luteus* by the somewhat longer propodeum, antennal club and ovipositor sheaths, and by the more numerous, somewhat more slender mesoscutal setae. The pupa of *notialis* is largely or entirely black. Although this species is evidently rather closely related to the **mytilaspidis** group, it cannot be considered a bona fide member of that group. *A. notialis* appears to be rather similar in general habitus to *immaculatus*, which is considered here a member of the **chrysomphali** group. However, it differs from *immaculatus* in the broader forewing, more numerous mesoscutal setae, longer and thicker antennal club, shorter propodeum, rounded crenulae and longer ovipositor sheaths, as well as in the position of the discoid sensilla on the scutellum.

**Potential Value for Biological Control.** This little-known species should be considered as an additional, potentially valuable natural enemy of the oleander scale.

## 56. **Aphytis taylori** Quednau

(Figures 843–852)

*Aphytis taylori* Quednau, 1964, J. Entomol. Soc. S. Afr., **27**: 104–106.

This South African species may be recognized by the yellow coloration, dusky thoracic sterna, rather dark thoracic setae, short antennal club, moderately short propodeum with low, flat crenulae forming a wavy ledge, and long ovipositor and sheaths; male antennal club with a partly cut-off sensory area on the ventral aspect.

*Female.* Eyes finely setose. Mandibles well developed, with 2 denticles and a dorsal truncation; maxillary palpi (Figure 843) 2-segmented, labial palpi 1-segmented. Antennal scape (Figure 844) slender, about $5\frac{1}{2}$ to $6\frac{1}{2}$ times as long as wide, a little longer than the club; pedicel about $1\frac{1}{2}$ to over $1\frac{4}{5}$ times as long as wide, $1\frac{1}{5}$ to nearly $1\frac{1}{2}$ times longer than the third segment of the funicle; first funicular segment very slightly trapezoidal, usually about $1\frac{1}{2}$ to $1\frac{2}{3}$ times as wide as long; second segment symmetrical, about as wide as the first segment but considerably shorter, 2 to $2\frac{1}{3}$ times as wide as long; third funicular segment relatively short, somewhat longer than wide, bearing 1–2 longitudinal sensilla; club $2\frac{2}{3}$ to 3 times as long as wide, about 3 times longer and considerably wider than the preceding segment, bearing 4–5 longitudinal sensilla.

Setae on thorax rather dark and coarse, more so than in *chrysomphali* but less than in *mytilaspidis*, visible under $\times 30$ magnification; setae on head paler, visible under $\times$ 120 magnification; those on sides of abdomen practically invisible even under $\times$ 120 magnification. Vertex with 2 pairs of long setae along occipital margin, in addition to numerous shorter setae. Mesoscutum (Figure 845) with 9–11 setae, each parapsis with 2, each axilla with 1 seta; scutellum with 4, the discoid sensilla closer to the anterior than to the posterior pair. Frontovertex, pronotum and mesonotal sclerites reticulate. Scutellum about $\frac{2}{3}$ to $\frac{4}{5}$ median length of mesoscutum. Metanotum reticulate except on the sides; anteromedian apodeme about as long as median length of metanotum.

Propodeum (Figures 845–847) usually about 4 times as long as the metanotum ($3\frac{3}{5}$ to nearly 5 times), about $\frac{2}{3}$ to over $\frac{4}{5}$ length of scutellum, faintly reticulate on the sides and centrally, the cells of the central reticulation broad; crenulae 4+4 to 6+7, minute, very low and wide, forming a wavy ledge on the posterior margin of the propodeum.

Second abdominal tergite (Figures 845, 847) with faint traces of transverse striation; tergites III–VII reticulate on the sides, with 2–5 setae in a short, transverse row on each reticulate area; third and seventh tergite with faint indications of transverse striation across center, the latter tergite bearing 2 setae submedially, between the lateral setiferous areas; eighth tergite (Figure 848) very faintly reticulate across, with a transverse row of 4 long setae between spiracles, often with an additional short seta just mesad of each spiracle; syntergum triangular, faintly punctate, bearing 5–7 setae in a transverse row. Cerci (Figure 848) about equidistant from posterior spiracles and tip of abdomen, with 2 long setae and 1 short seta. Ovipositor shaft relatively long, about twice as long as the middle tibia (1.84–2.15); ovipositor sheaths (Figure 849) long and narrow, about $\frac{2}{5}$ to $\frac{1}{2}$ length of middle tibia (0.42–0.50).

Mid-tibial spur somewhat shorter than the corresponding basitarsus.

Forewing (Figure 850) $2\frac{1}{3}$ to nearly $2\frac{1}{2}$ times as long as wide; marginal fringe not exceeding $\frac{1}{4}$ width of disk. Delta area with 31–71 setae in 4–7 rows (these numbers being in correlation with the size of specimens), these much longer and sparser than the setae distad of speculum, quite distinct from the row of setae along posterior margin of wing, widely separated from a few setae in a small group below distal part of submarginal vein; costal cell with several fine setae in a row along proximal half or three fifths, and 1 coarse seta near apex. Submarginal vein bearing 2 coarse setae, the proximal one about $\frac{3}{5}$ to $\frac{3}{4}$ length of the distal, and 15–22 bullae. Marginal vein bearing 7–12 prominent, subequal setae along anterior margin, these distinctly longer than the setae in a row along center of vein.

Forewing generally hyaline, distinctly infumated below distal portion of submarginal vein and at base of delta; a dusky streak along posterior margin of wing, distad of speculum.

General coloration yellow; central portion of posterior margin of scutellum narrowly lined with blackish; tegulae dusky; a short black streak near base of forewing, below tegula; thoracic sterna strongly infuscated. Antennal scape pale, pedicel and first funicular segment rather faintly dusky, rest of antenna somewhat more strongly so. Tibiae and tarsi stronger yellow but not distinctly dusky. Wing veins pale, lined with brownish.

Length 0.75–1.15 mm.

*Male*. Very similar to the female in structure, chaetotaxis, sculpture and coloration.

Antennal scape (Figure 851) pale, rest of antenna uniformly, faintly dusky; club nearly to fully 3 times as long as wide, $2\frac{1}{2}$ to nearly 3 times longer than the preceding segment, bearing 2–3 longitudinal sensilla; a partially cut-off sensory area, bearing short setae, is present on the ventral surface of the club (Figure 852). Propodeum $3\frac{3}{4}$ to $4\frac{1}{2}$ times as long as the metanotum (only $3\frac{1}{5}$ times in a minute specimen); crenulae 3+3 to 4+5, as in the female.

Genitalia about $\frac{2}{3}$ to nearly $\frac{4}{5}$ length of middle tibia; digital sclerites about $\frac{1}{4}$ the combined length of aedeagus and apodemes (0.25–0.27).

Length 0.62–0.84 mm.

Redescribed from 22♀♀ and 5♂♂, syntypes, reared by J. S. Taylor from *Aspidiotus capensis* Newstead on *Cycas* sp., Port Elizabeth, South Africa, February 1962; all mounted on 3 slides. The specimens are in rather poor shape; most of them lack complete antennae, some are partially distorted; the sculpture is hardly visible, and this may be due to the method of clearing or to the mounting medium; some of the above measurements may therefore not be accurate.

Type series in the collection of the Plant Protection Research Institute, Pretoria, South Africa.

**Notes.** *A. taylori* appears to occupy an intermediate position between the **mytilaspidis** and **chrysomphali** groups. The dark thoracic setae and the short antennal club would place it in the **mytilaspidis** group; however, the cut-off sensory area on the club of the male is characteristic of the **lingnanensis** group and several members of the **chrysomphali** group. The length of the propodeum may be regarded as intermediate between the two groups, whereas the peculiar crenulae appear to be similar to those of *testaceus*, which is an aberrant member of the **proclia** group.

Quednau (1964b) considered *taylori* to differ from *mytilaspidis* in having pigmented thoracic sterna. This, of course, was based on his misinterpretation of *mytilaspidis* as a species possessing unpigmented sterna. Quednau, however, did note the peculiar crenulae and long ovipositor sheaths of *taylori*, which are the main distinguishing characters of this species.

*A. taylori* appears to be a biparental species. It was originally described from a series of 40♀♀ and 7♂♂.

## 57. **Aphytis rolaspidis** DeBach and Rosen

(Figures 853–864)

*Aphytis flavus* Quednau (nec Nees), 1964, J. Entomol. Soc. S. Afr., **27**: 102, 104, 105; preoccupied.
*Aphytis rolaspidis* DeBach and Rosen, 1976, Ann. Entomol. Soc. Amer., **69**: 544.

This South African species may be recognized by the entirely yellow coloration, immaculate thoracic sterna, rather numerous, coarse, dark thoracic setae, short, stout antennal club, rather long ovipositor, broad forewing, long propodeum and few irregular, indistinct crenulae.

*Female*. Eyes finely setose. Mandibles well developed; maxillary palpi (Figure 853) 2-segmented, labial palpi 1-segmented. Antennal scape (Figure 854) about $4\frac{1}{5}$ to $4\frac{4}{5}$ times as long as wide, a little longer than the club; pedicel a little over $1\frac{1}{2}$ to $1\frac{2}{3}$ times as long as wide, distinctly longer ($1\frac{1}{8}$ to $1\frac{2}{5}$ times) than the third segment of the funicle; first funicular segment trapezoidal, $1\frac{1}{4}$ to $1\frac{2}{5}$ times as wide as long; second segment nearly symmetrical, considerably shorter and usually wider than the first segment, $1\frac{3}{5}$ to $2\frac{1}{4}$ times as wide as long; third funicular segment as wide as long to $1\frac{1}{5}$ times as long as wide, bearing 2–3 longitudinal sensilla; club $2\frac{3}{4}$ to nearly 3 times as long as wide, $2\frac{1}{4}$ to nearly 3 times longer and distinctly wider than the preceding segment, bearing 4–6 longitudinal sensilla.

Setae on thorax rather coarse, dark (less so and somewhat longer than in *mytilaspidis*), readily visible under $\times 30$ magnification; setae on head barely visible under $\times 60$ magnification; those on abdomen paler, slender, invisible even under $\times 120$ magnification. Vertex with 2 pairs of long setae along occipital margin, in addition to numerous shorter setae. Mesoscutum (Figure 855) with 12–15 setae, including 4 submedian pairs; each parapsis with 3 (rarely 4) shorter setae, each axilla with 1 seta; scutellum with 4, the discoid sensilla about equidistant from the two pairs. Frontovertex, pronotum, and mesonotal sclerites reticulate. Scutellum about $\frac{7}{10}$ to $\frac{4}{5}$ median length of mesoscutum. Metanotum (Figures 855, 856) rather short, reticulate except on the sides; anteromedian apodeme distinctly ($1\frac{1}{7}$ to $1\frac{2}{5}$ times) longer than median length of metanotum.

Propodeum (Figures 855–857) rather long, with a distinct median salient, $4\frac{2}{3}$ to $6\frac{1}{3}$ times as long as the metanotum, about $\frac{3}{5}$ to $\frac{7}{10}$ length of scutellum, faintly reticulate on the sides, reticulate with rather elongate cells on a trapezoidal central portion; crenulae low, very indistinct, irregular, apparently 3 to 5 on each side of median salient, usually minute but sometimes apparently larger (Figure 857) and crowded.

Second abdominal tergite with faint transverse striation anteriorly on both sides and posteriorly across center; tergites III–VI (Figure 858) reticulate on the sides, with 2–3 setae in a short transverse row on each reticulate area; third tergite also reticulate mesad of setiferous areas, transversely striated across center; tergites IV–VI transversely striate-punctate across center; seventh tergite bearing 2–6 setae on each side and 2 submedian setae, reticulate on setiferous areas, transversely striated across center between the submedian setae; eighth tergite delicately reticulate-punctate, this fading centrally, bearing 5–7 setae in a transverse irregular row between spiracles; syntergum (Figure 858) triangular, with a somewhat elongate cauda, distinctly reticulate-punctate, bearing 7–11 setae in 1–2 transverse rows. Cerci somewhat closer to posterior spiracles than to tip of abdomen, with 2 long setae and 1 short seta. Ovipositor rather long, the shaft $1\frac{3}{4}$ to $1\frac{4}{5}$ times as long as the middle tibia (1.76–1.80); sheaths rather long and narrow, a little over $\frac{2}{5}$ length of middle tibia (0.42–0.45).

Mid-tibial spur somewhat shorter than the corresponding basitarsus.

Forewing (Figure 859) relatively broad, $2\frac{3}{10}$ to a little over $2\frac{2}{5}$ times as long as wide; marginal fringe not exceeding $\frac{1}{5}$ width of disk (usually $\frac{1}{7}$–$\frac{1}{8}$). Delta area with 30–46 setae in 4–6 rows, these considerably longer and sparser than the setae distad of speculum, quite distinct from row of setae along posterior margin of wing, widely separated from a few setae in a small group below distal portion of submarginal vein; costal cell with 3–5 fine setae in a row along proximal two thirds, these widely separated from a single coarse seta near apex. Submarginal vein bearing 2 coarse setae, the proximal one appearing as long as, or considerably shorter than, the distal seta, and 14–19 bullae. Marginal vein bearing 8–13 prominent, subequal setae along anterior margin, these somewhat (up to $1\frac{1}{4}$ times) longer than the setae in a row along center of vein.

Forewing perfectly hyaline, faintly infumated below junction of submarginal and marginal veins (under phase contrast, the basal half of the wing appears very slightly infumated).

General coloration pale yellow, thoracic sterna immaculate; posterior margin of scutellum narrowly lined with blackish; a short black streak at base of forewing, below tegula. Antennal scape pale, rest of antenna very faintly, uniformly dusky. Tibiae and tarsi of all legs not distinctly dusky, appearing darker yellow.

Length 0.95–1.18 mm.

*Male.* Very similar to the female in structure, chaetotaxis, sculpture and coloration.

Antennal scape (Figure 860) wider, $3\frac{3}{5}$ times as long as wide, about $1\frac{1}{4}$ times as long as the club; club about $2\frac{7}{10}$ times as long as wide, bearing 2 longitudinal sensilla, without a cut-off sensory area on the ventral surface.

Mesoscutum (Figure 861) with 13 setae, each parapsis with 3; propodeum (Figures 861, 862) about $4\frac{3}{4}$ times as long as the metanotum, crenulae as in the female.

Genitalia (Figure 863) $\frac{7}{10}$ length of middle tibia; digital sclerites nearly $\frac{2}{5}$ the combined length of aedeagus and apodemes (0.38).

Forewing (Figure 864) essentially as in the female; delta with 28 setae.

Length 0.91 mm.

Redescribed from the entire syntype series: 5♀♀ and 1♂, reared by J. Munting from *Rolaspis chaetachmae* (Brain) on *Chaetachme aristata*, Durban, Natal, South Africa, November 1962. All cleared and mounted on one slide. A female specimen (2nd row, 3rd from right) is hereby designated lectotype. The single male is designated allotype; the other female specimens are paralectotypes.

Type series in the collection of the Plant Protection Research Institute, Pretoria, South Africa.

**Notes.** Two additional males from the same lot, determined by Quednau as syntypes and mounted on the same slide, apparently belong to another, probably undescribed, species. These males are superficially very similar to the syntypes of *rolaspidis*, but differ in having a very long propodeum, 9 to $9\frac{1}{5}$ times as long as the metanotum, with minute but very distinct, regular crenulae (Figures 865, 866). Additional material is required before the specific identity of these males can be determined. A mix-up might have occurred since at least two additional species, *A. funicularis*, and the type of *A. setosus*, were reared together with the original material of *rolaspidis* (Quednau, 1964b).

Although the status of *Eulophus flavus* Nees and *Trichogramma flavus* Ashmead is not clear, there is a strong possibility that at least one of these species was an *Aphytis* (see p. 732). It was therefore advisable to regard the name *flavus* as preoccupied in the genus *Aphytis*, and to consider *A. flavus* Quednau as a junior homonym. Accordingly, the species was renamed *rolaspidis* (DeBach and Rosen, 1976b).

*A. rolaspidis* appears to occupy an intermediate position between the **mytilaspidis** and **chrysomphali** groups; it has the coarse and short antennal club of the former, the long propodeum of the latter group. It appears to be rather closely related to *immaculatus*, but may be readily separated from this species by the darker thoracic setae, the indistinct, low, irregular propodeal crenulae, the longer ovipositor and the broader forewing. In fact, the darker thoracic setae are the main reason for not assigning *rolaspidis* to the **chrysomphali** group.

Because of the few specimens available (5♀♀, 1♂) it is uncertain whether this is a biparental species. However, with a uniparental species it would be very unusual to obtain 1♂ out of a total of 6 reared specimens.

## 58. **Aphytis capensis** DeBach and Rosen

(Figures 867–878)

*Aphytis capensis* DeBach and Rosen, 1976, Ann. Entomol. Soc. Amer., **69**: 544.

This large, biparental South African species appears to be very closely related to *rolaspidis*. It is similar to *rolaspidis* in the general yellow coloration, broad forewing, rather numerous dark thoracic setae, relatively short antennal club and long propodeum, but may be readily recognized by the short ovipositor and sheaths, minute but distinct propodeal crenulae, faintly dusky thoracic sterna and more numerous longitudinal sensilla on the antennae.

*Female*. Eyes finely setose. Mandibles (Figure 867) well developed, with 2 denticles and a dorsal truncation; maxillary palpi 2-segmented, labial palpi 1-segmented. Antennal scape (Figures 868, 869) slender, about $5\frac{1}{2}$ to $6\frac{1}{3}$ times as long as wide, a little longer than the club; pedicel rather slender, $1\frac{3}{4}$ times to fully twice as long as wide, a little longer than the third segment of the funicle; first funicular segment trapezoidal, ventral aspect longer than the dorsal, $1\frac{1}{3}$ to $1\frac{1}{2}$ times as wide as long; second segment symmetrical, nearly rectangular, considerably shorter and as wide as or slightly wider than the first segment, about 2 to $2\frac{1}{4}$ times as wide as long; third funicular segment $1\frac{1}{4}$ to $1\frac{2}{3}$ times as long as wide, bearing 4–6 longitudinal sensilla; club $2\frac{3}{4}$ to $3\frac{1}{6}$ times as long as wide, $2\frac{1}{4}$ to $2\frac{3}{5}$ times longer than the preceding segment, bearing 8–10 longitudinal sensilla.

Setae on thorax about as coarse and dark as in *rolaspidis*; somewhat more elongate, paler and more slender than in *mytilaspidis* but readily visible under ×30 magnification; setae on head paler, visible only under ×120 magnification, those on abdomen invisible even at ×120 magnification. Vertex with 2 pairs of long setae along occipital margin, in addition to numerous shorter setae. Mesoscutum (Figure 870) with 12–16 setae, the posterior pair and one seta at each antero-lateral corner considerably longer and coarser than the others; each parapsis with 3 (rarely 4) setae, each axilla with 1 or 2; scutellum with 4, the discoid sensilla about equidistant from the two pairs. Fronto-vertex, pronotum and mesonotal sclerites reticulate. Scutellum oval, about $\frac{3}{4}$ to $\frac{4}{5}$ median length of mesoscutum. Metanotum (Figure 871) somewhat arcuate, transversely striated anteriorly, reticulate centrally, smooth on the sides; anteromedian apodeme usually considerably longer than median length of metanotum (up to $1\frac{2}{5}$ times, but only $\frac{4}{5}$ length in one specimen).

Propodeum (Figure 871) rather long, about 4–7 times (usually 5–6 times) as long as the metanotum, $\frac{3}{4}$ to $\frac{4}{5}$ length of scutellum, posterior margin nearly straight, without a prominent median salient, reticulate centrally and on the side, faintly transversely striated along posterior margin; crenulae 4+4 to 6+7, minute, rounded or somewhat elongate, distinct, nonoverlapping.

Second abdominal tergite transversely striated anteriorly on both sides, more faintly so posteriorly, smooth centrally; tergites III–VII reticulate on the sides, with a few fine setae in 1–2 short transverse rows on each reticulate area; third tergite transversely reticulate also mesad of the lateral setiferous areas, the reticulation fading centrally, transversely striated along posterior margin; tergites IV–VI with faint indications of transverse striation centrally; seventh tergite very faintly reticulate across, bearing 2 submedian setae between the lateral rows; eighth tergite very faintly reticulate, with a transverse row of 4 setae between spiracles and 1 additional seta immediately below each spiracle; syntergum (Figure 872) triangular, faintly punctate, bearing 6–10 setae in a transverse row. Cerci about equidistant from posterior spiracles and tip of abdomen, with 2 long setae and 1 short seta. Ovipositor short, the shaft only about $1\frac{1}{4}$ to $1\frac{1}{3}$ times as long as the middle tibia (1.25–1.37), the sheaths about $\frac{1}{4}$ to $\frac{3}{10}$ length of middle tibia (0.23–0.29).

Mid-tibial spur nearly as long as the corresponding basitarsus.

Forewing (Figure 873) rather broad, about $2\frac{1}{4}$ to $2\frac{1}{2}$ times as long as wide, marginal fringe not exceeding $\frac{1}{7}$ width of disk. Delta area with 45–66 setae in 5–7 rows, these considerably longer and sparser than the setae distad of speculum, clearly separated from row of setae along posterior margin of wing, widely separated from a small group of 2–5 setae below distal portion of submarginal vein; costal cell with a row of fine setae along proximal three fifths or so, bare at the apex but with 1 coarse seta on the vein. Submarginal vein bearing 2 coarse setae, the proximal one about $\frac{3}{4}$ to $\frac{4}{5}$ length of the distal, and 17–22 bullae. Marginal vein bearing 10–14 prominent, subequal setae along anterior margin, these up to $1\frac{2}{5}$ times as long as the setae in a row along center of vein.

Forewing hyaline, proximal half very faintly infumated, more distinctly so below distal portion of submarginal vein and at base of delta; a fuscous streak along posterior margin of wing, distad of speculum.

General coloration yellow; posterior margin of scutellum conspicuously lined with black centrally; tegula faintly dusky; a short black streak at base of forewing, below tegula; sides of abdominal tergites very faintly dusky. Pro- and mesosternum faintly dusky, metasternum more distinctly so; plates around base of ovipositor faintly dusky. Antennal scape faintly infuscated, rest of antenna more distinctly, uniformly fuscous. Legs concolorous with body, except fore tibia and tarsus which are distinctly fuscous, middle tibia and tarsus more faintly so. Wing veins hyaline, lined with brownish.

Length 1.22–1.30 mm.

*Male*. Very similar to the female in structure, chaetotaxis, sculpture and coloration, differing slightly in the structure of the antennae.

Antennal scape (Figures 874, 875) wider than in the female, $3\frac{1}{2}$ to $4\frac{1}{5}$ times as long as wide, up to $1\frac{1}{4}$ times longer than the club; pedicel $1\frac{2}{3}$ to nearly twice as long as wide, considerably ($1\frac{1}{3}$ to $1\frac{2}{5}$ times) longer than the third segment of the funicle; third funicular segment up to $1\frac{1}{3}$ times as long as wide, bearing 2–3 longitudinal sensilla; club $2\frac{3}{5}$ to 3 times as long as wide, about $2\frac{1}{2}$ to $2\frac{4}{5}$ times longer and somewhat wider than the preceding segment, bearing 3–5 longitudinal sensilla, lacking a specialized sensory area on the ventral surface.

Mesoscutum with 14–17 setae. Propodeum (Figure 876) $4\frac{2}{3}$ to 5 times as long as the metanotum, $\frac{2}{3}$ to $\frac{3}{4}$ length of scutellum; crenulae 3+3 to 4+4, as in the female.

Genitalia (Figure 877) about $\frac{3}{5}$ to $\frac{3}{4}$ length of middle tibia (0.60–0.77); digital sclerites about $\frac{1}{4}$ to $\frac{1}{3}$ the combined length of aedeagus and apodemes (0.26–0.33).

Forewing (Figure 878) as in the female, $2\frac{1}{4}$ to $2\frac{1}{3}$ times as long as wide, marginal fringe not exceeding $\frac{1}{5}$ width of disk; delta with 31–45 setae; submarginal vein bearing 15–21 bullae; marginal vein bearing 9–13 prominent, subequal setae along anterior margin.

Length 0.80–1.19 mm.

Described from 15♀♀ and 13♂♂ (♀ holotype, ♂ allotype, and paratypes), reared by E. W. Rust from *Dentachionaspis margaritae* (Brain), Camp's Bay, Cape Province, South Africa, April 1923–July 1925. Holotype reared July 13, 1925, cleared and mounted in Canada balsam; allotype reared June 10, 1925, remounted in Hoyer's medium from an uncleared balsam slide.

Type series in the collection of the Division of Biological Control, University of California, Riverside.

**Additional Material:** 3♀♀, 2♂♂, reared by H. P. Insley from an undetermined armored scale insect on *Carissa bispinosa*, St. Winifred, South Africa, November 1970; 12♀♀, 4♂♂, reared by H. P. Insley from an undetermined armored scale insect, Oribi Gorge, South Africa, August 1972.

**Notes.** Like *taylori* and *rolaspidis*, *A. capensis* appears to occupy an intermediate position between the **mytilaspidis** and **chrysomphali** groups, but may be more closely related to the latter group, with which it shares the relatively long propodeum. It may be readily separated from both *taylori* and *rolaspidis* by the considerably shorter ovipositor and sheaths, as well as by the minute, distinct propodeal crenulae. The thoracic sterna of *capensis* are decidedly paler than in *taylori*, and the male of *capensis* also differs from the male of *taylori* in the absence of a specialized sensory area on the ventral surface of the antennal club.

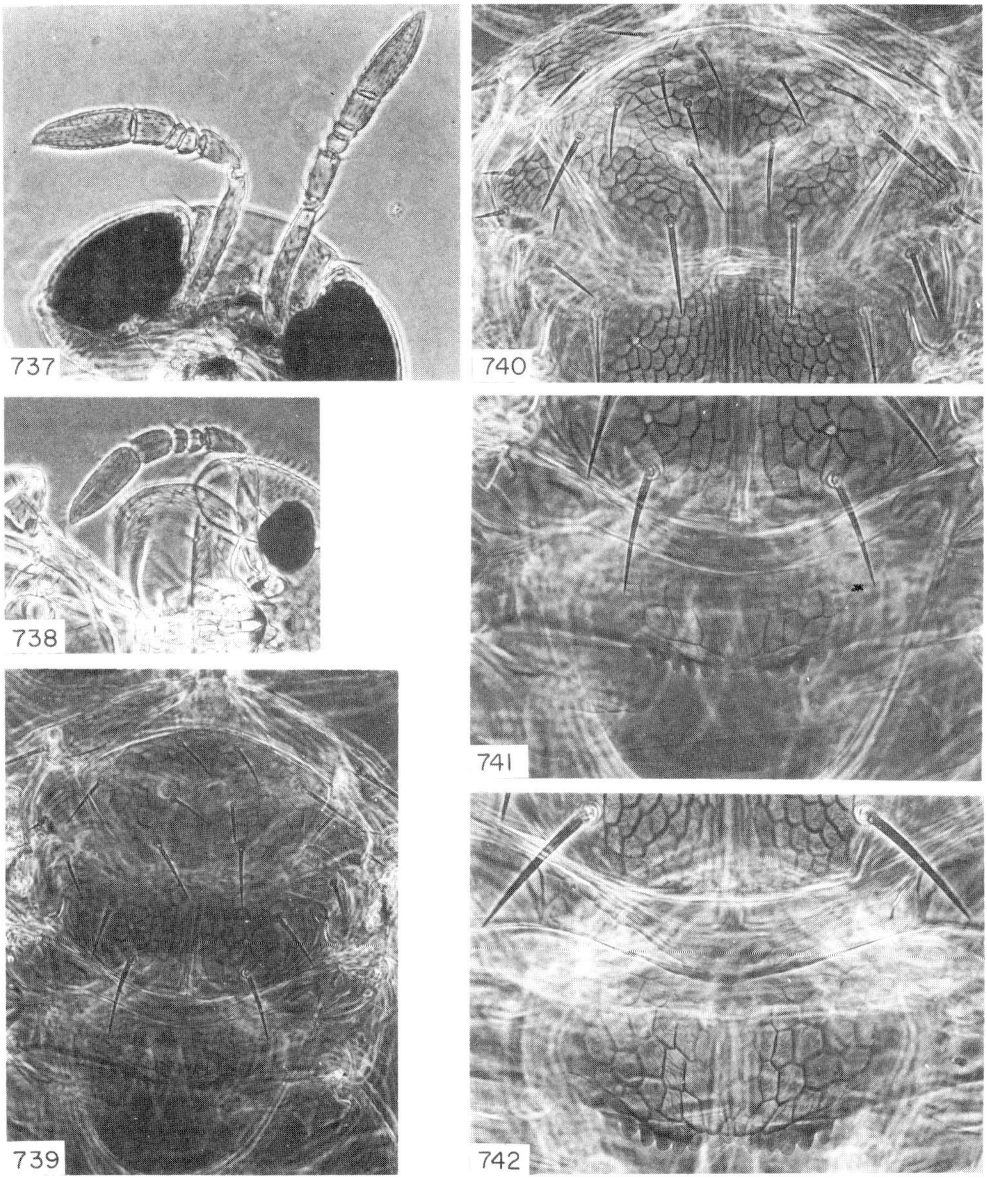


Figures 737–742. *Aphytis mytilaspidis* (Le Baron), ♀ (neotype series)
737, 738. Antennae. 739. Thorax and propodeum (small specimen, few mesoscutal setae). 740. Mesoscutum (large specimen, more numerous setae). 741, 742. Metanotum, propodeum and crenulae.

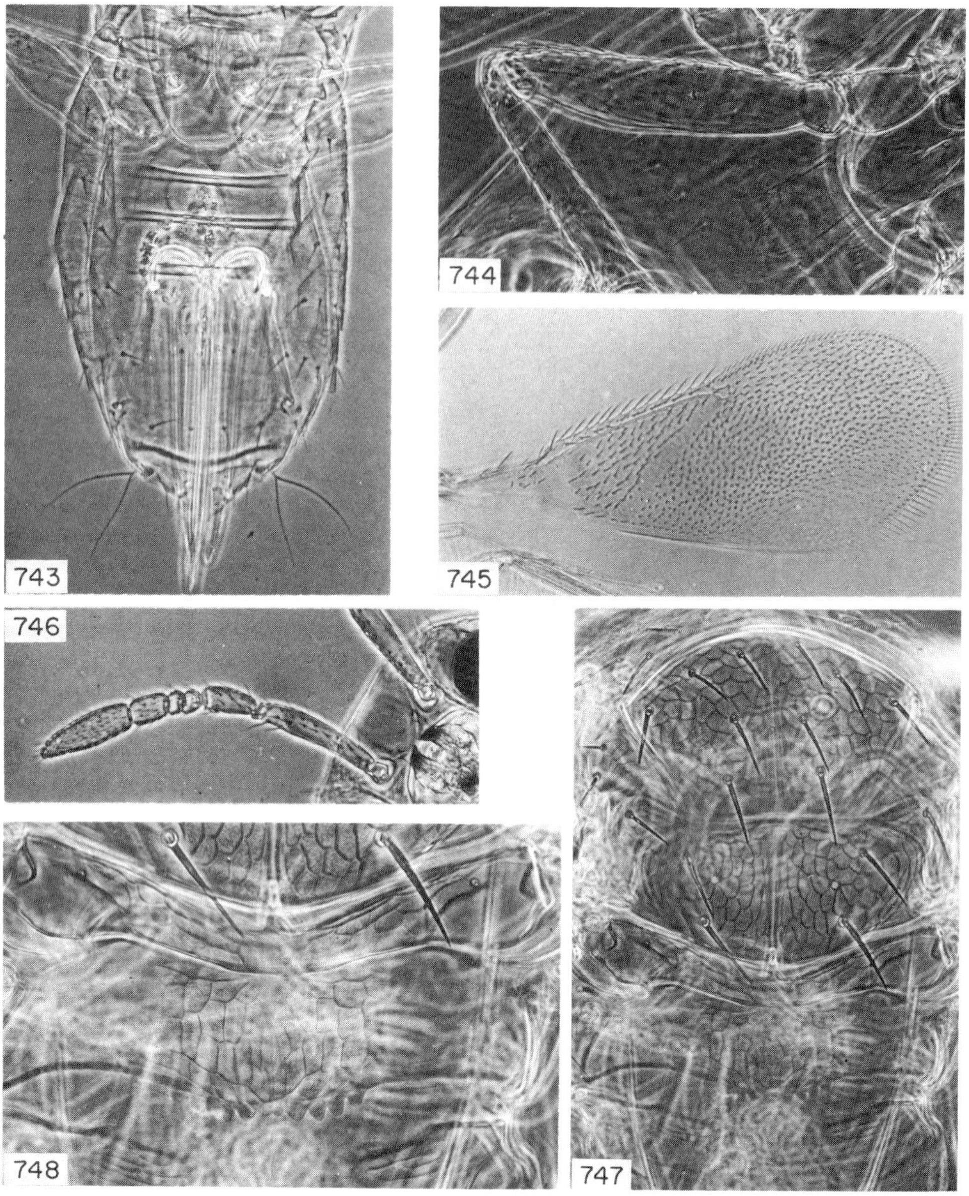


Figures 743–748. *Aphytis mytilaspidis* (Le Baron) (neotype series)
743. ♀: Abdomen. 744. ♀: Middle femur, showing femoral duct. 745. ♀: Forewing. 746. ♂: Antenna. 747. ♂: Thorax and propodeum. 748. ♂: Metanotum, propodeum and crenulae.

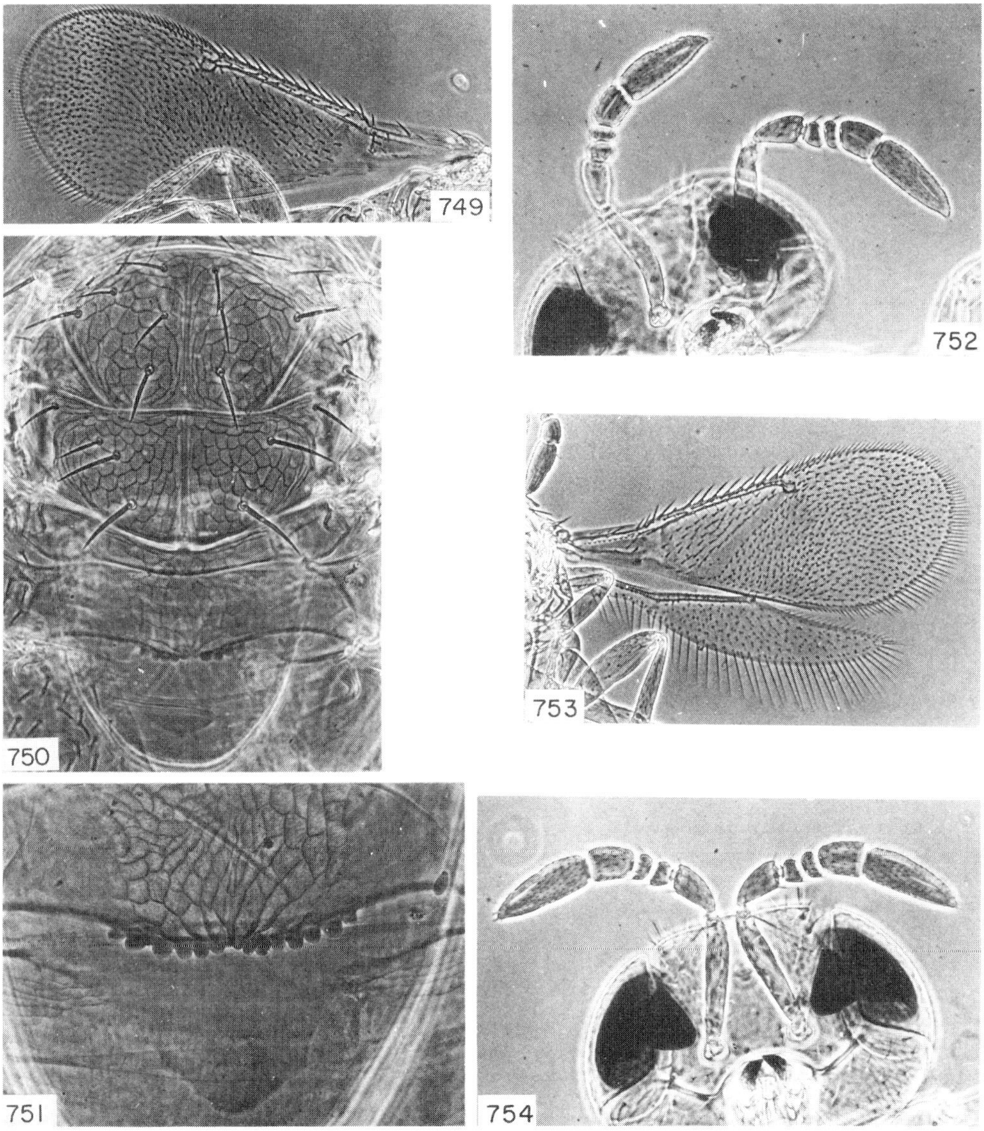


Figures 749–754. *Aphytis mytilaspidis* (Le Baron)
749. ♂: Forewing (neotype series). 750, 751. Uniparental form "3a", ♀: 750. Thorax, propodeum and base of gaster; note dusky spot on second abdominal tergite; this specimen has an abnormal seta arising from a scutellar sensillum. 751. Propodeal crenulae and second abdominal tergite. 752, 753. Uniparental form "3b", ♀: 752. Antennae. 753. Wings. 754. Uniparental form "3c", ♀: Antennae.

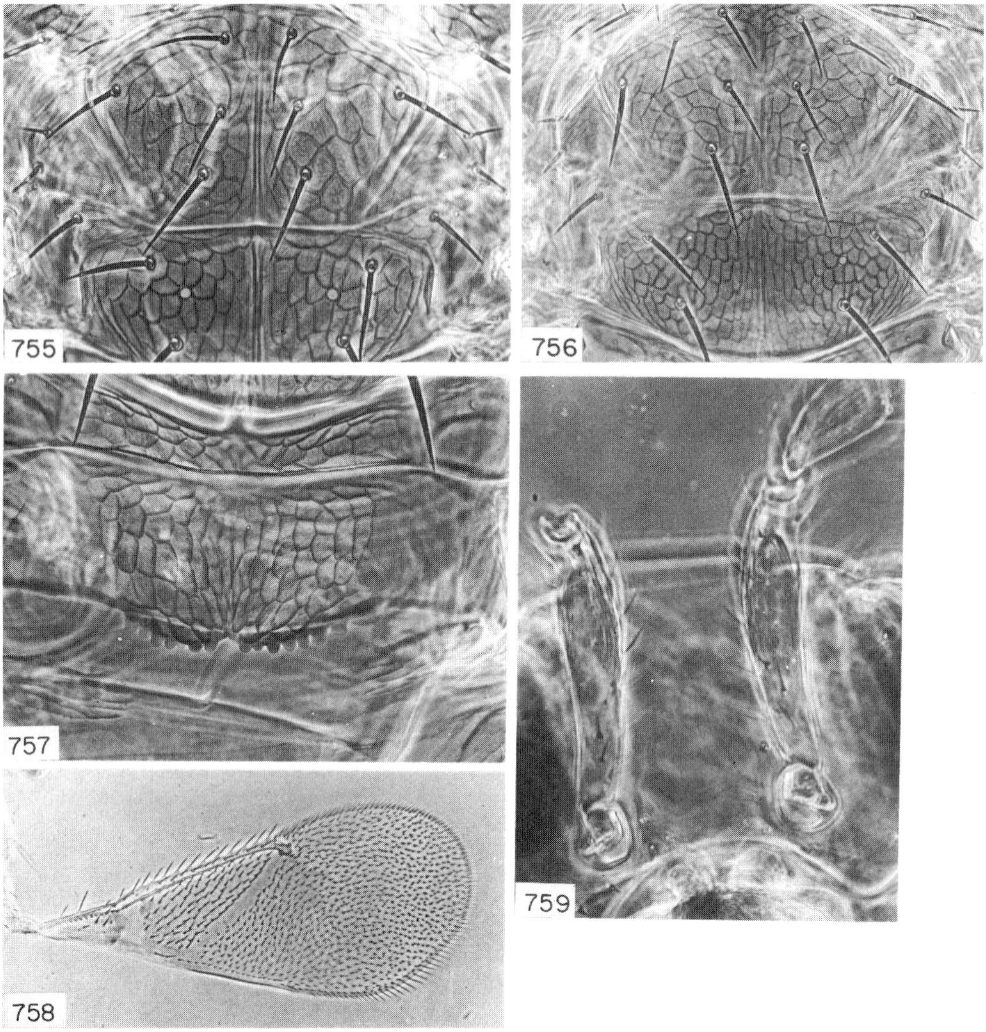


Figures 755–759. *Aphytis mytilaspidis* (Le Baron), uniparental form "3c"
755, 756. ♀: Mesonotal chaetotaxis. 757. ♀: Metanotum and propodeum; note dusky stripe on posterior margin of propodeum. 758. ♀: Forewing. 759. ♂ (rare): Antennal scape, showing ventral ridge.

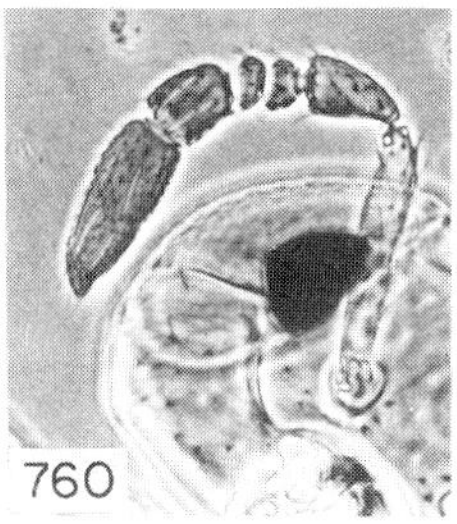


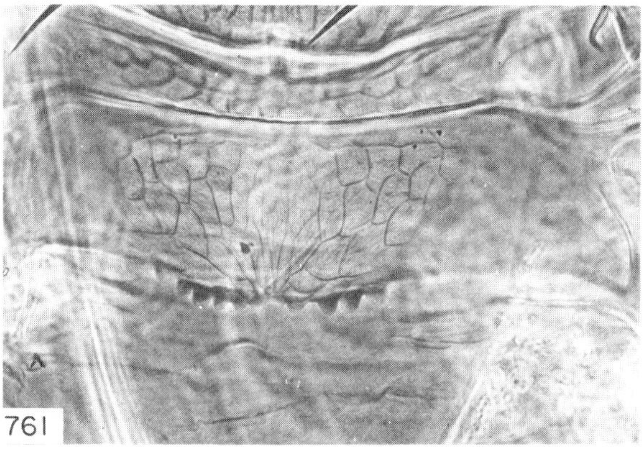


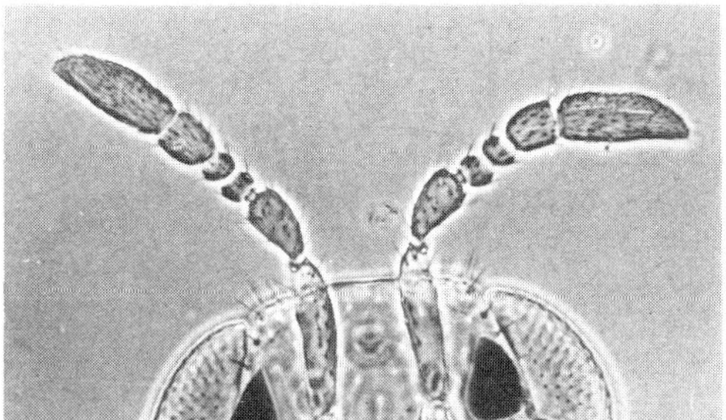


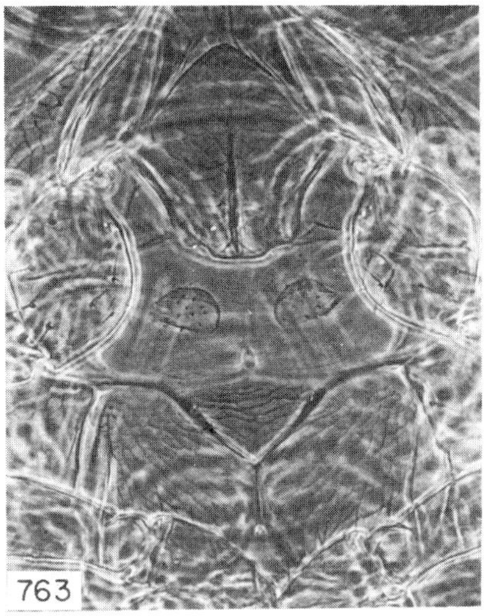


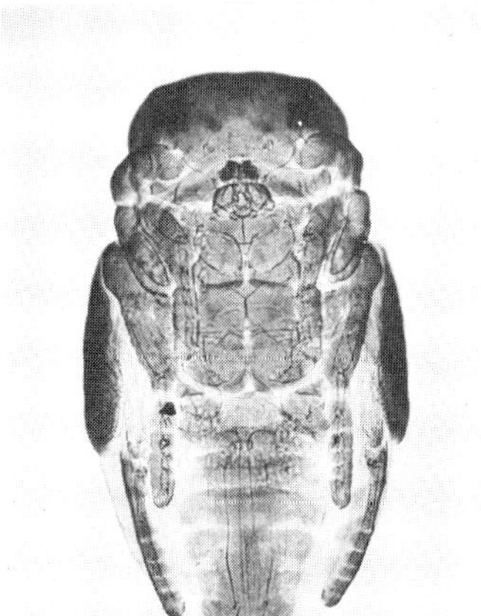


Figures 760–762. *Aphytis mytilaspidis* (Le Baron), biparental form "2"
760. ♀: Antenna. 761. ♀: Metanotum and propodeum. 762. ♂: Antennae.
Figure 763. *Aphytis mytilaspidis* (Le Baron), ♀: Venter of thorax, showing epicoxal pads and dusky sternites.
Figure 764. *Aphytis mytilaspidis* (Le Baron), ♀ pupa.

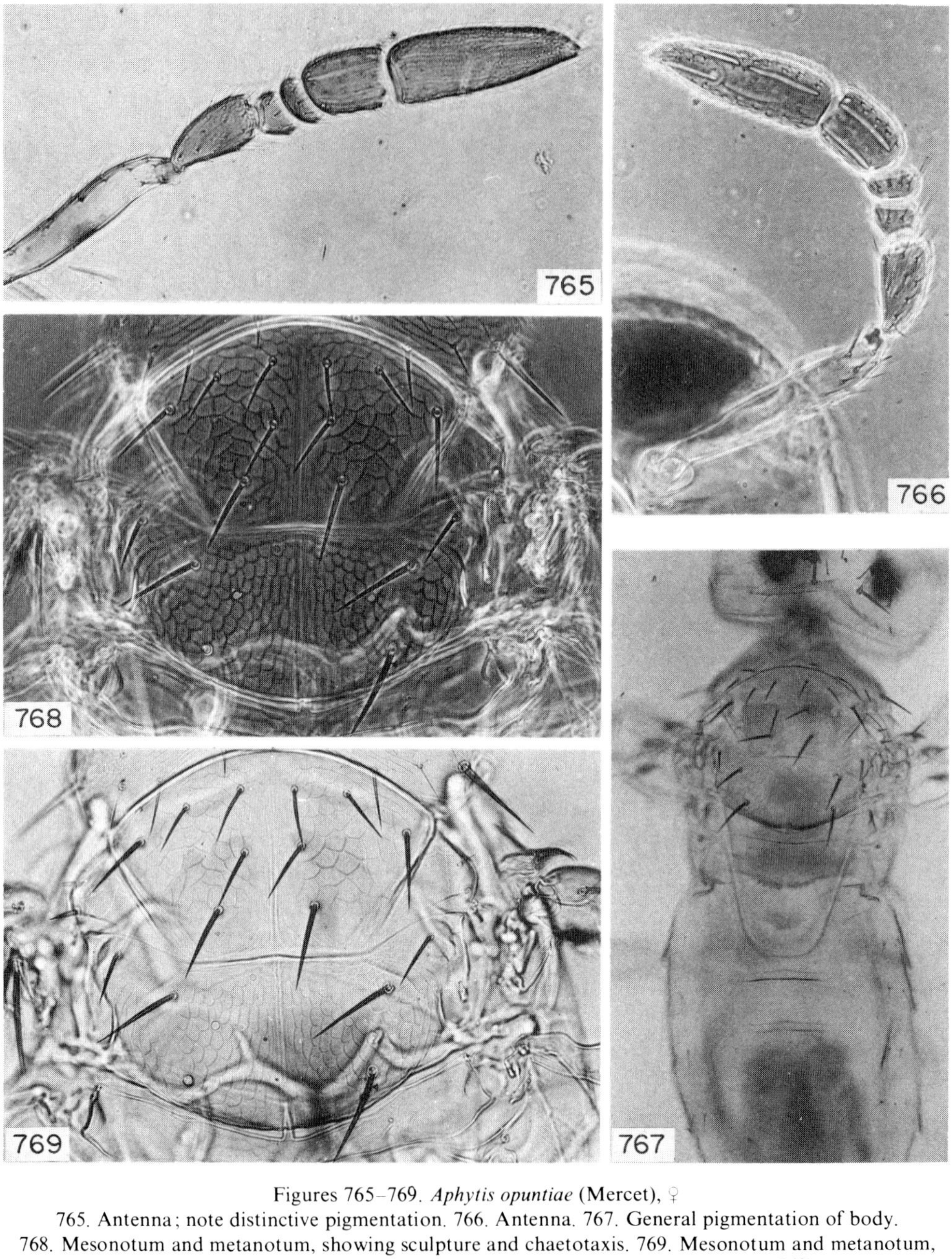


Figures 765–769. *Aphytis opuntiae* (Mercet), ♀
765. Antenna; note distinctive pigmentation. 766. Antenna. 767. General pigmentation of body. 768. Mesonotum and metanotum, showing sculpture and chaetotaxis. 769. Mesonotum and metanotum, showing pigmentation and chaetotaxis (same specimen as in Figure 768).

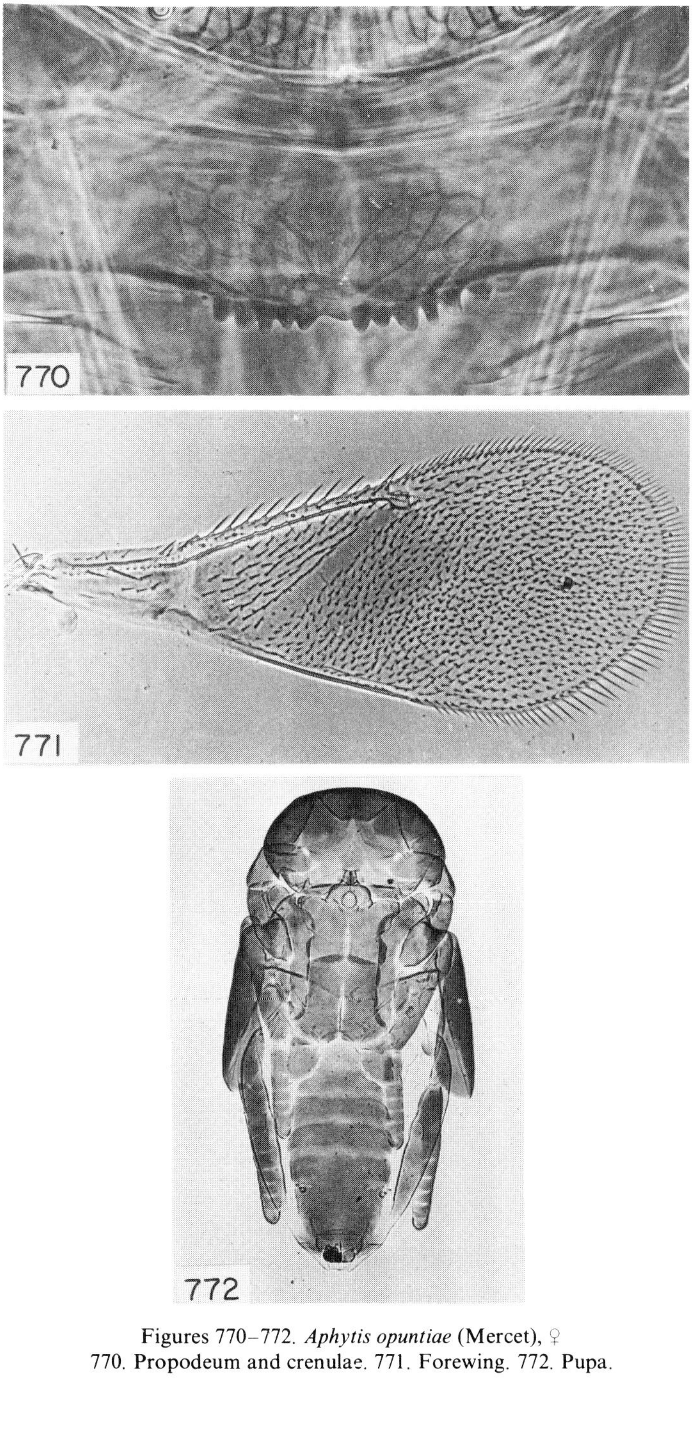


Figures 770–772. *Aphytis opuntiae* (Mercet), ♀
770. Propodeum and crenulae. 771. Forewing. 772. Pupa.

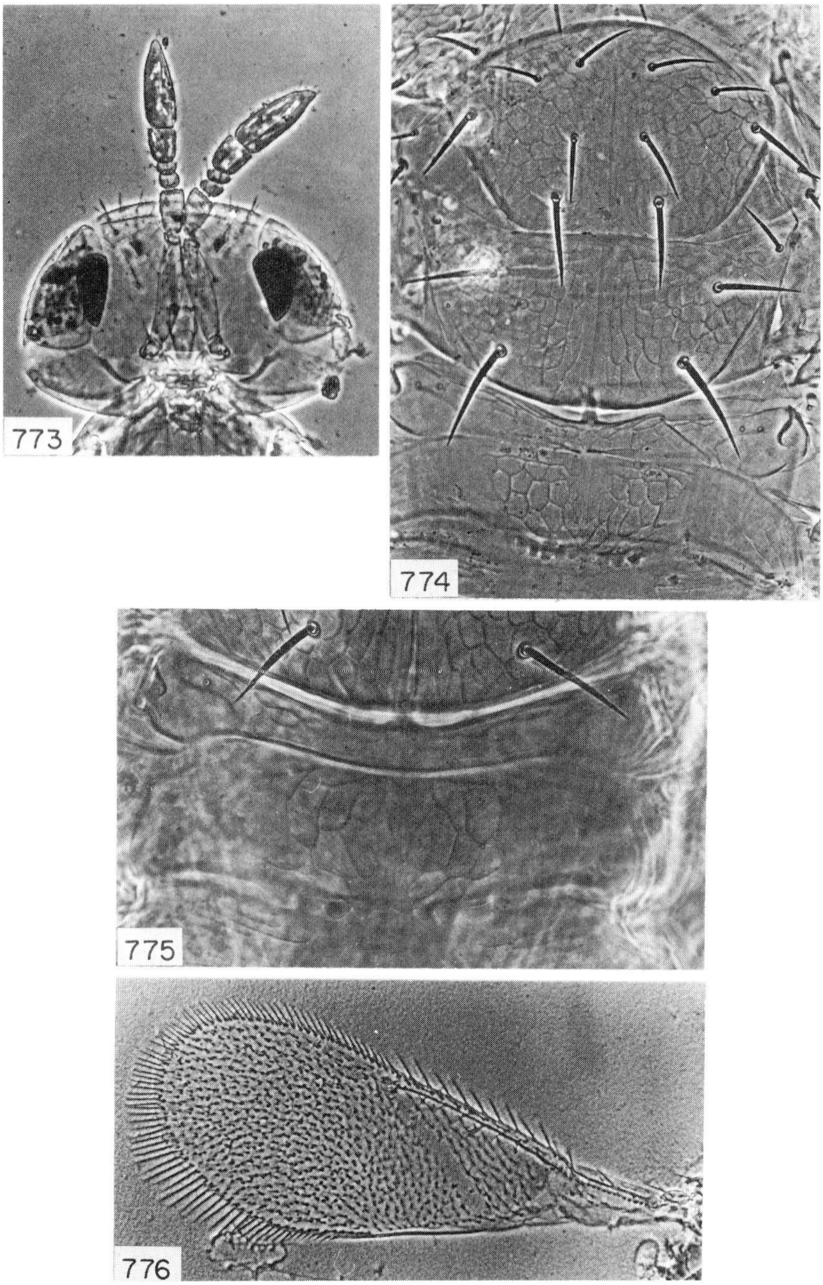


Figures 773–776. *Aphytis moldavicus* Jasnosh, ♀
773. Head and antennae. 774. Thorax and propodeum (paratype). 775. Metanotum, propodeum and crenulae (paratype). 776. Forewing (paratype).

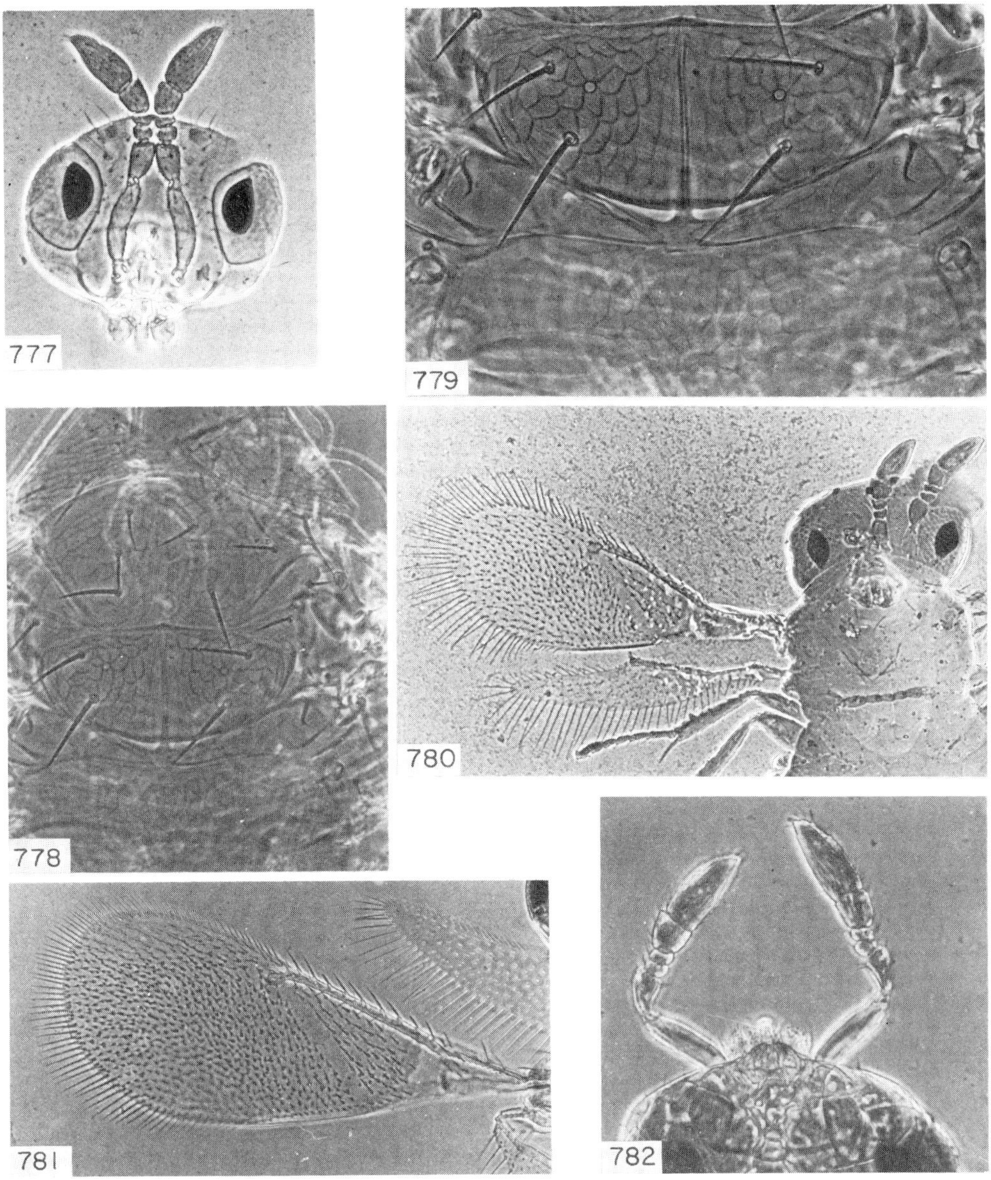


Figures 777–782. *Aphytis aonidiae* (Mercet), ♀
777–779. Holotype: 777. Head and antennae. 778. Thorax and propodeum. 779. Scutellum, metanotum and propodeum.
780. Mercet's specimen, ex *Aonidia lauri*, Madrid. 781. Holotype of *dubius* De Santis: Forewing.
782. Holotype of *dubius* var. *intermedia* De Santis: Antennae.

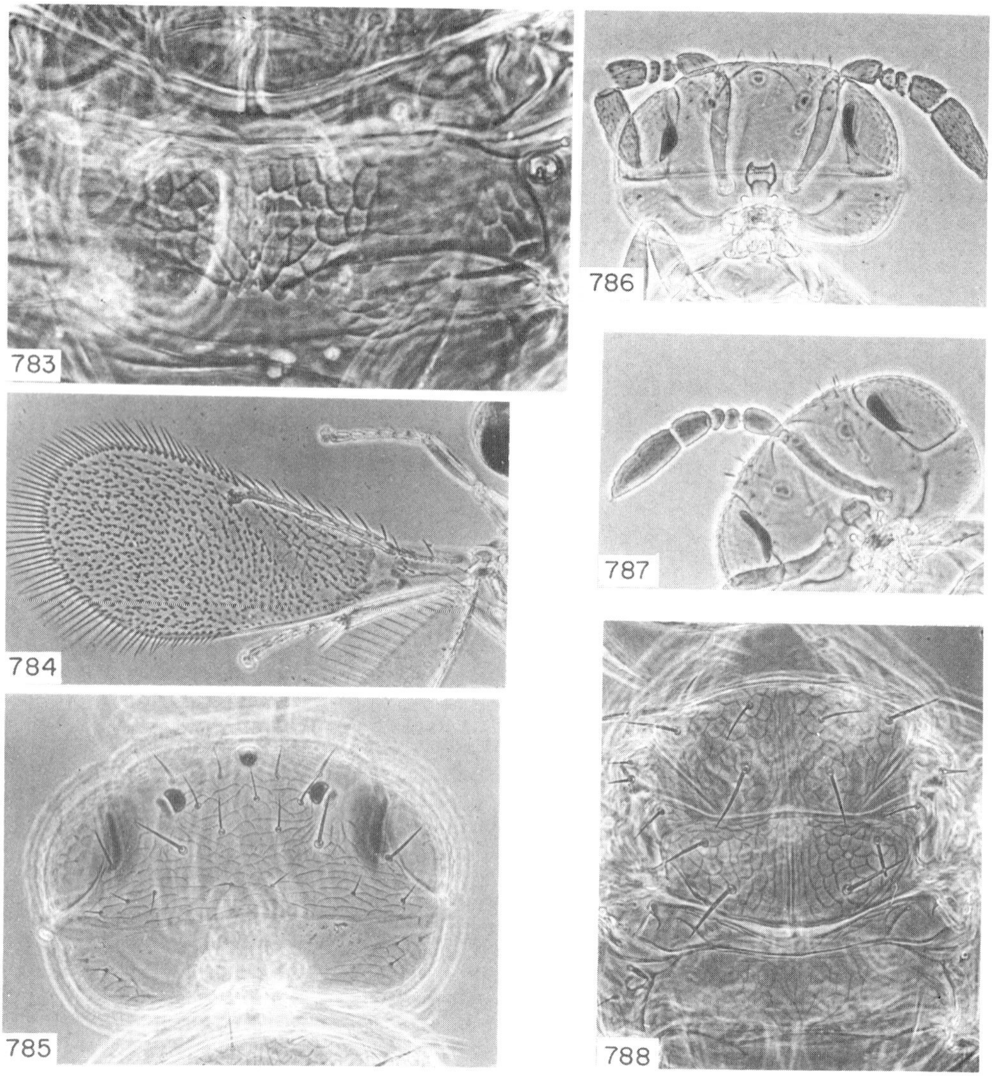


Figures 783–788. *Aphytis aonidiae* (Mercet), ♀
783, 784. Holotype of *dubius* var. *intermedia* De Santis: 783. Metanotum and propodeum. 784. Forewing. 785–788. Syntypes of *citrinus* Compere: 785. Head, showing chaetotaxis and sculpture (paralectotype). 786. Head and antennae (lectotype). 787. Head and antennae (paralectotype). 788. Thorax and propodeum (minute paralectotype specimen).

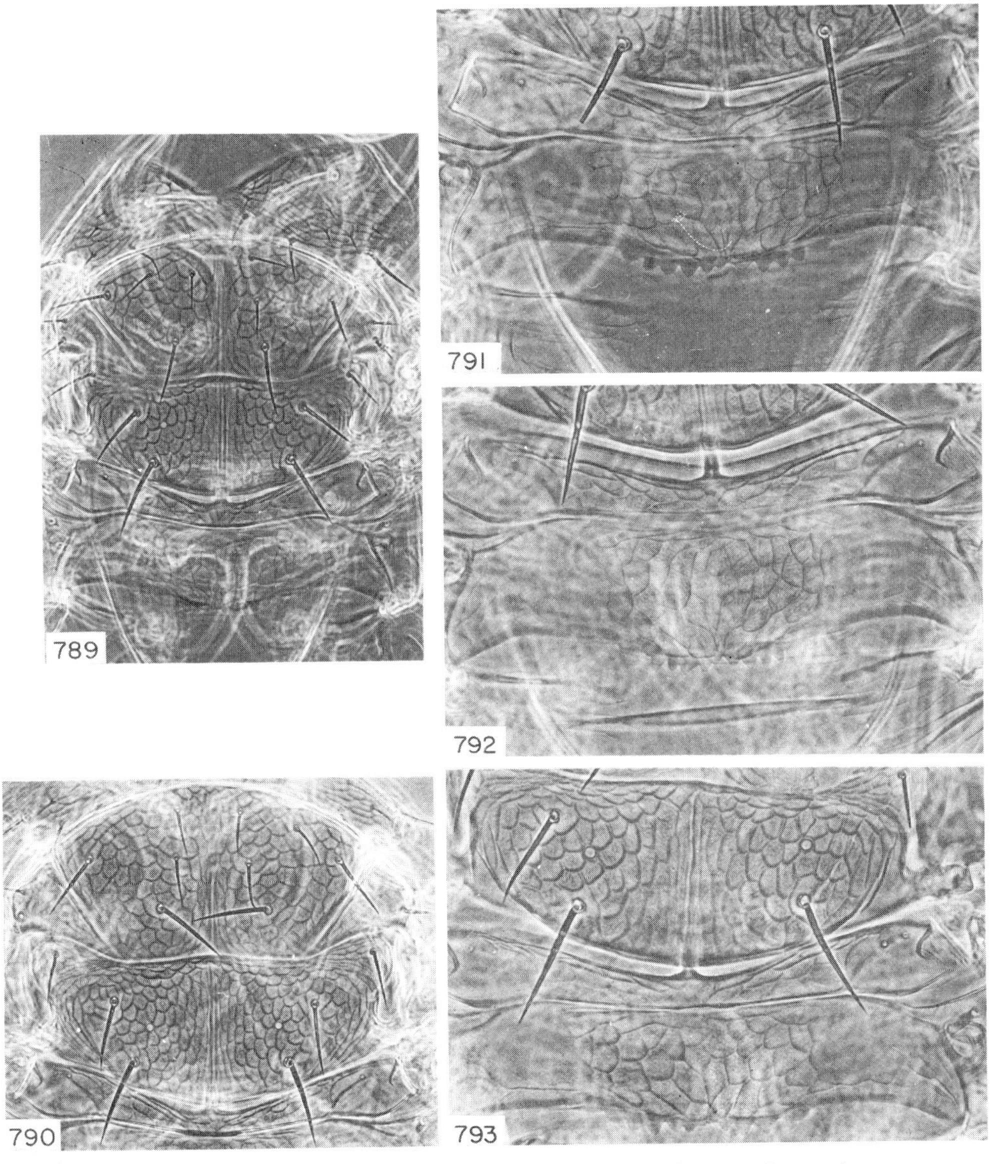


Figures 789–793. *Aphytis aonidiae* (Mercet), ♀ (syntypes of *citrinus* Compere)
789. Thorax and propodeum (large paralectotype specimen). 790. Mesonotum and metanotum (paralectotype). 791. Metanotum, propodeum and base of gaster (lectotype). 792. Metanotum and propodeum (paralectotype). 793. Scutellum (note abnormal chaetotaxis), metanotum and propodeum (paralectotype).

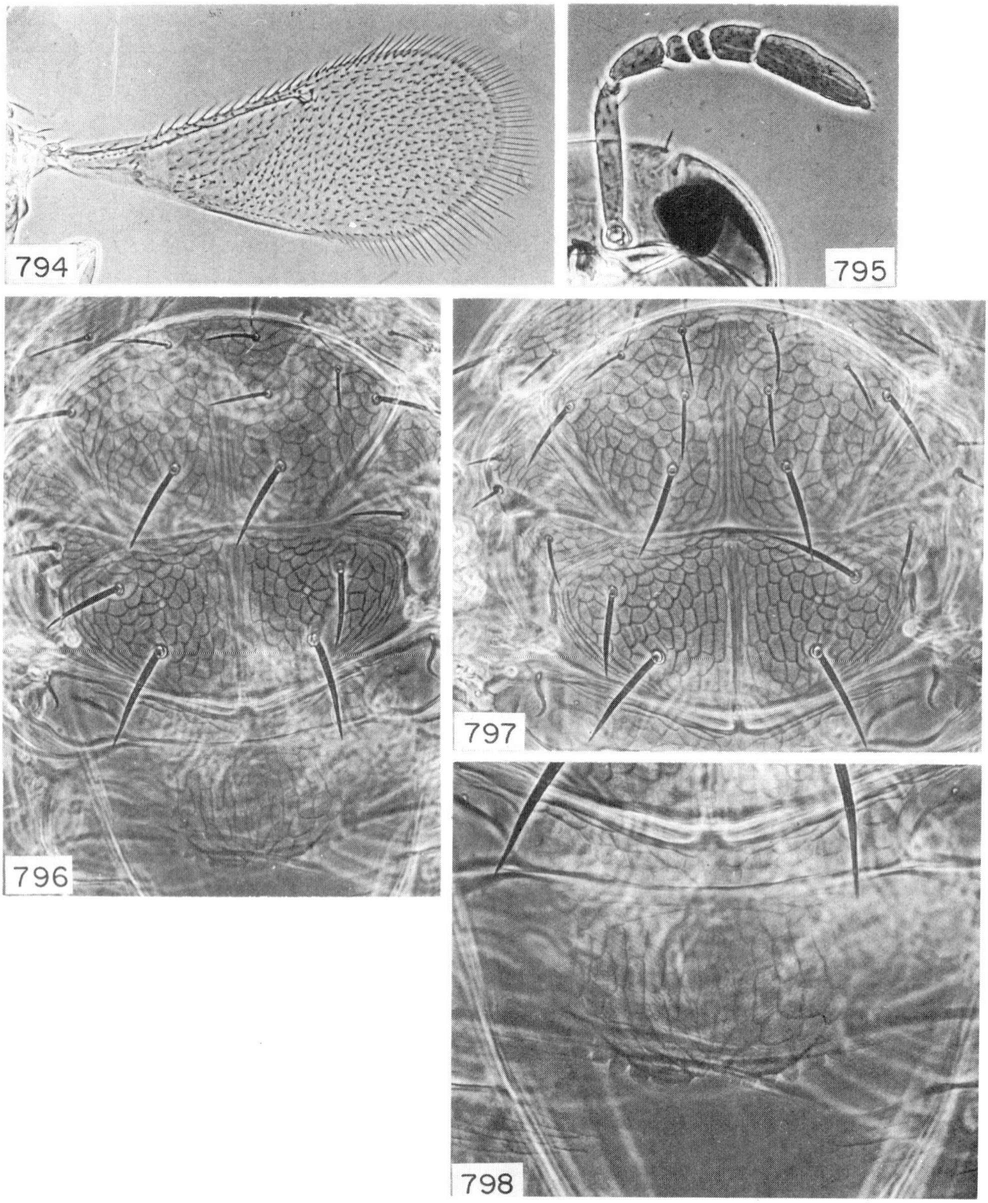


Figures 794–798. *Aphytis aonidiae* (Mercet), ♀
794. Paralectotype of *citrinus* Compere: Forewing.
795–798. California material ex *Quadraspidiotus perniciosus*: 795. Antenna. 796, 797. Thorax, showing size-linked variation in the number of mesoscutal setae. 798. Metanotum and propodeum.

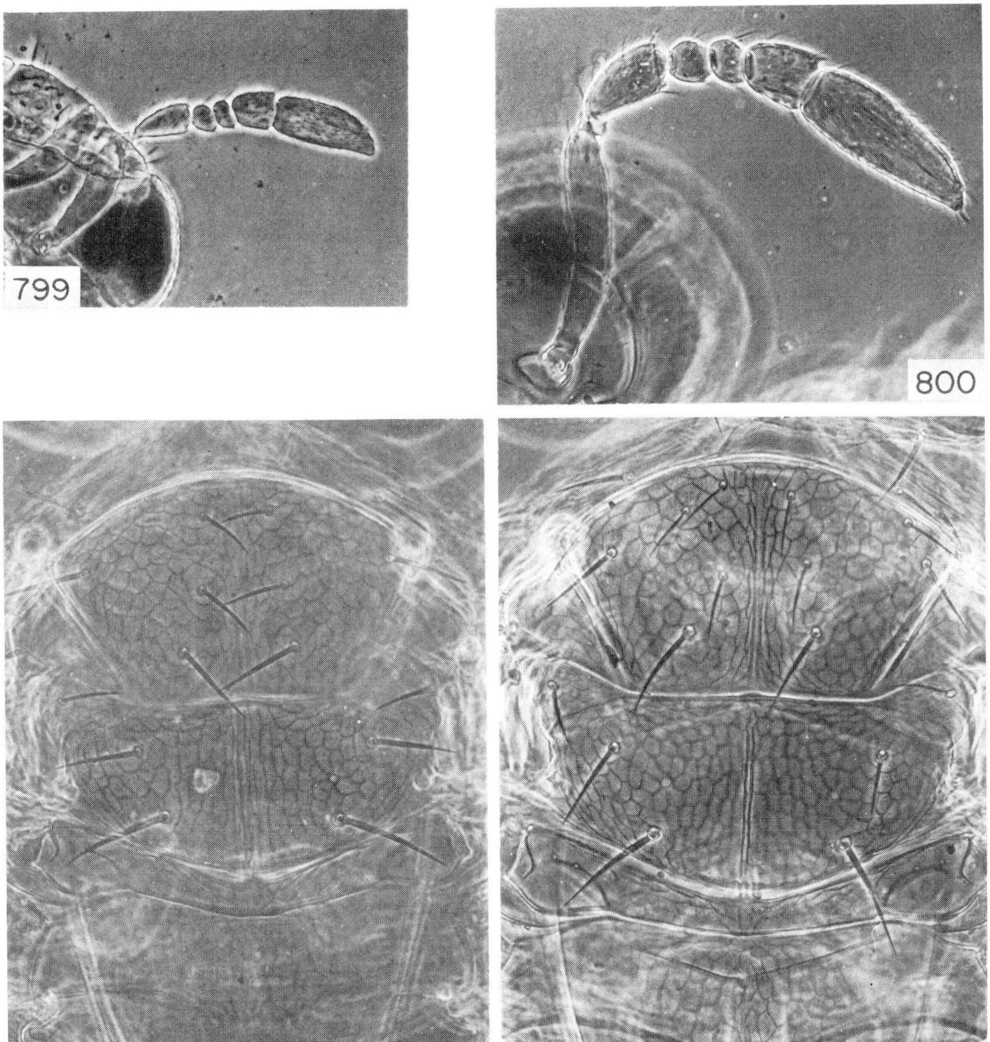


Figures 799–802. *Aphytis libanicus* Traboulsi, ♀ (Israel material)
799, 800. Antennae. 801, 802. Thorax and propodeum; note variation in number of mesoscutal setae.

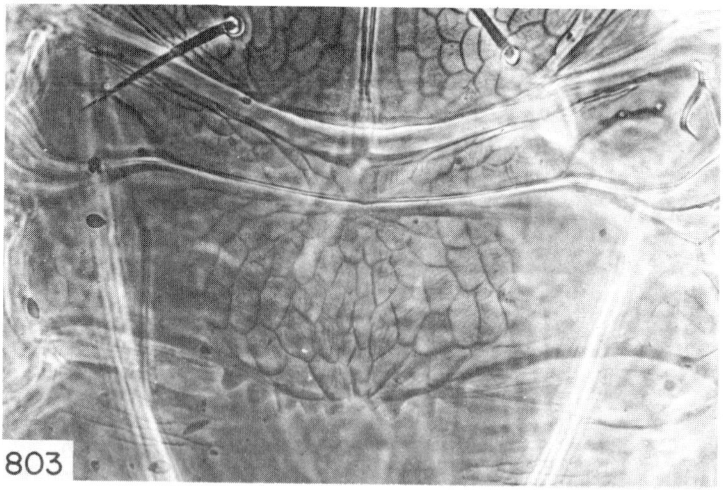


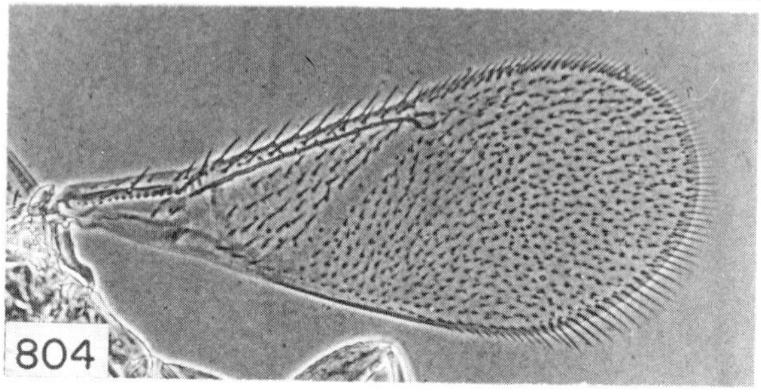


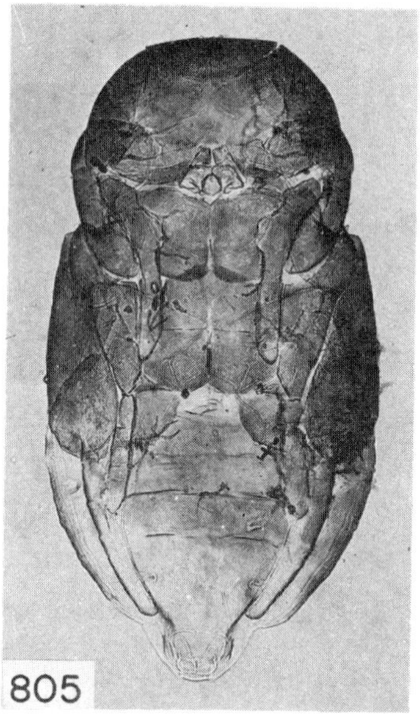


Figures 803–805. *Aphytis libanicus* Traboulsi, ♀ (Israel material)
803. Metanotum and propodeum. 804. Forewing. 805. Pupa.

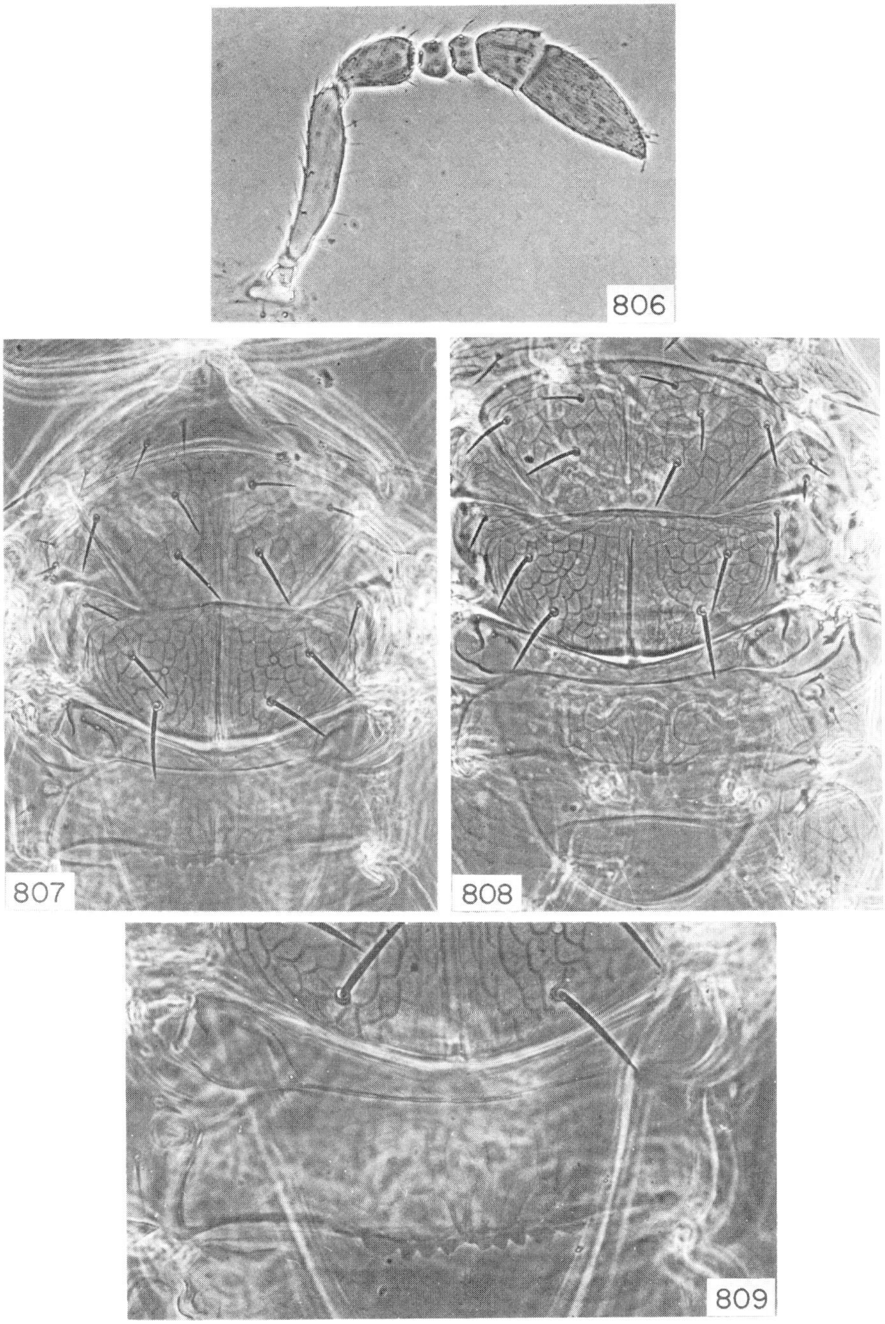


Figures 806–809. *Aphytis phoenicis* DeBach and Rosen, ♀
806. Antenna. 807, 808. Thorax and propodeum. 809. Metanotum and propodeum.

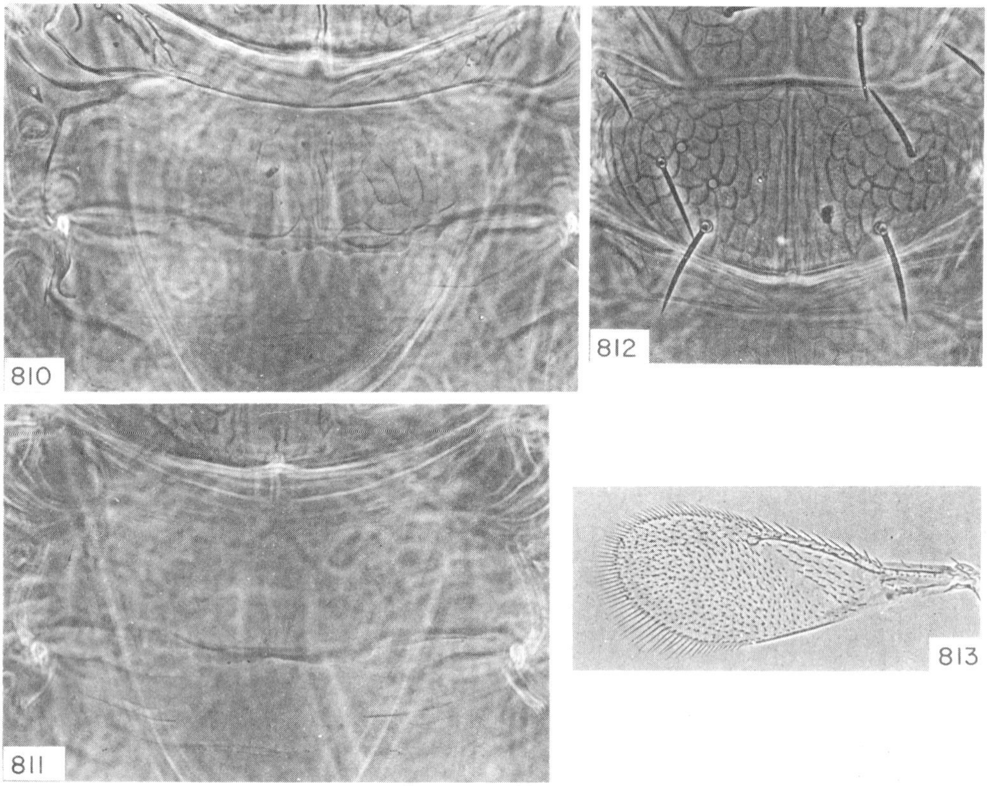


Figures 810–813. *Aphytis phoenicis* DeBach and Rosen, ♀
810, 811. Metanotum and propodeum; note irregular crenulae. 812. Scutellum with supernumerary sensillum. 813. Forewing.

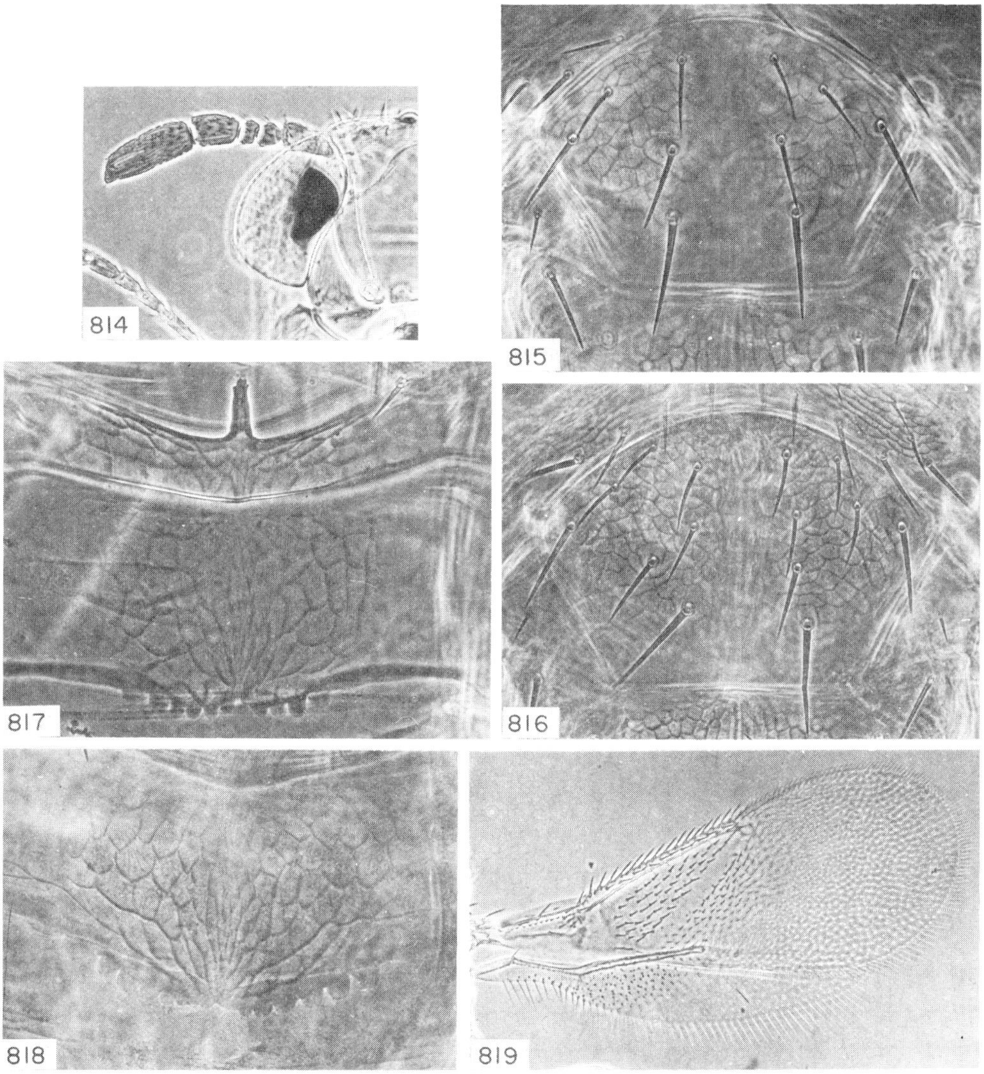


Figures 814–819. *Aphytis luteus* (Ratzeburg), ♀
814. Antenna. 815, 816. Mesoscutum, showing chaetotaxis and sculpture. 817. Metanotum and propodeum. 818. Propodeal crenulae. 819. Wings.

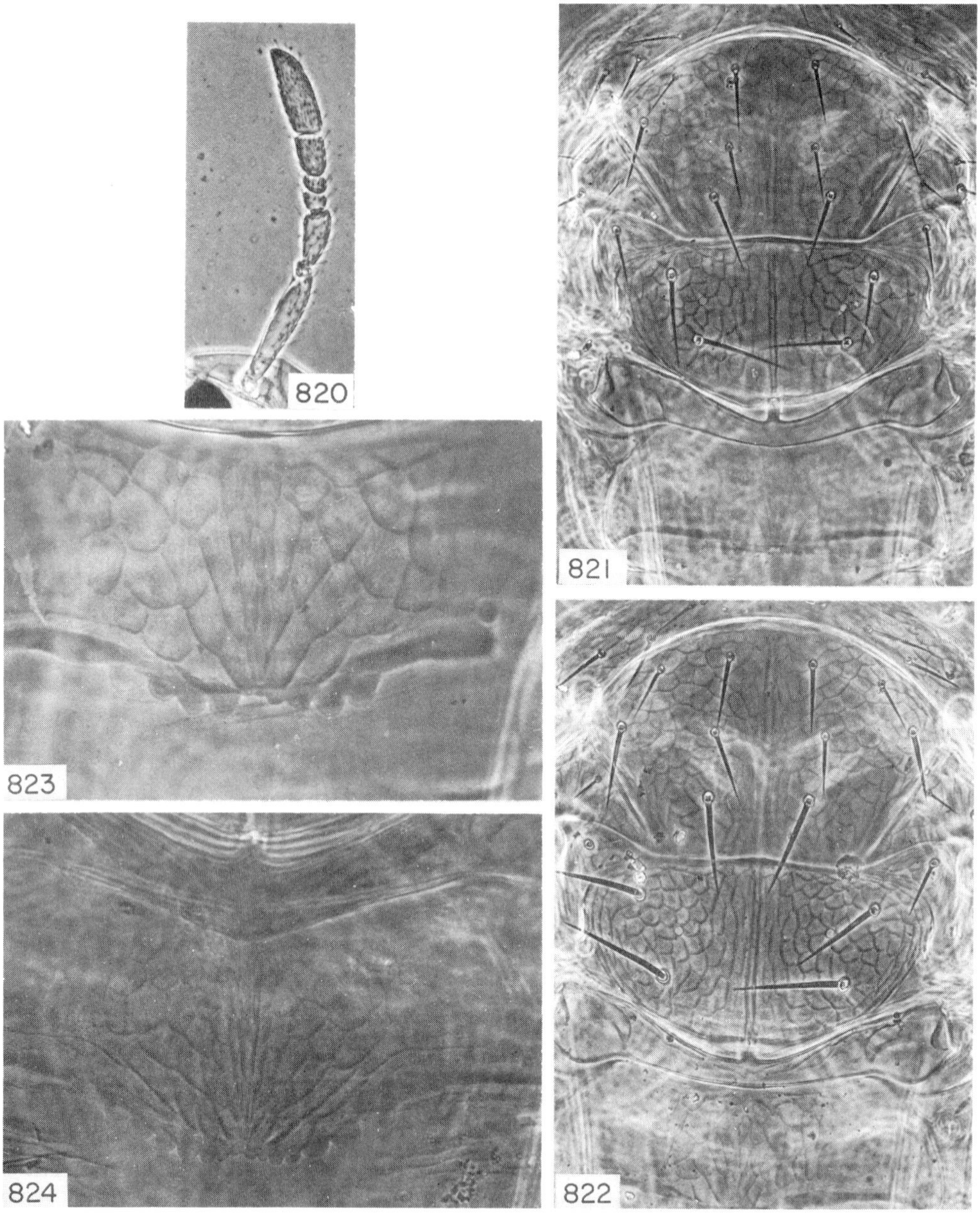


Figures 820–824. *Aphytis luteus* (Ratzeburg), ♂
820. Antenna; note ridge on scape. 821, 822. Thorax and propodeum. 823, 824. Propodeal crenulae.

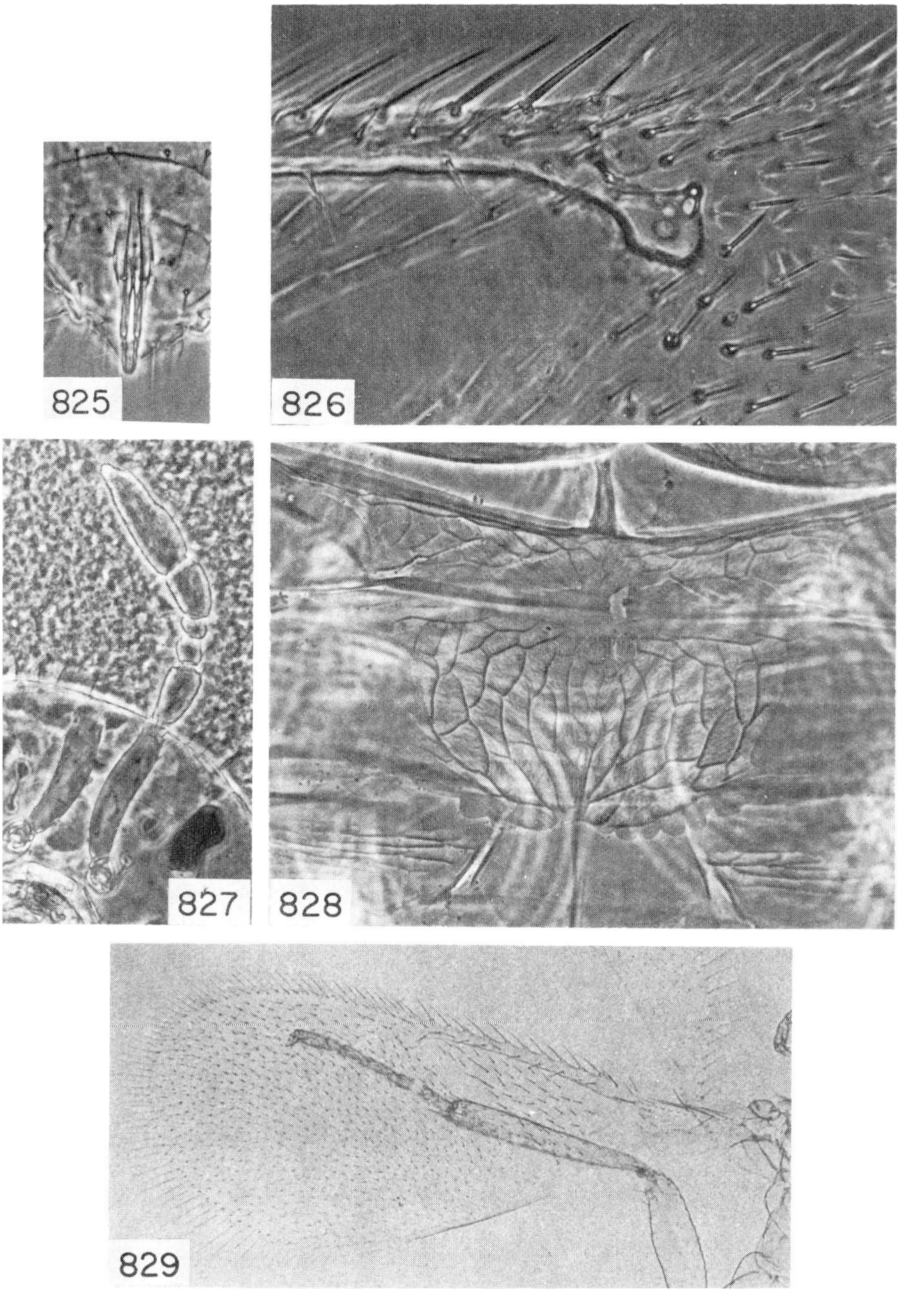


Figures 825–829. *Aphytis luteus* (Ratzeburg)
825. ♂: Genitalia. 826. ♂: Stigmal vein of forewing.
827–829. Holotype ♀: 827. Antenna. 828. Metanotum and propodeum. 829. Forewing.

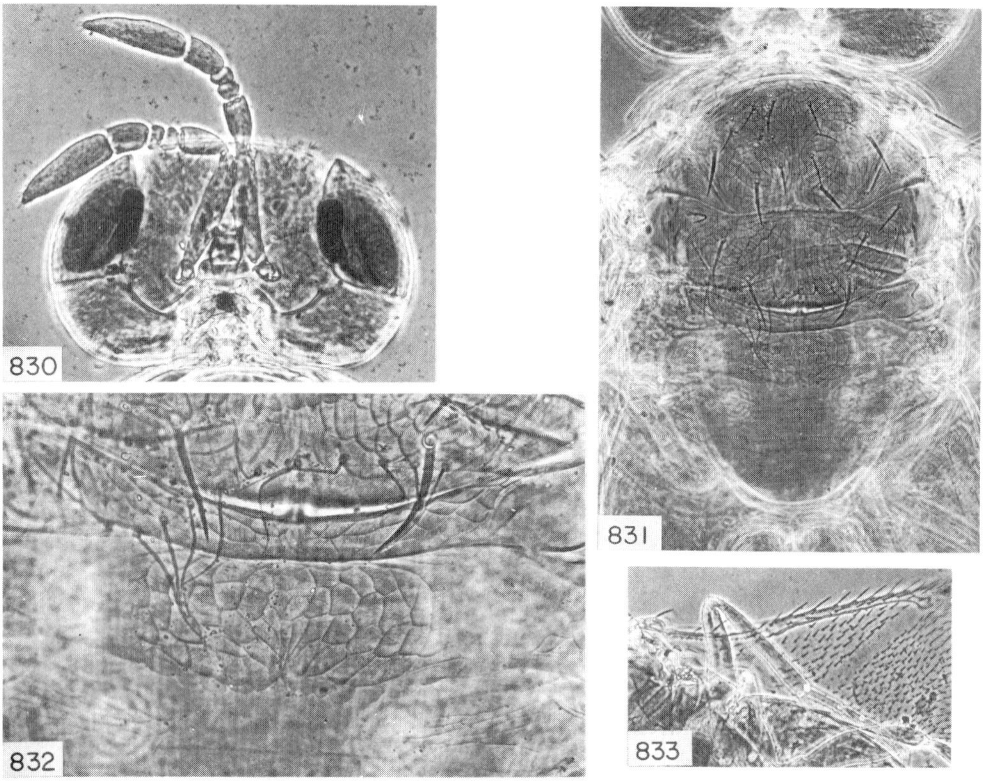


Figures 830–833. *Aphytis neuter* Jasnosh and Myartseva, ♀ (paratype)
830. Head and antennae. 831. Thorax, propodeum and base of gaster. 832. Metanotum, propodeum and crenulae. 833. Forewing (basal part).

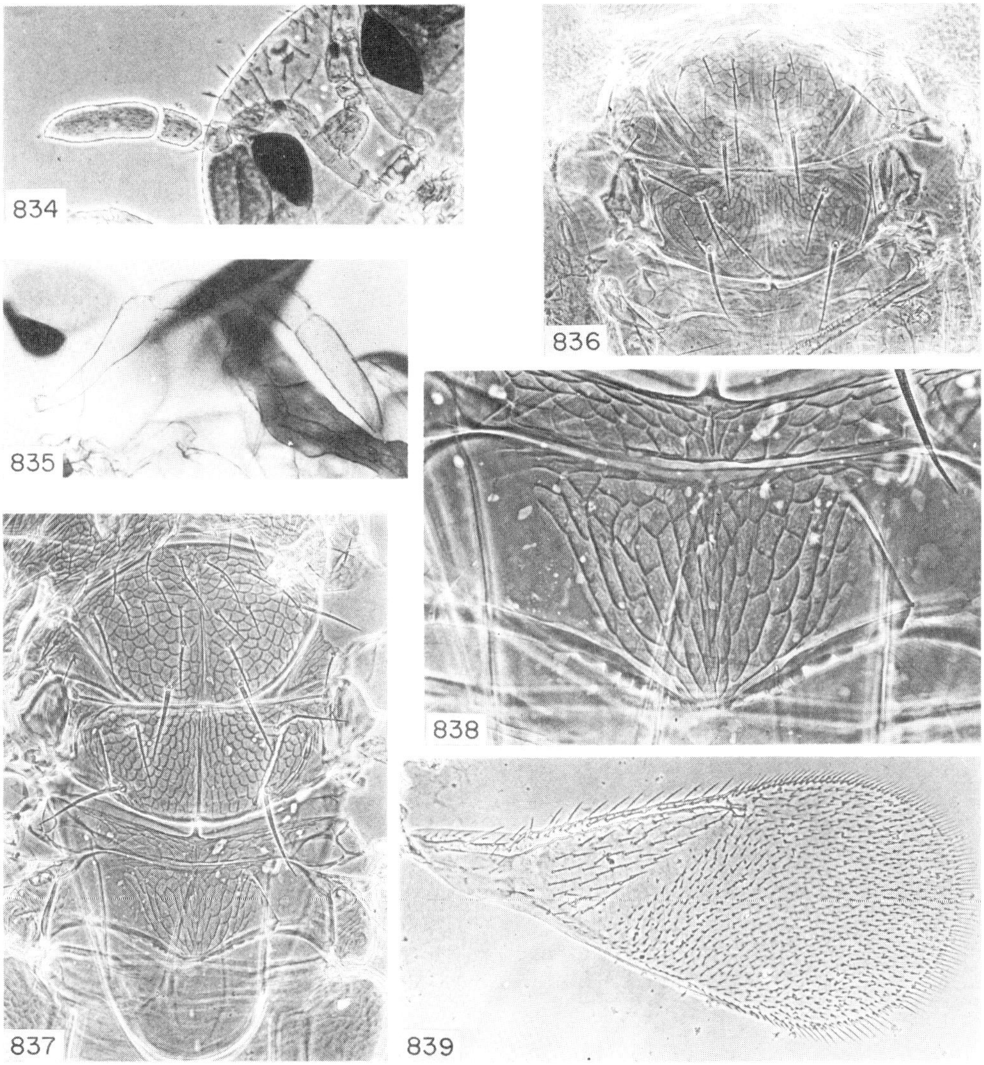


Figures 834–839. *Aphytis notalis* De Santis, ♀
834. Antenna. 835. Antenna (partly emerged from pupal case). 836. Thorax. 837. Thorax and propodeum. 838. Metanotum and propodeum. 839. Forewing.

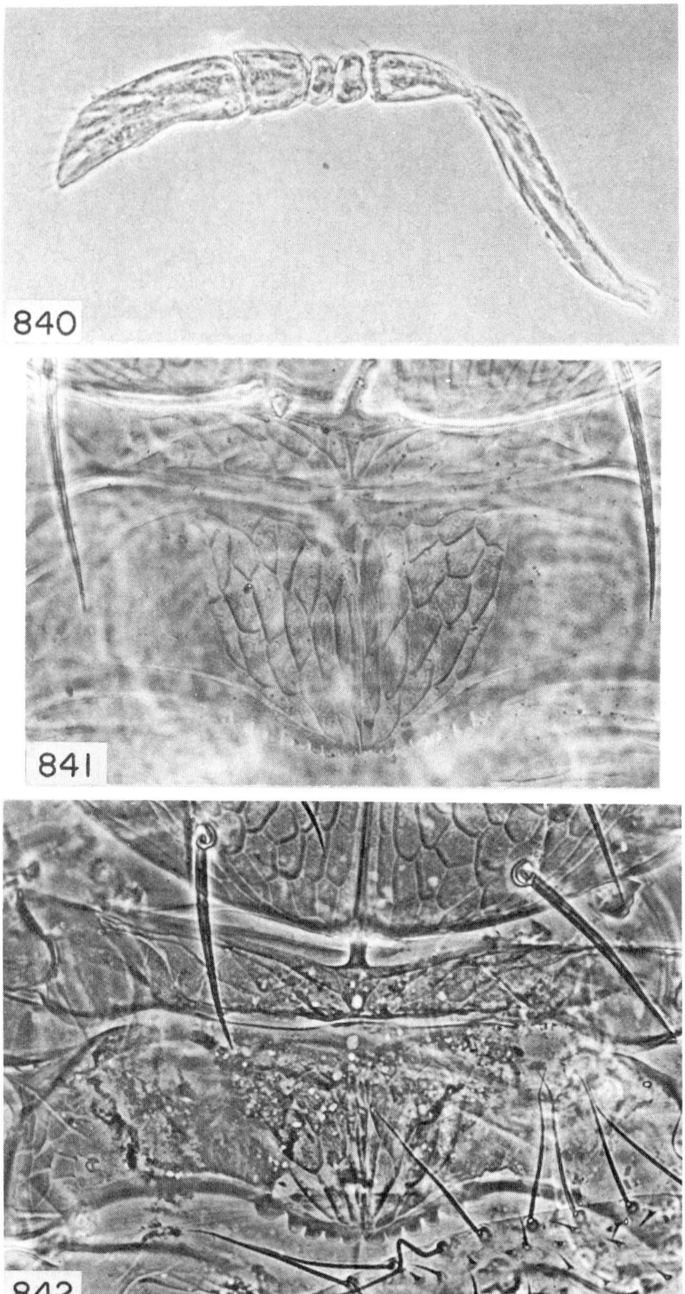

Figures 840–842. *Aphytis notialis* De Santis, ♂
840. Antenna (allotype). 841, 842. Metanotum and propodeum.

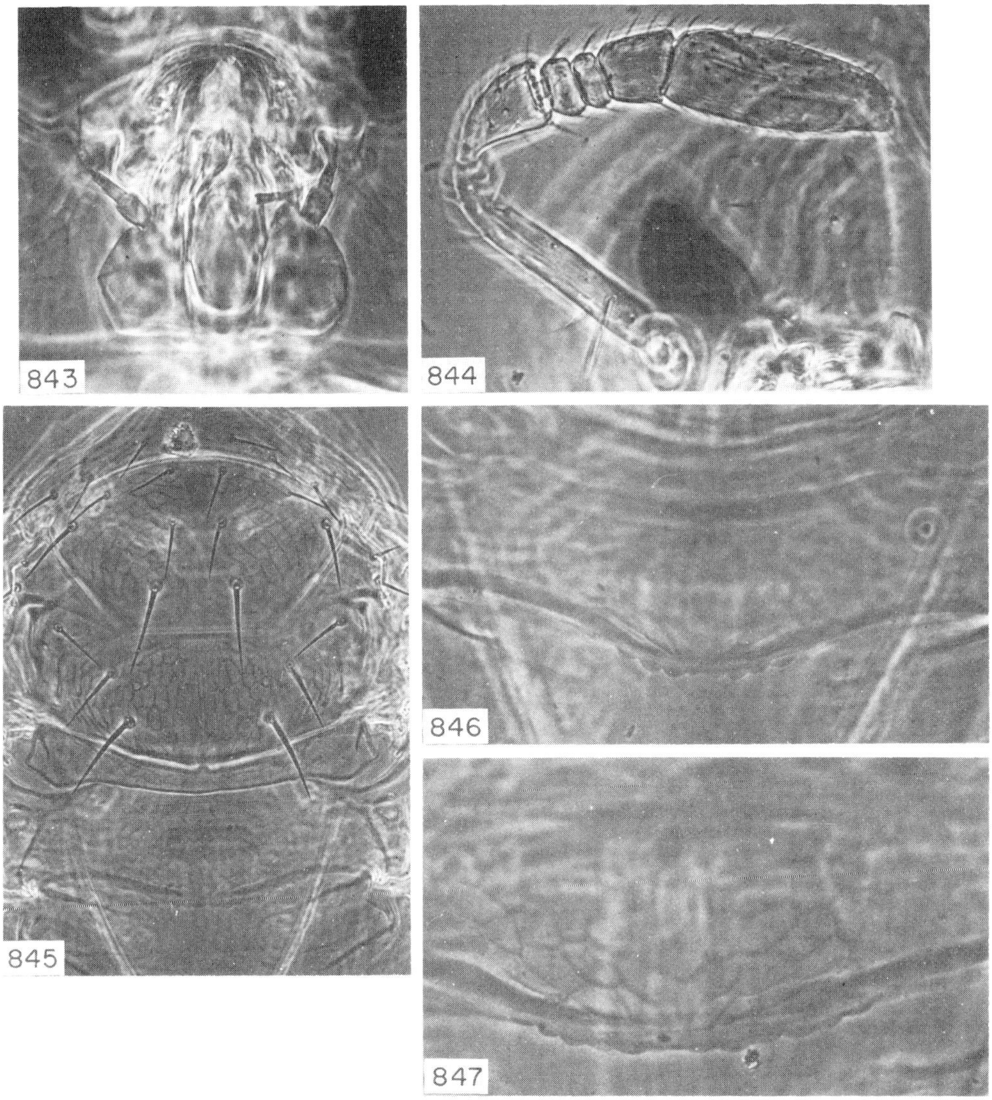


Figures 843–847. *Aphytis taylori* Quednau, ♀ (syntypes)
843. Mouthparts. 844. Antenna. 845. Thorax and propodeum. 846, 847. Propodeum.

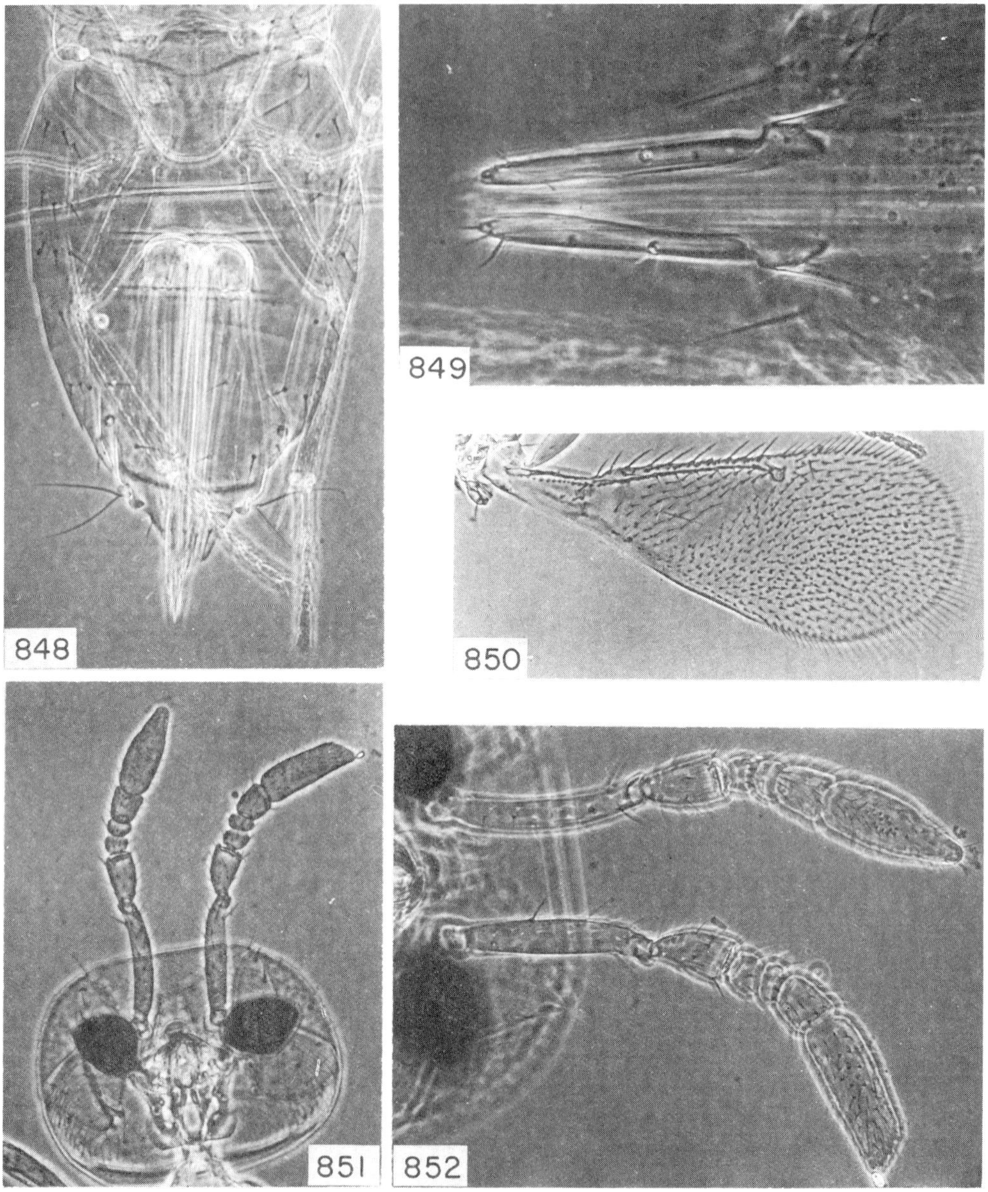


Figures 848–852. *Aphytis taylori* Quednau (syntypes)
848. ♀: Abdomen. 849. ♀: Ovipositor sheaths. 850. ♀: Forewing. 851. ♂: Head and antennae.
852. ♂: Antennae, showing specialized sensory area on ventral aspect of club.

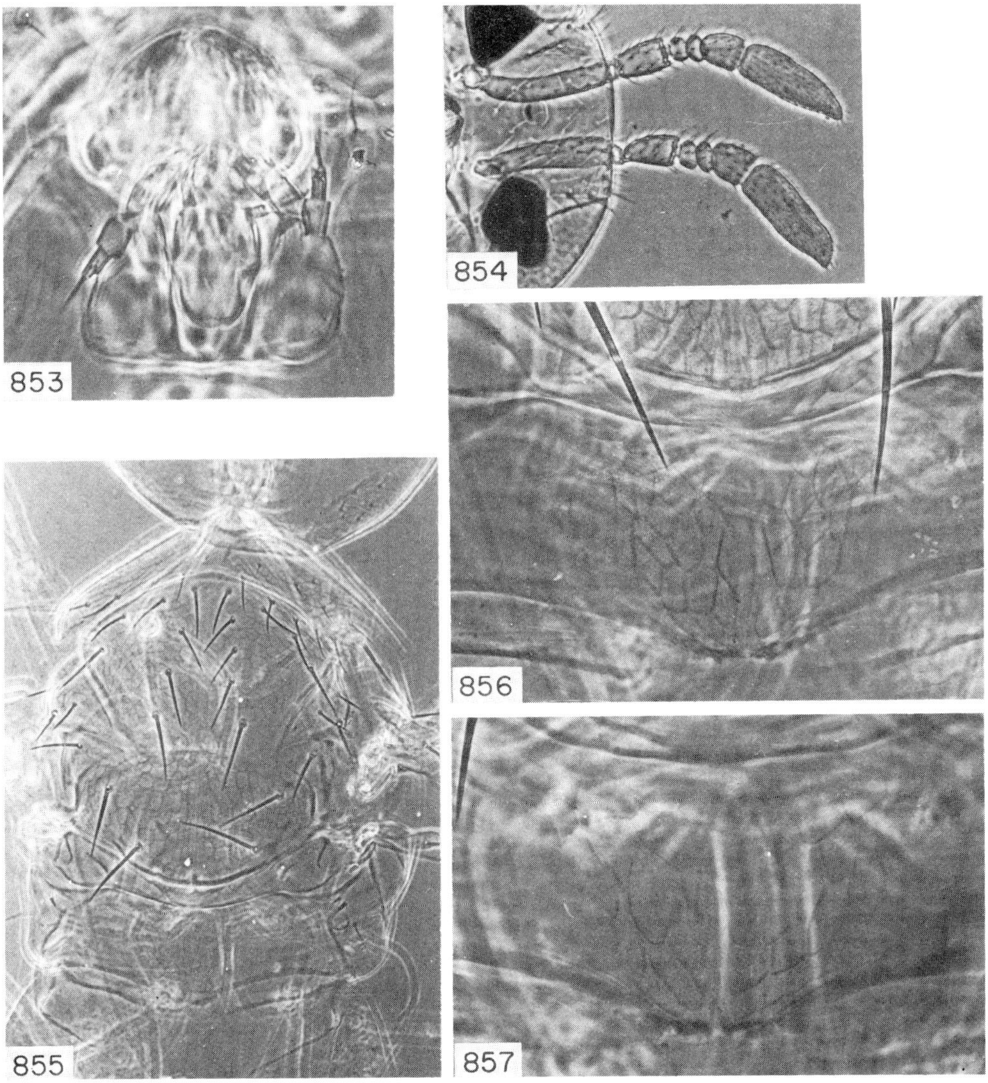


Figures 853–857. *Aphytis rolaspidis* DeBach and Rosen, ♀
853. Mouthparts (lectotypes). 854. Antennae (lectotype). 855. Thorax and propodeum (lectotype). 856. Metanotum and propodeum; note indistinct crenulae (paralectotype). 857. Propodeum; note irregular, rather large crenulae (paralectotype).

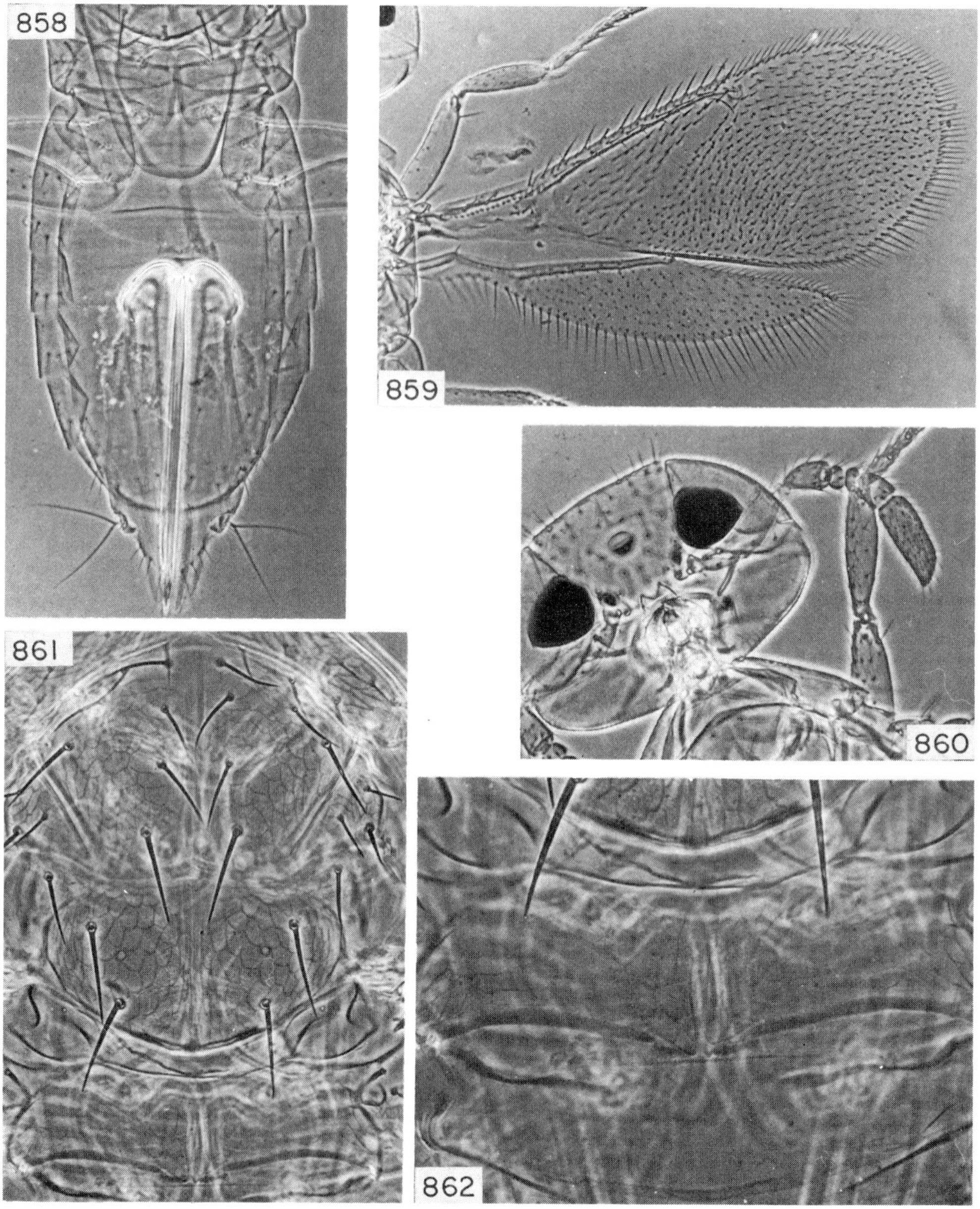


Figures 858–862. *Aphytis rolaspidis* DeBach and Rosen
858. ♀: Abdomen (paralectotype). 859. ♀: Wings (paralectotype). 860. ♂: Antenna (allotype). 861. ♂: Thorax and propodeum (allotype). 862. ♂: Metanotum and propodeum; note indistinct crenulae (allotype).

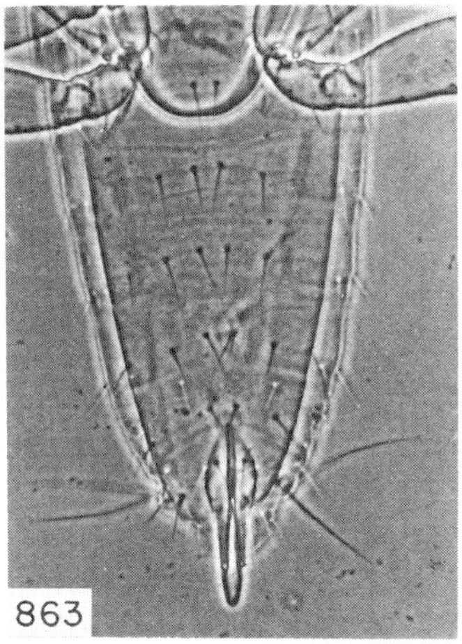


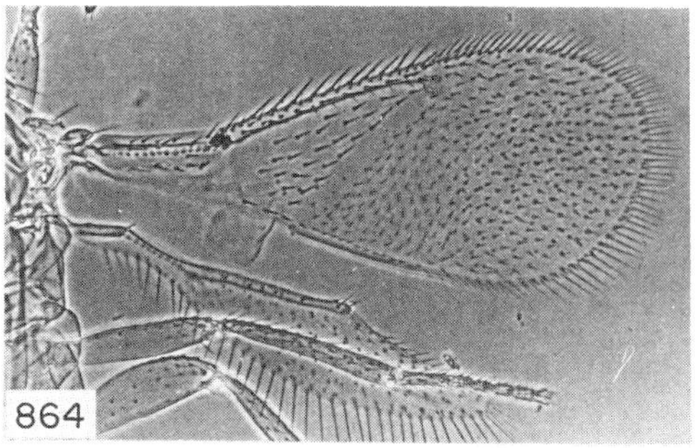


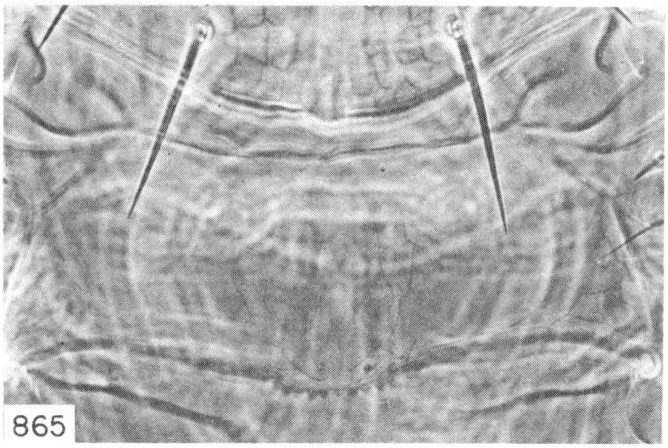


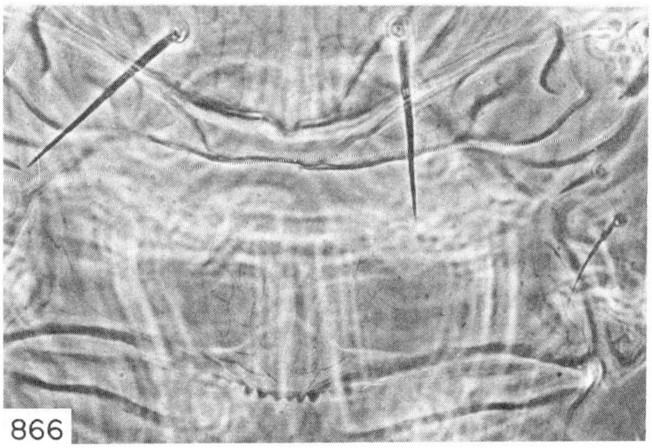


Figures 863–864. *Aphytis rolaspidis* DeBach and Rosen, ♂ (allotype)
863. Abdomen (ventral view) and genitalia. 864. Wings.
Figures 865, 866. Syntype ♂♂ of *Aphytis flavus* Quednau (= *A. rolaspidis* DeBach and Rosen)—apparently a different species: Metanotum and propodeum; note long propodeum and distinct crenulae.

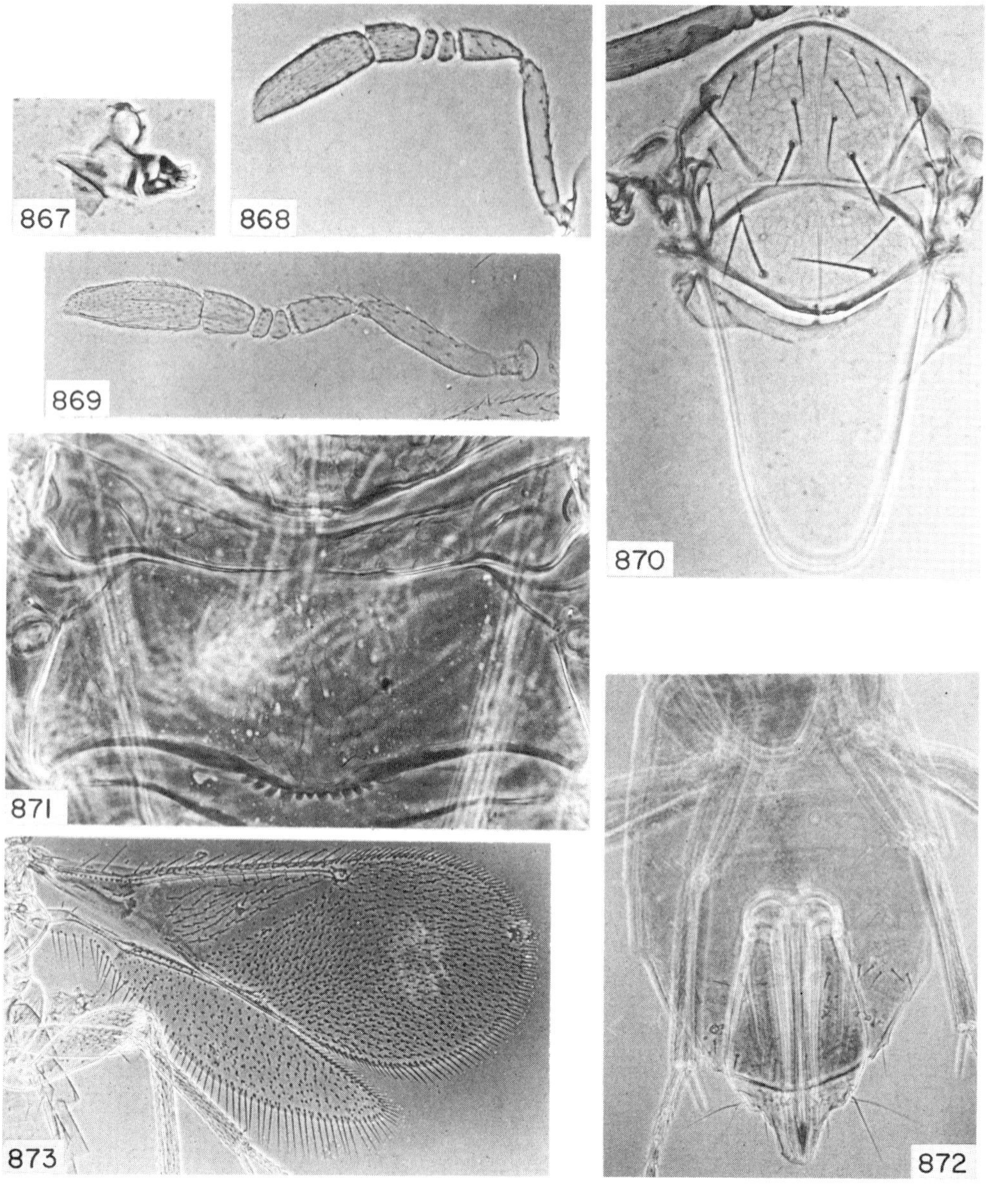


Figures 867–873. *Aphytis capensis* DeBach and Rosen, ♀
867. Mandible (paratype). 868, 869. Antennae (paratypes). 870. Mesonotum (dissected), including endophragma (paratype). 871. Metanotum and propodeum. 872. Abdomen; note relative length of ovipositor and middle tibia (paratype). 873. Wings (paratype).

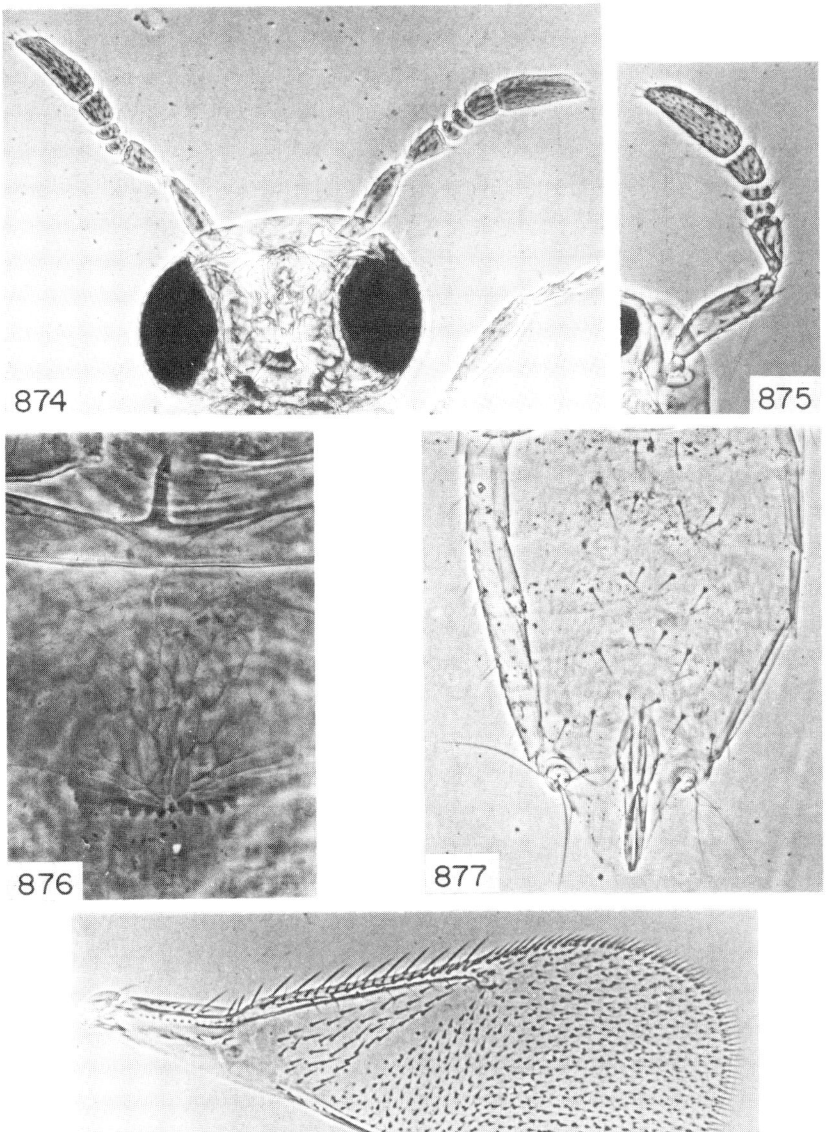


Figures 874–878. *Aphytis capensis* DeBach and Rosen, ♂
874. Head and antennae (allotype). 875. Antenna (paratype). 876. Central part of metanotum and propodeum, showing the anteromedian apodeme and crenulae (paratype). 877. Venter of abdomen and genitalia (paratype). 878. Forewing (paratype).

# V. THE LINGNANENSIS GROUP

This complex, highly uniform group is characterized by a relatively long propodeum with large, strongly overlapping crenulae. General coloration yellow; wings uniformly setose, hyaline or nearly so. Male abdomen with the posterior sternites rather conspicuously stippled; male antennal club usually with a partially cut-off sensory area on the ventral aspect. Body setae and thoracic sternal pigmentation are interspecifically diverse in this group, as is pupal pigmentation.

The **lingnanensis** group is believed to be closely related to the **chrysomphali** group, probably occupying an intermediate position between that highly advanced group and the more primitive **proclia** group.

This group at present includes the following 9 species: *africanus* Quednau, *coheni* DeBach, *equatorialis* n.sp., *fisheri* DeBach, *holoxanthus* DeBach, *lingnanensis* Compere, *margaretae* DeBach and Rosen, *melinus* DeBach, and *yasumatsui* Azim. It appears to be centered in the Far East: five members are of Oriental origin, with another one from the East Palearctic region (Japan); two are Ethiopian, and one occurs in the Western Hemisphere.

The Ethiopian members of the group, *africanus* and *equatorialis*, differ from the Far Eastern species in having a shorter club in the antenna of the female (the club in females of the Far Eastern members is quite long) and rather weak stippling on the abdominal sternites of the male. The American species, *margaretae*, is provisionally assigned to the **lingnanensis** group but differs from all other members in several obvious characters, such as the presence of specialized sense organs on the antennal scape of the male, and may not be a bona fide member of this group. It is, of course, quite possible that the **lingnanensis** group as currently understood is polyphyletic in origin.

Probably more is known about the biosystematic status of the forms comprising the **lingnanensis** group than about any of the other species groups in *Aphytis*. The group can be conveniently divided into two subgroups—one with dusky, the other with immaculate thoracic sterna. These subgroups also differ in the shape of the spermathecae and in details of courtship behavior. Extensive reciprocal crossing tests have revealed the existence of several complexes of closely related forms within the two subgroups, exhibiting various degrees of reproductive isolation and morphological distinctness (see Rao and DeBach, 1969a, 1969b; Gordh and DeBach, 1978). These are briefly discussed below, and will be presented in some more detail under the relevant species.

The *lingnanensis–coheni* complex actually consists of two sibling complexes of semispecies, all possessing dusky thoracic sterna and nearly indistinguishable from one another in imaginal as well as in pupal characters. The semispecies in each complex are only partially reproductively isolated from one another, but exhibit complete or near-complete isolation from the components of the other sibling complex.

*A. africanus* and *A. margaretae* have similarly pigmented thoracic sterna but differ in other morphological characters as mentioned above.

The *holoxanthus–melinus–yasumatsui–fisheri* complex includes sibling or near-sibling species possessing immaculate thoracic sterna, differing from one another mainly in pupal pigmentation. *A. melinus* and *A. fisheri* are nearly identical in imaginal characters, but are completely reproductively isolated from one another. On the other hand, *yasumatsui* shows only partial reproductive isolation from *melinus* and is regarded as a semispecies in relation to this species, but is morphologically more similar to *holoxanthus*, from which it is completely isolated reproductively. Thus, it appears that morphological divergence and reproductive isolation have evolved at a different pace in the various components of this baffling complex.

*A. equatorialis* also has immaculate thoracic sterna, but is otherwise more similar morphologically to *africanus*.

In the following discussion, the members of the **lingnanensis** group have been conveniently arranged according to sternal pigmentation, although it is quite obvious from the foregoing discussion that certain species with pigmented thoracic sterna may in fact be more closely related to species possessing immaculate thoracic sterna than to similarly pigmented forms. Separating the members of this complex group is often very difficult. Aside from sternal pigmentation, certain species can be separated from one another by slight differences in the coarseness of the body setae, in the antennal proportions, or in the chaetotaxis of the wings, etc. Patterns of pupal pigmentation frequently offer useful diagnostic characters in this group. In the final analysis, only reciprocal crossing tests may provide unequivocal evidence regarding the status of species in the **lingnanensis** group.

More than any other group, the members of the **lingnanensis** group have been successfully employed in biological control projects. Some of them, such as *holoxanthus*, *lingnanensis* and *melinus*, have in the last 20–30 years become established in numerous habitats in virtually all zoogeographical regions of the world in control of major citrus pests. Further study of this interesting group is likely to yield additional effective natural enemies.

## 59. **Aphytis lingnanensis** Compere

(Figures 108, 150, 157, 167, 181, 189, 190, 197, 203, 204, 216, 219, 233; and 879–898)

*Aphytis lingnanensis* Compere, 1955, Univ. Calif. Publ. Entomol., **10**: 303–305.
*Aphytis lingnanensis*: DeBach, 1959, Ann. Entomol. Soc. Amer., **52**: 357, 360; Quednau, 1964, J. Entomol. Soc. S. Afr., **27**: 107–110; Traboulsi, 1969, Ann. Soc. Entomol. Fr., N.S., **5**: 53–54.
For an extensive list of early references to *lingnanensis*, see Peck (1963).

This biparental Oriental species may be readily recognized by the generally yellow coloration, hyaline wings and long propodeum with large, strongly overlapping crenulae in both sexes, and by the conspicuous microscopic stippling of the posterior abdominal sternites in the male. It may be separated from other members of the **lingnanensis** group by the dusky thoracic sterna, relatively pale setae, moderately elongate antennal club (less than $3\frac{1}{2}$ times as long as wide) and extensively pigmented pupae.

*Female.* Eyes finely setose. Mandibles (Figures 879, 880) well developed, with a distinct ventral denticle and a central denticle merging into a dorsal truncation (see also Figure 108); maxillary palpi (Figure 881) 2-segmented, labial palpi 1-segmented. Antennal scape (Figures 882, 883) slender, 5–6 times as long as wide, distinctly (up to $1\frac{1}{5}$ times) longer than the club; pedicel about $1\frac{3}{5}$ times to fully twice as long as wide, considerably ($1\frac{1}{3}$ to $1\frac{2}{5}$ times) longer than the third segment of the funicle; first funicular segment somewhat trapezoidal, about $1\frac{1}{4}$ to $1\frac{3}{5}$ times as wide as long; second segment nearly symmetrical, considerably shorter and usually slightly wider than the first segment, usually $1\frac{4}{5}$ to $2\frac{2}{3}$ times as wide as long; third funicular segment varying from about as long as wide in small specimens to $1\frac{2}{5}$ times as long as wide in large ones, bearing 2 (rarely 3) longitudinal sensilla; club about 3 times as long as wide (2.55–3.40), $2\frac{3}{5}$ to 3 times longer and considerably wider than the preceding segment, bearing 5–7 longitudinal sensilla.

Setae on thorax moderately slender and pale, visible under ×60 magnification; those on head visible only under ×120 magnification; setae on sides of abdomen paler, usually invisible or barely discernible under high magnification (×120 under a stereoscopic microscope, on a white background). Vertex with 2 pairs of long setae along occipital margin, in addition to numerous shorter setae. Mesoscutum (Figure 885) with 9–13 (usually 10–12) setae, the posterior pair and 1 seta at each anterolateral corner distinctly longer than the others; each parapsis with 2 shorter setae, each axilla with 1 seta; scutellum with 4, the discoid sensilla about equidistant from the two pairs or somewhat closer to the anterior pair. Frontovertex, pronotum and mesonotal sclerites reticulate. Pronotum shown in Figure 884. Scutellum oval, about $\frac{3}{4}$ median length of mesoscutum (0.71–0.77). Metanotum (Figures 885–887) short, reticulate except on the sides, very slightly arcuate, posterior margin nearly straight; anteromedian apodeme (Figure 887) moderately robust, about as long as median length of metanotum, sometimes considerably (up to $1\frac{3}{10}$ times) longer.

Propodeum (Figures 885–887) long, $4\frac{3}{5}$ to $6\frac{1}{4}$ times as long as the metanotum, $\frac{7}{10}$ to $\frac{4}{5}$ length of scutellum, reticulate on the sides, broadly reticulate on a wide central area, without a distinct median salient; crenulae (Figure 181, p. 158) 3+4 to 7+8, large, elongate, strongly overlapping, the 2 sets clearly separated and forming 2 submedian lobes.

Second abdominal tergite (Figures 885–887) transversely striated anteriorly on both sides, rather extensively transversely reticulate centrally; tergites III–VII reticulate on the sides, with a few fine setae in a short transverse row on each reticulate area; third tergite (Figure 887) transversely reticulate-striated also mesad of the lateral setiferous areas, more narrowly so across center; tergites IV–VI (Figure 888) faintly transversely striated anteriorly across center, longitudinally striated posteriorly, this sculpture being more evident on the sixth tergite; seventh tergite transversely reticulate anteriorly across center, longitudinally striated posteriorly, bearing a pair of fine submedian setae (rarely 1 or 3 setae); eighth tergite delicately reticulate, with a transverse row of 3–6 (usually 4) setae between spiracles; syntergum (Figure 888) moderately long, triangular, very faintly reticulate, bearing 5–8 setae in a transverse row. Cerci situated somewhat closer to posterior spiracles than to tip of abdomen, with 2 long setae and 1 short seta. Ovipositor shaft about $1\frac{2}{3}$ times to nearly twice as long as the middle tibia (1.64–1.90); ovipositor sheaths about $\frac{2}{5}$ to $\frac{1}{2}$ length of middle tibia (0.43–0.51).

Strigil of foreleg shown in Figure 889. Mid-tibial spur (Figure 890) $\frac{3}{4}$ to nearly full length of the corresponding basitarsus.

Forewing (Figure 891) about $2\frac{2}{5}$ to $2\frac{2}{3}$ times as long as wide, marginal fringe usually not exceeding $\frac{1}{4}$ width of disk. Delta area with 30–51 setae in 4–5 rows, these considerably longer and sparser than the setae distad of speculum, not clearly separated from row of setae along posterior margin of wing, widely separated from 1–3 setae below distal portion of submarginal vein; costal cell with a row of fine setae along proximal half to three fifths or so, and 1 coarse seta near apex. Submarginal vein usually bearing 2 coarse setae (rarely 3), the proximal one about $\frac{3}{5}$ to $\frac{7}{10}$ length of the distal, and 12–23 bullae (see also Figure 233, p. 187). Marginal vein bearing 9–13 prominent, subequal setae along anterior margin, these $1\frac{1}{7}$ to nearly $1\frac{1}{2}$ times longer than the setae in a row along center of vein.

Forewing hyaline, faintly infumated below junction of submarginal and marginal veins and at base of delta; a fuscous streak along posterior margin of wing, distad of speculum.

General coloration yellow; thoracic sterna dusky, usually rather faintly so, the mesosternum often with just the longitudinal stem of the furca ("Y") marked with blackish; posterior margin of scutellum narrowly lined with blackish; a short black streak at base of forewing, below tegula; abdomen immaculate. Legs concolorous with body. Antennal scape pale, faintly marked with fuscous ventrally, rest of antenna uniformly faintly infumated. Wing veins hyaline.

Length 0.73–1.05 mm.

*Male*. Essentially similar to the female in structure, chaetotaxis, sculpture and coloration, differing mainly in details of antennal structure and abdominal sculpture.

Antennal scape (Figure 892) about $4\frac{1}{2}$ to 6 times as long as wide, up to $1\frac{1}{4}$ times as long as the club; pedicel $1\frac{3}{4}$ to 2 times as long as wide, $1\frac{1}{3}$ to $1\frac{2}{3}$ times longer than the third segment of the funicle; first funicular segment up to $1\frac{2}{3}$ times as wide as long, second segment $1\frac{1}{3}$ to 2 times as wide as long; third funicular segment up to $1\frac{1}{4}$ times as long as wide, bearing 2 longitudinal sensilla (rarely 1); club about $2\frac{3}{5}$ to 3 times as long as wide, about $2\frac{1}{2}$ to 3 times longer and considerably wider than the preceding segment, bearing 3–5 longitudinal sensilla, with a partly cut-off, specialized sensory area, bearing minute setae, on the ventral surface (Figure 893).

Pronotum shown in Figure 894. Mesoscutum (Figure 895) with 10–15 setae, each parapsis with 2, each axilla with 1, scutellum with 4 (Figure 895 shows an abnormal chaetotaxis of the scutellum, with 6 setae). Propodeum somewhat shorter than in the female, $3\frac{3}{5}$ to $4\frac{4}{5}$ times as long as the metanotum, $\frac{3}{5}$ to $\frac{4}{5}$ length of scutellum; crenulae 4+4 to 6+7, as in the female.

The 3 posterior abdominal sternites (Figure 896) are conspicuously stippled anteriorly, with several rows of minute spines (see also Figure 216, p. 177). Genitalia (Figure

896) about $\frac{2}{3}$ to $\frac{4}{5}$ length of middle tibia (0.66–0.78); digital sclerites $\frac{1}{5}$ to $\frac{1}{4}$ the combined length of aedeagus and apodemes (0.21–0.25).

Forewing (Figure 897) about $2\frac{3}{10}$ to $2\frac{3}{5}$ times as long as wide; delta with 20–37 setae in 3–5 rows. Submarginal vein with 14–21 bullae. Marginal vein with 7–12 prominent setae along anterior margin.

Pigmentation of thoracic sterna usually somewhat paler than in the female.

Length 0.69–0.96 mm.

Redescribed from numerous ♀♂ specimens, from a stock originally obtained from the California red scale, *Aonidiella aurantii* (Maskell), in China in 1947 and propagated on the oleander scale, *Aspidiotus nerii* Bouché [= *A. hederae* (Vallot)] in the insectary of the University of California, Riverside; some of these specimens were used by Compere (1955) in his original description of *lingnanensis*; also from a "selected strain", propagated on oleander scale at the Riverside insectary (see below).

Type series in the collection of the Division of Biological Control, University of California, Riverside.

Three slides bearing syntypes of *lingnanensis* were deposited at the British Museum (Natural History), London.

**Notes.** The pupa of *lingnanensis* is characterized by dark (shiny blackish), well-defined pigmentation on the mid-thoracic sterna and on the mid-ventral abdominal plates; head, wing pads and appendages usually yellow to yellow-brown; dorsal surface yellow (see Figure 247 and DeBach, 1959).

The spermathecal capsule of *lingnanensis* is reniform, averaging 0.05 mm in length and 0.04 mm in width (Rao and DeBach, 1969a).

Extensive laboratory mating tests have shown that *lingnanensis* is completely reproductively isolated from *lepidosaphes*, *africanus*, *holoxanthus*, *melinus*, *fisheri* and *coheni* (Rao and DeBach, 1969a; see further notes under "Related Forms" (below) and under *A. coheni*).

*A. lingnanensis* was originally imported into California from South China by J. L. Gressitt in 1947. It was first recognized as distinct from *chrysomphali*—then the only species of *Aphytis* parasitic upon the California red scale in California—on account of its distinctive pupal pigmentation. Early records referred to this species as *Aphytis* "A".

The following quotation from Compere (1955) may be of historical interest and illustrates the importance of accurate taxonomy for biological control: "The early history of this species cannot now be traced with certainty. Circumstantial evidence indicates that it was probably this species—not *chrysomphali*—that George Compere collected in South China in 1906 and attempted to introduce into California and Western Australia. At an early date the identity of the species was a matter of dispute between George Compere and L. O. Howard. The latter insisted that it was *mytilaspidis*, and the former that it was not. *A. lingnanensis* was collected in South China more than once by F. Silvestri in 1924–1925. However, no attempt was made to propagate the species or to colonize it in California at that time, owing to my mistaken belief that it was the same as *chrysomphali*. In 1932, I collected *lingnanensis* in South China but again did not make an attempt to introduce it into California, owing to the belief that it was *chrysomphali*. In fact, measures were taken in China to destroy a colony that developed accidentally in a breeding cage reserved for propagating forms of *Casca* and *Comperiella* on red scale."

**Utilization in Biological Control.** Imported from southern China in 1947, *A. lingnanensis* (then designated *Aphytis* "A") was first released in California against the California red scale in 1948, and establishment was evident by the end of that year. It was mass-reared and distributed throughout the infested citrus areas of the state during 1948–1954. Although its dispersal was rather slow, it soon became the dominant parasite of the scale, gradually displacing *A. chrysomphali* in the interior and intermediate climatic areas. By 1959, *A. chrysomphali* was virtually eliminated from all of southern California, except for a few small coastal pockets. Overall control of the California red scale was improved, but *A. lingnanensis* was not sufficiently effective in the interior citrus areas, where climatic extremes caused mortality of the parasites and prevented them from asserting their full biotic potential (see p. 62). Efforts were therefore made to augment this species by periodic releases (DeBach, Dietrich, Fleschner and Fisher, 1950): 100,000 to 400,000 parasites were colonized per acre each year, in 9 monthly allotments, usually from March through November. Results were good in intermediate climatic areas but unsatisfactory in interior areas (DeBach, Rosen and Kennett, 1971; Rosen and DeBach, 1978).

In order to further augment *A. lingnanensis* in interior areas, an attempt was made to select cold- and heat-resistant strains of this species. Subcultures of adult *A. lingnanensis* were exposed to various regimes of selective thermal pressure for over 100 generations. Quite striking and essentially permanent increases in tolerance of both high and low temperatures were obtained (White, DeBach and Garber, 1970). The various subcultures were eventually combined into one "selected strain", which was released in the interior areas of California. However, by that time *A. melinus*, a species even more adapted to extreme conditions, had become established in southern California and proceeded to displace *A. lingnanensis* from interior areas (see p. 73). At present, *A. lingnanensis* is the dominant parasite of the California red scale in the immediate coastal areas of California (DeBach, Rosen and Kennett, 1971).

Following its establishment in California, *A. lingnanensis* was introduced into several other countries. Establishment on the California red scale has been recorded in Turkey, Cyprus, Morocco and South Africa. Interestingly, *lingnanensis* had apparently become established by ecesis in Texas and Mexico even prior to its introduction into California (DeBach, 1971b). Likewise, this species must have entered Australia long before it was discovered and described. Specimens reared by A. A. Girault from "*Mytilaspis* and *Aspidiotus aurantii*" in Indooroopilly, Brisbane, as early as April 1930, and identified by him as *chrysomphali*, were found by us, upon remounting, to be *A. lingnanensis* (see Figure 898). Also, we collected *lingnanensis* throughout Queensland in 1971. To our knowledge it has never been purposely imported there. This raises the question whether Australia is not part of the original range of *lingnanensis*.

*A. lingnanensis* was recently discovered as a parasite of the purple scale, *Cornuaspis* [=*Lepidosaphes*] *beckii* (Newman) on citrus in Queensland. This interesting Australian form should be considered as a possible candidate for importation against that serious pest of citrus in other countries.

*A. lingnanensis* has been imported during the past several years into Florida from Hong Kong (ex snow scale, *Unaspis citri* (Comstock)) for control of the snow scale in that state. Dr. Robert Brooks (personal communication) reports good establishment and initial reduction in snow scale populations in test groves. We have confirmed the identity of this species by hybridization tests.

**Biology.** Information on the biology of *lingnanensis* was presented by DeBach and White (1960), Quednau (1964c) and various other authors and was recently summarized by Rosen and DeBach (1978). The species is an arrhenotokous parasite of the California red scale, *Aonidiella aurantii* (Maskell); the oleander scale, *Aspidiotus nerii* Bouché [= *A. hederae* (Vallot)], may also serve as a host. The sex ratio in the field is about 50 : 50. The life cycle from egg to adult takes 54 days at 15.6°C, 22 days at 21°C, 15 days at 26.7°C and 13 days at 32°C. The average longevity of the ovipositing female is 10 days and may be as long as 25 days. At 27°C and 60 % relative humidity, an average fecundity as high as 57 eggs per female has been obtained, and the female destroys an average of 46 scales by predatory host-feeding and mutilation. All stages of *A. lingnanensis* are adversely affected by low temperatures and low humidities (see also DeBach and Rao, 1968).

**Material Examined.** Specimens determined as *lingnanensis* in the collection of the Division of Biological Control, University of California, Riverside, were obtained from the California red scale, *Aonidiella aurantii* (Maskell), on citrus in China, Hong Kong, California, Texas, Mexico, Cyprus, Morocco and Australia, and on rose in India; from the yellow scale, *Aonidiella citrina* (Coquillett), on citrus in Pakistan; from the Florida red scale, *Chrysomphalus aonidum* (L.), on citrus in Taiwan; from (?) the dictyosperum scale, *Chrysomphalus dictyospermi* (Morgan), on breadfruit in Fiji; from the latania scale, *Hemiberlesia lataniae* (Signoret), on avocado in California; from the cactus scale, *Diaspis echinocacti* (Bouché), on *Opuntia* in Mexico; from the laurel scale, *Aonidia lauri* (Bouché), on *Laurus nobilis* in Spain; from *Pseudaonidia trilobitiformis* (Green) on *Laurus nobilis* in New Caledonia; from the rufous scale, *Selenaspidus articulatus* (Morgan), on citrus in Trinidad; from the coconut scale, *Temnaspidiotus destructor* (Signoret), on banana in Puerto Rico and on papaya in Trinidad; from the purple scale, *Cornuaspis* [= *Lepidosaphes*] *beckii* (Newman), on citrus in Queensland, Australia; from the citrus snow scale, *Unaspis citri* (Comstock), on citrus in Hong Kong and in Mexico; from undetermined hosts on citrus in Hong Kong and El Salvador, on *Ficus* in Taiwan, on rose in Pakistan, on live oak on Santa Cruz Island, California, on *Cycas revoluta* in Brazil and on *Colias* in Jamaica; from an undetermined host in Malaya; and from insectary cultures on California red scale in California and on oleander scale in California and Australia.

All these series cannot at present be distinguished from *lingnanensis*. Obviously, some of the host records may be erroneous, or the parasites may represent forms that are biologically distinct from *lingnanensis*. In fact, a biosystematic study by Rao and DeBach (1969a, 1969b) has indicated that certain biological forms of *Aphytis* which are morphologically indistinguishable from *lingnanensis* indeed exhibit various degrees of reproductive isolation from this species and must therefore be regarded as semispecies or even distinct species in relation to it. A discussion of two such forms follows.

**Related Forms.** A culture obtained from the coconut scale, *Temnaspidiotus destructor* (Signoret), on coconut palm in Puerto Rico, designated "2002", cannot be separated from *lingnanensis* by morphological criteria. However, reciprocal crossing tests have demonstrated that it is partially reproductively isolated from *lingnanensis*, although the occasional hybrids between them are fertile. Moreover, "2002" and *lingnanensis* show great differences in their crossability with other forms, such as "R-65-23" (see below). "2002" is therefore regarded as a semispecies in relation to *lingnanensis*.

Another culture, obtained in 1965 from the citrus snow scale, *Unaspis citri* (Comstock), on citrus in Florida and designated "R-65-23", poses an even more puzzling problem. Morphologically indistinguishable from both *lingnanensis* and "2002", it hybridizes fairly readily with "2002" and may be considered a semispecies in relation to that form. However, "R-65-23" exhibits strong reproductive isolation in relation to *lingnanensis*, with which it produces only sterile hybrids. On the basis of available information, therefore, "R-65-23" must be regarded as biologically distinct from *lingnanensis*, although some gene exchange may theoretically occur between them, with "2002" serving as an intermediary form.

The complex relations between some of the forms comprising the *lingnanensis* complex are presented in Figure 263 (p. 209).

A series obtained from the Glover scale, *Insulaspis* [= *Lepidosaphes*] *gloverii* (Packard), on citrus in Sinaloa, Mexico, appears to be somewhat more elongate than typical *lingnanensis*, with a somewhat narrower forewing, apparently somewhat more strongly infuscated thoracic sterna and a certain amount of duskiness at the base of the abdomen. These slight differences, as well as the unusual host record, seem to indicate that this may possibly be yet another distinct form. However, no biological information is available regarding it. A series obtained from the same host in Florida appears to be more similar to typical *lingnanensis*.

Further biosystematic investigation will undoubtedly reveal additional biologically distinct forms which are at present referred to *lingnanensis*. Although very confusing to the taxonomist, discovery of such forms may prove to be of great importance for biological control.

## 60. **Aphytis coheni** DeBach

(Figures 899–912)

*Aphytis coheni* DeBach, 1960, Ann. Entomol. Soc. Amer., **53**: 705.
*Aphytis coheni*: Quednau, 1964, J. Entomol. Soc. S. Afr., **27**: 110–112; Ferrière, 1965, Hymenoptera Aphelinidae d'Europe et du Bassin Méditerranéen, pp. 87–88.

This problematic biparental species is extremely closely related to *lingnanensis*, differing slightly from the latter species in the more pronounced pigmentation of the thoracic sterna, somewhat more dusky general coloration, more numerous setae in the delta of the forewing, somewhat more numerous, darker setae on the mesoscutum, and somewhat darker setae on the sides of the abdomen. In order to avoid unnecessary repetition, *coheni* is redescribed below relative to *lingnanensis*.

*Female*. Eyes and mouthparts as in *lingnanensis*. Antennae (Figures 899, 900) essentially as in *lingnanensis*; club about 3 times as long as wide.

Setae on thorax somewhat coarser and darker than in *lingnanensis*, readily visible under $\times$ 60 magnification; those on head paler but clearly visible under $\times$ 120 magnification; setae on sides of abdomen darker than in *lingnanensis*, usually distinct at high magnification ($\times$ 120 under a stereoscopic microscope, on a white background). Chaetotaxis of head as in *lingnanensis*. Mesoscutum (Figures 901–904) with 11–15 setae (usually 12–14); discoid sensilla on scutellum usually closer to the anterior than to the posterior pair of setae. Otherwise structure, chaetotaxis and sculpture of head and thorax as in *lingnanensis*.

Propodeum (Figures 901–903, 905, 906) about $4\frac{1}{5}$ to 5 times as long as the metanotum, about $\frac{7}{10}$ to $\frac{4}{5}$ length of scutellum; crenulae 5+5 to 8+8, as in *lingnanensis*.

Sculpture and chaetotaxis of abdominal tergites essentially as in *lingnanensis*, setae on sides of tergites sometimes in 2 short rows. Cerci and ovipositor as in *lingnanensis*.

Legs as in *lingnanensis*.

Forewing (Figure 907) essentially as in *lingnanensis*, but delta area with 50–70 setae in 5–6 rows; costal cell sometimes with 2 coarse setae near apex. Forewing hyaline, somewhat more distinctly infumated at base than in *lingnanensis*.

General coloration yellow, very similar to that of *lingnanensis*; thoracic sterna more strongly infuscated, including a distinctly fuscous triangle between the arms of the mesosternal furca; posterior margin of scutellum distinctly lined with black; a short black streak at base of forewing, below tegula; in cleared specimens, the basal abdominal sternites and sides of basal abdominal tergites are distinctly, although faintly, dusky. Fore and middle tibiae and tarsi faintly dusky. Antennal pedicel, funicle and club uniformly, more distinctly dusky than in *lingnanensis*.

Length 0.86–1.11 mm.

*Male*. Essentially similar to the female in structure, chaetotaxis, sculpture and coloration, and very similar to the male of *lingnanensis*.

Antennae essentially as in *lingnanensis*. Mesoscutum (Figures 908, 909) with 10–15 setae. Propodeum (Figures 908–910) over 4 times as long as the metanotum ($4\frac{1}{7}$ to $4\frac{3}{5}$ times), about $\frac{7}{10}$ to $\frac{3}{4}$ length of scutellum; crenulae 4+4 to 7+7, as in *lingnanensis*.

Abdominal sternites (Figure 911) as conspicuously stippled as in *lingnanensis*. Genitalia (Figure 911) as in *lingnanensis*, $\frac{2}{3}$ to $\frac{4}{5}$ length of middle tibia (0.66–0.81); digital sclerites about $\frac{1}{4}$ the combined length of aedeagus and apodemes (0.23–0.25).

Forewing (Figure 912) as in *lingnanensis*, but delta with 28–49 setae in 4–6 rows.

General coloration as in the female, somewhat duskier than *lingnanensis*.

Length 0.66–0.90 mm.

Redescribed from the following material: 1♀ (lectotype), reared by D. Nadel from the California red scale, *Aonidiella aurantii* (Maskell), on citrus, Ashkelon, Israel, January 18, 1960; 1♂ (paralectotype), same data (on same slide); numerous ♀♂ (paralectotypes), from an insectary stock reared on the oleander scale, *Aspidiotus nerii* Bouché [= *A. hederae* (Vallot)], Riverside, California, May 1960—originally from the California red scale on citrus in Israel. Also from numerous ♀♂, reared on the yellow scale, *Aonidiella citrina* (Coquillett), in the insectary at Riverside; additional ♀♂ reared from the California red scale on citrus in Israel and Cyprus and in the insectary in Cyprus; from the oleander scale on acacia in Israel; and from the latania scale, *Hemiberlesia lataniae* (Signoret), on acacia in Israel and on an ornamental vine in Cyprus.

Lectotype and paralectotypes in the collection of the Division of Biological Control, University of California, Riverside; additional type material in the U.S. National Museum of Natural History, Washington, and in the British Museum (Natural History), London.

**Notes.** The pupa of *coheni* is indistinguishable from that of *lingnanensis*. Also, the spermathecal capsule of *coheni* is similar to that of *lingnanensis* (Rao and DeBach, 1969a).

The origin of *coheni* is a mystery. It is known to occur naturally in the field only in Israel (DeBach, 1960; Rosen, 1965) and nearby Lebanon (Traboulsi, 1969) but, as pointed out by DeBach (1960), it is unlikely to have evolved there as it is not known to occur in any other Mediterranean country. Although *coheni* is not known at present to occur in the Orient, a form very closely related to it—the "khunti" *Aphytis* (see below)—was obtained in India (DeBach, 1959; see also Rao and DeBach, 1969a, 1969b). It may therefore be assumed that *coheni* has become established in Israel by ecesis from Southeast Asia.

Rivnay (1968) suggested that *coheni* was accidentally introduced into Israel from Hong Kong, together with *A. holoxanthus*. His suggestion was based on the alleged determination of *Aphytis* material from Hong Kong as *coheni* by Rosen, but this must be ascribed to some misunderstanding as we have never collected or seen any *coheni* material from Hong Kong. At any rate, positive identification of *coheni* can at present be made only by crossing tests with authenticated live material.

*A. coheni* and *A. lingnanensis* are at present regarded as sibling species. The morphological differences between them are at best very slight, and it may be argued that they are related to the somewhat larger size of *coheni*. With some experience, the greater number of setae in the delta of the forewing, the somewhat darker pigmentation, and the darker setae on the sides of the abdominal tergites may usually be used reliably to separate the two species; however, minute specimens of *coheni* are indeed almost indistinguishable from *lingnanensis*. Nevertheless, there is no question that the two species are reproductively isolated (DeBach, 1960; Rao and DeBach, 1969a, 1969b). Hybrids between *coheni* and *lingnanensis* are sterile, and the two species show highly significant sexual isolation, hence they can be considered to have acquired species status with respect to each other.

**Utilization in Biological Control.** *A. coheni* was recorded by DeBach (1960) as an important parasite of the California red scale in Israel. In the early 1960's, this species was the only parasite obtained from that scale in the inland valleys of Israel and the most abundant one along the coastal plain, where it was believed to be displacing *A. chrysomphali* (Rosen, 1965). Recent studies have shown that while *chrysomphali* remains the dominant parasite in coastal areas and *melinus* is spreading in intermediate areas, *coheni* is still the dominant parasite of the California red scale in the hot inland valleys of Israel (S. Kamburov, unpublished data, and Y. Rössler, personal communication).

Attempts were made to introduce *coheni* into California, Cyprus, Greece and South Africa. Initial establishment was reported from Cyprus by Wood (1963), but subsequent information indicated that *coheni* was largely displaced there by *A. melinus*. Initial recoveries were also reported in South Africa (Quednau, 1964b). Importation into California and Greece has failed (see Rosen and DeBach, 1978). Nevertheless, *A. coheni* should be considered as a potential candidate for introduction against the California red scale wherever this pest poses an economic problem, especially in more extreme climatic areas.

**Biology.** The life history of *coheni* is essentially similar to that of other members of the **lingnanensis** group. According to Avidov, Balshin and Gerson (1970), the life cycle takes 25 days at 20°C, 16 days at 24°C, 12 days at 28°C and 11 days at 32°C. Mated females laid an average of 26 eggs per female, unmated females 25. The species is arrhenotokous, females comprising only 44.6% of the population in field samples on citrus in Israel (Rosen, 1965).

females laid an average of 26 eggs per female, unmated females 25. The species is ar-

**Related Form.** A form of *Aphytis* was obtained by DeBach (1959) from *Aonidiella orientalis* (Newstead) on rose at Khunti, India (near New Delhi), and was at first designated "the khunti *lingnanensis*", although it would not cross with *lingnanensis*. This form (Figures 913–917) is morphologically indistinguishable from *coheni*. Reciprocal crossing tests reported by Rao and DeBach (1969a, 1969b) have demonstrated that "*Aphytis coheni* and 'khunti' hybridize readily in the laboratory, yielding fertile hybrids. However, they show different crossing relations with '2002', *lingnanensis*, and 'R-65-23', indicating that they are genetically rather distinct. They are therefore considered to be semispecies with respect to each other." Both *coheni* and "khunti" are considered good species with respect to *lingnanensis*, "2002" and "R-65-23" (see also Figure 263, p. 209).

Thus, whereas *lingnanensis*, "2002" and "R-65-23" form one complex of semispecies in the **lingnanensis** group, *coheni* and "khunti" form a sibling complex of semispecies that is reproductively isolated from the former complex.

## 61. **Aphytis africanus** Quednau

(Figures 918–930)

*Aphytis africanus* Quednau, 1964, J. Entomol. Soc. S. Afr., **27**:112–113.

This interesting biparental African species is rather similar to *lingnanensis*, but may be readily distinguished by the somewhat smaller, less elongate propodeal crenulae, the somewhat shorter antennal club and ovipositor sheaths, and by the very weak stippling on the abdominal sternites of the male. In order to avoid unnecessary repetition, *africanus* is redescribed here relative to *lingnanensis*.

*Female.* Eyes and mouthparts as in *lingnanensis.* Antennae (Figures 918, 919) essentially as in *lingnanensis*; third funicular segment varying from about as long as wide (rarely somewhat wider than long) to about $1\frac{1}{3}$ times longer than wide, bearing 1–2 longitudinal sensilla; club about $2\frac{3}{5}$ to 3 times as long as wide (usually less than 3 times), $2\frac{2}{3}$ to $3\frac{1}{4}$ times longer and considerably wider than the preceding segment, bearing 5–8 longitudinal sensilla.

Structure, chaetotaxis and sculpture of head and thorax essentially as in *lingnanensis.* Setae generally somewhat paler; those on thorax visible under $\times 60$ magnification, those on head only under $\times 120$ magnification; setae on sides of abdomen invisible under $\times 120$ magnification. Mesoscutum (Figures 920, 921) with 9–12 (usually 10) setae; each parapsis with 2 setae, each axilla with 1; scutellum with 4 setae, the discoid sensilla usually considerably closer to the anterior than to the posterior pair.

Propodeum (Figures 921–923) as in *lingnanensis*, $4\frac{4}{5}$ to $6\frac{1}{3}$ times as long as the metanotum, $\frac{3}{4}$ to $\frac{9}{10}$ length of scutellum; crenulae 3+4 to 6+6, distinctly overlapping but somewhat oblique, smaller, less elongate than in *lingnanensis.*

Sculpture and chaetotaxis of abdominal tergites (Figure 924) essentially as in *lingnanensis*, but the sculpture is generally fainter; eighth tergite with a transverse row of 4 setae between spiracles; syntergum bearing 5–6 setae in a transverse row. Cerci (Figure 924) as in *lingnanensis.* Ovipositor shaft about $1\frac{1}{2}$ to $1\frac{3}{4}$ times as long as the middle tibia, ovipositor sheaths shorter than in *lingnanensis*, $\frac{2}{5}$ length of middle tibia or nearly so (0.37–0.40).

Legs as in *lingnanensis.*

Forewing (Figure 925) about $2\frac{2}{5}$ to $2\frac{3}{5}$ times as long as wide; marginal fringe not exceeding $\frac{1}{3}$ width of disk in smallest specimen. Delta area with 32–67 setae in 4–7 rows; general chaetotaxis as in *lingnanensis.* Submarginal vein bearing 11–18 bullae. Marginal vein bearing 7–11 prominent, subequal setae along anterior margin.

Forewing hyaline, faintly infumated basally, as in *lingnanensis.*

General coloration as in *lingnanensis*; pigmentation of thoracic sterna less pronounced than in *coheni*; mesosternum usually with both the arms and the stem of the furca fuscous, the surrounding plates pale.

Length 0.61–0.89 mm.

*Male.* Very similar to the female in structure, chaetotaxis, sculpture and coloration; rather similar to the male of *lingnanensis.*

Antennal scape (Figures 926, 927) $4\frac{1}{2}$ to $5\frac{2}{5}$ times as long as wide, somewhat longer than the club; pedicel about $1\frac{3}{5}$ times as long as wide; third funicular segment about as long as wide or somewhat wider, bearing 1–2 longitudinal sensilla; club short, $2\frac{2}{5}$ to $2\frac{2}{3}$ times as long as wide, about 3 times longer than the preceding segment, bearing 2–3 longitudinal sensilla, with an oval area on the ventral aspect (Figure 927) bearing minute setae but not visibly cut off by a sensillumlike structure as in *lingnanensis.*

Mesoscutum with 10 setae. Propodeum (Figure 928) as in the female, $4\frac{1}{2}$ to a little over 6 times as long as the metanotum, about $\frac{2}{3}$ to $\frac{3}{4}$ length of scutellum; crenulae 3+3 to 5+6, as in the female.

Abdominal sternites (Figure 929) very weakly stippled, much less conspicuously so than in *lingnanensis*. Genitalia about $\frac{7}{10}$–$\frac{4}{5}$ length of middle tibia (0.69–0.83); digital sclerites about $\frac{1}{4}$ the combined length of aedeagus and apodemes (0.24–0.27).

Forewing (Figure 930) about $2\frac{1}{3}$ to $2\frac{2}{5}$ times as long as wide, marginal fringe not exceeding $\frac{1}{4}$ width of disk. Delta with 32–49 setae in 4–6 rows. Submarginal vein with 12–16 bullae. Marginal vein with 8–10 prominent, subequal setae along anterior margin.

Coloration as in the female.

Length 0.49–0.80 mm.

Redescribed from 20♀♀ and 14♂♂ (syntypes), reared from the California red scale, *Aonidiella aurantii* (Maskell), on orange at Letaba, Transvaal, South Africa, May 26 and October 1, 1961; also from numerous ♀♂, reared from *A. aurantii* on citrus in South Africa and Mozambique; from the Florida red scale, *Chrysomphalus aonidum* (L.), on citrus in South Africa; from the black araucaria scale, *Lindingaspis rossi* (Maskell), on *Phoenix* in South Africa; from the aloë red scale, *Separaspis capensis* (Walker), on *Aloë* in South Africa; and from a laboratory culture on *A. aurantii* in Riverside, originally from Rustenburg, South Africa. Quednau (1964b) listed a few additional hosts from South Africa.

Type series in the collection of the Plant Protection Research Institute, Pretoria, South Africa.

**Notes.** The fully developed pupa of *africanus* is indistinguishable from that of *lingnanensis* and *coheni*. However, in *africanus* the head begins to darken already at the beginning of the pupal stage, when the thorax is just dusky and the abdomen is still pale (Quednau, 1964b).

Some of the apparent differences between *africanus* and *lingnanensis*, especially in antennal proportions, may be partly due to flattening of the type specimens of *africanus* on the slides. However, the shape of the propodeal crenulae is distinctly different: they are smaller and somewhat oblique in *africanus*. Moreover, the lack of conspicuous stippling on the abdominal sternites of the male separates *africanus* from all other members of the **lingnanensis** group. This character, as well as the smaller crenulae and the apparent absence of a sensillumlike structure surrounding the specialized sensory area on the antennal club of the male, may be interpreted to indicate that *africanus* occupies an intermediate position between the **lingnanensis** and **chrysomphali** groups.

*A. africanus* shows complete reproductive isolation from *lingnanensis*, *coheni* and related forms in the **lingnanensis** group, such as "2002" and "khunti", as well as from *holoxanthus*, *melinus* and *fisheri* (Rao and DeBach, 1969a). It also differs markedly from these species in its host preferences, being unable to develop on the oleander scale, *Aspidiotus nerii* Bouché [=*A. hederae* (Vallot)], which is readily accepted by *lingnanensis* and serves as an excellent laboratory host for the latter species and its close allies (see Quednau, 1964b, 1965).

**Biology.** *A. africanus* is an arrhenotokous species. Data on the duration of its life cycle vary: development on the California red scale at 26.7°C takes 12 days according to Rao and DeBach (1969a), $13\frac{1}{2}$ days according to Quednau (1964b). Its fecundity on the California red scale—averaging about 20 eggs per female—is lower than that of several other members of the **lingnanensis** group (Quednau, 1964a, 1965). Obviously, such data cannot be relied upon for the separation of closely related species of *Aphytis*, unless a very high degree of standardization of experimental techniques is achieved.

**Utilization in Biological Control.** Known only from southern Africa, *A. africanus* is believed to have evolved on some native scale insect in that region. The California red scale, which is undoubtedly of Oriental origin, cannot be its original host. Nevertheless, *africanus* is an important parasite of the California red scale in South Africa. However, based upon rather large series of *Aphytis* reared from that host in South Africa and sent to us for identification, *A. melinus* seems to be in the process of displacing *africanus*, at least in some areas.

*A. africanus* had been introduced and released extensively against the California red scale in California and Israel, but establishment did not occur.

## 62. **Aphytis margaretae** DeBach and Rosen

(Figures 931–944)

*Aphytis margaretae* DeBach and Rosen, 1976, Ann. Entomol. Soc. Amer., **69**: 544.

This interesting American species superficially resembles *lingnanensis* in the large propodeal crenulae and pigmented thoracic sterna, but may be readily recognized by the rather peculiar shape of the elongate, conspicuously pigmented crenulae, the somewhat narrower wings, and by the distinctive male antennae, bearing sense organs on the scape and lacking a cut-off sensory area on the club; male abdominal sterna rather weakly stippled.

*Female.* Eyes finely setose. Mandibles well developed, with 2 denticles and a dorsal truncation; maxillary palpi 2-segmented, labial palpi 1-segmented. Antennal scape (Figure 931) slender, about $4\frac{1}{2}$–6 times as long as wide, somewhat longer than the club; pedicel about $1\frac{3}{5}$ to over $1\frac{4}{5}$ times as long as wide, considerably ($1\frac{1}{6}$ to $1\frac{2}{5}$ times) longer than the third segment of the funicle; first funicular segment somewhat trapezoidal, usually $1\frac{2}{5}$ to $1\frac{2}{3}$ times as wide as long; second segment more nearly symmetrical, distinctly shorter and wider than the first segment, about $1\frac{3}{4}$ to $2\frac{1}{2}$ times as wide as long; third funicular segment usually somewhat longer than wide (up to $1\frac{1}{4}$ times), bearing 2 longitudinal sensilla; club tapering, about $2\frac{3}{4}$ to a little over 3 times as long as wide, $2\frac{2}{3}$ to 3 times longer and somewhat wider than the preceding segment, bearing 5–7 longitudinal sensilla.

Setae on thorax rather long, moderately dark, visible under $\times 30$ magnification; those on head and sides of abdomen visible under $\times 60$ magnification. Vertex with 2 pairs of long setae along occipital margin, in addition to numerous shorter setae. Mesoscutum (Figures 932, 933) with 10–14 setae, the posterior pair and 1 seta at each antero-lateral corner distinctly longer and coarser than the others; each parapsis with

2 shorter setae, each axilla with 1 seta (rarely with a supernumerary seta as in Figure 933); scutellum with 4, the discoid sensilla somewhat closer to the anterior than to the posterior pair. Frontovertex, pronotum and mesonotal sclerites reticulate. Scutellum oval, about $\frac{2}{3}$ to $\frac{3}{4}$ median length of mesoscutum. Metanotum (Figures 932, 934) short, slightly arcuate, transversely reticulate except on the sides; anteromedian apodeme robust, usually considerably longer than median length of metanotum.

Propodeum (Figures 932, 934) long, 4–5 times as long as the metanotum, about $\frac{2}{3}$ to $\frac{4}{5}$ length of scutellum, reticulate on the sides, broadly reticulate on a wide central area; posterior margin usually rather deeply notched medially; crenulae 4+5 to 6+7, large, elongate, closely appressed and slightly overlapping, striated longitudinally, the 2 sets distinctly separated by the median notch.

Second abdominal tergite (Figure 934) transversely reticulate anteriorly on both sides and on a wide central area; tergites III–VII (Figure 935) reticulate on the sides, with a few fine setae in a short transverse row on each reticulate area; third tergite transversely reticulate-striated across center; tergites IV–VI with very faint indications of transverse striation; seventh tergite faintly transversely striated across center, bearing 2 (rarely 1 or 3) fine submedian setae; eighth tergite delicately reticulate, with a transverse row of 4 setae between spiracles; syntergum (Figure 935) triangular, rather broad, faintly reticulate, bearing 6–8 setae in a transverse row. Cerci situated considerably closer to posterior spiracles than to tip of abdomen, with 2 long setae and 1 short seta. Ovipositor shaft about $1\frac{3}{5}$ to $1\frac{3}{4}$ times as long as the middle tibia (1.57–1.77, usually about $1\frac{2}{3}$ times); ovipositor sheaths nearly $\frac{1}{2}$ length of middle tibia (0.45–0.50).

Mid-tibial spur somewhat shorter than the corresponding basitarsus.

Forewing (Figure 936) narrow, $2\frac{7}{10}$ to nearly 3 times as long as wide; anterior margin slightly indented beyond venation; marginal fringe not exceeding $\frac{1}{4}$ width of disk (usually about $\frac{1}{6}$ or $\frac{1}{7}$). Delta area with 28–60 setae in 4–6 rows, these considerably longer and sparser than the setae distad of speculum, not clearly separated from row of setae along posterior margin of wing, widely separated from 1–4 setae below distal portion of submarginal vein; costal cell with a row of fine setae along proximal half to three quarters or so, and 1–2 coarse setae near apex, the latter situated on the vein rather than in the cell proper. Submarginal vein bearing 2 coarse setae, the proximal one usually considerably shorter, $\frac{1}{2}$ to $\frac{3}{4}$ length of the distal, and 15–19 bullae. Marginal vein bearing 8–13 prominent, long, subequal setae along anterior margin, about $1\frac{1}{2}$ to $1\frac{3}{4}$ times longer than the setae in a row along center of vein.

Forewing hyaline, faintly infumated below apex of submarginal vein, at base of delta and below stigmal vein; a fuscous streak along posterior margin of wing, distad of speculum.

General coloration yellow; prosternum faintly dusky, mesosternum and metasternum strongly, rather conspicuously infuscated; posterior margin of scutellum narrowly lined with fuscous; a short black streak at base of forewing, below the faintly dusky tegula; posterior margin of propodeum and crenulae distinctly marked with blackish (Figure 934); abdomen immaculate. Legs concolorous with body, tibiae and tarsi faintly dusky. Antennal scape and pedicel faintly dusky, first 2 funicular segments somewhat paler, third funicular segment and club uniformly blackish. Wing veins hyaline.

Length 0.78–1.11 mm.

*Male.* Essentially similar to the female in structure, chaetotaxis, sculpture and coloration, differing mainly in details of antennal structure, sense organs and abdominal sculpture.

Antennal scape (Figures 937–939) somewhat wider than in the female, 4 to $4\frac{4}{5}$ times as long as wide, distinctly longer than the club, with an oval plate on the ventral aspect bearing 2 minute peglike sense organs; pedicel and funicular segments essentially as in the female; club as in the female, about $2\frac{2}{5}$ to $3\frac{2}{5}$ times longer than the preceding segment, bearing 2–4 longitudinal sensilla, lacking any specialized sensory area on the ventral surface.

Mesoscutum (Figures 940, 941) with 9–12 setae. Propodeum (Figure 941) somewhat shorter than in the female, $3\frac{2}{5}$ to 5 times as long as the metanotum, $\frac{3}{5}$ to $\frac{2}{3}$ length of scutellum; crenulae 3 + 3 to 6 + 8, as in the female.

Posterior abdominal sternites (Figures 942, 943) stippled, the stippling more diffuse, less conspicuous than in *lingnanensis*. Genitalia (Figures 942, 944) about $\frac{3}{5}$ to $\frac{3}{4}$ length of middle tibia (0.60–0.74); phallobase narrowly elongate proximally; digital sclerites about $\frac{1}{4}$ the combined length of aedeagus and apodemes (0.22–0.29).

Forewing somewhat wider than in the female, $2\frac{1}{2}$ to $2\frac{4}{5}$ times as long as wide, marginal fringe not exceeding $\frac{1}{6}$ width of disk; delta area with 23–39 setae in 4–5 rows. Submarginal vein with 13–19 bullae. Marginal vein with 6–10 prominent setae along anterior margin.

Pigmentation of thoracic sterna often somewhat paler than in the female, the prosternum varying from immaculate to faintly dusky, the mesosternum and metasternum distinctly infuscated.

Length 0.69–0.97 mm.

Described from numerous ♀♂ specimens, all reared by Paul DeBach as follows: from the cactus scale, *Diaspis echinocacti* (Bouché), on *Opuntia* sp., La Paz, Baja California Sur, Mexico, June 23, 1966 (♀ holotype, ♂ allotype, and paratypes), July 1966 (paratypes), December 28 (paratypes), and July 9, 1969 (paratypes); from the coconut scale, *Temnaspidiotus destructor* (Signoret), on coconut palm, La Paz, July 9, 1969 (paratypes); from *Pseudoparlatoria* sp. on honeysuckle, La Paz, July 9, 1969 (paratypes); from undetermined hosts on fig and on "ginger-like ornamental," La Paz, January 15, 1967 (paratypes); also from the following ♀♂ material, not included in the type series, reared by Paul DeBach: from undetermined hosts on mixed ornamentals, Guaymas, Sonora, Mexico, November 1965; from undetermined hosts on "canna or ginger," Terezopolis, Rio de Janeiro State, Brazil, April 7, 1962; from the cyanophyllum scale, *Abgrallaspis cyanophylli* (Signoret), on an ornamental tree, Copacabana, Rio de Janeiro, Brazil, April 20, 1962; and from undetermined hosts on "wild jungle plant," Acapa Territory, Brazil, July 25, 1962. All mounted in Hoyer's medium; holotype and allotype together on a separate slide, with one pupa.

Type series in the collection of the Division of Biological Control, University of California, Riverside. Paratypes have been deposited in the U. S. National Museum of Natural History and in the British Museum (Natural History).

**Notes.** The pupa of *margaretae* is entirely yellow, with dusky wing-pads. In a mature, cleared pupa the dusky meso- and metasternum of the imago show through the immaculate pupal integument.

*A. margaretae*, apparently an endemic American species, differs markedly from *lingnanensis* and its Oriental allies in the presence of sense organs on the antennal scape of the male and in the absence of a cut-off sensory area on the antennal club of the male. The crenulae of *margaretae* are elongate, often nearly parallel-sided, usually only slightly overlapping, and rather distinctively different from the crenulae of *lingnanensis*. In fact, although we have referred it for the time being to the **lingnanensis** group, *margaretae* may not be a bona fide member of this group.

This species was named by us in honor of Mrs. Margaret DeBach, who helped collect original material.

**Potential Value for Biological Control.** *A. margaretae* is a potentially important new natural enemy of the coconut scale, and should be considered for introduction into areas where this serious pest still causes economic damage.

## 63. **Aphytis holoxanthus** DeBach

(Figures 945–955)

*Aphytis holoxanthus* DeBach, 1960, Ann. Entomol. Soc. Amer., **53**:704–705.
*Aphytis holoxanthus*: Quednau, 1964, J. Entomol. Soc. S. Afr., **27**:115; Traboulsi, 1969, Ann. Soc. Entomol. Fr., **5**:54–56.

This important, biparental, Oriental parasite of the Florida red scale is closely related to *lingnanensis*, but even more so to *melinus* and *fisheri*. It differs markedly from *lingnanensis* in having immaculate thoracic sterna. It may be distinguished from *melinus* and *fisheri* by the relatively coarse, dark setae on the thorax and head, the long but more robust antennal club, the somewhat longer ovipositor and sheaths, and the broad forewing, with less than 50 setae in the delta. *A. holoxanthus* is described below relative to *lingnanensis*.

*Female.* Eyes and mouthparts as in *lingnanensis*. Antennal scape (Figures 945, 946) slender, about 5 to $6\frac{1}{2}$ times as long as wide, a little longer than the club; pedicel slender, $1\frac{3}{4}$ to 2 times as long as wide, considerably (up to $1\frac{2}{5}$ times) longer than the third segment of the funicle; first funicular segment usually about $1\frac{1}{4}$ to $1\frac{1}{2}$ times as wide as long; second segment usually slightly shorter and wider than the first, about $1\frac{3}{5}$ to $1\frac{7}{10}$ times as wide as long; third funicular segment about $1\frac{1}{5}$ to $1\frac{2}{5}$ times as long as wide, bearing 2 (sometimes 3) longitudinal sensilla; club rather long, robust, about $3\frac{2}{5}$ to $3\frac{7}{10}$ times as long as wide (usually not more than $3\frac{1}{2}$ times), nearly 3 to $3\frac{1}{3}$ times longer and somewhat wider than the preceding segment, bearing 5–6 longitudinal sensilla.

Setae on thorax rather coarse, dark, visible under $\times 30$ magnification; those on head paler but readily visible under $\times 120$ magnification; setae on sides of abdomen pale, slender, invisible even under $\times 120$ magnification. Vertex with 2 pairs of long setae along occipital margin, in addition to numerous shorter setae. Mesoscutum (Figure

947) with 10–13 (usually 10–12) setae, each parapsis with 2, each axilla with 1 seta; scutellum with 4 setae, the discoid sensilla somewhat closer to the anterior than to the posterior pair. Sculpture of head and thorax as in *lingnanensis*. Scutellum $\frac{7}{10}$ to over $\frac{4}{5}$ median length of mesoscutum (0.71–0.84). Metanotum as in *lingnanensis*.

Propodeum (Figures 947–949) $4\frac{1}{4}$ to $5\frac{1}{3}$ times as long as the metanotum, $\frac{3}{4}$ to over $\frac{4}{5}$ length of scutellum, sculptured as in *lingnanensis*; crenulae 5+5 to 7+8, large, elongate, strongly overlapping, as in *lingnanensis*.

Sculpture and chaetotaxis of abdominal tergites as in *lingnanensis*; seventh tergite with a pair of submedian setae; eighth tergite with a transverse row of 4 setae between spiracles; syntergum essentially as in *lingnanensis*, with 6 setae in a transverse row. Cerci as in *lingnanensis*. Ovipositor rather long, the shaft twice as long as the middle tibia or nearly so (1.90–2.01), the sheaths about $\frac{1}{2}$ length of middle tibia or somewhat longer (0.49–0.56).

Mid-tibial spur nearly as long as the corresponding basitarsus.

Forewing (Figure 950) rather broad, about $2\frac{1}{4}$ to somewhat less than $2\frac{1}{2}$ times as long as wide; marginal fringe not exceeding $\frac{1}{5}$ width of disk. General chaetotaxis as in *lingnanensis*. Delta area with 30–46 setae in 4–6 rows (approximately 35–50 according to DeBach's original description). Submarginal vein bearing 2 coarse setae, the proximal one about $\frac{3}{5}$ to $\frac{2}{3}$ length of the distal, and 14–18 bullae. Marginal vein bearing 9–11 prominent, subequal setae along anterior margin, these $1\frac{1}{6}$ to $1\frac{1}{3}$ times longer than the setae in a row along center of vein.

Forewing hyaline, faintly infumated basally, as in *lingnanensis*.

General coloration entirely yellow; thoracic sterna concolorous, immaculate; posterior margin of scutellum faintly lined with fuscous; as in all related species, a short black streak at base of forewing, below tegula. Legs yellow. Antennal scape fuscous ventrally, rest of antenna uniformly, rather strongly infumated. Wing veins hyaline.

Length 0.92–1.18 mm.

*Male*. Very similar to the female in structure, chaetotaxis, sculpture and coloration, differing mainly in the antennal proportions.

Antennal scape (Figure 951) about $4\frac{1}{3}$ to nearly 6 times as long as wide; pedicel $1\frac{1}{4}$ to $1\frac{1}{3}$ times as long as the third segment of the funicle; the latter segment about $1\frac{1}{7}$ to $1\frac{3}{10}$ times as long as wide, bearing 2 longitudinal sensilla; club usually $2\frac{3}{4}$ to $3\frac{3}{10}$ times as long as wide, $2\frac{3}{4}$ to 3 times longer than the preceding segment, bearing 2–3 longitudinal sensilla, with a partly cut-off sensory area on the ventral aspect as in *lingnanensis*.

Mesoscutum (Figure 952) with 11–12 setae. Propodeum (Figures 952, 953) 4–6 times as long as the metanotum, $\frac{7}{10}$ to $\frac{3}{4}$ length of scutellum; crenulae 4+4 to 6+7, as in the female.

Posterior abdominal sternites (Figure 954) conspicuously stippled, as in *lingnanensis*. Genitalia (Figure 954) about $\frac{3}{4}$ to $\frac{4}{5}$ length of middle tibia (0.72–0.79); digital sclerites usually a little over $\frac{1}{4}$ the combined length of aedeagus and apodemes (0.25–0.28).

Forewing (Figure 955) wider than in the female, $2\frac{1}{5}$ to $2\frac{1}{3}$ times as long as wide; marginal fringe not exceeding $\frac{1}{5}$ width of disk. Delta with 24–33 setae in 3–5 rows. Submarginal vein bearing 14–18 bullae. Marginal vein bearing 7–10 prominent, subequal setae along anterior margin.

Length 0.71–0.92 mm.

Redescribed from numerous ♀♂ (syntypes), from a laboratory culture reared on the oleander scale, *Aspidiotus nerii* Bouché [= *A. hederae* (Vallot)], in Riverside, California, originally obtained from the Florida red scale, *Chrysomphalus aonidum* (L.), on citrus in Israel.

Type series in the U. S. National Museum of Natural History, Washington, in the British Museum (Natural History), London, and in the collection of the Division of Biological Control, University of California, Riverside. 1♀ in the Riverside collection designated lectotype and clearly labeled as such.

**Additional Material.** Numerous ♀♂ specimens, reared from *C. aonidum* on citrus in Israel, Taiwan, Florida, Mexico and Trinidad, and on *Ficus* in Taiwan; also from laboratory cultures in California and Cyprus. Several ♀♂ specimens reared from the rufous scale, *Selenaspidus articulatus* (Morgan), on citrus in Trinidad also appear indistinguishable from *holoxanthus*.

**Notes.** The pupa of *holoxanthus* is practically indistinguishable in pigmentation from that of *lingnanensis*, perhaps with somewhat less pigment toward the tip of the abdomen.

The spermathecal capsule of *holoxanthus* is nearly spherical, averaging 0.03 mm in diameter (Rao and DeBach, 1969a).

*A. holoxanthus* is an Oriental species. For a while it was confused with *lingnanensis*, which was at the time believed to be a parasite of both the Florida red scale, *Chrysomphalus aonidum* (L.), and the California red scale, *Aonidiella aurantii* (Maskell), in the Orient, and was actually introduced under that name from Hong Kong into Israel against the Florida red scale. Only after the successful termination of that project was it discovered that the introduced species was not *A. lingnanensis*, which is a parasite of the California red scale, but the closely related, theretofore undescribed *A. holoxanthus*. The initial confusion was later explained when the cryptic history of *A. lingnanensis* was revealed. When this species was first introduced into California from southern China in 1947, it was obtained from a shipment of Florida red scale, and was successfully propagated on California red scale. It was therefore believed to be a parasite of both scales. In the light of present knowledge, it appears very likely that this original Florida red scale shipment to California also included a few California red scales, unnoticed at the time, from which several specimens of *lingnanensis* emerged. Since the parasites obtained from that shipment were confined with the California red scale, those that issued from the Florida red scale—presumably *holoxanthus*—were outcompeted by the few individuals of *lingnanensis* present for which that scale is a preferred host (DeBach, 1960; DeBach, Rosen and Kennett, 1971).

Interestingly, *A. holoxanthus* was apparently encountered several times in the past, but was not recognized as a distinct species. Several specimens, reared by A. M. Koebele from the Florida red scale on citrus in Hong Kong, apparently some time toward the end of the last century, were found in the collection of the U. S. National Museum. They had been identified as *mytilaspidis* and were subsequently ignored but they appear to be *holoxanthus*. Likewise, it is likely that the species of *Aphytis* described by Taylor (1935) as a "form" of *chrysomphali* with pigmented pupae, parasitizing the Florida red scale in Java, was actually *holoxanthus*. Needless to say, had the true identity of *holoxanthus* been known earlier, this effective parasite would have been available long ago for the biological control of Florida red scale in various countries.

**Utilization in Biological Control.** *A. holoxanthus* was introduced into Israel from Hong Kong in 1956–1957 and effected complete biological control of the Florida red scale throughout the coastal plain, the main citrus-growing area of Israel, within two or three years. The Florida red scale has been virtually eliminated from that area as a pest of economic importance. A few localized infestations of the scale still occur in the hot Jordan Valley, but are usually brought under satisfactory control by periodic releases of the parasite during fall and winter (DeBach, Rosen and Kennett, 1971; Rosen, 1974). This success has enabled the development of an effective program of integrated control of citrus pests in Israel (Harpaz and Rosen, 1971). Following this spectacular success, *A. holoxanthus* was introduced, with similar results, into Mexico, Florida, South Africa, Brazil and Peru (DeBach and Rosen, 1976a). It has recently been established also in Queensland, Australia (D. Smith, personal communication, 1976; Lukins and Snowball, 1977). Attempts to establish this species in California and Cyprus against the California red scale have, however, been unsuccessful.

**Biology.** Available information on the biology of *A. holoxanthus* was recently summarized by Rosen and DeBach (1978). It is an arrhenotokous, rather host-specific parasite of the Florida red scale, attacking all host stages except first-instar larvae and molt stages. In the laboratory at Riverside, it reproduced well on both the oleander scale, *Aspidiotus nerii* Bouché [= *A. hederae* (Vallot)], and the cactus scale, *Diaspis echinocacti* (Bouché), but not on the California red scale. The average life cycle on oleander scale ranges from 61 days at 15.6°C to 14 days at 27–32°C. An average of 33.8 progeny per female are produced at optimal temperatures.

*A. holoxanthus* is completely reproductively isolated from *lingnanensis, coheni* and all related forms, as well as from *melinus* and *fisheri*, to which it is much more closely related (DeBach, 1960; Rao and DeBach, 1969a).

**Related Forms.** *A. holoxanthus* has generally not been believed to attack the California red scale, *Aonidiella aurantii* (Maskell), and material obtained from Israel failed to develop on this host in California and Cyprus. However, Rosen (1965) recorded rearing a biparental form of *Aphytis*, indistinguishable from *holoxanthus*, from California red scale on citrus in the northern coastal plain of Israel. This had apparently occurred before *A. melinus*, with which *holoxanthus* might possibly be confused, was introduced into Israel. Traboulsi (1969) recorded a similar form apparently from adjacent citrus areas in Lebanon. Specimens at hand (Figures 956, 957) are indeed indistinguishable from *holoxanthus*. Whether they represent a distinct form cannot be determined at present, but it is possible that *A. holoxanthus* does occasionally develop on the California red scale. It is interesting to note in this context that Quednau (1964a) reported that *A. holoxanthus*, imported into South Africa, was capable of developing on California red scale in the laboratory. Recent laboratory studies at Riverside and Rehovot confirm this.

Two uniparental series, reared from *Chrysomphalus aonidum* on banana in the Dominican Republic and on citrus in Taiwan, also appear indistinguishable from *holoxanthus*. The specimens from the Dominican Republic seem to have somewhat paler thoracic setae, but this may be due to the fact that they have not been as perfectly cleared as the other series on hand. For the time being, these uniparental "forms" are referred to *holoxanthus*.

An interesting biparental series was reared by V. G. Ortega from *C. aonidum* on coconut palm in Cebu, the Philippines, during February–March 1964 and June 1965. It agrees with *holoxanthus* in pupal pigmentation and in most imaginal characters, but appears to differ slightly in having somewhat paler setae and antennae and a somewhat shorter antennal club. It is very likely that this form will eventually be recognized as a distinct species. However, in the absence of any biological information, and since the material at hand has been rather poorly preserved, it is for the time being referred to *holoxanthus*.

## 64. **Aphytis melinus** DeBach

(Figures 100, 102, 105, 106, 111, 125, 126, 132, 138–144, 149, 159, 160, 168, 179, 180, 186, 201, 211, 214, 217, 224, 232, 236, 237, 239, 240; and 958–977)

*Aphytis melinus* DeBach, 1959, Ann. Entomol. Soc. Amer., **52**: 361–362.
*Aphytis melinus*: Quednau, 1964, J. Entomol. Soc. S. Afr., **27**: 114–115.

This biparental Oriental parasite of the California red scale is very closely related to *holoxanthus*, differing from the latter species mainly in the long, slender antennal club and in the characteristic pigmentation of the mature pupa, with blackish thoracic sterna and unpigmented abdominal sterna. The shorter ovipositor and digital sclerites, slightly narrower wings with somewhat more numerous setae in the delta, and the somewhat paler setae and antennae may serve as additional diagnostic characters. *A. melinus* is redescribed below relative to *holoxanthus*.

*Female*. Eyes (see Figures 100, 105, 106) and mouthparts (Figure 111) as in *holoxanthus*. Antennae (Figures 958, 959) slender, with only a slight taper from scape to tip of club; scape as in *holoxanthus*; pedicel even more slender, $1\frac{4}{5}$ to $2\frac{1}{6}$ times as long as wide, considerably longer than the third segment of the funicle; first funicular segment usually only somewhat (up to $1\frac{1}{4}$ times) wider than long; second segment usually somewhat shorter and slightly wider than the first, as in *holoxanthus*; third funicular segment $1\frac{1}{5}$ to $1\frac{3}{5}$ times as long as wide, bearing 1–2 longitudinal sensilla; club long and slender, usually $3\frac{2}{5}$ to fully 4 times as long as wide (occasionally appearing wider when flattened on slide), $2\frac{3}{4}$ to $3\frac{1}{4}$ times longer and somewhat wider than the preceding segment, bearing 5–7 longitudinal sensilla.

Setae on thorax usually paler than in *holoxanthus*, though still visible under × 30 magnification; those on head paler, invisible or at most hardly visible under × 120 magnification; setae on sides of abdomen more slender, invisible even under × 120 magnification. Chaetotaxis and sculpture of head (Figure 100) and thorax (Figures 960, 961) essentially as in *holoxanthus*. Mesoscutum with 10–12 setae. Scutellum and metanotum as in *holoxanthus*; anteromedian apodeme (Figure 963) robust, about as long as median length of metanotum or nearly so. Epicoxal pads shown in Figures 201 and 962.

Propodeum (Figures 960, 961, 963, 964) about $4\frac{1}{4}$ to $5\frac{3}{4}$ times as long as the metanotum, $\frac{3}{4}$ to over $\frac{4}{5}$ length of scutellum, essentially as in *holoxanthus*; crenulae 4+4 to 7+8, large, elongate, strongly overlapping, as in *holoxanthus* and *lingnanensis* (see also Figures 179, 180).

Sculpture and chaetotaxis of abdominal tergites (Figures 960, 961, 965) as in *holoxanthus*; eighth tergite with a transverse row of 4 setae (occasionally 3 or 5) between the spiracles; syntergum and cerci as in *holoxanthus* and *lingnanensis*, syntergum bearing 6–8 setae in a transverse row. Ovipositor (Figure 966) somewhat shorter than in *holoxanthus*, the shaft about $1\frac{2}{3}$ times to somewhat less than twice as long as the middle tibia (1.68–1.93), the sheaths usually somewhat less than $\frac{1}{2}$ length of middle tibia (0.46–0.51).

Mid-tibial spur somewhat shorter than the corresponding basitarsus.

Forewing (Figures 967, 968) slightly narrower than in *holoxanthus*, $2\frac{2}{5}$ to $2\frac{3}{5}$ times as long as wide; marginal fringe not exceeding $\frac{1}{5}$ width of disk. General chaetotaxis as in *holoxanthus*, but delta area with somewhat more numerous setae, 39–63, usually in 5–7 rows. Submarginal vein as in *holoxanthus*, bearing 14–21 bullae (see also Figure 236). Marginal vein as in *holoxanthus*, bearing 7–11 prominent, subequal setae along anterior margin.

Forewing hyaline, faintly infumated basally, essentially as in *holoxanthus* or somewhat paler.

Hind wing (Figures 968, 969) rather narrow, width of disk about $1\frac{1}{5}$ to $1\frac{1}{2}$ times length of longest hairs in marginal fringe.

General coloration entirely yellow, as in *holoxanthus* or somewhat paler; thoracic sterna immaculate. Antennae more faintly infumated than in *holoxanthus*.

Length 0.83–1.21 mm (usually more than 1 mm).

*Male*. Essentially similar to the male of *holoxanthus*; very similar to the female in structure, chaetotaxis, sculpture and coloration, differing mainly in the antennal proportions.

Antennal scape (Figures 970, 971) about $4\frac{1}{5}$ to $5\frac{1}{2}$ times as long as wide; pedicel shorter than in the female, $1\frac{2}{3}$ to over $1\frac{4}{5}$ times as long as wide, about $1\frac{1}{4}$ to $1\frac{2}{5}$ times longer than the third segment of the funicle; the latter segment up to $1\frac{2}{5}$ times as long as wide, bearing 1–2 longitudinal sensilla; club nearly 3 to $3\frac{1}{3}$ times as long as wide, $2\frac{3}{5}$ to nearly 3 times longer than the preceding segment, bearing 2–4 longitudinal sensilla, with a partially cut-off sensory area on the ventral aspect as in *holoxanthus* and *lingnanensis* (Figures 143, 144).

Mesoscutum (Figures 972, 973) with 10–12 setae. Propodeum (Figures 973, 974) 4 to $5\frac{1}{2}$ times as long as the metanotum, $\frac{2}{3}$ to $\frac{3}{4}$ length of scutellum; crenulae 4+5 to 6+7, as in the female.

Posterior abdominal sternites (Figures 217, 975) conspicuously stippled, as in *holoxanthus* and *lingnanensis*. Genitalia (Figures 975, 976) about $\frac{2}{3}$ to $\frac{4}{5}$ length of middle tibia (0.67–0.83); digital sclerites somewhat shorter than in *holoxanthus*, a little less than $\frac{1}{4}$ the combined length of aedeagus and apodemes (0.21–0.24).

Forewing (Figure 977) somewhat wider than in the female but narrower than in male *holoxanthus*, about $2\frac{1}{4}$ to nearly $2\frac{2}{5}$ times as long as wide; marginal fringe not exceeding $\frac{1}{5}$ width of disk. Setae in delta somewhat more numerous than in *holoxanthus*,

26–41, in 4–5 rows. Submarginal vein bearing 13–18 bullae. Marginal vein bearing 8–10 prominent, subequal setae along anterior margin.

Length 0.73–1.01 mm.

Redescribed from numerous ♀♂ specimens (syntypes), from a mixed laboratory culture (see below), originally obtained from the California red scale, *Aonidiella aurantii* (Maskell) on rose in India and Pakistan, 1956, and from the yellow scale, *A. citrina* (Coquillett), on citrus in Pakistan, 1957, and reared on the oleander scale, *Aspidiotus nerii* Bouché [=*A. hederae* (Vallot)] in the insectary of the University of California, Riverside, October 1958; also from numerous ♀♂ specimens from the original shipments from India and Pakistan.

Type series in the U. S. National Museum of Natural History, Washington, in the British Museum (Natural History), London, and in the collection of the Division of Biological Control, University of California, Riverside.

**Notes.** The mature (green-eyed) pupa of *melinus* is characterized by dark (blackish) pigmentation on the midthoracic sterna which is less well defined than in *lingnanensis*; abdomen clear (yellowish) ventrally; head, wing-pads and appendages usually yellow; dorsal surface yellow (see Figure 247 and DeBach, 1959). This distinguishes it from the pupae of all other known species.

The spermathecal capsule of *melinus* is similar to that of *holoxanthus* (Rao and DeBach, 1969a).

Extensive laboratory mating tests have repeatedly shown that *melinus* is completely reproductively isolated from *lepidosaphes*, *africanus*, *lingnanensis*, *coheni* and all related forms, as well as from *holoxanthus* and *fisheri*, to which it is much more closely related morphologically. When mated with *holoxanthus* males with the aid of mating inducers, *melinus* females produced a few fertile hybrids. However, reciprocal tests yielded no hybrid progeny and it is very unlikely that these two species would be able to hybridize successfully in nature (DeBach, 1959, 1960; Rao and DeBach, 1969a; see further notes under *A. yasumatsui*).

*A. melinus* is almost a sibling of *holoxanthus*. Adult specimens of these two species are nearly indistinguishable from each other and may be easily confused. There is considerable variation in the antennal proportions in both species, resulting in some overlap between them. Moreover, the antennal proportions may be affected by mounting procedures, so that individual specimens of *melinus* may be misidentified as *holoxanthus* when determination is based on that character alone. However, a series of properly mounted specimens of *melinus* is bound to include at least a few females with very slender antennae, with the club fully 4 times as long as wide or nearly so, and with some experience correct identification can be made on that basis.

The degree of pigmentation of the body setae is practically impossible to quantify in *Aphytis*, and may be subject to considerable size-linked variation. The appearance of setae may also be affected by the degree of clearing of the body parts that serve as background to them when viewed under the microscope. Nevertheless, type and other original material of *melinus*, as well as abundant insectary-reared specimens at hand, do appear to have, as a rule, somewhat paler setae on the thorax and head than do specimens of *holoxanthus* of comparable size. This difficult character may therefore be of supportive value in the separation of these two species, as are other characters mentioned in the above redescription.

The characteristic pigmentation of the pupa can usually provide a shortcut to the identification of *melinus*. Although this character may sometimes vary under the effect of climatic conditions, it is usually quite reliable. Samples of mature (green-eyed) pupae should therefore accompany adult specimens whenever possible, to insure proper identification of *melinus*.

The distinctive host preferences of *holoxanthus* and *melinus* may also serve as a valid diagnostic character. Whereas the former species is an effective parasite of the Florida red scale and does not normally parasitize the California red scale, the latter species is an important parasite of the California red scale and the yellow scale, but is not known to attack the Florida red scale.

**History and Utilization in Biological Control.** "*Aphytis melinus* was first received in the insectary of the Department of Biological Control at Riverside on September 12, 1956. One hundred and eight females and twenty-seven males emerged from *Aonidiella aurantii* collected by G. W. Angalet on rose at New Delhi, India, September 5, 1956. Original data are recorded under S. & R. No. 1643 of the Department of Biological Control, Citrus Experiment Station, Riverside. Dr. T. W. Fisher successfully cultured this material in quarantine. Angalet made two more shipments of *Aonidiella aurantii* on rose, which also gave rise to cultures of *Aphytis melinus*. These came from Lahore, Pakistan, September 16, 1956 (S. & R. No. 1646), and from Gurgaon, India, October 26, 1956 (S. & R. No. 1651). A fourth culture was obtained from a shipment (S. & R. No. 1737) sent by [DeBach] from Saidpur Village (near Rawalpindi), West Pakistan, April 3, 1957. These parasites were collected on the yellow scale, *Aonidiella citrina*, on orange and pummelo.

"Mating tests showed that all four of these cultures (New Delhi, Lahore, Gurgaon, and Saidpur Village [=Culture "P"]) crossed readily and produced normal sex ratios and numbers of progeny. They were therefore combined into one mixed culture which has been mass produced in the insectary . . ." (DeBach, 1959).

Extensive releases on California red scale were started in California in the fall of 1957 and were continued for several years. About 2.5 million mated females were colonized during 1957–1959 in some 200 citrus plots all over southern California. Numerous field recoveries were soon made, and *melinus* became well established in California by the end of 1958, exhibiting the fastest rate of dispersal of any parasite of California red scale, especially in interior areas. It soon proceeded to displace *lingnanensis* in the vicinity of the original release sites. By 1964, *A. melinus* had completely displaced *lingnanensis* from interior and intermediate citrus areas in southern California and became the dominant natural enemy of California red scale in all but immediate coastal areas. A general decline of red scale was evident in California, and effective biological control by *A. melinus* was demonstrated in numerous untreated citrus plots (DeBach, Rosen and Kennett, 1971; Rosen and DeBach, 1978).

Following its successful establishment in California, *A. melinus* was transferred to many other citrus-growing countries of the world. In recent years it has become established on the California red scale in Greece, Sicily, Cyprus, Israel, Morocco, South Africa, Chile, Argentina and Australia, and has become an important parasite of that serious pest in all those countries. It has demonstrated its great capacity for dispersal by spreading from Greece to Turkey and from Argentina to Paraguay. *A. melinus* is now considered the most effective known natural enemy of the California red scale. In addition, it has proved an effective natural enemy of the yellow scale, the oleander scale and the dictyospermum scale in various areas, reducing the dictyospermum scale to low levels in Sicily and effecting complete biological control of the dictyospermum and oleander scales in Greece (DeBach and Rosen, 1976a).

*A. melinus* has been repeatedly released in Israel since 1961, but at first inexplicably failed to become firmly established. Late in 1967, what appeared to be this species was found to be the dominant parasite of *Aonidiella aurantii* in certain citrus areas on the West Bank of the river Jordan. Crossing tests with California material proved that this supposedly adapted strain was indeed conspecific with *melinus*, and it has since become established over large citrus areas, displacing other parasites (DeBach and Rosen, 1976a).

**Biology.** *A. melinus* is the only species of *Aphytis* known to attack commonly both the California red scale, *Aonidiella aurantii* (Maskell), and the yellow scale, *Aonidiella citrina* (Coquillett). It is also an effective parasite of the dictyospermum scale, *Chrysomphalus dictyospermi* (Morgan), and the oleander scale, *Aspidiotus nerii* Bouché [=*A. hederae* (Vallot)]; the latter species may serve as a convenient laboratory host for mass rearing of *melinus*. The species is biparental, arrhenotokous. At 26.7°C and 50% RH, the life cycle takes 12–13 days; oviposition takes about 12 minutes, and an average of 2.8 eggs per scale are laid on the oleander scale (Rosen and DeBach, 1978). Longevity of the female has been reported to average 54.1 days at 20°C, 29.8 days at 25°C, 18.0 days at 30°C. On the California red scale as host, the fecundity of *melinus* is relatively high, averaging 51.2 progeny per female at 20°C, 67.4 at 25°C, 55.5 at 30°C, each female destroying an average of 54.2, 61.6 and 45.0 host scales by oviposition at those temperatures, respectively, and an additional 41.5, 50.6 and 35.2 host scales, respectively, by mutilation and host-feeding. The theoretical threshold of development was calculated as approximately 11°C (Abdelrahman, 1974b). The behavior of *melinus* females in oviposition, mutilation and host-feeding was recently described in detail by Abdelrahman (1974c). Reports on the effects of extreme climatic conditions on *melinus* vary: DeBach and Argyriou (1966) reported that the adverse effects of cold on mating and reproduction are more severe in this species than in *lingnanensis*, whereas Abdelrahman (1974a) found that extreme temperatures had no effect on the fertility of mated females of *melinus* in laboratory tests. The latter author also reported *melinus* to be more tolerant of extreme heat, but less tolerant of extreme cold, than *A. chrysomphali*.

**Material Examined.** Specimens determined as *melinus* in the collection of the Division of Biological Control, University of California, Riverside, were obtained from the California red scale, *Aonidiella aurantii* (Maskell), on citrus in California, Chile, Argentina, Paraguay, India, Pakistan, Turkey, Israel, Cyprus, Morocco, South Africa and Australia, on *Moraea* and *Agave* in California, and on rose in India; from the yellow scale, *Aonidiella citrina* (Coquillett), on citrus in Pakistan and Turkey; from the dictyospermum scale, *Chrysomphalus dictyospermi* (Morgan), on citrus in California, Spain and Turkey, on rubber plant in California, on ligustrum and ivy in Turkey; from the latania scale, *Hemiberlesia lataniae* (Signoret), on avocado in California; from the coconut scale, *Temnaspidiotus destructor* (Signoret), on mango in Pakistan; from the oleander scale, *Aspidiotus nerii* Bouché [ = *A. hederae* (Vallot)], on oleander in Israel and on an undetermined plant in Chile; from the walnut scale, *Quadraspidiotus juglansregiae* (Comstock), on willow in California; from the San Jose scale, *Q. perniciosus* (Comstock), on peach and apple in California; from the cactus scale, *Diaspis echinocacti* (Bouché), on *Opuntia* in Baja California, Mexico; from chaff scale material, *Parlatoria pergandii* Comstock, on citrus in California; from undetermined scale insect hosts on citrus, rose and oleander in Pakistan, on citrus in India; and from laboratory cultures on the California red scale and oleander scale in California and Australia. All these series appear indistinguishable from *melinus*, and are at present referred to this species. It is, however, quite possible that some of the above-listed host records are erroneous, due to misidentification or contamination of samples (e.g., chaff scale in California).

**Related Forms.** A series of *Aphytis* specimens was reared by H. E. Martin from *Aonidiella orientalis* (Newstead) on citrus at Riyadh Erka and Hofuf, Saudi Arabia, January to March 1967. At hand are 12♀♀ and one mature ♀ pupa, as well as several cast pupal skins. The specimens appear very similar to *melinus*, entirely yellow with immaculate thoracic sterna and large overlapping propodeal crenulae, but the antennal club is somewhat shorter than in *melinus*, the thoracic setae appear somewhat darker, and the line on the posterior margin of the scutellum appears somewhat more pronounced than in typical *melinus*. Also, the pupa is very similar to *melinus*, but exhibits additional dark pigmentation on the appendages and top of head and some brownish at the base of the abdomen. The fact that only female specimens were obtained suggested that this might be a uniparental form.

This interesting form was successfully introduced from Saudi Arabia into California in 1977. It proved indeed to be uniparental, and was found to be capable of parasitizing both the California red scale and the oleander scale. Being adapted to the hot desert area near Riyadh, Saudia Arabia, it is considered a promising candidate for colonization in extreme climatic zones. It is now being released against the California red scale in the interior regions of California and in the hot San Joaquin and Imperial Valleys (DeBach, 1977). Although this form appears to represent a distinct species, it seems advisable to reserve judgment as to its taxonomic position in relation to *melinus* until more is known about it.

*A. yasumatsui* Azim is described below as distinct, although it may be regarded as a strong semispecies in relation to *melinus* (see discussion on p. 560).

## 65. **Aphytis yasumatsui** Azim

(Figures 978–988)

*Aphytis yasumatsui* Azim, 1963, J. Fac. Agric. Kyushu Univ., **12**:284–287.

This enigmatic Japanese parasite of the bifasciculate scale and the Asiatic red scale presents perplexing taxonomic and nomenclatural problems. It occupies an intermediate position between *holoxanthus* and *melinus* in pupal and some imaginal characters, but appears to be more closely related morphologically to the former species, from which it is completely isolated reproductively, than to the latter species, from which it is only partially isolated. *A. yasumatsui* is briefly redescribed below relative to *holoxanthus* and *melinus*.

*Female*. Eyes and mouthparts as in *holoxanthus* and *melinus*. Antennae (Figures 978, 979) more similar to those of *holoxanthus*; scape slender, about 5 to 6 times as long as wide; pedicel about $1\frac{2}{3}$ to 2 times as long as wide, up to $1\frac{1}{3}$ times longer than the third segment of the funicle; first funicular segment trapezoidal, $1\frac{1}{3}$ to $1\frac{2}{5}$ times as wide as long; second segment more nearly symmetrical, about $1\frac{1}{2}$ to $1\frac{3}{5}$ times as wide as long; third funicular segment about $1\frac{1}{4}$ to $1\frac{2}{5}$ times as long as wide, bearing 1–2 longitudinal sensilla; club comparatively robust, a little over 3 to $3\frac{1}{2}$ times as long as wide, $2\frac{2}{3}$ to 3 times longer and somewhat wider than the preceding segment, bearing 5–6 longitudinal sensilla.

Setae on thorax and head slightly more like those of *melinus*, somewhat paler than in *holoxanthus*, perhaps intermediate between the two. Structure, chaetotaxis and sculpture of head and thorax as in *holoxanthus* and *melinus*. Mesoscutum (Figure 980) with 10–12 setae.

Propodeum (Figures 980, 981) about 4 to nearly 5 times as long as the metanotum, a little over $\frac{4}{5}$ length of scutellum, essentially as in *holoxanthus* and *melinus*. Crenulae 5+5 to 8+9.

Abdomen essentially as in both *holoxanthus* and *melinus*. Ovipositor rather long, as in *holoxanthus*, the shaft about $1\frac{4}{5}$ times to fully twice as long as the middle tibia (1.83–2.03), the sheaths a little over $\frac{1}{2}$ length of middle tibia (0.51–0.57).

Forewing (Figure 982) as in *holoxanthus*, rather broad, $2\frac{2}{5}$ to $2\frac{1}{2}$ times as long as wide; delta area with 32–43 setae in 4–6 (usually 5) rows.

Wing pattern and general coloration essentially as in *holoxanthus*.

Length 0.81–1.18 mm.

*Male*. Essentially similar to the male of *holoxanthus*, but genitalia (Figure 984) more like those of *melinus*, about $\frac{3}{4}$ to $\frac{4}{5}$ length of middle tibia (0.76–0.84), with the digital sclerites usually somewhat less than $\frac{1}{4}$ the combined length of aedeagus and apodemes (0.21–0.26). Antennae shown in Figure 983.

Length 0.62–1.02 mm.

Redescribed from 1♂ (syntype), reared by A. Azim from the bifasciculate scale, *Chrysomphalus bifasciculatus* Ferris, on *Illicium religiosum*, Fukuoka, Kyushu, Japan, September 10, 1962; 10♀♀, 5♂♂, same data, May–June 1961 (apparently from the syntype series but not labeled as such); 3♀♀, reared by R. van den Bosch from the same host on *Laurus nobilis* and from the holly scale, *Dynaspidiotus britannicus* (Newstead), on ?*Cyclobalanopsis*, same locality, April 1964; and from numerous ♀♂ specimens, reared by T. Tachikawa and M. Sato from *C. bifasciculatus* on *Lithocarpis edulis* and from the Asiatic red scale, *Aonidiella taxus* Leonardi, on *Podocarpus macrophylla*, Saijo, Uwajima and Matsuyama, Ehime Ken, Shikoku, Japan, June–September 1973; also from ♀♂ specimens reared by T. Tachikawa from *C. bifasciculatus* on *Aspidistra elatior*, Tsukumi City, Kyushu, Japan, November 1973.

Type series in the collection of the Entomological Laboratory, Kyushu University, Fukuoka, Japan.

**Notes.** The pupa of *yasumatsui* (Figures 985–987) appears to be intermediate in pigmentation between the *holoxanthus–lingnanensis* type and the *melinus* type: mid-thoracic sterna with strong, well-defined dark pigmentation; abdominal sterna with paler, more diffuse, less clearly defined pigmentation; head, appendages and dorsal surface clear.

Figure 988 shows an abnormal male scutellum, with 3 pairs of setae and 1 pair of discoid sensilla.

Azim (1963b) described *yasumatsui* as "very similar to, if not identical with, *Aphytis chrysomphali* Mercet." Unfortunately, Azim's original description cannot be relied upon for a positive identification of this species. Thus, for instance, his rather confusing key describes the propodeal crenulae of *yasumatsui* as nonoverlapping, his text refers to them as being small and indistinct, whereas his Figure 10 shows the crenulae to be large, distinct and strongly overlapping, as in other members of the **lingnanensis** group. Likewise, Azim's key describes *yasumatsui* with "antennal club usually black at the apex," whereas his Figure 7 shows no such pigmentation and his text refers to "cleared antennae, especially pedicel, funicle and club, faintly brownish" (which they are not). The original description would therefore be better ignored.

In most diagnostic characters—shape and proportions of antennal segments, relative length of ovipositor, shape and chaetotaxis of forewing, pigmentation of pupa, etc.—*yasumatsui* appears to be more closely related to *holoxanthus* than to *melinus*. In fact, we at first considered it to be a synonym of *holoxanthus*. However, reciprocal mating tests at Riverside have shown that *yasumatsui* is completely reproductively isolated from *holoxanthus*, as well as from *fisheri*, *lingnanensis* and several closely related *lingnanensis* forms, but is incompletely isolated from *melinus*. Several stocks of *yasumatsui*, obtained in 1973 in Ehime, Japan, from *C. bifasciculatus* (R73-93, R73-114) and from *A. taxus* (R73-94, R73-112, R73-113) were found to cross freely with each other and were therefore determined as conspecific and combined into a single stock. Reciprocal mating tests with *holoxanthus*, *fisheri* and *lingnanensis* yielded no female progeny. Likewise, when females of *yasumatsui* were mated with males of *melinus*, no female progeny were produced. However, when males of *yasumatsui* were mated with females of *melinus*, about 50% female progeny were obtained. Thus, *yasumatsui* appears to represent a distinct biological species in relation to *holoxanthus*, *fisheri* and *lingnanensis*, but may be regarded as a semispecies in relation to *melinus* (see Gordh and DeBach, 1978).

Since *yasumatsui* is quite distinct morphologically from *melinus*, but is only partially reproductively isolated from this species, it seems that morphological divergence, as well as certain biological characteristics (e.g., host preferences), have evolved faster in this species complex than has reproductive isolation between its allopatric components.

Recognition of distinct, strong semispecies, especially ones possessing distinct host preferences or other biological attributes, may of course be of great practical importance to the biological control worker in search of new natural enemies. But where does it leave the taxonomist? The discovery of a taxon exhibiting strong semispecific status in relation to a valid named species presents a rather baffling nomenclatural problem. With evidence of only partial reproductive isolation in the laboratory, and (usually) with no information on the possible fate of hybrids in the field, should a strong semispecies be assigned a separate specific name? On the other hand, if a formal specific name already happens to be assigned to it in the literature, would it be justifiable to sink it into synonymy on the formal grounds of the occurrence of some introgression?

We cannot offer a general answer to this question. Since the degree of isolation between semispecies differs, each case should be evaluated on its own merits. Thus, although we have not proposed names for other strong semispecies discussed elsewhere in this revision (e.g., forms related to *lingnanensis* and *coheni*), it is our opinion that synonymizing *yasumatsui* with *melinus* would not be justifiable and would certainly not serve any useful purpose. We are suggesting this not only, and not even primarily, because the name *yasumatsui* is already available in the literature, but because the weight of both biological and morphological evidence indicates that *yasumatsui* is closer to a distinct species—biologically, ethologically and morphologically—than to a mere "strain" or "form" of *melinus*. We therefore propose to retain the name *yasumatsui* for this form, at least for the time being. It will be recalled that in the case of the forms related to *lingnanensis* or *coheni* there are no known morphological differences.

This species was named by Azim (1963b) in honor of Professor Keizo Yasumatsu of Kyushu University, Fukuoka, Japan.

**Additional Material.** Additional specimens of *yasumatsui* were reared by M. Tanaka from *C. bifasciculatus* on *Euonymus japonicus* in Miyazaki City, Kyushu, Japan, December 1968 (together with *A. japonicus* DeBach and Azim); by S. C. Warner from same host on *Aucuba japonica* in Taono-cho, Kagoshima Prefecture, Kyushu, Japan, June 1969; and by K. Nohara from *Dialeurodes citri* material (evidently from an unnoticed armored scale insect) on citrus, Hagi City, Yamaguchi Prefecture, Honshu, August 1973.

A slide bearing several specimens of *yasumatsui*, reared by F. Silvestri from "*Chrysomphalus aonidum*" at Kagoshima, Japan, apparently during his exploration of the Orient in 1924–1925, was received from the Filippo Silvestri Collection, Portici, Italy. We assume that Silvestri's specimens were in fact reared from *C. bifasciculatus*, which was generally confused with *C. aonidum* at that time. The latter species is not known to exist in Japan.

**Biology.** Azim (1963c) described the mating behavior, oviposition and development of *yasumatsui* in some detail. The species is biparental, arrhenotokous. Eggs are usually deposited on the ventral surface of the scale-insect host. At 30°C, development from egg to adult takes 15–18$\frac{1}{2}$ days.

In "no-choice" laboratory tests at Riverside, this species did better on the California red scale, *Aonidiella aurantii* (Maskell), than on any other available host, producing an average of 42.75 progeny per female.

**Potential Value for Biological Control.** According to Azim (1963c), *A. yasumatsui* is a quite important parasite of the bifasciculate scale in Japan. It also appears to be an abundant parasite of the Asiatic red scale in that country. It should be considered a promising candidate for utilization in biological control projects directed against these pests, and possibly other scale insects as well, in other countries.

## 66. Aphytis fisheri DeBach

(Figures 182, 207; and 989–1001)

*Aphytis fisheri* DeBach, 1959, Ann. Entomol. Soc. Amer., **52**:362.

This biparental parasite of the California red scale is a sibling species of *melinus*. It is virtually identical to *melinus* in all imaginal characters, except for some slight differences in the proportions of the hind wing and its marginal fringe, but differs markedly from that species in having entirely unpigmented pupae. In order to avoid unnecessary repetition, *fisheri* is briefly redescribed below relative to *melinus*.

*Female*. Eyes and mouthparts as in *melinus*. Antennae (Figures 989, 990) slender, virtually identical to those of *melinus*; pedicel even somewhat more slender, a little over $1\frac{4}{5}$ to $2\frac{1}{4}$ times as long as wide; club $3\frac{2}{5}$ to fully 4 times as long as wide.

Setae on thorax perhaps somewhat darker than in *melinus*; those on head and abdomen as in *melinus*. Structure, chaetotaxis and sculpture of head and thorax as in *melinus*; mesoscutum (Figure 991) with 9–12 (usually 10) setae.

Propodeum (Figures 991, 992) $3\frac{3}{4}$ to 5 times as long as the metanotum, about $\frac{3}{4}$ to $\frac{4}{5}$ length of scutellum, essentially as in *melinus*; crenulae 4+5 to 8+9, large, overlapping, as in *melinus*.

Structure, chaetotaxis and sculpture of abdominal tergites as in *melinus*. Ovipositor somewhat shorter, shaft about $1\frac{3}{5}$ to $1\frac{3}{4}$ times as long as the middle tibia (1.59–1.77), sheaths $\frac{2}{5}$ to somewhat less than $\frac{1}{2}$ length of middle tibia (0.39–0.49).

Forewing (Figures 993, 994) essentially as in *melinus*, about $2\frac{1}{3}$ to $2\frac{2}{3}$ times as long as wide; delta area with 35–59 setae; submarginal vein bearing 14–20 bullae; marginal vein bearing 6–12 prominent, subequal setae along anterior margin.

Forewing pattern identical to that of *melinus*.

Hind wing (Figures 994, 995) somewhat broader than in *melinus*, width of disk about $1\frac{3}{5}$ to $1\frac{4}{5}$ times length of longest hairs in marginal fringe.

General coloration as in *melinus*.

Length 0.68–1.24 mm (usually about 1 mm or longer).

*Male* (Figures 996–1001). Very similar to the female in structure, chaetotaxis, sculpture and coloration, differing mainly in antennal proportions; virtually identical to the male of *melinus*.

Setae on thorax perhaps somewhat darker than in *melinus*.

Genitalia (Figure 1001) about $\frac{2}{3}$ to $\frac{4}{5}$ length of middle tibia; digital sclerites a little less than $\frac{1}{4}$ the combined length of aedeagus and apodemes (0.19–0.25), as in *melinus*.

Length 0.61–0.86 mm.

Redescribed from numerous ♀♂ specimens (including syntypes), from a laboratory culture reared on the oleander scale, *Aspidiotus nerii* Bouché [= *A. hederae* (Vallot)] at the insectary of the University of California, Riverside, 1958–1973; originally obtained from a shipment (S. & R. No. 1673) of the California red scale, *Aonidiella aurantii* (Maskell), collected by P. DeBach on rose at Taunggyi and Heho, South Shan States, Burma, December 21–25, 1956.

Type series in the U. S. National Museum of Natural History, Washington, in the British Museum (Natural History), London, and in the collection of the Division of Biological Control, University of California, Riverside.

**Notes.** The mature pupa of *fisheri* is readily recognizable by the clear (yellowish) thoracic and abdominal sterna and head; wing-pads and appendages may vary from yellow to yellow-brown; dorsal surface yellow; no black lines or furcae whatsoever on body (see Figure 247 (p. 199) and DeBach, 1959).

The spermathecal capsule of *fisheri* is similar to that of *melinus* and *holoxanthus* (Rao and DeBach, 1969a).

*A. fisheri* appears to be very closely related to *melinus*. It resembles *melinus* in virtually all imaginal characters, including the few diagnostic characters separating *melinus* from *holoxanthus*. *A. fisheri* and *A. melinus* are therefore currently regarded as sibling species. As pointed out by DeBach (1959), "morphological distinction between the two is now more of an art than a science, and not a very well-developed art at that."

DeBach (op. cit.) offered the following characters to help differentiate between the two species:

| "*A. melinus* | *A. fisheri* |
|---|---|
| Hairs on edge of hind wing from hamulae to the wing-tip (to first long wing-tip hair) nearly equal in length. | Hairs on edge of hind wing from hamulae to wing-tip (to first long wing-tip hair) decreasing noticeably in length from hamulae to tip. |
| The two long hairs at apex of the hind wing about 2 times as long as hairs on leading edge just before the apex. | The two long hairs at apex of the hind wing about 2.5 to 3 times as long as hairs on leading edge just before the apex. |
| 6–8 pale spines evident on mesoscutum at 45 × magnification (dissection binocular). | 10 spines evident on mesoscutum at 45 × magnification (dissection binocular). |
| Greatest hind-wing width about 1.1 to 1.5 times that of the longest fringe hairs. | Greatest hind-wing width about 1.5 to 2 times that of the longest fringe hairs. |
| Spines on male notum indistinct at 45 × magnification. | Spines on male notum distinct at 45 × magnification." |

As can be seen from the above redescription of *fisheri*, we have been unable to find any additional reliable characters for the separation of *fisheri* and *melinus*. In fact, some of the characters offered by DeBach are very difficult to use. Thus, for instance, although the thoracic setae of *fisheri* do indeed usually appear to be somehat darker than those of *melinus* in both female and male specimens, this character is very difficult to quantify, is subject to considerable variation, and may be affected by mounting procedures, therefore requiring considerable experience. Likewise, the two long hairs at the apex of the hind wing are indeed somewhat longer in *fisheri* in relation to the hairs just before the apex, but this character, too, appears to be rather variable and the values for *fisheri* and *melinus* often overlap (hairs at apex 2.14–2.57 times longer than hairs just before apex in a random sample of *fisheri*, as compared with 1.78–2.33 in a random sample of *melinus*).

The only reliable imaginal character separating *fisheri* from *melinus* is the relatively broader hind wing of the former. In a random sample of female specimens of *fisheri*, the width of the disk of the hind wing was 1.58–1.78 times the length of the longest fringe hairs, as compared with 1.22–1.50 in a random sample of female specimens of *melinus*. Interestingly, the values for *holoxanthus* and *yasumatsui* are considerably more variable, their ranges overlapping those of both *fisheri* and *melinus*.

Thus, the relative width of the hind wing, and pupal pigmentation, at present appear to be the only reliable morphological characters separating *fisheri* from *melinus*. Nevertheless, extensive reciprocal mating tests have repeatedly shown that *fisheri* is completely isolated reproductively from *melinus*, as well as from *holoxanthus*, *yasumatsui*, *lingnanensis*, *coheni*, *africanus*, *lepidosaphes* and related forms (DeBach, 1959, 1960; Rao and DeBach, 1969a; Gordh and DeBach, 1978). *A. fisheri* and *A. melinus* are therefore considered valid, albeit sibling, species in relation to one another. It appears that, unlike in the case of *yasumatsui* (see above), reproductive isolation between *fisheri* and *melinus* has evolved faster than morphological divergence.

This species was named by DeBach in honor of Dr. Theodore W. Fisher, University of California, Riverside, who successfully cultured the original material obtained from Burma.

**Biology.** *A. fisheri* is a biparental, arrhenotokous species, females comprising about $\frac{3}{5}$ of its populations in laboratory cultures. When reared on oleander scale, development from egg to adult emergence takes 13–14 days at 26.7°C. Adult longevity is relatively long: 75 percent mortality was reached after 33 days (with honey as food and no oviposition). Average total fecundity is about 26 progeny per female, and an average of 28 scales per female are destroyed by host-feeding (DeBach and Sundby, 1963).

**Utilization in Biological Control.** *A. fisheri* was first noticed in the field by DeBach, December 21–22, 1956, on *Aonidiella aurantii* on pummelo and rose at Kalaw and Taunggyi, South Shan States, Burma, on the basis of pupal pigmentation. A shipment of *A. aurantii* on rose collected December 21–25, 1956 at Taunggyi and Heho yielded only 4♀♀ and 2♂♂ of *fisheri* in the insectary at Riverside. These few were successfully cultured in quarantine, and subsequently (1957–1958) this species was mass cultured and colonized throughout southern California. Although numerous field recoveries were initially made, and although laboratory tests had indicated *fisheri* to be a potentially effective parasite, competition by *melinus* and *lingnanensis* apparently precluded it from becoming established in California (DeBach, 1959; DeBach and Sundby, 1963; DeBach, Rosen and Kennett, 1971).

**Additional Material.** *A. fisheri* appears to be quite widespread in the Orient. DeBach (1959) recorded *fisheri*-type pupae of *Aphytis* on California red scale in Burma, Thailand, Taiwan and at Kagoshima (southern Kyushu), Japan. He obtained adult specimens of *fisheri* (apparently mixed with *yasumatsui*) by beating citrus trees at Kagoshima in 1956. *A. fisheri* (at least, a form indistinguishable from this species in both pupal and imaginal characters) was subsequently reared from California red scale on citrus at Kagoshima Prefecture, Japan, by P. DeBach on July 14, 1971 and by J. Yukawa on November 24, 1971. Some of this Kagoshima material was suspected at the time to be uniparental, but this should be verified by careful biological research. Some of the material at hand appears to be mixed with the closely related *yasumatsui*.

## 67. **Aphytis equatorialis** n.sp.

(Figures 1002–1013)

This interesting, small biparental parasite of the coconut scale from Equatorial West Africa is superficially rather similar to *africanus*, but may be readily separated from that species by the immaculate thoracic sterna. It differs markedly from other members of the **lingnanensis** group having unpigmented sterna in its small size, its short, robust antennae, less elongate propodeal crenulae and relatively short ovipositor, as well as by the very weak stippling on the abdominal sternites of the male.

*Female.* Eyes finely setose. Mandibles tridentate, rather slender, the ventral denticle sharply pointed, the median and dorsal denticles apparently more distinct than in other species; maxillary palpi 2-segmented, labial palpi 1-segmented. Antennae (Figures 1002, 1003) rather short, robust; scape about 4–5 times as long as wide, somewhat longer than the club; pedicel about $1\frac{3}{5}$ to $1\frac{3}{4}$ times as long as wide, considerably ($1\frac{1}{3}$ to $1\frac{3}{5}$ times) longer than the third segment of the funicle; first funicular segment semi-globular or somewhat trapezoidal, only $1\frac{1}{7}$ to $1\frac{2}{5}$ times as wide as long; second segment more symmetrical, shorter and somewhat wider than the first, about $1\frac{1}{2}$ to $1\frac{9}{10}$ times as wide as long; third funicular segment about as long as wide (ranging from somewhat wider than long to somewhat longer than wide), bearing 1 longitudinal sensillum; club short, only $2\frac{2}{3}$ to $2\frac{9}{10}$ times as long as wide, 3 to nearly $3\frac{1}{2}$ times longer and distinctly wider than the preceding segment, bearing 5–6 longitudinal sensilla.

Setae on thorax rather pale, moderately slender, the largest ones barely visible under × 60 magnification; setae on head paler, invisible or barely discernible under × 120 magnification; those on sides of abdomen invisible even under × 120 magnification. Vertex (Figure 1004) with 2 pairs of long setae along occipital margin, in addition to numerous short setae. Mesoscutum (Figure 1005) with 9–11 (usually 10) setae, the posterior pair and 1 seta at each antero-lateral corner considerably longer and coarser than the others; each parapsis with 2 shorter setae, each axilla with 1 seta; scutellum with 4 setae, the discoid sensilla somewhat closer to the anterior than to the posterior pair. Frontovertex, pronotum and mesonotal sclerites reticulate. Scutellum oval to nearly rectangular, transverse, about $\frac{2}{3}$ to $\frac{3}{4}$ median length of mesoscutum. Metanotum (Figures 1005, 1006) short, reticulate except on the sides, very lightly arcuate, posterior margin nearly straight; anteromedian apodeme rather slender, about as long as median length of metanotum or somewhat longer.

Propodeum (Figures 1005, 1006) $4\frac{1}{4}$ to $5\frac{1}{5}$ times as long as the metanotum, about $\frac{3}{4}$ to over $\frac{4}{5}$ length of scutellum, reticulate on the sides, broadly reticulate on a wide central area; crenulae 4+4 to 5+5, rather large, distinctly overlapping but broader and less elongate than in other known members of the **lingnanensis** group.

Second abdominal tergite (Figures 1005, 1006) transversely striated anteriorly on both sides, transversely reticulate-striated centrally; tergites III–VI reticulate on the sides, bearing 2 (rarely 3) fine setae in a short transverse row on each reticulate area; third tergite transversely reticulate-striated also mesad of the lateral setiferous areas, transversely striated across center; tergites IV–VI faintly transversely striated across center; seventh tergite (Figure 1007) reticulate on the sides, transversely striated and lightly punctate across center, usually bearing 3 (rarely 4) setae on each lateral reticulate area, and a pair of fine submedian setae; eighth tergite delicately reticulate-punctate, with a transverse row of 4 (rarely 5) setae between spiracles, the lateral pair considerably longer than the submedian pair; syntergum (Figure 1007) triangular, rather short and wide, faintly reticulate-punctate, bearing 5–7 (usually 6) setae in a transverse row. Cerci about equidistant from posterior spiracles and from tip of abdomen or somewhat closer to the former, with 2 long setae and 1 short seta. Ovipositor short, the shaft about $1\frac{1}{3}$ to $1\frac{1}{2}$ times as long as the middle tibia (1.35–1.50), the sheaths rather robust, only about $\frac{1}{3}$ to $\frac{2}{5}$ length of middle tibia (0.32–0.38, measured along outer margin).

Mid-tibial spur somewhat longer than the corresponding basitarsus.

Forewing (Figure 1008) $2\frac{2}{5}$ to nearly $2\frac{3}{5}$ times as long as wide, marginal fringe relatively long, not exceeding $\frac{1}{3}$ width of disk. Delta area with 32–42 setae in 4–6 rows, these considerably longer and sparser than the setae mesad of speculum, not clearly separated from row of setae along posterior margin of wing, widely separated from 1–2 dorsal setae below distal portion of submarginal vein; costal cell with a few fine setae in a row along proximal $\frac{3}{5}$ or so, and 1 coarse seta situated on the vein near apex of cell. Submarginal vein bearing 2 coarse setae, the proximal one about $\frac{2}{3}$ to $\frac{4}{5}$ length of the distal, and 11–16 bullae. Marginal vein bearing 6–9 prominent, subequal setae along anterior margin, these about $1\frac{1}{4}$ to $1\frac{1}{2}$ times longer than the setae in a row along center of vein.

Forewing entirely hyaline; a fuscous streak along posterior margin of wing, distad of speculum.

Hind wing relatively narrow, width of disk about equal to length of longest fringe hairs.

General coloration entirely yellow; thoracic sterna immaculate, concolorous with rest of body; posterior margin of scutellum faintly lined with fuscous; apex of tegula blackish; a short black streak at base of forewing, below tegula; antennae uniformly, faintly dusky; legs concolorous with body; wing veins hyaline.

Length 0.70–0.80 mm.

*Male*. Essentially similar to the female in structure, chaetotaxis, sculpture and coloration, differing mainly in the more robust antennae.

Antennal scape (Figure 1009) $3\frac{2}{3}$ to $4\frac{1}{5}$ times as long as wide, about as long as the club or somewhat longer; pedicel only about $1\frac{2}{5}$ to $1\frac{3}{5}$ times as long as wide, considerably longer than the third segment of the funicle; first 2 funicular segments as in the female, third segment short, somewhat (up to $1\frac{1}{5}$ times) wider than long, bearing 1 longitudinal sensillum; club only $2\frac{2}{5}$ to $2\frac{3}{4}$ times as long as wide, 3 to $3\frac{3}{5}$ times longer and distinctly wider than the preceding segment, bearing 2–3 longitudinal sensilla, with a partly cut-off sensory area on the ventral surface, comprising several rows of minute setae surrounding an oval bare area.

Mesoscutum (Figure 1010) with 9–11 (usually 10) setae, as in the female. Propodeum (Figures 1010, 1011) about $4\frac{1}{3}$ to $5\frac{1}{2}$ times as long as the metanotum, $\frac{2}{3}$ to $\frac{3}{4}$ length of scutellum; crenulae 3+3 to 4+5, as in the female.

Posterior abdominal sternites (Figure 1012) weakly, inconspicuously stippled centrally, with a few short transverse rows of minute spines. Genitalia (Figure 1012) $\frac{2}{3}$ to $\frac{3}{4}$ length of middle tibia (0.65–0.76); digital sclerites about $\frac{1}{4}$ the combined length of aedeagus and apodemes (0.23–0.28).

Forewing (Figure 1013) somewhat broader than in the female, $2\frac{1}{3}$ to $2\frac{1}{2}$ times as long as wide; delta with 25–35 setae in 4–6 rows; veins as in the female.

Length 0.60–0.74 mm.

Described from 24♀♀ and 46♂♂ (syntypes), reared by A. Vilardebo from the coconut scale, *Temnaspidiotus destructor* (Signoret), on avocado, Ivory Coast, Equatorial West Africa, February 1971. Specimens preserved in alcohol, subsequently mounted in Hoyer's medium on 2 slides.

Type series in the collection of the Division of Biological Control, University of California, Riverside.

**Notes.** Some of the differences between *equatorialis* and other members of the **lingnanensis** group, such as *holoxanthus* or *melinus*, may probably be ascribed to the minute size of *equatorialis*. Also, the fact that the specimens had been stored in alcohol prior to mounting doubtless affected their degree of cleaning and subsequent appearance. Nevertheless, *equatorialis* is undoubtedly a distinct species, and the short antennae, rather low propodeal crenulae, short ovipositor and weak stippling of the male abdominal sterna may serve as reliable diagnostic characters for its separation from other species. In the shape of crenulae and weak stippling, *equatorialis* is rather similar to *africanus*, an African member of the **lingnanensis** group with pigmented thoracic sterna.

In many of the female specimens at hand, the components of the ovipositor have become separated in preparation, with the stylets unnaturally extruded. Measurements were therefore taken only from the base to the tip of the central component.

In one female specimen, the mesonotal furca appears to be very faintly lined with brownish, whereas the sternal plates are immaculate. Since in all other specimens the furca as well as the sternal plates are immaculate, invisible under a dissecting microscope, we consider this to be an individual aberration, perhaps due to some quirk of preparation.

**Potential Value for Biological Control.** The coconut scale is an important pest of coconut, avocado and other fruit trees in various tropical countries. Although information on the biology of *equatorialis* is at present lacking, this interesting species should be considered a potential candidate for importation in biological control projects directed against that serious pest.

**Related Form.** A series of 7♀♀, reared from the masked scale, *Mycetaspis personata* (Comstock), on avocado in Cameroun, Equatorial West Africa, was received from A. Vilardebo in February 1971. These specimens rather closely resemble the syntypes of *equatorialis* in most diagnostic characters, but are somewhat larger (0.68–1.05 mm) and have a somewhat longer ovipositor (shaft 1.51–1.60 times as long as the middle tibia, sheaths 0.43–0.51 length of middle tibia). Since the small series consists of females only, it is possible that it represents a uniparental form. For the time being, this is regarded as a form closely related to *equatorialis*.

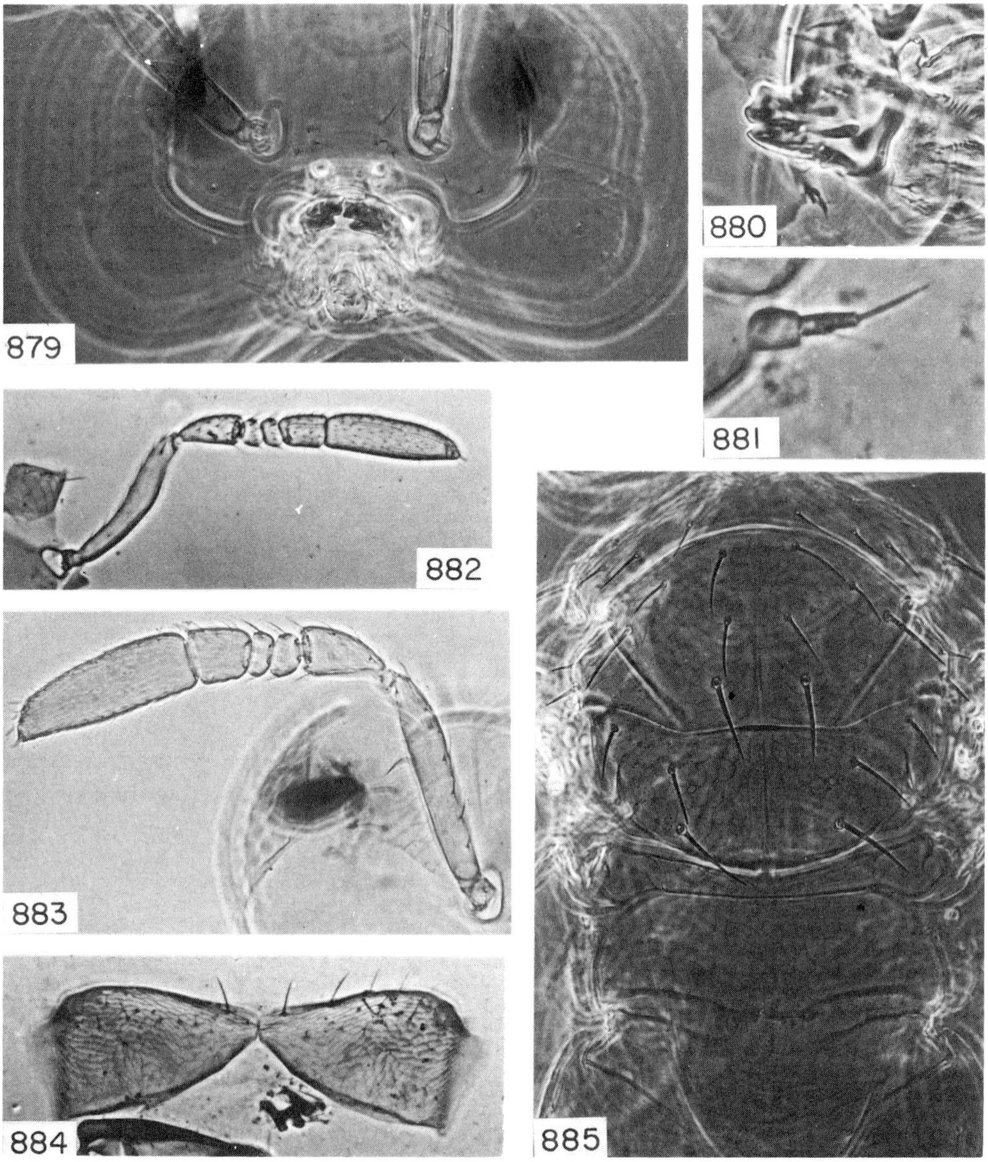


Figures 879–885. *Aphytis lingnanensis* Compere, ♀
879. Face and mouthparts. 880. Mandible. 881. Maxillary palpus. 882, 883. Antennae. 884. Pronotum. 885. Thorax, propodeum and base of gaster.

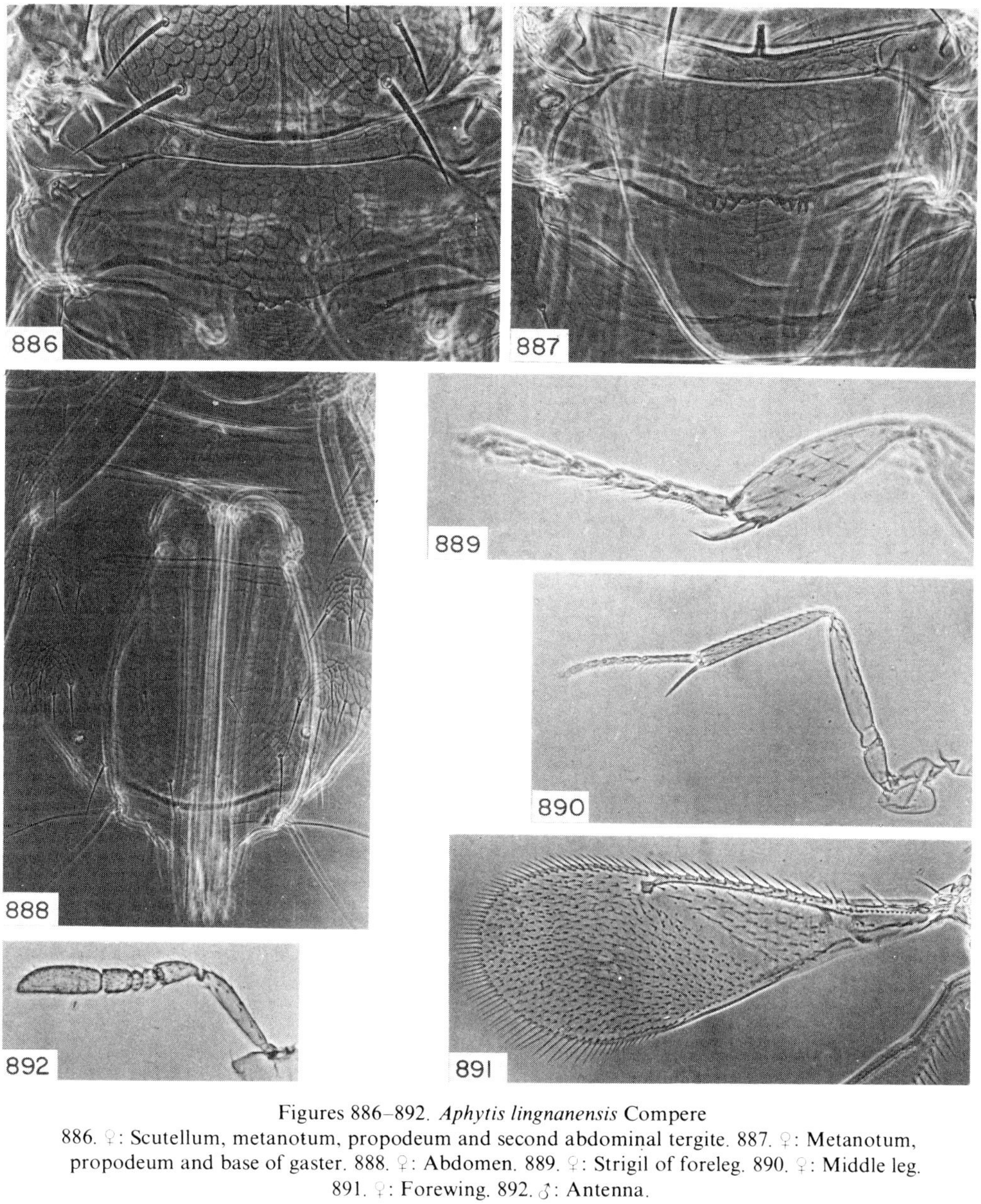


Figures 886–892. *Aphytis lingnanensis* Compere
886. ♀: Scutellum, metanotum, propodeum and second abdominal tergite. 887. ♀: Metanotum, propodeum and base of gaster. 888. ♀: Abdomen. 889. ♀: Strigil of foreleg. 890. ♀: Middle leg. 891. ♀: Forewing. 892. ♂: Antenna.

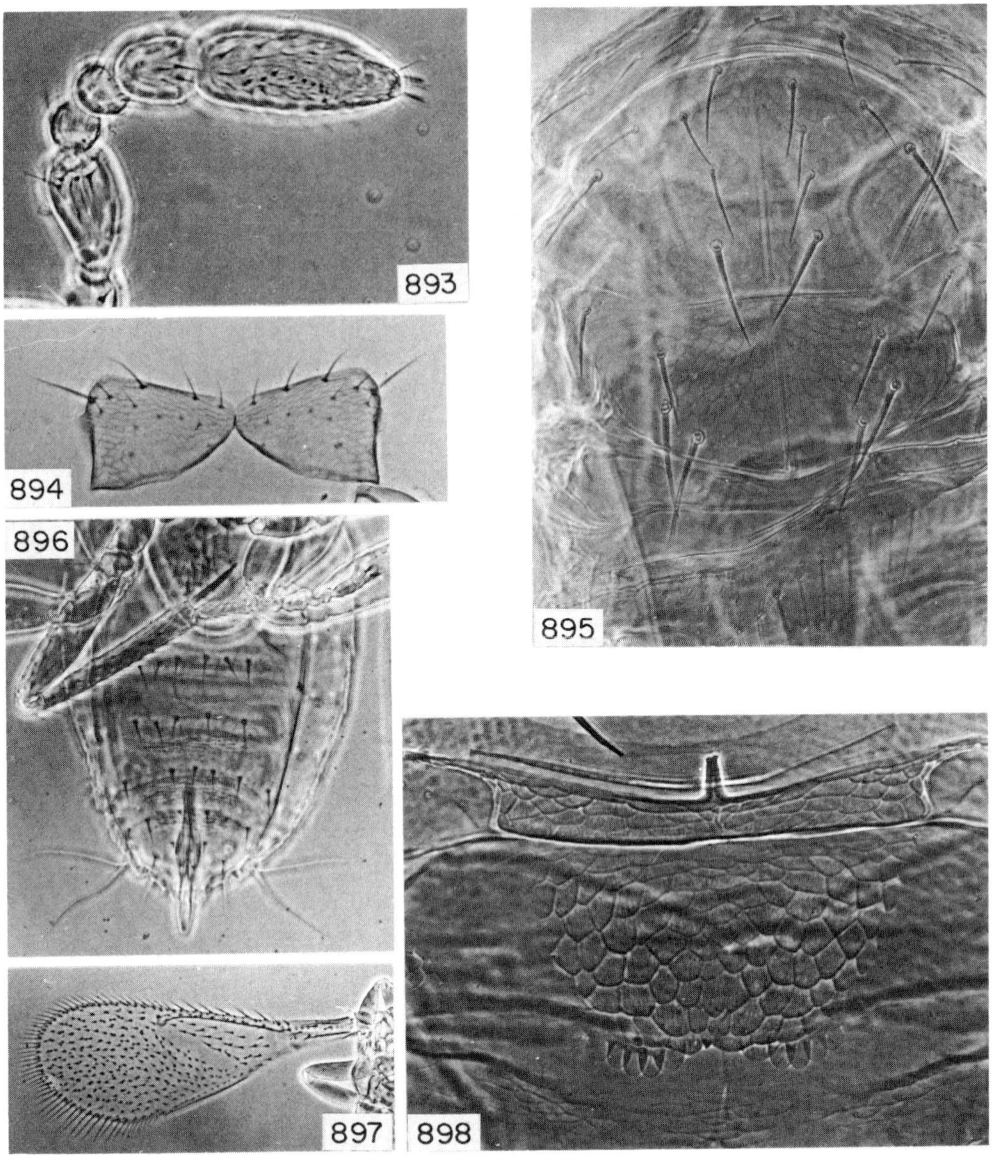


Figures 893–898. *Aphytis lingnanensis* Compere

893. ♂: Antennal club, showing specialized sensory area. 894. ♂: Pronotum. 895. ♂: Mesonotum; note abnormal chaetotaxis of scutellum (6 setae). 896. ♂: Abdomen, showing ventral stippling and genitalia (also middle tibia). 897. ♂: Forewing. 898. ♀ reared by Girault in Australia, misidentified as *chrysomphali*: Metanotum, propodeum and second abdominal tergite.

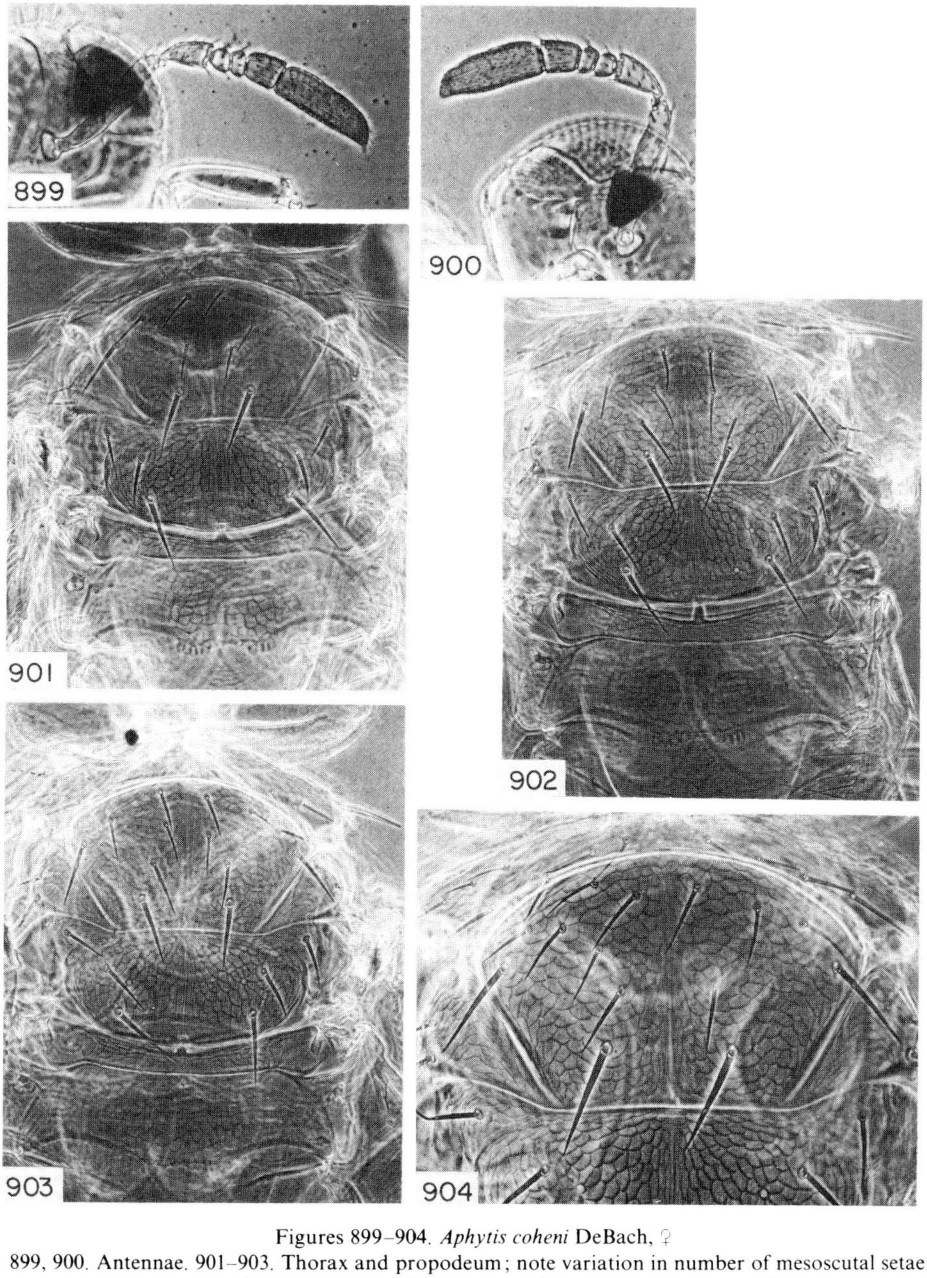


Figures 899–904. *Aphytis coheni* DeBach, ♀
899, 900. Antennae. 901–903. Thorax and propodeum; note variation in number of mesoscutal setae (paralectotypes). 904. Pronotum and mesonotum (14 mesoscutal setae; paralectotype).

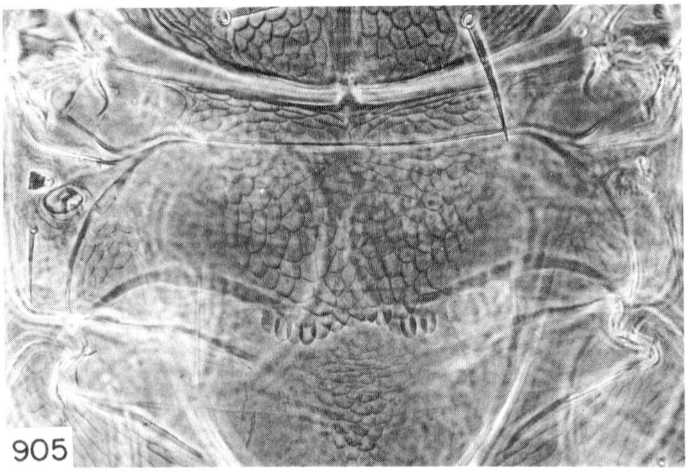


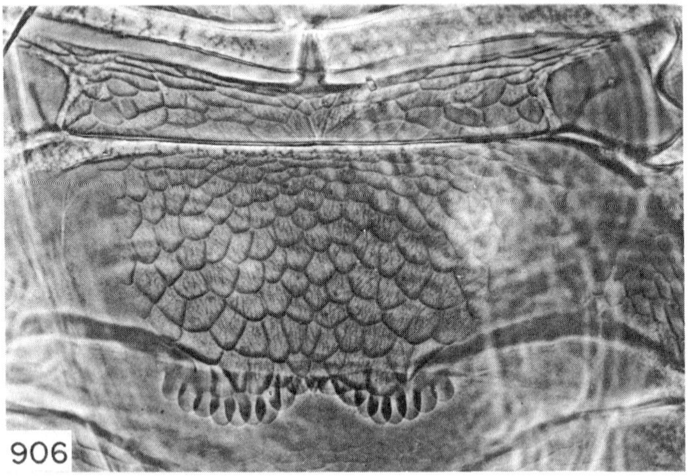


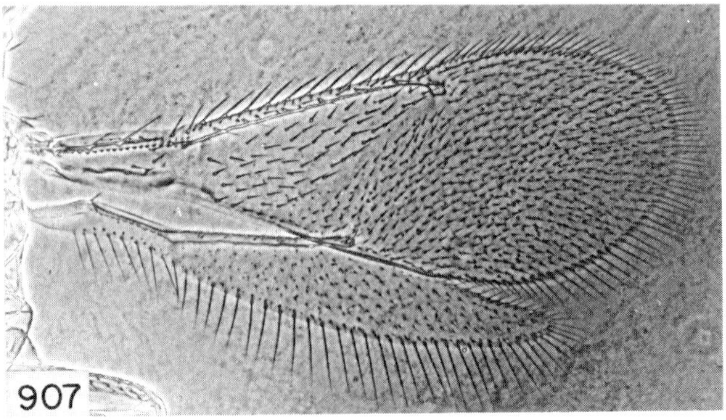


Figures 905–907. *Aphytis coheni* DeBach, ♀
905, 906. Metanotum and propodeum (paralectotypes). 907. Wings.

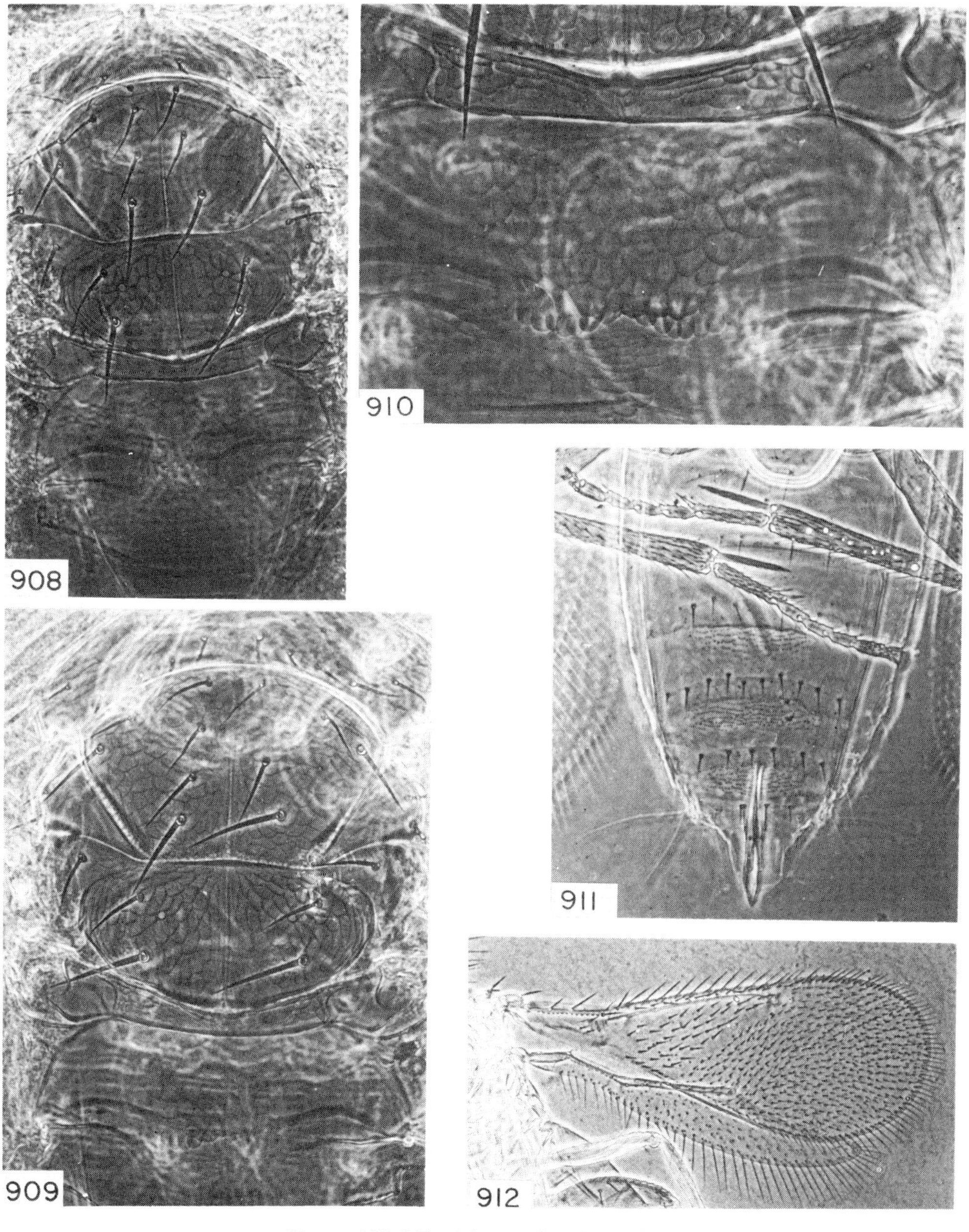


Figures 908–912. *Aphytis coheni* DeBach, ♂
908. Thorax and propodeum (minute specimen; paralectotype). 909. Thorax and propodeum (paralectotype). 910. Metanotum and propodeum (paralectotype). 911. Abdominal sterna and genitalia; note extensive stippling. 912. Wings.

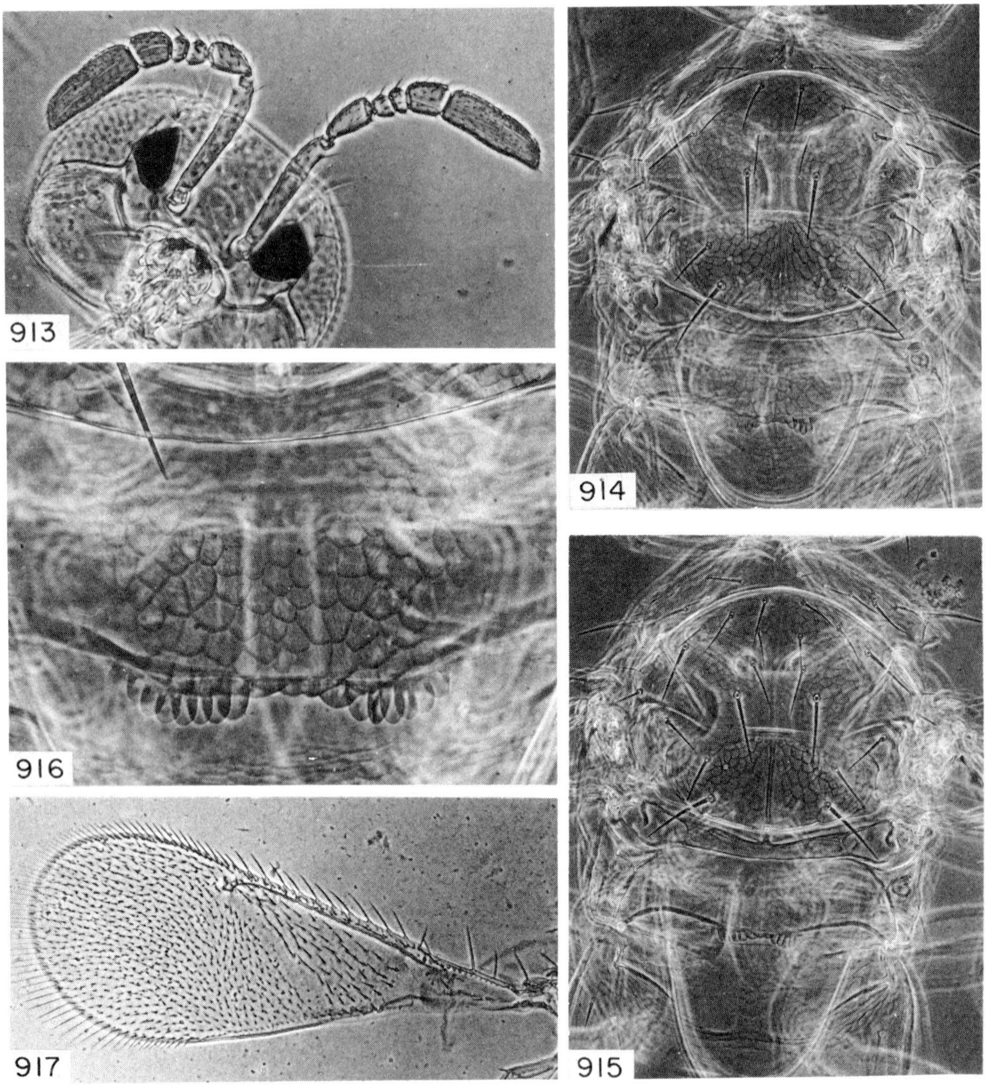


Figures 913–917. *Aphytis* "khunti" (a semispecies of *A. coheni* DeBach)
913. ♀: Head and antennae. 914. ♀: Thorax and propodeum. 915. ♂: Thorax and propodeum.
916. ♀: Propodeum and crenulae. 917. ♀: Forewing.

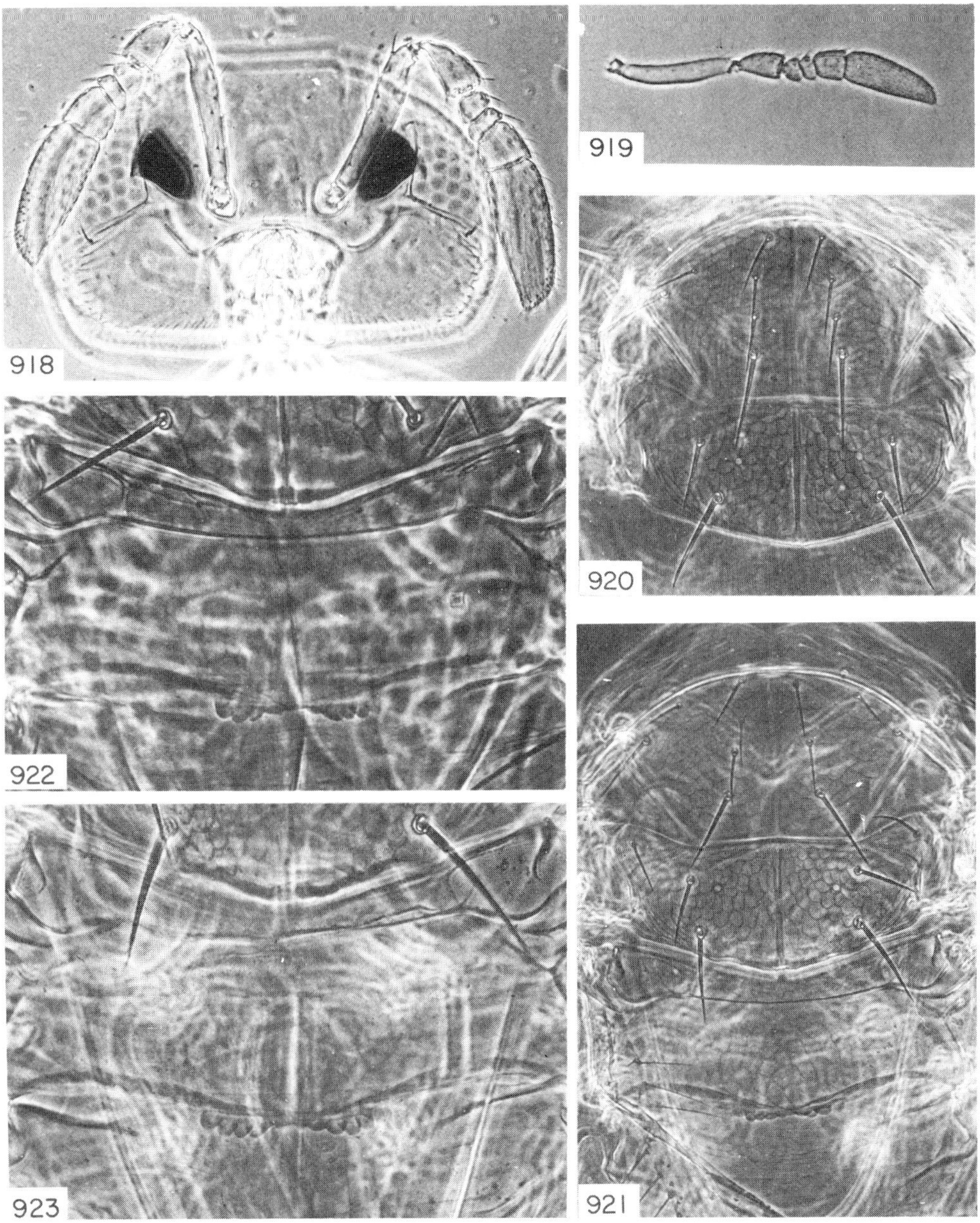


Figures 918–923. *Aphytis africanus* Quednau, ♀
918. Head and antennae (syntype). 919. Antenna. 920. Mesonotum. 921. Thorax, propodeum and base of gaster. 922. Metanotum and propodeum. 923. Propodeum (syntype).

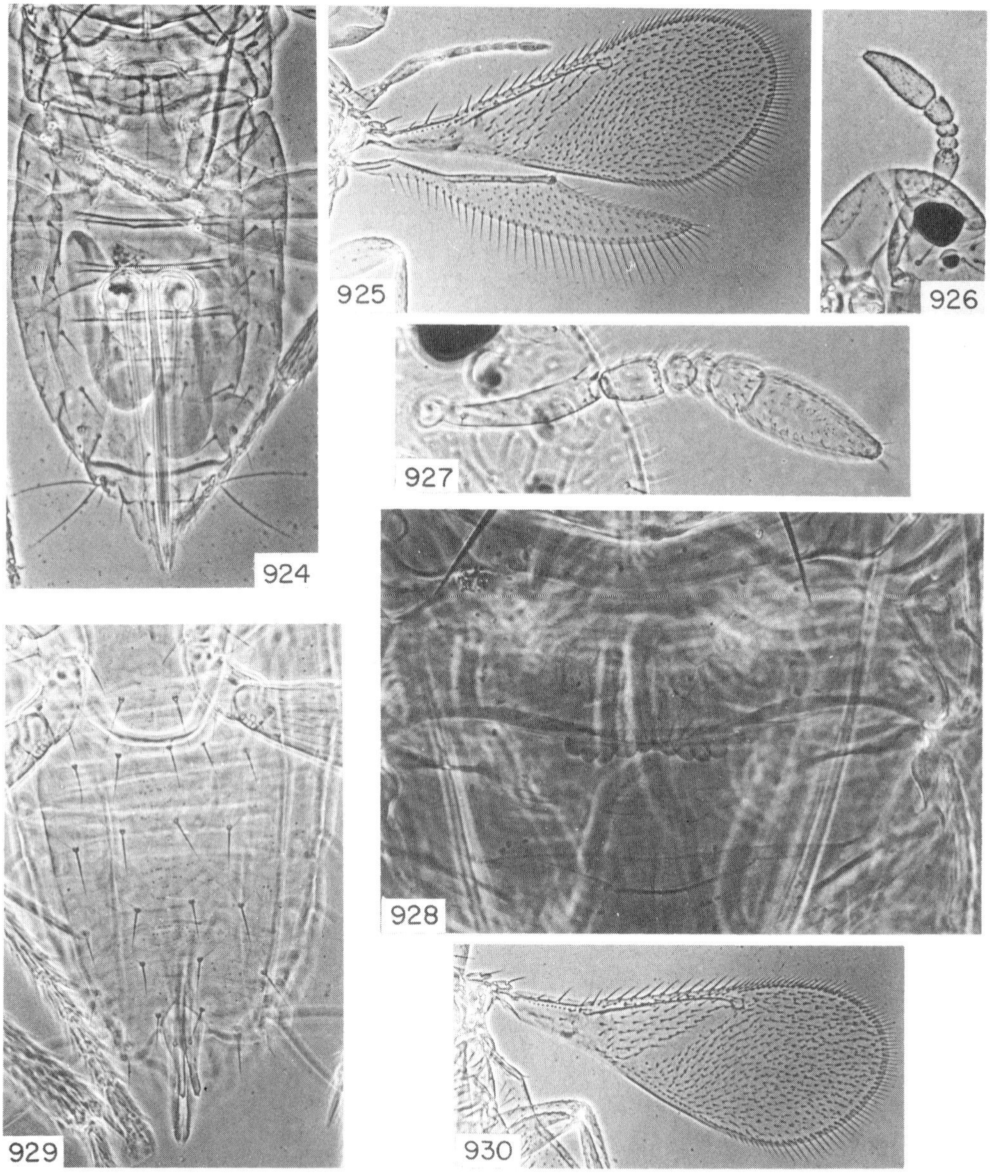

Figures 924–930. *Aphytis africanus* Quednau
924. ♀: Abdomen (syntype). 925. ♀: Wings (syntype). 926. ♂: Antenna. 927. ♂: Antenna, showing sensory area on club. 928. ♂: Propodeum and base of gaster (syntype). 929. ♂: Venter of abdomen and genitalia; note light stippling (syntype). 930. ♂: Forewing (syntype).

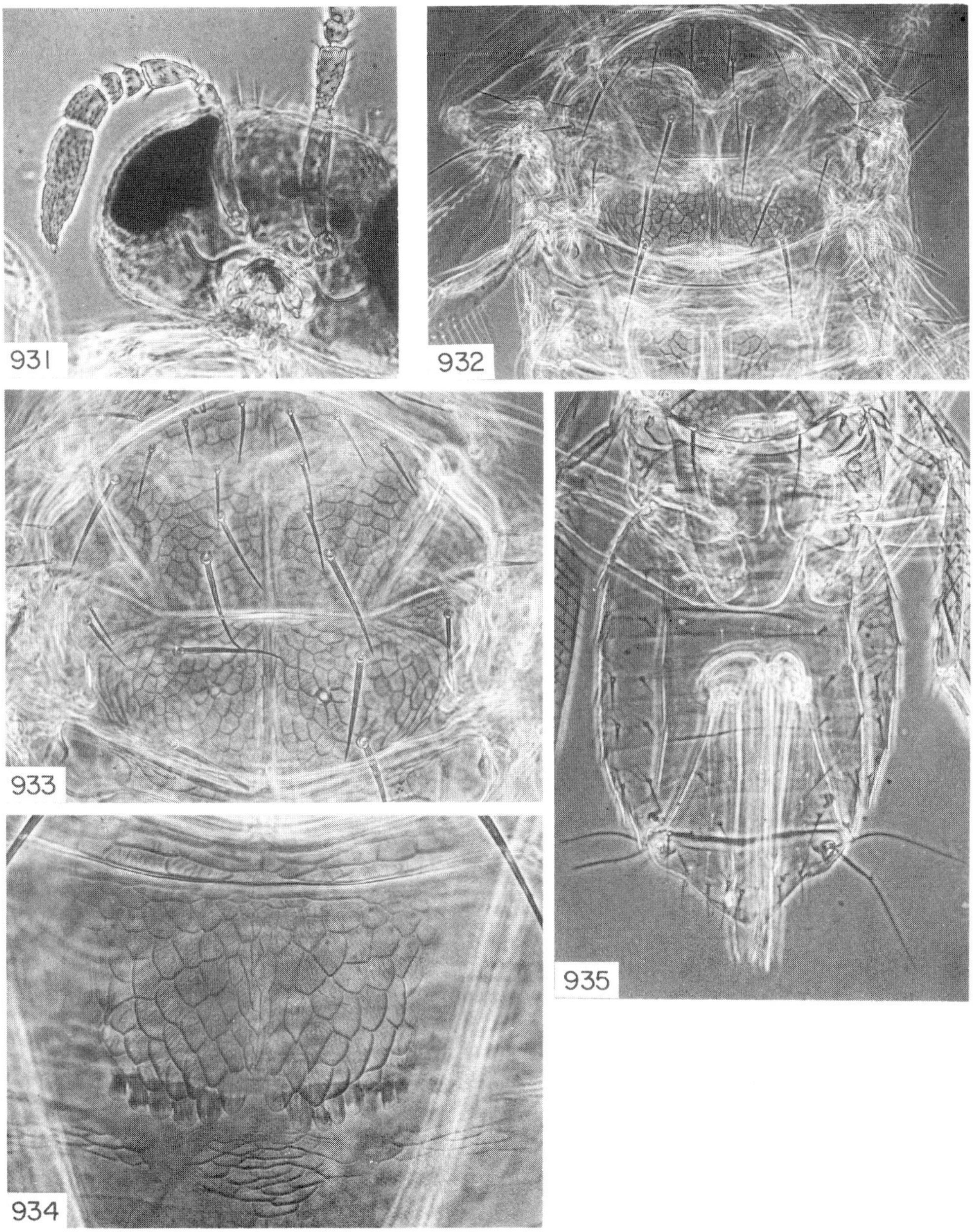


Figures 931–935. *Aphytis margaretae* DeBach and Rosen, ♀ (paratypes)
931. Antenna. 932. Thorax and propodeum. 933. Mesonotum; note supernumerary seta on left axilla. 934. Metanotum, propodeum and second abdominal tergite. 935. Abdomen.

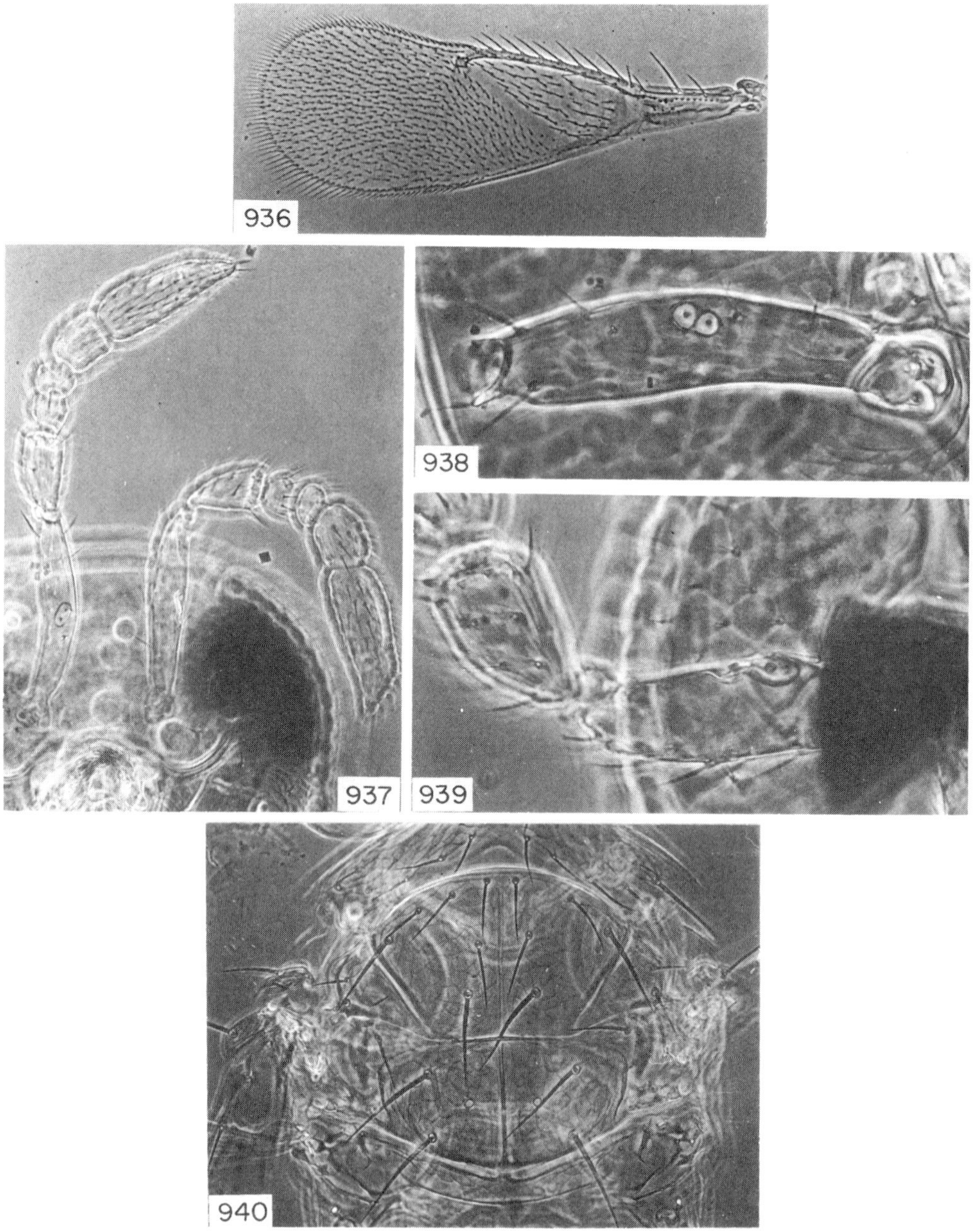


Figures 936–940. *Aphytis margaretae* DeBach and Rosen
936. ♀: Forewing. 937. ♂: Antennae; note sense organs on scape (paratype). 938. ♂: Antennal scape, showing sense organ in ventral view. 939. ♂: Antennal scape and pedicel, showing sense organ in lateral view. 940. ♂: Thorax.

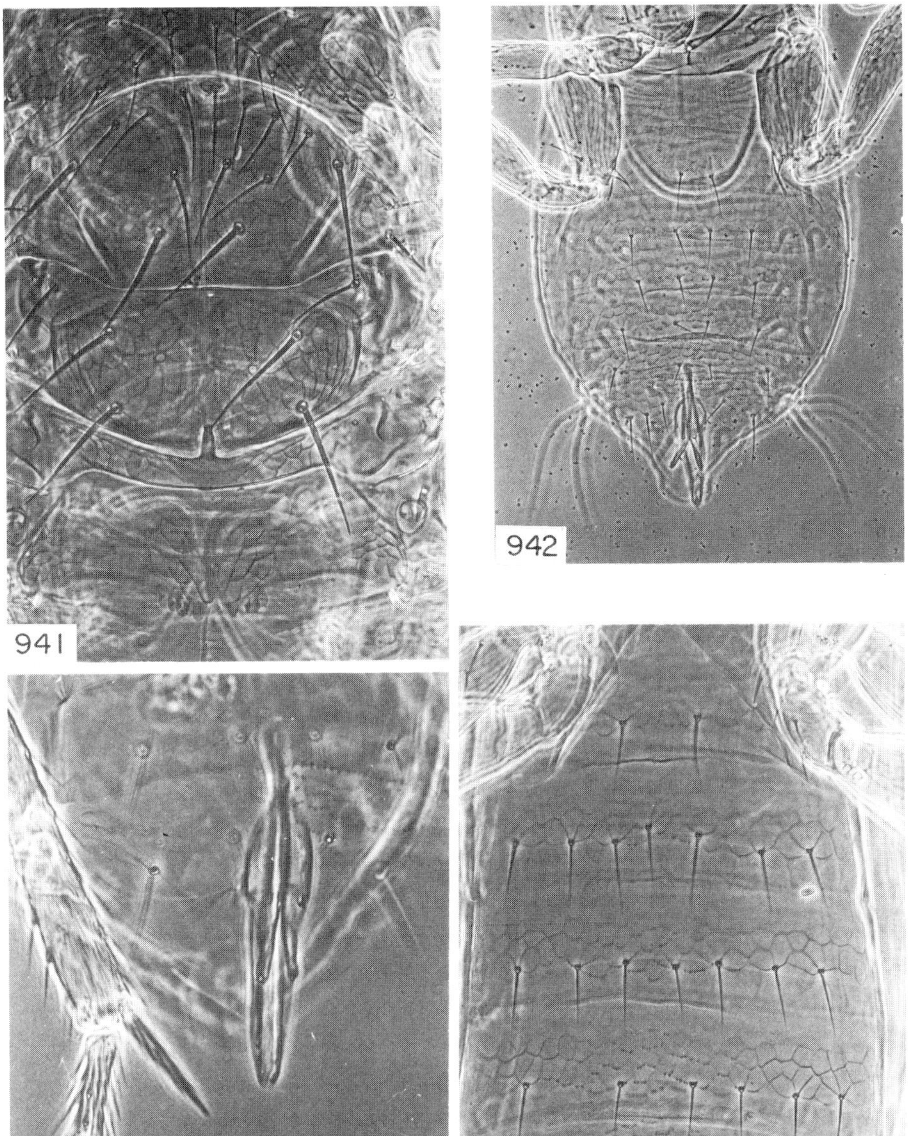


Figures 941–944. *Aphytis margaretae* DeBach and Rosen, ♂
941. Thorax and propodeum. 942. Venter of abdomen and genitalia. 943. Chaetotaxis and sculpture of abdominal sternites, showing light stippling (paratype). 944. Genitalia; note also mid-tibial spur (paratype).

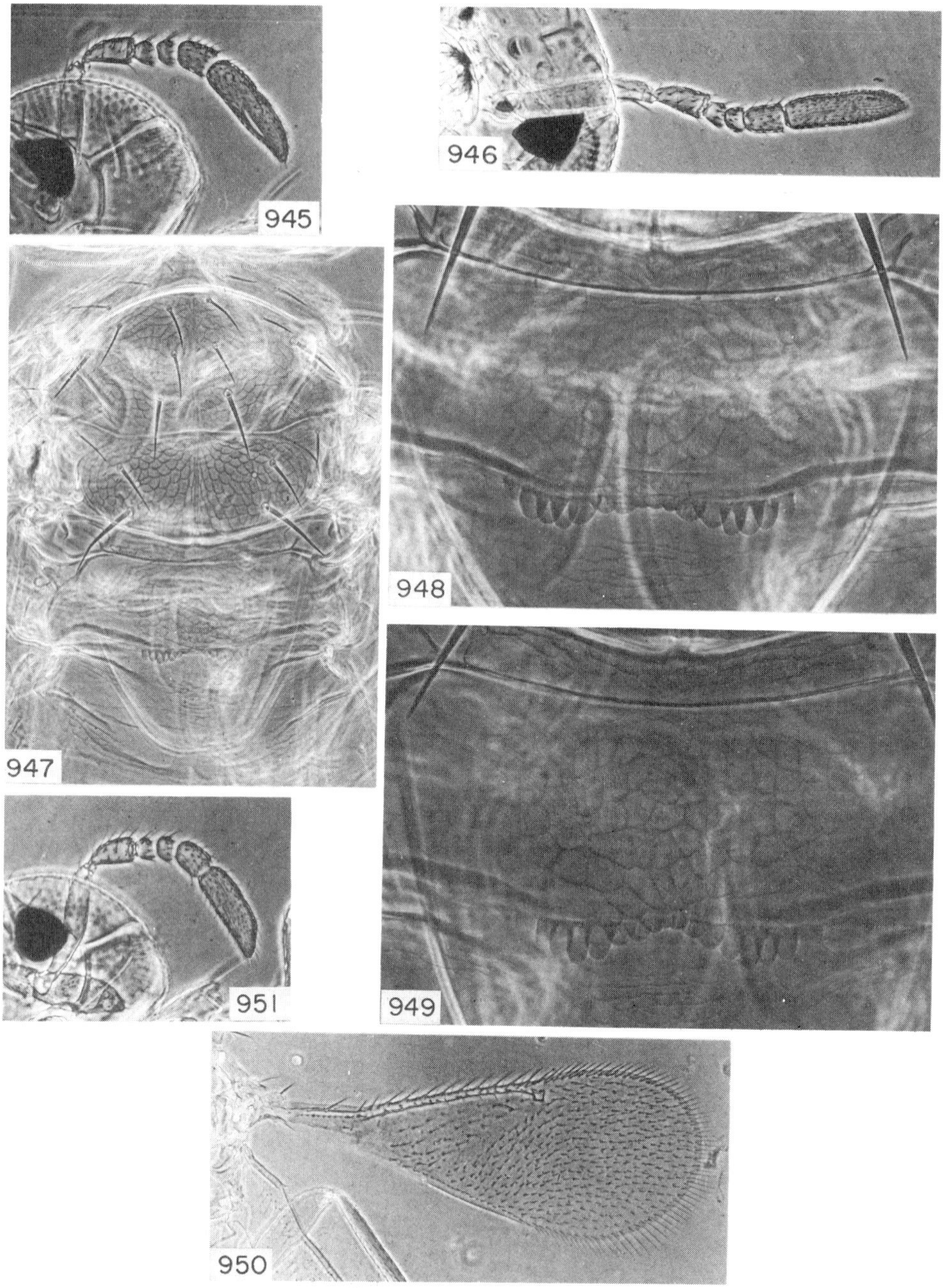


Figures 945–951. *Aphytis holoxanthus* DeBach (paralectotypes)
945, 946. ♀: Antennae. 947. ♀: Thorax, propodeum and base of gaster. 948, 949. ♀: Propodeum and crenulae. 950. ♀: Forewing. 951. ♂: Antenna.

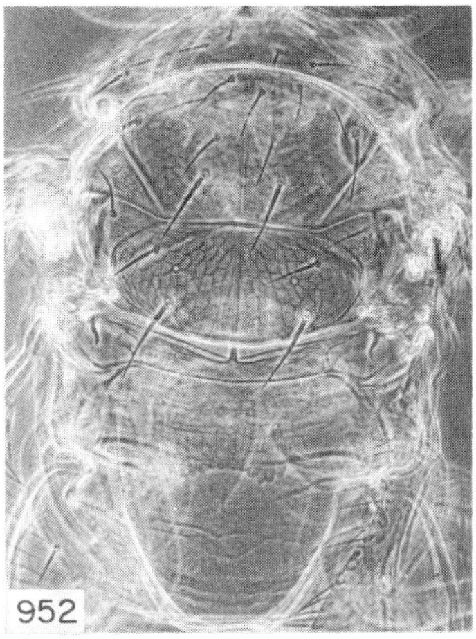


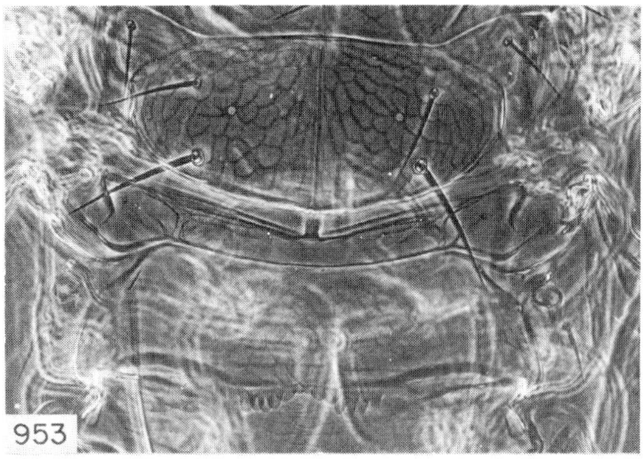


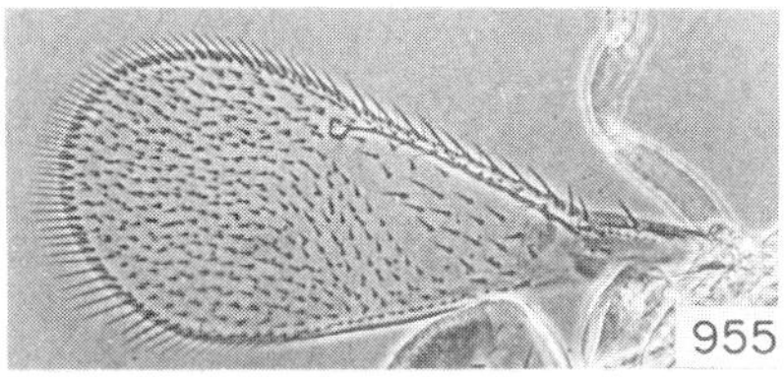


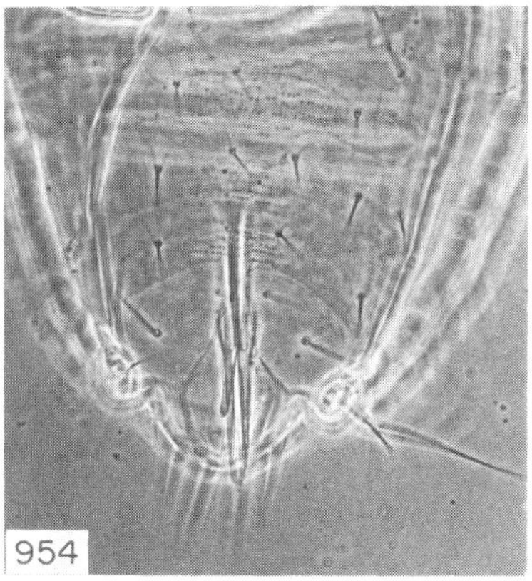


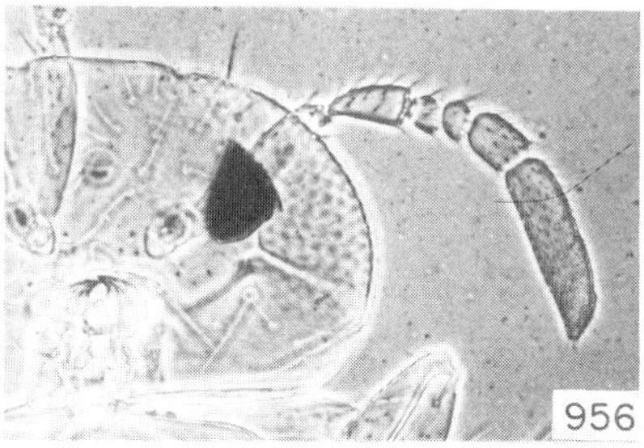


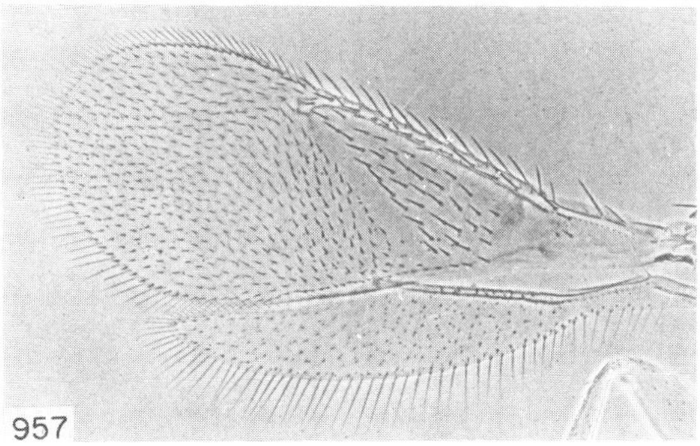


Figures 952–955. *Aphytis holoxanthus* DeBach, ♂
952. Thorax, propodeum and base of gaster (paralectotype). 953. Scutellum, metanotum and propodeum (paralectotype). 954. Posterior abdominal sternites and genitalia. 955. Forewing (paralectotype).
Figures 956, 957. *Aphytis ?holoxanthus* ex California red scale, Lebanon, ♀
956. Antenna. 957. Wings.

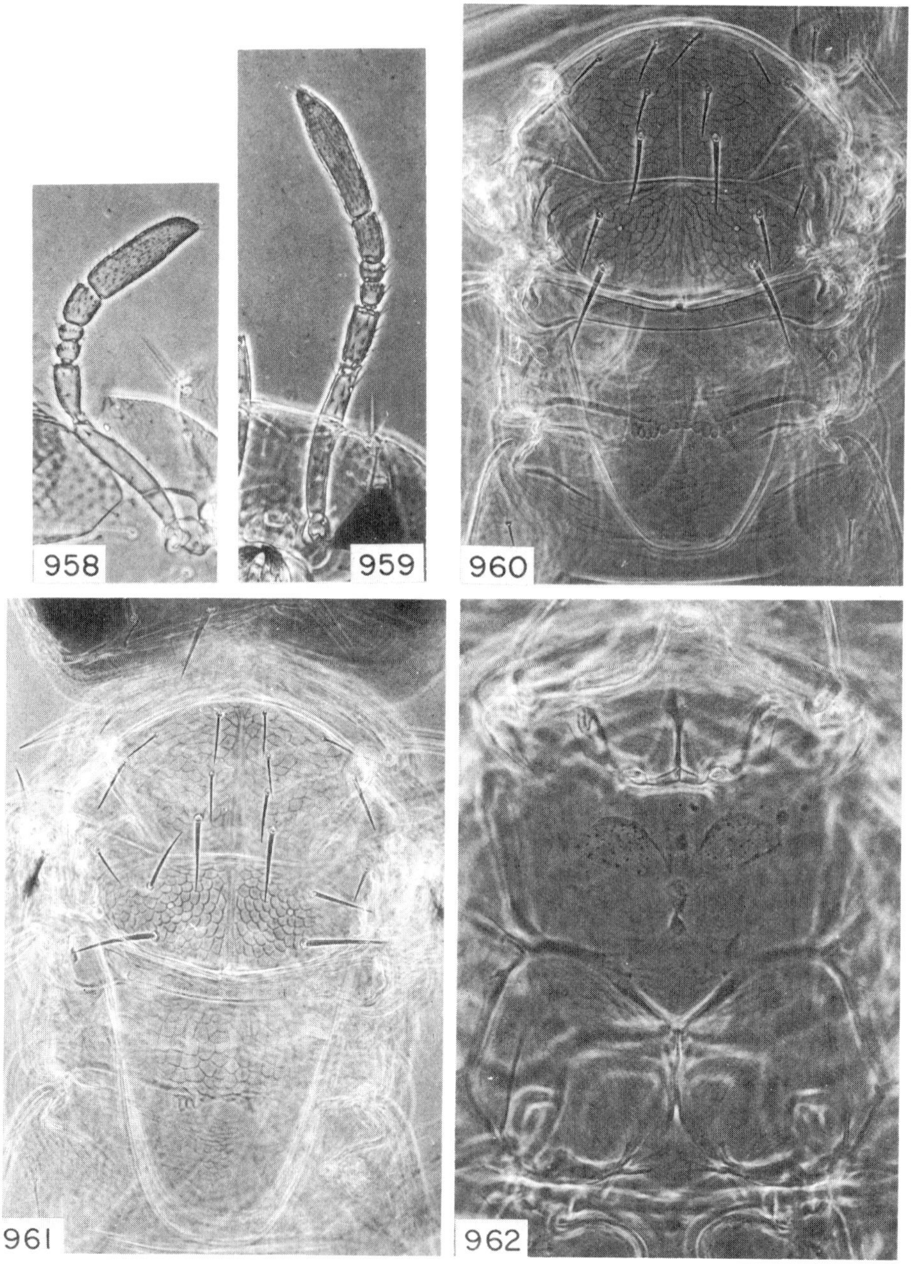


Figures 958–962. *Aphytis melinus* DeBach, ♀
958, 959. Antennae. 960, 961. Thorax, propodeum and base of gaster. 962. Prosternum, mesosternum and epicoxal pads.

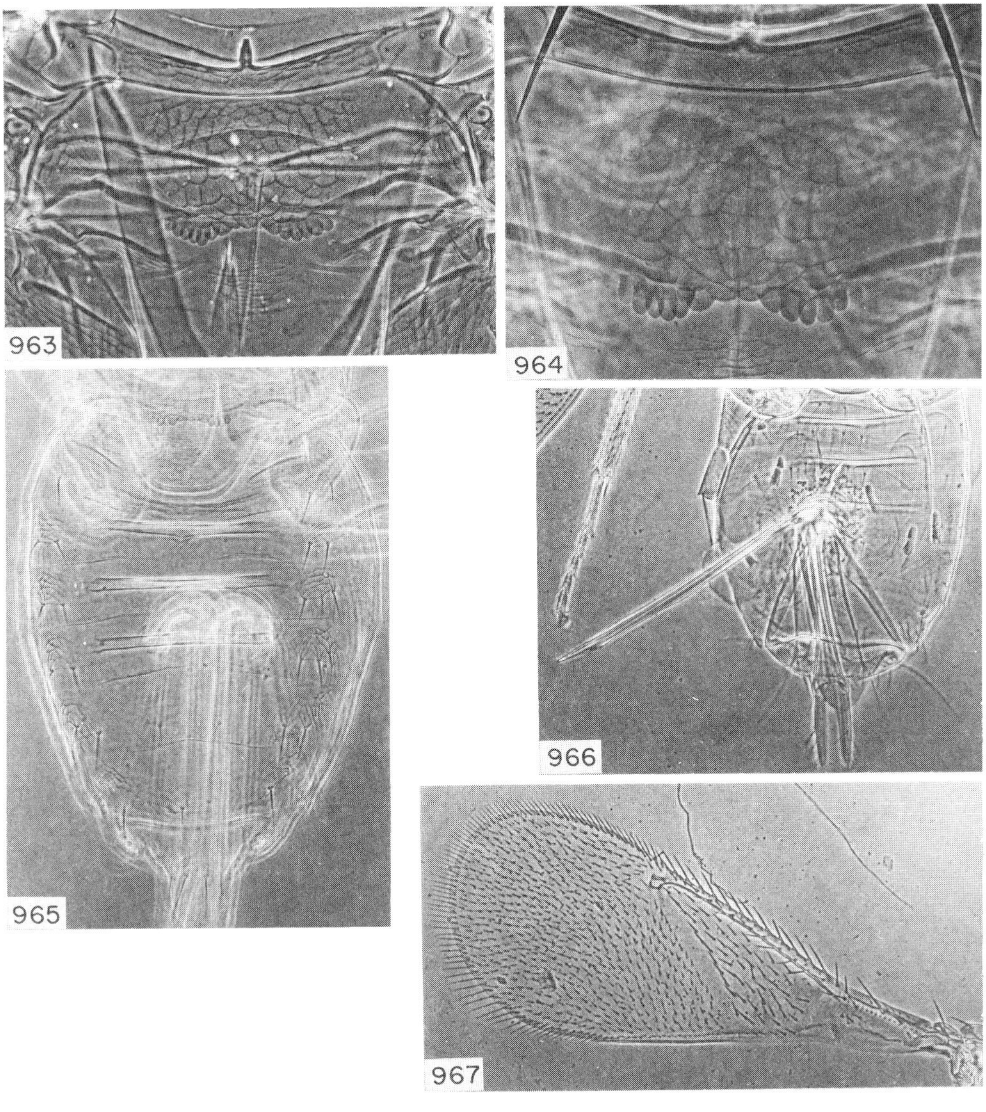


Figures 963–967. *Aphytis melinus* DeBach, ♀
963. Metanotum, propodeum and second abdominal tergite. 964. Propodeal crenulae. 965. Abdomen, dorsal view. 966. Abdomen, ventral view, showing exserted ovipositor. 967. Forewing.

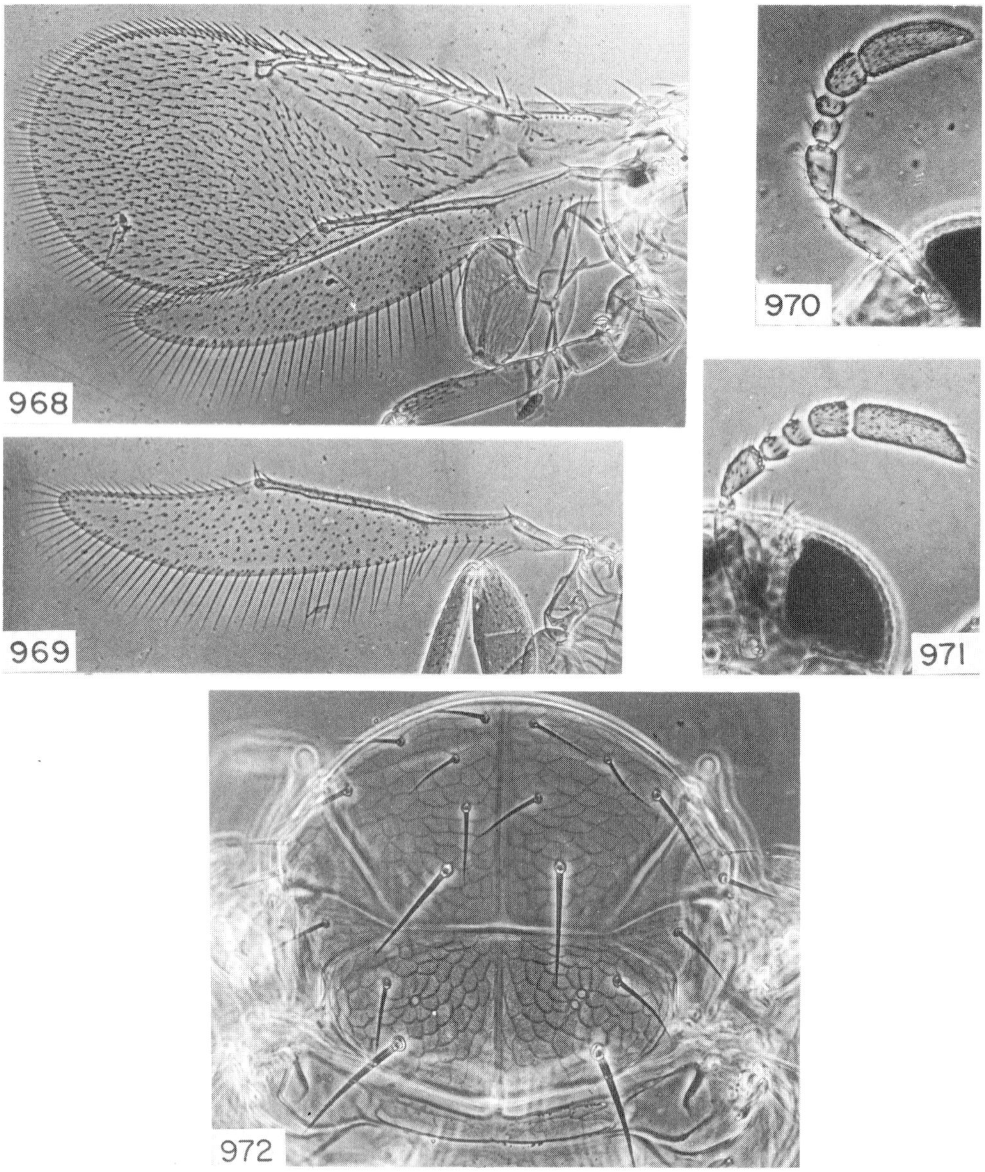

Figures 968–972. *Aphytis melinus* DeBach
968. ♀: Wings. 969. ♀: Hind wing. 970. ♂: Antenna (syntype). 971. ♂: Antenna. 972. ♂: Mesonotum and metanotum; note supernumerary discoid sensillum on scutellum.

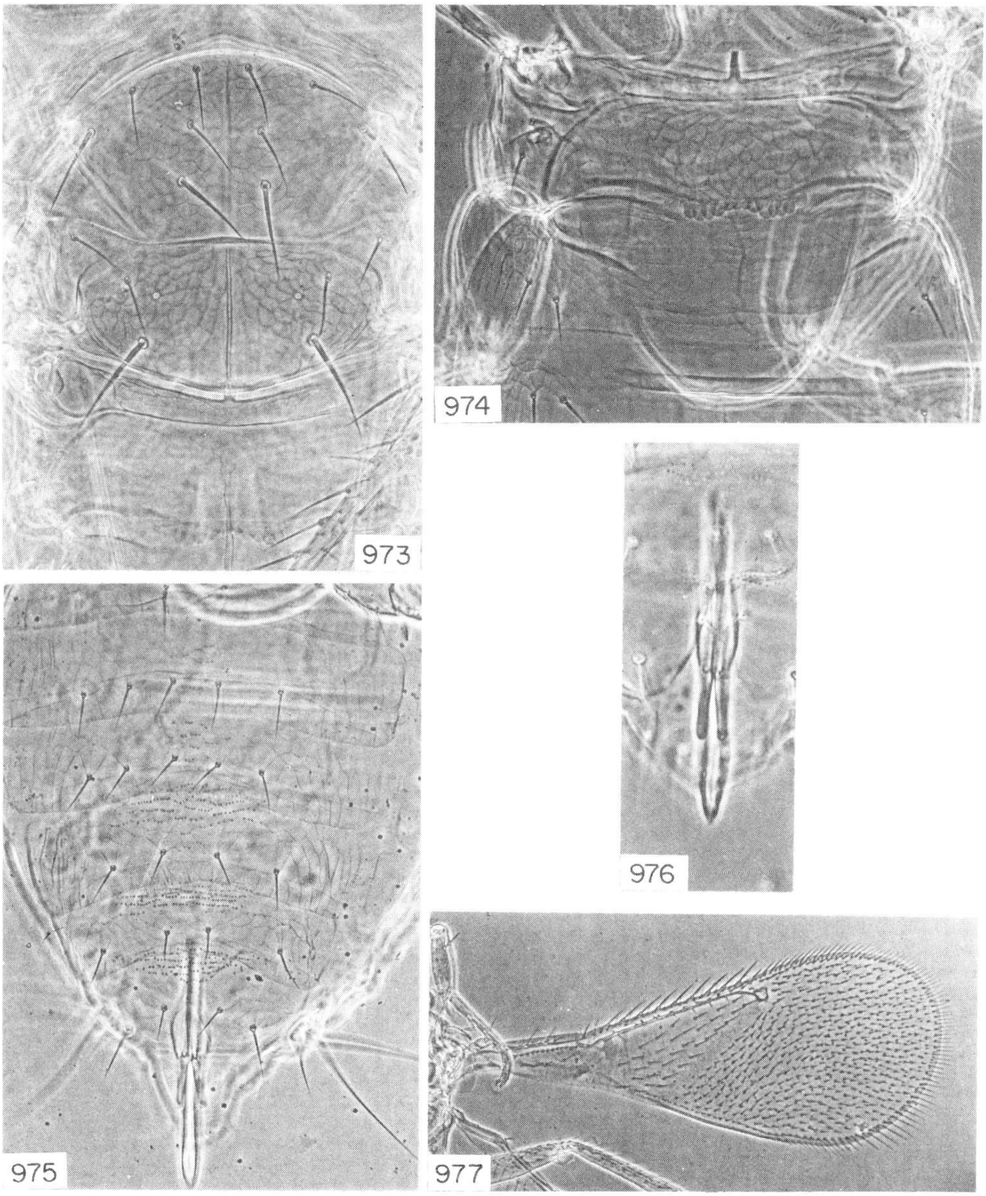


Figures 973–977. *Aphytis melinus* DeBach, ♂
973. Thorax and propodeum. 974. Metanotum, propodeum and base of gaster. 975. Posterior abdominal sternites, showing chaetotaxis and extensive stippling, and genitalia. 976. Genitalia. 977. Forewing.

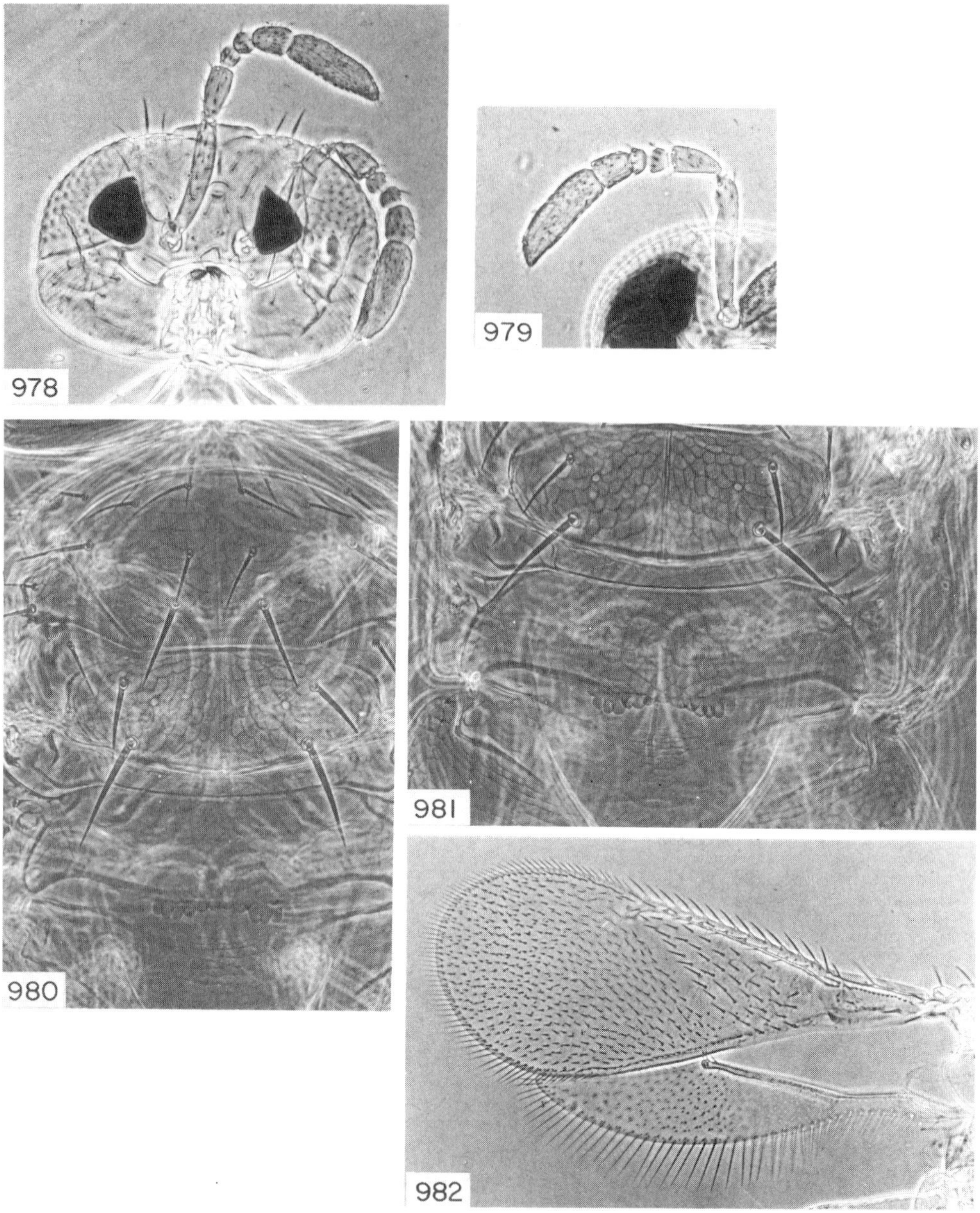


Figures 978–982. *Aphytis yasumatsui* Azim, ♀
978. Head and antennae (syntype series). 979. Antenna. 980. Thorax, propodeum and base of gaster (syntype series). 981. Scutellum, metanotum, propodeum and base of gaster (syntype series). 982. Wings.

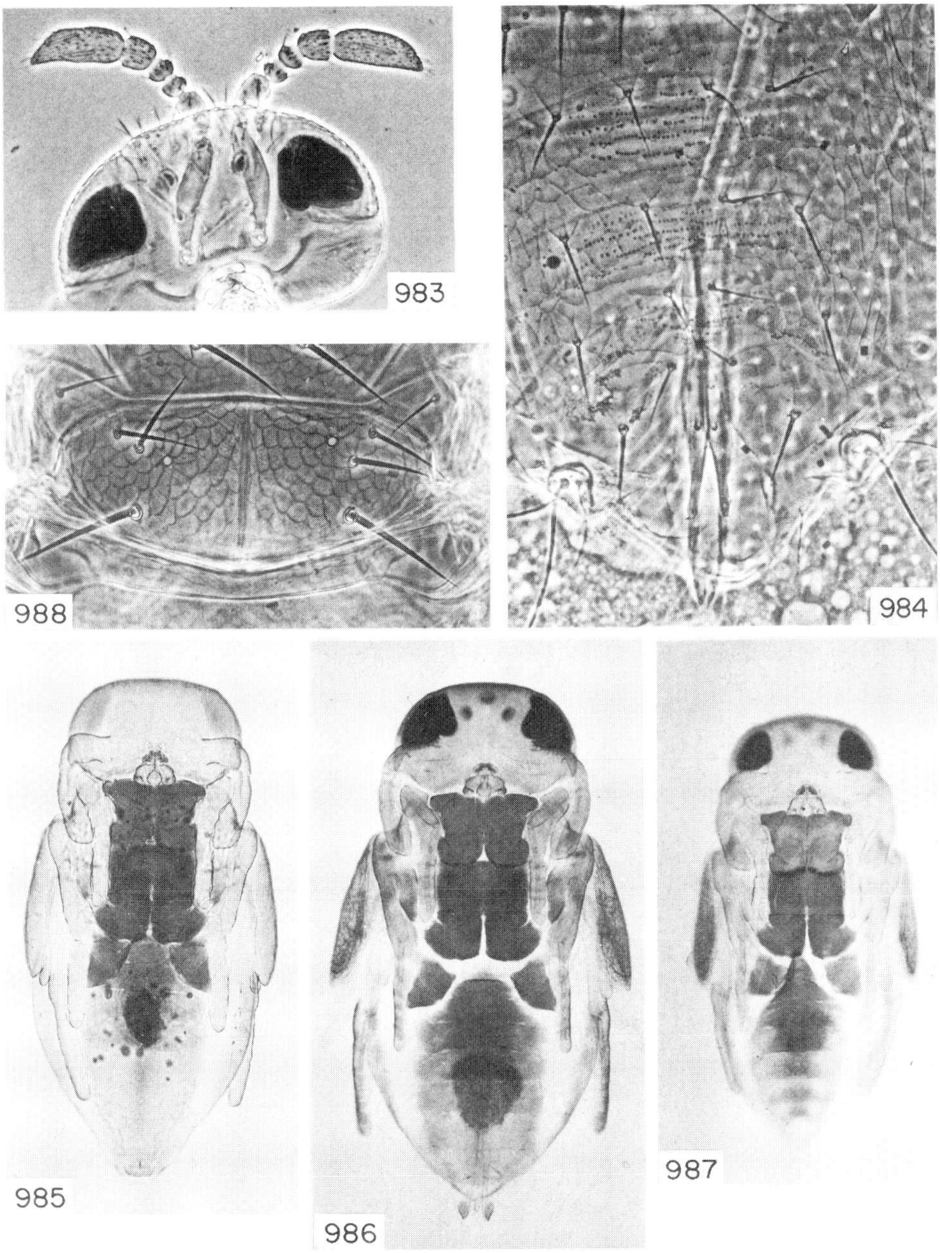


Figures 983–988. *Aphytis yasumatsui* Azim

983. ♂: Head and antennae. 984. ♂: Genitalia and stippling of gastral sternites (syntype). 985. Young ♀ pupa; abdominal pigmentation incomplete. 986. Mature ♀ pupa. 987. Mature ♂ pupa. 988. ♂: Abnormal scutellum with 6 setae and 2 discoid sensilla, showing also axillae and metanotum.

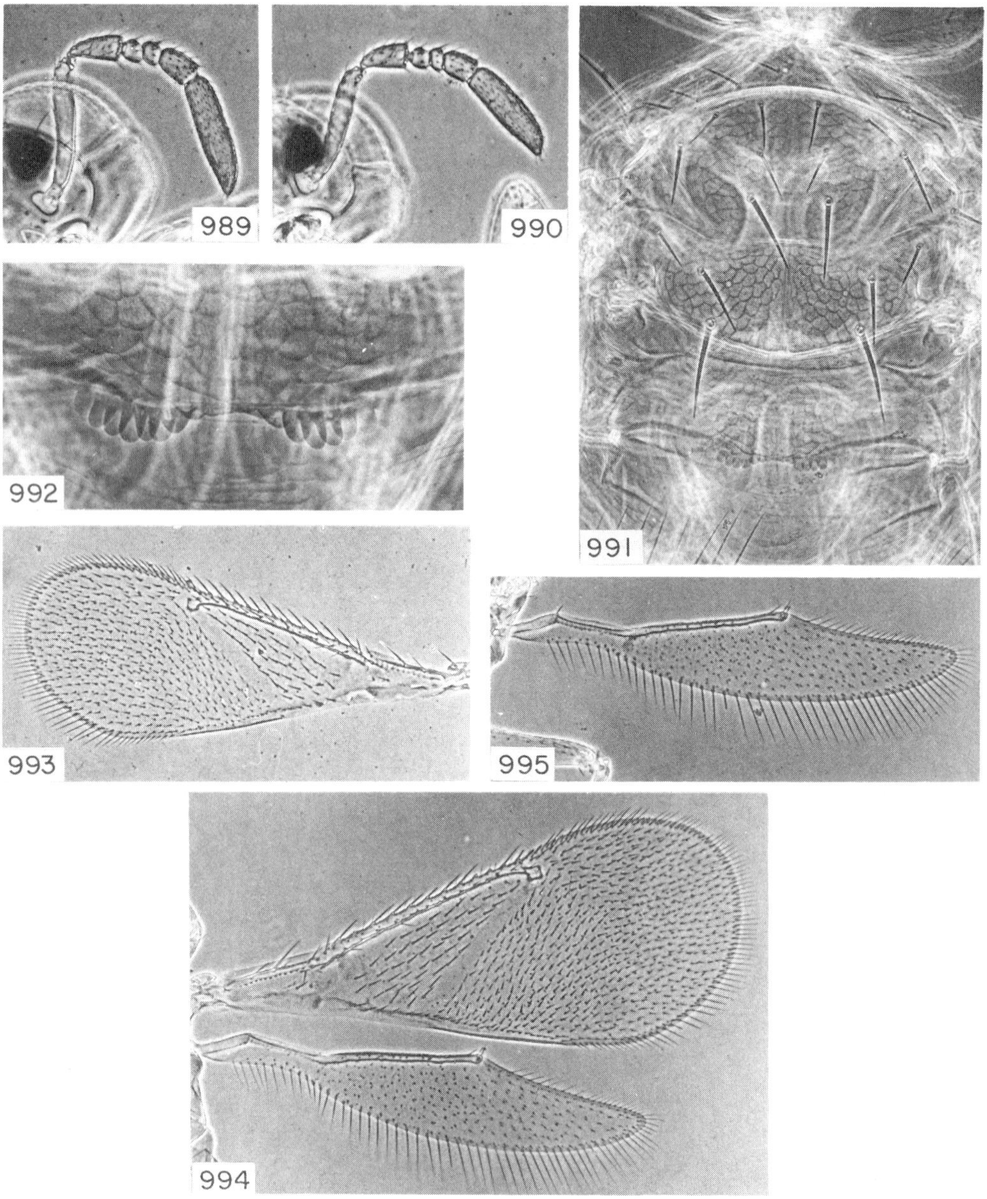


Figures 989–995. *Aphytis fisheri* DeBach, ♀
989, 990. Antennae. 991. Thorax and propodeum. 992. Propodeal crenulae. 993. Forewing. 994. Wings. 995. Hind wing.

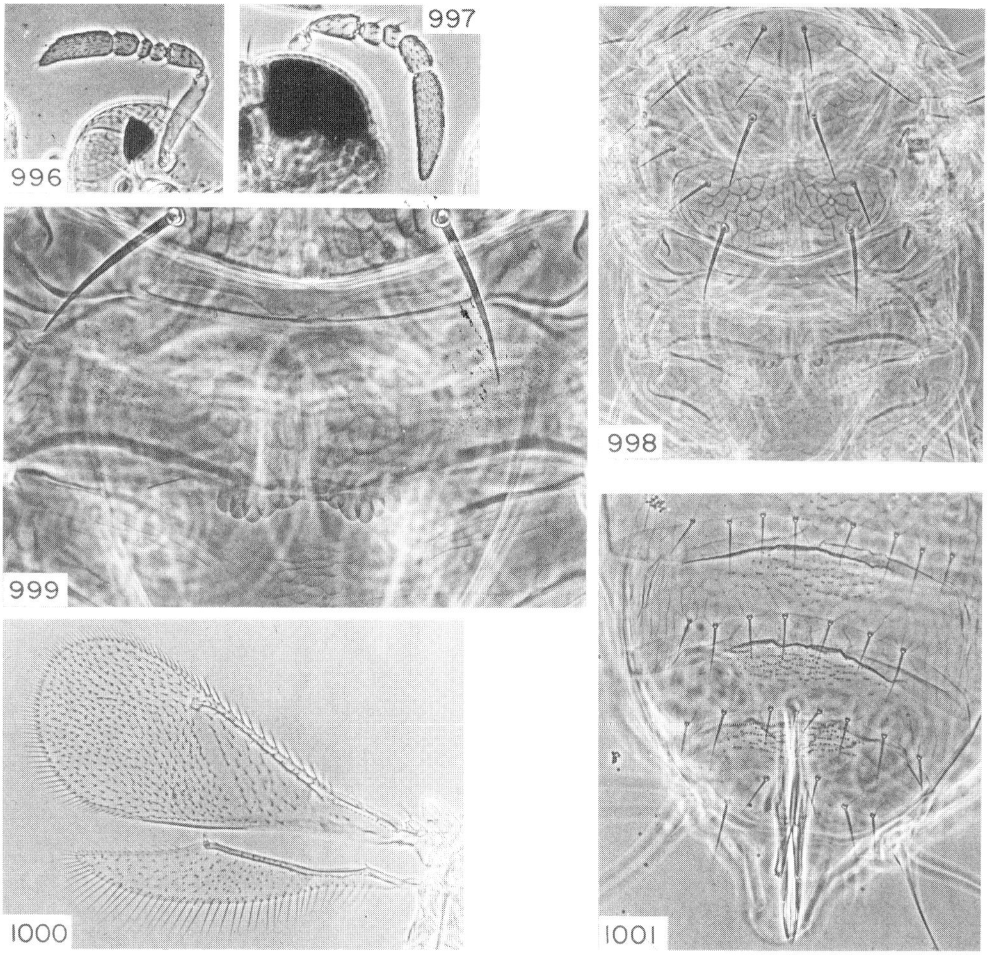


Figures 996–1001. *Aphytis fisheri* DeBach, ♂
996, 997. Antennae. 998. Thorax, propodeum and base of gaster. 999. Metanotum, propodeum and base of gaster. 1000. Wings. 1001. Posterior abdominal sternites, showing chaetotaxis and extensive stippling, and genitalia.

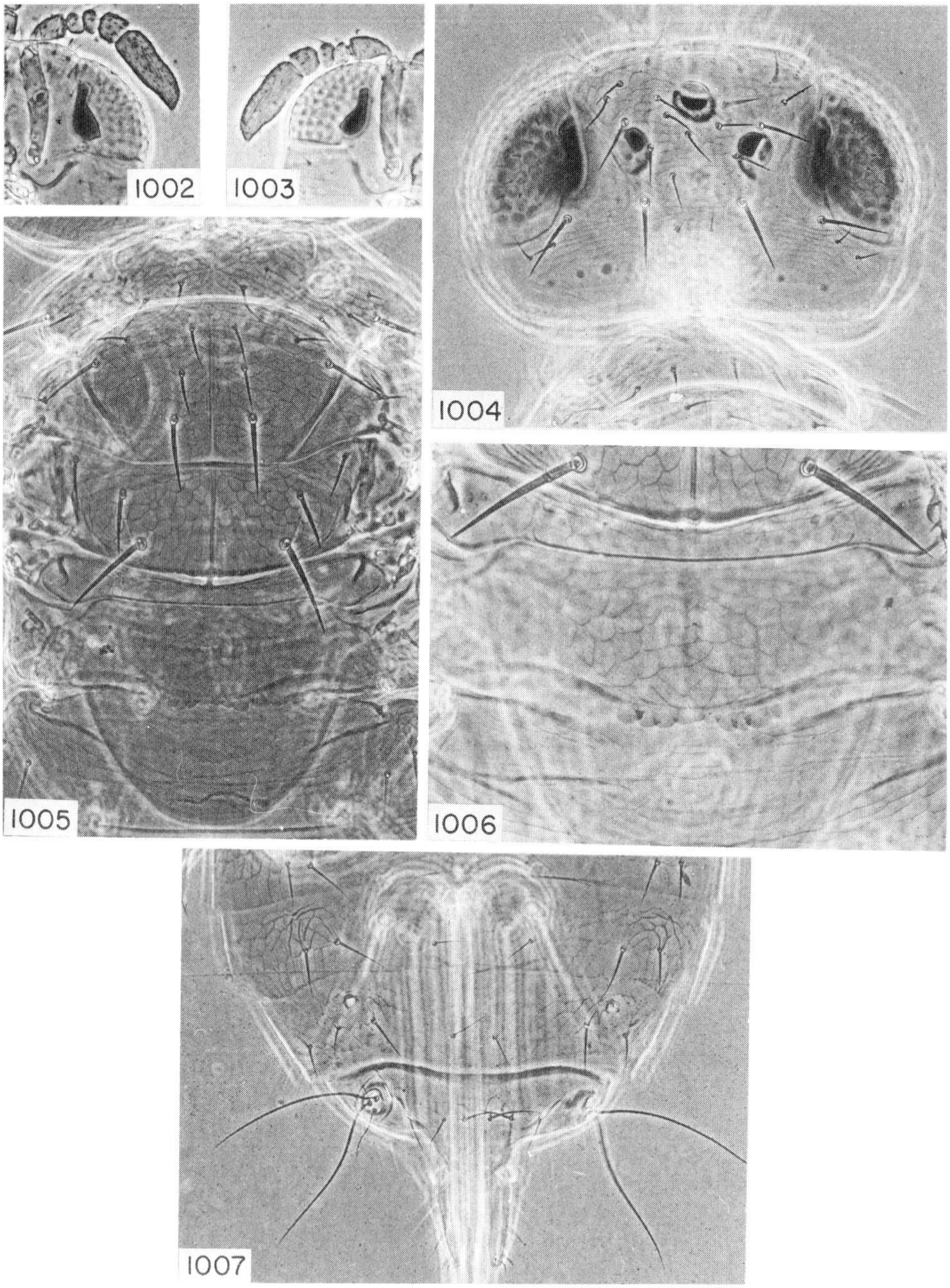


Figures 1002–1007. *Aphytis equatorialis* n.sp., ♀ (syntypes)
1002, 1003. Antennae. 1004. Head and pronotum, showing eyes, ocelli, chaetotaxis and sculpture. 1005. Thorax, propodeum and base of gaster. 1006. Metanotum, propodeum and base of gaster. 1007. Posterior abdominal tergites.

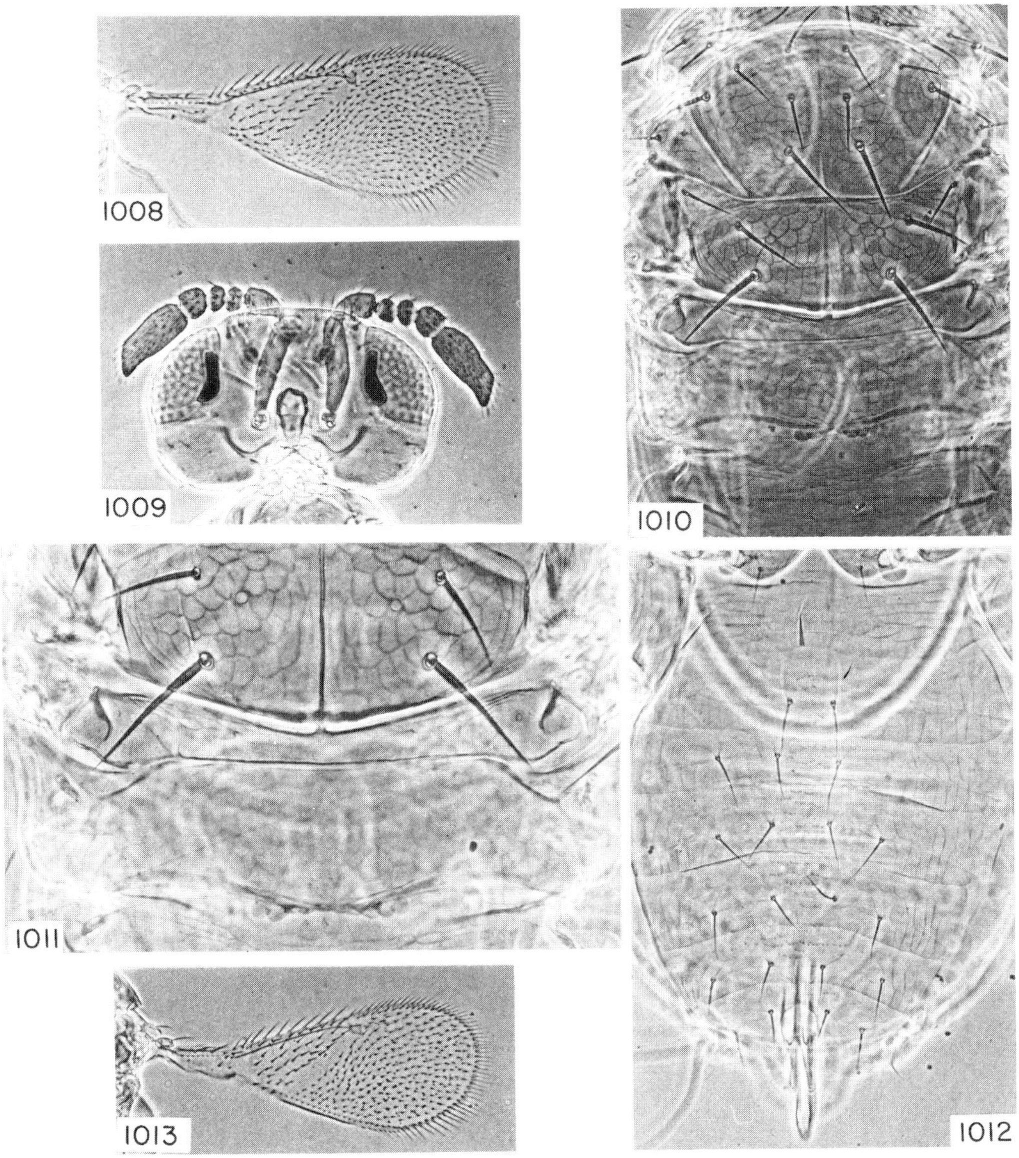


Figures 1008–1013. *Aphytis equatorialis* n.sp. (syntypes)
1008. ♀: Forewing. 1009. ♂: Head and antennae. 1010. ♂: Thorax, propodeum and base of gaster. 1011. ♂: Scutellum, metanotum and propodeum. 1012. ♂: Abdominal sternites (note weak stippling) and genitalia. 1013. ♂: Forewing.

# VI. THE CHRYSOMPHALI GROUP

This rather heterogenous, mainly Far Eastern group is characterized by a long propodeum—often as long as the scutellum—with small or minute crenulae, pale, slender thoracic setae, a generally yellow coloration and hyaline, uniformly setose wings. Male abdominal sterna not stippled. Pupae are generally yellow in this group. Although the propodeal crenulae of certain members are overlapping, they are always considerably smaller than in the **lingnanensis** group. Likewise, although the thoracic setae vary between the species, they are always considerably paler and more slender than in the **mytilaspidis** group. A specialized, partly cut-off sensory area is present on the antennal club of the male in some species, absent in most.

Of the forms originally referred by Compere (1955) to his **chrysomphali** group, apparently only *chrysomphali* is a valid species, the others representing either synonyms or unrecognizable species (see p. 596). Nevertheless, we feel that a group of species does exist, that is distinct from both the **mytilaspidis** and the **lingnanensis** groups. As currently understood, this group comprises the following 9 species: *anneckei* DeBach and Rosen, *chrysomphali* (Mercet), *cylindratus* Compere, *debachi* Azim, *immaculatus* Compere, *japonicus* DeBach and Azim, *lepidosaphes* Compere, *mazalae* DeBach and Rosen, and *sensorius* DeBach and Rosen. Seven of these species occur in the Oriental region or in the East Palearctic region (Japan), one is Ethiopian, and one, *chrysomphali*, is cosmopolitan, probably of Palearctic or Oriental origin.

The **chrysomphali** group appears to be the most advanced group of species in *Aphytis*. The pale coloration of adults and pupae, the pale, slender thoracic setae, the hyaline wings, the long propodeum and small crenulae, and the absence of a specialized sensory area on the antennal club of the male in most species, all these may be regarded as advanced characters, far removed from the ancestral type of *Aphytis* as represented by the **vittatus** group.

This group is probably polyphyletic in origin. Certain members, such as *debachi* or *cylindratus*, appear to be closely related to the **lingnanensis** group and may have evolved from it. As pointed out earlier, certain species of *Aphytis*, such as *africanus* or *equatorialis*, may be regarded as occupying an intermediate position between the **lingnanensis** and **chrysomphali** groups. On the other hand, other members of the **chrysomphali** group, such as *japonicus* or *lepidosaphes*, appear to exhibit more affinity to the **mytilaspidis** group, whereas species such as *rolaspidis* or *capenis* may be intermediate between the **mytilaspidis** and **chrysomphali** groups. In fact, *japonicus* was originally referred by DeBach and Azim (1962) to the **mytilaspidis** group. It is hoped that the discovery of additional species will eventually help elucidate the origin and affinities of the **chrysomphali** group.

The single Ethiopian member of this group, *anneckei*, shares a peculiar characteristic—reduced mandibles—with certain other Ethiopian species belonging to diverse groups (e.g., *merceti, funicularis, setosus, roseni,* etc.).

The **chrysomphali** group appears to be one of the most important groups of biological control agents in *Aphytis*. Most of its known members are important natural enemies of serious pest species, and some, like *lepidosaphes*, have been utilized in successful biological control projects.

The members of the **chrysomphali** group can be readily separated from one another by rather obvious structural characters such as the shape of the antennae, the propodeum or the crenulae, as well as by details of chaetotaxis and coloration.

## 68. **Aphytis chrysomphali** (Mercet)

(Frontispiece and Figures 1014–1031)

*Aphelinus chrysomphali* Mercet, 1912, Bol. R. Soc. Españ. Hist. Nat., **12**:135–140.

*Aphelinus chrysomphali*: Mercet, 1912, Trab. Mus. Nac. Cienc. Nat. Madrid, **10**:67–72; Malenotti, 1918, Redia, **13**:31–36.

?*Aphelinus (Prospaphelinus) silvestrii* De Gregorio, 1914, N. Ann. Agric. Sicil., **3**(4):227–230.

? *Aphelinus (Prospaphelinus) chrysomphali* var. *silvestrii* De Gregorio, 1915, Nat. Sicil., **22**:164–190.

*Aphelinus quaylei* Rust, 1915, Entomol. News, **26**:75.

*Aphytis chrysomphali*: Timberlake, 1926, Proc. Hawaii. Entomol. Soc., **6**:315.

*Aphytis (Prospaphelinus) chrysomphali*: Mercet, 1932, Eos, **8**:359–360; De Santis, 1948, Rev. Mus. La Plata (N.S.), **5**, Zool., 121–126.

?*Aphytis chrysomphali* var. *mazandaranica* Kiriukhin, 1946, Lab. Entomol. Phytopathol. Appl. Teheran, **2**:4–6, 13–22.

*Aphytis chrysomphali*: Compere, 1955, Univ. Calif. Publ. Entomol., **10**:314–316; Quednau, 1964, J. Entomol. Soc. S. Afr., **27**:113–114; Ferrière, 1965, Hymenoptera Aphelinidae d'Europe et du Bassin Méditerranéen, pp. 85–87; Nikol'skaya and Yasnosh, 1966, Aphelinids of the European Part of the USSR and the Caucasus, pp. 201–202 (in Russian); Traboulsi, 1969, Ann. Soc. Entomol. Fr. (N.S.), **5**:58–59.

For an extensive bibliography of *A. chrysomphali*, see De Santis (1948) and Peck (1963). However, many of the early records are probably erroneous. As pointed out earlier (p. 405), from 1894 to 1915 *chrysomphali* was commonly misidentified as *diaspidis*. Later, and until fairly recently, other species such as *lingnanensis, melinus* and *africanus* were misidentified as *chrysomphali*.

This distinctive, uniparental, nearly cosmopolitan species may be recognized by the following characters: General coloration yellow, thoracic sterna faintly dusky with a conspicuous black median line on the mesosternum; thoracic setae slender, pale, mesoscutum usually with 10 setae; antennae slender, club usually well over 3 times as long as wide; propodeum long, posterior margin broadly arcuate, crenulae small, rounded, nonoverlapping; the extremely rare males with very light stippling on the abdominal sternites, no specialized sensory area on the antennal club.

*Female*. Eyes finely setose. Mandibles (Figure 1014) well developed, with a sharp ventral denticle and a central denticle merging into a dorsal truncation; maxillary palpi 2-segmented, labial palpi 1-segmented. Antennae (Figures 1015, 1016) slender; scape slender, about 5–7 times as long as wide, about as long as the club or slightly longer; pedicel slender, usually about twice as long as wide ($1\frac{2}{3}$ times in minute specimens), considerably (usually about $1\frac{1}{4}$ times) longer than the third segment of the funicle; first 2 funicular segments subglobular, the first about $1\frac{1}{4}$ to $1\frac{1}{2}$ times as wide as long, the second considerably shorter and usually somewhat wider, about $1\frac{7}{10}$ times to a little over twice as wide as long; third funicular segment usually about $1\frac{1}{3}$ times as long as wide, bearing 2–3 longitudinal sensilla; club slender, usually $3\frac{1}{3}$ to $3\frac{4}{5}$ times as long as wide (sometimes fully 4 times, a little less than 3 times in minute specimens), usually 3 to $3\frac{1}{4}$ times longer and a little wider than the preceding segment, bearing 6–7 longitudinal sensilla.

Setae on thorax slender, pale, visible under × 60 magnification; setae on head paler, barely visible under × 120 magnification; those on abdomen invisible even under × 120 magnification. Vertex with 2 pairs of long setae along occipital margin, in addition to numerous shorter setae. Mesoscutum (Figure 1017) usually with 10 (rarely 11–13) setae, the posterior pair and 1 seta at each antero-lateral corner considerably longer and coarser than the others; each parapsis with 2 short setae, each axilla with 1 seta; scutellum with 4, the discoid sensilla usually somewhat closer to the anterior than to the posterior pair. Frontovertex, pronotum and mesonotal sclerites reticulate. Scutellum oval, about $\frac{2}{3}$ to $\frac{4}{5}$ median length of mesoscutum. Metanotum (Figures 1017, 1019) short, reticulate except on the sides, posterior margin nearly straight; antero-median apodeme as long as, or considerably longer than (up to $1\frac{3}{5}$ times), median length of metanotum.

Propodeum (Figures 1017, 1019–1021) long, 6 to $7\frac{4}{5}$ times as long as the metanotum, nearly as long as the scutellum or somewhat longer, reticulate on the sides, broadly reticulate on a wide central area, posterior margin broadly arcuate but without a distinct median salient; crenulae 3 + 3 to 6 + 7, small, rounded, usually nonoverlapping but sometimes slightly overlapping (Figure 1021), the 2 sets rather widely separated medially.

Second abdominal tergite transversely striated anteriorly on both sides, transversely reticulate centrally; tergites III–VII (Figure 1022) reticulate on the sides, bearing a few fine setae in a short transverse row on each reticulate area; third tergite transversely reticulate also mesad of the lateral setiferous areas, transversely striated across center; tergites IV–VI with faint indications of transverse striation across center; seventh tergite faintly transversely striated across center, bearing a pair of submedian setae (rarely 1 or 3 setae); eighth tergite delicately reticulate-striated, with a transverse row of 4 (rarely 3 or 5) setae between spiracles; syntergum (Figure 1022) triangular, faintly reticulate, bearing 6 (rarely 5) setae in a transverse row. Cerci situated somewhat closer to posterior spiracles than to tip of abdomen, with 2 long setae and 1 short seta. Ovipositor shaft $1\frac{2}{5}$ to $1\frac{2}{3}$ times as long as the middle tibia (1.40–1.65); ovipositor sheaths long, about $\frac{2}{5}$ to nearly $\frac{1}{2}$ the length of middle tibia (0.37–0.48).

Mid-tibial spur nearly as long as the corresponding basitarsus.

Forewing (Figures 1023, 1024) about $2\frac{1}{2}$ to $2\frac{2}{3}$ times as long as wide; marginal fringe not exceeding $\frac{1}{3}$ width of disk (usually about $\frac{1}{6}$). Delta area with 24–41 setae in 4–5 rows, these considerably longer and sparser than the setae distad of speculum, not clearly separated from row of setae along posterior margin of wing, widely separated from a single dorsal seta below distal portion of submarginal vein; costal cell with a row of fine setae along proximal three fifths or so, and 1 coarse seta near apex. Submarginal vein bearing 2 coarse setae, the proximal one varying from $\frac{3}{5}$ to full length of the distal, and 10–18 bullae. Marginal vein bearing 5–11 (usually 8–10) prominent, subequal setae along anterior margin, these $1\frac{1}{4}$ to $1\frac{4}{5}$ times longer than the setae in a row along center of vein.

Forewing hyaline, distinctly infumated below junction of submarginal and marginal veins; a faint fuscous streak along posterior margin of wing, distad of speculum.

General coloration yellow; thoracic sterna (Figure 1018) faintly dusky, with a conspicuous longitudinal median black line on the stem of the mesosternal furca ("Y"); posterior margin of scutellum narrowly lined with blackish; a short black streak at base of forewing, below tegula. Legs and antennal scape concolorous with body, middle tibia faintly dusky, antennal pedicel, funicle and club uniformly dusky. Wing veins hyaline.

Length 0.55–0.99 mm.

*Male*. Essentially similar to the female in structure, chaetotaxis, sculpture and coloration, differing mainly in the antennal proportions.

Antennal scape (Figure 1025) about 4 to $5\frac{1}{2}$ times as long as wide, somewhat longer than the club; pedicel $1\frac{3}{5}$ times to nearly twice as long as wide, $1\frac{1}{5}$ to $1\frac{1}{3}$ times longer than the third segment of the funicle; first two funicular segments about as in the female, third segment up to about $1\frac{1}{4}$ times as long as wide, bearing 1–2 longitudinal sensilla; club shorter than in the female, 3 to $3\frac{1}{4}$ times as long as wide, about $2\frac{3}{5}$ to $2\frac{4}{5}$ times longer and only very slightly wider than the preceding segment, bearing 3–4 longitudinal sensilla and lacking a specialized sensory area on the ventral aspect.

Mesoscutum (Figure 1026) with 10 setae. Anteromedian apodeme usually considerably longer (up to $1\frac{7}{10}$ times) than median length of metanotum. Propodeum (Figure 1026) $5\frac{2}{3}$ to $7\frac{2}{5}$ times as long as the metanotum, nearly as long as the scutellum; crenulae 3 + 3 to 4 + 5, as in the female.

Abdominal sternites (Figure 1027) with very light stippling centrally. Genitalia (Figure 1028) about $\frac{1}{2}$ to $\frac{2}{3}$ length of middle tibia (0.47–0.69), digital sclerites about $\frac{1}{4}$ to $\frac{1}{3}$ the combined length of aedeagus and apodemes (0.26–0.31).

Forewing delta with 21–27 setae in 3–4 rows; submarginal vein bearing 12–15 bullae; marginal vein bearing 6–8 prominent setae along anterior margin.

Length 0.56–0.86 mm.

**Material Examined.** Numerous specimens, reared from the California red scale, *Aonidiella aurantii* (Maskell), on citrus in California, Texas, Argentina, Chile, Israel,

Crete, Cyprus, Turkey, Morocco, South Africa, Japan, Taiwan, Hong Kong and Australia (Western Australia, New South Wales, Queensland and Victoria), on olive fruit in Australia, on *Acacia* in Crete, on rose and *Ligustrum* in Greece; from the yellow scale, *Aonidiella citrina* (Coquillett), on citrus in California, Turkey and Iran; from the dictyospermum scale, *Chrysomphalus dictyospermi* (Morgan), on citrus in Spain, Greece, Iran and Florida, on oleander and palm in Yugoslavia, on ?*Eleagnus* and *Ligustrum* in Italy, on false acacia and undetermined hedge in Iran; from the Florida red scale, *Chrysomphalus aonidum* (L.), on coconut in the Philippines and on an unrecorded plant in the Dominican Republic; from *Chrysomphalus* sp. on *Ficus* in Argentina; from the coconut scale, *Temnaspidiotus destructor* (Signoret), on coconut in Brazil and Tahiti, on plantain in the Dominican Republic, on banana in Costa Rica, Panama and Hawaii, on anona in Brazil, on *Ficus* in Taiwan; from the oleander scale, *Aspidiotus nerii* Bouché [= *A. hederae* (Vallot)] on oleander in Greece, in laboratory culture in Australia and South Africa; from *Pseudaonidia trilobitiformis* (Green) on citrus in Brazil; from the cactus scale, *Diaspis echinocacti* (Bouché), on *Opuntia* in Greece; from *Melanaspis inopinata* (Leonardi) on *Crataegus* in Greece; from *Fulaspis* sp. on *Acacia* in South Africa; from the rufous scale, *Selenaspidus articulatus* (Morgan), on *Ficus nitida* in Peru and on an ornamental plant in El Salvador; from ?*Hemiberlesia lataniae* (Signoret) on avocado in California; from an undetermined host on citrus in New Caledonia; and from an undetermined host in Puerto Rico.

All these series agree with the above redescription in all significant characters. All are uniparental, with males absent altogether or extremely rare. In the absence of any biological evidence to the contrary, we consider them all to be conspecific.

**Notes.** The pupa of *chrysomphali* is entirely yellow, except for a longitudinal black line on the mesosternum.

*A. chrysomphali* was originally described from material reared from "*Chrysomphalus dictyospermi* var. *pinnulifera*" in Spain and the Balearic Islands (Mercet 1912a, 1912b). As pointed out by Traboulsi (1969), *Chrysomphalus pinnulifer* (Maskell) is a distinct species, and the original host of *chrysomphali* may actually have been the dictyospermum scale, *Chrysomphalus dictyospermi* (Morgan).

Since Mercet's type material of *chrysomphali* was not available for study, we have followed Compere's (1955) interpretation of this species, which appears to be in close agreement with Mercet's descriptions.

Timberlake (1926) considered both *Aphelinus quaylei* Rust and *Aphelinus limonus* Rust as synonyms of *chrysomphali*. Mercet (1932) concurred, and added *Prospaphelinus silvestrii* De Gregorio as yet another synonym. Compere (1955), however, made the following statement: "Recent findings indicate that different species may now be concealed under the name *chrysomphali* (Mercet). Scientific names that have been applied to forms of the Chrysomphali group are: *chrysomphali* (Mercet), *silvestrii* (DeGregorio), *quaylei* (Rust), *limonus* (Rust), *chrysomphali* var. *manzardaranicus* [!] Kiriukhim [!]. How to recognize the units to which these names were applied remains an unsolved problem. Until biological evidence suggested the contrary, I believed that these five names applied to one species only. This opinion has been replaced by great uncertainty."

We have not seen the types of *silvestrii* and are unable to contribute any new information regarding the identity of that species, which was recorded by De Gregorio (1914, 1915) as a parasite of *Chrysomphalus dictyospermi* in Sicily. However, it should be noted here that Malenotti (1918a), who presented a detailed redescription of *chrysomphali*, considered *silvestrii* to be a distinct species.

*A. quaylei* was described by Rust (1915) from female specimens reared from *Selenaspidus articulatus* (Morgan), *Chionaspis minor* Maskell and "*Aspidiotus camelliae*" [= apparently *Hemiberlesia rapax* (Comstock)] on various host plants in Peru, and was recorded by him also from *Aonidiella aurantii* and *Aonidiella citrina* on citrus in California. We have examined several cleared syntype female specimens of *quaylei* (Figures 1029–1031) and have no doubt whatsoever that this species is indeed identical to *chrysomphali*.

*A. limonus* was described by Rust (1915) from female specimens reared from *Chionaspis minor* Maskell on "pigeon pea" in Hawaii. The type slide, deposited in the U.S. National Museum, includes 1♀ and 3♂♂; the specimens are not cleared and important characters cannot be seen. Although it superficially resembles *chrysomphali*, we are unable to accept *limonus* as a straight synonym and regard it, for the time being, as an unrecognizable species of *Aphytis* (see p. 735).

Kiriukhin (1946) described *A. chrysomphali* var. *mazandaranica* as a parasite of *Chrysomphalus dictyospermi* in northern Iran. We have not seen the type material. Nikol'skaya and Yasnosh (1966) found it to differ slightly from *chrysomphali*, but were unable to solve the problem of its identity due to the inadequacy of the material at their disposal. We collected typical specimens of *chrysomphali* from the same area on *C. dictyospermi* and *A. citrina* on citrus.

*A. chrysomphali* has been suggested as having evolved in the Mediterranean Basin, from which it has supposedly spread by ecesis to most other regions of the world (see Figure 67 and DeBach, 1971b). The above list of records, based only on authenticated specimens at hand, indicates that this species is nearly cosmopolitan. It is present in the Orient, including mainland China, where it is apparently rather rare. Since our present knowledge of the *Aphytis* fauna of China is far from adequate, and since several of its hosts are of Oriental origin, the possibility that *chrysomphali* is an Oriental species cannot be discounted; however, neither can its origin in any other region in which it occurs.

Taylor (1935) recorded "*Aphelinus chrysomphali*" as a parasite of the coconut scale, *Temnaspidiotus destructor* (Signoret), in Fiji. However, although we have recorded *chrysomphali* from that host in various localities (see above), the identity of Taylor's species is questionable. *A. chrysomphali* is a uniparental species, with very rare males, whereas Taylor noted that "the males are greatly outnumbered by the females but are nevertheless of common occurrence. This is noteworthy in view of the fact that in certain allied species of *Aphelinus* males are extremely rare." Thus, he was evidently referring to a biparental species. Also, the pupae of *chrysomphali* exhibit a characteristic longitudinal black streak, whereas Taylor described the pupae of his species as entirely yellow.

Taylor (op. cit.) also recorded a "form" of *chrysomphali*, parasitic upon the Florida red scale, *Chrysomphalus aonidum* (L.), in Java with distinctly pigmented pupae. Although his figure of the pupa is suggestive of *A. melinus*, in view of the host record we assume that this may have been *A. holoxanthus* (Taylor did not refer to the age of the pupa, and his figure may have represented a young pupa of *holoxanthus* with incomplete pigmentation). In any event, that "form" was certainly not *chrysomphali*.

**Biology.** Information on the biology of *A. chrysomphali* has been presented by several authors and was recently summarized by Rosen and DeBach (1978). The species is uniparental (thelytokous) and males are very rare. The life cycle takes 45 days at 15.6°C, 22 days at 21°C, 14 days at 26.7°C; all the eggs die at 32°C. The average longevity of the ovipositing female is 10 days, and oviposition is more or less continuous during life. An average of 13 eggs is laid at 26.7°C (maximum 44). At that temperature and 60 percent relative humidity, the female destroys an average of 22 hosts by predatory host-feeding. Thus, the total number of hosts destroyed by both oviposition and host-feeding is much lower than in *A. lingnanensis* (see DeBach and White, 1960). Abdelrahman (1974a, b) reported similar data from Australia, with higher longevity (30 days at 20°C, 18 at 25°C, 9 at 30°C) and fecundity (21.3, 25.1 and 6.6 eggs per female, respectively) and calculated the theoretical threshold of development of *chrysomphali* as 8.5°C. He found that the innate capacity for increase ($r_m$) of *chrysomphali* was higher than that of *melinus* at 20° and 25°C, but lower at 30°C, and that *chrysomphali* was more tolerant of extreme cold, but less tolerant of extreme heat, than *melinus*. (For a more detailed discussion of Abdelrahman's data, see the Ecology section, pp. 56, 58, and 77.)

**Role in Biological Control.** *A. chrysomphali* was apparently accidentally introduced into California around the beginning of the present century, and until 1948 was the only species of *Aphytis* parasitic upon the California red scale, *Aonidiella aurantii*, on citrus in that state. It has since been almost entirely displaced from the populations of that host in California by the recently introduced species *A. lingnanensis* and *A. melinus*. Similarly, *A. chrysomphali* was known in the past to attack immature males of the Florida red scale, *Chrysomphalus aonidum*, in Israel but has been entirely displaced from the populations of that pest by the more effective *A. holoxanthus*, introduced from Hong Kong in 1956–1957 (see DeBach, Rosen and Kennett, 1971). It has still persisted as the dominant parasite of the California red scale along the coastal plain of Israel.

## 69. **Aphytis anneckei** DeBach and Rosen

(Figures 1032–1047)

*Aphytis anneckei* DeBach and Rosen, 1976, Ann. Entomol. Soc. Amer., **69**: 544.

This is a large, generally yellow African species; thoracic sterna strongly infuscated, especially the stem of the mesosternal furca; mandibles small, without distinct denticles; antennal club long; propodeum very long; crenulae small, strongly overlapping, the 2 sets widely separated; ovipositor long. Male genitalia long, digital sclerites short.

*Female*. Eyes finely setose. Mandibles (Figure 1032) small, broad, nonoverlapping, without any distinct denticles; maxillary palpi 2-segmented, labial palpi 1-segmented. Antennal scape (Figure 1033) slender, varying from nearly 5 to $7\frac{1}{2}$ times as long as wide, about $1\frac{1}{4}$ times longer than the club; pedicel slender, $2\frac{1}{7}$ to $2\frac{1}{4}$ times as long as wide, $1\frac{1}{5}$ to nearly $1\frac{1}{2}$ times longer than the third segment of the funicle; first 2 funicular segments subrectangular; first segment as long as wide or a little wider than long (up to $1\frac{1}{5}$ times), sometimes apparently longer than wide; second segment a little shorter and wider than the first, varying from as long as wide to about $1\frac{2}{3}$ times as wide as long; third funicular segment $1\frac{3}{10}$ to $1\frac{1}{2}$ times as long as wide, about as long as the first two segments combined, bearing 2 longitudinal sensilla; club long, $3\frac{2}{5}$ to $3\frac{3}{5}$ times as long as wide, $2\frac{7}{10}$ to 3 times longer and somewhat wider than the preceding segment, bearing 6–7 longitudinal sensilla.

Setae on thorax moderately coarse and dark, the longest ones visible under $\times 30$ magnification in the largest specimen, under $\times 60$ magnification in smaller specimens; setae on head somewhat paler, those on abdomen invisible even under $\times 120$ magnification. Vertex with 2 pairs of very long setae along occipital margin, in addition to numerous shorter setae. Mesoscutum (Figure 1034) with 10–12 setae, the posterior pair and 1 seta at each antero-lateral corner considerably longer and coarser than the others; each parapsis with 2 setae, each axilla with 1 seta; scutellum with 4 setae, the discoid sensilla equidistant from the two pairs or somewhat closer to the anterior pair. Frontovertex, pronotum and mesonotal sclerites reticulate. Scutellum flat oval, about $\frac{3}{4}$ median length of mesoscutum. Metanotum (Figure 1035) reticulate except on the sides, posterior margin straight; anteromedian apodeme robust, about $\frac{2}{3}$ to $\frac{4}{5}$ median length of metanotum.

Propodeum (Figures 1034–1036) very long, $5\frac{2}{3}$ to $6\frac{3}{10}$ times as long as the metanotum, somewhat longer than the scutellum, reticulate on the sides and on a broad trapezoidal central portion, the cells somewhat wider than long, posterior margin broadly rounded; crenulae 7+7 to 7+9, small, rounded, strongly overlapping, the 2 sets widely separated.

Second abdominal tergite (Figure 1035) rather extensively reticulate on a wide, trapezoidal central portion, extending antero-laterally; tergites III–VII (Figure 1037) reticulate on the sides, bearing a few setae in a short transverse row on each reticulate area; third tergite extensively, transversely reticulate-striated anteriorly across center, longitudinally striated posteriorly but this fading toward center; tergites IV–VI with indications of transverse striation anteriorly, longitudinally striated posteriorly across center; seventh tergite obliquely reticulate-striated across center, bearing 2 fine, submedian setae between the lateral setiferous areas; eighth tergite delicately reticulate-punctate, with a transverse row of 4 setae (3 in one specimen) between the spiracles, the lateral pair considerably longer than the submedian setae; syntergum triangular, with a rather elongate cauda, faintly reticulate-punctate, with 5–6 setae in an arcuate transverse row. Cerci about equidistant from posterior spiracles and from tip of abdomen, with 2 long setae and 1 short seta. Ovipositor shaft a little over $1\frac{4}{5}$ times as long as the middle tibia (1.83–1.89); ovipositor sheaths long, narrow, nearly $\frac{1}{2}$ length of middle tibia (0.47–0.49).

Mid-tibial spur about $\frac{4}{5}$ length of the corresponding basitarsus.

Forewing (Figure 1038) broad, $2\frac{3}{10}$ to $2\frac{2}{5}$ times as long as wide, marginal fringe not exceeding $\frac{1}{8}$ width of disk ($\frac{1}{10}$ in large specimens). Delta area with 40–57 setae in 5–6 rows, these considerably longer, coarser and sparser than the setae distad of speculum, not quite distinct from row of setae along posterior margin of wing, widely separated from just 2 dorsal setae below apical portion of submarginal vein; costal cell with a row of fine setae along proximal three fifths or so, with no setae near the apex except for one coarse seta arising from the vein itself. Submarginal vein bearing 2 coarse setae, the proximal one about $\frac{3}{5}$ length of the long distal seta, and 16–18 bullae. Marginal vein bearing 10–14 prominent, subequal setae along anterior margin, these $1\frac{1}{3}$ to $1\frac{3}{5}$ times longer than the setae in a row along center of vein.

Forewing faintly, rather uniformly infumated, more strongly so below junction of submarginal and marginal veins and at base of delta; a fuscous streak along posterior margin of wing, distad of speculum.

General coloration yellow; thoracic sterna with small but strongly infuscated areas, especially the stem of the mesosternal furca ("Y") (Figure 1039) which is black, and a black margin of the metasternum surrounding a pale central spot (Figure 1040); posterior margin of scutellum narrowly lined with black; a short black streak near base of forewing, below the faintly dusky tegula. Antennal scape pale, lined with fuscous ventrally; rest of antenna uniformly, rather faintly dusky. Femora of all legs concolorous with body, tibiae and tarsi stronger yellow or faintly dusky. Wing veins hyaline.

Length 1.09–1.45 mm.

*Male.* Essentially similar to the female in structure, chaetotaxis, sculpture and coloration.

Antennal scape wider than in the female, $3\frac{2}{3}$ times as long as wide; third funicular segment (Figure 1041) only a little longer than wide, bearing 2 longitudinal sensilla; club nearly 3 times as long as wide, nearly 3 times longer than the preceding segment, bearing 4 longitudinal sensilla, with an elongate, vaguely defined ventral plate bearing short setae.

Setae on thorax paler than in the female, barely visible under $\times 60$ magnification. Thorax (Figure 1042) and propodeum (Figure 1043) as in the female.

Abdomen delicately stippled ventrally on a rather narrow, median longitudinal stripe (Figure 1044). Genitalia (Figure 1045) long, narrow, with a distinct "bottleneck" at base, about $\frac{7}{10}$ length of middle tibia; digital sclerites short, $\frac{1}{5}$ the combined length of aedeagus and apodemes.

Strigil of foreleg shown in Figure 1046. Forewing a little broader than in the female, $2\frac{1}{4}$ times as long as wide; delta with 40–41 setae.

Venter of thorax (Figure 1047) as in the female.

Length 1.04 mm.

Described from 1♀ (holotype) and 1♂ (allotype), reared by B. R. Bartlett from *Chrysomphalus* sp. on *Trichilia emetica*, Pretorius Kop, South Africa, July 10, 1958;

also 2♀♀ (paratypes) reared by D. Gerling from *Paraselenaspidus madagascariensis* (Mamet) on *Dovyalis caffra*, Kenya ("Shipment 7"), March 31, 1969.

Type series in the collection of the Division of Biological Control, University of California, Riverside.

**Notes.** This species is rather closely related to *chrysomphali*. It shares with *chrysomphali* the relatively long antennal club, the general shape of the propodeum, the relatively few bullae and conspicuously unequal setae on the submarginal vein of the forewing, as well as many other characters, but differs markedly in the arrangement of the propodeal crenulae, in details of sternal pigmentation, and in the shape of the mandibles and male genitalia. It is also rather closely related to *debachi*, from which it differs in the shape of the propodeal crenulae and in the length of the antennal club.

The number of bullae on the submarginal vein of the forewing, as well as the number of mesoscutal setae, is usually correlated with the size of specimens in *Aphytis*. When these numbers are as low as they are, in specimens as large as those of *anneckei*, these appear to be valid specific characters.

This African species was named by us in honor of Dr. D. P. Annecke of the Plant Protection Research Institute, Pretoria, South Africa, in recognition of his outstanding contribution to the systematics of the Aphelinidae and in gratitude for the abundant African material of *Aphytis* that he kindly put at our disposal during the preparation of this monograph.

## 70. **Aphytis debachi** Azim

(Figures 1048–1055)

*Aphytis debachi* Azim, 1963, J. Fac. Agr. Kyushu Univ. **12**:287–290.

This little-known Far Eastern species may be recognized by the following combination of characters: Generally yellow, thoracic sterna fuscous; mesonotal setae dark; mandibles small, distinctly tridentate; antennal club short; propodeum long, crenulae nonoverlapping, the 2 sets widely separated by a wavy margin; ovipositor and sheaths long.

*Female.* Eyes finely setose. Mandibles (Figure 1048) small, nonoverlapping, with 3 distinct denticles and no truncation; maxillary palpi 2-segmented, labial palpi 1-segmented. Antennal scape (Figure 1049) slender, $5\frac{1}{3}$ to $6\frac{2}{3}$ times as long as wide, up to $1\frac{1}{4}$ times longer than the club; pedicel about $1\frac{2}{3}$ to $1\frac{4}{5}$ times as long as wide, $1\frac{1}{4}$ to $1\frac{1}{2}$ times longer than the third segment of the funicle; first funicular segment trapezoidal, about $1\frac{1}{3}$ to $1\frac{2}{3}$ times as wide as long; second segment nearly symmetrical, somewhat shorter and wider than the first, about $1\frac{2}{3}$ times to twice as wide as long; third funicular segment a little longer than wide, bearing 1–2 longitudinal sensilla; club short, $2\frac{1}{2}$ to nearly 3 times as long as wide, about 3 times longer and somewhat wider than the preceding segment, bearing 5–8 longitudinal sensilla.

Setae on thorax rather coarse, dark, the longest ones visible under ×30 magnification; setae on head visible under ×60 magnification, those on abdomen visible under ×120 magnification. Vertex with 2 pairs of long setae along occipital margin, in addition to numerous shorter setae. Mesoscutum (Figure 1050) usually with 10 setae, rarely with 11 (9–11 according to Azim's original description), the posterior pair considerably longer and coarser than the others; each parapsis with 2 setae, each axilla with 1 seta; scutellum with 4 setae, the discoid sensilla considerably closer to the anterior than to the posterior pair. Frontovertex, pronotum and mesonotal sclerites reticulate. Scutellum flattened oval, about $\frac{2}{3}$ to $\frac{4}{5}$ median length of mesoscutum. Metanotum (Figures 1050–1053) reticulate except on the sides, the posterior margin nearly straight; anteromedian apodeme robust, as long as or shorter than median length of metanotum.

Propodeum (Figures 1050–1053) long, $4\frac{3}{5}$ to nearly 7 times as long as the metanotum, about as long as the scutellum, reticulate on the sides and on a broad central area, the cells about as wide as long; crenulae 3 + 3 to 5 + 6, rather small, rounded and somewhat elongate, nonoverlapping, the 2 sets widely separated by a wavy portion of the posterior margin of the propodeum which may sometimes look like additional, large, overlapping crenulae (compare Figures 1051 and 1052).

Second abdominal tergite (Figures 1051, 1052) transversely striated anteriorly on each side, reticulate-striated centrally; tergites III–VII (Figure 1054) reticulate on the sides, with 2–4 setae in a short transverse row on each reticulate area; third tergite transversely striated across center, between the lateral reticulate areas; tergites IV–VI transversely striated anteriorly, longitudinally striated posteriorly, this more or less fading centrally; seventh tergite obliquely reticulate-striated across, with 2 fine submedian setae; eighth tergite reticulate-punctate, with a transverse row of 4 (rarely 3) setae between the spiracles, the lateral pair considerably longer than the submedian pair; syntergum (Figure 1055) triangular, faintly reticulate-punctate, with 5–8 (usually 6–7) setae in a transverse row. Cerci situated somewhat closer to posterior spiracles than to tip of abdomen, with 2 long setae and 1 short seta. Ovipositor long, the shaft about twice as long as the middle tibia (1.88–2.06), the sheaths long and narrow, nearly $\frac{3}{5}$ length of middle tibia (0.54–0.59).

Mid-tibial spur usually somewhat shorter than the corresponding basitarsus.

Forewing about $2\frac{1}{3}$ to $2\frac{3}{5}$ times as long as wide; marginal fringe usually not exceeding $\frac{1}{5}$ width of disk. Delta area sparsely setose, with 26–37 setae in 4 rows, these considerably longer and sparser than the setae distad of speculum, quite distinct from row of setae along posterior margin of wing, widely separated from 1 or 2 dorsal setae below distal portion of submarginal vein; costal cell with a few fine setae in a row along proximal three fifths or two thirds, widely separated from 1 coarse seta on the vein near apex. Submarginal vein bearing 2 coarse setae, the proximal one $\frac{2}{3}$ to $\frac{4}{5}$ length of the distal, and 12–16 bullae. Marginal vein bearing 7–11 prominent, subequal setae along anterior margin, these about $1\frac{2}{5}$ to $1\frac{3}{5}$ times longer than the setae in a row along center of vein.

Forewing nearly hyaline, proximal half faintly infumated, strongly so below junction of submarginal and marginal veins and at base of delta; a fuscous streak along posterior margin of wing, distad of speculum.

General coloration yellow; thoracic sterna strongly infuscated; posterior margin of scutellum narrowly lined with black; a short black streak near base of forewing, below tegula; plates around base of ovipositor faintly dusky. Antennal scape pale, faintly lined with fuscous ventrally; rest of antenna uniformly, rather faintly dusky. Tarsi of all legs faintly dusky. Wing veins lined with brownish.

Length 0.77–0.98 mm.

*Male*. Not studied. According to Azim (1963b), "males resemble females in general but are slightly smaller and exhibit the usual sex differences."

Redescribed from 1♀ (headless "cotype" (= syntype)), reared by A. Azim from *Parlatoria* sp., Fukuoka, Kyushu, Japan, August 27, 1962; 2♀♀, same data but not marked as types, July 15, 1961; also from the following specimens, all reared from the snow scale, *Unaspis citri* (Comstock), on citrus in Hong Kong: 1♀, A. G. Selhime, August 1971; 3♀♀, S. K. Cheng, December 5, 1971; 1♀, S. K. Cheng, September 16, 1972; 1♀, S. K. Cheng, September 24, 1972 (reared in the insectary at Riverside, California, on the cactus scale, *Diaspis echinocacti* (Bouché)); 1♀, S. K. Cheng, September 8, 1973; 3♀♀, S. K. Cheng, October 13, 1973; also 1♀, reared by A. G. Selhime from an undetermined armored scale insect on citrus, Hong Kong, October 1971.

**Notes.** Azim (1963b, 1963c) recorded *debachi* as a primary parasite of *Parlatoria* sp. on *Camellia japonica*; his material was collected in Fukuoka and Kumamoto, Kyushu, Japan, between June 1961 and October 1962.

Unfortunately, Azim's original description cannot be relied upon for a proper identification of this species. His key describes *debachi* with a black-tipped antennal club and nonoverlapping propodeal crenulae, whereas his specific description states that the antenna is faintly dusky and that "the crenulation on the posterior margin of the propodeum and the structure of the propodeum of this new species is like *Aphytis lingnanensis*"; the crenulae are further described as "large indistinct overlapping." Azim's figure of the popodeum (1963b: fig. 13) apparently does not show any crenulae. No mention is made of the dusky thoracic sterna, an important diagnostic character of *debachi*; as a matter of fact, Azim even states that *debachi* differs from *lingnanensis* in the absence of such pigmentation. Thus the original description appears to be self-contradictory and inconsistent with the material upon which it was supposedly based. It would therefore be better ignored. Additional material, especially males, should be studied in order to augment this redescription.

*A. debachi* can be easily recognized by a number of good distinguishing characters, especially the shape of the mandibles and popodeum and the long oivpositor. The species appears to be a bona fide member of the **chrysomphali** group.

This species was named by Azim (op. cit.) in honor of Dr. Paul DeBach, Division of Biological Control, University of California, Riverside.

**Biology.** Azim (1963c) reported *debachi* to be a biparental, arrhenotokous species and described its mating behavior in some detail. Surprisingly, only female specimens of this species have been obtained in Hong Kong so far. It seems that the mode of reproduction of *debachi* merits further investigation.

According to Azim (1963c), *debachi* prefers to lay its eggs singly on the ventral surface of the body of its *Parlatoria* host. Under room conditions in summer in Fukuoka, Japan, the duration of development was 18–21$\frac{1}{2}$ days, and average adult longevity was 13 days.

**Potential Value for Biological Control.** Azim (1963b, 1963c) recorded *debachi* as an effective natural enemy of *Parlatoria* in Kyushu. Unfortunately, the specific identity of its host in Japan is not known. Since several species of *Parlatoria* are regarded as serious agricultural pests in various countries, it would certainly be worth while to study the host range of *debachi* in Japan.

*A. debachi* had not been previously known to attack the snow scale, *Unaspis citri*. Although the few specimens obtained from this host in Hong Kong were largely outnumbered by *A. lingnanensis*, this new record may prove to be of great potential value for biological control efforts directed against this serious pest of citrus in Florida and elsewhere. Importation of *A. debachi* from Hong Kong against the snow scale is highly recommended.

## 71. **Aphytis cylindratus** Compere

(Figures 1056–1065)

*Aphytis cylindratus* Compere, 1955, Univ. Calif. Publ. Entomol., **10**:303.
*Aphytis cylindratus*: Azim, 1963a, Mushi, **37**:53–54; Azim, 1963b, J. Fac. Agric. Kyushu Univ., **12**:275–276.

This large, distinctive, biparental parasite of *Pseudaonidia* spp. may be readily recognized by the long, cylindrical antennal club and third funicular segment of the female, the long, slender thoracic setae, small, overlapping propodeal crenulae, broad forewing and dusky thoracic sterna; male antennal club somewhat shorter than in the female, bearing a partly cut-off sensory area on the ventral aspect.

*Female.* Eyes finely setose, the inter-ommatidial setae relatively long. Mandibles (Figure 1056) well developed, with 2 denticles and a broad dorsal truncation; maxillary palpi 2-segmented, labial palpi 1-segmented. Antennal scape (Figure 1057) slender, about 4–6 times as long as wide; pedicel 1$\frac{3}{5}$ times to nearly twice as long as wide; first funicular segment trapezoidal, about 1$\frac{3}{5}$ to 1$\frac{2}{3}$ times as wide as long; second segment symmetrical, considerably shorter and somewhat wider than the first segment, twice to nearly 3 times as wide as long; third funicular segment cylindrical, 1$\frac{1}{2}$ to 1$\frac{3}{4}$ times as long as wide, up to 1$\frac{1}{4}$ times longer than the pedicel, bearing 2–6 longitudinal sensilla; club long, thick, cylindrical, 3$\frac{1}{2}$ to 4$\frac{1}{3}$ times as long as wide (usually 4 times or longer), up to 1$\frac{1}{3}$ times longer than the scape, 2$\frac{2}{3}$ to 3$\frac{1}{3}$ times longer and somewhat wider than the preceding segment, bearing 10–17 longitudinal sensilla.

Setae on thorax long, slender, moderately dark, barely visible under ×30 magnification; setae on head and sides of abdomen paler, visible under ×60 magnification. Vertex with 2 pairs of long setae along occipital margin, in addition to numerous shorter setae. Mesoscutum (Figure 1058) with 10–11 rather long setae, the posterior pair and 1 seta at each antero-lateral corner considerably longer and coarser than the others; each parapsis with 2 shorter setae, each axilla with 1 seta, scutellum with 4, the discoid sensilla about equidistant from the two pairs or somewhat closer to the anterior pair. Frontovertex, pronotum and mesonotal sclerites reticulate. Scutellum oval, about $\frac{3}{4}$ to $\frac{4}{5}$ median length of mesoscutum. Metanotum (Figures 1059, 1060) reticulate centrally, transversely striated anteriorly on the sides; anteromedian apodeme slender, usually about as long as median length of metanotum ($\frac{4}{5}$ to $1\frac{1}{2}$ times).

Propodeum (Figures 1059, 1060) about 4–5 times as long as the metanotum ($6\frac{4}{5}$ times in one specimen), $\frac{3}{5}$ to $\frac{3}{4}$ length of scutellum, reticulate on the sides and on a broad central area, with faint indications of transverse striation along posterior margin; crenulae 4 + 4 to 10 + 11, small, rounded, somewhat elongate, closely appressed, usually somewhat overlapping.

Second abdominal tergite transversely striated anteriorly on both sides, smooth centrally; tergites III–VII reticulate on the sides, with a few setae in a short transverse row on each reticulate area; third tergite transversely reticulate also mesad of the lateral setiferous areas, smooth or with just faint indications of transverse striation centrally; tergites IV–VI with faint indications of transverse striation across center; seventh tergite faintly reticulate mesad of the lateral setiferous areas, faintly transversely striated across center, bearing a pair of submedian setae; eighth tergite delicately reticulate, with a transverse row of 4 setae between spiracles; syntergum triangular, faintly punctate, bearing 4–7 setae in a transverse row. Cerci about equidistant from posterior spiracles and from tip of abdomen, with 2 long setae and 1 short seta. Ovipositor shaft about $1\frac{3}{5}$ to $1\frac{4}{5}$ times as long as the middle tibia (1.58–1.83); ovipositor sheaths usually about $\frac{2}{5}$ length of middle tibia (0.38–0.44; 0.49 in a minute specimen).

Mid-tibial spur long, usually about as long as the corresponding basitarsus.

Forewing (Figures 1061, 1062) broad, $2\frac{1}{4}$ to $2\frac{1}{3}$ times as long as wide; marginal fringe not exceeding $\frac{1}{5}$ width of disk. Delta area with 22–52 setae in 4–6 rows, these considerably longer and sparser than the setae distad of speculum, not clearly separated from row of setae along posterior margin of wing, widely separated from a small group of setae below distal portion of submarginal vein; costal cell with a row of fine setae along proximal three fifths or so, and 1 coarse seta near apex. Submarginal vein bearing 2 coarse setae, the proximal one varying from $\frac{2}{5}$ to $\frac{4}{5}$ length of the distal, and 15–23 bullae. Marginal vein bearing 8–14 (usually 10–11) prominent, subequal setae along anterior margin, these about $1\frac{3}{5}$ to $1\frac{2}{3}$ times as long as the setae in a row along center of vein.

Forewing hyaline, faintly infumated below junction of submarginal and marginal vein and at base of delta; a fuscous streak along posterior margin of wing, distad of speculum.

General coloration yellow; thoracic sterna strongly infuscated; posterior margin of scutellum narrowly lined with blackish; a short black streak at base of forewing, below tegula; abdominal tergites III–VII faintly dusky on sides. Antennal scape pale, narrowly lined with fuscous ventrally; rest of antenna uniformly, faintly dusky. Legs concolorous with body. Wing veins hyaline, faintly lined with brownish.

Length 0.76–1.36 mm (usually over 1 mm).

*Male.* Essentially similar to the female in structure, chaetotaxis, sculpture and coloration, differing mainly in the antennal proportions.

Antennal scape (Figure 1063) $3\frac{3}{5}$ to $5\frac{1}{2}$ times as long as wide; pedicel about $1\frac{1}{2}$ to $1\frac{3}{4}$ times as long as wide, distinctly (up to $1\frac{1}{3}$ times) longer than the third segment of the funicle; the latter segment shorter than in the female, about $1\frac{1}{5}$ to $1\frac{3}{5}$ times as long as wide, bearing 1–2 longitudinal sensilla; club somewhat shorter than in the female, about 3–4 times as long as wide, a little longer than the scape, $2\frac{1}{2}$ to $3\frac{1}{2}$ times longer and distinctly wider than the preceding segment, bearing 2–8 longitudinal sensilla, with a partly cut-off sensory area on the ventral aspect.

Mesoscutum (Figure 1064) with 10 setae (13 in one specimen). Propodeum (Figure 1065) usually 4 to $4\frac{2}{5}$ times as long as the metanotum, $\frac{3}{5}$ to $\frac{7}{10}$ length of scutellum, bearing 5 + 5 to 7 + 7 crenulae.

Abdominal sternites faintly punctate medially. Genitalia about $\frac{2}{3}$ to $\frac{3}{4}$ length of middle tibia (0.66–0.76); digital sclerites about $\frac{1}{4}$ to $\frac{1}{3}$ the combined length of aedeagus and apodemes (0.26–0.32).

Forewing as in the female; delta with 20–33 setae in 4 rows; submarginal vein bearing 15–20 bullae; marginal vein bearing 7–11 prominent, subequal setae along anterior margin.

Length 0.66–0.85 mm.

Redescribed from the following material: 2♀♀ and 3♂♂ (syntypes), reared by C. P. Clausen from the camphor scale, *Pseudaonidia duplex* (Cockerell), on persimmon, Osaka, Japan, January 5, 1921; 11♀♀, 2♂♂, reared by C. P. Clausen (apparently from the same host) from *Ceroplastes rubens* (Maskell) material, Nagasaki, Japan, April 17, 1916; 19♀♀, 12♂♂, reared by C. P. Clausen from *P. duplex*, Nagasaki, Japan, May 16, 1916; 2♀♀, 5♂♂, reared by A. Azim from *P. duplex* on camelia and citrus, Fukuoka, Japan, May 1961; 2♀♀, reared by P. DeBach from *P. trilobitiformis* (Green) on *Citrus maxima*, Rural University, Rio de Janeiro State, Brazil, March 25, 1962; 1♂, reared by P. DeBach from the same host on citrus, Goiana, Pernambuco, Brazil, April 10, 1962; 7♀♀, 2♂♂, reared by P. DeBach from the same host on cashew tree leaves, Floresta Belém de São Francisco, Pernambuco, Brazil, April 13, 1962; 16♀♀, 4♂♂, reared by P. DeBach from the same host on citrus, Sete Lagoas, Minas Gerais, Brazil, May 1, 1962; 11♀♀, 6♂♂, reared by P. DeBach from an undetermined armored scale insect on an ornamental tree, Rio, Brazil, June 29, 1962; 1♀, 1♂, reared by E.J. Rankin from *P. trilobitiformis* on cashew, La Brea, Trinidad, October 8, 1969; 2♀♀, reared by E. J. Rankin from the same host on *Ixora*, Tacarigua, Trinidad, May 24, 1970.

All the specimens listed above are deposited in the collection of the Division of Biological Control, University of California, Riverside.

**Notes.** The mature pupa of *cylindratus* is light golden yellow, with the thoracic sterna, wing pads, antennae and legs brownish.

*A. cylindratus* appears to be a quite unique member of the **chrysomphali** group. It probably exhibits rather close affinity to the **lingnanensis** group.

Compere (1955) compared *cylindratus* with *ignotus*. As pointed out in the discussion of the latter species (p. 701), we do not consider *ignotus* to be closely related to *cylindratus*.

**Biology.** The biology of *cylindratus* was reported in detail by Azim (1963a). This species is the most effective parasite of the camphor scale in Japan. It appears to be highly host-specific, having been recorded only from the genus *Pseudaonidia* in the field; attempts by Azim to rear it on various other armored scale insects, including *Chrysomphalus bifasciculatus* Ferris and *Aonidiella taxus* Leonardi, were unsuccessful. *A. cylindratus* is a gregarious, arrhenotokous species. Progeny averaged about 80% females in the field in Japan. Under unspecified summer conditions, the life cycle was found to take about 17 days, and the longevity of the female averaged 14 days.

**Potential Value for Biological Control.** *A. cylindratus* is known to attack *Pseudaonidia* spp. in Japan, Brazil and Trinidad, and is considered an effective natural enemy of *P. duplex* in Japan. We assume that this species is of Far Eastern origin, and has followed its hosts in their spread to the Neotropical region. It should be regarded as a promising candidate for introduction into other regions against the camphor scale.

## 72. **Aphytis lepidosaphes** Compere

(Figures 152, 177, 178, 193, 194, 209, 241; and 1066–1078)

*Aphytis lepidosaphes* Compere, 1955, Univ. Calif. Publ. Entomol., **10**:307.
*Aphytis lepidosaphes*: DeBach and Landi, 1961, Hilgardia, **31**:465–467; Quednau, 1964, J. Entomol. Soc. S. Afr., **27**:107.

This distinctive biparental Oriental parasite of the purple scale, *Cornuaspis* [= *Lepidosaphes*] *beckii* (Newman), may be readily recognized by the following characters: General coloration yellow, thoracic sterna strongly infuscated; antennal club about 3 times as long as wide or somewhat longer; mesoscutum usually with 12 setae; propodeum about 4 to $4\frac{1}{2}$ times as long as the metanotum, crenulae small, nonoverlapping; male antennal club lacking a specialized sensory area on the ventral aspect.

*Female.* Eyes finely setose. Mandibles well developed, with 2 denticles and a dorsal truncation; maxillary palpi 2-segmented, labial palpi 1-segmented. Antennal scape (Figures 1066, 1067) slender, $4\frac{1}{2}$ to $5\frac{3}{4}$ times as long as wide, only somewhat longer than the club; pedicel $1\frac{2}{3}$ to $1\frac{4}{5}$ times as long as wide, usually somewhat longer than the third segment of the funicle; first funicular segment trapezoidal, $1\frac{2}{5}$ to $1\frac{4}{5}$ times as wide as long; second segment nearly symmetrical, usually shorter and wider than the first, about $1\frac{3}{4}$ to $2\frac{2}{5}$ times as wide as long; third funicular segment about $1\frac{1}{5}$ to $1\frac{2}{5}$ times as long as wide, bearing 2–4 longitudinal sensilla; club $2\frac{4}{5}$ to $3\frac{2}{5}$ times as long as wide (usually over 3 times), about $2\frac{3}{5}$ to $2\frac{3}{4}$ times longer and somewhat wider than the preceding segment, bearing 7–9 longitudinal sensilla.

Setae on thorax rather coarse, dark, readily visible under $\times 30$ magnification (but considerably paler than in *mytilaspidis*); setae on head paler, visible under $\times 60$ magnification, those on abdomen invisible even under $\times 120$ magnification. Vertex with 2 pairs of long setae along occipital margin, in addition to numerous shorter setae. Mesoscutum (Figures 152, 1068, 1069) with 10–15 (usually 12) setae, including 4 submedian pairs, the posterior pair and 1 seta at each antero-lateral corner considerably longer and coarser than the others; each parapsis with 2 setae, each axilla with 1 seta; scutellum with 4 setae, the discoid sensilla about equidistant from the two pairs or slightly closer to the anterior pair. Frontovertex, pronotum and mesonotal sclerites reticulate. Scutellum oval, $\frac{3}{4}$ to $\frac{4}{5}$ median length of mesoscutum. Metanotum (Figures 152, 1070, 1071) reticulate except on the sides; anteromedian apodeme robust, about $\frac{4}{5}$ to $\frac{9}{10}$ median length of metanotum.

Propodeum (Figures 1069–1072) about 4 to $4\frac{1}{2}$ times as long as the metanotum, about $\frac{3}{4}$ length of scutellum, reticulate on the sides, broadly reticulate centrally, sometimes with faint indications of transverse striation along posterior margin, with a distinct median salient; crenulae 5 + 6 to 7 + 8, small, rounded, somewhat elongate, nonoverlapping (see also Figures 177, 178).

Second abdominal tergite transversely striated anteriorly on both sides, faintly striated centrally; tergites III–VII reticulate on the sides, bearing a few setae in a transverse row on each reticulate area; third tergite transversely reticulate also mesad of the lateral setiferous areas, faintly striated centrally; tergites IV–VI with faint indications of transverse striation across center; seventh tergite faintly reticulate across, bearing a pair of submedian setae; eighth tergite (see Figure 209) delicately reticulate, with a transverse row of 4 (rarely 5) setae between spiracles, usually with an additional seta immediately below each spiracle; syntergum triangular, faintly punctate, bearing 6 (rarely 7 or 8) setae in a transverse row. Cerci about equidistant from posterior spiracles and from tip of abdomen, with 2 long setae and 1 short seta. Ovipositor shaft $1\frac{3}{4}$ to 2 times as long as the middle tibia (1.75–1.99); ovipositor sheaths about $\frac{2}{5}$ length of middle tibia (0.37–0.43).

Mid-tibial spur $\frac{3}{4}$ to $\frac{4}{5}$ length of the corresponding basitarsus.

Forewing (Figure 1073) about $2\frac{1}{2}$ times as long as wide (2.43–2.56); marginal fringe not exceeding $\frac{1}{6}$ width of disk. Delta area with 32–48 setae in 4–5 rows, these considerably longer and sparser than the setae distad of speculum, not so clearly separated from row of setae along posterior margin of wing, widely separated from a few setae below distal portion of submarginal vein; costal cell with a row of fine setae along proximal two thirds or so and 1 coarse seta near apex, the latter usually situated on the border of the vein. Submarginal vein bearing 2 coarse setae, the proximal one $\frac{3}{4}$ to $\frac{9}{10}$ length of the distal, and 15–21 bullae. Marginal vein bearing 8–12 prominent, subequal setae along anterior margin, these only about $1\frac{1}{7}$ to $1\frac{1}{3}$ times as long as the setae in a row along center of vein.

Forewing hyaline, faintly infumated below junction of submarginal and marginal veins and at base of delta; a fuscous streak along posterior margin of wing, distad of speculum.

General coloration yellow; thoracic sterna strongly infuscated; posterior margin of scutellum lined with black; a short black streak at base of forewing, below tegula. Antennal scape pale, with a fuscous stripe ventrally, rest of antenna faintly, uniformly dusky. Legs concolorous with body. Wing veins pale.

Length 0.90–1.15 mm.

*Male.* Essentially similar to the female in structure, chaetotaxis, sculpture and coloration.

Antennal scape (Figure 1074) about $3\frac{2}{5}$ to 5 times as long as wide, up to $1\frac{1}{5}$ times longer than the club; pedicel $1\frac{1}{2}$ to $1\frac{4}{5}$ times as long as wide, up to $1\frac{1}{3}$ times longer than the third segment of the funicle; first 2 funicular segments as in the female, third segment up to $1\frac{1}{4}$ times as long as wide, bearing 2 longitudinal sensilla; club $2\frac{3}{4}$ to $3\frac{1}{5}$ times as long as wide, bearing 3–5 longitudinal sensilla, lacking a specialized sensory area on the ventral aspect.

Mesoscutum (Figure 1075) with 10–13 (usually 12) setae. Propodeum (Figures 1076, 1077) as in the female, about 4 to $4\frac{1}{2}$ times as long as the metanotum, $\frac{3}{5}$ to $\frac{2}{3}$ length of scutellum; crenulae 3 + 3 to 5 + 6.

Genitalia (Figure 1078) about $\frac{2}{3}$ to $\frac{3}{4}$ length of middle tibia (0.66–0.76); digital sclerites nearly $\frac{1}{3}$ the combined length of aedeagus and apodemes (0.30–0.32).

Forewing somewhat wider than in the female, about $2\frac{3}{10}$ to $2\frac{2}{5}$ times as long as wide; delta with 20–31 setae in 3–4 rows; submarginal vein bearing 15–19 bullae; marginal vein bearing 8–10 prominent, subequal setae along anterior margin.

Length 0.70–0.93 mm.

Redescribed from numerous ♀♂ specimens, reared from the purple scale, *Cornuaspis* [= *Lepidosaphes*] *beckii* (Newman), including the syntype series: "Thirty unstained specimens under one cover slip on one slide, and approximately 20 or more stained and dissected specimens under two cover slips, both on one slide. . . . The unstained specimens were propagated on *Lepidosaphes beckii* in the insectary at Riverside; the stained specimens were propagated on the same host in the insectary at Anaheim" (Compere, 1955).

Type series in the collection of the Division of Biological Control, University of California, Riverside.

**Notes.** The pupa of *lepidosaphes* is predominantly yellow, with the wing-pads slightly tinted with brown.

This interesting species was first recognized as distinct upon its introduction from China and Taiwan into California in 1948–1949. An initial breeding stock of 86 specimens was received at Riverside from J. L. Gressitt and Y. W. Djou. This material was reared from purple scale that had been collected on oranges at Lo-Kong-Tung, near Canton, China, and shipped to the quarantine insectary at Riverside on November 28, 1948. Earlier than that, Silvestri (1929) had recorded observing a purple scale parasite on citrus in southern China in 1924–1925, which in all probability was *lepidosaphes* (Compere, 1955; DeBach and Landi, 1961). Early reports by Flanders (1950) and DeBach (1953) referred to this species as "*Aphytis* X".

*A. lepidosaphes* is known to occur throughout the Orient (China, Taiwan, Thailand, Burma, India, Pakistan and the Philippines). Following its establishment in California, this species has been successfully introduced into numerous countries, including Texas, Mexico, Peru, Brazil, Chile, Spain, Greece, Crete, Cyprus and South Africa. Additionally, it has been gradually moving accidentally around the world and has become established by ecesis in Hawaii, Louisiana, Florida, Puerto Rico, Guadaloupe, Jamaica,

El Salvador, Argentina, Turkey, Israel, Australia, New Caledonia and Fiji (DeBach, 1971). It was recently recorded also from Lebanon (Traboulsi, 1969).

Material at hand includes specimens reared from *Cornuaspis* [= *Lepidosaphes*] *beckii* in China, Taiwan, Burma, California, Texas, Louisiana, Florida, Hawaii, Puerto Rico, Jamaica, Trinidad, Peru, Chile, Israel, Turkey, Cyprus and Australia.

Although *A. lepidosaphes* is regarded here as a bona fide member of the **chrysomphali** group, it shows some resemblance to the **mytilaspidis** group and in certain characters appears to occupy an intermediate position between the two groups. Thus, the propodeum of *lepidosaphes* appears to be shorter, and the thoracic setae darker, than in any other member of the **chrysomphali** group.

**Biology.** The biology of *A. lepidosaphes* was reported in detail by DeBach and Landi (1961). This species appears to be highly host-specific, and is not known to occur on scales other than *Cornuaspis* [= *Lepidosaphes*] *beckii* in the field. Attempts to rear it in the laboratory on other scale insects, including *Aonidiella aurantii* (Maskell) and *Aspidiotus nerii* Bouché, have been unsuccessful. Freshly laid eggs and early larval stages of *A. lepidosaphes* failed to complete development when transferred to *A. nerii* and *A. aurantii*.

*A. lepidosaphes* is generally gregarious under both laboratory and field conditions (see Figure 38). Under insectary conditions of 26.7 ± 1°C and 50% relative humidity, the average life cycle is 17 days, and the average longevity of nonproducing adult females with access to fresh food (honey) and in the absence of hosts is 11 days. Average total oviposition was 32.1 eggs per female. The species is arrhenotokous; the progeny reared under laboratory conditions averaged 52.9% females, whereas under field conditions the progeny averaged 80.8% females. Ovipositional behavior and developmental stages were described in detail by DeBach and Landi (1961).

**Utilization in Biological Control.** *A. lepidosaphes* is an effective parasite of the purple scale. Following its introduction from China, it brought about partial control of this serious pest in California, which enabled the development of an effective program of integrated control by strip treatment (DeBach and Landi, 1961). It has recently been reported by Meyerdirk (1972) to regulate purple scale populations on citrus in California below the economic injury level, especially in climatically mild coastal areas.

*A. lepidosaphes* now appears to be responsible for substantial to complete biological control of purple scale in every one of the countries into which it has gained entry (see DeBach, 1971; DeBach and Rosen, 1976a; Rosen and DeBach, 1978). The complete biological control of purple scale effected by this species in Texas was recently reported by Dean (1975).

## 73. **Aphytis immaculatus** Compere

(Figures 1079–1087)

*Aphytis immaculatus* Compere, 1955, Univ. Calif. Publ. Entomol., **10**:307–308.

This biparental Oriental parasite of *Lepidosaphes* spp. appears to be rather closely related to *lepidosaphes*, but may be readily separated from the latter species by the immaculate thoracic sterna and pale thoracic setae. It may be further recognized by the shorter antennal club, somewhat longer propodeum, triangular crenulae and shorter ovipositor; the antennal club of the male lacks a specialized sensory area on the ventral aspect.

*Female.* Eyes finely setose. Mandibles well developed, with 2 denticles and a dorsal truncation; maxillary palpi 2-segmented, labial palpi 1-segmented. Antennal scape (Figures 1079, 1080) slender, about 4–6 times as long as wide, somewhat longer than the club; pedicel $1\frac{1}{2}$ to $1\frac{4}{5}$ times as long as wide, considerably ($1\frac{3}{10}$ to $1\frac{1}{2}$ times) longer than the third segment of the funicle; first funicular segment somewhat trapezoidal, about $1\frac{2}{5}$ to $1\frac{2}{3}$ times as wide as long; second segment nearly symmetrical, usually considerably shorter and somewhat wider than the first segment, usually twice as wide as long or wider ($1\frac{9}{10}$ to $2\frac{2}{5}$ times); third funicular segment usually about as long as wide (sometimes wider than long, rarely longer than wide), bearing 2 longitudinal sensilla; club usually $2\frac{2}{3}$ to $2\frac{4}{5}$ times as long as wide, about 3 to $3\frac{1}{2}$ times longer and distinctly wider than the preceding segment, bearing 5–7 longitudinal sensilla.

Setae on thorax rather slender, pale, visible only under $\times 120$ magnification; those on head and abdomen paler, invisible even under $\times 120$ magnification. Vertex with 2 pairs of long setae along occipital margin, in addition to numerous shorter setae. Mesoscutum (Figure 1081) with 12–15 setae, including 4 submedian pairs, the posterior pair and 1 seta at each antero-lateral corner considerably longer and coarser than the others; each parapsis with 2 setae, each axilla with 1 seta; scutellum with 4 setae, the discoid sensilla about equidistant from the two pairs. Frontovertex, pronotum and mesonotal sclerites reticulate. Scutellum oval, about $\frac{3}{4}$ to $\frac{4}{5}$ median length of mesoscutum. Metanotum (Figures 1081–1083) short, reticulate except on the sides; anteromedian apodeme usually considerably (up to $1\frac{1}{2}$ times) longer than median length of metanotum.

Propodeum (Figures 1081–1083) long, about 5–7 times as long as the metanotum, $\frac{3}{4}$ to over $\frac{4}{5}$ length of scutellum, reticulate on the sides, rather broadly reticulate centrally, lightly transversely striated along posterior margin, with a distinct median salient; crenulae 3 + 4 to 6 + 6, rather small but very distinct, usually triangular, closely spaced, sometimes slightly overlapping at base (see Figure 1083).

Second abdominal tergite transversely striated anteriorly on both sides, smooth centrally; tergites III–VII reticulate on the sides, bearing 1–4 setae in a short transverse row on each reticulate area; third tergite strongly transversely striated across center, tergites IV–VI more faintly so; seventh tergite faintly reticulate across center, bearing a pair of submedian setae; eighth tergite delicately reticulate, with a transverse row of 4 setae along posterior margin between spiracles; syntergum triangular, faintly punctate, bearing 6 (rarely 7) setae in a transverse row. Cerci situated somewhat closer to posterior spiracles than to tip of abdomen, with 2 long setae and 1 short seta. Ovipositor shaft about $1\frac{1}{2}$ to $1\frac{2}{3}$ times as long as the middle tibia (1.45–1.67); ovipositor sheaths about $\frac{1}{3}$ length of middle tibia (0.31–0.37).

Mid-tibial spur about as long as the corresponding basitarsus.

Forewing (Figure 1084) about $2\frac{1}{2}$ to $2\frac{3}{4}$ times as long as wide; marginal fringe not exceeding $\frac{1}{4}$ width of disk. Delta area with 26–44 setae in 4 (rarely 5) rows, these considerably longer and sparser than the setae distad of speculum, not clearly separated from row of setae along posterior margin of wing, widely separated from 1–3 setae below distal portion of submarginal vein; costal cell with a row of 4–5 fine setae and 1 coarse seta near apex, the latter apparently arising from the vein. Submarginal vein bearing 2 coarse setae, the proximal one $\frac{2}{3}$ to $\frac{3}{4}$ length of the distal, and 17–22 bullae. Marginal vein bearing 7–10 prominent, subequal setae along anterior margin, these about $1\frac{2}{5}$ to $1\frac{2}{3}$ longer than the setae in a row along center of vein.

Forewing entirely hyaline, very faintly infumated below junction of submarginal and marginal veins and at base of delta; a fuscous streak along posterior margin of wing, distad of speculum.

General coloration entirely yellow; thoracic sterna immaculate; posterior margin of scutellum narrowly lined with fuscous; a short blackish streak at base of forewing, below tegula. Antennae faintly dusky yellowish. Legs concolorous with body. Wing veins pale, faintly lined with brownish.

Length 0.84–1.03 mm.

*Male.* Essentially similar to the female in structure, chaetotaxis, sculpture and coloration.

Antennal scape (Figure 1085) about 4 to $4\frac{2}{3}$ times as long as wide, about $1\frac{1}{5}$ times longer than the club; pedicel and funicle as in the female, third funicular segment bearing 1–2 longitudinal sensilla; club shorter than in the female, $2\frac{2}{5}$ to $2\frac{1}{2}$ times as long as wide, $2\frac{2}{3}$ to $3\frac{1}{4}$ times longer than the preceding segment, bearing 2 longitudinal sensilla (5 in one specimen); no specialized sensory area is evident on the club in the specimens at hand.

Mesoscutum with 12–15 setae, as in the female. Propodeum (Figure 1086) as in the female, 4 to $6\frac{3}{4}$ times as long as the metanotum, about $\frac{2}{3}$ length of scutellum; crenulae 3+3 to 4+5.

Genitalia (Figure 1087) about $\frac{2}{3}$ to $\frac{3}{4}$ length of middle tibia (0.63–0.76); digital sclerites less than $\frac{1}{3}$ the combined length of aedeagus and apodemes (0.27–0.30).

Length 0.73–1.04 mm.

Redescribed from numerous ♀♂ specimens (including syntypes), reared by T. C. Maa from *Lepidosaphes* sp. on *Agalma* sp., Taiwan, November–December 1952. According to Compere (1955), "the samples of *immaculatus* were obtained from receiving cages containing three different host insects and three different species of *Aphytis*. It was impossible to distinguish among the three species of *Aphytis* before they were mounted and examined microscopically. Two or possibly three species may be represented on some slides. For this reason, the type series of *immaculatus* is limited to 11 females and 1 male, all under three cover slips on one slide. Six females and 1 male were cleared and stained and are under one cover slip, except for the propodea of four specimens that were detached and flattened under a separate cover slip. Four females were mounted before shrinkage without clearing and staining. These are now badly shriveled and decolorized." The latter specimens were subsequently remounted in Hoyer's medium, which has since deteriorated, and are in rather poor shape.

Type series in the collection of the Division of Biological Control, University of California, Riverside.

**Notes and Additional Material.** According to Compere (1955), "the species *Aphytis immaculatus* traces back to some shipments imported from Formosa in October, November and December, 1952 (S&R numbers 1017, 1018, 1033 and 1034). These shipments contained parasitized *Aonidiella*, *Chrysomphalus*, and *Lepidosaphes* species, collected on *Agalma* species at Peitou, Peh-Tan, Wai-Suan-Chi, and Yuang-Shan, Formosa. The material was collected by T. C. Maa and received by Flanders. . . . three species of *Aphytis* issued from the imported scales, namely, *chrysomphali* (Mercet), *lingnanensis* (new species), and *immaculatus* (new species). The species under consideration here, *immaculatus*, issued from the *Lepidosaphes* species. Attempts to propagate this species in the insectary at Riverside on *Lepidosaphes beckii* (Newman), *Aspidiotus camelliae* Signoret, and *Hemiberlesia lataniae* (Signoret) were unsuccessful. A colony of sixty was given to the Orange County Insectary for trial there, and another small colony was released by DeBach on citrus trees infested with *Lepidosaphes gloverii* (Pack.) at San Juan Capistrano" (designated as *Aphytis* "Y" on the release card).

A series reared by J. L. Gressitt and Y. W. Djou from *Lepidosaphes* sp. on wild shrub (?*Breynia*), 60 miles WNW of Canton, China, December 12, 1948, includes specimens that appear to be conspecific with *immaculatus*.

Some of the apparent differences between *immaculatus* and *lepidosaphes* may be artifacts of different mounting procedures. This may be especially true for the apparently shorter antennal club of *immaculatus*, which may be due to some flattening of specimens on the slides. However, the striking difference in the coloration of the thoracic sterna, as well as the different shape of the propodeal crenulae and other structural differences leave no doubt that these are two distinct species, attacking closely related hosts in the Orient.

**Potential Value for Biological Control.** In spite of previous failures, there is a possibility that *A. immaculatus* is a parasite of the purple scale, *Cornuaspis* [= *Lepidosaphes*] *beckii* (Newman). In view of the economic importance of purple scale as a serious pest of citrus in many regions of the world, efforts should be continued to secure and use *immaculatus* in further trials against that pest. In any event, this little-known species may prove useful in the biological control of one species or another of *Lepidosaphes*.

## 74. **Aphytis japonicus** DeBach and Azim

(Figures 1088–1099)

*Aphytis japonicus* DeBach and Azim, 1962, Mushi, **36**: 1–8.
*Aphytis japonicus*: Azim, 1963, J. Fac. Agr. Kyushu Univ., **12**: 283–284.

This small, biparental East Palearctic species may be readily recognized by the entirely yellow coloration, the immaculate thoracic sterna, the rather pale, slender thoracic setae, the short antennal club, the long propodeum with minute but distinct crenulae, and the rather narrow forewing. The antennal club of the male lacks a specialized sensory area on the ventral surface.

*Female.* Eyes finely setose. Mandibles well developed, with 2 distinct denticles and a dorsal truncation, sometimes appearing tridentate; maxillary palpi 2-segmented, labial palpi 1-segmented. Antennal scape (Figure 1088) slender, 5 to $6\frac{3}{4}$ times as long as wide, considerably (up to $1\frac{2}{5}$ times) longer than the club; pedicel about $1\frac{1}{2}$ to $1\frac{4}{5}$ times as long as wide, $1\frac{1}{6}$ to $1\frac{2}{5}$ times as long as the third segment of the funicle; first 2 funicular segments usually subglobular, the first somewhat trapezoidal, $1\frac{1}{7}$ to $1\frac{2}{5}$ times as wide as long; second segment nearly symmetrical, somewhat shorter and slightly wider than the first, usually $1\frac{1}{2}$ to $1\frac{9}{10}$ times as wide as long; third funicular segment relatively short, as long as wide to about $1\frac{1}{4}$ times longer than wide, bearing 2 longitudinal sensilla (rarely 1); club short, about $2\frac{1}{3}$ to $2\frac{4}{5}$ times as long as wide, $2\frac{1}{5}$ to nearly 3 times longer and a little wider than the preceding segment, bearing 4–5 longitudinal sensilla.

Setae on thorax rather pale, relatively long and slender, barely visible under $\times 30$ magnification; setae on abdomen slender, pale, invisible even under $\times 120$ magnification. Vertex with 2 pairs of long setae along occipital margin, in addition to numerous shorter setae. Mesoscutum (Figure 1089) usually with 10 setae, rarely with 9; each parapsis with 2 setae, these shorter than the other mesonotal setae but still relatively long; each axilla with 1 seta; scutellum with 4, the discoid sensilla equidistant from the two pairs or somewhat closer to the anterior pair. Frontovertex, pronotum, and mesonotal sclerites reticulate. Scutellum about $\frac{3}{4}$ to $\frac{4}{5}$ median length of mesoscutum. Metanotum (Figures 1089, 1090) rather short, somewhat curved, reticulate except on the sides; anteromedian apodeme considerably (1.45–1.67 times) longer than median length of metanotum.

Propodeum (Figures 1089–1091) distinctly trapezoidal, relatively long, $4\frac{3}{4}$ to $6\frac{1}{4}$ times as long as the metanotum, $\frac{2}{3}$ to $\frac{9}{10}$ length of scutellum, faintly reticulate on the sides, rather broadly reticulate centrally, the cells being as long as wide anteriorly, longer than wide posteriorly; crenulae $4+5$ to $7+7$, minute but distinct, slightly elongate, nonoverlapping.

Second abdominal tergite transversely striated anteriorly on both sides, reticulate centrally; tergites III–VII reticulate on the sides, bearing 2–5 setae in a short transverse row on each reticulate area; third tergite distinctly transversely striated across center; tergites IV–VI with faint indications of transverse striation and punctation; seventh tergite faintly transversely striated across center, bearing 2 submedian setae between the lateral setiferous areas; eighth tergite delicately reticulate-punctate, bearing 4 setae in a transverse row between spiracles; syntergum triangular, punctate, bearing 6–7 setae in a transverse row. Cerci situated somewhat closer to posterior spiracles than to tip of abdomen, with 2 long setae and 1 short seta. Ovipositor shaft (Figure 1092) varying from about $1\frac{1}{2}$ to $1\frac{9}{10}$ times as long as the middle tibia (1.52–1.89); ovipositor sheaths long, narrow, usually over $\frac{2}{5}$ length of middle tibia (0.40–0.46).

Mid-tibial spur relatively long, slender, nearly as long as or somewhat longer than the corresponding basitarsus.

Forewing (Figures 1093, 1094) rather narrow, $2\frac{3}{5}$ to $2\frac{4}{5}$ times as long as wide; marginal fringe relatively long, about $\frac{1}{3}$ to $\frac{1}{4}$ width of disk. Delta area with 26–41 setae in 4–5 rows, these much longer and sparser than the setae distad of speculum, not quite distinct from row of setae along posterior margin of wing, widely separated from 2–3 setae below distal portion of submarginal vein; costal cell with a row of fine setae along proximal three fifths and 1 coarse seta near apex. the latter apparently on the vein rather than in the cell itself. Setae on submarginal and marginal veins rather long. Submarginal vein bearing 2 setae, the proximal one about $\frac{1}{2}$ to $\frac{3}{4}$ length of the distal, and 12–16 bullae. Marginal vein bearing 6–10 prominent, subequal setae along anterior margin, these $1\frac{1}{2}$ times to nearly twice longer than the setae in a row along center of vein.

Forewing almost perfectly hyaline, very faintly infumated below junction of submarginal and marginal veins and at base of delta; a fuscous streak along posterior margin of wing, distad of speculum; under phase contrast, the basal half of the wing appears very faintly infumated.

General coloration entirely yellow; thoracic sterna concolorous, immaculate; posterior margin of scutellum sometimes faintly, narrowly lined with fuscous; a short black streak near base of forewing, below tegula; plates around base of ovipositor very faintly dusky. Antennal scape pale, rest of antenna uniformly, faintly dusky. Legs concolorous with body, tibiae and tarsi darker yellow. Wing veins colorless.

Length 0.73–0.89 mm.

*Male.* Essentially similar to the female in structure, chaetotaxis, sculpture and coloration, differing mainly in the shape of the antennae.

Antennal scape (Figures 1095, 1096) a little less slender than in the female, $4\frac{1}{3}$ to $5\frac{1}{2}$ times as long as wide, $1\frac{1}{5}$ to $1\frac{1}{2}$ times as long as the club; third funicular segment more elongate than in the female, somewhat asymmetrical, $1\frac{1}{5}$ to $1\frac{1}{2}$ times as long as wide, bearing 1–2 longitudinal sensilla; club $2\frac{1}{5}$ to $2\frac{3}{4}$ times as long as wide, only about twice as long as the preceding segment, bearing 2–3 longitudinal sensilla and lacking a cut-off sensory area on the ventral surface.

Thoracic setae (Figure 1097) even paler than in the female. Posterior margin of scutellum not lined with fuscous. Propodeum (Figure 1097) $3\frac{2}{3}$ to $5\frac{1}{4}$ times as long as the metanotum, $\frac{2}{3}$ to $\frac{3}{4}$ length of scutellum; crenulae 3 + 5 to 6 + 6, as in the female.

Genitalia usually about $\frac{3}{4}$ length of middle tibia (0.63–0.87); digital sclerites $\frac{1}{3}$ to $\frac{2}{5}$ the combined length of aedeagus and apodemes (0.32–0.40).

Wings (Figures 1098, 1099) as in the female.

Length 0.64–0.83 mm.

Redescribed from the following material: Numerous specimens, reared by A. Azim from *Chrysomphalus bifasciculatus* Ferris on *Euonymus japonicus*, Fukuoka, Japan (14♀♀, 6♂♂—♀ holotype, ♂ allotype, and paratypes—May 1961; 4♀♀, June 28, 1960; 72♀♀, 43♂♂, October 1960; 26♀♀, 8♂♂, November 1960; 40♀♀, 79♂♂, May 1961); 5♀♀, 9♂♂, reared by A. Azim from *Aonidiella taxus* Leonardi, Fukuoka, Japan, July 15, 1961; 44♀♀, 1♂, collected by net by A. Azim, Fukuoka, Japan, April 30 and May 1, 1961; and numerous ♀♂ specimens, reared by H. Uematsu from *Chrysomphalus bifasciculatus* on *Euonymus japonicus*, Miyazaki City, Japan, December 27, 1968.

Holotype, allotype and paratypes in the collection of the Division of Biological Control, University of California, Riverside. Additional paratypes—probably labeled as "syntypes"—in the British Museum (Natural History), in the U.S. National Museum of Natural History, and in the collection of the Entomological Laboratory, Kyushu University, Fukuoka, Japan.

**Notes.** The mature pupa of *japonicus* is entirely yellow. Although originally included in the **mytilaspidis** group, we now consider this species to be a bona fide member of the **chrysomphali** group, as understood here. It differs from the members of the **mytilaspidis** group in having pale, slender setae, a long propodeum, narrow wings and a yellow pupa; from *chrysomphali* in having pale thoracic sterna and a short antennal club. The somewhat asymmetrical third funicular segment of the male (Figure 1096) may also serve as a diagnostic character.

According to DeBach and Azim (1962), "*Aphytis japonicus* is very similar to, if not identical with, the so-called 'Kagoshima form of *Aphytis citrinus*' discussed by DeBach (1959). He collected specimens by beating citrus trees heavily infested with *Aonidiella aurantii* (Mask.) at Kagoshima, Japan, November 15, 1956." Material collected by DeBach on citrus at Kagoshima in 1956 includes both *japonicus* and *mazalae*.

**Biology.** Azim (1961, 1963c) reported in some detail on the biology and developmental stages of *japonicus*. This species is biparental, the females outnumbering males by a ratio of about 2:1 in field samples. Under unspecified room conditions during summer in Fukuoka, Japan, egg development took 4–5 days, larval development 5–6 days, the prepupal stage 1–2 days and the pupal stage 6–7 days, the entire life cycle thus totaling 16–20 days. Average adult longevity was about 15 days in summer, 30.5 days in autumn. The parasite was reported as solitary, although superparasitism may occur.

**Potential Value for Biological Control.** *A. japonicus* is apparently an abundant parasite of *Chrysomphalus bifasciculatus* in southern Japan. It should be considered as a promising candidate for introduction into other areas for biological control of this and related species of scale insects.

## 75. **Aphytis mazalae** DeBach and Rosen

(Figures 1100–1109)

*Aphytis mazalae* DeBach and Rosen, 1976, Ann. Entomol. Soc. Amer., **69**: 544.

This small, biparental Far Eastern species is rather closely related to *japonicus*, but may be readily recognized by the distinctive pigmentation of the middle tibiae, which are conspicuously tipped with black in the male, faintly tipped with brownish in the female; also, third funicular segment strongly asymmetrical in the male, somewhat asymmetrical in the female. General coloration yellow; thoracic sterna immaculate; antennae strongly infuscated, tip of club blackish; propodeum rather long, crenulae small, nonoverlapping; forewing rather narrow, hyaline.

*Female*. Eyes finely setose. Mandibles well developed, with 2 denticles and a dorsal truncation; maxillary palpi 2-segmented, labial palpi 1-segmented. Antennal scape (Figures 1100, 1109) slender, 5 to $5\frac{1}{2}$ times as long as wide, distinctly (up to $1\frac{1}{5}$ times) longer than the club; pedicel about $1\frac{3}{4}$ times as long as wide, about as long as the third segment of the funicle (varying from slightly shorter to slightly longer); first funicular segment subglobular, $1\frac{1}{4}$ to $1\frac{1}{3}$ times as wide as long; second segment shorter and somewhat wider than the first, about $1\frac{3}{4}$ to $1\frac{4}{5}$ times as wide as long; third funicular segment slightly but distinctly asymmetrical (see Figure 1109), the inner or ventral aspect longer than the outer or dorsal aspect, $1\frac{1}{4}$ to a little over $1\frac{2}{5}$ times as long as wide, bearing 1–2 longitudinal sensilla; club $2\frac{3}{4}$ to 3 times as long as wide, $2\frac{1}{5}$ to $2\frac{3}{10}$ times longer and somewhat wider than the preceding segment, bearing 4–5 longitudinal sensilla.

Setae on thorax (Figure 1101) rather coarse, dark, visible under $\times 30$ magnification; setae on head visible under $\times 120$ magnification; those on abdomen slender, invisible even under $\times 120$ magnification. Vertex with 2 pairs of long setae along occipital margin, in addition to numerous shorter setae. Mesoscutum with 10 rather long setae, the posterior pair and 1 seta at each antero-lateral corner noticeably longer than the others; each parapsis with 2 shorter setae, each axilla with 1 seta; scutellum with 4, the discoid sensilla about equidistant from the two pairs or somewhat closer to the anterior pair. Frontovertex, pronotum and mesonotal sclerites reticulate. Scutellum oval, about $\frac{3}{4}$ median length of mesoscutum. Metanotum (Figure 1102) reticulate except on the sides; anteromedian apodeme distinctly longer than median length of metanotum.

Propodeum (Figures 1101, 1102) rather long, $4\frac{1}{3}$ to fully 5 times as long as the metanotum, $\frac{3}{4}$ to $\frac{9}{10}$ length of scutellum, with a distinct median salient, reticulate on the sides and on a trapezoidal central area, narrowly striated along posterior margin; crenulae $4 + 5$ to $7 + 7$, small but distinct, rounded, nonoverlapping.

Second abdominal tergite transversely reticulate anteriorly on each side, faintly broadly reticulate centrally, vaguely transversely striated posteriorly; tergites III–VII reticulate on the sides, bearing a few setae in a short transverse row on each reticulate area; third tergite distinctly transversely striated across center, between the lateral reticulate areas; tergites IV–VI with indications of transverse striation; seventh tergite indistinctly reticulate-striated, bearing 1–2 submedian setae between the lateral setiferous areas; eighth tergite delicately reticulate-punctate, bearing a transverse row of 4 setae between the spiracles; syntergum triangular, with a distinct, rather elongate cauda, delicately reticulate-punctate, with 6–7 setae in a transverse row. Cerci situated somewhat closer to posterior spiracles than to tip of abdomen, with 2 long setae and 1 short seta. Ovipositor shaft about $1\frac{2}{5}$ to $1\frac{1}{2}$ times as long as the middle tibia (1.41–1.56); ovipositor sheaths about $\frac{2}{5}$ length of middle tibia (0.38–0.41).

Mid-tibial spur about as long as the corresponding basitarsus, or somewhat shorter.

Forewing (Figure 1103) rather narrow, about $2\frac{3}{4}$ times as long as wide (2.69–2.79); marginal fringe long, not exceeding $\frac{1}{3}$ width of disk. Delta area with 29–60 setae in 4–6 rows, these considerably longer and sparser than the setae distad of speculum, not clearly separated from row of setae along posterior margin of wing, widely separated from 2–3 setae below distal part of submarginal vein; costal cell with a row of fine setae along proximal two thirds, and 1 coarse seta near apex. Submarginal vein bearing 2 coarse setae, the proximal one about $\frac{2}{3}$ to $\frac{3}{4}$ length of the distal, and 15–21 bullae. Marginal vein bearing 7–10 long, prominent, subequal setae along anterior margin, these about $1\frac{3}{5}$ times to twice as long as the setae in a row along center of vein.

Forewing nearly hyaline, faintly dusky on basal half, more distinctly infumated below submarginal vein and at base of delta; a narrow fuscous streak along posterior margin of wing, distad of speculum.

General coloration yellow; thoracic sterna immaculate; posterior margin of scutellum narrowly lined with blackish centrally; a short black streak at base of forewing, below tegula. Antennal scape pale, rest of antenna rather uniformly, strongly infuscated, tip of club blackish. Tibiae and tarsi of all legs faintly dusky, middle tibia with a small brownish spot at apex, fore tibia sometimes faintly tipped with brownish, apical tarsal segment sometimes faintly fuscous. Wing veins colorless.

Length 0.79–0.82 mm.

*Male*. Essentially similar to the female, differing mainly in the coloration of the middle tibia and in the structure and coloration of the antennae.

Antennal scape and pedicel (Figures 1104, 1105) essentially as in the female; first funicular segment about $1\frac{1}{5}$ to $1\frac{3}{10}$ times as wide as long; second segment about $1\frac{3}{5}$ to $1\frac{2}{3}$ times as wide as long; third funicular segment strongly asymmetrical, obliquely truncate, ventral aspect considerably longer than the dorsal, asymmetrically joined to the club, in its longest aspect about $1\frac{3}{4}$ times to twice as long as wide and about $1\frac{1}{3}$ to $1\frac{2}{5}$ times longer than the pedicel, bearing 1–2 longitudinal sensilla; club strongly asymmetrical at base, in its longest aspect about $2\frac{3}{5}$ to $2\frac{2}{3}$ times as long as wide, only about $1\frac{1}{2}$ times longer and slightly wider than the preceding segment, bearing 2–3 longitudinal sensilla, lacking any specialized sensory area on the ventral surface.

Mesoscutum with 10–11 setae. Propodeum (Figure 1106) 4–5 times as long as the metanotum, about $\frac{3}{4}$ length of scutellum; crenulae 5 + 6 to 6 + 7, as in the female.

Genitalia about $\frac{3}{5}$ to $\frac{3}{4}$ length of middle tibia (0.62–0.73), digital sclerites $\frac{1}{3}$ to nearly $\frac{2}{5}$ the combined length of aedeagus and apodemes (0.33–0.38).

General coloration similar to that of the female, differing as follows: infuscation on posterior margin of scutellum very faint or absent; antennal scape pale, pedicel very faintly dusky, funicular segments and base of club strongly infuscated, rest of club rather faintly dusky; middle tibia (Figures 1107, 1108) conspicuously marked with black at apex; middle basitarsus blackish on basal half.

Length 0.65–0.79 mm.

Described from 3♀♀ and 1♂ (♀ holotype, ♂ allotype, and paratypes), reared by P. Lin-moo Peng from a host recorded as *Chrysomphalus aonidum* (L.) (but considered questionable at the time) on citrus, near Taipei, Taiwan, February 14, 1966; 1♀ and 2♂♂ (paratypes), reared by E. I. Schlinger from *Aulascaspis murrayae* Takahashi on *Murraya funiculara*, Taipei, Taiwan, March 26, 1961.

Type series in the collection of the Division of Biological Control, University of California, Riverside.

**Notes and Additional Material.** The pupa of *mazalae* is entirely yellow.

One ♀ and 2♂♂, obtained by P. DeBach by beating citrus trees at Kagoshima, Japan, November 15, 1956 (together with *japonicus* and ?*fisheri*), are inseparable from the type series of *mazalae*, with which they are evidently conspecific. The female is a little larger than the females of the Taiwan series, measuring 0.88 mm in length, but is otherwise very similar to them; the males possess all the striking characteristics of *mazalae*.

Two ♀♀, reared by M. A. Ghani from the yellow scale, *Aonidiella citrina* (Coquillett), on *Citrus sinensis*, Mardan, West Pakistan, December 9, 1964 (together with *melinus*), also appear to be conspecific with the type series of *mazalae*. They, too, possess the slightly but distinctly pigmented tibiae, the slightly asymmetrical third funicular segment (Figure 1109), and all other characteristics of *mazalae*.

Additional material from Pakistan presents a taxonomic problem. A series of 4♀♀ and 3♂♂, reared by M. A. Ghani from *Pinnaspis* sp. on *Ficus palmata*, Saidpur, West Pakistan, May 26, 1965, and another series of 33♀♀ and 22♂♂, reared by M. A. Ghani from *Pinnaspis strachani* (Cooley) on *Ficus palmata*, Saidu Sharif, West Pakistan, November 18, 1965, are considerably smaller and generally paler than the specimens of the type series. The females (Figure 1110), measuring 0.59–0.75 mm in length, lack the pigmentation on the posterior margin of the scutellum and have somewhat paler thoracic setae and somewhat narrower forewings than in the type series (up to 3 times as long as wide). In some female specimens, slight brownish pigmentation is distinctly evident at the apex of the middle tibia, but this is very faint or entirely absent in other specimens. The slight asymmetry of the third funicular segment is evident in larger female specimens, but not in others. The males do not differ significantly from those of the type series, except for being smaller and generally somewhat paler. As in the type series, these males possess a conspicuously black-tipped middle tibia and a strongly asymmetrical third funicular segment (Figure 1111). Like the type series, they do not have abnormally enlarged sensilla on the first two funicular segments (see below). As far as can be determined from the slide-mounted material at hand, these two series from *Pinnaspis* can be considered to represent small specimens of *mazalae*.

The identity of other series from Pakistan, reared from *Aonidiella* spp., is even more questionable. These include the following material: 1♀, reared by M. A. Ghani from *A. citrina* on citrus, Rawalpindi, January 25, 1961; 1♂, reared by S. K. Kazimi from *Aonidiella* sp. on citrus, Rawalpindi, February 7–9, 1961; 2♀♀, 4♂♂, reared by S. K. Kazimi from *A. citrina* on citrus, Rawalpindi, September 10–12, 1961 (Figure 1112); 11♀♀, 1♂, reared by M. A. Ghani from *A. orientalis* (Newstead) on *Dalbergia sissoo*, Wah, November 10, 1965 (Figures 1113–1115); 2♀♀, 6♂♂, reared by R. Ahmad and M. A. Ghani from *Aonidiella* spp. on citrus, Rawalpindi, October 4, 1968 (Figure 1116); 7♀♀, 6♂♂, reared by R. Ahmad from *A. aurantii* (Maskell) and *A. citrina* on citrus, Rawalpindi, November 18, 1968 (Figure 1117). The females in these series are small and generally pale, with rather pale thoracic setae, without any trace of pigmentation on the tibiae, and with no indication of asymmetry of the third funicular segment. The males have the conspicuously pigmented middle tibia and strongly asymmetrical third funicular segment of *mazalae*, but most of them possess conspicuously elongate, spindle-shaped or spatulate sensilla on the first two funicular segments (Figure 1117), similar to those of *sensorius* (see p. 620). A thorough biological study is required in order to determine whether some of these specimens represent yet another undescribed species of *Aphytis*.

*A. mazalae* can be considered an aberrant species in the **chrysomphali** group, related to *japonicus* but differing in the pigmented middle tibiae of both sexes, in the coarse black thoracic setae and in the strongly asymmetrical third funicular segment of the male. Small female specimens may be very hard to separate from *japonicus*. The species appears to be widespread in the Far East (Taiwan, Japan and Pakistan) but some material referred to it may eventually prove to represent a complex of closely related species.

This distinctive species was named by us in honor of Mrs. Mazal Rosen.

**Potential Value for Biological Control.** *A. mazalae* appears to be associated with economically important species of *Aonidiella*, *Pinnaspis*, possibly *Chrysomphalus*, etc. It should be studied further in order to ascertain its role and potential value as a natural enemy of these and other serious pests.

## 76. **Aphytis sensorius** DeBach and Rosen

(Figures 1118–1128)

*Aphytis sensorius* DeBach and Rosen, 1976, Ann. Entomol. Soc. Amer., **69**: 544–545.

This small, biparental Oriental species is rather closely related to *japonicus* and *mazalae*. It may be recognized by the peculiar spindle-shaped sensillum on each of the first two funicular segments of the male, by the strongly asymmetrical third funicular segment of the male, as well as by the entirely yellow coloration, dusky tibiae and tarsi, long propodeum, elongate crenulae, long endophragma and narrow forewing.

*Female.* Eyes finely setose. Mandibles well developed, with 2 denticles and a dorsal truncation; maxillary palpi 2-segmented, labial palpi 1-segmented. Antennal scape (Figure 1118) slender, about $4\frac{3}{4}$ to $5\frac{3}{4}$ times as long as wide, $1\frac{1}{5}$ to $1\frac{3}{10}$ times longer than the club; pedicel about $1\frac{2}{5}$ to $1\frac{2}{3}$ times as long as wide, distinctly ($1\frac{1}{7}$ to $1\frac{1}{3}$ times) longer than the third segment of the funicle; first funicular segment subglobular, as long as wide or a little (up to about $1\frac{1}{3}$ times) wider than long; second segment considerably shorter and somewhat wider than the first, symmetrical, about $1\frac{2}{5}$ to $1\frac{4}{5}$ times as wide as long; third funicular segment slightly asymmetrical, ventral aspect somewhat longer than the dorsal, somewhat longer than wide (up to $1\frac{1}{3}$ times), bearing 1 longitudinal sensillum; club $2\frac{1}{4}$ to $2\frac{1}{2}$ times as long as wide, a little over twice to $2\frac{3}{5}$ times longer and somewhat wider than the preceding segment, bearing 3–5 longitudinal sensilla.

Setae on thorax rather long, slender, pale, the longest ones barely visible under $\times 60$ magnification; setae on head and abdomen invisible even under $\times 120$ magnification. Vertex with 2 pairs of long setae along occipital margin, in addition to numerous shorter setae. Mesoscutum (Figure 1119) with 8–10 setae, each parapsis with 1–2, each axilla with 1 seta; scutellum with 4 setae, the discoid sensilla about equidistant from the two pairs, occasionally somewhat closer to the anterior pair. Frontovertex, pronotum, and mesonotal sclerites reticulate. Scutellum oval, about $\frac{7}{10}$ median length of mesoscutum. Metanotum (Figures 1119–1121) reticulate except on the sides, rather short centrally; anteromedian apodeme long, about $1\frac{2}{5}$ to $1\frac{2}{3}$ times median length of metanotum.

Propodeum (Figures 1119–1121) relatively long, with a distinct median salient, $4\frac{4}{5}$ to $6\frac{1}{2}$ times as long as the metanotum, nearly as long as the scutellum (0.88–0.98), faintly reticulate on the sides, delicately striated along posterior margin, reticulate on a trapezoidal central area, the cells longer than wide; crenulae 3 + 4 to 5 + 5, small, distinctly elongate, nonoverlapping.

Second abdominal tergite slightly transversely reticulate anteriorly on each side, smooth or very slightly sculptured centrally; tergites III–VII reticulate on the sides, with a few setae in a short transverse row on each reticulate area; third tergite transversely striated across center; tergites IV–VI with some indications of transverse striation; seventh tergite longitudinally striated between the lateral reticulate areas, fading toward center, with 2 (rarely 3) submedian setae; eighth tergite longitudinally striated, this fading toward center, with a transverse row of 4 setae between spiracles; syntergum triangular, faintly punctate, with 6–7 setae in a transverse row. Cerci about equidistant from posterior spiracles and from tip of abdomen, with 2 long setae and 1 short seta. Ovipositor shaft about $1\frac{2}{3}$ times as long as the middle tibia (1.60–1.70); ovipositor sheaths usually a little over $\frac{2}{5}$ length of middle tibia (0.40–0.44). Endophragma (Figure 1122) elongate, usually about $1\frac{1}{2}$ times as long as wide (in most species of *Aphytis* it is about as long as wide at base).

Mid-tibial spur a little shorter than the corresponding basitarsus.

Forewing (Figure 1123) very narrow, a little over 3 times as long as wide (up to $3\frac{1}{4}$ times); marginal fringe relatively long, ranging from about $\frac{1}{3}$ width of disk to nearly $\frac{1}{2}$ in minute specimens. Delta area with 24–39 setae in 4–5 rows, these not distinct from row of setae along posterior margin of wing, widely separated from a few setae below distal half of submarginal vein; costal cell with a few fine setae in a row along proximal three fifths or so, and 1 coarse seta near apex (the latter most often on the vein rather than in the cell itself). Submarginal vein bearing 2 coarse, rather long setae, the proximal one $\frac{3}{4}$ length of the distal or more, and 11–14 bullae. Marginal vein bearing 5–7 prominent, subequal, rather long, widely spaced setae along anterior margin, these about $1\frac{3}{5}$ times to fully twice as long as the setae in a row along center of vein.

Forewing nearly hyaline, proximal half very faintly infumated, more distinctly so below submarginal vein; a narrow fuscous streak along posterior margin of wing, distad of speculum.

General coloration entirely yellow; thoracic sterna immaculate; tegula faintly dusky; a short black streak at base of forewing, below tegula; plates around base of ovipositor faintly dusky. Antennal scape pale, rest of antenna rather uniformly dusky. Tibiae and tarsi of all legs rather strongly, contrastingly dusky, femora concolorous with body. Wing veins colorless.

Length 0.52–0.74 mm.

*Male*. Essentially similar to the female in structure, chaetotaxis, sculpture and coloration.

Antennal scape (Figure 1124) about $1\frac{1}{4}$ to $1\frac{2}{5}$ times longer than the club; pedicel sometimes distinctly shorter than the third segment of the funicle, not more than $1\frac{1}{5}$ times longer; each of the first two funicular segments bearing an elongate, spindle-shaped sensillum (Figures 1125, 1126); third funicular segment (Figure 1124) rather strongly asymmetrical, more so than in *japonicus* but less than in *mazalae*, $1\frac{1}{10}$ to $1\frac{3}{5}$ times as long as wide, bearing 1 longitudinal sensillum; club about 2 to $2\frac{1}{2}$ times as long as wide, about $1\frac{4}{5}$ to $2\frac{1}{5}$ times longer than the preceding segment, bearing 2–3 longitudinal sensilla, lacking any specialized sensory area on the ventral surface.

Metanotal apodeme up to nearly twice median length of metanotum. Propodeal crenulae 3 + 3 to 4 + 5, as in the female.

Genitalia (Figure 1127) about $\frac{4}{5}$ length of middle tibia (0.74–0.84); digiti 0.32–0.39 the combined length of aedeagus and apodemes.

Forewing (Figure 1128) about 3 times as long as wide; marginal vein bearing 4–8 prominent, subequal setae along anterior margin.

Length 0.43–0.67 mm.

Described from 28♀♀ and 25♂♂ (♀ holotype, ♂ allotype, and paratypes), reared by M. A. Ghani from *Anamaspis coniferarum* (Hall and Williams) on *Pinus roxburghii*, Murree Hills, Rawalpindi, Pakistan, August 23, 1966. Some of these specimens are headless or otherwise in poor shape.

Type series in the collection of the Division of Biological Control, University of California, Riverside.

**Notes.** *A. sensorius* is a minute species, similar in many respects to minute specimens of *aonidiae*, and was at first regarded as a member of the **mytilaspidis** group. However, it differs greatly from *aonidiae* in having a long propodeum, and therefore clearly belongs to the **chrysomphali** group. It can be best recognized by the peculiar sensilla on the first two funicular segments of the male, by the long endophragma, and by the distinctive pigmentation of the legs.

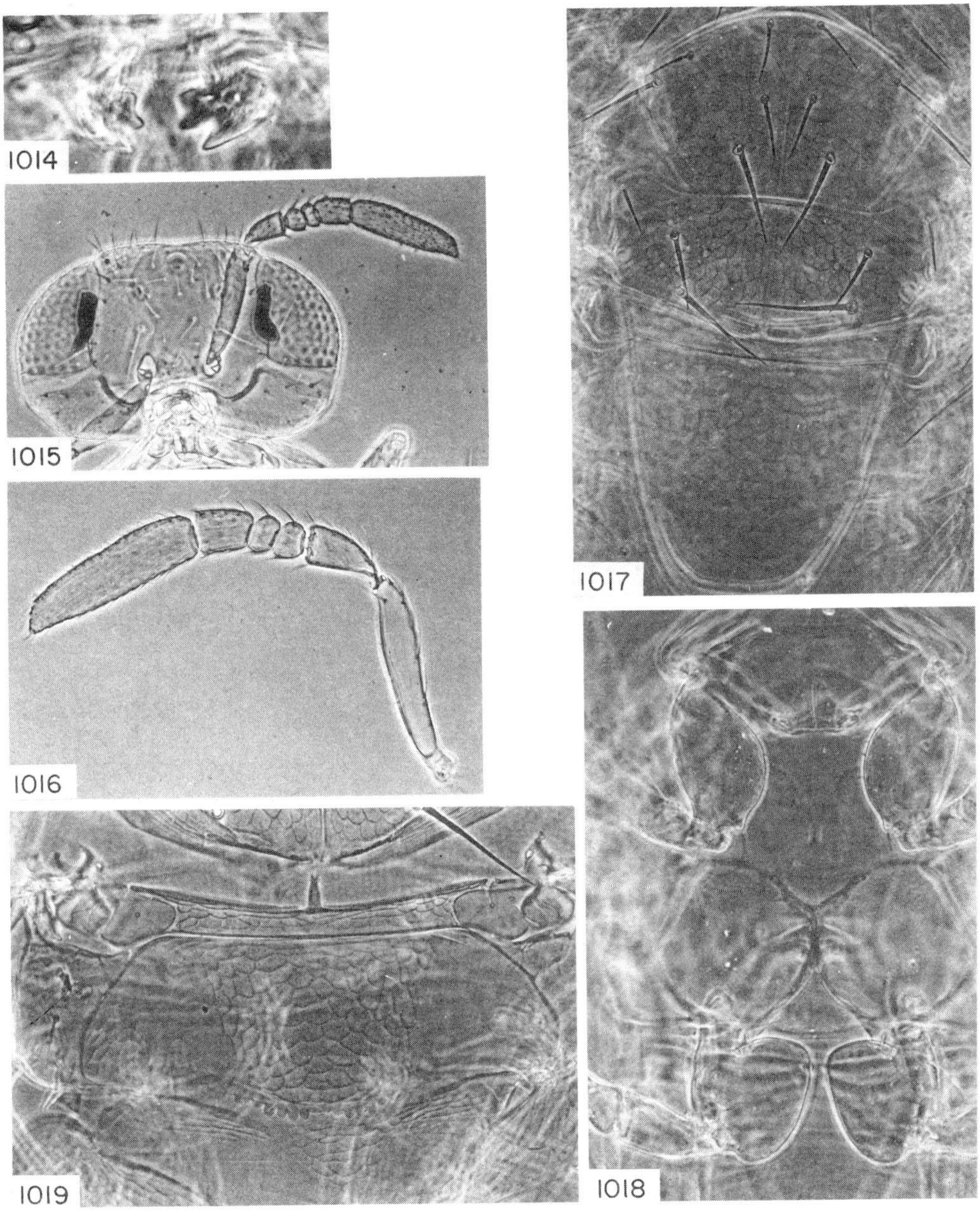


Figures 1014–1019. *Aphytis chrysomphali* (Mercet), ♀
1014. Mandibles. 1015. Head and antenna. 1016. Antenna. 1017. Thorax, propodeum and endophragma. 1018. Thoracic sterna. 1019. Metanotum, propodeum and second abdominal tergite.

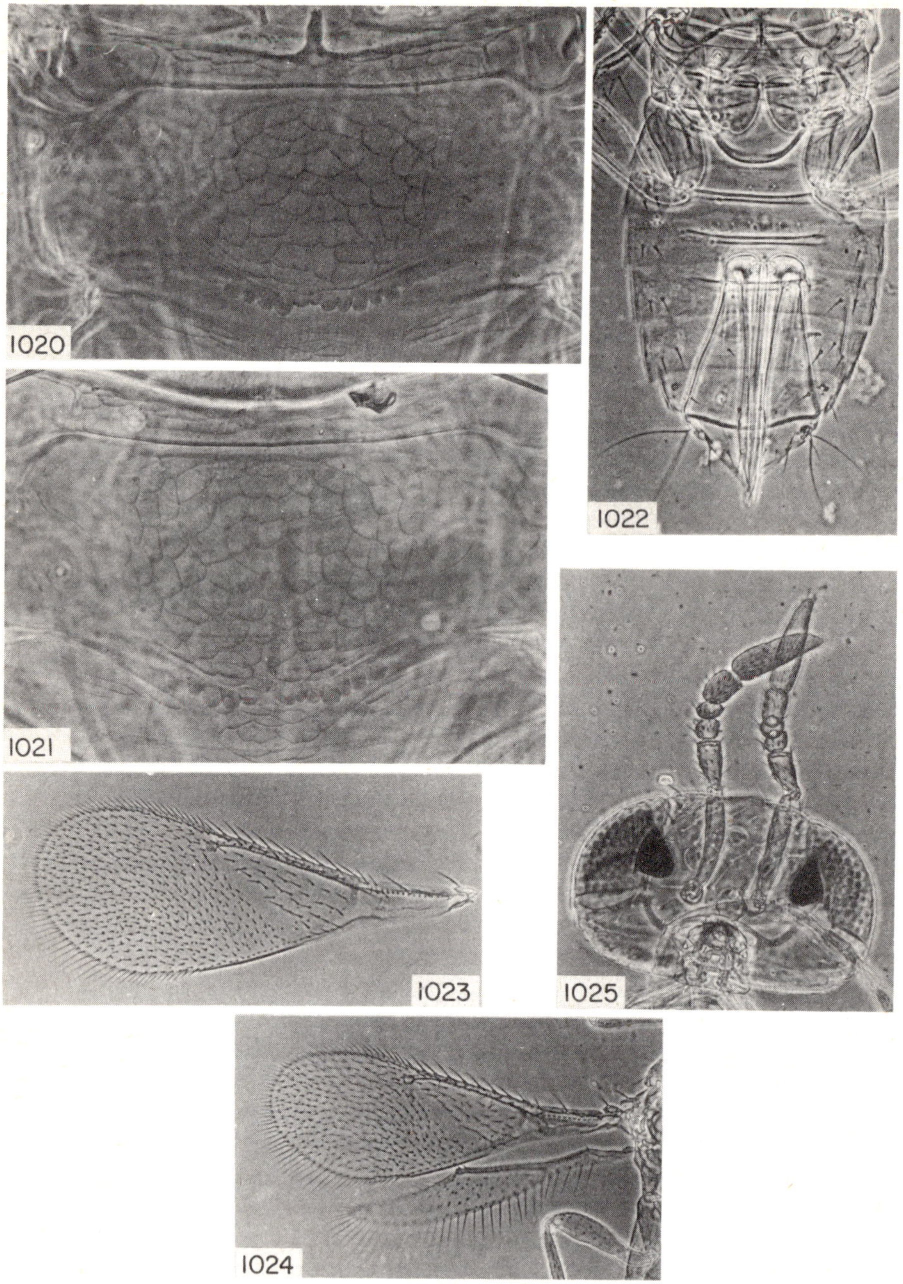


Figures 1020–1025. *Aphytis chrysomphali* (Mercet)
1020. ♀: Metanotum and propodeum. 1021. ♀: Propodeum; note slightly overlapping crenulae.
1022. ♀: Abdomen. 1023. ♀: Forewing. 1024. ♀: Wings. 1025. ♂: Head and antennae.

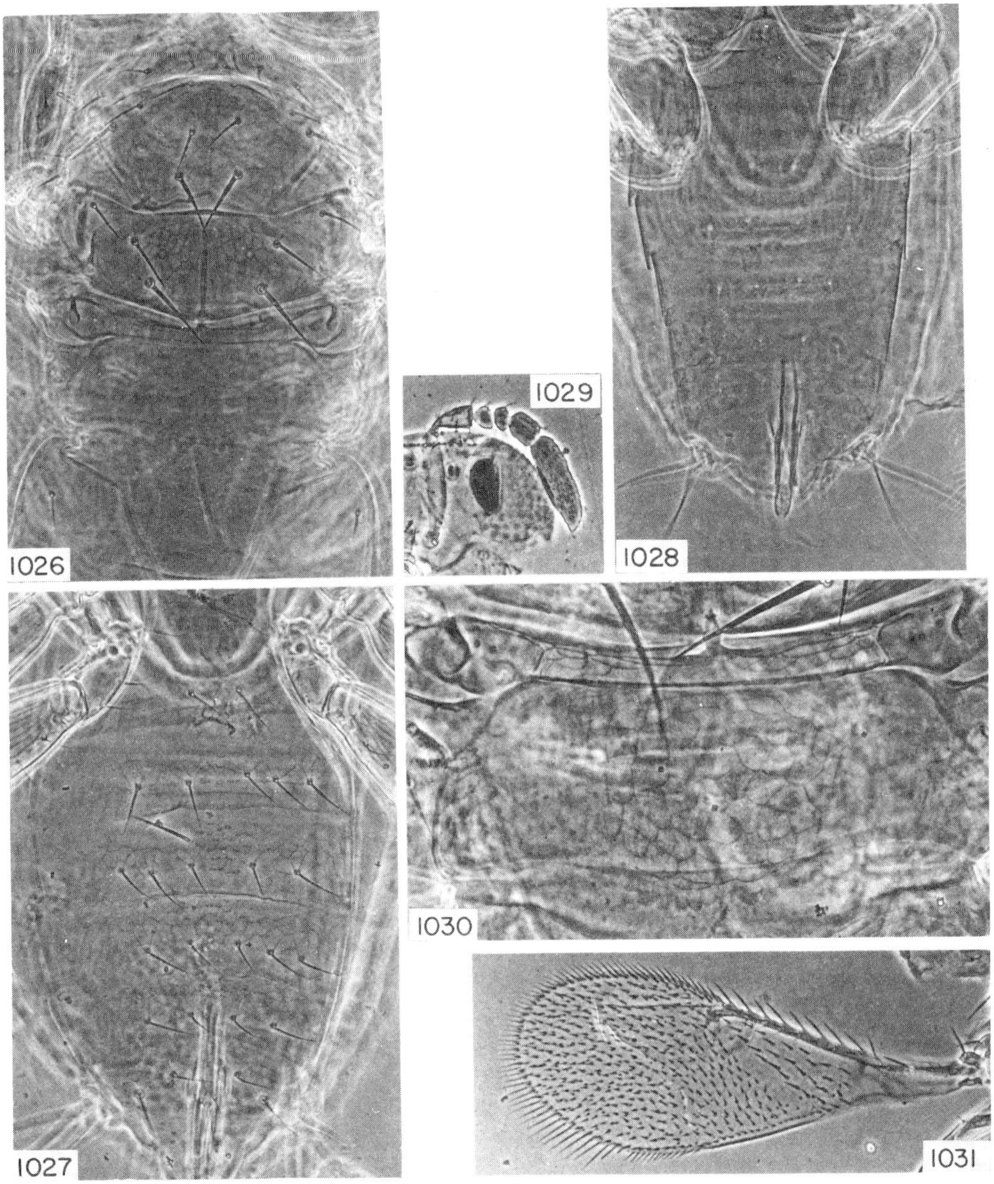


Figures 1026–1031. *Aphytis chrysomphali* (Mercet)
1026. ♂: Thorax, propodeum and base of gaster. 1027. ♂: Abdominal sterna, showing chaetotaxis and light stippling. 1028. ♂: Genitalia. 1029–1031. ♀ syntype of *quaylei* (Rust): 1029. Antenna. 1030. Metanotum and propodeum. 1031. Forewing.

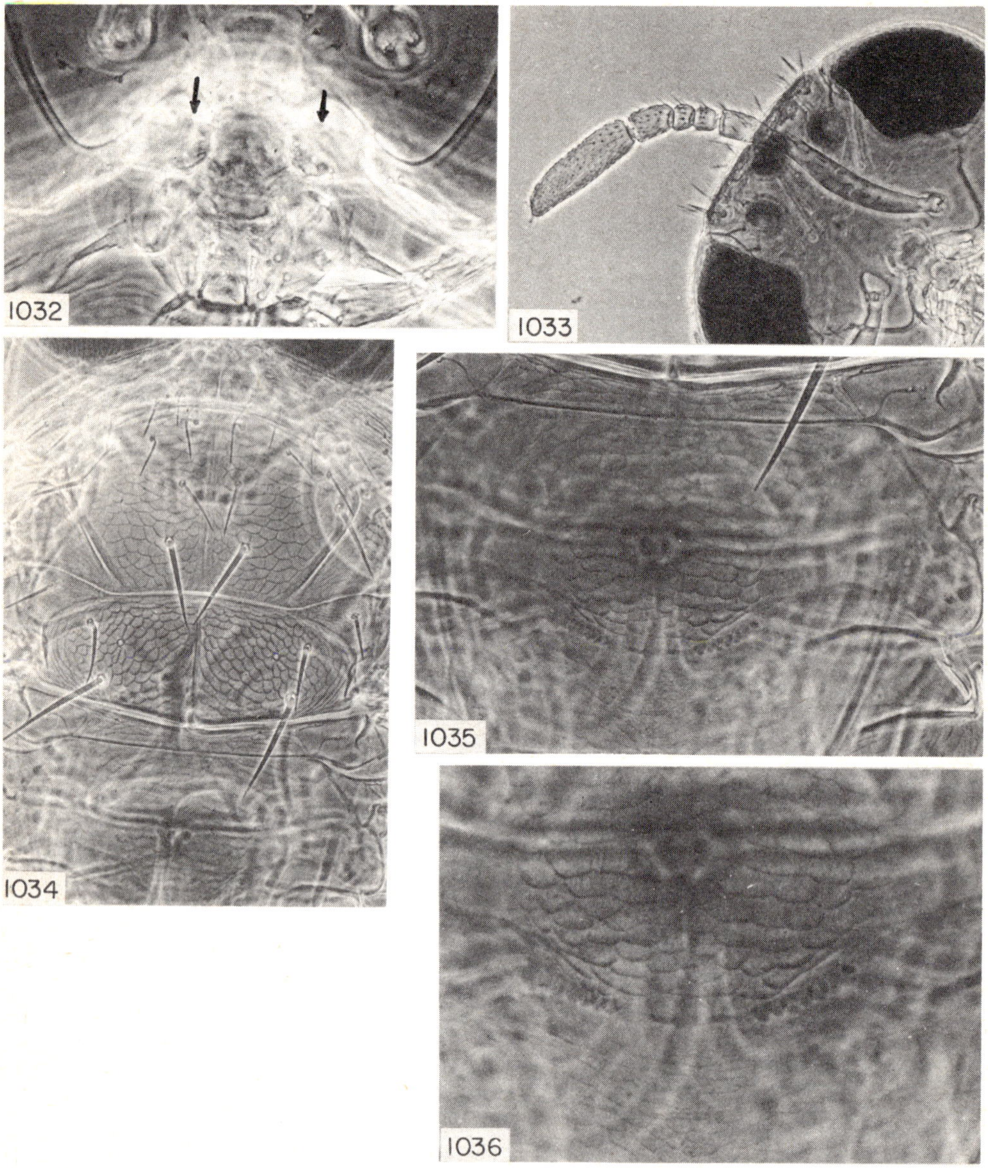


Figures 1032–1036. *Aphytis anneckei* DeBach and Rosen, ♀ (paratypes)
1032. Mandibles; note absence of distinct denticles. 1033. Antenna. 1034. Thorax and propodeum. 1035. Metanotum and propodeum. 1036. Propodeal crenulae.

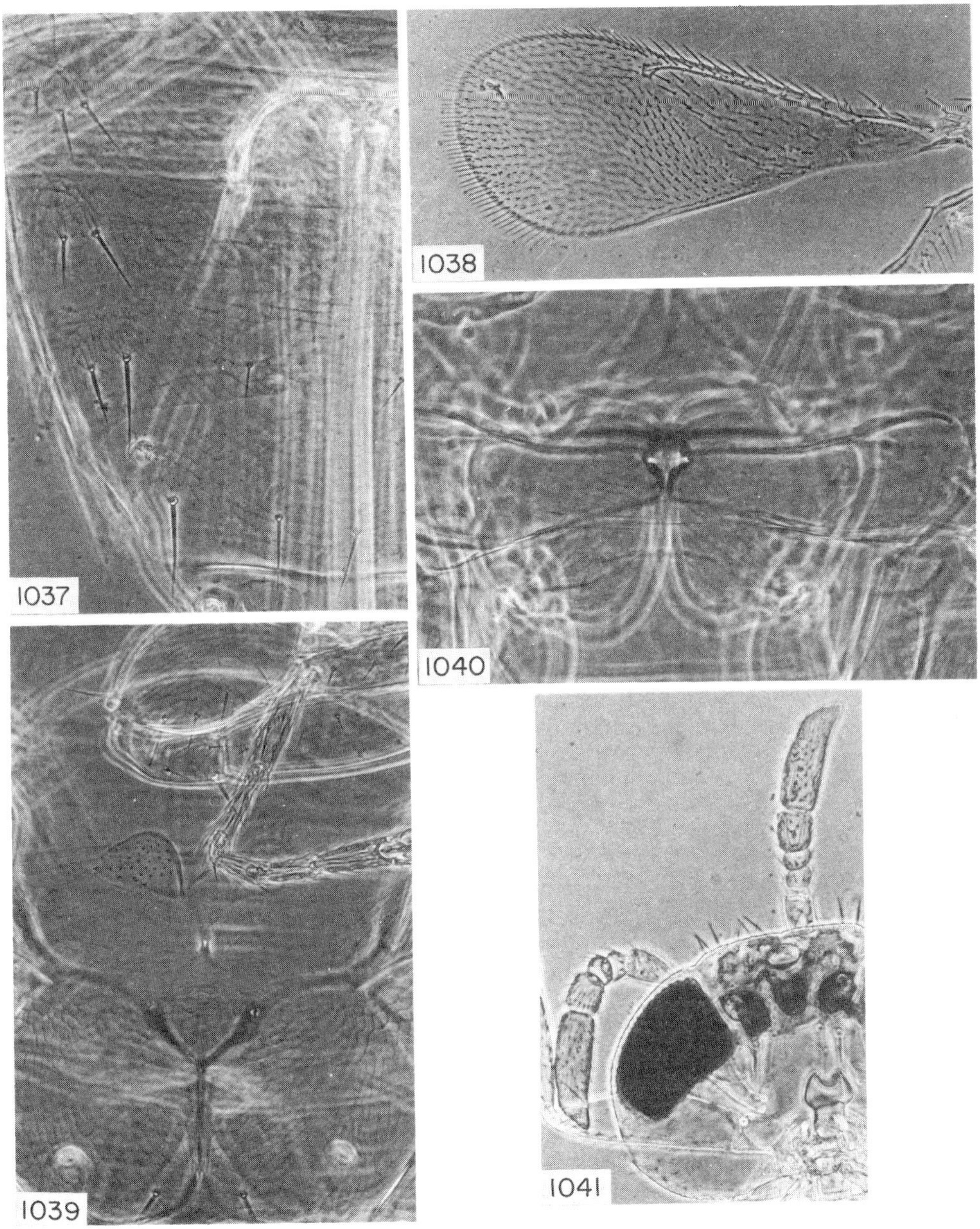


Figures 1037–1041. *Aphytis anneckei* DeBach and Rosen
1037. ♀: Abdominal tergites III–VII, showing sculpture and chaetotaxis (holotype). 1038. ♀: Forewing (holotype). 1039. ♀: Mesosternum (paratype). 1040. ♀: Metasternum (paratype). 1041. ♂: Antennae (allotype).

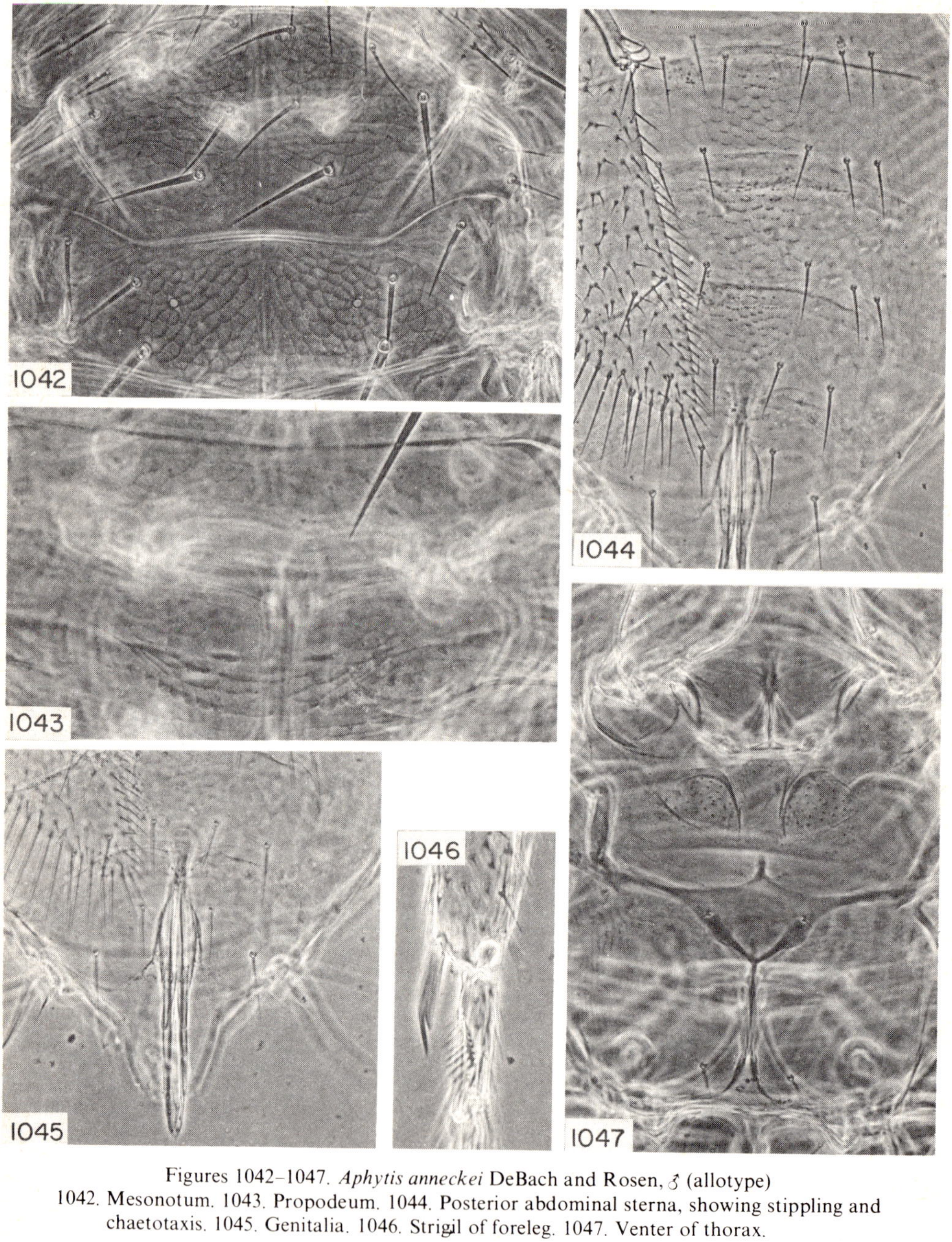


Figures 1042–1047. *Aphytis anneckei* DeBach and Rosen, ♂ (allotype)
1042. Mesonotum. 1043. Propodeum. 1044. Posterior abdominal sterna, showing stippling and chaetotaxis. 1045. Genitalia. 1046. Strigil of foreleg. 1047. Venter of thorax.

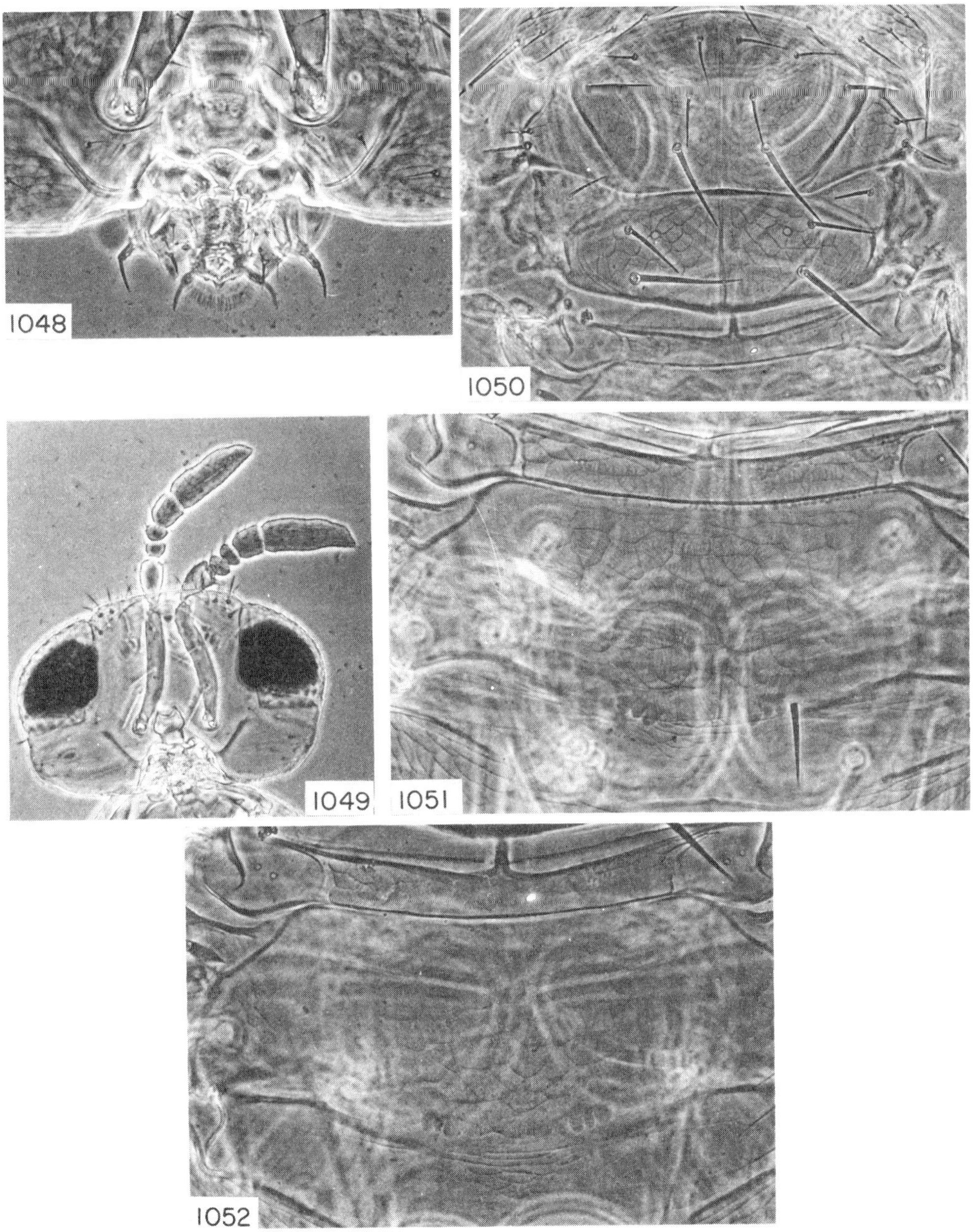


Figures 1048–1052. *Aphytis debachi* Azim, ♀
1048. Face and mouthparts; note triangular lower denticle of mandible. 1049. Head and antennae. 1050. Thorax. 1051. Metanotum and propodeum (syntype). 1052. Metanotum and propodeum.

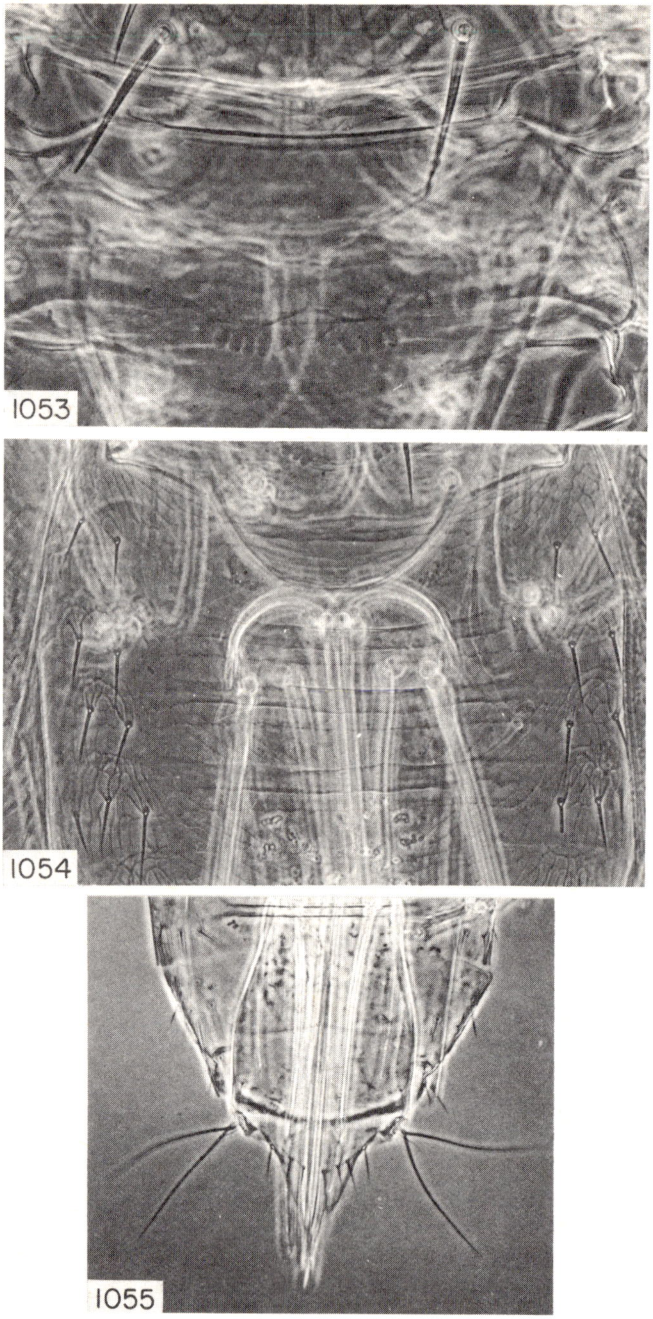


Figures 1053–1055. *Aphytis debachi* Azim, ♀
1053. Metanotum and propodeum. 1054. Abdominal tergites II–VI. 1055. Posterior abdominal tergites.

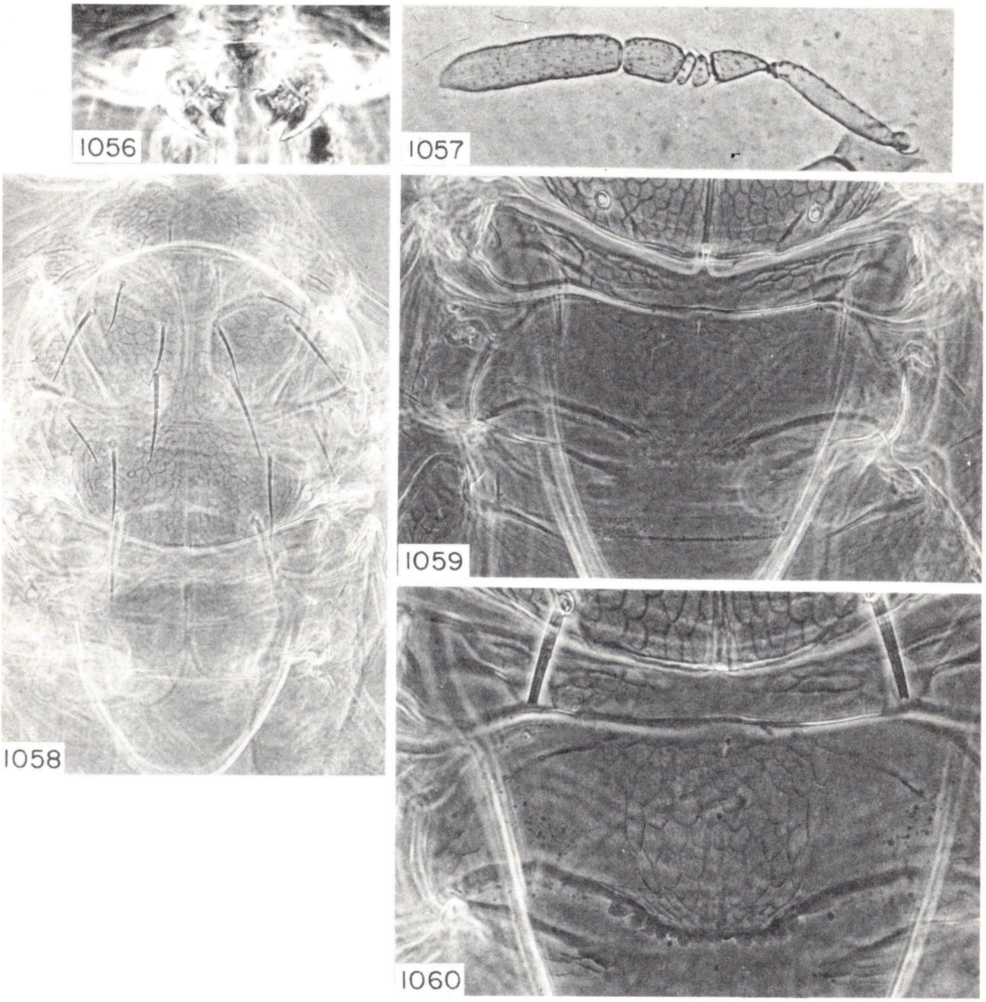


Figures 1056–1060. *Aphytis cylindratus* Compere, ♀
1056. Mandibles. 1057. Antenna (syntype). 1058. Thorax and propodeum. 1059. Metanotum, propodeum and base of gaster. 1060. Metanotum and propodeum.

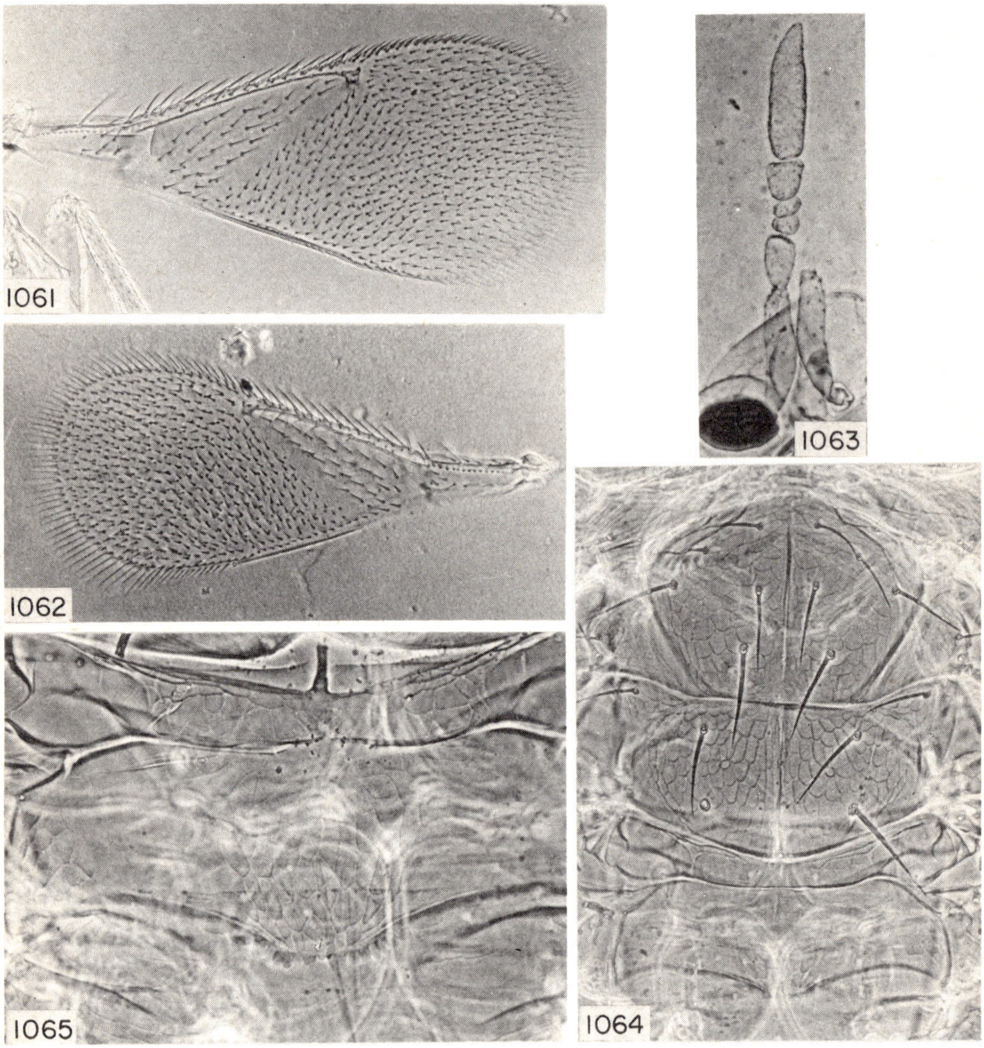


Figures 1061–1065. *Aphytis cylindratus* Compere
1061, 1062. ♀: Forewings. 1063. ♂: Antenna (syntype). 1064. ♂: Thorax and propodeum.
1065. ♂: Metanotum and propodeum.

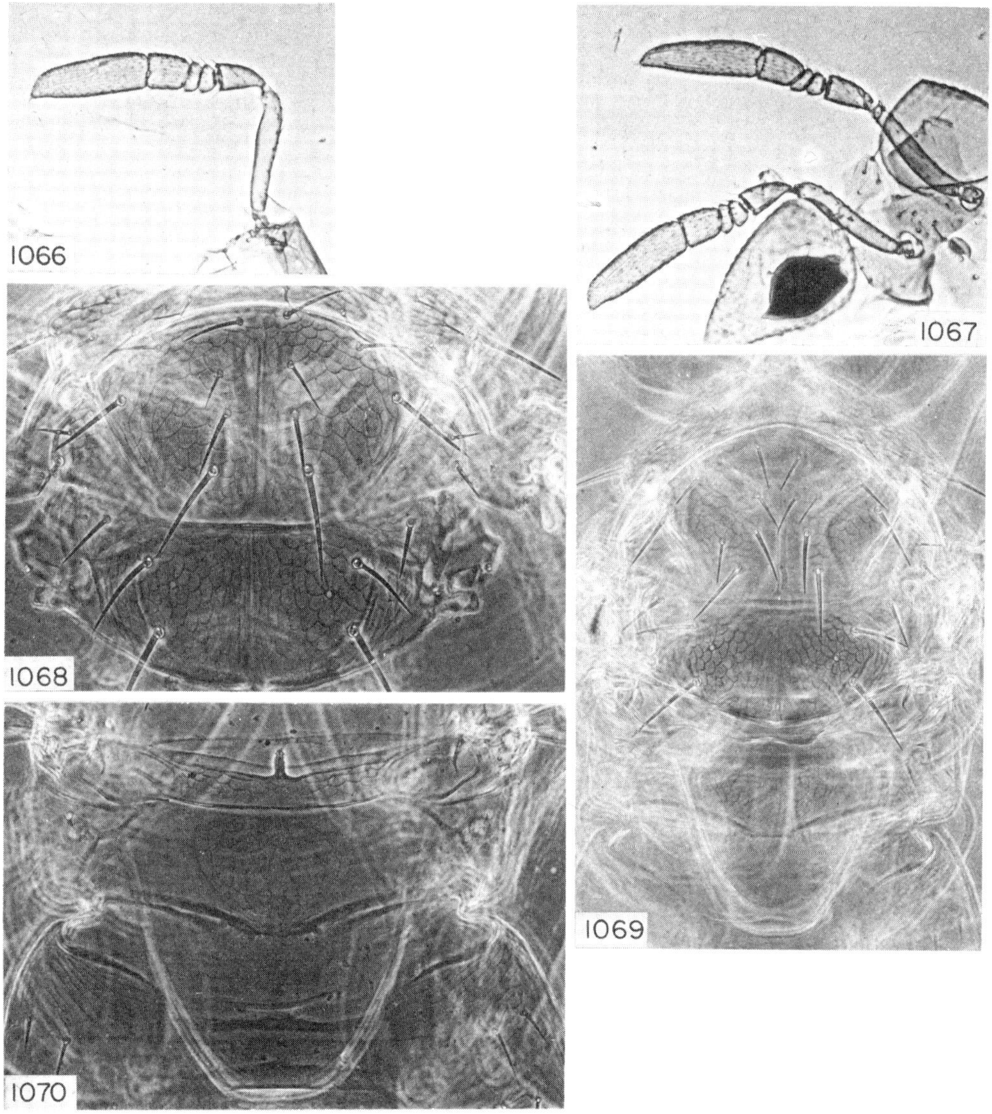


Figures 1066–1070. *Aphytis lepidosaphes* Compere, ♀
1066, 1067. Antennae. 1068. Mesontoum; note 4 pairs of submedian setae on mesoscutum. 1069. Thorax and propodeum; note 14 mesoscutal setae. 1070. Metanotum, propodeum and base of gaster.

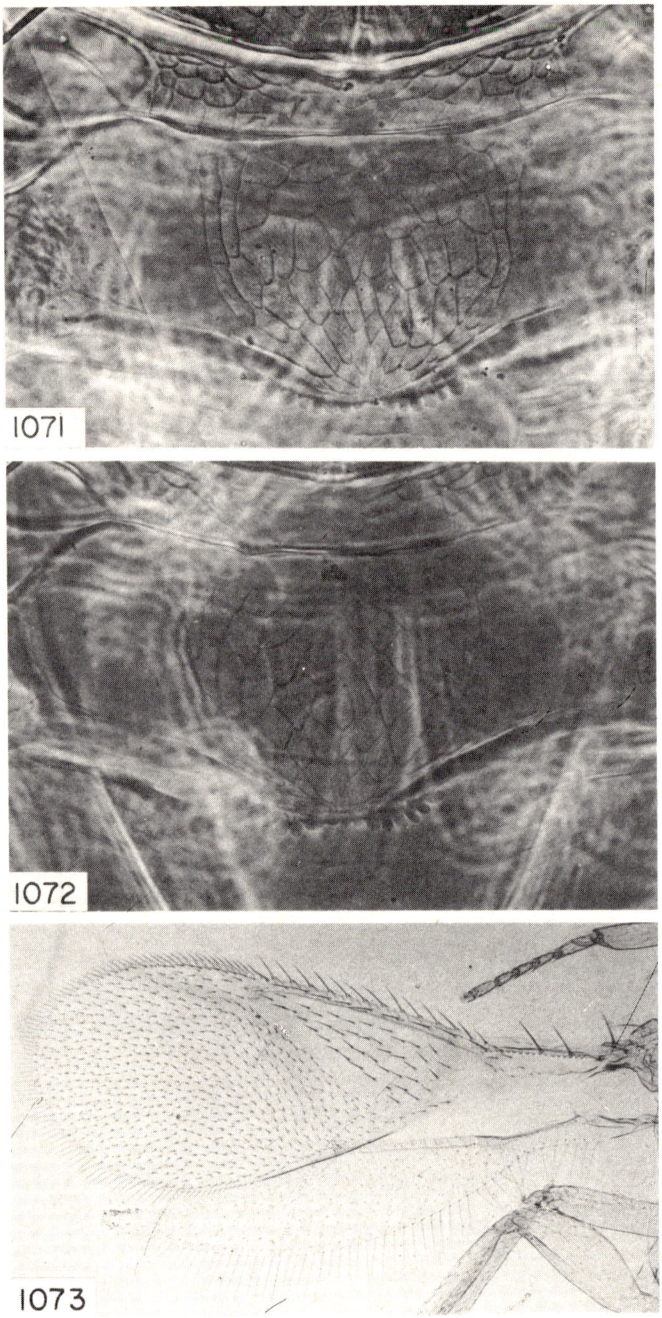


Figures 1071–1073. *Aphytis lepidosaphes* Compere, ♀
1071, 1072. Metanotum and propodeum. 1073. Wings.

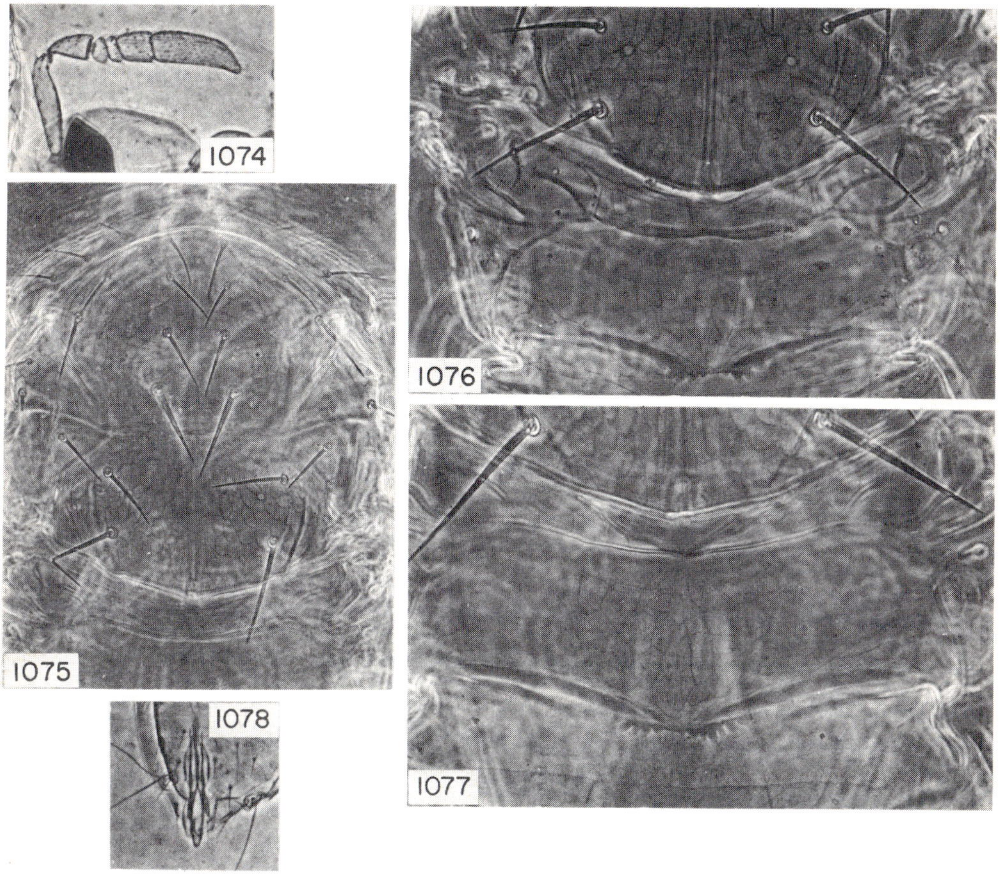


Figures 1074–1078. *Aphytis lepidosaphes* Compere, ♂
1074. Antenna (syntype). 1075. Thorax. 1076. Scutellum, metanotum and propodeum. 1077. Metanotum and propodeum. 1078. Genitalia (syntype).

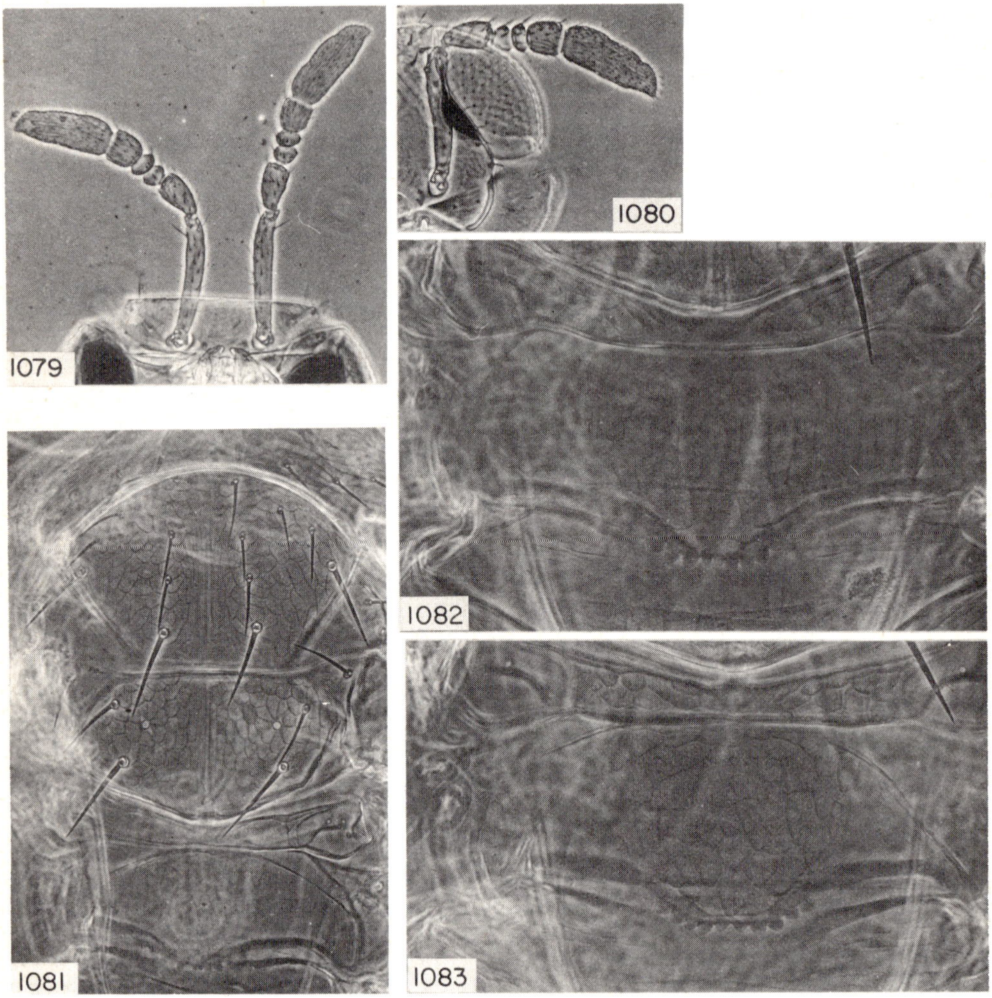


Figures 1079–1083. *Aphytis immaculatus* Compere, ♀
1079, 1080. Antennae. 1081. Thorax and propodeum. 1082, 1083. Metanotum and propodeum.

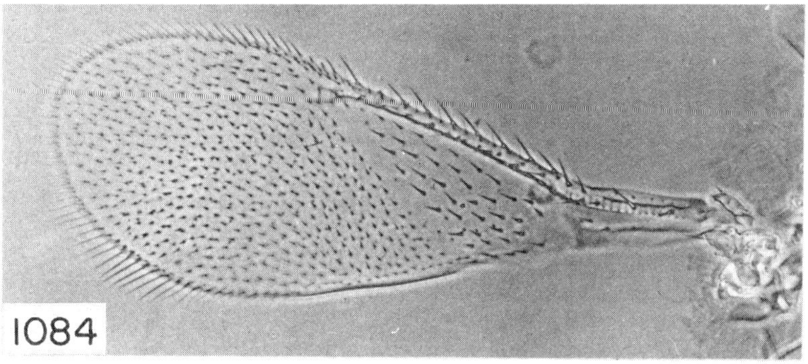


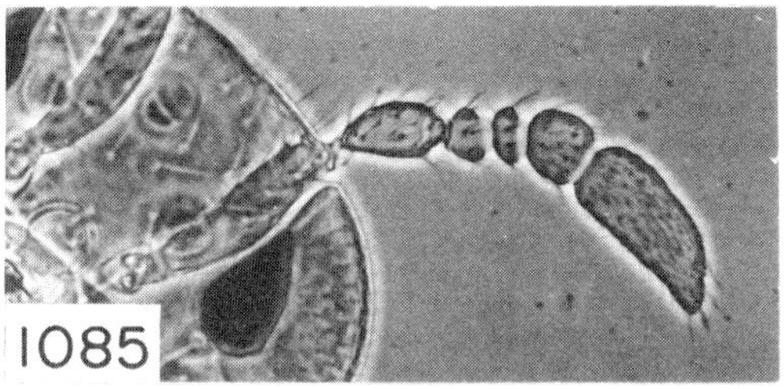


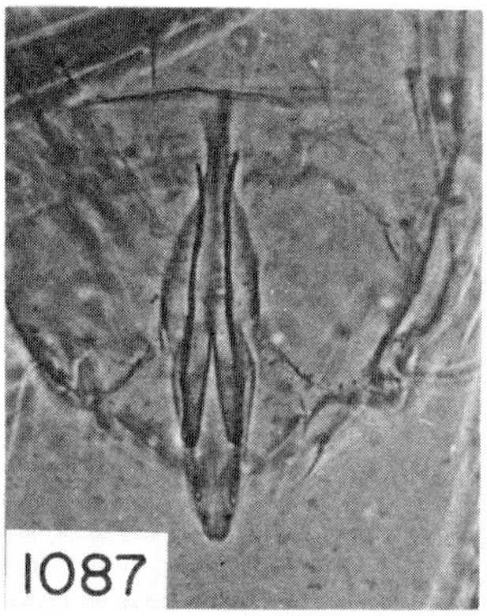


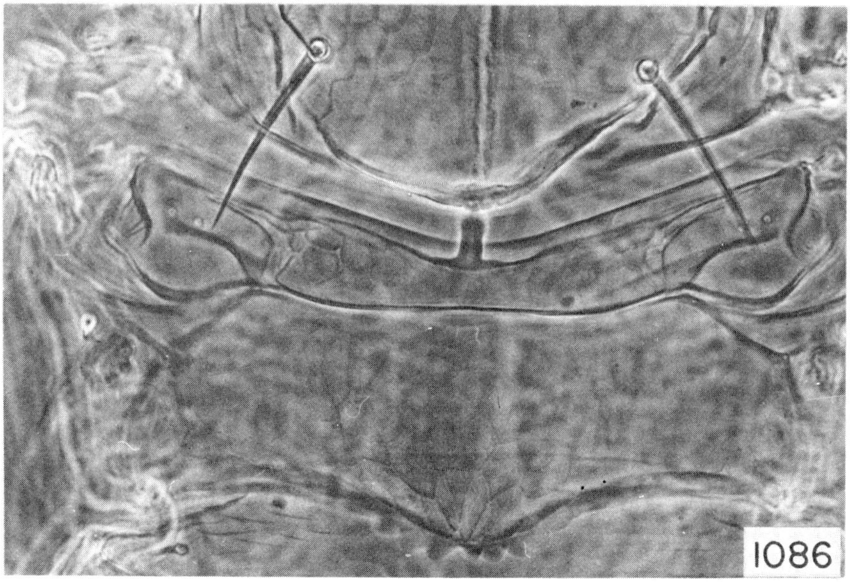


Figures 1084–1087. *Aphytis immaculatus* Compere
1084. ♀: Forewing. 1085. ♂: Antenna. 1086. ♂: Metanotum and propodeum. 1087. ♂: Genitalia.

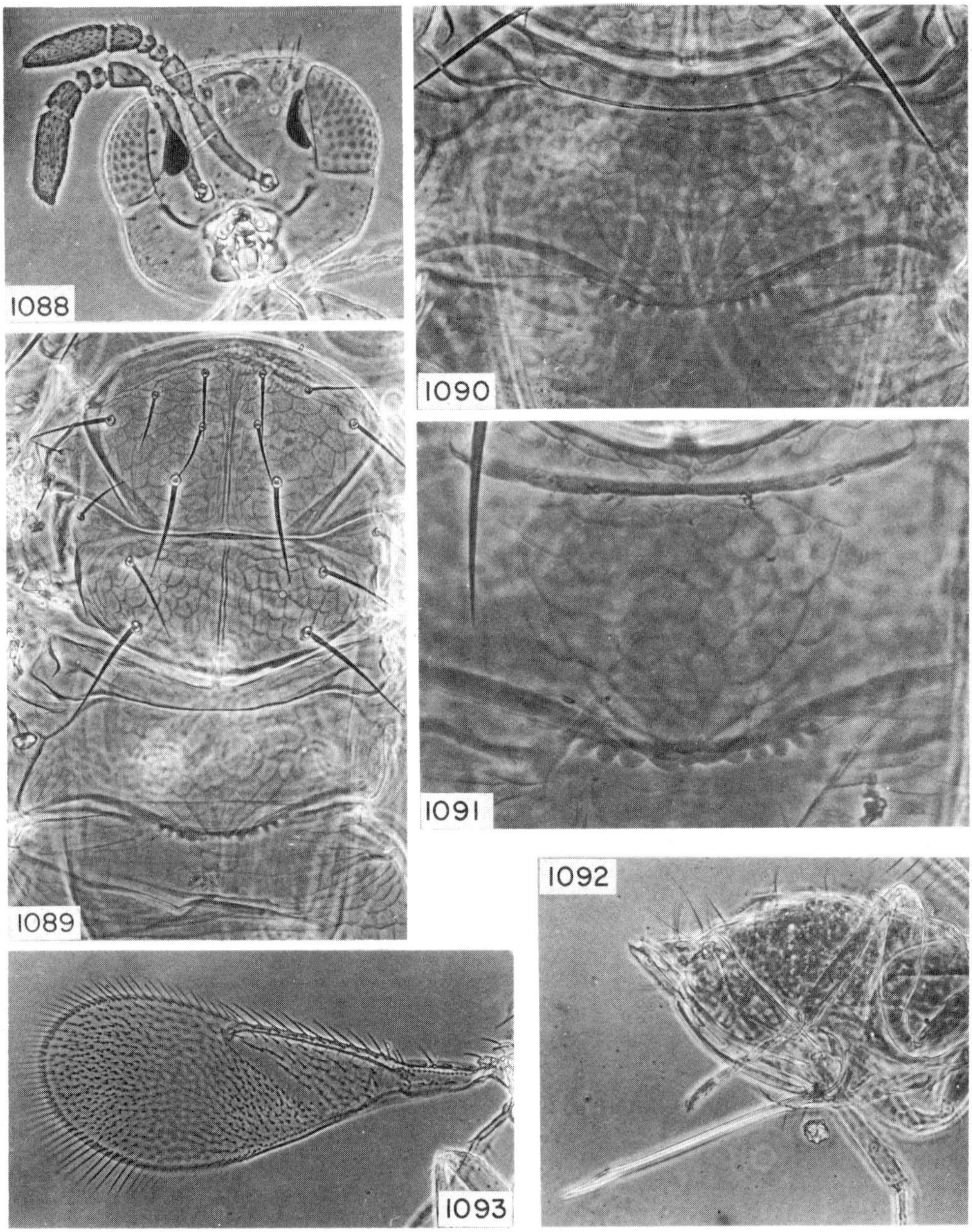


Figures 1088–1093. *Aphytis japonicus* DeBach and Azim, ♀
1088. Head and antennae. 1089. Thorax, propodeum and base of gaster. 1090, 1091. Metanotum and propodeum. 1092. Gaster, lateral view, with ovipositor shaft exserted. 1093. Forewing.

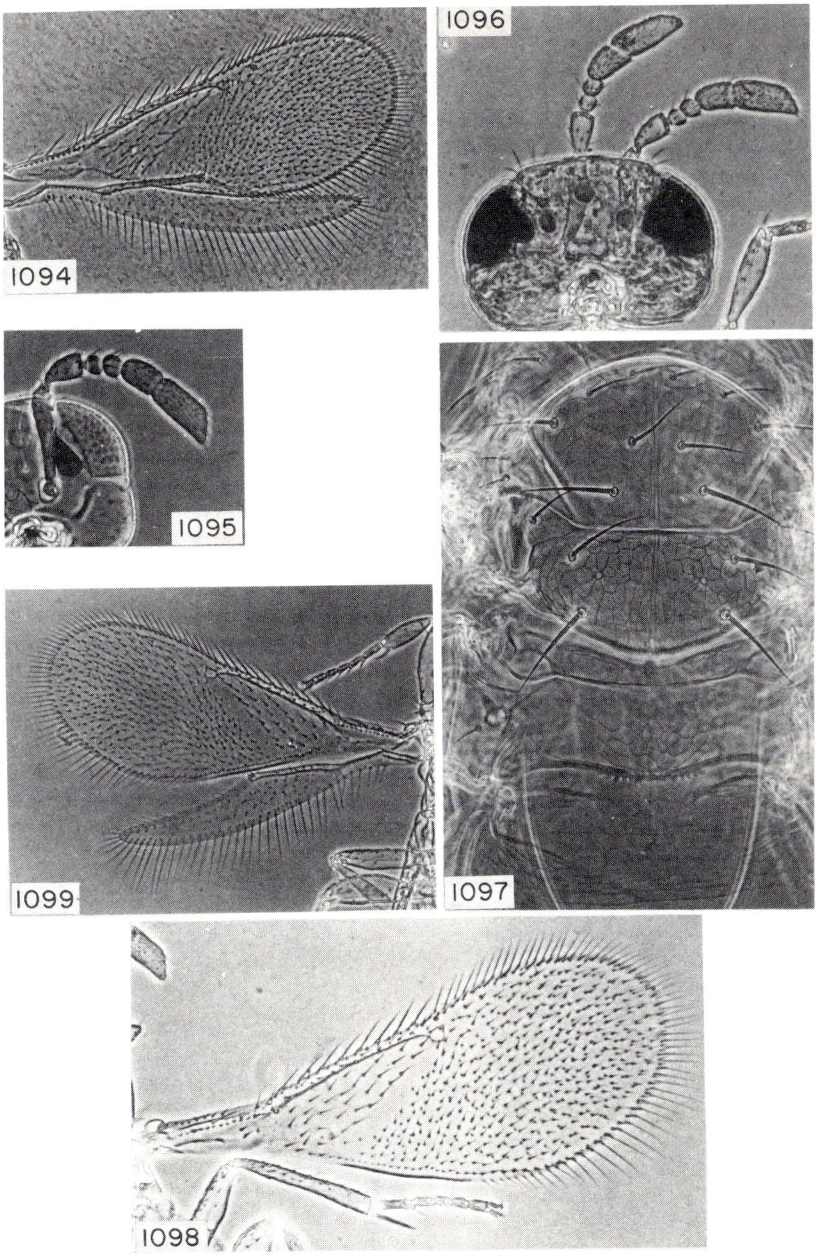


Figures 1094–1099. *Aphytis japonicus* DeBach and Azim
1094. ♀: Wings. 1095. ♂: Antenna. 1096. ♂: Head and antennae; note the somewhat asymmetrical third funicular segment. 1097. ♂: Thorax, propodeum and base of gaster. 1098. ♂: Forewing and middle leg; note long mid-tibial spur. 1099. ♂: Wings.

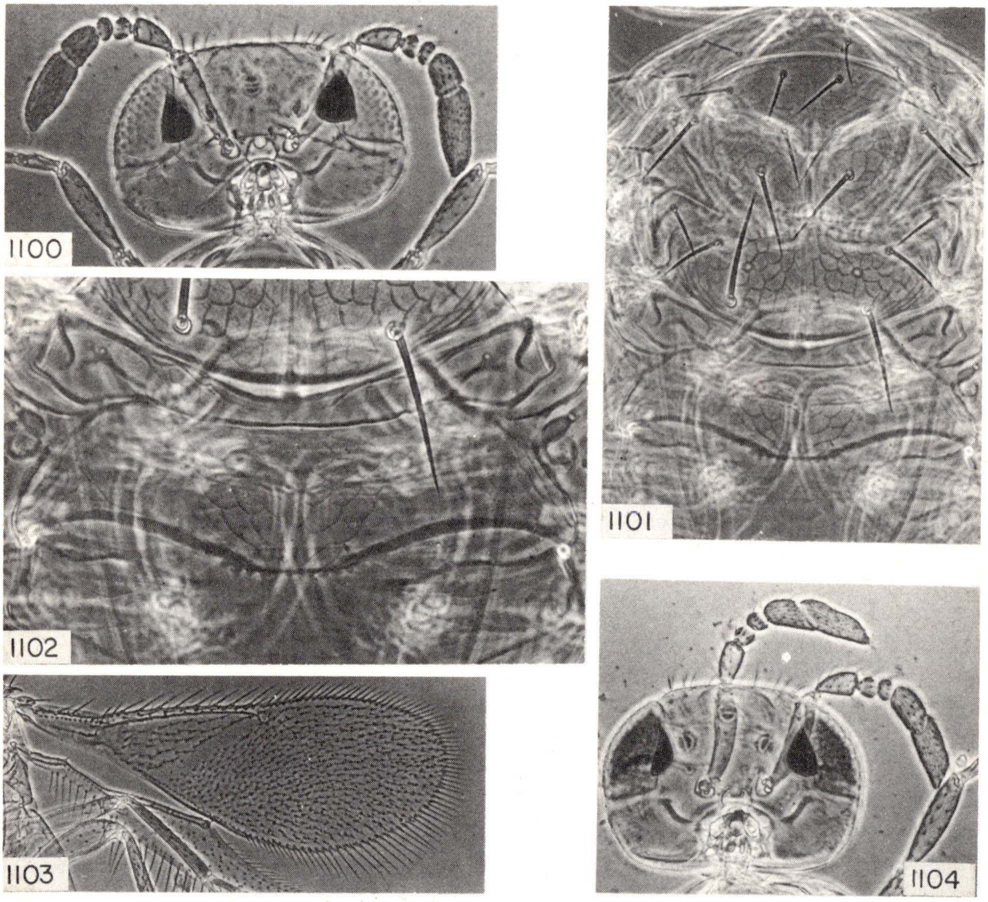


Figures 1100–1104. *Aphytis mazalae* DeBach and Rosen
1100. ♀: Head and antennae (holotype). 1101. ♀: Thorax and propodeum (holotype). 1102. ♀: Metanotum and propodeum (holotype). 1103. ♀: Forewing (holotype). 1104. ♂: Head and antennae; note asymmetrical third funicular segment (allotype).

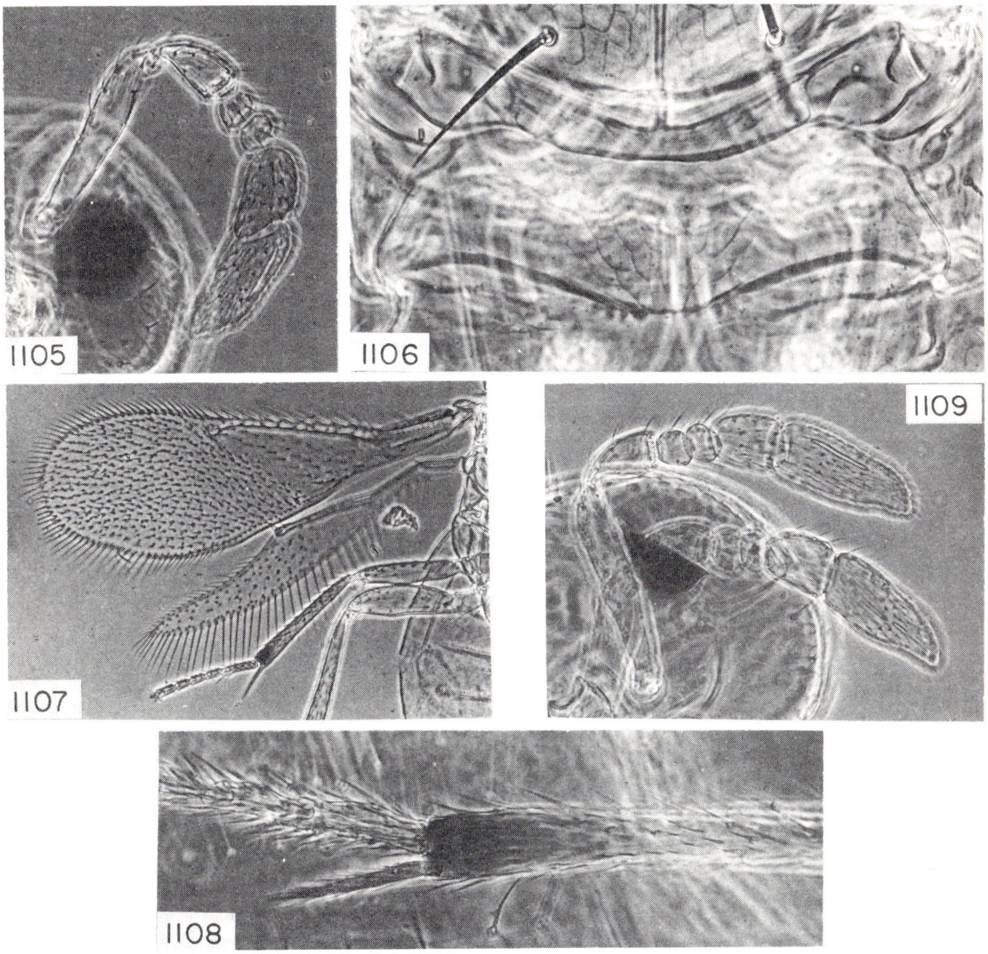


Figures 1105–1109. *Aphytis mazalae* DeBach and Rosen
1105. ♂: Antenna, showing pronounced asymmetry of third funicular segment (paratype). 1106. ♂: Metanotum and propodeum (allotype). 1107. ♂: Wings and middle leg; note distinctive pigmentation of middle tibia (allotype). 1108. ♂: Middle tibia, showing pigmentation and spur (paratype). 1109. ♀ (ex *Aonidiella aurantii*, Pakistan): antenna, showing slight asymmetry of third funicular segment.

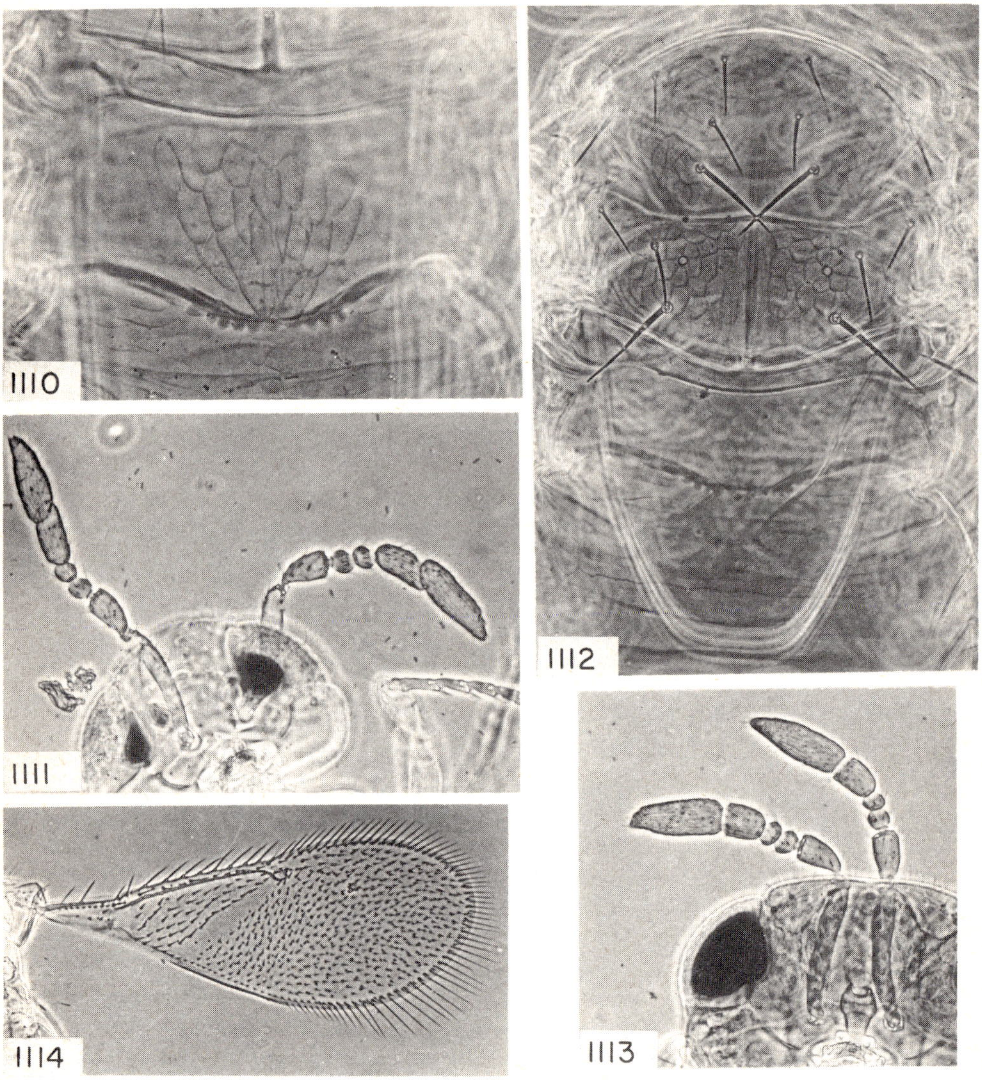


Figures 1110–1114. *Aphytis* ?*mazalae*, Pakistan material
1110. ♀ (ex *Pinnaspis strachani*): Metanotum, propodeum and base of gaster. 1111. ♂ (ex *Pinnaspis strachani*): Antennae, showing asymmetry of third funicular segment and club. 1112. ♀ (ex *Aonidiella citrina*): Thorax, propodeum and base of gaster. 1113. ♀ (ex *Aonidiella orientalis*): Head and antennae. 1114. ♀ (ex *Aonidiella orientalis*): Forewing.

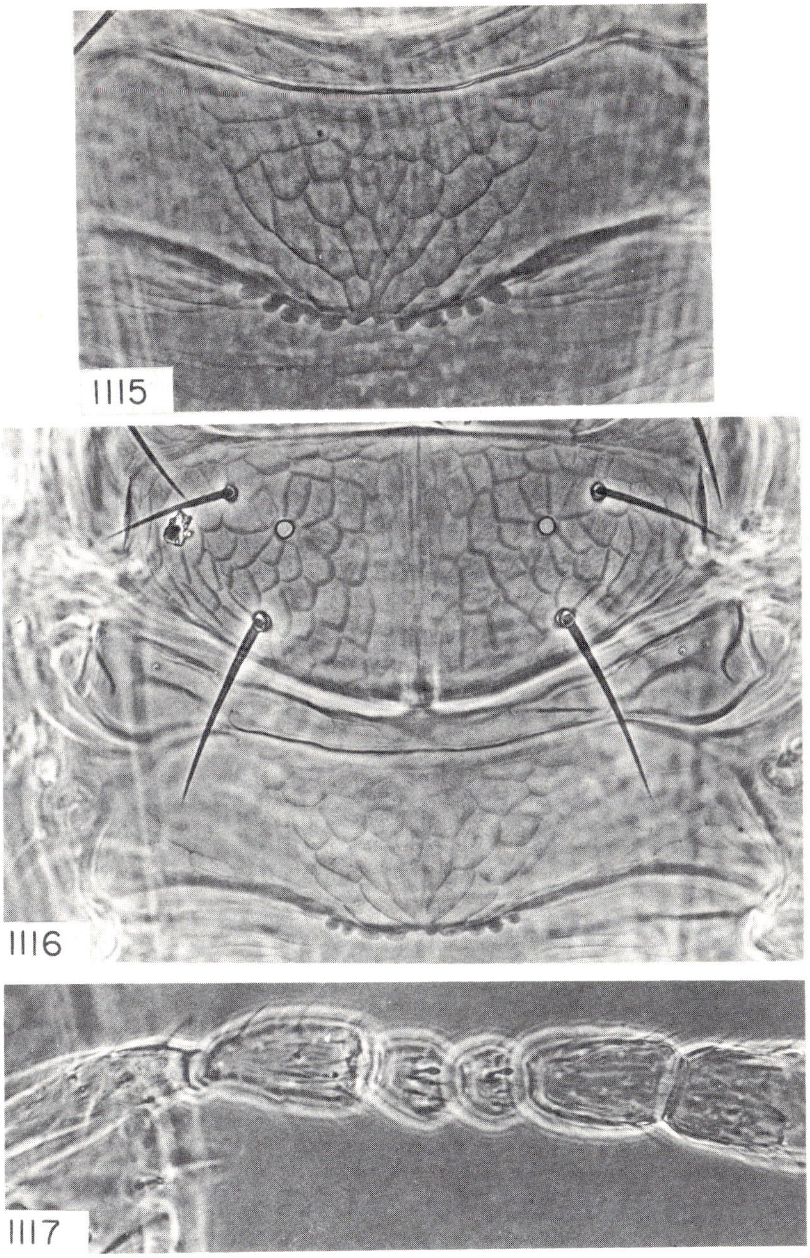


Figures 1115–1117. *Aphytis ?mazalae*, Pakistan material
1115. ♀ (ex *Aonidiella orientalis*): Propodeum. 1116. ♀ (ex *Aonidiella* sp.): Scutellum, metanotum and propodeum. 1117. ♂ (ex *Aonidiella* sp.): Antenna, showing spatulate sensilla on first two funicular segments; such organs are not present in the type series of *mazalae*.

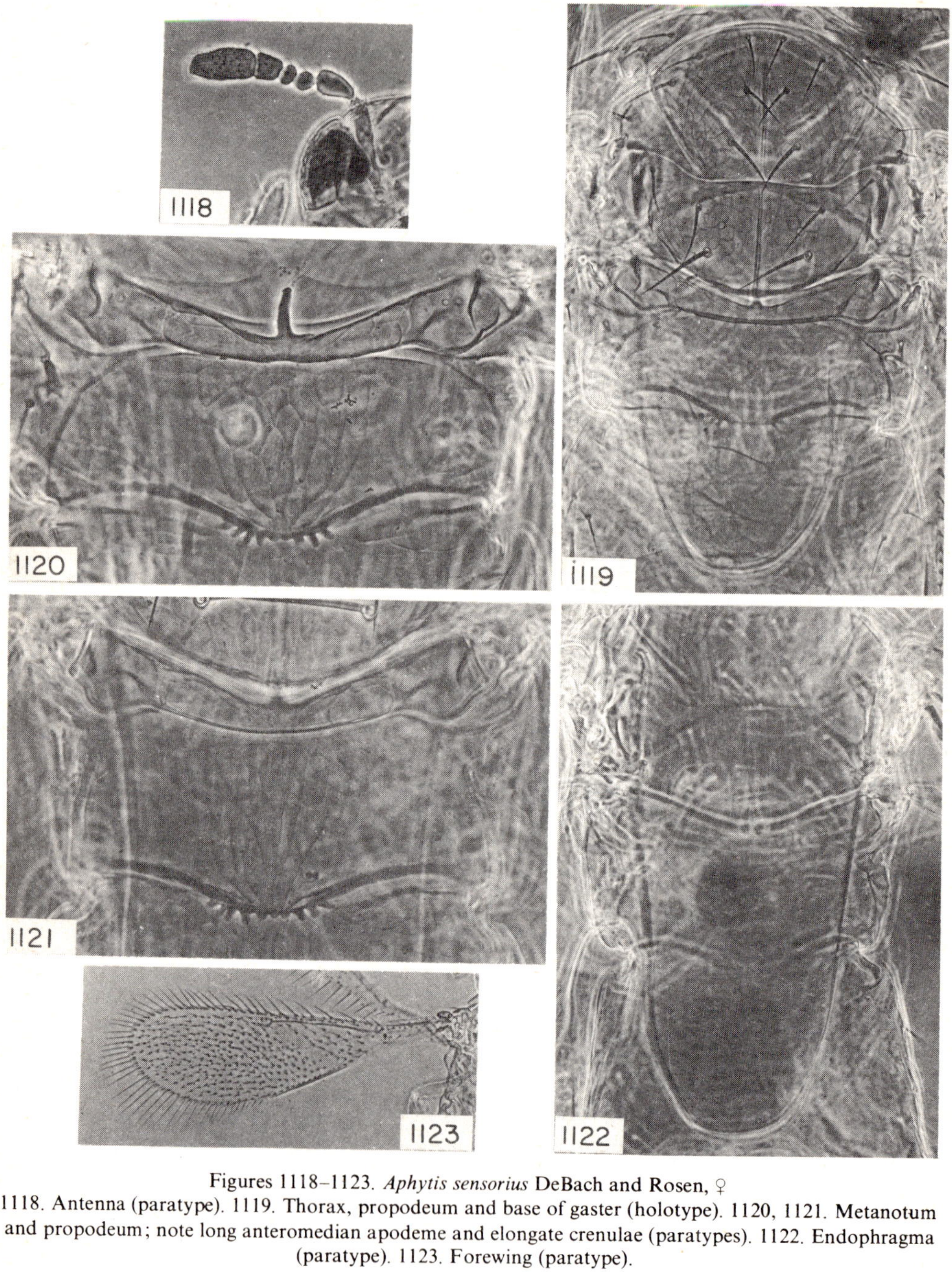


Figures 1118–1123. *Aphytis sensorius* DeBach and Rosen, ♀
1118. Antenna (paratype). 1119. Thorax, propodeum and base of gaster (holotype). 1120, 1121. Metanotum and propodeum; note long anteromedian apodeme and elongate crenulae (paratypes). 1122. Endophragma (paratype). 1123. Forewing (paratype).

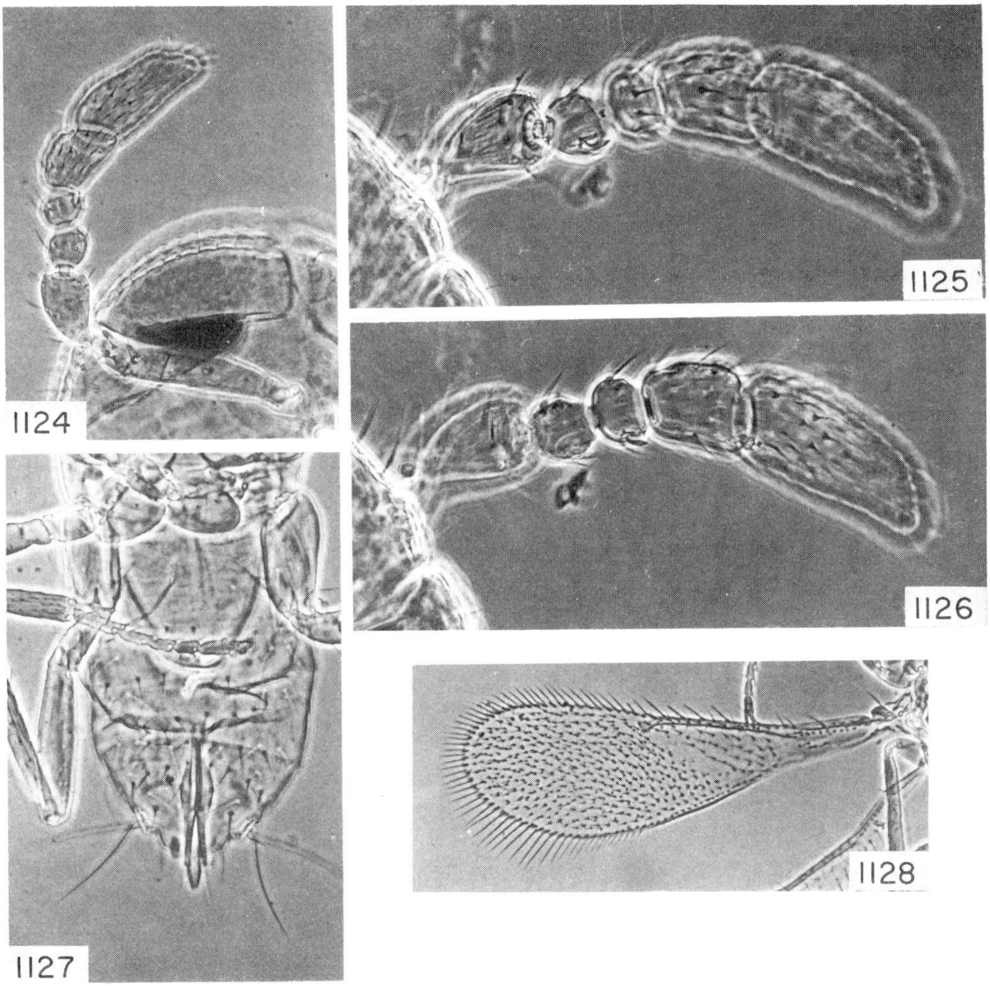


Figures 1124–1128. *Aphytis sensorius* DeBach and Rosen, ♂
1124. Antenna; note asymmetrical third funicular segment (paratype). 1125. Antenna; note sensillum on first funicular segment (paratype). 1126. Antenna; note sensillum on second funicular segment (same specimen as in Figure 1125, different focal plane). 1127. Venter of abdomen and genitalia (allotype). 1128. Forewing (allotype).

# VII. THE FUNICULARIS GROUP

This small, aberrant group of small species is characterized by the reduction of antennal segmentation and mouthparts in both sexes: Antennae 5-segmented (1121), lacking the third funicular segment; mandibles minute or absent; maxillary palpi 1-segmented. General coloration yellow; wings uniformly setose, hyaline or nearly so. The propodeum is short, not exceeding 4 times length of the metanotum or $\frac{3}{4}$ length of scutellum; the shape of the propodeal crenulae and other morphological characters vary considerably among the members of this group.

The **funicularis** group at present comprises the following 4 species: *funicularis* Compere, *gordoni* DeBach and Rosen, *theae* (Cameron) and *ulianovi* Girault. Surprisingly, it appears to be quite widely distributed, with two Oriental members, one Ethiopian and one Australian member.

Girault (1932) referred *ulianovi*—an Australian species with 5-segmented antennae—to the genus *Aphytis* because he apparently believed this genus to have 5-segmented antennae, as originally described by Howard (1900) (see p. 8). We have examined the types of *ulianovi*, as well as abundant recent material referable to this species, and consider it to be a bona fide member of *Aphytis*, as understood here.

Compere (1955) described a similar species from Africa, *funicularis*, in the genus *Aphytis*, and made the following comments: "In conventional systematics this species is out of place in *Aphytis*, having only two segments in the funicle instead of the usual three. However, in my opinion, *funicularis* is more closely related to the common yellow species of *Aphytis* than are many species that cannot be distinguished by diagnostic characters of generic value." We agree with Compere's conclusion. Not only does *funicularis* possess all the main diagnostic characters of *Aphytis*, including the well-developed, crenulate propodeum, but the female has a vestigial transverse suture on the antennal club which, if complete, would have separated a normal third funicular segment, thus clearly demonstrating an affinity to the usual species of *Aphytis* even in this respect.

DeBach and Rosen (1976b) recently described an Oriental species, *gordoni*, as another aberrant member of *Aphytis*, closely related to *funicularis*. Finally, with the resurrection of *theae* as yet another Oriental species of *Aphytis* with 5-segmented antennae, Rosen and DeBach (1977a) established the **funicularis** group to accommodate these four aberrant species.

This aberrant group differs markedly from all other species groups in *Aphytis* and does not exhibit any obvious affinities to any particular group. Not much can be said at present regarding the possible origin or phylogenetic status of this peculiar group. The **chilensis** group, with aberrant male antennae exhibiting marked reduction of the first two funicular segments and 6-segmented female antennae, is obviously unrelated. Perhaps the fact that three of the four known members of the **funicularis** group possess specialized sensilla on the antennal scape of the male—a primitive characteristic found in other aphelinine genera and in the primitive species groups but not in the more advanced groups of *Aphytis*—may be interpreted to indicate that this group split from the mainstream of the genus at an early evolutionary stage. Reduction of mouthparts is obviously not a primitive characteristic and must have occurred subsequently. The host range of the **funicularis** group, limited as it is to armored scale insects of the subfamily Diaspidinae (see p. 115), tends to support the hypothesis that this is a primitive group. Interestingly, two Oriental species, *malayensis* and *mandalayensis*, have 5-segmented antennae similar to those of the **funicularis** group but differ markedly in being heavily pigmented and in having well-developed mouthparts, and are currently listed as related to the **vittatus** group. It is possible that they represent an ancestral form from which the **funicularis** group may have evolved. Discovery of additional species will hopefully shed more light on this interesting topic.

Very little is known about the biology of the four members of the **funicularis** group. In view of the absence of functional mandibles, it would be interesting to know whether these species are capable of gnawing an exit hole in the covering scale of the host, like other species of *Aphytis*, or are obliged to emerge by pushing their way from underneath the edge of the covering scale. At least two members of this group appear to show some promise as natural enemies of important scale insect pests: *gordoni*, parasitic upon the citrus snow scale, and *theae*, parasitic upon the tea scale.

The members of the **funicularis** group may be readily separated from one another by obvious morphological characters such as the shape of the propodeal crenulae, male genitalia or sensilla on the antennal scape of the male, the number of mesonotal setae, the length of the ovipositor, or the pigmentation of the thoracic sterna. The main differences between the 4 known members of this group are summarized in Table 3 (p. 660).

## 77. **Aphytis funicularis** Compere

(Figures 1129–1145)

*Aphytis funicularis* Compere, 1955, Univ. Calif. Publ. Entomol., **10**: 279, 282–283.
*Aphytis funicularis*: Quednau, 1964, J. Entomol. Soc. S. Afr., **27**: 94; Rosen and DeBach, 1977, Fla. Entomol., **60**: 3–5.

This interesting, biparental African species can be readily recognized by its 5-segmented antennae and reduced mouthparts in both sexes. It may be separated from other members of the **funicularis** group by the following combination of characters: Thoracic sterna strongly infuscated; thoracic setae numerous, relatively long; propodeal crenulae minute, triangular, nonoverlapping; forewing with numerous setae in delta; female ovipositor relatively long; male antennal scape bearing an elongate plate with several sensilla, the tip of the club blackish; male genitalia relatively long, with elongate papillae.

*Female.* Eyes finely setose. Mandibles (Figure 1129) reduced, pointed, nonoverlapping, not sclerotized; both maxillary and labial palpi 1-segmented. Scrobes (Figure 1130) long, straight, very distinct. Antennae (Figures 1131, 1132) 5-segmented (1121), with 2 small funicular segments, the third segment missing; club usually with a rudimentary, incomplete transverse suture (Figure 1133) at about $\frac{1}{5}$ or $\frac{1}{4}$ length dorsally (if complete, this suture would separate a third funicular segment from the club); scape about 4 to $5\frac{1}{2}$ times as long as wide, a trifle longer than the club; pedicel $1\frac{2}{3}$ times to nearly twice as long as wide, considerably longer than the 2 segments of the funicle combined; first funicular segment trapezoidal, usually $1\frac{1}{2}$ to $1\frac{3}{4}$ times as wide as long; second segment symmetrical, somewhat shorter and wider than the first segment, considerably over twice as wide as long; club about 3 times as long as wide or nearly so (wider when flattened on slide), about $2\frac{1}{4}$ times longer and considerably wider than the pedicel, bearing 4–6 longitudinal sensilla.

Setae on head, thorax and abdomen relatively slender, rather pale; only the longest setae on the thorax are visible under × 60 magnification, whereas those on the abdomen are invisible even under × 120 magnification. Vertex with 2 pairs of long setae along occipital margin, in addition to numerous medium-long setae. Mesoscutum (Figure 1134) with 15–26 (usually 20 or more) setae, the posterior pair and one seta at each antero-lateral corner considerably longer than the others; each parapsis with 3 setae, these about as long as the shorter setae on the mesoscutum; each axilla with 1 long seta; scutellum with 4, the discoid sensilla about equidistant from the two pairs or somewhat closer to the posterior pair; setae on tegulae relatively long. Frontovertex, pronotum, and mesonotal sclerites reticulate, with a faint, delicate infrasculpture within the cells. Scutellum about $\frac{2}{3}$ to $\frac{4}{5}$ median length of mesoscutum. Metanotum (Figure 1134) reticulate except on the sides; anteromedian apodeme robust, usually somewhat shorter than median length of metanotum.

Propodeum (Figures 1134–1136) about 3 to $3\frac{2}{3}$ times as long as the metanotum, $\frac{2}{3}$ to $\frac{3}{4}$ length of scutellum, rather widely reticulate centrally, faintly reticulate on the sides; posterior margin with a weak median salient; crenulae 4 + 5 to 6 + 7, small, triangular, nonoverlapping, forming one continuous row; the most distal one on each side sometimes considerably wider than the others.

Second abdominal tergite faintly transversely striated on the sides, smooth centrally; tergites III–VII reticulate on the sides, bearing 2–6 fine setae in a transverse row on each reticulate area, with faint indications of transverse striation across center; the row of setae on each side of the seventh tergite considerably wider than on the preceding tergites; eighth tergite reticulate across, with a transverse row of 6–7 setae between spiracles; syntergum (Figure 1137) with a more deeply incised cauda than in other species, faintly reticulate-punctate, bearing 5–8 setae in a transverse row; a distinct, dense group of short setae above each cercal plate. Cerci (Figure 1137) apparently situated more posterad and considerably closer to each other than in other species of *Aphytis*, closer to tip of abdomen than to posterior spiracles, with 2 long setae and 1 short seta. Ovipositor shaft $1\frac{2}{3}$ to over $1\frac{4}{5}$ times as long as the middle tibia (1.68–1.88); ovipositor sheaths nearly $\frac{1}{2}$ length of middle tibia (0.44–0.51); ovipositor plates elongate, relatively narrow.

Mid-tibial spur about as long as the corresponding basitarsus.

Forewing (Figure 1138) broad, usually $2\frac{1}{4}$ to $2\frac{1}{3}$ times as long as wide ($2\frac{2}{3}$ times in a minute specimen); marginal fringe short, not exceeding $\frac{1}{4}$ width of disk (usually $\frac{1}{6}$–$\frac{1}{7}$). Delta area with 51–97 setae in 6–9 rows, these sparser and considerably longer than the setae distad of speculum, clearly separated from row of setae along posterior margin of wing, widely separated from a few setae below distal half of submarginal vein; costal cell with a longitudinal row of 3–4 fine setae centrally, and sometimes 1 coarse seta near apex. Submarginal vein bearing 2 long, subequal setae, the proximal one only a little shorter than the distal, and 12–19 bullae. Marginal vein bearing 7–10 prominent, subequal setae along anterior margin, these up to $1\frac{1}{2}$ times longer than the setae in a row along center of vein.

Forewing faintly dusky on proximal half or so, hyaline beyond venation, more strongly clouded below stigma and below base of marginal vein; a fuscous streak along posterior margin of wing, distad of speculum.

General coloration pale yellowish; anterior margin of mesoscutum and posterior margin of scutellum faintly, narrowly lined with fuscous; posterior margin of propodeum fuscous, crenulae blackish; a short black streak at base of forewing, below the dusky tegula. Thoracic sterna strongly infuscated, especially the meso- and metasternum; a pair of sublateral sutures on mesosternum lined with blackish. Antennal scape lined with fuscous ventrally, pedicel and funicle fuscous, club more strongly, uniformly dusky. Hind femora with a fuscous streak along ventral margin, all tibiae and tarsi faintly dusky. Wing veins faintly lined with brownish.

Length 0.71–1.09 mm (usually less than 1 mm).

*Male.* Essentially similar to the female in structure, chaetotaxis, sculpture and coloration, differing mainly in antennal characters.

Antennal scape (Figures 1139–1141) about 4–5 times as long as wide, $1\frac{1}{5}$ to $1\frac{1}{4}$ times longer than the club, bearing an elongate plate with several minute tuberculous sensilla ventrally near apex; second funicular segment apparently less than twice as wide as long; club nearly 3 times as long as wide, nearly twice longer than the pedicel, bearing 3 longitudinal sensilla.

Mesoscutum with 15–19 setae. Propodeum (Figures 1142, 1143) as in the female. Dense group of setae above cerci missing.

Posterior abdominal sternites (Figure 1144) with some fine stippling. Genitalia (Figure 1145) relatively long, usually nearly as long as middle tibia (0.89–0.97); papillae on phallobase elongate; digital sclerites narrowing on distal half, sclerotized on distal half and along inner margin, a little over $\frac{1}{4}$ the combined length of aedeagus and apodemes (0.22–0.31).

Forewing as in the female.

Antennal club distinctly blackish at apex.

Length 0.61–0.93 mm.

Redescribed from the following material: 1♀, 1♂ (paratypes), reared by E. W. Rust from the cycad scale, *Furchadiaspis zamiae* (Morgan) (recorded at the time as *Diaspis rhusae* Brain), Camp's Bay, Cape Province, South Africa, July 12–17, 1925; 4♀♀, 4♂♂ (♀ holotype, ♂ allotype, and paratypes), reared by E. W. Rust from *Rolaspis chaetach-*

*mae* (Brain) on *Chaetachme aristata*, Durban, Natal, South Africa, August 28, 1925; 6♂♂ (paratypes), same data but reared September 9, 1925; 1♀, captured on window by D. P. Annecke, Plettenberg Bay, Cape Province, South Africa, December 1960; 9♀♀, 1♂, reared by M. Naude from *Rolaspis incisa* Munting on *Euclea crispa*, Orighstad, Cape Province, South Africa, January 1965; 1♀, 2♂♂, reared by J. Munting from *Rolaspis* sp. near *incisa* on *Colpoon compressum*, Tradown Pass, Cape Province, South Africa, January 1969. Type series in balsam, except 6♂♂ remounted in Hoyer's medium; all other specimens in Hoyer's medium.

Type series in the collection of the Division of Biological Control, University of California, Riverside.

**Notes.** *A. funicularis* is the darkest member of the **funicularis** group. Its mandibles, although greatly reduced, are still larger than those of other members. It is the only member of the group with a vestigial suture on the antennal club. These characters may be interpreted to indicate that this species may be the most primitive member of the **funicularis** group.

The main differences between *funicularis* and other members of the group are summarized in Table 3 (p. 660).

Compere (1955) recorded an "Australian form" of *funicularis*, differing from the typical South African form in the paler thoracic sterna and hyaline wings. This was apparently *A. ulianovi* Girault, which is redescribed below as a distinct species.

## 78. **Aphytis ulianovi** Girault

(Figures 1146–1158)

*Aphytis ulianovi* Girault, 1932, New Lower Hymenoptera from Australia and India. Priv. Publ., Brisbane, 6 pp., page 3.

*Aphytis ulianovi*: Rosen and DeBach, 1977, Fla. Entomol., **60**: 3–5.

This little-known Australian species is very closely related to *funicularis*, differing from the latter species mainly in its larger, overlapping propodeal crenulae and paler thoracic sterna. The long, slender metanotal apodeme and mid-tibial spur, the somewhat shorter ovipositor and male genitalia, and the absence of specialized sense organs on the antennal scape of the male may serve as additional distinguishing characters. *A. ulianovi* is redescribed below relative to *funicularis*.

*Female.* Eyes finely setose. Mandibles minute, even more reduced in size than in *funicularis*, not sclerotized, not visible in most cleared specimens; both maxillary and labial palpi (Figures 1146, 1147) 1-segmented. Antennal scrobes shorter, less distinct than in *funicularis*. Antennae (Figure 1148) 5-segmented, essentially as in *funicularis* but no vestigial suture evident on club; scape 4–5 times as long as wide, about as long as the club or somewhat longer; pedicel bulbous, about $1\frac{3}{5}$ times as long as wide; both funicular segments trapezoidal, the first segment about $1\frac{1}{4}$ to $1\frac{1}{2}$ times as wide as long, the second about as long as but considerably wider than the first segment, about $1\frac{1}{2}$

to $2\frac{1}{4}$ times as wide as long; club thick, about $2\frac{4}{5}$ to 3 times as long as wide, $2\frac{1}{3}$ to $2\frac{4}{5}$ times longer and considerably wider than the pedicel, bearing 4–6 longitudinal sensilla.

Setae on head, thorax and abdomen usually considerably shorter, more slender and paler than in *funicularis*; only the longest setae on the head and thorax are visible under $\times 120$ magnification. Chaetotaxis and sculpture of head and thorax as in *funicularis*. Mesoscutum (Figure 1149) with 16–32 (usually 20 or more) setae, each parapsis with 3, each axilla with 1 seta; scutellum with 4 setae, the discoid sensilla about equidistant from the two pairs. Scutellum about $\frac{2}{3}$ to $\frac{3}{4}$ median length of mesoscutum. Anteromedian apodeme of metanotum long, slender, up to $1\frac{2}{5}$ times longer than median length of metanotum.

Propodeum (Figures 1149–1152) $2\frac{1}{2}$ to $3\frac{3}{5}$ times as long as the metanotum, about $\frac{1}{2}$ to $\frac{2}{3}$ length of scutellum; sculpture as in *funicularis*; posterior margin with a weak median salient; crenulae 7 + 7 to 9 + 9, considerably larger than in *funicularis*, forming two more or less distinct sets, the lateral 3 or 4 on each side much larger than the others, rounded, overlapping.

Sculpture and chaetotaxis of abdominal tergites essentially as in *funicularis*; eighth tergite with a transverse row of 6 setae, the most lateral one on each side situated directly below the spiracle; syntergum moderately incised. Cerci about equidistant from posterior spiracles and from tip of abdomen, with 2 long setae and 1 short seta. Ovipositor shorter than in *funicularis*, shaft about $1\frac{2}{5}$ to nearly $1\frac{3}{5}$ times as long as the middle tibia (1.41–1.57), sheaths about $\frac{1}{3}$ length of middle tibia (0.29–0.37).

Mid-tibial spur (Figure 1153) long, slender, up to $1\frac{1}{4}$ times as long as the corresponding basitarsus.

Forewing (Figure 1154) essentially as in *funicularis*. Delta area with 56–88 setae in 6–9 rows; costal cell with a row of 4–5 fine setae centrally, and 1 coarse seta near apex. Submarginal vein bearing 14–19 bullae. Marginal vein bearing 8–13 prominent, subequal setae along anterior margin. Forewing pattern very similar to that of *funicularis* but somewhat fainter.

Coloration essentially as in *funicularis* but generally paler. Anterior margin of mesoscutum and posterior margins of scutellum and propodeum very faintly lined with dusky, considerably paler than in *funicularis*. Thoracic sterna faintly dusky. Antennae pale, club uniformly dusky, sometimes darker at tip.

Length 0.50–1.02 mm.

*Male*. Similar to the female in structure, chaetotaxis and sculpture, differing mainly in the generally darker coloration.

Antennal scape (Figure 1155) as long as the club or a trifle longer, not bearing any specialized sense organs; funicular segments as in the female; club elongate, $3\frac{1}{4}$ to nearly 4 times as long as wide, about $2\frac{1}{4}$ to $2\frac{3}{5}$ times longer than the pedicel, bearing 3–4 longitudinal sensilla.

Mesoscutum (Figure 1156) with 10–26 setae. Metanotal apodeme long and slender, as in the female. Propodeum (Figures 1156, 1157) $2\frac{1}{2}$ to $3\frac{1}{5}$ times as long as the metanotum; crenulae 5 + 5 to 9 + 9, as in the female.

Stippling on posterior abdominal sternites (Figure 1158) considerably coarser than in *funicularis*. Genitalia (Figure 1158) somewhat shorter than in *funicularis*, about

$\frac{3}{4}$ length of middle tibia (0.70–0.83); papillae on phallobase minute, not elongate; digital sclerites about $\frac{1}{4}$ the combined length of aedeagus and apodemes (0.24–0.28).

Mid-tibial spur long and slender, as in the female. Forewing essentially as in the female; delta with 33–75 (usually about 50 or more) setae.

General coloration somewhat darker than in the female; a fuscous blotch on back of head, surrounding foramen; pronotum, anterior margin of mesoscutum, tegulae, and posterior margins of scutellum and propodeum more distinctly marked with fuscous; thoracic sterna more conspicuously fuscous than in the female, but less so than in the male of *funicularis*; abdominal tergites rather faintly suffused with dusky. Antennae pale, apical $\frac{2}{5}$ of club blackish.

Length 0.41–0.87 mm.

Redescribed from the following material: 3♀♀ and 1♂ (types), reared from "Diaspinae on Gum," Indooroopilly (Brisbane), Queensland, Australia, April 1931; 4♀♀, 3♂♂ (paratypes), no data except January 1933; 2♀♀, 2♂♂ (paratypes), no data; 13♀♀, 16♂♂, Indooroopilly, January 1933. All these specimens had been mounted in balsam by A. A. Girault; the paratypes were cleared and remounted by us in Hoyer's medium. Also 2♀♀, 2♂♂, reared by P. DeBach from an unidentified armored scale insect on *Croton*, Innisfail, Queensland, August 6, 1971; 26♀♀, 22♂♂, reared by P. DeBach from an unidentified armored scale insect on *Pandanus*, 50 miles south of Ayr, Queensland, August 7, 1971.

Type series in the Queensland Museum, Brisbane. Other specimens in the collection of the Division of Biological Control, University of California, Riverside.

**Notes.** All the available specimens of *ulianovi* are in rather poor condition.

Girault (1932) apparently referred *ulianovi* to *Aphytis* because he believed this peculiar species to be congeneric with *chilensis*, the generotype of *Aphytis*, which had been originally described by Howard (1900) with 5-segmented antennae. Girault must have been unaware of Timberlake's (1924) correction of Howard's original error (see p. 10). We now consider *ulianovi* to be a bona fide member of the **funicularis** group of *Aphytis*. The main differences between *ulianovi* and other members of the group are summarized in Table 3 (p. 660).

Compere (1955) recorded a so-called "Australian form" of *funicularis*, differing from the typical South African form of that species in having pallid thoracic sterna and apophyses and hyaline wings. We obtained Compere's Australian specimens (3♀♀, 2♂♂, reared by H. Compere from an unidentified coccid on *Doryantes palmeri*, Botanical Gardens, Sydney, Australia, January 8, 1928, uncleared, in balsam on one slide) on loan from the British Museum (Natural History), London, and remounted 1♀. As far as can be ascertained from the poorly preserved material, this "Australian form" is identical to *ulianovi*.

## 79. **Aphytis gordoni** DeBach and Rosen

(Figures 1159–1169)

*Aphytis gordoni* DeBach and Rosen, 1976, Ann. Entomol. Soc. Amer., **69**: 545.
*Aphytis gordoni*: Rosen and DeBach, 1977, Fla. Entomol., **60**: 3–5.

This interesting, minute, biparental Oriental species is closely related to both *funicularis* and *ulianovi*, sharing with these species the peculiar 5-segmented antennae and reduced mouthparts. It resembles *funicularis* in its minute, nonoverlapping propodeal crenulae and *ulianovi* in its faintly dusky thoracic sterna, but may be readily separated from both species by its much fewer mesonotal setae, more sparsely setose forewing, and short ovipositor. The male antennal scape bearing 2 distinct sensilla also separates it from *ulianovi*. *A. gordoni* is described below relative to *funicularis*.

*Female*. Eyes finely setose. Mandibles (see Figure 1163) minute, apparently more reduced than in *funicularis*, not sclerotized, usually hardly visible; both maxillary and labial palpi 1-segmented. Scrobes shorter, less distinct than in *funicularis*. Antennae (Figure 1159) 5-segmented, essentially as in *funicularis*, the club without any trace of a transverse suture; scape slender, about $4\frac{3}{4}$ to 6 times as long as wide, as long as the club or a trifle longer; pedicel about $1\frac{3}{5}$ to $1\frac{4}{5}$ times as long as wide, considerably longer than the 2 segments of the funicle combined; both funicular segments trapezoidal, the first one $1\frac{1}{5}$ to $1\frac{2}{5}$ times as wide as long, the second somewhat more symmetrical, shorter and wider than the first segment, $1\frac{2}{5}$ to $1\frac{4}{5}$ times as wide as long; club $2\frac{3}{4}$ to $3\frac{1}{3}$ times as long as wide, about twice to $2\frac{3}{5}$ times longer and considerably wider than the pedicel, bearing 4–7 longitudinal sensilla.

Setae on head, thorax and abdomen relatively slender, rather pale, as in *funicularis*. Chaetotaxis of head as in *funicularis*. Mesoscutum (Figure 1160) with 10–12 setae, the posterior pair and 1 seta at each antero-lateral corner considerably longer and coarser than the others; each parapsis with 1 short seta, each axilla with 1 long seta; scutellum with 4 setae, the discoid sensilla considerably closer to the anterior than to the posterior pair. Sculpture of head and thorax as in *funicularis*. Scutellum about $\frac{3}{4}$ median length of mesoscutum. Anteromedian apodeme of metanotum slender, usually considerably longer than median length of metanotum.

Propodeum (Figures 1160, 1161) about $2\frac{3}{4}$ to 4 times as long as the metanotum, $\frac{2}{3}$ to $\frac{3}{4}$ length of scutellum; sculpture as in *funicularis*; posterior margin straight; crenulae 6+7 to 8+8, small, more rounded than in *funicularis*, nonoverlapping, forming one continuous row; the 1 or 2 most distal ones on each side usually considerably wider than the others.

Sculpture of gastral tergites similar to that in *funicularis* but fainter; second abdominal tergite smooth; tergites III–VII bearing 2–4 fine setae in a transverse row on each reticulate lateral area; seventh tergite with 2 submedian setae; eighth tergite with a transverse row of 4 setae between spiracles; syntergum normal, triangular, less deeply incised than in *funicularis*, bearing 6 (rarely 7) setae in a transverse row. Cerci situated closer to posterior spiracles than to tip of abdomen, with 2 long setae and 1 short seta. Ovipositor considerably shorter than in *funicularis*, the shaft about $1\frac{1}{6}$ to $1\frac{1}{3}$ times as long as the middle tibia (1.17–1.36), the sheaths about $\frac{1}{3}$ length of middle tibia (0.29–0.35).

Mid-tibial spur distinctly longer than the corresponding basitarsus.

Forewing (Figure 1162) narrower than in *funicularis*, about $2\frac{1}{2}$ to nearly $2\frac{2}{3}$ times as long as wide; marginal fringe relatively long, not exceeding $\frac{1}{3}$ width of disk (usually about $\frac{1}{4}$). Delta area more sparsely setose than in *funicularis*, with 24–41 setae in 4–5 rows, these considerably sparser and longer than the setae distad of speculum, not clearly separated from row of setae along posterior margin of wing, widely separated from 1–2 setae below distal portion of submarginal vein; costal cell with a longitudinal row of 1–4 setae centrally. Submarginal vein bearing 2 long setae, the proximal one over $\frac{4}{5}$ length of the distal, and 12–17 bullae. Marginal vein bearing 7–9 prominent, subequal setae along anterior margin, these about $1\frac{1}{3}$ to $1\frac{2}{5}$ times longer than the setae in a row along center of vein.

Forewing paler than in *funicularis*, nearly entirely hyaline, very faintly dusky on proximal half or so, with a trace of fuscous cloud below junction of submarginal and marginal veins; a faint fuscous streak along posterior margin of wing, distad of speculum.

General coloration pale yellowish, considerably paler than *funicularis*. A short black streak at base of forewing, below tegula; posterior margin of scutellum very faintly, narrowly lined with fuscous; posterior margin of propodeum almost imperceptibly lined with faint fuscous. Thoracic sterna faintly dusky. Antenna faintly, uniformly dusky. Legs concolorous with body; a small black spot at dorso-distal apex of fore tibia. Wing veins pale, faintly lined with brownish.

Length 0.65–0.83 mm.

*Male.* Essentially similar to the female in structure, chaetotaxis, sculpture and coloration.

Mandibles seen in Figure 1163. Antennal scape (Figures 1164, 1165) $4\frac{1}{5}$ to $4\frac{1}{2}$ times as long as wide, as long as the club or a trifle longer, bearing 2 distinct discoid sensilla, each with a central tubercle, ventrally at about midlength; club nearly 3 to $3\frac{2}{5}$ times as long as wide, $2\frac{1}{3}$ to $2\frac{2}{5}$ times longer than the pedicel, bearing 3–5 longitudinal sensilla.

Mesocutum (Figure 1166) with 8–11 setae. Propodeum (Figures 1166, 1167) $2\frac{3}{4}$ to $3\frac{1}{2}$ times as long as the metanotum; crenulae 5 + 6 to 9 + 9, as in the female.

Posterior abdominal sternites (Figure 1168) stippled, the minute spines considerably coarser than in *funicularis*. Genitalia (Figure 1168) about $\frac{3}{4}$ length of middle tibia (0.71–0.78); papillae on phallobase somewhat elongate, as in *funicularis*; digital sclerites about $\frac{1}{4}$ the combined length of aedeagus and apodemes (0.25–0.29).

Forewing (Figure 1169) $2\frac{1}{2}$ to $2\frac{3}{4}$ times as long as wide; delta with 19–38 setae; submarginal vein bearing 13–15 bullae; marginal vein bearing 6–9 prominent setae along anterior margin.

Thoracic sterna more distinctly infuscated than in the female but considerably less so than in *funicularis*. Antenna faintly, uniformly dusky. Black spot at apex of fore tibia somewhat more conspicuous than in the female.

Length 0.55–0.77 mm.

Redescribed from the following material: 8♀♀, 7♂♂ (paratypes), obtained by A. G. Selhime in Florida via the USDA Quarantine Laboratory at Moorestown, New Jersey, from citrus snow scale material, *Unaspis citri* (Comstock), collected by S. K. Cheng on citrus, Hong Kong, September–October 1971; 2♀♀ (paratypes), reared in Riverside, California from *U. citri* collected by S. K. Cheng on citrus, Hong Kong, November 29, 1971; 2♀♀ (paratypes), reared in Riverside from *U. citri* collected by S. K. Cheng on tangerine, New Territories, Hong Kong, July 21, 1972; 1♀ (paratype), same data, August 7, 1972; 4♀♀, 2♂♂ (paratypes), same data, September 24, 1972; 2♀♀, 3♂♂ (♀ holotype, ♂ allotype, and paratypes), reared in Riverside from *U. citri* collected by S. K. Cheng on lemon, Hong Kong, October 8, 1972; 1♀, 1♂ (paratypes), reared in Riverside from *U. citri* collected by S. K. Cheng on citrus, Hong Kong, October 1, 1972; 1♀, 1♂ (paratypes), same data, November 15, 1972; 1♀ (paratype), insectary-reared in Riverside on the euonymus scale, *U. euonymi* (Comstock), from material originally obtained from *U. citri* collected by S. K. Cheng on citrus in Hong Kong, October 31, 1972. All specimens mounted in Hoyer's medium; most of them are in rather poor condition, mounted together with specimens of *lingnanensis* and *debachi*.

Type series in the collection of the Division of Biological Control, University of California, Riverside.

**Notes.** Whereas *gordoni* resembles *funicularis* in certain diagnostic characters such as the shape of the thoracic setae, propodeal crenulae and papillae of the male genitalia, it appears to be more similar to *ulianovi* in certain other characters, such as the shape of the antennal scrobes, metanotal apodeme and mid-tibial spur, the length of the ovipositor and male genitalia, and the paler coloration. The main differences between *gordoni* and other members of the **funicularis** group are summarized in Table 3 (p. 660). Although some of the apparent differences between this species and *funicularis* or *ulianovi* may be ascribed to the generally smaller size of *gordoni*, we still consider these forms to represent 3 distinct species.

*A. gordoni* was named by us in honor of Dr. Gordon Gordh, now at the Division of Biological Control, University of California, Riverside.

**Potential Value for Biological Control.** *A. gordoni* was discovered during an Integrated Pest Management (IPM) project of parasite importation for biological control of the citrus snow scale, *Unaspis citri* (Comstock), funded by the U.S. National Science Foundation. Although snow-scale material obtained from Hong Kong yielded only a few specimens of *gordoni*, and attempts to start a live culture for subsequent colonization in Florida have so far failed, *gordoni* appears to be a potentially valuable natural enemy of this very serious pest of citrus.

## 80. **Aphytis theae** (Cameron)

(Figures 1170–1187)

*Aphelinus theae* Cameron, 1891, Mem. Proc. Manchester Lit. Phil. Soc., 4th Ser., **4**:183–184; Pl. 1, Figures 5, 5a.

*Aphytis theae*: Compere, 1955, Univ. Calif. Publ. Entomol., **10**:317; Rosen and DeBach, 1977, Fla. Entomol., **60**:1–9.

This minute, biparental Oriental species differs from all other members of the **funicularis** group in its entirely yellow coloration, immaculate thoracic sterna and perfectly hyaline wings. It can be further recognized by the following characters: Mesoscutum with an intermediate number of setae (13–19 in the female, 10–14 in the male); propodeal crenulae minute, triangular, nonoverlapping; ovipositor short; male antennal scape bearing an oval plate with 2 tuberculous sensilla.

*Female*. Eyes finely setose. Mandibles (Figure (1172) reduced to minute, unsclerotized rudiments, not visible in most cleared specimens; both maxillary and labial palpi 1-segmented. Antennal scrobes rather indistinct. Antennae (Figure 1173) 5-segmented (1121); each of the two small funicular segments separated from the following segment by a normal constriction ventrally but apparently only by a suture dorsally, the club without any evidence of a transverse suture; scape slender, $4\frac{2}{3}$ to $5\frac{1}{5}$ times as long as wide, up to $1\frac{1}{4}$ times longer than the club; pedicel usually $1\frac{3}{5}$ to $1\frac{2}{3}$ times as long as wide (varying from $1\frac{1}{2}$ to $1\frac{4}{5}$ times), considerably longer than the 2 segments of the funicle combined; first funicular segment trapezoidal or subrectangular, $1\frac{1}{2}$ times to twice as wide as long; second segment more symmetrical, subrectangular, usually somewhat longer and wider than the first segment, about $1\frac{3}{4}$ times to twice as wide as long; club rather robust, truncate, about $2\frac{3}{5}$ to 3 times as long as wide, about twice to $2\frac{3}{4}$ times longer and considerably wider than the pedicel, bearing 4–5 longitudinal sensilla.

Setae on thorax slender, the largest ones dark, visible under $\times 30$ magnification, the others paler, barely visible under $\times 120$ magnification; setae on head and abdomen pale, invisible even under $\times 120$ magnification. Vertex with 2 pairs of long setae along occipital margin, in addition to numerous medium-long setae. Mesoscutum (Figure 1174) with 13–19 setae, the posterior pair and one seta at each antero-lateral corner longer and somewhat coarser than the others; each parapsis with 3 short setae, each axilla with 1 long seta; scutellum with 4 setae, the discoid sensilla about equidistant from the two pairs. Frontovertex, pronotum and mesonotal sclerites reticulate, with a delicate infrasculpture within the cells. Scutellum $\frac{3}{4}$ to $\frac{4}{5}$ median length of mesoscutum. The two epicoxal pads on the mesosternum, bearing minute spines, appear to be united by a median stippled area (Figure 1175). Metanotum (Figure 1176) arcuate, reticulate except on the sides; anteromedian apodeme slender, straight, distinctly longer (up to $1\frac{2}{5}$ times) than median length of metanotum.

Propodeum (Figures 1174, 1176, 1177) short, varying from a little over twice to fully 3 times as long as the metanotum, about $\frac{1}{2}$ to $\frac{2}{3}$ length of scutellum (usually about $\frac{3}{5}$ or less), reticulate centrally, the cells considerably longer than wide, faintly reticulate on the sides; posterior margin nearly straight, with a weak median salient; crenulae 5+5 to 8+8, minute, triangular, distinctly elongate, nonoverlapping, usually forming one continuous row but the median crenulae may be smaller or irregular so that two sets are sometimes evident.

Second abdominal tergite (Figures 1176, 1177) transversely striated on the sides, reticulate on a small central area; tergites III–VII (Figure 1178) reticulate on the sides, bearing 2–4 fine setae in a transverse row on each reticulate area; third tergite transversely reticulate mesad of the lateral setiferous areas, transversely striated across center; tergites IV–VI transversely striated across center; seventh tergite longitudinally reticulate mesad of the lateral setiferous areas, transversely striated across center, bearing a pair of submedian setae; eighth tergite reticulate-punctate across, with a transverse row of 4 setae between spiracles; syntergum triangular, rather broad, the cauda weakly incised and considerably wider than long, faintly punctate, bearing 6–7 setae in a transverse row. Cerci situated considerably closer to posterior spiracles than to tip of abdomen, with 2 long setae and 1 short seta. Ovipositor short, the shaft about $1\frac{1}{4}$ times as long as the middle tibia (1.15–1.29), the sheaths $\frac{1}{3}$ to $\frac{2}{5}$ length of middle tibia (0.33–0.40).

Mid-tibial spur considerably (up to $1\frac{2}{5}$ times) longer than the corresponding basitarsus.

Forewing (Figure 1179) rather broad, about $2\frac{1}{3}$ to $2\frac{1}{2}$ times as long as wide; marginal fringe not exceeding $\frac{1}{5}$ width of disk. Delta area with 61–85 setae in 7–8 rows, these somewhat sparser and distinctly longer than the setae distad of speculum, not clearly separated from row of setae along posterior margin of wing, widely separated from 1 or 2 dorsal setae below distal part of submarginal vein; costal cell with a longitudinal row of 2–4 fine setae centrally, and sometimes 1 coarse seta near apex. Submarginal vein bearing 2 long, subequal setae, the proximal one only a little shorter than the distal, and 11–14 bullae. Marginal vein bearing 8–11 prominent, subequal setae along anterior margin, these usually about $2\frac{1}{2}$ times longer than the setae in a row along center of vein.

Forewing perfectly hyaline; a faint fuscous streak along posterior margin, distad of speculum.

General coloration entirely pale yellowish, without any dark markings on head or body; even the short black streak at the base of the forewing, common to all other yellow species of *Aphytis*, is absent in this species. Thoracic sterna immaculate, concolorous with rest of body. Antennal scape pale, pedicel dusky, funicle and club uniformly, faintly dusky. All legs concolorous with body, except middle tibiae, which are distinctly dusky. Wing veins hyaline.

Length 0.53–0.77 mm.

*Male*. Essentially similar to the female in structure, chaetotaxis, sculpture and coloration, differing mainly in antennal characteristics.

Antennal scape (Figures 1180–1183) somewhat widening anteriorly, $3\frac{3}{4}$ to $4\frac{1}{2}$ times as long as wide, a little longer than the club, bearing on ventral aspect an oval, partly divided plate with 2 minute tuberculous sensilla, bordered by a row of setae; pedicel as in the female; first funicular segment about $1\frac{1}{2}$ to 2 times as wide as long, second segment $1\frac{2}{3}$ to $2\frac{1}{4}$ times as wide as long; club as in the female, bearing 2–3 longitudinal sensilla.

Mesoscutum (Figure 1184) with 10–14 setae. Propodeum (Figures 1184–1186) about $2\frac{1}{3}$ to $2\frac{2}{3}$ times as long as the metanotum, $\frac{1}{2}$ to $\frac{3}{5}$ length of scutellum; crenulae 5+5 or 5+6, as in the female.

Abdominal sterna not stippled, faintly punctate. Genitalia (Figure 1187) about $\frac{3}{4}$ length of middle tibia (0.72–0.76); phallobase well defined, tapering anteriorly; papillae minute, not elongate; digital sclerites robust, with a strong apical claw, a little less than $\frac{1}{3}$ the combined length of aedeagus and apodemes (0.28–0.31).

Forewing as in the female; delta with 49–62 setae in 5–6 rows; submarginal vein bearing 11–13 bullae; marginal vein bearing 7–9 prominent, subequal setae along anterior margin.

Coloration as in the female; antennae somewhat paler.

Length 0.55–0.78 mm.

Redescribed from numerous ♀♂ specimens, obtained from the tea scale, *Fiorinia theae* Green, on tea plants, Jorhat (Assam), India. Some of these specimens were reared in India by T. Sankaran of the CIBC in December, 1975, and sent to the authors at Riverside for identification; other material was reared by T. M. Manjunath and shipped as pupae to the USDA Biological Control Laboratory at Gainesville, Florida, where it was cultured in the quarantine laboratory and insectary; specimens were later sent to us. A ♀ neotype and a ♂ allotype were selected from a series reared by Frederick A. Collins III from tea scale on *Camellia japonica* at the insectary of the USDA Biological Control Laboratory, Gainesville, May 27, 1976.

Neotype and allotype in the collection of the Division of Biological Control, University of California, Riverside; a few specimens from the neotype series were deposited at the Florida State Collection of Arthropods, Gainesville.

**Notes.** The pupa of *theae* is generally yellow, with the top of the head, antennal cases and wing cases blackish, the leg cases dusky.

Cameron (1891) described *Aphelinus theae* from a single specimen, reared from the tea scale at Janygo, India. Unfortunately his original description, quoted in full below, is hopelessly inadequate:

> "Yellow; the legs pallid, with apex of the hinder tibiae and tarsal joints infuscated. Head dilated behind the eyes. The second antennal joint small; the third large, thicker than the second or fourth; the fourth and fifth not half the size of the third, and equal in length; the club abrupt, longer than the preceding four joints united; the last joint conical, apparently thinner than the penultimate. Hinder tarsal spur as long as the metatarsus. Wings with a long hair fringe.
> Length not $\frac{1}{2}$ millim.
> The only specimen I have seen is mounted in balsam and has got flattened, so that its exact shape cannot be seen satisfactorily. Apparently there are two broad triangular processes projecting from the thorax to near the middle of the abdomen; but their precise relationship or structure cannot be correctly made out. They seem to proceed from the base of the meso-

> notum. The sutures of the thorax cannot be observed. I am not aware of any similar structure being known in the family. It is so peculiar that I have no doubt that it will be proved, on further examination, from fresh specimens, of generic value—that the species forms the type of a new genus. Otherwise the species compares fairly well with *Aphelinus*."

Cameron's two original figures of *theae* are reproduced here (Figures 1170, 1171). The original description is, of course, very confusing. Cameron was obviously referring to the antennae of *theae* as being 8-segmented. However, a close study of his fig. 5a readily reveals that he must have imagined the pedicel to be divided into two segments—undoubtedly an error. Considering the poor quality of the specimen, the drawbacks of balsam as a mounting medium for *Aphytis*, and the optical equipment available to him, we assume that his subdivision of the club into three segments was also erroneous. In fact, as pointed out by Rosen and DeBach (1977a), no known genus in the Aphelinidae has the antennal configuration depicted by Cameron, with 2 small funicular segments and a large, 3-segmented club. On the other hand, if the subdivisions of the pedicel and club are ignored, the outline of the antenna drawn by Cameron is very similar to that of the 5-segmented antennae of the **funicularis** group. Cameron's fig. 5 appears to be a male. The "triangular processes" described by him must have been the endophragma, a common feature of *Aphytis* and other aphelinids.

The type specimen of *theae* could not be found either in the Manchester Museum, where Cameron worked, or in the British Museum (Natural History), and has apparently been lost (R. R. Askew and Z. Bouček, personal communications, 1976). Nevertheless, we concur with Compere (1955) who, on the basis of the original description, correctly recognized *theae* as a species of *Aphytis*. Furthermore, we have no hesitation in referring the species recently obtained from tea scale in India to Cameron's old species, which was obtained from the same host in the same country.

*A. theae* differs from the three other known members of the **funicularis** group in its entirely yellow coloration, immaculate thoracic sterna and dusky middle tibiae, as well as the absence of conspicuous stippling on the posterior abdominal sterna of the male. The unusual length of the setae along the anterior margin of the marginal vein, in comparison with those along the center of the vein, also seems to be a valid diagnostic character. *A. theae* resembles the other Oriental member of this group, *gordoni*, in the short ovipositor and in the shape of the female syntergum, as well as in the pigmentation of the antennae and in the shape of the sensilla on the antennal scape of the male. It differs from *gordoni* in having more numerous mesonotal setae and broader, more densely setose wings, more elongate, triangular propodeal crenulae and minute papillae on the male genitalia. The main differences between *theae* and other members of the **funicularis** group are summarized in Table 3 (p. 660).

The integument of *theae* appears to be thinner than in other species of *Aphytis*. Most of the specimens at hand have therefore become more or less distorted during mounting.

**Utilization in Biological Control.** The tea scale, *Fiorinia theae* Green, is a pest of

TABLE 3. Comparison of the members of the **funicularis** group.

| | *A. funicularis* Compere | *A. ulianovi* Girault | *A. gordoni* DeBach and Rosen | *A. theae* (Cameron) |
|---|---|---|---|---|
| Known distribution | Ethiopian | Australian | Oriental | Oriental |
| ♀♂ | | | | |
| Mandibles | Small, pointed | Minute or absent | Minute | Minute |
| Antennal scrobes | Long, distinct | Shorter, less distinct | Shorter, less distinct | Shorter, less distinct |
| Metanotal apodeme | Stout, shorter than metanotum | Slender, longer than metanotum | Slender, longer than metanotum | Slender, longer than metanotum |
| Propodeal crenulae | Minute, triangular, nonoverlapping, forming 1 row | Larger, rounded, overlapping, forming 2 sets | Minute, rounded, nonoverlapping, forming 1 row | Minute, triangular, nonoverlapping, forming 1 row |
| Mid-tibial spur | As long as basitarsus | Longer than basitarsus | Longer than basitarsus | Longer than basitarsus |
| ♀ | | | | |
| Antennal club | Vestigial transverse suture present | Transverse suture absent | Transverse suture absent | Transverse suture absent |
| Setae on meoscutum | Numerous (15–26) | Numerous (16–32) | Few (10–12) | Intermediate (13–19) |
| Setae on parapsis | 3 | 3 | 1 | 3 |
| Syntergum | Cauda deeply incised, cerci closer to tip of abdomen | Cauda moderately incised, cerci about equidistant from posterior spiracles and tip of abdomen | Cauda weakly incised, cerci closer to posterior spiracles | Cauda weakly incised, cerci closer to posterior spiracles |
| Ovipositor shaft | Long | Intermediate | Short | Short |
| Ovipositor sheaths | Long | Medium | Medium | Medium |
| Forewing | Broad | Broad | Narrow | Broad |
| Setae in delta | Numerous (51–97 in 6–9 rows) | Numerous (56–88 in 6–9 rows) | Sparse (24–41 in 4–5 rows) | Numerous (61–85 in 7–8 rows) |
| Thoracic sterna | Strongly infuscated | Faintly dusky | Faintly dusky | Immaculate |
| ♂ | | | | |
| Antennal scape | Plate bearing several sensilla | Sensilla absent | 2 distinct sensilla | Plate bearing 2 sensilla |
| Setae on mesocutum | Numerous (15–19) | Numerous (10–26) | Few (8–11) | Intermediate (10–14) |
| Posterior abdominal sterna | Finely stippled | Coarsely stippled | Coarsely stippled | Faintly punctate, not stippled |
| Genitalia | Long, papillae elongate | Medium, papillae minute | Medium, papillae elongate | Medium, papillae minute |
| Cephalic pigmentation | Head immaculate, tip of antennal club blackish | Fuscous blotch around foramen, tip of antennal club blackish | Head immaculate, antennal club uniformly dusky | Head immaculate, antennal club uniformly dusky |
| Thoracic sterna | Strongly infuscated, as in ♀ | Lightly infuscated, darker than in ♀ | Lightly infuscated, darker than in ♀ | Immaculate |

tea and other economic crops in the Orient, and an important pest of ornamental *Camellia* spp. in Florida. *A. theae* appears to be a promising parasite of this pest, and efforts have been made to introduce it from India into Florida for biological control (Rosen and DeBach, 1977a).

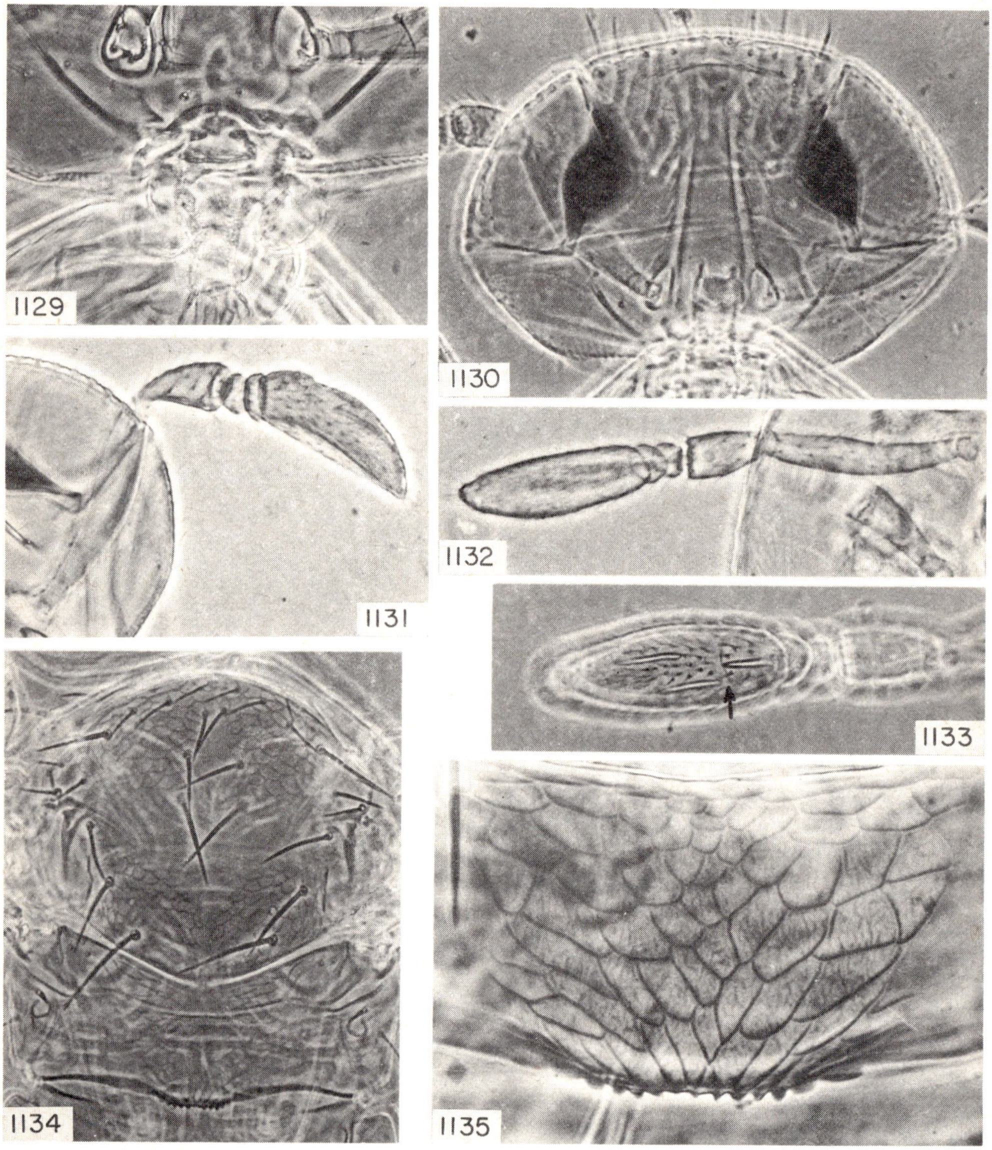

Figures 1129–1135. *Aphytis funicularis* Compere, ♀

1129. Lower part of face, showing reduced mandibles. 1130. Head, frontal view; note distinct scrobes. 1131. Antenna, lateral view. 1132. Antenna, dorso-ventral view. 1133. Antennal club; vestigial suture indicated by arrow. 1134. Thorax and propodeum. 1135. Propodeal crenulae.

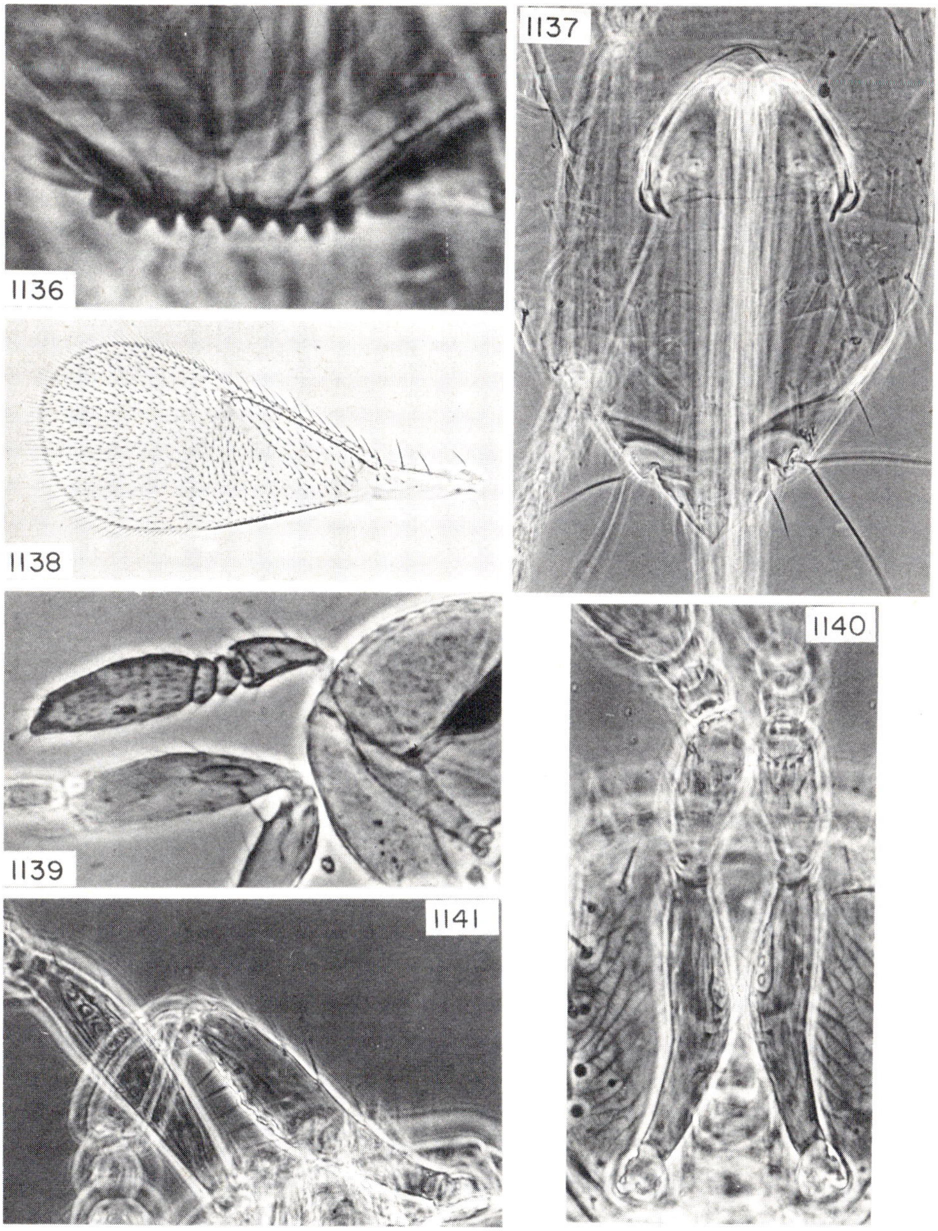


Figures 1136–1141. *Aphytis funicularis* Compere
1136. ♀: Propodeal crenulae. 1137. ♀: Gaster; note incised syntergum. 1138. ♀: Forewing.
1139. ♂: Antenna. 1140, 1141. ♂: Antennal scape, showing sense organs.

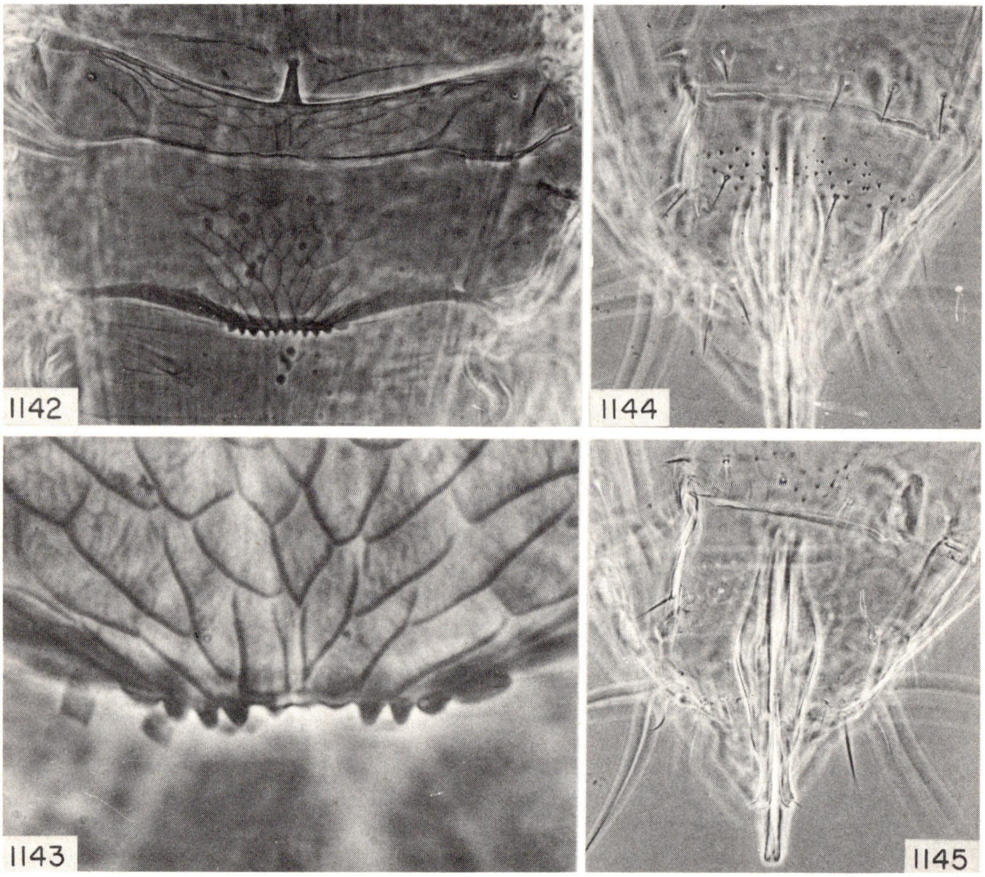


Figures 1142–1145. *Aphytis funicularis* Compere, ♂
1142. Metanotum and propodeum. 1143. Propodeal crenulae. 1144. Stippling on posterior gastral sternites; note also elongate genital papillae. 1145. Genitalia.

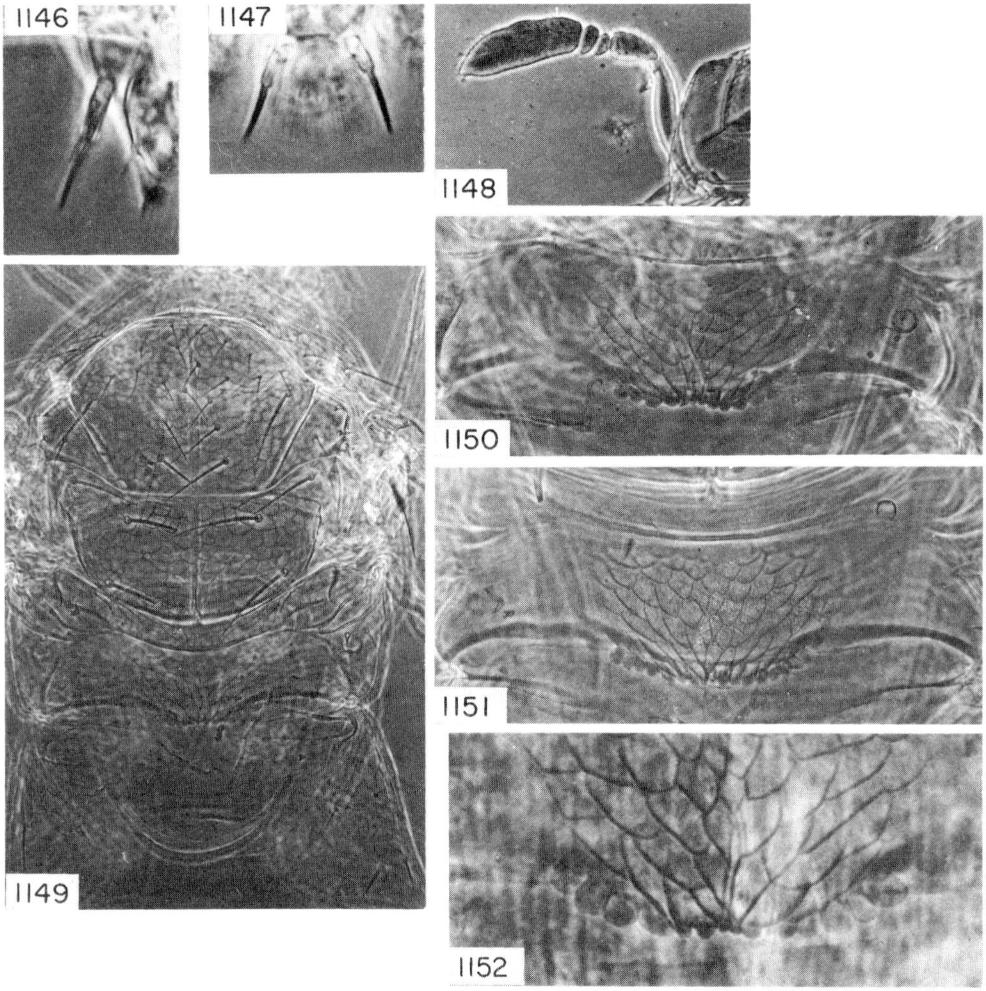


Figures 1146–1152. *Aphytis ulianovi* Girault, ♀
1146. Maxilla and maxillary palpus (paratype). 1147. Labial palpi (paratype). 1148. Antenna. 1149. Thorax, propodeum and base of gaster. 1150, 1151. Metanotum and propodeum; note large crenulae. 1152. Propodeal crenulae (paratype).

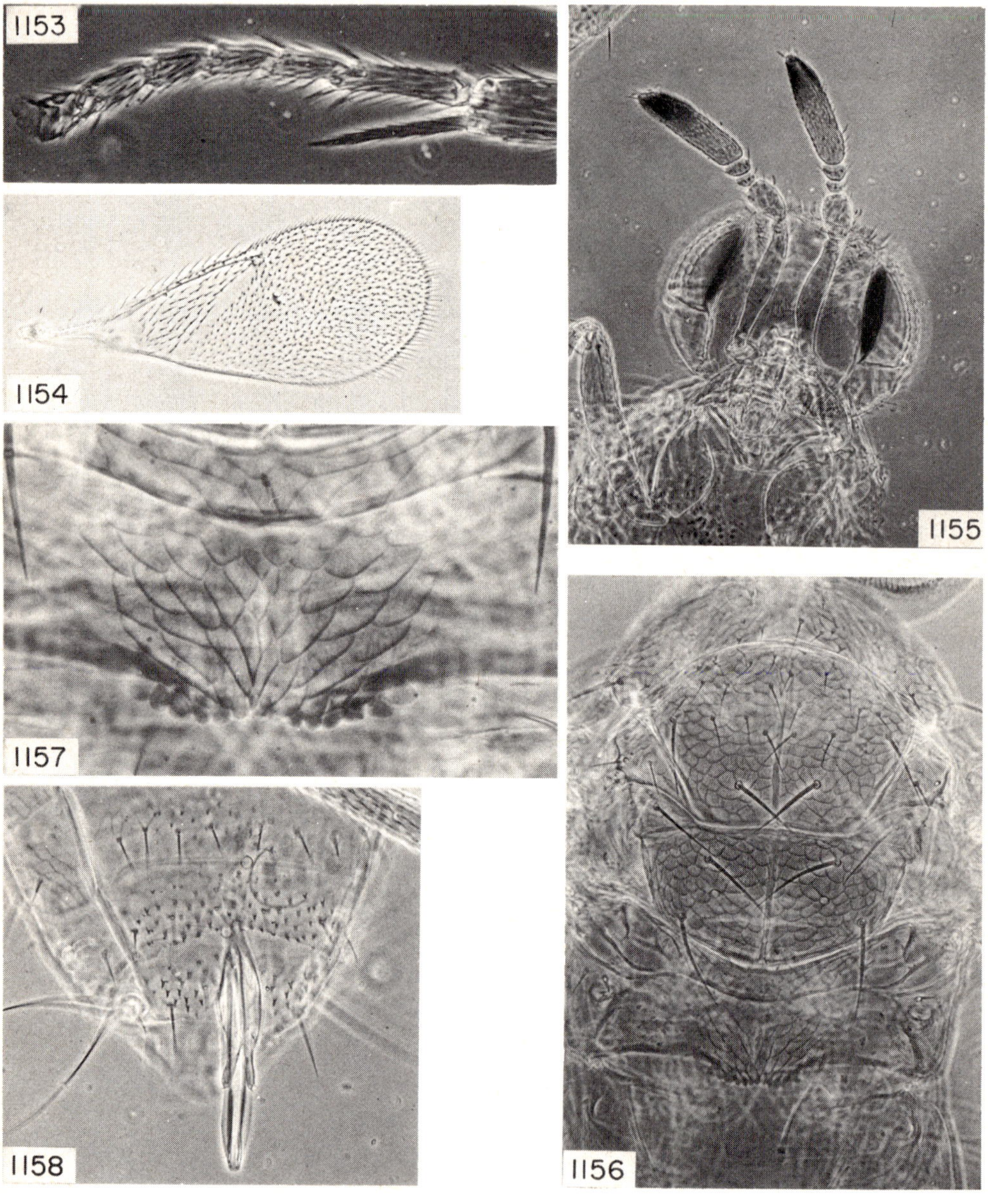


Figures 1153–1158. *Aphytis ulianovi* Girault
1153. ♀: Mid-tibial spur and middle tarsus. 1154. ♀: Forewing (paratype). 1155. ♂: Head and antennae. 1156. ♂: Thorax, propodeum and base of gaster. 1157. ♂: Metanotum, propodeum and crenulae. 1158. ♂: Posterior gastral sternites (note stippling) and genitalia.

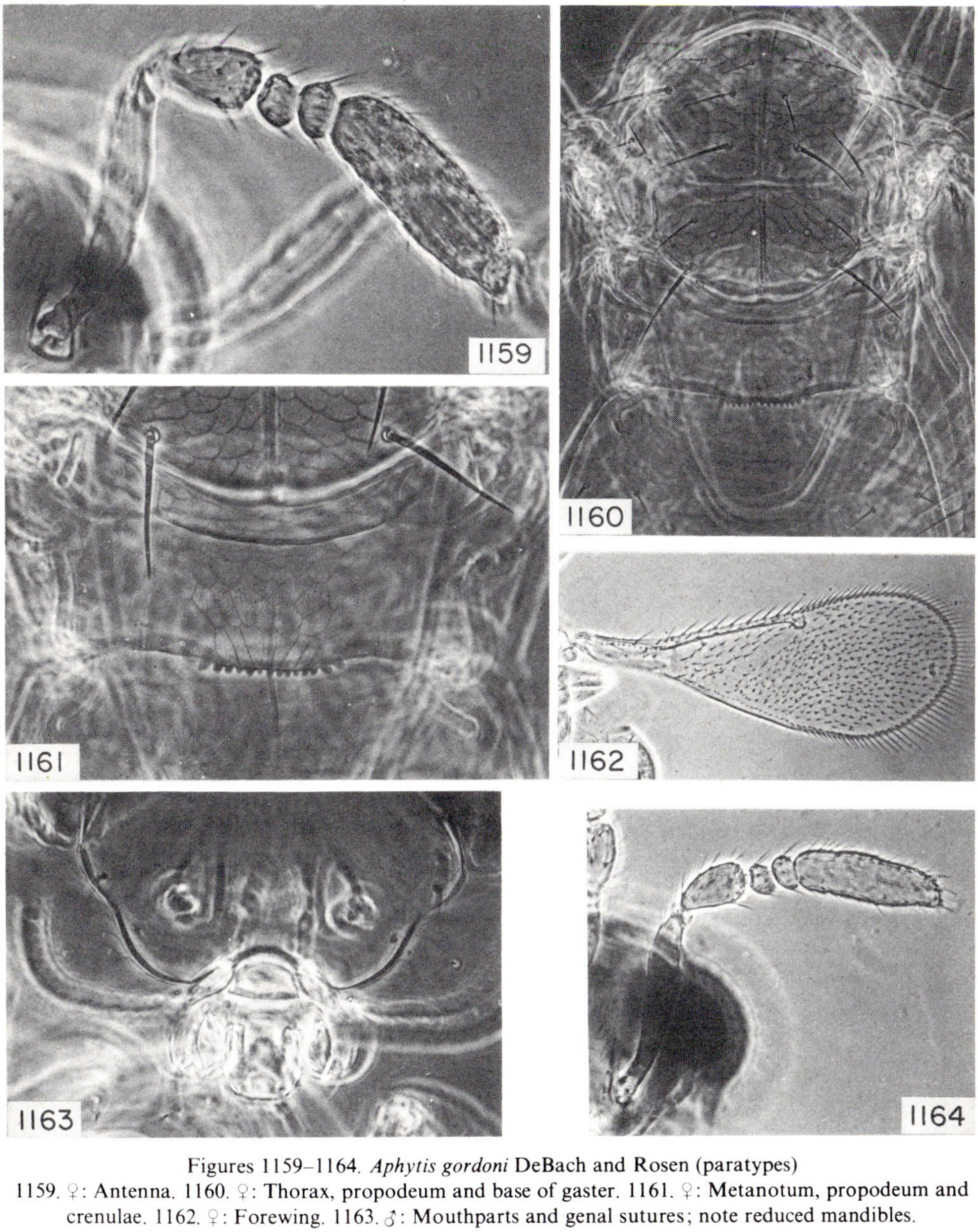


Figures 1159–1164. *Aphytis gordoni* DeBach and Rosen (paratypes)
1159. ♀: Antenna. 1160. ♀: Thorax, propodeum and base of gaster. 1161. ♀: Metanotum, propodeum and crenulae. 1162. ♀: Forewing. 1163. ♂: Mouthparts and genal sutures; note reduced mandibles. 1164. ♂: Antenna.

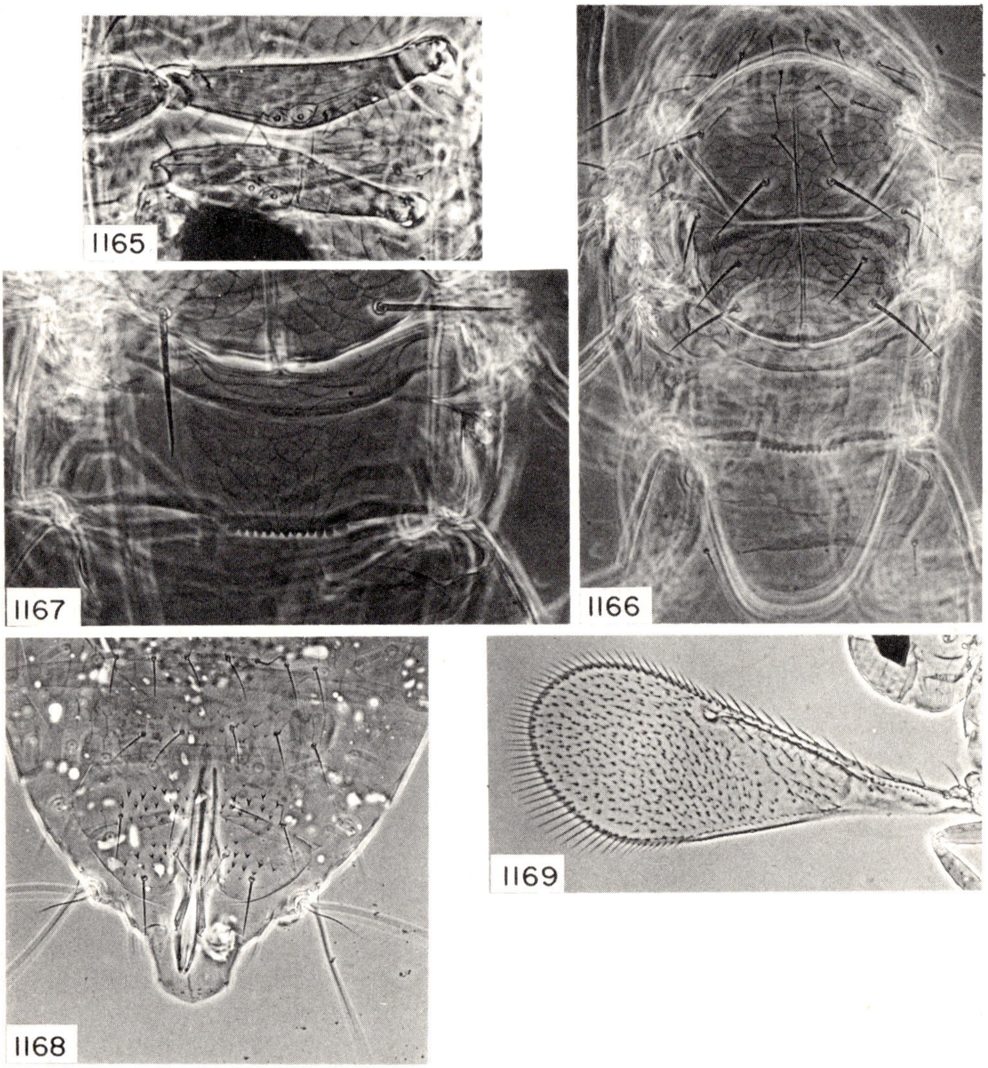


Figures 1165–1169. *Aphytis gordoni* DeBach and Rosen, ♂ (paratypes)
1165. Antennal scape, showing sense organs. 1166. Thorax, propodeum and base of gaster. 1167. Metanotum, propodeum and crenulae. 1168. Posterior gastral sternites, showing chaetotaxis and stippling, and genitalia. 1169. Forewing.

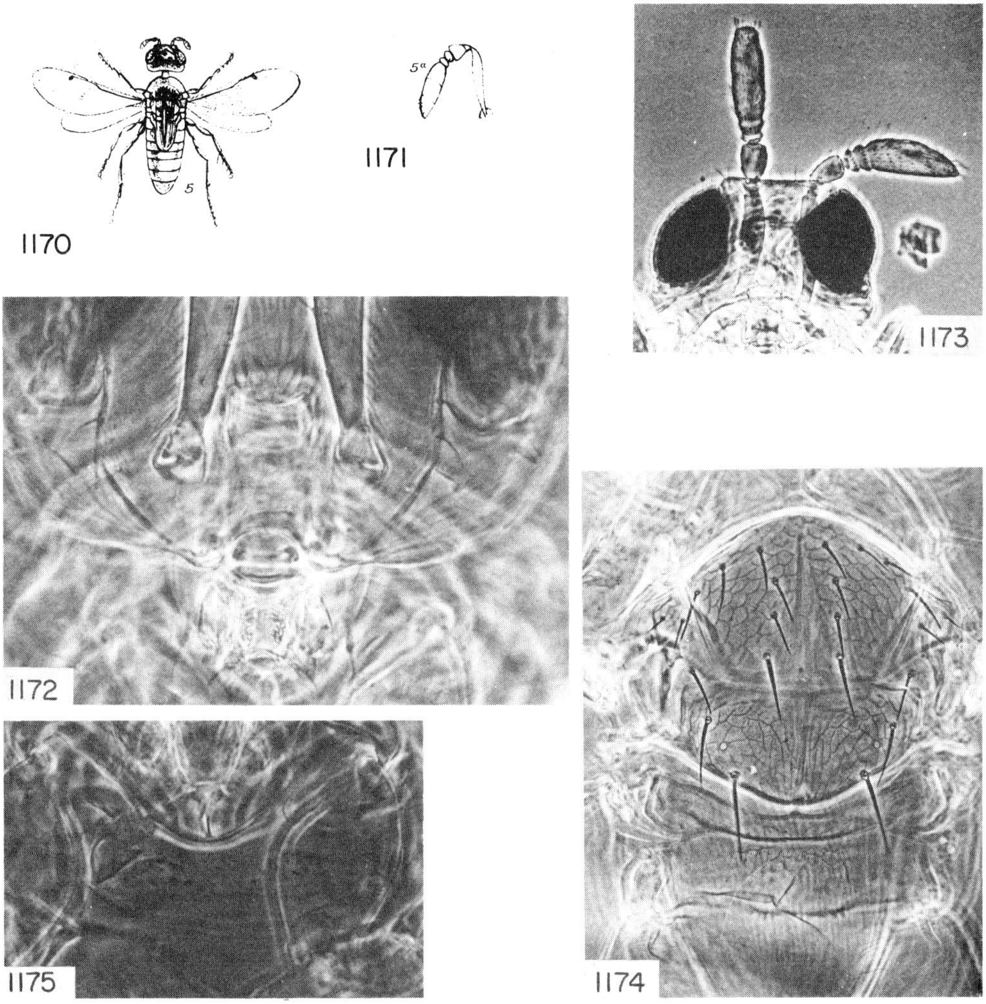


Figures 1170–1175. *Aphytis theae* (Cameron)
1170, 1171. Cameron's original figures of *Aphelinus theae* (from Cameron, 1891, figures 5, 5a): 1170. Female. 1171. Antenna.
1172–1175. Neotype series, ♀: 1172. Lower face, showing rudimentary mandibles. 1173. Head and antennae. 1174. Thorax and propodeum. 1175. Prosternum and epicoxal pads; note median stippled area. (All but Figure 1175 from Rosen and DeBach, 1977a.)

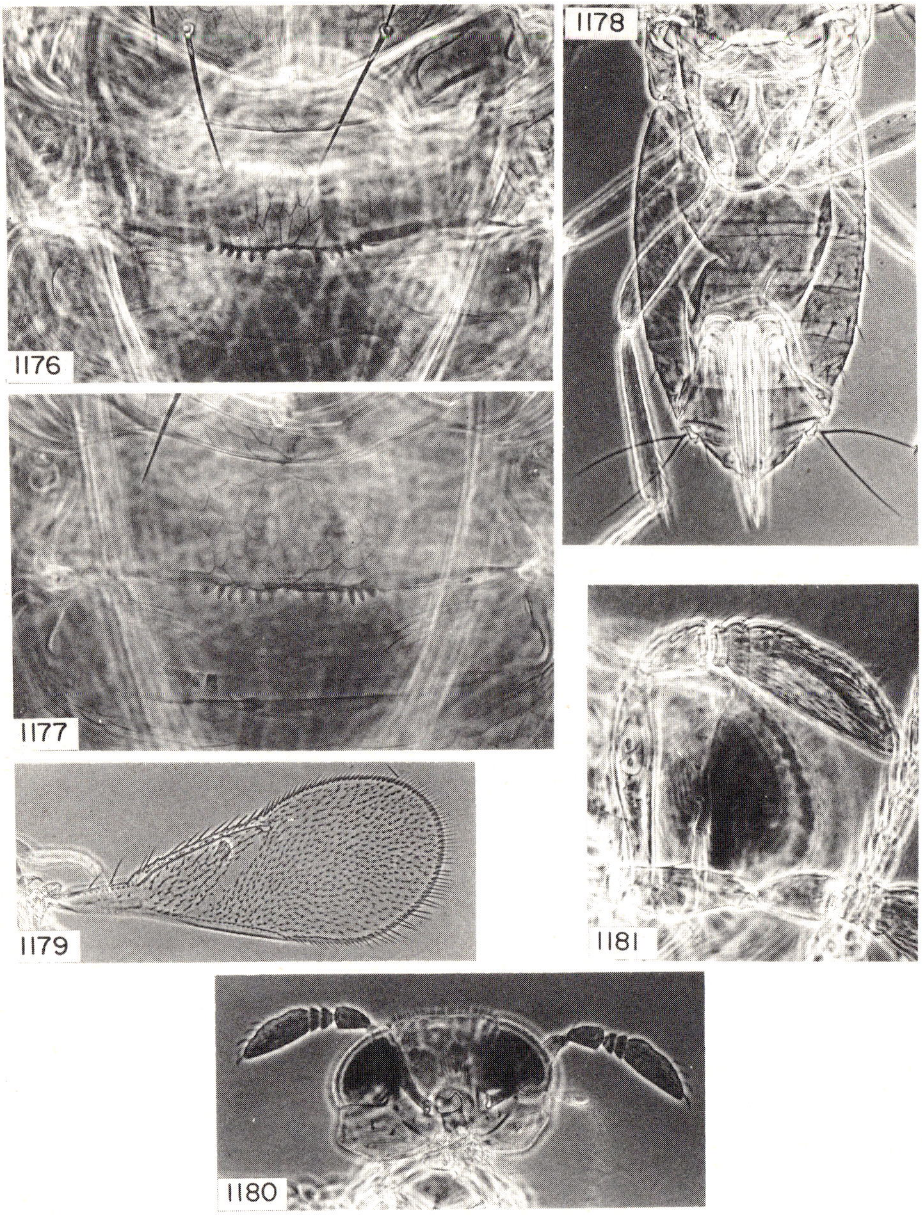


Figures 1176–1181. *Aphytis theae* (Cameron) (neotype series)
1176, 1177. ♀: Metanotum, propodeum and base of gaster. 1178. ♀: Abdomen. 1179. ♀: Forewing (dark spot below marginal vein is an artifact). 1180. ♂: Head and antennae (allotype). 1181. ♂: Antenna, showing sense organs on scape.
(From Rosen and DeBach, 1977a.)

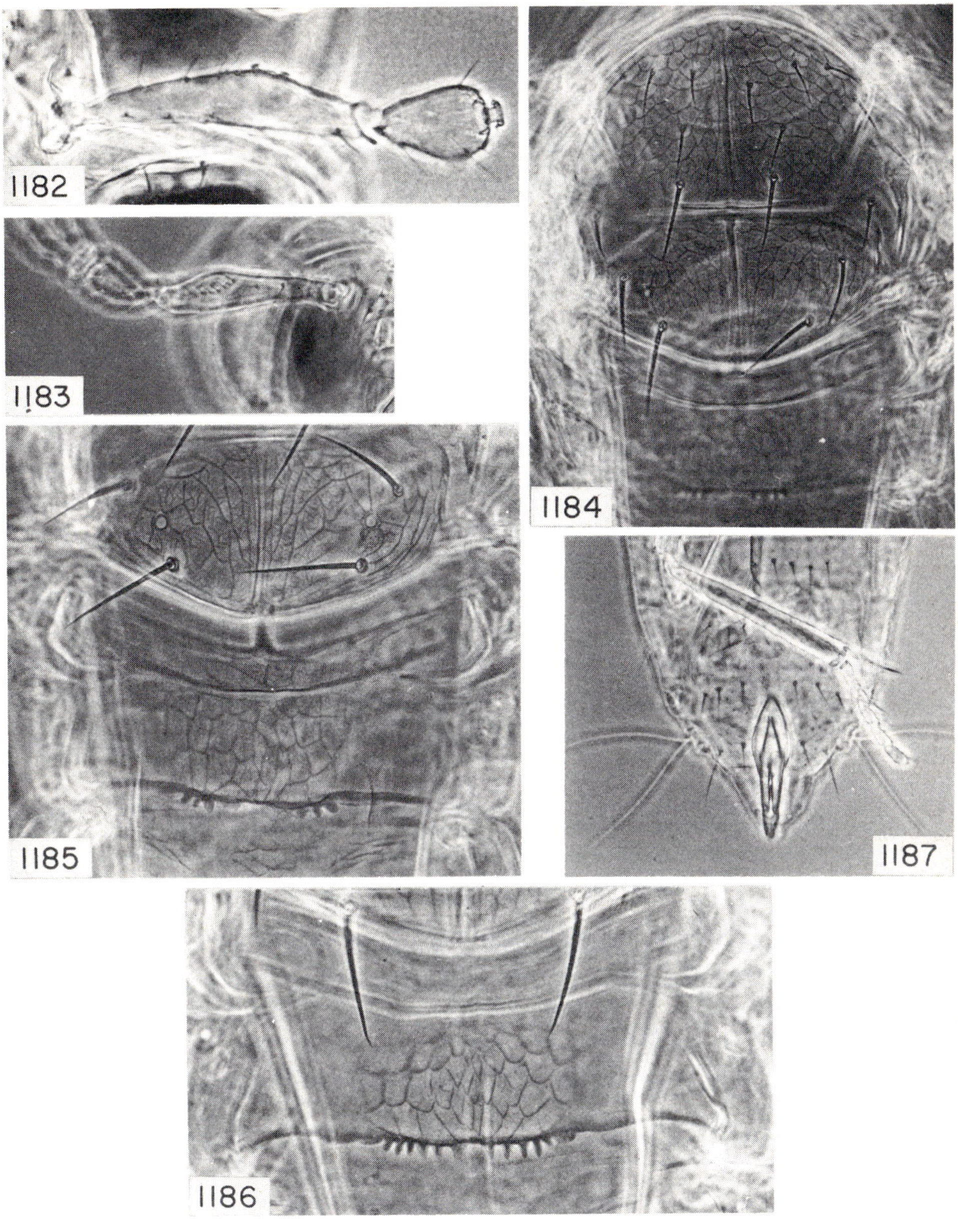


Figures 1182–1187. *Aphytis theae* (Cameron), ♂ (neotype series)
1182. Antennal scape and pedicel, showing tuberculous sensilla in lateral view. 1183. Antennal scape and pedicel, showing setae bordering plate on scape. 1184. Thorax and propodeum. 1185, 1186. Metanotum and propodeum. 1187. Posterior gastral sternites, genitalia and middle tibia.
(From Rosen and DeBach, 1977a.)

# VIII. UNASSIGNED SPECIES

Ten bona fide species of *Aphytis* cannot be assigned to any of the seven existing species groups and are assembled below. With the discovery of additional species, new groups may eventually be formed around some of them.

*A. secundus* (Compere), an African species with 4-segmented antennae and greatly reduced mouthparts, has been transferred to *Aphytis* from the genus *Marlattiella* Howard. This aberrant species is apparently related to the **funicularis** group, but obviously cannot be regarded as a member of it.

Four African species: *erythraeus* (Silvestri), *pilosus* DeBach and Rosen, *roseni* DeBach and Gordh, and *setosus* DeBach and Rosen, and one Oriental species, *longicaudus* n.sp., have peculiar male antennae, bearing long setae on the club, sometimes also on the third funicular segment. They share this unusual characteristic with three Neotropical members of the **vittatus** group, viz. *costalimai* (Gomes), *maculatipennis* (Dozier), and *perplexus* n.sp. Although these eight species share such a distinctive character in the male sex, they differ markedly from one another in the female sex and do not seem to form a cohesive group but rather a heterogenous, polyphyletic assemblage. Thus, *costalimai*, *maculatipennis* and *perplexus* are heavily pigmented, with mottled wings, and are therefore regarded as members of the **vittatus** group, whereas the other five species are generally yellow, with unmottled wings. The female of *roseni* could be accepted as a member of the **chrysomphali** group, whereas the female of *longicaudus* differs from all other species of *Aphytis* in that the abdominal syntergum is produced into a very long cauda. In the males of *erythraeus*, *longicaudus* and *roseni*, the third funicular segment is distinctly wider than long, whereas in the other five species this segment is distinctly longer than wide. A key for the separation of these eight species is given on p. 690.

Four other species: *bedfordi* n.sp., *ignotus* Compere, *mimosae* DeBach and Rosen and *salvadorensis* n.sp. possess unique combinations of characters and do not appear to be closely related to any of the existing species groups.

### 81. **Aphytis secundus** (Compere) n.comb.

(Figures 1188–1192)

*Marlattiella secunda* Compere, 1936, Univ. Calif. Publ. Entomol., **6**:280–281.
*Marlattiella secunda*: Rosen and DeBach, 1970, Mushi, **43**:41–42.

This peculiar African species differs from all other known species of *Aphytis* in its 4-segmented antennae, with a single small funicular segment. It may be further recognized by its entirely yellow coloration and hyaline wings. Male antennal scape bearing a specialized sense organ. The following redescription is based on the uncleared, slide-mounted type series.

*Female*. Eyes finely setose. Mandibles very small, unsclerotized, hardly visible in the uncleared specimens; both maxillary and labial palpi 1-segmented. Antennae (Figures 1188, 1189) 4-segmented (1111); scape slender, about $5\frac{1}{2}$ to $6\frac{1}{2}$ times as long as wide, about as long as the club or somewhat longer; pedicel about $1\frac{3}{4}$ times to fully twice as long as wide, considerably ($2\frac{2}{5}$ to $3\frac{1}{3}$ times) longer than the funicle; the single funicular segment small, trapezoidal, about $1\frac{1}{4}$ times as wide as long; club rather robust, about $3\frac{1}{4}$ to $3\frac{2}{5}$ times as long as wide, up to 3 times longer than the pedicel, sensilla not seen clearly.

Setae on head, thorax and sides of abdomen slender, pale, practically invisible in the uncleared specimens even under $\times 120$ magnification. Mesoscutum (Figure 1188) with numerous setae (about 26 according to the original description); scutellum with 4 setae, not 6 as described by Compere (1936), discoid sensilla unseen. Sculpture of head and thorax unseen. Scutellum about $\frac{3}{4}$ median length of mesoscutum. Anteromedian apodeme of metanotum unseen.

Propodeum apparently at least 3 times as long as the metanotum; sculpture and crenulae unseen in the uncleared material at hand.

Sculpture and chaetotaxis of abdominal tergites unseen. Syntergum triangular; cerci about equidistant from posterior spiracles and tip of abdomen, with 2 long setae and 1 short seta. Ovipositor rather long, the shaft somewhat over $1\frac{4}{5}$ times as long as the middle tibia (1.84–1.85), the sheaths about $\frac{1}{2}$ length of middle tibia (0.47–0.51).

Mid-tibial spur about $\frac{4}{5}$ length of the corresponding basitarsus.

Forewing (Figure 1188; see also Figure 1191) rather broad, about $2\frac{1}{3}$ times as long as wide; marginal fringe not exceeding $\frac{1}{5}$ width of disk. Delta area with 70–77 setae in 7 rows, these somewhat sparser and longer than the setae distad of speculum, rather clearly separated from row of setae along posterior margin of wing, widely separated from a few setae below submarginal vein; costal cell with a row of fine setae along central portion, and 1 coarse seta near apex. Submarginal vein bearing 2 coarse, subequal setae, sometimes with 1 shorter seta distad, and 16–18 bullae. Marginal vein bearing 9–10 prominent, subequal setae along anterior margin, these only up to $1\frac{1}{5}$ times longer than the setae in a row along center of vein.

Forewing hyaline, very lightly infumated below base of marginal vein and in a small spot below stigma; a fuscous streak along posterior margin of wing, distad of speculum.

General coloration apparently entirely yellow, no darker pigmentation visible in the uncleared balsam mounts, except a short black streak at base of forewing, below tegula. Antennae and legs apparently concolorous with body.

Length 0.88–1.01 mm.

*Male*. Apparently similar to the female in all but sexual and antennal characters. Antennae (Figure 1190) 4-segmented; scape bearing an oval plate on ventral aspect, with 2 sensilla, the exact shape of which cannot be ascertained in the material at hand; funicular segment $1\frac{2}{3}$ times to twice as wide as long; club about $2\frac{4}{5}$ times as long as wide, up to $2\frac{1}{2}$ times longer than the pedicel.

Genitalia about $\frac{3}{4}$ length of middle tibia; digital sclerites apparently about $\frac{1}{4}$ the combined length of aedeagus and apodemes.

Forewing (Figure 1191) essentially as in the female, nearly perfectly hyaline, the small spot below stigma almost imperceptible. Hind wing as in Figure 1192.

General coloration apparently entirely yellow, as in the female; abdominal tergites seemingly somewhat dusky. Antenna uniformly dusky. Legs paler than body.

Length about 0.71 mm.

Redescribed from 2♀♀ and 2♂♂ (syntypes), reared by H. Compere "from an unknown host, probably a coccid, on *Olea chrysophylla*, Nefasit, Eritrea," April 4, 1930. All specimens uncleared, mounted in balsam on 1 slide; the females intact, the males partly dissected.

Type series in the U.S. National Museum of Natural History, Washington, D.C.

**Notes.** As we pointed out in an earlier paper (Rosen and DeBach 1970), this species cannot be considered congeneric with *Marlattiella prima* Howard, the type species of *Marlattiella*. Although the poor condition of the uncleared type specimens and the fact that they are mounted in balsam preclude adequate examination of the propodeum and various other important diagnostic characters, it is readily obvious that this species possesses many characteristics of *Aphytis*: the head is relatively short and wide, ocelli situated in an obtuse triangle; mesoscutum with numerous setae, including 2 submedian rows; forewing chaetotaxis as in *Aphytis*; outer ovipositor plates with a longitudinal ridge apparently starting near the apex. Even the propodeum appears similar to that of uncleared *Aphytis* (Rosen and DeBach 1970). We therefore formally transfer this species from *Marlattiella* to *Aphytis*. This transfer necessitates a change of gender, from *secunda* to *secundus* (see p. 12).

Unlike in *faurei*, in which the 4-segmented antenna of the female is very similar to the peculiar antennae of the males in the **chilensis** group, with an extremely elongate club, in *secundus* the antennae of both female and male are quite normal for *Aphytis*, except for the absence of the second and third funicular segments.

*A. secundus* is obviously very closely related to the **funicularis** group, which is characterized by similarly reduced mouthparts and 5-segmented antennae.

## 82. **Aphytis erythraeus** (Silvestri)

(Figures 1193–1206)

*Aphelinus erythraeus* Silvestri, 1915, Boll. Lab. Zool. Gen. Agr. Portici, **9**:315–317.
*Aphytis erythraeus*: Compere, 1955, Univ. Calif. Publ. Entomol., **10**:299.

This little-known African species may be recognized by the following combination of characters: Generally yellow; wings broad, hyaline; mandibles reduced, bidentate;

numerous black, coarse setae on mesoscutum; propodeum trapezoidal, crenulae small, rounded, slightly overlapping; ♀: third funicular segment short; ovipositor sheaths long; ♂: third funicular segment wider than long; club long, bearing long setae; genitalia with elongate papillae; thoracic sterna pigmented; middle tibia tipped with black.

The following redescription is based on the few, partly dissected type specimens; some of the characters had to be reconstructed with the aid of Silvestri's original description and figures.

*Female.* Eyes not seen. Mandibles (Figure 1193, see also Figure 1198) greatly reduced, with 2 minute denticles at apex; maxillary palpi 2-segmented, labial palpi 1-segmented. Antennal scape slender; according to Silvestri's figure it is about $4\frac{3}{4}$ times as long as wide and distinctly shorter than the club; pedicel, according to Silvestri's figure, a little less than twice as long as wide, about $1\frac{1}{5}$ times longer than the third segment of the funicle; first funicular segment (Figure 1194) trapezoidal, about $1\frac{2}{3}$ times as wide as long; second segment about as long as the first segment but wider, nearly twice as wide as long; third funicular segment trapezoidal, dorsal margin the longest, about as wide as long, bearing 2 longitudinal sensilla; club robust, about $2\frac{4}{5}$ times as long as wide, fully 3 times longer and somewhat wider than the preceding segment, bearing 6 longitudinal sensilla.

Setae on thorax black, coarse, the largest ones visible under × 30 magnification; those on abdomen slender, invisible even under × 120 magnification. Setae on head not seen; according to Silvestri, vertex with 2 pairs of long, coarse setae, in addition to numerous short setae. Mesoscutum (Figure 1195) with 19 setae (not 21 as figured by Silvestri), as follows: 13 short, relatively slender setae anteriorly; 1 long, coarse seta at each antero-lateral corner; a pair of long, coarse setae postero-submedially; a pair of somewhat shorter, weaker setae submedially, anterad of the posterior pair; each parapsis with 3 setae (not 2 as figured by Silvestri), each axilla with 1 seta; scutellum (Figure 1196) with 4 setae, the discoid sensilla about equidistant from the two pairs, or considerably closer to the anterior pair (in one specimen). Frontovertex, pronotum, and mesonotal sclerites reticulate. Scutellum about $\frac{3}{4}$ median length of mesoscutum. Metanotum (Figure 1196) reticulate except on the sides; anteromedian apodeme robust, about $\frac{2}{3}$ median length of metanotum.

Propodeum (Figure 1197) rather long, about $4\frac{1}{4}$ times as long as the metanotum, over $\frac{4}{5}$ length of scutellum, distinctly trapezoidal, the median salient forming a rather sharp angle apically; propodeum faintly reticulate on the sides, rather coarsely reticulate on a trapezoidal central portion; crenulae 7 + 9, rather small but distinct, rounded, crowded, slightly overlapping.

Second abdominal tergite reticulate centrally, transversely striated anteriorly on each side; tergites III–VII reticulate on the sides, bearing a few setae on each reticulate area; third tergite transversely reticulate-striated across; tergites IV–VII transversely striated anteriorly, longitudinally striated posteriorly; seventh tergite bearing 3 setae in a transverse row between the lateral setiferous areas; eighth tergite reticulate, bearing a transverse row of 4 setae between the spiracles; syntergum reticulate-punctate, bearing 7 setae in a transverse row. Cerci with 2 long setae and 1 short seta. Ovipositor shaft $1\frac{2}{3}$ times as long as the middle tibia (1.67); ovipositor sheaths long, narrow, over $\frac{2}{5}$ length of middle tibia (0.45).

Mid-tibial spur about $\frac{3}{4}$ length of the corresponding basitarsus.

Forewing very broad, only $2\frac{1}{6}$ times as long as wide; marginal fringe short, not exceeding $\frac{1}{10}$ width of disk. Delta area with 54–58 setae in 6 rows, these considerably longer, coarser and sparser than the setae distad of speculum, not distinctly separated from row of setae along posterior margin of wing, widely separated from a few setae in a small group below distal two fifths of submarginal vein; costal cell with a few fine setae in a longitudinal row, and 1 coarser seta near apex. Submarginal vein bearing 2 coarse, long setae, the proximal one only about $\frac{3}{5}$ length of the distal, and 14–17 bullae. Marginal vein bearing 11–12 prominent, subequal setae along anterior margin, these only about $1\frac{1}{4}$ times longer than the setae in a row along center of vein.

Forewing perfectly hyaline, with a narrow fuscous streak on posterior margin centrally, but no trace of infumation.

General coloration yellow, thoracic sterna concolorous; posterior margin of scutellum narrowly lined with black; a short black streak near base of forewing, below tegula. Antennae not seen. Tibiae and tarsi of all legs apparently faintly dusky. Wing veins pale brownish.

Length according to Silvestri 0.98 mm.

*Male*. Similar to the female in general structure, chaetotaxis, sculpture and coloration, differing mainly in the structure of the antennae and in the coloration of the thoracic sterna and middle tibia.

Eyes finely setose. Mandibles (Figure 1198) as in the female. Antennal scape (Figure 1199) rather slender, about 4 to $4\frac{1}{3}$ times as long as wide; pedicel about $1\frac{3}{5}$ times as long as wide, considerably ($1\frac{2}{5}$ to $1\frac{4}{5}$ times) longer than the third segment of the funicle; first funicular segment a little over $1\frac{4}{5}$ times as wide as long; second segment considerably longer and wider than the first, also a little over $1\frac{4}{5}$ times as wide as long; third funicular segment short, shorter than the 2 preceding segments combined, over $1\frac{1}{2}$ to nearly $1\frac{3}{4}$ times as wide as long, bearing 1–2 longitudinal sensilla; club (Figures 1199, 1200) $3\frac{1}{3}$ to nearly 4 times as long as wide, $5\frac{1}{2}$ to $6\frac{2}{3}$ times longer than but about as wide as the preceding segment, bearing 3–4 longitudinal sensilla and rather long setae, the latter distinctly shorter than width of club; there is an indication of a vaguely cut-off area, bearing short setae, on the ventral surface of the club.

Mesoscutum with 18 setae. Propodeum (Figures 1201, 1202) as in the female, $4\frac{3}{10}$ times as long as the metanotum, $\frac{7}{10}$ length of scutellum; crenulae 7 + 7.

Posterior abdominal sternites extensively stippled centrally. Genitalia (Figures 1203, 1204) rather long, nearly $\frac{3}{4}$ length of middle tibia (0.70–0.74); papillae on phallobase (Figure 1204) elongate, longer than in any other known species of *Aphytis*; digital sclerites (Figure 1203) nearly $\frac{1}{3}$ the combined length of aedeagus and apodemes (0.31–0.32).

Mid-tibial spur (Figure 1205) over $\frac{4}{5}$ length of the corresponding basitarsus.

Forewing (Figure 1206) as broad as in the female in a large specimen, about $2\frac{1}{2}$ times as long as wide in a smaller specimen. Delta area with 37–45 setae in 5 rows. Submarginal vein as in the female. Marginal vein bearing 10–13 prominent, subequal setae along anterior margin.

General coloration yellow; posterior margin of scutellum narrowly lined with black; a short black streak near base of forewing, below tegula; prosternum faintly marked with fuscous posteriorly; mesosternum with a distinct, v-shaped blackish mark. Antennae apparently faintly, uniformly dusky. Legs apparently concolorous with body, middle tibia conspicuously marked with black at apex (Figure 1205). Wing veins pale brownish.

Length about 0.90 mm.

Redescribed from 1♀ and 2♂♂ (syntypes), reared by F. Silvestri from *Aspidiotus elaeidis* Marchal (recorded as *A. oppugnatus* Silvestri) on olive, Nefasit, Eritrea, August-September 1914. Remounted in Hoyer's medium on 1 slide, from the badly deteriorated original glycerin slide. The female specimen with the head dissected, the only antenna (Figure 1194) unfortunately lost in remounting; one male with the head separate, the other male dissected.

Types in the collection of the Istituto di Entomologia Agraria "Filippo Silvestri," Portici, Italy.

**Notes.** *A. erythraeus* is a peculiar, distinctive species. The reduced mandibles may indicate a possible relationship with the aberrant **funicularis** group, although several other African species also appear to share this peculiar characteristic. The long setae on the male antenna resemble those of *roseni*, *longicaudus*, *pilosus* and *setosus*, as well as those of *maculatipennis*, *costalimai* and *perplexus*. The striking pigmentation of the middle tibia of the male is similar to that of *melanostictus* and *mazalae*. The elongate papillae on the male genitalia appear to be a unique character, although *funicularis* and *gordoni* also have somewhat elongate papillae. *A. erythraeus* is also unique in that the thoracic sterna of the male are distinctly pigmented, whereas those of the female appear to be immaculate. In other species of *Aphytis*, the sexes are usually more uniform in this respect.

The above redescription may have been affected by the poor condition of the type series. The specimens appear dusky, or dirty yellow, probably due to the accumulation of internal artifacts, and the coloration of certain parts cannot be ascertained. The specimens also appear to be flattened, and this might have affected the apparent antennal proportions; i.e., the segments appear relatively wider than in Silvestri's original figure. Enough good characters are evident to make *erythraeus* an easily recognizable species. However, additional material is required in order to better understand the morphology, variation, and affinities of this peculiar species.

*A. erythraeus* was described by Silvestri (1915) as a parasite of females and male larvae of *Aspidiotus elaeidis*.

An additional slide in Silvestri's collection was labeled "*Aphelinus erythraeus* Silv. ♀ e ♂, Nefasit, *Aspidiotus*, olivo." It contained 1♀ and 1♂, apparently reared together with the types of *erythraeus*. Upon remounting, it was found that these specimens do not belong to *erythraeus*, but to another—possibly undescribed—species, apparently in the **lingnanensis** group. They have rather large, strongly overlapping crenulae (Figure 1207), very faintly pigmented thoracic sterna, a long antennal club in the female, and a weakly stippled abdomen in the male. They may be conspecific with *africanus* (p. 542), but their poor condition precludes a definite identification.

Silvestri (1915) also recorded a "smaller form" of *erythraeus*, which he figured with a considerably shorter antennal club, and noted that it may represent a different species. A single female specimen (Figures 1208, 1209) was remounted from Silvestri's slide labeled "*Aphelinus erythraeus* Silv. var.?, ♀, Nefasit, olivo" and was found to belong to *A. mimosae* (see p. 696).

Thus, *erythraeus* is now known only from the partly dissected 1♀ and 2♂♂ of the syntype series.

## 83. **Aphytis roseni** DeBach and Gordh

(Figures 1210–1231)

*Aphytis roseni* DeBach and Gordh, 1974, Entomophaga, **19**:260–265.

This peculiar biparental African species, now established in South America, may be readily recognized by a unique combination of characters: General coloration yellow, thoracic sterna dusky with a longitudinal black line on the mesosternum; wings nearly hyaline; mandibles reduced; pronotal plates broadly joined; propodeum very long, crenulae small, indistinct, overlapping; female antennal club, ovipositor and sheaths long; male antennal club elongate, bearing long setae, third funicular segment wider than long.

*Female*. Eyes finely setose. Mandibles (Figure 1210) greatly reduced, nonoverlapping, nearly truncate, weakly sclerotized at apex, with 2 minute denticles and a small dorsal truncation; maxillary palpi 2-segmented, labial palpi 1-segmented. Antennal scape (Figures 1211, 1212) slender, about $5\frac{1}{2}$ to $6\frac{1}{2}$ times as long as wide; pedicel slender, usually over twice as long as wide, considerably longer than the third segment of the funicle (up to about $1\frac{1}{3}$ times); first funicular segment slightly trapezoidal, usually somewhat wider than long (up to $1\frac{1}{4}$ times); second segment subquadrate, somewhat shorter and wider than the first segment, up to about $1\frac{3}{5}$ times as wide as long; third funicular segment about $1\frac{2}{5}$ to $1\frac{4}{5}$ times as long as wide, bearing 1–4 longitudinal sensilla; club elongate, $3\frac{2}{3}$ to $4\frac{1}{5}$ times as long as wide, about as long as the scape (varying from somewhat shorter to somewhat longer), about $2\frac{2}{3}$ to $3\frac{3}{5}$ times longer and considerably wider than the preceding segment, bearing 7–11 longitudinal sensilla.

Setae on thorax rather slender, pale, barely visible under $\times 60$ magnification; those on head and sides of abdomen visible only under $\times 120$ magnification. Vertex with 2 pairs of long setae along occipital margin, in addition to numerous shorter setae. Mesoscutum (Figures 1213–1215) with 10–13 setae, the posterior pair and 1 seta at each antero-lateral corner considerably longer than the others; each parapsis with 2 (rarely 3) setae, each axilla with 1 seta; scutellum with 4 setae, the discoid sensilla somewhat closer to the anterior than to the posterior pair. Frontovertex, pronotum and mesonotal sclerites reticulate. Pronotum (Figure 1213) composed of 2 broadly joined plates. Scutellum (Figures 1214, 1215) oval, about $\frac{2}{3}$ to $\frac{4}{5}$ median length of mesoscutum. Metanotum (Figures 1214–1217) short, transversely reticulate except on the sides, posterior margin straight or nearly so; anteromedian apodeme rather short, usually shorter than median length of metanotum (sometimes somewhat longer).

Propodeum (Figures 1214–1218) very long, about 7 to $9\frac{1}{2}$ times as long as the metanotum, $1\frac{1}{6}$ to $1\frac{1}{3}$ times longer than the scutellum, broadly reticulate on a wide central area, faintly reticulate on the sides, lacking a distinct median salient but the posterior margin broadly arcuate; crenulae 5+5 to 9+9, small, nearly indistinct, rounded, strongly overlapping, the 2 sets separated by an indistinct median ledge.

Gastral tergites extensively, delicately sculptured; second abdominal tergite (Figures 1216–1218) transversely reticulate centrally and along anterior margin; tergites III–VII (Figure 1219) reticulate on the sides, with a few fine setae in a short transverse row on each reticulate area; third tergite reticulate also mesad of the lateral setiferous areas, transversely reticulate across center; tergites IV–VI transversely reticulate anteriorly, longitudinally striated posteriorly; seventh tergite delicately reticulate across, bearing a pair of fine submedian setae; eighth tergite (Figure 1220) relatively long, delicately reticulate-striated, with a transverse row of 4 setae between spiracles; syntergum triangular, with an elongate cauda, faintly reticulate-punctate, bearing 5–10 setae in 1–3 short rows. Cerci about equidistant from posterior spiracles and tip of abdomen, or somewhat closer to the former, with 2 long setae and 1 short seta. Ovipositor long, the shaft twice to $2\frac{1}{4}$ times as long as the middle tibia (2.02–2.28), the sheaths about $\frac{3}{5}$ length of middle tibia (0.57–0.61).

Mid-tibial spur about $\frac{3}{4}$ to $\frac{4}{5}$ length of the corresponding basitarsus.

Forewing (Figure 1221) about $2\frac{1}{3}$ to $2\frac{1}{2}$ times as long as wide; marginal fringe short, not exceeding $\frac{1}{8}$ width of disk. Delta area with 38–64 setae in 4–6 (usually 5) rows, these considerably longer, coarser and sparser than the setae distad of speculum, not clearly separated from row of setae along posterior margin of wing, widely separated from 2–4 setae below distal portion of submarginal vein; costal cell with a row of fine setae along central part. Submarginal vein bearing 2 coarse setae, the proximal one about $\frac{2}{3}$ to $\frac{3}{4}$ length of the distal, and 14–20 bullae. Marginal vein bearing 9–14 prominent, subequal setae along anterior margin, these about $1\frac{1}{4}$ to $1\frac{2}{5}$ times longer than the setae in a row along center of vein.

Forewing nearly hyaline, very faintly, uniformly dusky, slightly more noticeably so basally; a fuscous streak along posterior margin of wing, distad of speculum.

General coloration yellow; thoracic sterna rather faintly dusky, stem of mesosternal furca ("Y") conspicuously black; posterior margin of scutellum narrowly lined with black; a short black streak at base of forewing, below tegula. Antennal scape marked with fuscous ventrally, rest of antenna uniformly dusky. Legs concolorous with body, all tibiae and tarsi appearing darker yellow in cleared specimens. Wing veins lined with brownish.

Length 1.18–1.59 mm.

*Male.* Essentially similar to the female in structure, chaetotaxis, sculpture and coloration, differing mainly in the shape of the antennae.

Antennal scape (Figure 1222) about $4\frac{1}{2}$ to $5\frac{1}{2}$ times as long as wide; pedicel $1\frac{2}{5}$ to $1\frac{3}{5}$ times as long as wide, up to $2\frac{3}{5}$ times longer than the third segment of the funicle; first 2 funicular segments small, symmetrical, transverse, the first one about $1\frac{2}{3}$ to $2\frac{1}{5}$ times as wide as long, the second segment subequal in length but somewhat wider, about $2\frac{1}{7}$ to $2\frac{3}{4}$ times as wide as long; third funicular segment asymmetrical, ventral aspect longer than the dorsal, somewhat longer and wider than the second segment, about $1\frac{4}{5}$ to $2\frac{3}{10}$ times as wide as long, bearing 1–3 (usually 2) oblique longitudinal sensilla; club obliquely attached to the preceding segment, elongate, about $4\frac{2}{3}$ to $5\frac{4}{5}$ times as long as wide, $1\frac{1}{3}$ to $1\frac{2}{3}$ times longer than the scape, about 8 to 11 times longer and somewhat narrower than the third segment of the funicle, bearing 2–5 longitudinal sensilla and numerous long setae, the latter usually about as long as width of club; ventral aspect of club (Figure 1223) bearing a distinct group of short sensory setae.

Pronotum as in the female. Mesoscutum (Figure 1224) with 10–12 setae; scutellum $\frac{3}{5}$ to $\frac{3}{4}$ median length of mesoscutum. Propodeum (Figures 1224–1226) somewhat shorter than in the female, about 6 to $7\frac{4}{5}$ times as long as the metanotum, as long as the scutellum or up to $1\frac{1}{6}$ times longer; crenulae 5+5 to 8+8, as in the female.

Abdominal tergites II–VIII (Figure 1228) about as in the female; syntergum shorter, bearing 4–7 setae in a single transverse row; sternites IV–VIII (Figure 1229) reticulate, appearing lightly stippled centrally, each bearing a transverse row of 2–7 (usually 4) setae. Genitalia (Figures 1229, 1230) about $\frac{2}{3}$ to over $\frac{4}{5}$ length of middle tibia (0.69–0.85); digital sclerites short, about $\frac{1}{5}$ the combined length of aedeagus and apodemes (0.18–0.21).

Forewing (Figure 1227) somewhat broader than in the female, about $2\frac{1}{9}$ to $2\frac{1}{4}$ times as long as wide; delta with 24–42 setae in 4–5 rows; submarginal vein with 14–18 bullae; marginal vein with 7–12 prominent, subequal setae along anterior margin.

Thoracic sterna (Figure 1231) considerably more strongly infuscated than in the female. Antennal scape pale, the club somewhat darker than in the female. Forewing more nearly hyaline.

Length 0.68–1.16 mm.

Redescribed from 13♀♀ and 8♂♂ (♀holotype, ♂ allotype, and paratypes), reared by Oscar Beingolea from the rufous scale, *Selenaspidus articulatus* (Morgan), on citrus, Huaru Valley, Peru, May 8, 1973; 36♀♀, 7♂♂ (paratypes), same data, August 4, 1972; also 11♀♀, 11♂♂, collected on Natal plum and (?) *Dovyalis caffra*, Nairobi, Kenya, April 14, 1970.

Holotype, allotype and numerous paratypes in the collection of the Division of Biological Control, University of California, Riverside.

**Notes.** *A. roseni* is a peculiar species. The female resembles *chrysomphali* in the elongate antennal club and propodeum, as well as in general coloration and habitus. In fact, we would have unhesitatingly assigned this species to the **chrysomphali** group were it not for the peculiar structure of the male antennae, with an elongate club bearing long setae and a short third funicular segment, very similar to the antenna of male *erythraeus*. The reduced mandibles of *roseni* also resemble those of *erythraeus*, whereas the broadly joined pronotal plates appear to set this species apart from most other species of *Aphytis*. At the present state of our knowledge, *roseni* appears to be rather closely related to *erythraeus*. However, the African fauna of *Aphytis* will probably have to be far better understood before any definite conclusions can be reached as to the true affinities of this interesting species.

This species was named by DeBach and Gordh (1974) in honor of Dr. David Rosen of the Hebrew University, Faculty of Agriculture, Rehovot, Israel.

**Biology.** According to Bartra (1974), the life cycle of *A. roseni* in the laboratory was 15.4 days under 26°C and 85% RH, 21.4 days under 17°C and 65% RH. Longevity under 25°C and 80% RH averaged 21 days for females but only 9 days for males. Fecundity was 40–61 eggs per female.

**Utilization in Biological Control.** The rufous scale, *Selenaspidus articulatus* (Morgan), used to be a very serious pest of citrus in Peru and in the West Indies. Inasmuch as virtually no effective natural enemies had been known, special efforts were initiated in 1970 to ascertain the native home of that important pest and to discover and import parasites or predators. A survey of the literature strongly indicated *S. articulatus* to have originated in Africa, and this led to its collection in Uganda. Shipments made from Uganda to Peru in 1970 yielded a species of *Aphytis* that was subsequently described as *roseni*. The parasite was successfully cultured, colonized and established in Peru, and by 1972 was showing considerable promise of effecting biological control of the rufous scale. By early 1973, one previously heavily infested citrus grove had been cleaned up by *roseni*. By June 1975 the project was reported as a complete success (DeBach and Gordh, 1974; Bartra, 1974; DeBach and Rosen, 1976a; and O. Beingolea, personal communication).

## 84. **Aphytis longicaudus** n.sp.

(Figures 1232–1244)

This large, distinctive Oriental species can be readily separated from all other species of *Aphytis* by the fact that the syntergum of the female is produced into an elongate cauda, whereas the male resembles *erythraeus* and *roseni* in antennal structure, with the third funicular segment wider than long and the club bearing long setae. The species may be further recognized by the following characters: Coloration entirely yellow, thoracic sterna immaculate; wings hyaline; mandibles well developed; mesonotal setae dark, very numerous; propodeal crenulae elongate, overlapping; ♀: antennae and ovipositor elongate; ♂: antennal scape bearing specialized sensilla, digital sclerites short.

*Female*. Eyes finely setose. Mandibles well developed, with 2 denticles and a dorsal truncation; maxillary palpi 2-segmented, labial palpi 1-segmented. Antennal scape (Figures 1232, 1233) rather slender, about $4\frac{3}{4}$ to $5\frac{1}{5}$ times as long as wide; pedicel slender, twice to $2\frac{1}{3}$ times as long as wide, up to $1\frac{1}{4}$ times longer than the third segment of the funicle; first 2 funicular segments subequal, the first one somewhat trapezoidal when viewed in profile, only up to $1\frac{1}{4}$ times as wide as long; second segment nearly symmetrical, usually only slightly shorter and wider than the first, up to $1\frac{1}{3}$ times as wide as long; third funicular segment elongate, $1\frac{3}{5}$ times to a little over twice as long as wide, bearing 3–4 longitudinal sensilla; club elongate, $3\frac{1}{2}$ to nearly $4\frac{1}{2}$ times as long as wide, up to $1\frac{1}{5}$ times as long as the scape, about $2\frac{3}{4}$ to 3 times longer and distinctly wider than the preceding segment, bearing 9–12 longitudinal sensilla.

Setae on head and thorax rather long, coarse, dark, the larger ones readily visible under $\times 30$ magnification; those on sides of abdomen paler, barely visible under $\times 120$ magnification. Vertex with 2 pairs of long, coarse setae along occipital margin, the submedian pair considerably larger than the lateral pair, and numerous shorter setae. Mesoscutum (Figure 1234) with 29–36 setae, the posterior pair and 1 seta at each antero-lateral corner considerably longer and coarser than the others; each parapsis with 3 shorter setae, each axilla with 1 long seta; scutellum with 4, the discoid sensilla about equidistant from the two pairs or somewhat closer to the anterior pair. Frontovertex, pronotum and mesonotal sclerites reticulate, with a distinct infrastructure within the cells. Scutellum about $\frac{3}{4}$ median length of mesoscutum. Metanotum (Figures 1234, 1235) rather long, reticulate except on the sides; anteromedian apodeme rather robust, usually somewhat shorter than median length of metanotum.

Propodeum (Figures 1234, 1235) about $3\frac{1}{2}$ to $3\frac{4}{5}$ times as long as the metanotum, $\frac{2}{3}$ length of scutellum, reticulate centrally, reticulate-dentate on the sides, faintly transversely striated along posterior margin, with a moderately produced median salient; crenulae 7+8 to 9+11, elongate, overlapping, becoming progressively larger and more strongly overlapping from center to sides, the most distal one on each side considerably wider than long.

Second abdominal tergite faintly striated transversely on the sides, smooth centrally; tergites III–VII reticulate-punctate on the sides, with a few fine setae in several irregular transverse rows on each reticulate area; third and seventh tergites reticulate also mesad of the lateral setiferous areas, this fading into faint transverse striation centrally; tergites IV–VI with faint indications of transverse striation across center; seventh tergite bearing 2–4 fine setae in a transverse row across center; eighth tergite rather large, reticulate-punctate, with 9–14 setae in several irregular rows, the lateral pair considerably longer and coarser than the others; syntergum (Figures 1236, 1237) rather narrow, with a very elongate, densely setose cauda. Cerci situated rather close to one another, considerably closer to the posterior spiracles than to tip of cauda, with 2 long setae and 1 short seta. Ovipositor long, the shaft $2\frac{1}{6}$ to $2\frac{1}{4}$ times as long as the middle tibia (2.16–2.23), the sheaths $\frac{7}{10}$ to $\frac{4}{5}$ length of middle tibia (0.71–0.80).

Mid-tibial spur long, about as long as the corresponding basitarsus.

Forewing (Figure 1238) about $2\frac{2}{5}$ to a little over $2\frac{1}{2}$ times as long as wide; marginal fringe short, not exceeding $\frac{1}{9}$ width of disk. Discal setae relatively long; delta area with 46–66 setae in 5–7 rows, these considerably longer and sparser than the setae distad of speculum, not clearly separated from row of setae aong posterior margin of wing, widely separated from a few setae below distal two thirds of submarginal vein, the latter about as long as the setae in delta; costal cell with a row of fine setae along proximal two thirds or so, and 1 coarse seta near apex. Submarginal vein bearing 3–4 long, coarse setae and 19–22 bullae. Marginal vein bearing 12–15 prominent, subequal setae along anterior margin, these only up to $1\frac{1}{5}$ times longer than the setae in a row along center of vein.

Forewing entirely hyaline; even the fuscous streak along the posterior margin is very pale.

General coloration entirely yellow; thoracic sterna immaculate; no dark markings whatsoever, except dark setae and a short black streak at base of forewing, below tegula. Antennae pale or faintly dusky. Legs concolorous with body. Wing veins pale.

Length 1.45–1.61 mm.

*Male*. Essentially similar to the female in structure, chaetotaxis, sculpture and coloration, differing mainly in the structure of the antennae and tip of abdomen.

Antennal scape (Figures 1239–1241) considerably thicker than in the female, about $3\frac{1}{5}$ to $3\frac{3}{4}$ times as long as wide, bearing an oval plate with 2 minute tuberculous sensilla on the ventral aspect; pedicel about $1\frac{3}{4}$ times as long as wide, considerably ($2\frac{2}{5}$ to 3 times) longer than the third segment of the funicle; all funicular segments considerably wider than long; first segment somewhat trapezoidal, about $1\frac{4}{5}$ to $2\frac{1}{3}$ times as wide as long; second segment nearly symmetrical, somewhat shorter and wider than the first, about $2\frac{1}{2}$ to 3 times as wide as long; third funicular segment trapezoidal, ventral aspect longer than the dorsal, shorter than the first 2 segments combined, about $1\frac{3}{4}$ times to nearly twice as wide as long, bearing 1 longitudinal sensillum; club robust, about 3 times as long as wide, $1\frac{2}{5}$ to $1\frac{1}{2}$ times as long as the scape, about $2\frac{3}{4}$ to 3 times longer and considerably wider than the preceding segment, bearing 5–7 longitudinal sensilla and numerous long setae, the latter distinctly shorter than width of club.

Mesoscutum (Figure 1242) with 25–31 setae. Propodeum (Figures 1242, 1243) apparently shorter than in the female, only somewhat over $2\frac{1}{2}$ times as long as the metanotum, a little over $\frac{1}{2}$ length of scutellum; crenulae 6+6, as in the female.

Syntergum wider than long, without an elongate cauda; cerci about equidistant from posterior spiracles and tip of abdomen. Abdominal sternites lightly stippled centrally. Genitalia (Figure 1244) about $\frac{3}{4}$ length of middle tibia (0.71–0.76); digital sclerites short, only $\frac{1}{9}$ the combined length of aedeagus and apodemes (0.11).

Forewing broader than in the female, only up to $2\frac{1}{6}$ times as long as wide; marginal fringe not exceeding $\frac{1}{8}$ width of disk. Delta area with 19–25 setae in 3–4 rows. Marginal vein bearing 7–12 prominent, subequal setae along anterior margin. Otherwise as in the female.

Length 0.77–0.96 mm.

Described from 2♀♀ and 1♂ (♀ holotype, ♂ allotype, and ♀ paratype), reared by S. E. Flanders from *Pseudaonidia trilobitiformis* (Green) on citrus, Taipo, Hong Kong, December 11, 1953; also 2♀ pupae (1 partly emerged), same data; 2♀♀, 1♂ (paratypes), reared by F. Silvestri from the camphor scale, *Pseudaonidia duplex* (Cockerell), Kowloon, Hong Kong, presumably during his 1924–1925 collecting trip to the Orient for the University of California.

Type series in the collection of the Division of Biological Control, University of California, Riverside.

**Notes.** The pupa of *longicaudus* is entirely black.

The few available specimens of this species are not in very good condition, probably somewhat flattened on the slides, and their condition may have affected the accuracy of the above description.

It is very difficult to evaluate the affinities of this peculiar Oriental species. The female can, of course, be readily separated from all other known species of *Aphytis* by the elongate abdominal cauda. On the other hand, the peculiar antennae of the male appear to place it in a rather heterogenous assemblage, including *erythraeus*, *roseni*, *pilosus* and *setosus*. The male of *longicaudus* differs from these 4 African species in the presence of specialized sensilla on the antennal scape, as well as in the peculiar shape of the genitalia, with very short digital sclerites. Like *erythraeus* and *roseni*, it differs from both *pilosus* and *setosus* in having the third funicular segment distinctly wider than long. It differs from *erythraeus* in pigmentation (especially the thoracic sterna and middle tibia) and in having well-developed mandibles, from *roseni* in having numerous thoracic setae and well-developed mandibles, as well as in the shape of the propodeal crenulae.

**Potential Value for Biological Control.** Both *Pseudaonidia trilobitiformis* and *P. duplex* are widely distributed, polyphagous pests of citrus and various other tropical and subtropical fruit trees and ornamentals (see Borkhsenius, 1966). *A. longicaudus* should be included in biological control projects directed against these species.

## 85. **Aphytis pilosus** DeBach and Rosen

(Figures 1245–1258)

*Aphytis pilosus* DeBach and Rosen, Ann. Entomol. Soc. Amer., **69**: 545.

This unusual biparental African species may be recognized as follows: Generally yellow, thoracic sterna dusky; forewing rather narrow, nearly hyaline; mandibles well developed; propodeal crenulae distinct, rounded, nonoverlapping, forming a continuous row; female antennae bearing a single long seta on third funicular segment; male antennae with both the third funicular segment and club elongate, bearing numerous long setae.

*Female.* Eyes finely setose. Mandibles well developed, with 2 denticles and a dorsal truncation; maxillary palpi 2-segmented, labial palpi 1-segmented. Antennal scape (Figures 1245, 1246) slender, about 5–6 times as long as wide, somewhat longer than the club (up to $1\frac{1}{4}$ times); pedicel $1\frac{3}{5}$ times to nearly twice as long as wide, somewhat longer than the third segment of the funicle; first funicular segment trapezoidal, $1\frac{1}{2}$ to $1\frac{4}{5}$ times as wide as long; second segment nearly symmetrical, shorter and somewhat wider than the first segment, 2 to $2\frac{1}{3}$ times as wide as long; third funicular segment (Figures 1245–1247) asymmetrical, dorsal aspect the longest, about $1\frac{1}{7}$ to $1\frac{1}{2}$ times as long as wide, bearing 1 longitudinal sensillum (rarely 2 sensilla) and a single long seta (Figures 1246, 1247), which is considerably longer than the width of the segment, in addition to the usual short setae; club robust, $2\frac{2}{5}$ to $2\frac{3}{4}$ times as long as wide, $2\frac{1}{7}$ to $2\frac{3}{4}$ times longer and $1\frac{1}{4}$ to $1\frac{1}{3}$ times wider than the preceding segment, bearing 3–5 longitudinal sensilla.

Setae on thorax dark, moderately coarse, readily visible under $\times 60$ magnification; setae on head and sides of abdomen barely visible, or invisible, even under $\times 120$ magnification. Vertex with 2 pairs of longer setae along occipital margin, in addition to numerous short setae. Mesoscutum (Figure 1248) with 8–15 (usually 10–11) setae, each parapsis with 2 short setae, each axilla with 1 seta; scutellum (Figures 1248, 1249) with 4, the discoid sensilla considerably closer to the anterior than to the posterior pair. Frontovertex, pronotum, and mesonotal sclerites reticulate. Scutellum about $\frac{4}{5}$ median length of mesoscutum (0.77–0.91). Metanotum (Figures 1248–1251) reticulate except on the sides; anteromedian apodeme varying from $\frac{2}{3}$ length to nearly as long as the metanotum.

Propodeum (Figures 1248, 1250, 1251) $3\frac{3}{5}$ to $4\frac{3}{5}$ times as long as the metanotum, $\frac{2}{3}$ to $\frac{3}{4}$ length of scutellum, very faintly reticulate on the sides, rather coarsely reticulate on a trapezoidal central portion, the cells usually longer than wide; crenulae 3+4 to 5+5, rather large, very distinct, rounded, nonoverlapping, forming a continuous row along median salient of posterior margin.

Second abdominal tergite (Figures 1248, 1251) rather coarsely reticulate across on about anterior half; tergites III–VII reticulate on the sides, with several setae in a short transverse row on each reticulate area; third tergite transversely reticulate-striated across, tergites IV–VI with some transverse striation between the lateral reticulate areas; seventh tergite reticulate across anteriorly, longitudinally striated posteriorly, bearing 2 submedian setae between the lateral setiferous areas; eighth tergite distinctly reticulate across, with a transverse row of 4–5 setae between the spiracles, usually with an additional shorter seta just mesad of each spiracle; syntergum triangular, relatively narrow, with a well-defined, rather deeply incised cauda, delicately reticulate-punctate, bearing 5–9 setae in 1–2 rows. Cerci about equidistant from posterior spiracles and tip of abdomen, with 2 long setae and 1 short seta. Ovipositor rather long, shaft $1\frac{3}{5}$ to $1\frac{9}{10}$ times as long as the middle tibia (1.60–1.87), sheaths $\frac{2}{5}$ to over $\frac{1}{2}$ length of middle tibia (0.40–0.54).

Mid-tibial spur nearly as long as (rarely somewhat longer than) the corresponding basitarsus.

Forewing (Figure 1252) rather narrow, $2\frac{1}{2}$ to fully 3 times as long as wide; marginal fringe usually not exceeding $\frac{1}{3}$ width of disk ($\frac{1}{6}$ in a large specimen). Delta area with 39–91 setae in 4–7 rows, these sparser and considerably longer than the setae distad of speculum, quite distinct from row of setae along posterior margin of wing, widely separated from a few setae below distal half of submarginal vein; costal cell with a row of fine setae along proximal half, and 1–2 coarse setae near apex. Submarginal vein bearing 2–3 coarse setae and 12–18 bullae. Marginal vein bearing 6–9 prominent, subequal setae along anterior margin, these about $1\frac{1}{3}$ to $1\frac{4}{5}$ times longer than the setae in a row along center of vein.

Forewing nearly hyaline, faintly infumated below junction of submarginal and marginal veins and at base of delta; a fuscous streak along posterior margin of wing, distad of speculum.

General coloration yellow; thoracic sterna dusky; occiput and pronotum usually faintly dusky; posterior margin of scutellum narrowly lined with fuscous; crenulae and central part of posterior margin of propodeum usually distinctly fuscous (Figures 1248, 1250, 1251); apical half of ovipositor sheaths fuscous. Antennal scape faintly, rest of antenna rather strongly and uniformly dusky. Legs concolorous with body, tarsal segments dusky. Wing veins colorless.

Length 0.65–0.97 mm.

*Male*. Essentially similar to the female in structure, chaetotaxis and sculpture, differing mainly in the structure of the antennae and in the generally more extensive pigmentation.

Antennal scape (Figures 1253–1255) slender, about 5 times as long as wide; pedicel $1\frac{1}{2}$ to $1\frac{2}{3}$ times as long as wide, somewhat shorter than the third segment of the funicle; first 2 funicular segments strongly compressed, the first one triangular, $1\frac{7}{10}$ to $1\frac{9}{10}$ times as wide as long; second segment saucer-shaped, considerably shorter and wider than the first segment, $2\frac{3}{5}$ to $3\frac{1}{2}$ times as wide as long; third funicular segment and club bearing long setae which are distinctly longer than width of club; third funicular segment rather long, asymmetrical, dorsal aspect the longest, $1\frac{1}{2}$ to $1\frac{2}{3}$ times as long as wide, slightly over twice as long as the 2 preceding segments combined, bearing 1 longitudinal sensillum; club long, narrowing toward a truncate apex, $3\frac{1}{5}$ to over $3\frac{4}{5}$ times as long as wide, distinctly longer than the scape, $2\frac{2}{5}$ to $2\frac{4}{5}$ times longer and somewhat wider than the preceding segment, bearing 2–3 longitudinal sensilla; there is a small, distinct, elongate plate bearing short, fine setae on the ventral surface of the club (Figure 1255).

Mesocutum (Figure 1256) with 10–15 setae; discoid sensilla on scutellum considerably closer to the anterior than to the posterior pair of setae. Propodeum (Figures 1256, 1257) as in the female; crenulae 4+4 to 5+6, forming a continuous row.

Second abdominal tergite (Figures 1256, 1257) rather coarsely reticulate, as in the female; posterior abdominal sternites weakly stippled. Genitalia $\frac{2}{3}$ to $\frac{3}{4}$ length of middle tibia (0.65–0.80); digital sclerites $\frac{1}{5}$ to nearly $\frac{1}{4}$ the combined length of aedeagus and apodemes (0.20–0.23).

Forewing (Figure 1258) rather narrow, $2\frac{1}{2}$ to $2\frac{3}{5}$ times as long as wide; delta with 43–60 setae in 4–6 rows; pattern essentially as in the female, but somewhat more strongly infumated at base.

General coloration dusky yellow; thoracic sterna more strongly infuscated than in the female; occiput and pronotum distinctly fuscous, genal sutures narrowly lined with blackish; anterior margin of mesoscutum centrally, and posterior margin of scutellum, lined with blackish; a short black streak near base of forewing, below tegula; posterior margin of propodeum and crenulae fuscous; second abdominal tergite fuscous anteriorly, subsequent abdominal tergites dusky. Antennal coloration similar to that of female but somewhat darker. Tibiae and tarsi of all legs dusky.

Length 0.73–0.80 mm.

Described from 15♀♀ and 6♂♂ (♀ holotype, ♂ allotype, and paratypes), reared by G. J. Snowball from "*Gascardia mimosae* material" on *Acacia karroo*, Pienaarspoort, Transvaal, South Africa, April 1966 (together with the types of *A. mimosae*).

Type series in the collection of the Division of Biological Control, University of California, Riverside.

**Additional Material:** 5♀♀, 9♂♂, reared by E. C. G. Bedford from "*Fulaspis* n.sp." on *Acacia karroo*, Pienaarspoort, Transvaal, South Africa, February 1954; 7♀♀, 4♂♂, reared by H. P. Insley from an undetermined armored scale insect, same host plant and locality, September 1971.

**Notes.** In the rather coarse shape and the arrangement of the propodeal crenulae, *pilosus* appears to be related to *wallumbillae*, *mimosae*, and probably to some bona fide members of the **vittatus** group. On the other hand, the peculiar male antennae, bearing long setae, are similar to those of *erythraeus*, *roseni*, *longicaudus* and *setosus*, as well as to those of *maculatipennis*, *costalimai* and *perplexus*. (Males of *pilosus* and *setosus* may be readily separated from males of *erythraeus*, *roseni* and *longicaudus* by the shape of the third segment of the funicle, which is longer than wide in the former, wider than long in the latter; *maculatipennis*, *costalimai* and *perplexus* are of course members of the **vittatus** group.) The peculiar long seta on the third funicular segment of the female sets *pilosus* apart from all other known species of *Aphytis*, except *setosus*.

The host record suggested for the type series—a soft scale—is undoubtedly erroneous. We assume that the species of *Fulaspis*, collected on the same host plant in the same locality and recorded as the host of the 1954 series, is indeed the host of *pilosus*.

## 86. **Aphytis setosus** DeBach and Rosen

(Figures 1259–1280)

*Aphytis ciliatus* Quednau (nec Dodd), 1964, J. Entomol. Soc. S. Afr., **27**: 102, 103.
*Aphytis setosus* DeBach and Rosen, 1976, Ann. Entomol. Soc. Amer., **69**: 545.

This interesting biparental African species was hitherto known only from the male, which is rather similar to the male of *pilosus*. Recent discovery of additional material has enabled us to provide a more complete description of both sexes. *A. setosus* may be readily recognized by the following combination of characters: Generally yellow; thoracic sterna dusky; forewing broad, nearly hyaline; mandibles reduced; pronotal plates broadly joined; propodeum long, with small, nonoverlapping crenulae forming 2 widely separated sets; male antennae with both the third funicular segment and the club elongate, bearing numerous long setae; male genitalia with short digital sclerites.

*Female*. Eyes finely setose. Mandibles (Figure 1259) reduced, quadrate, nearly truncate apically, with minute denticles; maxillary palpi 2-segmented, labial palpi 1-segmented. Antennal scape (Figures 1260, 1261) slender, 5 to $5\frac{4}{5}$ times as long as wide, somewhat longer than the club; pedicel $1\frac{3}{5}$ to $1\frac{3}{4}$ times as long as wide, somewhat longer than the third segment of the funicle; first funicular segment trapezoidal, about $1\frac{1}{4}$ to $1\frac{1}{2}$ times as wide as long; second segment more nearly symmetrical, somewhat shorter and wider than the first, up to twice as wide as long; third funicular segment asymmetrical, dorsal aspect the longest, about $1\frac{1}{4}$ to $1\frac{2}{5}$ times as long as wide, bearing 2 longitudinal sensilla and a single long seta which is about as long as the width of the segment, in addition to the usual short setae; club about 3 to $3\frac{1}{2}$ times as long as wide, about $2\frac{3}{4}$ to $2\frac{4}{5}$ times longer and somewhat wider than the preceding segment, bearing 5–7 longitudinal sensilla.

Setae on thorax dark, moderately coarse, the larger ones visible under $\times$ 30 magnification; setae on head and sides of abdomen paler, barely visible under $\times$ 120 magnification. Vertex with 2 pairs of longer setae along occipital margin, in addition to numerous shorter setae. Mesoscutum (Figure 1262) with 10 setae (occasionally with 9 or 11), the posterior pair and 1 seta at each antero-lateral corner considerably longer and coarser than the others; each parapsis with 2 setae, each axilla with 1 seta; scutellum with 4 setae, the discoid sensilla about equidistant from the two pairs. Frontovertex, pronotum, and mesonotal sclerites reticulate. Pronotum (Figures 1262, 1263) composed of 2 broadly joined plates. Scutellum about $\frac{3}{4}$ to $\frac{4}{5}$ median length of mesoscutum. Metanotum (Figures 1262, 1264, 1265) short, posterior margin nearly straight, transversely reticulate except on the sides; anteromedian apodeme considerably ($1\frac{1}{4}$ to nearly $1\frac{1}{2}$ times) longer than median length of metanotum.

Propodeum (Figures 1262, 1264, 1265) long, 6 to $7\frac{3}{4}$ times as long as the metanotum, about $\frac{3}{4}$ to $\frac{4}{5}$ length of scutellum, reticulate with some punctation on the sides, reticulate on a broad central portion, the cells usually wider than long; crenulae 3+4 to 5+6, very small, rounded, nonoverlapping, forming 2 widely separated sets on the arcuate posterior margin.

Gastral tergites (Figures 1262, 1264, 1265) extensively sculptured; second abdominal tergite transversely reticulate along entire anterior margin as well as centrally; tergites III–VII reticulate on the sides, with 2–4 setae in a short transverse row on each side, the central portion transversely striated and slightly punctate anteriorly, longitudinally striated posteriorly; seventh tergite also with 2 fine submedian setae; eighth tergite transversely reticulate-striated and punctate across, bearing 4 setae in a transverse row between the spiracles; syntergum relatively narrow, with a well-defined cauda, delicately striated transversely and rather extensively punctate, bearing 6 setae in a transverse row. Cerci about equidistant from posterior spiracles and tip of abdomen, with 2 long setae and 1 short seta. Ovipositor shaft about $1\frac{3}{4}$ to $1\frac{4}{5}$ times as long as the middle tibia (1.72–1.81), ovipositor sheaths about $\frac{1}{2}$ length of middle tibia or nearly so (0.44–0.51).

Mid-tibial spur about as long as the corresponding basitarsus.

Forewing (Figure 1266) broad, about $2\frac{1}{9}$ to $2\frac{1}{4}$ times as long as wide; marginal fringe not exceeding $\frac{1}{5}$ width of disk. Delta area with 29–38 setae in 4–5 rows, these considerably longer and sparser than the setae distad of speculum, not clearly separated from row of setae along posterior margin of wing, widely separated from a single dorsal seta below distal half of submarginal vein; costal cell with a row of fine setae along proximal three fifths or so, and 1 coarse seta near apex, the latter situated on the vein rather than in the cell itself. Submarginal vein bearing 2 coarse setae, the proximal one only about $\frac{1}{2}$ to $\frac{3}{5}$ length of the distal, and 12–16 bullae. Marginal vein bearing 7–9 prominent, subequal setae along anterior margin, these about $1\frac{2}{5}$ times longer than the setae in a row along center of vein.

Forewing nearly entirely hyaline, faintly infumated below junction of submarginal and marginal veins and at base of delta; a fuscous streak along posterior margin of wing, distad of speculum.

General coloration yellow; thoracic sterna (Figure 1267) rather strongly infuscated, stem of mesosternal furca lined with black; posterior margin of scutellum narrowly lined with blackish; tegula fuscous; a short black streak at base of forewing, below tegula; abdominal tergites faintly dusky. Antennal scape pale, longitudinally lined with fuscous ventrally; pedicel and first 2 funicular segments faintly, third funicular segment and club more noticeably, uniformly dusky. Legs concolorous with body, posterior femora lined with fuscous ventrally. Wing veins hyaline.

Length 0.62–0.92 mm.

*Male*. Essentially similar to the female in structure, chaetotaxis, sculpture and coloration, differing mainly in the shape of the antennae.

Mandibles (Figure 1268) reduced, as in the female. Antennal scape (Figures 1269, 1270) slender, about 5 to $5\frac{2}{3}$ times as long as wide; pedicel about $1\frac{2}{5}$ to $1\frac{1}{2}$ times as long as wide, somewhat longer than the third segment of the funicle; first 2 funicular segments strongly compressed, the first one triangular, about $1\frac{1}{2}$ to $1\frac{4}{5}$ times as wide as long; second segment symmetrical, saucer-shaped, shorter and considerably wider than the first, varying from $2\frac{2}{5}$ to $3\frac{2}{3}$ times as wide as long; third funicular segment and club bearing numerous long setae, these distinctly longer than width of club, with large round bases; third funicular segment asymmetrical, dorsal aspect the longest, about $1\frac{1}{5}$ to $1\frac{2}{5}$ times as long as wide, considerably longer than the 2 preceding segments combined, bearing 1 longitudinal sensillum; club distinctly narrowing toward a truncate apex, about $3\frac{1}{5}$ to $3\frac{2}{3}$ times as long as wide, distinctly longer than the scape, $2\frac{2}{3}$ to 3 times longer and somewhat wider than the preceding segment, bearing 2–3 longitudinal sensilla, with a partly cut-off sensory area bearing minute setae on the ventral aspect (Figure 1271).

Mesoscutum (Figures 1272, 1273) with 10 (occasionally 11) setae. Propodeum (Figures 1272–1274) long, $5\frac{3}{4}$ to $6\frac{1}{2}$ times as long as the short metanotum; crenulae 2+2 to 3+4, as in the female.

Abdominal tergites (Figures 1272–1275) extensively sculptured, as in the female; venter of abdomen (Figure 1276) lightly stippled posteriorly. Genitalia (Figures 1276, 1277) about $\frac{3}{4}$ to $\frac{4}{5}$ length of middle tibia (0.75–0.81); digital sclerites short, sometimes unequal, about $\frac{1}{6}$ the combined length of aedeagus and apodemes (0.16–0.18).

Mid-tibial spur (Figure 1278) up to $1\frac{1}{5}$ times as long as the corresponding basitarsus.

Forewing (Figure 1279) broad, essentially as in the female; delta with 21–27 setae in 3–4 rows; pattern as in the female.

General coloration as in the female; thoracic sterna (Figure 1280) strongly infuscated; antennal club and third funicular segment more strongly, uniformly dusky.

Length 0.58–0.73 mm.

Described from 1♂ (holotype), reared by J. Munting from *Rolaspis chaetachmae* (Brain) on *Chaetachme aristata*, Durban, Natal, South Africa, November 1962 (together with *funicularis* and *rolaspidis*). Also from 3♀♀ and 3♂♂, reared by H. P. Insley from an undetermined armored scale insect on *Chaetachme aristata*, Umkomaas, Natal, South Africa, November 1970; 20♀♀, 22♂♂, reared by G. L. Prinsloo from an undetermined armored scale insect on *Bequaertiodendron megalismontanum*, Graskop, Transvaal, South Africa, April 1972. One of the 1970 female specimens (top right on slide) is designated allotype.

Holotype in the collection of the Plant Protection Research Institute, Pretoria, South Africa (mounted on a slide together with several specimens of *funicularis*). Allotype in the collection of the Division of Biological Control, University of California, Riverside.

**Notes.** The name *ciliatus* is preoccupied in the genus *Aphytis* by *A. ciliatus* (Dodd), of which Quednau's 1964 name is a junior homonym. We have therefore renamed this species *setosus* (DeBach and Rosen, 1976b).

Quednau (1964b) described this species from a single male specimen. With the additional material at hand, including the hitherto unknown female, it is obvious that *setosus* is indeed a valid species, possessing some unique characteristics and quite distinct from *pilosus*. Thus, although it resembles *pilosus* in the general shape of the male antennae and in the peculiar long seta on the third funicular segment of the female, *setosus* differs markedly from that species in the shape and arrangement of the propodeal crenulae, in the reduced mandibles, in the broad forewing, in the short digital sclerites of the male, and in numerous details of sculpture and coloration. In fact it is difficult to assess the affinities of *setosus* at the present state of our knowledge: It appears to be rather similar to *roseni* in the reduced mandibles, peculiar shape of the pronotal plates, long propodeum and broad forewing; however, it can be readily distinguished from *roseni* by obvious diagnostic characters such as the shorter antennal club and ovipositor of the female, the elongate third funicular segment of the male, and the nonoverlapping crenulae.

*A. setosus* is one of 8 known species of *Aphytis* with peculiar male antennae bearing long setae. As pointed out above (p. 672), these species differ markedly from one another in most other respects, and cannot be considered to constitute a cohesive species group. The following key will help in distinguishing them.

## KEY TO THE SPECIES OF *APHYTIS* WITH HAIRY MALE ANTENNAE

### FEMALES

1. Thorax and abdomen extensively pigmented; forewing with patches of coarse, dark setae against a background of fine, hyaline setae; eyes coarsely setose; cerci with 3 long setae and 1 short seta . . . . . . . . . . . . . . . . . . . . . . . 2

— Thorax and abdomen yellow; forewing hyaline or nearly so; eyes finely setose; cerci with 2 long setae and 1 short seta . . . . . . . . . . . . . . . . . . . . . . . 4

2 (1). Ovipositor relatively short, less than twice length of middle tibia; dark patches on forewing relatively small; setae along anterior margin of marginal vein subequal in length; submedian dark stripes present on both mesoscutum and scutellum . . . . . . . . . . . . . . . . . . . . . . . . . *maculatipennis*

— Ovipositor longer, at least twice length of middle tibia; dark patches on forewing more extensive; setae along anterior margin of marginal vein decreasing in length toward apex of vein; submedian dark stripes present only on scutellum . . . 3
3 (2). Abdomen dark brown basally and distally, pale centrally; propodeal crenulae distinct, rounded, overlapping . . . *costalimai*
— Abdomen with a continuous dark brown blotch centrally; propodeal crenulae indistinct, more elongate, nonoverlapping . . . *perplexus*
4 (1). Mandibles well developed . . . 5
Mandibles reduced . . . 6
5 (4). Thoracic sterna dusky; antennal club short; third funicular segment bearing a single long seta; propodeal crenulae rounded, nonoverlapping; syntergum normal, cauda not exceedingly long . . . *pilosus*
Thoracic sterna immaculate; antennal club long; third funicular segment not bearing a long seta; propodeal crenulae elongate, overlapping; syntergum produced into a very long cauda . . . *longicaudus*
6 (4). Thoracic sterna dusky; mesoscutum with 10–13 setae; mandibles nearly truncate . . . 7
— Thoracic sterna immaculate; mesoscutum with 19 setae; mandibles distinctly bidentate . . . *erythraeus*
7 (6). Antennal club 3 to $3\frac{1}{2}$ times as long as wide; propodeum shorter than scutellum, crenulae nonoverlapping; ovipositor less than twice length of middle tibia . . . *setosus*
— Antennal club considerably more than $3\frac{1}{2}$ times as long as wide; propodeum longer than scutellum, crenulae overlapping; ovipositor more than twice length of middle tibia . . . *roseni*

## MALES

1. Third funicular segment distinctly longer than wide . . . 2
— Third funicular segment distinctly wider than long . . . 6
2 (1). Thorax and abdomen with dark brown markings; eyes coarsely setose . . . 3
— Thorax and abdomen yellow; eyes finely setose . . . 5
3 (2). Abdomen entirely dark brown; antennal scape with 1 row of sensilla; setae along anterior margin of marginal vein decreasing in length toward apex of vein . . . 4
— Abdomen dark brown basally and distally, pale centrally; antennal scape with 3 rows of sensilla; setae along anterior margin of marginal vein subequal in length . . . *maculatipennis*
4 (3). Mesonotal sclerites largely yellow; propodeal crenulae overlapping . . *costalimai*
— Mesonotal sclerites uniformly suffused with brownish; propodeal crenulae nonoverlapping . . . *perplexus*
5 (2). Mandibles well developed; propodeal crenulae forming a continuous row . . . *pilosus*
— Mandibles reduced; propodeal crenulae forming 2 widely separated sets . . . *setosus*

6 (1). Thoracic sterna immaculate; mandibles well developed; antennal scape bearing 2 sensilla; digital sclerites very short, $\frac{1}{9}$ the combined length of aedeagus and apodemes . . . . . . . . . . . . . . . . . . . . . . . . . . . . . . . *longicaudus*

— Thoracic sterna dusky; mandibles reduced; antennal scape not bearing specialized sensilla; digital sclerites longer . . . . . . . . . . . . . . . . . . . . . . . . . . 7

7 (6). Middle tibia contrastingly tipped with black; propodeum shorter than scutellum; papillae on phallobase elongate . . . . . . . . . . . . . . . . . . . *erythraeus*

— Middle tibia not tipped with black; propodeum as long as scutellum or longer; papillae on phallobase minute . . . . . . . . . . . . . . . . . . . . . . . . . *roseni*

## 87. **Aphytis bedfordi** n.sp.

(Figures 1281–1285)

This South African species may be readily recognized by the peculiar, small, rather irregular crenulae that form a wavy ledge on the posterior margin of the propodeum. The short antennal club, the short, dark thoracic setae (15–16 on the mesoscutum, 3 on each parapsis), the broad, nearly hyaline forewing, and the generally yellow coloration with immaculate thoracic sterna, may serve as additional distinguishing characters.

*Female.* Eyes finely setose. Mandibles well developed; maxillary palpi 2-segmented, labial palpi 1-segmented. Antennal scape (Figures 1281, 1282) slender, $5\frac{1}{4}$ to $5\frac{4}{5}$ times as long as wide, considerably ($1\frac{1}{3}$ to $1\frac{2}{5}$ times) longer than the club; pedicel $1\frac{3}{4}$ to $1\frac{4}{5}$ times as long as wide, considerably ($1\frac{1}{4}$ to $1\frac{2}{5}$ times) longer than the third segment of the funicle; first funicular segment trapezoidal, about $1\frac{1}{5}$ times as wide as long; second segment symmetrical, shorter and somewhat wider than the first, about $1\frac{4}{5}$ times as wide as long; third funicular segment rather short, about $1\frac{1}{6}$ times as long as wide, bearing 1 longitudinal sensillum; club short, thick, $2\frac{2}{5}$ to $2\frac{1}{2}$ times as long as wide, about $2\frac{2}{5}$ to $2\frac{3}{5}$ times longer and somewhat wider than the preceding segment, bearing 5 longitudinal sensilla.

Setae on thorax relatively short, moderately coarse, dark, readily visible under $\times 30$ magnification; setae on head paler, the largest ones barely visible under $\times 120$ magnification; those on sides of abdomen invisible even under $\times 120$ magnification. Vertex with 2 pairs of longer setae along occipital margin, in addition to numerous short setae. Mesoscutum (Figure 1283) with 14–16 setae, each parapsis with 3, each axilla with 1 seta; scutellum with 4 setae, the discoid sensilla considerably closer to the anterior than to the posterior pair. Frontovertex, pronotum, and mesonotal sclerites reticulate, with a distinct infrasculpture within the cells. Scutellum oval, about $\frac{4}{5}$ median length of mesoscutum. Metanotum (Figures 1283, 1284) arcuate, reticulate except on the sides; anteromedian apodeme moderately robust, varying from $\frac{2}{3}$ to full median length of metanotum.

Propodeum (Figures 1283, 1284) rather short, $3\frac{3}{5}$ to 4 times as long as the metanotum, $\frac{2}{3}$ length of scutellum, faintly reticulate on the sides, rather coarsely reticulate on a trapezoidal central area, the cells longer than wide; crenulae 4 + 4, small, irregular, rather low and wide, nonoverlapping, forming a wavy ledge on posterior margin of propodeum.

Second abdominal tergite faintly reticulate on the sides, transversely reticulate on a small central area; tergites III–VII reticulate on the sides, with a few setae in a short transverse row on each reticulate area; third tergite transversely striated across center, tergites IV–VI less extensively so; seventh tergite transversely striated anteriorly, with faint indications of longitudinal striation posteriorly, bearing 2–3 submedian setae between the lateral setiferous areas; eighth tergite reticulate on the sides, faintly punctate centrally, with a transverse row of 4 subequal setae between the spiracles and an additional shorter seta just mesad of each spiracle; syntergum triangular, finely punctate, bearing 6 setae in a transverse row. Cerci situated somewhat closer to tip of abdomen than to posterior spiracles, with 2 long setae and 1 short seta. Ovipositor shaft about $1\frac{4}{5}$ times as long as the middle tibia (1.79–1.81); ovipositor sheaths nearly $\frac{1}{2}$ length of middle tibia (0.48).

Mid-tibial spur somewhat shorter than the corresponding basitarsus.

Forewing (Figure 1285) broad, $2\frac{1}{4}$ to $2\frac{1}{3}$ times as long as wide, marginal fringe not exceeding $\frac{1}{6}$ width of disk. Delta area with 56–62 setae in 6–7 rows, these sparser and considerably longer than the setae distad of speculum, rather well separated from row of setae along posterior margin of wing, widely separated from a few setae below distal half of submarginal vein; costal cell with 4–5 fine setae in a row along proximal half or three fifths, and 1–2 coarser setae near apex. Submarginal vein bearing 2 coarse, nearly subequal setae, the proximal one about $\frac{9}{10}$ length of the distal, and 18–19 bullae. Marginal vein bearing 9–10 prominent, subequal setae along anterior margin, these about $1\frac{1}{4}$ to $1\frac{2}{5}$ times longer than the setae in a row along center of vein.

Forewing nearly hyaline, faintly infumated below junction of submarginal and marginal veins and at base of delta; a fuscous streak along posterior margin of wing, distad of speculum.

General coloration yellow; thoracic sterna immaculate; posterior margin of scutellum narrowly lined with blackish; tip of tegula blackish; a short black streak at base of forewing, below tegula. Antennal scape pale, rest of antenna uniformly dusky. Legs concolorous with body, tibiae and tarsi somewhat darker yellow. Wing veins nearly hyaline, lined with brownish.

Length 1.11–1.12 mm.

*Male.* Unknown.

Described from 2♀♀ (holotype and paratype), reared by M. J. Mynhardt from an undetermined armored scale insect host collected on *Rhus queinzii*, Pienaarspoort, Transvaal, South Africa, September 1971. Both specimens cleared and mounted in Hoyer's medium on one slide.

Types in the collection of the Division of Biological Control, University of California, Riverside.

**Notes.** *A. bedfordi* superficially resembles the female of *pilosus*, but differs markedly from this species in the peculiar shape of its rather undeveloped propodeal crenulae, as well as in its immaculate thoracic sterna and various other characters. On the other hand, it may be related to the **mytilaspidis** group. In the absence of males, it is impossible to ascertain the affinities of this interesting species.

*A. bedfordi* is at present known only from 2 female specimens, which are nearly identical in size and in most morphological characteristics. It can be assumed that the range of morphological variation within this species will eventually prove to be considerably wider than that represented in the above description.

This interesting species is named in honor of Dr. Eric C. G. Bedford, of the Citrus and Subtropical Fruit Research Institute, Nelspruit, Transvaal, South Africa.

## 88. **Aphytis mimosae** DeBach and Rosen

(Figures 1286–1296)

*Aphytis mimosae* DeBach and Rosen, 1976, Ann. Entomol. Soc. Amer., **69**:545.

This rather distinctive biparental African species can be recognized by the generally yellow coloration with dusky thoracic sterna, the hyaline wings, the very short, thick antennal club, the coarse, black mesonotal setae (12–17 on mesoscutum, 3 on each parapsis) and the well-developed, rounded, nonoverlapping propodeal crenulae.

*Female.* Eyes finely setose. Mandibles well developed; maxillary palpi 2-segmented, labial palpi 1-segmented. Antennal scape (Figure 1286) slender, about $5\frac{2}{3}$ times as long as wide, $1\frac{3}{10}$ to $1\frac{2}{5}$ times longer than the short club; pedicel $1\frac{1}{2}$ to $1\frac{3}{4}$ times as long as wide, distinctly ($1\frac{1}{8}$ to $1\frac{3}{10}$ times) longer than the third segment of the funicle; first funicular segment trapezoidal, $1\frac{1}{3}$ to $1\frac{2}{5}$ times as wide as long; second segment nearly symmetrical, considerably shorter and wider than the first segment, about twice as wide as long; third funicular segment distinctly trapezoidal, dorsal margin the longest, rather short, only slightly longer than wide, bearing 1 longitudinal sensillum; club short and thick, about twice as long as wide, 2 to $2\frac{1}{4}$ times longer and somewhat wider than the preceding segment, bearing 4–5 longitudinal sensilla.

Setae on thorax coarse, black, readily visible under $\times 30$ magnification; setae on head short, coarse, visible under $\times 60$ magnification; those on abdomen more slender, invisible even under $\times 120$ magnification. Vertex with 2 pairs of medium-long setae along occipital margin, in addition to numerous short setae. Mesoscutum (Figure 1287) with 12–17 setae, each parapsis with 3, each axilla with 1 seta; scutellum with 4 setae, the discoid sensilla considerably closer to the anterior than to the posterior pair. Frontovertex, pronotum and mesonotal sclerites reticulate. Scutellum about $\frac{4}{5}$ median length of mesoscutum. Metanotum (Figures 1288, 1289) relatively long, somewhat bilobed posteriorly, reticulate except on the sides; anteromedian apodeme rather stout, $\frac{2}{3}$ to $\frac{4}{5}$ median length of metanotum.

Propodeum (Figures 1288, 1289) about $3\frac{1}{2}$ times as long as the metanotum, $\frac{7}{10}$ median length of scutellum, faintly striated on the sides, rather coarsely reticulate on a median trapezoidal area; crenulae 5+5 to 5+6, rather large, rounded, crowded, usually nonoverlapping but sometimes slightly overlapping.

Second abdominal tergite transversely reticulate centrally, reticulate-striated anteriorly on each side; tergites III–VII reticulate on the sides, bearing a few setae in a short transverse row on each reticulate area; third segment distinctly, transversely striated across center between the reticulate areas, tergites IV–VII more faintly so; seventh tergite sometimes with faint indications of longitudinal striation posteriorly, bearing 2–3 setae between the lateral setiferous areas; eighth tergite reticulate-punctate across, with a transverse row of 4 setae between the spiracles and an additional, shorter seta just mesad of each spiracle, anterad of the transverse row; syntergum triangular, faintly reticulate-punctate, bearing 6–8 setae in 1–2 transverse rows. Cerci about equidistant from posterior spiracles and tip of abdomen, with 2 long setae and 1 short seta. Ovipositor shaft about $1\frac{4}{5}$ times as long as the middle tibia (1.76–1.85); ovipositor sheaths relatively long, $\frac{2}{5}$ to nearly $\frac{1}{2}$ length of middle tibia (0.42–0.48).

Mid-tibial spur somewhat shorter than the corresponding basitarsus.

Forewing (Figure 1291) relatively broad, about $2\frac{1}{3}$ times as long as wide; marginal fringe short, not exceeding $\frac{1}{6}$ width of disk. Delta area with 63–85 setae in 6–8 rows, these considerably sparser and longer, somewhat coarser, than the setae distad of speculum, not quite distinct from row of setae along posterior margin of wing, widely separated from a small group of about 6 setae below distal half of submarginal vein; costal cell with a longitudinal row of 3–5 setae, widely separated from 1–2 coarse setae near apex. Submarginal vein bearing 2–3 coarse setae and 15–18 bullae. Marginal vein bearing 9–11 prominent, subequal setae along anterior margin, these $1\frac{1}{4}$ to $1\frac{2}{5}$ times longer than the setae in a row along center of vein.

Forewing nearly entirely hyaline, faintly infumated below junction of submarginal and marginal veins and at base of delta; under phase contrast, the basal half appears very faintly infumated; a narrow fuscous streak along posterior margin of wing, distad of speculum.

General coloration yellow; posterior margin of scutellum narrowly lined with blackish; tegula dusky; a short blackish streak near base of forewing, below tegula; prosternum pale, meso- and metasternum distinctly dusky (see Figure 1290); plates around base of ovipositor faintly dusky. Antennae faintly, rather uniformly dusky. Tibiae and tarsi of all legs faintly dusky. Wing veins colorless, faintly lined with brownish.

Length 0.81–0.85 mm.

*Male.* Essentially similar to the female in structure, chaetotaxis, sculpture and coloration.

Antennal scape (Figures 1292, 1293) about $4\frac{3}{5}$ to a little over 5 times as long as wide, considerably longer than the short club; pedicel $1\frac{3}{5}$ to $1\frac{7}{10}$ times as long as wide, about $1\frac{1}{3}$ times longer than the third segment of the funicle; first funicular segment about $1\frac{3}{5}$ times as wide as long, second segment twice as wide as long; third funicular segment as long as wide or slightly longer than wide, bearing 1 longitudinal sensillum; club short, 2 to $2\frac{1}{5}$ times as long as wide, $2\frac{3}{10}$ to $2\frac{3}{5}$ times longer than the preceding segment, bearing 2 longitudinal sensilla, with a distinct, cut-off area bearing short setae on the ventral surface (Figure 1293).

Mesoscutum (Figure 1294) with 14–20 setae, each parapsis with 2–3. Propodeum (Figure 1295) $3\frac{2}{5}$ to $3\frac{3}{5}$ times as long as the metanotum, $\frac{3}{5}$ to $\frac{2}{3}$ length of scutellum; crenulae 4+4 to 5+6, as in the female.

Genitalia about $\frac{2}{3}$ length of middle tibia (0.67–0.69); digital sclerites about $\frac{1}{5}$ to $\frac{1}{4}$ the combined length of aedeagus and apodemes (0.22–0.25).

Forewing (Figure 1296) broad, $2\frac{1}{5}$ to $2\frac{1}{3}$ times as long as wide; marginal fringe not exceeding $\frac{1}{7}$ width of disk; delta with 49–65 setae; marginal vein bearing 8–10 prominent, subequal setae along anterior margin.

Length 0.74–0.83 mm.

Described from 4♀♀ and 3♂♂ (♀ holotype, ♂ allotype, and paratypes), reared by G. J. Snowball from "*Gascardia mimosae* material" on *Acacia karroo*, Pienaarspoort, Transvaal, South Africa, April 1966, together with the types of *A. pilosus*. Also from 11♀♀, 11♂♂, reared by H. P. Insley from an undetermined host on *Scutia myrtina*, Oribi Gorge, South Africa, August 1972.

Type series in the collection of the Division of Biological Control, University of California, Riverside.

**Notes.** *A. mimosae* cannot be assigned to any of the seven species groups of *Aphytis*. Although the short antennal club, relatively short propodeum and dark thoracic setae may superficially appear to place this species in proximity to the **mytilaspidis** group, *mimosae* differs markedly in general habitus and does not seem to be closely related to this group. The rather large, coarse propodeal crenulae of *mimosae* are similar to those of *pilosus* and *wallumbillae*, and probably to those of some bona fide members of the **vittatus** group. However, *mimosae* differs from *pilosus* in the absence of long setae on the male antennae and a single long seta on the third funicular segment of the female, from *wallumbillae* in its general coloration.

The host record for the type series is presumably erroneous: *Gascardia* is, of course, a wax scale (Homoptera: Coccidae); we assume that *mimosae*, like all other species of *Aphytis*, develops as a parasite of armored scale insects and must have emerged from a host that went unnoticed in the sample of *Gascardia*.

**Additional Material.** 1♀ from the collection of F. Silvestri, labeled "*Aphelinus erythraeus* Silv. Var.?, ♀, Nefasit, olivo," was apparently reared by Silvestri from *Aspidiotus elaeidis* Marchal (recorded as *A. oppugnatus* Silvestri) on olive, Nefasit, Eritrea, 1914, and was recorded by him (Silvestri, 1915) as a "form" of *erythraeus* (Silvestri's fig. LXVII, 6 apparently refers to this specimen). The specimen (Figures 1208 and 1209, p. 705) was remounted in Hoyer's medium from the original, badly deteriorated glycerin slide, and appears to be conspecific with the type series of *mimosae*. It is a little larger (0.99 mm in length), and has a slightly longer antennal club ($2\frac{1}{4}$ times as long as wide), 17 mesoscutal setae, a slightly longer propodeum ($4\frac{3}{10}$ times as long as the metanotum) with fewer crenulae (3 + 4), and a narrower forewing ($2\frac{2}{3}$ times as long as wide), but is otherwise very similar to *mimosae*. For the time being, this specimen is referred to *mimosae*.

## 89. **Aphytis salvadorensis** n.sp.

(Figures 1297–1315)

This peculiar, biparental Central American species may be readily recognized by the following unique combination of characters: Coloration entirely yellow, immaculate;

genal sutures parallel; each parapsis with 1 seta, axillae asetose; propodeal crenulae rounded, overlapping; syntergum broad, arcuate; forewing hyaline, sparsely setose, setae along marginal vein decreasing in length toward apex of vein. Male with a specialized sense organ on antennal scape, genitalia truncate basally, without apodemes, middle tibia contrastingly tipped with black.

*Female.* Eyes finely setose. Mandibles (Figure 1297) well developed, with 2 denticles and a rounded dorsal truncation; maxillary palpi 2-segmented, labial palpi 1-segmented. Genal sutures (Figure 1298, see also Figure 1307) curving at an acute angle, then running parallel to each other. Antennal scape (Figure 1299) slender, 5 to over 6 times as long as wide, $1\frac{1}{4}$ to $1\frac{1}{3}$ times longer than the club; pedicel about $1\frac{1}{2}$ to $1\frac{4}{5}$ times as long as wide, $1\frac{1}{4}$ to $1\frac{2}{5}$ times longer than the third segment of the funicle; first funicular segment trapezoidal, $1\frac{1}{3}$ to $1\frac{1}{2}$ times as wide as long; second segment somewhat more symmetrical, somewhat shorter and wider than the first, about $1\frac{3}{4}$ times to twice as wide as long; third funicular segment rather short, usually somewhat wider than long, bearing 2–3 longitudinal sensilla; club $2\frac{1}{2}$ to a little over $2\frac{4}{5}$ times as long as wide, $2\frac{3}{4}$ to $3\frac{1}{3}$ times longer and somewhat wider than the preceding segment, bearing 6–8 longitudinal sensilla.

Setae on head and thorax moderately coarse, dark, visible under × 30 magnification; those on abdomen slender, paler, invisible even under × 120 magnification. Vertex with 2 pairs of long setae along occipital margin, in addition to numerous short setae. Mesoscutum (Figure 1300) with 10 setae, the posterior pair and 1 seta at each antero-lateral corner considerably longer than the others; each parapsis with 1 seta, axillae asetose; scutellum with 4 setae, the discoid sensilla considerably closer to the anterior than to the posterior pair. Frontovertex, pronotum and mesonotal sclerites reticulate. Scutellum almost rectangular, about $\frac{3}{4}$ to $\frac{4}{5}$ median length of mesoscutum. Metanotum (Figures 1301, 1302) reticulate except on the sides; anteromedian apodeme very slender, usually considerably shorter than median length of metanotum.

Propodeum (Figures 1302, 1303) about $2\frac{4}{5}$ to $3\frac{4}{5}$ times as long as the metanotum, $\frac{3}{4}$ to $\frac{4}{5}$ length of scutellum, reticulate on a broad trapezoidal central area, coarsely sculptured mesad of spiracles, with a rather prominent median salient; crenulae 3+4 to 6+6, moderately large, rounded, overlapping.

Second abdominal tergite faintly transversely striated; tergites III–VII reticulate on the sides, bearing a few fine setae on each reticulate area, with faint indications of transverse striation across center; seventh tergite asetose centrally; eighth tergite (Figure 1304) large, delicately reticulate-strigose, with a transverse arcuate row of 4 setae between the spiracles, along the posterior margin; syntergum (Figure 1304) arcuate, v-shaped, bearing 14–18 setae in a transverse, arcuate row. Cerci situated very close to the posterior spiracles, with 2 long setae and 1 short seta. Ovipositor shaft about $1\frac{2}{3}$ to $1\frac{4}{5}$ times as long as the middle tibia (1.69–1.83); ovipositor sheaths about $\frac{2}{5}$ length of middle tibia (0.37–0.43).

Mid-tibial spur somewhat shorter than the corresponding basitarsus.

Forewing (Figure 1305) about $2\frac{2}{5}$ times as long as wide; marginal fringe short, not exceeding $\frac{1}{7}$ width of disk; anterior margin of wing not in a straight line, indented in region of stigma. Delta area with 31–39 setae in 5–6 rows, these longer and considerably sparser than the setae distad of speculum, not clearly separated from sparse row of setae along posterior margin of wing, widely separated from 2–4 setae below distal half of submarginal vein; costal cell with 3–4 fine setae along central portion, and 1–2 coarse setae near apex. Submarginal vein bearing 2 coarse, long setae, the proximal one about $\frac{2}{3}$ to $\frac{4}{5}$ length of the distal, followed by a considerably shorter coarse seta, and 14–21 bullae. Marginal vein (Figure 1306) bearing 8–11 prominent, unequal setae along anterior margin which decrease considerably in length toward the apex of the vein, the proximal setae $1\frac{3}{5}$ times to nearly twice as long as the distal ones and about $1\frac{1}{4}$ to $1\frac{1}{3}$ times longer than the setae in a row along center of vein.

Forewing perfectly hyaline; a faintly fuscous streak along posterior margin, distad of speculum.

Length 0.98–1.12 mm.

*Male*. Essentially similar to the female in structure, chaetotaxis, sculpture and general coloration, differing mainly in antennal characteristics and in the pigmentation of the middle tibia.

Genal sutures (Figure 1307) as in the female. Antennal scape (Figures 1307–1309) $4\frac{1}{3}$ to $5\frac{1}{4}$ times as long as wide, with an oval plate on the ventral aspect bearing a pair of minute tuberculous sensilla; club $2\frac{1}{2}$ to nearly 3 times as long as wide, bearing 4–5 longitudinal sensilla.

Thorax (Figure 1310) and propodeum (Figure 1311) as in the female; crenulae 3+3 to 4+5. Abdominal tergites (Figure 1312) as in the female. Genitalia (Figure 1313) about $\frac{2}{3}$ to nearly $\frac{3}{4}$ length of middle tibia (0.64–0.72); phallobase narrow, truncate basally, aedeagus without distinct apodemes, digital sclerites slender, about $\frac{1}{5}$ length of genitalia (0.18–0.23).

Middle tibia (Figure 1314) swollen apically, contrastingly marked with black at apex; mid-tibial spur somewhat shorter than the corresponding basitarsus. Forewing (Figure 1315) as in the female; delta with 25–31 setae in 4–5 rows; marginal vein bearing 7–10 prominent, unequal setae along anterior margin, these decreasing in length toward apex of vein.

Length 0.75–0.92 mm.

Described from 6♀♀ and 8♂♂ (♀ holotype, ♂ allotype, and paratypes), reared by P. A. Berry from an undetermined scale insect, San Miguel, El Salvador, May 30, 1957; originally mounted in balsam (some specimens not cleared, others cleared in KOH, some dissected), subsequently remounted in Hoyer's medium.

Type series in the collection of the Division of Biological Control, University of California, Riverside.

**Notes.** *A. salvadorensis* is a unique species, presenting a peculiar combination of "primitive" characters, such as the unequal setae along the marginal vein of the forewing, the slender metanotal apodeme, and the presence of specialized sensilla on the antennal scape of the male, with presumably "advanced" characters such as the immaculate coloration and reduced chaetotaxis. It resembles *melanostictus* (p. 291) in several important characters, including the chaetotaxis of the mesonotum (asetose axillae), the shape of the syntergum, the presence of sensilla on the antennal scape of the male, and the black-tipped middle tibia of the male. However, it differs markedly from *melanostictus* in coloration and wing pattern, in the sparsely setose forewing, in the unequal setae along the marginal vein, and in the shape of the antennae and male genitalia, the latter being aberrant in the genus *Aphytis*. *A. salvadorensis* cannot be assigned to any of the seven species groups at present recognized in *Aphytis*.

## 90. **Aphytis ignotus** Compere

(Figures 1316–1328)

*Aphytis ignotus* Compere, 1955, Univ. Calif. Publ. Entomol., **10**:300.

This large, peculiar, biparental Australian species may be readily separated from all other species of *Aphytis* by the extensive stippling of the abdominal sterna in both sexes. It may be further recognized by the following combination of characters: Mandibles reduced; antennal club long; mesonotal setae numerous, dark; propodeum moderately long, crenulae numerous, small, nonoverlapping; ovipositor long; coloration entirely yellow, thoracic sterna immaculate; forewing broad, hyaline.

*Female*. Eyes finely setose. Mandibles (Figure 1316) reduced, nonoverlapping, weakly sclerotized, with 1 minute dorsal denticle and a broad, low ventral truncation; maxillary palpi 2-segmented, labial palpi 1-segmented (Figure 1316, see also Figure 1324). Antennal scape (Figure 1317) rather slender, usually 5 to 6 times as long as wide, as long as the club or somewhat longer; pedicel elongate, $1\frac{3}{4}$ times to fully twice as long as wide, considerably (about $1\frac{1}{3}$ to $1\frac{1}{2}$ times) longer than the third segment of the funicle; first funicular segment trapezoidal, up to $1\frac{1}{3}$ times as wide as long; second segment nearly symmetrical, subquadrate, somewhat shorter and wider than the first segment, $1\frac{2}{5}$ to $1\frac{3}{4}$ times as wide as long; third funicular segment rather short, about $1\frac{1}{10}$ to $1\frac{3}{10}$ times as long as wide, somewhat trapezoidal with the dorsal aspect longer than the ventral, bearing 1–2 longitudinal sensilla; club robust, cylindrical, nearly truncate, $3\frac{1}{5}$ to $3\frac{2}{3}$ times as long as wide, about $3\frac{1}{5}$ to $3\frac{3}{5}$ times longer and considerably wider than the preceding segment, bearing 5–8 longitudinal sensilla.

Setae on head and thorax rather long, coarse, dark, readily visible under ×30 magnification; those on sides of abdomen somewhat paler, visible in properly cleared specimens under ×60 magnification. Vertex with 2 pairs of long, coarse setae along occipital margin, in addition to numerous shorter setae. Mesoscutum (Figure 1318) with 15–20 setae, the posterior pair and 1 seta at each antero-lateral corner considerably longer and coarser than the others; each parapsis usually with 3 setae, rarely with 2, each axilla with 1 seta; scutellum with 4 setae, the discoid sensilla considerably closer to the anterior than to the posterior pair. Frontovertex, pronotum and mesonotal sclerites reticulate. Scutellum about $\frac{3}{4}$ to $\frac{4}{5}$ median length of mesoscutum. Metanotum (Figures 1318–1320) reticulate except on the sides; anteromedian apodeme rather robust, usually considerably shorter than median length of metanotum.

Propodeum (Figures 1318–1321) about $3\frac{1}{4}$ to $4\frac{1}{4}$ times as long as the metanotum,

$\frac{3}{5}$ to $\frac{2}{3}$ length of scutellum, reticulate on the sides, reticulate-strigose on a central trapezoidal area, with a distinct median salient; crenulae 7+7 to 11+11, small, distinct, oval, nonoverlapping or slightly overlapping at base, forming a continuous row without a medial gap.

Second abdominal tergite (Figures 1319–1321) transversely reticulate-striated on the sides, transversely striated on a small central area; tergites III–VII reticulate on the sides, bearing a few fine setae on each reticulate area, extensively sculptured across center, between the lateral setiferous areas; third tergite (Figure 1321) transversely reticulate-striated anteriorly, reticulate-dentate posteriorly; tergites IV–VII similar, with indications of longitudinal striation posteriorly; seventh tergite usually bearing 2 fine submedian setae; eighth tergite reticulate, with an arcuate, irregular row of 7–10 setae between spiracles; syntergum elongate, relatively narrow, faintly reticulate-punctate, bearing 11–12 setae in several irregular rows. Cerci situated rather close to one another, about equidistant from posterior spiracles and tip of abdomen, with 2 long setae and 1 short seta. Abdominal sternites III–VII smooth anteriorly, reticulate posteriorly, extensively stippled centrally (Figure 1322): the reticulate areas bear transverse rows of robust denticles; on sternites V–VII, the anterior smooth areas bear dense rows of minute, elongate spines. Ovipositor long, the shaft twice as long as the middle tibia or nearly so (1.89–2.02), the sheaths about $\frac{1}{2}$ to $\frac{3}{5}$ length of middle tibia (0.52–0.58); ovipositor plates narrow.

Mid-tibial spur (Figure 1323) long, usually only somewhat shorter than the corresponding basitarsus.

Forewing (Figure 1323) broad, about $2\frac{1}{4}$ to $2\frac{1}{3}$ times as long as wide, marginal fringe short, not exceeding $\frac{1}{8}$ width of disk. Delta area rather sparsely setose, with 38–52 setae in 4–5 rows, these considerably longer and sparser than the setae distad of speculum, not clearly separated from row of setae along posterior margin of wing, widely separated from a few setae below distal half of submarginal vein; costal cell with a row of fine setae along proximal two thirds or so, and 1–2 coarse setae near apex. Submarginal vein bearing 2 coarse setae, the proximal one about $\frac{2}{3}$ to $\frac{4}{5}$ length of the distal, and 16–24 bullae. Marginal vein bearing 10–14 prominent, subequal setae along anterior margin, these as long as or up to $1\frac{1}{5}$ times longer than the setae in a row along center of vein.

Forewing nearly hyaline, faintly infumated on basal half, slightly more noticeably so below junction of submarginal and marginal veins and at base of delta; a fuscous streak along posterior margin of wing, distad of speculum.

General coloration yellow; thoracic sterna immaculate; posterior margin of scutellum narrowly lined with black; tegula dusky distally; a short black streak at base of forewing, below tegula. Antennal scape and pedicel pale, funicle and club uniformly dusky. Legs concolorous with body. Wing veins hyaline.

Length 1.23–1.45 mm.

*Male.* Essentially similar to the female in structure, chaetotaxis, sculpture and coloration.

Palpi shown in Figure 1324. Antennal scape (Figure 1325) about $4\frac{3}{4}$ to $5\frac{2}{5}$ times as long as wide, slightly longer than the club; pedicel $1\frac{2}{3}$ times to twice as long as wide, $1\frac{2}{5}$ to $1\frac{1}{2}$ times longer than the third segment of the funicle; first funicular segment $1\frac{1}{4}$ to $1\frac{3}{5}$

times as wide as long, second segment $1\frac{2}{5}$ to $2\frac{2}{5}$ times as wide as long; third funicular segment only up to $1\frac{1}{5}$ times as long as wide, bearing 1–2 longitudinal sensilla; club about $2\frac{4}{5}$ to $3\frac{3}{4}$ times as long as wide, up to $3\frac{2}{5}$ times longer than the preceding segment, bearing 3–4 longitudinal sensilla, with an extensive area on the ventral aspect bearing numerous minute sensory setae, not cut off from surrounding areas.

Mesoscutum with 15–22 setae. Propodeum (Figure 1326) somewhat shorter than in the female, $2\frac{3}{4}$ to $3\frac{4}{5}$ times as long as the metanotum, about $\frac{1}{2}$ to $\frac{3}{5}$ length of scutellum; crenulae 6+6 to 7+10, as in the female.

Sculpture of abdominal tergites and sternites (Figures 1326, 1327) as in the female; eighth sternite stippled like sternites V–VII. Genitalia (Figure 1328) about $\frac{2}{3}$ length of middle tibia (0.61–0.72); digital sclerites $\frac{1}{5}$ to $\frac{1}{4}$ the combined length of aedeagus and apodemes (0.20–0.26).

Mid-tibial spur as long as the corresponding basitarsus, or somewhat longer. Forewing even broader than in the female, up to $2\frac{1}{5}$ times as long as wide; delta with 28–43 setae; submarginal vein bearing 16–19 bullae; marginal vein bearing 9–14 prominent, subequal setae along anterior margin.

Length 0.62–1.22 mm.

Redescribed from the following material: 3♀♀, 2♂♂ (paratypes), collected by H. Compere on *Phoenix canariensis*, the females observed ovipositing in the black araucaria scale, *Lindingaspis rossi* (Maskell), Botanical Gardens, Sydney, N.S.W., Australia, September 17, 1927; 1♀ (paratype), collected by H. Compere on *Kentia*, Marrickville, Sydney, Australia, October 6, 1927; 1♂, collected by H. Compere in Centennial Park, Sydney, Australia, October 25, 1927; 2♀♀, 1♂ (holotype ♀, allotype ♂ and paratype ♀), reared by S.E. Flanders from *L. rossi*, Sydney, Australia, April 21, 1931; 1♀, reared by S. E. Flanders from an undetermined host, Sydney, Australia, May 26, 1931; 7♀♀, 2♂♂, reared by E. W. Valentine from *L. rossi* on *Phebalium*, New Plymouth, New Zealand, October 16, 1962; 17♀♀, 1♂, reared by E. W. Valentine from *L. rossi* on *Phebalium*, Nelson, New Zealand, October 19, 1962.

Type series in the collection of the Division of Biological Control, University of California, Riverside.

**Notes.** *A. ignotus* is an unusual species. It resembles the **chrysomphali** group in the shape of the antennae and propodeum, the **mytilaspidis** group in the dark thoracic setae; it differs from all known species of *Aphytis* in the very extensive stippling of the abdominal sternites. Although the long, cylindrical antennal club of *ignotus* is quite similar to that of *cylindratus* (p. 604), we do not agree with Compere (1955) that these two species are very closely related. *A. ignotus* differs markedly from *cylindratus* in the extensive abdominal stippling, as well as in the reduced mandibles, numerous mesonotal setae and long ovipositor. As pointed out by Compere (1955), *ignotus* is rather similar to *erythraeus* in the female sex, but differs markedly from this species in the male sex. At the present state of our knowledge, the affinities of this peculiar Australian species cannot be determined with any degree of certainty.

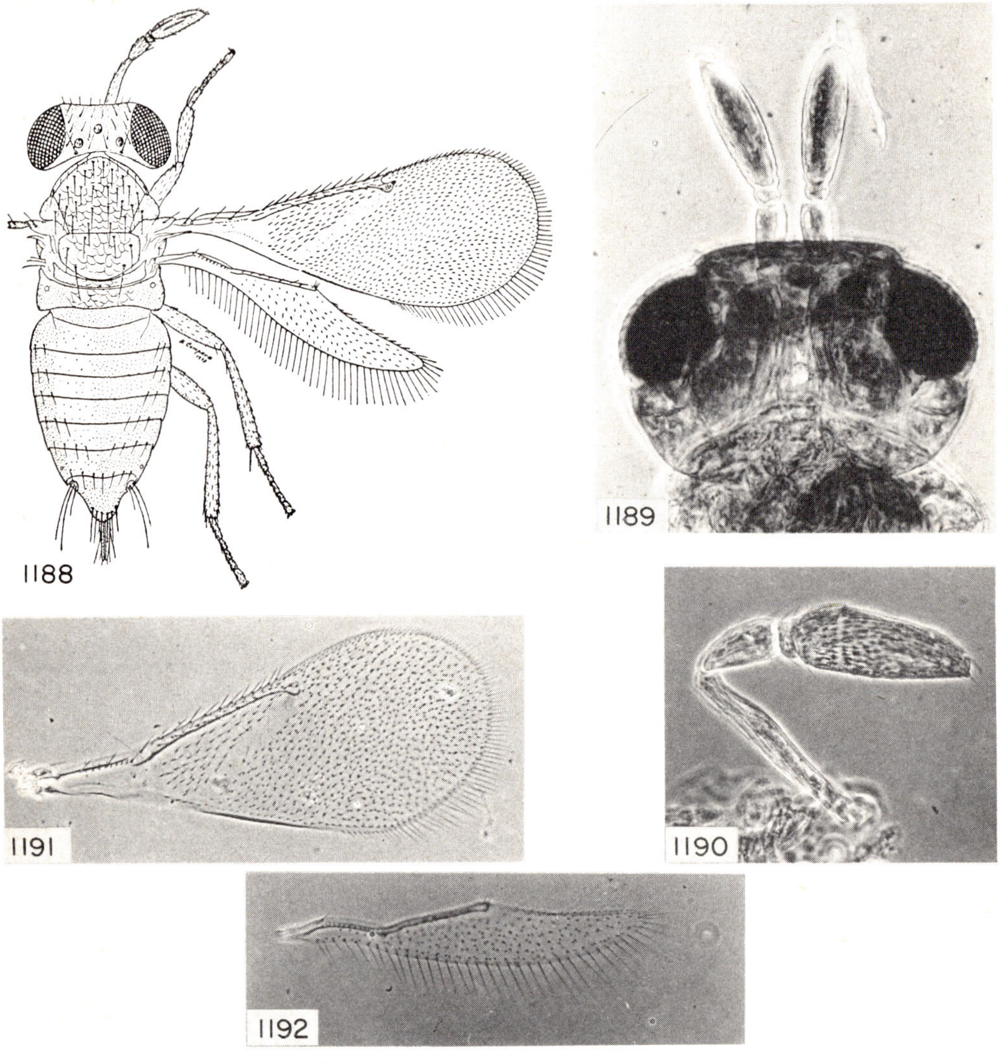


Figures 1188–1192. *Aphytis secundus* (Compere) (syntypes)
1188. ♀ (from Compere, 1936); note: the scutellum bears only 2 pairs of setae, not 3 pairs as shown here. 1189. ♀: Head and antennae (from Rosen and DeBach, 1970). 1190. ♂: Antenna. 1191. ♂: Forewing (from Rosen and DeBach, 1970). 1192. ♂: Hind wing.

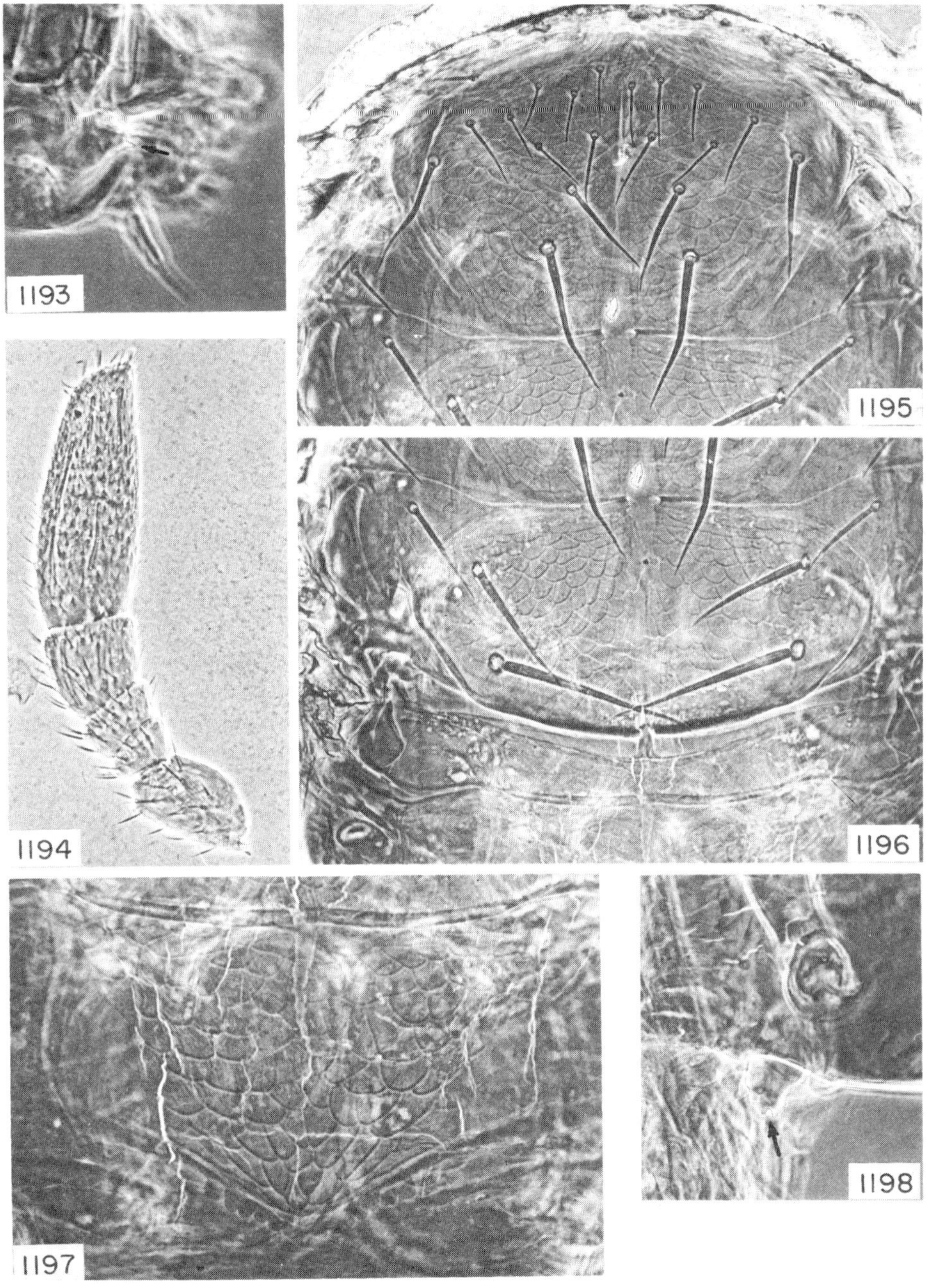


Figures 1193–1198. *Aphytis erythraeus* (Silvestri) (syntypes)
1193. ♀: Mandible. 1194. ♀: Antennal pedicel, funicle and club (apparently flattened on slide).
1195. ♀: Mesoscutum. 1196. ♀: Scutellum and metanotum. 1197. ♀: Propodeum. 1198. ♂: Mandible.

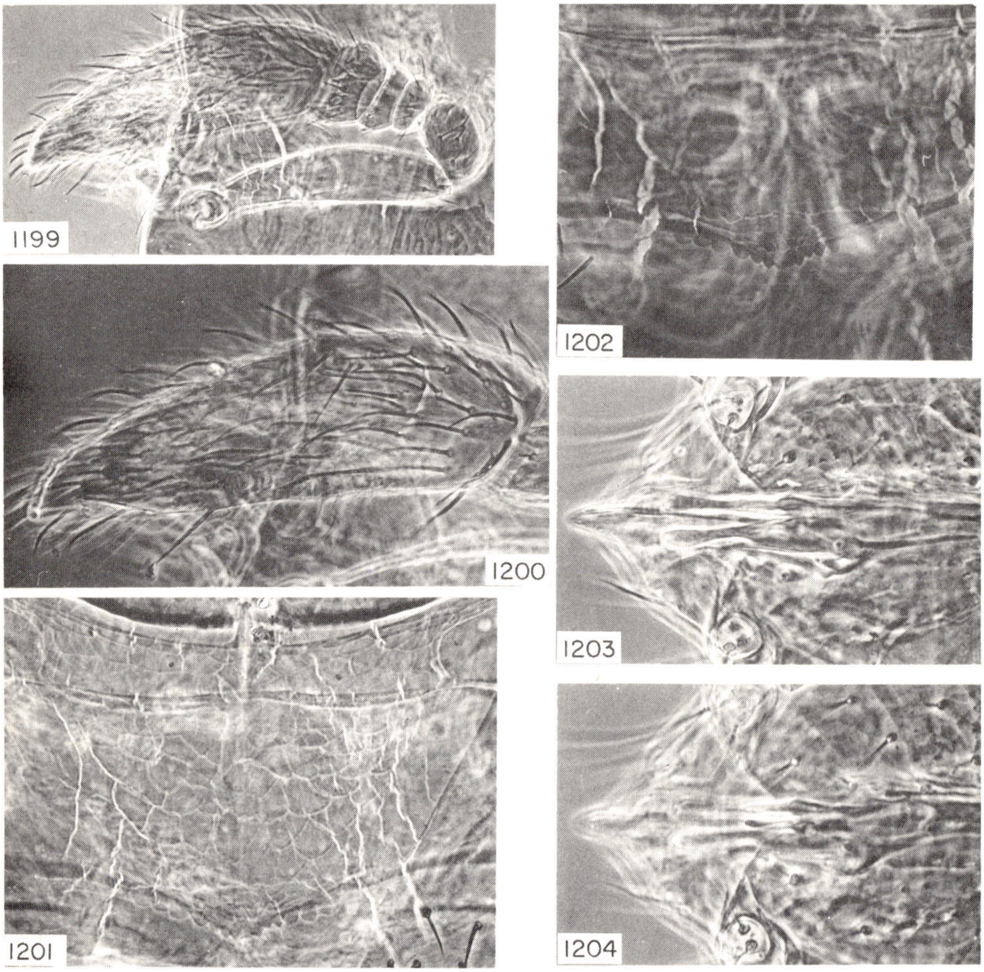


Figures 1199–1204. *Aphytis erythraeus* (Silvestri), ♂ (syntypes)
1199. Antenna. 1200. Antennal club, showing long setae. 1201. Metanotum and propodeum. 1202. Propodeum. 1203. Genitalia, focus on digital sclerites. 1204. Genitalia (same as in Figure 1203), focus on elongate papillae.

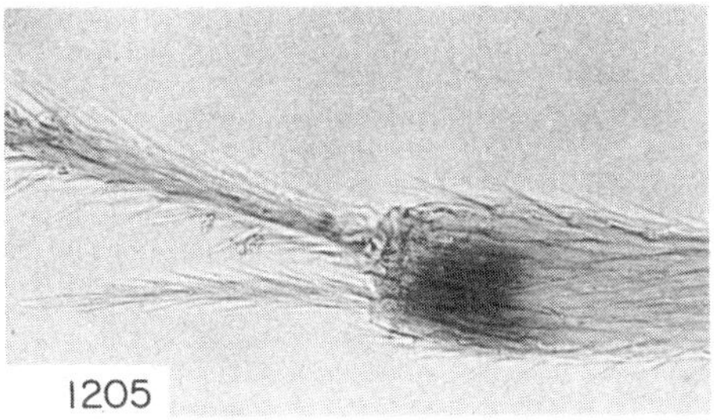


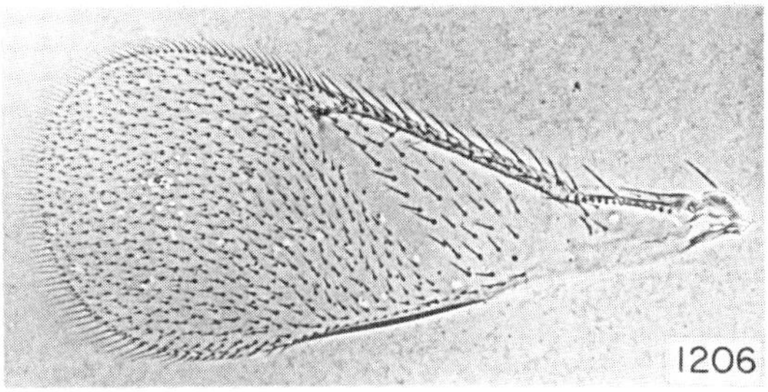


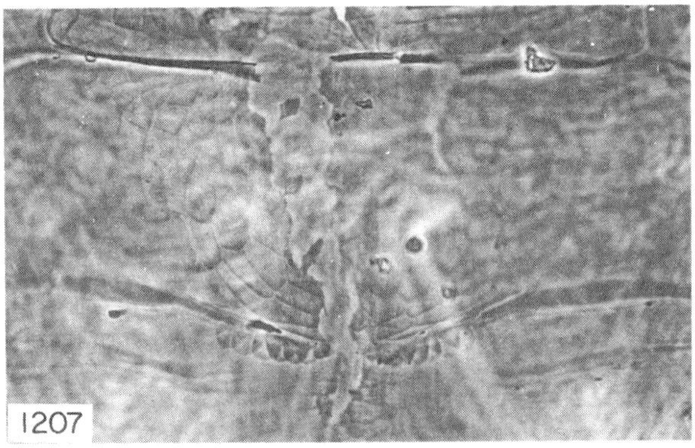


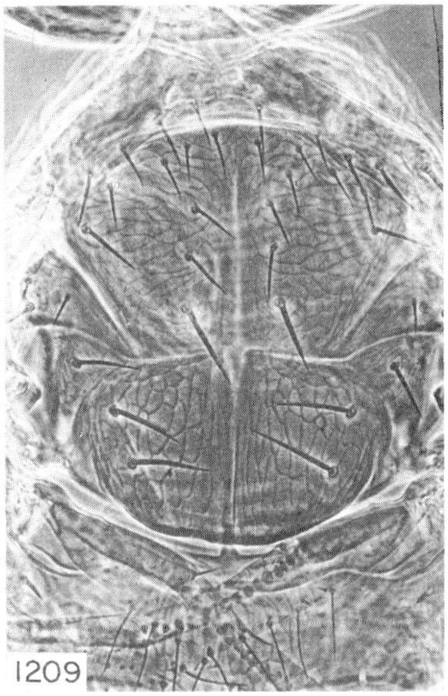


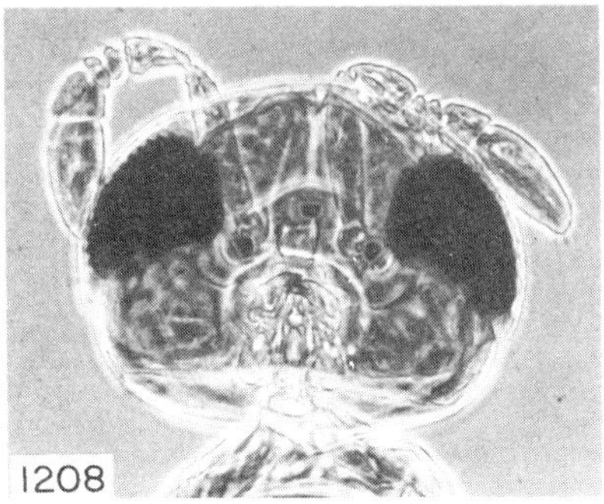


Figures 1205, 1206. *Aphytis erythraeus* (Silvestri), ♂ (syntypes)
1205. Mid-tibial spur and basitarsus; note pigmentation of middle tibia. 1206. Forewing.

Figure 1207. *Aphytis ?africanus*, misidentified by Silvestri as *erythraeus*, ♀: Propodeum; note overlapping crenulae.

Figures 1208, 1209. *Aphytis mimosae*, recorded by Silvestri as a "form" of *erythraeus*, ♀
1208. Head and antennae. 1209. Thorax.

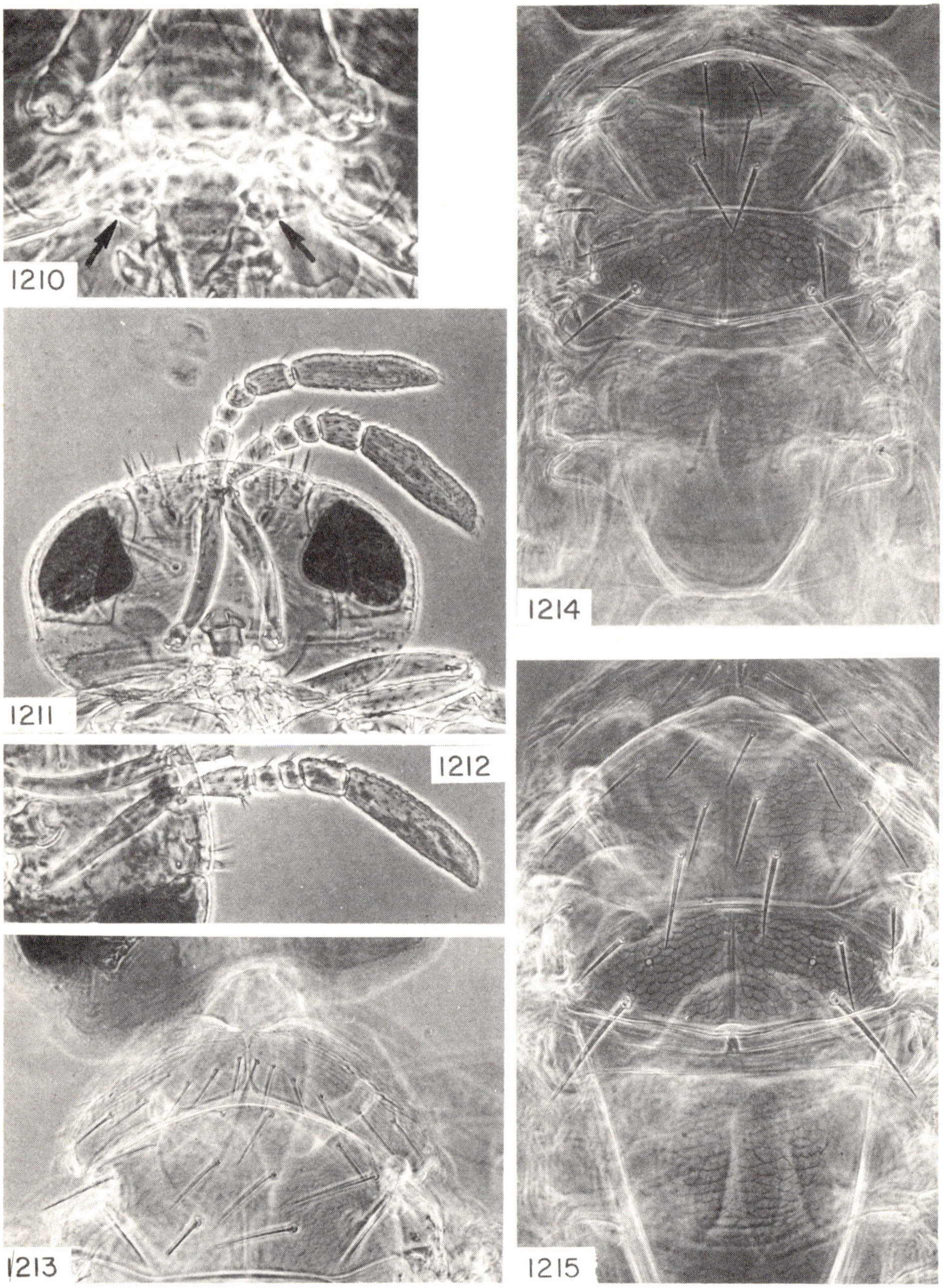


Figures 1210–1215. *Aphytis roseni* DeBach and Gordh, ♀
1210. Mandibles (paratype). 1211. Head and antennae (holotype). 1212. Antenna (paratype).
1213. Pronotum, showing broadly joined plates (paratype). 1214. Thorax and propodeum.
1215. Thorax and propodeum (paratype).

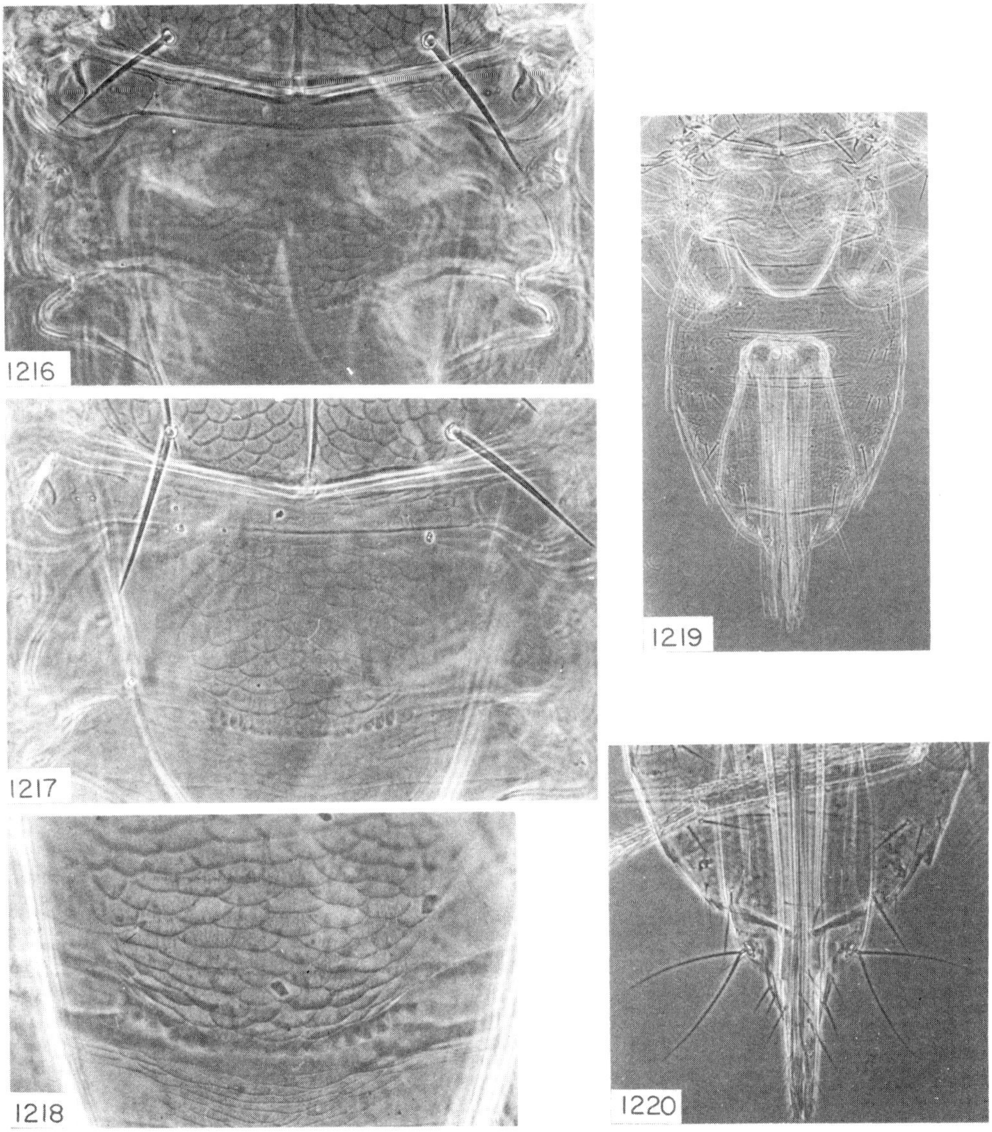


Figures 1216–1220. *Aphytis roseni* DeBach and Gordh, ♀
1216, 1217. Metanotum and propodeum. 1218. Propodeal crenulae (paratype). 1219. Abdomen.
1220. Posterior abdominal tergites, showing cerci and posterior spiracles (paratype).

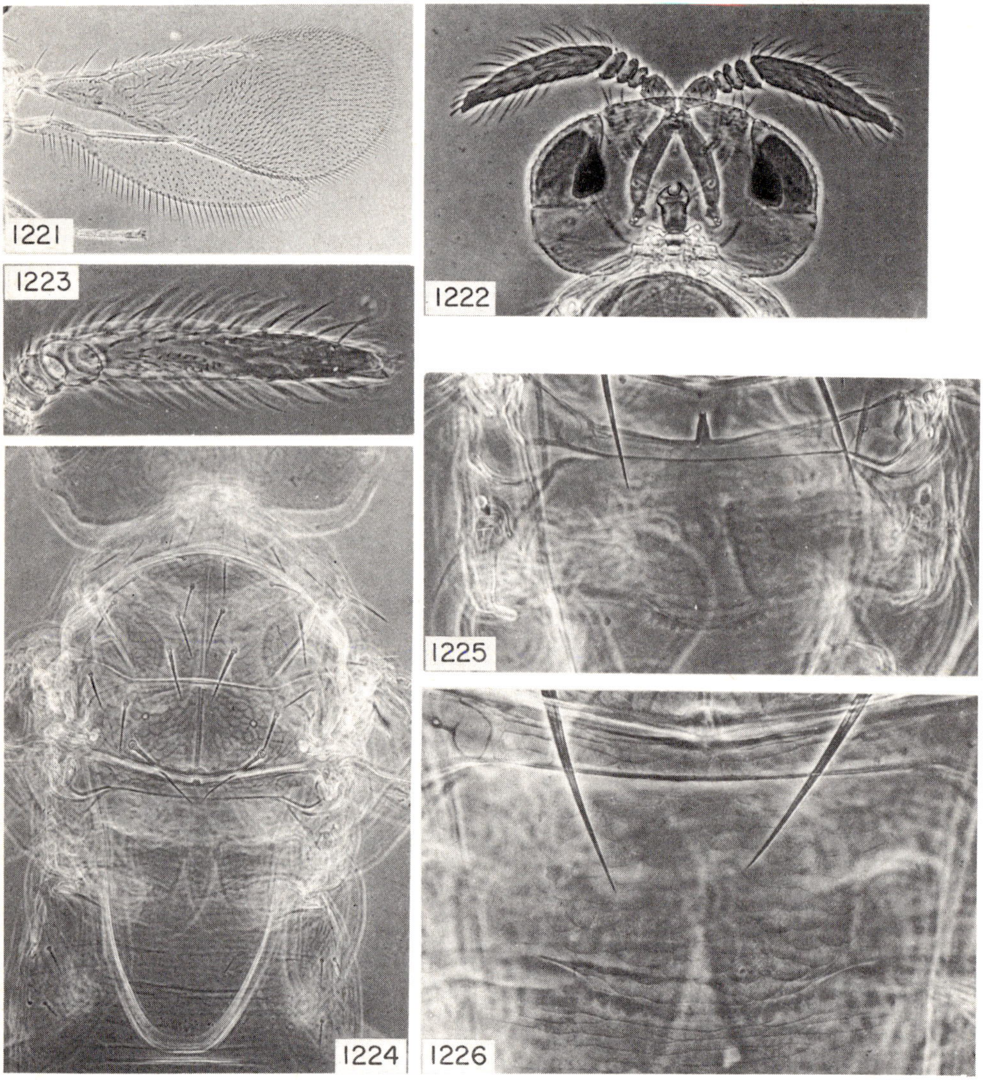


Figures 1221–1226. *Aphytis roseni* DeBach and Gordh
1221. ♀: Wings (paratype). 1222. ♂: Head and antennae (allotype). 1223. ♂: Sensory area on venter of antennal club (paratype). 1224. ♂: Thorax and propodeum (paratype). 1225. ♂: Metanotum and propodeum (paratype). 1226. ♂: Propodeum and crenulae (paratype).

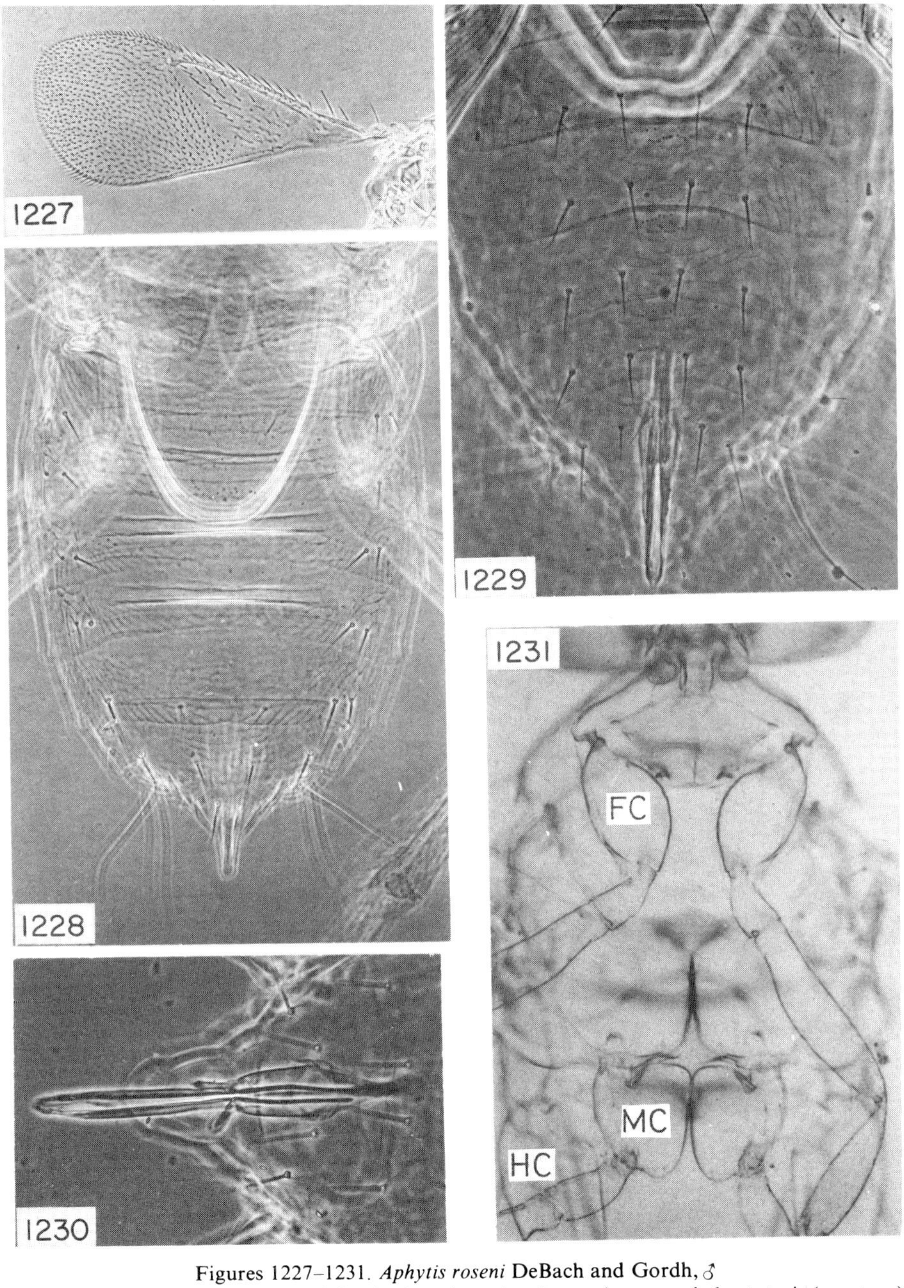


Figures 1227–1231. *Aphytis roseni* DeBach and Gordh, ♂
1227. Forewing (paratype). 1228. Abdominal tergites, showing sculpture and chaetotaxis (paratype). 1229. Abdominal sternites, showing sculpture and chaetotaxis, and genitalia. 1230. Genitalia (paratype). 1231. Thoracic sterna, showing pattern of pigmentation (paratype).
*FC* = fore coxa; *HC* = hind coxa; *MC* = middle coxa.

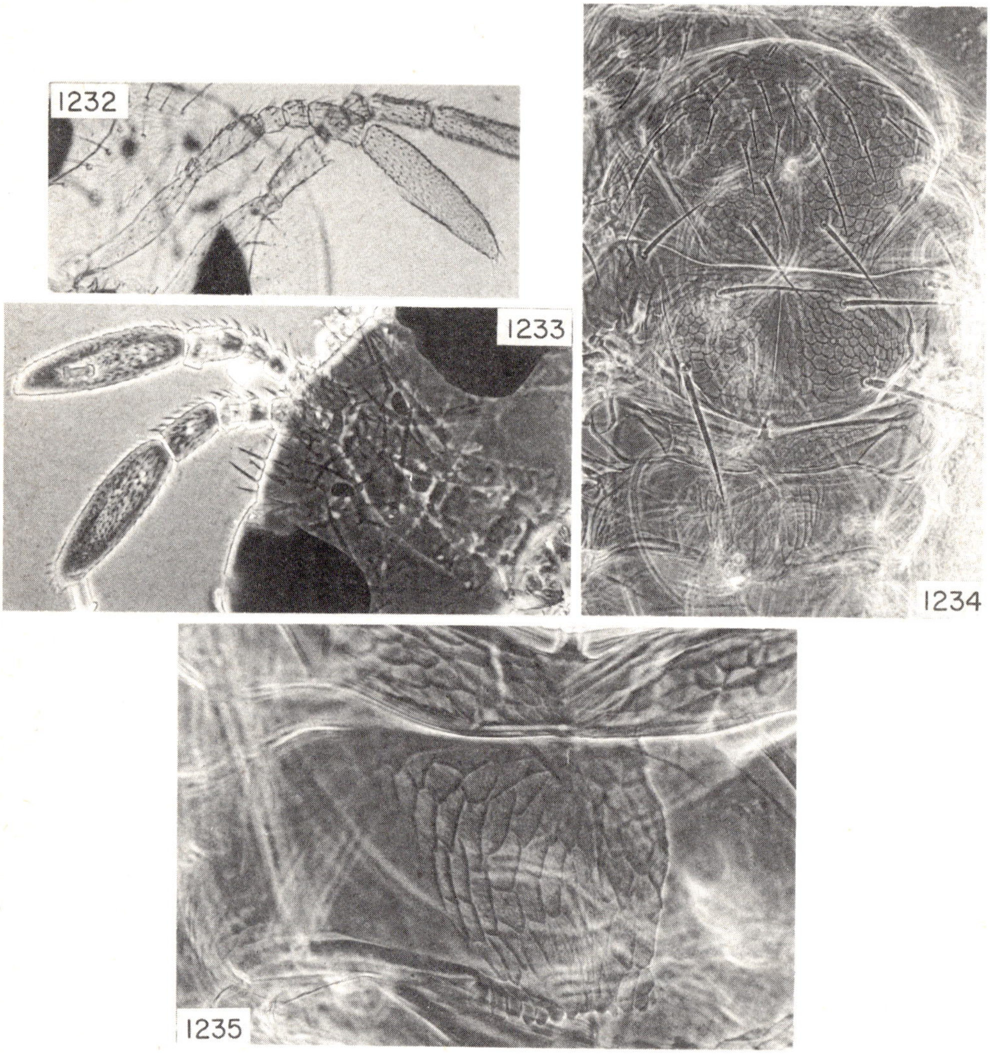


Figures 1232–1235. *Aphytis longicaudus* n.sp., ♀
1232. Antenna (holotype). 1233. Antennae (paratype; some artifacts present). 1234. Thorax and propodeum (holotype). 1235. Metanotum and propodeum (paratype).

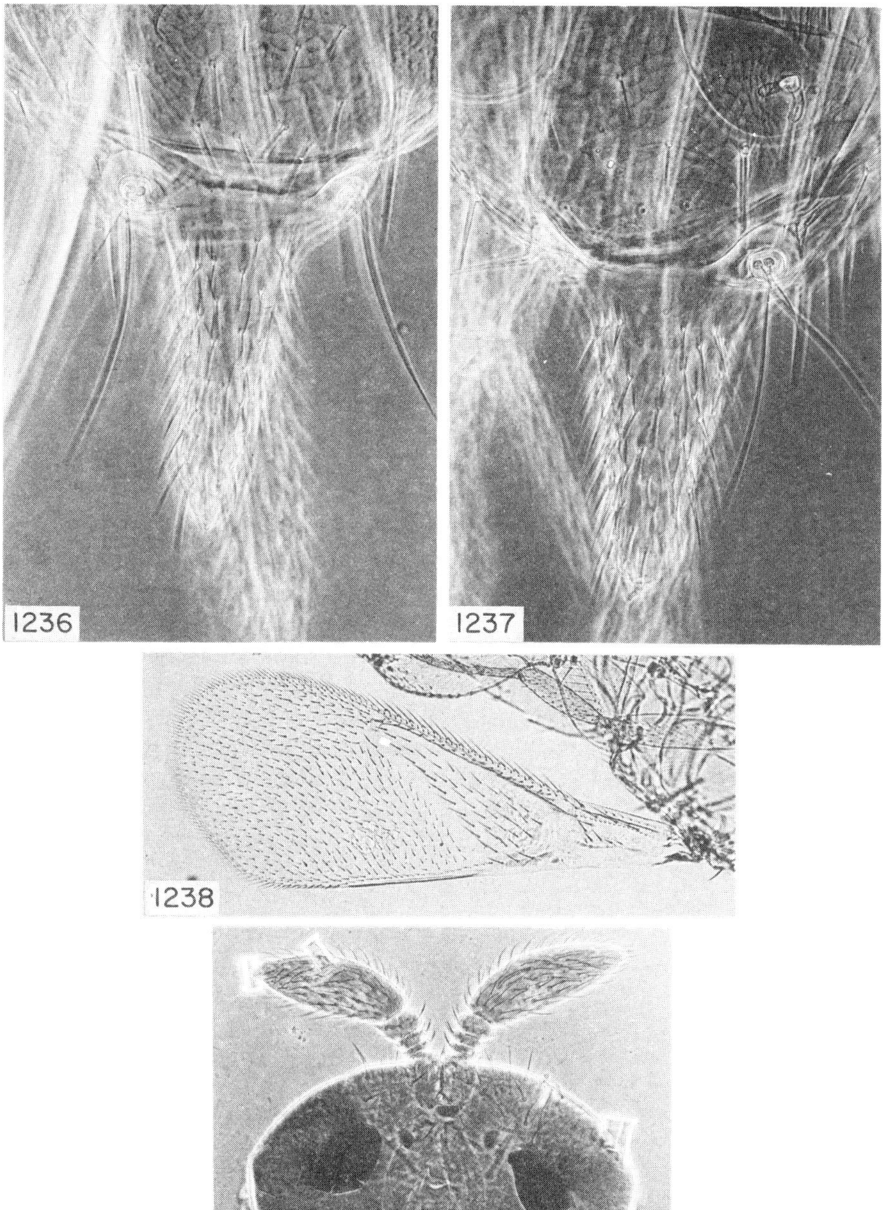


Figures 1236–1239. *Aphytis longicaudus* n.sp.
1236. ♀: Syntergum; note elongate cauda (holotype). 1237. ♀: Eighth abdominal tergite and syntergum (paratype). 1238. ♀: Forewing (holotype). 1239. ♂: Head and antennae (paratype; some artifacts present).

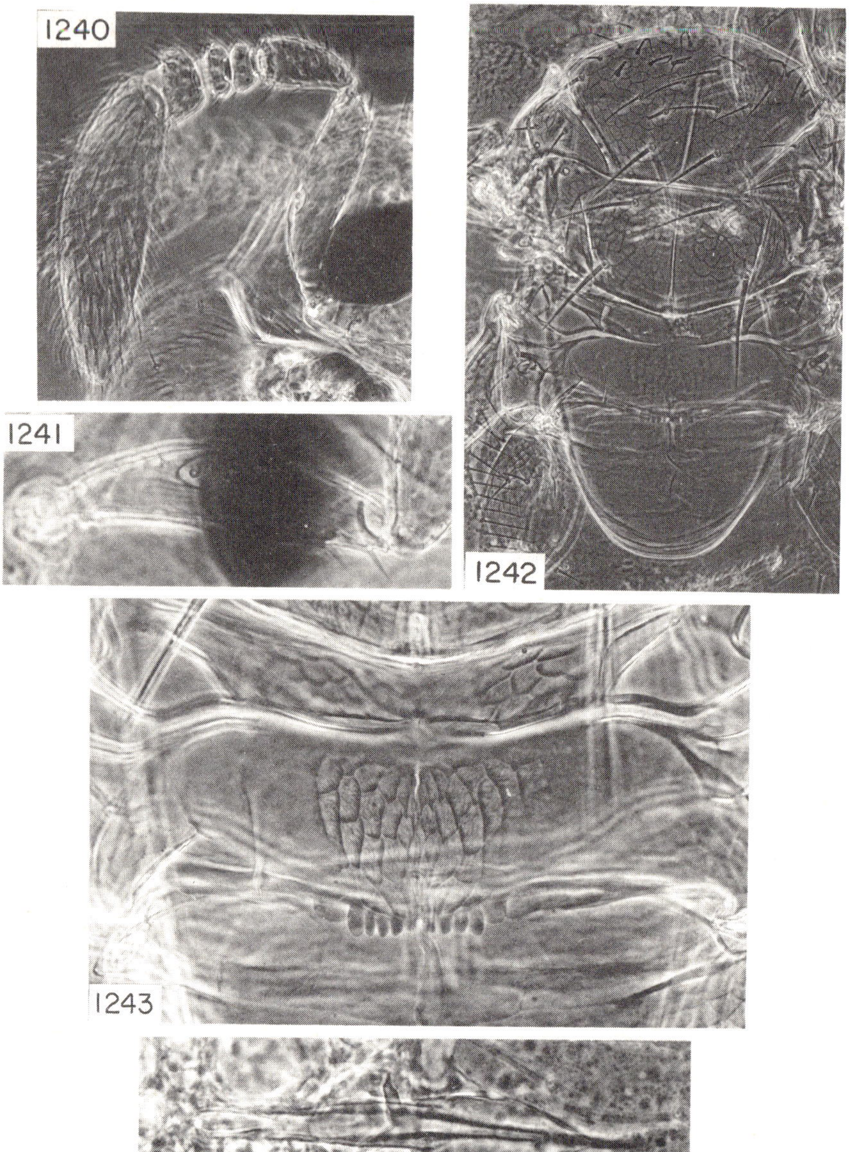


Figures 1240–1244. *Aphytis longicaudus* n.sp., ♂ (allotype)
1240. Antenna; note sensilla on scape. 1241. Antennal scape, showing sensilla. 1242. Thorax, propodeum and endophragma. 1243. Metanotum and propodeum. 1244. Genitalia.

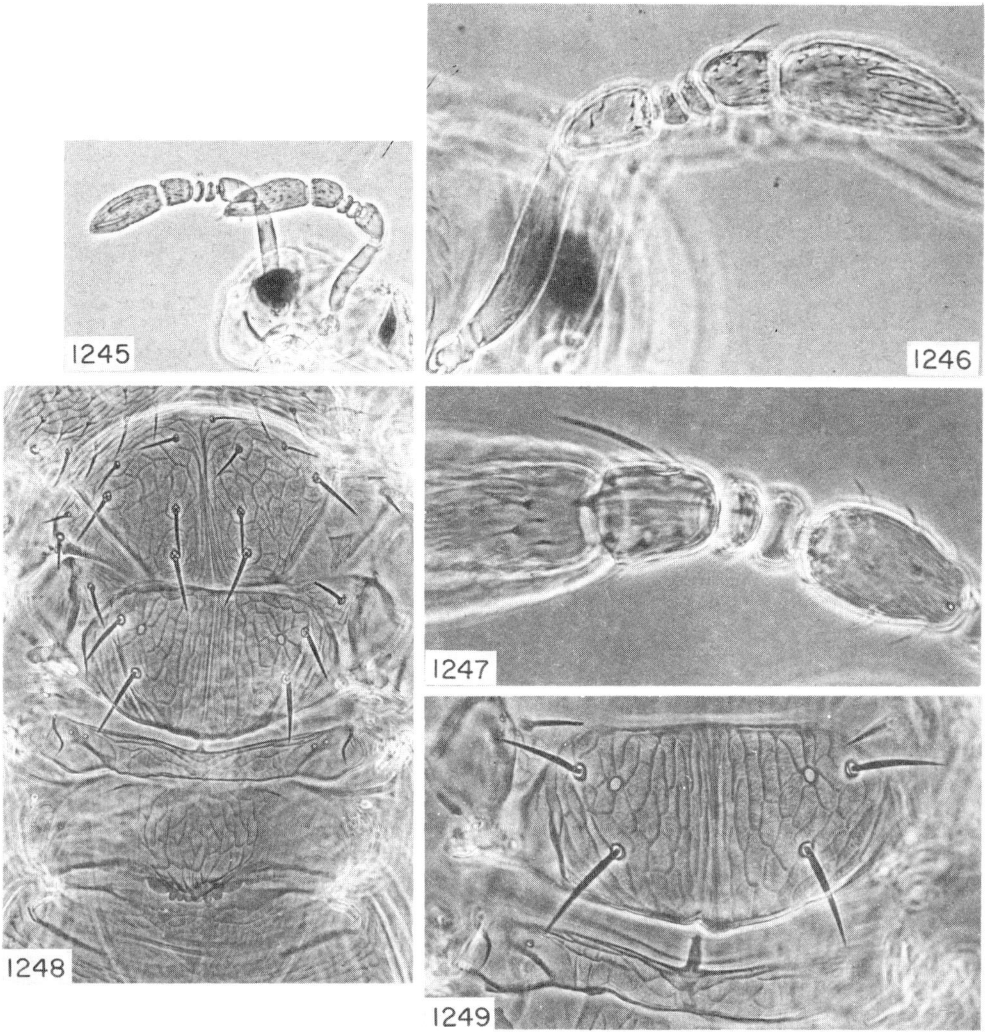


Figures 1245–1249. *Aphytis pilosus* DeBach and Rosen, ♀
1245. Antennae (holotype). 1246. Antenna; note long seta on third funicular segment (paratype).
1247. Antennal pedicel, funicle and base of club, showing long seta on third funicular segment (paratype).
1248. Thorax, propodeum and base of gaster (paratype). 1249. Scutellum and metanotum (paratype).

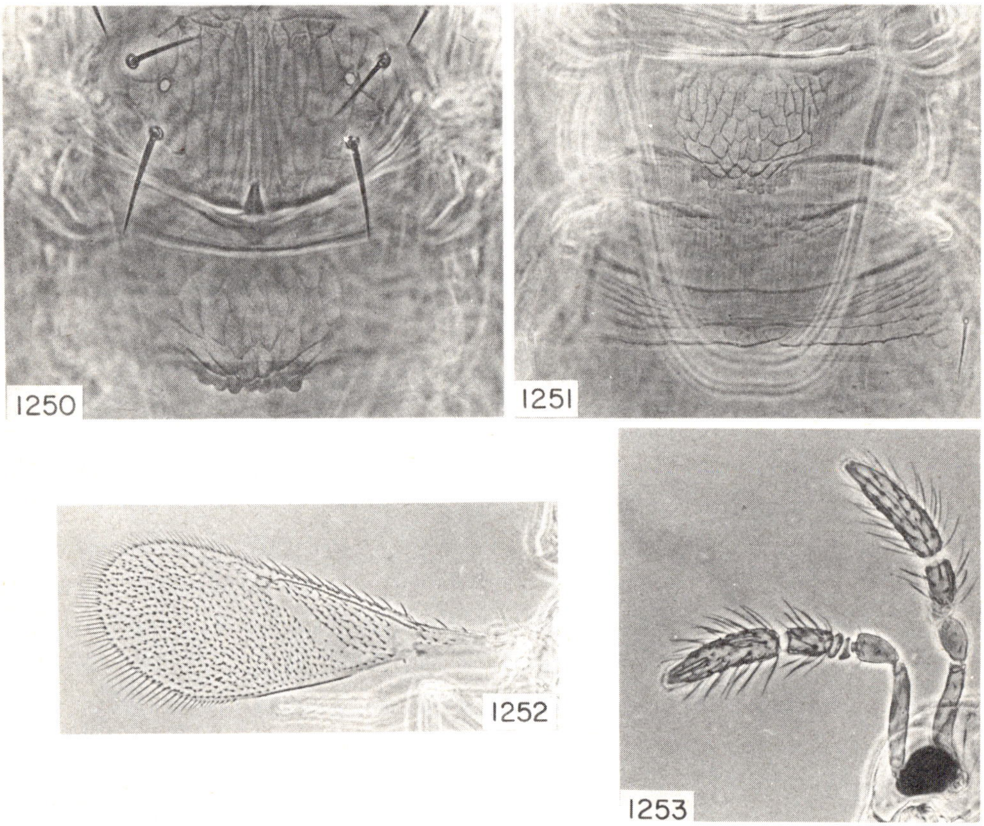


Figures 1250–1253. *Aphytis pilosus* DeBach and Rosen (paratypes)
1250. ♀: Scutellum, metanotum and propodeum. 1251. ♀: Metanotum, propodeum and base of gaster.
1252. ♀: Forewing. 1253. ♂: Antennae.

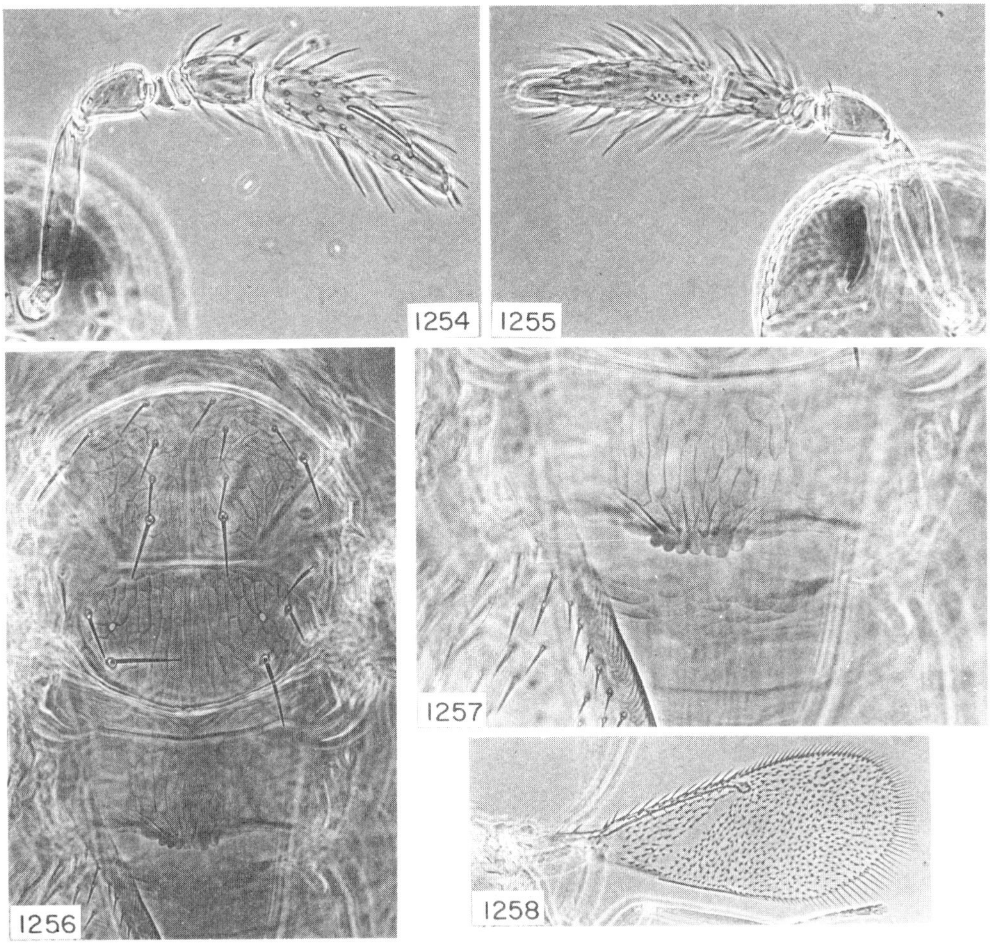


Figures 1254–1258. *Aphytis pilosus* DeBach and Rosen, ♂
1254. Antenna (allotype). 1255. Antenna, showing specialized sensory area on club (allotype).
1256. Thorax and propodeum (allotype). 1257. Propodeum and second abdominal tergite (allotype).
1258. Forewing (paratype).

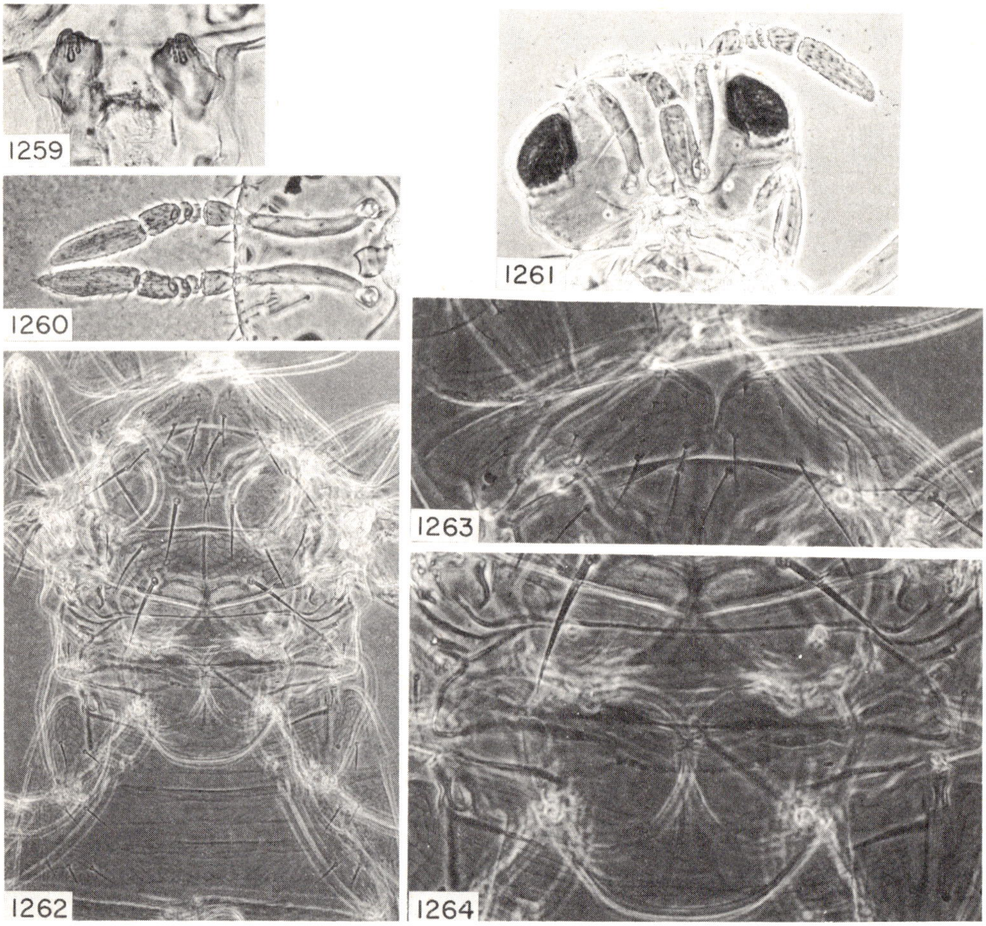


Figures 1259–1264. *Aphytis setosus* DeBach and Rosen, ♀
1259. Mandibles (allotype). 1260, 1261. Antennae. 1262. Thorax, propodeum and base of gaster (allotype). 1263. Pronotum; note broadly joined plates (allotype). 1264. Metanotum, propodeum and base of gaster (allotype).

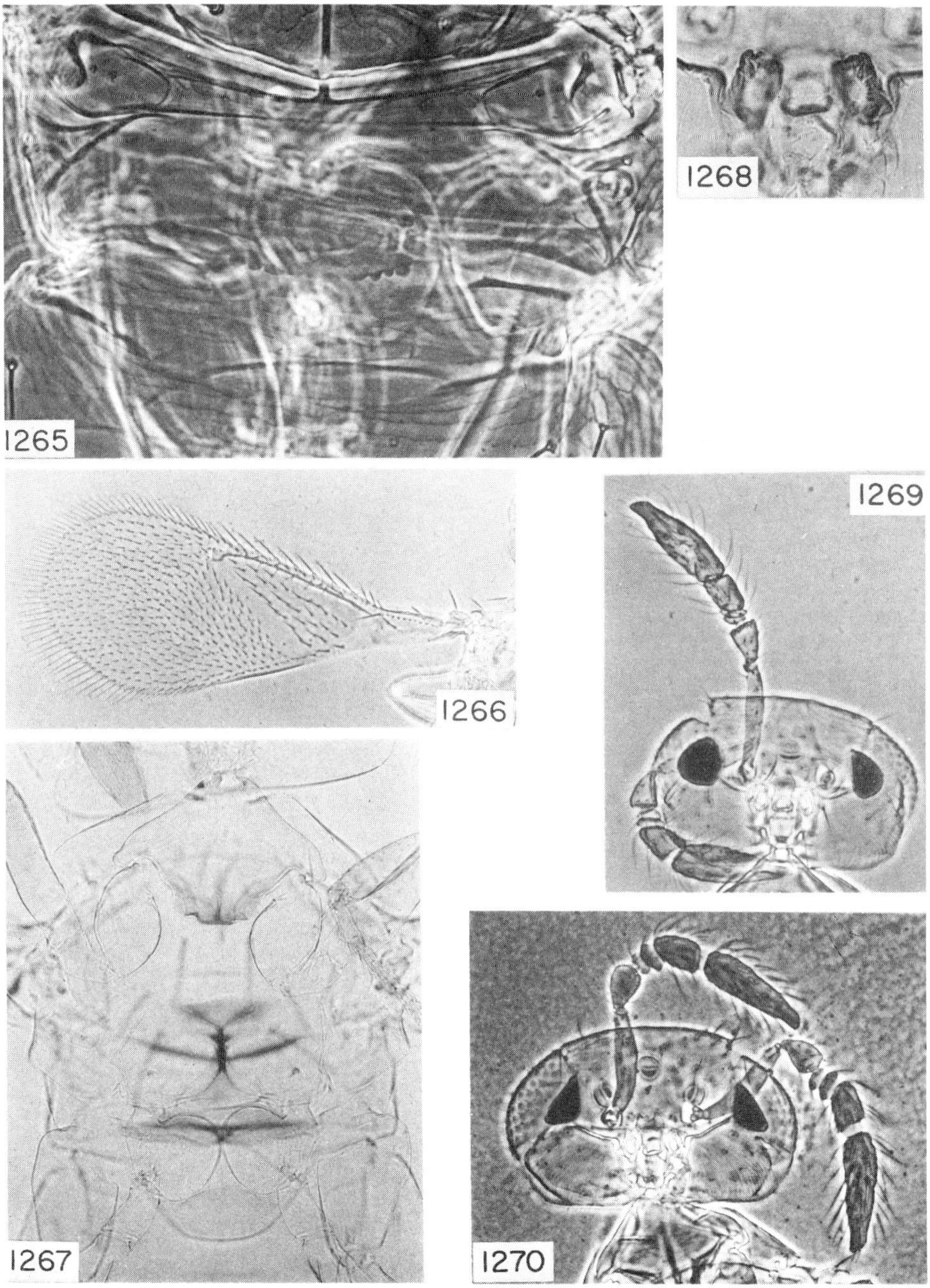


Figures 1265–1270. *Aphytis setosus* DeBach and Rosen
1265. ♀: Metanotum, propodeum and base of gaster. 1266. ♀: Forewing. 1267. ♀: Venter of thorax, showing sternal pigmentation (allotype). 1268. ♂: Mandibles. 1269. ♂: Head and antennae (holotype). 1270. ♂: Head and antennae.

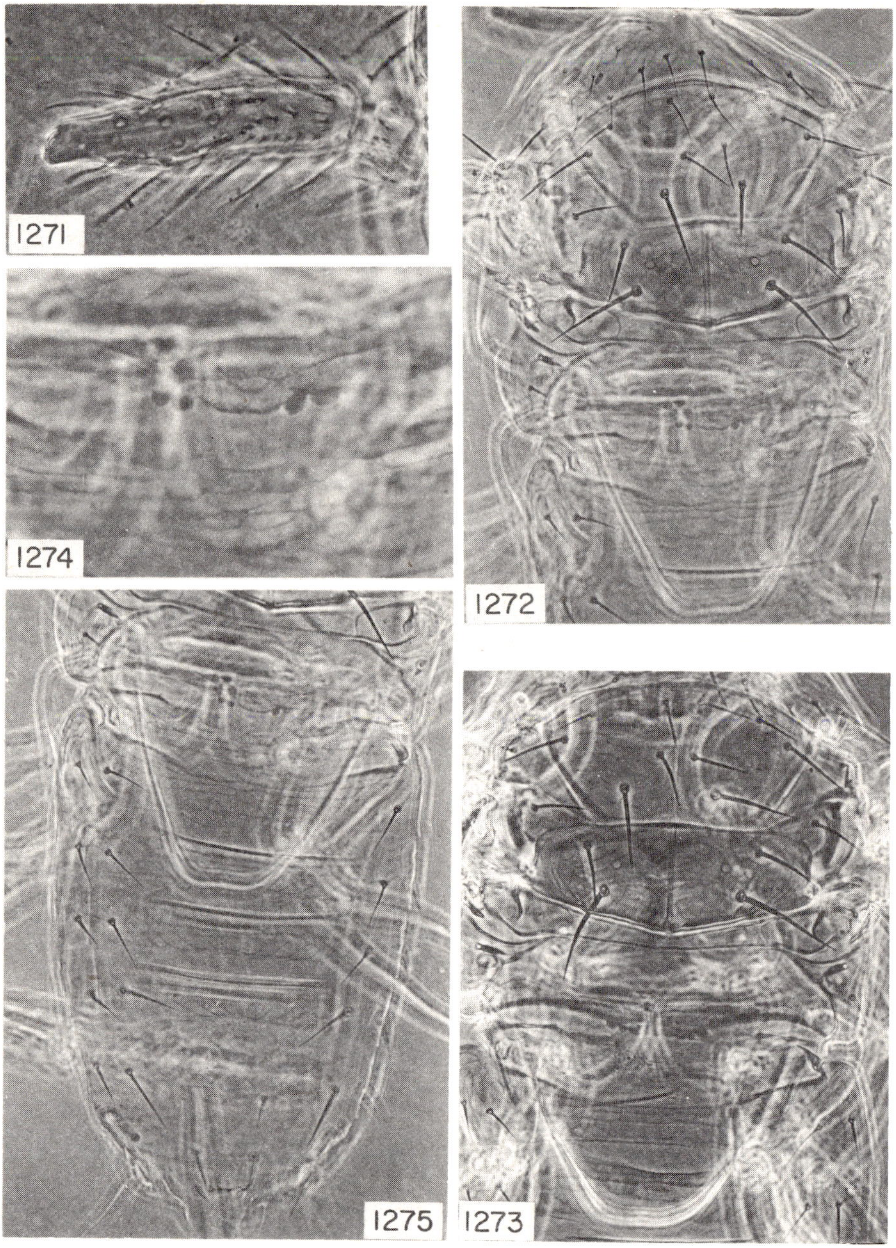


Figures 1271–1275. *Aphytis setosus* DeBach and Rosen, ♂
1271. Ventral aspect of antennal club, showing sensory area partly cut off by a sensillum-like structure. 1272. Thorax, propodeum and base of gaster (holotype). 1273. Thorax, propodeum and base of gaster. 1274. Propodeal crenulae (holotype). 1275. Propodeum and gastral tergites, showing sculpture and chaetotaxis (holotype).

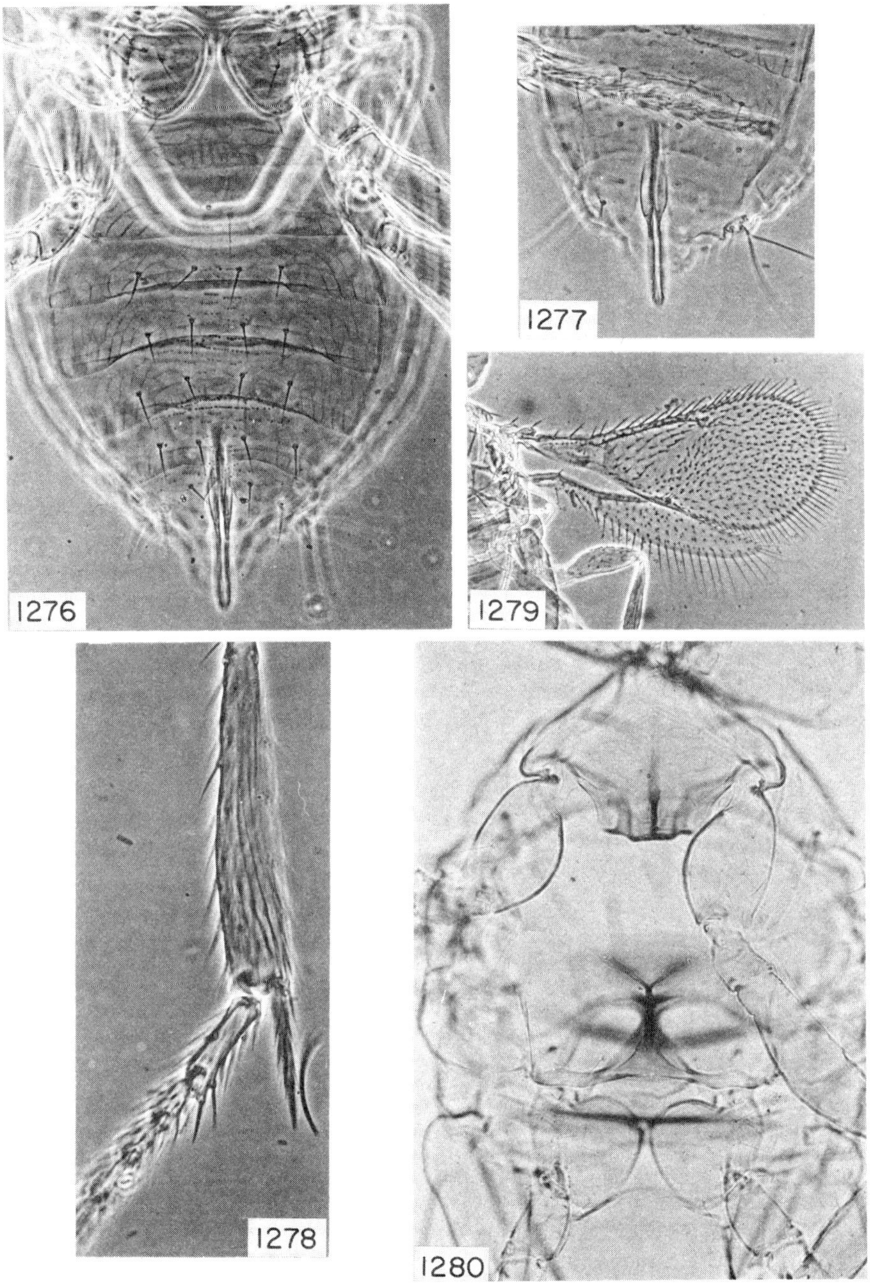


Figures 1276–1280. *Aphytis setosus* DeBach and Rosen, ♂
1276. Venter of abdomen, showing genitalia and light stippling of posterior sternites. 1277. Genitalia (holotype). 1278. Middle tibia, spur and basitarsus (holotype). 1279. Wings. 1280. Venter of thorax, showing sternal pigmentation.

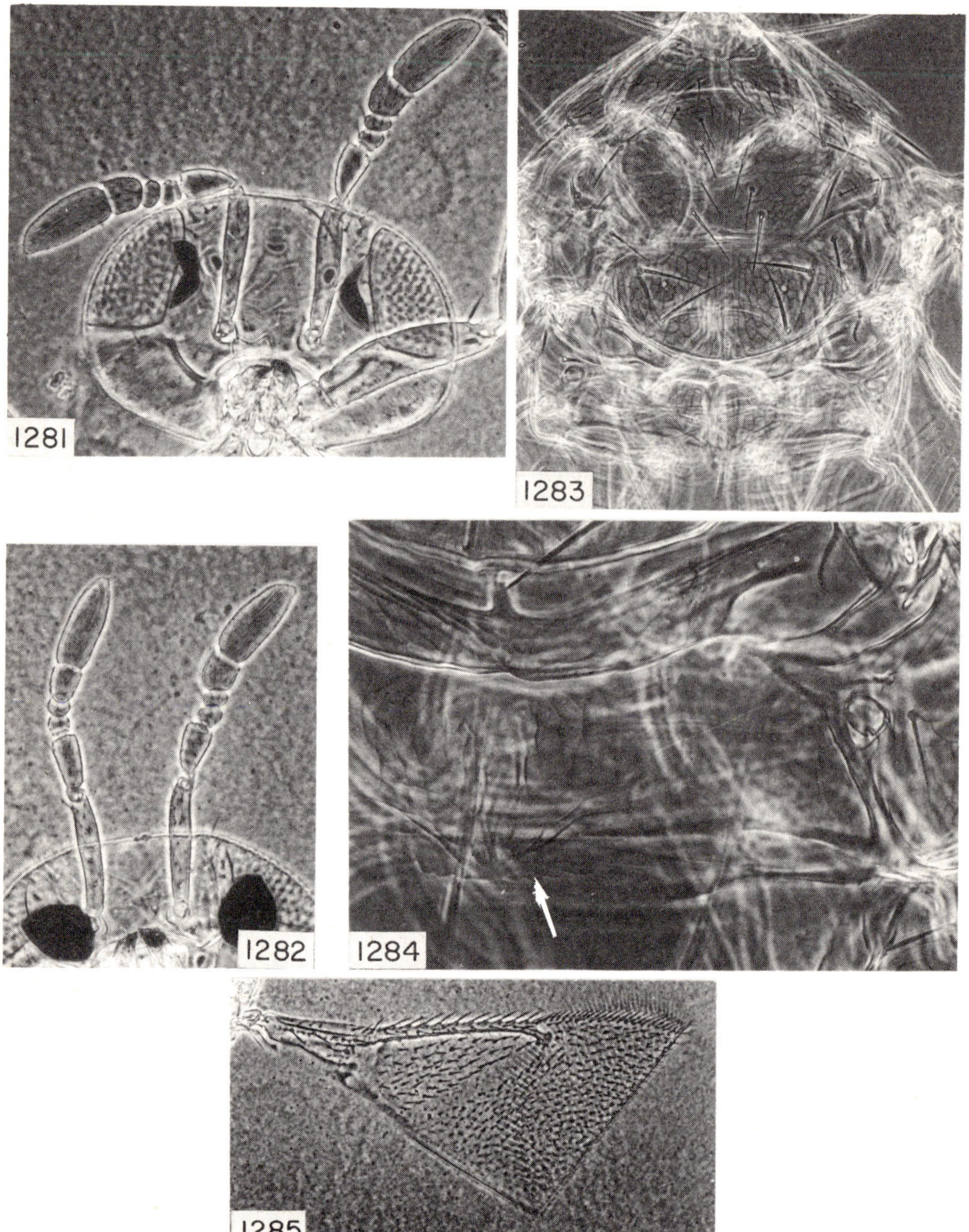


Figures 1281–1285. *Aphytis bedfordi* n.sp., ♀
1281. Head and antennae (holotype). 1282. Antennae (paratype). 1283. Thorax, propodeum and base of gaster (holotype). 1284. Metanotum and propodeum; note the small, irregular crenulae, indicated by an arrow (paratype). 1285. Forewing, folded (paratype).

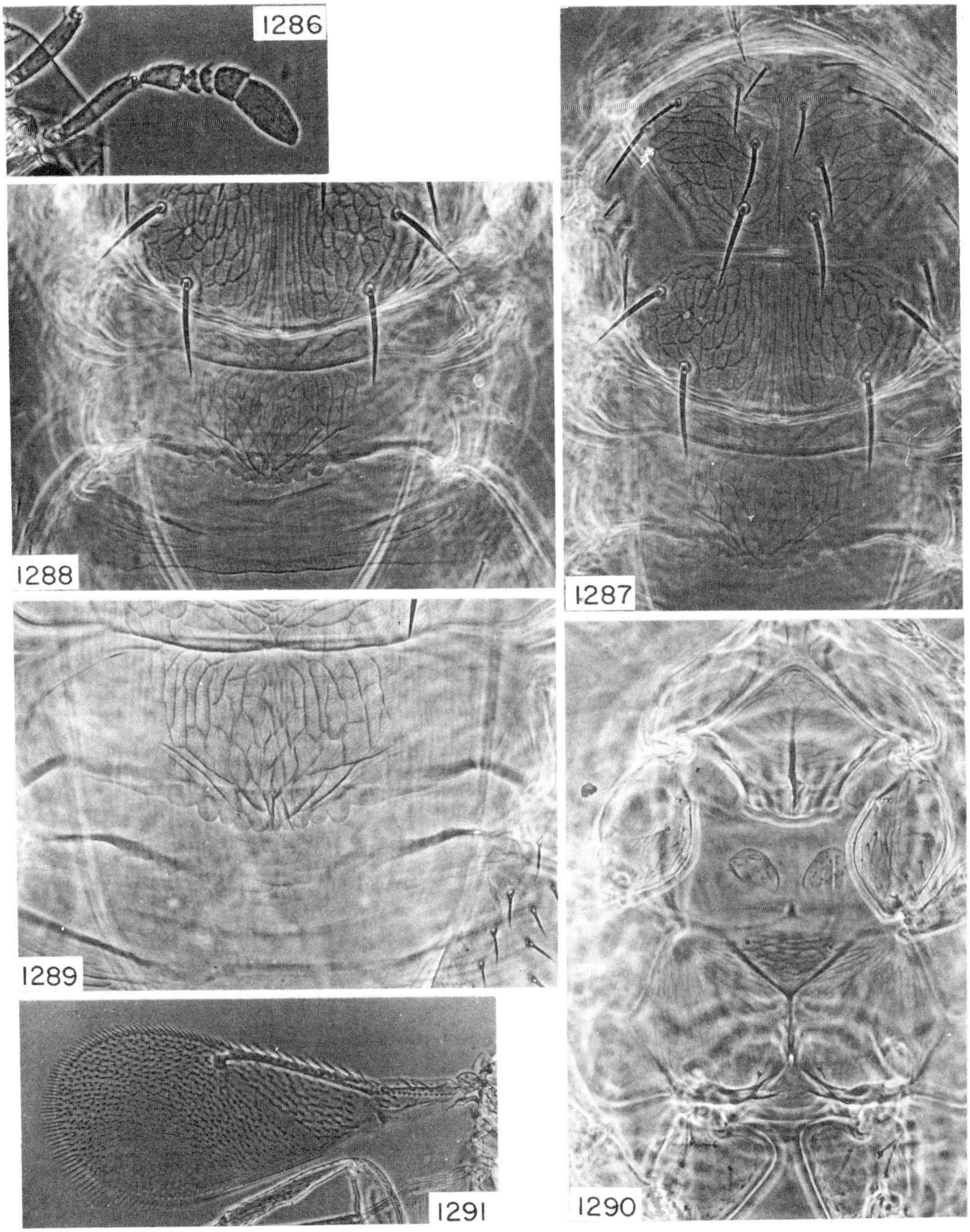


Figures 1286–1291. *Aphytis mimosae* DeBach and Rosen, ♀ (paratypes)
1286. Antenna. 1287. Thorax and propodeum. 1288. Scutellum, metanotum, propodeum and base of gaster. 1289. Metanotum, propodeum and base of gaster. 1290. Venter of thorax, showing epicoxal pads and sternal pigmentation. 1291. Forewing.

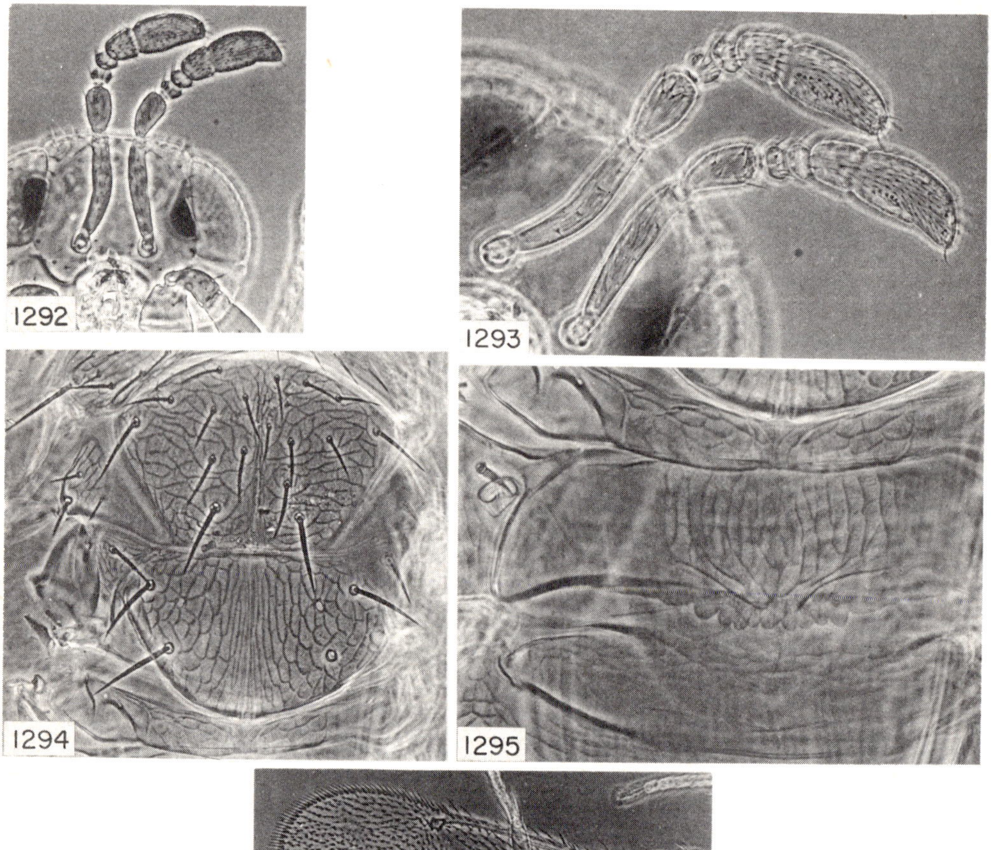


Figures 1292–1296. *Aphytis mimosae* DeBach and Rosen, ♂ (allotype)
1292. Antennae. 1293. Antennae, showing specialized sensory area on club. 1294. Mesonotum and metanotum. 1295. Metanotum and propodeum. 1296. Forewing.

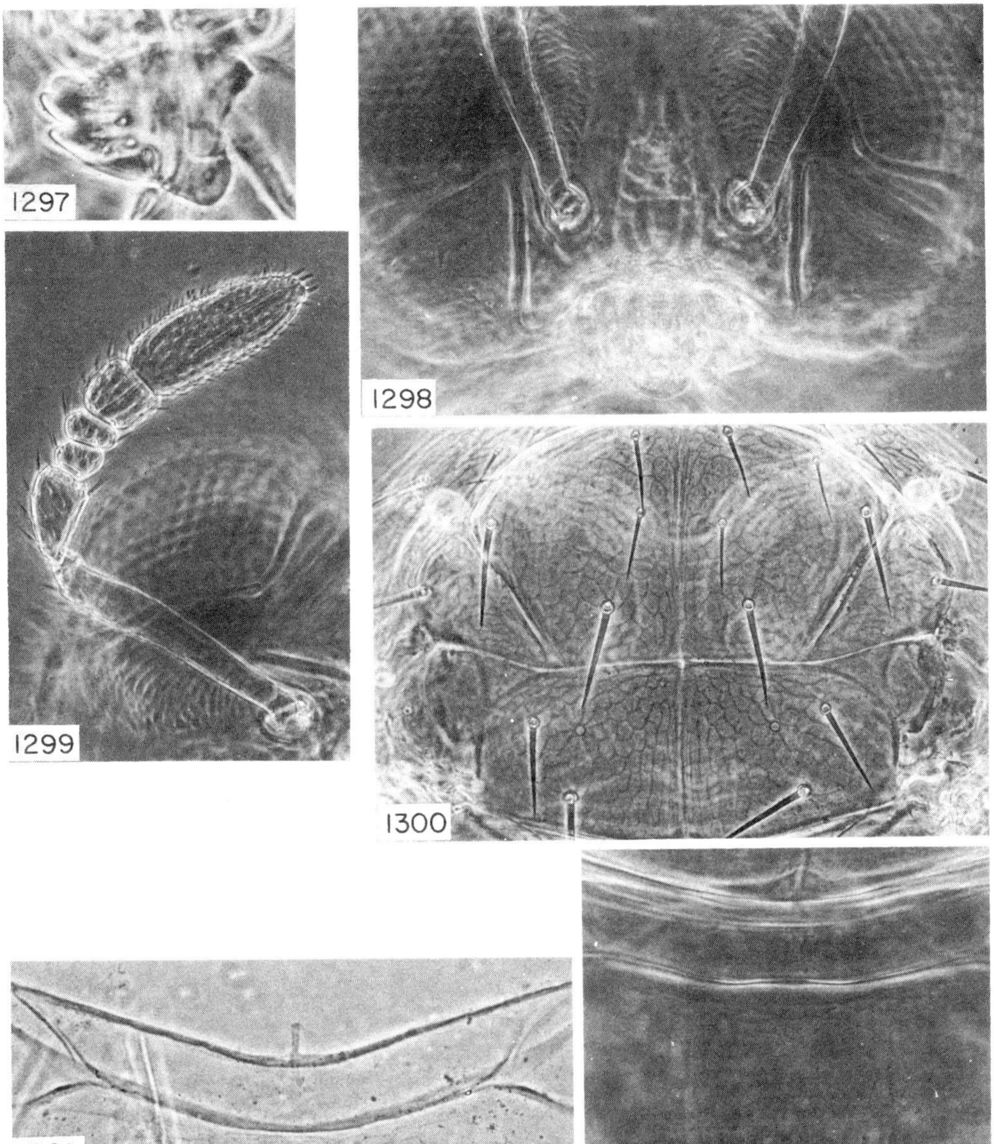


Figures 1297–1302. *Aphytis salvadorensis* n.sp., ♀
1297. Mandible (paratype). 1298. Face, showing genal sutures (paratype). 1299. Antenna (paratype). 1300. Mesonotum; note asetose axillae (paratype). 1301. Metanotum, showing slender apodeme (paratype). 1302. Metanotum and propodeum (holotype).

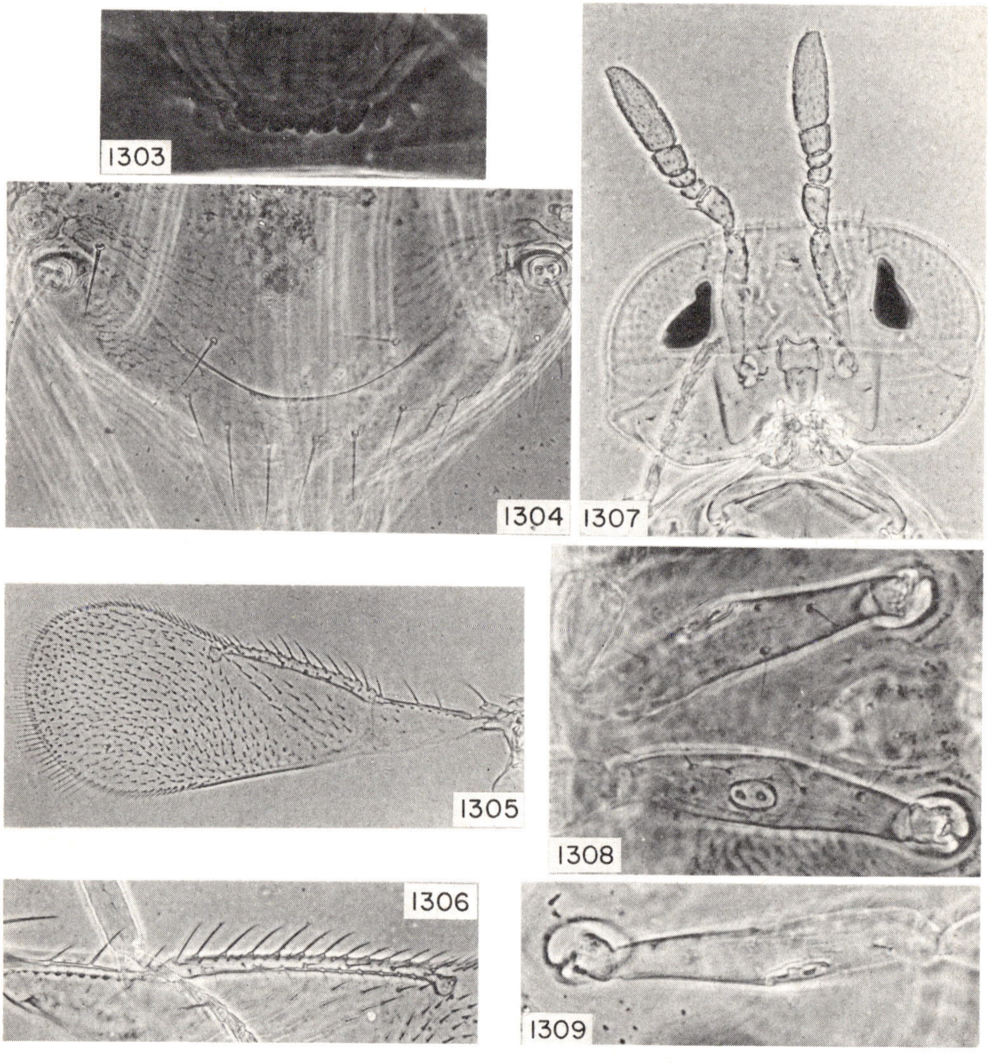


Figures 1303–1309. *Aphytis salvadorensis* n. sp. (paratypes)
1303. ♀: Propodeal crenulae. 1304. ♀: Eighth abdominal tergite and syntergum. 1305. ♀: Forewing. 1306. ♀: Venation of forewing; note unequal setae along marginal vein. 1307. ♂: Head and antennae; note genal sutures. 1308, 1309. Antennal scape, showing sensilla.

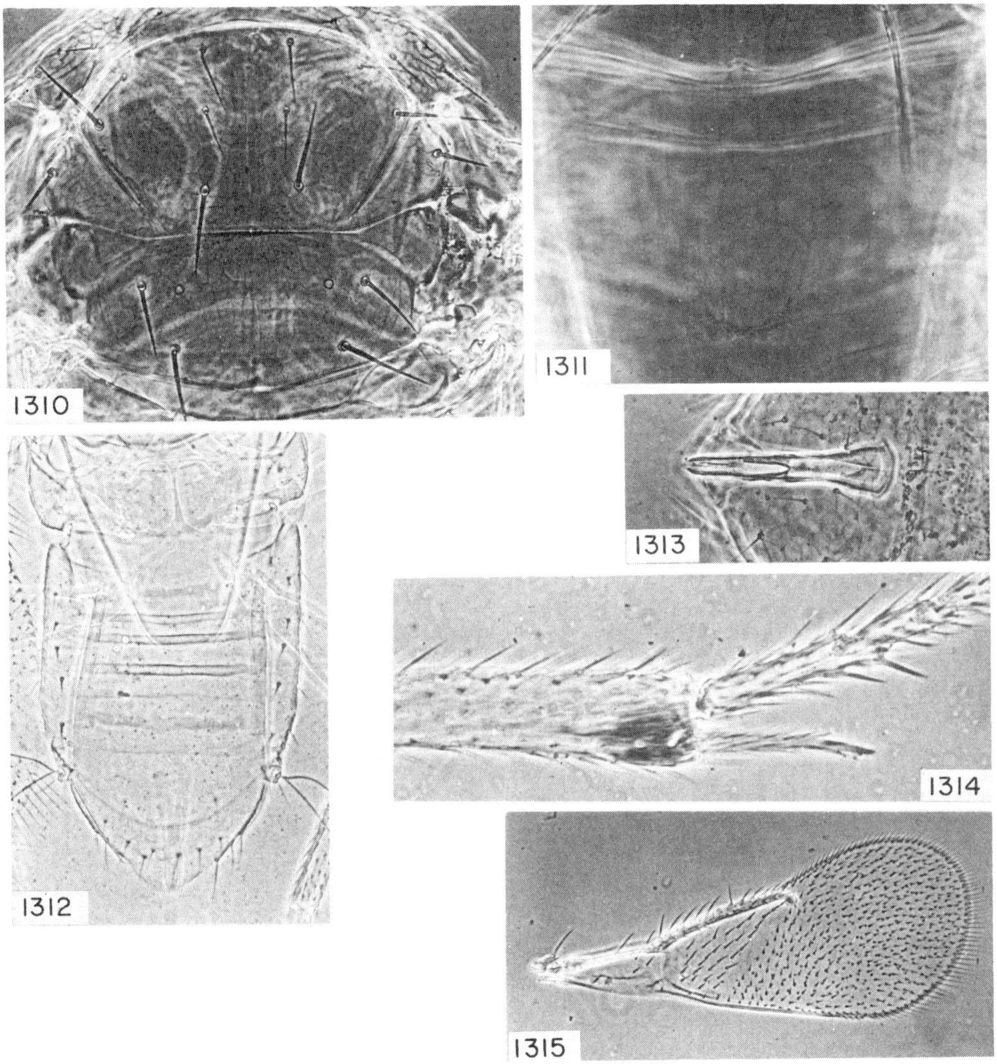


Figures 1310–1315. *Aphytis salvadorensis* n. sp., ♂
1310. Mesonotum and metanotum; note asetose axillae (paratype). 1311. Metanotum and propodeum (allotype). 1312. Abdominal tergites (paratype). 1313. Genitalia (paratype). 1314. Middle tibia, spur and basitarsus (paratype). 1315. Forewing (allotype).

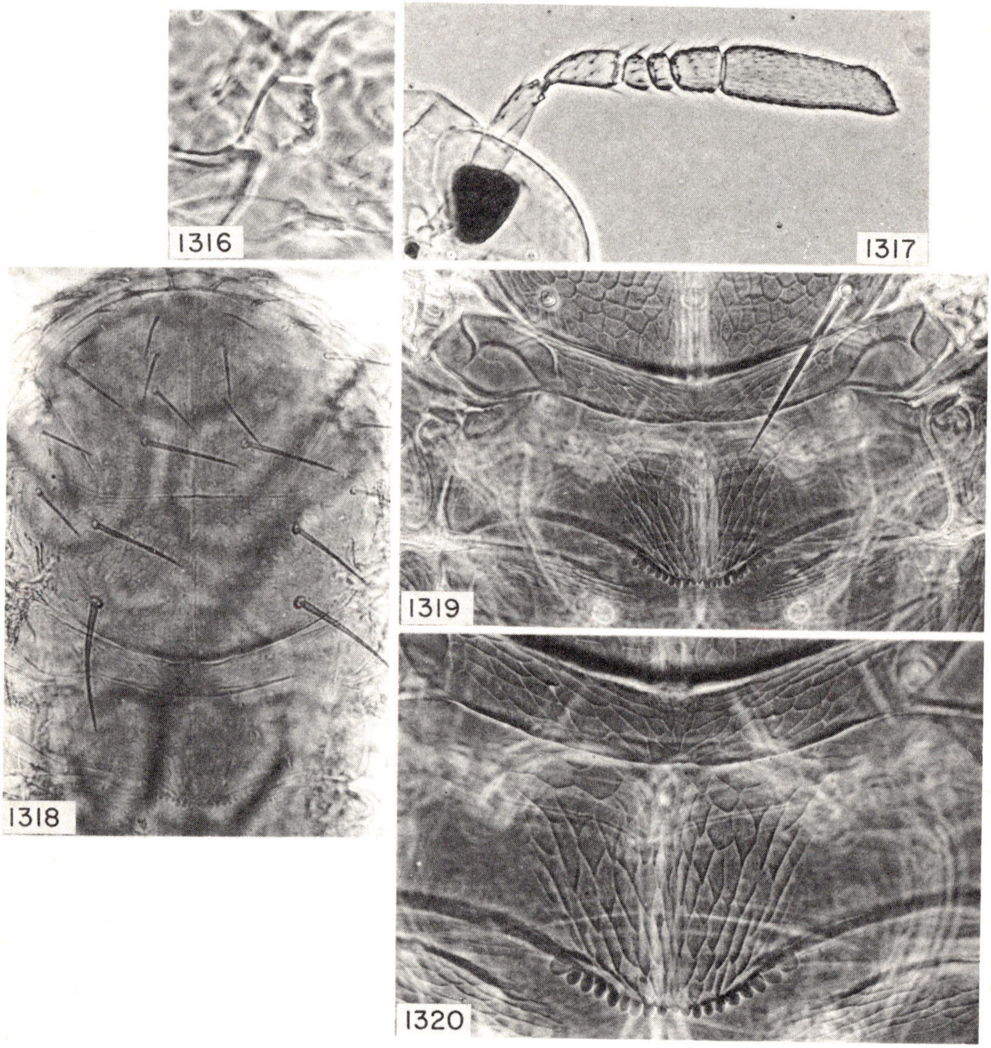


Figures 1316–1320. *Aphytis ignotus* Compere, ♀
1316. Mandible and maxillary palpus (paratype). 1317. Antenna (paratype). 1318. Thorax and propodeum. 1319. Metanotum and propodeum (holotype). 1320. Metanotum and propodeum (paratype).

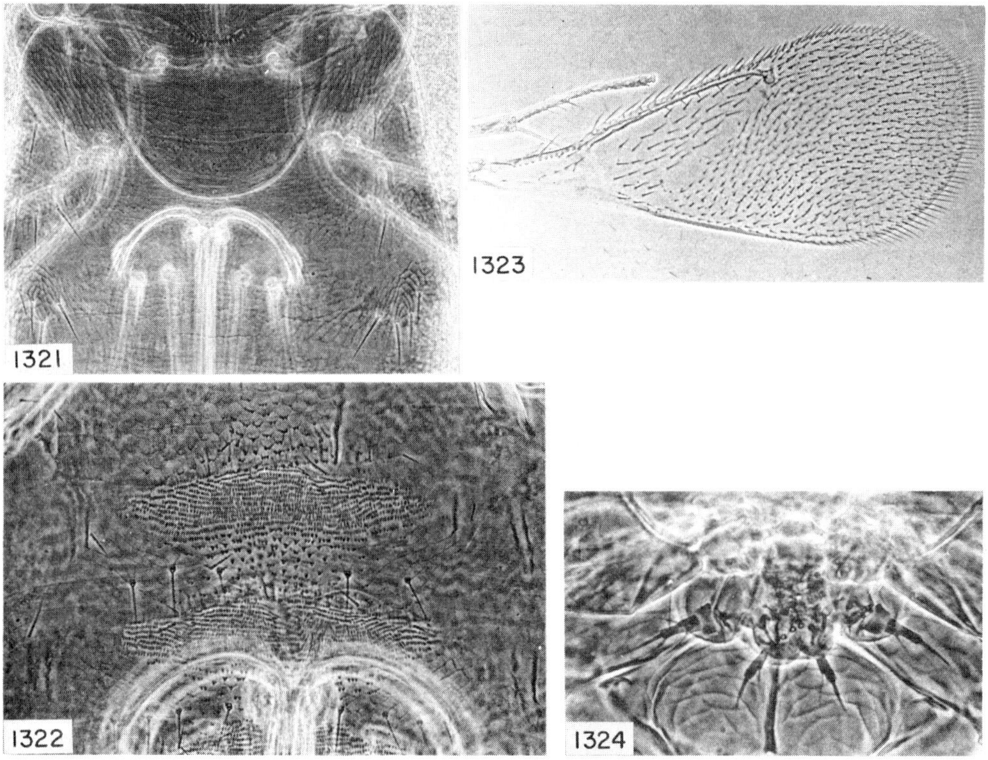


Figures 1321–1324. *Aphytis ignotus* Compere
1321. ♀: Propodeum and abdominal tergites II–V, showing sculpture and chaetotaxis (holotype). 1322. ♀: Abdominal sternites IV–VII, showing stippling. 1323. ♀: Forewing, with mid-tibial spur and middle tarsus. 1324. ♂: Maxillary and labial palpi.

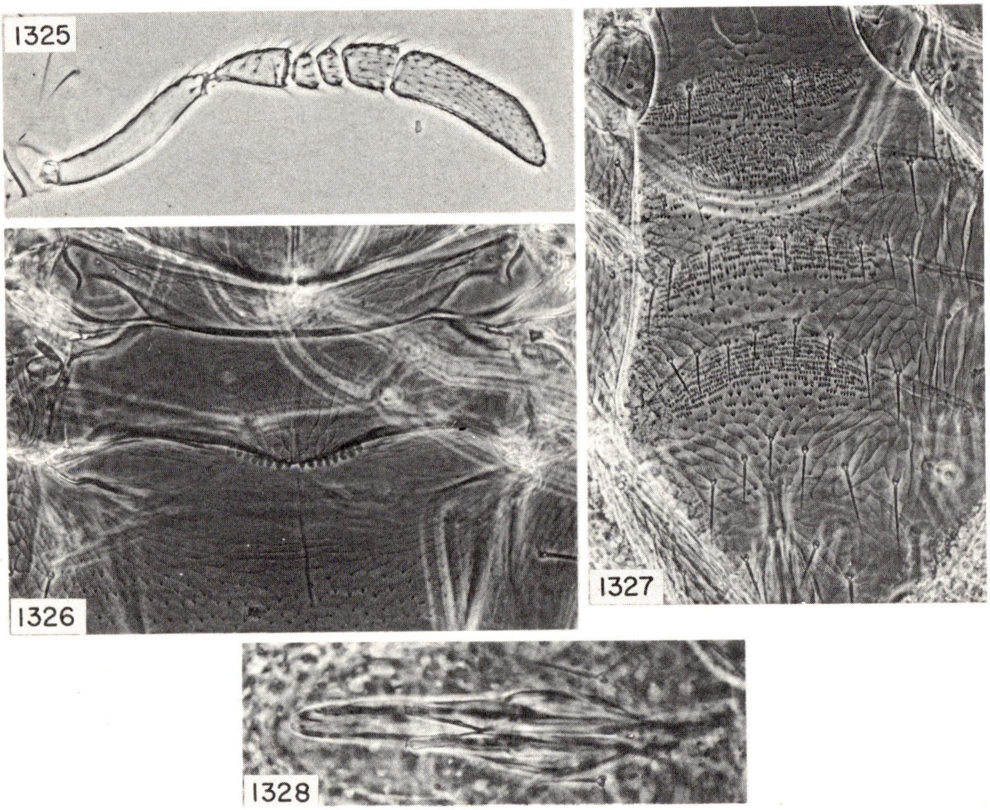

Figures 1325–1328. *Aphytis ignotus* Compere, ♂
1325. Antenna (paratype). 1326. Metanotum, propodeum and base of gaster. 1327. Abdominal sternites V–VIII, showing stippling and chaetotaxis (allotype). 1328. Genitalia.

# UNRECOGNIZABLE, DUBIOUS AND UNKNOWN SPECIES

The following 18 species cannot be recognized at present. Some are known to us only from the inadequate original descriptions, while others are represented only by poorly preserved type specimens. All these species appear to be members of the genus *Aphytis*, although even the generic position of some of them is rather doubtful. They are listed here in alphabetical order.

### **Aphytis abnormis** (Howard)

*Aphelinus abnormis* Howard, 1881, Ann. Rep. Comm. Agr. for 1880, p. 355.
*Aphelinus abnormis*: Howard, 1895, U.S. Dept. Agr. Div. Entomol. Tech. Ser., **1**:26.
*Aphytis abnormis*: Compere, 1955, Univ. Calif. Publ. Entomol., **10**:317.

Howard (1881) originally described *abnormis* as follows:

"Length, 0.55 mm; wing expanse, 1.4 mm; greatest width of forewing, 1.23 mm. Proportions as in the preceding species [=*mytilaspidis*]. Scutellum sharply pointed anteriorly (in this respect differing from all other species of *Aphelinus* with which we are acquainted). General color, light lemon yellow; antennae dusky, eyes blackish, ocelli reddish, legs with yellowish femora and dusky tibiae and tarsi. Wings perfectly clear, veins transparent.

"Described from 1♀ specimen.

"Parasitic upon *Mytilaspis* sp. on *Salix caprea* (District of Columbia).

"The peculiarity of the scutellum may ultimately cause this species to be referred to a new genus, but since it is so evidently closely related to *Aphelinus* in other respects, it seems best to place it here."

Subsequently, Howard (1895) added: "General proportions as with *A. mytilaspidis*, from which it differs only in the following respects: The scutellum is pointed anteriorly, seemingly entering a reentering angle on the base of mesoscutum; antennal club three times as long as penultimate joint. . . ." He recorded the host as *Mytilaspis pomorum* (Bouché) which is a synonym of *Lepidosaphes ulmi* (L.), and added: "No other specimen of this peculiar form has ever been found. There is a possibility that it may be simply a dwarfed and distorted specimen of *A. mytilaspidis*."

We have studied the type specimen, preserved in the U.S. National Museum. The specimen is uncleared and mounted in balsam. The propodeum, crenulae, and other important characters cannot be seen. The peculiar folding of the scutellum and mesoscutum is clearly visible; it was obviously caused by abnormal shrinkage of the specimen, and should not be regarded as a diagnostic character.

*A. abnormis* is most probably a synonym of *mytilaspidis*, but this cannot be definitely established unless the specimen is cleared and remounted. For the time being, *abnormis* stands as an unidentifiable species, probably related to the **mytilaspidis** group.

Mercet (1931) resurrected the name *abnormis* for a species of *Aphytis* parasitic upon *Leucaspis signoreti* Targioni-Tozzetti on pine in Corsica. His identification was based mainly on the details of coloration as described by Howard. We consider the species discussed by Mercet to be identical to *A. luteus* (Ratzeburg) (see p. 489).

## **Aphytis alami** Agarwal

*Aphytis alami* Agarwal, 1964, Proc. Ind. Acad. Sci., **60**: 316–319.

This species is known to us from the description only. Agarwal (1964) recorded *alami* as an "endoparasite on *Pseudococcus citri* Risso on *Citrus medica* L." at Aligarh, India. He referred this species to the **mytilaspidis** group, but compared it to *chrysomphali* rather than to any member of that group.

According to the original description, the female of *alami* is yellow, but "apex of scutellum with blackish vertical lines"; antennal club about 3 times as long as wide, $2\frac{1}{2}$ times longer than the preceding segment; mesoscutum with 10 setae, each parapsis with 1 seta; forewing "hyaline with a very light smoky background, especially at the base of the submarginal vein, at the junction of the submarginal and marginal veins and around the stigmal vein"; marginal fringe $\frac{1}{4}$ width of disk; delta with 5 rows of setae. Length 0.543 mm (although *A. alami* was originally described from the holotype and several paratype specimens, no range was given for this or any other measurement). The presence—or absence—of males was not mentioned.

Unfortunately, Agarwal's description does not disclose important diagnostic characters such as the pigmentation of the thoracic sterna and setae, the shape of the propodeum and crenulae (other than "propodeum of uniform width") etc., and is therefore inadequate for proper identification of the species, or even for its correct assignment to any of the species groups. However, the original description and figures do indicate that this is an *Aphytis*. Since Agarwal's type material was not available for study, we have to consider *alami* as an unrecognizable species. The host record—a mealybug—is presumably erroneous.

## **Aphytis angeloni** (Girault) n.comb.

*Marietta angeloni* Girault, 1932, New Lower Hymenoptera from Australia and India. Priv. Publ., Brisbane, 6pp., page 2.

Girault's original description of *angeloni* was very brief: "Cilia proximad hairless line spotted (4 dark areae), legs immaculate. Pale yellow, abdomen with transverse lateral marks. Wing with 3 marginal bows (reciprocals) and between these on disc 2 triangles of 3 spots each." No data on host, locality or date of collection of type material were recorded.

The type slide, preserved in the Queensland Museum, Brisbane, Australia, is labeled as follows: "Type ♀ *Perrissopterus* [!]: 2. *argenticorpus* Gir. 1. *angeloni* Girault. *Ufensia pretiosa*. Qsld."; *angeloni* is represented by a single female specimen, labeled "3763". The type is dissected but not cleared, mounted in balsam; the cover glass is crushed, and the poor condition of the specimen precludes remounting. Not much can be added to the original description; the species is evidently an *Aphytis*, apparently of the **vittatus** group, but is practically unrecognizable.

## ?**Aphytis caucasicus** Tshumakova

*Aphytis caucasicus* Tshumakova, 1964, Trud. Vsesoyuz. Inst. Zashchit. Rast., **21**:25–26.
*Aphytis caucasicus*: Nikol'skaya and Yasnosh, 1966, Aphelinids of the European Part of the USSR and the Caucasus, pp. 284–285 (in Russian).
*Aphytis caucasica*: Nikol'skaya and Yasnosh, 1968, Trud. Vsesoyuz. Entomol. Obshchest., **52**:15.

*A. caucasicus* was described by Chumakova (1964) from 2♀♀, reportedly reared from *Lepidosaphes ulmi* (L.) on poplar in the Caucasus, August 1953 and September 1954. According to the original description and drawings, the female is entirely yellow, with a decidedly narrow body. Mandibles tridentate. Antennae 6-segmented, robust; the club short, somewhat less than twice as long as the third funicular segment. Mesoscutum with 10 setae. Mid-tibial spur slender, as long as the corresponding basitarsus. Forewing narrow, 3 times as long as wide; marginal fringe about $\frac{1}{5}$ width of disk; rows of setae in delta incomplete, obliterated posteriorly; marginal vein distinctly shorter than the submarginal; a short postmarginal vein is indicated. Length 0.7 mm. The shape of the propodeum is not mentioned.

The types of *caucasicus* were reportedly deposited in the collection of the Zoological Institute, USSR Academy of Sciences, Leningrad. However, Dr. V. A. Yasnosh (personal communication, 1975) informed us that the holotype could not be found in that collection, whereas the paratype, mounted in balsam and kept in Dr. Chumakova's collection, was in poor shape and could not be redescribed.

In fact, the original description raises some doubts as to the generic position of *caucasicus*, some of the described characters being rather suggestive of *Centrodora*. For the time being we have to regard *caucasicus* as an unrecognizable, questionable species of *Aphytis*.

## **Aphytis diaspidioti** Tshumakova

*Aphytis diaspidioti* Tshumakova, 1957, Zool. Zhur., **36**:542–543.

Chumakova (1957) described *diaspidioti* as a parasite of *Quadraspidiotus gigas* (Thiem and Gerneck) in the Soviet Far East. According to her original description, this species is very closely related to *mytilaspidis*, differing from the latter only in the number and relative length of the setae along the anterior margin of the marginal vein (7–8, over $1\frac{1}{2}$ times as long as the setae along center of vein in *diaspidioti*, as against 12 and less than $1\frac{1}{2}$ times in *mytilaspidis*). These are known to be unreliable characters. Ferrière (1965) and Nikol'skaya and Yasnosh (1966) listed *diaspidioti* as a synonym of *mytilaspidis*. However, *diaspidioti* was described as entirely yellow, and no reference was made to sternal pigmentation—a key character in the **mytilaspidis** group. If it lacks sternal pigmentation, it could be a synonym of *aonidiae*. We therefore cannot accept the proposed synonymy with *mytilaspidis*. Since the types of *diaspidioti* were not available for study, we have to regard this species, for the time being, as an unrecognizable member of the **mytilaspidis** group.

## ?**Aphytis flavus** (Ashmead)

*Trichogramma flavus* Ashmead, 1880, Orange Insects, pp. 33–34.
*Trichogramma flavum*: Peck, 1963, Canad. Entomol., Suppl. **30**:903–904.

Ashmead (1880) described *Trichogramma flavus* as follows:
"TRICHOGRAMMA FLAVUS, N. SP.—Female.—Length .04 of an inch. Head wider than thorax, brownish, three ocelli triangularly arranged, with two smaller red ones back of these. Eyes reddish, excepting dark spot on side nearest ocelli. Antennae, five-jointed, yellowish red, first joint longer than two and three combined and narrower than joint two. Joint two not as long as first, but wider—as long as joints three and four together. Joints three and four equal, narrower than second. Fifth, or apical joint, as long as second, third and fourth combined and much wider, claviform. Thorax and abdomen a bright yellow, reddish along hinder part of thorax, where it joins abdomen. Abdomen brownish on segments one to five around the spiracles, also a few hairs issuing therefrom. Ovipositor long, surrounded at base with short hairs. Wings hyaline, fore wings rather long and well rounded, with fringing of short fine ciliae. Hind wings narrow, curving into a sharp point from the middle, also ciliated. Under surface uniform yellow. Legs thin, paler but uniform in color, sparsely covered with hair, a short tibial spur, tarsi five jointed. . . . . . Described from four specimens."

Ashmead's type material could not be found in the U.S. National Museum of Natural History and has apparently been lost (Peck, 1963; G. Gordh, personal communication, 1976). Peck (1951, 1963) and Burks (1958) listed *flavus* as an unplaced species of Chalcidoidea. However, it should be noted that Girault (1911) regarded this species as "an unknown Aphelinine", whereas Timberlake (unpublished Catalogue of Hymenoptera) referred it to *Aphytis*. The original description is indeed entirely inadequate by modern standards. Ashmead apparently mistook the pigment spots, commonly seen in dry or slide-mounted specimens of small Chalcidoidea, for supernumerary ocelli. His reference to the presence of spiracles on abdominal [= ?gastral] tergites I–V was presumably also erroneous, inasmuch as in all known species of Chalcidoidea, spiracles are present only on the propodeum and eighth abdominal tergite. Nevertheless, Ashmead's (1880) original description and figure leave little doubt that *flavus* was an aphelinid (with 5-segmented tarsi, it certainly could not have been a trichogrammatid). As far as can be surmised from the scant evidence presented by Ashmead, *flavus* may have been a species of *Aphytis*, or possibly *Eretmocerus*. Specific identification is, of course, at present impossible.

The type specimens of *flavus* were collected on orange leaves in Florida. One of them was found under the brown soft scale, *Coccus hesperidum* L., leading Ashmead (1880) to presume that it was parasitic upon that species. If our interpretation of *flavus* as an *Aphytis* is correct, it was probably parasitic upon an armored scale insect.

## ?**Aphytis flavus** (Nees)

*Eulophus flavus* Nees ab Esenbeck, 1834, Hymenopterorum Ichneumonibus Affinium Monographiae, Genera Europaea et Species Illustrantes. Vol. **2**, Stuttgart, p. 167.
*Aphelinus flavus*: Mercet, 1912, Los Enemigos de los Parasitos de las Plantas. Los Afelininos. Trab. Mus. Cienc. Natur. Madrid, **10**:99–100.
*Aphytis flavus*: DeBach and Rosen, 1976, Ann. Entomol. Soc. Amer., **69**:544.

The history of this species is rather confusing. The original description is given here in full:

"?18. *Eulophus flavus*.

E. flavus, oculis stemmatibusque nigris, abdomine conico-subulato. *Fem.* Long. lin. $\frac{1}{3}$. Refero hunc in locum insectum quoddam, paucis verbis in schedis notatum, sed prius, quam accuratius examinari potuerit, casu mihi ereptum neque postea unquam obvium. Proprii generis esse suspicor.

*Antennae* septemarticulatae videbantur (sed fortiori lente adhuc explorandae erant). Totum corpus laeve, nitidum, *flavum*, solis *oculis stemmatibus*que nigris. *Ungues* fusci. *Alae* hyalinae, ramulo stigmatico punctiformi *basi propiori*. *Abdomen* angustissimum, subulatum, dorso planum, ut in *Cleonymo bimaculato*.

Cepi unam *feminam* 2. Octobris a.1809 sub foliis aridis directis prope *Sickershausen*.

Adnot. Ob radium stigmaticum ante median alam enatum *Encyrtis* accedere videtur, sed habitus repugnat, et antennae paucioribus potius quam pluribus articulis componi videbantur."

The type has apparently been lost. Several authors have made references to what they believed to be *flavus* Nees. Mercet (1912b) presented the following synonymy:

"*Aphelinus flavus* Nees

*Eulophus flavus* Nees ab Esenbeck, Hym. Ichn. affin., Monogr. II, 1834.
*Aphelinus flavus* Walker, Monogr. Chalcid., I, 1839.
*Myina flava* Walker, List. Hymen. Brit. Mus. Chalcid., II, 1848.
*Encyrtus pallidus* Ratzeburg, Ichneum. d. Förstins, II, 1848.
*Coccobius pallidus* Ratzeburg, Ichneum, d. Förstins, III, 1852.
——— *flavus* Reinhard, Berl. Ent. Zeitschr., II, 1858.
*Aphelinus flavus* Thomson, Hymen. Scandin., IV, 1875."

Thomson (1875: 186) described the species *Aphelinus flavus*, which he considered identical to *Eulophus flavus* Nees, and recorded the host as *Coccus*. Kurdjumov (1913) listed *flavus* Nees in his key to the European species of *Aphelinus* parasitic in aphids.

Nowicki (1930:179) discussed "*Aphelinus flavus* Nees" as follows (translated): "It would be best to designate this species as "incertae sedis." The very imperfect original description makes the impression that we are dealing with a species of another genus, or even another family. The yellow species of *Aphytis* and *Aphelinus* are numerous in Europe; therefore, the identity of *flavus* of Nees, Walker, Reinhard, Thomson and Kurdjumov, and the synonymy of *Coccobius pallidus* Ratz., appear to me to be very questionable . . . ."

According to Ferriere (1965), *Aphelinus flavus* Walker is a synonym of *Mesidiopsis subflavescens* (Westwood), *Aphelinus flavus* Kurdjumov is a synonym of *Mesidia annulipes* (Walker), and *Aphelinus flavus* Thomson is a true *Aphelinus* and a parasite of aphids. "*Aphelinus flavus* Nees, although the type is lost, is not an *Aphelinus* as thought by Thomson, but probably an Encyrtid" (Ferrière 1965: 62, translated).

On the other hand, Bouček (1964: 667) examined a female specimen of "*pallidum* Ratzeburg" in the Ratzeburg collection at Eberswalde, and determined it as an *Aphytis*. He agreed with Nowicki (op. cit.) that the synonymy of this species with *Eulophus flavus* Nees was doubtful.

A single specimen preserved in the U.S. National Museum of Natural History is labeled "*Aphelinus (Coccobius) flavus* ♀ Nees." Although it is mounted on a pin (actually pierced by a micro pin!), this specimen clearly belongs to a yellow species of *Aphytis*.

Thus, the identity and generic position of *Eulophus flavus* Nees remain uncertain. However, the possibility that it may have been an *Aphytis* cannot be discounted. In view of the uncertain position of both *flavus* (Nees) and *flavus* (Ashmead) (see above), it seems advisable to regard the name *flavus* as preoccupied in the genus *Aphytis*. *A. flavus* Quednau is therefore considered here to be a junior synonym (see DeBach and Rosen, 1976b, and p.499 herein).

The story of *flavus* illustrates the difficulties encountered by students of the Aphelinidae during the last century and the beginning of the present one, when inadequately mounted specimens were examined with poor optical equipment. Unfortunately, the modern student of the Chalcidoidea cannot ignore this burdensome legacy of misinterpretation, but has to cope with it to the best of his ability.

## ?**Aphytis grotiusi** (Girault)

*Aphelinus grotiusi* Girault, 1913, Mem. Queensl. Mus., **2**:181–182.
*Aphytis grotiusi*: Compere, 1955, Univ. Calif. Publ. Entomol., **10**:316.

Girault (1913) described *grotiusi* "from one female captured from a window in a building on a sugar-cane farm, December 18, 1911 . . . Nelson (Cairns), Queensland," Australia. His original description was very brief:

"*Female*:—Length, 0.75 mm.

"Almost exactly like *australiensis* but the second funicle joint is wider than long, subequal to the first, distinctly less than half the length of the first club joint which is somewhat longer than wide. Also the general colour is lemon yellow."

Girault's description is, of course, inadequate. It should be noted that *Paraphelinus australiensis* Girault, with which *grotiusi* was compared, evidently belongs in *Centrodora*, and that in the same publication Girault (1913: 180) synonymized *Paraphelinus* (=*Centrodora*) with *Aphelinus*.

Compere (1955) made the following comments: "The single type female, number 3772, Queensland Museum, is poorly preserved—the body setae are not discernible, the antennae are detached and partly folded, and one front wing is folded over the abdomen. The following descriptive notes made in 1932 contribute little if anything of value toward distinguishing this species.

"*Description*:—Antennal club almost four times as long as the third funicular segment (11:3), the latter slightly longer than wide (6:5). It is now impossible to ascertain the exact proportions of the first and second funicular segments because of oblique position. Costal margin of marginal vein with 5 strong setae; 1 seta of similar size on the stigmal vein and another, similar seta on the upward bend of the submarginal vein. Stigmal vein separated from the speculum by one irregular row of hairs. Disc of front wings with moderately sparse hairs."

The holotype of *grotiusi*, a female specimen deposited in the Queensland Museum, Brisbane, Australia, is indeed very poorly preserved as described by Compere. It is mounted in balsam, uncleared, together with the type of *newtoni* and a trichogrammatid specimen, and the cover glass is cracked just above it. Although most diagnostic characters cannot be clearly seen, enough is visible to put even the generic status of this species in serious doubt: The first two funicular segments appear to be rather elongate, the propodeum appears to be only slightly longer than the metanotum, and the marginal vein of the forewing appears to be shorter than the submarginal vein. In fact, *grotiusi* may belong in *Centrodora*. At best, it is an unrecognizable species which is only questionably referable to *Aphytis*.

## Aphytis limonus (Rust)

(Figures 1329–1331)

*Aphelinus limonus* Rust, 1915, Entomol. News, **26**:76–77.
*Aphytis limonus*: Timberlake, 1924, Proc. Hawaii. Entomol. Soc., **5**:412.

Rust (1915) described *limonus* "from fourteen female specimens received from E. M. Ehrhorn in a lot of *Hemichionaspis minor* on 'pigeon-pea,' which he sent to this office from Honolulu, Territory of Hawaii, in June, 1911," and stated that the male of this species was unknown. However, the type slide (No. 40223, U.S. National Museum of Natural History) includes 1♀ and 3♂♂, and the specimen designated by Rust as the type is a male.

Rust described *limonus* as very closely related to *quaylei*, which is recognized here as a synomym of *chrysomphali*. He listed several diagnostic characters for separating the two species, most of which are insignificant (e.g., color of the eyes in balsam mounts), but noted that in *limonus* the wings are entirely hyaline and the antennal club is stouter than in *quaylei*.

Mercet (1932: 359) listed *limonus* as a synonym of *chrysomphali*, and this synonymy was accepted by Ferrière (1965: 85) and by Nikol'skaya and Yasnosh (1966: 201). However, Compere (1955 : 314), while including *limonus* in the **chrysomphali** group, expressed some doubts regarding the synonymy of this species with *chrysomphali*.

Rust's type specimens were mounted in balsam without clearing, and important diagnostic characters such as the shape of the propodeum and crenulae cannot be seen (Figures 1330, 1331). In the only female available for study, the antennal club (Figure 1329) does appear stouter than in *chrysomphali*, only slightly over 3 times as long as wide; the forewing is perfectly hyaline, the 2 coarse setae on the submarginal vein more nearly subequal than in *chrysomphali*; the propodeum appears to be shorter than in *chrysomphali*. *A. limonus* is a yellow species, thoracic sterna faintly dusky, the stem of the mesosternal furca blackish but apparently fainter than in *chrysomphali*.

In view of these apparent differences, and the apparent abundance of males, we cannot accept the synonymy of *limonus* with *chrysomphali*. For the time being, therefore, we consider *limonus* to be an unrecognizable species of *Aphytis*.

The host of *limonus* was apparently the lesser snow scale, *Pinnaspis strachani* (Cooley), which was recorded in the past as "*Hemichionaspis minor*" (see Borkhsenius, 1966).

## **Aphytis maculatipes** (Girault) n.comb.

(Figure 1332)

*Marietta maculatipes* Girault, 1917, Insecut. Inscit. Menstr., **5**:32.
*Marietta maculatipes*: Compere, 1936, Univ. Calif. Publ. Entomol., **6**:315.

This species is known only from the type series: 5♀♀ and 5♂♂, collected by G. Compere on *Hakea* sp., Ravensthorpe, Western Australia (No. 990 in Compere's original ledger; U.S. National Museum slide No. 25688). The specimens are in poor shape, apparently mounted in balsam after partial clearing, rather crumpled, most diagnostic characters not visible.

As far as can now be determined, *maculatipes* appears to agree with *A. capillatus* in general coloration and all structural characters, including the peculiar shape and sense organs of the male antennae. The forewing of the female (Figure 1332) is nearly hyaline, with a faint, almost imperceptible pattern of somewhat contrasting coarse and fine setae, very indistinct but apparently similar to the *capillatus* pattern; this is somewhat more readily visible in the delta area.

Although the structure of the propodeum, male genitalia and other key characters cannot be seen, it is assumed from the general habitus that *maculatipes* is indeed an *Aphytis*. Whether it is just another variant of *capillatus*, an artifact of poor mounting, or a distinct species cannot be determined at present. For the time being, due to the very poor condition of the available material, *maculatipes* has to be listed as an unrecognizable species, apparently belonging to the *capillatus* complex in the **vittatus** group.

## **Aphytis minutissimus** (Girault)

(Figures 1333–1335)

*Aphelinus minutissimus* Girault, 1913, Mem. Queensl. Mus., **2**:182.
*Aphytis minutissimus*: Compere, 1955, Univ. Calif. Publ. Entomol., **10**:316.

Girault (1913) described this Australian species "from a single female reared from a *Chionaspis* on foliage of cockatoo apple, forest, December 18, 1911 . . . . Nelson (Cairns), Queensland." The original description refers to *minutissimus* as "pale lemon yellow," does not mention most of the essential diagnostic characters, and is inadequate for proper identification of this species.

We are in complete agreement with the following comments, made by Compere (1955) regarding *minutissimus*: "The type specimen, No. 3768, Queensland Museum, is a male, not a female as described. On the same page (loc. cit.), where *minutissimus* is described but with reference to the species identified as *fuscipennis* (Howard), Girault wrote: "Five females reared from a *Chionaspis* on cockatoo apple with *minutissimus*, forest, Nelson (Cairns), Queensland." Since the single male *minutissimus* was reared from the same lot of scales as the so-called *fuscipennis* females, there is a possibility that the two forms are opposite sexes of the same species."

The holotype of *minutissimus* is a minute male specimen (length about 0.40 mm), showing the characteristic cephalic pigmentation and infuscated thoracic sterna of the **proclia** group. The specimen is partly cleared (propodeal crenulae unseen), slightly distorted (especially the antennae, Figure 1333), apparently due to immersion in KOH prior to mounting in balsam, but is otherwise in quite good condition (Figure 1334); the forewing (Figure 1335) exhibits characteristics of a minute specimen: rather narrow (nearly 3 times as long as wide), marginal fringe long (nearly $\frac{1}{3}$ width of disk), marginal vein bearing only 4 setae along the anterior margin.

Obviously, the above description would fit almost any member of the **proclia** group. The identity of Girault's ♀ specimens of "*fuscipennis*" has not been determined by us. *A. fuscipennis* (Howard) now stands as a synonym of *A. diaspidis* (Howard) (see p. 407), a nearly cosmopolitan species recorded also from Australia. *A. minutissimus* is for the present unidentifiable.

## **Aphytis newtoni** (Girault)

(Figures 1336, 1337)

*Aphelinus newtoni* Girault, 1913, Mem. Queensl. Mus., **2**: 182–183.
*Aphytis newtoni*: Compere, 1955, Univ. Calif. Publ. Entomol., **10**:316.

Girault (1913) compared *newtoni* with *minutissimus*, noting some differences in the shape and chaetotaxis of the forewing. His brief original description lacks most essential diagnostic characters and is entirely worthless. He described this species "from one female captured from a window, March 13, 1912 . . . Thursday Island, Torres Strait and Nelson, Queensland," Australia. However, Compere (1955) subsequently pointed out that "the single type specimen, number 3769, Queensland Museum, is a male, not a female as described."

The holotype of *newtoni* is indeed a male. It is mounted in balsam, uncleared, together with the holotype of *grotiusi* and a trichogrammatid, and is fairly well preserved, with one forewing detached. This species is undoubtedly an *Aphytis*. It is entirely yellow, with the mesosternal furca ("Y") and the metasternum rather faintly dusky. The forewing (Figures 1336, 1337) is hyaline, sparsely setose. Length 0.48 mm.

Until the type specimen is adequately cleared and remounted, *newtoni* will remain an unrecognizable species of *Aphytis*.

## **Aphytis pallidus** (Ratzeburg)

*Encyrtus pallidus* Ratzeburg, 1848, Die Ichneumonen der Forstinsecten, **2** : 149.
*Coccobius pallidus*: Ratzeburg, 1852, Die Ichneumonen der Forstinsecten, **3** : 195.
*Aphytis pallidus*: Bouček, 1964, Beitr. Entomol., **14**:667.

This species is known to us from the description only. Ratzeburg (1848) described *pallidus* in the genus *Encyrtus*, but subsequently (1852) he transferred it to his newly described genus *Coccobius*. Reinhard (1858) synonymized this species with *Eulophus flavus* Nees, but Nowicki (1930) regarded this synonymy as questionable (see quotation on p. 733 herein). Bouček (1964), who examined a female specimen labeled "pallidum Rtz." in the remnants of the Ratzeburg collection at Eberswalde, determined it as belonging to the genus *Aphytis* and agreed with Nowicki (op. cit.) regarding the questionable synonymy. The available descriptions of *pallidus* do not permit its specific identification.

Ratzeburg (1848) described *pallidus* from a specimen reared by Bouché from "*Coccus (Aleyrodes) Aceris* Bé"; later (1852) he recorded the host as "*Coccus Tiliae.*" That species, now known as *Eulecanium tiliae* (L.), is a soft scale insect (Homoptera: Coccidae). However, if *pallidus* had indeed been an *Aphytis*, we assume that, like all its known congeners, it was parasitic upon an armored scale insect.

## **Aphytis perissoptroides** (Girault) n.comb.

(Figure 1338)

*Aphelinus perissoptroides* Girault, 1915, Mem. Queensl. Mus., **4**:46.
*Marietta perissoptroides*: Compere, 1936, Univ. Calif. Publ. Entomol., **6**:311.

This species was described "from one female captured in forest, April 17, 1914 (A. P. Dodd)," Cloncurry, Queensland, Australia. In his original description, Girault (1915) inexplicably compared it to *grotiusi*, an entirely yellow species listed here as unrecognizable and of doubtful generic status (see p. 734). He noted that *perissoptroides* differs from *grotiusi* mainly in that the third segment of the funicle "widens distad and is somewhat wider than long while the discal ciliation of the forewings is broken into naked areas forming a colorless pattern somewhat as in *Perissopterus* except that the ciliated spaces are not infuscated deeply and the hairless ones less numerous than usual." He also made the following comments: "This species has all the characters of *Perissopterus* except the small triangular postscutellum . . . The genus *Perissopterus* should be carefully scrutinized because its general resemblance to *Aphelinus* is most striking and we have just seen a species of the latter which has the fore wings which approach the peculiar type of wing pattern common to species of *Perissopterus.*"

Compere (1936) referred *perissoptroides* to the genus *Marietta*, but commented as follows: "If one may judge by the description, this species seems to be more closely related to *Aphytis* Howard than is any of the other species. The penultimate joint of the antenna is wider than long, and the metanotum is smooth across the center. This species is described as having seven rows of cilia basad of the speculum, and in this respect is similar to an *Aphytis.*"

The female holotype of *perissoptroides*, No. 3754 in the Queensland Museum, is mounted in balsam, uncleared, with the head and one forewing separated from the body and both antennae broken. Although important diagnostic characters such as the shape of the propodeum cannot be seen, this is undoubtedly an *Aphytis*, apparently related to the *capillatus* complex in the **vittatus** group. The forewing pattern (Figure 1338) is very faint but quite extensive, similar to that of *nigripes*, the apical cloud merging with the two subapical clouds. General coloration yellow, faintly marked with brownish on mesonotal margins and sutures, more conspicuously so on antero-lateral margins of endophragma and on sides of propodeum. If there are any crossbands on the abdominal tergites, they cannot be seen in this specimen. Assuming that this is not a mere artifact of mounting, *perissoptroides* appears to be the palest member of the *capillatus* complex. However, inasmuch as the shape of the antennae, propodeum and crenulae, the chaetotaxis of the thorax, and various other diagnostic characters are not available for study in the single specimen at hand, this species is at present unrecognizable.

### ?**Aphytis punctaticorpus** (Girault) n.comb.

*Perissopterus punctaticorpus* Girault, 1917, Descriptiones stellarum novarum. Priv. Publ., 22 pp., pages 2–3.
*Marietta punctaticorpus*: Compere, 1936, Univ. Calif. Publ. Entomol., **6**:315.

Girault (1971b) described the coloration and wing pattern of *punctaticorpus* in some detail, but made no reference to most of the structural characters. His original description is quoted here in full:

"Somewhat larger than usual. Honey yellow, the vertex orange and with numerous blackish dots. Dorsum of abdomen (except six marginal spots, 2–5 dots, 1 wider than long, twice the size of 2, 6 triangular, large, on each side of the apex), propodeum, a distinct streak along each side of a narrow median sulcus on the scutellum, a somewhat narrower and similar one on the scutum but not quite complete cephalad, numerous dots on the scutum, two larger dots in a longitudinal line on each side of the scutellum, caudal margin (caudo-lateral border broader) of scutellum and an obscure dot center of each axilla, blackish. Middle and cephalic tibiae with three cincti, the third at apex; caudal tibiae with four (at least from dorsal aspect but the first two dorsad only and irregular). Femora immaculate or nearly. Fore wings with the pattern of spots—proximad of the hairless line an ovate ring which nearly opens caudodistad onto the hairless line; a long, thick curved dash from the stigmal vein to about the middle of the wing and from the hairless line, proximad of this dash, a small, longer than wide area; caudad of this but not touching the hairless line, a large ovate area; caudo-distad of this and caudad of the apex (nearly) of the long stigmal dash, a small round area not half the size of the preceding; some distance distad (in a straight line) of the venation, a minute area or dot; then two "eye-spots" about half way to apex from the venation (the cephalic one may be connected very narrowly with the dot straight out from venation), one on each side of the middle line; these are smaller than the large area near the hairless line but larger than the small one opposite the stigmal dash. Stigmal vein with a small brown substigmal spot. Proximal half of first joint of caudal tibiae brown. Caudal femur with a spot at knee and another farther proximad but not distant from knee. First four antennal joints white, club 1 yellow, 2 black and the longest antennal joint, 1 subequal to the scape, over twice longer than wide. Pedicel barely longer than wide. Scape with a narrow blackish stripe along about the middle of its side. Mexico (A. Koebele)."

This species is represented in the U.S. National Museum of Natural History (No. 20080) by 2♀♀ (syntypes?), both pasted on one point; one specimen headless and with crumpled wings, the other with one antenna and the wings missing.

Not much can be added to Girault's original description. The mandibles are well developed. The first two segments of the antennal funicle, if in fact present, are minute and cannot be seen clearly, whereas the third funicular segment (Girault's "club 1") is elongate as described, though apparently not quite as long as the scape. Mesoscutum with numerous setae. Metanotum without the diamond-shaped sculpture characteristic of *Marietta*. The shape of the propodeum cannot be seen clearly, but this tergite appears to be several times longer than the metanotum. Cerci with 3 long setae and 1 short seta.

It is impossible to determine the correct generic position of this peculiar Neotropical species from the poor material available for study. The shape of the metanotum and propodeum and the general habitus are suggestive of *Aphytis*, and the wing pattern as described by Girault is rather similar to that of members of the *capillatus* complex, as well as other members of the **vittatus** group. The peculiar antennae, on the other hand, may be those of a true *Marietta*. For the time being we regard *punctaticorpus* as an unrecognizable species of highly questionable status, probably related to the **vittatus** group.

### **Aphytis ruskini** (Girault) n.comb.

(Figures 1339–1341)

*Aphelinus ruskini* Girault, 1915, Mem. Queensl. Mus., **4**:46–47.

This Australian species was described "from one female taken by sweeping in forest, January 4, 1912," Capeville (Pentland), Queensland. The original description is not very helpful: Girault (1915) compared *ruskini* to *grotiusi*, listed here as an unrecognizable species of doubtful generic status, to *fuscipennis*, and to *miltoni*, which is evidently a species of *Centrodora*.

The holotype specimen, a female, deposited in the Queensland Museum, is in very poor shape: partly cleared, badly distorted, mounted in balsam with the head separated from the body, together with the type of *Coccophagus filius* Girault and a few undetermined Aphelinidae. Important diagnostic characters such as the shape of the propodeum and crenulae cannot be seen. The following notes were made from the holotype specimen.

Mandibles well developed; maxillary palpi 2-segmented. Antennae (Figure 1339) distorted, the club apparently robust, relatively short. Mesoscutum (Figure 1340) apparently with 12 coarse setae, the largest ones seemingly composed of distinct filaments. Cerci apparently with 2 long setae and 1 short seta. Ovipositor shaft about $1\frac{1}{2}$ times as long as the middle tibia; ovipositor sheaths about $\frac{1}{3}$ length of middle tibia. Forewing (Figure 1341) perfectly hyaline, uniformly setose, about $2\frac{2}{5}$ times as long as wide; marginal fringe short, not exceeding $\frac{1}{12}$ width of disk; delta area with 7 rows of setae, these longer and somewhat sparser than the setae distad of speculum; costal cell with 3–4 fine setae along proximal half or so, and 2 coarse setae near apex; submarginal vein bearing 5 coarse setae and 18 bullae; marginal vein bearing 9 prominent, subequal setae along anterior margin, these only slightly longer than the setae in a row along center of vein. General coloration apparently yellow (Girault's reference to *ruskini* as "dull honey yellow tinged with reddish" apparently reflected the unnatural coloration due to coagulation of body contents in the poorly preserved specimen); parapsidal sutures, posterior margins of mesoscutum and scutellum, anterior and posterior margins of propodeum laterally, and lateral margins of endophragma anteriorly, lined with blackish; pigmentation of head, thoracic sterna and abdomen, if any, unseen. Length about 1.10 mm.

From the above description it appears that *ruskini* may be rather similar, if not identical, to *A. wallumbillae* (Girault) (see p. 287). However, inasmuch as essential aspects of structure and pigmentation cannot be seen in the poorly preserved type specimen of *ruskini*, and whereas the condition of the type precludes its safe remounting, there is no sufficient basis for synonymizing *wallumbillae* with *ruskini*. Thus, *ruskini* remains an unrecognizable species of *Aphytis*.

## **Aphytis simplex** (Zehntner)

*Aphelinus simplex* Zehntner, 1897, Arch. Java-Suikerindust., **4**:19–20; Pl. I, Fig. 18–19.
*Aphytis simplex*: Compere, 1955, Univ. Calif. Publ. Entomol., **10**:317.

*A. simplex* was described as a parasite of *Duplachionaspis saccharifolii* (Zehntner) on sugarcane in Java. Unfortunately the original description, though rather detailed, failed to mention some of the most important diagnostic characters and is inadequate by modern standards.

We have examined what appears to be the type series, deposited in the U.S. National Museum: 4♀♀ and 2♂♂, mounted in balsam, without clearing, on a single slide. The specimens are all in rather poor condition, opaque, partly distorted. All that can be said now is that *simplex* is an entirely yellow species, apparently without any dark markings except for a short blackish streak at the base of the forewing, below the tegula; mandibles well developed; antennae normal, 6-segmented in both sexes (according to the original description, the pedicel is twice as long as wide, longer than the third segment of the funicle; the first 2 funicular segments are wider than long, and the club is truncate, $2\frac{1}{2}$ times longer than the preceding segment); setae on head, thorax and abdomen pale; mesoscutum apparently with 10–11 setae; forewing perfectly hyaline. Length: ♀ 0.63–0.72 mm; ♂ 0.51–0.56 mm.

This could be a small member of the **chrysomphali** group. For the time being, *simplex* stands as an unrecognizable species of *Aphytis*.

## **Aphytis variolosum** Alam

*Aphytis variolosum* Alam, 1956, Trans. R. Entomol. Soc. Lond., **108**:373–374.
*Aphytis variolosus*: Ferrière, 1965, Hymenoptera Aphelinidae d'Europe et du Bassin Méditerranéen, pp. 90, 91; syn. of *mytilaspidis*.

Alam (1956a) described *variolosum* from 2♀♀, reared July 3, 1953, and recorded as ectoparasites of *Asterolecanium variolosum* (Ratzeburg) on *Quercus robur* at Silwood Park, England. The species was described as closely related to *mytilaspidis*, and the differences between the two were tabulated as follows:

| "*A. variolosum* sp.n. | *A. mytilaspidis* Le Baron |
|---|---|
| (1) Entire antenna yellow. | Scape and pedicel dusky. |
| (2) Length of antenna greater than width of head. | Length of antenna equals width of head. |
| (3) Scutum with four transverse rows of setae. | Scutum with three transverse rows of setae. |
| (4) Coxae, trochanters and femora whitish-yellow; tibiae and tarsi brown. | Entire legs yellow. |
| (5) Body whitish-yellow. | Body bright lemon-yellow." |

In the original description, *variolosum* was compared to *A. zonatus* Alam, a synonym of *proclia* (see p. 381). The description is rather inadequate.

Ferrière (1965) noted that the type was in very poor shape, and listed *variolosum* as a synonym of *mytilaspidis*.

Upon a visit to the British Museum (Natural History) in 1963, DeBach made the following notes: "Examined the holotype female (B. M. Type Hym. 5-1678) of *Aphytis variolosum* Alam . . . which is mounted on a slide. The specimen is badly shriveled and distorted and not at all cleared. It must have been mounted fresh in balsam. In its present condition it is nearly worthless for detailed analytic study. . . . There appear to be discrepancies in the description, and the characters used to separate it from *mytilaspidis* are open to question. The host record is questionable. Only 2 specimens were obtained by Alam. In my opinion *A. variolosum* should be considered a synonym of *A. mytilaspidis* for the present. . . . Boratynski's group studying the ecology and populations of *Asterolecanium variolosum* have never reared or seen an external parasite on *Asterolecanium*. I checked 60 specimens reared in July 1963 and none were *Aphytis*."

And again in 1968: "I would draw the same conclusions. Additionally, note that this form of *mytilaspidis* does not show any pigmentation of sternal apophyses or elsewhere whatsoever—it appears to be completely yellowish."

It should be noted that at the time it was not known whether the thoracic sterna of *mytilaspidis* were pigmented or clear. (See p. 462 for a discussion of sternal pigmentation and other diagnostic characters in the **mytilaspidis** group.)

The type of *variolosum* was received on loan from the British Museum (Natural History) and was restudied for the present revision. It is indeed in very poor shape, mounted in balsam without clearing, distorted, with the propodeum and crenulae unseen and most other parts unfit for measurement. The synonymy with *mytilaspidis* cannot be accepted in the light of present knowledge, since there is no trace of pigmentation on the thoracic sterna.

*A. variolosum* must be regarded as an unrecognizable species. Even its association with the **mytilaspidis** group is open to question: the thoracic setae appear to be pale (7 are barely visible on one side of the mesoscutum), and with the propodeum invisible, it may as well be related to the **chrysomphali** group. In the **mytilaspidis** group, *variolosum* appears to be related to *aonidiae* and *libanicus*.

The host of *variolosum* was most probably an armored scale insect which went unnoticed in a rearing sample of *Asterolecanium variolosum*.

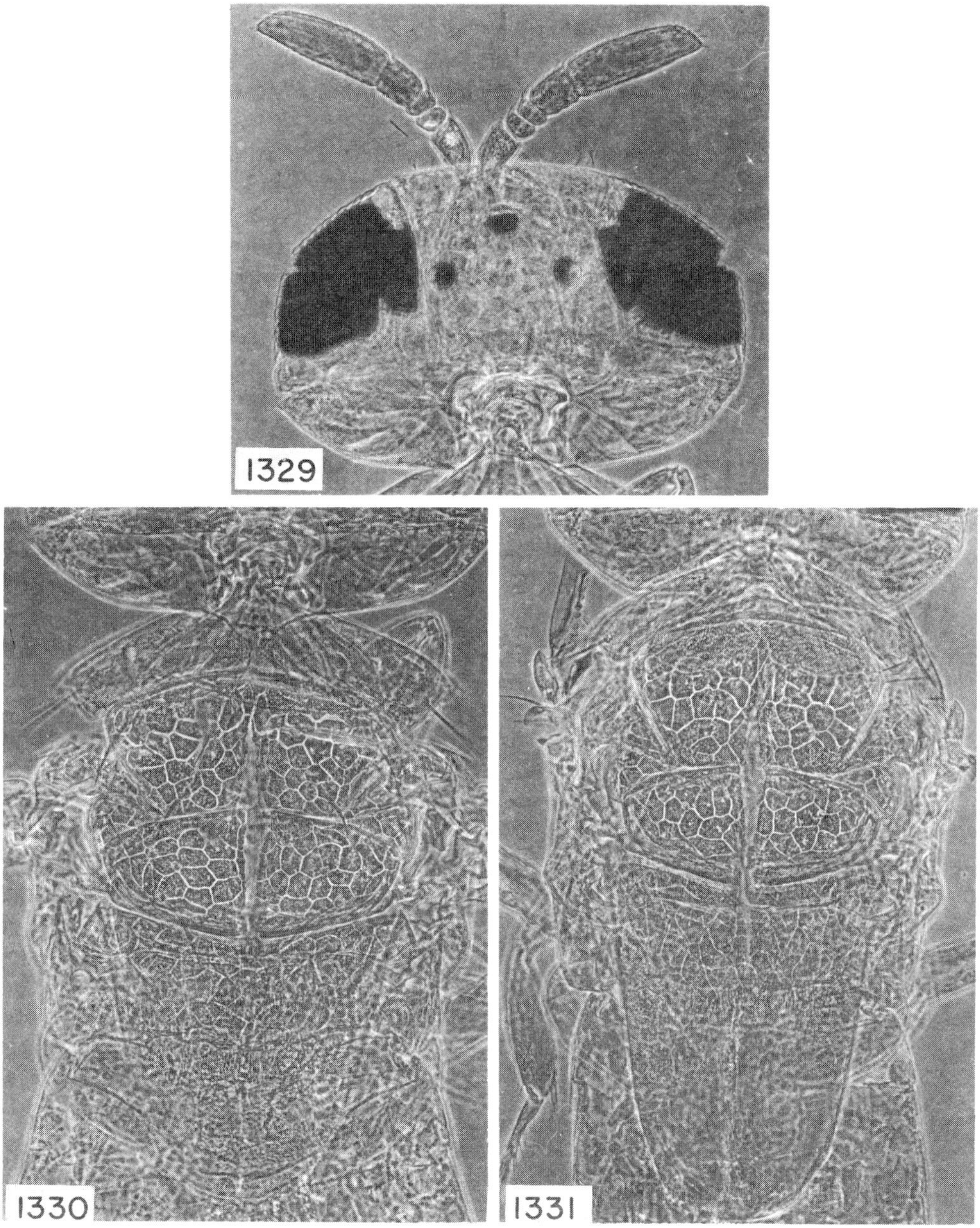


Figures 1329–1331. *Aphytis limonus* (Rust)
1329. ♀: Head and antennae (paratype). 1330. ♀: Thorax, propodeum and base of gaster (paratype). 1331. ♂: Thorax, propodeum and base of gaster (holotype).

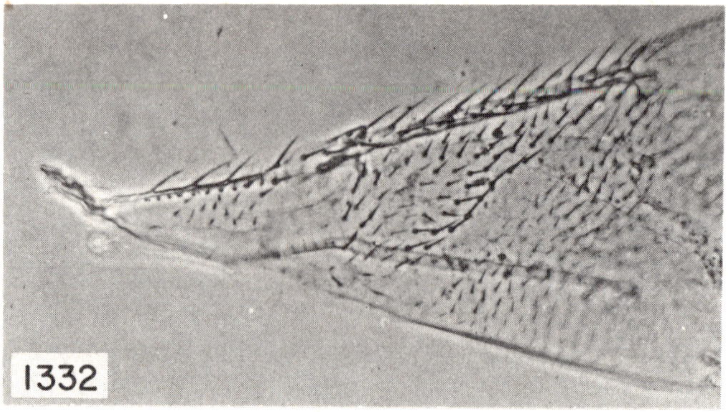

Figure 1332. *Aphytis maculatipes* (Girault), ♀ (syntype): Base of forewing.

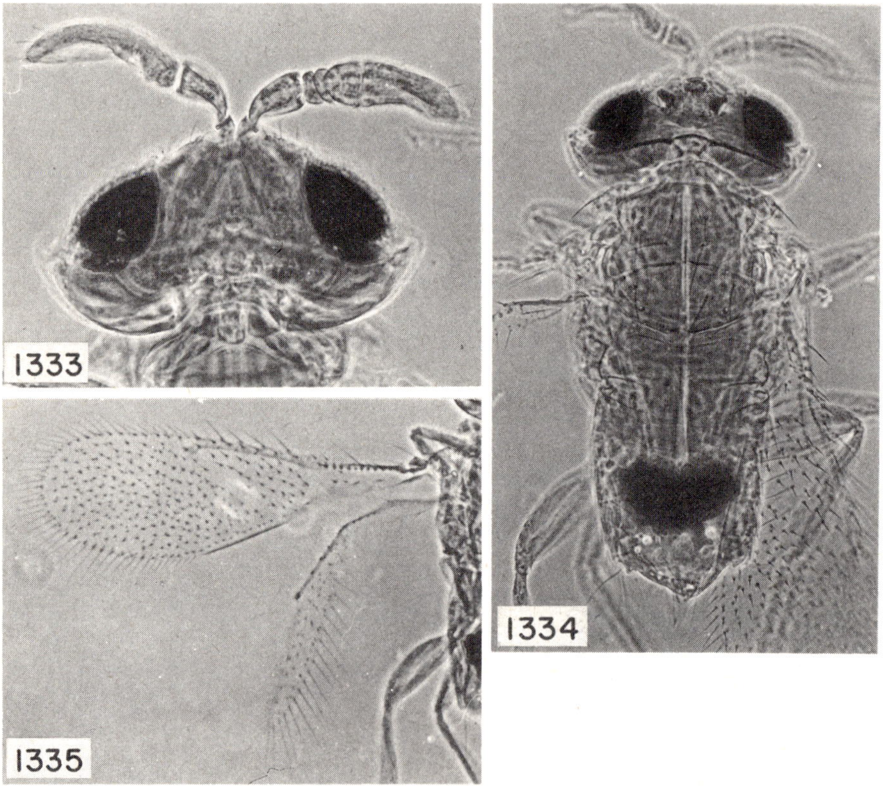

Figures 1333–1335. *Aphytis minutissimus* (Girault), ♂ (holotype)
1333. Head and antennae. 1334. Entire body. 1335. Wings.

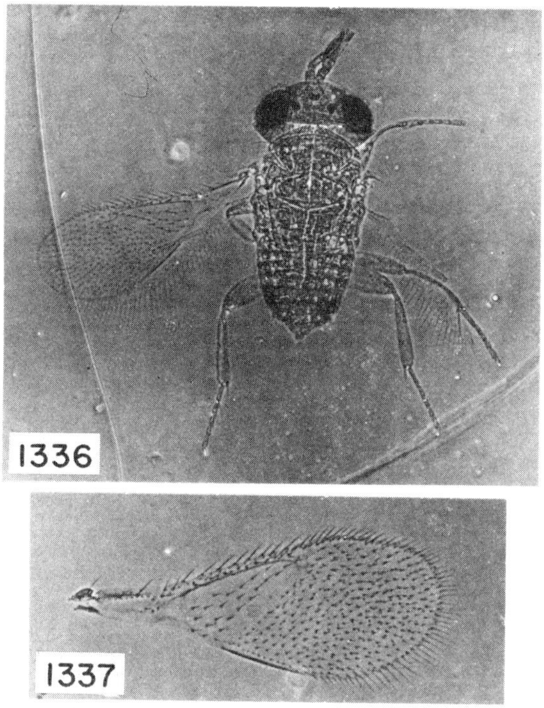


Figures 1336, 1337. *Aphytis newtoni* (Girault), ♂ (holotype)
1336. Entire specimen. 1337. Forewing.

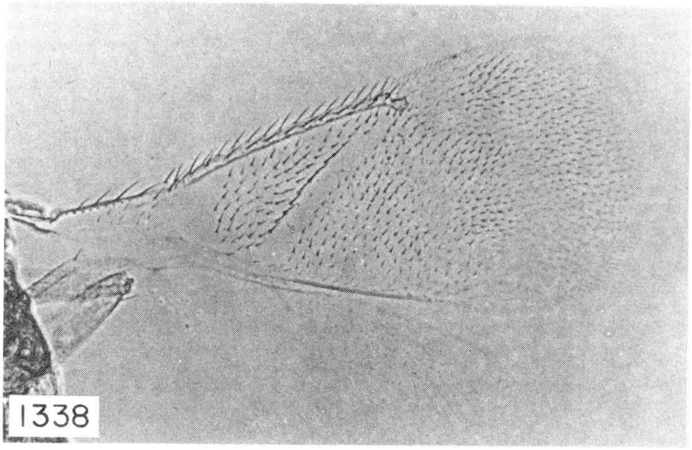


Figure 1338. *Aphytis perissoptroides* (Girault), ♀ (holotype): Forewing.

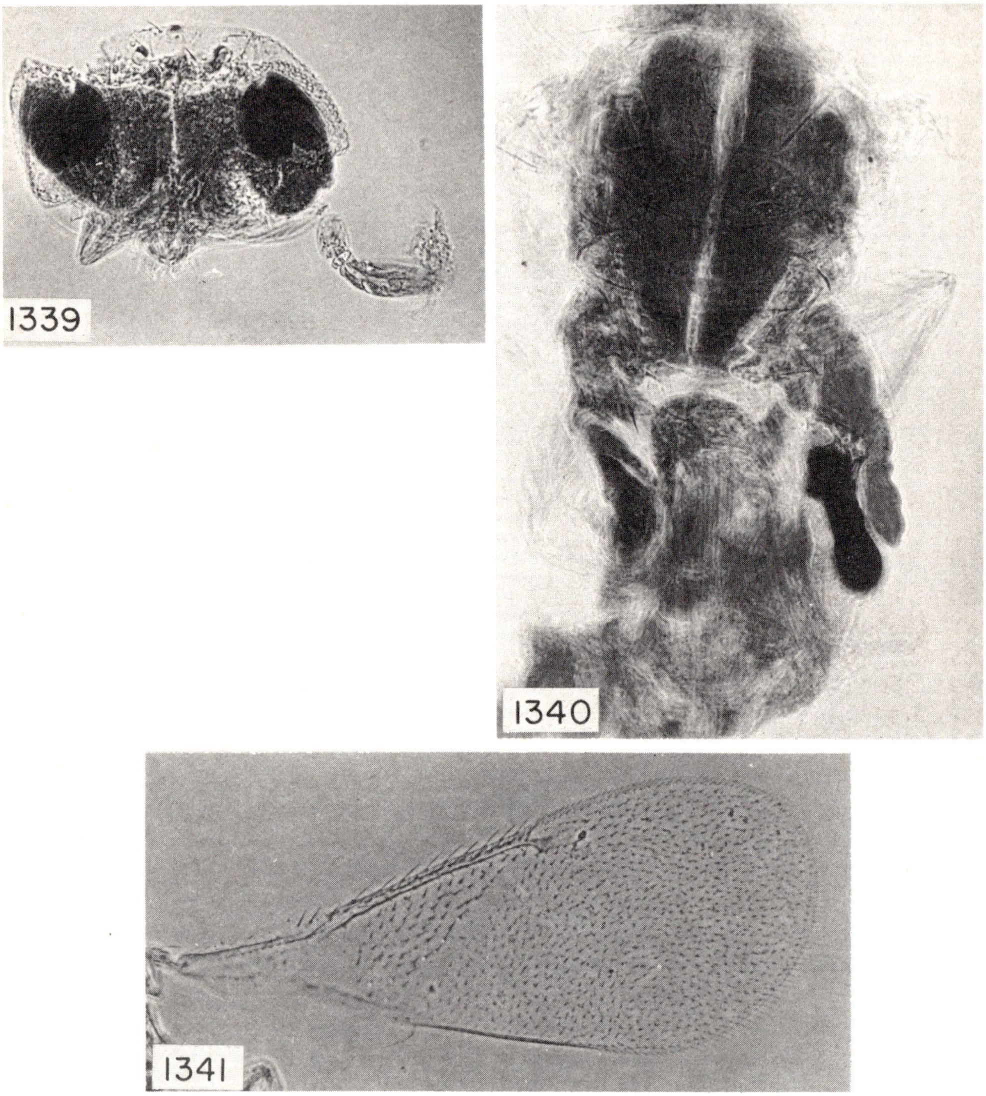


Figures 1339–1341. *Aphytis ruskini* (Girault), ♀ (holotype)
1339. Head and antennae. 1340. Thorax and abdomen. 1341. Forewing.

# APPENDICES

# APPENDIX 1. DISTRIBUTION OF *APHYTIS* SPP. IN THE MAIN ZOOGEOGRAPHICAL REGIONS OF THE WORLD

The following table presents available data on the distribution of the known species of *Aphytis* in the six main zoogeographical regions of the world. Only authenticated records of valid species have been included; species regarded here as unrecognizable have been excluded from the table. The species are arranged by species groups, and alphabetically within the groups. The zoogeographical regions have been defined according to Wallace (1876). Thus, the Australian region includes New Caledonia and New Zealand, the Palearctic region includes the Soviet Far East, Japan, and Africa north of the Sahara, and the Nearctic region includes Baja California. An asterisk (*) denotes establishment of a deliberately introduced species. For each group, the total number of species recorded from each region is given, excluding deliberate introductions.

TABLE 4. Zoogeographical distribution of the species of *Aphytis*

| | Australian | Neotropical | Ethiopian | Oriental | Palearctic | Nearctic |
|---|---|---|---|---|---|---|
| The **vittatus** group | | | | | | |
| *acutaspidis* | | + | | | | |
| *angustus* | | | | + | | |
| *anomalus* | | + | | | | |
| *argenticorpus* | + | | | | | |
| *australiensis* | + | | | | | |
| *capillatus* | + | | | | | |
| *ciliatus* | + | | | | | |
| *cochereaui* | + | | | | | |
| *costalimai* | | + | | | | |
| *fabresi* | + | | | | | |
| *haywardi* | | + | | | | |
| *maculatipennis* | | + | | | | |
| *nigripes* | + | | | | | |
| *noumeaensis* | + | | | | | |
| *obscurus* | | + | | | | |
| *peculiaris* | | | | + | | |
| *perplexus* | | + | | | | |
| *vittatus* | | | | + | | |
| | 8 | 7 | 0 | 3 | 0 | 0 |
| Related species | | | | | | |
| *dealbatus* | | | | | | + |
| *hyalinipennis* | + | | | | | |
| *malayensis* | | | | + | | |
| *mandalayensis* | | | | + | | |
| *melanostictus* | | | | | | + |
| *wallumbillae* | + | | | | | |
| | 2 | 0 | 0 | 2 | 0 | 2 |

TABLE 4 (continued)

| | Australian | Neotropical | Ethiopian | Oriental | Palearctic | Nearctic |
|---|---|---|---|---|---|---|
| The **chilensis** group | | | | | | |
| *cercinus* | | | + | | | |
| *chilensis* | + | + | + | | + | + |
| *columbi* | + | | | | | |
| *faurei* | | | + | | | |
| *merceti* | | | + | | | |
| | 2 | 1 | 4 | 0 | 1 | 1 |
| Related species | | | | | | |
| *antennalis* | | | | + | | |
| | 0 | 0 | 0 | 1 | 0 | 0 |
| The **proclia** group | | | | | | |
| *comperei* | | + | + | + | | + |
| *confusus* | | | + | | | |
| *diaspidis* | + | + | + | + | + | + |
| *hispanicus* | | + | | + | + | + |
| *maculicornis* | | * | | | + | * |
| *paramaculicornis* | | | | + | + | * |
| *philippinensis* | | | | + | | |
| *pinnaspidis* | | + | | | | + |
| *proclia* | | + | | + | + | + |
| *testaceus* | | | | | + | |
| *vandenboschi* | | | | | + | |
| | 1 | 5 | 3 | 6 | 7 | 5 |
| Related species | | | | | | |
| *acrenulatus* | | | | + | | |
| *amazonensis* | | + | | | | |
| *desantisi* | | + | | | | |
| *griseus* | | | + | | | |
| *tucumani* | | + | | | | |
| | 0 | 3 | 1 | 1 | 0 | 0 |
| The **mytilaspidis** group | | | | | | |
| *aonidiae* | | + | | | + | + |
| *libanicus* | | | | | + | |
| *luteus* | | | | | + | |
| *moldavicus* | | | | | + | |
| *mytilaspidis* | | | | | + | + |
| *opuntiae* | | | | | + | |
| *phoenicis* | | | | | + | |
| | 0 | 1 | 0 | 0 | 7 | 2 |
| Related species | | | | | | |
| *capensis* | | | + | | | |
| *neuter* | | | | | + | |
| *notialis* | | + | | | | |
| *rolaspidis* | | | + | | | |
| *taylori* | | | + | | | |
| | 0 | 1 | 3 | 0 | 1 | 0 |

TABLE 4 (continued)

| | Australian | Neotropical | Ethiopian | Oriental | Palearctic | Nearctic |
|---|---|---|---|---|---|---|
| The **lingnanensis** group | | | | | | |
| *africanus* | | | + | | | |
| *coheni* | | | | + | * | |
| *equatorialis* | | | + | | | |
| *fisheri* | | | | + | + | |
| *holoxanthus* | * | * | * | + | * | * |
| *lingnanensis* | + | + | * | + | * | * |
| *margaretae* | | + | | | | + |
| *melinus* | * | * | * | + | * | * |
| *yasumatsui* | | | | | + | |
| | 1 | 2 | 2 | 5 | 2 | 1 |
| The **chrysomphali** group | | | | | | |
| *anneckei* | | | + | | | |
| *chrysomphali* | + | + | + | + | + | + |
| *cylindratus* | | + | | | + | |
| *debachi* | | | | + | + | |
| *immaculatus* | | | | + | | |
| *japonicus* | | | | | + | |
| *lepidosaphes* | + | * | * | + | * | * |
| *mazalae* | | | | + | + | |
| *sensorius* | | | | + | | |
| | 2 | 2 | 2 | 6 | 5 | 1 |
| The **funicularis** group | | | | | | |
| *funicularis* | | | + | | | |
| *gordoni* | | | | + | | |
| *theae* | | | | + | | |
| *ulianovi* | + | | | | | |
| | 1 | 0 | 1 | 2 | 0 | 0 |
| Unassigned species | | | | | | |
| *bedfordi* | | | + | | | |
| *erythraeus* | | | + | | | |
| *ignotus* | + | | | | | |
| *longicaudus* | | | | + | | |
| *mimosae* | | | + | | | |
| *pilosus* | | | + | | | |
| *roseni* | | * | + | | | |
| *salvadorensis* | | + | | | | |
| *secundus* | | | + | | | |
| *setosus* | | | + | | | |
| | 1 | 1 | 7 | 1 | 0 | 0 |

# APPENDIX 2. HOST SPECIES AND THEIR KNOWN *APHYTIS* PARASITES

Of the 90 species of *Aphytis* recognized as valid in the present revision, there are no definite host records for the following 14 species:

| | | | | |
|---|---|---|---|---|
| *amazonensis* | *argenticorpus* | *faurei* | *nigripes* | *ulianovi* |
| *angustus* | *bedfordi* | *hyalinipennis* | *salvadorensis* | *wallumbillae* |
| *antennalis* | *fabresi* | *malayensis* | *secundus* | |

The armored scale-insect hosts of the remaining 76 species are listed here in alphabetical order. (Host records of other insect groups, such as soft scale insects, are presumed to be erroneous and have not been included.) The nomenclature and classification of the Diaspididae proposed by Borkhsenius (1966) have usually been followed. Under each host species, the species of *Aphytis* recorded from it are listed in alphabetical order, along with lists of countries where such records were made (including established introductions). Unrecognizable species of *Aphytis* are in parentheses. The countries are arranged according to the main zoogeographical regions.

In view of the great confusion that existed almost as a rule until recently regarding the identity of many *Aphytis* species, we have not included host records from the older literature for which we could not obtain first-hand corroboration. The list is therefore based mainly on our own records, represented by positively identified *Aphytis* specimens in the collection of the University of California, Riverside, and on some of the more reliable recent publications. However, although we are reasonably sure about the identity of the parasites, we realize that some mistakes in the identity of host species may not have been avoided.

The information compiled in the list is summarized in Tables 5, 6 and 7 (pp. 766–772).

In Table 5, the genera of armored scale insects are listed according to the classification of Borkhsenius (1966). All the major groups (subfamilies, tribes and subtribes) of the Diaspididae are represented, but genera are listed only if known to be attacked by *Aphytis*. The species of *Aphytis* are arranged by species groups, and alphabetically within each group. The numbers indicate how many species of a given scale-insect genus are known to be attacked by a given *Aphytis* species (*A. chilensis*, for instance, has been recorded from one species of *Diaspis*, three species of *Parlatoria*, two species of *Aspidiotus*, etc.). An asterisk (*) denotes a laboratory host. For each species of *Aphytis*, the total number of field hosts is given in the right-hand column. The total number of *Aphytis* species recorded from each scale-insect genus is given at the end of the table (unrecognizable species were not counted except when they constitute the only record; laboratory hosts were not included); a dash indicates no record.

Table 6 presents the number of armored scale-insect species, arranged by subfamilies, known to be attacked by members of the main species groups of *Aphytis*.

Table 7 presents the number of *Aphytis* species, arranged by species groups, recorded as parasitic upon members of the five subfamilies of the Diaspididae.

**Abgrallaspis cyanophylli** (Signoret) CYANOPHYLLUM SCALE
*costalimai*: Brazil
*margaretae*: Brazil

**Acutaspis agavis** (Townsend and Cockerell)
*diaspidis*: Bermuda

**Acutaspis albopicta** (Cockerell)
*acutaspidis*: Brazil

**Acutaspis scutiformis** (Cockerell)
*hispanicus*: Brazil

**Africaspis chionaspiformis** (Newstead)
*confusus*: South Africa

**Africaspis terminaliae** Munting
*cercinus*: South Africa

**Anamaspis coniferarum** (Hall and Williams)
*sensorius*: Pakistan

**Aonidia lauri** (Bouché) LAUREL SCALE
*aonidiae*: Spain, Greece
*chilensis*: Greece
*lingnanensis*: Spain
*mytilaspidis*: Spain

**Aonidiella aurantii** (Maskell) CALIFORNIA RED SCALE
*africanus*: South Africa, Mozambique
*aonidiae*: USA (California)
*chilensis*: USA (California)
*chrysomphali*: USA (California, Texas), Argentina, Chile, Israel, Greece, Crete, Cyprus, Turkey, Morocco, South Africa, Japan, Taiwan, Hong Kong, Australia
*coheni*: Israel, Cyprus
*desantisi*: Argentina
*diaspidis*: USA (California), South Africa
*fisheri*: Japan, Burma
*holoxanthus*: Israel, Lebanon (also in laboratory in California, Israel and South Africa)
*lingnanensis*: USA (California, Texas), Mexico, Cyprus, Morocco, China, Hong Kong, India, Australia

*?mazalae*: Pakistan
*melinus*: USA (California), Chile, Argentina, Paraguay, Israel, Cyprus, Turkey, Morocco, South Africa, India, Pakistan, Australia
*mytilaspidis*: USA (California)
*philippinensis*: India
*pinnaspidis*: Mexico (Baja California Sur)
*proclia*: Burma

**Notes.** Some of these records obviously reflect rare or occasional instances of parasitism. *A. chilensis*, for instance, is a very rare and insignificant parasite of the California red scale in California. On the other hand, species such as *A. desantisi* and *A. philippinensis* may be potentially important natural enemies and should receive further attention in biological control projects. *A. melinus* is currently considered to be the most effective available natural enemy of the California red scale.

**?Aonidiella aurantii** (Maskell)
*capillatus*: Australia
*comperei*: USA (Texas), China
*mandalayensis*: Burma
*paramaculicornis*: USA (California)

**Notes.** In these cases, the host was recorded as doubtful. *A. comperei*, for instance, was reared from "California red scale material," which may have included other scale-insect species, whereas the *capillatus* record is based on a single male specimen, reared "with" California red scale material.

**Aonidiella citrina** (Coquillett) — YELLOW SCALE
*aonidiae*: USA (California)
*chrysomphali*: USA (California), Turkey, Iran
*coheni*: (laboratory in California)
*lingnanensis*: Pakistan
*mazalae*: Pakistan
*?mazalae*: Pakistan
*melinus*: Turkey, Pakistan

**Aonidiella orientalis** (Newstead) — ORIENTAL SCALE
*aonidiae*: Iran
*coheni* (the "khunti" semispecies): India
*?mazalae*: Pakistan
*?melinus*: Saudi Arabia
*mytilaspidis*: Iran

**Aonidiella ?orientalis** (Newstead)
*peculiaris*: India

**Aonidiella** sp.
*?mazalae*: Pakistan

**Aonidiella taxus** Leonardi — ASIATIC RED SCALE
*japonicus*: Japan
*yasumatsui*: Japan

**Aspidaspis densiflorae** (Bremner) — TAN OAK SCALE
*aonidiae*: USA (California)

**Aspidiella zingiberi** Mamet
*acrenulatus*: Mauritius

**Aspidiotus capensis** Newstead
*taylori*: South Africa

**Aspidiotus elaeidis** Marchal
*erythraeus*: Eritrea
*mimosae*: Eritrea

**Aspidiotus hedericola** Leonardi
*chilensis*: Spain
*mytilaspidis*: Greece, Lebanon

**Aspidiotus nerii** Bouché [=*A. hederae* (Vallot)] — OLEANDER SCALE
*aonidiae*: Argentina, Israel
*capillatus*: Australia
*chilensis*: USA (California), Mexico, Chile, Europe, Middle East, South Africa
*chrysomphali*: Greece (also in laboratory in South Africa and Australia)
*coheni*: Israel (also in laboratory in California and Israel)
*diaspidis*: USA (California), Chile, Israel, Greece, Crete, South Africa
*fisheri*: (laboratory in California)
*hispanicus*: USSR (Caucasus)
*holoxanthus*: (laboratory in California)
*lingnanensis*: (laboratory in California)
*maculicornis*: Mexico, Greece
*melinus*: USA (California), Chile, Israel (also in laboratory in California, Israel and Australia)
*mytilaspidis*: Yugoslavia
*notialis*: Argentina, Chile
*paramaculicornis*: (laboratory in California)
*proclia*: (laboratory in California)
*vandenboschi*: (laboratory in California)

**Notes.** The oleander scale serves as an excellent laboratory host for quite a few species of *Aphytis* and has been used extensively for this purpose. *A. chilensis* is usually its most abundant parasite in the field, but species such as *A. melinus*, introduced against other hosts, seem to be displacing it in certain areas.

**Aspidiotus** sp.
*cercinus*: South Africa
*ciliatus*: Australia
*noumeaensis*: New Caledonia

**Aulacaspis murrayae** Takahashi
*mazalae*: Taiwan

**Aulacaspis rosae** (Bouché) ROSE SCALE
*aonidiae*: Crete
*columbi*: Australia
*diaspidis*: USA (California, Florida, Hawaii), Ruanda
*mytilaspidis*: Crete

**Carulaspis caruelii** (Targioni-Tozzetti) [=*C. minima* (Targioni-Tozzetti)] MINUTE CYPRESS SCALE
*chilensis*: USA (California)
*mytilaspidis*: USA (California)

**Carulaspis visci** (Schrank)
*mytilaspidis*: USA (Ohio), Greece, Morocco

**Chionaspis americana** Johnson ELM SCURFY SCALE
*proclia*: USA (Indiana)

**Chionaspis salicis** (L.)
*proclia*: England, USSR, Germany, Austria

**Chionaspis salicisnigrae** (Walsh) BLACK WILLOW SCALE
*proclia*: USA (Ohio)

**Chionaspis** sp.
*australiensis*: Australia
*capillatus*: Australia
*diaspidis*: Sri Lanka
*griseus*: South Africa
(*minutissimus*): Australia
*mytilaspidis*: Israel

**Chrysomphalus aonidum** (L.) FLORIDA RED SCALE
*africanus*: South Africa
*chrysomphali*: Dominican Republic, Philippines
*columbi*: Australia
*costalimai*: Brazil
*diaspidis*: South Africa
*holoxanthus*: USA (Florida), Mexico, Brazil, Peru, Trinidad, Dominican Republic, Israel, Hong Kong, Taiwan, Australia

*?holoxanthus*: Philippines
*lingnanensis*: Taiwan
*merceti*: South Africa
*philippinensis*: Philippines

**Notes.** *A. holoxanthus* is by far the most effective available natural enemy of the Florida red scale. However, such little-known species as *costalimai* or *philippinensis* should definitely be included in importation programs.

**?Chrysomphalus aonidum** (L.)
*comperei*: China, South Africa
*mazalae*: Taiwan
*pinnaspidis*: El Salvador

**Notes.** These are uncertain or doubtful records, reared from "Florida red scale material" or from questionably determined hosts.

**Chrysomphalus bifasciculatus** Ferris — BIFASCICULATE SCALE
*japonicus*: Japan
*vandenboschi*: Japan
*yasumatsui*: Japan

**Chrysomphalus dictyospermi** (Morgan) — DICTYOSPERMUM SCALE
*anomalus*: Brazil
*aonidiae*: Spain, Greece
*chilensis*: Argentina
*chrysomphali*: USA (Florida), Spain, Italy, Greece, Yugoslavia, Iran
*diaspidis*: Brazil
*hispanicus*: Brazil, Barbados, Spain
*melinus*: USA (California), Spain, Greece, Turkey
*mytilaspidis*: USSR (Georgia), Morocco
*proclia*: Mexico, El Salvador

**?Chrysomphalus dictyospermi** (Morgan)
*lingnanensis*: Fiji

**Chrysomphalus** sp.
*anneckei*: South Africa
*chrysomphali*: Argentina

**Clavaspis ?subsimilis** (Cockerell)
*melanostictus*: Mexico (Baja California Sur)

**Cornuaspis** [=*Lepidosaphes*] **beckii** (Newman) — PURPLE SCALE
*cochereaui*: New Caledonia
*columbi*: Australia

*lepidosaphes*: USA (California, Texas, Louisiana, Florida, Hawaii), Puerto Rico, Guadeloupe, Jamaica, Mexico, El Salvador, Argentina, Brazil, Chile, Peru, Trinidad, Israel, Lebanon, Cyprus, Greece, Crete, Spain, Turkey, South Africa, China, Taiwan, Thailand, Burma, India, Pakistan, Philippines, Fiji, New Caledonia, Australia
*lingnanensis*: Australia

**Notes.** Although *A. lepidosaphes* is currently by far the best-known natural enemy of the purple scale, the other species of *Aphytis* listed here should also be considered in biological control projects.

**?Cornuaspis** [=*Lepidosaphes*] **beckii** (Newman)
*comperei*: USA (Florida)

**Notes.** A doubtful record, reared from "purple scale material" that probably contained other hosts such as the chaff scale.

**Dentachionaspis margaritae** (Brain)
*capensis*: South Africa

**Diaspidiotus** sp.
*maculatipennis*: Haiti
*proclia*: Hungary

**Diaspis bromeliae** (Kerner) PINEAPPLE SCALE
*diaspidis*: Hawaii

**Diaspis echinocacti** (Bouché) CACTUS SCALE
*chilensis*: USA (California)
*chrysomphali*: Greece
*cochereaui*: (laboratory in California)
*debachi*: (laboratory in California)
*diaspidis*: USA (California), Mexico, Peru, Greece
*holoxanthus*: (laboratory in California)
*lingnanensis*: Mexico
*margaretae*: Mexico
*melinus*: Mexico
*mytilaspidis*: USA (California), Spain, Greece, Lebanon (also in laboratory in California)
*obscurus*: Argentina
*opuntiae*: Spain
*paramaculicornis*: (laboratory in California)
*vandenboschi*: (laboratory in California)

**Notes.** The cactus scale may serve as a convenient laboratory host for mass rearing of several *Aphytis* species.

**Diaspis ?echinocacti** (Bouché)
*tucumani*: Argentina

**Duplachionaspis saccharifolii** (Zehntner)
*(simplex)*: Java

**Dynaspidiotus britannicus** (Newstead) HOLLY SCALE
*aonidiae*: Greece
*libanicus*: Lebanon, Israel
*yasumatsui*: Japan

**Epidiaspis leperii** (Signoret) ITALIAN PEAR SCALE
*maculicornis*: USSR (Georgia)
*moldavicus*: USSR (Moldavia)
*mytilaspidis*: USA (California), USSR (Georgia)
*proclia*: USSR

**Fiorinia theae** Green TEA SCALE
*theae*: India (also in greenhouse in Florida)

**Fulaspis** sp.
*chrysomphali*: South Africa
*pilosus*: South Africa

**Furchadiaspis zamiae** (Morgan) CYCAD SCALE
*chilensis*: USA (California)
*funicularis*: South Africa

**Hemiberlesia lataniae** (Signoret) LATANIA SCALE
*aonidiae*: Israel, Greece, Crete (also in laboratory in California)
*chilensis*: USA (California), Greece
*coheni*: Israel, Cyprus
*diaspidis*: USA (California), Brazil, Greece, Crete, Cyprus, Lebanon (also in laboratory in California and Israel)
*hispanicus*: (laboratory in Israel)
*lingnanensis*: USA (California)
*maculicornis*: Greece
*melinus*: USA (California)
*mytilaspidis*: Crete
*paramaculicornis*: (laboratory in California)
*perplexus*: Brazil
*proclia*: Morocco (also in laboratory in California)
*vandenboschi*: (laboratory in California)

**Notes.** Like the oleander and cactus scales, the latania scale is a convenient laboratory host for several *Aphytis* species.

**?Hemiberlesia lataniae** (Signoret)
*chrysomphali*: USA (California)

**Hemiberlesia rapax** (Comstock) GREEDY SCALE
*chilensis*: USA (California), Mexico
*diaspidis*: USA (California)
*proclia*: Morocco

**Hemiberlesia** sp.
*diaspidis*: Greece, Crete

**Insulaspis** [=*Lepidosaphes*] **gloverii** (Packard) GLOVER SCALE
*?lingnanensis*: USA (Florida), Mexico

**?Insulaspis** [=*Lepidosaphes*] **gloverii** (Packard)
*diaspidis*: Iran

**Insulaspis juniperi** (Lindinger)
*aonidiae*: USSR (Georgia)
*mytilaspidis*: USSR (Georgia)

**Insulaspis pallida** (Green)
*hispanicus*: USSR (Caucasus)

**Ledaspis distincta** (Leonardi)
*confusus*: South Africa

**Lepidosaphes espinosai** Porter
*haywardi*: Argentina

**Lepidosaphes pallens** (Maskell)
*capillatus*: Australia

**Lepidosaphes pinnaeformis** (Bouché) [=?*Cornuaspis beckii* (Newman)]
*?proclia*: South Africa

**Lepidosaphes** sp.
*aonidiae*: Greece
*diaspidis*: USA (California, Florida, District of Columbia)
*immaculatus*: China, Taiwan
*mytilaspidis*: USSR (Kabardino-Balkaria)
*paramaculicornis*: Pakistan
*?proclia*: China
*vandenboschi*: Japan

**Lepidosaphes** sp. [=?*Paralepidosaphes tubulorum* (Ferris)]
*?proclia*: China

**Lepidosaphes ulmi** (L.) OYSTERSHELL SCALE
(*abnormis*): USA (District of Columbia)
(*caucasicus*): USSR (Caucasus)
*dealbatus*: USA (California)
*diaspidis*: Canada, Greece
*mytilaspidis*: Canada (Quebec), USA (California, Illinois), Hungary, Greece, Crete, Cyprus, Yugoslavia, USSR (Moldavia)
*proclia*: USA (Ohio)
*testaceus*: USSR

**Notes.** *A. mytilaspidis* is by far the most effective parasite of the oystershell scale.

**Leucaspis pini** (Hartig)
*luteus*: Germany, Greece

**Leucaspis riccae** Targioni-Tozzetti
*libanicus*: Israel
*mytilaspidis*: Israel

**Leucaspis signoreti** Targioni-Tozzetti
*luteus*: Corsica

**Leucaspis** sp.
*luteus*: Hungary

**Lindingaspis rossi** (Maskell) BLACK ARAUCARIA SCALE
*africanus*: South Africa
*ignotus*: Australia, New Zealand
*merceti*: South Africa

**Lopholeucaspis japonica** (Cockerell)
*hispanicus*: USSR (Caucasus)

**Melanaspis corticosa** (Brain)
*confusus*: South Africa
*merceti*: South Africa

**Melanaspis inopinata** (Leonardi)
*chrysomphali*: Greece

**Mycetaspis personata** (Comstock) MASKED SCALE
nr. *equatorialis*: Cameroun

**Mytilaspis conchiformis** (Gmelin) [=*Lepidosaphes ficus* (Signoret)] FIG SCALE
*hispanicus*: USSR (Caucasus)
*mytilaspidis*: Lebanon, Italy, USSR (Georgia)
*?opuntiae*: Italy

**Nelaspis exalbida** (Cockerell)
*griseus*: South Africa

**Neomorgania eucalypti** (Maskell)
*capillatus*: Australia

**Paralepidosaphes tubulorum** (Ferris)
*vittatus*: China

**Paraselenaspidus madagascariensis** (Mamet)
*anneckei*: Kenya

**Parlatoreopsis longispinus** (Newstead)
*?maculicornis*: Egypt

**Parlatoria blanchardi** (Targioni-Tozzetti) PARLATORIA DATE SCALE
*paramaculicornis*: Saudi Arabia
*phoenicis*: Saudi Arabia, Israel

**Parlatoria cinerea** Doane and Hadden TROPICAL GRAY CHAFF SCALE
*hispanicus*: Trinidad, Israel

**Parlatoria oleae** (Colvée) OLIVE SCALE
*chilensis*: Italy, Greece
*hispanicus*: USSR (Caucasus)
*maculicornis*: USA (California), Israel, Greece, Italy, Spain, Algeria, Egypt, USSR (Tadzhikistan, Armenia)
*mytilaspidis*: Greece
*paramaculicornis*: USA (California), Iraq, Iran, India, Pakistan

**Notes.** *A. paramaculicornis* is unquestionably the most effective natural enemy of the olive scale.

**Parlatoria pergandii** Comstock CHAFF SCALE
*chilensis*: Lebanon
*comperei*: USA (Texas, Florida), Mexico, Hong Kong
*hispanicus*: USA (California, Texas), Mexico, Argentina, Brazil, Trinidad, Israel, Turkey, Spain, USSR (Caucasus)
*paramaculicornis*: USA (California)

**Notes.** Although *A. hispanicus* is by far the most common and abundant parasite of the chaff scale, it is not always effective. Both *A. comperei* and the California "strain" of *A. paramaculicornis* should therefore be considered for introduction into areas where the chaff scale is a serious pest of citrus.

**?Parlatoria pergandii** Comstock
*aonidiae*: Spain
*melinus*: USA (California)

**Notes.** Questionable records, reared from "chaff scale material," presumably from other hosts.

**Parlatoria pittospori** Maskell PITTOSPORUM SCALE
*chilensis*: New Zealand
*diaspidis*: USA (California)

**Parlatoria** sp.
*debachi*: Japan

**Phenacaspis pinifoliae** (Fitch) PINE NEEDLE SCALE
*chilensis*: USA (California)
*mytilaspidis*: USA (California, Ohio)

**Pinnaspis** sp.
*?mazalae*: Pakistan

**Pinnaspis strachani** (Cooley) LESSER SNOW SCALE
(*limonus*): Hawaii
*?mazalae*: Pakistan
*pinnaspidis*: Mexico, Brazil

**Pseudaonidia duplex** (Cockerell) CAMPHOR SCALE
*cylindratus*: Japan
*longicaudus*: Hong Kong

**Pseudaonidia trilobitiformis** (Green)
*chrysomphali*: Brazil
*costalimai*: Brazil
*cylindratus*: Brazil, Trinidad
*lingnanensis*: New Caledonia
*longicaudus*: Hong Kong

**Pseudaulacaspis pentagona** (Targioni-Tozzetti) WHITE PEACH SCALE
*diaspidis*: Argentina, South Africa, Madagascar
*proclia*: Italy, France, Switzerland
*vandenboschi*: Japan

**Pseudoparlatoria** sp.
*margaretae*: Mexico

**Quadraspidiotus forbesi** (Johnson) FORBES SCALE
*proclia*: USA (Ohio)

**Quadraspidiotus gigas** (Thiem and Gerneck)
(*diaspidioti*): USSR (Maritime Territory)

**Quadraspidiotus juglansregiae** (Comstock) WALNUT SCALE
*aonidiae*: USA (California)
*diaspidis*: USA (California)
*maculicornis*: USA (California)
*melanostictus*: USA (California)
*melinus*: USA (California)
*mytilaspidis*: USA (California)

**Quadraspidiotus ostreaeformis** (Curtis) EUROPEAN FRUIT SCALE
*aonidiae*: Argentina
*proclia*: Morocco
*testaceus*: USSR (Moldavia), Hungary

**Quadraspidiotus perniciosus** (Comstock) SAN JOSE SCALE
*aonidiae*: USA (California, Ohio), Argentina, USSR (Armenia)
*diaspidis*: USA (California, Ohio, Georgia), Argentina, Chile, Pakistan, New Zealand
*melinus*: USA (California)
*mytilaspidis*: USA (California)
*paramaculicornis*: Pakistan
*proclia*: USA (Connecticut), USSR (Caucasus, Moldavia), France, North Africa
*vandenboschi*: Japan

**Quadraspidiotus pyri** (Lichtenstein)
*moldavicus*: USSR
*proclia*: USSR

**Quadraspidiotus slavonicus** (Green)
*neuter*: USSR (Central Asia)

**Quadraspidiotus zonatus** (Frauenfeld)
*mytilaspidis*: Greece
*proclia*: England

**Rolaspis chaetachmae** (Brain)
*funicularis*: South Africa
*rolaspidis*: South Africa
*setosus*: South Africa

**Rolaspis incisa** Munting
*funicularis*: South Africa

**Rolaspis lounsburyi** (Cooley)
*griseus*: South Africa

**Rolaspis** sp. nr. **incisa** Munting
*chilensis*: South Africa
*funicularis*: South Africa

**Selenaspidus articulatus** (Morgan) RUFOUS SCALE
*chrysomphali*: El Salvador, Peru
*diaspidis*: Peru
*holoxanthus*: Trinidad
*lingnanensis*: Trinidad
*roseni*: Peru, Kenya

**Notes.** *A. roseni* appears to be by far the most effective available natural enemy of the rufous scale.

**Separaspis capensis** (Walker) ALOË RED SCALE
*africanus*: South Africa

**Tecaspis visci** (Brain)
*cercinus*: South Africa

**Temnaspidiotus destructor** (Signoret) COCONUT SCALE
*chrysomphali*: Hawaii, Brazil, Panama, Costa Rica, Dominican Republic, Tahiti, Taiwan
*equatorialis*: Ivory Coast
*lingnanensis*: Puerto Rico, Trinidad
*margaretae*: Mexico
*melinus*: Pakistan

**Unaspis citri** (Comstock) CITRUS SNOW SCALE
*debachi*: Hong Kong
*gordoni*: Hong Kong
*lingnanensis*: USA (Florida), Mexico, Hong Kong

**Unaspis euonymi** (Comstock) EUONYMUS SCALE
*gordoni*: (laboratory in California)

TABLE 5. HOST RECORDS OF THE SPECIES OF APHYTIS

| | SUBFAMILY XANTHOPHTHALMINAE | SUBFAMILY DIASPIDINAE | | | | | | | | | | | | | | | | | | | | | | |
|---|---|---|---|---|---|---|---|---|---|---|---|---|---|---|---|---|---|---|---|---|---|---|---|---|
| | | ANTAKASPIDINI | LEPIDOSAPHIDINI | | | | | | | ANCEPASPIDINI | CHIONASPIDINI | | | | | | | | | | | | | |
| | | | COCCOMYTILINA | LEPIDOSAPHIDINA | | | | | | | SCLOPETASPIDINA | AUGULASPIDINA | | | KUWANASPIDINA | CHIONASPIDINA | | | | PHENACASPIDINA | | | | |
| | | | | Fulaspis | Paralepidosaphes | Lepidosaphes | Mytilaspis | Cornuaspis | Insulaspis | | | Rolaspis | Tecaspis | Dentachionaspis | | Chionaspis | Unaspis | Pinnaspis | Africaspis | Phenacaspis | **Duplachionaspis** | Ledaspis | Nelaspis | Aulacaspis |
| The **vittatus** group | | | | | | | | | | | | | | | | | | | | | | | | |
| *acutaspidis* | | | | | | | | | | | | | | | | | | | | | | | | |
| *angustus* | | | | | | | | | | | | | | | | | | | | | | | | |
| *anomalus* | | | | | | | | | | | | | | | | | | | | | | | | |
| *argenticorpus* | | | | | | | | | | | | | | | | | | | | | | | | |
| *australiensis* | | | | | | | | | | | | | | | | 1 | | | | | | | | |
| *capillatus* | | | | | | 1 | | | | | | | | | | 1 | | | | | | | | |
| *ciliatus* | | | | | | | | | | | | | | | | | | | | | | | | |
| *cochereaui* | | | | | | | | 1 | | | | | | | | | | | | | | | | |
| *costalimai* | | | | | | | | | | | | | | | | | | | | | | | | |
| *fabresi* | | | | | | | | | | | | | | | | | | | | | | | | |
| *haywardi* | | | | | | 1 | | | | | | | | | | | | | | | | | | |

| | SUBFAMILY DIASPIDINAE (cont.) | | | | | | | | | | SUBFAMILY LEUCASPIDINAE | | | | | |
|---|---|---|---|---|---|---|---|---|---|---|---|---|---|---|---|---|
| | FIORINIINI | DIASPIDINI | | | | | | | | | PARLATORIINI | | | LEUCASPIDINI | | |
| | | RUGASPIDIOTINA | HOWARDIINA | DIASPIDINA | | | | | | XEROPHILASPIDINA | PARLATORIINA | PARLATOREOPSIDINA | GYMNASPIDINA | | | |
| | Fiorinia | | | Furchadiaspis | Carulaspis | Pseudoparlatoria | Diaspis | **Pseudaulacaspis** | Epidiaspis | | Parlatoria | Parlatoreopsis | | Leucaspis | Lopholeucaspis | Anamaspis |
| The **vittatus** group | | | | | | | | | | | | | | | | |
| *acutaspidis* | | | | | | | | | | | | | | | | |
| *angustus* | | | | | | | | | | | | | | | | |
| *anomalus* | | | | | | | | | | | | | | | | |
| *argenticorpus* | | | | | | | | | | | | | | | | |
| *australiensis* | | | | | | | | | | | | | | | | |
| *capillatus* | | | | | | | | | | | | | | | | |
| *ciliatus* | | | | | | | | | | | | | | | | |
| *cochereaui* | | | | | | | * | | | | | | | | | |
| *costalimai* | | | | | | | | | | | | | | | | |
| *fabresi* | | | | | | | | | | | | | | | | |
| *haywardi* | | | | | | | | | | | | | | | | |

| | SUBFAMILY ODONASPIDINAE | SUBFAMILY ASPIDIOTINAE | | | | | | | | | | | | | | | | | | | | | | | | |
|---|---|---|---|---|---|---|---|---|---|---|---|---|---|---|---|---|---|---|---|---|---|---|---|---|---|---|
| | | COMSTOCKIELLINI | PSEUDAONIDIINI | | | TARGIONIINI | | ASPIDIOTINI | | | | | | | | | | | | | | | | | | |
| | | | PSEUDAONIDIINA | | FURCASPIDINA | ASPIDIELLINA | TARGIONIINA | SELENASPIDINA | | ASPIDIOTINA | | | | | | | | | | | MELANASPIDINA | | | | AONIDIINA | |
| | | | Pseudaonidia | Neomorgania | Separaspis | Aspidiella | | Selenaspidus | Paraselenaspidus | Aspidiotus | Temnaspidiotus | Dynaspidiotus | Chrysomphalus | Aonidiella | Aspidaspis | Hemiberlesia | Abgrallaspis | **Clavaspis** | Diaspidiotus | Quadraspidiotus | Lindingaspis | Melanaspis | Acutaspis | Mycetaspis | Aonidia | TOTAL HOSTS |
| The **vittatus** group | | | | | | | | | | | | | | | | | | | | | | | | | | |
| *acutaspidis* | | | | | | | | | | | | | | | | | | | | | | | 1 | | | 1 |
| *angustus* | | | | | | | | | | | | | | | | | | | | | | | | | | — |
| *anomalus* | | | | | | | | | | | | | 1 | | | | | | | | | | | | | 1 |
| *argenticorpus* | | | | | | | | | | | | | | | | | | | | | | | | | | — |
| *australiensis* | | | | | | | | | | | | | | | | | | | | | | | | | | 1 |
| *capillatus* | | | | 1 | | | | | | 1 | | | | | | | | | | | | | | | | **4** |
| *ciliatus* | | | | | | | | | | 1 | | | | | | | | | | | | | | | | 1 |
| *cochereaui* | | | | | | | | | | | | | | | | | | | | | | | | | | 1 |
| *costalimai* | | | | 1 | | | | | | | | | 1 | | | | | 1 | | | | | | | | 3 |
| *fabresi* | | | | | | | | | | | | | | | | | | | | | | | | | | — |
| *haywardi* | | | | | | | | | | | | | | | | | | | | | | | | | | 1 |

|||||||||||||||||||||||||||||||||||||||||||||||||||||||||||||||||||||
|---|---|---|---|---|---|---|---|---|---|---|---|---|---|---|---|---|---|---|---|---|---|---|---|---|---|---|---|---|---|---|---|---|---|---|---|---|---|---|---|---|---|---|---|---|---|---|---|---|---|---|---|---|---|---|---|---|---|---|---|---|---|---|---|---|---|---|
| *maculatipennis* ||||||||||||||||||||||||||||||||||||||||||||||||||||||||||||1|||||||1|
| *nigripes* ||||||||||||||||||||||||||||||||||||||||||||||||||||||||||||||||||—|
| *noumeaensis* ||||||||||||||||||||||||||||||||||||||||||||||||||1||||||||||||||||1|
| *obscurus* |||||||||||||||||||||||||||||||1||||||||||||||||||||||||||||||||||1|
| *peculiaris* ||||||||||||||||||||||||||||||||||||||||||||||||||||||1||||||||||||1|
| *perplexus* ||||||||||||||||||||||||||||||||||||||||||||||||||||||||1||||||||||1|
| *vittatus* |||||1|||||||||||||||||||||||||||||||||||||||||||||||||||||||||||||1|
| Related species |||||||||||||||||||||||||||||||||||||||||||||||||||||||||||||||||||
| *dealbatus* ||||||1||||||||||||||||||||||||||||||||||||||||||||||||||||||||||||1|
| *hyalinipennis* ||||||||||||||||||||||||||||||||||||||||||||||||||||||||||||||||||—|
| *malayensis* ||||||||||||||||||||||||||||||||||||||||||||||||||||||||||||||||||—|
| *mandalayensis* ||||||||||||||||||||||||||||||||||||||||||||||||||||||1||||||||||||1|
| *melanostictus* ||||||||||||||||||||||||||||||||||||||||||||||||||||||||||1||||||||1|
| *wallumbillae* ||||||||||||||||||||||||||||||||||||||||||||||||||||||||||||||||||—|
| The **chilensis** group |||||||||||||||||||||||||||||||||||||||||||||||||||||||||||||||||||
| *cercinus* |||||||||||||1||||||1|||||||||||||||||||||||||||||||1||||||||||||||||3|
| *chilensis* ||||||||||||1||||||||1||||||||1|1||1||||3|||||||||||||||2|||1|1||2|||||||||1|15|
| *columbi* ||||||||1||||||||||||||||1|||||||||||||||||||||||||||||1|||||||||||||3|
| *faurei* ||||||||||||||||||||||||||||||||||||||||||||||||||||||||||||||||||—|
| *merceti* |||||||||||||||||||||||||||||||||||||||||||||||||||||1||||||||1|1||||3|
| Related species |||||||||||||||||||||||||||||||||||||||||||||||||||||||||||||||||||
| *antennalis* ||||||||||||||||||||||||||||||||||||||||||||||||||||||||||||||||||—|
| The **proclia** group |||||||||||||||||||||||||||||||||||||||||||||||||||||||||||||||||||
| *comperei* |||||||||||||||||||||||||||||||||||1|||||||||||||||||||||||||||||||1|
| *confusus* |||||||||||||||||||1|||1||||||||||||||||||||||||||||||||||||||||1||||3|
| *diaspidis* ||||||1|||1|||||||1||||||||1|||||||2|1|||1|||||||||||||1||1|||2|1||2||||2|||1|||18|
| *hispanicus* |||||||1||1|||||||||||||||||||||||||||3||||1|||||||||||1|||1|||*|||||||1|||9|

TABLE 5. (CONT.)

| | SUBFAMILY XANTHOPHTHALMINAE | SUBFAMILY DIASPIDINAE | | | | | | | | | | | | | | | | | | | | | | | | | | | | | | | | SUBFAMILY LEUCASPIDINAE | | | | | | SUBFAMILY ODONASPIDINAE | SUBFAMILY ASPIDIOTINAE | | | | | | | | | | | | | | | | | | | | | | | | TOTAL HOSTS |
|---|---|---|---|---|---|---|---|---|---|---|---|---|---|---|---|---|---|---|---|---|---|---|---|---|---|---|---|---|---|---|---|---|---|---|---|---|---|---|---|---|---|---|---|---|---|---|---|---|---|---|---|---|---|---|---|---|---|---|---|---|---|---|---|---|---|
| | | ANTAKASPIDINI | LEPIDOSAPHIDINI | | | | | | | ANCEPASPIDINI | CHIONASPIDINI | | | | | | | | | | | | | | FIORINIINI | DIASPIDINI | | | | | | | | | PARLATORIINI | | | LEUCASPIDINI | | | | COMSTOCKIELLINI | PSEUDAONIDIINI | | | TARGIONIINI | | ASPIDIOTINI | | | | | | | | | | | | | | | | | | |
| | | | COCCOMYTILINA | LEPIDOSAPHIDINA | | | | | | | SCLOPETASPIDINA | AUGULASPIDINA | | | KUWANASPIDINA | CHIONASPIDINA | | | | PHENACASPIDINA | | | | | | RUGASPIDIOTINA | HOWARDIINA | DIASPIDINA | | | | | | XEROPHILASPIDINA | PARLATORIINA | PARLATOREOPSIDINA | GYMNASPIDINA | | | | | | PSEUDAONIDIINA | | FURCASPIDINA | ASPIDIELLINA | TARGIONIINA | SELENASPIDINA | | ASPIDIOTINA | | | | | | | | | | | MELANASPIDINA | | | | AONIDIINA | |
| | | | | Fulaspis | Paralepidosaphes | Lepidosaphes | Mytilaspis | Cornuaspis | Insulaspis | | | Rolaspis | Tecaspis | Dentachionaspis | | Chionaspis | Unaspis | Pinnaspis | Africaspis | Phenacaspis | **Duplachionaspis** | Ledaspis | Nelaspis | Aulacaspis | Fiorinia | | | Furchadiaspis | Carulaspis | Pseudoparlatoria | Diaspis | **Pseudaulacaspis** | Epidiaspis | | Parlatoria | Parlatoreopsis | | Leucaspis | Lopholeucaspis | Anamaspis | | | Pseudaonidia | Neomorgania | Separaspis | Aspidiella | | Selenaspidus | Paraselenaspidus | Aspidiotus | Temnaspidiotus | Dynaspidiotus | Chrysomphalus | Aonidiella | Aspidaspis | Hemiberlesia | Abgrallaspis | Clavaspis | Diaspidiotus | Quadraspidiotus | Lindingaspis | Melanaspis | Acutaspis | Mycetaspis | Aonidia | |
| *maculicornis* | | | | | | | | | | | | | | | | | | | | | | | | | | | | | | | | | 1 | | 1 | 1 | | | | | | | | | | | | | | 1 | | | | | | 1 | | | | 1 | | | | | | 6 |
| *paramaculicornis* | | | | | | 1 | | | | | | | | | | | | | | | | | | | | | | | | | * | | | | 3 | | | | | | | | | | | | | | | * | | | | 1 | | * | | | | 1 | | | | | | 6 |
| *philippinensis* | | | | | | | | | | | | | | | | | | | | | | | | | | | | | | | | | | | | | | | | | | | | | | | | | | | | | 1 | 1 | | | | | | | | | | | | 2 |
| *pinnaspidis* | | | | | | | | | | | | | | | | | | 1 | | | | | | | | | | | | | | | | | | | | | | | | | | | | | | | | | | | 1 | 1 | | | | | | | | | | | | 3 |
| *proclia* | | | | | | 2 | | | | | | | | | | 3 | | | | | | | | | | | | | | | | 1 | 1 | | | | | | | | | | | | | | | | | * | | | 1 | 1 | | 2 | | | 1 | 5 | | | | | | 17 |
| *testaceus* | | | | | | 1 | | | | | | | | | | | | | | | | | | | | | | | | | | | 1 | | | | | | | | | | | | | | | | | | | | | | | | | | | 1 | | | | | | 3 |
| *vandenboschi* | | | | | | 1 | | | | | | | | | | | | | | | | | | | | | | | | | * | 1 | | | | | | | | | | | | | | | | | | * | | | 1 | | | * | | | | 1 | | | | | | 4 |
| Related species | | | | | | | | | | | | | | | | | | | | | | | | | | | | | | | | | | | | | | | | | | | | | | | | | | | | | | | | | | | | | | | | | | |
| *acrenulatus* | | | | | | | | | | | | | | | | | | | | | | | | | | | | | | | | | | | | | | | | | | | | | | 1 | | | | | | | | | | | | | | | | | | | | 1 |
| *amazonensis* | | | | | | | | | | | | | | | | | | | | | | | | | | | | | | | | | | | | | | | | | | | | | | | | | | | | | | | | | | | | | | | | | | — |
| *desantisi* | | | | | | | | | | | | | | | | | | | | | | | | | | | | | | | | | | | | | | | | | | | | | | | | | | | | | | 1 | | | | | | | | | | | | 1 |

| | | | | | | | | | | | | | | | | | | | | | | | | | | | | | | | | | | | | | | | | | | | | | | | | | | | | | | | | | | | | | | | | | | | |
|---|---|---|---|---|---|---|---|---|---|---|---|---|---|---|---|---|---|---|---|---|---|---|---|---|---|---|---|---|---|---|---|---|---|---|---|---|---|---|---|---|---|---|---|---|---|---|---|---|---|---|---|---|---|---|---|---|---|---|---|---|---|---|---|---|---|---|
|*griseus*|||||||||||1||||1|||||||1|||||||||||||||||||||||||||||||||||||||||||3|
|*tucumani*||||||||||||||||||||||||||||||1||||||||||||||||||||||||||||||||||1|
|The **mytilaspidis** group|||||||||||||||||||||||||||||||||||||||||||||||||||||||||||||||||||
|*aonidiae*||||||1|||1||||||||||||||1|||||||||||||||||||||||||||1||1|1|3|1|1||||3|||||1|15|
|*libanicus*||||||||||||||||||||||||||||||||||||||1|||||||||||||1|||||||||||||2|
|*luteus*||||||||||||||||||||||||||||||||||||||2|||||||||||||||||||||||||||2|
|*moldavicus*|||||||||||||||||||||||||||||||||1||||||||||||||||||||||||||1||||||2|
|*mytilaspidis*||||||1|1||1|||||||1||||1||||1|||||2||1||1||1|||1||||||||||||2|||1|2||1||||3|||||1|22|
|*opuntiae*|||||||1||||||||||||||||||||||1|||||||||||||||||||||||||||||||||||2|
|*phoenicis*|||||||||||||||||||||||||||||||||||1||||||||||||||||||||||||||||||1|
|Related species|||||||||||||||||||||||||||||||||||||||||||||||||||||||||||||||||||
|*capensis*||||||||||||||1||||||||||||||||||||||||||||||||||||||||||||||||||||1|
|*neuter*||||||||||||||||||||||||||||||||||||||||||||||||||||||||||||1||||||1|
|*notialis*||||||||||||||||||||||||||||||||||||||||||||||||||1||||||||||||||||1|
|*rolaspidis*||||||||||||1||||||||||||||||||||||||||||||||||||||||||||||||||||||1|
|*taylori*||||||||||||||||||||||||||||||||||||||||||||||||||1||||||||||||||||1|
|The **lingnanensis** group|||||||||||||||||||||||||||||||||||||||||||||||||||||||||||||||||||
|*africanus*||||||||||||||||||||||||||||||||||||||||||||||1||||||||1|1|||||||1|||||4|
|*coheni*||||||||||||||||||||||||||||||||||||||||||||||||||1||||3||1||||||||||5|
|*equatorialis*|||||||||||||||||||||||||||||||||||||||||||||||||||1|||||||||||||1||2|
|*fisheri*||||||||||||||||||||||||||||||||||||||||||||||||||*||||1||||||||||||1|
|*holoxanthus*|||||||||||||||||||||||||||||||*|||||||||||||||||1||*|||1|1||||||||||||3|
|*lingnanensis*||||||||1|1||||||||1||||||||||||||1||||||||||||1|||||1||*|1||2|2||1|||||||||1|13|
|*margaretae*||||||||||||||||||||||||||||||1|1||||||||||||||||||||1||||||1|||||||||4|
|*melinus*|||||||||||||||||||||||||||||||1|||||||||||||||||||1|1||1|3||1||||2||||||10|
|*yasumatsui*||||||||||||||||||||||||||||||||||||||||||||||||||||1|1|1||||||||||||3|

TABLE 5. (CONT.)

| | SUBFAMILY XANTHOPHTHALMINAE | SUBFAMILY DIASPIDINAE | | | | | | | | | | | | | | | | | | | | | | |
|---|---|---|---|---|---|---|---|---|---|---|---|---|---|---|---|---|---|---|---|---|---|---|---|---|
| | | ANTAKASPIDINI | LEPIDOSAPHIDINI | | | | | | | ANCEPASPIDINI | CHIONASPIDINI | | | | | | | | | | | | | |
| | | | COCCOMYTILINA | LEPIDOSAPHIDINA | | | | | | | SCLOPETASPIDINA | AUGULASPIDINA | | | KUWANASPIDINA | CHIONASPIDINA | | | | PHENACASPIDINA | | | | |
| | | | | Fulaspis | Paralepidosaphes | Lepidosaphes | Mytilaspis | Cornuaspis | Insulaspis | | | Rolaspis | Tecaspis | Dentachionaspis | | Chionaspis | Unaspis | Pinnaspis | Africaspis | Phenacaspis | **Duplachionaspis** | Ledaspis | Nelaspis | Aulacaspis |
| The **chrysomphali** group | | | | | | | | | | | | | | | | | | | | | | | | |
| *anneckei* | | | | | | | | | | | | | | | | | | | | | | | | |
| *chrysomphali* | | | | 1 | | | | | | | | | | | | | | | | | | | | |
| *cylindratus* | | | | | | | | | | | | | | | | | | | | | | | | |
| *debachi* | | | | | | | | | | | | | | | | | 1 | | | | | | | |
| *immaculatus* | | | | | | 1 | | | | | | | | | | | | | | | | | | |
| *japonicus* | | | | | | | | | | | | | | | | | | | | | | | | |
| *lepidosaphes* | | | | | | | | 1 | | | | | | | | | | | | | | | | |
| *mazalae* | | | | | | | | | | | | | | | | | | 1 | | | | | | 1 |
| *sensorius* | | | | | | | | | | | | | | | | | | | | | | | | |

| | SUBFAMILY DIASPIDINAE | | | | | | | | | | SUBFAMILY LEUCASPIDINAE | | | | | | SUBFAMILY ODONASPIDINAE |
|---|---|---|---|---|---|---|---|---|---|---|---|---|---|---|---|---|---|
| | FIORINIINI | DIASPIDINI | | | | | | | | | PARLATORIINI | | | LEUCASPIDINI | | | |
| | | RUGASPIDIOTINA | HOWARDIINA | DIASPIDINA | | | | | | XEROPHILASPIDINA | PARLATORIINA | PARLATOREOPSIDINA | GYMNASPIDINA | | | | |
| | Fiorinia | | | Furchadiaspis | Carulaspis | Pseudoparlatoria | Diaspis | **Pseudaulacaspis** | Epidiaspis | | Parlatoria | Parlatoreopsis | | Leucaspis | Lopholeucaspis | Anamaspis | |
| The **chrysomphali** group | | | | | | | | | | | | | | | | | |
| *anneckei* | | | | | | | | | | | | | | | | | |
| *chrysomphali* | | | | | | | 1 | | | | | | | | | | |
| *cylindratus* | | | | | | | | | | | | | | | | | |
| *debachi* | | | | | | | 1 | | | | 1 | | | | | | |
| *immaculatus* | | | | | | | | | | | | | | | | | |
| *japonicus* | | | | | | | | | | | | | | | | | |
| *lepidosaphes* | | | | | | | | | | | | | | | | | |
| *mazalae* | | | | | | | | | | | | | | | | | |
| *sensorius* | | | | | | | | | | | | | | | | 1 | |

| | SUBFAMILY ASPIDIOTINAE | | | | | | | | | | | | | | | | | | | | | | | | TOTAL HOSTS |
|---|---|---|---|---|---|---|---|---|---|---|---|---|---|---|---|---|---|---|---|---|---|---|---|---|---|
| | COMSTOCKIELLINI | PSEUDAONIDIINI | | | TARGIONIINI | | ASPIDIOTINI | | | | | | | | | | | | | | | | | | |
| | | PSEUDAONIDIINA | | FURCASPIDINA | ASPIDIELLINA | TARGIONIINA | SELENASPIDINA | | ASPIDIOTINA | | | | | | | | | | | MELANASPIDINA | | | | AONIDIINA | |
| | | Pseudaonidia | Neömorgania | Separaspis | Aspidiella | | Selenaspidus | Paraselenaspidus | Aspidiotus | Temnaspidiotus | Dynaspidiotus | Chrysomphalus | Aonidiella | Aspidaspis | Hemiberlesia | Abgrallaspis | Clavaspis | Diaspidiotus | Quadraspidiotus | Lindingaspis | Melanaspis | Acutaspis | Mycetaspis | Aonidia | |
| The **chrysomphali** group | | | | | | | | | | | | | | | | | | | | | | | | | |
| *anneckei* | | | | | | | | 1 | | | | 1 | | | | | | | | | | | | | 2 |
| *chrysomphali* | | 1 | | | | | 1 | | 1 | 1 | | 2 | 2 | | 1 | | | | | | 1 | | | | 12 |
| *cylindratus* | | 2 | | | | | | | | | | | | | | | | | | | | | | | 2 |
| *debachi* | | | | | | | | | | | | | | | | | | | | | | | | | 3 |
| *immaculatus* | | | | | | | | | | | | | | | | | | | | | | | | | 1 |
| *japonicus* | | | | | | | | | | | | 1 | 1 | | | | | | | | | | | | 2 |
| *lepidosaphes* | | | | | | | | | | | | | | | | | | | | | | | | | 1 |
| *mazalae* | | | | | | | | | | | | 1 | 3 | | | | | | | | | | | | 6 |
| *sensorius* | | | | | | | | | | | | | | | | | | | | | | | | | 1 |

| | | | | | | | | | | | | | | | | | | | | | | | | | | | | | | | | | |
|---|---|---|---|---|---|---|---|---|---|---|---|---|---|---|---|---|---|---|---|---|---|---|---|---|---|---|---|---|---|---|---|---|---|
| The **funicularis** group | | | | | | | | | | | | | | | | | | | | | | | | | | | | | | | | | |
| *funicularis* | | | | | | | | | | | | 2 | | | | | | | | | | | | | | | | 1 | | | | | |
| *gordoni* | | | | | | | | | | | | | | | | | 1 | | | | | | | | | | | | | | | | |
| *theae* | | | | | | | | | | | | | | | | | | | | | | | | | 1 | | | | | | | | |
| *ulianovi* | | | | | | | | | | | | | | | | | | | | | | | | | | | | | | | | | |
| Unassigned species | | | | | | | | | | | | | | | | | | | | | | | | | | | | | | | | | |
| *bedfordi* | | | | | | | | | | | | | | | | | | | | | | | | | | | | | | | | | |
| *erythraeus* | | | | | | | | | | | | | | | | | | | | | | | | | | | | | | | | | |
| *ignotus* | | | | | | | | | | | | | | | | | | | | | | | | | | | | | | | | | |
| *longicaudus* | | | | | | | | | | | | | | | | | | | | | | | | | | | | | | | | | |
| *mimosae* | | | | | | | | | | | | | | | | | | | | | | | | | | | | | | | | | |
| *pilosus* | | | | 1 | | | | | | | | | | | | | | | | | | | | | | | | | | | | | |
| *roseni* | | | | | | | | | | | | | | | | | | | | | | | | | | | | | | | | | |
| *salvadorensis* | | | | | | | | | | | | | | | | | | | | | | | | | | | | | | | | | |
| *secundus* | | | | | | | | | | | | | | | | | | | | | | | | | | | | | | | | | |
| *setosus* | | | | | | | | | | | | 1 | | | | | | | | | | | | | | | | | | | | | |
| Unrecognizable species | | | | | | | | | | | | | | | | | | | | | | | | | | | | | | | | | |
| *abnormis* | | | | | | 1 | | | | | | | | | | | | | | | | | | | | | | | | | | | |
| *caucasicus* | | | | | | 1 | | | | | | | | | | | | | | | | | | | | | | | | | | | |
| *diaspidioti* | | | | | | | | | | | | | | | | | | | | | | | | | | | | | | | | | |
| *limonus* | | | | | | | | | | | | | | | | | | 1 | | | | | | | | | | | | | | | |
| *minutissimus* | | | | | | | | | | | | | | | | 1 | | | | | | | | | | | | | | | | | |
| *simplex* | | | | | | | | | | | | | | | | | | | | | 1 | | | | | | | | | | | | |
| Total number of *Aphytis* species | — | — | — | 2 | 1 | 11 | 3 | 4 | 5 | — | — | 5 | 1 | 1 | — | 6 | 3 | 2 | 2 | 2 | 1 | 1 | 1 | 5 | 1 | — | — | 2 | 2 | 1 | 11 | 3 | 5 |

| | | | | | | | | | | | | | | | | | | | | | | | | | | | | | | | | |
|---|---|---|---|---|---|---|---|---|---|---|---|---|---|---|---|---|---|---|---|---|---|---|---|---|---|---|---|---|---|---|---|---|---|
| The **funicularis** group | | | | | | | | | | | | | | | | | | | | | | | | | | | | | | | | | |
| *funicularis* | | | | | | | | | | | | | | | | | | | | | | | | | | | | | | | | | 3 |
| *gordoni* | | | | | | | | | | | | | | | | | | | | | | | | | | | | | | | | | 1 |
| *theae* | | | | | | | | | | | | | | | | | | | | | | | | | | | | | | | | | 1 |
| *ulianovi* | | | | | | | | | | | | | | | | | | | | | | | | | | | | | | | | | — |
| Unassigned species | | | | | | | | | | | | | | | | | | | | | | | | | | | | | | | | | |
| *bedfordi* | | | | | | | | | | | | | | | | | | | | | | | | | | | | | | | | | — |
| *erythraeus* | | | | | | | | | | | | | | | | | 1 | | | | | | | | | | | | | | | | 1 |
| *ignotus* | | | | | | | | | | | | | | | | | | | | | | | | | | | | 1 | | | | | 1 |
| *longicaudus* | | | | | | | | | | 2 | | | | | | | | | | | | | | | | | | | | | | | 2 |
| *mimosae* | | | | | | | | | | | | | | | | | 1 | | | | | | | | | | | | | | | | 1 |
| *pilosus* | | | | | | | | | | | | | | | | | | | | | | | | | | | | | | | | | 1 |
| *roseni* | | | | | | | | | | | | | | | 1 | | | | | | | | | | | | | | | | | | 1 |
| *salvadorensis* | | | | | | | | | | | | | | | | | | | | | | | | | | | | | | | | | — |
| *secundus* | | | | | | | | | | | | | | | | | | | | | | | | | | | | | | | | | — |
| *setosus* | | | | | | | | | | | | | | | | | | | | | | | | | | | | | | | | | 1 |
| Unrecognizable species | | | | | | | | | | | | | | | | | | | | | | | | | | | | | | | | | |
| *abnormis* | | | | | | | | | | | | | | | | | | | | | | | | | | | | | | | | | 1 |
| *caucasicus* | | | | | | | | | | | | | | | | | | | | | | | | | | | | | | | | | 1 |
| *diaspidioti* | | | | | | | | | | | | | | | | | | | | | | | | | | | 1 | | | | | | 1 |
| *limonus* | | | | | | | | | | | | | | | | | | | | | | | | | | | | | | | | | 1 |
| *minutissimus* | | | | | | | | | | | | | | | | | | | | | | | | | | | | | | | | | 1 |
| *simplex* | | | | | | | | | | | | | | | | | | | | | | | | | | | | | | | | | 1 |
| Total number of *Aphytis* species | — | 9 | 1 | — | 3 | 1 | 1 | — | — | 5 | 1 | 1 | 1 | — | 5 | 1 | 17 | 5 | 3 | 22 | 21 | 1 | 11 | 2 | 1 | 2 | 11 | 3 | 3 | 3 | 1 | **4** | |

TABLE 6. Number of species of Diaspididae attacked by the members of the main species groups of *Aphytis**

| | XANTHOPHTHALMINAE | DIASPIDINAE | LEUCASPIDINAE | ODONASPIDINAE | ASPIDIOTINAE |
|---|---|---|---|---|---|
| The **vittatus** group | 0 | 6 | 0 | 0 | 9 |
| Related species | 0 | 1 | 0 | 0 | 2 |
| The **chilensis** group | 0 | 3 | 1 | 0 | 4 |
| The **proclia** group | 0 | 9 | 5 | 0 | 10 |
| Related species | 0 | 2 | 0 | 0 | 2 |
| The **mytilaspidis** group | 0 | 4 | 4 | 0 | 4 |
| Related species | 0 | 2 | 0 | 0 | 3 |
| The **lingnanensis** group | 0 | 3 | 0 | 0 | 9 |
| The **chrysomphali** group | 0 | 5 | 2 | 0 | 5 |
| The **funicularis** group | 0 | 3 | 0 | 0 | 0 |
| Unassigned species | 0 | 2 | 0 | 0 | 5 |
| TOTAL | 0 | 40 | 12 | 0 | 53 |

* The numbers represent host-parasite associations; thus, if a diaspidid species has been recorded as host to several members of a given group, it was counted several times.

TABLE 7. Number of *Aphytis* species attacking members of the five subfamilies of the Diaspididae

| | XANTHOPHTHALMINAE | DIASPIDINAE | LEUCASPIDINAE | ODONASPIDINAE | ASPIDIOTINAE |
|---|---|---|---|---|---|
| The **vittatus** group | 0 | 7 | 0 | 0 | 12 |
| Related species | 0 | 1 | 0 | 0 | 2 |
| The **chilensis** group | 0 | 9 | 3 | 0 | 12 |
| The **proclia** group | 0 | 25 | 11 | 0 | 37 |
| Related species | 0 | 4 | 0 | 0 | 2 |
| The **mytilaspidis** group | 0 | 16 | 6 | 0 | 24 |
| Related species | 0 | 2 | 0 | 0 | 3 |
| The **lingnanensis** group | 0 | 7 | 0 | 0 | 38 |
| The **chrysomphali** group | 0 | 8 | 2 | 0 | 20 |
| The **funicularis** group | 0 | 5 | 0 | 0 | 0 |
| Unassigned species | 0 | 2 | 0 | 0 | 6 |
| TOTAL | 0 | 86 | 22 | 0 | 156 |

# REFERENCES


ABDELRAHMAN, I. (1973). Toxicity of malathion to the natural enemies of California red scale, *Aonidiella aurantii* (Mask.) (Hemiptera: Diaspididae). *Austral. J. Agric. Res.*, **24**: 119–133.

ABDELRAHMAN, I. (1974a). The effect of extreme temperatures on California red scale, *Aonidiella aurantii* (Mask.) (Hemiptera: Diaspididae), and its natural enemies. *Austral. J. Zool.*, **22**: 203–212.

ABDELRAHMAN, I. (1974b). Growth, development and innate capacity for increase in *Aphytis chrysomphali* Mercet and *A. melinus* DeBach, parasites of California red scale, *Aonidiella aurantii* (Mask.), in relation to temperature. *Austral. J. Zool.*, **22**: 213–230.

ABDELRAHMAN, I. (1974c). Studies in ovipositional behaviour and control of sex in *Aphytis melinus* DeBach, a parasite of California red scale, *Aonidiella aurantii* (Mask.). *Austral. J. Zool.*, **22**: 231–247.

AGARWAL, M. M. (1964). Studies of forms of Aphelinidae (Hymenoptera—Chalcidoidea) collected at Aligarh (India)—II. *Proc. Ind. Acad. Sci.*, **60**: 315–335.

ALAM, S. M. (1956a). The taxonomy of some British aphelinid parasites (Hymenoptera) of scale insects (Coccoidea). *Trans. Roy. Entomol. Soc. London*, **108**: 357–384.

ALAM, S. M. (1956b). Life-cycle with notes on host-parasite relationship in the field and the anatomy of fully-grown larva of *Aphytis diaspidis* (Silwood material). *Proc. Nat. Acad. Sci. India*, **26**: 69–74.

ANNECKE, D. P. (1963). New and little known species of South African Aphelinidae (Hymenoptera: Aphelinidae). *Ann. Mag. Nat. Hist.*, Ser. 13, **6**: 337–352.

ANNECKE, D. P. (1965). A new species of *Centrodora* Foerster (Hymenoptera: Aphelinidae) parasitic in the eggs of a cicadellid injurious to citrus in South Africa. *S. Afr. J. Agric. Sci.*, **8**: 1133–1137.

ANNECKE, D. P., and H. P. INSLEY (1970). New and little known species of *Azotus* Howard, *Ablerus* Howard and *Physcus* Howard (Hym., Aphelinidae) from Africa and Mauritius. *Bull. Entomol. Res.*, **60**: 237–251.

ANNECKE, D. P., and H. P. INSLEY (1971). Catalogue of Ethiopian Encyrtidae and Aphelinidae (Hymenoptera: Chalcidoidea). *Dept. Agric. Tech. Serv. S. Afr., Entomol. Mem.* **23**, 53 pp.

ANNECKE, D. P., and H. P. INSLEY (1972). The species of *Marietta* and a new *Centrodora* from South Africa (Hymenoptera: Aphelinidae). *J. Entomol. Soc. S. Afr.*, **35**: 1–15.

ARGYRIOU, L. C. (1974). Data on the biological control of citrus scales in Greece. *Bull. SROP* 1974/3: 89–94.

ASHMEAD, W. H. (1880). *Orange Insects; a Treatise on the Injurious and Beneficial Insects Found on the Orange Trees of Florida*. Ashmead Bros., Jacksonville, XV + 78 pp.

ASHMEAD, W. H. (1904). Classification of the chalcid flies or the superfamily Chalcidoidea, with descriptions of new species in the Carnegie Museum, collected in South America by Herbert H. Smith. *Mem. Carnegie Mus.*, **1**(4): XI + 225–551, pl. XXXI–XXXIX.

AVIDOV, Z., M. BALSHIN, and U. GERSON (1970). Studies on *Aphytis coheni*, a parasite of the California red scale, *Aonidiella aurantii*, in Israel. *Entomophaga*, **15**: 191–210.

AZIM, A. (1961). Mass-production of *Chrysomphalus bifasciculatus* Ferris and its hymenopterous parasites. *Mushi,* **35**: 97–109.

AZIM, A. (1963a). *Aphytis cylindratus* Compere (Hymenoptera, Aphelinidae), an effective parasite of *Pseudaonidia duplex* Cockerell. *Mushi,* **37**: 53–63.

AZIM, A. (1963b). Systematic and biological studies on the genus *Aphytis* Howard (Hymenoptera, Aphelinidae) of Japan. Part 1. Taxonomy. *J. Fac. Agric. Kyushu Univ.,* **12**: 265–290.

AZIM, A. (1963c). Systematic and biological studies on the genus *Aphytis* Howard (Hymenoptera, Aphelinidae) of Japan. Part 2. Biology and mass production. *J. Fac. Agric. Kyushu Univ.,* **12**: 291–321.

BAKER, J. L. (1976). Determinants of host selection for species of *Aphytis* (Hymenoptera: Aphelinidae), parasites of diaspine scales. *Hilgardia,* **44**: 1–25.

BARTLETT, B. R. (1951). The action of certain "inert" dust materials on parasitic Hymenoptera. *J. Econ. Entomol.,* **44**: 891–896.

BARTLETT, B. R. (1962). The ingestion of dry sugars by adult entomophagous insects and the use of this feeding habit for measuring the moisture needs of parasites. *J. Econ. Entomol.,* **55**: 749–753.

BARTLETT, B. R. (1963). The contact toxicity of some pesticide residues to hymenopterous parasites and coccinellid predators. *J. Econ. Entomol.,* **56**: 694–698.

BARTLETT, B. R., and T. W. FISHER (1950). Laboratory propagation of *Aphytis chrysomphali* for release to control California red scale. *J. Econ. Entomol.,* **43**: 802–806.

BARTRA, C. E. (1974). Biologia de *Selenaspidus articulatus* Morgan y sus principales controladores biologicos. *Rev. Peru. Entomol.,* **17**: 60–68.

BEARD, R. L. (1971). Production and use of venom by *Bracon brevicornis* (Wesm.), pp. 181–190 in: A. de Vries and E. Kochva (Editors), *Toxins of Animal and Plant Origin,* Vol. 1. Gordon and Breach, New York, London and Paris.

BEDFORD, E. C. G. (1975). The present status of the integrated control of citrus pests in South Africa. *Proc. I Congr. Entomol. Soc. S. Afr.* (1974): 203–215.

BEDFORD, E. C. G. (1976). Citrus pest management in South Africa. *Proc. Tall Timbers Conf. Ecol. Anim. Contr. Habitat Mgmt.,* **6** (1974): 19–42.

BENASSY, C. (1955). Remarques sur deux Aphelinidés: *Aphytis mytilaspidis* Le Baron et *Aphytis proclia* Walker. *Ann. Epiphyt.,* **6**: 11–17.

BEN-DOV, Y., and D. ROSEN (1969). Efficacy of natural enemies of the California red scale on citrus in Israel. *J. Econ. Entomol.,* **62**: 1057–1060.

BORKHSENIUS, N. S. (1965). Principles of classification of the armored scale insects (Homoptera, Coccoidea, Diaspididae). *Entomol. Obozr.,* **44**: 362–376 (in Russian). (English transl.: *Entomol. Rev.,* **44**: 208–214).

BORKHSENIUS, N. S. (1966). *A Catalogue of the Armored Scale Insects (Diaspidoidea) of the World.* Nauka, Moscow and Leningrad, 449 pp. (in Russian).

BOUČEK, Z. (1964). Proctotrupoidea und Chalcidoidea aus den Resten der Ratzeburg-Sammlung. *Beitr. Entomol.,* **14**: 663–673.

BRÈTHES, J. (1913). Himenópteros de la América meridional. *An. Mus. Nac. Hist. Nat. Buenos Aires,* **24**: 35–165.

BRÈTHES, J. (1916). Hyménoptères parasites de l'Amérique méridionale. *An. Mus. Nac. Hist. Nat. Buenos Aires,* **27**: 401–430.

BROWN, S. W., and H. L. MCKENZIE (1962). Evolutionary patterns in the armored scale insects and their allies (Homoptera: Coccoidea: Diaspididae, Phoenicococcidae and Asterolecaniidae). *Hilgardia,* **33**: 141–170A.

BURKS, B. D. (1958). Superfamily Chalcidoidea, pp. 62–84 in: K. V. Krombein (Editor), *Hymenop-*

*tera of America North of Mexico—Synoptic Catalog.* Agric. Monogr. **2**, first supplement. U.S. Dept. Agric., Washington, D.C.

BURKS, B. D. (1967). Superfamily Chalcidoidea, pp. 213–282 in: K. V. Krombein and B. D. Burks (Editors), *Hymenoptera of America North of Mexico—Synoptic Catalog.* Agric. Monogr. **2**, second supplement. U.S. Dept. Agric., Washington, D.C.

CAMERON, P. (1891). Hymenopterological notices. *Mem Proc. Manchester Lit. Phil. Soc.*, 4th Ser., **4**: 182–194, Pl. 1.

CHUMAKOVA, B. M. (1957). Parasites of ostreaeform scale insects in the Maritime Territory. *Zool. Zhur.*, **36**: 533–547 (in Russian).

CHUMAKOVA, B. M. (1961). Parasites of injurious coccids from Kabardino-Balkaria (Hymenoptera, Chalcidoidea). *Entomol. Obozr.*, **40**: 315–338 (in Russian).

CHUMAKOVA, B. M. (1964). A review of the species of the family Aphelinidae (Hymenoptera)—parasites of coccid pests of tree cultures on the Black Sea coast of the RSFSR. *Trud. Vsesoyuz. Inst. Zashchit. Rast.*, **21**: 14–39 (in Russian).

CILLIERS, C. J. (1971). Observations on circular purple scale *Chrysomphalus aonidum* (Linn.), and two introduced parasites in western Transvaal citrus orchards. *Entomophaga*, **16**: 269–284.

CLAUSEN, C. P. (1942). The relation of taxonomy to biological control. *J. Econ. Entomol.*, **35**: 744–748.

COMPERE, H. (1925). A new genus and species of Aphelinidae (Hymenoptera) from China. *Trans. Amer. Entomol. Soc.*, **51**: 129–134.

COMPERE, H. (1936). Notes on the classification of the Aphelinidae with descriptions of new species. *Univ. Calif. Publ. Entomol.*, **6**: 277–321.

COMPERE, H. (1955). A systematic study of the genus *Aphytis* Howard (Hymenoptera, Aphelinidae) with descriptions of new species. *Univ. Calif. Publ. Entomol.*, **10**: 271–319.

COMPERE, H. (1961). The red scale and its insect enemies. *Hilgardia*, **31**: 173–278.

COMPERE, H. (1969). The role of systematics in biological control: a backward look. *Israel J. Entomol.*, **4**: 5–10.

COMPERE, H., and D. P. ANNECKE (1961). Descriptions of parasitic Hymenoptera and comments (Hymenopt.: Aphelinidae, Encyrtidae, Eulophidae). *J. Entomol. Soc. S. Afr.*, **24**: 17–71.

COMPERE, H., and D. ROSEN (1970). The prescutum in Hymenoptera. *Proc. Roy. Entomol. Soc. London* (A), **45**: 91–97.

DALLA TORRE, C. G. DE (1898). *Catalogus hymenopterorum hucusque descriptorum systematicus et synonymicus.* Vol. V: *Chalcididae et Proctotrupidae.* Engelmann, Lipsiae, 598 pp.

DALMAN, J. W. (1820). Försök till uppställning af insect-familjen Pteromalini, i synnerhet med afseende på de i Sverige funne arter. *K. Svenska VetenskAkad. Handl.*, **41**: 123–174.

DEAN, H. A. (1965). An *Aphytis* complex (Hymenoptera: Eulophidae) of chaff scale. *Ann. Entomol. Soc. Amer.*, **58**: 142–145.

DEAN, H. A. (1975). Complete biological control of *Lepidosaphes beckii* on Texas citrus with *Aphytis lepidosaphes. Environ. Entomol.*, **4**: 110–114.

DEBACH, P. (1951). The necessity for an ecological approach to pest control on citrus in California. *J. Econ. Entomol.*, **44**: 443–447.

DEBACH, P. (1953). Purple scale parasites in southern California. *Calif. Citrog.* **38**: 219–222.

DEBACH, P. (1954). Relative efficacy of the red scale parasites *Aphytis chrysomphali* Mercet and *Aphytis* "A" on citrus trees in southern California. *Boll. Lab. Zool. Gen. Agr. "Filippo Silvestri"*, Portici, **33**: 135–151.

DEBACH, P. (1958a). The role of weather and entomophagous species in the natural control of insect populations. *J. Econ. Entomol.*, **51**: 474–484.

DeBach, P. (1958b). Application of ecological information to control of citrus pests in California. *Proc. 10th Internat. Congr. Entomol.* (Montreal, 1956), **3**: 187–194.

DeBach, P. (1958c). Selective breeding to improve adaptations of parasitic insects. *Proc. 10th Internat. Congr. Entomol.* (Montreal, 1956), **4**: 759–768.

DeBach, P. (1959). New species and strains of *Aphytis* (Hymenoptera, Eulophidae) parasitic on the California red scale, *Aonidiella aurantii* (Mask.), in the Orient. *Ann. Entomol. Soc. Amer.*, **52**: 354–362.

DeBach, P. (1960). The importance of taxonomy to biological control as illustrated by the cryptic history of *Aphytis holoxanthus* n.sp. (Hymenoptera: Aphelinidae), a parasite of *Chrysomphalus aonidum*, and *Aphytis coheni* n.sp., a parasite of *Aonidiella aurantii*. *Ann. Entomol. Soc. Amer.*, **53**: 701–705.

DeBach, P. (1963). *"Aphytis costa-limai"* (Gomes, 1941) comb. n., a parasite of *"Chrysomphalus aonidum"* (L.) in Brazil (Chalcidoidea, Aphelinidae). *Rev. Brasil. Biol.*, **23**: 35–38.

DeBach, P. (1964). Some species of *Aphytis* Howard (Hymenoptera: Aphelinidae) in Greece. *Ann. Inst. Phytopathol. Benaki*, N.S., **7**: 5–18.

DeBach, P. (1965a). Weather and the success of parasites in population regulation. *Canad. Entomol.*, **97**: 848–863.

DeBach, P. (1965b). Some biological and ecological phenomena associated with colonizing entomophagous insects. pp. 287–306 in: H. G. Baker and G. L. Stebbins (Editors), *The Genetics of Colonizing Species*. Academic Press, New York and London.

DeBach, P. (1966). The competitive displacement and coexistence principles. *Ann. Rev. Entomol.*, **11**: 183–212.

DeBach, P. (1969a). Biological control of diaspine scale insects on citrus in California. *Proc. 1st. Internat. Citrus Symp.* (Riverside, 1968), **2**: 801–822.

DeBach, P. (1969b). Uniparental, sibling and semi-species in relation to taxonomy and biological control. *Israel J. Entomol.*, **4**: 11–28.

DeBach, P. (1971a). The use of imported natural enemies in insect pest management ecology. *Proc. Tall Timbers Conf. Ecol. Anim. Contr. Habitat Mgmt.*, **3**: 211–233.

DeBach, P. (1971b). Fortuitous biological control from ecesis of natural enemies. pp. 293–307 in: *Entomological Essays to Commemorate the Retirement of Professor K. Yasumatsu*. Hokuryukan Publ. Co., Tokyo, 389 pp.

DeBach, P. (1971c). The theoretical basis of importation of natural enemies. *Proc. 13th Internat. Congr. Entomol.* (Moscow, 1968), **2**: 140–142.

DeBach, P. (1974). *Biological Control by Natural Enemies*. Cambridge Univ. Press, London and New York, 323 pp.

DeBach, P. H. (1976). The *Aphytis* story: an illustration of principles and practices in biological control. *25th Ann. Fac. Res. Lecture, Univ. Calif., Riverside*, 23pp.

DeBach, P. (1977). New red scale parasite imported from Saudi Arabia. *Citrograph*, **62**: 321–322.

DeBach, P., and L. C. Argyriou (1966). Effects of short-duration sub-optimum pre-oviposition temperatures on progeny production and sex ratio in species of *Aphytis* (Hymenoptera: Aphelinidae). *Res. Popul. Ecol.*, **8**: 69–77.

DeBach, P., and L. C. Argyriou (1967). The colonization and success in Greece of some imported *Aphytis* spp. (Hym. Aphelinidae) parasitic on citrus scale insects (Hom. Diaspididae). *Entomophaga*, **12**: 325–342.

DeBach, P., and A. Azim (1962). *Aphytis japonicus* n.sp., a parasite of *Chrysomphalus bifasciculatus* Ferris in Japan (Hymenoptera: Aphelinidae). *Mushi*, **36**: 1–8.

DeBach, P., and B. R. Bartlett (1964). Methods of colonization, recovery and evaluation. Chapter 14 in: P. DeBach (Editor), *Biological Control of Insect Pests and Weeds*. Chapman and Hall, London, pp. 402–426.

DeBach, P., E. J. Dietrick, and C. A. Fleschner (1951). Ants vs. biological control of citrus pests. *Calif. Citrogr.*, **36**: 312, 347–348.

DeBach, P., E. J. Dietrick, C. A. Fleschner, and T. W. Fisher (1950). Periodic colonization of *Aphytis* for control of the California red scale. Preliminary tests, 1949. *J. Econ. Entomol.*, **43**: 783–802.

DeBach, P., T. W. Fisher, and J. Landi (1955). Some effects of meteorological factors on all stages of *Aphytis lingnanensis*, a parasite of the California red scale. *Ecology*, **36**: 743–753.

DeBach, P., C. A. Fleschner, and E. J. Dietrick (1948). Natural control of the California red scale on citrus. *Calif. Citrogr.*, **34**: 6, 38–39.

DeBach, P., C. A. Fleschner, and E. J. Dietrick (1951). A biological check method for evaluating the effectiveness of entomophagous insects. *J. Econ. Entomol.*, **44**: 763–766.

DeBach, P., C. A. Fleschner, and E. J. Dietrick (1953). Natural control of the California red scale in untreated citrus orchards in southern California. *Proc. 7th Pacif. Sci. Congr.*, **4**: 255–259.

DeBach, P., and G. Gordh (1974). A new species of *Aphytis* that attacks important armored scale insects. *Entomophaga*, **19**: 259–265.

DeBach, P., and K. S. Hagen (1964). Manipulation of entomophagous species. Chapter 15 in: P. DeBach (Editor), *Biological Control of Insect Pests and Weeds*. Chapman and Hall, London, pp. 429–458.

DeBach, P., and C. B. Huffaker (1971). Experimental techniques for evaluation of the effectiveness of natural enemies. Chapter 5 in: C. B. Huffaker (Editor), *Biological Control*. Plenum Press, New York and London, pp. 113–140.

DeBach, P., C. B. Huffaker, and A. W. MacPhee (1976). Evaluation of the impact of natural enemies. Chapter 11 in: C. B. Huffaker and P. S. Messenger (Editors), *Theory and Practice of Biological Control*. Academic Press, New York and London, pp. 255–285.

DeBach, P., and J. Landi (1961). The introduced purple scale parasite, *Aphytis lepidosaphes* Compere, and a method of integrating chemical with biological control. *Hilgardia*, **31**: 459–497.

DeBach, P., J. Landi, and L. Jeppson (1959). Integrated control measures. *Calif. Agric.*, **13**(7): 12, 15.

DeBach, P., and S. V. Rao (1968). Transformation of inseminated females of *Aphytis lingnanensis* into factitious virgins by low-temperature treatment. *Ann. Entomol. Soc. Amer.*, **61**: 332–337.

DeBach, P., and M. Rose (1977). Environmental upsets caused by chemical eradication. *Calif. Agric.*, **31**(7): 8–10.

DeBach, P., M. Rose, and D. Rosen (1978). Biological and systematic studies of developmental stages in *Aphytis* (Homoptera: Aphelinidae). III. Meconia as a possible systematic tool in *Aphytis*. *Hilgardia*, **46**: 110–112.

DeBach, P., and D. Rosen (1976a). Armoured scale insects. Chapter 6 in: V. L. Delucchi (Editor), *Studies in Biological Control. Internat. Biol. Prog. Vol. 9*, Cambridge Univ. Press, pp. 139–178.

DeBach, P., and D. Rosen (1976b). Twenty new species of *Aphytis* (Hymenoptera: Aphelinidae) with notes and new combinations. *Ann. Entomol. Soc. Amer.*, **69**: 541–545.

DeBach, P., D. Rosen, and C. E. Kennett (1971). Biological control of coccids by introduced natural enemies. Chapter 7 in: C. B. Huffaker (Editor), *Biological Control*. Plenum Press, New York and London, pp. 165–194.

DeBach, P., and P. Sisojević (1960). Some effects of temperature and competition on the distribution and relative abundance of *Aphytis lingnanensis* and *A. chrysomphali* (Hymenoptera: Aphelinidae). *Ecology*, **41**: 153–160.

DeBach, P., and R. A. Sundby (1963). Competitive displacement between ecological homologues. *Hilgardia*, **34**: 105–166.

DeBach, P., and E. B. White (1960). Commercial mass culture of the California red scale parasite *Aphytis lingnanensis*. *Calif. Agr. Expt. Sta. Bull.* **770**, 58 pp.

De Gregorio, A. (1914). Nuovo *Aphelinus* parassita del *Chrysomphalus dictyospermi*. *N. Ann. Agric. Sicil.*, **3**(4): 224–230.

De Gregorio, A. (1915). Caratteri e biologia del *Chrysomphalus dictyospermi* Morg. auctorum (an potius *Aspidiotus agrumincola* De Greg.?) e del suo parassita distruttore *Aphelinus chrysomphali* Gar. Merc. Var. *Silvestrii* De Greg. con cenni di due ragni submicroscopici (*Licosa*). *Natur. Sicil.* (Ann. 22, 1914), N. Ser., **2**: 125–190; Tav. XV–XXIV.

Delucchi, V. (1964). Une nouvelle espèce d'*Aphytis* du groupe *chilensis* Howard (Hym., Chalcidoidea, Aphelinidae). *Rev. Pathol. Veg. Entomol. Agr. Fr.*, **43**: 135–140.

Delucchi, V., D. Rosen, and E. I. Schlinger (1976). Relationship of systematics to biological control. Chapter 4 in: C. B. Huffaker and P. S. Messenger (Editors), *Theory and Practice of Biological Control*. Academic Press, New York and London, pp. 81–91.

De Santis, L. (1941). Lista de himenopteros parasitos y predatores de los insectos de la Republica Argentina. *Bol. Soc. Brasil. Agron.*, **4**: 1–66.

De Santis, L. (1946). Taxonomía de la familia Aphelinidae (Hymenoptera, Chalcidoidea). *Rev. Mus. La Plata, Zool.* (N. Ser.), **5**: 1–21.

De Santis, L. (1948). Estudio monográfico de los afelínidos de la Republica Argentina (Hymenoptera, Chalcidoidea). *Rev. Mus. La Plata, Zool.* (N. Ser.), **5**: 23–280.

De Santis, L. (1965). Adiciones a la fauna argentina de afelinidos. VI (Insecta Hymenoptera). *Rev. Soc. Entomol. Argent.*, **27**: 75–79 (1964).

De Santis, L. (1967). *Catálogo de los himenópteros argentinos de la serie Parasitica, incluyendo Bethyloidea*. Prov. Buenos Aires Gober., Com. Invest. Cient., La Plata, 337 pp.

De Santis, L., and L. Esquivel (1966 (1967)). Tercera lista de himenopteros parasitos y predatores de los insectos de la Republica Argentina. *Rev. Mus. La Plata, Zool.* (N. Ser.), **9**: 47–215.

Dodd, A. P. (1917). Records and descriptions of Australian Chalcidoidea. *Trans. Roy. Soc. S. Austral.*, **41**: 344–368.

Doutt, R. L. (1954). An evaluation of some natural enemies of the olive scale. *J. Econ. Entomol.*, **47**: 39–43.

Doutt, R. L. (1964). Biological characteristics of entomophagous adults. Chapter 6 in: P. DeBach (Editor), *Biological Control of Insect Pests and Weeds*. Chapman and Hall, London, pp. 145–167.

Dozier, H. L. (1933). Miscellaneous notes and descriptions of chalcidoid parasites (Hymenoptera). *Proc. Entomol. Soc. Wash.*, **35**: 85–100.

Eady, R. D. (1968). Some illustrations of microsculpture in the Hymenoptera. *Proc. Roy. Entomol. Soc. London* (A), **43**: 66–72, 31 figs.

Eliraz, A., and D. Rosen (1978). Biological and systematic studies of developmental stages in *Aphytis* (Hymenoptera: Aphelinidae). II. Larval criteria in the systematics of *Aphytis*. *Hilgardia*, **46**: 96–101.

Enghoff, H. (1976). Taxonomic problems in parthenogenetic animals. *Zool. Scripta*, **5**: 103–104.

Erdös, J. (1964). Hymenoptera II. Chalcidoidea III. *Fauna Hungar.*, **73**, 372 pp.

Fabres, G. (1973). Particularités biologiques d'*Aphytis cochereaui* Rosen (Hym. *Aphelinidae*), parasite de *Lepidosaphes beckii* Newman (Homopt. *Coccoidea Diaspididae*) en Nouvelle-Calédonie. *C. R. Acad. Sci. Paris*, **277**: 1159–1160.

Fabres, G. (1974). Contribution à l'étude d'*Aphytis cochereaui* et d'*A. lepidosaphes (Hym. Aphelinidae)*, parasites de *Lepidosaphes beckii (Hom. Diaspididae)* en Nouvelle-Calédonie. *Ann. Soc. Entomol. Fr.* (N.S.), **10**: 371–379.

Fabres, G. (1975). Intervention du parasite *Aphytis cochereaui (Hym.: Aphelinidae)* lors d'une pullulation de son hôte *Lepidosaphes beckii (Hom.: Diaspididae)* en Nouvelle-Calédonie. *Entomophaga*, **20**: 81–92.

FABRES, G. (1977). Élevage au laboratoire d'*Aphytis cochereaui* et précision de ses paramètres biologiques (Hym. Aphelinidae). *Ann. Soc. Entomol. Fr.* (N.S.), **13**: 237–242.

FERRIÈRE, C. (1965). *Hymenoptera Aphelinidae d'Europe et du Bassin Méditerranéen. Faune de l'Europe et du Bassin Méditerranéen,* **1**. Masson, Paris, 206 pp.

FISHER, T. W. (1952). *Comparative Biologies of Some Species of Aphytis (Hymenoptera: Chalcidoidea) with Particular Reference to Racial Differentiation.* Ph.D. Dissertation, Univ. Calif., V + 106 pp., 4 pl. (mimeogr.).

FLANDERS, S. E. (1945). The bisexuality of uniparental Hymenoptera, a function of the environment. *Amer. Natur.,* **79**: 122–141.

FLANDERS, S. E. (1950). An enemy of purple scale recently established in California. *Calif. Citrogr.,* **36**: 64–65.

FLANDERS, S. E. (1965). On the sexuality and sex ratios of hymenopterous populations. *Amer. Natur.,* **99**: 489–494.

FLANDERS, S. E. (1971). Multiple parasitism of armored coccids (Homoptera) by host-regulative aphelinids (Hymenoptera); ectoparasites versus endoparasites. *Canad. Entomol.,* **103**: 857–872.

FOERSTER, A. (1856). *Hymenopterologische Studien.* 2. Heft. *Chalcidiae und Proctotrupii.* Ernst ter Meer, Aachen, 152 pp.

GAHAN, A. B. (1919). Notes on some genera and species of chalcid-flies belonging to the Aphelininae with description of a new species. *Proc. U.S. Nat. Mus.,* **55**: 403–407.

GAHAN, A. B. (1933). The serphoid and chalcidoid parasites of the Hessian fly. *U.S. Dept. Agric. Misc. Publ.,* **174**: 1–147.

GAHAN, A. B., and M. M. FAGAN (1923). The type species of the genera of Chalcidoidea or chalcid flies. *Bull. U.S. Nat. Mus.,* **124**, 173 pp.

GERSON, U. (1967). *Aphytis riadi*—a synonym of *A. chilensis* (Hymenoptera: Aphelinidae). *Ann. Entomol. Soc. Amer.,* **60**: 1116–1118.

GERSON, U. (1968). The comparative biologies of two hymenopterous parasites of the chaff scale, *Parlatoria pergandii. Entomophaga,* **13**: 163–173.

GIRAULT, A. A. (1911). Hosts of insect egg-parasites in North and South America. II. *Psyche,* **18**: 146–153.

GIRAULT, A. A. (1913). Australian Hymenoptera Chalcidoidea—IV. The family Eulophidae with descriptions of new genera and species. *Mem. Queensl. Mus.,* **2**: 140–334.

GIRAULT, A. A. (1915). Australian Hymenoptera Chalcidoidea—VII. The family Encyrtidae with descriptions of new genera and species. *Mem. Queensl. Mus.,* **4**: 1–184.

GIRAULT, A. A. (1917a). Some new Australian chalcid-flies, mostly of the family Encyrtidae. *Insecut. Inscit. Menstr.,* **5**: 29–37.

GIRAULT, A. A. (1917b). *Descriptiones stellarum novarum.* Priv. Publ., 22pp.

GIRAULT, A. A. (1919). *Hymenoptera Chalcidoidea Nova Australiensis.* Priv. Publ., Brisbane, 3 pp.

GIRAULT, A. A. (1923). New Encyrtidae from Australia—II (Hymenoptera). *Insecut. Inscit. Menstr.,* **11**: 141–148.

GIRAULT, A. A. (1924). *Homo perniciosus and New Hymenoptera.* Priv. Publ., Brisbane, 4 pp.

GIRAULT, A. A. (1932). *New Lower Hymenoptera from Australia and India.* Priv. Publ., Brisbane, 6 pp.

GIRAULT, A. A. (1933). *Some Beauties Inhabitant Not of Commercial Boudoirs but of Nature's Bosom, Notably New Insects.* Priv. Publ., Brisbane, 5 pp.

GOMES, J. G. (1942). Subsídios à sistemática dos calcidídeos brasileiros. *Bol. Esc. Nac. Agron.,* Rio de Janeiro, **2** (1941): 9–45.

GORDH, G., and P. DEBACH (1976). Male inseminative potential in *Aphytis lingnanensis* (Hymenoptera: Aphelinidae). *Canad. Entomol.,* **108**: 583–589.

GORDH, G., and P. DEBACH (1978). Courtship behavior in the *Aphytis lingnanensis* group, its potential usefulness in taxonomy, and a review of sexual behavior in the parasitic Hymenoptera (Chalcidoidea: Aphelinidae). *Hilgardia,* **46**:37–75.

GORDH, G., and D. ROSEN (1973). A new genus and species of Aphelinidae from Greece (Hymenoptera: Chalcidoidea). *J. Entomol. Soc. S. Afr.,* **36**:275–279.

GRAHAM, M. W. R. DE V. (1976). The British species of *Aphelinus* with notes and descriptions of other European Aphelinidae (Hymenoptera). *Syst. Entomol.,* **1**:123–146.

GRESSITT, J. L. (1974). Insect biogeography. *Ann. Rev. Entomol.,* **19**:293–321.

GRISWOLD, G. H. (1925). A study of the oyster-shell scale, *Lepidosaphes ulmi* (L.), and one of its parasites, *Aphelinus mytilaspidis* Le B. Part II. Biology of a parasite of the oyster-shell scale. *Cornell Univ. Agric. Expt. Sta. Mem.* **93**:57–67.

GULMAHAMAD, H., and P. DEBACH (1978). Biological studies on *Aphytis aonidiae* (Mercet) (Hymenoptera: Aphelinidae), an important parasite of the San Jose scale, *Quadraspidiotus perniciosus* (Comstock), in southern California. *Hilgardia* (in press).

HAFEZ, M., and R. L. DOUTT (1954). Biological evidence of sibling species in *Aphytis maculicornis* (Masi) (Hymenoptera, Aphelinidae). *Canad. Entomol.,* **86**:90–96.

HAGEN, K. S. (1964). Developmental stages of parasites. Chapter 7 in: P. DeBach (Editor), *Biological Control of Insect Pests and Weeds.* Chapman and Hall, London, pp. 168–246.

HARPAZ, I., and D. ROSEN (1971). Development of integrated control programs for crop pests in Israel. Chapter 20 in: C. B. Huffaker (Editor), *Biological Control.* Plenum Press, New York and London, pp. 458–468.

HAYAT, M. (1972). A new aphelinid genus *Eriaphytis* (Hymenoptera, Chalcidoidea) reared from *Cerococcus* spp. *Polsk. Pismo Entomol.,* **42**:151–156.

HAYAT, M. (1973). Two new Indian species of *Centrodora* Foerster, 1878 (Hymenoptera: Aphelinidae) parasitic in the eggs of membracids (Homoptera: Membracidae). *Entomophaga,* **18**:41–46.

HAYAT, M. (1974). On the types of *Marietta peculiaris* Girault, 1932 and *M. javensis indi* Girault, 1932 (Hymenoptera: Aphelinidae). *J. Nat. Hist.,* **8**:345–347.

HAYWARD, K. J. (1941). Insectos de importancia económica en la región de Concordia (Entre Ríos). *Rev. Soc. Entomol. Argent.,* **11**:68–109.

HOWARD, L. O. (1881). Report on the parasites of the Coccidae in the collection of this department. Report of the Entomologist, Part III, in: *Ann. Rep. Commissioner Agric. for 1880,* Washington, D.C., pp. 350–371, pl 23–24.

HOWARD, L. O. (1895). Revision of the Aphelininae of North America, a subfamily of hymenopterous parasites of the family Chalcididae. *U.S. Dept. Agric. Div. Entomol. Tech. Ser.,* **1**:44 pp.

HOWARD, L. O. (1900). A new genus of Aphelininae from Chile. *Canad. Entomol.,* **32**:167–168.

HOWARD, L. O. (1907). New genera and species of Aphelininae, with a revised table of genera. *U.S. Dept. Agric. Bur. Entomol. Tech Ser.* **12** (4):69–88.

HOWARD, L. O. (1908). Upon the aphis-feeding species of *Aphelinus. Entomol. News,* **19**:365–366.

HUFFAKER, C. B., and C. E. KENNETT (1966). Studies of two parasites of olive scale, *Parlatoria oleae* (Colvée). IV. Biological control of *Parlatoria oleae* (Colvée) through the compensatory action of two introduced parasites. *Hilgardia,* **37**:283–335.

HUFFAKER, C. B., C. E. KENNETT, and G. L. FINNEY (1962). Biological control of olive scale, *Parlatoria oleae* (Colvée), in California by imported *Aphytis maculicornis* (Masi) (Hymenoptera: Aphelinidae). *Hilgardia,* **32**:541–636.

HUXLEY, J. S. (Editor) (1940). *The New Systematics.* Clarendon Press, Oxford, 583 pp.

ICZN (1961). *International Code of Zoological Nomenclature Adopted by the XV International Congress of Zoology.* N. R. Stoll et al. (Editors). Internat. Trust Zool. Nomencl., London, 176 pp.

IMMS, A. D. (1916). Observations on the insect parasites of some Coccidae. I—On *Aphelinus mytilaspidis* Le Baron, a chalcid parasite of the mussel scale (*Lepidosaphes ulmi* L.). *Quart. J. Microsc. Sci.*, N.S., **61**: 217–274, pl. 19–20.

KEHAT, M., E. SWIRSKI, D. BLUMBERG, and S. GREENBERG (1974). Integrated control of date palm pests in Israel. *Phytoparasitica*, **2**: 141–149.

KHASIMUDDIN, S., and P. DEBACH (1975). Mating behavior and evidence of a male sex pheromone in species of the genus *Aphytis*. *Ann. Entomol. Soc. Amer.*, **68**: 893–896.

KHASIMUDDIN, S., and P. DEBACH (1976a). Hybridization tests: a method for establishing biosystematic statuses of cryptic species of some parasitic Hymenoptera. *Ann. Entomol. Soc. Amer.*, **69**: 15–20.

KHASIMUDDIN, S., and P. DEBACH (1976b). Biosystematic status of three allopatric populations of *Aphytis maculicornis* (Masi) (Hym.: Aphelinidae). *Entomophaga*, **21**: 81–92.

KHASIMUDDIN, S., and P. DEBACH (1976c). Biosystematic and evolutionary statuses of two sympatric populations of *Aphytis mytilaspidis* (Hym.: Aphelinidae). *Entomophaga*, **21**: 113–122.

KIRIUKHIN, G. (1946). Les parasites du *Chrysomphalus dictyospermi* Morg. au nord de l'Iran. *Lab. Entomol. Phytopathol. Appl. Teheran*, **2**: 4–6, 13–22 (in Persian, with a summary in French).

KURDJUMOV, N. V. (1913). Notes on European species of the genus *Aphelinus* Dalm. (Hymenoptera, Chalcidoidea), parasitic upon the plant-lice. *Rev. Russ. Entomol.*, **13**: 266–270.

LE BARON, W. (1870). The chalcideous parasite of the apple-tree bark-louse (*Chalcis* [*Aphelinus*] *mytilaspidis*, n. sp.). *Amer. Entomol. Bot.*, **2**: 360–362.

LEVENE, H. (1949). A new measure of sexual isolation. *Evolution*, **3**: 315–321.

LIZER Y TRELLES, C. A. (1939). Catálogo sistemático razonado de los Cóccidos ("Hom. Sternor.") vernáculos de la Argentina. *Physis*, **17**: 157–209.

LORD, F. T. (1947). The influence of spray programs on the fauna of apple orchards in Nova Scotia. II. Oystershell scale *Lepidosaphes ulmi* (L.). *Canad. Entomol.*, **79**: 196–209.

LORD, F. T., and A. W. MACPHEE (1953). The influence of spray programs on the fauna of apple orchards in Nova Scotia. VI. Low temperatures and the natural control of the oystershell scale, *Lepidosaphes ulmi* (L.) (Homoptera: Coccidae). *Canad. Entomol.*, **85**: 282–291.

LUKINS, R. G., and G. J. SNOWBALL (1977). Biological control of circular black scale. *CSIRO Div. Entomol. Ann. Rep. for 1976–77*, pp. 5–6.

MACKAUER, M. (1972). The aphid-attacking genera of Aphelinidae (Hymenoptera), including the description of a new genus. *Canad. Entomol.*, **104**: 1771–1779.

MALENOTTI, E. (1918a). I nemici naturali della "Bianca-Rossa" (*Chrysomphalus dictyospermi* Morg.). *Redia*, **13**: 17–53.

MALENOTTI, E. (1918b). Nuovi calciditi. *Redia*, **13**: 77–92.

MARCHAL, P. (1909). La ponte des *Aphelinus* et l'intérêt individuel dans les actes liés à la conservation de l'espèce. *C. R. Acad. Sci. Paris*, **148**: 1223–1225.

MASI, L. (1911). Contribuzioni alla conoscenza dei Calcididi Italiani. *Boll. Lab. Zool. Gen. Agr. Portici*, **5**: 140–171.

MAYR, E. (1942). *Systematics and the Origin of Species*. Columbia Univ. Press, New York, 334 pp.

MAYR, E. (1963). *Animal Species and Evolution*. Belknap Press, Harvard Univ. Press, 797 pp.

MAYR, E. (1969). *Principles of Systematic Zoology*. McGraw-Hill, New York, 428 pp.

MAYR, E., E. G. LINSLEY, and R. L. USINGER (1953). *Methods and Principles of Systematic Zoology*. McGraw-Hill, New York, Toronto and London, 336 pp.

MCCONNELL, W. R. (1916). Notes on the biology of *Paraphelinus speciosissimus* Girault. *Ann. Entomol. Soc. Amer.*, **9**: 97–102.

Mercet, R. G. (1911a). Las plagas del naranjo en Valencia. *Asoc. Españ. Prog. Cienc., Congr.* (Valencia, 1910), **5** (reprint, 17 pp., 3 pl.).

Mercet, R. G. (1911b). Los Calcídidos parásitos de Cóccidos. *Bol. R. Soc. Españ. Hist. Nat.,* **11**:507–515.

Mercet, R. G. (1912a). Un parásito del "poll-roig". *Bol. R. Soc. Españ. Hist. Nat.,* **12**:135–140.

Mercet, R. G. (1912b). Los enemigos de los parásitos de las plantas. Los Afelíninos. *Trab. Mus. Cienc. Nat. Madrid,* **10**, 306 pp.

Mercet, R. G. (1921). Notas sobre afelininos (Hym. Chalc.). *R. Soc. Espan. Hist. Nat., 50th Aniv. Vol.,* Madrid, pp. 299–309.

Mercet, R. G. (1928). La fauna afelinina española. *Asoc. Españ. Prog. Cienc., Congr.* (Cadiz, 1927), **4**:235–243.

Mercet, R. G. (1929). Los afelínidos de España. *Rev. Biol. Forest. Limnol.,* Ser. B., **1**:3–28.

Mercet, R. G. (1930). Los afelinidos de España. Seguñda parte. *Rev. Biol. Forest. Limnol.,* Ser. B., **2**:29–106.

Mercet, R. G. (1931). Afelínidos paleárticos (Hym. Chalc.) 6.ª nota. *Eos,* **7**:5–7.

Mercet, R. G. (1932). Notas sobre afelínidos (Hym. Chalc.) 6.ª nota. *Eos,* **8**:353–365.

Meyerdirk, D. E. (1972). *The Parameters Affecting the Distribution and Regulating the Abundance of Purple Scale, Lepidosaphes beckii (Newman) (Homoptera: Diaspididae).* Ph.D. Thesis, University of California, Riverside, 108 pp. (mimeogr.).

Michener, C. D. (1956). 23. Hymenoptera. pp. 131–140 in: S. L. Tuxen (Editor), *Taxonomist's Glossary of Genitalia in Insects.* Ejnar Munksgaard, Copenhagen.

Nakayama, S. (1921). An enumeration of the Japanese Aphelininae, with descriptions of two new species. *Philip. J. Sci.,* **18**:97–100.

Nees ab Esenbeck, C. G. (1834). *Hymenopterorum Ichneumonibus affinium monographiae, genera Europaea et species illustrantes,* **2**. Stuttgart and Tubingen, 448 pp.

Nikol'skaya, M. N. (1952). *The Chalcid Fauna of the USSR. Opred. Faun. SSSR,* **44**. Akad. Nauk SSSR, Moscow and Leningrad, 574 pp. (in Russian). (English transl.: Israel Prog. Sci. Transl., Jerusalem 1963, 593 pp.).

Nikol'skaya, M. N. (1963). Two new genera of Aphelinidae (Hymenoptera, Chalcidoidea), bred from the bamboo scale *Odonaspis secreta* (Ckll.) in the Caucasus. *Entomol. Obozr.,* **42**: 186–189 (in Russian). (English transl.: *Entomol. Rev.,* **42**: 102–103).

Nikol'skaya, M. N., and V. A. Yasnosh (1966). *Aphelinids of the European Part of the USSR and the Caucasus. Opred. Faun. SSSR,* **91**. Nauka, Moscow and Leningrad, 296 pp. (in Russian).

Nikol'skaya, M. N., and V. A. Yasnosh (1968). On the aphelinid fauna of the Caucasus. *Trud. Vsesoyuz. Entomol. Obshchest.,* **52**:3–42 (in Russian).

Novitzky, S. (1961). Note on *Agonioneurus subflavescens* Westwood and *A. albidus* Westwood (Hym., Aphelinidae). *Entomol. Mon. Mag.,* **97**:195.

Nowicki, S. (1930). Bemerkungen zu den europäischen Apheliniden-Gattungen (Hym., Chalc.). *Zeitschr. Wiss. Insektenbiol., Neue Beitr. Syst. Insektenkunde,* **4**:153–160, 177–181.

Parker, H. L. (1924). Recherches sur les formes post-embryonnaires des chalcidiens. *Ann. Soc. Entomol. Fr.* **93**:261–379, pl. 2–39.

Peck, O. (1951). Superfamily Chalcidoidea. pp. 410–594 in: C. F. W. Muesebeck, K. V. Krombein, and H. K. Townes (Editors), *Hymenoptera of America North of Mexico—Synoptic Catalog.* Agric. Monogr. **2**, U.S. Dept. Agric., Washington, D.C.

Peck, O. (1963). A catalogue of the Nearctic Chalcidoidea (Insecta: Hymenoptera). *Canad. Entomol.,* Suppl. **30**, 1092 pp.

PECK, O., Z. BOUČEK, and A. HOFFER (1964). Keys to the Chalcidoidea of Czechoslovakia (Insecta: Hymenoptera). *Mem. Entomol. Soc. Canada,* **34**:7–120.

QUAYLE, H. J. (1910). *Aphelinus diaspidis* Howard. *J. Econ. Entomol.,* **3**:398–401.

QUEDNAU, F. W. (1964a). Experimental evidence of differential fecundity on red scale [*Aonidiella aurantii* (Mask.)] in six species of *Aphytis* (Hymenoptera, Aphelinidae). *S. Afr. J. Agr. Sci.,* **7**:335–340.

QUEDNAU, F. W. (1964b). A contribution on the genus *Aphytis* Howard in South Africa (Hymenoptera: Aphelinidae). *J. Entomol. Soc. S. Afr.,* **27**:86–116.

QUEDNAU, F. W. (1964c). An evaluation of fecundity, host-mutilation and longevity on three species of diaspine scales in *Aphytis lingnanensis* Compere (Hymenoptera, Aphelinidae). *S. Afr. J. Agr. Sci.,* **7**:521–530.

QUEDNAU, F. W. (1965). A technique for identifying mixed populations of six species of *Aphytis* parasitic on red scale [*Aonidiella aurantii* (Mask.)], for recognition after recovery from scales collected in citrus orchards (Hymenoptera, Aphelinidae). *S. Afr. J. Agr. Sci.,* **8**:43–56.

QUEDNAU, F. W., and H. M. HÜBSCH (1964). Factors influencing the host-finding and host-acceptance pattern in some *Aphytis* species (Hymenoptera: Aphelinidae). *S. Afr. J. Agr. Sci.,* **7**:543–554.

QUEZADA, J. R., P. DEBACH, and D. ROSEN (1973). Biological and taxonomic studies of *Signiphora borinquensis,* new species, (Hymenoptera: Signiphoridae), a primary parasite of diaspine scales. *Hilgardia,* **41**:543–601.

RAO, S. V., and P. DEBACH (1969a). Experimental studies on hybridization and sexual isolation between some *Aphytis* species (Hymenoptera: Aphelinidae). I. Experimental hybridization and an interpretation of evolutionary relationships among the species. *Hilgardia,* **39**:515–553.

RAO, S. V., and P. DEBACH (1969b). Experimental studies on hybridization and sexual isolation between some *Aphytis* species (Hymenoptera: Aphelinidae). II. Experiments on sexual isolation. *Hilgardia,* **39**:555–567.

RAO, S. V., and P. DEBACH (1969c). Experimental studies on hybridization and sexual isolation between some *Aphytis* species (Hymenoptera: Aphelinidae). III. The significance of reproductive isolation between interspecific hybrids and parental species. *Evolution,* **23**:525–533.

RATZEBURG, J. T. C. (1848). *Die Ichneumonen der Forstinsecten in forstlicher und entomologischer Beziehung; ein Anhang zur Abbildung und Beschreibung der Forstinsecten,* **2**. Berlin, VII + 238 pp., 3 tab., 3 pl.

RATZEBURG, J. T. C. (1852). *Die Ichneumonen der Forstinsecten in forstlicher und entomologischer Beziehung; ein Anhang zur Abbildung und Beschreibung der Forstinsecten,* **3**. Berlin, III–XVIII + 272 pp., 3 tab.

REINHARD, H. (1858). Beiträge zur Geschichte und Synonymie der Pteromalinen (Schluss). *Berlin. Entomol. Zeitschr.,* **2**:10–23.

RICHARDS, O. W. (1956). Hymenoptera: Introduction and Key to Families. Roy. Entomol. Soc., London, *Handbooks for the Identification of British Insects,* **6** (1), 94 pp.

RIEK, E. F. (1970). Hymenoptera. Chapter 37 in: *The Insects of Australia.* Div. Entomol. CSIRO, Melbourne Univ. Press, pp. 867–959.

RISBEC, J. (1952). Contribution à l'étude des chalcidoïdes de Madagascar. *Mem. Inst. Sci. Madagascar,* Ser. E., **2**:1–449.

RIVNAY, E. (1968). Biological control of pests in Israel (a review 1905–1965). *Israel J. Entomol.* **3** (1):1–156.

ROSEN, D. (1965). The hymenopterous parasites of citrus armored scales in Israel (Hymenoptera: Chalcidoidea). *Ann. Entomol. Soc. Amer.,* **58**:388–396.

ROSEN, D. (1966). Keys for the identification of the hymenopterous parasites of scale insects, aphids and aleyrodids on citrus in Israel. *Scripta Hierosolym.*, **18**: 43–79, 267 figs.

ROSEN, D. (1967a). Effect of commercial pesticides on the fecundity and survival of *Aphytis holoxanthus* (Hymenoptera: Aphelinidae). *Israel J. Agric. Res.*, **17**: 47–52.

ROSEN, D. (1967b). Biological and integrated control of citrus pests in Israel. *J. Econ. Entomol.*, **60**: 1422–1427.

ROSEN, D. (1969). The parasites of coccids, aphids and aleyrodids on citrus in Israel: some zoogeographical considerations. *Israel J. Entomol.*, **4**: 45–53.

ROSEN, D. (1973). Methodology for biological control of armored scale insects. *Phytoparasitica*, **1**: 47–54.

ROSEN, D. (1974). Current status of integrated control of citrus pests in Israel. *EPPO Bull.*, **4**: 363–368.

ROSEN, D., and P. DEBACH (1970). Notes on the genus *Marlattiella* Howard (Hymenoptera: Aphelinidae). *Mushi*, **43**: 39–44.

ROSEN, D., and P. DEBACH (1973). Systematics, morphology and biological control. *Entomophaga*, **18**: 215–222.

ROSEN, D. and P. DEBACH (1976). Biosystematic studies on the species of *Aphytis* (Hymenoptera: Aphelinidae). *Mushi*, **49**: 1–17.

ROSEN, D., and P. DEBACH (1977a). Resurrection of *Aphytis theae* (Hymenoptera: Aphelinidae), a parasite of tea scale, with notes on a new group of species. *Fla. Entomol.*, **60**: 1–10.

ROSEN, D., and P. DEBACH (1977b). Use of scale-insect parasites in Coccoidea systematics. *Va. Polytech. Inst. & State Univ. Res. Div. Bull.* **127**: 5–21.

ROSEN, D., and P. DEBACH (1978). Diaspididae. pp. 78–128 in: C. P. Clausen (Editor), *Introduced Parasites and Predators of Arthropod Pests and Weeds: a World Review*. Agric. Handbook **480**, U.S. Dept. Agric., Washington, D.C.

ROSEN, D., and A. ELIRAZ (1978). Biological and systematic studies of developmental stages in *Aphytis* (Hymenoptera: Aphelinidae). I. Developmental history of *Aphytis chilensis* Howard. *Hilgardia*, **46**: 77–95.

RÖSSLER, Y., and P. DEBACH (1972a). The biosystematic relations between a thelytokous and an arrhenotokous form of *Aphytis mytilaspidis* (Le Baron) (Hymenoptera: Aphelinidae) 1. The reproductive relations. *Entomophaga*, **17**: 391–423.

RÖSSLER, Y., and P. DEBACH (1972b). The biosystematic relations between a thelytokous and an arrhenotokous form of *Aphytis mytilaspidis* (Le Baron) (Hymenoptera: Aphelinidae), 2. Comparative biological and morphological studies. *Entomophaga*, **17**: 425–435.

RÖSSLER, Y., and P. DEBACH (1973). Genetic variability in a thelytokous form of *Aphytis mytilaspidis* (Le Baron) (Hymenoptera: Aphelinidae). *Hilgardia*, **42**: 149–175.

RUST, E. W. (1915). Three new species of *Aphelinus* (Hym.). *Entomol. News*, **26**: 73–77.

SABROSKY, C. W. (1955). The interrelations of biological control and taxonomy. *J. Econ. Entomol.*, **48**: 710–714.

SCHLINGER, E. I., and R. L. DOUTT (1964). Systematics in relation to biological control. Chapter 8 in: P. DeBach (Editor), *Biological Control of Insect Pests and Weeds*. Chapman and Hall, London, pp. 247–280.

SCHMIEDEKNECHT, O. (1909). Hymenoptera fam. Chalcididae. Fasc. **97** in: P. Wytsman, *Genera Insectorum*. Verteneuil and Desmet, Bruxelles, 550 pp., 8 pl.

SILVESTRI, F. (1915). Contributo alla conoscenza degli insetti dell' olivo dell' Eritrea e dell' Africa meridionale. *Boll. Lab. Zool. Gen. Agr. Portici*, **9**: 240–334.

SILVESTRI, F. (1929). Preliminary report on the citrus scale-insects of China. *Fourth Internat. Congr. Entomol.*, Ithaca, N.Y. (1928), II. Trans., pp. 897–904.

SMITH, J. M. (1957). Effects of the food plant of California red scale, *Aonidiella aurantii* (Mask.) on reproduction of its hymenopterous parasites. *Canad. Entomol.*, **89**: 219–230.

SNODGRASS, R. E. (1935). *Principles of Insect Morphology*. McGraw-Hill, New York and London, 667 pp.

SNODGRASS, R. E. (1941). The male genitalia of Hymenoptera. *Smithson. Misc. Coll.*, **99** (14), 86 pp., 33 pl.

STERNLICHT, M. (1973). Parasitic wasps attracted by the sex pheromone of their coccid hosts. *Entomophaga*, **18**: 339–342.

STEYN, J. J. (1958). The effect of ants on citrus scales at Letaba, South Africa. *Proc. 10th Internat. Congr. Entomol.* (Montreal, 1956), **4**: 589–594.

TACHIKAWA, T. (1962). Discovery of *Marlattiella prima* Howard, a parasite of the pear white scale, from Japan (Hymenoptera: Aphelinidae). *Trans. Shikoku Entomol. Soc.*, **7**: 74–77.

TAYLOR, T. H. C. (1935). The campaign against *Aspidiotus destructor*, Sign., in Fiji. *Bull. Entomol. Res.*, **26**: 1–102.

THOMSON, C. G. (1875). *Scandinaviens Hymenoptera*, **4**: e Delen, Innehållande Slägtet *Pteromalus* Sverderus. Lund, 259 pp. (Actual date of publication apparently 1876.)

TIMBERLAKE, P. H. (1924). Descriptions of new chalcid-flies from Hawaii and Mexico (Hymenoptera). *Proc. Hawaii. Entomol. Soc.*, **5**: 395–417.

TIMBERLAKE, P. H. (1926). New species of Hawaiian chalcid-flies (Hymenoptera). *Proc. Hawaii. Entomol. Soc.*, **6**: 305–320.

TRABOULSI, R. (1968). Prédateurs et parasites d'*Aphytis* (Hym., Aphelinidae). *Entomophaga*, **13**: 345–355.

TRABOULSI, R. (1969). Contribution à l'étude des *Aphytis* Howard du Liban (Hym. Chalcidoidea, Aphelinidae). *Ann Soc. Entomol. Fr.* (N.S.), **5**: 5–72.

TURNBULL, A. L., and D. A. CHANT (1961). The practice and theory of biological control of insects in Canada. *Canad. J. Zool.*, **39**: 697–753.

VAN DEN BOSCH, R., and A. D. TELFORD (1964). Environmental modification and biological control. Chapter 16 in: P. DeBach (Editor), *Biological Control of Insect Pests and Weeds*. Chapman and Hall, London, pp. 459–488.

WALKER, F. (1839). *Monographia Chalciditum*, **1**. Hyppolitus Bailliere, London, 333 pp.

WALLACE, A. R. (1876). *The Geographical Distribution of Animals*, Vol. **1**. Reprinted 1962, Hafner Publ. Co., New York and London, 503 pp.

WATERSTON, J. (1917). A new species of *Paraphelinus*, Perk., from British Guiana, with a discussion of the genus and the allied *Aphelinus*, Dalm. *Bull. Entomol. Res.*, **8**: 43–58.

WATT, K. E. F. (1965). Community stability and the strategy of biological control. *Canad. Entomol.*, **97**: 887–895.

WESTWOOD, J. O. (1837). Descriptions of some new British species of hymenopterous insects. *Phil. Mag.*, Third Ser., **10**: 440–442.

WHITE, E. B., P. DEBACH, and M. J. GARBER (1970). Artificial selection for genetic adaptation to temperature extremes in *Aphytis lingnanensis* Compere (Hymenoptera: Aphelinidae). *Hilgardia*, **40**: 161–192.

WHITE, M. J. D. (1964). Cytogenetic mechanisms in insect reproduction. pp. 1–12 in: K. C. Highnam, (Editor), *Insect Reproduction*. Symposium No. 2, Roy. Entomol. Soc., London.

WOOD, B. J. (1963). Imported and indigenous natural enemies of citrus coccids and aphids in Cyprus, and an assessment of their potential value in integrated control programmes. *Entomophaga*, **8**: 67–82.

Yasnosh, V. A. (1972). On the biosystematic characteristics of species of the genus *Aphytis* Howard (Chalcidoidea, Aphelinidae)—parasites of scale insects in the USSR. *Entomol. Obozr.*, **51**: 240–253 (in Russian). (English transl.: *Entomol. Rev.*, **51**: 146–152).

Yasnosh, V. A. (1976). Classification of the parasitic Hymenoptera of the family Aphelinidae (Hymenoptera, Chalcidoidea). *Entomol. Obozr.*, **55**: 159–168 (in Russian). (English transl.: *Entomol. Rev.*, **55** (1): 114–120.

Yasnosh, V. A., and E. N. Mindiashvili (1971). Biology and importance of *Aphytis maculicornis* (Masi) (Chalcidoidea, Aphelinidae)—a parasite of the olive scale *Parlatoria oleae* Colvée (Homoptera, Coccoidea). *Trud. Inst. Zashchit. Rast. Gruz. SSR*, **23**: 73–79 (in Russian).

Yasnosh, V. A., and S. N. Myartseva (1971). Two new aphelinid species (Chalcidoidea, Aphelinidae)—parasites of armored scale insects (Coccoidea, Diaspididae) in Central Asia. *Izv. Akad. Nauk Turkmen. SSR, Ser. Biol. Nauk*, **6**: 35–41 (in Russian).

Zehntner, L. (1897). De plantenluizen van het suikerriet op Java. II. *Arch. Java-Suikerindust.*, N.S., **4** (36): 193–218, pl. III–IV.

# INDEX TO ANIMAL NAMES

In the following index, both the scientific and common names of animal species and genera are listed in alphabetical order. The authors of all specific and generic names are included, even if not mentioned in the text. Page references to a major section dealing with a given taxon are presented in boldface, whereas references to pages including figures are in italics.

# INDEX TO PLANT NAMES